Rock Mechanics and Engineering

Rock Mechanics and Engineering

Volume 3: Analysis, Modeling & Design

Editor

Xia-Ting Feng

*Institute of Rock and Soil Mechanics, Chinese Academy of Sciences
State Key Laboratory of Geomechanics and Geotechnical
Engineering, Wuhan, China*

CRC Press
Taylor & Francis Group
Boca Raton London New York

CRC Press is an imprint of the
Taylor & Francis Group, an **informa** business

A BALKEMA BOOK

First published in paperback 2024

First published 2017
by CRC Press/Balkema
4 Park Square, Milton Park, Abingdon, Oxon, OX14 4RN

and by CRC Press/Balkema
2385 NW Executive Center Drive, Suite 320, Boca Raton FL 33431

CRC Press/Balkema is an imprint of the Taylor & Francis Group, an informa business

© 2017, 2024 Taylor & Francis Group, LLC

Publisher's Note
The publisher has gone to great lengths to ensure the quality of this reprint but points out that some imperfections in the original copies may be apparent.

Library of Congress Cataloging-in-Publication Data

ISBN: 978-1-138-02761-9 (hbk)
ISBN: 978-1-03-291740-5 (pbk)
ISBN: 978-1-315-36424-7 (ebk)
ISBN: 978-1-315-70813-3 (eBook+)

Typeset by Integra Software Services Private Ltd

Visit the Taylor & Francis Web site at
http://www.taylorandfrancis.com

and the CRC Press Web site at
http://www.crcpress.com

Contents

Foreword

Although engineering activities involving rock have been underway for millennia, we can mark the beginning of the modern era from the year 1962 when the International Society for Rock Mechanics (ISRM) was formally established in Salzburg, Austria. Since that time, both rock engineering itself and the associated rock mechanics research have increased in activity by leaps and bounds, so much so that it is difficult for an engineer or researcher to be aware of all the emerging developments, especially since the information is widely spread in reports, magazines, journals, books and the internet. It is appropriate, if not essential, therefore that periodically an easily accessible structured survey should be made of the currently available knowledge. Thus, we are most grateful to Professor Xia-Ting Feng and his team, and to the Taylor & Francis Group, for preparing this extensive 2017 "Rock Mechanics and Engineering" compendium outlining the state of the art—and which is a publication fitting well within the Taylor & Francis portfolio of ground engineering related titles.

There has previously only been one similar such survey, "Comprehensive Rock Engineering", which was also published as a five-volume set but by Pergamon Press in 1993. Given the exponential increase in rock engineering related activities and research since that year, we must also congratulate Professor Feng and the publisher on the production of this current five-volume survey. Volumes 1 and 2 are concerned with principles plus laboratory and field testing, i.e., understanding the subject and obtaining the key rock property information. Volume 3 covers analysis, modelling and design, i.e., the procedures by which one can predict the rock behaviour in engineering practice. Then, Volume 4 describes engineering procedures and Volume 5 presents a variety of case examples, both these volumes illustrating 'how things are done'. Hence, the volumes with their constituent chapters run through essentially the complete spectrum of rock mechanics and rock engineering knowledge and associated activities.

In looking through the contents of this compendium, I am particularly pleased that Professor Feng has placed emphasis on the strength of rock, modelling rock failure, field testing and Underground Research Laboratories (URLs), numerical modelling methods—which have revolutionised the approach to rock engineering design—and the progression of excavation, support and monitoring, together with supporting case histories. These subjects, enhanced by the other contributions, are the essence of our subject of rock mechanics and rock engineering. To read through the chapters is not only to understand the subject but also to comprehend the state of current knowledge.

I have worked with Professor Feng on a variety of rock mechanics and rock engineering projects and am delighted to say that his efforts in initiating, developing and seeing through the preparation of this encyclopaedic contribution once again demonstrate his flair for providing significant assistance to the rock mechanics and engineering subject and community. Each of the authors of the contributory chapters is also thanked: they are the virtuosos who have taken time out to write up their expertise within the structured framework of the "Rock Mechanics and Engineering" volumes. There is no doubt that this compendium not only will be of great assistance to all those working in the subject area, whether in research or practice, but it also marks just how far the subject has developed in the 50+ years since 1962 and especially in the 20+ years since the last such survey.

John A. Hudson, Emeritus Professor, Imperial College London, UK
President of the International Society for Rock Mechanics (ISRM) 2007–2011

Introduction

The five-volume book "Comprehensive Rock Engineering" (Editor-in-Chief, Professor John A. Hudson) which was published in 1993 had an important influence on the development of rock mechanics and rock engineering. Indeed the significant and extensive achievements in rock mechanics and engineering during the last 20 years now justify a second compilation. Thus, we are happy to publish 'ROCK MECHANICS AND ENGINEERING', a highly prestigious, multi-volume work, with the editorial advice of Professor John A. Hudson. This new compilation offers an extremely wide-ranging and comprehensive overview of the state-of-the-art in rock mechanics and rock engineering. Intended for an audience of geological, civil, mining and structural engineers, it is composed of reviewed, dedicated contributions by key authors worldwide. The aim has been to make this a leading publication in the field, one which will deserve a place in the library of every engineer involved with rock mechanics and engineering.

We have sought the best contributions from experts in the field to make these five volumes a success, and I really appreciate their hard work and contributions to this project. Also I am extremely grateful to staff at CRC Press / Balkema, Taylor and Francis Group, in particular Mr. Alistair Bright, for his excellent work and kind help. I would like to thank Prof. John A. Hudson for his great help in initiating this publication. I would also thank Dr. Yan Guo for her tireless work on this project.

Editor
Xia-Ting Feng
President of the International Society for Rock Mechanics (ISRM) 2011–2015
July 4, 2016

Numerical Modeling Methods

Coupled THMC modeling for safety assessment of geological disposal of radioactive wastes: The DECOVALEX project (1992–2015)

John A. Hudson[1], Chin-Fu Tsang[2] & Lanru Jing[3]
[1]*Imperial College London, UK*
[2]*Lawrence Berkeley National Laboratory, Berkeley, CA, USA*
[3]*Royal Institute of Technology, Stockholm, Sweden (lanru@kth.se)*

Abstract: This Chapter describes a long-term research effort (1992 to the present) on coupled THMC processes in geological systems in the context of the safe geological disposal of radioactive wastes: the DECOVALEX project (DEvelopment of COupled models and their VAlidation through EXperiments). This project is a unique international co-operative research project which was initiated in 1991, officially started in 1992, has continued through a number of phases without interruption since then and is still continuing in the time of preparing this chapter by the authors. The overall objective of this research has been the development, validation and application of numerical modeling methods and techniques for the performance and safety assessments of geological disposal of radioactive waste (GDRW) in underground repositories. The cooperation has been financed by national waste management organizations, regulatory bodies and national research institutes and individual universities in Canada, China, Czech Republic, European Commission, Finland, France, Germany, Republic of Korea, Spain, Japan, Sweden, UK and USA. Over the period of 23 years, the project has made impressive advanced researches in the field of coupled THMC (Thermo-Hydro-Mechanical-Chemical) processes in geological systems, especially in fractured crystalline and sedimentary rocks and buffer/backfill materials, through integrated numerical modeling and laboratory and field experiments. The experiments cover scales ranging from laboratory-sized samples to in situ experiments in underground research laboratories (URLs) in different host rocks in different countries. The work has resulted in an impressive number of major developments, as reported in scientific publications and helped to educate and train younger generations of researchers in this field. This Chapter presents the goals, structure, contents and approaches of the project, as well as achievements and lessons learned during this long-term project, at both the fundamental and application levels.

I INTRODUCTION

The subject of couplings between the thermal (heat transfer), hydrological (fluid flow) and mechanical (stress, deformation, damage and failure) processes in fractured rocks has become an important subject in rock mechanics and engineering since the early 1980s (Tsang, 1987, 1991), mainly due to the modeling requirements for the design and performance assessment of underground radioactive waste repositories, and other engineering fields in which heat transfer and fluid flow play important roles, such as gas/oil recovery, geothermal energy extraction, contaminant migration control and environment impact evaluation in general. In fact, the coupling can be extended to include geological, chemical and biological processes, but the coupled THM (thermo-hydro-mechanical) and coupled THMC (thermo-hydro-mechanical and chemical) processes are the ones most often required in coupled models in rock engineering. The need to couple the processes in geological systems is a reflection of the fact that the processes affect each other, which occurs mainly in two ways. The first is direct coupling in which one process induces the development of another process, representable by a cross term involving both processes in the governing equations. The second way is indirect coupling in which one process changes the property parameters controlling the progress of another process.

In general, the impact of natural or man-made perturbations on rock masses, such as tectonic events, glaciation cycles, drilling, excavations and injections of fluids or solid particles, on the energy and mass transport in rocks cannot be predicted with adequate reliability by considering each process independently without consideration of couplings among the processes involved.

Underground repositories in either crystalline rocks (mostly granites) or sedimentary rocks (mostly clayey rocks) have been considered in many countries worldwide as a potential solution for disposal of radioactive waste. Figure 1 shows a conceptual illustration of a Geological Disposal of Radioactive Waste (GDRW) repository concept developed by SKB, Sweden, based on the concept of multi-barrier system composed of canisters, bentonite buffer, backfill materials (mostly crushed rocks and bentonite mixtures) and the host rock. The host rock represents the geosphere of the system and is often referred to as the far-field of the repository. The canister, bentonite buffer and backfill, placement deposition hole and the transportation tunnel are often referred to as the near-field. The coupled THM/THMC processes in the above conceptual

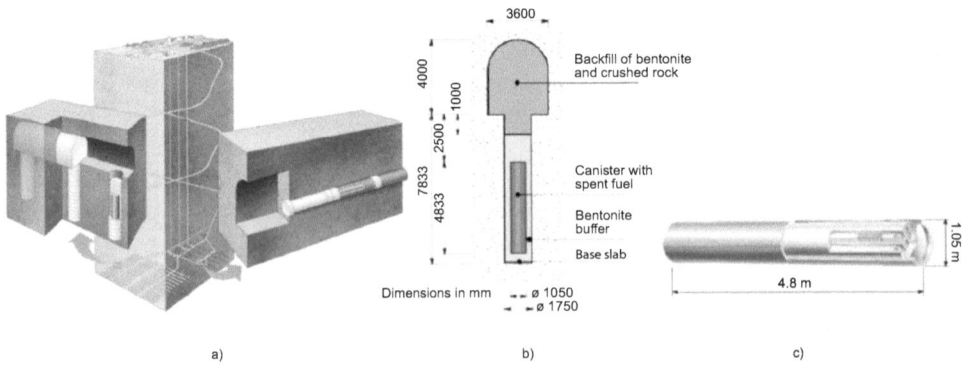

Figure 1 a) The SKB's KBS-3 concept of a Geological Disposal of Radioactive Waste (GDRW) repository in crystalline rocks with alternative vertical or horizontal placement of cast-iron canisters containing the spent fuel surrounded by a bentonite buffer and the host rock; b) the near-field of the repository with canister, bentonite buffer in the placement borehole, and backfill in the transportation tunnels; and c) the canister containing the spent fuel.

GDRW repositories are composed of interactions among mechanisms of energy and mass transport due to: (a) heat generation (radioactive decay by spent fuel) and transfer (by conduction, advection, diffusion and radiation in excavations and materials in the near- and far-fields); (b) flows of geo-fluids (groundwater, gases, steam-vapor and their mixtures) in the porous rock matrix and fracture networks; (c) stress, deformation, damage and failure of all materials in the near- and far-fields; and (d) geochemical processes that affect the generation, evolution and transport of radioactive nuclides in the near- and far-fields. Radionuclide transport (path, time and amount) is the key concern for the safety of the repository concerned.

The coupled THM/THMC processes of geological systems are one of the key issues for the study of GDRW repositories, due to the two most important aspects of any GDRW project.

1. Environmental safety is required for tens of thousands or millions of years, which means that any current laboratory or field experiment lacks the capability for observing and establishing the long-term behavior of the repository system.
2. The multi-scale complexity of the fracture systems, and heterogeneity in general, in rock masses cannot be measured in detail and is difficult to be reliably characterized mathematically in order to support proper analysis of large scale in situ experiments.

These two factors make the predictive mathematical models and associated computational methods the only quantitative means to assess GDRW long-term safety based on fundamental understanding about the interactions among processes, properties and parameters of the whole system over such long time periods before, during and after the repository construction and operation. Owing to the fact that the verification of the reliability of such predictions is possible only for short time behavior of repository design, construction, operation and post-closure monitoring systems. There is a great need that such predictions will be conducted as much as possible carefully and completely without bias because of researchers' background and prior concepts. This is the main motivation for the international co-operative research project DECOVALEX that was initiated in 1991, launched in 1992 and has been continuously extended since then.

Based on a series of initial discussions during 1990–1992, the DECOVALEX project was launched in Stockholm, Sweden, and managed by the Swedish Nuclear Power Inspectorate (SKI). It was originally planned for the period of 1992–1995. The overall objective of the DECOVALEX project was "to increase the understanding of various thermohydromechanical processes of importance for radionuclide release and transport from a repository to the biosphere and how these could be described by mathematical models". The modeling and testing of the coupled THM processes in geological media, which were commonly used in safety assessment of GDRW repositories at that time, was not a mature field of science and technology. The practical objectives were focused on increasing basic understanding of the coupled THM processes, advancing modeling capacities and tools for simulating THM processes in fractured hard rocks with validations against well controlled, small scale laboratory experiments, especially on rock fracture behavior, and exchanges of test data among the funding organizations.

The achievements and lessons learnt through this first DECOVALEX project led the funding organizations to decide on an extension to a second phase named

DECOVALEX II (with the original DECOVALEX project called DECOVALEX I), for the period 1996–1999. The success of the work continued, leading to the subsequent phases of DECOVALEX III (2000–2003), DECOVALEX–THMC (2004–2007), DECOVALEX–2011 (2008–2011) and DECOVALEX–2015 (2012–2015) at the time of writing this Chapter. Such continued research is a rare phenomenon in international research co-operations in the field of radioactive waste management or in geosciences in general. The main reasons for such a long-term research co-operation are:

1. the main research topics are important for performance and safety assessments of GDRW repositories;
2. the project takes an integrated research approach combining numerical modeling, small scale laboratory tests and larger scale in situ experiments in different host rocks to conduct the verification, validation and reliability assessments of the application of numerical methods and techniques in support of long-term predictive modeling required by the safety assessments;
3. the research topics of THM and THMC processes in rock fractures and continuum geological porous media over periods of 3–4 years per project phase have been suitable for educating Ph.D. students. Over 30 doctoral students have conducted research on the DECOVALEX project's problems and successfully obtained their Ph.D. degrees—many of whom continued on to work in the field of radioactive waste management or similar fields concerning energy and mass transport;
4. research results have been widely published in the form of journal and conference papers, technical reports, Ph.D. theses, and edited volumes; and finally and most importantly,
5. DECOVALEX project provides an international platform for supporting the funding organizations' own R&D programs in reasonable time intervals, from multidisciplinary points of view, and from different countries with different backgrounds and emphasis.

2 PROJECT ORGANIZATION AND RESEARCH MANAGEMENT

Figure 2 illustrates the organization of the DECOVALEX I project, which started with DECOVALEX I and has continued, with only minor changes, up to the present

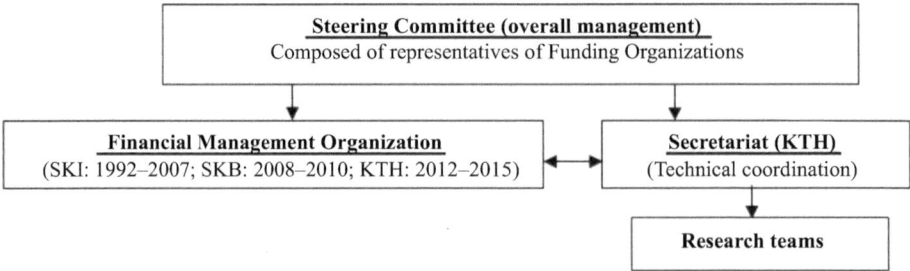

Figure 2 Organization structure of the DECOVALEX Project.

day. The Steering Committee is composed of one representative from each funding organization, and has the overall role of deciding project aims, content, tasks, duration, reporting/publications and the finances. The financial managing organization is responsible for managing the project economy, mainly issues of budget and expenditures. The Secretariat is responsible for administrative and technical management of the project, including arrangements of meetings, task coordination, communications, publications, reporting, and archives. The full names and the acronyms of the funding organizations and research teams from 1992 to the present day are listed in Tables 1 and 2. (Note that the participants in the planned 2016–2019 DECOVALEX phase are not included here because the arrangements for this phase are in the process of being developed as this Chapter is being written in 2015—but can be found at www.DECOVALEX.org). See Table 1 for the acronyms of the funding organizations, where letter 'D' indicates 'DECOVALEX', and research teams in Table 2.

For each DECOVALEX phase, the funding organizations propose tasks that are of importance to their respective R&D program. The final selection of tasks to be studied is decided by the Steering Committee through discussion and voting. The finally selected proposals are then further developed, including definition of objectives, coupled THM or THMC processes involved, input data (initial and boundary conditions, material properties), output specification requirements and time schedule. Usually, one problem is studied by several research teams sponsored by different funding organizations from different countries, with different modeling approaches, tools and different understanding of the physics involved—which is the key advantage that motivated continuation of the DECOVALEX project phases. Cycles of blind predictions followed by model calibrations with measured data, if available, have been the main approach to evaluate scientific quality and reliability of the results.

The selected proposals are classified as a Test Case (TC) or a Benchmark-Test (BMT) problem. The BMTs are mostly problems without (or with limited) data support from laboratory or in situ experiments. However, they can identify the need to establish more comprehensive conceptual understanding or key relevant parameters on coupled THM or THMC processes in porous fractured or continuum media in specific geological conditions, *e.g.* issues of complex fracture network geometry, upscaling and homogenization, special effects of certain physical processes such as heat advection by groundwater in fractures, impacts of glaciation and deglaciation cycles over long periods of time, etc. The TCs are defined based on lab or in situ experiments with measured data for testing, verifying, validating and supporting the development of constitutive models and computer codes applied. Often this helps to establish their shortcomings for further development to improve their capability for predictive modeling.

Both categories, BTMs and TCs, are necessary for planning, developing, validating and confidence building of numerical modeling techniques and are useful not only for GDRW repositories, but also for all fields involving energy and mass transport in geological systems. We use the term 'verifying' to relate to the process of comparing the computer modeling output of different teams, and 'validating' to compare the modeling output with measured lab and/or in situ data.

Table 1 Funding organizations for the DECOVALEX project (1992–2015).

Funding Organization	Acronym	Country	Phases participated
National Agency for Radioactive Waste Management	ANDRA	France	DI, DII, DIII
Atomic Energy Commission	CEA	France	DI, DIII
Institute for Nuclear Protection and Safety	IRSN/IPSN	France	DI–D2015
Atomic Energy of Canada Ltd.	AECL	Canada	DI
Ontario Hydro Co.	OH	Canada	DII
Ontario Power Generation Co.	OPG	Canada	DIII
Nuclear Waste Management Organization (former OPG)	NWMO	Canada	D-THMC
Atomic Energy Control Board/ Canadian Nuclear Safety Commission	AECB/CNSC	Canada	DI, DII, DIII, D-THMC
Radioactive Waste Repository Authority	RAWRA	Czech	D-2011, D-2015
Chinese Academy of Sciences, Institute of Rock and Soil Mechanics	CAS	China	D-THMC, D-2011, D-2015
Wuhan University	WHU	China	D-2011
European Commission (through the BENCHPAR project)	EU		DIII
Federal Institute for Geoscience and Natural Resources	BGR	Germany	DIII, D-THMC, D-2015
Helmholtz Centre for Environmental Research	UFZ	Germany	D-2015
Power Reactor and Nuclear Fuel Development Co.	PNC	Japan	DI
Japan Nuclear Cycle Development Institute (former PNC)	JNC	Japan	DII, DIII,
Japan Atomic Energy Agency (former JNC)	JAEA	Japan	D-THMC, D-2011, D-2015
Korean Atomic Energy Research Institute	KAERI	Korea (Republic of)	D-2011
Empresa Nacional de Residuos Radioactivos, S. A.	ENRESA	Spain	DII, DIII
United Kingdom Nirex Ltd.	NIREX	UK	DI, DII, DIII
Nuclear Decommissioning Authority	NDA	UK	D-2011, D-2015
Radioactive Waste Management	RWM	UK	D-2015
Environmental Agency	EA	UK	DII
Nuclear Regulatory Commission	NRC	USA	DI, DIII, D-2015
Department of Energy	DOE	USA	DIII, D-THMC, D-2015
Swedish Nuclear Fuel and Waste Management Co., Sweden	SKB	Sweden	DI, DII, DIII, D-THMC, D-2011
Swedish Nuclear Power Inspectorate, Sweden	SKI	Sweden	DI, DII, DIII, D-THMC
Centre for Radiation and Nuclear Safety	STUK	Finland	DI, DII, DIII, D-THMC
Posiva Oy	POSIVA	Finland	D-2011
Swiss Federal Nuclear Safety Inspectorate	ENSI	Switzerland	D-2015

Table 2 Research teams and their sponsors.

Research team	Acronym	Funded by
Institute of Rock and Soil Mechanics, Chinese Academy of Sciences	CAS	CAS
INERIS-LAEGO, Ecole des Mines de Nancy, France	INERIS	CEA, ANDRA
Ecole Polytechnique, G3S, France	ANG	ANDRA
Ecole Nationale Supérieure des Mines de Paris, France	ENSMP	CEA, IPSN
CEA/DM25/SEMT, France	CEA	IPSN
AECL, Canada	AECL	AECL, OH, OPG, NWMO
Atomic Energy Control Board(AECB), Canada	AECB	AECB
Kyoto University, Japan,	KPH, JNC, KIPH, JAEA	PNC, JNC, JAEA
Hazama Corporation, Japan	KPH, JNC, KIPH, JAEA	PNC, JNC, JAEA
Tokai Works, PNC(JNC, JAEA), Japan	KPH, JNC, KIPH, JAEA	PNC, JNC, JAEA
Iwate University, Japan	KIPH	JNC
Centre for Nuclear Waste Regulatory Analysis, SWRI, USA	CNWRA	NRC
Itasca Geomechanics AB, Sweden	ITASCA, ITAs	SKB
ITASCA Consultants s.a., France	ITAf	ANDRA
Lund University of Technology, Sweden	LTH	SKB
Clay Technology AB, Sweden	CLAY	SKB
Chalmers University of Technology, Sweden	CTH	SKB
Lawrence Berkeley National Laboratory, USA	DOE, SKI, NDA	SKI, DOE
Royal Institute of Technology, Sweden	KTH	SKI, SKB
Chalmers University of Technology, Sweden	CTH	SKB
VTT Technical Research Centre of Finland	VTT	STUK
Uppsala University, Sweden	UU	STUK
Helsinki University of Technology, Finland	HUT	STUK
University of Edinburgh, UK	EU, EoU	NDA, RWM
University of Newcastle, UK	UNEW	EA
University of Birmingham, UK	UoB	EA
AEA Technology, UK	AEA	NIREX
Norwegian Geotechnical Institute, Norway	NGI	NIREX
Atomic Energy Control Board, Canada	AECB	AECB
Canadian Nuclear Safety Commission, Canada	CNS, CNSC	CNSC
Universitat Politècnica de València, Spain	UPV	ENRESA
Universitat Politècnica de Catalunya, Spain	UPC	ENRESA
University of Tübingen, Germany	BGR	BGR
Institute of Fluid Mechanics and Computer Applications in Civil Engineering, University of Hanover, Germany	ISEB-BGR	BGR
Federal Institute for Geosciences and Natural Resources, Germany	BGR	BGR
Sandia National Laboratory, USA	DOE, Sandia	DOE
Institute of Geonics AS CR, Czech Republic	UGN	RAWRA(SURAO)

Table 2 (Cont.)

Research team	Acronym	Funded by
Technical University of Liberec, Czech Republic	TUL	RAWRA (SURAO)
Korea Atomic Energy Research Institute, Republic of Korea	KAERI	KAERI
Inha University, Republic of Korea	Inha	KAERI
University of Alberta, Canada	UoA, SKB	SKB
Imperial College London, UK	ICL	NDA
Serco Ltd., UK	ICL	NDA
Helmholtz Centre for Environmental Research, Germany	BGR/UFZ	BGR
Quintessa Ltd., UK	Quintessa	NDA(RWM)
Swiss Federal Nuclear Safety Inspectorate, Switzerland	ENSI	ENSI

3 PROBLEMS, SCIENTIFIC ISSUES AND OUTCOMES OF THE DECOVALEX PHASES

Over the 23 years of the DECOVALEX project phases, a large number of BMTs and TCs have been studied. A summary review for the phases during 1992–2007, from DECOVALEX I to DECOVALEX-THMC, was presented in Tsang *et al.* (2009) with the details and links to publications; so only a brief description is presented here of the research topics, processes concerned, publications and outstanding issues for these phases. A slightly more detailed summary of the problems defined, physics involved, experiments, modeling tools and results, is included for the latter two phases during 2008–2015, *i.e.*, DECOVALEX-2011 and DECOVALEX-2015. Details can be found in the publications listed in the references at the end of this Chapter.

3.1 DECOVALEX I (1992–1995)

The tasks of the DECOVALEX I project studied three BMTs (BMT1–BMT3) and six TCs (TC1–TC6).

BMT1 was defined for a repository located at 500 m depth in a generic far-field 2D model of fractured crystalline rock of two orthogonal sets of parallel fractures (Durin *et al.*, 1995).

These fractures are assigned three alternative fracture spacing (100 m, 50 m and 25 m) with each fracture following the cubic law governing water flow. The main objective was to evaluate the performance of the computational methods and computer codes and to evaluate the state-of-the-arts of modeling methods and codes for simulating coupled THM processes and understanding their impact on repository performance.

BMT2 was defined as a near-field 2D model of fractured rock of size 0.75 m of horizontal length and 0.5 m in vertical height, containing four orthogonal fractures that allow water flow under heating. An initial temperature $t_0 = 15°C$ and initial horizontal and vertical stress components $= -4$ MPa were specified, with the factures having a constant initial aperture of 300 μm. The aim was to investigate the interactive behavior of the fractures and intact rock matrix during fully coupled THM processes, especially the forced heat convection along the fractures (Tin *et al.*, 1995).

BMT3 was a model of the near-field of a hypothetical repository set up in a 2D space, with a size of 50 m in height and 50 m in width, located 500 m below the ground surface. The model contains 6580 fractures that were created from a 2D simplification of one of the 3D realizations generated based on site mapping data from the Stripa Mine, Sweden. The aim of the BMT3 problem was to evaluate numerical methods and approaches, material characterization and computational capacities for coupled THM processes of fractured rocks with a large number of fractures—for a near-field study.

For the three BMTs, besides the achieved aims for comparing performance and relative verification of computer codes and models for such near- and far-field problems with explicit representation of fractures, the main scientific findings were: (a) the significant difference caused by whether or not the thermal convection by water flow in the fractures; (b) the importance of thermal expansion inside the host rock far from the repository or near ground surface; (c) the dominating fracture transmissivity under confined thermo-mechanical conditions during heating and thermal expansion processes; and (d) critically inadequate knowledge on the importance of the representative mechanical behavior generated by discrete and continuum models for the fractured rocks. This indicated the need for further study on the fundamental aspects of numerical modeling using different material characterizations of fractured rocks. An understanding of these issues was also important for modeling mass transport in fractured rocks, and served as the main motivation for the extension of the project to the DECOVALEX II, III and D-2011 phases, based on conceptual understanding developed from BMT3 with alternative model geometry and material behavior/properties.

TC1, TC4, TC5 and TC6 were lab tests of single rock fractures under HM or THM loading conditions, aimed at understanding the constitutive behavior and models for THM processes in rock fractures. They play a dominating role on energy and mass transfer in fractured rock masses, especially the hard rocks like granite. However, due to time limitations and unexpected challenges in the laboratory testing process, TC4 and TC5 could not be studied during the DECOVALEX I project duration.

TC1 was based on a laboratory experiment conducted at NGI for coupled stress-flow processes in a rock fracture. The fracture sample was created from a large diameter core containing a single fracture and was subjected to multiple-step loading-unloading cycles while water was conducted through the fracture. The normal and shear stresses and displacements, as well as water flowrate, were measured during the test.

The results demonstrated that numerical models did not have adequate ability to enable proper representation of the stress-path dependence which led to inadequate agreement between the measured and calculated hydraulic aperture, and the measured and calculated shear dilatancy. These findings caused significant attention in the project so that coupled THM processes of single rock fractures became a continued issue of research and discussion in the subsequent DECOVALEX project phases, see Makurat *et al.* (1995) for more details.

TC6 involved numerical modeling of an in situ field injection experiment at a depth of 356–7 m on a sub-horizontal fracture isolated by pressurized packers in a borehole of 56 mm in diameter at Luleå University of Technology, Sweden. A series of pressure pulse tests, step pressure tests (hydraulic jacking) and constant pressure injection tests were conducted. These tests were simulated in the TC6 work, with the objectives mainly related to the calibration of numerical models against the measured field data, with targeted calibration variables being (a) Young's modulus of the host rock, (b) initial hydraulic aperture of the tested fracture, and (c) initial and boundary (stress) conditions. The modeling results were found to agree well in general with the measured data, and demonstrated the importance of coupled hydro-mechanical coupling behavior of rock fractures under high fluid pressure. The key parameters identified were hydraulic aperture, normal stiffness and effective normal stress of the facture—an important indicator of the important role of fractures at the field scale (Claesson *et al.*, 1995; Rutqvist, 1995).

TC2 was developed to simulate an intermediate-scale in situ heating test on a host rock at the Fanay-Augères test site, France, considering coupled thermo-mechanical processes under an unsaturated in situ condition and low permeability of the fractures at the test site. The objective of TC2 was to partially validate the abilities of numerical modeling of 3D fractured rocks under combined heating and mechanical loading/unloading conditions. It was found that both modeling approaches, of equivalent continuum (FEM model) and discrete system (DEM model), were able to model the coupled TM processes in a fractured rock mass. The main challenges were to develop an adequate 3D model setup, and obtain reliable constitutive models of the fractures. However, despite these challenges, acceptable results were obtained for the heat transfer.

TC3 was defined based on a large-scale laboratory experiment concerning the coupled THM processes of the engineered barrier system for GDRW, using the BIG-BEN (Big-Bentonite) facility at Tokai Works, PNC, Japan. The aim of TC3 was to test and compare the capabilities of different numerical models and computer codes for simulating one of the test pits. Results showed that predicting the stress field was more challenging than predicting fluid flow and heat transport processes. This finding indicated the importance of further fundamental study on the swelling pressure phenomenon of the bentonite, which became a key issue of research for the following DECOVALEX phases of DII, DIII and D-THMC.

In summary, the TCs and BMTs served well for establishing current state-of-the-capabilities of the existing computer codes and numerical modeling approaches available at the time. They also highlighted outstanding issues of importance for the national R&D program and the need for continued development and verification/validation (Jing *et al.*, 1995; Stephansson *et al.*, 1996). Thus, DECOVALEX I was extended to DECOVALEX II.

3.2 DECOVALEX II (1996–1999)

The DECOVALEX II project was initiated by the recognition of the need for a proper evaluation of the current capabilities of the numerical modeling of coupled THM processes when applied to fractured geological media that were less characterized, more complex and realistic as commonly found in in situ experiments at larger scales. This was in contrast to the small scale, well characterized and relatively much simpler BMTs for fractured rock masses or laboratory experiments on single fractures investigated during the DECOVALEX I project. Thus the overall objective of DECOVALEX II was to increase the understanding of various coupled THM processes of importance for radionuclide release and transport from a repository to the biosphere and how they could be described by mathematical models, with focus on large scale in situ experiments for more complex and hence more realistic test conditions. An additional objective was to evaluate how studies conducted in the DECOVALEX I and DECOVALEX II projects could be used for design and performance assessment of radioactive waste repositories.

Four tasks were defined for the DECOVALEX II project:

- Task 1: numerical study of the then-planned Nirex's RCF Shaft excavation at Sellafield, UK, with numerical simulations of the coupled hydro-mechanical processes in the RCF3 pumping test at the site and the responses of the rock mass to the shaft excavation, including study of the excavation disturbed zone (EDZ);
- Task 2: numerical study of JNC's in situ THM experiment at the Kamaishi Mine, Japan, which was an integrated investigation on a complete rock-buffer-heater system under in situ conditions over an extended period of heating-cooling time;
- Task 3: Review of the state-of-the-art of the constitutive relations of rock joints; and
- Task 4: A report on the current understanding of the coupled THM processes related to design and performance assessment of radioactive waste repositories. The management style was the same as that for the DECOVALEX I project.

Task 1 consisted of three sub-phases: Task 1A: blind prediction of the flowrate at the source zone located in borehole RCF 3 from which a pump test had been carried out by Nirex, and the pressure responses in 18 monitoring zones in nearby boreholes, considering the effects of coupled hydro-mechanical processes on the results; Task 1B: calibration of the numerical model for Task 1A against measured data; and Task 1C: prediction of coupled hydro-mechanical responses in a specified sector of the host rock (Borrowdale Volcanic Group, BVG) at the Sellafield site due to a planned shaft excavation, with EDZ effects as one of the main research issues. Task 1 was successfully carried out by a diverse combination of different modeling approaches, numerical methods and computer codes, and material characterization schemes—which was possible due to large amount of site characterization data supplied by NIREX, which enabled more reliable numerical modeling, with reasonable results of pressure field and the general trend of the flowrate data.

However, significant uncertainties still existed, which were related to flow properties, characterization of field conditions, and reliability of model conceptualization concerning fracture systems. The blind predictions performed for Task 1C helped in enhancing understanding of the effects of the coupled stress–flow processes of the BVG caused by shaft sinking, especially on the EDZ phenomenon at a time when the EDZ issue was just started to be considered more widely and comprehensively in the international community of GDRW. The main outstanding issue of Task 1 of the DECOVAEX II project was the need for more in-depth research on characterization of the site geology and the physical and chemical processes involved—as the basis for more successful and reliable numerical modeling of coupled THM or THMC processes, especially for fractured crystalline rocks. Details of the research results can be found in (Knight *et al.*, 2001; Gómez-Hernández *et al.*, 2001; Kobayashi *et al.*, 2001; Hakami *et al.*, 2001; Rajeb & Bruel, 2001).

The aim of Task 2 of the DECOVALEX II project was to study the multiple barrier system in the near-field of a repository with special focus on the bentonite behavior, constitutive models and parameters. This was accomplished by modeling a comprehensive in situ THM experiment at the abandoned Kamaishi Mine, Japan, where PNC (later renamed as JNC) ran an underground laboratory. Considering the complexity of the in situ experiment, Task 2 was divided into three sub-tasks: Task 2A, Task 2B and Task 2C. Numerical modelings were conducted in the subtasks, through cycles of blind prediction–model calibration in steps with the progress of the experiments and measured data availability.

– Task 2A: pure blind predictive modeling of the effect of the test pit excavation with four steps: calculation of fracture distribution on the wall of the test pit from surface fracture mapping, mechanical effects of the test pit excavation, water inflow into the test pit and pore pressure changes in the rocks due to excavation of the test pit. One of the aims of Task 2A to was to use the modeling results to assist in the in situ THM experimental design, such as locations of monitoring sensors, measurement techniques and field mapping of the fractures in the near-field of the test pit, as part of a prediction–design–testing–improvement cycle conducted by PNC.

– Task 2B: model calibration against measured data during the excavation of the test pit, with the blind prediction results obtained during Task 2A as a starting point.

– Task 2C: predictive modeling of the T-H-M behavior of both rock and buffer caused by the buffer emplacement and heating experiment. Four laboratory tests were carried out during the project for improving the material characterization of the bentonite. They were suction test, water infiltration test, thermal gradient test and swelling pressure test—which were modeled by the teams in steps of blind predictions and calibrations using measured data.

Task 2C was a major accomplishment in terms of testing and modeling the near-field multi-barrier system behavior of GDRW repositories not only through continuous cycles of blind-prediction-followed-by-model-calibration practice in the DECOVALEX project, but also as an integrated modeling and experiment design and implementation with DECOVALEX researchers working closely with the PNC scientists. The specially designed and executed laboratory tests on the bentonite during Task 2C proved to be important for improving material characterization as identified during Task 2B. This demonstrated the advantage of integrated testing and modeling approaches. The main open issues that emerged were the less accurate results (both measurements and modeling) for the mechanical behavior of the host rock near the test pit, due mainly to: (a) the largely unknown fracture system inside the rock around the test pit locally and at the site globally; (b) unclear initial and boundary hydraulic conditions of the site model surrounding the test area; and (c) EDZ effects on the surrounding rocks by the test pit excavation. Detailed results can be found in (Chijimatsu *et al.*, 2001; Nguyen *et al.*, 2001; Börgesson *et al.*, 2001; Rutqvist *et al.*, 2001a, 2001b).

Based on the lessons learnt from DECOVALEX I regarding the importance of the fundamental behavior of rock fractures and the importance of coupled THM processes for the performance assessment of GDRW repositories, a forum of lectures on rock fractures and preparation of a comprehensive review on the state-of-the-art of coupled THM processes in the international GDRW community was implemented. These two activities were defined as Task 3 and Task 4, respectively and proved to be very effective for promoting research at both the fundamental science level and for under-standing the applicability of research results for GDRW management practices world-wide. The successful modeling results. The impressive outcome of the fracture forum and THM process review motivated the extension of the project as a useful research and communication platform, thus leading to DECOVALEX III.

3.3 DECOVALEX III (2000–2003)

The DECOVALEX III project was initiated with two main objectives. The first was the further verification/validation of computer codes, either self-developed or commercial codes, used by the research teams. Two large-scale in situ experiments were simulated: the FEBEX T H M experiment in the URL at Grimsel, Switzerland, designated as Task 1; and the drift scale heater test at Yucca Mountain URL, Nevada, USA, designated as Task 2. The second objective was to determine the relevance of THM processes on the safety of a repository by numerical modeling of three BMTs to examine the relevance of THM processes in the context of performance and safety assessments. Task 3 was defined to achieve this objective by using three BMTs of a generic nature but supported by realistic material properties and a proper model setup. An additional Task 4 was also organized to present the state of the current understanding of the impacts and treatments of the THM issues on the PA (Performance Assessment) and SA (Safety Assessment) of nuclear waste repositories, from the views of the Funding Organizations of the DECOVALEX project, through compilation of answers to a questionnaire prepared for this purpose (Stephansson et al., 2004; Tsang et al., 2005).

Task 1: The international FEBEX (Full-Scale Engineered Barriers Experiment in Crystalline Host Rock) project was in operation from 1994 to 2003, and aimed at the study of the various processes occurring in the near-field of a high activity radio-active waste storage in crystalline rock. The experiment was installed at the Grimsel Test Site with a multiple Engineered Barrier System (EBS), supported by detailed information on geological and hydrogeological characterization of the Grimsel Test Site, comprehensive characterization of the bentonite used to fabricate the engineered barrier, and the monitoring undertaken during the drilling of the FEBEX tunnel, as well as during the test.

The modeling work of Task 1 was divided into three parts: Part A – hydro-mechan-ical modeling of the Grimsel granite host rock; Part B – thermo-hydro-mechanical modeling of the bentonite behavior; and Part C – thermo-hydro-mechanical modeling of the Grimsel granite rock. A cyclical blind numerical prediction and model calibra-tion with measured data approach was executed through the modeling process for the three parts.

The FEBEX test was one of the few large-scale tests available two decades ago enabling an integrated perspective of the behavior of the then current concepts for nuclear waste disposal in crystalline rock. The test provided valuable measurement data support for verifying, validating and enabling further appropriate developments of the capabilities of a number of numerical codes used in the DECOVALEX III for project teams to handle coupled problems in geological porous media. The main scientific findings and developments from the integrated cycles of blind predictions were: model calibrations and code developments in the subject areas of: water inflow into the excavated tunnel with a calibrated hydrogeological model; specific idealization of the rock mass (equivalent porous media, discrete fractures); development and dissipation of excess pore water pressures in the vicinity of the advancing tunnel (at the time of the FEBEX tunnel excavation); predicting the behavior of the buffer under combined heating and wetting actions; identification of the necessary physical processes controlling the bentonite behavior; influence of the host rock on hydration of the bentonite buffer; and the significance of rock stress alteration in the vicinity of the FEBEX tunnel by heating.

The entire project was a learning process experienced by the involved participants—despite the fact that a true blind prediction was difficult to make, even for a highly controlled and documented experiment such as the FEBEX in situ test. The modeling could be improved to be more accurate and reliable through gradual comparison with the real field measurements made available to modelers. Identification of the relevant couplings was a key issue in the modeling for Task 1, such as full HM coupling, 3D initial stress field, and excavation process for Part A; phase change of water and vapor flow, saturation dependence of water permeability and thermal conductivity and suction-induced deformation for Part B; and thermal dilation of water and skeleton during the HM coupling process for Part C. A number of outstanding issues were identified, such as the long-term behavior of the system, applicability of modeling results and knowledge to the real scale for a real repository with much more complexities in terms of the repository design and processes involved, and reliable characterization of the mechanical response of the buffer and the blocky nature of the buffer blocks used in the FEBEX test. The achievements and outstanding issues were reported in (Alonso et al., 2005; Nguyen et al., 2005).

Task 2: The Drift Scale Test (DST) at Yucca Mountain in Nevada, USA was a large scale, long-term, field thermal test being conducted for the United States Department of Energy (DOE). The heating phase of the DST was initiated in December 1997 and terminated in January 2002, which ushered in the cooling phase of the test, continuing for four further years. The overarching objective of the DST was to study coupled thermal, hydrological, mechanical, and chemical processes caused by the decay heat from high level radioactive wastes (HLW) and spent nuclear fuel (SNF) emplaced in an underground geological repository.

Task 2 was divided into four sub-tasks: Task 2A, Task 2B, Task 2C and Task 2D, with close links among them. Task 2A was to mathematically simulate and study the thermal-hydrological (TH) responses of the rock mass in the DST. The modeling objectives included numerical predicting of the time-evolution of temperature distribution and saturation changes in the test block at suitable time intervals, with gradual developments of blind prediction and model calibration with measured data support, and further modeling and results comparison. Tasks 2B and 2C dealt with the mechanical processes in the DST during the heating and cooling phases of the test. The difference between Task 2B and 2C was that, in 2B, the mechanical effects were simulated using modeled temperature distributions as inputs; while, in 2C, the mechanical effects were calculated using measured temperature distributions as inputs. The provided data were basically the same as that for Task 2A, but the objectives were predicting the time-evolution of the displacements in the test block measured in the MPBX holes and changes in the (fracture) permeability of the rock due to thermal-mechanical processes at suitable time intervals. The general objective of Task 2D was to develop a thorough understanding of the coupled TMHC processes in the rock mass immediately surrounding the proposed repository because of the decay heat from the nuclear waste. Task 2D included a series of specific studies involving the four principal thermal, hydraulic, mechanical and chemical processes regarding measurements, evaluations, investigations, monitoring, property evaluations, observations, data collections and analyzes.

The main achievements were made by developments of various conceptual models verified by comparing the simulated and measured temperatures by applying different numerical methods and material models for the fractured porous tuff rock, regarding material properties and parameters, problem dimensions, alternative process coupling combinations, and their impacts on performance assessments. The major achievements and outstanding issues were published in (Hsiung et al., 2005; Olivella & Gens, 2005; Rutqvist et al., 2005a; Sonnenthal et al., 2005).

Task 3: Task 3 included 3 BMT problems and was proposed to examine the relevance of THM processes in the context of performance and safety assessments for the geological disposal of radioactive waste repositories regarding: the impact of THM processes in the near-field of a hypothetical repository in fractured hard rock; homogenization and upscaling of hydro-mechanical properties of fractured rocks and their impact on far-field performance and safety assessments; and impact of glaciation process on far-field performance and safety assessments.

BMT1 of Task 3 was aimed at using scoping calculations to estimate the influence of THM processes on the flow pattern, as well as the structural integrity, of the geological and engineered barriers in the near-field of a typical repository. The technical definition was based on a conceptual design of a hypothetical nuclear waste repository in a granitic rock used for the Kamaishi THM experiment in Japan, and measured rock mass properties based on typical Canadian Shield granite data. The components of the research included temperature evolution in the near-field, buffer re-saturation time, stress states in buffer and canisters, permeability and water flow in the near-field, rock failure, and uncertainties and rock property variability. Three steps were used to enable progressive research in order to: a) Step 1: identify the coupling mechanisms of importance for construction, performance and safety of the repository through scoping calculations considering varying degrees of THM coupling; b) Step 2: to simplify the calculation process and focus on the physics of the problem instead of computational efforts; and c) Step 3: to perform scoping calculations with different coupling combinations based on the same conceptual repository design but with added fractures in order to evaluate the impact of fracture flow on the variables of importance for Performance Assessment. Another model of a highly fractured rock mass with two orthogonal sets of fractures was also studied by one team. It was found that the following results had a strong influence on the long term safety and performance of the repository: the maximal temperature created by the thermal loading from the emplaced wastes; the time for re-saturation of the buffer; the maximal swelling stress developed in the buffer; the structural integrity of the rock mass; the permeability evolution in the rock mass. The results can be found in (Chijimatsu et al., 2005; Millard et al., 2005; Rutqvist et al., 2005b).

The BMT 2 of Task 3 was concerned with the impact of upscaling of the THM processes in a fractured rock mass and its significance for large-scale repository performance assessment of repositories in fractured host rocks. A key issue of the work was to explore the extent to which fractured rock properties can be up-scaled

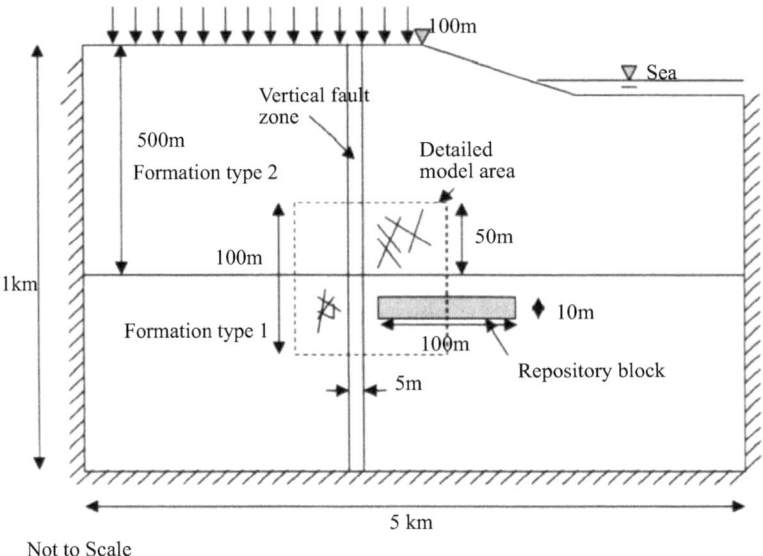

Figure 3 Conceptual model geometry and boundary conditions (not to scale) for the reference problem of the BMT2 for Task 3, DECOVALEX III project.

and represented by 'equivalent' continuum properties for appropriate PA calculations. The aims of the BMT2 were to understand the impact of upscaling of coupled processes on the approach to performance assessment modeling—and the analysis of the model results, together with the impact on uncertainty analysis. The problem was established as a numerical study of a realistic large-scale reference problem which was concerned with a hypothetical heat producing repository located at depth in a hypothetical fractured rock (Fig.3). However, in order to obtain realistically complex and spatially varying geological data, the hydrogeological and hydromechanical input data were loosely based on the basement geology at the Sellafield site, UK, and on the site characterization data acquired by the then Nirex UK. Although the geometry analyzed was purely fictional and did not represent the actual conditions at the Sellafield site. The main research undertaken concerned with the far-field groundwater flow and solute transport for a situation where a heat producing repository was to be located in a fractured rock mass. A vertical fracture zone cuts both rock units but lies beyond the end of the repository tunnel. The main findings have been published in (Cassiraga *et al.*, 2005; Min *et al.*, 2005; Blum *et al.*, 2005; Öhman *et al.*, 2005).

The BMT3 of Task 3 was also a generic numerical exercise developed from the geological and hydrogeological characteristics of a Northern Hemisphere site subjected to a prescribed time sequence of climatically driven glaciation/deglaciation events. A generic spent-fuel repository was assumed to be located in a crystalline rock mass, which consists of low-permeability, low-porosity, sparsely fractured, intact rock matrix (with 10-19 m^2 or lower permeability), traversed by an interconnected network of fracture zones. These conditions were representative of those that would be encountered in a geological Shield setting. The objectives were:

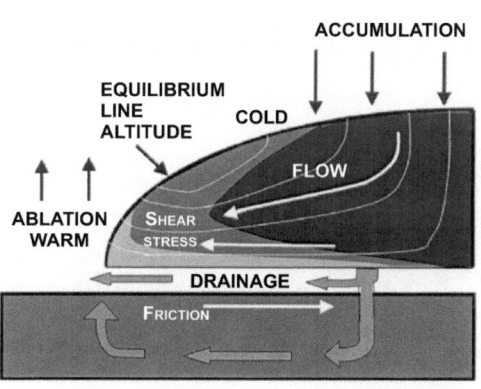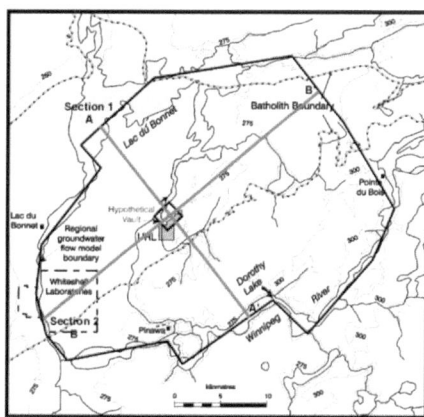

Figure 4 a) HM processes associated with a glaciation/deglaciation processes; b) Simplified map of the Whiteshell Research Area (WRA), Canada, for the model setup.

– to study, by analytical and/or numerical modeling the long-term evolution of a fractured rock mass in which a generic repository is located, as it undergoes a glaciation/deglaciation cycle in a time frame of 100,000 years (Fig. 4a);
– to assess the impact of the coupled hydromechanical responses of the repository system to a glaciation/deglaciation cycle on its long-term performance in waste isolation; and,
– to improve the scientific basis for supporting the Safety Case for a deep geological repository.

Simplified data from a specific Canadian Shield research area were utilized to make the simulations realistic. Site attributes were largely based on those of the Whiteshell Research Area (WRA) in eastern Manitoba, Canada, at 255–290 masl (meters above sea level). (Fig. 4b)

The numerical study was undertaken in three phases: Phase I aimed at enhancing numerical tools for simulations of the climate driven, ice-sheet loading and basal thermal and hydrological regime, including permafrost phenomenon; in Phase II, the modeling covered the 3D ice-sheet/drainage process in order to provide boundary conditions for 2D and 3D modeling, and the effects of groundwater salinity on the development of permafrost and perennially frozen ground; in Phase III, 3D ice-sheet/drainage simulations were conducted to generate spatially and temporally variable 2D mechanical and hydraulic boundary conditions to be used directly by other teams for coupled HM site-scale rock mass modeling in 2D and 3D, including studies on ice-sheet/drainage and permafrost models thermodynamically at the ice-bed interfaces.

The transient, coupled, hydro-mechanical modeling in this study represented a major step forward in advancing the state of the science for modeling the geosphere response to glaciation. Although models of glacier–groundwater, glacier–permafrost–groundwater and glacier–groundwater–shallow failure systems had been presented previously, this BMT was one of the first attempts to assess impacts at repository depths using site-specific (though simplified) data, and the study probably represented the first successful attempt at scaling down an ice-sheet/drainage model with 10 km resolution to a 200 m resolution to interface with site-scale subsurface modeling. The results provide valuable insights into the magnitude and rate of change of site-specific hydro-geological and geomechanical responses to external, transient climate forcing. They clearly demonstrate the importance of glaciation scenarios in performance assessments and the reality of effects that result from H-M coupling, and underline the need for transient analyses of these coupled phenomena. Although no enhancements to the computer codes were undertaken within the BMT3 study, it was no small achievement to demonstrate for the first time a capability to simulate the coupled H-M rock mass response to glaciation using realistic time-dependent glacial boundary conditions, along with reasonably realistic site-specific hydraulic and mechanical attributes. A number of outstanding issues were also identified regarding uncertainties in problem setup, mechanisms and processes, and their implications for performance assessment (Chan *et al.*, 2005; Chan & Stanchell, 2005)[34, 35].

Task 4: from the accumulated knowledge and experience of the two DECOVALEX project phases, it had been found that the following important issues needed to be continuously addressed: clarifying the role of THM processes for PA; demonstration analysis of disposal system stability; scale-dependent properties relevant to repository design and PA; and a technical auditing demonstration of specific numerical codes and the overall modeling. Thus, Task 4 was developed to produce conclusions and recommendations on practices for addressing THM issues in Performance and Safety Assessment applications, based on the findings of the DECOVALEX III project. More specifically the Task intended:

– to provide concrete examples on when T-H-M couplings may need to be considered in a quantitative fashion for post-closure performance assessment of nuclear waste repositories in hard rock formations and when T-H-M couplings do not need to be included in such assessments; and
– to provide a practical approach to the general problem of simplifications of T-H-M analyses such that they can be properly incorporated in Performance Assessment analysis, and to evaluate uncertainties introduced through such simplifications.

In order to address the issues as bulleted above, Task 4 teams had: 1) analyzed two major T-H-M experiments (Task 1 and Task 2) and three different Bench Mark Tests (Task 3) setups to explore the significance of T-H-M in some potentially important safety assessment applications; 2) to compile and evaluate the use of T-H-M modeling in SA at the time (the year 2000) using a questionnaire and answers from the concerned funding organizations in the DECOVALEX III project; and 3) to organize a forum of interchange between PA-analysts and THM-modelers at each DECOVALEX III workshop.

The main general conclusions of Task 4 were that: 1) the full development of T-H-M modeling was still at an early stage during the DECOVALEX III project time, and it was not evident whether the current computational methods could provide sufficient information as was required for PA/SA; 2) the current situation (at that time) did not directly imply that all existing coupled mechanisms must be represented numerically in PA/SA. Modeling for specific purposes can be conducted to match the required confidence level of the modeling objective and results; 3) coupled THM modeling must incorporate uncertainties, in both the conceptual model and in the data assessment; 4) the emphasis on the need for THM modeling differs among disciplines. For geological radioactive waste disposal in crystalline and other similar hard rock formations, it is essential to understand the stress-permeability couplings when interpreting stress and permeability field data, to understand the impacts of the coupled processes involved on the re-saturation of the near-field, the Excavated Disturbed Zones (EDZ), and large-scale and significant climatic events, such as glaciations and permafrost. The approaches and achievements of the technical auditing work completed for Task 4 are given in (Hudson *et al.*, 2005).

3.4 DECOVALEX-THMC project (2003–2007)

The DECOVALEX-THMC project was the fourth phase of the ongoing DECOVALEX project for the period 2003–2007, which extended the scope of research into chemical coupling, a major step forward. The general objective was to characterize and evaluate the coupled THMC processes in the near-field and far-field of a geological repository and to assess their impact on PA/SA during the three phases of repository development: excavation phase, operational phase and post-closure phase—for three different rocks types, crystalline, argillaceous and tuff, with specific focus on the EDZ issues, permanent property changes in the rock masses, and glaciation and permafrost phenomena. The project contained 5 Tasks: Task A: influence of near-field coupled THM phenomena on performance assessment; Task B: MHC studies of the EDZ; Task C: EDZ at the argillaceous rock at the Tournemire site; Task D: permanent permeability/porosity changes due to THC and THM processes; and Task E: THM Processes associated with long-term climate change: a glaciations case study. The project was sponsored and guided by 10 Funding Organizations with research conducted by 15 international research teams.

Task A: The objective of Task A was to assess the implications of coupled THM processes in the near-field of a typical repository on its long-term performance, with special concerns about the impacts of buffer material behavior and rock damage on performance and safety assessments. Task A was divided into two subtasks: 1) Subtask A-1: Preliminary THM analysis of the near-field—develop a preliminary THM model of the near-field of the repository, including one room and one pillar, and perform one set of preliminary calculations; and 2) Sub-task A-2: model development and calibration considering rock damage and unsaturated bentonite buffer materials, using laboratory and in situ experiments of the CEA MX-80 buffer test and the TSX experiment, at the URL, Whiteshell, Canada.

For Subtask A-1, the repository being considered was based on one of the several Canadian case studies of different repository concepts. With the base scenario for Canadian HLW repository concepts as the foundation, some issues of a higher degree of uncertainty for PA/SA were considered, such as temperature field, buffer swelling pressure as loads on containers, repository re-saturation rate and time, influence of interfaces between the repository components, and impact of THM coupling on PA/SA, among many other issues. Task A-1 was then defined to address as much as possible the uncertainties in the subject areas as listed above, with the THM calculations performed for a common numerical model though three steps: 1) simple thermal calculations to help finalising parameters for repository design with a temperature limit of 100 °C for the container; 2) a steady state HM analysis to establish pre-excavation conditions, followed by a transient HM analysis for 30 years to simulate the operational phase of the repository; 3) starting after a time of 30 years, assuming that all emplacement boreholes were instantaneously filled with container and bentonite. A THM analysis representing a repository evolution time of 1000 years was undertaken for evaluating and addressing key parameters of importance for PA/SA and identifying outstanding uncertainties.

In view of the results and outstanding issues identified from the Subtask A-1 results, a series of laboratory tests on bentonite and rock damage (at both laboratory and field scales) were conducted and modeled, in order to establish a better understanding of the constitutive behavior of the bentonite and rock damage mechanisms. The bentonite models were calibrated against results from four laboratory experiments: swelling pressure tests, water uptake tests, thermal gradient tests and a mock-up THM test, run by SKB and CEA, respectively. The damage behavior of rocks, as measured in both lab and in situ test site, in lab samples and at tunnel scales, in Canada, were made available by the CNSC as the experimental basis for the assessment of damage of granite samples subjected to triaxial test conditions around tunnel excavations, and for the development of numerical models for rock damage and EDZ simulations. The lab test data from cyclic triaxial tests performed on Lac du Bonnet granite of the Canadian Shield were collected and the evolution of the elastic modulus, the anisotropy ratio, the permeability and the strength parameters of the sample were traced. Using irrecoverable deformation as a measure of damage, the evolution of the above mechanical and hydraulic properties could be expressed as simple mathematical functions.

Calibrated numerical modeling (against measured data) of bentonite tests yielded reliable results, in general, leading to better prediction capability by the constitutive models of bentonite in the teams' models, with much improved reliability for future modeling work. Some uncertainty issues were identified with the required experimental work for increasing the reliability of constitutive models of both bentonite and rock damage, and followed by successful modeling, thus increasing confidence in the numerical modeling ability for identifying the key safety features in the near-field of a hypothetical repository, using coupled THM scoping calculations (Tsang, 2009; Tsang *et al.*, 2009; Chijimatsu *et al.*, 2009; Nguyen *et al.*, 2009; Rutqvist *et al.*, 2009a).

Task B: the overall objective of Task B was to improve the current understanding of the EDZ evolution and to enhance the ability of modeling the HMC mechanisms of the EDZ in crystalline rocks. The overall approach was a combination of laboratory tests on intact rock samples saturated with fluids having different salinity in order to calibrate and improve the damage models; and a series of BMT models of different sizes and fracturing conditions, in order to evaluate the EDZ effects on rock permeability fields. The coupled processes concerned were the stress and deformation, heat transfer and thermal stress/deformation, chemical effects of salinity fluids of rock samples and time. The research performed included the following issues: understanding and characterizing the complete failure of intact rock in uniaxial compression using different numerical models; physical testing with chemical effects; characterizing the failure of intact rock with fractures; benchmark test (BMT) modeling of the EDZ with heat, water, chemical effects and failure, with and without pre-existing fractures; methodology for dealing with uncertainties; using the in situ experimental work at Äspö HRL, Sweden, with tunnel wall sampling, crack mapping and all relevant information for more realistic modeling; and developing a method to characterize and measure the EDZ in a new crystalline rock situation. SKB performed the laboratory experiments on intact rock samples of Ävrö granite saturated with fluids of different salinities, plus the fracture mapping for the Bench-Mark analyses.

The results obtained have demonstrated how widely different modeling approaches can be adapted to simulate the evolution of EDZ around a heat-releasing nuclear waste emplacement tunnel in fractured rock. Important role of fractures and scale-dependent (laboratory to tunnel scale) properties in evaluating the EDZ evolution was identified. Scale-dependent properties and how to capture critical changes occurring in fractures around a drift (laboratory versus tunnel scale) were identified as major outstanding issues, together with time dependent strength degradation over long-term periods. See detailed results in (Hudson *et al.*, 2009; Pan *et al.*, 2009; Rutqvist *et al.*, 2009b).

Task C: This Task was based on an on-going research program of IRSN at its Tournemire URL site, France, for studying the site properties, fluid transport processes and the EDZ effects in an argillaceous rock. The objective was to understand the physical phenomena induced by excavation in an argillaceous medium, and to develop numerical models for interpretation of observed damaged zones around openings. At this site, three openings were constructed at three times in the past (100, 10, 3 years), and research teams modeled the evolution of the EDZ with time and compared the results with measurements of the EDZ phenomena in the tunnels (Fig.5) with the material data and in situ EDZ measurement support through an integrated modeling and experimental approach. Modeling was focused on three issues: (a) the extent of the EDZ around openings; (b) the EDZ properties; and (c) advancement of modeling capabilities for the EDZ extent and properties. Therefore Task C was divided into two subtasks: Subtask C1 concerned with modeling the EDZ development around the old tunnel by investigations into the failure mechanism of the host rock, the in situ stress field, effects of bedding planes with respect to the axis of the tunnel; and Subtask C2, which dealt with the time evolution of the EDZ around three openings excavated at three different times.

Research was conducted with an adaptive step-by-step modeling strategy, supported by two complementary experiments: a mine-by-test and an unsaturated zones

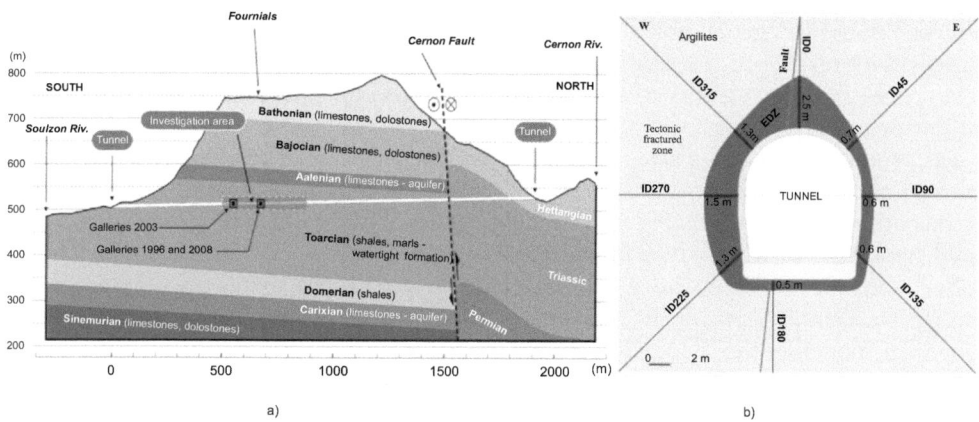

Figure 5 (a) The Tournemire site geology profile, and (b) the measured extent of the EDZ around the old tunnel.

experiment. The results showed that simulation of the response of argillite to excavation required the use of coupled hydro-mechanical models for unsaturated conditions, with special focus on effects of the excavation and natural ventilation in the openings. The unsaturated zone experiments and mine-by-test provided an outstanding opportunity to identify the in situ hydro-mechanical properties of the Tournemire argillite. The hydro-mechanical approach presented in this study confirmed the assumption that the damage observed in the field was completely governed by desaturation/ re-saturation processes, and the models reproduced correctly the hydro-mechanical response of the argillite to the excavation of the 2003 gallery. Also, results explained the tensile failure along the bedding planes and thus the EDZ around the new galleries. The main lesson learned was that the EDZ problem in the case of Tournemire site was complex because of the delayed failure mechanisms. In fact, these mechanisms, induced by various closely-coupled processes, were only partially understood. In contrast, the instantaneous failure mechanism, induced by the excavation of the rock mass, was understood well and predicted properly. To improve this EDZ modeling, it was recommended that more analyses of the stresses induced by excavation and desaturation processes should be performed, in addition to the failure criteria. The main results can be found in (Maβmann et al., 2009; Millard et al., 2009).

Task D: The research program developed for Task D of DECOVALEX-THMC involved both geomechanical and geochemical research areas. The geomechanical task case, referred to as D-THM, was built on the knowledge gained from modeling the short-term in situ DST heater experiments at Yucca Mountain Project of DOE, USA, in the DECOVALEX-III project, in order to conduct the evaluation of long-term THM processes in two generic conceptual geological repositories for radioactive waste. The geochemical task case, referred to as D-THC, addressed long-term THC effects and their relevance through consideration of the same two generic conceptual repositories as studied in D-THM. The objective of D-THC task case was to model the THC processes in the fractured rock close to representative emplacement tunnels as a function of time, and predict the changes in water and gas chemistry, mineralogy, and hydrological properties. The ultimate research topic in Task D was to evaluate and predict long-term changes in the near-field hydrological properties caused by heat-driven geomechanical and geochemical alterations. Such changes in hydrological properties (mostly with respect to fracture porosity and permeability) affect the flow and transport processes in the vicinity of emplacement tunnels and can thus be very important for performance assessment.

The generic conceptual waste repositories evaluated in Task D represented simplified versions of two possible repository sites and emplacement conditions considered by the participating organizations (Fig. 6). The first repository (Fig.6a) was located in a saturated crystalline rock, in which emplacement tunnels were backfilled with a bentonite buffer material. This repository was referred to as a FEBEX type, since many of its features were similar to the FEBEX field test setting. The second repository (Fig.6b) was a simplified model of the Yucca Mountain Project site, featuring a deep, unsaturated, volcanic rock formation with emplacement in open gas-filled tunnels (Yucca Mountain type). For each generic repository, geomechanical and geochemical simulations were conducted separately; i.e., a fully coupled THMC model analysis was not performed. Since two different generic repository settings are considered in the FEBEX type and Yucca Mountain type, there are two sub-tasks each for D-THM and D-THC:

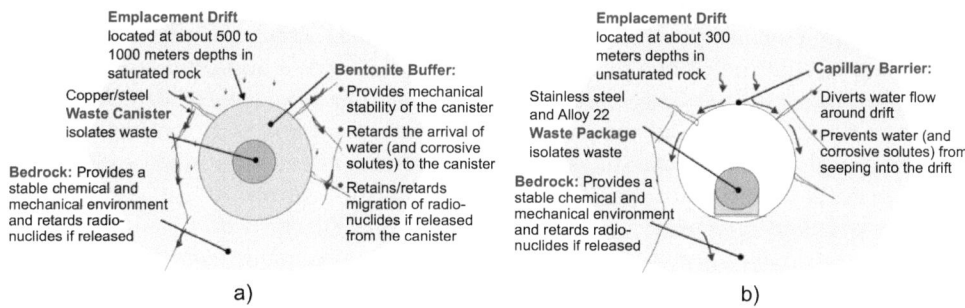

Figure 6 Schematic showing the two repository types evaluated in tasks D-THM and D-THC: (a) bentonite-back-filled repository in saturated rock (FEBEX type), and (b) open-drift repository in unsaturated rock (Yucca Mountain type).

– Task D-THM1: Geomechanical simulations for a generic repository located in saturated crystalline rock, where emplacement tunnels were backfilled with buffer material (FEBEX type).
– Task D-THM2: Geomechanical simulations for a generic repository located in unsaturated volcanic rock, with emplacement in open gas-filled tunnels (Yucca Mountain type).
– Task D-THC1: Thermal-hydrological-chemical simulations for a generic repository located in saturated crystalline rock, where emplacement tunnels were backfilled with buffer material (FEBEX type).
– Task D-THC2: Thermal-hydrological-chemical simulations for a generic repository located in unsaturated volcanic rock, with emplacement in open gas-filled tunnels (Yucca Mountain type).

The teams working on D-THM task case conducted a comparative analysis of coupled thermal, hydrological, and mechanical interactions in both the backfilled and open repository-drift designs, using five different numerical models with varying degrees of complexity. Comparison of the results indicated a good overall agreement in the basic THM processes, in part because the prediction of thermo-mechanical changes was relatively insensitive to the detailed near-field thermal-hydrological processes. Most challenging proved to be the complex heat-induced flow processes at the Yucca Mountain site, which ideally would require modeling approaches capable of handling multi-phase processes and complex interactions between fractures and rock matrix. Research teams identified several outstanding issues in the THM field study, *e.g.* stress-induced changes in hydrological properties, the potential of shear-induced permeability changes, the complexity of heterogeneity in mechanical or hydrological properties, and the importance of chemical processes affecting the THM response. The teams participating in the geochemical D-THC task, using in-house reactive transport simulators, focused on the FEBEX type repository with simplified models for conducting sensitivity analyses to a variety of model parameters, leading to a good quantitative agreement of the predicted aqueous species concentrations and mineral alterations. The simplified modeling exercise provided an ideal starting point for future project phases, where additional complexities should be addressed.

In the D-THM case, coupled THM processes could be reasonably well approximated without a detailed description of flow, this was not the D-THC case that required an accurate representation of reactive transport processes with a good representation of flow in the fractures and the rock matrix. Additional challenges were identified from the strong thermally-induced driving forces in such a system, with boiling of pore water, vapor-phase transport, and where subsequent condensation might induce significant mineral precipitation and dissolution. Details of the results and outstanding issues can be found in (Rutqvist *et al.*, 2009c; Liu *et al.*, 2009).

Task E: the primary purpose of Task E was to provide a reasoned basis to support the treatment of long-term climate change in performance assessment and an overall Safety Case for a deep geological repository of spent nuclear fuel in a crystalline shield setting, as in Canada. A key goal of Task E was to derive a complementary geoscientific basis to address the implications of long-term climate change, especially glaciation and de-glaciation process, on groundwater flow system dynamics as it could potentially affect repository performance. In this regard, a finite-element approach, considering surface and subsurface conditions of the coupled THMC (salinity) simulations, was adopted on a sub-regional scale to address the following flow system/geosphere responses associated with a glacial event: 1) infiltration of glacial meltwaters to the subsurface; 2) anomalous hydraulic head, defined as the difference between the present-day equivalent freshwater head and the corresponding topographically driven head; and 3) evolution of the state of stress. The influence on modeled responses to various model parameters, including the degree of coupling of THMC (salinity) processes and model dimensionality (2D vs. 3D), as well as surface boundary conditions including alternative representations of ice-sheet topography, and two alternative glacial scenarios were investigated. In addition, a limited numerical study was conducted to simulate groundwater flow dynamics under permafrost conditions. Discrete fracture network (DFN) realization models were developed for the site with the site investigation results being used for simulation model setup, plus the results of the variation of pressure-adjusted freezing point with horizontal distance at various times. The modeling was conducted by AECL sponsored by NWMO.

A number of key findings of importance for PA and SA of radioactive waste repositories located in cold regions, such as in Canada, Sweden, Finland and UK, were obtained in the subject areas were investigated with impressive results of freshwater head evolution, residual glacially-induced anomalous hydraulic head, changes and differences in hydraulic head, effects of large fracture zones' features, impact of density effects from salinity and thermal processes, and the mechanical factor of safety change during glaciation and deglaciation cycles. A number of outstanding issues pertaining to THM and THMC modeling of sub-surface response to glaciation were identified, in terms of THM vs THMC modeling of transient evolution of permafrost, the influence of repository heat on permafrost evolution, the need for fully-coupled THMC modeling, stress and depth-dependent fracture or fracture zone permeability, etc. See (Chan & Stanchell, 2009) for more details.

3.5 DECOVALEX-2011 project (2008–2011)

The DECOVALEX-2011 project was the fifth phase of the DECOVALEX project, and took place during the period of 2008–2011. Its general aims were: to characterize and evaluate the impact of coupled THMC processes on PA/SA of geological repositories in fractured rocks (crystalline and argillaceous) and buffer materials; investigating and verifying the predictive capabilities of different codes to simulate field experiments; to improve the understanding of the constitutive behavior of crystalline and argillaceous rock masses and buffer materials; to perform THMC calculations in a PA context; and to review the state-of-the-art in coupled THMC issues in PA. To achieve the above general aims and associated objectives, three tasks were defined: Task A was concerned with the impact of HMC processes in argillaceous rocks; Task B related to TM modeling of fracture initiation and propagation, rock spalling in excavations and its consequences; and Task C for assessment of coupled THMC processes in single fractures and fractured rocks. The modeling defined for the three Tasks was divided into different phases or steps so that the progress could be monitored and the achievements documented in project reports and other publications.

Task A: This task was defined to investigate the effects of ventilation on Opalinus clay at the Mont Terri URL site, Switzerland, based on the Ventilation Experiment (VE) installed and executed in a non-lined micro-tunnel of 10 m in length and 1.3 m in diameter, for the period of 2003–2006 (Fig. 7). The issues of importance that were identified for investigations included desaturation/resaturation of the rock; phase change at air/rock interface; damage/micro-cracking of the host rock due to hydro-mechanical and/or chemical effects; and EDZ evolution. The main items of investigation were identification of the relevant processes and parameters of Opalinus Clay on the basis of the laboratory drying test, hydromechanical modeling of the VE up to the end of the first drying phase (on the basis of the laboratory parameters), calibration of the models, and advanced hydromechanical modeling of the VE, including blind prediction of the second drying period. Five research teams sponsored by four funding organizations (CAS, IRSN, NDA and JAEA) conducted the research by performing hydro-mechanical and geochemical modeling of the different phases of the VE tests and

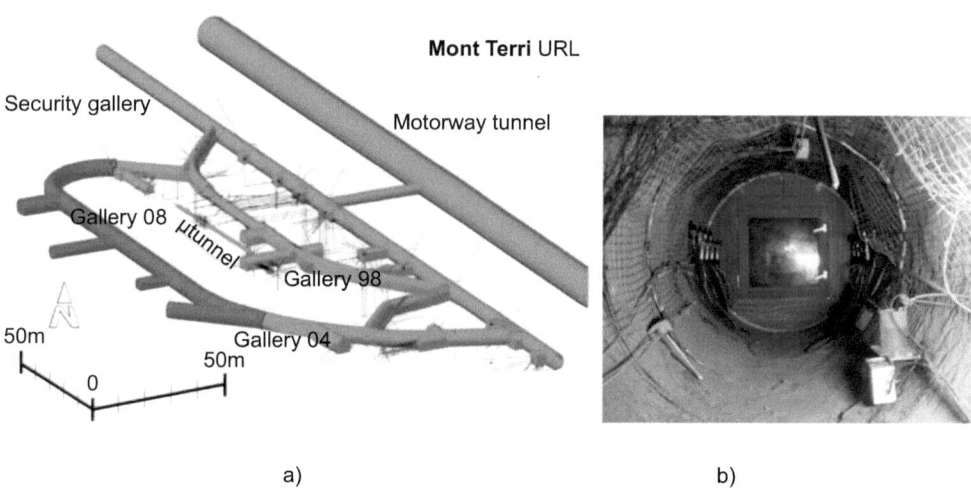

a)   b)

Figure 7 a) Layout of the Mont Terri URL; b) the micro-tunnel where the Ventilation Experiment (VE) was installed and implemented.

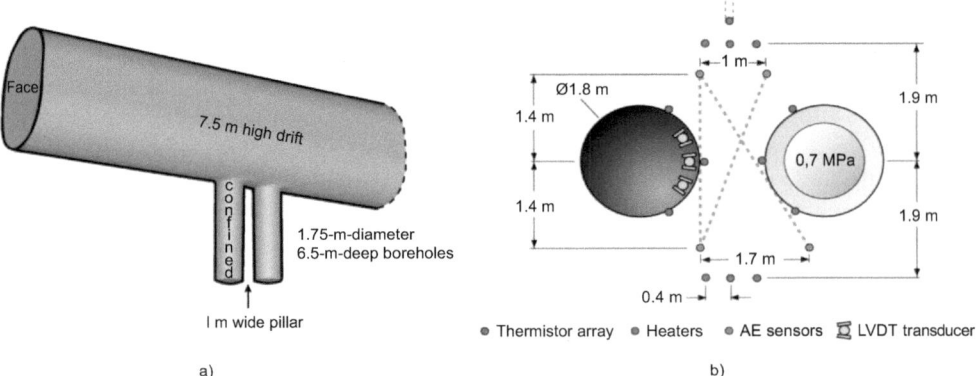

Figure 8 Layout of the APSE experiment. a) Relative positions of the horizontal APSE tunnel, two vertical deposition holes and the pillar between them; b) Plan layout of deposition holes, dimensions and instrumentation borehole locations.

comparing calculated results with experimental observations. This approach allowed checking the capabilities of the various modeling tools and advancing the understanding of ventilation effects on an argillaceous host rock. More detailed presentation of this task is given in Section 5.2 as an example and in published papers (Garitte *et al.*, 2013; Bond *et al.*, 2013a, 2013b; Millard *et al.*, 2013; Zhang *et al.*, 2013).

Task B: SKB conducted the Äspö Pillar Stability Experiment (APSE) in order to determine the spalling/yielding strength of a granitic rock and to test how small confining pressures may initiate the crack initiation strength (the stress state at which the spalling process initiates). The APSE was located at the 450 m level of the Äspö Hard Rock Laboratory in Sweden. The experimental layout consisted of a tunnel with arched roof and floor, in which two boreholes, which were separated by a 1 m thick pillar of Äspö diorite containing fractures, were drilled in order to generate spalling in the pillar which was encouraged by artificial heating (Fig. 8). The APSE work was used for Task B in the DECOVALEX-2011 project, and its purpose aimed to enhance our knowledge and modeling ability in the context of the challenging issue of rock spalling— for more reliable and realistic performance and safety assessments of radioactive waste disposal in crystalline rocks. The hydraulic and chemical processes were excluded due to the fact that very little water was observed to be present during the APSE experiment, plus the test period was not long enough and the temperature was not high enough to cause any significant influences from hydraulic and geochemical processes. The problem was therefore a coupled thermo-mechanical one with the challenges of understanding and modeling the rock spalling processes at both the micro- and macroscopic levels.

The modeling of the APSE experiment in Task B was divided into three stages. Stage 1 aimed to calibrate computer codes in order to identify and model the properties and failure mechanisms of the Äspö Diorite; Stage 2 aimed to simulate the spalling processes during the APSE as time-dependent coupled TM processes through back-analysis; Stage 3 aimed to simulate the entire excavation and heating phases of the experiment in order to determine the spalling strength, deformations and the spalling notch geometry, and to analyze result sensitivity among the teams' models.

The computer model and code calibration against the measured properties and failure mechanisms of the Äspö diorite during Stage 1 were completed successfully by most of the teams. The final modeling results of Stage 2 matched the measured data well, but there was a general over estimation of the rock temperature in the results. For Stage 3, the estimation of spalling strength via back-calculations showed close agreement with the measured data, with a small standard deviation of 7 MPa, compared with the spalling strength of 120 MPa of the pillar, based on elastic models. These results were comparable with earlier findings on the spalling strength of granitic rock. Research also found that modeling the mechanical behavior of crystalline rocks was a difficult challenge, as evidenced by the past experience during the earlier DECOVALEX project phases, with rocks containing fractures of varying sizes, shapes, water-bearing status, mineral filling status, and displacement/damage histories. The main difficulty, and the outstanding issue, was the challenge of characterization of the rock volumes to be modeled, since the details of the fracture system geometry and their mechanical behavior could not be known beforehand, or even after the tests. This challenge was understood as a general feature of fractured crystalline rocks, not a special issue limited only to Task B. The detailed results were described in (Kwon *et al.*, 2013; Koyama *et al.*, 2013; Rinne *et al.*, 2013; Blaheta *et al.*, 2013; Pan & Feng, 2013).

Task C: Task C of the DECOVALEX-2011 project was developed to research the stress effects on flow and transport processes in fractured crystalline rocks. The task included two numerical modeling problems:

–   a Test Case (Task C1): an essentially hydrogeological problem at the Bedrichov Tunnel in the Czech Republic (Fig. 9a), defined as a three-dimensional Test Case involving the characterization, parameterization and numerical modeling of the hydrogeological features, fluid flow and tracer transport processes at the tunnel site, with the challenging tasks of data collection, model setup, parameterization, and uncertainty evaluation;

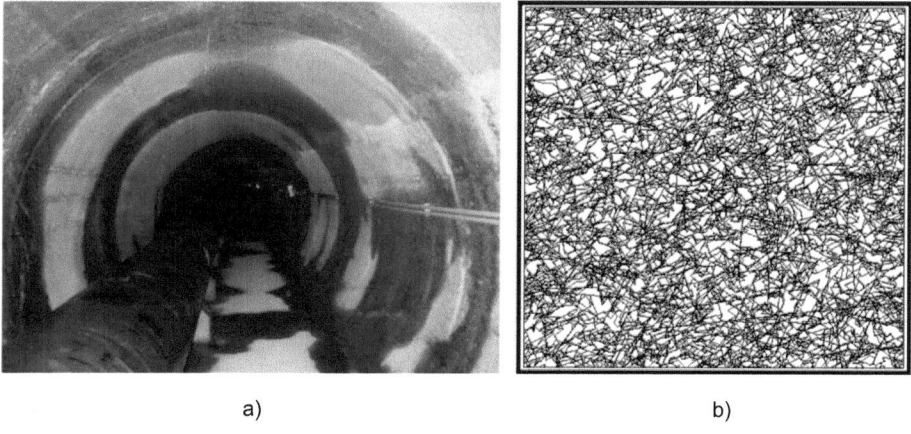

a)   b)

Figure 9 a): The Bedrichov Tunnel for water transport for Task C1; b): The fracture system, model dimension (20 m × 20 m in size) for the 2D BMT problem definition of Task C2.

–   a generic Bench-Mark Test (BMT) problem (Task C2) involving coupled HMC processes in a discrete fracture network model defined in two dimensions, considering the stress effect on fluid flow and solute transport with the retardation mechanisms of advection and matrix diffusion, in order to improve the understanding of the fundamental behavior of interactions between stress and transport processes, and the representation and characterization of a system of 7797 fractures having different lengths and hydraulic apertures (Fig. 9b).

Study of coupled processes in fracture networks has been an on-going fundamental subject in the DECOVALEX project due to its importance for PA and SA of GDRW repositories in fractured crystalline rocks such as granite, which has been adopted as the host rock in Sweden and Finland, and also potentially in Canada, China and some other countries. Much simplified fracture system models were investigated in DECOVALEX I (BMT1, BMT2 and BMT3), and BMT2 of Task 3 of DECOVALEX II, where coupled THM processes were the main concern. Inclusion of solute transport process in fracture networks in Task C2 was therefore a necessary effort in this direction despite its generic nature. Due to this important feature, Task C2 was chosen as an example as presented in the section of examples below.

The model development and initial modeling results of the Task C1 illustrated the importance of understanding, handling and predicting the impact of uncertainties in the site characterization and reliability of predictive modeling for even such apparently 'simple' hydrogeological cases that are in reality the corner stones of understanding and modeling the groundwater flow and solute transport processes for GDRW problems. This outstanding issue was the basis for its continuation to the next phase of the DECOVALEX project, DECOVALEX-2015.

The modeling of the interactions among stress, fluid flow and solute/particle transport in a complex fracture network was conducted in Task C2, considering advection and matrix diffusion retardation mechanisms, and the outcome served as a good demonstration of the importance of the stress effect on flow and transport in fracture systems, thus being an important issue in PA and SA. The major outstanding issues related to identification and management of the uncertainties in the input data, such as fracture system geometry and fracture properties (hydraulic and mechanical); reliability of the in situ measured data; development of a systematic approach for identification, quantification and monitoring of the sources and evolution of the uncertainties with the progress of modeling steps; and how to use the current numerical tools for investigating uncertainty issues. The main results of Task C2 can be seen in (Rutqvist *et al.*, 2013; Zhao *et al.*, 2013).

3.6 DECOVALEX-2015 project (2012–2015)

The current phase of the DECOVALEX project, at the time of writing this Chapter (D-2015, *i.e.*, the phase that ends in 2015) concerned with modeling coupled THMC processes considered in five tasks of large scale *in situ* experiments in URLs or site investigations in different countries and lab tests on single fractures. Five modeling tasks were defined to cover the range of argillaceous, sedimentary and crystalline rocks: Task A: The Sealex *in situ* experiment, Tournemire Site, France; Task B1: HE-E heater test, Mont Terri, Switzerland; Task B2: EBS experiment, Horonobe, Japan; Task C1: THMC of single rock fractures; and Task C2: water inflow, Bedrichov Tunnel, Czech Republic.

Figure 10 Work on the in situ experimental studies at the Tournemire site in France.

Task A: This test case involves study of large-diameter bentonite cores and bentonite–rock interfaces with the objective of evaluating the impact of the seals (bentonite and concrete plugs) on the performance and safety assessment functions (Fig.10). The knowledge gained from the experiments can also be applicable to other host rocks, such as crystalline and sedimentary rocks (the main instrumented component is bentonite which is needed for repositories in crystalline rocks, such as granite). Also, the experiment has a direct impact on design, implementation, evaluation and monitoring of the sealing systems (bentonite and concrete plug) of radioactive waste repositories, especially on the post-closure issues of safety assessment after the sealing system installation in different host rocks, for both near- and far-field safety cases.

The main testing features are: placement of pre-fabricated bentonite cores (seals) into horizontal boreholes (600 mm in diameter) in the wall of galleries of the URL; forced saturation of the bentonite cores with watertight sealing of the boreholes; intra-core instrumentation with wireless systems installed in the bentonite core to prevent potential preferential fluid flow pathways. The participating research teams are performing numerical simulations of the saturation phase of the Sealex in situ tests for different testing conditions and modeling the coupled hydro-mechanical behavior of the bentonite–rock interfaces.

Task B1: This work is linked to an experiment in the Opalinus clay at the Mont Terri URL in Switzerland and is a Test Case based on the NAGRA PEBS (Long-term Performance of Engineered Barrier Systems) program, Figure 11. The objective is the evaluation of the sealing and barrier performance of the EBS (Engineered Barrier System) with time during the heating phase of a repository. Similarly to Task A, the knowledge gained from experiments can be applicable to other host rocks. The work involves a combination of blind prediction, model calibration with measured data and long-term impact evaluation of the EBS, important for PA and SA.

The main scientific issues being considered are the processes of thermal evolution, buffer (bentonite) re-saturation, *in situ* determination of thermal conductivity of the

Figure 11 The research tunnel at the Mont Terri URL, Switzerland.

bentonite and its dependency on saturation, pore water pressure in the near-field, swelling pressure evolution of the bentonite, and water input from rock to the EBS. The heating started in 2011 and was continued for a period of three years with a designed heater surface temperature of 135°C. Laboratory tests are also being performed to characterize the behavior and parameters of bentonite blocks, granular bentonite particles and the Opalinus clay rock.

Task B2: This task is based on the in situ EBS (Engineered Barrier System) experiment that is being conducted in a sedimentary rock at the Horonobe URL in Japan, with the main instrumented components being bentonite and sand, Figure 12. The objectives are to validate the coupled THMC models and to obtain data on an engineered barrier system and in the surrounding rock. Additional objectives are to confirm both the applicability of the measurement techniques for confirming the performance of EBS and its 'setup technology', including the backfill of the tunnel using practical techniques given by the in situ environment. A similar *in situ* near-field THM experiment was conducted in the Kamaishi Mine, Japan and was investigated as Task 2 of DECOVALEX II (*cf.* Section 3.2) but the host rock in that case was granite with different behavior and test design.

As with the Tasks A and B1, the knowledge gained from the experiments and modeling of Task B2 can also be applicable to other host rocks because the main instrumented experimental component is the EBS (*i.e.* bentonite and plug). The work involves a combination of blind prediction, model calibration with measured data and long-term prediction for the impact of the EBS on PA/SA. The scientific issues include thermal evolution, buffer (bentonite) re-saturation process, backfill effects, pore water pressure evolution in the near-field, swelling pressure evolution of the bentonite, water

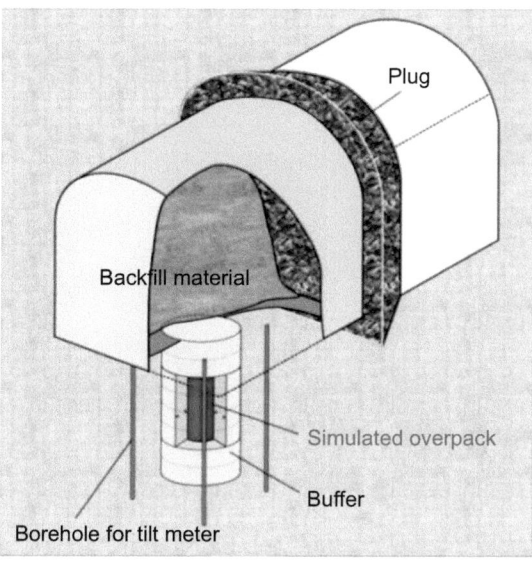

Figure 12 The Engineered Barrier System at the Horonobe URL in Japan.

input from rock to the EBS (involving characterization of rock saturation surrounding the EBS), and possible chemical issues, with model development and validation. Confidence building is also one of the major objectives.

Task C1: The work for Task C1 involves the study of laboratory sample-sized rock fractures for understanding and modeling the fully coupled THMC processes as fluids flow through the fractures, Figure 13. Investigating the fully-coupled THMC processes of a single rock fracture has not been attempted before within the DECOVALEX project phases, but these processes are issues of dominating importance for the PA/SA of repositories in fractured crystalline rocks, such as granite, and may also have important reference values for repositories in other types of host rocks. This research subject is one of the key issues of theoretical and practical importance on the PA/SA of GDRW repositories in crystalline rocks, and the main reason for its continued concern in all past DECOVALEX phases.

The work is aimed at modeling the fully-coupled THMC processes of rock fractures based on data from the laboratory experiments on fracture samples of novaculite and granite rocks (Yasuhara *et al.*, 2004, 2006, 2008, 2011). The overall objective of the Task is to use the published experimental results to build and refine conceptual and physical process models for the single fracture system and to provide useful input for underpinning science in radioactive waste disposal and safety case development. The novaculite fracture experiment is being studied first, together with considering the impacts of different approaches adopted by the research teams. The work will then move on to consider the granite fracture case based on learnings from the fracture surface evolution in the novaculite fracture experiment—through moving from a well hydraulically constrained system (the novaculite fracture) to a less constrained system (the granite fracture).

Task C2: This Test Case problem is the extension of Task C1 of DECOVALEX-2011 phase (*cf.* Section 3.5). The work for this Task involves predicting the water flow into

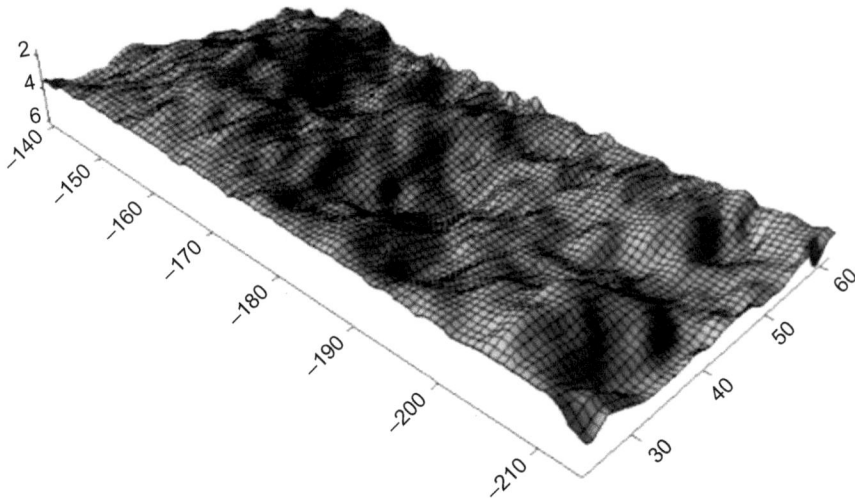

Figure 13 Illustrative example of an irregular rock fracture surface.

the Bedrichov water transfer tunnel located in the Bohemian granite massif in the Czech Republic. The objective is to evaluate the groundwater flow and tracer transport processes at the site scale and compare the result with the recorded data, and to consider the treatment of uncertainties for site characterization for GDRW repositories. The experiment and research will interact with the site characterization practices by considering the challenging fracture system characterization issues that play an important role in the reliability and uncertainty issues of PA/SA. A comprehensive database was already established containing the available data on site geology, fracture mapping (inside the tunnel), resistivity profiles, water inflow, water chemistry, stable isotope sampling and results, and fracture displacements.

The main scientific issues being considered are: hydro-geological characterization of the test site; stress measurement and interpretation of results at the site scale; groundwater flow and reactive tracer transport study; measurements and numerical modeling, considering discontinuity and heterogeneity issues; impact of uncertainty of the fracture system geometry and hydro-mechanical behavior on water flow and tracer transport.

Further information on the history and current activities of the DECOVALEX program can be found at: www.DECOVALEX.org, where an internal interim report (Jing *et al.*, 2014) containing detailed descriptions of DECOVALEX-2015 can be reached.

4 EXAMPLES

4.1 BMT models for stress effect on flow and transport processes in fractured rocks

For the GDRW problems in crystalline rocks, one of the main rock mechanics issues is the stress effect on fluid flow and transport processes in fracture networks—which requires characterization of fracture system geometry, scaling and homogenization for deriving equivalent continuum properties based on the REV concept (Representative Elementary Volume), and numerical prediction of the stress effect on nuclide transport processes for PA/SA of a GDRW repository. This concern led to continued research as undertaken in DECOVALEX I (BMT1 and BMT 2), DECOVALEX III (BMT2 of Task C), and DECOVALEX-2011 (Task C1), during the period of 1992–2011. The reason for the chosen BMT problems is the fact that large scale (tens or hundreds of meters) field tests for proper evaluation of REVs containing thousands or tens of thousands of fractures of effectively randomly distributed geometrical properties of size, orientation, location and hydro-mechanical properties (aperture/transmissivity, stiffness, frictional properties, etc.) has not been possible. Therefore, a DEM solution is so far the most suitable approach to derive the equivalent properties at their REV scales and the stress effect upon them, therefore supporting large scale and long-term simulations for PA/SA using equivalent continuum approaches, at conceptual level (predictive modeling) or application level (with measured data support).

As mentioned before, BMTs in the DECOVALEX I project were mainly designed for testing modeling approaches and the applicability of computer codes for simple (BMT1 and BMT2) to complex (BMT3) fractured system models. The research was then extended to scaling, homogenization and REV issues, studied in BMT2 of Task C in the DECOVALEX III project, focusing mainly on hydrological and mechanical property identification approaches. Besides the derivations of equivalent elastic compliance and permeability tensors, the shear stress caused fluid flow channeling was identified as an important issue for modeling flow and solute transport processes (Min & Jing, 2003; Min *et al.*, 2004) (Fig. 14). The work was then extended to consider the effects of correlation between size and hydraulic aperture of the fractures, which led to findings of stress-dependent REV variations and related permeability changes (Baghbanan & Jing, 2007, 2008), as shown in Fig. 15.

The findings in DECOVALEX I and III established the mathematical basis for modeling the stress effect on flow and solute transport processes conducted in Task C1 of the DECOVALEX-2011 project, with the DFN model at the REV size of 20m×20m containing nearly 8000 fractures with apertures correlated to their lengths (*cf.* Fig.9b), as a generic study. Figure 16 shows the different breakthrough curves for solute transport at different ratios (K) of the vertical stress value divided by the horizontal stress value (Zhao *et al.*, 2013). The difference in orders of magnitude for travel time between stressed and unstressed results vividly indicate the importance of

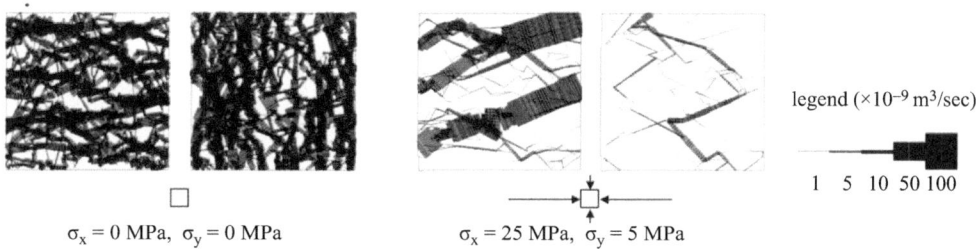

$\sigma_x = 0$ MPa, $\sigma_y = 0$ MPa $\sigma_x = 25$ MPa, $\sigma_y = 5$ MPa

legend ($\times 10^{-9}$ m³/sec)

1   5   10 50 100

Figure 14 Comparison of fluid pathways with and without applied stress in the horizontal and vertical hydraulic pressure gradient directions. One line indicates the flow rate of 10-9 m3/sec, as shown in the legend to the right (Min *et al.*, 2004).

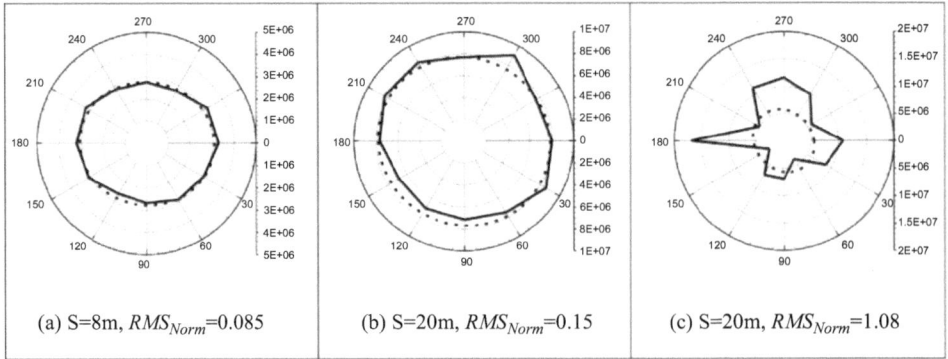

Figure 15 Directional permeability contours for DEM model sizes of 0.5m, 5m, 8m and 20m, and mean square error for the models with (a) constant aperture of 65 µm (b=0); (b) correlated aperture-length with b=1.0, and (c) correlated aperture-length with b=3.0. Scales at the right-hand side of the plots are values of $1/K1/2$, with K being permeability (Baghbanan & Jing (2007, 2008)).

the stress conditions on PA/SA, especially when significant stress field changes occur, such as during glaciation and deglaciation cycles—which is one of the most important scenarios in Canada, Finland, Sweden and the UK.

4.2 Modeling THM processes of Opalinus Clay as influenced by tunnel excavation

Task A of the DECOVALEX-2011 project aimed at identification, understanding and quantification of mechanisms taking place during the ventilation of a tunnel in an argillaceous host rock by investigating the capacity of modeling approaches and codes to reproduce these processes, Fig.17. The four years long in situ ventilation experiment took place in a 1.3 m diameter unlined micro-tunnel (*cf.* Fig.7b) and included two re-saturation-desaturation cycles. A laboratory drying experiment was designed to mimic the *in situ* conditions and was carried out before the VE. The main result from the modeling showed that the response of the host rock to ventilation in an argillaceous rock was mainly governed by hydraulic processes (advective Darcy flow and non-advective vapor diffusion) and that the mechanical-hydraulic back coupling was relatively weak. Therefore, results are useful for understanding and measuring the hydraulic conductivity of the rock mass for GDRW repositories in clayey rock.

The multi-step progressive modeling of this Task led to a satisfactory reproduction of relevant measurement results (water mass balance in the experiment, relative humidity, pore water pressure, water content and deformation) generated by teams using different codes and approaches (Figs. 18 and 19). The achievements validated the numerical tools used, and calibrated the reliability of the modeling approaches and results—and were therefore useful for assessing the relevance of a number of desaturation processes, permeability, partial saturation related issues, anisotropic rock properties (permeability and deformability), hydrological effects on stress redistribution and EDZ, among

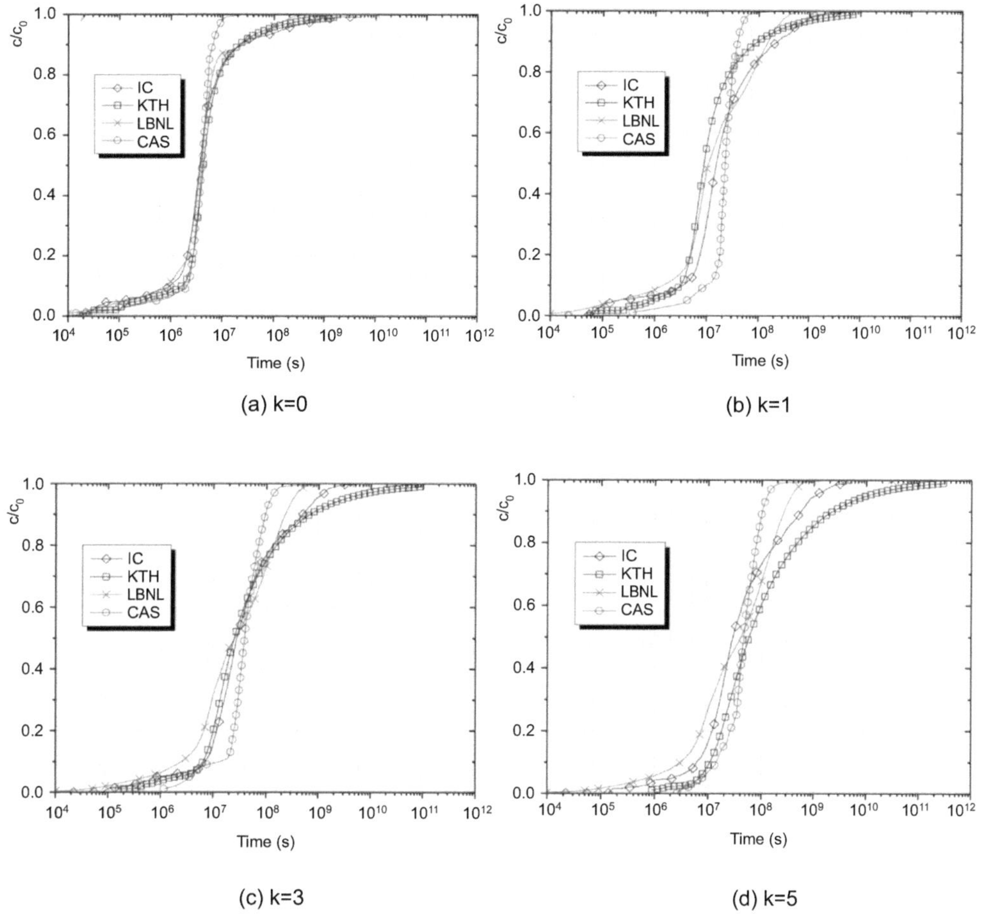

(a) k=0

(b) k=1

(c) k=3

(d) k=5

Figure 16 Comparison of breakthrough curves for interacting tracers exiting from all the three outlet boundaries with increasing stress ratio, under a horizontal hydraulic gradient of 10 Pa/m (0.001 m/m) (Zhao *et al.*, 2013).

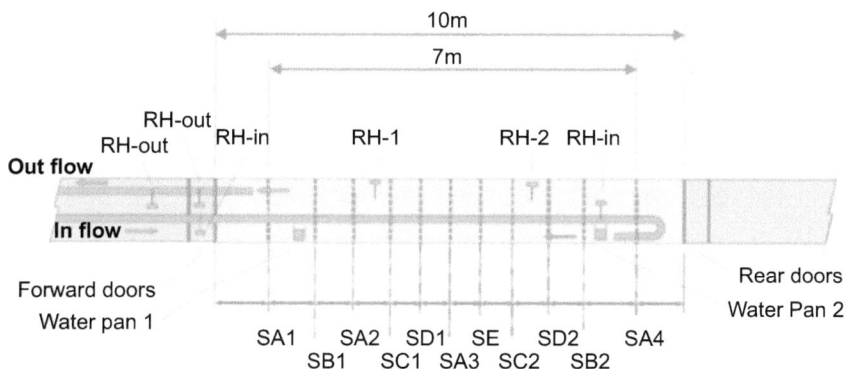

Figure 17 Layout of the Ventilation Experiment (Garitte *et al.*, 2013). RH indicated relative humidity; SA, SB, SC and SD are locations of instrumentation sections.

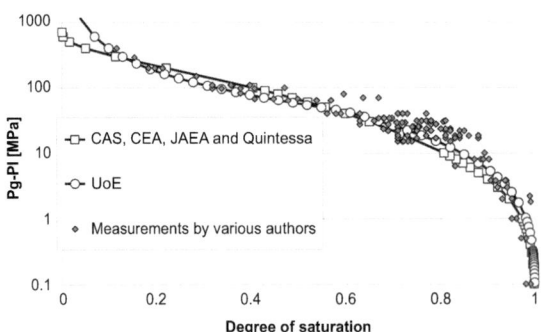

Figure 18 Water retention curve of Opalinus Clay: measurements compared with the functions used by the different modelers (Garitte *et al.*, 2013). The vertical axis represents suction.

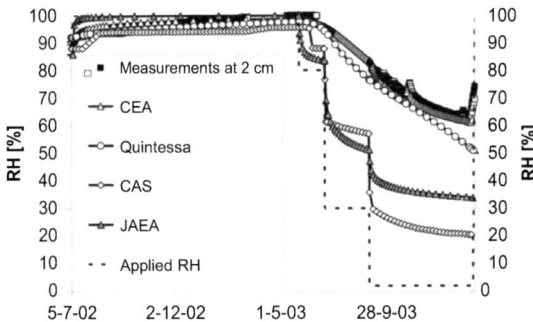

Figure 19 Measured and simulated relative humidity 20 mm inside the rock mass (Garitte *et al.*, 2013).

others. The main outstanding issue was inadequate depth of research on chemical coupling and uncertainties of the effects of fracture systems in the host rock on both flow and transport processes, as well as a lack of extended research on the impact of performance and safety assessments over a longer time period, which was mainly due to time limitations of this phase of the DECOVALEX-2011 project, and similarly with previous phases.

5 CONCLUDING REMARKS

The DECOVALEX project has been and continues to be an unusually long-term international co-operational research project, at both the fundamental and application levels, aimed at supporting the safe management of radioactive wastes in the international communities of rock mechanics and radioactive waste management. To date, the project has yielded an impressive number of research articles of comprehensive scientific originality, novelty and applicability for, not only GDRW repositories, but for other aspects of general rock engineering and projects such as CO_2 sequestration and geothermal energy modeling.

From its start, the DECOVALEX project has played an important role in the development of the coupled THM/THMC models and computational tools for geo-system evaluations applicable not only to engineering for GDRW repositories but also to geothermal and hydrocarbon reservoir engineering, environmental engineering and resources engineering, such as mining and hydropower, due to the fact that the environmental safety has become a key issue in all engineering practice worldwide today. During this project, coupled THM/THMC models and computing tools have been initiated, developed and enhanced continuously through researches performed for solving the problems involving different scales, variable processes and host rocks, with the uncertainties evaluated and data requirements established. The work has incorporated a variety of scenarios from the detailed geometry and behavior of single fractures at centimeter scales to geosphere responses to glaciation and permafrost at the multiple kilometer scale over millions of years.

The success of the project has been the result of the DECOVALEX framework that has provided, and is continuing to provide, integration of numerical modeling and laboratory/in situ experiments, intensive interactions among research teams representing different disciplines and with different modeling approaches, integrated PhD studies and project problem solutions. In particular, the support provided by a considerable number of large-scale field tests of coupled processes, conducted in Canada, Czech Republic, Japan, France, Switzerland, Sweden, United Kingdom and the United States, has been extremely valuable. In fact, the results of these efforts have been of major benefit to the advancement of knowledge on waste disposal in general, and to the funding organizations of the DECOVALEX project in particular. The achievements obtained in such an integrated and co-operative research environment would have been less significant if the teams had worked independently over short project timespans, because some of the important issues can only be properly investigated by joint researches from multiple disciplinary viewpoints over considerable period of time, as demonstrated by the continued research on THM and THMC processes in rock fractures, fracture networks, and the EBS systems.

ACKNOWLEDGMENTS

The authors wish to express their sincere gratitude to all funding organizations of the DECOVALEX project (Table 1) for their outstanding financial and technical support, experts for their valuable advices, and, especially, the research teams (Table 2) for their extensive contributions and significant achievements that have made the project successful. Especially, we would like to thank the Swedish Nuclear Fuel and Waste Management Co. (SKB) and Swedish Nuclear Power Inspectorate (SKI, now reorganized as the Swedish Radiation Safety Authority, SSM) for their sustained support and encouragement throughout the major part of this successful and unusually long-term international cooperative project.

REFERENCES

Alonso, E.E., Alcoverro, J., Coste, F., Malinsky, L., Merrien-Soukatchoff, V., Kadiri, I., Nowak, T., Shao, H., Nguyen, T.S., Selvadurai, A.P.S., Armand, G., Sobolik, S.R., Itamura, M., Stone, C.M., Webb, S.W., Rejeb, A., Tijani, M., Maouche, Z., Kobayashi, A., Kurikami, H., Ito, A.,

Sujita, Y., Chijimatsu, M., Bergesson, L., Hernelind, J., Rutqvist, J., Nguyen, T.S., Selvadurai, A.P.S. & Armand, G. (2005) *Modelling the FEBEX THM experiment using a state surface approach*. International Journal of Rock Mechanics and Mining Sciences, 42(5–6), 639–651.

Baghbanan, A. & Jing, L. (2007) *Hydraulic properties of fractured rock masses with correlated fracture length and aperture*. International Journal of Rock Mechanics and Mining Sciences, 44(5), 704–719.

Baghbanan, A. & Jing, L. (2008) *Stress effects on permeability in fractured rock mass with correlated fracture length and aperture*. International Journal of Rock Mechanics and Mining Sciences, 45(8), 1320–1334.

Blaheta, R., Byczanski, P., Čermák, M., Hrtus, R., Kohut, R., Kolcun, A., Malík, J. & Sysala, S. (2013) *Analysis of Äspö Pillar Stability Experiment: Continuous thermo-mechanical model development and calibration*. Journal of Rock Mechanics and Geotechnical Engineering, 5(2), 124–135.

Blum, P., Mackay, R., Riley, M.S. & Knight, L. (2005) *Performance assessment of a nuclear waste repository: Upscaling coupled hydro-mechanical properties for far-field transport analysis*. International Journal of Rock Mechanics and Mining Sciences, 42(5–6), 781–792.

Bond, A., Benbow, S., Wilson, J., Millard, A., Nakama, S., English, M., McDermott, C. & Garitte, B. (2013a) *Reactive and non-reactive transport modelling in partially water saturated argillaceous porous media around the ventilation experiment, Mont Terri*. Journal of Rock Mechanics and Geotechnical Engineering, 5(1), 44–57.

Bond, A., Millard, A., Nakama, S., Zhang, C.-Y. & Garritte, B. (2013b) *Approaches for representing hydro-mechanical coupling between sub-surface excavations and argillaceous porous media at the ventilation experiment, Mont Terri*. Journal of Rock Mechanics and Geotechnical Engineering, 5(2), 85–96.

Börgesson, L., Chijimatsu, M., Fujita, T., Nguyen, T.S., Rutqvist, J. & Jing, L. (2001) *Thermo-hydro-mechanical characterisation of a bentonite-based buffer material by laboratory tests and numerical back analyses*. International Journal of Rock Mechanics and Mining Sciences, 38(1), 95–104.

Cassiraga, E.F., Fernàndez-Garcia, D. & Gómez-Hernández, J.J. (2005) *Performance assessment of solute transport upscaling methods in the context of nuclear waste disposal*. International Journal of Rock Mechanics and Mining Sciences, 42(5–6), 756–764.

Chan, T., Khair, K., Jing, L., Ahola, M., Noorishad, J. & Vuillod, E. (1995) *International comparison of coupled thermo-hydro-mechanical models of a multiple-fracture bench mark problem: DECOVALEX phase I, bench mark test 2*. International Journal of Rock Mechanics and Mining Sciences & Geomechanics Abstracts, 32(5), 435–452.

Chan, T., Christiansson, R., Boulton, G.S., Ericsson, L.O., Hartikainen, J., Jensen, M.R., Mas Ivars, D., Stanchell, F.W., Vistrand, P. & Wallroth, T. (2005) *DECOVALEX III BMT3/BENCHPAR WP4: The thermo-hydro-mechanical responses to a glacial cycle and their potential implications for deep geological disposal of nuclear fuel waste in a fractured crystalline rock mass*, International Journal of Rock Mechanics and Mining Sciences, 42(5–6), 805–827.

Chan, T. & Stanchell, F.W. (2005) *Subsurface hydro-mechanical (HM) impacts of glaciation: Sensitivity to transient analysis, HM coupling, fracture zone connectivity and model dimensionality*. International Journal of Rock Mechanics and Mining Sciences, 42(5–6), 828–849.

Chan, T. & Stanchell, F.W. (2009) *Implications of subsurface thermal-hydraulic-mechanical processes associated with glaciation on shield flow system evolution and performance assessment*. Journal of Environmental Geology, 57(6), 1371–1389.

Chijimatsu, M., Fujita, T., Sugita, Y., Amemiya, K. & Kobayashi, A. (2001) *Field experiment, results and THM behaviour in the Kamaishi mine experiment*. International Journal of Rock Mechanics and Mining Sciences, 38(1), 67–78.

Chijimatsu, M., Nguyen, T.S., Jing, L., De Jonge, J., Kohlmeier, M., Millard, A., Rejeb, A., Rutqvist, J., Souley, M. & Sugita, Y. (2005) *Numerical study of the THM effects on the near-field safety of a hypothetical nuclear waste repository—BMT1 of the DECOVALEX III project. Part 1: Conceptualization and characterization of the problems and summary of results.* International Journal of Rock Mechanics and Mining Sciences, 42(5–6), 720–730.

Chijimatsu, M., Börgesson, L., Fujita, T., Jussila, P., Nguyen, T.S., Rutqvist, J. & Jing, L. (2009) *Model development and calibration for the coupled thermal, hydraulic and mechanical phenomena of the bentonite,* Journal of Environmental Geology, 57(6), 1255–1261.

Claesson, J., Follin, S., Hellström, G. & Wallin, N.-O. (1995) *On the use of the diffusion equation in Test Case 6 of DECOVALEX.* International Journal of Rock Mechanics and Mining Sciences & Geomechanics Abstracts, 32(5), 525–528.

Durin, M., Stietel, A., Thoraval, A., Vuillod, E., Baroudi, H., Plas, F., Bougnoux, A., Vouille, G., Kobayashi, A., Hara, K., Fujita, T. & Ohnishi, Y. (1995) *Discrete and continuum approaches to simulate the thermo-hydro-mechanical couplings in a large, fractured rock mass.* International Journal of Rock Mechanics and Mining Sciences & Geomechanics Abstracts, 32(5), 409–434.

Garitte, B., Bond, A., Millard, A., Zhang, C.-Y., Mcdermott, C., Nakama, S. & Gens, A. (2013) *Analysis of hydro-mechanical processes in a ventilated tunnel in an argillaceous rock on the basis of different modelling approaches.* Journal of Rock Mechanics and Geotechnical Engineering, 5(1), 1–17.

Gómez-Hernández, J.J., Hendricks, H.J., Franssen, A. & Cassiraga, E.F. (2001) *Stochastic analysis of flow response in a three-dimensional fractured rock mass block.* International Journal of Rock Mechanics and Mining Sciences, 38(1), 31–44.

Hakami, H. (2001) *Rock characterisation facility (RCF) shaft sinking—numerical computations using FLAC.* International Journal of Rock Mechanics and Mining Sciences, 38(1), 59–65.

Hsiung, S.M., Chowdhury, A.H. & Nataraja, M.S. (2005) *Numerical simulation of thermal–mechanical processes observed at the Drift-Scale Heater Test at Yucca Mountain, Nevada, USA.* International Journal of Rock Mechanics and Mining Sciences, 42(5–6), 652–666.

Hudson, J.A., Stephansson, O. & Andersson, J. (2005) *Guidance on numerical modelling of thermo-hydro-mechanical coupled processes for performance assessment of radioactive waste repositories.* International Journal of Rock Mechanics and Mining Sciences, 42(5–6), 850–870.

Hudson, J.A., Bäckström, A., Rutqvist, J., Jing, L. Backers, T., Chijimatsu, M., Christiansson, R., Feng, X.-T., Kobayashi, A., Koyama, T., Lee, H.-S., Neretnieks, I., Pan, P.-Z., Rinne, M. & Shen, B.-T. (2009). *Characterising and modelling the excavation damaged zone in crystalline rock in the context of radioactive waste disposal,* Journal of Environmental Geology, 57(6), 1275–1297.

Jing, L., Hudson, J.A. & Birkholzer, J. (eds.) (2014) DECOVALEX-2015 Project: Interim Report. Internal report of DECOVALEX-2015 project.

Jing, L., Tsang, C.-F. & Stephansson, O. (1995) *DECOVALEX – an International co-operative research project on mathematical models of coupled T-H-M processes for safety analysis of radioactive waste repositories.* International Journal of Rock Mechanics and Mining Sciences & Geomechanics Abstracts, 32(5), 389–398.

Knight, L. (2001) *Prediction of the hydro-mechanical response during shaft sinking for the proposed Nirex Rock Characterisation Facility near Sellafield, Cumbria, United Kingdom,* International Journal of Rock Mechanics and Mining Sciences, 38(1), 5–16.

Kobayashi, A., Fujita, T. & Chijimatsu, M. (2001) *Continuous approach for coupled mechanical and hydraulic behaviour of a fractured rock mass during hypothetical shaft sinking at Sellafield, UK.* International Journal of Rock Mechanics and Mining Sciences, 38(1), 45–57.

Koyama, T., Chijimatsu, M., Shimizu, H., Nakama, S., Fujita, T., Kobayashi, A. & Ohnishi, Y. (2013) *Numerical modelling for the coupled thermo-mechanical processes and spalling phenomena in Äspö Pillar Stability Experiment (APSE).* Journal of Rock Mechanics and Geotechnical Engineering, 5(1), 58–72.

Kwon, S., Lee, C., Jeon, S. & Choi, H.-J. (2013) *Thermo-mechanical coupling analysis of APSE using submodels and neural networks.* Journal of Rock Mechanics and Geotechnical Engineering, 5(1), 32–43.

Liu, X., Zhang, C.-Y., Liu, Q.-S. & Birkholzer, J. (2009) *Multiple-point statistical prediction on fracture networks at Yucca Mountain.* Journal of Environmental Geology, 57(6), 1361–1370.

Makurat, A., Ahola, M., Khair, A., Noorishad, J., Rosengren, L. & Rutqvist, J. (1995) *The DECOVALEX Test—Case One.* International Journal of Rock Mechanics and Mining Sciences & Geomechanics Abstracts, 32(5), 399–408.

Maßmann, J., Uehara, S.-I., Rejeb, A. & Millard, A. (2009) *Investigation of desaturation in an old tunnel and new galleries at an argillaceous site.* Journal of Environmental Geology, 57(6), 1337–1345.

Millard, A., Rejeb, A., Chijimatsu, M., Jing, L., De Jonge, J., Kohlmeier, M., Nguyen, T.S., Rutqvist, J., Souley, M. & Sugita, Y. (2005) *Numerical study of the THM effects on the near-field safety of a hypothetical nuclear waste repository—BMT1 of the DECOVALEX III project. Part 2: Effects of THM coupling in continuous and homogeneous rocks.* International Journal of Rock Mechanics and Mining Sciences, 42(5–6), 731–744.

Millard, A., Maßmann, J., Rejeb, A. & Uehara, S. (2009) *Study of the initiation and propagation of excavation damaged zones around openings in argillaceous rock.* Journal of Environmental Geology, 57(6), 1325–1335.

Millard, A., Bond, A., Nakama, S., Zhang, C.-Y., Barnichon, J. & Garitte, B. (2013) *Accounting for anisotropic effects in the prediction of the hydro-mechanical response of a ventilated tunnel in an argillaceous rock.* Journal of Rock Mechanics and Geotechnical Engineering, 5(2), 97–109.

Min, K.-B. & Jing, L. (2003) *Numerical determination of the equivalent elastic compliance tensor for fractured rock masses using the distinct element method.* International Journal of Rock Mechanics and Mining Sciences, 40(6), 795–816.

Min, K.-B., Jing, L. & Stephansson, O. (2004a) *Determining the equivalent permeability tensor for fractured rock masses using a stochastic REV approach: Method and application to the field data from Sellafield, UK.* Hydrogeology Journal, 12(5), 497–510.

Min, K.-B., Rutqvist, J., Tsang, C.-F. & Jing, L. (2004b) *Stress-dependent permeability of fractured rock mass: A numerical study.* International Journal of Rock Mechanics and Mining Sciences, 41(7), 1191–1210.

Min, K.-B., Rutqvist, J., Tsang, C.-F. & Jing, L. (2005) *Thermally induced mechanical and permeability changes around a nuclear waste repository—a far-field study based on equivalent properties determined by a discrete approach.* International Journal of Rock Mechanics and Mining Sciences, 42(5–6), 765–780.

Nguyen, T.S., Börgesson, L., Chijimatsu, M., Rutqvist, J., Fujita, T., Hernelind, J., Kobayashi, A., Ohnishi, Y., Tanaka, M. & Jing, L. (2001) *Hydro-mechanical response of a fractured granitic rock mass to excavation of a test pit—the Kamaishi Mine experiment in Japan.* International Journal of Rock Mechanics and Mining Sciences, 38(1), 79–94.

Nguyen, T.S., Börgesson, L., Chijimatsu, M., Hernelind, J., Jing, L., Kobayashi, A. & Rutqvist, J. (2009) *A case study on the influence of THM coupling on the near-field safety of a spent fuel repository in sparsely fractured granite.* Journal of Environmental Geology, 57(6), 1239–1254.

Öhman, J., Niemi, A. & Tsang, C.-F. (2005) *Probabilistic estimation of fracture transmissivity from Wellbore hydraulic data accounting for depth-dependent anisotropic rock stress.* International Journal of Rock Mechanics and Mining Sciences, 42(5–6), 793–804.

Olivella, S. & Gens, A. (2005) *Double structure THM analyses of a heating test in a fractured tuff incorporating intrinsic permeability variations.* International Journal of Rock Mechanics and Mining Sciences, 42(5–6), 667–679.

Pan, P.-Z., Feng, X.-T., Huang, X.-H., Cui, Q. & Zhou, H. (2009) *Coupled THM processes in EDZ of crystalline rocks using an elasto-plastic cellular automaton.* Journal of Environmental Geology, 57(6), 1299–1311.

Pan, P. & Feng, X. (2013) *Numerical study on coupled thermo-mechanical processes in Äspö Pillar Stability Experiment.* Journal of Rock Mechanics and Geotechnical Engineering, 5(2), 136–144.

Rejeb, A. & Bruel, D. (2001) *Hydromechanical effects of shaft sinking at the Sellafield site.* International Journal of Rock Mechanics and Mining Sciences, 38(1), 17–29.

Rinne, M., Shen, B. & Backers, T. (2013) *Modelling fracture propagation and failure in a rock pillar under mechanical and thermal loadings.* Journal of Rock Mechanics and Geotechnical Engineering, 5(1), 73–83.

Rutqvist, J. (1995) *Determination of hydraulic normal stiffness of fractures in hard rock from well testing.* International Journal of Rock Mechanics and Mining Sciences & Geomechanics Abstracts, 32(5), 513–523.

Rutqvist, J., Börgesson, L., Chijimatsu, M., Kobayashi, A., Jing, L., Nguyen, T.S., Noorishad, J. & Tsang, C.-F. (2001a) *Thermohydromechanics of partially saturated geological media: governing equations and formulation of four finite element models.* International Journal of Rock Mechanics and Mining Sciences, 38(1), 105–127.

Rutqvist, J., Börgesson, L., Chijimatsu, M., Nguyen, T.S., Jing, L., Noorishad, J. & Tsang, C.-F. (2001b) *Coupled thermo-hydro-mechanical analysis of a heater test in fractured rock and bentonite at Kamaishi Mine—comparison of field results to predictions of four finite element codes.* International Journal of Rock Mechanics and Mining Sciences, 38(1), 129–142.

Rutqvist, J., Barr, D., Datta, R., Gens, A., Millard, A., Olivella, S., Tsang, C.-F., Tsang, Y. (2005a) *Coupled thermal–hydrological–mechanical analyses of the Yucca Mountain Drift Scale Test—Comparison of field measurements to predictions of four different numerical models.* International Journal of Rock Mechanics and Mining Sciences, 42(5–6), 680–697.

Rutqvist, J., Chijimatsu, M., Jing, L., Millard, A., Nguyen, T.S., Rejeb, A., Sugita, Y. & Tsang, C.-F. (2005b) *A numerical study of THM effects on the near-field safety of a hypothetical nuclear waste repository—BMT1 of the DECOVALEX III project. Part 3: Effects of THM coupling in sparsely fractured rocks.* International Journal of Rock Mechanics and Mining Sciences, 42(5–6), 745–755.

Rutqvist, J., Börgesson, L., Chijimatsu, M., Hernelind, J., Jing, L., Kobayashi, A. & Nguyen, T.S. (2009a) *Modelling of damage, permeability changes and pressure responses during excavation of the TSX tunnel in granitic rock at URL, Canada.* Journal of Environmental Geology, 57(6), 1263–1274.

Rutqvist, J., Bäckström, A., Chijimatsu, M., Feng, X.-T., Pan, P.-Z., Hudson, J.A., Jing, L., Kobayashi, A., Koyama, T., Lee, H.-S., Huang, X.-H., Rinne, M. & Shen, B.T. (2009b) *A multiple-code simulation study of the long-term EDZ evolution of geological nuclear waste repositories.* Journal of Environmental Geology, 57(6), 1313–1324.

Rutqvist, J., Barr, D., Birkholzer, J.T., Fujisaki, K., Kolditz, O., Liu, Q.-S., Fujita, T., Wang, W. & Zhang, C.-Y. (2009c) *A comparative simulation study of coupled THM processes and their effect on fractured rock permeability around nuclear waste repositories.* Journal of Environmental Geology, 57(6), 1347–1360.

Rutqvist, J., Leung, C., Hoch, A., Wang, Y. & Wang, Z. (2013) *Linked multicontinuum and crack tensor approach for modelling of coupled geomechanics, fluid flow and transport in fractured rock.* Journal of Rock Mechanics and Geotechnical Engineering, 5(1), 18–31.

Sonnenthal, E., Ito, A., Spycher, N., Yui, M., Apps, J., Sugita, Y., Conrad, M., Kawakami, S. (2005) *Approaches to modelling coupled thermal, hydrological, and chemical processes in the drift scale heater test at Yucca Mountain.* International Journal of Rock Mechanics and Mining Sciences, 42(5–6), 698–719.

Stephansson, O., Jing, L. & Tsang, C.-F. (eds.) (1996). Coupled thermo-hydro-mechanical processes of fractured media-mathematical and experimental studies: Recent development of DECOVALEX project for radioactive waste repositories. Elsevier Development in Geotechnical Engineering Series, Volume 79. Elsevier, Oxford.

Stephansson, O., Hudson, J.A. & Jing, L. (eds.) (2004). Coupled thermo-hydro-mechanical-chemical processes in geo-systems: Fundamentals, modelling, experiments and applications. Elsevier Geo-Engineering Book Series, Volume 2. Elsevier, Oxford.

Tsang, C.-F. (ed.) (1987) Coupled processes associated with nuclear waste repositories. Academic Press, New York.

Tsang, C.-F. (1991) *Coupled hydromechanical-thermomechanical processes in rock fractures.* Review of Geophysics, 29(4), 537–551.

Tsang, C.-F. (2009) *Introductory editorial to the special issue on the DECOVALEX-THMC Project.* Journal of Environmental Geology, 57(6), 1217–1219.

Tsang, C.-F., Jing, L., Stephansson, O. & Kautsky, F. (2005) *The DECOVALEX III project: A summary of activities and lessons learned.* International Journal of Rock Mechanics and Mining Sciences, 42 (5–6), 593–610.

Tsang, C.-F. & Jussila, P. (2005) *The FEBEX benchmark test: Case definition and comparison of modelling approaches.* International Journal of Rock Mechanics and Mining Sciences, 42 (5–6), 611–638.

Tsang, C.-F., Stephansson, O., Jing, L. & Kautsky, F. (2009) *DECOVALEX Project: From 1992 to 2007.* Journal of Environmental Geology, 57(6), 1221–1237.

Yasuhara, H., Elsworth, D. & Polak, A. (2004) *Evolution of permeability in a natural fracture: Significant role of pressure solution.* Journal of Geophysical Research, 109, B03204.

Yasuhara, H., Polak, A., Mitani, Y., Grader, A., Halleck, P. & Elsworth, D. (2006) *Evolution of fracture permeability through fluid-rock reaction under hydrothermal conditions.* Earth and Planetary Science Letters, 244, 186–200.

Yasuhara, H. & Elsworth, D. (2008) *Compaction of a rock fracture moderated by competing roles of stress corrosion and pressure solution.* Pure and Applied Geophysics, 165, 1289–1306.

Yasuhara, H., Kinoshita, N., Ohfuji, H., Lee, D.S., Nakashima, S. & Kishida, K. (2011) *Temporal alteration of fracture permeability in granite under hydrothermal conditions and its interpretation by coupled chemo-mechanical model.* Applied Geochemistry, 26, 2074–2088.

Zhang, C., Liu, X. & Liu, Q. (2013) *A thermo-hydro-mechano-chemical formulation for modelling water transport around a ventilated tunnel in an argillaceous rock.* Journal of Rock Mechanics and Geotechnical Engineering, 5 (2), 145–155.

Zhao, Z., Rutqvist, J., Leung, C., Hokr, M., Liu, Q., Neretnieks, I., Hoch, A., Havlíček, J., Wang, Y., Wang, Z., Wu, Y. & Zimmerman, R.W. (2013) *Impact of stress on solute transport in a fracture network: A comparison study.* Journal of Rock Mechanics and Geotechnical Engineering, 5(2), 110–123.

Chapter 2

Dynamic relaxation applied to continuum and discontinuum numerical models in geomechanics

Peter Cundall & Christine Detournay
Itasca Consulting Group Inc., Minneapolis, Minnesota, USA

Abstract: Rock often exhibits nonlinear, localized and unstable behavior, and rock masses contain discontinuities at all scales. This Chapter presents the 3D formulation of a numerical solution scheme that follows the dynamic response, even for quasi-static problems, so as to follow in a realistic manner nonlinear and unstable behavior, as well as localization such as fracturing. Fluid coupling in continua and discontinua is also described. Several examples are provided for diverse continuous and discontinuous systems, including hydro-fracturing.

1 INTRODUCTION

In geomechanics, and rock mechanics in particular, there are many examples of highly nonlinear, localized and unstable behavior. Rock samples in the laboratory display peak strengths and subsequent softening, and even snap-back. In the field, mine pillars exhibit spalling and progressive failure, and initially intact rock can fracture and develop localized features. Most rock contains many existing discontinuities, from micro-cracks to faults, which may slip, open and extend in complex, and often unstable, ways. Numerical schemes that were originally developed for structural analysis, such as matrix-based methods, are not ideally suited to modeling rock because the assumption of linearity inherent in the solution of matrix equations must somehow be circumvented to treat the behaviors noted above, for example, by using iterative procedures. Further, the notion of finding a "solution" using matrix-based schemes, even if nonlinearities are included, may not be realistic for processes that are path-dependent. For these reasons, we prefer to use a simulation scheme that follows the dynamic behavior of a system, even for quasi-static response. By solving the equations with an explicit, time-marching scheme, we also avoid the use of large matrices (Suave & Metzger, 1995) and can follow even softening behavior directly without needing iteration.

The numerical procedures described in this chapter have a common ancestor in Dynamic Relaxation (Day, 1965; Otter *et al.*, 1966), which embodies the explicit solution of the equations of motion and constitutive equations for application to static problems. Independently, Mark Wilkins and his colleagues were using similar methods for dynamic simulations (*e.g.*, Wilkins, 1969; Wilkins *et al.*, 1971), based on earlier work by Von Neumann and Richtmeyer (1950). Dynamic Relaxation (DR) replaces the differential equations of motion and elasticity by central difference expressions, using (in its original form) interlaced meshes in which stress components and velocity components are located at different points in space, so that central differences arise naturally.

In this chapter, we introduce the DR scheme in Section 1.1 by way of its implementation for a one-dimensional bar and an examination of the static and dynamic solutions for the bar. Then in Section 2, the method is generalized to a three-dimensional continuum. Section 3 describes the formulation applied to discontinuous media such as a rock mass, with a three-dimensional example of application to a large-scale engineering project. In Section 4, the application to fracture propagation is described, together with a mention of the synthetic rock mass (SRM), in which both new and existing fractures are modeled. Finally in Section 5, the formulations for fluid coupling in both continua and discontinua are presented with an example of 3D hydraulic fracturing. It is important to note that the same explicit dynamic solution scheme is embodied in all these seemingly diverse formulations.

1.1 A one-dimensional illustration of the solution scheme

The explicit DR scheme can be illustrated by a numerical model of the one-dimensional elastic bar shown in Figure 1, discretized into elements of equal width, Δx, where x increases from left to right.

The constitutive equation (of elasticity, in this example)

$$\sigma = E\frac{\partial u}{\partial x} \tag{1}$$

is replaced by the difference equations

$$\sigma_i^t = E\frac{u_{i+1}^t - u_i^t}{\Delta x} \tag{2}$$

And the equation of motion

$$\rho\ddot{u} = \frac{\partial \sigma}{\partial x} + \rho g \tag{3}$$

is replaced by the difference equations

$$\dot{u}_i^{t+\Delta t/2} = \dot{u}_i^{t-\Delta t/2} + \frac{(\sigma_i^t - \sigma_{i-1}^t)\Delta t}{\rho\,\Delta x} + g\Delta t$$

$$u_i^{t+\Delta t} = u_i^t + \dot{u}_i^{t+\Delta t/2}\Delta t \tag{4}$$

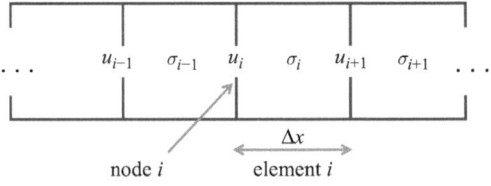

Figure 1 Location of stress (σ) and displacement (u) variables in a one-dimensional bar.

The time step is Δt, the Young's modulus is E, the density is ρ and the acceleration due to gravity is g. Subscripts denote sequential locations in space and superscripts denote the time at which the variable exists. A single dot denotes velocity and a double dot acceleration. Note that (a) central differences are used in both time and space (leading to second-order accuracy), and (b) all quantities on the right-hand sides of the difference expressions are known (*i.e.*, have already been evaluated); this is the essence of the explicit solution scheme. An explicit solution has the advantage that the constitutive behavior can be nonlinear without any change to the logic. Thus, expression (2) can simply be replaced by one in which stress is an arbitrary function of strain rather than the linear one shown; no iteration is necessary to follow a nonlinear relation. The disadvantage of the explicit method is that there is a strict limit on time step:

$$\Delta t < \Delta t_{crit} = \frac{\Delta x}{c} \tag{5}$$

where c is the maximum sound velocity in the medium. This limit is equivalent to the requirement that information cannot be communicated between neighboring elements in one time step. As a positive side to this limitation, it is not necessary to assemble and solve matrices because there is no connection between remote elements during the update from one time step to the next. Accordingly, the storage requirements for a DR solver are small.

1.2 Damping formulation for static solutions

The above equations are for the case with no energy dissipation (apart from dissipation that may occur within a nonlinear constitutive relation). When DR is used to solve static elastic problems, Otter *et al.* (1966) add a viscous term η to (3), which becomes $\rho\ddot{u} + \eta\dot{u} = \partial\sigma/\partial x + \rho g$, in order for the solution to converge to that of static equilibrium. However, this implies that viscous body forces act on nodes during the simulation. This is not realistic for common cases in geomechanics in which there is steady yielding, because the body forces will affect the solution, *e.g.*, when evaluating the factor of safety. An alternative damping formulation, applied to the equation of motion in difference form (4), is

$$\dot{u}_i^{t+\Delta t/2} = \dot{u}_i^{t-\Delta t/2} + \frac{\Delta t}{\rho\Delta x}\left[F_i^t - \alpha|F_i^t|\operatorname{sgn}\left(\dot{u}_i^{t-\Delta t/2}\right)\right] \tag{6}$$

where the unbalanced force, $F_i^t = \sigma_{i-1}^t - \sigma_i^t + \rho g\Delta x$ (for unit area cross-section), α is a non-dimensional constant and $\operatorname{sgn}(\dot{u}_i^t)$ is the sign of the velocity (where $\operatorname{sgn}(a) = 1$ if $a \geq 0$ and $\operatorname{sgn}(a) = -1$ if $a < 0$). This formulation has three advantages: first, the damping is zero for steady flow; second, it is controlled by a non-dimensional factor (in contrast to the viscous damping formulation, in which the viscosity coefficient has the dimension of $MT^{-1}L^{-3}$); and third, the damping does not need to be adjusted according to the frequencies present in the system being simulated. The damping scheme is called local damping, because the damping force is proportional to the local unbalanced force; α is commonly set to 0.8 for good convergence in quasi-static simulations.

1.2.1 An example that illustrates the operation of local damping and the solution cycle

To illustrate the operation of this form of damping, consider a gravity-loaded column of 10 elements that is fixed at its base (node 1), with $g = -10$, $E = 10^7$, $\rho = 10^3$, $\Delta x = 1$, $\alpha = 0.8$ and $\Delta t = 5 \times 10^{-3}$. The time step is chosen here to be half the critical value of $\Delta t_{crit} = \Delta x/c = 10^{-2}$, where the maximum wave speed in this example is $c = \sqrt{E/\rho} = 100$. Note that one solution step consists of applying Equation (2) to *all* elements, followed by the application of (4) to *all* nodes. This cycle is repeated 500 times, and the resulting displacement history of the top of the column (node 11) is shown in Figure 2, which shows that the motion converges to the static solution after a brief transient. The final displacement profile is shown in Figure 3, compared to the analytical solution given by the following formula.

$$u = \frac{g\rho}{E} \left\{ hx - \frac{x^2}{2} \right\} \tag{7}$$

where h is the height of the column, which is 10 units in this example.

Although local damping is good for quasi-static and plasticity problems, it is not appropriate for systems that involve free fall under gravity, dynamic simulations or impact problems; other formulations, such as strain-rate viscosity or a hysteretic constitutive law (*e.g.*, Cundall, 2006), should be used in these cases.

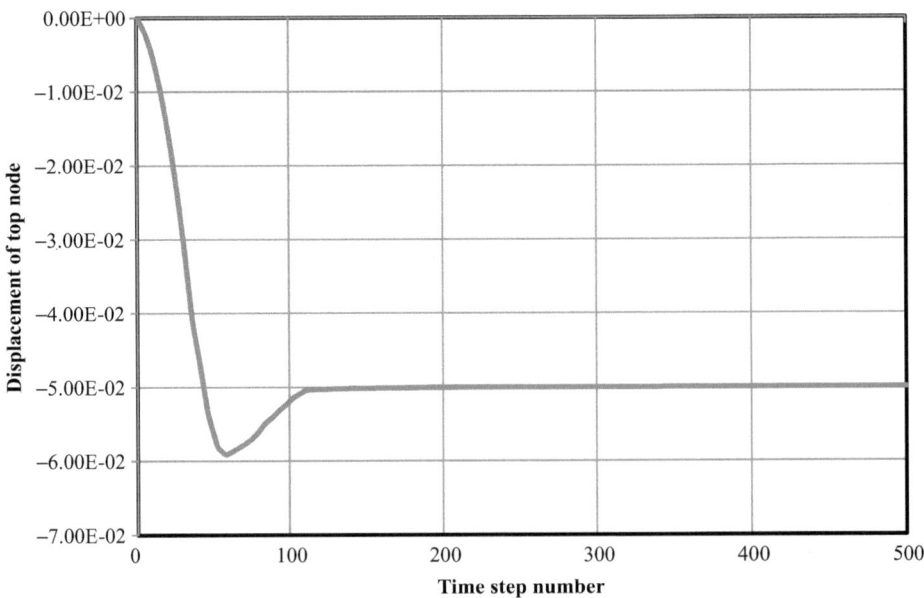

Figure 2 Displacement of top node versus time step for gravity-loaded column.

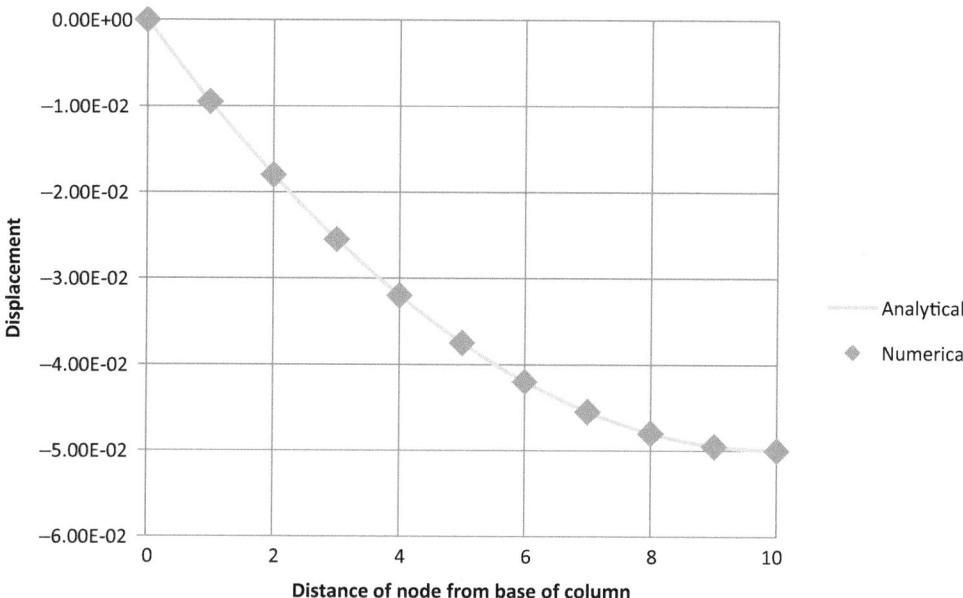

Figure 3 Profile of displacement: numerical results compared to analytical.

1.3 Mass-scaling for quasi-static problems

The simulation of quasi-static problems may be made by using the full dynamic equations of motion (as exemplified by Equations (3), (4) and (6)), and setting the damping coefficient α to an appropriate non-zero value. However, such solutions of systems of two- and three-dimensional grids (as used by the codes *FLAC* and *FLAC3D* [Itasca, 2011, 2012], for example) are more efficiently performed by scaling inertial masses of nodes. The reason for doing this is that, in general, both the grid geometry and the material properties may be non-uniform. Thus, regions that contain stiff and/or small elements have short natural periods, while regions that contain soft and/or large elements have long natural periods. Since the time step must be chosen to respect the smallest natural period, T_{min}, of the system ($\Delta t < T_{min}/\pi$), the response of the long-period regions will be sluggish and will prolong the overall time needed to reach final equilibrium. In an attempt to equalize the local response periods throughout the system, inertial masses may be adjusted so that the apparent time step for all elements is the same. (But note that gravitational body forces are unaffected). The formulation for mass scaling is described for the three-dimensional grid in Section 2.3.3.

1.4 Fully dynamic solutions

Fully dynamic solutions may be obtained by setting $\alpha = 0$ with no mass-scaling. Since the numerical scheme is an approximation to the governing differential equations, it is important to evaluate the accuracy of the solution. Note, however, that momentum is conserved exactly (to machine accuracy), because Equation (4) directly derives the change in momentum from the unbalanced force.

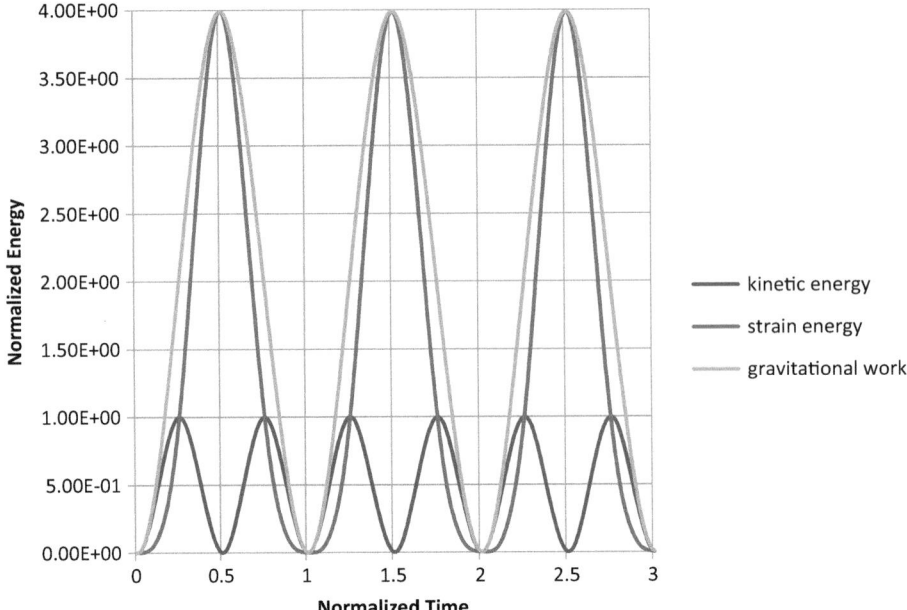

Figure 4 Time histories of energy components for undamped, gravity-loaded column.

1.4.1 Energy conservation

Although energy conservation is not directly enforced, the second-order, time-marching scheme ensures good accuracy in conserving energy. This assertion may be tested by considering the same example of a column (Section 1.2.1). In the dynamic solution ($\alpha = 0$), the sudden application of gravity causes the column to oscillate indefinitely at its natural period of $T = 4h/c$, as shown by the histories of energy quantities in Figure 4, in which the time scale is normalized by dividing the model time by T, and using $\Delta t = 0.5 T_{crit}$ with the simulation being carried out for three complete periods of oscillation. The energy components are kinetic energy, $E_k/\overline{E}_s$, strain energy, $E_s/\overline{E}_s$, and work done by gravity, $W_g/\overline{E}_s$, where the normalizing factor $\overline{E}_s$ is the strain energy of the column under equilibrium conditions: $\overline{E}_s = (g\rho)^2 h^3/6E$. The components are evaluated as follows, ensuring that all variables are evaluated at the same simulation time.

$$E_k = \frac{1}{2} \sum_{nodes} m\dot{u}^2$$

$$E_s = \frac{1}{2} \sum_{elements} E\varepsilon^2 \qquad (8)$$

$$W_g = \sum_{nodes} mgu$$

where the strain for element i is $\varepsilon_i = (u_{i+1} - u_i)/\Delta x$.

Table 1 Excess energy for various values of time step.

Fraction of critical time step	Maximum excess energy, E_t
0.5	1.77×10^{-3}
0.1	7.04×10^{-5}
0.05	1.76×10^{-5}
0.01	7.04×10^{-7}

To determine how well energy is conserved, we track the normalized excess energy E_t, where

$$E_t = (E_k + E_s - W_g)/\overline{E}_s \tag{9}$$

which should be zero for perfect energy conservation. The results are shown in Table 1 for various values of time step.

Even for the largest time step, the maximum error over the whole simulation period is less than 0.2%. To evaluate the long-term stability of the solution, the simulation (for the case of $\Delta t = 0.5\Delta t_{crit}$) was continued for one million time steps, and the maximum error over the whole simulation period was found to be 1.79×10^{-3}, which indicates that there is essentially no long-term drift (comparing this value to the value in row 1 of Table 1).

1.4.2 Wave propagation

The explicit solution scheme for continuum simulations embodies a diagonal mass matrix obtained by mass lumping (see Section 2.2). Given this approximation, it is instructive to evaluate the accuracy of wave propagation. Accordingly, the one-dimensional grid used in the previous examples (with the same properties and element dimensions without damping) was extended to allow a pulse to propagate without reflection. The pulse was introduced at one end of the grid by setting the nodal velocity to the value $\dot{u} = (1 - \cos(2\pi t/T_p))$ for time $t < T_p$ and $\dot{u} = 0$ for $t \geq T_p$, where T_p is the duration of the pulse. Since the wave speed is 100, the wavelength, λ_p, of the pulse is $100T_p$. The numerical wave speed was calculated by noting the times at which the half-amplitude (0.5) of the pulse passed locations $x = 100$ and $x = 300$. The wave amplitude was evaluated as the mean of maximum velocities at the same two locations. Table 2 records the errors in wave speed and pulse amplitude. Clearly the wavelength

Table 2 Results from wave-propagation tests.

$\lambda_p = N\Delta x$ where $N =$	$\Delta t = 0.5\Delta t_{crit}$		$\Delta t = 0.25\Delta t_{crit}$	
	Velocity error %	Amplitude error %	Velocity error %	Amplitude error %
10	−0.47	18.0	−0.56	−21.4
20	−0.26	−0.28	−0.29	−0.027
30	−0.22	0.11	−0.25	0.38
40	−0.10	0.098	−0.15	−0.089
50	−0.018	−0.046	−0.035	−0.026

determines the accuracy to a large extent. A common rule of thumb in dynamic analysis is that the shortest wavelength should be greater than ten times the element size, but the pulse used here contains higher frequency components than that implied by $1/T_p$, as would be shown by a Fourier analysis of the pulse (noting that only an infinitely extended sinusoidal wave contains a single frequency). Thus, a better criterion might be $\lambda_p = 20\,\Delta x$, which ensures reasonable accuracy both in wave speed and amplitude fidelity, as indicated in Table 2.

1.5 Practical implementation of the DR scheme

Several computer codes will be used in example applications presented in the remainder of this chapter. These codes are *FLAC* and *FLAC3D* (Itasca, 2011, 2012) for 2D and 3D continuum solutions, *UDEC* and *3DEC* (Itasca, 2004, 2008a) for 2D and 3D distinct element simulations of angular blocks, and *PFC* (Itasca, 2008b) for distinct element simulations of circular particle assemblies in 2D and 3D. All codes embody the same explicit, time-stepping algorithm presented above. Note that the codes also include many nonlinear constitutive models, structural coupling, thermal analysis, automatic rezoning and creep analysis, but the corresponding formulations will not be described further due to lack of space.

1.6 Notation

Throughout the rest of this Chapter, index notation will be used, whereby indices i,j,k take the values 1–3 and denote components of a vector or tensor in the global coordinate system. The summation convention applies for repeated indices in one term. The permutation tensor is e_{ijk}.

2 CONTINUUM: FINITE VOLUME METHOD

The basic equations used in the numerical scheme (including the equations of motion and the damping term) were originally developed using the conceptual model of a discrete assembly of nodes (carrying mass) connected by springs. However, it is also possible to derive identical discrete numerical equations starting from a continuum conceptual model.

2.1 Basic equations

The approach is documented in this section using *FLAC3D* as an example. The governing equations are listed in Section 2.1, the numerical formulation is summarized in Section 2.2, and the calculation scheme is outlined in Section 2.3.

2.1.1 Cauchy's equations of motion

As noted earlier, an important aspect of the model is the inclusion of the equations of motion, although *FLAC3D* is primarily concerned with the state of stress and deformation of the medium near the state of equilibrium. The continuum form of the momentum principle yields Cauchy's equation of motion:

$$\sigma_{ij,j} + \rho b_i = \rho \ddot{u}_i \qquad (10)$$

where ρ is the mass-per-unit volume of the medium, b_i is the body force per unit mass, and $\ddot{u}_i$ is the material derivative of the velocity. Note that, in the case of static equilibrium of the medium, the acceleration $\ddot{u}_i$ is zero, and Equation (10) reduces to the equations of equilibrium:

$$\sigma_{ij,j} + \rho b_i = 0 \qquad (11)$$

2.1.2 Rate of strain and rate of rotation

Let the particles of the medium move with velocity $\dot{u}_i$. In an infinitesimal time, dt, the medium experiences an infinitesimal strain determined by the translations $\dot{u}_i dt$, and the corresponding components of the strain-rate tensor may be written as

$$\dot{\varepsilon}_{ij} = \frac{1}{2}(\dot{u}_{i,j} + \dot{u}_{j,i}) \qquad (12)$$

where partial derivatives are taken with respect to components of the current position vector x_i (updated Lagrange approach).

Aside from the rate of deformation characterized by the tensor $\dot{\varepsilon}_{ij}$, a volume element experiences an instantaneous rigid-body displacement, determined by the translation velocity $\dot{u}_i$, and a rotation with angular velocity

$$\Omega_i = -\frac{1}{2} e_{ijk} \omega_{jk} \qquad (13)$$

where e_{ijk} is the permutation symbol, and ω_{ij} is the rate of rotation tensor whose components are defined as

$$\omega_{ij} = \frac{1}{2}(\dot{u}_{i,j} - \dot{u}_{j,i}) \qquad (14)$$

2.1.3 Constitutive equations

The equations of motion (10) together with the definitions (12) for the rates of strain constitute nine equations for fifteen unknowns (the 6 + 6 components of the stress- and strain-rate tensors and the three components of the velocity vector). Six additional relations are provided by the constitutive equations that define the nature of the particular material under consideration. They are usually given in the form

$$\dot{\sigma}_{ij}^c = H_{ij}(\sigma_{kl}, \dot{\varepsilon}_{kl}, \kappa) \qquad (15)$$

in which $\dot{\sigma}_{ij}^c$ is the co-rotational stress-rate tensor, H_{ij} is the functional form of the material constitutive law, and κ is a parameter that takes into account the history of loading. The co-rotational stress rate is equal to the material derivative of the stress as it would appear to an observer in a frame of reference attached to the material point and rotating with it at an angular velocity equal to the instantaneous value of the angular velocity Ω_{ij} of the material. The stress rate components referred to the fixed frame of reference are defined as:

$$\dot{\sigma}_{ij} = \dot{\sigma}^c_{ij} + \left(\omega_{ik}\sigma_{kj} - \sigma_{ik}\omega_{kj}\right) \tag{16}$$

in which the terms in parentheses accounts for the finite body rotations.

2.2 Numerical formulation

The DR method of solution used in *FLAC3D* has the following components:

1. Finite volume approach which eliminates the restriction of using regular grids. (First-order space derivatives of velocity are calculated by application of the Gauss divergence theorem, assuming linear variation of the variable over tetrahedral elements.)
2. Diagonal mass matrix obtained by mass lumping and diagonal damping matrix. (The continuous medium is effectively replaced by a discrete equivalent, one in which all forces involved—applied and interactive—and masses are concentrated at the nodes of a three-dimensional mesh used in the medium representation.)
3. Dynamic-solution approach. (The inertial terms in the equations of motion are used as numerical means to reach the equilibrium state of the system under consideration.)
4. Mixed discretization technique, based on low order constant strain-rate elements, which has the advantage of effectively increasing the order of the element without requiring the use of an element stiffness matrix, and in addition is capable of addressing volumetric locking in problems involving plastic deformation.

The laws of motion for the continuum are, by means of these approaches, transformed into discrete forms of Newton's law at the nodes, which are damped to provide static or quasi-static (non-inertial) solutions. The resulting system of ordinary differential equations is then solved numerically using an explicit finite difference formulation (second-order accurate), centered in time.

2.2.1 Nodal formulation of the equation of motion

To formulate the numerical scheme, the medium is discretized into constant strain-rate elements of tetrahedral shape whose vertices are the nodes of a three-dimensional mesh.

The forces $f_i^{(n)}$ acting on the nodes of a single tetrahedron in quasi-static equilibrium with the tetrahedron stresses and body forces are derived by application of the theorem of virtual work:

$$f_i^{(n)} = \frac{T_i^{(n)}}{3} + \frac{\rho b_i V}{4} - \int_V \rho N^{(n)} \ddot{u}_i dV \tag{17}$$

where the superscript (n) indicates the node, $N^{(n)}$ is a linear shape function, and

$$T_i^{(n)} = \sigma_{ij} n_j^{(n)} S^{(n)} \tag{18}$$

Also, $n_i^{(n)}$ is the outside unit normal to the tetrahedron face opposite node n, and $S^{(n)}$ is the area of the face.

The quasi-static equilibrium conditions are established by requiring that at each node, the sum of the statically equivalent forces f_i of all contributing tetrahedral and nodal contributions p_i of applied loads and concentrated forces be zero. This gives:

$$\Sigma \int_V \rho N \ddot{u}_i dV = \Sigma \frac{T_i}{3} + \Sigma \frac{\rho b_i V}{4} + \Sigma p_i \qquad (19)$$

A diagonal mass matrix is obtained by mass lumping:

$$m\ddot{u}_i = \Sigma \frac{T_i}{3} + \Sigma \frac{\rho b_i V}{4} + \Sigma p_i \qquad (20)$$

where $m = \Sigma \rho V / 4$.

2.2.2 Local non-viscous damping

The equations of motion must be damped to provide static or quasi-static (non-inertial) solutions. A local non-viscous damping term, similar to that described in Section 1.2 (Cundall, 1987) is added to Equation (20), such that the damped equations of motion can be written as:

$$m\ddot{u}_i = F_i \left(1 - \alpha \, \text{sgn}(F_i) \text{sgn}(\dot{u}_i) \right) \qquad (21)$$

Where α is a constant (equal to 0.8); sgn(x) is 1 if x > 0, –1 if x < 0, and 0 if x = 0; and the *unbalanced force F* is:

$$F_i = \Sigma \frac{T_i}{3} + \Sigma \frac{\rho b_i V}{4} + \Sigma p_i \qquad (22)$$

The local non-viscous damping formulation is similar to hysteretic damping, in which the energy loss per cycle is independent of the rate at which the cycle is executed. The advantages of this form of damping were discussed in Section 1.

2.2.3 Time integration method

The solution of (21) is obtained using an explicit time integration method. A central difference scheme is used because it has the largest stability limit for second-order accurate multistep integration formula (Krieg, 1973). With it, velocities are calculated at the mid-point of the time step considered for displacements and forces, and the approximations for the time derivatives are as follows:

$$\ddot{u}_i^t = \frac{\dot{u}_i^{t+\Delta t/2} - \dot{u}_i^{t-\Delta t/2}}{\Delta t}$$

$$\dot{u}_i^{t+\Delta t/2} = \frac{u_i^{t+\Delta t} - u_i^t}{\Delta t} \qquad (23)$$

The equations for advancing the velocity and displacement to the next time step are obtained by substituting Equation (23) in (21):

$$\dot{u}_i^{t+\Delta t/2} = \dot{u}_i^{t-\Delta t/2} + \frac{\Delta t}{m} F_i^t \left(1 - \alpha \, \text{sgn}\left(F_i^t\right) \text{sgn}\left(\dot{u}_i^{t-\Delta t/2}\right)\right) \tag{24}$$

$$u_i^{t+\Delta t} = u_i^t + \Delta t \dot{u}_i^{t+\Delta t/2} \tag{25}$$

where (see (22)):

$$F_i^t = \Sigma \frac{T_i^t}{3} + \Sigma \frac{\rho b_i^t V}{4} + \Sigma p_i^t \tag{26}$$

and (see (18)):

$$T_i^t = \sigma_{ij}^t n_j^t S_j^t \tag{27}$$

2.2.4 Strain rate calculation

For a constant strain rate tetrahedron, the components of the strain-rate tensor are computed using a finite volume formulation, derived by application of the Gauss divergence theorem to the tetrahedron:

$$\dot{\varepsilon}_{ij} = -\frac{1}{6V} \sum_{l=1}^{4} \left(\dot{u}_i^{(l)} n_j^{(l)} + \dot{u}_j^{(l)} n_i^{(l)}\right) S^{(l)} \tag{28}$$

where, by convention, $n_i^{(n)}$ is the outside unit normal to the tetrahedron face opposite node n, $S^{(n)}$ is the area of the face opposite node n, and the summation is on the four nodes of the tetrahedron.

2.2.5 Time step

The difference equations (24) to (27) will not provide valid answers unless the numerical scheme is stable, see Section 1.1. In theory, the stability of the central-difference integrator is expressed based on the Courant-Friedrichs-Lewy condition (Courant *et al.*, 1928) as:

$$\Delta t < \frac{2}{\sqrt{\lambda_m}} \tag{29}$$

where λ_m is the largest eigenvalue of the matrix represented by the product of the inverse of the (diagonal) mass matrix, M and the global tangent linear stiffness matrix, k (see Oakley *et al.*, 1995). An upper bound for λ_m may be obtained from Gerschgorin's theorem (Strang, 1976):

$$\lambda_m = \max_i \frac{\Sigma_j |k_{ij}|}{m_i} \tag{30}$$

where k_{ij} has units of [force/displacement]. In the quasi-static application of the DR method, the inertial terms in the equations of motion are used as numerical means to reach the equilibrium state of the system under consideration. Equations (29) and (30) show that the time step and nodal mass are not independent. The value of one of those two parameters can be fixed and the other one determined from the relationship. In *FLAC3D*, the time step is assigned a value of 1, and the nodal mass is calculated locally by adding the mass contribution of all tetrahedra meeting at the node. The expression for the tetrahedron mass contribution at node l is as follows:

$$m^{(l)} = \frac{K + 4G/3}{9V} (\Delta t)^2 \max \left(\left[n_i^{(l)} S^{(l)} \right]^2, i = 1, 3 \right) \qquad (31)$$

where K and G are the bulk and shear material modulus, respectively, V is the nodal volume, and $\Delta t = 1$. In the dynamic formulation, *FLAC3D* also uses lumped masses and a diagonal mass matrix, with "real" masses used at nodes instead of the fictitious masses used for optimum convergence in the static solution scheme.

2.2.6 Mixed discretization

Among three-dimensional constant strain-rate elements, tetrahedra have the advantage of *not* generating hourglass deformations (*i.e.*, deformation patterns created by combinations of nodal velocities producing no strain rate, and thus no nodal force increments). However, when used in the framework of plasticity, these elements do not provide for enough modes of deformation (see Nagtegaal *et al.*, 1974). In particular situations, for example, they cannot deform individually without change of volume as required by certain important constitutive laws. In those cases, the elements are known to exhibit an overly stiff response compared to what is expected from theory. To overcome this problem, a process of *mixed discretization* is applied in *FLAC3D*.

The principle of the mixed discretization technique is to give the element more volumetric flexibility by proper adjustment of the first invariant of the tetrahedron strain-rate tensor. Following the approach pioneered by Marti and Cundall (1982), a coarser discretization in hexahedral elements is superposed to the tetrahedral discretization, and the first strain-rate invariant of a particular tetrahedron in an element is evaluated as the volumetric-average value over all tetrahedron in the element. In situations where the volume of the element is to remain constant, for example, application of the mixed discretization process allows each individual tetrahedron in an element to reflect this behavior.

2.3 Main calculation steps

FLAC3D uses an explicit, time-marching, finite difference solution scheme. For every time step, the calculation sequence can be summarized by the following:

1. New strain rates are derived from nodal velocities for all elements.
2. Constitutive equations are used to calculate new stresses from the strain rates and stresses at the previous time.
3. The equations of motion are invoked to derive new nodal velocities and displacements from stresses and forces for all nodes.

The sequence is repeated at every time step, and the maximum unbalanced force in the model is monitored. This force will either approach zero, indicating that the system is reaching an equilibrium state or steady motion (such as plastic flow), or it will approach a constant, nonzero value, indicating that a portion (or all) of the system is accelerating, thus indicating a physical instability. It is important to note that physical instability does not lead to numerical instability in the explicit DR method. The calculation may be interrupted at any point in order to analyze the solution.

A summary of equations that are used in the calculation sequence is presented below. In the notation, superscripts are used to particularize a node or tetrahedron only when needed in the local context.

2.3.1 Strain rate calculation

For increased accuracy, a typical *FLAC3D* mesh uses hexahedral elements, composed of two overlays of five tetrahedra each.

Starting from a known velocity field, the incremental components of the strain-rate tensor are computed, for each tetrahedron in an element, using Equation (28):

$$\Delta \varepsilon_{ij} = -\frac{\Delta t}{6V} \sum_{l=1}^{4} \left(\dot{u}_i^{(l)} n_j^{(l)} + \dot{u}_j^{(l)} n_i^{(l)} \right) S^{(l)} \tag{32}$$

For the hexahedral grid, a procedure of mixed discretization is applied, and new diagonal incremental strain-rate tensor components are calculated. For a particular tetrahedron, now locally labeled l in the element, these equations yield

$$\dot{\varepsilon}_{ij}^{[l]} = \dot{\eta}_{ij}^{[l]} + \frac{\dot{\varepsilon}^z}{3} \delta_{ij} \tag{33}$$

where $\dot{\eta}_{ij}$ is the deviatoric strain rate tensor, and $\dot{\varepsilon}^z$, the average value for the element of the first strain-rate invariant, is calculated from

$$\dot{\varepsilon}^z = \frac{\sum_{k=1}^{n_t} \dot{\varepsilon}^{[k]} V^{[k]}}{\sum_{k=1}^{n_t} V^{[k]}} \tag{34}$$

The total number of tetrahedra involved in the element computation is n_t, and the volume of a tetrahedron labeled k is $V^{[k]}$.

2.3.2 Stress calculation

The constitutive equations (see Equations (15) and (16)) are used in their incremental form, to calculate stress increments from strain increments (see Equation (32)) for each tetrahedron in an element:

$$\Delta \sigma_{ij} = H_{ij} \left(\sigma_{kl}, \dot{\varepsilon}_{kl}, \kappa \right) \Delta t + \Delta \sigma_{ij}^R \tag{35}$$

where

$$\Delta \sigma_{ij}^R = \left(\omega_{ik} \sigma_{kj} - \sigma_{ik} \omega_{kj} \right) \Delta t \tag{36}$$

and

$$\omega_{ij} = -\frac{1}{6V} \sum_{l=1}^{4} \left(\dot{u}_i^{(l)} n_j^{(l)} - \dot{u}_j^{(l)} n_i^{(l)} \right) S^{(l)} \tag{37}$$

The stress correction, $\Delta\sigma_{ij}^R$ is taken into consideration in large-strain mode.

New stress values are then derived by addition of the stress increments.

For the hexahedral grid, a technique of mixed discretization is applied, and the diagonal components of the stress-tensor components are adjusted accordingly. For a particular tetrahedron, locally labeled l in the element, these equations are

$$\sigma_{ij}^{[l]} = s_{ij}^{[l]} + \sigma^z \delta_{ij} \tag{38}$$

where s_{ij} is the deviatoric stress tensor, and

$$\sigma^z = \frac{\sum_{k=1}^{n_t} \sigma^{[k]} V^{[k]}}{\sum_{k=1}^{n_t} V^{[k]}} \tag{39}$$

2.3.3 Nodal mass calculation

In quasi-static analyses, the tetrahedron contribution $m^{(l)}$ to the nodal mass at local node l is evaluated using:

$$m^{(l)} = \frac{K + 4G/3}{9V} \max \left(\left[n_i^{(l)} S^{(l)} \right]^2, i = 1,3 \right) \tag{40}$$

noting that the magnitude of the time step is set equal to unity.

The nodal mass $M^{(l)}$ at global node l is computed as the sum of contributions at the node of all tetrahedra having that node in common:

$$M^{(l)} = \sum m^{(l)} \tag{41}$$

In the small-strain mode, the nodal mass calculation is only carried out once before cycling. In large-strain mode, values are updated every 10 steps, or less for rapidly-changing geometry.

2.3.4 Unbalanced force calculation

The tetrahedron contribution p_i^l to the unbalanced force at local node l is computed from the stresses and body forces using

$$p_i^l = \frac{1}{3} \sigma_{ij} n_j^{(l)} S^{(l)} + \frac{1}{4} \rho b_i V \tag{42}$$

The sum of contributions (42) at global node l of all tetrahedra having that node in common is evaluated. Averaging over two overlays is carried out when appropriate, and the contributions $P_i^{(l)}$ of applied loads and concentrated forces are added, thus providing the expression for the unbalanced force $F_i^{(l)}$ at global node l:

$$F_i^{(l)} = \sum p_i^{(l)} + P_i^{(l)} \tag{43}$$

The unbalanced force is monitored to detect whether the system has reached a state of equilibrium or steady flow. The damping term is evaluated as

$$D_i^{(l)} = -\alpha \left| F_i^{(l)} \right| \mathrm{sgn}\left(\dot{u}_i^{(l)} \right) \tag{44}$$

and is added to the unbalanced force.

2.3.5 Velocity and displacement calculations

Equation (24) is invoked, taking the damping into account, to derive new nodal velocities from known unbalanced forces:

$$\dot{u}_i^{t+\Delta t/2} = \dot{u}_i^{t-\Delta t/2} + \frac{\Delta t}{M^{(l)}} \left(F_i^{(l)} + D_i^{(l)} \right) \tag{45}$$

In turn, nodal displacements are evaluated using:

$$u_i^{t+\Delta t} = u_i^t + \dot{u}_i^{t+\Delta t/2} \Delta t \tag{46}$$

2.3.6 Geometry update calculation

In large-strain mode, node locations are updated, and the grid then deforms accordingly. For global node l, we have

$$x_i^{t+\Delta t} = x_i^t + u_i^{t+\Delta t/2} \Delta t \tag{47}$$

In small-strain mode, this geometry adjustment is not performed.

2.4 Applications

Actual slopes are not infinitely long and straight; usually, they are curved in both plan and elevation. The effect of slope curvature can really only be analyzed with a three-dimensional model.

Hoek and Bray (1981) observed that the lateral restraint provided by material on either side of a potential slope failure will increase as the slope becomes more concave. They recommend that when the radius of curvature of the slope is less than the height of the slope, the allowed slope angle can be 10° steeper than the angle suggested by conventional two-dimensional stability analyses. Further, for radii of curvature greater than twice the slope height, the maximum slope angle given by a two-dimensional analysis should be used.

The model shown in Figure 5 represents a quarter-section of an open pit. The height of the slope is 25 m, and the slope angle is 2 vertical to 1 horizontal (approximately 63°). It is expected that plane-strain conditions will prevail along the plane shown on the front to the right in Figure 5 ($y = -30$m), while axisymmetric conditions will be predominant at the plane on the front to the left in the same figure ($x = 0$).

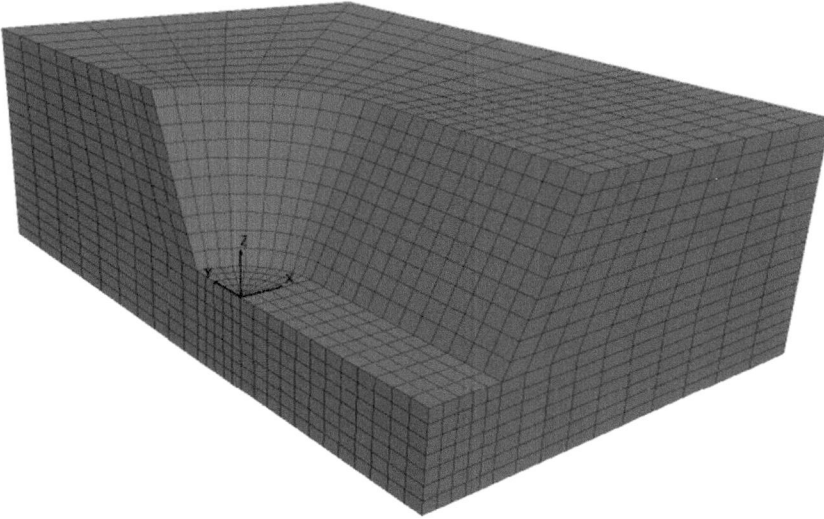

Figure 5 FLAC3D grid to evaluate stability of a slope.

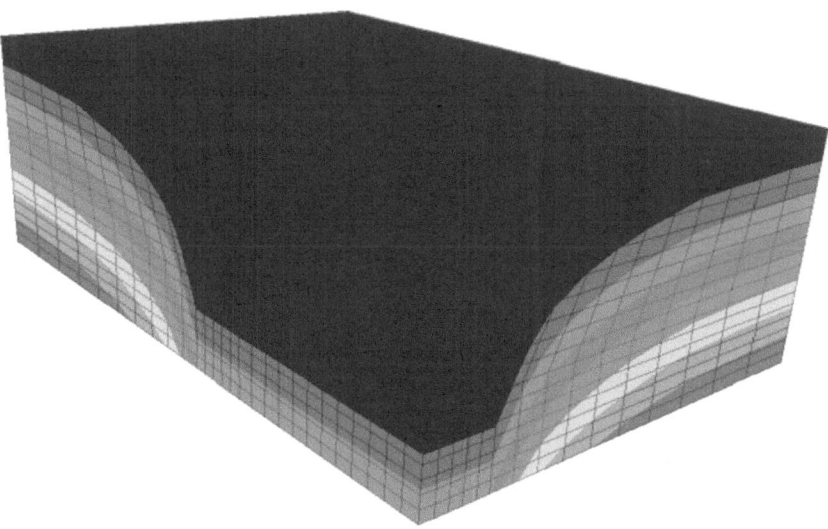

Figure 6 Pore pressure contours.

The free water surface imposed in this problem intersects the top of the model 50 m behind the toe of the slope, and there is seepage on the bottom half of the slope face. This water table, under steady-state conditions, will lead to the pore-pressure distribution shown in Figure 6.

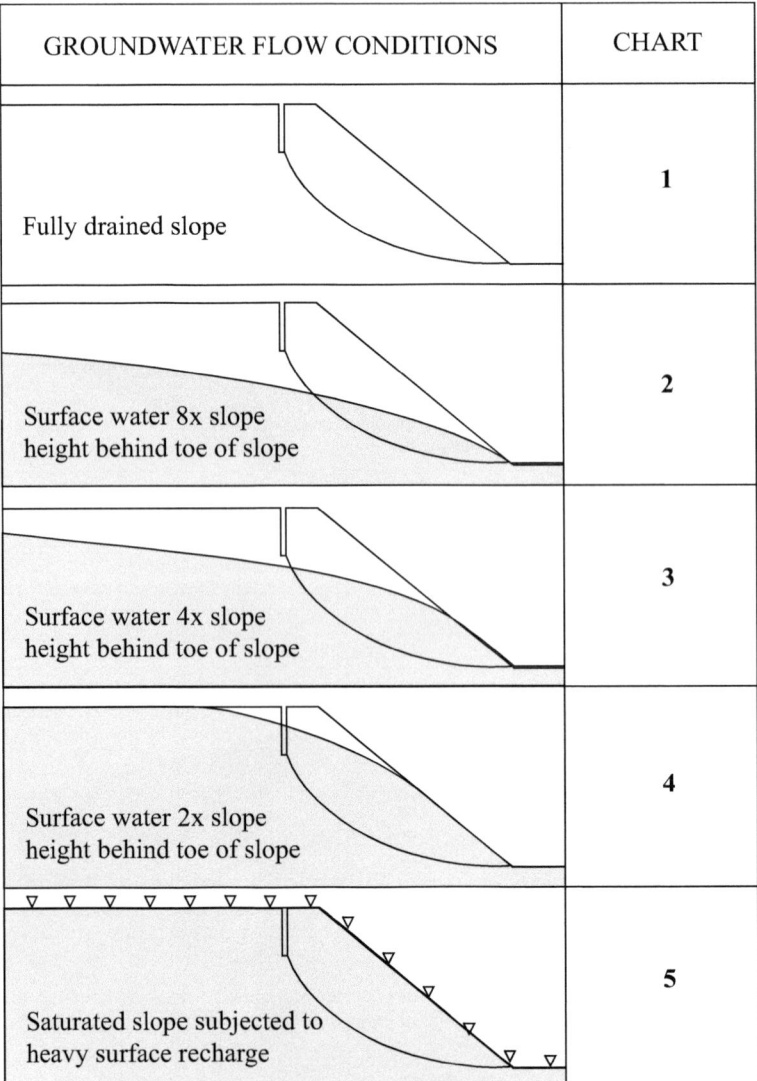

GROUNDWATER FLOW CONDITIONS	CHART
Fully drained slope	1
Surface water 8x slope height behind toe of slope	2
Surface water 4x slope height behind toe of slope	3
Surface water 2x slope height behind toe of slope	4
Saturated slope subjected to heavy surface recharge	5

Figure 7 Chart number as a function of groundwater conditions (adapted from Wyllie & Mah, 2004).

The strength parameters chosen for this model are selected for comparison of *FLAC3D* results to circular 2D (plane strain) failure charts published by Hoek and Bray (1981). Figure 7 shows which chart to use as a function of the groundwater flow conditions. In our case, the chart used is number 4. For example, if we assume a friction angle of 45° (tan $\phi = 1$) and a factor of safety $F = 1$, then we can draw a horizontal line in chart number 4 (see Figure 8) until we intersect the slope angle of 63°. If we draw a vertical line, we obtain a value of 0.06 for $c/\gamma HF$. For a specific weight, γ, of 25,000 N/m^3 and a height, H, of 25 m, we obtain a cohesion of 37.5 kPa.

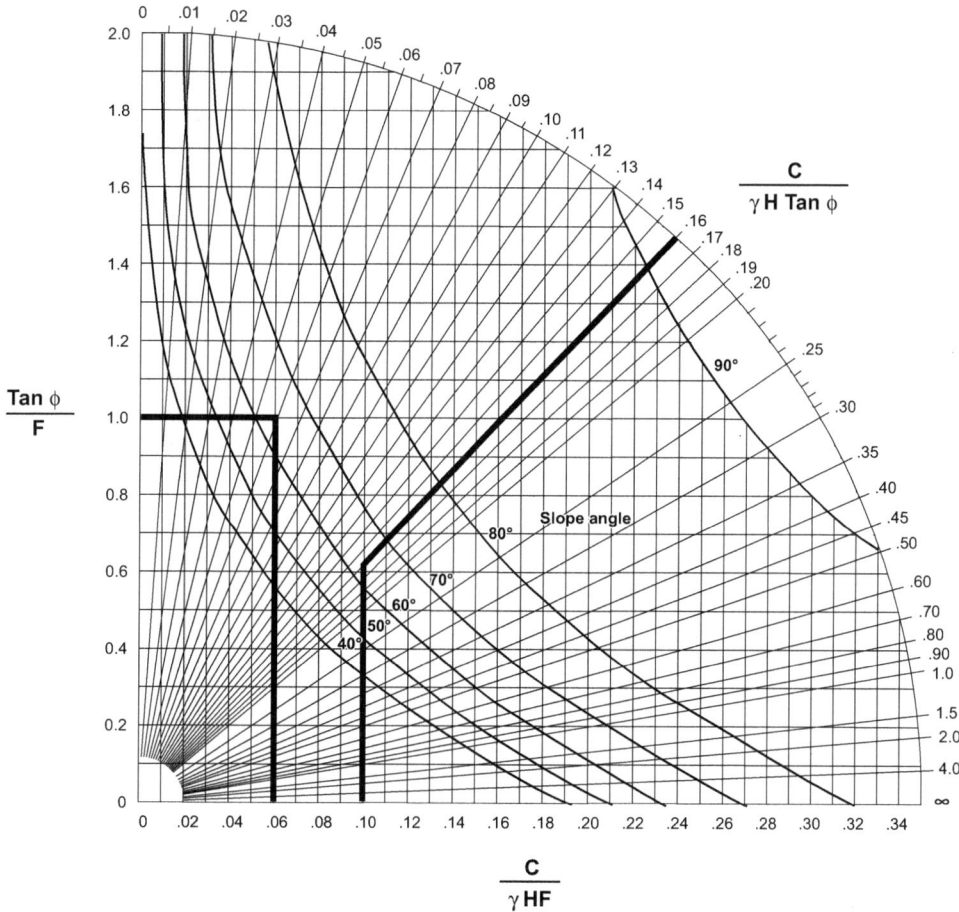

Figure 8 Circular failure chart number 4 (Wyllie & Mah, 2004).

For our analysis, we select a cohesion value of 100 kPa in order to start with a stable slope. The value for $c/(\gamma H \tan\phi)$ is then 0.16 and, using Figure 8, the value for $c/(\gamma HF)$ is 0.1 and F is 1.61.

The factor of safety for the *FLAC3D* model is calculated by the strength reduction method (Dawson *et al.*, 1999). A value of 1.70 is calculated for F. This is slightly higher than the factor of safety produced by the circular failure chart, which suggests that there is a slight effect of slope curvature on the stability. The resulting failure surface is depicted by the shear strain rate contour plot shown in Figure 9. This plot shows that a "scoop-shaped" failure surface develops along the long side of the open pit, but the slope is stable at the curved end.

This problem was also run with the two-dimensional program *FLAC* in both plane-strain mode and axisymmetry mode. The model geometry was created to match that in the vertical section through the *FLAC3D* model.

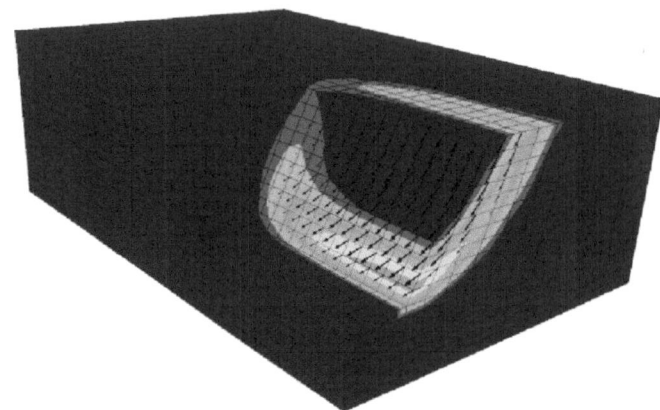

Figure 9 Shear strain rate contours and velocity in the *FLAC3D* model at the failure state.

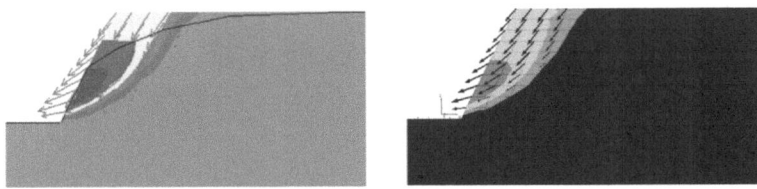

Figure 10 Displacement contours and vectors for plane-strain *FLAC* model (left) and for *FLAC3D* model along a plane at y = −30 (right).

The calculation for factor of safety in the plane-strain model matches that from the circular failure chart, $F = 1.61$. The displacement contour and vector plot at failure show a similar failure surface to that from *FLAC3D*. Compare the left and right graphics in Figure 10, which plots displacement contours and vectors on a vertical plane through the *FLAC3D* model at $y = -30$.

The factor-of-safety calculation for the axisymmetric model produces a value for $F = 2.35$. This further indicates that the greater curvature produces a more stable slope.

3 DISCONTINUUM: DISTINCT ELEMENT METHOD

3.1 Introduction and background

The response of a rock mass is often dominated by discontinuities that cut through the rock, because they are usually much weaker and more deformable than the intact rock. Thus, as a first approximation, the behavior of a rock mass under light loads (when intact deformation is small) can be treated as an assemblage of rigid, angular blocks that interact at their points and planes of contact. Cundall (1971) described such a numerical model and applied it to the toppling failure of a rock slope. Subsequently, the method became known as the distinct element method (Cundall & Strack, 1979) or the discrete element method (DEM) and it has been extended to deformable rock blocks and applied to such diverse systems as granular material, masonry structures and hydraulic fracturing. Although rock discontinuities had previously been introduced (as specialized elements) into the finite element method (Goodman *et al.*, 1968), the DEM is different because it characterizes a joint as a nonlinear boundary condition, rather than an element, and it allows arbitrary displacement and rotation of rock blocks and unlimited freedom for any object to interact with any other object.

Any viable DEM code requires an underlying process that continuously identifies pairs of neighboring blocks and the specific entities (corners, edges and faces) that may interact between each pair. These tasks must execute in linear time for the code to be efficient for simulating systems with thousands of blocks and complex block shapes. One such scheme is described by Cundall (1988), but there are other algorithms that also avoid the polynomial-time searches implied by a brute-force approach.

3.2 General formulation in three dimensions

The mechanical formulation of a DEM simulation follows the general scheme of dynamic relaxation presented in Section 1, but with the stress-strain equation of a continuum element replaced by a force-displacement equation of a spring that is presumed to exist at a contact point between neighboring blocks. Also, the equations of translational motion are supplemented by a rotation (or spin) equation. Following the analysis of Hart *et al.* (1988), the relative contact velocity, $\dot{u}_i^R$, at a contact point between block B and block A is:

$$\dot{u}_i^R = \dot{u}_i^B + e_{ijk}\omega_j^B\left(x_k^C - x_k^B\right) - \dot{u}_i^A - e_{ijk}\omega_j^A\left(x_k^C - x_k^A\right) \tag{48}$$

where x_i^C is the position vector of the contact point, x_i^A and x_i^B are the position vectors of the centroids of blocks A and B, $\dot{u}_i^A$ and $\dot{u}_i^B$ are their translational velocity vectors and ω_i^A and ω_i^B are their corresponding angular velocity vectors. Note that all terms in Equation (48) are at the point $(t + \Delta t/2)$ in time. At each contact point, a unit normal vector n_i is established, either by the "common-plane" logic described by Cundall (1988) or by some other scheme. This enables the contact force vector, F_i to be decomposed into normal (scalar F^n) and shear (vector F_i^s) components:

$$\begin{aligned} F^n &= F_i n_i \\ F_i^s &= F_i - F^n n_i \end{aligned} \tag{49}$$

The force components are updated from the relative velocity vector derived from Equation (48):

$$\begin{aligned} F^{n,t+\Delta t} &= F^{n,t} - k^n \dot{u}_i^{R,t+\Delta t/2} n_i^{t+\Delta t/2} \Delta t \\ F_i^{s,t+\Delta t} &= F_i^{s,t} - k^s\left(\dot{u}_i^{R,t+\Delta t/2} - \dot{u}_k^{R,t+\Delta t/2} n_k^{t+\Delta t/2} n_i^{t+\Delta t/2}\right)\Delta t \end{aligned} \tag{50}$$

where k^n, k^s are the apparent contact stiffnesses in units of force/displacement. For a point contact, these stiffnesses are specified directly, but for a planar contact (*e.g.*, a rock joint), the stiffness are $k^n = K^n A^c$ and $k^s = K^s A^c$, where K^n, K^s are the joint stiffnesses (in units of stress/displacement) and A^c is the contact area. The second superscripts to the forces and velocities in (50) denote the times at which the variables exist. Nonlinear contact stiffnesses may be substituted with no essential change to the logic, because the explicit solution scheme is used. In particular, tensile breaking of a bonded contact is modeled by setting F^n to zero if $F^n < -T$, where T is the tensile strength (in force units). Similarly, a shear strength may be specified, such that sliding does not occur when the shear force is below the given strength. Sliding of a frictional contact (assuming any shear bond is broken) may be achieved by adjusting the shear force as follows:

$$\text{if } \left| F_i^s \right| > F^n \tan \phi$$
$$F_i^s \leftarrow F_i^s \left(F^n \tan \phi / \left| F_i^s \right| \right) \tag{51}$$

where ϕ is the friction angle, the symbol $\leftarrow$ denotes "replaced by", and a compressive normal force is taken as positive. Once the normal and shear forces have been obtained, the total contact force vector is given by $F_i = F^n n_i + F_i^s$, and the force and moment sums acting on the two blocks are calculated as follows:

$$F_i^A \leftarrow F_i^A + F_i$$
$$M_i^A \leftarrow M_i^A + e_{ijk}(x_j^C - x_j^A)F_k$$
$$F_i^B \leftarrow F_i^B - F_i$$
$$M_i^B \leftarrow M_i^B - e_{ijk}(x_j^C - x_j^B)F_k \tag{52}$$

where the force F_i^D and moment M_i^D acting on block D are the sums of contributions from all contacts on the block for the current time $(t \leftarrow t + \Delta t)$. The forces are reset to zero after the equations of motion have been applied for each time step. The central difference expressions for the translational equations of motion of the block centroid are:

$$\dot{u}_i^{D,t+\Delta t/2} = \dot{u}_i^{D,t-\Delta t/2} + F_i^{D,t} \Delta t / m^D$$
$$x_i^{D,t+\Delta t} = x_i^{D,t} + \dot{u}_i^{D,t+\Delta t/2} \Delta t \tag{53}$$

for block D, where m^D is the mass of block D. The rotational motion of an undamped rigid body is described by Euler's equations:

$$I_1 \dot{\omega}_1 + (I_3 - I_2)\omega_3 \omega_2 = M_1$$
$$I_2 \dot{\omega}_2 + (I_1 - I_3)\omega_1 \omega_3 = M_2$$
$$I_3 \dot{\omega}_3 + (I_2 - I_1)\omega_2 \omega_1 = M_3 \tag{54}$$

where I_i is a component of the moment of inertia tensor about principal axis i, and the subscripts here refer to principal axes. For a fully dynamic simulation, the above equations need to be solved numerically, which is complicated by the presence of non-linear coupling terms. For quasi-static simulations, these second-order terms are small and the inertia tensor can be approximated by a scalar, I^D, for block D. In this case,

$$\omega_i^{D,t+\Delta t/2} \approx \omega_i^{D,t-\Delta t/2} + M_i^{D,t} \Delta t / I^D \tag{55}$$

Following the centroid updates of Equations (53) and (55), block vertices are updated using an expression similar to (48), in which the contact location is replaced by the vertex location. Of course, if the body is a sphere, then no such update is necessary. For fully dynamic simulations of systems of angular blocks, the Euler equations (54) are solved in an iterative manner.

Local damping may be added to equations (53) and (55) for quasi-static simulations, as described in Section 1, although for simulations involving impact of bodies, other forms of energy dissipation are more appropriate and realistic, such as hysteretic damping at contacts.

The above set of equations relates to rigid blocks or particles, which is a good simplification when contact forces are low enough that the deformation of intact material is small compared to relative displacements of bodies by sliding or separation at contacts. The assumption is reasonable for circular particles representing granular material, and the code *PFC* (Itasca, 2008b) uses the above formulation directly for modeling disks and spheres. When intact block deformations need to be taken into account for systems of polygons or polyhedra, the blocks are discretized, and the Finite Volume method described in Section 2 is used to compute the response of the intact material. The code *3DEC* (Itasca, 2008a) is a three-dimensional DEM package that embodies the equations set out above, as well as the continuum formulation for deformable blocks and full fluid/mechanical interaction and flow in joints. Some examples of the use of *3DEC* are provided below.

3.3 Example of large 3D engineering project studied with the DEM

3DEC has been used to model many aspects of the Baihetan hydropower structures in Western China (Meng & Zhu, 2009; Jiang *et al.*, 2010). The underground structures in one of the two dam abutments are shown in Figure 11.

The major discontinuities were included in the model (see Figure 12) together with *in-situ* stresses that were first studied by a preliminary *3DEC* simulation of the regional tectonic processes. After simulating the excavation of the various underground structures shown in Figure 11, the displacement field could be studied, with particular emphasis on the influence of the various discontinuities. For example, Figure 13 shows contours of displacements on selected surfaces. It was found that the shear zones (labeled C3, C4 and C5) had a significant impact.

The above studies were made with the deformable block option of *3DEC*. In order to determine the potential for structurally controlled failure, the rigid-block option of *3DEC* was used to construct a model composed of many thousands of blocks that are formed by the intersections of all types of discontinuities, such as bedding planes, joints and faults. A view of the blocky system around the caverns is shown in Figure 14, which also includes the stereonet upon which the statistical generation of joint planes was based. Figure 15 shows potentially unstable blocks calculated by *3DEC* for the surge chamber.

The example illustrates how the 3D DEM can be used to model deformations and failure modes in large-scale and complex engineering work. The method can handle many hundreds of discontinuities and thousands of discrete blocks of arbitrary shapes.

Note that the *3DEC* model documented in this example represents the 2011 facility layout of the Baihetan hydropower plant. This particular plan has been changed to the final facility layout with eight chambers. The *3DEC* analysis has played a supporting role for the current cavern layout design. In particular, numerical comparison studies

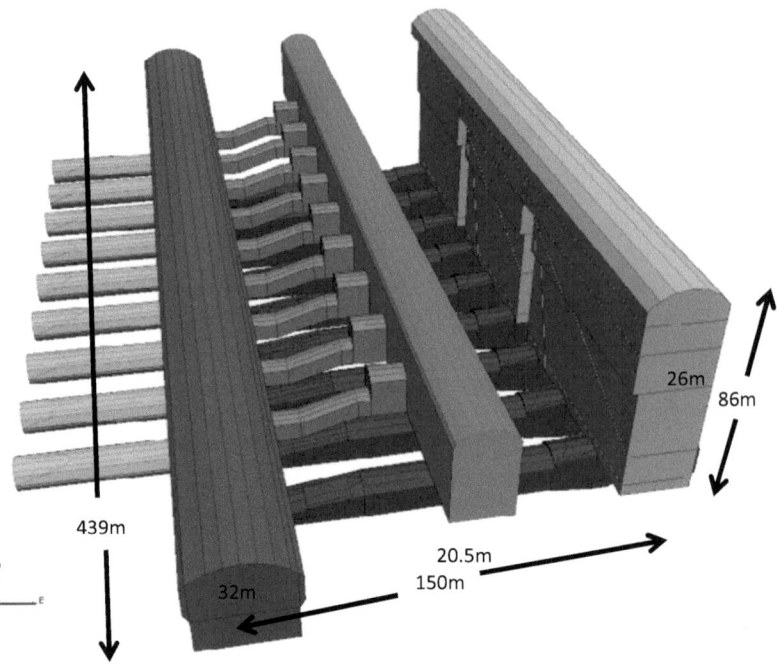

Figure 11 View of underground structures in one abutment of Baihetan site, including the powerhouse (left chamber), the transformer chamber (center chamber) and the tailrace surge tank (right chamber).

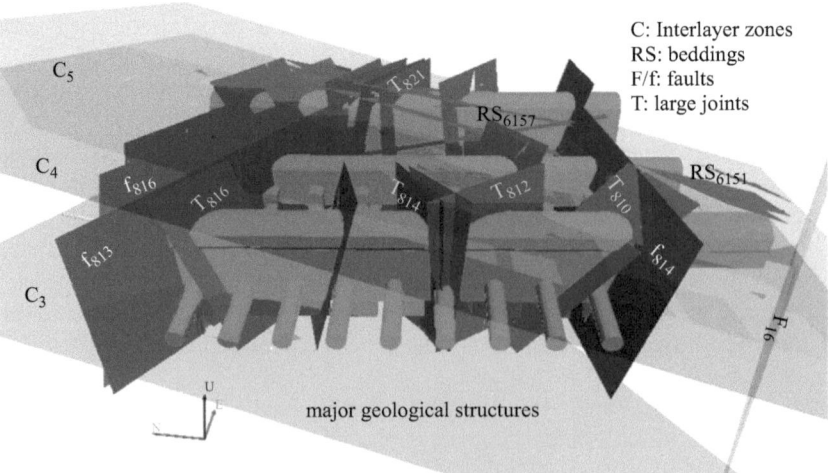

Figure 12 Major geological discontinuities. Courtesy G. Meng (Figures 11–14).

Figure 13 Contours of displacement on selected surfaces.

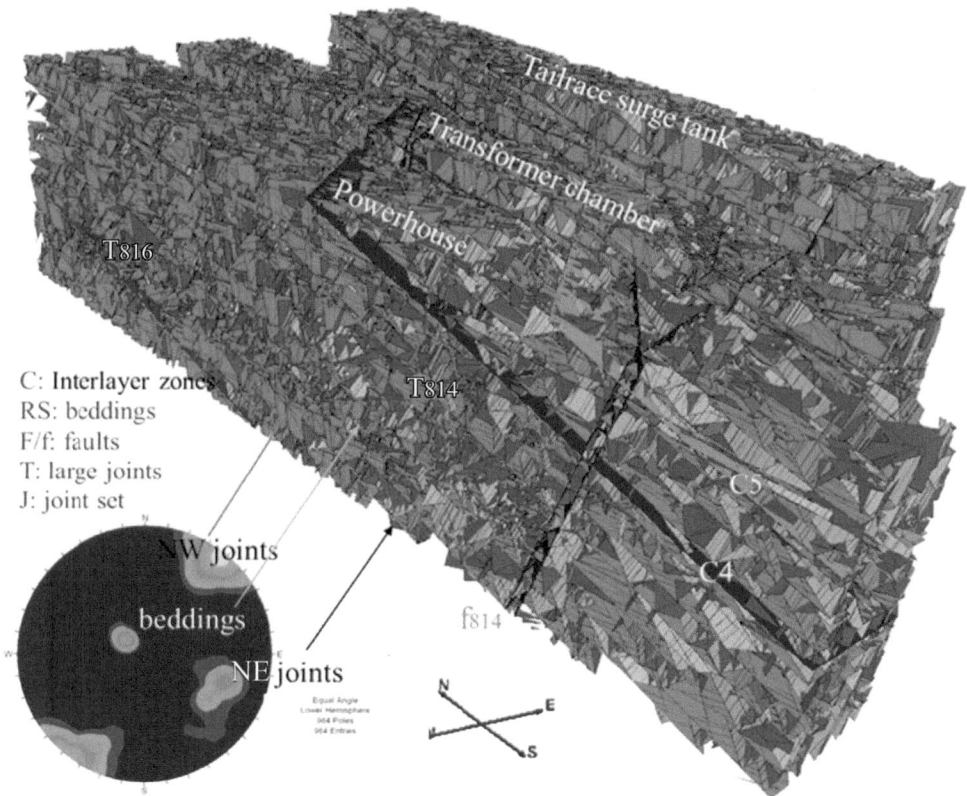

Figure 14 Joint planes, bedding and faults surrounding the three main caverns.

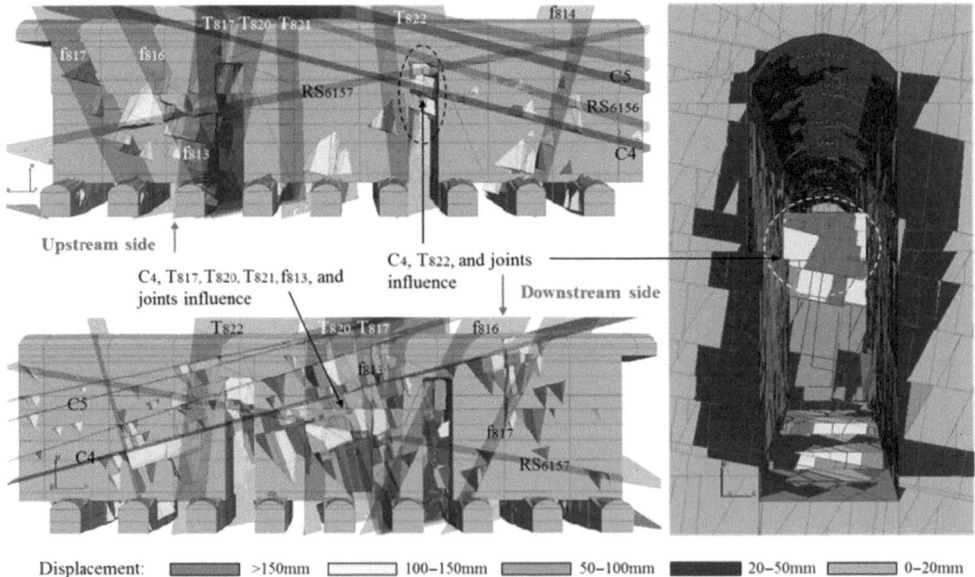

Displacement: ▣ >150mm ▢ 100–150mm ▢ 50–100mm ■ 20–50mm ▢ 0–20mm

Figure 15 Potentially unstable blocks simulated by *3DEC* for the surge chamber.

have shown that the rock mass surrounding the chamber was more stable for a cylindrical chamber compared to a chamber in the shape of a long corridor.

4 FRACTURES AND DISCONTINUITIES

4.1 Localization in continuum models

Existing (*i.e.*, *in-situ*) discontinuities may be modeled as discussed in Section 3. In rock, new fractures or localized features may appear in response to loading. There are several approaches to modeling the genesis of such fractures, depending on the nature of the process. Shear bands (such as faults) may evolve as strain localizations within a continuum. Although the thickness of a shear band depends on the grid resolution, it is generally not necessary to perform remeshing; rather, the band simply manifests as a local distortion of the mesh. As an example, Figure 16 shows a family of shear bands that forms in a sample that is distorted by tilting part of a loading frame. The left-hand sketch shows experimentally observed bands (Mandl, 1988) and the right-hand plot is of strain rate contours from a corresponding simulation with *FLAC*, using a constitutive model with weak strain-softening (Cundall, 1990).

4.2 Bonded particle models to represent tensile fracture

In contrast with shear bands, tensile cracks are difficult to resolve in a continuum grid without breaking or reforming the grid. An alternative approach is to represent the rock material as a bonded particle model (BPM), which consists of a closely packed assembly of particles (simulated by the distinct element method: see Section 3 for the

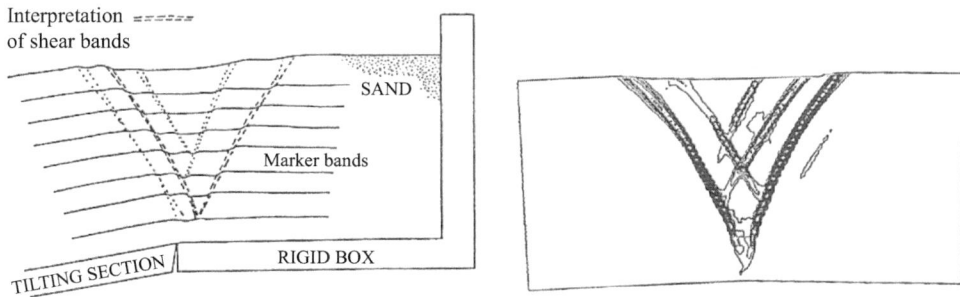

Figure 16 Left: Experimentally observed shear bands. Right: Numerical simulation with *FLAC*.

formulation) with bonds between neighbors that break at a specified magnitude of tensile force. Such bond breaks are regarded as micro-cracks, which may link up to form macro-cracks, depending on boundary conditions and the kinematics of deformation. Potyondy and Cundall (2004) developed this idea as a model for brittle rock. The micro-properties (contact stiffness and contact strength) are calibrated such that the given elasticity and tensile strength of the target material are reproduced for the particle ensemble. The apparent fracture toughness of the synthetic material is then related to the mean particle radius, R:

$$K_{Ic} = \sigma_t \sqrt{\pi a R} \qquad (56)$$

where σ_t is the tensile strength of the ensemble and a is a dimensionless factor that depends on particle packing, bond-strength heterogeneity and other things. The relation between fracture toughness and internal length scale (ILS) of a material appears to have been first formulated by Eringen (1977), who took the ILS to be a molecular dimension, but the result is independent of the nature of the ILS. Huang and Detournay (2008) further developed the connection between fracture toughness and the ILS and verified its applicability to particle models. The formulation is derived on the basis of forces, rather than energy, which is more usual in fracture mechanics. As Eringen remarks: "No poorly known constant such as surface energy appears…" (in reference to the existence of surface energy in the Griffith criterion, for example). The BPM reproduces all the results of linear elastic fracture mechanics (LEFM), as well as the size effect (*e.g.*, Bazant, 1998), whereby the strength of samples change from an inverse square-root relation at large sample sizes to a constant-strength asymptote at small sizes. The existence of an internal length scale explains this transition, because the stress amplification at the bond located at the tip of a crack depends on the ratio between the crack length and the particle size (ILS). The use of a BPM to simulate compression tests is illustrated in Figure 17, reproduced from a paper by Potyondy (2014), in which fragments are depicted by distinct shades, where a fragment is defined as a set of grains joined by contacts that have non-zero remaining bond strength. In this example, the bond model is called the "flat-jointed model," which allows moments to develop at contacts. Note that predominantly axial splitting is seen for the unconfined test (left-hand plots), while shear fractures are seen for the confined test (right-hand plots).

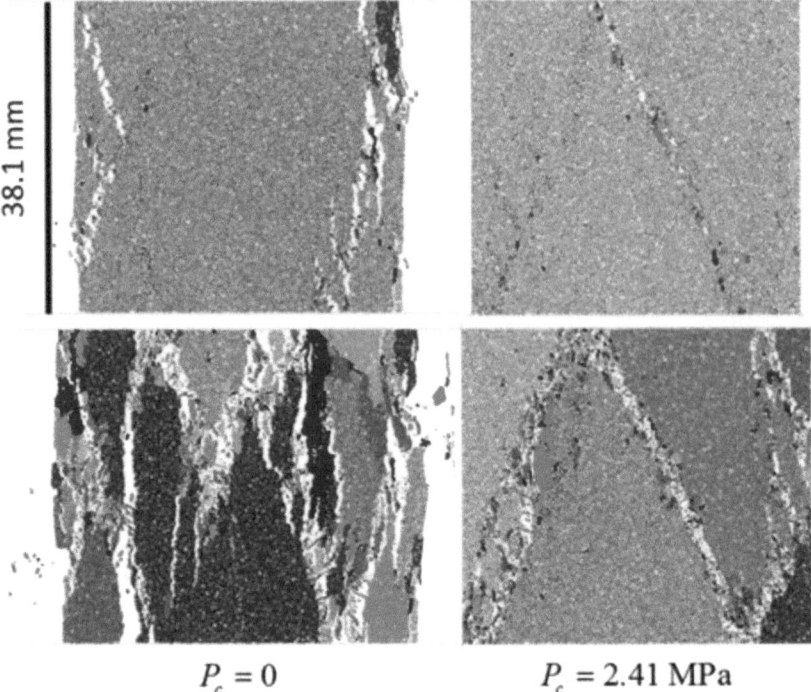

38.1 mm

$P_c = 0$ $P_c = 2.41$ MPa

Figure 17 Fragmentation of 2D sample at postpeak (top) and residual (bottom) stages of compression tests, for two different confining stresses (P_c): after Potyondy (2015).

4.3 Example of coupled particle model and continuum model

The bonded particle model (BPM) may be coupled to the continuum model described in Section 2 when both brittle and ductile materials are in contact. In this example of rock overlying salt, the simulation is performed over geological time scales. Salt is mechanically weak and flows like a fluid, even at geologically high strain rates, and is hence inherently unstable under a wide range of geologic conditions (Jackson *et al.*, 1994; Hudec & Jackson, 2007). Gravitational loading is produced by a combination of the weight of rocks overlying the salt and the gravitational body forces within the salt. During the course of a company-funded project conducted at the University of Vienna, mechanical coupling of *PFC2D* and *FLAC* was utilized to model the very large strains characteristic for salt tectonics, which required automatic rezoning of the *FLAC* grid. Mechanical coupling is achieved by exchanging forces and velocities across a common interface (*i.e.*, the top boundary of the salt layer) in each time-step. After rezoning, the interface particles are automatically remapped onto the new interface segments. Synsedimentary deformation is modeled by adding particle layers that embody a certain geological time interval, such as 0.1 Ma (where Ma = million years), and letting the salt flow until the creep-time that the layer represents has elapsed. Repeated addition of particle layers and creep lead to quasi-continuous sedimentation. In the example shown in Figure 18, a flat-topped delta with a linear slope of 2 degrees is prograding horizontally

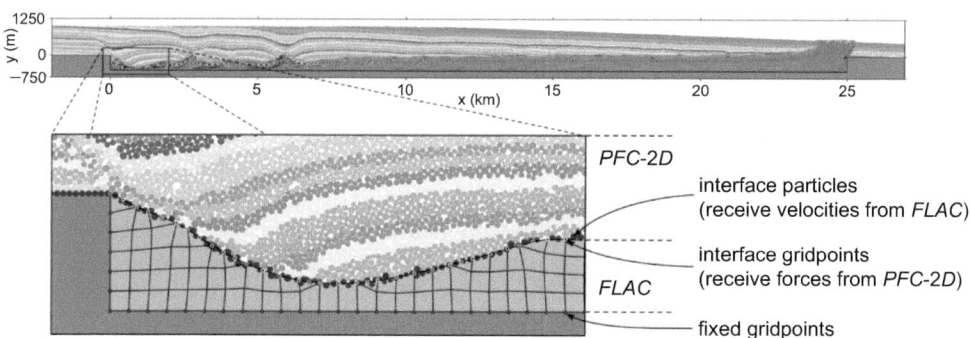

Figure 18 Coupled *PFC2D-FLAC* simulation of salt tectonics. The linear viscous salt (lower layer; black lines in top figure are initially vertical passive marker lines) is modeled with *FLAC* and was initially tabular (500 m thick). The actual *FLAC* grid is shown in the enlargement. The frictional-plastic sedimentary overburden is modeled with *PFC2D* (particles shaded according to depositional age). The total elapsed creep time is 4 Ma. Courtesy Martin Schöpfer.

at a constant rate of 1 cm/year. The model was developed and run by Martin Schöpfer, and the figure is used with permission from Statoil ASA. The differential load of the sediments produces a horizontal pressure gradient in the salt, driving a lateral squeeze-flow (Lehner, 2000) that leads to the formation of fault-bound extensional basins and salt expulsion in the proximal part and horizontal compression above a salt swell in the distal part (toe) of a delta.

4.4 Schemes to overcome particle size limitations

The representation of brittle material of given fracture toughness by the BPM can be used as a model for the fracture process in rock systems, provided that the particle size is chosen to satisfy Equation (56). For many rock types, this implies that the particle size should be of the order of 1 cm. Unfortunately, many rock systems encompass dimensions that are many orders of magnitude larger, and it would be computationally infeasible to model the entire system with small particles. Several strategies have been tried. One approach was the implementation of a J-integral-based criterion in the BPM. The J integral is a measure of energy release rate during a small fracture extension. It was originally formulated as a surface integral using classical continuum variables such as stress, displacement gradient and strain (Rice, 1968). For application to the BPM, the domain expression of the J integral was used, and the integral was reformulated in terms of variables of a discrete system. The J-integral-based criterion is non-local and requires time-consuming integration within a domain around a contact. The radius of the integration domain should be equal to at least a couple of particle diameters. Also, this criterion did not provide additional accuracy in resolution of fracture propagation mainly because of non-uniformity of forces (dispersion of local strain and stress) in contacts between particles in the BPM. The local, force-based criterion for fracture propagation is simpler and faster than non-local criteria such as one based on the J integral.

In the force-based criterion, the contact bond strength can be calculated from Equation (56) to match required toughness for any particle radius, R. Clearly, if the bond strength is rescaled to match material toughness, such a synthetic material will not match the tensile strength of the material (*i.e.*, the strength of material in absence of any fractures). The tensile strength of the material will typically be underestimated. In addition, non-uniformity of the contact forces will result in further weakening of the material. Under certain conditions, such a material can exhibit diffuse cracking even under an all-compressive stress state that has a significant deviatoric component. The contact force dispersion can be significantly reduced if the contacts between the particles correspond to interfaces between domains in a Voronoi lattice (see Section 4.6) and the contact shear and normal stiffnesses are assumed to be equal.

Another possible solution to matching fracture toughness is to employ a "dual-mesh" scheme, whereby the main extent of the system is represented by large-diameter particles, while regions surrounding crack tips are also represented by small particles, with dimensions that respect Equation (56). Figure 19 illustrates the concept.

The motions (velocities) of the coarse particles are imposed on the corresponding boundary particles in the fine-grid region, which is allowed to deform independently of the main (coarse-particle) grid. Cracks that develop in the fine-grid region are transmitted to the coarse-grid assembly by breaking bonds at contacts that correspond to

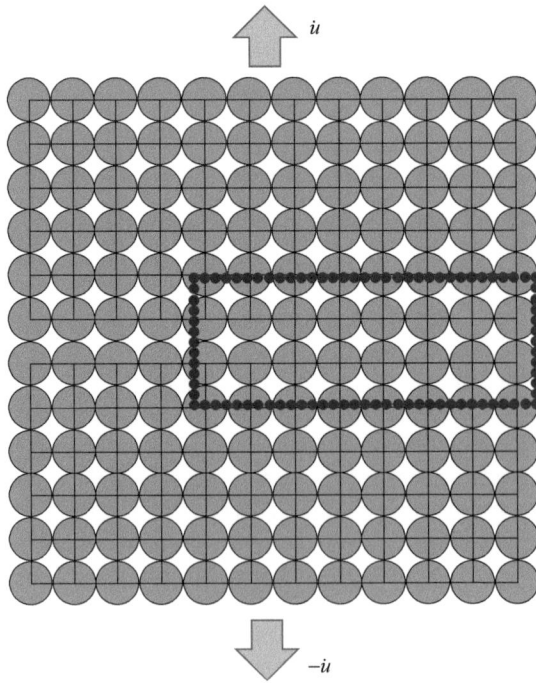

Figure 19 Tensile test sample composed of large particles with an initial crack extending across half the sample width (note the absence of bonds, which are denoted by black lines). A fine-grid region is superimposed on the coarse grid, so as to encompass the crack tip. (Only the boundary of the fine-grid region is shown, in dark grey).

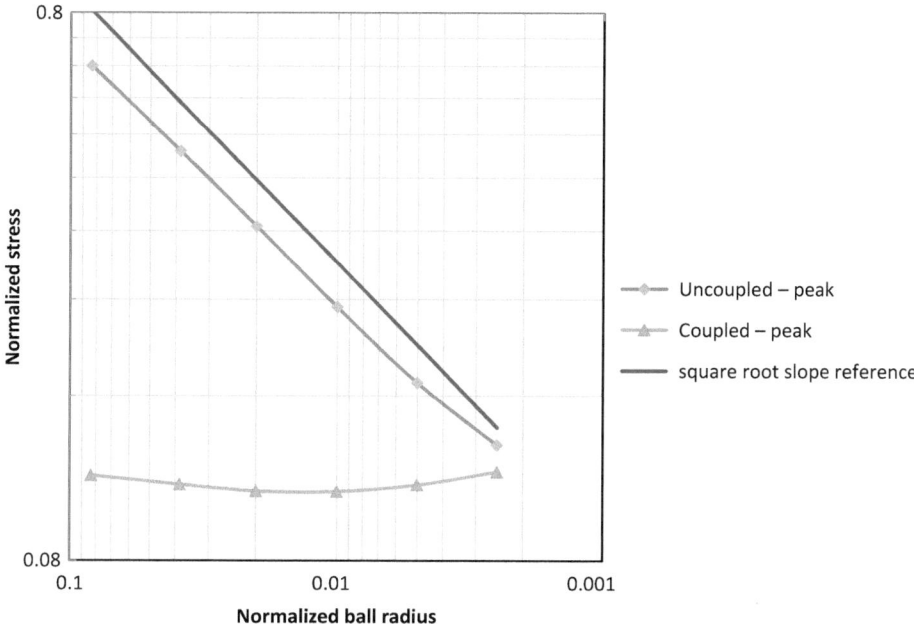

Figure 20 Strength (peak stress) of cracked sample for different conditions.

cracks in the fine grid. It is important to note that the coupling is only partial—both grids operate independently, with no forces being transmitted between the fine and coarse grids. Only boundary velocities are transmitted from the coarse grid to the fine grid, and only fine-grid cracks are transmitted to the coarse grid. As a test of the dual-mesh idea, several tensile tests were performed with different coarse-grid particle sizes, both with and without the coupling logic. Figure 20 shows the results. Without coupling, the center curve shows that sample strength trends according to $R^{-1/2}$, where R is the particle radius (comparing the slope with the upper reference slope). When the fine-grid coupling (according to the prescription noted above, using a fine-grid radius of 0.002) is active, the response is shown as the lower curve. Now, the strength is almost constant, which is consistent with an apparent fracture toughness corresponding to the internal length scale of the fine grid, rather than that of the coarse grid. These results are preliminary, but later developments have shown that the technique represents a robust way to achieve a given fracture toughness in a BPM independently of the particle radii. In practice, an adaptive scheme can be used, such that fine grids are placed when required around areas of high strain gradients. Note that the extent of the fine grid (away from a fracture tip) is not particularly critical, given that strain concentrations decay rapidly away from the tip.

4.5 Synthetic rock mass model

A rock mass typically contains pre-existing fractures at all scales. The yield and failure of the rock mass involves failure and yield on both the pre-existing fractures and new

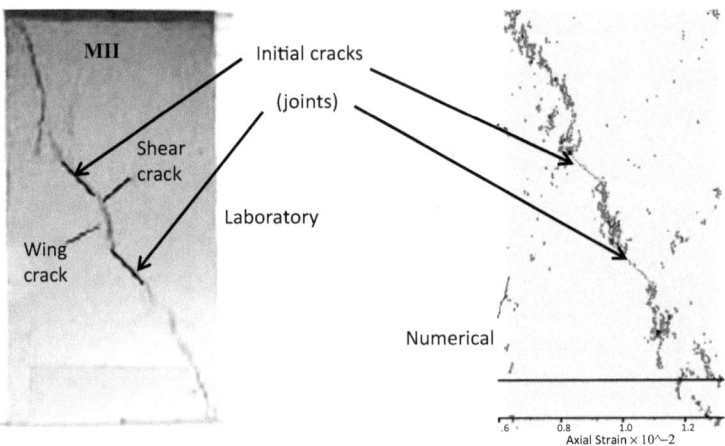

Figure 21 DEM simulation of the compressive response of a pre-cracked sample compared to a similar test performed in the laboratory (Wong et al., 2001). New cracks are seen to link existing cracks.

fractures. A simulation technique called the synthetic rock mass (SRM) allows both of these mechanisms to take place by using the bonded particle model (Section 4.2) with embedded planar fractures. These existing fractures are introduced by a special contact model that allows planar sliding that effectively cuts through the circular DEM particles, avoiding the "bumpy road" effect of particles sliding over one another. Figure 21 shows a validation test in which a numerical simulation is compared to a laboratory test (Wong *et al.*, 2001) in which several initial cracks were introduced into a compressed sample. In a complete rock mass, there are many *in-situ* fractures, which are generated according to a discrete fracture network (DFN) that conforms statistically to observations in the field. The application of the SRM consists of a calibrated BPM to represent the intact rock and an overlay of discontinuities that conforms to a given DFN. The properties of the discontinuities may be determined by physical tests done on samples of joints. Thus, the response of the synthetic material should simulate that of a large-scale rock mass, including size effects. The SRM has been described many times (*e.g.*, Pierce *et al.*, 2007), and that material will not be repeated here.

4.6 Lattice models for fracture

The bonded-particle models described thus far have involved circular particles interacting at their contact points. A simplification is possible whereby finite particles are replaced by point masses and contacts are replaced by springs. In small-strain mode, all spring constants and geometrical data may be pre-calculated, thus removing the overhead of contact detection and geometry updates. This type of scheme is termed a lattice model, and has been used by several people to simulate the fracture process in brittle materials: *e.g.*, Schlanger and Garboczi (1996). Bolander and Sukumar (2005) use a Voronoi tessellation and stiffness-scaling to generate springs that connect to the array of point masses. This approach provides a significant improvement in strain uniformity

Figure 22 Flexural toppling failure seen in a model test: after Adhikari *et al.* (1997).

compared to simply reproducing the nearest-neighbor contact logic of the DEM with uniform-stiffness springs. The formulation is identical to that given in Section 3, except that a "contact" is replaced by a spring, with the notional location of the contact point being the center of the spring. Joint sets are superimposed on the lattice in the same way as described in Section 4.5. When a joint crosses a spring, the axial spring vector is replaced by the unit normal vector to the joint plane, allowing sliding and opening to take place with respect to the actual joint direction.

As an application of the lattice model, a centrifuge test on flexural toppling was reproduced numerically. The test was reported by Adhikary *et al.* (1997), and the case examined here is shown in Figure 22, in which the failure of columns of brittle material separated by joints can be seen. The properties used in the simulation are given in Table 3, but the fracture toughness of the test material was not provided. For the simulation, it was necessary to use a small lattice resolution, which implies (according to Equation (56)) a fracture toughness that is probably smaller than the actual one. Accordingly, the observed onset of failure was matched by increasing the given tensile strength by a factor of 1.5.

Figure 23 presents a summary of the numerical results compared to the experimentally observed trace of the line of cracks that act as hinge points for the toppling columns. The two results match well, giving confidence that the lattice formulations for both fracturing and joint sliding are realistic. The experiment is a good test of the numerical model because the failure mode involves both sliding on joints and fracture of intact material; both mechanisms are necessary for the particular failure mode to develop.

Table 3 Properties used in numerical simulation of Adhikari's test.

Ratio of slope height to joint spacing	30
Joint spacing	12 mm
Slope angle	70°
Joint dip angle	69°
Tensile strength	1070–1140 KPa
Density	23.8 KN/m³
Joint friction angle	22°

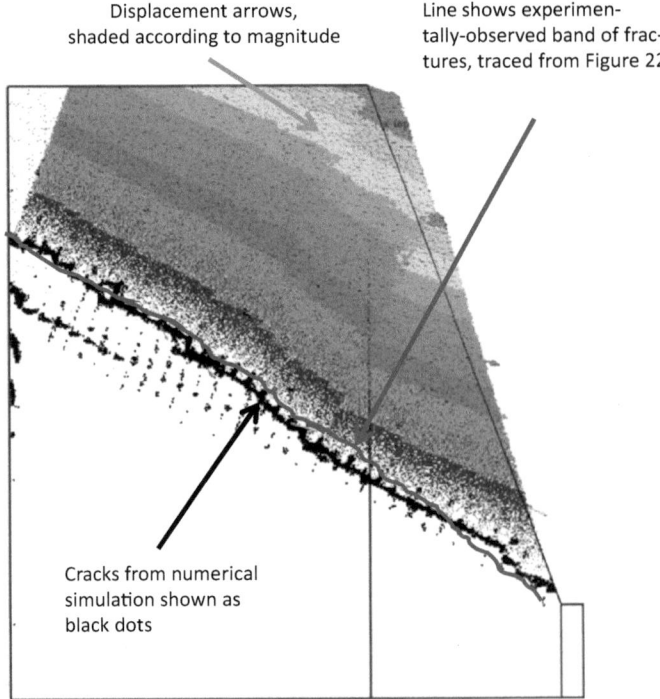

Displacement arrows, shaded according to magnitude

Line shows experimentally-observed band of fractures, traced from Figure 22

Cracks from numerical simulation shown as black dots

Figure 23 Numerical results (cracks and displacement contours) with superimposed experimentally observed fracture trace.

5 FLUID-MECHANICAL COUPLING

The presence of fluid has a profound impact on deformation and integrity of a soil or rock mass. Also, the withdrawal or injection of fluid from the earth subsurface, where pre-existing fractures often exist, may create major geo-mechanical issues, including severe surface subsidence or heave, earthquake generation, and water contamination. These processes are associated with industries on which most people rely on a daily basis, such as water and oil and gas production.

5.1 Theory outline

The logic for fluid flow and its coupling to mechanical deformation is integrated in both the continuum and the discontinuum dynamic relaxation procedures described in previous sections. The fluid flow logic can be exercised on its own, or coupled to the mechanical logic. Fluid flow coupling involves two mechanical effects. First, changes in pore pressure cause changes in stresses, which affect the response of the solid (for example, a reduction in effective stress may induce plastic yield in a continuum and/or slip along a joint). Second, the fluid reacts to mechanical volume changes by a change in pore pressure. In addition, a change in aperture for a joint in a discontinuum impacts its conductivity.

Several methods for incorporating fluid flow in the Discrete Element Method have been developed over the years, with specific applications in mind; a comprehensive review is presented by Furtney *et al.* (2013). In this section, we will review in some detail the approach taken in a code called *XSite* based on the lattice formulation (Cundall, 2011) and in the continuum codes.

5.1.1 Lattice code

In the lattice implementation (mentioned in Section 4.6), the mechanical behavior of the fractured rock mass is described using an assembly of nodes connected by normal and shear springs, called a lattice. The connecting springs are broken when the failure criteria is met in order to simulate the creation of new fractures, or when they intersect the DFN (Discrete Fracture Network) set, to account for the presence of pre-existing joints.

Fluid nodes, which are defined at broken spring locations, are regarded as penny-shaped cracks with apertures that depend on their fluid content. Flow of fluid in the lattice is simulated using the network of fluid nodes connected dynamically by one-dimensional pipe elements. The pipes inherit the mean properties of the associated fluid nodes. Also, the flow resistance of the pipes is evaluated from an equivalent parallel plate model with the same aperture and extent:

$$q = \beta s^2 (3 - s) \frac{a^3}{12\mu} \left[P^A - P^B + \rho_f g (z^A - z^B) \right] \tag{57}$$

where a is hydraulic aperture, μ is viscosity of the fluid, P^A and P^B are fluid pressures at fluid nodes "A" and "B" respectively, z^A and z^B are the node elevations, ρ_f is fluid density, and s is saturation. The dimensionless number β is a calibration parameter, a function of fluid resolution, used to match the conductivity of a pipe network to the conductivity of a joint represented by parallel plates with aperture a.

5.1.2 Explicit time step

For future reference, the stable explicit fluid time step for a numerical scheme, based on the pipe network, that uses the true fluid bulk modulus as a relaxation parameter is given by:

$$\Delta t = \min \left(\frac{R^2}{c_F} \right) \tag{58}$$

where R is the lattice resolution, and c_F is the pipe diffusivity. The pipe diffusivity is the product of the fluid bulk modulus, K_F (the inverse of fluid compressibility), and the pipe mobility coefficient, k_m, *i.e.*,

$$c_F = K_F k_m \tag{59}$$

and

$$k_m = \frac{a^2}{12\mu} \tag{60}$$

The final expression is:

$$\Delta t = \min\left(\frac{a}{K_F} \frac{R^2 12\mu}{a^3}\right) \tag{61}$$

5.1.3 Incompressible fluid-mechanical scheme

A new coupled fluid-mechanical scheme is used in the *HF-simulator* to model the mechanisms associated with hydraulic fracturing and/or fluid injection in pre-existing joints. The scheme (called MIF for Mechanical Incompressible Fluid) is applicable to situations in which the rock compressibility is much larger than that of the fluid. One of the main advantages of this explicit scheme is that the stable time step is proportional to the product of rock compressibility and discretization length. This time step is typically orders of magnitude greater than the explicit time step, proportional to the product of fluid compressibility and fracture aperture used in approaches that use the true fluid bulk modulus as the relaxation parameter. (The rock and fluid compressibilities are roughly of the same order of magnitude, but the fracture aperture is typically orders of magnitude smaller than the discretization length.) The scheme and its effects on time steps are described briefly in this section.

Consider an element of rock that includes a single joint. The element has the dimension of the lattice resolution, R; Figure 24 illustrates the mechanical arrangement

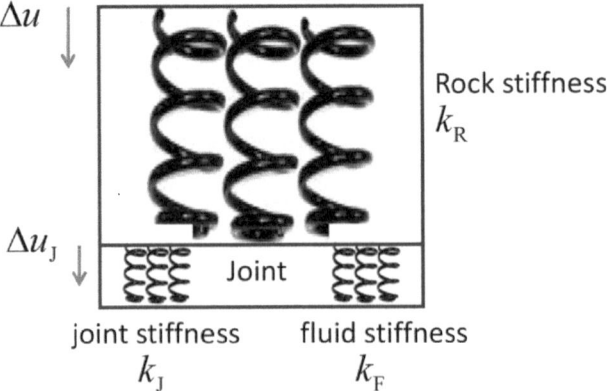

Figure 24 Schematic of lattice element with embedded fluid-filled joint, normal direction.

in the normal direction. The discussion in this section is in terms of one-dimensional stiffnesses having units of [force/displacement]. Thus, $k_F = K_F A/a$, where K_F is fluid bulk modulus, A is the apparent area of the joint element (of the order of R^2), and a is the joint aperture, and $k_R = K_R A/R$, where K_R is rock bulk modulus. Because $R \gg a$, typically, $k_F \gg k_R$.

The combined joint normal stiffness, k_C, is $(k_J + k_F)$, and the total element stiffness, k_T, is

$$k_T = \frac{k_R k_C}{k_R + k_C} \tag{62}$$

Due to an incremental total relative displacement, Δu, the increment in force, ΔF, for the element is

$$\Delta F = k_T \Delta u \tag{63}$$

giving rise to a joint displacement of

$$\Delta u_J = \frac{\Delta F}{k_C} \tag{64}$$

The increment in fluid force is $\Delta F_F = \Delta u_J k_F$, which (after some substitution) becomes

$$\Delta F_F = \frac{\Delta u\, k_R\, k_F}{k_R + k_J + k_F} \tag{65}$$

If $k_F \gg k_R$, then the fluid pressure increment approaches

$$\Delta P \to \Delta u\, k_R/A \tag{66}$$

Thus, in the MIF scheme, the fluid pressure increments induced by mechanical deformation do not depend on the fluid stiffness k_F, but instead depend only on lattice spring stiffness. This implies that the stable mechanical time step for hydro-mechanical simulation (which is inversely proportional to maximum stiffness in the model) is practically the same as for uncoupled simulations. This allows for a large decrease in computational time, compared to schemes that use the true fluid bulk modulus as a relaxation parameter.

The apparent fluid modulus (used in the flow part of the model and affecting the stable flow time step) also is changed as a result of the new algorithm, as follows. Consider a fluid "displacement" increment, Δu_F (derived from a flow imbalance, ΣQ, in the joint), which is imagined to be injected in series with the fluid spring, k_F, at each fluid time step. The increment in fluid force is then

$$\Delta F_F = \frac{(k_J + k_R) k_F}{k_J + k_R + k_F} \Delta u_F \tag{67}$$

The pressure increment is $\Delta P = \Delta F_F/A$, which approaches

$$\Delta P \to \frac{\Delta u_F (k_R + k_J)}{A} \tag{68}$$

when $k_F \gg k_R$. Thus, the fluid "sees" an apparent bulk modulus of

$$\overline{K}_F = \frac{a}{A}(k_R + k_J) \tag{69}$$

in the limit of $k_F \gg k_R$. If we ignore k_J, for simplicity, $\overline{K}_F \propto K_R a/R$. The stable fluid time step in the MIF scheme is obtained after substitution of $\overline{K}_F$ for K_F in (61):

$$\Delta t = \min\left(\frac{R}{K_R} \frac{R^2 12\mu}{a^3}\right) \tag{70}$$

Therefore, the apparent fluid bulk modulus being much smaller than the fluid bulk modulus $\overline{K}_F \ll K_F$, allows for much longer fluid time steps.

Finally, the lattice spring force is incremented as a result of fluid displacement arising from the flow calculation, as follows:

$$\Delta F = \frac{\Delta u_F k_R k_F}{k_R + k_J + k_F} \tag{71}$$

In the limit $k_F \gg k_R$,

$$\Delta F \rightarrow \Delta u_F k_R \tag{72}$$

The joint spring carries the total force, which affects the force balance and the motion. However, the effective stress is considered in assessment of slip or opening of the joint.

If, in the MIF scheme, the aperture is increased by 2, the time step is divided by 8, see Equation (70). To prevent a prohibitive decrease of fluid flow time step, there is an upper bound to the apparent fracture aperture. Note that this limit applies to the flow resistance—not the fluid storage volume.

The stable mechanical step is typically smaller than the stable fluid step. In the MIF scheme, uniform density scaling is applied by giving the mechanical time step a value equal to that of the fluid. (In this technique, inertial masses are multiplied by a factor such that the computed mechanical time step is equal to the fluid time step. Gravitational masses are unaffected.)

5.1.4 Continuum code

The formulation of coupled fluid-deformation mechanisms in *FLAC/FLAC3D* is based on the Biot theory of poro-mechanics; it can be applied to problems involving Darcy flow in a porous medium. The governing equations are as follows.

Transport law

Fluid transport is described by Darcy's law,

$$q_i = -k_{ij}\hat{k}(s)\frac{\partial}{\partial x_j}\left(p - \rho_f g_k x_k\right) \tag{73}$$

where q is the specific discharge vector, k is the mobility coefficient (a tensor), $\hat{k}$ is the relative mobility coefficient, which is a function of the saturation s, p is the fluid pressure, ρ_f is the mass density of the fluid, and g is the gravity vector.

Balance law

The fluid mass balance relation is

$$\dot{\zeta} = -\frac{\partial q_i}{\partial x_i} + q_v \tag{74}$$

where ζ is the variation of fluid volume per unit volume of porous material, and q_v is the volumetric fluid source intensity.

The balance of momentum has the form

$$\frac{\partial \sigma_{ij}}{\partial x_j} + \rho g_i = \rho \ddot{u}_i \tag{75}$$

where $\rho = (1 - n)\rho_s + ns\rho_f$ is the solid bulk density; ρ_s and ρ_f are the densities of the solid and fluid phase, respectively, and n is porosity.

Constitutive laws

The response equation for the pore fluid depends on the value of the saturation. At full saturation, $s = 1$, $\hat{k} = 1$, and the fluid can sustain a tension up to a limit $P = T_f$; the response equation is

$$\dot{p} = M(\dot{\zeta} - \alpha \dot{\varepsilon}) \tag{76}$$

where M is the Biot modulus, α is the Biot coefficient, and ε is the volumetric strain.

The Biot modulus is related to the drained bulk modulus of the porous medium, K, and the fluid bulk modulus, K_f as $M = K_f/[n + (\alpha - n)(1 - \alpha)K_f/K]$.

For saturation less than one, the constitutive equation of the pore fluid is described by the saturation equation

$$\dot{s} = \frac{1}{n}(\dot{\zeta} - \alpha \dot{\varepsilon}) \tag{77}$$

The pressure is zero in the unsaturated zone, and unsaturated flow is governed solely by gravity. Also, the relation between relative permeability and saturation is given by a cubic law

$$\hat{k}(s) = s^2(3 - 2s) \tag{78}$$

The logic is applicable to coarse materials when capillary effects can be neglected, and to large scale problems when the height of the capillary zone can be neglected compared to the representative length in the problem.

The constitutive response of the porous solid is described by

$$\dot{\sigma}_{ij} + \alpha \dot{p} \delta_{ij} = H_{ij}(\sigma_{kl}, \dot{\varepsilon}_{kl}, \kappa) \tag{79}$$

where H_{ij} is the functional form of the material constitutive law, and κ is a history parameter.

Compatibility equation

The relationship between strain rate and velocity gradient is

$$\dot{\varepsilon}_{ij} = \frac{1}{2}\left(\frac{\partial \dot{u}_i}{\partial x_j} + \frac{\partial \dot{u}_j}{\partial x_i}\right) \tag{80}$$

The solid weight, the buoyancy and the drag or seepage force are automatically taken into account in the formulation. This can be shown as follows. By definition of effective stress, we have

$$\sigma_{ij} = \sigma'_{ij} - p\delta_{ij} \tag{81}$$

After substitution of Equation (81) in the equilibrium equation $\partial\sigma_{ij}/\partial x_j + \rho g_i = 0$, see Equation (75), and using the definition of the solid bulk density ρ, we obtain, after some manipulations:

$$\frac{\partial \sigma'_{ij}}{\partial x_j} + (1-n)\rho_s g_i - (1-n)\frac{\partial p}{\partial x_i} - n\gamma_f \frac{\partial \phi}{\partial x_i} = 0 \tag{82}$$

where we have used the definition of fluid unit weight, $\gamma_f = \rho_f|g|$, and piezometric head, $\phi = p/(\rho_f|g|) - g_k x_k/|g|$.

In Equation (82), the second term is associated with solid weight, the third with buoyancy and the last one with seepage force (drag).

Substitution of the fluid mass balance, Equation (74), in the constitutive equation for the pore fluid (76) yields the following expression for the fluid continuity equation for saturated conditions:

$$\dot{p} = M\left(-q_{i,i} + q_v - \alpha\dot{\varepsilon}\right) \tag{83}$$

The numerical implementation of the governing equations for the fluid uses the same discretization in elements as the one used to solve the mechanical equations. The pore pressure is defined at the nodes of the element, and two overlays of constant discharge tetrahedra in 3D (or triangles in 2D) are used (assuming linear pore pressure variation in a tetrahedral or triangle). A finite volume approach, described in Section 2, is used to derive the numerical fluid flow formulation, in which the numerical scheme is equivalent to the mechanical constant stress formulation of Section 2 that leads to the nodal form of Newton's law. The fluid formulation is obtained by substituting the pore pressure, specific discharge vector and pore pressure gradient for velocity vector, stress tensor and strain-rate tensors, respectively, in the mechanical formulation. Both explicit and implicit schemes are considered for the integration in time of the resulting system of ordinary differential equations. The current implicit scheme uses a Gauss

Seidel approximation, with a Jacobi iterative procedure (valid for fully saturated domains).

Starting from a state of mechanical equilibrium, a fully coupled hydro-mechanical (static) simulation involves a series of calculation steps. Each step includes one or more flow steps (flow loop) followed by enough mechanical steps to maintain quasi-static equilibrium.

The increment of pore pressure due to fluid flow is evaluated in the flow loop; the contribution from mechanical strain is evaluated in the mechanical loop as an element value, which is then distributed to the nodes.

For the effective stress calculation, the total stress increment due to pore pressure change arising from mechanical volume strain is evaluated in the mechanical loop, and that arising from fluid flow, in the flow loop.

In the continuum codes, all material models are expressed in terms of Terzaghi effective stresses (*i.e.*, effective stresses are used to detect yielding in plastic materials). In this context, the pore pressure field may originate from different sources: a fluid flow analysis; a coupled fluid/mechanical simulation; or an initialization of the pore pressure field.

5.2 Example of fluid-mechanical simulation with the lattice model

Hydraulic fracturing is a technique used in the oil and gas industry to stimulate production. In this process, a fluid (usually water with additives) is injected into a borehole and the rock mass via injection clusters, see sketch in Figure 25. The fluid pressure is purposely high enough to generate fractures in the rock formation. The induced fractures are desirable for production because they generally afford an increase in rock permeability and an enhanced contact area with the fluid being produced.

Sets of large parallel fractures are viewed as an ideal outcome of the technique. However, hydraulic fracturing is a complex (tightly coupled) process, and the target is not always achieved because fractures interact with each other in non-obvious ways, depending on a number of parameters, including rock mass properties, initial and induced stresses in the rock, distance between fractures, and fluid viscosity.

The example in this section considers fluid injection in a horizontal borehole with up to six clusters drilled in a homogeneous medium composed of a top layer and a bottom layer of higher toughness.

A fracture in a homogeneous medium has a tendency to grow as a penny-shaped feature in the direction normal to the minimum in-situ compressive stress. Also, the fluid in the fracture exerts a pressure on the surrounding rock and thus impacts the local stress state, and this in turn may potentially influence the growth of a fracture in its vicinity. This mechanism, called the 'stress shadow' effect in the literature, is illustrated in this example.

The conditions for the problem are outlined in Figure 26.

The impact of distance between clusters on fracture growth is illustrated in three cases in Figure 27 for the conditions outlined in Figure 26. In Case 1, a borehole injection in two clusters that are 15 m apart produces simulated fractures that grow slightly away from each other. When the two clusters are 30 m apart, as in Case 2, the induced fractures do not seem to interfere. In Case 3, injection is made via three clusters

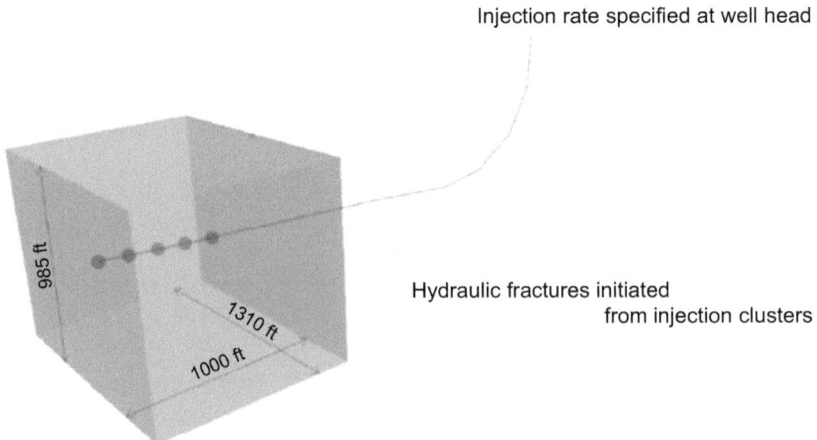

Injection rate specified at well head

Hydraulic fractures initiated
from injection clusters

985 ft

1310 ft

1000 ft

Figure 25 Model of a horizontal borehole with injection clusters.

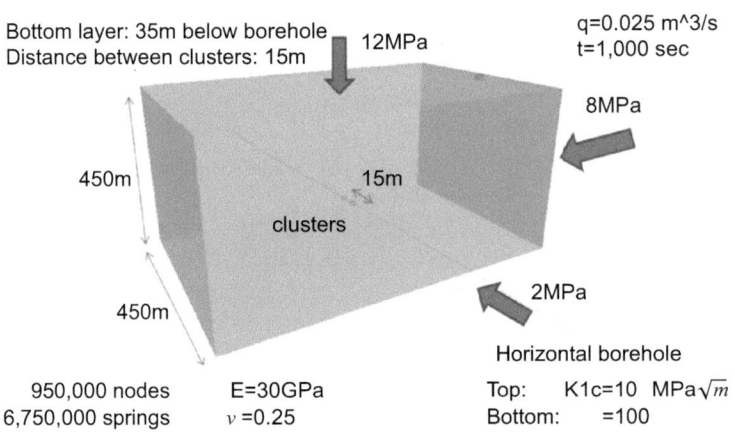

Bottom layer: 35m below borehole
Distance between clusters: 15m

12MPa

q=0.025 m^3/s
t=1,000 sec

8MPa

450m

15m

clusters

450m

2MPa

Horizontal borehole

950,000 nodes   E=30GPa
6,750,000 springs v =0.25

Top: K1c=10 MPa√m
Bottom: =100

Figure 26 Problem conditions.

15 m apart. The simulations show two fractures evolving from the clusters located at the extremities. Also, the growth of the fracture from the middle cluster is inhibited by the two others in this case.

Simulation results obtained for injection in a borehole with six clusters are shown in Figure 28.

The plots in Figure 28 show a vertical asymmetric fracture growth. This occurs because the fracture does not penetrate in the bottom rock layer, which is of higher toughness. Also, the fractures originating from the clusters at the extremities (with only one neighbor) are well developed in the simulation. This is in contrast with the fractures originating from the middle clusters, whose paths appear to be strongly influenced by each other.

The example illustrates the mechanism whereby the growth of a particular fracture can be inhibited by the propagation of fractures in its vicinity.

Cluster spacing: 15m

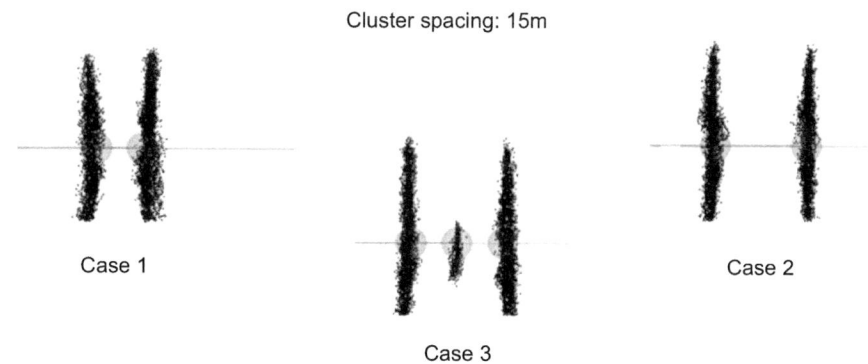

Case 1   Case 2

Case 3

Figure 27 Simulation of fracture growth from two and three clusters.

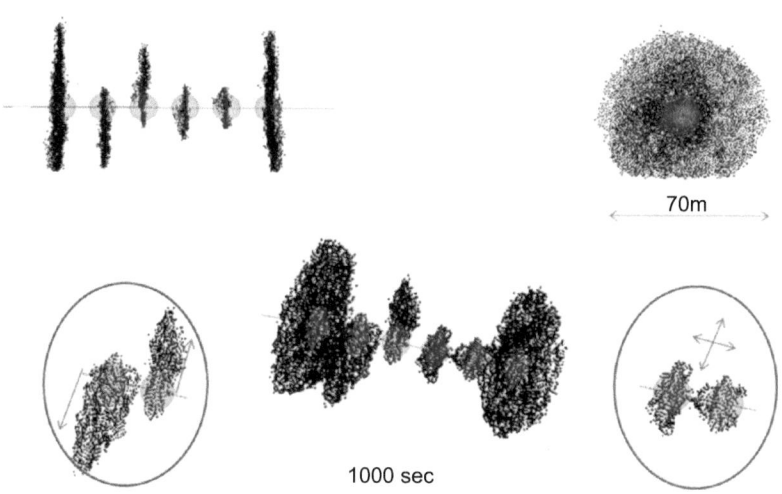

70m

1000 sec

Figure 28 Simulation of fracture growth from six clusters.

REFERENCES

Adhikary, D. P., Dyskin, A. V., Jewell, R. J. & Stewart, D. P. (1997) A study of the mechanism of flexural toppling failure of rock slopes. *Rock Mech. Rock Eng.*, 30 (2), 75–93.

Bazant, Z. P. & Planas, J. (1998) *Fracture and Size Effect in Concrete and Other Quasibrittle Materials*. Boca Raton, Florida, CRC Press.

Bolander, J. E. & Sukumar, N. (2005) Irregular lattice model for quasistatic crack propagation. *Phys. Rev. B*, 71 (9), 1–12.

Courant, R., Friedrichs, K. & Lewy, H. (1928) On the partial difference equations of mathematical physics. *Mathematische Annalen*, 100, 32–74. English translation in IBM J. (March, 1967), 215–234.

Cundall, P. A. (1971) A computer model for simulating progressive large scale movements in blocky rock systems. In: *Proceedings of the Symposium of the International Society for Rock Mechanics, September 1971, Nancy, France.* pp. 2–8.

Cundall, P. A. & Strack, O. D. L. (1979) A discrete numerical model for granular assemblies. *Géotechnique*, 29 (1), 47–65.

Cundall, P. A. (1987) Distinct Element Models of Rock and Soil Structure, in Brown, E. T. (ed.) *Analytical and Computational Methods in Engineering Rock Mechanics*, London: Allen & Unwin. pp. 129–163.

Cundall, P. A. (1988) Formulation of a three-dimensional distinct element model – Part I. A Scheme to Detect and Represent Contacts in System Composed of Many Polyhedral Blocks. *Int. J. Rock Mech. & Mining Sci. & Geomechanics Abstracts*, 25 (3), 107–116.

Cundall, P.A. (1990) Numerical modelling of jointed and faulted rock. In Rossmanith, H.P. (ed.), Mechanics of Jointed and Faulted Rock. A.A. Balkema: Rotterdam.

Cundall, P. A. (2006) A simple hysteretic damping formulation for dynamic continuum simulations. In: Varona, P. and Hart, R. (eds.) *Proceedings of the 4th International FLAC Symposium*, 29–31 May 2006, *Madrid, Spain*, Minneapolis, Itasca Consulting Group. pp. 359–364.

Cundall, P. A. (2011) Lattice method for modeling brittle, jointed rock. In: Sainsbury, D., Hart, R., Detournay, C. & Nelson, M. (eds.) *Continuum and Distinct Element Numerical Modeling in Geomechanics, Proceedings of the 2nd International FLAC/DEM Symposium 14–16 February 2011, Melbourne, Australia.* Minneapolis, Itasca Consulting Group. pp. 11–19.

Dawson, E., Roth, W. & Drescher, A. (1999) Slope stability analysis by strength reduction. *Géotechnique*, 49 (6), 835–840.

Day, A. S. (1965) An introduction to dynamic relaxation, *The Engineer*, 219, 218–221.

Eringen, A. C. (1977) Continuum mechanics at the atomic scale. *Crystal Lattice Defects*, 7, 109–130.

Furtney, J., Zhang, F. & Han, Y. (2013) Review of methods and applications for incorporating fluid flow in the Discrete Element Method. In: Detournay, C., Hart, R. & Nelson, M. (eds.) *Continuum and Distinct Element Numerical Modeling in Geomechanics, Proceedings of the 3rd International FLAC/DEM Symposium, 22–24 October 2013, Hangzhou, China.* Minneapolis, Itasca Consulting Group. Paper: 10–01.

Goodman, R. E., Taylor, R. L. & Brekke, T. L. (1968) A model for the mechanics of jointed rock. *J. Soil Mech. & Foundations Division, ASCE*, 94 (SM 3), 637–659.

Hart, R., Cundall, P. A. & Lemos, J. (1988). Formulation of a Three-dimensional Distinct Element Model – Part II. Mechanical Calculations for Motion and Interaction of a System Composed of Many Polyhedral Blocks. *Int. J. Rock Mech. & Mining Sci. & Geomechanics Abstracts*, 25 (3), 117–125.

Hoek, E. & Bray, J. W. (1981). *Rock Slope Engineering, 3rd Ed.* London: The Institute of Mining and Metallurgy.

Huang, H. & Detournay, E. (2008) Intrinsic length scales in tool-rock interaction 1. *Int. J. Geomech*, 8 (Special Issue), 39–44.

Hudec, M. R. & Jackson, M. P. A. (2007). Terra infirma: Understanding salt tectonics. *Earth-Science Reviews*, 82, 1–28.

Itasca Consulting Group, Inc. (2004) *UDEC – Universal Distinct Element Code*, Ver. 4.0. Minneapolis: Itasca.

Itasca Consulting Group, Inc. (2008a) *3DEC – Three-Dimensional Distinct Element Code*, Ver. 4.1. Minneapolis: Itasca.

Itasca Consulting Group, Inc. (2008b) *PFC – Particle Flow Code in Two and Three Dimensions*, Ver. 4.0. Minneapolis: Itasca.

Itasca Consulting Group, Inc. (2011) *FLAC — Fast Lagrangian Analysis of Continua*, Ver. 7.0. Minneapolis: Itasca.

Itasca Consulting Group, Inc. (2012) *FLAC3D — Fast Lagrangian Analysis of Continua in Three-Dimensions*, Ver. 5.0. Minneapolis: Itasca.

Jackson, M. P. A., Vendeville, B. C. & Schultz-Ela, D. D. (1994) Structural dynamics of salt systems. *Annu. Rev. Earth and Pl. Sc.*, 22, 93–117.

Jiang, Y., Xu, J., Meng, G., Xu, J. Q., Zhu, H. C., Li, H., & Wang, Y. F. (2010) Numerical evaluation of cavern layout design for the Baihetan Hydropower Project in China, In *the 44th US Rock Mechanics Symposium and 5th U.S.-Canada Rock Mechanics Symposium*, 27–30 June 2010, *Salt Lake City, Utah*. Paper no. ARMA-10-119.

Krieg, R. (1973) Unconditional stability in numerical time integration methods. *ASME. J. Appl. Mech.*, 40 (2), 417–421.

Lehner, F. K. (2000) Approximate Theory of Substratum Creep and Associated Overburden Deformation in Salt Basins and Deltas. In Lehner, F. K. & Urai, J. L. (eds.) *Aspects of Tectonic Faulting*, New York: Springer Berlin Heidelberg. pp. 21–47.

Marti, J. & Cundall, P. (1982) Mixed Discretization Procedure for Accurate Modelling of Plastic Collapse. *Int. J. Num. & Analy. Methods in Geomech.*, 6 (1), 129–139.

Meng, G. & Zhu, H. (2009) Report on stability analysis of the underground powerhouse surrounding rock mass for Baihetan hydropower station, Itasca Consulting China Ltd. (In Chinese).

Nagtegaal, J. C., Parks, D. M. & Rice, J. R. (1974) On Numerically Accurate Finite Element Solutions in the Fully Plastic Range, *Comp. Meth. Appl. Mech. & Eng.*, 4, 153–177.

Oakley, D. R. & Knight, N. F. K. Jr. (1995) Adaptive Dynamic Relaxation algorithm for nonlinear hyperelastic structures. Part I. Formulation. *Comput. Meth. Appl. Mech. & Eng.*, 126 (1–2), 67–89.

Otter, J. R., Cassell, A. C. & Hobbs, R. E. (1966) Dynamic Relaxation. *Proc ICE*, 35 (4), 633–656.

Pierce, M., Mas Ivars, D., Cundall, P. A. & Potyondy, D. O. (2007) A synthetic rock mass model for jointed rock. In: Eberhardt E. (ed.) *Rock Mechanics: Meeting Society's Challenges and Demands In Proceedings of the 1st Canada-U.S. Rock Mechanics Symposium, Vancouver*, 27–31 May 2007, Vol. 1: Fundamentals, New Technologies & New Ideas. London: Taylor & Francis Group. pp. 341–349.

Potyondy, D. O. (2015). The bonded-particle model as a tool for rock mechanics research and application: Current trends and future directions. *Geosystem Eng.*, 18 (1), 1–28.

Potyondy, D. O. & Cundall, P. A. (2004). A bonded-particle model for rock. *Int. J. Rock Mech. & Min. Sci.*, 41 (8), 1329–1364.

Press, W. H., Flannery, B. P., Teukolsky, S. A. & Vetterling, W. T. (1986) *Numerical Recipes: The Art of Scientific Computing*. Cambridge: Cambridge University Press.

Rice, J. R. (1968) A path independent integral and the approximate analysis of strain concentration by notches and cracks. *J. Appl. Mech.*, 35 (2), 379–386.

Sauvé, R. G. & Metzger, D. R. (1995) Advances in Dynamic Relaxation Techniques for Nonlinear Finite Element Analysis. *Journal of Pressure Vessel Technology*, 117 (2), 170–176.

Strang, G. (1976) *Linear Algebra and Its Applications*. Academic Press: New York.

Underwood, P. (1983) Dynamic relaxation. In: Belytschko, T. and Hughes, T. J. R. (eds.) *Computational Methods for Transient Dynamic Analysis*. Amsterdam: North Holland. pp. 246–265.

Von Neumann, J. & Richtmeyer, R. D. (1950) A Method for the Numerical Calculation of Hydrodynamic Shocks. *J. Appl. Phys.*, 21, 232.

Wilkins, M. L. (1969) *Calculation of Elastic-Plastic Flow*. Livermore, California, Lawrence Radiation Laboratory.

Wilkins, M. L., French, S. J. & Sorem, M. (1971) Finite difference scheme for calculating problems in three space dimensions and time. In: *Proceedings 2nd International Conference on Numerical Methods in Fluid Dynamics: Lecture Note in Physics*, 15–19 September, 1970, *Berkeley, California*. 8, pp. 30–33.

Wong, R. H. C., Chau, K. T., Tang, C. A. & Lin, P. (2001) Analysis of crack coalescence in rock-like materials containing three flaws: Part I: experimental approach. *Int. J. Rock Mech. & Min. Sci.* 38 (7), 909–924.

Wyllie, D.C. & Mah, C.W. (2004) Rock Slope Engineering: Civil and Mining, 4th edition. CRC Press / Taylor & Francis Group: Abingdon.

Chapter 3

The numerical Discontinuous Deformation Analysis (DDA) method: Benchmark tests

G. Yagoda-Biran[1] & Y.H. Hatzor[2]

[1]*Engineering Geology and Geological Hazards, Geological Survey of Israel, Jerusalem, Israel*
[2]*Geological and Environmental Sciences, Ben-Gurion University of the Negev, Beer-Sheva, Israel*

Abstract: This chapter reviews benchmarking studies of 2D and 3D-DDA performed by the rock mechanics group at Ben-Gurion University of the Negev over the past decade. DDA fundamentals are briefly reviewed first, followed by a comprehensive review of verification and validation studies. We conclude by presenting some limitations of 3D-DDA in its current formulation.

I INTRODUCTION

Discontinuous Deformation Analysis (DDA) has been in use by the professional rock mechanics community for three decades, and has been successful in modeling discontinuous rock masses and masonry structures. DDA is an implicit, discrete element, numerical method proposed in the late 1980s (Shi, 1988, 1993; Shi & Goodman, 1985, 1989), that provides a useful tool for investigating the dynamics of blocky rock masses and systems composed of multiple blocks, such as masonry structures.

In the 1980s Shi (Shi, 1988) proposed the two-dimensional DDA (2D-DDA) as a PhD thesis at UC Berkeley, and later on, the 3D-DDA was published (Shi, 2001). A review of the essentials of DDA is provided by Jing (1998). Reviews of DDA within the scope of other numerical methods used today to solve problems in rock mechanics and rock engineering are provided by Jing (2003), Jing and Hudson (2002) and Jing and Stephansson (2007). MacLaughlin and Doolin (2006) published a comprehensive review of 2D-DDA validations. The 3D-DDA, being a more recent development, has not been verified extensively; some new and useful validations of 3D-DDA are presented in this chapter.

Many research groups have made modifications to the original code developed by Dr. Shi (Shi, 1988, 1993; Shi & Goodman, 1985) in an attempt to better address some of the fundamental issues in DDA.

The contact algorithm has been the subject of many research groups. Lin *et al.* (1996) modified the original contact model of DDA, which is based on the penalty method, by adopting the Lagrange type approach. Ning *et al.* (2010) modified the contact algorithm of DDA by adopting the Augmented Lagrangian method. Later on Bao *et al.* (2014b) introduced another version of the augmented Lagrangian method, which makes use of advantages of both the Lagrangian multiplier method and the penalty method so as not to alter the structure of the system of equations. Bao and Zhao (2010, 2012) have made some enhancements to the vertex to vertex contact. In the 3D-DDA front, Beyabanaki *et al.* (2009) verified the 3D-DDA with the analytical solution for the case of a block on an incline, and modified the edge-to-edge contact constraints by using the augmented Lagrangian method instead of the penalty method. By doing so, they claim to retain the simplicity of the penalty method, and reduce its disadvantages. Ahn and Song (2011) proposed a new contact-definition algorithm for 3D-DDA using a virtually inscribed sphere installed in every contacting vertex, and reported an increase in the efficiency and stability of the code, using their suggested algorithm. Mikola and Sitar (2013) developed a 3D-DDA formulation using an explicit time integration procedure, and a different contact detection algorithm. Wu *et al.* (2014) addressed the issue of contact-detection algorithm, claimed to be the most computation-consuming stage in 3D-DDA simulations. They use a novel multi-shell cover system, and claim that this method greatly reduces the contact detection volume and iterations. They report improved efficiency when their algorithm is implemented in the 3D-DDA code.

In an attempt to overcome the DDA simply deformable blocks assumption and therefore uniform distribution of stresses within blocks, Shi (1997) developed the Numerical Manifold Method, using superposition of a mathematical cover over the physical mesh of the blocks. Bao and Zhao (2013) have integrated the advantages of both DDA and the finite element methods (FEM), and developed the hybrid nodal based DDA (NDDA), thus improving the accuracy of stress distribution and allowing for crack propagation within blocks. Jiao *et al.* (2012) developed a two-dimensional contact constitutive model to simulate the fragmentation of jointed rock. Several other research groups have developed higher order DDA codes to address this issue (*e.g.* Grayeli & Mortazavi, 2006).

Considering friction and cohesion of block interfaces, Zhang *et al.* (2014a) added an additional contact type determination for the edge-to-edge contact, in order to better address the issue of discontinuities with both cohesion and friction. They verified and validated their improved code, and found it to be accurate. Zhang *et al.* (2015) replaced the Mohr-Coulomb joint failure criterion in the original 3D-DDA code, which has a constant friction coefficient, with rate-and-state friction laws. They tested their code by modeling slide-hold-slide and velocity stepping tests, and reported that their numerical results compared well with experimental results. Wang *et al.* (2013), in an attempt to overcome the cohesion treatment in the original DDA code, developed displacement-dependent interface shear strength. They report that landslides, simulated using the modified algorithm, exhibit a two-stage failure pattern: a relatively slow, downward progressive failure stage followed by a rapid, massive run-out failure stage.

Jiao *et al.* (2007) developed a viscous boundary for DDA based on the standard viscous boundary condition provided in the original DDA formulation, in order to deal with dynamic wave propagation problems. Later on Bao *et al.* (2012) implemented new viscous boundary conditions to the 2D-DDA, in order to improve the absorbing efficiency. Zhang *et al.* (2014b) made two modifications to the original DDA code, in order to investigate the seismic response of underground houses. They incorporated viscous boundary conditions, and applied the seismic input as stress input from the bottom of the model. They report that the numerical solution of the modified DDA is close to the theoretical solution. Ning and Zhao (2012b) coupled the superposition boundary algorithm, to obtain a non-reflecting boundary, with the DDA, for wave propagation modeling in infinite media.

Jiang *et al.* (2012) introduced the softening block approach for the simulation of stage-wise sequential excavations in jointed rock masses. Their proposed modification to the original code eliminates the need to remove the excavated blocks from the calculation model. They present validations for their method, and report of accurate results. Tal *et al.* (2014) improved the original numerical manifold method capability to model stability of underground openings embedded in discontinuous rock masses by implementing an algorithm which models the excavation sequence during simulations, starting with a domain with no opening at all and progressively adding openings according to the planned construction phases. Later on this modification was implemented in the original DDA code and was applied in stability analysis of deep tunnels excavated in the columnar jointed basalts of the Baihetan hydroelectric project in South West China (Hatzor *et al.*, 2015).

Kim *et al.* (1999) and Jing *et al.* (2001) have made a modification where they compute water pressure and seepage through rock mass, this way coupling fluid flow in fractures. More recent implementations of hydro-mechanical coupling in DDA was developed by Chen *et al.* (2013) and Ben *et al.* (2011). Koyama *et al.* (2011) combined the DDA and the finite element method for fluid flow simulation to model the interaction between solid particles movement and fluid flow. Jiao *et al.* (2015) have made modifications to the original code in order to simulate the hydraulic fracturing process. In their proposed approach, they first perform the calculation of fluid mechanics to obtain seepage pressure near the tips of existing cracks, and then treat the fluid pressure as linearly distributed loads on corresponding block boundaries, by adding these components to the force matrix of the global equilibrium equation. They verified their new approach and found that it effectively simulates hydraulic rock fracturing. Morgan and Aral (2015) have also approached the subject of hydro-mechanical coupling for hydraulic fracturing. They combined the DDA with finite volume fracture network model for simulation of compressible fluid flow in fractures. The authors claim to improve the fluid, contact and coupling components to increase the accuracy of the solution, and report successful verification of their coupled H-M model against analytical and semi-analytical solutions.

Due to the first order displacement function used in DDA, the error, when computing block rotations, may become excessive if the blocks are undergoing large rotations (Ohnishi *et al.*, 1995). Wu *et al.* (2005) developed a post-contact adjustment method to overcome rotation errors when addressing rock fall problems in the original code. Liu *et al.* (2012) coupled the 3D-DDA with tetrahedron finite elements to overcome the problem of block expansion under rigid body rotation. They validated their method with a physical model of a wedge, and got a very good agreement.

A model for cable bolt-rock mass interaction was integrated with DDA by Moosavi and Grayeli (2006). Other useful developments and applications of DDA are summarized in a series of ICADD proceedings (International Conference on Analysis of Discontinuous Deformation) published biannually since 1995.

In this chapter illustrative benchmark tests performed by members of the rock mechanics group at Ben-Gurion University of the Negev (BGU) during the past decade are reviewed. These tests could be performed in every development of the DDA code, for verification purposes. We compare all benchmark tests reported here with analytical solutions, some developed at BGU and some adopted from existing publications. The authors do acknowledge other verification studies performed by other research groups, such as the extensive study of sliding blocks performed at Nanyang Technological University in Singapore (Ning & Zhao, 2012a), slope stability kinematics (MacLaughlin *et al.*, 2001) and analysis of three-hinged beams (Yeung, 1996) performed at U.C. Berkeley, and more, but here only verifications resulting from BGU research are discussed, for brevity.

2 DDA FUNDAMENTALS: A BRIEF REVIEW

DDA considers both statics and dynamics using a time-step marching scheme and an implicit algorithm formulation. The static analysis assumes the velocity of the different block elements is zero at the beginning of each time step, while the dynamic analysis assumes the velocity at the beginning of a time step is inherited from the previous one. The criterion for convergence in DDA is that there will be neither tension nor penetration between the blocks at the end of a time step in the entire block system. These two constraints are applied using a penalty method, where stiff springs are attached to the contacts. Extension or compression of the springs are energy consuming, therefore the minimum energy solution utilized in DDA assures no penetration or tension between the blocks.

In the original DDA code a damping submatrix was not included in the equilibrium equations. There are two ways to introduce damping in the original code: the time step marching scheme introduces algorithmic damping (Doolin & Sitar, 2004), that is determined by the time step size used, and will be briefly discussed later, and kinetic damping. The latter can be applied by assigning a number lower than 1 for the dynamic control parameter: a value of zero means the analysis is static and the velocity is zeroed at the beginning of each time step, a value of unity means the analysis is fully dynamic and the velocity at the beginning of a time step is inherited from the previous one, and any value between zero and one corresponds to the percentage of the velocity that is inherited from one time step to the next. For example, a value of 0.98 corresponds to 2% kinetic damping.

For the sake of brevity, the DDA basic equations will not be reviewed here. The interested reader is encouraged to refer to basic DDA references (Shi, 1988, 1993; Shi & Goodman, 1985).

3 DDA VERIFICATIONS

3.1 Sliding block on an inclined plane

A block sliding on an inclined plane is a classic problem in rock mechanics, as it is a simple and intuitive model for some cases of rock slopes, and has a straightforward analytical solution. The DDA has been verified by several researchers with the analytical solution for a block on an inclined plane. Tsesarsky *et al.* (2005) and Kamai and Hatzor (2008) have verified the 2D-DDA with the analytical solution for static and dynamic input, where in the latter a horizontal acceleration was added. Ning and Zhao (2012a) have verified the 2D-DDA for a block sliding on an inclined plane under both horizontal and vertical accelerations, and under different mechanisms of loading. In this subsection the analytical solution for a block sliding on an incline is reviewed, and

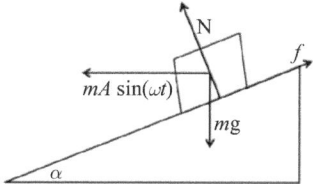

Figure 1 Schematics of the block on an inclined plane problem.

the results of former verification studies of the 2D-DDA, as well as new verifications of 3D-DDA are presented.

3.1.1 Single face sliding

The model of a block on an inclined plane is presented in Figure 1. The inclination angle of the slope is α, and the friction angle of the sliding interface is ϕ. The forces acting on the block are its self-weight mg, the normal from the plane N and the frictional force at the interface between the block and the plane, f. In the most generalized case, where an external force is applied on the block in the form of a harmonic function (Figure 1), the downslope displacements of the block can be calculated by:

$$d(t) = \frac{1}{2}g(\sin\alpha - \cos\alpha\tan\phi)t^2 - \frac{A}{\omega^2}\sin(\omega t)(\cos\alpha + \sin\alpha\tan\phi) + \dot{d}_0 t + d_0 \quad (1)$$

where g is the acceleration of gravity, $\tan\phi$ is the friction coefficient of the sliding interface, A and ω are the amplitude and angular frequency of the harmonic input acceleration, respectively, $\dot{d}_0$ is the initial velocity of the sliding block and d_0 is its initial displacement.

3.1.1.1 2D-DDA

Kamai (2006) performed a 2D-DDA verification study of a block sliding on an inclined plane subjected to gravity ($A = 0$), starting at rest ($\dot{d}_0 = 0$, $d_0 = 0$), for an inclination angle of $\alpha = 28°$, and different values of friction angle. She obtained a good agreement between the two solutions (Figure 2), demonstrated by the relative numerical error, E_N, defined as:

$$E_N = \left|\frac{d_A - d_N}{d_A}\right| \cdot 100\% \quad (2)$$

where d_A and d_N are the analytical and numerical displacements, respectively. She found that the relative numerical error increases with increasing friction angle, and was lower than 8% for $\phi = 25°$, and at the order of 0.05% for $\phi = 5°$.

Kamai and Hatzor (2008) then proceeded to loading the block with a sinusoidal horizontal acceleration, as in Figure 1. In this case, downslope sliding will initiate only when the yield acceleration, $a_y = g\tan(\phi - \alpha)$, is exceeded, at time

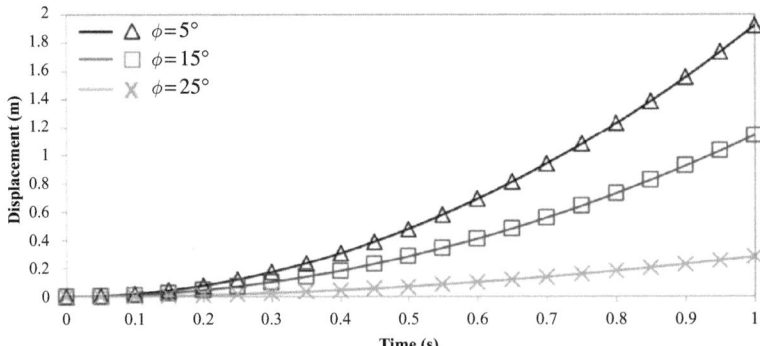

Figure 2 Block displacement vs. time for the case of a block on an incline – gravitational loading only. Comparison between analytical (curves) and DDA (symbols) solutions. After Kamai (2006).

$\theta = \sin^{-1}(\tan(\phi - \alpha)g/A) * (1/\omega)$, a model proposed by Newmark (1965) and Goodman and Seed (1966), and today largely referred to as 'Newmark type' analysis. In this case, Eq. 1 becomes (assuming that $d(\theta) = \dot{d}(\theta) = 0$):

$$d(t) = g\left[(\sin \alpha - \cos \alpha \tan \phi)\left(\tfrac{1}{2}t^2 - \theta t + \tfrac{1}{2}\theta^2\right)\right]$$
$$+ \frac{A}{\omega^2}\left[(\cos \alpha + \sin \alpha \tan \phi)(\omega\cos(\omega\theta)(t - \theta) - \sin(\omega t) + \sin(\omega\theta))\right] \tag{3}$$

The downslope displacements, $d(t)$, are calculated while a_y is exceeded for the first time at θ_1, or the block's velocity is positive. If neither condition is fulfilled, the block is at rest, and will initiate sliding only once a_y is exceeded again, at θ_2, and so on.

Kamai and Hatzor (2008) have used this solution to verify the 2D-DDA. The inclination angle they used for the slope was $\alpha = 20°$, with different friction angles (for the entire set of results, please refer to their original paper). They obtained an excellent agreement between the analytical and numerical solutions, with relative numerical error lower than 1% for most of the simulation time (Figure 3), with a friction angle of 30°.

3.1.1.2 3D-DDA

(1) One direction of motion

The 3D-DDA mesh for the model of the block on an incline is similar to the one in Figure 1, and is constructed of a triangular prism base block, fixed in space, serving as the incline, and a box as the sliding block.

This verification study of the block on an incline is performed in three steps: first the response of the block when subjected to gravity, starting at rest, is examined, then the block is given initial horizontal velocity, and finally the block is subjected to one-dimensional horizontal sinusoidal acceleration. When subjecting the 3D block to initial velocity and acceleration in one direction only, the problem is basically reduced to a 2D problem.

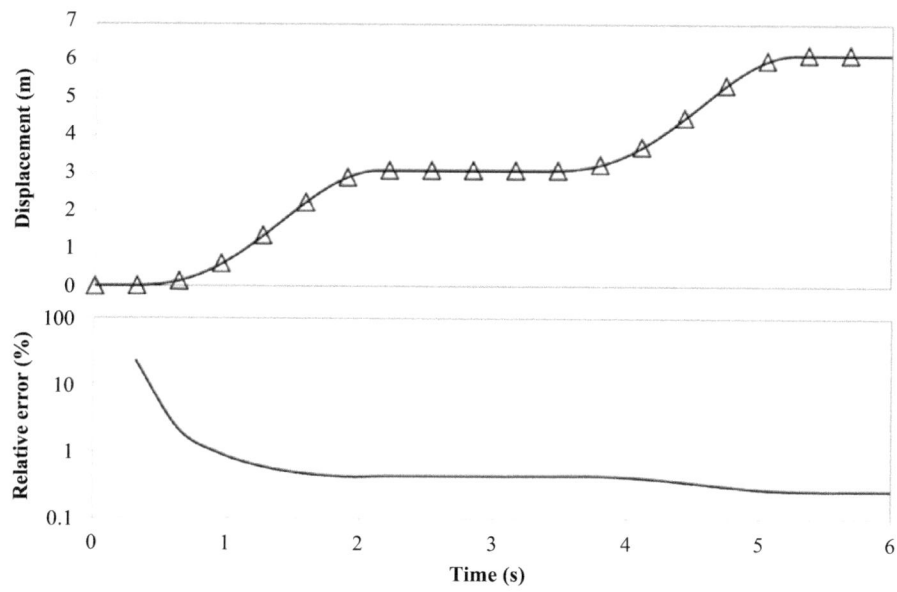

Figure 3 Verification of the dynamic case of a block on an incline for interface friction angle of 30°: top – comparison between analytical (curve) and DDA (symbols); bottom—relative error. After Kamai and Hatzor (2008).

(a) Block starting at rest

In this step the block is subjected to gravity alone ($A = 0$), starting at rest ($\dot{d}_0 = 0$, $d_0 = 0$). In this case, Eq. 1 becomes:

$$d(t) = \frac{1}{2}g(\sin \alpha - \cos \alpha \tan \phi)t^2 \tag{4}$$

The numerical and physical parameters used can be found in (Yagoda-Biran & Hatzor, 2016). The downslope displacement history is compared for friction angle of 20° (remembering the inclination angle of the slope is 45°). Note the excellent agreement between the analytical and numerical solutions (Figure 4a), further demonstrated by the low relative error in Figure 4b: after 0.2 s the numerical error drops to below 1%.

(b) Block starting with initial velocity

The next step of the verification study is applying initial velocity $\dot{d}_0$ to the sliding block. In this case, where no external forces are applied on the block, Eq. 1 becomes:

$$d(t) = \frac{1}{2}g(\sin \alpha - \cos \alpha \tan \phi)t^2 + \dot{d}_0t \tag{5}$$

An initial velocity of 1 m/s is applied horizontally in the dip direction. The agreement between the analytical and the numerical solution is good (Figure 5a), as demonstrated by the relative numerical error plotted in Figure 5b: less than 1% after 0.5 s of the analysis.

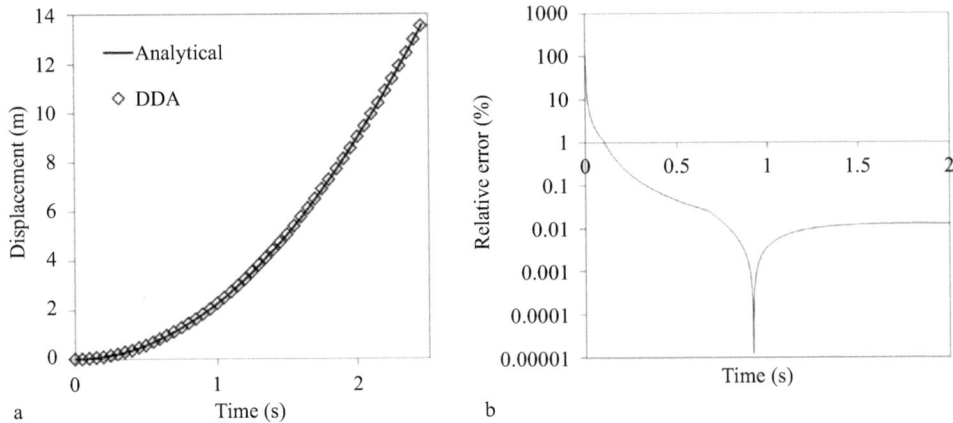

Figure 4 a) Downslope displacement histories of a block on an inclined plane subjected to gravity alone, with interface friction of 20°. b) The relative numerical error.

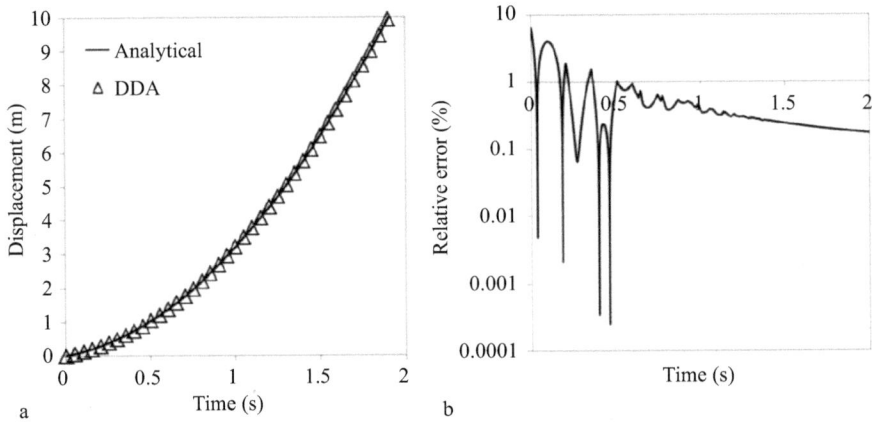

Figure 5 a) Downslope displacement histories of a block on an inclined plane subjected to gravity and initial velocity. b) The relative numerical error.

(c) Block subjected to sinusoidal acceleration input

The third step of the verification study is subjecting the block to one-dimensional horizontal sinusoidal acceleration, as in Figure 1. The amplitude and frequency used for the input acceleration are 2 m/s² and 1 Hz, respectively.

Here the 'Newmark' type analysis is used as the analytical solution, as in the 2D-DDA verification explained earlier. The friction angle of the interface between the slope and the sliding block is set to $\phi = 50°$, higher than the inclination angle $\alpha = 45°$, so block sliding will initiate only when the yield acceleration is exceeded. In Figure 6a the downslope displacement histories calculated by the Newmark analysis and the

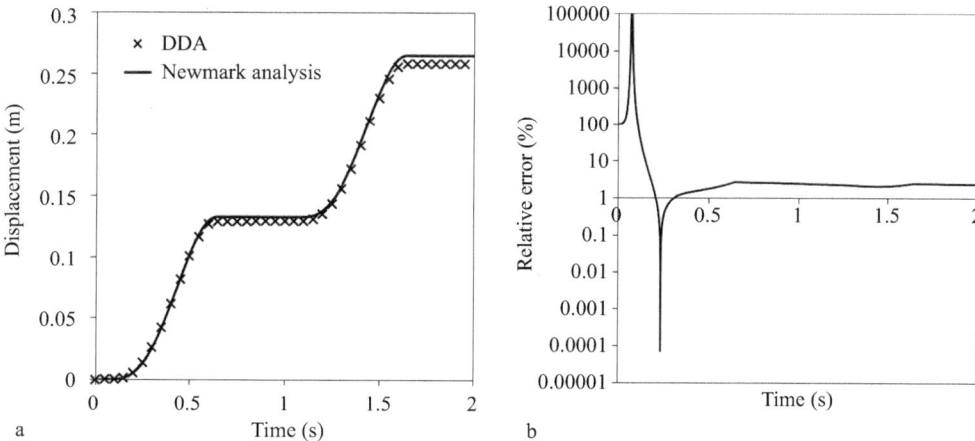

Figure 6 a) Downslope displacement histories of a block on an inclined plane subjected to gravity and
1-D sinusoidal input function. b) The relative numerical error.

3D-DDA code are presented. The agreement between the two is good, and can again be
expressed in terms of relative error, presented in Figure 6b, which during most of the
analysis remains below 3%.

(2) Block sliding on an incline – loading in two directions

Bakun-Mazor *et al.* (2009, 2012) have derived a semi-analytical 3D-formulation
for solving dynamic three dimensional displacements of single and double plane
sliding. In their paper, Bakun-Mazor *et al.* (2009) presented an analytical formulation which they called Vector Analysis (VA), based on the limiting equilibrium
equations of vector forces acting on a block on an inclined plane. The dynamic
equations of motion of their analytical solution have a discrete nature, therefore
the solution is considered semi-analytical. The formulation is explained in details
in (Bakun-Mazor *et al.*, 2009, 2012). They compared the 3D-DDA numerical
results to the results they obtained with their 3D solution, when subjecting the
block to sinusoidal accelerations in the horizontal dip and strike directions, with
different amplitudes and frequencies, and obtained a good agreement, where the
numerical error in the final position of the block after 6 seconds was approximately 8% (Figure 7).

3.1.2 Double face sliding with 3D-DDA

Double face sliding refers to the classic problem in rock mechanics – the wedge failure.
Bakun-Mazor *et al.* (2009) verified the 3D-DDA with their VA solution for a wedge
failure (Figure 8a). The input acceleration to the wedge was a sine function acting
parallel to the line of intersection between the boundary planes along which the wedge

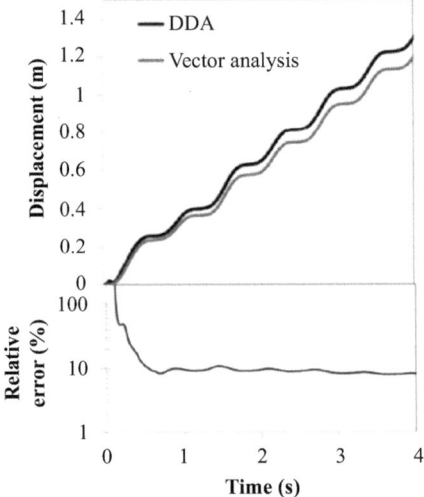

Figure 7 Comparison between vector analysis and 3D-DDA for 2D horizontal input motion simultaneously. The relative error is plotted in the lower panel where the VA is used as a reference.
After Bakun-Mazor *et al.* (2009).

slides. The plunge of the line of intersection was 30° below the horizon, and the friction
of the sliding interfaces was set to 20°. They obtained a good agreement between the
3D-DDA and their VA formulation where the relative numerical error remained below
7% for the entire simulation (Figure 8b).

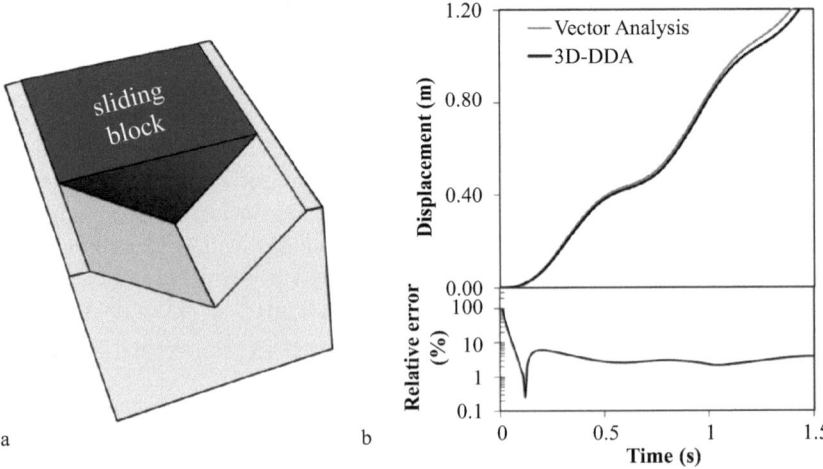

Figure 8 Dynamic sliding of a wedge: comparison between 3D-DDA and VA solutions. (a) The wedge
model in the 3D-DDA. (b) Wedge response to one component of horizontal sinusoidal input
motion and self-weight. Lower panel presents the relative error. After Bakun-Mazor *et al.*
(2009).

3.2 Failure mode mapping for a block on an inclined plane

A block on an incline, of which sliding failure was reviewed in the previous section, has actually four possible modes: it can stay at rest, it can slide, it can topple, or it can slide and topple simultaneously. The actual failure mode is controlled by three factors, when the block is subjected to gravity: the inclination of the slope α, the slenderness of the block δ, and the friction angle of the interface, ϕ (see Figure 9a). The boundaries between the modes were modified over the years (Ashby, 1971; Bray & Goodman, 1981; Hoek & Bray, 1981; Yeung, 1991) (see Figure 9b), and recently mapping the failure mode when the block is subjected to an external pseudo-static earthquake inertia force F (see Figure 9a), has been demonstrated (Yagoda-Biran & Hatzor, 2013). Yagoda-Biran and Hatzor (2013) have found that when adding a horizontal pseudo-static force, with its resultant with the weight vector forming an angle β with the vertical direction (see Figure 9a), the mode of the block is now controlled by the slenderness of the block δ, the friction angle of the interface ϕ, and a new angle $\psi = \alpha + \beta$, rather than α.

3.2.1 2D-DDA

The first comparison between 2D-DDA and analytically derived failure mode chart for toppling and sliding was performed by Yeung (1991) at U.C. Berkeley, and the results are discussed extensively in his PhD dissertation. Inconsistencies found in his comparison actually led him to develop a new equation for the boundary between toppling and sliding and toppling, turning it from static to dynamic in nature.

In this section, the agreement between the DDA and the mode chart can't be discussed in terms of relative error, since in this case only the first motion of the block is of interest, so the degree of agreement can be thought of as binary – either the solutions agree, or they don't.

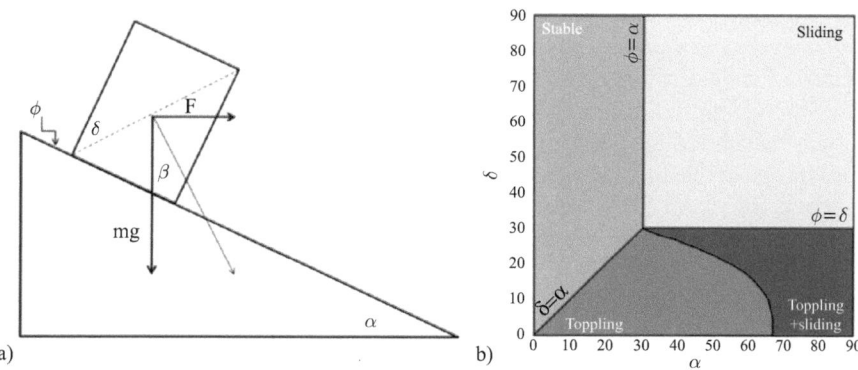

Figure 9 a) Schematics of the block on an incline and the angles controlling its mode. b) An example for the failure mode chart for gravitational loading, at $\phi = 30°$. After Yagoda-Biran & Hatzor (2013).

3.2.1.1 Comparison between 2D-DDA and modified failure mode chart under pseudo-static loading

Yagoda-Biran and Hatzor (2013) used the 2D-DDA to verify their newly developed failure mode chart when a pseudo-static inertia force is considered. They put forth a set of rules to determine the first motion of the block as obtained with DDA, and compared it to the prediction of their modified mode chart. The slope angle α used was 10° for most of the simulations, and they used a wide range of slenderness and friction. The change in the angle ψ was controlled by changing the magnitude of the applied pseudo-static force. An excellent agreement was obtained between the two solutions, with the DDA returning the modes as predicted by the analytical mode chart for 106 out of the 110 simulations.

3.2.2 3D-DDA

In the case of 3D-DDA, the numerical solution was compared for both static and pseudo-static cases.

3.2.2.1 Gravitational loading

Since the 3D-DDA has not been verified many times in the past, Yagoda-Biran and Hatzor (2013) first verified the 3D-DDA with the previously published failure mode chart as derived by (Ashby, 1971; Bray & Goodman, 1981; Hoek & Bray, 1981; Yeung, 1991), using the same criteria for determining the first motion of the block as used for the 2D comparison. The range for α was between 14 and 50°, ϕ was 20° in most of the simulations, and δ was between 11 and 50°. An excellent agreement was obtained between the two solutions; 49 out of the 51 DDA simulations produced the failure mode predicted by the analytical mode chart.

3.2.2.2 Pseudo-static loading

After verifying the 3D-DDA with the original failure mode chart for gravitational loading, Yagoda-Biran and Hatzor (2013) proceeded with comparing the 3D-DDA and their newly developed mode chart incorporating a pseudo-static force, using the same criteria for determining the first motion of the block as used for the 2D comparison. The slope angle α used was 10° for most of the simulations. The block slenderness range δ was between 6 and 70°, and the range of input friction angle ϕ between 6 and 80°. The change in the angle ψ was controlled by changing the magnitude of the applied pseudo-static force. All 89 simulations performed with the 3D-DDA returned failure modes as predicted by the analytical mode chart.

3.3 Block rotation under dynamic excitation

Makris & Roussos (2000) studied the problem of the dynamic rocking of a free-standing column subjected to a sinusoidal input acceleration and their analytical solution can be reviewed in (Makris & Roussos, 2000; Yagoda-Biran & Hatzor, 2010). The free body diagram for the problem is shown in Figure 10a. The analytical solution assumes that no sliding occurs at the base of the rocking block.

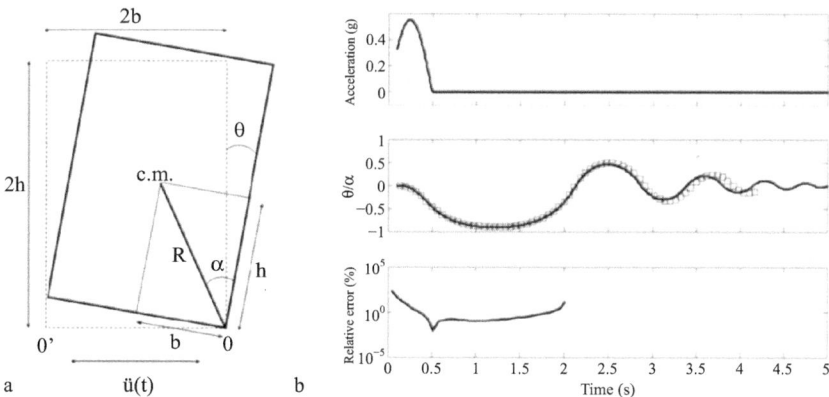

Figure 10 a) Free body diagram and sign convention used in this paper for the rocking block problem. b) upper panel: input accleration. Middle panel: comparison of analytical (curve) and DDA (symbols) solution for dynamic column rotation (b=0.2 m, h=0.6 m). lower panel: relative numerical error. After Yagoda-Biran & Hatzor (2010).

Yagoda-Biran and Hatzor (2010) compared between results from 2D-DDA simulations and the Makris and Roussos (2000) solution. They selected a geometry of a block where b = 0.2 and h = 0.6m, and used an acceleration input function of the form $\ddot{u}_g(t) = a_p \sin(\omega_p t + \psi)$, with changing amplitude a_p and $\omega = 2\pi$ from $t = 0$ to $t = 0.5$ s. For this specific geometry and frequency of motion, the analytical solution shows that the block will not topple with $a_p = 5.43 m/s^2$, but will topple with $a_p = 5.44 m/s^2$. Yagoda-Biran and Hatzor (2010) found the value of contact spring stiffness (*i.e.* the optimal penalty parameter) that will give the same results, in terms of stability-failure, in the 2D-DDA, and then compared the rotation time histories of the column calculated by the analytical and DDA solutions. They obtained an excellent agreement between the two solutions, where the relative error drops below 10% after about 0.3 s of simulation, and below 1% after about 0.5 s (Figure 10b). Yagoda-Biran and Hatzor (2010) found that the error grows larger and the DDA deviates from the analytical solution as soon as the first impact between the rocking column and the fixed base occurs. They explained it by the way damping is implemented in the two solutions. While in the analytical solution the motion during impact is energetically damped due to conservation of angular momentum following the constant value of the coefficient of restitution (Makris & Roussos, 2000), in DDA oscillations at contact points are restrained due to inherent algorithmic damping (Doolin & Sitar, 2004; Ohnishi *et al.*, 2005b).

3.4 Block response to shaking foundation

In the preceding sections the dynamic response of the blocks was a result of direct dynamic input to the blocks centroid. In this sub-section dynamic loadings will be induced by displacement input to a foundation block, a state that resembles the true loading mechanism during an earthquake.

3.4.1 2D-DDA

Kamai and Hatzor (2008) verified the 2D-DDA with a semi-analytical solution for a block responding to induced displacements at its foundation. The model comprises of three blocks as follows (**Figure 11**): a stationary base block (0), an intermediate block (1) to which the input displacements are applied, and an overlying very flat block (2) which responds to the induced displacements in Block 1.

Block 1 is subjected to a horizontal displacement input function in the form of a cosine, starting from zero:

$$d(t) = D\left(1 - \cos(2\pi f t)\right) \tag{6}$$

The only force acting on Block 2, other than its weight and the normal from Block 1, is the frictional force, which determines the acceleration of block 2. For full derivation of equations please refer to Kamai and Hatzor (2008). Kamai and Hatzor (2008) defined a set of inequalities and boundary conditions that determine the magnitude and direction of the acceleration of Block 2, as a function of the acceleration of block 1 and the relative velocity between the blocks. Kamai and Hatzor (2008) compared the displacements computed with 2D-DDA and the analytical solution for changing values of friction angle of the interface between blocks 1 and 2, and for changing amplitudes of input displacement. They found that generally the relative numeric error stayed below 5%, and was more sensitive to friction coefficient than amplitude changes (Kamai & Hatzor, 2008). Example of the response of block 2 is presented in Figure 12.

3.4.2 3D-DDA

The verification of the case of a responding block to moving foundation in three dimensions is based on the one-dimensional verification described by Kamai and Hatzor (2008), with the exception that here the displacements, velocities and accelerations are vectors.

3.4.2.1 The semi-analytical solution

The model used for the verification study is presented in **Figure 11**. Each of the two moving blocks, blocks 1 and 2, has time dependent displacements $\mathbf{d}(t)$, velocities $\dot{\mathbf{d}}(t)$ and accelerations $\ddot{\mathbf{d}}(t)$. The displacement induced to block 1, $\mathbf{d}_1$, is in the form of a cosine function:

Figure 11 The model used in the DDA verification of block response to induced displacements.

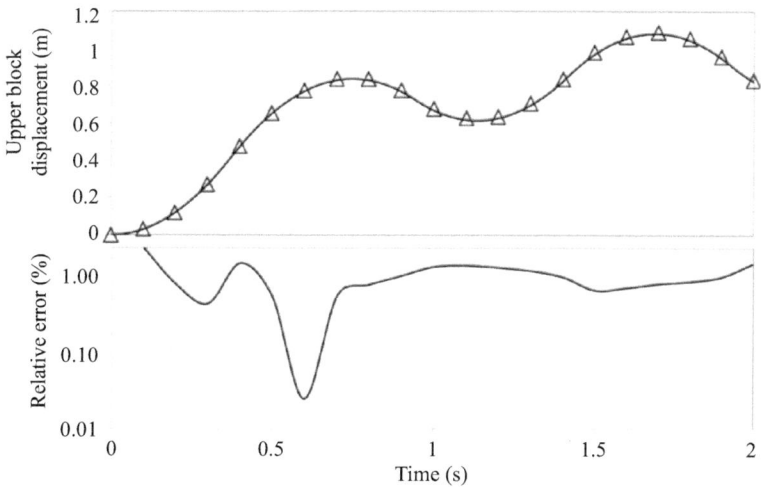

Figure 12 Response of Block 2 to displacement input of f=1Hz, and amplitude of 0.5 m. Curve – analytical, symbols – DDA. After Kamai & Hatzor (2008).

$$\mathbf{d}_1(t) = \mathbf{A}\left(1 - \cos\left(2\pi \bar{f} t\right)\right) \tag{7}$$

where $\mathbf{A}$ and $\bar{f}$ are the amplitude and frequency of motion, respectively. The forces acting on block 2 are its weight, $m_2 g$, the normal from block 1, $N = m_2 g$, and the frictional force between the two blocks, $\mu^* m_2 g$, where μ is the friction coefficient. Newton's second law of motion yields that the acceleration of block 2 is $|\ddot{\mathbf{d}}_2| = \mu * g$. Following Kamai and Hatzor (2008), the direction of the frictional force, and therefore of $\ddot{\mathbf{d}}_2$, is determined by the direction of the relative velocity between the two blocks, $\dot{\mathbf{d}}^* \equiv \dot{\mathbf{d}}_1 - \dot{\mathbf{d}}_2$, defined by the unit vector of the relative velocity, $\hat{\mathbf{d}}^*$, When $|\dot{\mathbf{d}}^*| = 0$, the acceleration of block 2 ($\ddot{\mathbf{d}}_2$) is determined by the acceleration of block 1 ($\ddot{\mathbf{d}}_1$). When the acceleration of block 1 exceeds the yield acceleration $\mu^* g$, over which block 2 no longer moves in harmony with block 1, the frictional force direction is determined by the direction of $\hat{\mathbf{d}}^*$, but the magnitude of $\ddot{\mathbf{d}}_2$ is equal to $\mu^* g$. This rationale can be formulated as follows:

$$\text{If } \left|\dot{\mathbf{d}}^*\right| = 0 \ldots \ldots \ldots \text{ and } \left|\ddot{\mathbf{d}}_2\right| \leq \mu * g \ldots \ldots \text{ then } \quad \ddot{\mathbf{d}}_2 = \ddot{\mathbf{d}}_1$$

$$\text{and } \left|\ddot{\mathbf{d}}_2\right| > \mu * g \ \ldots \ldots \text{ then } \quad \ddot{\mathbf{d}}_2 = (\mu * g) \cdot \hat{\ddot{\mathbf{d}}}_1 \tag{8}$$

If $|\hat{\mathbf{d}}^*| \neq 0 \ \ldots \ldots$ then $\ddot{\mathbf{d}}_2 = (\mu * g) \cdot \hat{\mathbf{d}}^*$

This set of conditions and inequalities was applied with a time step of 0.0001 s. Since the analytical solution is calculated numerically, it is actually a semi-analytical solution.

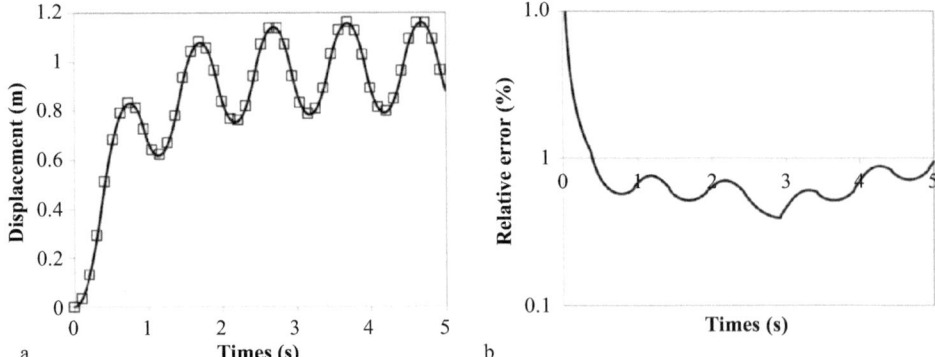

Figure 13 a) Comparison between the analytical solution (curve) and 3D-DDA solution (symbols), for amplitude of A = 0.2 m, friction coefficient of 0.6, and frequency of 2 Hz for the input motion. b) relative numerical error.

3.4.2.2 The numerical model

The actual model used for the 3D-DDA is shown in **Figure 11**. Block 2 was designed to be very flat, so as to avoid rotations during motion. The physical and numerical control parameters used in the verification analyses can be found in (Yagoda-Biran & Hatzor, 2016).

3.4.2.3 One direction of motion

The first step was inducing displacements to block 1 in the x-direction (see **Figure 11**), similar to the work reported by Kamai and Hatzor (2008). This was done with changing amplitudes, frequencies and friction angles, which can be viewed in (Yagoda-Biran & Hatzor, 2016). One example can be seen in Figure 13, where the input motion is of amplitude of 0.5 m, frequency of 2 Hz and friction coefficient of 0.6. The two solutions compare very well (Figure 13a), and the relative numerical error stays below 1% after about 0.3 s into the simulation.

3.4.2.4 Two directions of motion

In the second step of the verification study, displacements were induced to block 1 in the x and y directions (**Figure 11**), each with different amplitude and frequency. Some results are presented in Figure 14. where the resultant horizontal displacement vs. time is presented in a). Figure 14c is a 3D plot of the x and y displacements vs. time, presented as the vertical axis. Again, the agreement between the 3D-DDA and the analytical solution is good, as expressed by the numerical error in Figure 14b, which remains below 10% for the entire simulation. A deviation between the numerical and analytical solutions is observed with increasing simulation time.

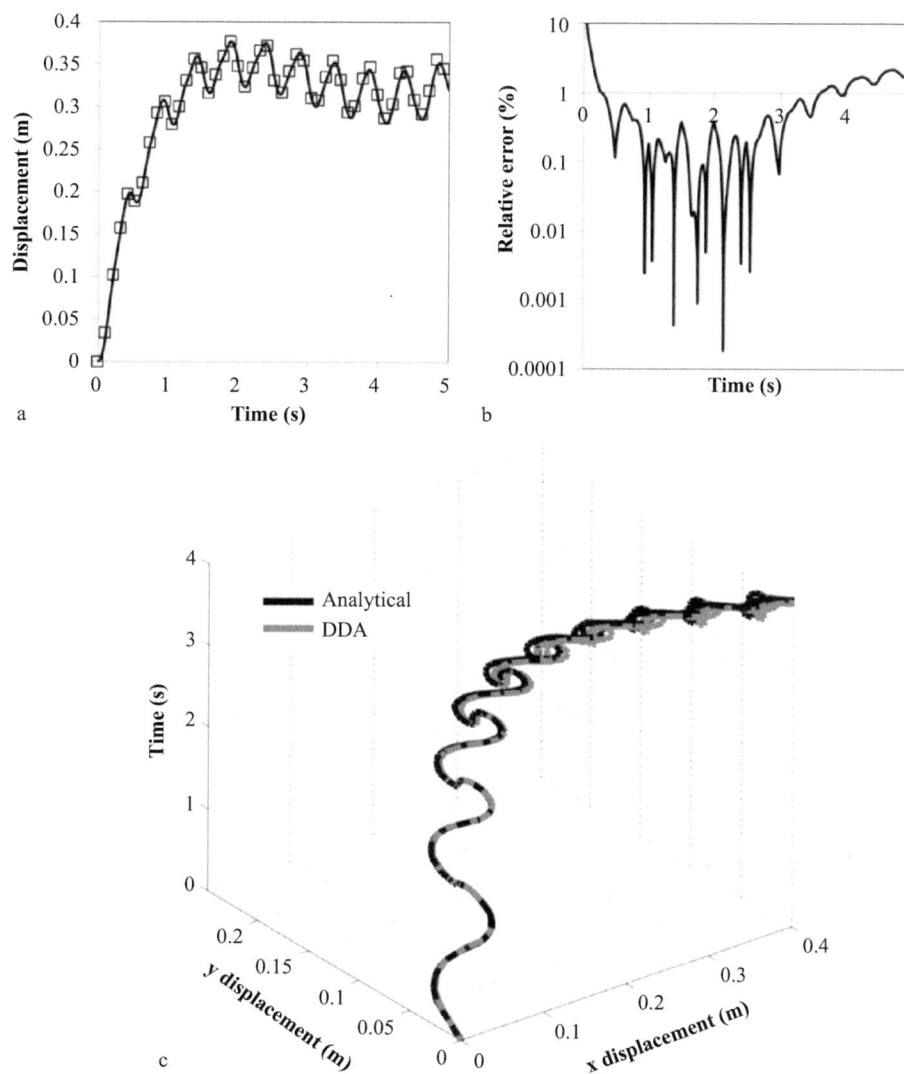

Figure 14 a) Comparison between analytical (curve) and 3D-DDA (symbols) solutions. The amplitude and frequency for the input displacements are 0.3 m and 2 Hz, 0.2 m and 4Hz for the x and y directions, respectively. b) relative numerical error. c) x and y displacements as a function of time (vertical axis).

3.2.4.5 Three directions of motion

The third verification step was subjecting block 1 to sinusoidal displacements in all three directions: x, y and z. Adding input in the vertical direction affects the response of block 2 as it changes the normal force between the two blocks, and therefore the frictional force between them. This in turn changes the acceleration of block 2, $\ddot{\mathbf{d}}_2$. Applying time-dependent displacements in the z direction is actually equivalent to time-dependent changes in g: when block 1 has positive z acceleration $(\ddot{d}_1 \hat{k} > 0)$, it is added to g. When $\ddot{d}_1 \hat{k}$ is negative, it is subtracted from g. The analytical solution in this case assumes no other effect of the vertical displacement of block 1 on the horizontal displacement of block 2. The induced displacement function is now:

$$\mathbf{d}_1(t) = 0.1\left(1 - \cos(2\pi 2t)\right) \cdot \hat{i} + 0.1\left(1 - \cos(2\pi 4t)\right) \cdot \hat{j} + 0.1\left(1 - \cos(2\pi t)\right) \cdot \hat{k}$$

$$(9)$$

In Figure 15 results of the verification study with three components of induced displacements are presented. Figure 15a presents the resultant horizontal (x-y plane) displacement vs. time, for different values of the k-normal contact spring stiffness. The black curve is the analytical solution, while the other curves are the 3D-DDA numerical solutions for different values of k. The range of contact spring stiffness that best fits the analytical solution is between $1*10^7$ and $1*10^9$ N/m, with stiffness of $k = 1*10^7$ N/m, or $0.0003\ E*L$, being the optimal selection, where E is the Young's modulus of the block and L is the length of the line across which the contact springs are attached. When considering 3D-DDA, it might be more relevant to compare k to $E*A$, where A is the area across which the contact springs are attached. In this case, $k = 1*10^7$ is ~$0.0001\ E*A$, not much different from $E*L$. Figure 15b demonstrates this with the relative numerical error. Note that for the results obtained with $k = 1*10^7$ N/m, the relative error stays below 3% for the entire analysis, and the error is well below 10% for $k = 1*10^8$ and $1*10^9$ N/m as well.

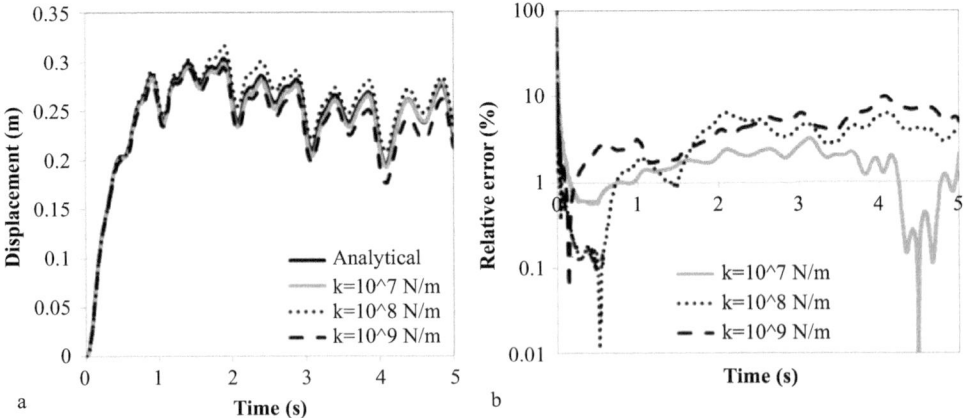

Figure 15 a) Comparison between analytical and 3D-DDA solutions. The best fit is obtained with contact spring stiffness of $k = 1*10^7$ N/m, but the overall trend of the analytical solution is maintained for all values of stiffness. b) Relative numerical error for the different solutions presented in a).

3.5 Shear wave propagation

Bao *et al.* (2014a) tested the ability of the 2D-DDA to correctly simulate wave propagation. They modeled a set of stacked horizontal layers (Figure 16a), generated a shear wave by inducing horizontal displacements at the base, and compared the waveforms at the measurement point in the model (Figure 16a). Because the DDA blocks are linear elastic and un-damped, the analytical amplitude in the tests is the same amplitude as that of the incident wave at the rigid base. After performing sensitivity analysis to the block size and time step size, they managed to successfully preserve the wave form in the DDA model (Figure 16b).

Bao *et al.* (2014a) also validated the DDA with SHAKE program for site response applications. SHAKE (Lysmer *et al.*, 1972; Schnabel *et al.*, 1972) is a program that analyzes the 1-dimensional response of a stack of linear-elastic layers in the frequency domain. The stack can be composed of several layers with varying properties, and it is subjected to seismic motion through its base. Although in this chapter we review verification studies, we choose to present this validation part, because the SHAKE program has been verified many times, its accuracy is well established for the underlying assumptions and boundary conditions, and validating the DDA with SHAKE for wave propagation applications seems to be an important step. Bao *et al.* (2014a) generated a stack of 15 layers in 2D-DDA, each with different mechanical properties (Figure 17a), and applied a real earthquake time history at its base. They modeled the same sequence of layers in SHAKE, and compared the spectral amplifications obtained with the two approaches at the top of the layer stack, with respect to its base. A very good agreement between the two methods was obtained (Figure 17b), both for homogeneous and inhomogeneous media, both for frequency and amplitude. Their results suggest the DDA can be used to model wave propagation through discontinuous media, provided that the numerical control parameters are well conditioned.

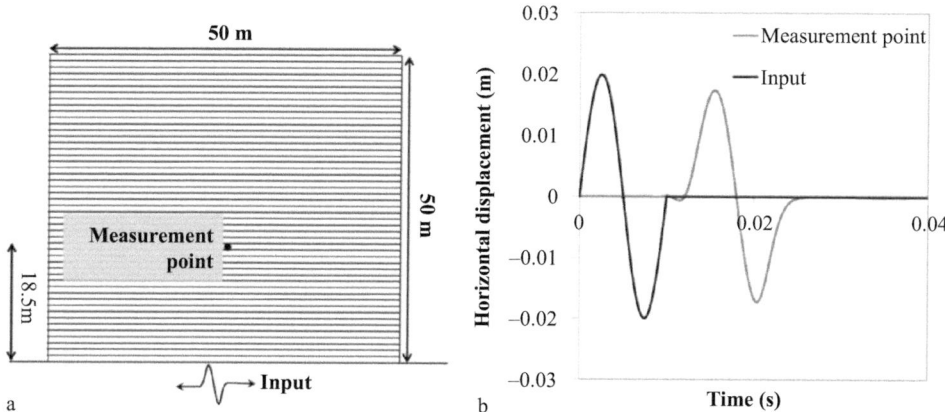

Figure 16 Wave propagation through a stack of layers with DDA. a) DDA model. b) One-cycle sinusoidal incident wave time history used to induce vertical shear wave propagation in the DDA, and wave forms as obtained at the measurement point. After Bao *et al.* (2014a).

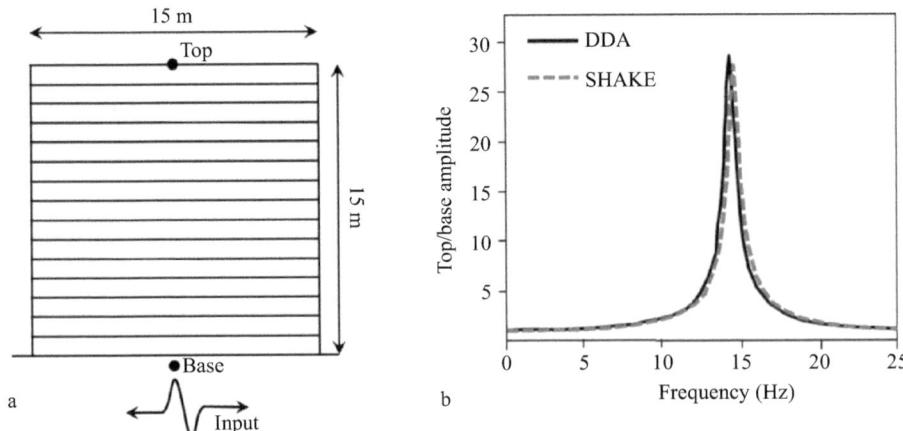

Figure 17 a) The model of the stacked layers in DDA. b) Spectral ratios as obtained with DDA and SHAKE. After Bao *et al.* (2014a).

4 DISCUSSION

4.1 Numerical control parameters

The user-defined numerical control parameters have a significant effect on the results of DDA simulations. In this section a discussion of these parameters and suggestions as to optimal selection of them will be made. It is however important to stress that although some general guidance for optimal selection of the parameters is presented here, given the effect they have on the results of the simulation, their calibration should be performed whenever possible, and routine sensitivity analyzes are highly recommended.

4.1.1 Normal contact spring stiffness

The normal contact spring stiffness, k, is the stiffness of the virtual springs assigned at the dynamically formed contacts. In many sensitivity analyzes it was found that the value of k significantly affects the results of the simulation.

Shi (1996), in his user manual, recommended that as a rule of thumb, $k = E * L$, where E is the Young's modulus of intact block material and L is the average block diameter. However, it is sometimes reported that the optimal value does not follow this rule. In their study of wave propagation with DDA, Bao *et al.* (2014a) found that a k value lower by 1.5 orders of magnitude than Shi's rule of thumb is optimal. They suggested that the condition of the interface between the blocks might have an effect on the value selected: a weathered interface might effectively lower the Elastic modulus, therefore lower the optimal stiffness value. This observation is supported by Yagoda-Biran and Hatzor (2010) who modeled a physical problem somewhat similar to the one modeled by Bao *et al.* (2014a) and concluded that a k value of about 2 orders of magnitude lower than Shi's rule of thumb (Shi, 1996) would be optimal. In other cases however, it seems that the optimal stiffness value even further deviates from Shi's rule of thumb, such as the case presented in section 3.4.2. In this case, the optimal stiffness that results in the smallest numerical error is 2 to 4 orders of magnitude lower than Shi's recommendation. In this case however the simulation is in 3D-DDA, while Shi's recommendations were given as a guide for 2D-DDA, therefore generalization might not be appropriate in this case.

4.1.2 Time step interval

Selecting an appropriate time step interval is an important issue with DDA simulations, and should be a balance between accuracy and computational efficiency. As reviewed in Bao *et al.* (2014a), various authors proposed different rules of thumb for optimal time step size in relation to the wave period: less than 1% of the primary wave period in finite element simulations of nonlinear sound wave propagation (Kagawa *et al.*, 1992), 5% of the shortest period of incident waves in Newmark time integration scheme (Moser *et al.*, 1999), or smaller than $2/\pi$ of the un-damped period of vibration of the system, in order to avoid bifurcation in the DDA solution (Doolin & Sitar, 2004).

In order to ensure the stability of the numerical solution, the time step interval should be smaller than the fraction of the period that is equal to the ratio between the side length of the element along the direction of wave propagation path, and the wavelength, according to the Courant–Friedrichs–Levy condition (Courant *et al.*, 1967). This condition however does not ensure accuracy. In the next section we discuss how the choice of time step interval is reflected in the systems' damping.

4.1.3 Damping

In the original DDA code a damping submatrix was not incorporated in the equilibrium equations. Therefore, if the original code is to be used without modifications, damping can be introduced artificially by means of either kinetic damping or algorithmic damping. Kinetic damping is applied when the transferred velocity to the consecutive time step is reduced by some measure. Any value between 0 and 1 of the user defined dynamic control parameter would correspond to the percentage of velocity transferred from one time step to the following, that is, a dynamic parameter of 0.97 corresponds to 3% damping. When studying dynamic deformation of jointed rock slopes, Hatzor *et al.* (2004) reported that a 2% kinetic damping is required to obtain stable solution with the 2D-DDA version they used at the time. But if true and accurate displacements are required, then no kinetic damping should be introduced at all. This can be done provided that all other numerical control parameters are properly conditioned.

Algorithmic damping (Doolin & Sitar, 2004) is associated with the time integration scheme used for integrating second order systems of equations over time. Numerical damping stabilizes the numerical integration scheme by damping out the unwanted high frequency modes. For the Newmark time integration scheme used in DDA (with $\beta = 0.5$ and $\gamma = 1.0$, see Ohnishi *et al.*, 2005a), it also affects the lower modes and reduces the accuracy of integration scheme to first order. In DDA, the numerical damping that is associated with the time integration scheme increases with increasing time step size. If the time step is small enough, the numerical damping phenomenon is insignificant. Bao *et al.* (2014a) suggested a way to utilize this time step size dependence of algorithmic damping, and obtained an equivalent damping ratio by seeking the time step size that will result in exactly the same damping ratio that would have been assumed otherwise in the structural analysis. They inspected the damped oscillations of the free end of a cantilever beam modeled with DDA with different time step intervals, and obtained an equivalent damping ratio, using the algorithmic damping in DDA as a function of time step interval. Then they modeled a stack of horizontal layers in DDA, subjected to earthquake displacements at the foundation, with a time step size of 0.001s, which corresponded in that case to 2.3% damping. They compared the amplification and resonance frequency obtained with the DDA model, to those of an equivalent SHAKE (Lysmer *et al.*, 1972; Schnabel *et al.*, 1972) model with an input of 2.3% damping, and obtained an extremely good agreement between the two methods (see section 3.5).

4.2 Limitations of 3D-DDA

4.2.1 Constructing a 3D-DDA mesh

Modeling three dimensional multi-block structures in 3D-DDA is an elaborate and challenging task. The block cutting code in 3D-DDA does not have a user friendly graphic interface, and does not accept three-dimensional blocks as input, but rather two-dimensional triangles, of which the blocks are constructed. For example, in order to build a rectangular face of a box, two triangles will be required, with three vertices each. Therefore, to form a simple box, one needs to input the vertices for 12 triangles. The accuracy of the input coordinates is of great importance as well: if the coordinates of two adjacent triangles do not exactly coincide, the code will not be able to cut a block. This requires sophisticated pre-processing with suitable software. When modeling problems involving only of a few blocks, the process of building the mesh might be tolerable, but when constructing meshes which consist of many blocks (see section 4.2.2) the task becomes difficult and exhausting. It is therefore recommended to use computer aided design (CAD) software to construct a 3D-DDA mesh, and use a function of the CAD software to export the model. In the work presented here in section 4.2.2 we built the model in a CAD software, exported the nodes, and then wrote the model's nodes to a file readable by the DDA. The scope of this chapter does not allow for a full presentation of the process, but the interested reader is welcome to contact the corresponding author for more information.

4.2.2 The L'Aquila case study

The DDA in its two dimensional formulation has been used several times as a tool to estimate historical seismic hazard (Kamai & Hatzor, 2008; Yagoda-Biran & Hatzor, 2010). While attempting to use the 3D-DDA in a similar manner, we have stumbled upon limitations that suggest the 3D-DDA, at its current formulation, is still not ready for solving reliably dynamic problems involving a large number of blocks. We use the case study of L'Aquila to illustrate this problem.

The city of L'Aquila, the capital of the Abruzzo region, Italy, suffered strong ground motions during the M_w 6.3 earthquake of April 6, 2009. Over 300 people were killed, and many of the buildings in the old city were severely damaged and evacuated. Since there are several strong ground motion accelerographs in the city, this seemed as an

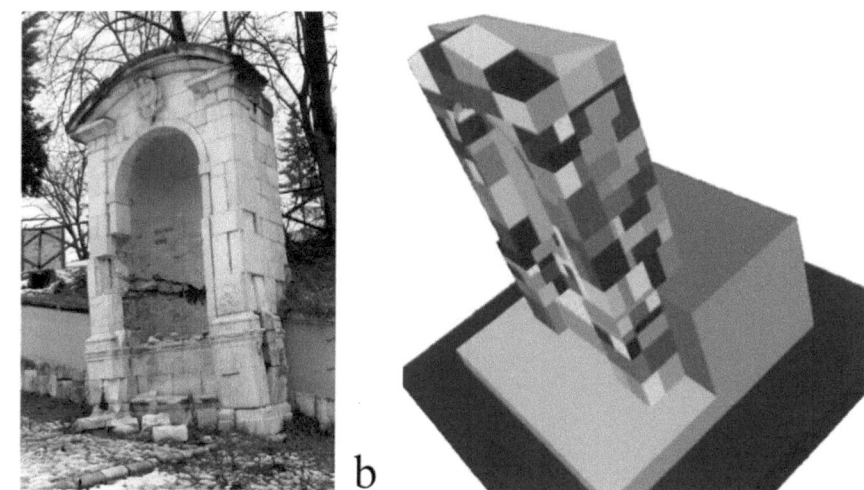

Figure 18 a) The damaged masonry structure in the city of L'Aquila, Italy following the Mw 6.3 earthquake that struck the region, b) snapshot of the corresponding 3D-DDA mesh.

excellent case study to check the validity of the DDA for solving complicated dynamic problems in three dimensions.

We searched the old city of L'Aquila for small, simple buildings that can be easily modeled with the 3D-DDA. We found a small masonry structure that was damaged by the earthquake, but did not collapse (Figure 18). Naturally, the observed damage cannot be modeled correctly with 2D-DDA, and a 3D approach is required. The model of the structure in 3D-DDA was comprised of 197 blocks.

When trying to run forward analyzes with the model, subjecting it to the acceleration time series as recorded in a station located about 500 meters away from the structure, the solution converged only when kinetic damping of at least 3% was used, and a time step size no larger than 10^{-5} s. With less kinetic damping and longer time steps a stable solution could not be obtained, probably due to limitations of the contact algorithm in its current form, which does not allow the system to converge when a large number of contacts must be solved in every iteration. The choice of such a small time step inevitably leads to extremely long CPU time, especially when running simulations with long "real" run time of tens of seconds. The use of small time step intervals in the 3D-DDA, smaller than the ones used in similar simulations in 2D-DDA, was observed many times. For example, the 3D simulations in Yagoda-Biran and Hatzor (2013) used a time step size two orders of magnitude smaller than the time step size used in similar 2D-DDA simulations. For the case of the responding block to induced displacements, the time step size used in the 3D-DDA case in section 3.4 is one order of magnitude smaller than the one used in 2D-DDA (Kamai, 2006). For the case of a block on an incline, the time step size used in the 2D-DDA case is 0.002 s (Kamai & Hatzor, 2008), while the 3D-DDA simulation in section 3.1.1.2 used a time step size of 0.0001 s.

Furthermore, we noticed that the displacements of the blocks were several orders of magnitude smaller than expected. These results led us to investigate what effect does the coupling of kinetic damping and small time step has on the numerically obtained

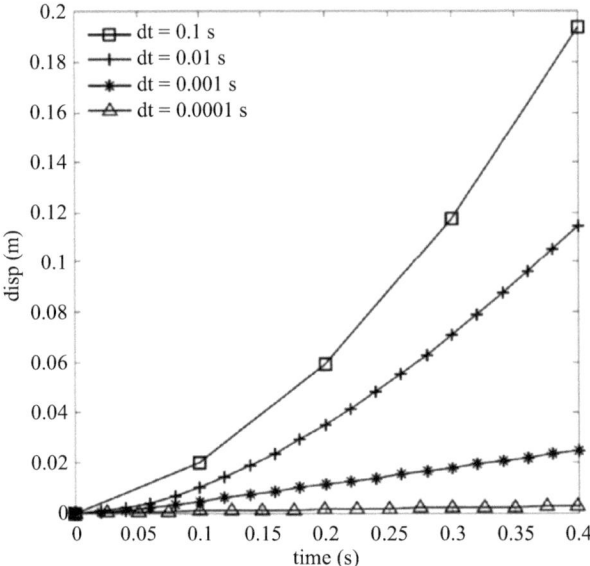

Figure 19 Cumulative displacement of a mass subjected to constant force, under 3% kinetic damping and different time steps, as calculated semi-analytically. After Yagoda-Biran & Hatzor (in review).

cumulative displacements. We conducted a numerical experiment, where the time dependent displacements of a mass driven by a constant force were computed, with kinetic damping of 3% and different time step sizes. As observed in Figure 19, when using kinetic damping, the cumulative displacement decreases with decreasing time step, an effect also observed in the same experiment performed with the 3D-DDA (see Yagoda-Biran & Hatzor, 2016). This effect was not observed when no kinetic damping was applied: the time step size had no effect on the cumulative displacements. Furthermore, decreasing the kinetic damping by even 2%, down to 1% damping, did not change the results significantly: the displacements were still highly restrained.

It is thus evident that a combination of a very small time step interval and kinetic damping of a small percentage significantly decreases the cumulative displacement during the simulation rendering the numerical results inaccurate and unrealistic. It is also evident that a small increase in kinetic damping coefficient will not make a great difference when very small time steps are used. Ideally, it would be preferable to use zero kinetic damping in dynamic DDA simulations, as the displacements per time step are reduced with increasing time step size anyhow due to the inherent algorithmic damping in DDA.

In cases such as these, where displacements are the desirable output, these limitations of the current version of the 3D-DDA code are not tolerable, and this effect makes the 3D-DDA, in its current form, ill-suited for dynamic simulations of multi-block systems. A completely new contact algorithm has recently been proposed by Shi (2013), but has not yet been implemented in executable codes which are available to us for testing. After the new contact algorithm is implemented 3D-DDA should be tested again for its applicability for multi-block systems and multiple contacts using benchmark tests similar to those presented in this chapter.

5 SUMMARY AND CONCLUSIONS

In this chapter the validity of dynamic analysis with 2D and 3D-DDA is verified by reviewing published verification studies. As the numerical discrete element DDA method is becoming more popular in rock mechanics and engineering geology research worldwide, it becomes supremely important to verify the accuracy and applicability of the method before it is accepted and established as a standard analytical approach in the practice. The DDA has been proven to accurately solve problems involving block translations (block on an incline, double face sliding) and rotations (rocking of a free standing column) when the loading is applied at the center of the block, as well as translations when the loading is applied at the foundation (responding block). It has been proven to accurately solve large displacements, as well as wave propagation problems, that involve small displacements, despite the simply deformable blocks assumption and the first order approximation. This accuracy however is highly dependent on an educated selection of the numerical control parameters, first and foremost the penalty parameter otherwise known as the contact spring stiffness, as well as the time step size. A wise selection of the time step size would balance between small computation times and high accuracy. A wise contact spring stiffness selection would ensure accuracy of the solution; the block size becomes an issue primarily when dealing with wave propagation problems.

Naturally, 2D-DDA cannot be used in cases where out-of-plane deformations are expected in the physical problem. In such cases using the 3D-DDA would seem more appropriate although as much as we have experimented with 3D-DDA in its current formulation we have concluded that obtaining a stable solution to dynamic problems involving a large number of blocks is a very challenging task.

ACKNOWLEDGMENTS

We wish to thank very sincerely Dr. Gen-hua Shi for his support of our research over the years by coming to visit us in our laboratory and by providing access always to most updated versions of the 2D and 3D-DDA codes, his ongoing code modifications that many times resulted from our fruitful interaction, and for stimulating discussions. We also wish to thank sincerely the graduates of our rock mechanics group Drs. Michael Tsesarsky, Ronnie Kamai, Dagan Bakun-Mazor and Huirong Bao for their friendship, collaboration, stimulating discussions and for sharing their ideas and data with us. Financial support of Israel Science Foundation through grant No. ISF-2201, Contract No. 556/08 is thankfully acknowledged.

REFERENCES

Ahn, T.Y. & Song, J.J., 2011. New Contact-Definition Algorithm Using Inscribed Spheres for 3d Discontinuous Deformation Analysis. *International Journal of Computational Methods*, 8, 171–191.
Ashby, J.P., 1971. Sliding and Toppling Modes of Failure in Models and Jointed Rock Slopes: University of London, Imperial College.

Bakun-Mazor, D., Hatzor, Y.H. & Glaser, S.D., 2009. 3D DDA vs. analytical solutions for dynamic sliding of a tetrahedral wedge, in: Ma, G., Zhou, Y. (Eds.), Proceedings of ICADD-9: The 9th International Conference on Analysis of Discontinuous Deformation. Singapore: Research Publishing, pp. 193–200.

Bakun-Mazor, D., Hatzor, Y.H. & Glaser, S.D., 2012. Dynamic sliding of tetrahedral wedge: The role of interface friction. *International Journal for Numerical and Analytical Methods in Geomechanics*, 36, 327–343.

Bao, H.R., Hatzor, Y.H. & Huang, X., 2012. A New Viscous Boundary Condition in the Two-Dimensional Discontinuous Deformation Analysis Method for Wave Propagation Problems. *Rock Mechanics and Rock Engineering*, 45, 919–928.

Bao, H.R., Yagoda-Biran, G. & Hatzor, Y.H., 2014a. Site response analysis with two-dimensional numerical discontinuous deformation analysis method. *Earthquake Engineering & Structural Dynamics*, 43, 225–246.

Bao, H.R. & Zhao, Z.Y., 2010. An alternative scheme for the corner-corner contact in the two-dimensional Discontinuous Deformation Analysis. *Advances in Engineering Software*, 41, 206–212.

Bao, H.R. & Zhao, Z.Y., 2012. The vertex-to-vertex contact analysis in the two-dimensional discontinuous deformation analysis. *Advances in Engineering Software*, 45, 1–10.

Bao, H.R. & Zhao, Z.Y., 2013. Modeling brittle fracture with the nodal-based Discontinuous Deformation Analysis. *International Journal of Computational Methods*, 10(6).

Bao, H.R., Zhao, Z.Y. & Tian, Q., 2014b. On the Implementation of augmented Lagrangian method in the two-dimensional discontinuous deformation Analysis. *International Journal for Numerical and Analytical Methods in Geomechanics*, 38, 551–571.

Ben, Y.X., Xue, J., Miao, Q.H. & Wang, Y., 2011. Coupling fluid flow with discontinuous deformation analysis, in: Zhao, J., Ohnishi, Y., Zhao, G.F. (Eds.), 10th International Conference on Advances in Discontinuous Numerical Methods and Applications in Geomechanics and Geoengineering (ICADD). Honolulu, HI: CRC Press, Taylor & Francis Group, pp. 107–112.

Beyabanaki, S.A.R., Ferdosi, B. & Mohammadi, S., 2009. Validation of dynamic block displacement analysis and modification of edge-to-edge contact constraints in 3-D DDA. *International Journal Rock Mechanics and Mining Sciences*, 46, 1223–1234.

Bray, J.W. & Goodman, R.E., 1981. The Theory of Base Friction Models. *International Journal of Rock Mechanics and Mining Sciences.*, 18, 453–468.

Chen, H.M., Zhao, Z.Y. & Sun, J.P., 2013. Coupled hydro–mechanical model for fracture rock masses using the Discontinuous Deformation Analysis. *Tunnelling and Underground Space Technology*, 38, 506–516.

Courant, R., Friedrichs, K. & Lewy, H., 1967. On the partial difference equations of mathematical physics. *IBM Journal of Research and Development*, 11, 215–234.

Doolin, D.M. & Sitar, N., 2004. Time integration in discontinuous deformation analysis. *Journal of Engineering Mechanics ASCE*, 130, 249–258.

Goodman, R.E. & Seed, H.B., 1966. Earthquake-induced displacements in sand embankments. *Journal of Soil Mechanics and Foundation Division, ASCE*, 90, 125–146.

Grayeli, R. & Mortazavi, A., 2006. Discontinuous deformation analysis with second-order finite element meshed block. *International Journal for Numerical and Analytical Methods in Geomechanics*, 30, 1545–1561.

Hatzor, Y.H., Arzi, A.A., Zaslavsky, Y. & Shapira, A., 2004. Dynamic stability analysis of jointed rock slopes using the DDA method: King Herod's Palace, Masada, Israel. *International Journal of Rock Mechanics and Mining Sciences*, 41, 813–832.

Hatzor, Y.H., Feng, X.-T., Li, S., Yagoda-Biran, G., Jiang, Q. & Hu, L., 2015. Tunnel reinforcement in columnar jointed basalts: The role of rock mass anisotropy. *Tunnelling and Underground Space Technology*, 46, 1–11.

Hoek, E. & Bray, J.W., 1981. *Rock Slope Engineering*, 3rd revised ed. London: E & FN SPON.

Jiang, Q.H., Zhou, C.B., Li, D.Q. & Yeung, M.C.R., 2012. A softening block approach to simulate excavation in jointed rocks. *Bulletin of Engineering Geology and the Environment*, 71, 747–759.

Jiao, Y.Y., Zhang, H.Q., Zhang, X.L., Li, H.B. & Jiang, Q.H., 2015. A two-dimensional coupled hydromechanical discontinuum model for simulating rock hydraulic fracturing. *International Journal for Numerical and Analytical Methods in Geomechanics*, 39, 457–481.

Jiao, Y.Y., Zhang, X.L. & Zhao, J., 2012. Two-Dimensional DDA Contact Constitutive Model for Simulating Rock Fragmentation. *Journal of Engineering Mechanics-ASCE*, 138, 199–209.

Jiao, Y.Y., Zhang, X.L., Zhao, J. & Liu, Q.S., 2007. Viscous boundary of DDA for modeling stress wavve propagation in jointed rock. *International Journal of Rock Mechanics and Mining Sciences*, 44, 1070–1076.

Jing, L., 2003. A review of techniques, advances and outstanding issues in numerical modeling for rock mechanics and rock engineering. *International Journal of Rock Mechanics and Mining Sciences*, 40, 283–353.

Jing, L. & Hudson, J.A., 2002. Numerical methods in rock mechanics. *International Journal of Rock Mechanics and Mining Sciences*, 39, 409–427.

Jing, L., Ma, Y. & Fang, Z., 2001. Modeling of fluid flow and solid deformation for fractured rocks with discontinuous deformation analysis (DDA) method. *International Journal of Rock Mechanics of Mining Sciences Geomechanics Abstracts*, 343–355.

Jing, L. & Stephansson, O., 2007. Fundamentals of Discrete Element Methods for Rock Engineering: Theory and Applications: Elsevier, Amsterdam

Jing, L.R., 1998. Formulation of discontinuous deformation analysis (DDA) – an implicit discrete element model for block systems. *Engineering Geology*, 49, 371–381.

Kagawa, Y., Tsuchiya, T., Yamabuchi, T., Kawabe, H. & Fujii, T., 1992. Finite-element simulation of nonlinear sound-wave propagation. *Journal of Sound and Vibration*, 154, 125–145.

Kamai, R., 2006. Estimation of Historical Seismic Ground-Motions Using Back Analysis of Structural Failures in Archaeological Sites, MSc thesis, The Department of Geological and Environmental Sciences. Beer-Sheva: Ben-Gurion University of the Negev, p. 127.

Kamai, R. & Hatzor, Y.H., 2008. Numerical analysis of block stone displacements in ancient masonry structures: a new method to estimate historic ground motions. *International Journal for Numerical and Analytical Methods in Geomechanics*, 32, 1321–1340.

Kim, Y., Amadei, B. & Pan, E., 1999. Modelling the effect of water, excavation sequence and rock reinforcement with discontinuous deformation analysis. *International Journal of Rock Mechanics and Mining Science & Geomechanics Abstracts*, 36, 949–970.

Koyama, T., Nishiyama, S., Yang, M. & Ohnishi, Y., 2011. Modeling the interaction between fluid flow and particle movement with discontinuous deformation analysis (DDA) method. *International Journal for Numerical and Analytical Methods in Geomechanics*, 35, 1–20.

Lin, C.T., Amadei, B., Jung, J. & Dwyer, J., 1996. Extension of discontinuous deformation analysis for jointed rock masses. *International Journal of Rock Mechanics and Mining Science & Geomechanics Abstracts*, 33, 671–694.

Liu, J., Nan, Z. & Yi, P., 2012. Validation and application of three-dimensional discontinuous deformation analysis with tetrahedron finite element meshed block. *Acta Mechanica Sinica*, 28, 1602–1616.

Lysmer, J., Seed, H.B. & Schanable, P.B., 1972. SHAKE – A computer program for earthquake response analysis for horizontally layered sites: Berkeley.

MacLaughlin, M., Sitar, N., Doolin, D. & Abbot, T., 2001. Investigation of slope-stability kinematics using discontinuous deformation analysis. *International Journal of Rock Mechanics and Mining Sciences*, 38, 753–762.

MacLaughlin, M.M. & Doolin, D.M., 2006. Review of validation of the discontinuous deformation analysis (DDA) method. *International Journal for Numerical and Analytical Methods in Geomechanics*, 30, 271–305.

Makris, N. & Roussos, Y.S., 2000. Rocking response of rigid blocks under near-source ground motions. *Geotechnique*, 50, 243–262.

Mikola, R.G. & Sitar, N., 2013. Explicit three dimensional Discontinuous Deformation Analysis for blocky system 47th US Rock Mechanics / Geomechanics Symposium. San Francisco: American Rock Mechanics Association.

Moosavi, M. & Grayeli, R., 2006. A model for cable bolt-rock mass interaction: Integration with discontinuous deformation analysis (DDA) algorithm. *International Journal of Rock Mechanics and Mining Sciences*, 43, 661–670.

Morgan, W.E. & Aral, M.M., 2015. An implicitly coupled hydro-geomechanical model for hydraulic fracture simulation with the discontinuous deformation analysis. *International Journal of Rock Mechanics and Mining Sciences*, 73, 82–94.

Moser, F., Jacobs, L.J. & Qu, J.M., 1999. Modeling elastic wave propagation in waveguides with the finite element method. *Ndt & E Int*, 32, 225–234.

Newmark, N., 1965. Effects of earthquakes on dams and embankments. *Geotechnique*, 15, 139–160.

Ning, Y.J., Yang, J., Ma, G.W. & Chen, P.W., 2010. Contact Algorithm Modification of DDA and Its Verification, in: Ma, G., Zhou, Y. (Eds.), Analysis of Discontinuous Deformation: New Developments and Applications, Research Publishing, Singapore, pp. 73–81.

Ning, Y. & Zhao, Z., 2012a. A detailed investigation of block dynamic sliding by the discontinuous deformation analysis. *International Journal for Numerical and Analytical Methods in Geomechanics*, 37, 2373–2393.

Ning, Y.J. & Zhao, Z.Y., 2012b. Nonreflecting boundaries for the discontinuous deformation analysis, in: Advances in Discontinuous Numerical Methods and Applications in Geomechanics and Geoengineering, Takeshi Sasaki (Ed.), CRC Press 2012, pp. 147–154.

Ohnishi, Y., Chen, G. & Miki, S., 1995. Recent devloment of DDA in rock mechanics practice, in: Li, J.C., Wang, C.-Y., Sheng, J. (Eds.), The first international conference on analysis of discontinuous deformation. Changli, Taiwan, pp. 26–47.

Ohnishi, Y., Nishiyama, S., Akao, S., Yang, M. & Miki, S., 2005a. DDA for Elastic Elliptic Element, in: MacLaughlin, M., Sitar, N. (Eds.), Proceedings of ICADD-7. Honolulu, Hawaii: CRC Press, Taylor & Francis Group, pp. 103–112.

Ohnishi, Y., Nishiyama, S., Sasaki, T. & Nakai, T., 2005b. The application of DDA to practical rock engineering problems: issues and recent insights, in: Sitar, M.M.a.N. (Ed.), Proceedings of the 7th International Conference on the Analysis of Discontinuous Deformation. Honolulu, Hawaii, pp. 277–287.

Schnabel, P.B., Lysmer, J. & Seed, H.B., 1972. SHAKE: A computer program for earthquake response analysis of horizontally layered sites. Berkeley: Earthquake Engineering Research Center, University of California, p. 102.

Shi, G., 2013. Basic theory of two dimensional and three dimensional contacts, in: Chen, G.Q., Ohnishi, Y., Zheng, L., Sasaki, T. (Eds.), ICADD -11: The 11th International Conference on Analysis of Discontinuous Deformation. Fukuoka, Japan: Taylor and Francis, London, UK, pp. 3–14.

Shi, G.-h., 1988. Discontinuous Deformation Analysis–a new numerical method for the statics and dynamics of block system [Ph.D. thesis], Department of Civil Engineering. Berkeley: Berkeley, University of California, p. 378

Shi, G.-h., 1993. Block System Modeling by Discontinuous Deformation Analysis. Southampton, UK: Computational Mechanics Publication.

Shi, G.-h., 1996. Discontinuous Deformation Analysis Programs Version 96 User's Manual. Revised by Man-Chu Ronald Young, 1996. Unpublished Report.

Shi, G.-h., 1997. The Numerical Manifold Method and Simplex Integration. Vicksburg. MS: US Army Corps of Engineers: Waterways Experiment Station.

Shi, G.-h., 2001. Three Dimensional Discontinuous Deformation Analyses, Proceedings of the 38th US Rock Mechanics Symposium. Washington, DC, pp. 1421–1428.

Shi, G.-h. & Goodman, R.E., 1985. Two dimensional discontinuous deformation analysis. *International Journal for Numerical and Analytical Methods in Geomechanics*, 9, 541–556

Shi, G.-h. & Goodman, R.E., 1989. Generalization of two dimensional discontinuous deformation analysis for forward modeling. *International Journal for Numerical and Analytical Methods in Geomechanics*, 13, 359–380.

Tal, Y., Hatzor, Y.H. & Feng, X.T., 2014. An improved numerical manifold method for simulation of sequential excavation in fractured rocks. *International Journal of Rock Mechanics and Mining Sciences*, 65, 116–128.

Tsesarsky, M., Hatzor, Y.H. & Sitar, N., 2005. Dynamic displacement of a block on an inclined plane: Analytical, experimental and DDA results. *Rock Mechanics and Rock Engineering*, 38, 153–167.

Wang, L.Z., Jiang, H.Y., Yang, Z.X., Xu, Y.C. & Zhu, X.B., 2013. Development of discontinuous deformation analysis with displacement-dependent interface shear strength. *Computers and Geotechnics*, 47, 91–101.

Wu, J.H., Ohnishi, Y. & Nishiyama, S., 2005. A development of the discontinuous deformation analysis for rock fall analysis. *International Journal for Numerical and Analytical Methods in Geomechanics*, 29, 971–988.

Wu, W., Zhu, H.H., Zhuang, X.Y., Ma, G.W. & Cai, Y.C., 2014. A multi-shell cover algorithm for contact detection in the three dimensional discontinuous deformation analysis. *Theoretical and Applied Fracture Mechanics*, 72, 136–149.

Yagoda-Biran, G. & Hatzor, Y.H., 2010. Constraining paleo PGA values by numerical analysis of overturned columns. *Earthquake Engineering and Structural Dynamics*, 39, 462–472.

Yagoda-Biran, G. & Hatzor, Y.H., 2013. A new failure mode chart for toppling and sliding with consideration of earthquake inertia force. *International Journal for Rock Mechanics and Mining Sciences*, 64, 122–131.

Yagoda-Biran, G. & Hatzor, Y.H., 2016. Benchmarking the Numerical Discontinuous Deformation Analysis method. *Computers and Geotechnics*, 71, 30–46.

Yeung, M.R., 1991. Application of Shi's Discontinuous Deformation Analysis to the Study of Rock Behavior, Department of Civil Engineering. Berkeley: University of California, Berkeley, p. 339.

Yeung, M.R., 1996. Analysis of three-hinged beam using DDA, in: Salami, MR and Banks, D. (Eds.), Proceedings of the First International Forum on Discontinuous Deformation Analysis (DDA) and Simulations of Discontinuous Media. Berkeley, CA: TSI Press: Albuquerque, NM, pp. 462–469.

Zhang, L., Jiang, Z.S., Wu, Y.Q., Zou, Z.Y., Liu, X.X. & Wei, W.X., 2015. Development of the 3D DDA method with rate- and state-friction laws. *Chinese Journal of Geophysics-Chinese Edition*, 58, 474–480.

Zhang, Y.B., Xu, Q., Chen, G.Q., Zhao, J.X. & Zheng, L., 2014a. Extension of discontinuous deformation analysis and application in cohesive-frictional slope analysis. *International Journal of Rock Mechanics and Mining Sciences*, 70, 533–545.

Zhang, Y.H., Fu, X.D. & Sheng, Q., 2014b. Modification of the discontinuous deformation analysis method and its application to seismic response analysis of large underground caverns. *Tunnelling and Underground Space Technology*, 40, 241–250.

Chapter 4

Continuum–discontinuum element method

S.H. Li, D. Zhou, J. Wang & C. Feng
Institute of Mechanics, Chinese Academy of Sciences, Beijing, China

Abstract: The continuum–discontinuum element method is a dynamic explicit algorithm based on element fracture under the Lagrange system. This method has combined the advantages of the continuous model and the discrete element method. The finite element method and the spring element method can be applied to calculate the internal force of an element and to determine if the element has reached the fracture condition. The discrete element method can be used to detect contact and determine contact force on the fracture plane or on the boundary of a block. It is a unified expression for the whole failure process of a material from continuous state to discontinuous state. A series of new theories, concepts and methods have been merged into this method, which include a spring element method based on the new concept of Poisson spring and pure shear spring, structure layer model with specific physical meaning for the interface springs, a new algorithm which allows elements to fracture during calculation, and a new contact detection method based on edge and face. With the basic models mentioned above, dynamic deformation; fracture and motion of the material; and structure under complex conditions can be simulated with continuum–discontinuum element method.

1 INTRODUCTION

The continuum–discontinuum coupling method has been a hot topic in the area of computational mechanics in recent years. Goodman (1968) proposed an interface model in finite element method to simulate the initiation and propagation of the crack in materials. An extended finite element method introduced by Belytschko *et al.* (1999, 2001) can describe a discontinuous displacement field and the fracture of continuum media with step function. With this method, simulation of crack growth can be realized without mesh regeneration. Owen *et al.* (2004) investigated the mechanical behavior of brittle material under impact force with a combined finite/discrete element approach. A topology updating scheme is employed in this method to introduce new physical cracks. Munjiza *et al.* (2004) proposed a FEM/DEM coupling method in which the fracture theory is introduced into the finite element method and the contact of blocks is calculated with the discrete element method. The discontinuous deformation analysis method and the numerical manifold method proposed by Shi (1992, 1993) provided strict mathematical expressions for a unified scheme of continuum–discontinuum simulation.

This chapter introduces a new numerical method called the continuum–discontinuum element method (CDEM), which establishes a unified expression for continuum and discontinuum under the Lagrange system. The whole failure process of material from continuous deformation, crack initiation, multi-crack propagation, fragmentation, and discrete motion can be simulated with this method. The basic concept, theory and models of this method are introduced.

2 BASIC CONCEPTS AND DEFINITIONS

The continuum–discontinuum element method is a kind of numerical approach which is a combination of a continuum element model and a discrete element method. The algorithm is based on a time-dependent explicit iteration using dynamic relaxation. The deformation of elements or blocks is calculated with a continuum model, such as the finite element method, finite volume method or spring element method. While the interfaces between the blocks are calculated with a bond spring model or a contact spring model. The progressive failure process of materials or structures from continuous deformation, through crack growth, to totally failure can be modelled with this method. Some basic concepts and models are shown in Figure 1.

An element is a polyhedron which is cut by a mathematical mesh. A block is a polyhedron which is cut by physical faces in rock mass or in materials. The main difference between elements and blocks is that there are no joint faces or joint layers between neighboring blocks. A block can consist of several elements. The interfaces between blocks are called joints. The interface can be calculated with a spring model.

The initial geometrical model could be continuous, with initial joints, or totally discrete, as shown in Figure 2. If a crack passes through an element, the element will be divided into two sub-elements by the crack path as shown in Figure 3. The interactions at the crack are described with a contact model.

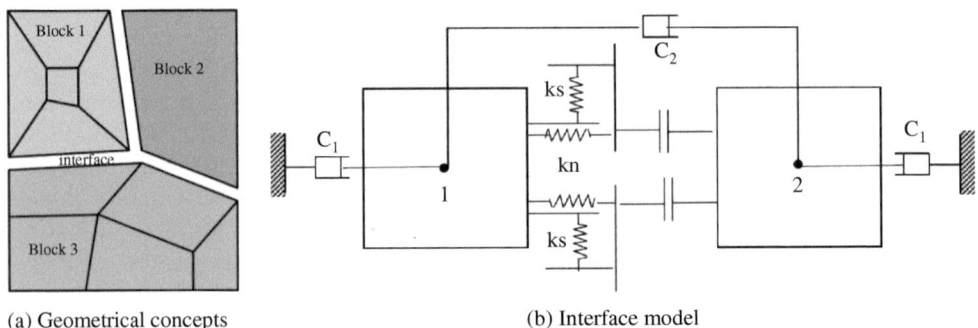

(a) Geometrical concepts   (b) Interface model

Figure 1 Basic concept and models.

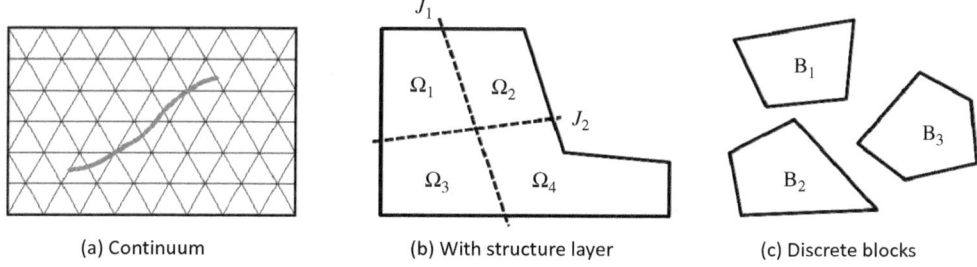

(a) Continuum   (b) With structure layer   (c) Discrete blocks

Figure 2 The initial geometrical model of CDEM.

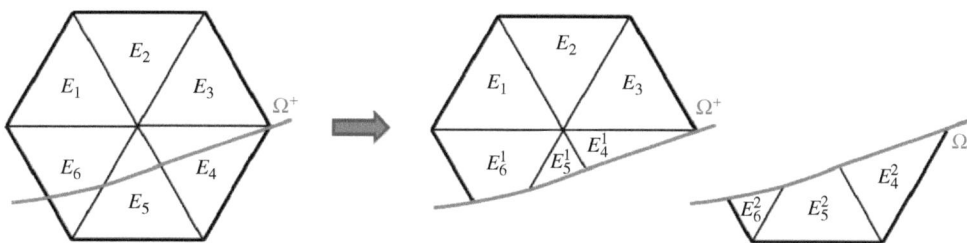

Figure 3 Division of the elements at the crack.

3 FUNDAMENTAL EQUATIONS AND COMPUTATIONAL SCHEME

3.1 Governing equation for the element

The governing equation is established with a Lagrange equation

$$\frac{d}{dt}\left(\frac{\partial L}{\partial \dot{u}_i}\right) - \frac{\partial L}{\partial u_i} = Q_i \tag{1}$$

where Q_i is the unconventional force, L is the Lagrange function which could be written as

$$L = \Pi_m + \Pi_e + \Pi_f \tag{2}$$

where Π_m, Π_e and Π_f represent kinetic energy, elastic energy and potential energy.
    The energy function for an element could be further written as

$$L = \frac{1}{2}\int_V \rho \dot{u}_i^2 dV + \int_V \frac{1}{4}\sigma_{ij}(u_{i,j} + u_{j,i})dV - \int_V f_i u_i dV \tag{3}$$

where ρ is density, u_i is the displacement of nodes, σ_{ij} and $u_{i,j}$ are the stress and strain tensor of the element, f_i is the body force, V is the volume of the element.
    The unconventional force includes a damping force and the external boundary force, which can be expressed as

$$Q_\mu = \int_V \mu \dot{u}_i dV, \quad Q_{\overline{T}} = -\int_S \overline{T}_i dS \tag{4}$$

where μ is the damping coefficient, $\overline{T}_i$ is the surface force on the boundary.

Then Equation (1) can be written as

$$-\left(\int_V \rho \ddot{u}_i dV + \int_V \sigma_{ij} \frac{\partial u_{i,j}}{\partial u_i} dV - \int_V f_i dV\right) = \int_V \mu \dot{u}_i dV - \int_S \overline{T}_i dS \tag{5}$$

According to the method of integration by part

$$\int_V \sigma_{ij} \frac{\partial u_{i,j}}{\partial u_i} dV = \int_S \sigma_{ij} n_j dS - \int_V \sigma_{ij,j} dV \tag{6}$$

Equation (5) can be written as

$$\int_V \left(\sigma_{ij,j} + f_i - \rho \ddot{u}_i - \mu \dot{u}_i\right) dV + \int_S \left(\overline{T}_i - \sigma_{ij} n_j\right) dS = 0 \tag{7}$$

Considering boundary condition

$$\sigma_{ij} n_j = \overline{T}_i \tag{8}$$

Equation (7) can be written as

$$\sigma_{ij,j} + f_i - \rho \ddot{u}_i - \mu \dot{u}_i = 0 \tag{9}$$

Equation (9) is the kinetic equilibrium differential equation. Because of the partial derivative form of stress, it could not directly be used for a numerical solution. The internal force can be calculated with a finite element method or other continuum model, which can be written as

$$F_i^e = \frac{\partial \Pi_e}{\partial u_i} = K_{ij}^e u_j \tag{10}$$

where K_{ij}^e is the stiffness coefficient.

Thus, Equation (7) can be written as

$$\int_V \rho \ddot{u}_i dV + \int_V \mu \dot{u}_i dV + F_i^e = \int_V f_i dV + \int_S \overline{T}_i dS \tag{11}$$

The governing equation is finally obtained as

$$\mathbf{M}\ddot{u}(t) + \mathbf{C}\dot{u}(t) + \mathbf{K}u(t) = \mathbf{F}(t) \tag{12}$$

where $\ddot{u}(t)$, $\dot{u}(t)$, $u(t)$ are the acceleration vector, velocity vector and displacement vector of the nodes. $\mathbf{M}$, $\mathbf{C}$, $\mathbf{K}$ and $\mathbf{F}(t)$ are the mass matrix, damping matrix, stiffness matrix and external nodal force of the element, respectively.

3.2 Computational scheme

The whole computational field is calculated with a dynamic relaxation method. The basic solving procedure is listed as follows.

1. Fix all the nodes at the beginning of each step.
2. Calculate the internal force F_i^e of each element, and combined with the external force F_i^B determine the resultant force of the node.

$$F_n = F_n^e + F_n^B \tag{13}$$

3. According to Equation (12), calculate the unbalance force of the nodes $\{F_n^r\}$.

$$F_n^r = F_n - C\dot{u}_n - Ku_n \tag{14}$$

4. Calculate the acceleration of the nodes.

$$a_{n+1} = M^{-1}F_n^r \tag{15}$$

5. Calculate the velocity and displacement of the nodes.

$$\dot{u}_{n+1} = \dot{u}_n + a_{n+1}\Delta t, \ u_{n+1} = u_n + \dot{u}_{n+1}\Delta t \tag{16}$$

6. Fix all the nodes in the new position, and jump to step (1).

3.3 Time step

Time step is dependent on the frequency of the external load and the natural frequency of the material. The size of the element is a characteristic scale. Wave propagation should not exceed an element during a time step to ensure the convergence condition.

The natural period of the element can be written as

$$T_e = 2\pi\sqrt{\frac{m}{K}} \tag{17}$$

where m and K are the mass and stiffness of the element.

When the frequency of the external load is high, the period of the load T_f should also be considered. The time step should satisfy the following equation

$$\Delta t \le \min(T_e, \ T_f) \tag{18}$$

3.4 Damping coefficient

In static problems, the damping coefficient is used to accelerate the convergence process, while in realistic dynamic problems, the damping coefficient represents the speed of energy dissipation. It would greatly influence the result of the kinetic iteration.

The governing equation of the node is

$$m\ddot{u}_i + c\dot{u}_i + F_i^e = F \tag{19}$$

considering the characteristic period in Equation (17), use variable substitution

$$t = T_e t' \tag{20}$$

Equation (19) can be written as

$$\ddot{u}_i + \frac{2\pi c}{\sqrt{mK}} \dot{u}_i + \frac{4\pi^2}{K} F_i^e = \frac{4\pi^2}{K} F \tag{21}$$

Then, define the dimensionless damping ratio as

$$\zeta = \frac{2\pi c}{\sqrt{mK}} \tag{22}$$

and so then the damping coefficient can be written as

$$c = \frac{1}{2\pi} \zeta \sqrt{mK} \tag{23}$$

4 BASIC MODELS AND TECHNIQUES

4.1 Spring element model

The spring element model (SEM) (Li *et al.*, 2010; Zhang *et al.*, 2013) is a new approach to calculate the inner force and describe the crack of an element. Deformation of an element can be described with four types of springs, which are normal spring, tangential spring, Poisson spring and pure shear spring. The computational result of SEM is exactly consistent with the finite element method, but the efficiency is greatly improved because of the less independent stiffness parameters used.

4.1.1 Three-node triangular element

For a triangle of any shape, the system includes six springs as shown in Figure 4.

The element is composed of the normal spring S_{xx} and the tangential spring S_{xy} along the bottom edge, the normal spring S_{yy} and the tangential spring S_{yx} along the height direction of the bottom edge, and a Poisson spring S_{xyp} related to the first two normal springs and a pure shear spring S_{xys} related to the first two tangential springs. The normal and the tangential springs follow Hooke's Law, and the direction of force is opposite to the relative displacement. The direction of the Poisson spring force is

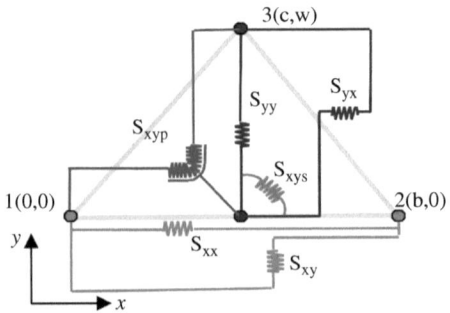

Figure 4 Spring system of three-node triangular element.

vertical to the one of displacement increment, and the energy of the Poisson spring is the product of its force and its vertical displacement. The energy of the pure shear spring is the product of shear stress and its vertical shear displacement.

The stiffness coefficient of each spring can be determined by the energy equivalence principle. The elastic strain energy of the element can be written as

$$\Pi = \int \frac{1}{2} \sigma_{ij}\varepsilon_{ij}dV(i,j = x,y) \tag{24}$$

which can be further written as

$$\Pi = \int \frac{1}{2}\left(\lambda\varepsilon_{kk}\delta_{ij} + 2G\varepsilon_{ij}\right)\varepsilon_{ij}dV$$

$$= \frac{abt}{4}\left(\begin{array}{c} (\lambda + 2G)\left(\dfrac{u_2^x - u_1^x}{a}\right)^2 + (\lambda + 2G)\left(\dfrac{u_3^y - u_{33}^y}{b}\right)^2 \\[2mm] + G\left(\dfrac{u_2^y - u_1^y}{a}\right)^2 + G\left(\dfrac{u_3^x - u_{33}^x}{b}\right)^2 \\[2mm] + 2\lambda\left(\dfrac{u_2^x - u_1^x}{a}\right)\left(\dfrac{u_3^y - u_{33}^y}{b}\right) + G\left(\dfrac{u_2^y - u_1^y}{a}\right)\left(\dfrac{u_3^x - u_{33}^x}{b}\right) \end{array} \right) \tag{25}$$

where λ and G are Lamé constants, u_i^x and u_i^y ($i = 1,2,3$) are the displacement of the node, u_{33}^x and u_{33}^y are the displacement of the intersection point of spring S_{xx} and spring S_{yy} in x and y direction, a and b are the length of spring S_{xx} and spring S_{yy}, t is the thickness of the element.

The spring force of spring S_{xx} and spring S_{yy} can be obtained as

$$F_x^x = \frac{\partial\Pi}{\partial\left(u_2^x - u_1^x\right)} = \frac{(\lambda + 2G)bt}{2a}\left(u_2^x - u_1^x\right) + \frac{\lambda t}{2}\left(u_3^y - u_{33}^y\right)$$

$$F_x^y = \frac{\partial\Pi}{\partial\left(u_2^y - u_1^y\right)} = \frac{btG}{2a}\left(u_2^y - u_1^y\right) + \frac{Gt}{4}\left(u_3^x - u_{33}^x\right)$$

$$F_y^x = \frac{\partial\Pi}{\partial\left(u_3^x - u_{33}^x\right)} = \frac{Gat}{2b}\left(u_3^x - u_{33}^x\right) + \frac{Gt}{4}\left(u_2^y - u_1^y\right) \tag{26}$$

$$F_y^y = \frac{\partial\Pi}{\partial\left(u_3^y - u_{33}^y\right)} = \frac{(\lambda + 2G)at}{2b}\left(u_3^y - u_{33}^y\right) + \frac{\lambda t}{2}\left(u_2^x - u_1^x\right)$$

The energy of the spring system can be expressed as

$$\Pi^e = \left(\begin{array}{c} \dfrac{1}{2}K_x^x\left(U_x^x\right)^2 + \dfrac{1}{2}K_x^y\left(U_x^y\right)^2 + \dfrac{1}{2}K_y^x\left(U_y^x\right)^2 + \dfrac{1}{2}K_y^y\left(U_y^y\right)^2 \\[2mm] + K_p^{xy}U_x^xU_y^y + K_s^{xy}U_x^yU_y^x \end{array} \right) \tag{27}$$

$$U_x^x = u_2^x - u_1^x \quad U_x^y = u_2^y - u_1^y \quad U_y^x = u_3^x - u_{33}^x \quad U_y^y = u_3^y - u_{33}^y$$

where $U_i^j(i,j = x,y)$ is the deformation of the spring, K_i^j is the stiffness coefficient.

The spring force of spring S_{xx} and spring S_{yy} can also be written as

$$F_x^x = \frac{\partial \prod^e}{\partial \left(u_2^x - u_1^x\right)} = K_x^x \left(u_2^x - u_1^x\right) + K_p^{xy}\left(u_3^y - u_{33}^y\right)$$

$$F_x^y = \frac{\partial \prod^e}{\partial \left(u_2^y - u_1^y\right)} = K_x^y \left(u_2^y - u_1^y\right) + K_s^{xy}\left(u_3^x - u_{33}^x\right)$$

$$F_y^x = \frac{\partial \prod^e}{\partial \left(u_3^x - u_{33}^x\right)} = K_y^x \left(u_3^x - u_{33}^x\right) + K_s^{xy}\left(u_2^y - u_1^y\right)$$

$$F_y^y = \frac{\partial \prod^e}{\partial \left(u_3^y - u_{33}^y\right)} = K_y^y \left(u_3^y - u_{33}^y\right) + K_p^{xy}\left(u_2^x - u_1^x\right)$$

$$(28)$$

According to the energy equivalence principle, Equation (26) and Equation (28) should be the same. So the stiffness of the springs is obtained as

$$K_i^j = \begin{cases} \dfrac{(\lambda + 2G)L_j t}{2L_i} & i = j \\[3mm] \dfrac{GL_j t}{2L_i} & i \neq j \end{cases}$$

$$K_p^{xy} = \frac{\lambda t}{2}$$

$$K_s^{xy} = \frac{Gt}{4} \tag{29}$$

where $L_i(i=x,y)$ is the length of spring S_{xx} and spring S_{yy}.

The nodal force can be further written as

$$\left.\begin{array}{ll} F_1^x = \dfrac{\partial \prod}{\partial u_1^x} = -F_x^x - (1-n)F_y^x & F_1^y = \dfrac{\partial \prod}{\partial u_1^y} = -F_x^y - (1-n)F_y^y \\[3mm] F_2^x = \dfrac{\partial \prod}{\partial u_2^x} = F_x^x - nF_y^x & F_2^y = \dfrac{\partial \prod}{\partial u_2^y} = F_x^y - nF_y^y \\[3mm] F_3^x = \dfrac{\partial \prod}{\partial u_3^x} = F_y^x & F_3^y = \dfrac{\partial \prod}{\partial u_3^y} = F_y^y \end{array}\right\} \tag{30}$$

4.1.2 Four-node tetrahedral element

By extending the method to a three dimensional problem, the spring system of a tetrahedral element is built. The structure of a tetrahedral spring system is shown as Figure 5. The stiffness of the spring can be determined in the same way, see Equation (31).

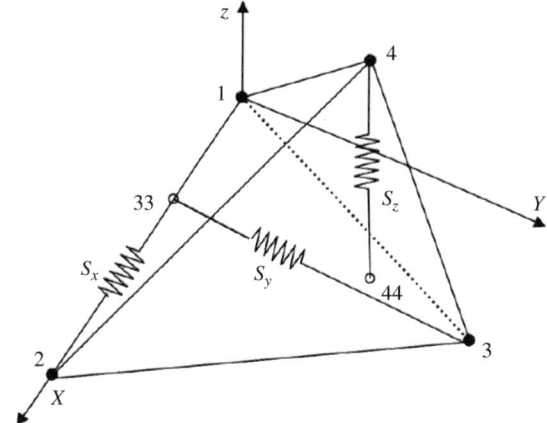

Figure 5 Spring system of the four-node tetrahedral element.

$$K_i^j = \begin{cases} \dfrac{(2G + \lambda)V}{(L_i)^2} & i = j \\[3mm] \dfrac{GV}{(L_i)^2} & i \neq j \end{cases}$$

$$K_p^{km} = K_p^{mk} = \frac{\lambda V}{L_k L_m} \quad (k \neq m)$$

$$K_s^{km} = K_s^{mk} = \frac{GV}{L_k L_m} \quad (k \neq m) \tag{31}$$

4.1.3 Other types of spring element

According to the concept, different types of spring element can be constructed, such as the four-node plane element, eight-node hexahedron element, etc. The spring system for different types of element are shown in Figure 6.

4.1.4 Validation of the spring element model

A different constitutive equation could be easily realized in SEM. One type of non-linear coefficients associated with the strain of spring is shown in Equation (32). According to this type of non-linear coefficients, spring force could be calculated (shown in Equation (33)). The comparison between the numerical and theoretical results of a uniaxial compression test is shown in Figure 7.

$$C_{ij} = 1 - e^{-10^4 |x_{ij}/L_i|} \quad (i, j = x, y) \tag{32}$$

$$\begin{cases} F_{ii} = C_{ii} K_{ii} \Delta u_{ii} + C_{jj} K_{ijp} \Delta u_{jj} \\ F_{ij} = C_{ij} K_{ij} \Delta u_{ij} + C_{ji} K_{ijs} \Delta u_{ji} \end{cases} \quad (i, j = x, y; i \neq j) \tag{33}$$

where, L_i represents the length and the height.

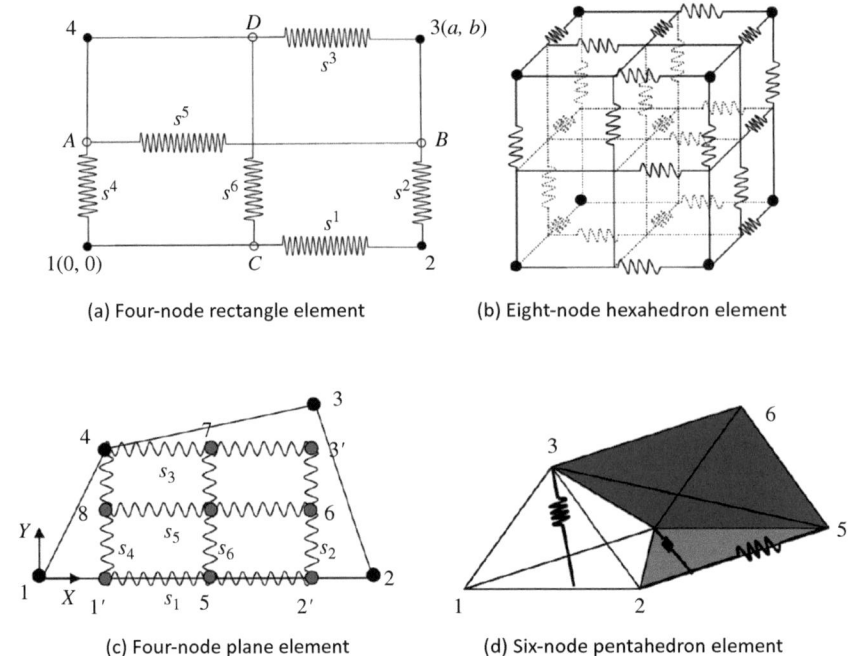

(a) Four-node rectangle element   (b) Eight-node hexahedron element

(c) Four-node plane element   (d) Six-node pentahedron element

Figure 6 Different types of spring element.

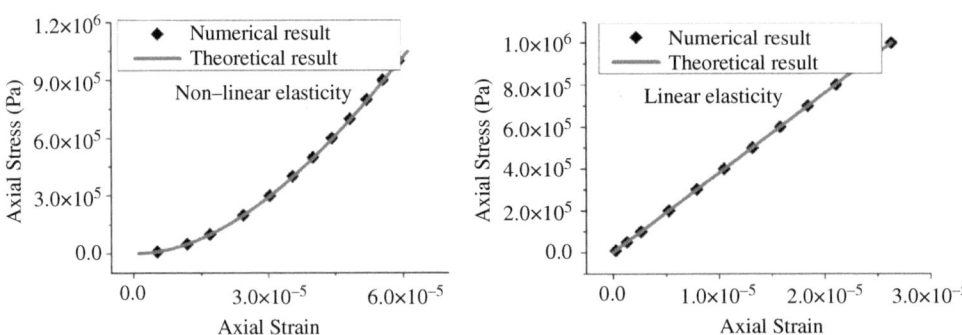

Figure 7 Comparison between numerical and theoretical result.

The numerical results agree well with the theoretical results, which validates the effectiveness of this model. SEM is more than twice as fast as FEM and the memory usage of SEM is about one-fifth compared to FEM (as shown in Table 1).

4.2 Structure layer model

The structure layer (Zhou, 2013) is composed of spring elements and adopted in the continuum–discontinuum element method (CDEM). It has thickness but is without

Table 1 Comparison between spring element and finite element in time cost.

	SEM	FEM	ratio	effect
Time cost (CPU clock)	792	1920	1 : 2.4	More than 2 times faster
RAM cost (Storage number)	15	78	1:5.2	Memory cost less than 1/5

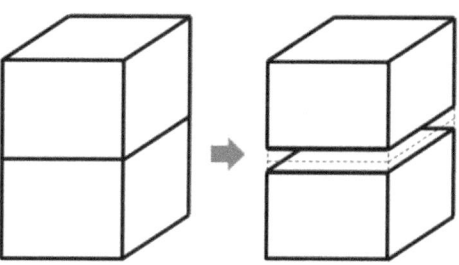

Figure 8 Indentation of element face.

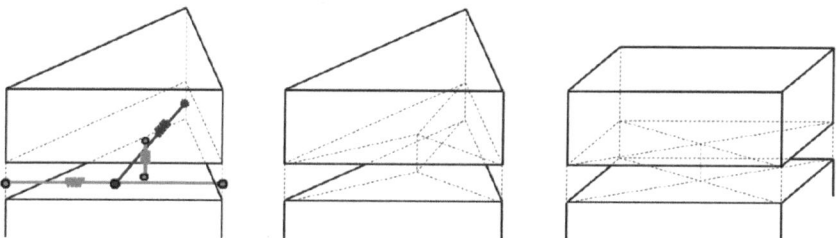

Figure 9 Spring element in structure layer model.

mass. Thin intercalation in the discontinuous geologic bodies, as well as the interface between elements in the continuous problem, can be modeled by structure layer. Different from the traditional interface spring model in discrete element method, the stiffness of the structure layer has definite physical meaning, because the layer has real thickness and all the parameters are determined by the factual material. With this model, no extra energy was added in the system, and a relatively large time step could be adopted without loss of the accuracy of displacement.

4.2.1 Construction of structure layer model

A structure layer model is established to simulate a joint. Two steps are adopted, including face indent and the construction of a thin layer spring element. The joint layer will be indented to construct a thin layer with its thickness according to the practical problem. The indented face should be parallel to the original face. Then a thin layer is formed between two elements, as shown in Figure 8. The thin indented layer will be filled by the spring element to calculate the deformation. The geometry of the element is shown in Figure 9.

The thin layer spring element is composed of a triangle spring element and a thickness spring. The stiffness of each spring is determined according to the material. The Poisson and pure shear effect of each spring are considered. A structure layer can consist of only one spring element, three spring elements, or four spring elements, according to accuracy demand and the shape of the layer.

Structure layer element can break itself when its strength is exceeded, and only three basic springs are left including the normal spring and two shear spring to describe contact and friction.

4.2.2 Validation of the structure layer model

4.2.2.1 Single structure layer

In this simple case, tension, compression and shear load are tested. Cloud maps of displacements are shown in Figure 10. Parameters used in this model are listed in Table 2. With structure layer, numerical results and theoretical results are highly consistent, as shown in Table 3.

4.2.2.2 Multi-structure layer

Multi-structure layer cases are tested to verify the adaptability of the model. Structure layers with different thickness and directions are set in the numerical model. Cloud maps of displacements are shown in Figure 11. Parameters are listed in Table 4 and numerical results are listed in Table 5. Numerical results agree with the theoretical results.

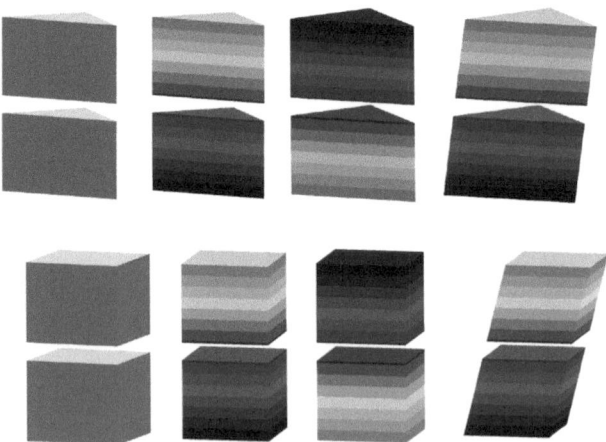

Figure 10 Cloud map of displacement under different load conditions.

Table 2 Parameters used for the single layer test.

E	v	Pressure	Length	Width	Height	Indent ratio
10 GPa	0.25	1 MPa	1 m	1 m	2 m	0.2

Table 3 Results of a single structure layer.

	Tension	Compression	Shear
Numerical (pentahedron)	2e-4m	−2e-4m	5e-4m
Numerical (hexahedron)	2e-4m	−2e-4m	5e-4m
Theoretical	2e-4m	−2e-4m	5e-4m

Table 4 Parameters used for the multi-layer test.

E	v	Pressure	Length	Width	Height	Indent ratio (I)	Indent ratio (II)
1.0e10 Pa	0.25	1 MPa	10 m	10 m	12 m	0.2	0.01

Table 5 Results of multi-structure layer.

	Tension	Compression	Shear
Numerical	1.20e-3m	−1.20e-3m	3.00e-3m
Theoretical	1.20e-3m	−1.20e-3m	3.00e-3m

Figure 11 Cloud map of displacement for a multi-structure layer case.

4.3 Element fracture model

Different failure criteria are used in CDEM to evaluate the initiation and propagation of a crack, such as strain strength distribution criterion (Li & Zhou, 2013), the mixed criterion of Mohr–Coulomb and maximum tensile strength, etc. The element fracture model (Wang *et al.*, 2013) introduced here can determine the fracture state and the propagation direction of the crack according to corresponding fracture criterion. The crack could pass through the element, and the fractured element will be cut into two sub-elements by the inner crack.

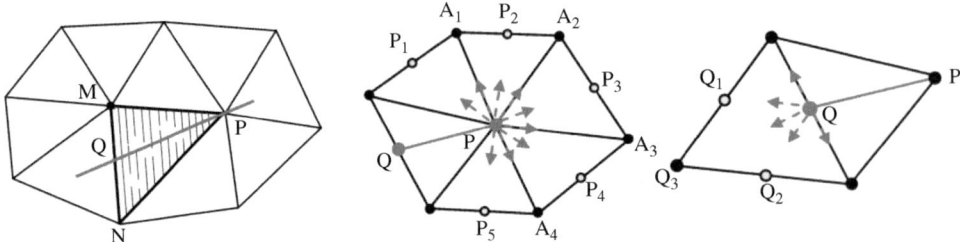

Figure 12 Initial crack and the possible propagation direction.

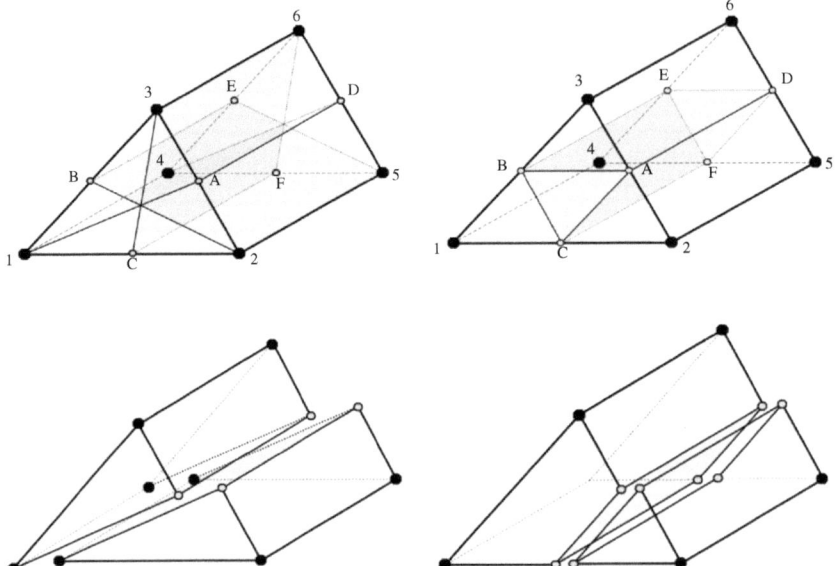

Figure 13 Possible fracture modes of the element.

4.3.1 Fracture of the element

Assume the line PQ is an initial crack. At the crack tip, P, crack propagation could be the boundaries of the elements PA_1–PA_4, or pass through the elements PP_1–PP_5, which depends on the stress state and the failure criterion used (as shown in Figure 12). New nodes will generate when the element has been divided into two sub-elements, as shown in Figure 13.

4.3.2 Validations and applications

4.3.2.1 Three point beam

Carpinteri & Colombo (1989) investigated the crack propagation of a three point beam with node relaxation technique in FEM. The model is shown in Figure 14, where

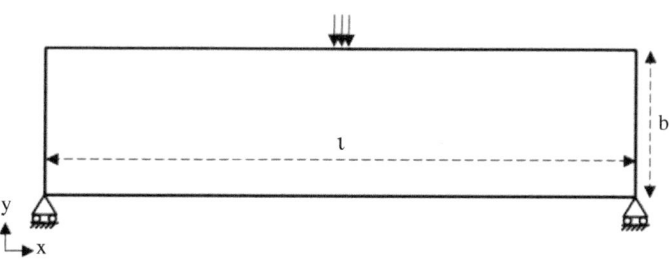

Figure 14 Possible fracture modes of the element.

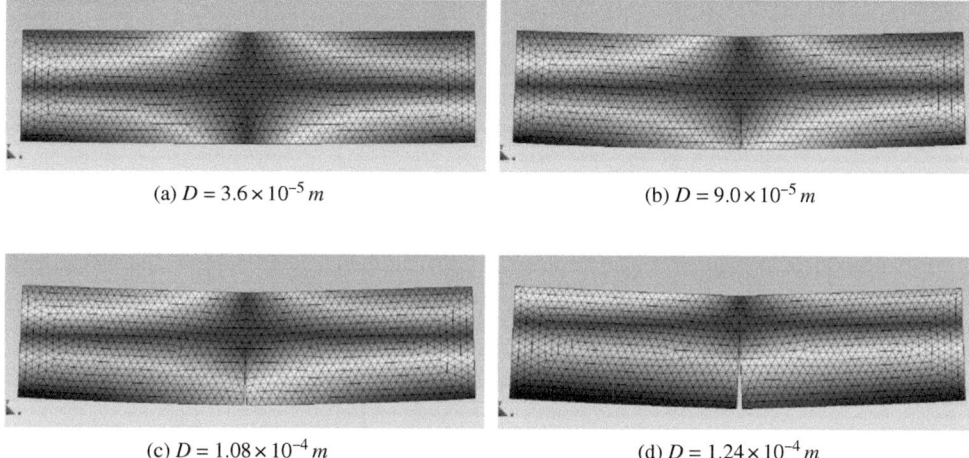

(a) $D = 3.6 \times 10^{-5}\,m$ (b) $D = 9.0 \times 10^{-5}\,m$

(c) $D = 1.08 \times 10^{-4}\,m$ (d) $D = 1.24 \times 10^{-4}\,m$

Figure 15 Displacement nephogram in different loading conditions.

$l = 0.6\,m$, $b = 0.15\,m$, $t = 0.01\,m$. t is the thickness of the beam. The material parameters are: $E = 36.5Gpa$, $v = 0.1$, the tensile strength $\sigma_u = 3.19Mpa$. The loading rate at the middle top is $3 \times 10^{-10}\,m/s$.

The result of the crack propagation process is shown in Figure 15. The crack initiates from the middle button of the beam where the tensile stress is the largest. The crack penetrates the element and the original element is divided into two new elements. The force–displacement curve is shown in Figure 16. With the same parameters and boundary conditions, the numerical results agree well with the results given by Carpinteri & Colombo (1989).

4.3.2.2 Failure of a slope

A generalized model for rock slope is shown in Figure 17. Parameters used are listed in Table 6.

Crack propagation within the slope is investigated with overload gravity. With increasing gravity, the cracks grow rapidly, and the deformation of the slope becomes

Table 6 Parameters of the slope.

E (GPa)	v	ρ (kg/m³)	c (kPa)	φ (°)	T (kPa)
30	0.22	2500	430	24	210

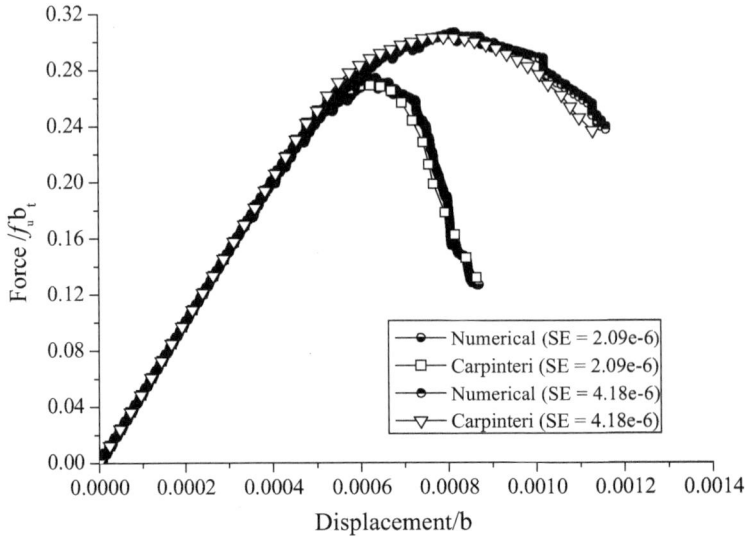

Figure 16 Force–displacement curve of the three point beam.

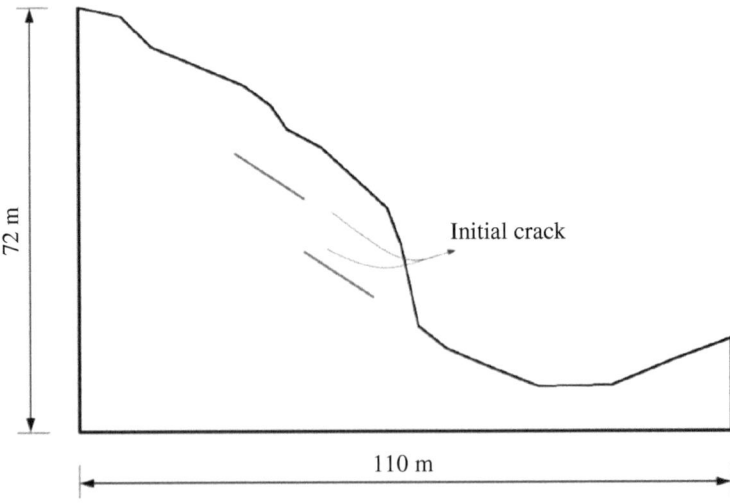

Figure 17 A generalized model for rock slope.

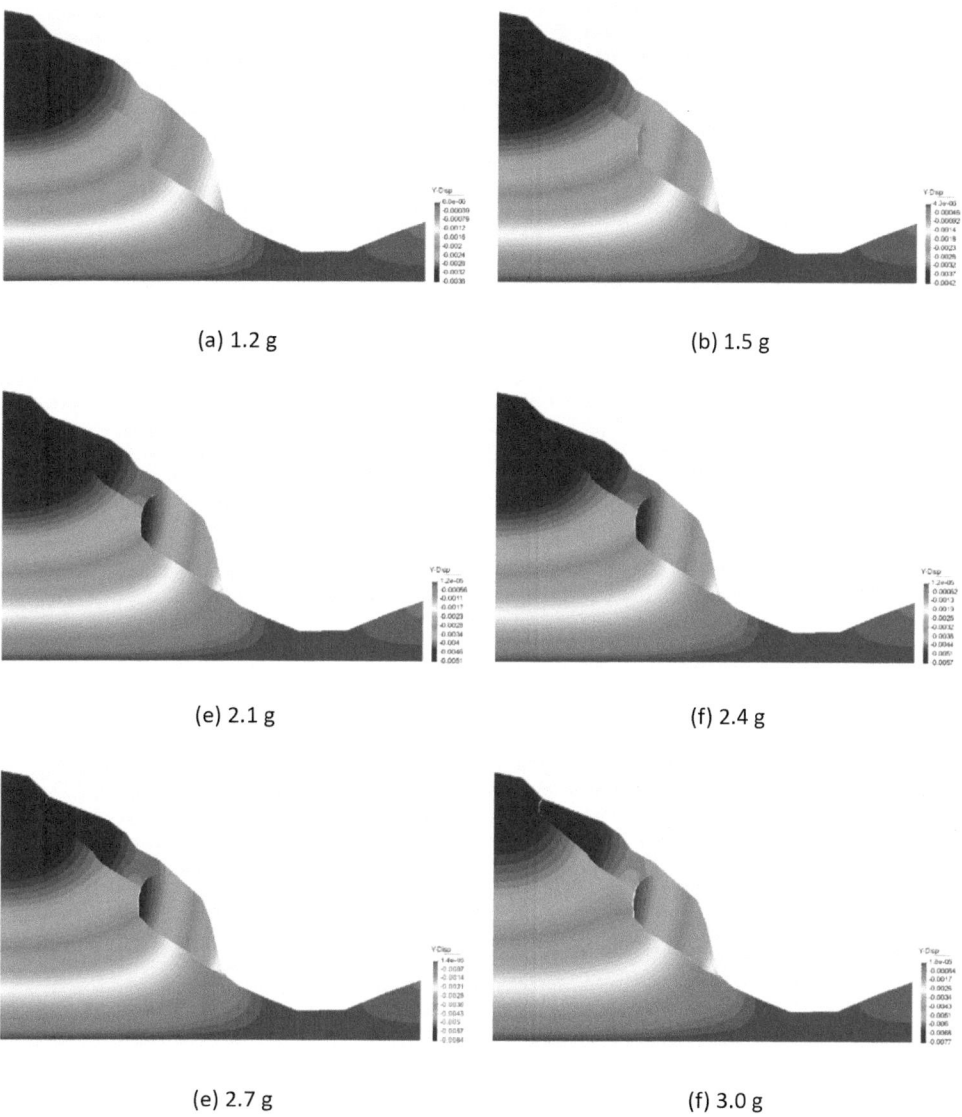

Figure 18 Crack propagation and failure of the slope under gravity overload.

much larger. After the crack is totally running through, the slope is unstable and begins to slide along the fracture surface. The crack propagation process is shown in Figure 18.

4.4 Techniques for contact detection

A new algorithm for the detection of contacts between arbitrary convex polyhedra with planar boundaries is introduced in this section (Wang, 2015). In this algorithm, a pair of contacting blocks is identified as (1) a main block and (2) a target block. The concept

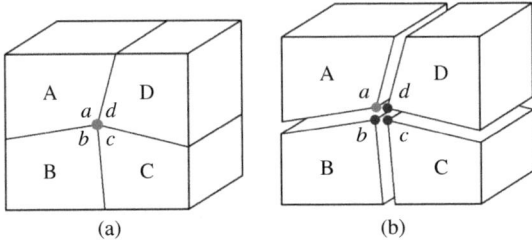

Figure 19 (a) Vertices located at the same position (b) Vertex-to-vertex contact is difficult to identify.

of a shrunken edge is introduced in this paper. First, each vertex of the main block is shrunk toward the center of the neighboring faces. The shrinkage is infinitesimal, yet useful for contact detection. Shrunken edges parallel to the original edges on the main block are established by connecting the shrunken points. Contact detection is then performed by determining the geometric relationship between a shrunken edge and its approaching face on the target block. From the three possible geometric relationships, all six contact types in three dimensions can be identified precisely, which allows for an easy and efficient detection process.

4.4.1 The concept of a shrunken edge

In hybrid continuous–discontinuous numerical methods, such as FDEM and NMM, blocks are initially neatly stacked, forming a continuous medium. Additionally, vertices and edges are shared by neighboring faces. In this condition, it is difficult to identify contacts as vertex-to-vertex or edge-to-edge. In some cases, the contact condition cannot be determined correctly. As an example, Figure 19(a) shows four blocks (A, B, C and D) with four vertices (a, b, c and d) located at the same position. In vertex-to-vertex contact detection, it is difficult to precisely identify contact between vertex a and vertices b, c and d (Figure 19(b)). Similarly, contact detection will be difficult if more than two parallel edges are located together. These conditions are commonly found in hybrid continuous–discontinuous methods; however, few effective solutions have been proposed. To resolve this problem, the shrunken edge model is proposed for contact detection.

A shrunken edge, unlike a conventional edge, is located on a face. It is assumed that each vertex of a block shrinks toward the center of the neighboring faces. The shrunken distance is infinitesimal but still useful for contact detection. Then, a shrunken edge parallel to the initial edge is established by connecting the shrunken points. As an example, a hexahedral block is presented in Figure 20(a). V_1 is one vertex of the block, and P_1 is the shrunken point corresponding to V_1 on face a. The position of point P_1 will depend on the shrinkage distance, which can be expressed as

$$|P_1 V_1| = \lambda |OV_1| \tag{34}$$

where λ is the shrinkage coefficient, which can be 0.1–1.0% (1.0% is used in this paper). To find the correct contact, the search tolerance should be less than the shrinkage distance. All of the shrunken points in the block are obtained from the

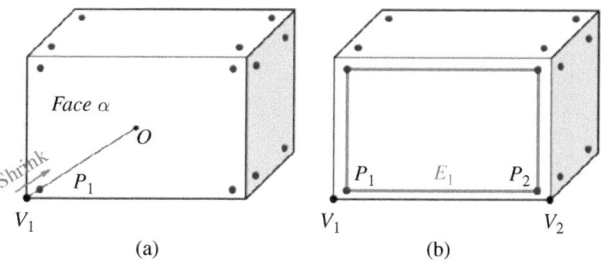

Figure 20 (a) Shrunken points and (b) shrunken edges of a hexahedral block.

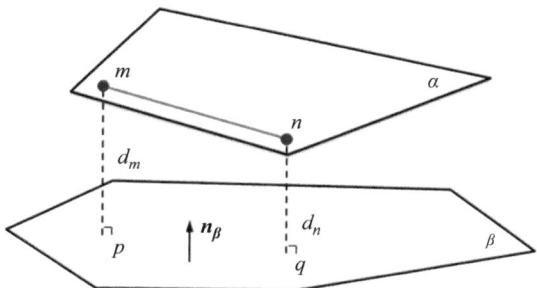

Figure 21 Contact detection between the indented edge and target face.

vertices in the same manner, and the shrunken edges are established. In Figure 20(b), points P_1 and P_2 are shrunken points and E_1 is a shrunken edge.

Here, the shrunken edge is used only for the geometric resolution phase of contact detection, and the shrinkage coefficient λ can be regarded as a way to set the tolerance for contacts. The initial configuration of the block should be used when calculating the motion and deformation of a block. Therefore, there is no reduction in the block volume in this algorithm.

There are several benefits of the shrunken edge concept. First, it is easier to establish the initial contact conditions in a continuous block system using shrunken edges because the edges used for detection shrink into the neighboring faces. Second, the seamless transition of a geological body from a continuum to discrete description can be achieved using shrunken edges.

4.4.2 Contact types

For two blocks which may have possible interactions, let A be the mother block and B be the target block. The contact detection is to judge the geometric relationship between indented edges on block A and target faces on block B. As shown in Figure 21, distances for two end points of indented edge mn to target face α, i.e. d_m and d_n, respectively, can be calculated. On the basis of relationships between d_m, d_n and d_{tol} which is the assumed tolerance for contact, four possibilities between indented edge and target face can be derived as follows:

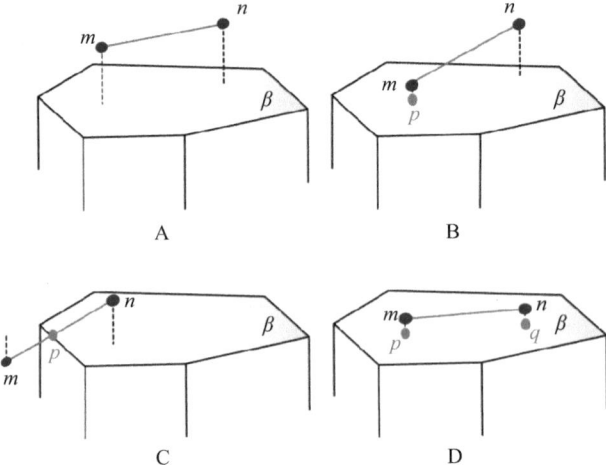

Figure 22 Possible geometric relationships between a shrunken edge and an approaching face.

1. $|d_m| > d_{tol}$, $|d_n| > d_{tol}$ and $d_m \cdot d_n > 0$ (Figure 22a). No contact would occur between indented edge mn and target face α.

2. $|d_m| > d_{tol}$, $|d_n| < d_{tol}$ or $|d_m| < d_{tol}$, $|d_n| > d_{tol}$ (Figure 22b). Let m' be the projection of point m on target face α. If m' is located in α, the contact point would be m' and semi-spring is located on point m. The corresponding contact type could be vertex-to-face contact, vertex-to-edge contact or vertex-to-vertex contact. In the vertex-to-face case, the contact normal is simply the target face normal. In the vertex-to-edge case, the contact normal can be computed by taking the average between the normal vectors of the faces neighboring the edge. In the vertex-to-vertex case, the contact normal can be computed from average normal vectors of the faces neighboring one of those vertexes.

3. $|d_m| > d_{tol}$, $|d_n| > d_{tol}$ and $d_m \cdot d_n < 0$ (Figure 22c). There must be point p on indented edge mn and its distance to target face α is zero. If point p is located in target face α, the contact point would be p and semi-springs on both ends of the indented edge would slide to p. The corresponding contact type is edge-to-edge contact. In this case, the contact normal is the cross product of the edge direction vectors.

4. $|d_m| < d_{tol}$, $|d_n| < d_{tol}$ (Figure 22d). Let m' and n' be the projection of point m and n on target face α, respectively. Segment $m'n'$ will be cut by each edge of the target face, thus there will be corresponding updates to $m'n'$ from time to time. If the length of final segment $m'n'$ is greater than zero, the contact would occur, and semi-springs would slide to m' and n'. The corresponding contact type could be parallel edge-to-edge contact, edge-to-face contact or face-to-face contact. In the parallel edge-to-edge case, the contact normal can be computed by taking the average between the normal vectors of the faces neighboring this edge. In the edge-to-face and face-to-face cases, the contact normal is the normal of target face.

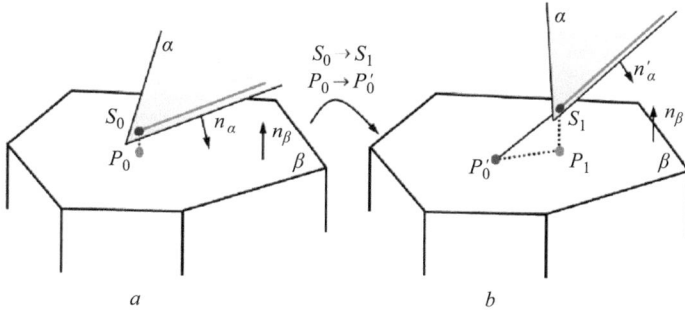

Figure 23 Position of the shrunken edge and its approaching face, (a) before a displacement increment and (b) after a displacement increment.

4.4.3 Contact forces

4.4.3.1 Normal contact force

In Figure 23, $S_0(x_s, y_s, z_s)$ is a point on a shrunken edge, and its projection on the approaching face β is $P_0(x_p, y_p, z_p)$ is the contact point. S_1 and P'_0 represent the locations of S_0 and P_0 after displacement increments (u_s, v_s, w_s) and (u_p, v_p, w_p), respectively. The normal displacement Δd_n of the contact can be determined from

$$\Delta d_n = \boldsymbol{n_\beta} \cdot \boldsymbol{P'_0 S_1} - \boldsymbol{n_\beta} \cdot \boldsymbol{P_0 S_0}$$

$$= \boldsymbol{n_\beta} \cdot \begin{bmatrix} x_s + u_s - x_p - u_p \\ y_s + v_s - y_p - v_p \\ z_s + w_s - z_p - w_p \end{bmatrix} - \boldsymbol{n_\beta} \cdot \begin{bmatrix} x_s - x_p \\ y_s - y_p \\ z_s - z_p \end{bmatrix} = \boldsymbol{n_\beta} \cdot \begin{bmatrix} u_s - u_p \\ v_s - v_p \\ w_s - w_p \end{bmatrix} \tag{36}$$

where $\boldsymbol{n_\beta}$ is the unit normal vector of face β. The displacement increments of S_0 and P_0 can be written in the following forms

$$\begin{Bmatrix} u_s \\ v_s \\ w_s \end{Bmatrix} = N_i(x_s, y_s, z_s)\boldsymbol{u_i} \tag{37}$$

$$\begin{Bmatrix} u_p \\ v_p \\ w_p \end{Bmatrix} = N_j(x_s, y_s, z_s)\boldsymbol{u_j} \tag{38}$$

where N_i and N_j are the shape functions of faces α and β, respectively. $\boldsymbol{u_i}$ and $\boldsymbol{u_j}$ are the displacement increment vectors of the vertices on faces α and β, respectively. Equation (36) can then be written as

$$\Delta d_n = \boldsymbol{n_\beta} \cdot N_i(x_s, y_s, z_s)\boldsymbol{u_i} - \boldsymbol{n_\beta} \cdot N_j(x_p, y_p, z_p)\boldsymbol{u_j} \tag{39}$$

Using the penalty method, a mathematical spring with stiffness K_n is placed between points S_0 and P_0 in the contact normal direction. The normal force increment, taking compressive force as positive, is calculated as

$$\Delta F_c^n = -K_n \Delta d_n \tag{40}$$

The total normal force is updated as

$$F_c^n(\tau + \Delta\tau) = F_c^n(\tau) + \Delta F_c^n \tag{41}$$

4.4.3.2 Shear contact force

The point P_1 in Figure 23 is the projection of S_1 onto the approaching face, after the application of the displacement increment. The shear displacement Δd_s of the contact can be determined from

$$\Delta d_s = |P_0' P_1| = \sqrt{|P_0 S_1|^2 - \Delta d_n^2} \tag{42}$$

A shear contact spring is activated when the shear force is less than the shear resistance of a discontinuity. A mathematical spring with stiffness K_s is placed between points S_0 and P_0 in the direction parallel to the contact face. The shear force vector increment is calculated as

$$\Delta F_{ci}^s = -K_s \Delta d_{si} \tag{43}$$

The total shear force vector is updated as

$$F_{ci}^s(\tau + \Delta\tau) = F_{ci}^s(\tau) + \Delta F_{ci}^s \tag{44}$$

4.4.4 Validation of the contact model

4.4.4.1 Sliding of a block

The example shown in Figure 24 simulates the sliding of a block on an inclined plane at an angle θ to the horizontal. In Figure 24, the numerical results are compared with the time-dependent displacement predicted by the analytical solution. The dots denote the numerical results, and the solid lines represent the corresponding analytical solution. The maximum relative error is less than 0.001% for all friction angles. The numerical results display satisfactory agreement with the analytical solutions.

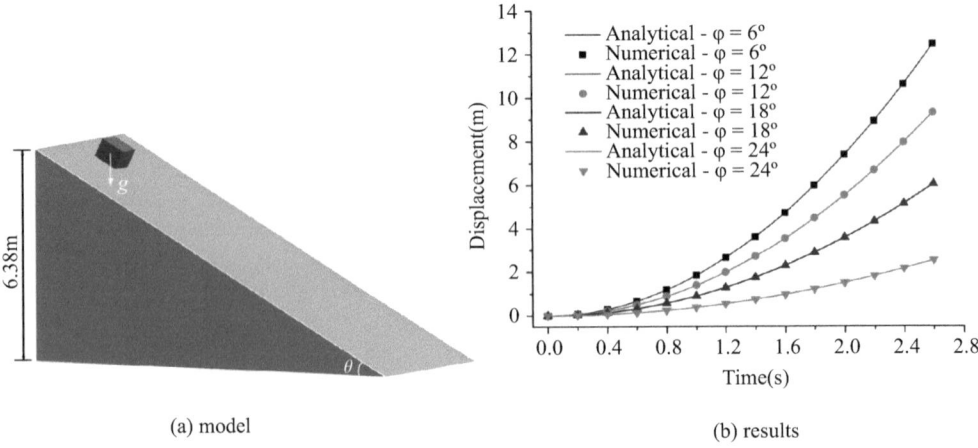

(a) model   (b) results

Figure 24 Model and results.

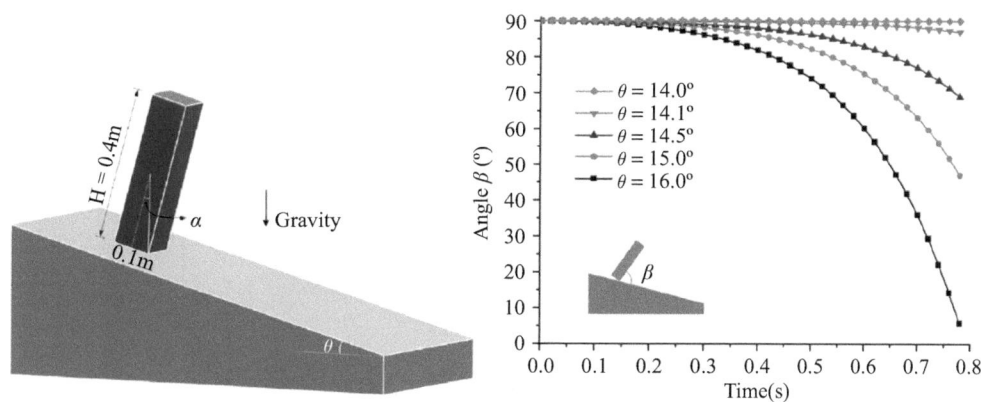

Figure 25 Model and results.

4.4.4.2 Block toppling

As shown in Figure 25, for a block on an inclined plane, if the slope angle θ is less than the friction angle φ, then the block will remain still if rotation is not considered. However, if the slope angle θ is greater than the angle α, the block will begin to rotate. The angle α can be calculated from the dimensions of the block, as expressed by

$$\alpha = \frac{180}{\pi} \times \arctan\left(\frac{W}{H}\right) \tag{45}$$

Where W and H are the width and height of the block, respectively. In the figure, angle α equals 14° when W and H are 0.1 m and 0.4 m, respectively. We assume that the density is 2,500 kg/m³, that the Young's modulus is 30 GPa and that the Poisson's ratio is 0.22 for each block. The stiffnesses of the normal and shear contact springs are 150 GN/m and 123 GN/m, respectively.

Slope angles θ of 14.0°, 14.1°, 14.5° and 16.0° are investigated. The numerical results for these cases are shown in Figure 25. Let β denote the intersection angle between the block and the inclined plane. Figure 25 shows angle β during the toppling motion. From this figure, the critical value for angle θ is 14.1°, which corresponds well with the analytical solution. The horizontal displacement of the block after 0.4s, 0.6s, 0.8s and 0.9s for $\theta = 15.0°$ is shown in Figure 25. In the process, the modeling of face-to-face and edge-to-face contacts is verified.

5 APPLICATION OF CDEM IN JIWEISHAN SLOPE

5.1 Numerical model

The deformation of Jiweishan slope has a long history, according to field survey and local people's accounts. A professional geological reconnaissance team found a crack on the back of the east surface in 1999. After that, so much work – such as investigation and evacuation – has been carried out by a geological teams and local government.

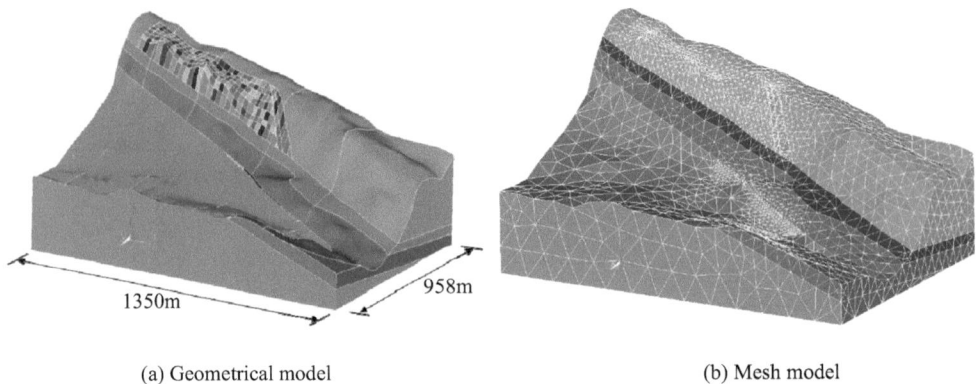

(a) Geometrical model   (b) Mesh model

Figure 26 Geometrical and numerical model of the source area.

However, there are still many important deviations between prediction and practical consequence, including failure mode, sliding direction, volume of unstable rock mass, and the distance. In this situation, is very important to analyze the start-up mechanism and kinetic features of the Jiweishan rockslide.

Consequently, two numerical models of Jiweishan – the source area and the entire area – were employed in the numerical experiment. In order to avoid large-scale calculation, the source area model was used to analyze the failure mechanism (start-up mechanism) of the Jiweishan slope. When analyzing the kinetic feature and runout distance of the rockslide, a full area model was used. Particular attention was devoted to the geometry of joints, friction, and cohesion, each of which contribute strong constraints on mechanical behavior during sliding.

Using the geological information, a three-dimensional geometrical model of the source area was first built, as shown in Figure 26a. Both the karst area and the excavated ore body are considered in this model. As mentioned before, the rock mass of this slope was cut into distinct blocks by several groups of joints. The dimension of the sliding rock mass is about 780 m (length) × 260 m (width) × 60 m (thickness), and the spacing of joints in this model is about 20 m. Based on the geometrical model, the numerical model was obtained as Figure 26b. The model contains 12,677 nodes and 68,350 tetrahedrons.

5.2 Parameters of rock mass and the monitored points

On the basis of *in situ* observation and laboratory experiment, the adopted material parameters of the rock mass are showed in Table 7. The strength of joints is obtained by the reduction of intact rock, and the reduction coefficient ranges from 0.2~0.3, with a uniform distribution. Roller boundary conditions are assumed along the lateral faces of the model such that no displacement is allowed in the normal direction. At the bottom face of the numerical model, the boundary is fixed such that no movement is allowed in the z-direction. The numbers in Figure 27 show monitored points where displacements and stress in the slope are recorded throughout the numerical simulation.

Table 7 Material parameters of rock mass for the simulations.

Parameters	E/GPa	v	ρ/ kg·m-3	C/ MPa	φ/°	T/ MPa
Plm	82	0.14	2710	17.3	45	90
karst zone	23	0.36	1750	5.0	25	15
slip band	34	0.21	1750	1.2	8	10
P_1q	66	0.32	2670	10.7	43	80
P_1l	59	0.28	2640	4.0	28	70
S_3hj	46	0.23	2630	4.5	25	20

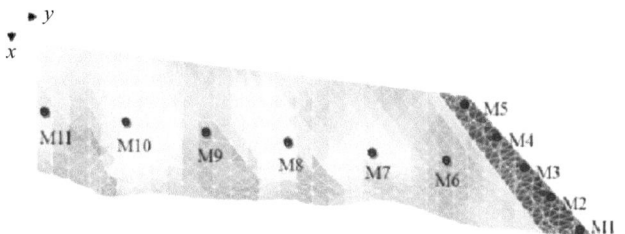

Figure 27 The position of monitored points.

The results from numerical simulation of the jointed rock slope were described for three stages. Initially, the rock slope was equilibrated under gravity using the input parameters described earlier, but the joints cannot be broken. After that, the Mohr–Coulomb criterion with a tensile cut off was used to obtain the fracture state of all joints under gravity. Lastly, the influence of excavation to the slope was analyzed by setting the ore body as an empty model. Considering the long-time creep of the Jiweishan slope, we adopted variable values for friction coefficient of the slip band as the major parameter to study in the numerical experiments.

5.3 Results and discussions

Figure 28 shows the displacement results of different monitor points (MP1–MP5) with the friction coefficient μ = 0.27. In this figure, the variations of displacement in three directions of the three stages can be analyzed. Based on the fact that all the curves tend to be stable, we can conclude that the block could not slide down to the cliff in this situation. For a friction coefficient of μ = 0.14, the displacement curves are shown in Figure 29. Compared to the result of the stable case, the displacement of monitor points (MP1–MP11) are increased with the calculation time, and the value of displacement is about 80 m. This result can indicate that rock mass of the slope is unstable in this case.

Figure 30 shows the results of the failure process at different calculation steps, which can be used to analyse the failure mechanism of the Jiweishan slope. Due to high compressibility of the karst area and creep characteristics of the slip band, the whole rock mass slips along the apparent dip direction (y direction) (Figure 30a–b), and the rock in the front edge is in the state of compression. If the shear stress on

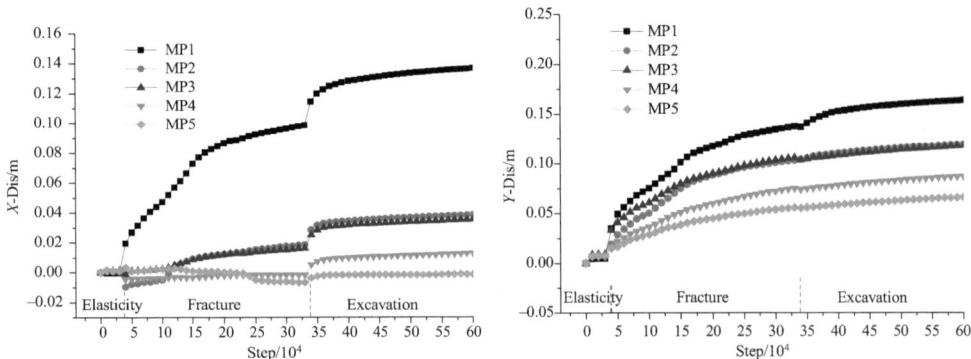

Figure 28 Displacement of monitor points in karst zone ($\mu = 0.27$).

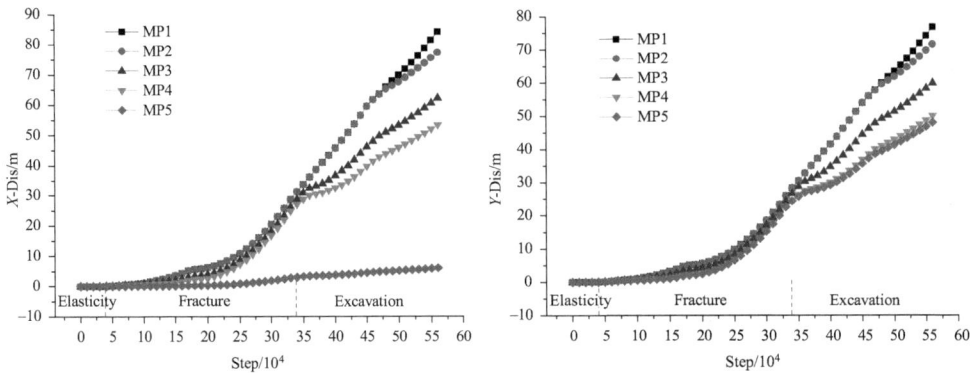

Figure 29 Displacement of monitor points (MPI–MP5) in karst zone ($\mu = 0.14$).

joint faces is greater than shear strength, the fracture of joints will happen (Figure 30c–d). After that, a key block in the front part of the slope is formed. If pushed by back rock masses continuously, the key block will move toward the cliff. When the key block falls down the cliff, the back rock mass of the slope will slide and the disaster will happen.

According to the field investigation and remote sensing interpretation results, the sliding mass changed its direction and traveled a further 2.2 km. In order to analyze the kinetic feature of this long runout rockslide, the entire area model is established. Similar to the source area model, the rock mass is also cut by three sets of joints, and the spacing of joints is about 20 m. As shown in Figure 31, the numerical model contains 18,264 nodes and 97,664 tetrahedrons. After the three stages introduced earlier, another stage – dynamic calculation – is added to analyze the movement of the rockslide.

Figure 32 shows the process of movement of the rockslide at different times, with the friction coefficient determined above, that is $\mu = 0.14$. The forming process of the key block was analyzed in detail by the use of the source area model. After that, the

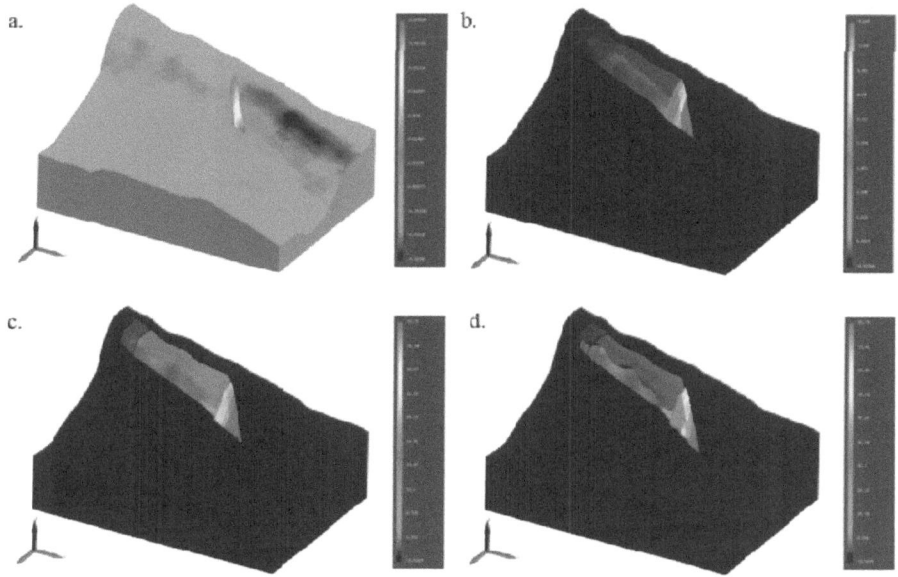

Figure 30 Failure process of the key block: (a) 40k step; (b) 200k step; (c) 350k step; (d) 550k step.

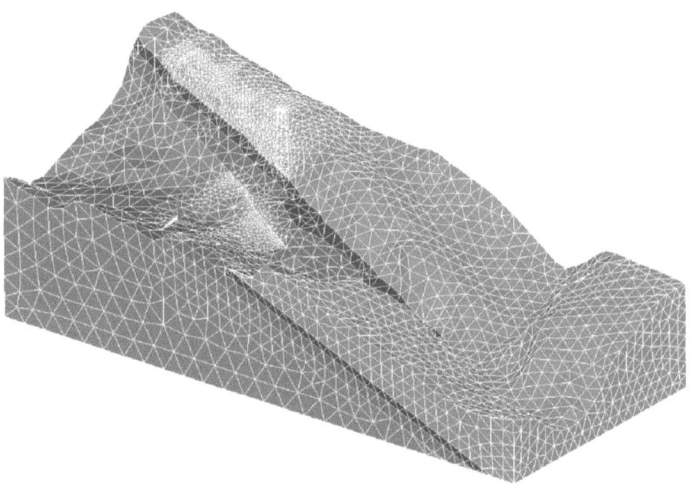

Figure 31 Numerical model of the entire area.

key block changed its moving direction and slid toward the cliff (Figure 32(a)). The key block first crossed over the Tiejiang creek, climbed up the opposite creek slope accompanied by a heavy collision with this slope (Figure 32b). Then, these blocks changed their direction of movement once more, traveling along the valley (Figure 32(c–d)).

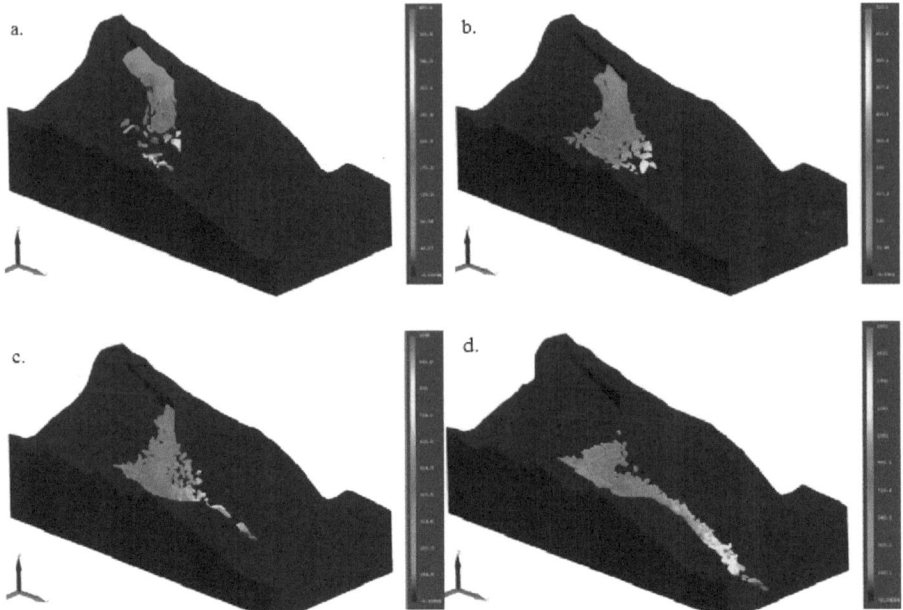

Figure 32 The kinetic characteristics of the Jiweishan rockslide at different times: (a) t = 25s; (b) t = 35s; (c) t = 45s; (d) t = 80s.

At the initial stage of deformation, the existence of the karst band had an important effect on changing the direction of movement of the slope rock mass. Actually, the phenomenon of a change of direction during movement down a slope is very common. As shown in Figure 33, there are six blocks, signified by colored points, and the dotted line is the reference mark. It is obvious that the initial direction of movement of these blocks is north, but changed as shown in Figure 33f.

The depositing characteristics of the Jiweishan rockslide are introduced in detail by Xu *et al.* (2009), in which the deposition area can be categorized into the scratching zone, the main deposition zone, the debris deposition zone, and the sprinkling zone. A comparison of the deposition area between the numerical results and actual field conditions is showed in Figure 34. It can be seen that the characteristics of the deposition area obtained from the numerical result are in accord with the field investigation.

The failure mechanism and the kinetic characteristics of the Jiweishan rockslide can be obtained by analyzing the source area model and the entire area model. The rock mass of the slope was cut into distinct blocks by several set of joints. As there exists karst zone and slip band, the slope slid toward the north constantly. Under high compressive stress, shear destruction of the slope in the front part occurred, and a key block was thus formed. The key block and the rock mass changed their sliding direction and fell down the cliff. Blocked by the opposite steep creek wall, the sliding mass changed its direction second time and traveled a further 2.2 km. Numerical results are in accordance with the field investigation, which is meaningful for the prediction of disaster range on this kind of rockslides.

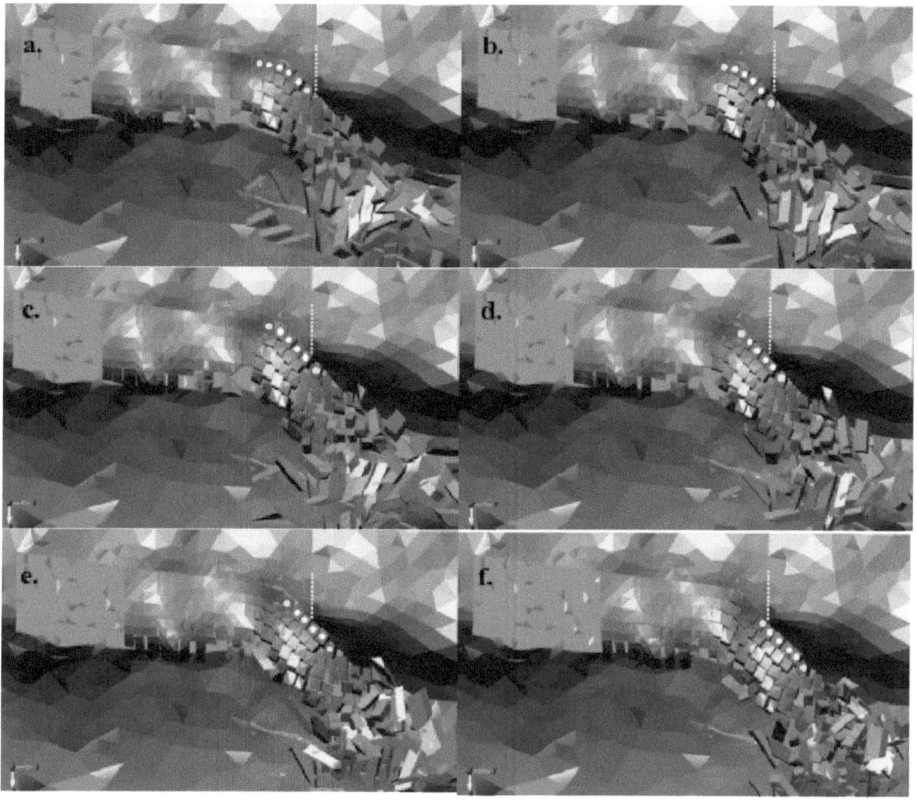

Figure 33 The change of sliding direction during the movement of slope mass: (a) t = 30s; (b) t = 31s; (c) t = 32s; (d) t = 33s; (c) t = 34s; (d) t = 36s.

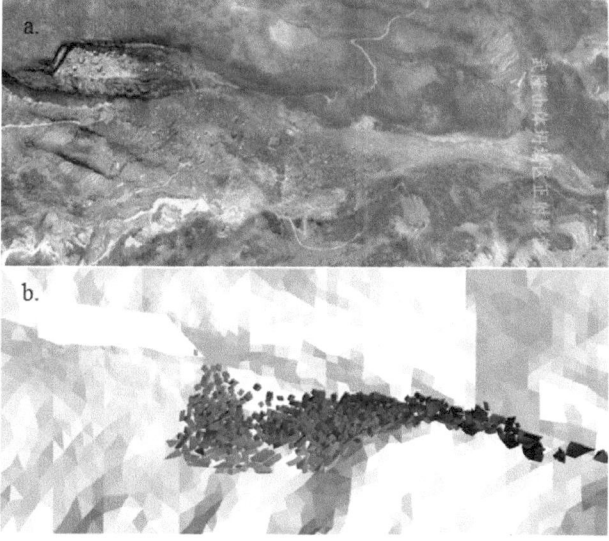

Figure 34 Comparison of the deposition area between (a) actual field conditions and (b) the numerical results.

6 SUMMARY

In this chapter the concept, governing equation, computational scheme and basic models of the continuum–discontinuum element method are briefly introduced. This method is a unified expression for continuum and discontinuum problems, which is very suitable for the whole process simulation for the failure of rock materials and structures. The spring element model, considering Poisson and pure shear effects, can properly describe the mechanical behavior of continuum media. The structure layer model is a useful tool to simulate the joints in rock material. The element fracture model has improved the accuracy of the crack propagation during the computation. The contact detection technique introduced in this chapter simplifies the detection procedure, and improved its efficiency.

REFERENCES

Belytschko, T. & Black, T. (1999). Elastic crack growth in finite elements with minimal remeshing. *International Journal for Numerical Methods in Engineering*, 45(5), 601–620.

Belytschko, T., Moës, N., Usui, S. & Parimi, C. (2001). Arbitrary discontinuities in finite element. *International Journal for Numerical Methods in Engineering*, 50(4), 993–1013.

Carpinteri, A. & Colombo, G. (1989). Numerical analysis of catastrophic softening behavior (snap-back instability). *Computer & Structures*, 31(4), 607–636.

Goodman, R. E., Taylor, R. L. & Brekke, T. (1968). A model for the mechanics of jointed rock. *Journal of the Soil Mechanics and Foundations Division*, ASCE, 94(3), 637–659.

Li, S. H., Zhang, Y. N. & Feng, C. (2010). A spring system equivalent to continuum model, 5th International Conference on Discrete Element Methods (DEM5), London: Queen Mary's University.

Li, S. H. & Zhou, D. (2013). Progressive failure constitutive model of fracture plane in geomaterial based on strain strength distribution, *International Journal of Solids and Structures*, 50 (3–4), 570–577.

Munjiza, A. (2004). *The combined finite-discrete element method*. New York, John Wiley and Sons Publishing.

Owen, D. R. J., Feng, Y.T., de Souza Neto, E.A., Cottrell, M.G., Wang, F., Andrade Pires, F.M. & Yu, J. (2004). The modeling of multi-fracturing solids and particulate media. *International Journal for Numerical Methods in Engineering*, 60(1), 317–339.

Shi, G. H. (1993). *Block system modeling by discontinuous deformation analysis*. Southampton, Computational Mechanics Publications.

Shi, G. H. (1992). Manifold Method of Material Analysis. In: *Transactions of the Ninth Army Conference on Applied Mathematics and Computing. Minneapolis, Minnesota, USA*, 51–76.

Wang, J., Li, S. H. & Zhou, D. (2013). A block-discrete-spring model to simulate failure process of rock. *Rock and Soil Mechanics*, 34(8), 2355–2362.

Wang, J., Li, S. H. & Feng, C. (2015). A shrunken edge algorithm for contact detection between convex polyhedral blocks, *Computers and Geotechnics*, 63, 315–330.

Zhang, Q. B. Li, S.-H., Feng, C. & Wang, J. (2013). Study of deformable block discrete element method based on SEM. *Rock and Soil Mechanics*, 34(8), 2385–2392.

Zhou, D. & Li, S. H. (2013). Structure layer model in CDEM. *Proceedings of the 6th International Conference on Discrete Element Methods. Colorado U.S.A, 5–6 August 2013*. pp. 445–450.

Chapter 5

Discrete element modeling: Principle and application in weak cemented particulate materials

T.G. Sitharam[1] & S.D. Anitha Kumari[2]
[1]*Professor, Department of Civil Engineering, Indian Institute of Science, Bangalore, India*
[2]*Associate Professor, Department of Civil Engineering, Ramaiah University of Applied Sciences, Bangalore, India*

Abstract: Discrete Element Method (DEM) is considered as one of the best tools to explain the physics behind the experimentally observed facts of granular materials. This numerical technique considers the discrete nature of the granular materials /rock mass and adopts the use of force-displacement law and Newton's laws of motion alternatively. It adopts an explicit finite difference method, which necessitates the use of extremely small time steps to ensure accuracy and numerical stability. The fundamental unit of DEM is the contact between two adjacent particles. The forces and displacements developed at the contacts help to characterize the micro-mechanical behavior, which is otherwise impossible to capture. The averaging of this behavior over the entire model helps to figure out the macroscopic response of the modeled assembly. By introducing a bonding concept, particles can be bonded together at contacts to reproduce the cementation effect similar to the cohesive strength of the rock masses. The study of the strength and behavior of weak cemented rock is one of the major thrust areas in rock mechanics and engineering. Generally, sedimentary rocks belong to the group of weak rock mass. They consist of rock fragments and minerals held together by natural cementing materials. The low strength and high deformability of these materials result in unexpected material behavior including yielding, squeezing, swelling etc. But, due to the practical difficulty in understanding the mechanical behavior of rock masses through physical experiments, numerical modeling has evolved as an efficient and practical tool in the field of rock mechanics. In this paper, the fundamental principles of DEM and the various micro-parameters associated with it are explained. The various issues related to the modeling of the particulate material are highlighted along with its solutions. Following an outlook into the methodology and numerical modeling steps, an underground weak cemented rock mass is modeled. The behavior of a single underground tunnel subjected to various structural and loading conditions are also added to furnish the utility of this numerical method in understanding and explaining the various mechanical phenomena from the fundamentals. The case study of a vertical cliff modeled adopting bonded systems and is subjected to non-uniform weathering is also presented to understand the application of DEM in cemented systems.

1 INTRODUCTION

Granular material behavior is highly influenced by the discrete nature of the particles constituting it. In such materials, the external loads are distributed among the particles in the assembly through the contacts between them. Since these physical processes are happening at a grain scale level due to the inherent discrete nature of the particles, a satisfactory explanation of the mechanisms using continuum theories does not seem apt. In addition, a proper analysis of the different modes of failure and deformation may also be affected if continuum approach is followed. This emphasizes the need for understanding the constitutive behavior of granular materials by adopting a suitable method that considers the discrete nature of the assembly. The basic element of interest in a granular media is the contact between particles as the load transfer occurs through these contacts. Any deformation or contact force developed in these contacts can be related to the macroscopic behavior of the granular material. Hence the Discrete element method (DEM) provides an apt method for studying the response of granular materials by understanding the underlying fundamental processes which controls the behavior of an assembly as a whole. Cundall (1971) introduced the concept of DEM to understand the behavior of rock slope, which was later extended by Cundall & Strack (1979a, 1979b) to study the granular media too. Several researchers viz. Rothenburg & Bathrust (1989), Bardet & Proubet (1991), Thornton & Sun (1993), Ng & Dobry (1994), Sitharam & Nimbkar (1997, 2000), Thornton & Antony (2000), Sitharam (2000, 2003), Sitharam & Dinesh (2002, 2003), Sitharam et al. (2002), Cheng et al. (2003, 2004), Dinesh et al. (2004), Sitharam & Vinod (2008, 2009, 2010), Sitharam et al. (2008, 2009), Soroush & Ferdowsi (2011) and Anitha & Sitharam (2011, 2012a, 2013) have successfully used DEM to understand the various aspects of granular materials which make their behavior unique. However in case of weak cemented rocks like sandstone, the introduction of cementation for properly assessing the behavior is very important. This is because the mechanical response of such cemented soils/ weak rocks is considerably different from the unbonded granular materials. The initial study of bonded materials adopting Bonded Particle Model (BPM) was suggested by Potyondy and Cundall (2004). Several researchers like Jiang et al. (2005), Jiang et al. (2006), Taboada et al. (2006), Utili & Nova (2008), Jiang et al. (2011), Estrada et al. (2010), Jiang et al. (2013) and Jiang et al. (2014) have studied cemented granular materials adopting DEM. In these studies referred to above, tensile strength and torque transmission at the contacts are considered to introduce the cementation effects. This paper is aimed at giving a brief overview on the discrete element method and the steps adopted for properly simulating a weak cemented granular mass. Following this, numerical simulations of a tunnel system in a representative weak cemented granular system are also presented.

2 DISCRETE ELEMENT METHOD (DEM)

Discrete element method is an explicit finite-difference method, which considers the discrete nature of the particles by taking into account the particle to particle interactions of the assembly. The fundamental unit/element in DEM is the contact between the particles. When an assembly of particles is subjected to an external load, particle rearrangement occurs to remain in equilibrium. The equilibrium contact forces and displacements in this stressed assembly are found out by monitoring the movement of individual particles. During each calculation cycle, a force-displacement law is applied to each contact, which calculates the contact force developed. Applying Newton's second law of motion, the new positions due to the contact force can be found out. This implies that DEM calculations alternate between the application of Newton's second law and force-displacement law at the contacts. To ensure that the disturbance arising from the contact force will not propagate beyond the immediate neighbors of the considered particle, the time step selected for the calculation should also be very small. This helps to keep the velocities and accelerations constant within the considered time step and the disturbances will not propagate beyond the particles that are in immediate contact. As a result, the resultant forces on a particle at any time are controlled exclusively by its interaction with particles that are in contact with it thereby maintaining conditions close to static equilibrium.

2.1 Fundamental principles of DEM

A typical calculation cycle in DEM is shown in Figure 1. The motion of individual particles is tracked and the positions and contact forces are updated applying Newton's second law and force-displacement laws respectively. The integration of the laws of the motion gives the displacements at the contacts and their new positions. From the displacements, the new contact forces are calculated using the force-displacement law. In the next step, these calculated contact forces will be applied to the particles and the cycle continues. Generally the loading rates applied are very low in order to keep the inertial forces very small and to maintain the condition of equilibrium.

A brief overview of the mathematical formulations associated with the DEM calculations are given below by considering two particles A and B as shown in Figure 2. This figure shows two particles A and B subjected to a movement due to the velocity v applied to the walls. After a short time Δt, overlap exists at the meeting points of the wall and the particles, whose magnitude is given by $\Delta n = v\Delta t$. Knowing this, by applying the contact-displacement law, the increment in the normal force $\Delta F_n = k_n (\Delta n)_{t1} = k_n v\Delta t$ (Cundall & Strack, 1979a) where k_n represents the stiffness of the contact. The force exerted by particle A and particle B in the x-direction can be represented as $F_{(A)x} = k_n(\Delta n)_{t1}$ and $F_{(B)x} = -k_n(\Delta n)_{t1}$.

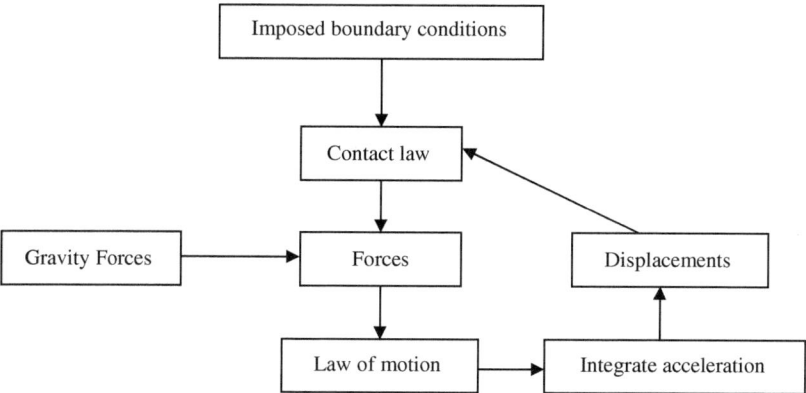

Figure 1 Typical DEM calculation cycle.

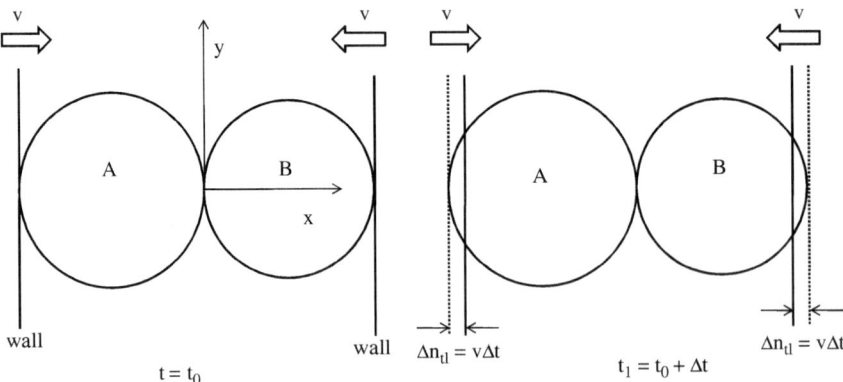

Figure 2 Movement of two particles due to an applied velocity (Ref : Cundall and Strack, 1979a).

Once these contact forces are known, applying Newton's second law, the accelerations of the particles can be obtained as $\ddot{x}_A = F_{A(x)}/m_A$ and $\ddot{x}_B = F_{B(x)}/m_B$ where $\ddot{x}_A$ and $\ddot{x}_B$ are the accelerations of the particles A and B respectively. After obtaining the accelerations, the corresponding velocities are obtained by integrating the accelerations. Integrating these velocities, the relative displacements during the next time step can be calculated as $\Delta n_{(t2)} = (v - [F_{(A)x}/m_A]\Delta t)\Delta t$. Thus, an alternate application of the force displacement law at the contacts to find out the forces and the Newton's law to the particles to find out the displacements forms the basis of the calculation cycle of DEM.

Figure 3 shows the representation of the contact between balls A and B. x_i^A and x_i^B represent the coordinates of the centre of the particles and x_i^C represent the contact

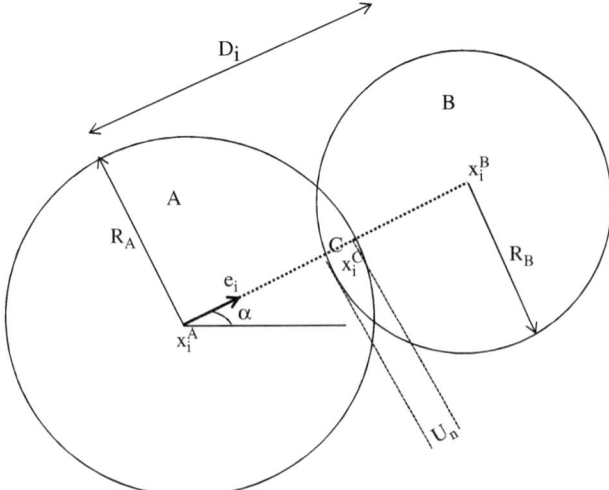

Figure 3 Representation of Contact between balls (Cundall and Strack, 1979a).

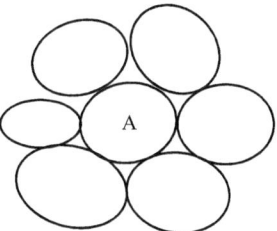

Figure 4 Particles in contact.

point. D_i represents the distance between the centre of the particles A and B. R_A and R_B are the radii of the particles A and B respectively. The overlap of the contact which is the relative contact displacement is denoted by U_n. The unit normal e_i defined along the line joining the centers of the spheres can be defined as $e_i = \frac{x_i^A - x_i^B}{D_i}$ and the overlap $U_n = R_A + R_B - D_i$. Knowing the relative contact displacement, the coordinates of the contact point C can be calculated as $x_i^C = x_i^A + \left(R_A - \frac{1}{2}U_n\right)e_i$.

In Figure 4, the particle A at equilibrium is in contact with several other particles. The number of contacts per particle is termed as the coordination number of the particle. Average coordination number represents the total number of contact points within the considered volume to the total number of particles and is an indicator of the density of the assembly. The average coordination number (γ) of an assembly is given as $\gamma = \frac{M}{N}$ where M is total number of contact points and N is total number of particles. Contact density (m_v) describes the number of contacts per unit volume of the assembly and is given by $m_v = M/V$ where V is the volume of assembly. In addition to these parameters, the orientation of the contacts is very important as the intergranular forces are dependent on it. Figure 5 shows the other descriptors associated with the micromechanics. When two particles are in contact the interparticle load transfer occurs through the contact points and is represented by the contact force f^c. The unit vector orthogonal to the contact plane is termed as contact normal (n_c). Contact vector (l_c) is the vector joining the centroid of the contacting particle and the contact point. For spherical particles contact normal coincides with the contact vector.

DEM follows an explicit algorithm scheme as against implicit methods. This is because for each time step, the number of iterations required for the implicit methods are very high especially in systems where contacts break and form in large numbers. However in explicit algorithms, for each particle the motion equation is integrated completely assuming that it is not in contact. The contact conditions are imposed later, after identifying the contact points at the end of the time step. Since a large number of particles are involved which result in innumerable number of contacts forming and breaking, explicit scheme provides a much more realistic simulation. This also signifies the importance of a calculation time step in DEM simulations. A very short/small time step is absolutely important as the stability of the system is dependent on it. However it should not be too small as it may affect the computational efficiency of the simulations. When particles move in a granular system generally the disturbances propagate in the form of a Rayleigh wave as reported by Timoshenko & Goodier (2007), Johnson (2004). The simulation time step is related to the time taken by the energy to transverse

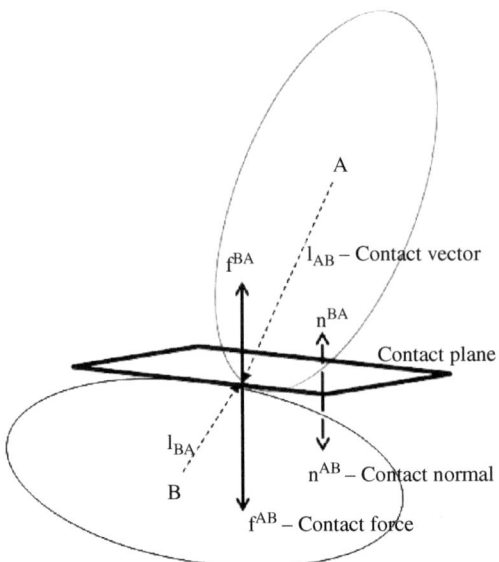

Figure 5 Contact force, normal and vector.

the smallest element in a considered system. Hence it should be so small that the disturbance due to the motion should not propagate beyond the particles which are immediate neighbors. Also, during a particular time step the velocity and acceleration are assumed to be constant. Different approaches have been adopted by various researchers for calculating the time step (Cundall & Strack, 1979, Itasca, 1998, O'Sullivan & Bray, 2004). For a single degree of freedom system, the critical time step as given by Bathe & Wilson (1976) is $t_{critical} = 2\sqrt{\frac{m}{k}}$ where m is the mass and k is the stiffness. Itasca, 1998 used a critical time step $t_{critical} = \sqrt{\frac{m}{k}}$ considering infinite series of point masses and springs for translational motion. However O'Sullivan & Bray (2004), suggested critical time steps for various configurations of 2D and 3D, but were more restrictive in nature compared to the time step suggested by Itasca 1998. They also reported that the critical time step is a function of packing configuration and number of contacts per particle.

2.2 Significance of DEM in cemented materials

The physical and engineering properties of naturally occurring rock masses are very difficult to establish when compared with artificially manufactured materials like steel and plastics. These rock masses are highly discontinuous, anisotropic and porous. Moreover, it is very difficult to capture the behavior under different loading conditions through experiments. This necessitates the importance of numerical modeling, which helps to study the fundamental mechanical behavior of rock masses. Lightly cemented sands are a common occurrence in different parts of the world. The constitutive behavior of cemented materials is considerably different from that of unbonded granular materials. The behavior of unbonded granular materials is controlled by the frictional component whereas both cohesion and friction play a role in the response of cemented granular materials. The presence of cementation results in the introduction of tensile force at the contact level. When the tensile force exceeds and the bond strength crosses the yield point, it starts degrading and finally the entire cementation will be lost and the material behaves as an unbonded granular material. In addition, due to mechanical loading, the load bearing capacity of the bonds gradually degrade resulting in different responses like global yielding, dilation and strain softening in cemented materials (Leroueil & Vaughan, 1990, Cuccovillo & Coop, 1997, Yun & Santamarina, 2005). Several studies have reported the ability of DEM to successfully reproduce the behavior of cemented granular materials (Utili & Nova, 2008, Zeghal & Shamy, 2008, Estrada et al., 2010, Jiang et al., 2011). This is because DEM can capture the bond degradation/damage at a local level and correlate it with the global failure in terms of macro response. Such a capturing of data at the local and global level simultaneously is extremely difficult in case of experiments. Several studies were also conducted to understand the behavior of weak and highly jointed rock masses (Tannant & Wang, 2004, Jiao et al., 2005, Chen & Zhao, 2002) which suggested that DEM is a better tool to understand the behavior of the discontinuous masses.

2.3 Bonded Particle Model (BPM)

The Bonded Particle model (BPM) suggested by Potyondy & Cundall (2004) represents rock as a cemented granular material as it assumes that the grains as well as the cement are deformable and breakable. In this model, the external load is carried by the grains and the cement that fills the voids in between the grains. Similar to granular material, force chains are formed which propagate through grain contacts and the cement filling in between them. These cementations can resist compression, tension and shear forces and may carry bending moments also. This results in a non-homogeneous load transfer between contacts and consequently some grains will be carrying loads much higher than the applied loads. These unusually high grain forces result in local loading leading to the initiation of bond breakage. Once the bonds start breaking, loads will be redistributed to the intact bonded grains thus subjecting them to high grain level forces leading to its breakage. Numerical modeling of cemented material can be done adopting indirect or direct approaches. Indirect methods consider the material as continuum whereas direct methods take into account the discontinuity/discrete nature of the particles. The BPM adopts a direct approach where the modeled material is considered as a collection of independent materials with cementation in between. The most significant advantage of BPM is that it does not make inherent assumptions like rock is an elastic continuum with a series of cracks. Instead, cracks are allowed to form, join and grow into fractures as the loading progresses. The deformations occurring in a granular assembly are due to the rotations and sliding of the particles. When bonds are added to this system, the assembly behaves as if cement is added between the grains of the particles and thus the behavior should replicate that of weak sedimentary rocks like sandstone. In both cases as the loading increases, there is gradual breaking of the bonds and the assembly slowly progresses toward the unbonded granular state. The cementation between the particles can be modeled in two ways. If the bonding is represented over a very small contact area, it is termed as contact bond. However, if more cementation is present in between the grains such that it covers a significant area surrounding the contact point, it is termed as parallel bond. The contact bond breaks if the tensile normal or contact shear force exceeds the limiting strength. Since parallel bonds extend over a larger area, they are able to resist moments too. The grains can transfer only forces. Hence the rolling of particles is resisted in an assembly cemented with parallel bonds.

In this research, the particle flow code 3D (PFC^{3D}) is adopted as a tool for the numerical simulation. For the current study, weak cemented materials are represented by a dense packing of circular or spherical particles that are bonded together at their contact points. The strength of the bonds represents the cementation of the particles. Generally for modeling weak material like sandstone, contact bonds are preferred. Contact bonds act at the contact point and are capable of transmitting force alone. The contact bond will have a tensile normal strength and shear contact-force strength. These bonds allow tensile forces to develop at a contact. The strength of the material is dependent upon the contact bond strength also. The contact bonding model supported in PFC^{3D} can transmit a force which acts only at the contact points and it consists of shear and normal strength components. If the magnitude of the tensile normal or shear contact force exceeds the normal or shear strength of the contact bond, the bond breaks. As long as the bond is intact, no slip is possible. This contact bond prevents the possibility of slip as the magnitude of the shear contact force is limited by the shear contact bond strength. These bonds allow tensile forces to develop at a contact. But rolling can occur between particles that are attached with contact bonds since these bonds cannot resist the moment acting at the contact point. Several researchers have conducted studies on weak cemented rocks adopting the bonding mechanism to represent the cementation (Funatsu *et al.*, 2008, Utili & Nova, 2008, Anitha & Sitharam, 2010a, 2010b, 2012b, Anitha *et al.*, 2011). The studies indicated that the damage caused to the bonded particulate systems can be represented by the broken bonds and the physical behavior underlying these mechanisms can be clearly understood from the particle scale.

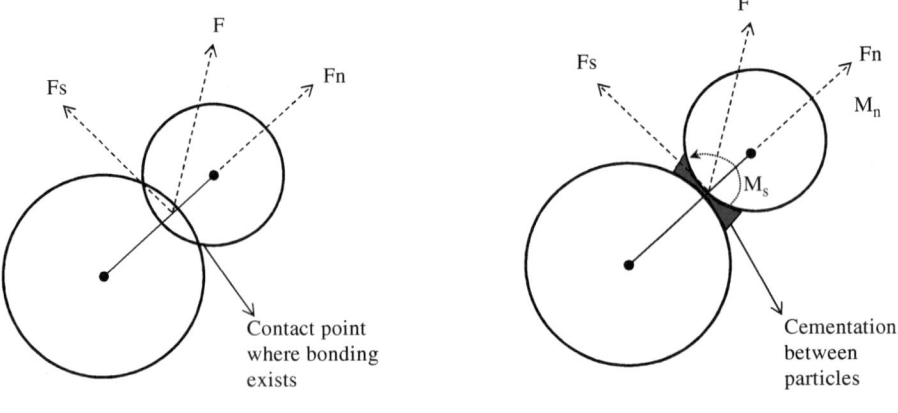

Figure 6 Typical bonding (Potyondy & Cundall, 2004, PFC^{3D} Manual).

3 CONSTITUTIVE BEHAVIOR OF BONDED AND UNBONDED SYSTEMS

A set of triaxial shear testing is performed on bonded and unbonded systems to understand the constitutive behavior of these systems. The simulations are performed on numerical samples under identical confining conditions, the only difference being in the bonding strength. Figure 7 shows the cylindrical assembly used for the triaxial testing and the contact bond representation adopted. The height to diameter ratio of the sample is kept as 2 similar to a laboratory sample. The sample consists of 3000 particles. The influence of bonding/cementation from experimental observation in terms of microparameters is very difficult. To understand the constitutive behavior of such bonded systems and its comparison with the numerical modeling, a qualitative comparison of the experimental test data reported by Marri *et al.* (2012), in sand bonded with various quantities of Portland cement is presented in Figure 8. Even though the purpose of their test was to understand the particle breakage, a qualitative comparison of the constitutive behavior of the assemblies with varying degrees of cementation is possible. The comparison between the 0% cement content and

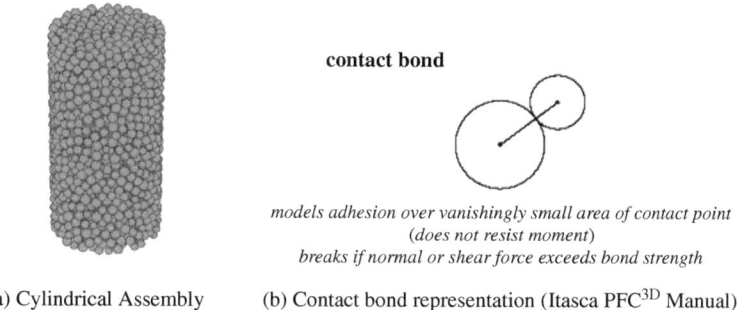

contact bond

models adhesion over vanishingly small area of contact point
(does not resist moment)
breaks if normal or shear force exceeds bond strength

(a) Cylindrical Assembly   (b) Contact bond representation (Itasca PFC^{3D} Manual)

Figure 7 Sample and contact bond representation.

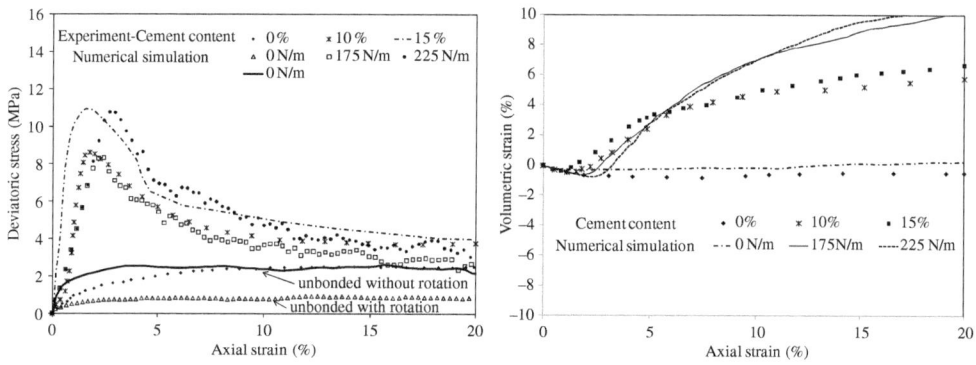

Figure 8 Comparison of the behavior of cemented sand and DEM simulations.

unbonded material shows qualitatively similar behavior in terms of strain hardening. The difference in the peak strength values obtained in the experimental and numerical simulations is attributed to the influence of restricted rotation of the irregular shaped sand particles resulting in more interlocking and higher resistance for the assembly as against the numerical model where all the particles are modeled as spheres. The stress-strain behavior of the numerical model also underlines the significance of the shape of the particle on the shear strength of the assembly. In the case of unbonded granular systems, the peak strength attained by the assembly was very low. This is due to the fact that the spherical particles which are used for modeling the assembly is free to rotate and hence not developing full resistance. So to bring out the influence of rotation, a simulation with restricted rotation was also performed on the same assembly. The results indicate that the strength gained in due course of loading is similar to the experimental observation underlining the significance of particle rotation which in turn can be related to particle shape. A bond model adopting this type of contact model is presented as a case study of an underground tunnel in the following section.

4 CASE I: NUMERICAL MODELING OF TUNNELS IN WEAK ROCK MASSES

An ever increasing need of developing additional infrastructure is the present day challenge as the urban areas continue to keep growing in terms of population and job opportunities. This demands the development of additional transportation facilities and other infrastructure in existing cities. The most feasible way of developing these amenities in the already congested cities is by properly utilizing the underground area. Generally tunneling is associated with all underground constructions. As far as a geotechnical engineer is concerned, tunneling in weak cemented soils/rocks pose unusual challenges since the deformation and forces/stresses around the openings may result in catastrophes. The selection of a most suitable tunneling method which is also economical is very significant. The generation of stresses/strains takes place as soon as the excavation starts. Hence to understand the various aspects of these cemented materials during tunneling, DEM is adopted in this study. However the accurate simulation of any numerical model greatly depends on the selection of the various parameters used for the study. The calibration studies are reported in detail by Anitha & Sitharam (2010a). Based on these studies, appropriate selection for the microparameters are done for this model study.

4.1 Modeling the assembly

The assembly is modeled in a section of weathered rock having a dimension of 25m × 10m × 25m and is shown in Figure 9. A total of 50000 particles are used for the simulation of the assembly. Based on the studies done on the influence of micro parameters on the macro properties, the values adopted for the particles used in the simulation are presented in Table 1. The particles are allowed to settle down under gravity. After applying the gravitational force, the system is subjected to a confining stress of 0.8MPa corresponding to the overburden pressure. This is to develop an

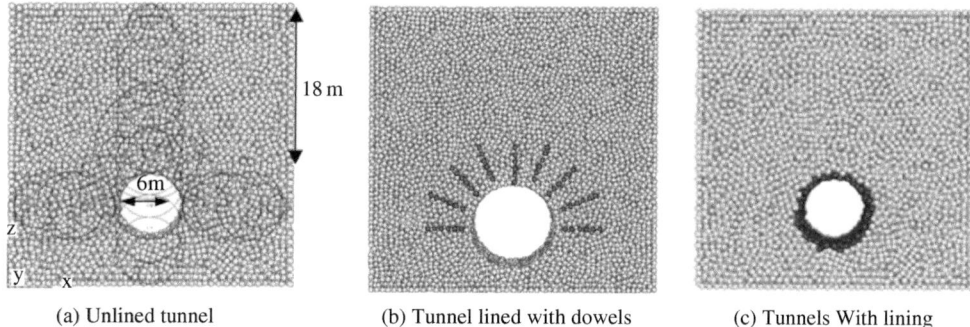

(a) Unlined tunnel   (b) Tunnel lined with dowels   (c) Tunnels With lining

Figure 9 Sample (25m x 10m x 25m) with excavated tunnel and Volumes (Measurement spheres) considered for averaging the stresses and strains.

Table 1 Properties of the particles used.

Property	Value
Normal stiffness of particles	100 MN/m
Shear stiffness of particles	100 MN/m
Wall stiffness	1000 MN/m
Density of particles	2600 kg/m^3
Normal Contact bond strength	20 kN
Shear Contact bond strength	20 kN
Porosity	0.4
No of particles	50000
Inter particle friction	0.45
Particle size	0.075–0.1m

isotropic loading condition within the model rock. Contact bonds are installed within the sample to represent the rock mass.

Following this, a tunnel of radius 3m is excavated in the soil mass at a depth of 18m from the ground surface as shown in Figure 9. Unlined and lined tunnels are shown in this figure. Tunnels can be lined either using a concrete liner or dowels can be provided which act as a passive reinforcement to support the crown of the tunnel. Once the tunnel is excavated, additional stresses and displacements are developed in the assembly. In order to extract the stresses and strains along the various sections of the model, measurement spheres are defined. Measurement spheres are used to measure various parameters over a specified spherical volume such that it represents a large number of discrete particles. These spherical volumes help to average the stresses and strains from the contact forces and relative displacements between the particles. Typical measurement spheres defined at various sections along the model are shown in Figure 9. The effect of the excavation of the tunnel is studied with respect to the contact force and circumferential stress variation. The variation is studied for two different cases (a) the tunnel is left unlined/without any support and (b) tunnel is lined. The collapse of the

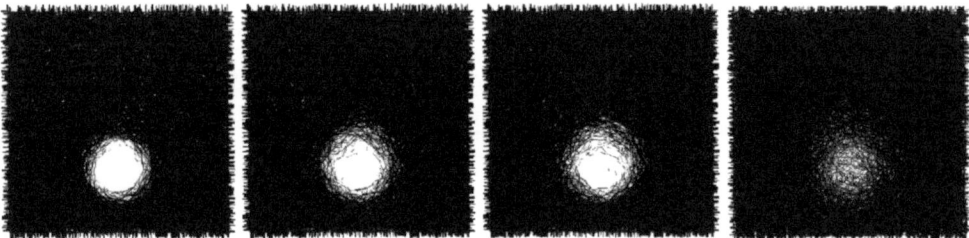

Figure 10 Contact forces at various stages of tunnel excavation – Unlined tunnel.

tunnel in terms of the contact force is shown in Figure 10. The thickness of the lines represents the magnitude of contact forces. Greater the thickness, greater will be the magnitude of contact force. It can be clearly observed that there is a significant reduction in the contact force once the tunnel is excavated leading to the ultimate failure of the tunnel. However if the tunnel is lined it is observed that the system is much more stable when compared to that of an unlined tunnel.

To make the effect of lining clear, the variation of circumferential stresses around the tunnel is calculated and the distribution is shown in Figure 11. This distribution indicates that the values are higher for the structure with lining. Moreover it can be seen that the stresses are almost uniformly distributed around the tunnel. This may be attributed to the fact that the presence of lining results in a confinement of the area surrounding it. Due to this, the particle movements are restricted. This increases the stability of the structure compared to that of the unlined structure. The stress

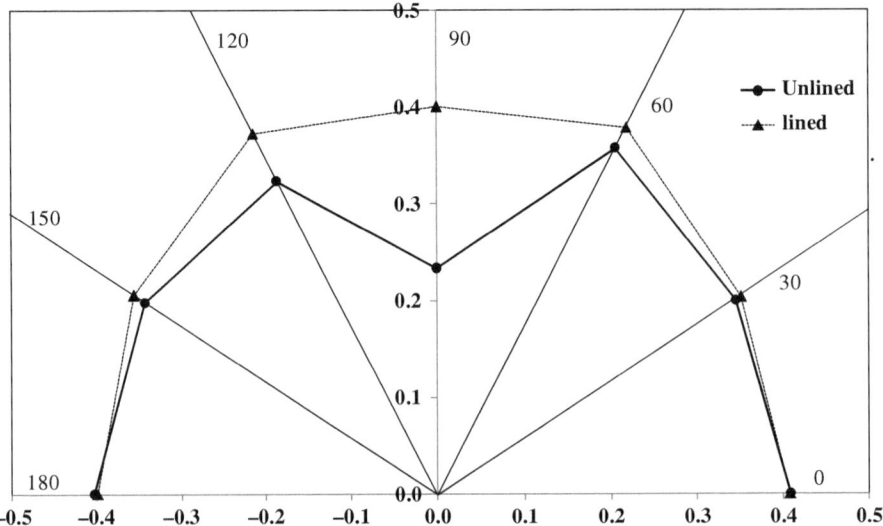

Figure 11 Distribution of circumferential stresses around the tunnel.

distribution of an unlined tunnel indicates that the stresses are comparatively low near the crown of the tunnel. This may be due to the loosening of the adjacent ground. However the progress of failure in the case of an unlined tunnel is almost the same in all directions except the top. This can be seen from Figure 10 where the tunnel collapses from all sides, more being from the top.

4.2 Finite element modeling

In addition to the discrete element simulation of the tunnel assembly, a series of finite element modeling was also done to understand the behavior from a continuum point of view. Finite element method can be applied to soil/rock mechanics problems irrespective of the complexities in geometry, loading or material models and are widely used in rock mechanics applications. In finite element method, the assembly is modeled as a continuum consisting of elements which are connected at discrete points called nodes. In 2-dimensional analysis, the assembly is discretized into triangle or quadrilateral shapes and all the forces are transmitted through nodes. The analysis of the problem is basically done in terms of these nodal forces and nodal displacements.

In order to generate the tunnel assembly and the subsequent analysis, the commercial software PLAXIS is used. While modeling the assembly, it is assumed that the material is isotropic and homogeneous. Similar to the tunnel assembly modeled using DEM, the geometry of the problem is maintained and gravity loading has to be defined. The assembly is modeled with a width of 25m and depth of 32m. A tunnel of diameter 6m is excavated at a depth of 18m from the ground surface. Once the geometry is defined, it is divided into a number of 15-noded triangular elements. After forming the meshes, it is important to provide proper material properties. A proper constitutive model has to be specified since the response of the material to a loading condition is dependent on the constitutive model. In this study, Mohr-Coulomb elasto-plastic model has been used for the modeled material. After providing the material properties and constitutive model, the boundary/initial conditions for the modeled assembly are specified. The top boundary is subjected to a natural condition of zero force and free displacement. Hence no explicit boundary condition is specified for the top boundary. The vertical and horizontal displacements of the bottom boundary are restricted due to the surrounding confinement and hence the boundary conditions are fixed full. The properties of the model used for the finite element study is given in Table 2. Gravity loading is applied to the assembly as initial conditions following the K_0 procedure since a

Table 2 Properties used for the FEM simulation of the tunnel assembly.

Model	Mohr-Coulomb
Behavior	Drained
Unit weight	26.5 kN/m^3
Young's modulus	5e6 kN/m^2
Poisson's ratio	0.2
Cohesion	20 kN/m^2
Angle of internal friction	24

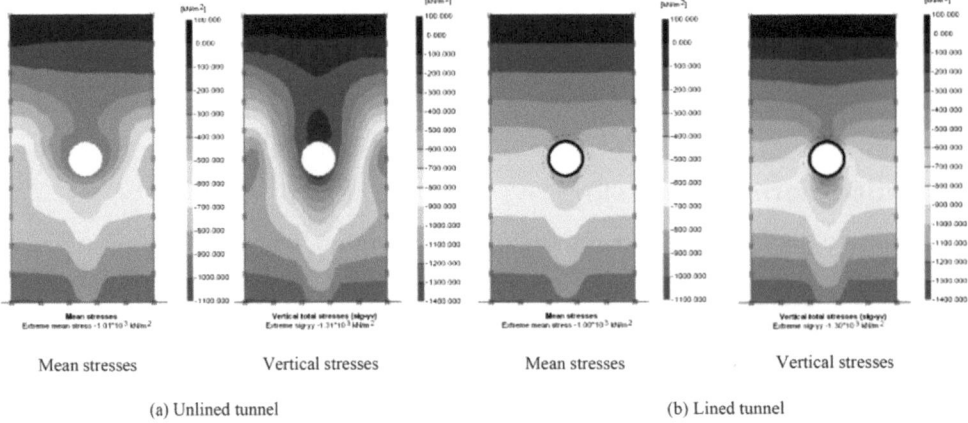

Mean stresses	Vertical stresses	Mean stresses	Vertical stresses
(a) Unlined tunnel		(b) Lined tunnel	

Figure 12 Stress distribution in unlined and lined tunnels.

horizontal surface is assumed. After applying the initial stress conditions, loading phase is initiated through automatic load stepping.

Similar to the discrete element simulations, it is observed that the deformations are very high near the crown of the tunnel indicating a complete collapse of the system. The stress distribution along the depth of the section indicates the variation in stresses due to the excavation. The complete collapse of the assembly is clearly visible from the stress contours shown in Figure 12(a), which are comparable to the observations made in discrete element simulations. Figure 12(b) shows the stress distribution for the lined assembly. An analysis of the displacement vectors proved that the magnitudes are less compared to that of the unlined tunnel and the stress contours indicate that they are almost uniform around the tunnel indicating a stable ground with more confinement. Also it is observed that the deformation is non-radial but almost uniform which can be attributed to the presence of the lining.

4.2.1 *Effect of particle aspect ratio on tunnel behavior*

After modeling and analyzing the results of the response of a tunnel assembly consisting of spherical particles, a series of runs were performed on assemblies consisting of different particle shapes to understand the effect of aspect ratio (AR) on the tunnel behavior. Three sets of assemblies in addition to the assembly consisting of only spherical particles were modeled to understand the effect of particle shape on the response of the structure. The first set of assembly consists of only spherical particles (aspect ratio AR 1), whereas the second model consists of clumped particles with aspect ratio 1.5. The third assembly consists of particles with aspect ratio 1.75 and fourth set consists of particles whose aspect ratio is 2.0. as shown in Figure 13. The clumped particles were added to the assembly by replacing particles each of which having the same volume as the replaced spherical ball. Clump logic used in PFC^{3D} is used for forming the clumps. The three sets of assemblies are subjected to the same conditions as that of the spherical assembly to understand the effect of excavation on the assemblies.

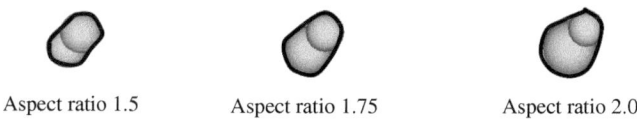

Aspect ratio 1.5   Aspect ratio 1.75   Aspect ratio 2.0

Figure 13 Shape of the particles used for modeling various tunnel assemblies.

An analysis into the variation of the circumferential and radial stresses around the various assemblies formed by particles of different aspect ratios, it is observed that the stresses are more uniform for non-spherical particle assembly. This is attributed to the fact that the non-spherical particles are able to make a large number of contacts compared to the spherical particles. Also, among the particles with higher aspect ratios, as the asperity increases it is observed that their strength developing property is reduced due to the difficulty in rearranging themselves to a better dense packing as the asperities are not allowed to break apart.

The distribution of radial and circumferential stresses above the crown of the lined tunnel is shown in Figure 14(a). The radial stresses indicated as σ_{rr} shows a very small value near the crown of the tunnel and increases gradually as the distance from the crown increases. The circumferential stresses ($\sigma_{\theta\theta}$) also increase as it moves away from the crown of the tunnel for all the assemblies. These results suggest that even though the shape changes from spheres to clumps with different aspect ratios, the stress distribution above the tunnel is not affected much. Deformations are also very important in the case of tunnel sections as it is the most important cause leading to the failure of surface or subsurface structures. Majority of the studies deal with ground deformations, but numerical modeling helps to monitor internal strains more effectively. Hence strains were also extracted for the different assemblies and are plotted in Figure 14(b). The amount of vertical strains increases toward the crown of the tunnel and the magnitude of the vertical strain is maximal for the assembly consisting of particles with aspect

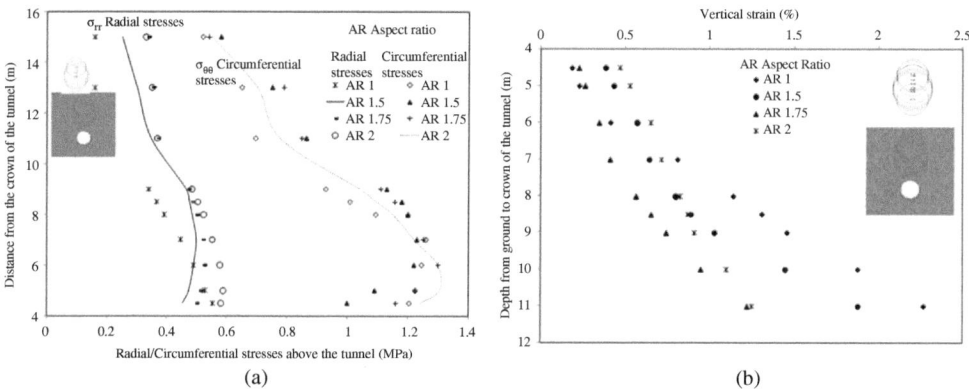

Figure 14 Distribution of radial/circumferential stresses and vertical strains above the tunnel.

ratio 1.0. As the aspect ratio increases, the amount of vertical strains near the tunnel crown decreases from 2.27% in the case of AR 1 to 1.25% for AR 1.75 and AR 2. The vertical strains plot indicates that the subsurface strains are significant as their values are comparatively higher than the surface settlements. These results match with those published by Ahmed & Iskander (2011) about subsurface settlement using plane strain approach.

5 CASE 2: NUMERICAL SIMULATION OF A VERTICAL CLIFF

While dealing with cohesive frictional material it is important to introduce the cohesion component in the modelling to simulate the accurate mechanical behavior. This is because in a real scenario the strength exhibited by any assembly can be considered as a combined contribution from cohesion as well as interparticle friction. The bond model shown in Figure 7(b) adopted in PFC^{3D} considers the bond to be fragile where it is assumed that as soon as the limit values of the normal or shear strengths are reached, the bonds break. To overcome this, Utili and Nova (2008) introduced a new bond model which considers the contributions from cohesion and friction separately. The bond model can be schematically represented as shown in Figure 15 where F^S and F^N represents shear and normal forces respectively.

According to this model, fragile behavior is considered when the contact shear force F^S reaches the contact strength and the contact breaks and the shear strength is reduced to $F^N \tan\phi$. In this model, the bonds in the normal direction are considered fragile whereas bonds in the shear direction are ductile. Also, it is clear that the failure surface of the broken contact reduces to the failure surface of a non-bonded material. However, when the shear force exceeds the shear strength, it is reduced to $F^N \tan\phi +$ c_μ and the shear strength is unaltered in the case of ductile response and the bond does not break. Hence, plastic yielding of the bond and the accumulation of the shear displacements can be modeled in this bond model. They reported that a cohesive

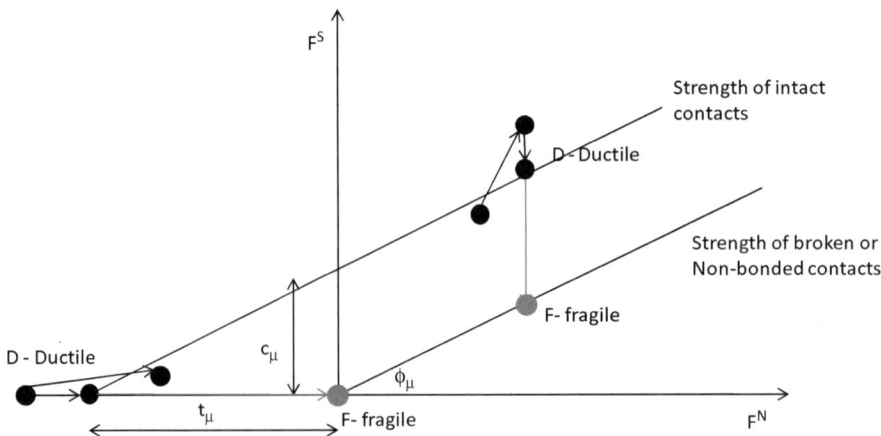

Figure 15 Representation of the bond model which gives a fragile and ductile behavior (Utili & Nova, 2008).

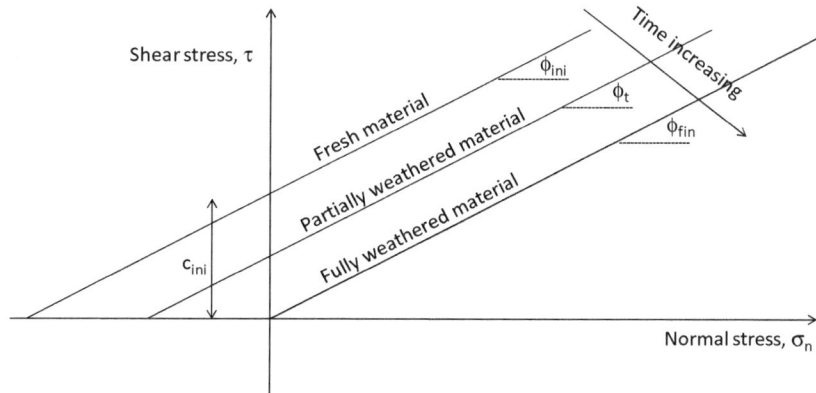

Figure 16 Progressive degradation of soil strength parameters due to continuous weathering (Utili & Nova, 2008).

frictional soil can very well be reproduced by a ductile fragile model which give a global ductile behavior. This model as implemented in the stability of a vertical cliff by Utili & Nova (2008) is shown in the next section along with its 3D simulations. The microparameters like contact bond strength and inter particle friction are obtained through a calibration procedure of macroparameters cohesion and friction angle respectively in this model. More information of the calibration procedure adopted can be found in Utili & Nova (2008). A series of calibration studies are performed to establish the relations between the microparameters and macroparameters. The progressive weathering occurring in a cliff as represented by the above researchers is given in Figure 16. When the shear stresses exceed the limiting shear strength of the slope, it becomes unstable and results in the movement of the soil mass toward the toe of the slope.

When such a weathering happens, the surface of the slope will degrade resulting in a new profile. A vertical slope of height 40m with properties consistent with the values taken for the 2D analysis (Utili & Crosta, 2008) is used to show the bonded systems subjected to weathering. Figure 17 shows the comparison of the weathering of a vertical cliff adopting 2D and 3D. In the case of cliffs, the weathering process is accelerated due to various reasons. The continuous exposure of the slope face to various natural weathering agents lead to non-uniform weathering of the slope. The degradation of the cohesion and friction values will be more intense on these exposed surface and it progresses inward as the slope failure occurs. Unlike continuum and limit analysis, in DEM the progressive failure of the slope can be precisely modeled. This slow degradation or lose of cohesive strength results in a gradual change in the slope profile over a long period of time. The step by step evolution of this process can be clearly explained adopting DEM and can also be used to identify the amount of debris that may get deposited at the base of the slope and forming a new sloped profile which is stable. The difference in the slope profile during the initial and final stages in Figure 17 clearly shows how the cementation are broken due to weathering and how the broken debris gets aligned to the new stable profile as time progresses. Also during the initial stages of the landslide, soil wedges are seen which has bonded material. But during the later stages, the failure is observed to be a scratch or ravelling pattern which can be

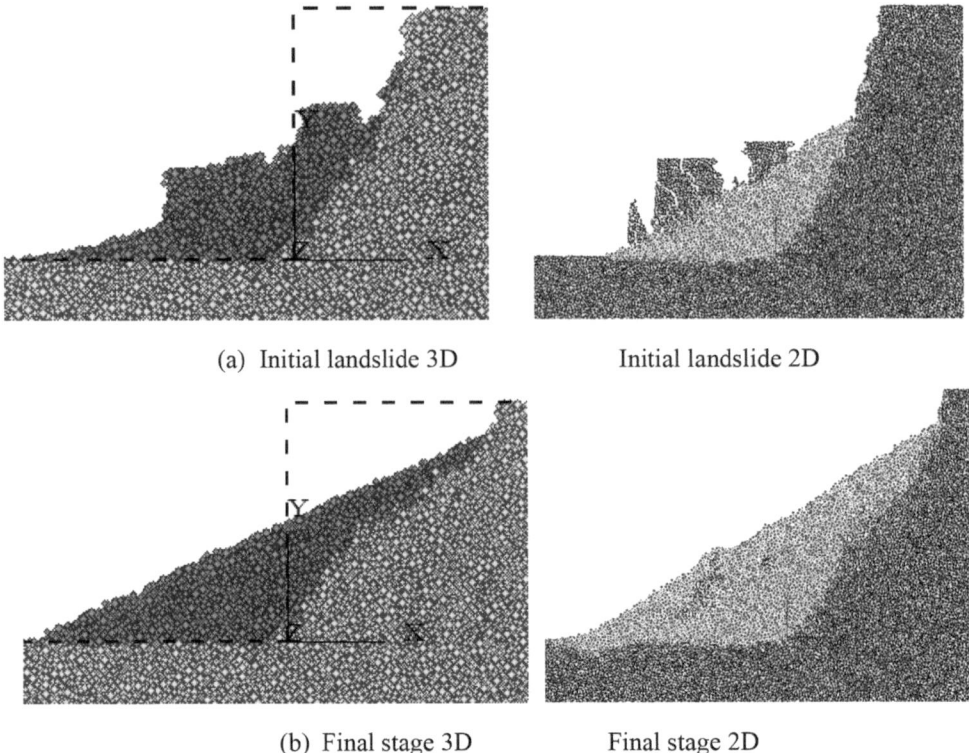

(a) Initial landslide 3D   Initial landslide 2D

(b) Final stage 3D   Final stage 2D

Figure 17 Weathering effects of a vertical bonded cliff 2D and 3D simulations. Broken line represents the initial cliff (Red colored particles are those which displaced more than 2m).

easily recognizable from the number of bonds broken and the unbonded material deposition. The inclination of the unbonded material forming the debris is nearly equal to the global friction angle of the material and it is also to be noted that the bonded material is now completely protected by a layer of this unbonded material.

6 CONCLUSIONS

DEM is a very powerful tool in simulating and explaining the fundamental mechanisms behind various processes involving discrete materials. The aptness of adopting DEM in analyzing the behavior of a weak cemented weak soil mass like sandstone is presented in this work. The cementation effects of these rocks are introduced in terms of various bonding models and the reviews indicate the efficiency of this method in reproducing the responses and also to explain these behaviors from the microparameters and the forces/displacements occurring to the individual particles due to the external loading. A numerical model is also developed to show how DEM modeling can be used for practical applications. A series of simulations are done to understand how DEM can be used to study the behavior of underground structures. The tunnel assembly was modeled with lined and unlined conditions. The response of the unlined assembly indicates that the system is unstable due to the low bonding capabilities of the weak soil leading to complete collapse of the structure with huge displacements and low stresses around the tunnel. But when lining is provided, it is observed that the ground surrounding the tunnel was confined resulting in a more uniform distribution of circumferential stresses around the tunnel leading to more stable situation compared to that of the unlined tunnel. The same trend of results was observed when the tunnels were simulated using continuum approach for both unlined and lined tunnels. The loosening of the ground and the formation of the arch action was well captured in the case of DEM compared to the finite element method. It also emphasizes the point that considering the discontinuous nature of the media will help us to model the tunnel behavior accurately and hence can be used to predict the responses of the structure. This also helps to understand the real phenomena from a microscale during various loading conditions. The studies to understand the effect of the aspect ratio on the tunnel behavior indicate that as the shape of the particle changes from sphere to clumps, there is an increase in the circumferential stresses around the tunnel leading to a more stable assembly. The non-spherical particles are having a larger number of contacts compared to spherical particles and results in higher contact forces and hence strong contact force chains leading to higher strengths. But when comparing the behavior of particles with aspect ratios 1.5, 1.75 and 2.0, there is not much variation in the strength and strain magnitudes. It may be due to the fact that as the asperity increases, the particles may find it difficult to rearrange themselves to a dense packing. An analysis into the weathering of a vertical cliff consisting of cemented/bonded systems indicates that the gradual degradation of the cementation leads to the slope failure. The analysis adopting DEM also indicates how the first significant landslide is triggered and a huge soil mass is falling off and creating a scree/talus. The amount of mass involved in the subsequent landslides is much less and the evolution of a new sloping stable profile is aptly captured by DEM studies. In addition to this, the subsequent erosion from the deposited material can also be studied adopting this method.

REFERENCES

Ahmed, M. & Iskander, M. (2011), Analysis of tunneling-induced ground movements using transparent soil models, Journal of Geotechnical & Geoenvironmental Engineering, 137(5), 525–535.

Anitha Kumari, S.D. & Sitharam, T.G. (2010a), Effect of contact bonds on the constitutive behaviour of cemented rocks: Simulations using DEM, International Symposium and 6th Asian Rock Mechanics Symposium (Advances in Rock Engineering), 2010, Delhi, India pp 77, 1–6 (CD Proceedings).

Anitha Kumari, S.D. & Sitharam, T.G. (2010b), Stability of tunnels – Effect of ground supports and reinforcement: 3-D Discrete element simulations, International Symposium and 6th Asian Rock Mechanics Symposium (Advances in Rock Engineering), 2010, Delhi, India, pp 146, 1–7 (CD Proceedings).

Anitha Kumari, S.D. & Sitharam, T.G. (2011), Liquefaction and post liquefaction behaviour of granular materials: Particle shape effect, Indian Geotechnical Journal, 41(4), 186–195.

Anitha Kumari, S.D., Vipin, K.S. & Sitharam, T.G. (2011), Stress distribution and deformations around tunnels: A study based on discrete and finite element methods, International Conference on Underground Space Technology and the 8th Asian Regional Conference of IAEG, 2011, Bangalore, India, pp DA03-01–DA03-08.

Anitha Kumari, S.D. & Sitharam, T.G. (2012a), Liquefaction and Dynamic properties of assemblies with particles of spherical and ellipsoidal shapes: A discrete element approach, International Journal of Geotechnical Earthquake Engineering, 3(1), 18–33.

Anitha Kumari, S.D. & Sitharam, T.G. (2012b), Tunnels in weak ground: Discrete element simulations, International Society of Rock Mechanics (India) Journal, 1(1), 15–22.

Anitha Kumari, S.D. & Sitharam, T.G. (2013), Effect of aspect ratio on the monotonic shear behavior: Micromechanical interpretations, Journal of Geotechnical and Geological Engineering, 31(5), 1543–1553.

Bardet, J.P. & Proubet, J. (1991), A numerical investigation of the structure of persistent shear bands in granular media, Geotechnique, 41(4), 599–613.

Bathe, K.J. & Wilson, E.L. (1976), Numerical Methods in Finite Element Analysis. Prentice-Hall, Englewood Cliffs, NJ.

Cheng, Y.P., Nakata, Y. & Bolton, M.D. (2003), Discrete element simulation of crushable soil, Geotechnique, 53(7), 633–641.

Cheng, Y.P., Bolton, M.D. & Nakata, Y. (2004) Crushing and particle deformation of soils simulated using DEM, Geotechnique, 54(2), 131–141.

Cuccovillo, T. & Coop, M.R. (1997), Yielding and pre-failure deformation of structured sands, Geotechnique, 47(3), 491–508.

Estrada, N., Lizcano, A. & Taboada, A. (2010), Simulation of cemented granular materials. I. Macroscopic stress-strain response and strain localization, Physical Review E82, 011303, 1–11. DOI: 10.1103/PhysRevE.82.011303.

Funatsu, T., Hoshino, T., Sawae, H. & Shimizu, N. (2008), Numerical analysis to better understand the mechanism of the effects of ground supports and reinforcements on the stability of tunnels using the distinct element method, Tunnelling and Underground Space Technology, 23, 561–573.

Holt, R.M., Kjølaas, J., Larsen, I., Li, L., Gotusso Pillitteri, A. & Sønstebø, E.F. (2005), Comparison between controlled laboratory experiments and discrete particle simulations of the mechanical behaviour of rock, International Journal of Rock Mechanics & Mining Sciences, 42, 985–995.

ISRM. (1983), Suggested methods for determining the strength of rock materials in triaxial compression: Revised version, International Journal of Rock Mechanics & Mining Sciences Geomechanics Abstracts, 20, 291–295.

Itasca Consulting Group. (1998), PFC^{3D} Particle Flow Code in Three Dimensions, Itasca Consulting Group.

Jiang, M., Leroueil, S. & Konrad, J. (2005), Yielding of microstructured geomaterial by distinct element method analysis, Journal of Engineering Mechanics, 131(11), 1209–1213.

Jiang, M., Yu, H.-S. & Leroueil, S. (2006), A simple and efficient approach to capturing bonding effect in naturally microstructured sands by discrete element method, International Journal for Numerical Methods in Engineering, 69(6), 1158–1193.

Jiang, M.J., Yan, H.B., Zhu, H.H. & Utili, S. (2011), Modeling shear behaviour and strain localization in cemented sands by two-dimensional distinct method analyses, Computers and Geotechnics, 38(1), 14–29.

Jiang, M., Zhang, W., Sun, Y. & Utili, S. (2013), An investigation on loose cemented granular materials via DEM analyses, Granular Matter, 15, 65–84. DOI: 10.1007/s10035-012-0382-8.

Jiang, M., Liu, F. & Zhou, Y. (2014), A bond failure criterion for DEM simulations of cemented geomaterials considering variable bond thickness, International Journal for Numerical and Analytical Methods in Geomechanics. DOI: 10.1002/nag.2282.

Johnson K.L. (2004), Contact Mechanics, 9th ed., Cambridge University Press, Cambridge, UK, p. 343.

Leroueil, S. & Vaughan, P.R. (1990), The general and congruent effects of structure in natural soils and weak rocks, Geotechnique, 40(3), 467–488.

Marri, A., Wanatowski, D. & Yu, H.S. (2012), Drained behaviour of cemented sand in high pressure triaxial compression tests, Geomechanics and Geoengineering, 7, 159–174.

O'Sullivan, C. & Bray, J.D. (2004), Selecting a suitable time step for discrete element simulations that use the central difference time integration scheme, Engineering Computations, 21 (2/3/4), 278–303.

Particle Flow Code (PFC^{3D}). (2004), User's manual, Itasca consulting group.

Potyondy, D.O. & Cundall, P.A. (2004), A bonded-particle model for rock, International Journal of Rock Mechanics & Mining Sciences, 41(8), 1329–1364.

Rothenburg, L. & Bathrust, R.J. (1989), An analytical study of induced anisotropy in idealized granular materials, Geotechnique, 39(4), 601–614.

Sitharam, T.G. (2000), Numerical simulation of particulate materials using discrete element modeling, Current Science, 78(7), 876–886.

Sitharam, T.G. & Nimbkar, M.S. (1997), Numerical modeling of the micromechanical behavior of granular media by discrete element method, Geotechnical Engineering International Resources Center, ISSN 0858-4869, 6(4), 261–283.

Sitharam, T.G. & Nimbkar, M.S. (2000), Micromechanical Modeling of Granular Materials: Effect of Particle Size and Gradation, Geotechnical and Geological Engineering, 18, 91–117.

Sitharam, T.G. & Dinesh, S.V. (2002), Micromechanical modeling of granular materials using three dimensional discrete element modeling, Key Engineering Materials, 227, 73–78.

Sitharam, T.G. & Dinesh, S.V. (2003), Numerical Simulation of Liquefaction Behaviour of Granular Materials Using Discrete Element Method, Proc. Indian Academy of Sciences. (Earth and planetary sciences), 112(3), 479–484.

Sitharam, T.G., Dinesh, S.V.& Shimizu, N. (2002), Micromechanical modeling of monotonic drained and undrained shear behaviour of granular media using three-dimensional DEM, International Journal for Numerical and Analytical Methods in Geomechanics, 26, 1167–1189.

Sitharam, T.G. & Vinod, J.S. (2008), Numerical simulation of Liquefaction and Pore pressure response of granular material using DEM, International Journal of Geotechnical Engineering, 2, 103–113.

Sitharam, T.G. & Vinod, J.S. (2009), Critical state behaviour of Granular materials from isotropic compression and rebound paths: DEM simulations, Granular Matter, 11(1), 33–42.

Sitharam, T.G. & Vinod, J.S. (2010), Shear modulus and damping ratio of granular materials: A discrete element approach, Geotechnical and Geological Engineering: An International Journal, 28(5), 591–601.

Sitharam, T.G., Vinod, J.S. & Ravishankar, B.V. (2008), Evaluation of undrained response from drained triaxial shear tests: DEM simulations and Experiments, Geotechnique, 58(7), 605–608.

Sitharam, T.G., Vinod, J.S. & Ravishankar, B.V. (2009), Post liquefaction undrained monotonic behaviour of sands: Experiments and DEM simulations, Geotechnique, 29(9), 739–749.

Taboada, A. Estrada, N. & Radjaï, F. (2006), Additive decomposition of shear strength in cohesive granular media from grain-scale interactions, Physical Review Letters, 97, 098302, 1–4. DOI: 10.1103/PhysRevLett.97.098302.

Timoshenko, S.P. & Goodier, J.N. (1970) Theory of Elasticity, 3rd ed., International Student Edition, Tokyo, Japan, p. 505.

Yun, T.S. & Santamarina, J.C. (2005), Decementation, softening, and collapse: Changes in small-strain shear stiffness in k(0) loading. Journal of Geotechnical and Geoenvironmental Engineering, 131(3), 350–358.

Chapter 6

Practical equivalent continuum analyses of jointed rockmass: Experiments, numerical modeling, validation and field case studies

T.G. Sitharam
Department of Civil Engineering, Indian Institute of Science, Bangalore, India

Abstract: The paper presents summary of the work carried out by the author on the modeling of jointed rock mass with some field applications. The practical equivalent continuum analyses approach presented attempts to use statistical relations, which are simple and obtained after analyzing a large data from the literature on laboratory test results of jointed rock masses. Systematic investigations were done including laboratory experiments to develop the methodologies to determine the equivalent material properties of rock mass and their stress-strain behavior, using a hyperbolic approach. Present study covers the development of equivalent continuum model for rock mass, implementation of the model in FLAC3D for 3-dimensional applications and subsequently verification leading to real field application involving jointed rocks. The model was rigorously validated by simulating jointed rock specimens. Element tests were conducted for both uniaxial and triaxial cases and then compared with the respective experimental results. The numerical test program includes laboratory tested cylindrical rock specimens of different rock types, from plaster of Paris representing soft rock to granite representing very hard rock. The results of the equivalent continuum modeling were also compared with explicit modeling results where joints were incorporated in the model as interfaces. Several case studies have been presented with the developed model.

1 INTRODUCTION

Rock is distinguished from other engineering materials by the presence of inherent discontinuities such as joints, bedding planes and faults that control its behavior. Hence, the prediction of the response of rocks and rock masses derives largely from their discontinuous and variable nature. Reliable characterization of the strength and deformation behavior of jointed rocks is important for the safe and economic design of civil and mining structures such as arch dams, bridge piers, tunnels, slopes and large caverns. Realistic evaluation of the shear strength and deformation characteristics presents formidable theoretical and experimental difficulties due to the complex geometry and behavior of jointed rock. A numerical approach to treating the rock mass with equivalent material properties for obtaining the overall response has been advocated in recent years (Sitharam, 2007, Sitharam *et al.*, 2001, Sitharam & Latha, 2002). Several numerical methods have been developed by various researchers to model jointed rock masses using various techniques. Singh (1973a,b) has presented continuum characterization methods for jointed rock masses and expressions were presented to estimate the elastic moduli of the equivalent continuum anisotropic rock mass. Zienkiewicz et al. (1977) has used the equivalent continuum approach, referred to as a multi-laminate model to simulate a discontinuous rock mass. Gerrard (1982) has also used an equivalent continuum approach by expressing the compliance of an element as the sum of the compliances of the intact rock and that of the individual joint sets. Cai & Horii (1992) proposed a constitutive model that presents the effects of density, orientation and connectivity of joints, as well as the property of the joints themselves. The constitutive equation is derived from the relation between the average stress and strain over a representative volume, which consists of many fractures, on the basis of micromechanics. This analysis is valid for small displacements before the failure of the joints. In the other micromechanics based model of Yoshida & Horii (2000), the jointed rock mass is replaced with an equivalent continuum body whose constitutive equation is obtained from the relation between the average stress and strain over a representative volume element. The constitutive equations directly reject the orientation and spacing of the dominant joints and can incorporate strong induced anisotropy. In the crack tensor model of Oda et al. (1993) the compliance tensor of the jointed rock mass is given as the sum of the elastic compliance tensor of the base rock and that corresponding to the crack deformation. The crack tensor is defined in terms of the size of the joints and components of the unit vector normal to the joints. The damage tensor model developed by Kawamoto et al. (1988) is based on the concept that the effective cross section of the material is reduced due to the damage. When the number of joints in the rock mass is many and it is not possible to obtain information about all of them, thus, it is not possible to deal with each joint individually. It is necessary to replace the jointed rock mass with an equivalent continuum body for analysis with an appropriate associated constitutive model. However, the constitutive models which are discussed above and many others, are very complex and need much input data from experimental or field testing in order to carry out the analysis. So, there is a need for a simple technique where the equivalent continuum method can capture sufficiently well the behavior of a jointed rock mass using minimal input from the field or from tests and experiments. Considering the inherently inhomogeneous nature of rock masses, the practical equivalent continuum approach attempts to use statistical relations, which are simple and obtained after analyzing a large set of data from the literature on laboratory test results relating to jointed rock masses. The material properties are represented by a set of relations, which express the properties of the jointed rock mass as a function of a joint factor and the properties of the intact rock. The joint factor for a given jointed rock is estimated based on the joint fabric. The tangent elastic modulus of the intact rock is represented by a confining stress-dependent hyperbolic relation. The jointed rock has been represented as an equivalent continuum using the statistical relations arrived at from the analysis of the experimental data of Roy (1993), Arora (1987), Yaji (1984), Brown & Trollope (1970), Vidya Bhushan Maji (2007) and Einstein & Hirschfield (1973). These relations express the tangent modulus of the jointed medium as a function of the joint factor and the tangent modulus of the intact rock. The results have been presented in the form of stress–strain curves for the jointed rocks and compared with the experimental results. These results are also compared with explicit modeling of rock mass where the joints are modeled using interface element. The developed model called as "Practical Equivalent Continuum Model" has been applied to the analyses of well-documented field case studies of large rock excavations and slopes.

2 EMPIRICAL RELATIONSHIPS

An effort has been made to arrive at empirical relations, which express the strength and deformation of jointed rock as a function of intact rock properties and a joint factor. These relations are determined by statistical analysis of a large amount of experimental data from Brown & Trollope (1970), Einstein & Hirschfeld (1973), Yaji (1984), Arora (1987), Vidya Bhushan Maji (2007), and Roy (1984). The experimental data covers a wide range of rocklike material and rocks namely plaster of Paris, different kinds of sandstone, granite and gypsum plaster with parallel and non-parallel joints with different joint fabric for different confining pressures. Thus, it is hoped that the developed equations for jointed rock mass may have a reasonably wide applicability. Based on the statistical analysis, the uniaxial compressive strength and elastic modulus obtained from uniaxial compressive tests and triaxial tests of jointed rock at different confining pressures are expressed as the function of the joint factor and intact rock properties. So, knowing the intact rock properties and the joint factor, the jointed rock properties can be estimated. Statistical analyses were carried out to arrive at possible empirical relations for the tangent modulus at different confinements and uniaxial compressive strength. A large amount of experimental data of uniaxial compressive strength ratio and elastic modulus ratio versus joint factor of the jointed rock specimens was digitally filtered to reduce the scatter in the data. Linear and nonlinear relationships between the uniaxial compressive strength, tangent elastic modulus at different confinements, and joint factor have been arrived at by using least-squares fitting for linear relationships and Lorentzian minimization for nonlinear relationships. Least-squares minimization assumes that the x values are accurately determined and that an error exists only in the dependent variable y. The errors are assumed to map a Gaussian profile and are normally distributed. Lorenzian minimization is very robust when the data is noisy and also converges quite rapidly. The correlation coefficient and the standard error of the fitted relationship are defined below. The correlation coefficient, r^2 of a relationship fitted to x,y, data is expressed as

$$r^2 = 1 - \frac{\sum_{i=1}^{n} [y_i - (y_p)_i]^2}{\sum_{i=1}^{n} (y_i - \bar{y})^2} \tag{1}$$

Where $y_i = y$ value for a given x; y data and the $(y_p)_i =$ value of y computed using the relationship fitted; and $\bar{y} =$ mean of y values. The standard error of the relation fitted is computed as

$$S = \sqrt{\frac{\sum_{i=1}^{n} [y_i - (y_p)_i]^2}{n - 1}} \tag{2}$$

In the present analysis the variable "y" is either uniaxial compressive strength ratio (σ_{cr}) or modulus ratio (E_r) and variable "x" is joint factor J_f. Both linear and nonlinear relations are obtained for σ_{cr} and E_r as a function of a joint factor J_f using the procedure described above. The experimental data is compiled in the form of a joint factor J_f, the uniaxial compressive strength ratio σ_{cr}, and elastic modulus ratio E_r. The experimental data used in the analysis covers a wide range of rocks like plaster of Paris, different kinds of sandstone, granite, and gypsum plaster with filled and unfilled joints with different joint fabric for different confining pressures. The experimental data is digitally filtered to reduce the scatter in the data for a better fit. The values of correlation coefficient for the fitted equations are given for the digitally filtered data set and the standard error of the fit is calculated (Sridevi, 2003) with respect to the actual experimental data. Rank of the relationship fitted is in the order of minimum standard error. The relation for which the standard error is minimum is ranked as 1 and the next relation is ranked 2, 3, and 4 based on the increase in the standard error.

Based on the statistical analysis of the data, empirical relationships for the uniaxial compressive strength ratio as a function of joint factor (J_f) are derived. Figure 1 shows the plot of the uniaxial compressive strength ratio versus the joint factor of the experimental data with the empirical relationship fitted for the equation with rank 1. The elastic modulus ratio from unconfined compressive strength test, from triaxial tests, with different confining pressures is considered in the analysis. Linear and non-linear relationships between the elastic modulus ratio at the three different confining pressures and the joint factor are derived using the experimental data. Much more unconfined compressive strength test data is available compared to triaxial test data. Hence the relationships arrived for elastic modulus using unconfined compressive test data are more reliable than those obtained using triaxial test data. Sample plot of the experimental data with the fitted relationship is shown in Figure 2. All the statistical relations arrived in the analysis are for $0 < J_f < 800$. For intact rock, $i.e.$, for $J_f = 0$, σ_{cr} and $E_r = 1$.

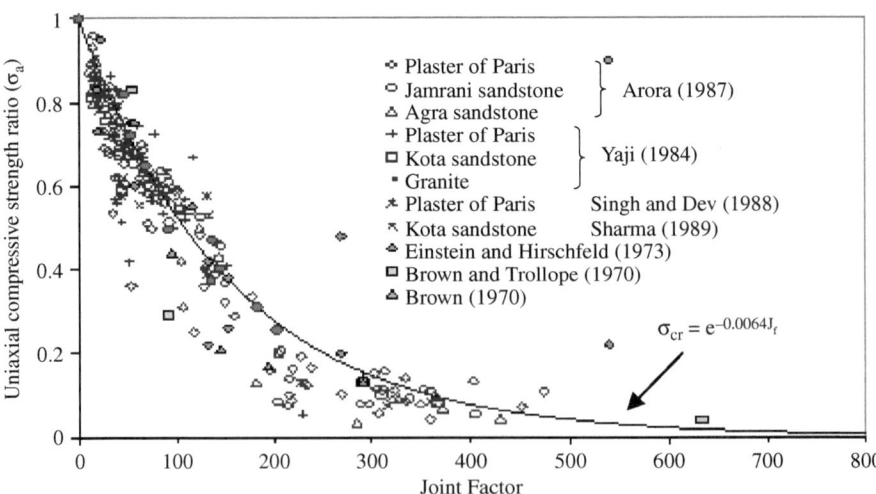

Figure 1 Uniaxial compressive test data and the fitted relation between J_f and σ_{cr}. Data from Arora (1987), Yaji (1984), Einstein and Hirschfeld (1973), Brown and Trollope (1970), Brown (1970) and Roy (1993).

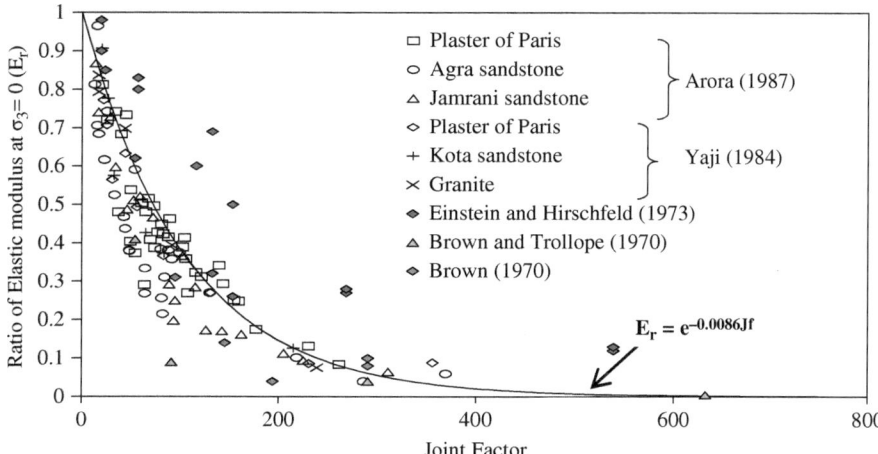

Figure 2 Experimental data and the fitted relation between J_f and E_r. Data from Arora (1987), Yaji (1984), Einstein and Hirschfeld (1973), Brown and Trollope (1970), Brown (1970), and Roy (1993).

Extensive laboratory testing of intact and jointed specimens of different grades of plaster of Paris, sandstone and granite in uniaxial and triaxial compression revealed that the important factors, which influence the strength and modulus values of the jointed rocks are (i) joint frequency, J_n, (ii) joint orientation, β, with respect to the major principal stress direction and (iii) joint strength. The joint factor (Ramamurthy, 1993) for a given jointed rock is estimated using the following equation:

$$J_f = \frac{J_n}{n.r} \tag{3}$$

where, J_n is number of joints per meter depth, 'n' is the inclination parameter depending on the orientation of the joint β, 'r' is the roughness or joint strength parameter depending on the joint condition. The value of 'n' is obtained by taking the ratio of log (strength reduction) at $\beta = 90°$ to log (strength reduction) at the desired value of β. The joint strength parameter 'r' is obtained from a shear test along the joint and is given as $r = \tau_j/\sigma_{nj}$ where τ_j is the shear strength along the joint and σ_{nj} is the normal stress on the joint. The inclination parameter, 'n' is independent of joint frequency. The values of 'n' for various orientation angles and the joint strength parameter 'r' for various uniaxial compressive strengths of intact rock are presented by Ramamurthy (1993), based on extensive laboratory testing of rocks. So, knowing the intact rock properties and the joint factor, the jointed rock properties can be estimated using the following groups of empirical/ statistical relations.

Group 1: The jointed rock is modeled using the statistical relations given as below.

$$\sigma_{cr} = \frac{\sigma_{cj}}{\sigma_{ci}} = 0.04 + 0.89\exp\left(\frac{-J_f}{161.0}\right) \tag{4}$$

Table I Values of empirical constant a, b, c for different confining pressures.

Confining Pressure (MPa)	Value of 'a' in Equation 5	Value of 'b' in Equation 5	Value of 'c' in Equation 5
0.0	0.035	0.879	92.69
1.0	0.332	0.706	61.67
5.0	0.345	0.707	43.14

$$E_r = \frac{E_j}{E_i} = a + b\exp\left(-\frac{J_f}{c}\right) \qquad (5)$$

These relationships arrived are based on the statistical analysis of a large amount of experimental data. E_r is the tangent modulus ratio, E_j is the tangent modulus of jointed rock, E_i is the tangent modulus of intact rock, J_f is the joint factor and a, b, and c are statistical constants. The values of these constants are given in Table 1 for 0, 1.0 and 5.0MPa confining pressures. For the confining pressures other than listed in the Table 1 the values of a, b, and c have been linearly interpolated or extrapolated based on the value of a, b, and c for 0–5.0MPa confining pressure.

Group 2: The jointed rock is modeled using the following empirical relations given by Ramamurthy (1993).

$$\sigma_{c\,r} = \frac{\sigma_{cj}}{\sigma_{ci}} = \exp(-0.008\ J_f) \qquad (6)$$

$$E_r = \frac{E_j(\sigma_3 = 0)}{E_i(\sigma_3 = 0)} = \exp(-1.15 \times 10^{-2}\ J_f) \qquad (7)$$

Tangent elastic modulus of jointed rock for any other σ_3 is derived from the tangent elastic modulus of jointed rock at $\sigma_3 = 0$ using the formula given below.

$$E_j\ (\sigma_3) = \frac{E_j\ (\sigma_3 = 0)}{1 - \exp\left[-0.1\ \left(\frac{\sigma_{cj}}{\sigma_3}\right)\right]} \qquad (8)$$

Equation 7 is valid for $\sigma_3 = 0$. For different confining pressures, the elastic modulus of jointed rock is calculated using Equation 8, where σ_{3j} is obtained from Equation 6 and E_j at $\sigma_3 = 0$ is obtained from Equation 7. These relations were given by Ramamurthy and Arora (1994) based on laboratory studies on numerous artificial joints.

Group 3: The jointed rock is modeled by a set of statistical relations after Sitharam (20) as given below.

$$\sigma_{c\,r} = \frac{\sigma_{cj}}{\sigma_{ci}} = \exp(-0.0065\ J_f) \qquad (9)$$

$$E_r = \frac{E_j(\sigma_3 = 0)}{E_i(\sigma_3 = 0)} = \exp(a\ \times J_f) \qquad (10)$$

These relations are arrived at (Equations 9, 10) by fitting the equations of the same form as Ramamurthy (1993) to the experimental data using statistical analysis. The value of the empirical constant 'a' is given in tabular form (Table 2) for different confining pressures. For confining pressure other than those listed in the Table 2 the value of 'a'

Table 2 Value of empirical constant 'a' for different confining pressures.

Confining Pressure (MPa)	Value of 'a'
0.0	−0.0113
1.0	−0.0064
5.0	−0.0103
7.0	−0.0082

can be interpolated or extrapolated based on the value of 'a' for 0 to 7.0 MPa confining pressure.

3 HYPERBOLIC STRESS-STRAIN RELATIONSHIPS

The constitutive relations for the intact and jointed rocks used in the equivalent continuum analysis are described in detail in this section. The stress–strain behavior of rocks over a wide range of stress field is nonlinear and dependent upon confining pressure. The nonlinear elastic confining stress-dependent model following a hyperbolic relation proposed by Duncan and Chang (1970) is used in the present study. The material behavior of intact rock is modeled using the following non-linear relation:

$$E_r = \frac{d(\sigma_1 - \sigma_3)}{d(\varepsilon_a)} = \frac{\frac{1}{E_{ti}}}{\left[\frac{1}{E_{ti}} + \frac{\varepsilon_a R_f}{(\sigma_1 - \sigma_3)f}\right]^2} = \left(1 - R_f\left(\frac{(\sigma_1 - \sigma_3)}{(\sigma_1 - \sigma_3)f}\right)\right)^2 E_{ti} \tag{11}$$

$$E_{ti} = K P_a \left(\frac{\sigma_3}{P_a}\right)^n \tag{12}$$

$$R_f = \frac{(\sigma_1 - \sigma_3)_f}{(\sigma_1 - \sigma_3)_{ult}} \tag{13}$$

Where, E_t is the tangent modulus computed at 50 percent of failure stress for the intact rock, E_{ti} is the initial tangent modulus, σ_1 is the major principal stress, σ_3 is the confining pressure, ϵ_a is axial strain, K is the modulus number, n is modulus exponent and P_a is atmospheric pressure. $(\sigma_1 - \sigma_3)_f$ is the failure stress and $(\sigma_1 - \sigma_3)_{ult}$ is the asymptotic value of stress. The value of modulus exponent 'K' and modulus number 'n' are determined from the plots of tangent modulus versus confining pressure for the intact rock. For brittle failure in rocks the failure stress and the asymptotic value of stress are almost same so $R_f = 1$. For heavily jointed rocks, which have a similar stress-strain relationship of soils, the asymptotic value of stress and failure stress are slightly different and so R_f is taken as 0.9. For the linear stress-strain relation an $R_f = 0.0$ is taken.

4 EQUIVALENT CONTINUUM MODEL

First implementation of this equivalent continuum model has been done with finite element (Sitharam, 2001), and the numerical model has been developed from an existing finite element code for a non-linear soil–structure interaction program (NLSSRIP). As mentioned, the jointed rock is represented as an equivalent continuum whose properties have been derived from the intact rock properties and a joint factor based on the joint fabric. Nonlinearity in the finite element analysis has been incorporated in the form of material non-linearity of both the intact and jointed rocks. The incremental method is used for the solution of the non-linear problem by the finite element method. The load is increased in a series of steps or increments. Each increment is analyzed twice: the first time using the moduli values for the elements based on the stresses at the beginning of the increment; and the second time using moduli value based on the average stresses during the increment. The changes in stress and strain of the elements and changes in the nodal displacements during each increment are added to the values at the beginning of the increment. At the beginning of each new increment of loading, an appropriate modulus value is selected for each element on the basis of the values of stress or strain in that element. The non-linear stress–strain behavior is approximated by a series of straight lines. The displacements at each increment of loading are accumulated to give the total displacement at any stage of loading, and the incremental process is repeated until the total load is applied. The principal advantages of this procedure are its complete generality and its ability to provide a relatively complete description of the load deformation behavior. Initial stresses may be readily accounted for as the tangent modulus is expressed in terms of the stresses only.

In the equivalent continuum approach, the discontinuous rock body is modeled using four-node quadrilateral elements, the properties of each element being defined in terms of some combination of the properties of the intact rock and those of the joints. Non-linearity in the finite element analysis has been incorporated in the form of material non-linearity of both the intact and jointed rocks. During loading, if a rock element is found to fail in shear, this is noted, but no changes are effected, and the element is allowed to follow the hyperbolic relation as before, in keeping with the non-linear elastic formulation of the problem. During analysis, it has been found that some elements which fail at one stage recoup at a subsequent higher loading stage. If the elements fail in tension, they are assigned very small values of the elastic modulus for the subsequent loads.

In the case of explicit representation of joints, the intact rock is represented by four-node quadrilateral elements and the joints are represented by a two dimensional gap and friction elements. This element is a two-node non-linear interface element used to model node-to-node contact between two bodies with or without friction. It is represented by a pair of coupled non-linear orthogonal springs in the normal and tangential directions to the interface, which are assumed to be very stiff, relative (with stiffness, K_n and K_t) to the bodies they are attached to. The Coulomb law is used for friction. Frictionless contact may be modeled by specifying a zero coefficient of friction. The element may assume open or a closed status on relative displacement in the normal direction. The closed status may be sticking or sliding depending on whether the friction limit $\mu |f_n|$ is reached, where μ is the coefficient of friction and f_n is the normal compressive force in the gap.

Later the model has also been incorporated in the commercial finite difference code First Langrangian Analysis of Continua FLAC (Sitharam & Latha, 2002). A FISH function has been written to incorporate the joint factor model with Duncan-Chang non-linear hyperbolic relationships in FLAC. The work verifies the validity of the proposed model for different field case studies, namely two large power station caverns, one in Japan and the other in the Himalayas and the Kiirunavara mine in Sweden. Sequential excavation was simulated in the analysis by assigning the null model available in FLAC to the excavated rock mass in each stage. The settlement and failure observations reported from field studies for these different cases were compared with the predicted observations from the numerical analysis in this study. The results of numerical modeling applied to these different cases are systematically analyzed to investigate the efficiency of the numerical model in estimating the deformations and stress distribution around the excavations. Results indicated that the model is capable of predicting the settlements and failure observations made in the field fairly well. Results from this study confirmed the effectiveness of the practical equivalent continuum approach and the joint factor model used together for solving various problems involving excavations in jointed rocks.

Recently, the applicability of the model has been extended by incorporating the model in the FLAC-3D (Sitharam, 2005). Initially, it has been validated with simple element tests *i.e.* by simulating 3-D triaxial testing of rock specimens and is compared with experimental results as well as explicit modeling. Later, the applicability of the model for field problems is investigated by undertaking numerical modeling of the Nathpa Jhakri powerhouse cavern, India, considering 3-dimensional geometry and stresses along with stages of excavation.

4.1 Validation of the model

4.1.1 Equivalent continuum model

For validation purposes, the finite element analysis has been carried out for single-, multiple- and block jointed rocks (Figure 3) of sandstone, granite, Agra sandstone and gypsum plaster using the proposed model. The jointed rock properties are expressed using Equations 6–8. The properties of the intact rocks used for the simulation are presented in Table 3. The results have been plotted in the form of stress–strain curves and compared with the experimental results. The stress–strain curves compare well with the experimental results for single-, multiple- and block jointed specimens with different confining pressures and joint inclination angles. Some sample stress–strain

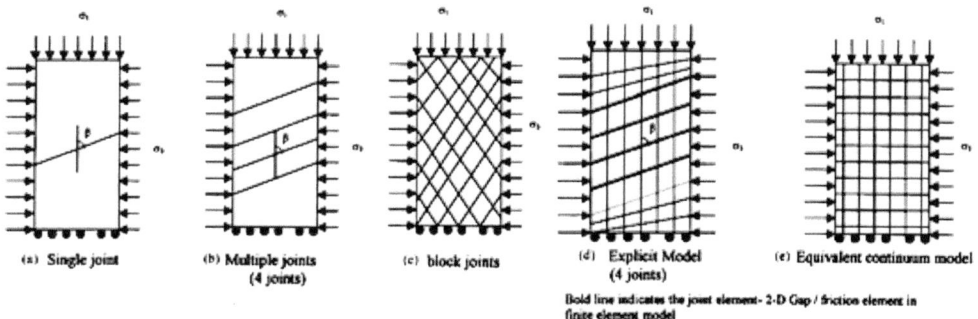

Figure 3 (a-e) Jointed rock specimens and corresponding finite element.

Table 3 Properties of intact rocks used for the numerical modeling.

Property	Jamrani sandstone	Agra Sandstone	Gypsum Plaster
Mass densitytext (KN/m^3)	22.5	22.17	1568
UCS (MN/m^2)	70	110	21
Mod. No. (K)	45000	43000	9000
Mod. Exp. (n)	0.115	0.243	0.5
Cohesion (MN/m^2)	14.0	25.5	4.0
Friction (°)	44.0	51.0	30
Classification	Hard Rock	very hard rock	Soft Rock

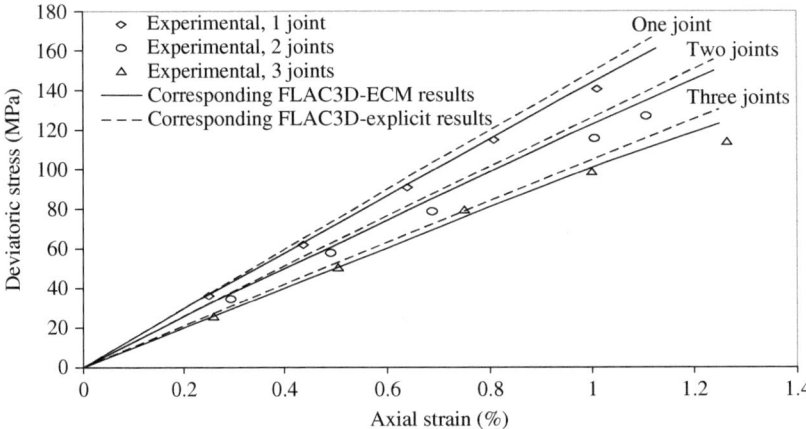

Figure 4 Stress–strain plot for different confining pressures (experimental data after Arora, 1987).

plots for multiple-jointed specimens of Agra Sandstones are shown in Figure 4 and block-jointed specimens of gypsum plaster for different confining pressures are shown in Figure 5 along with the experimental results. It can be seen from Figures 4 and 5 that the equivalent continuum model provides a good approximation of the jointed rock behavior.

4.1.2 Comparison of the equivalent continuum model with explicit modeling of joints

The results obtained using explicit modeling of joints for single-jointed specimens of sandstone and granite and multiple-jointed specimens of Agra sandstone are compared with the results obtained for the same specimens using equivalent continuum analysis. The comparison of stress–strain curves for single-jointed specimens of sandstone is given in Figure 7. The experimental results of Yaji (1984) and Arora (1987) are also plotted in the figure for comparison. Though only a few comparison plots are shown here, the trend of the results is consistent for different rocks, different joint inclination angles and confining pressures.

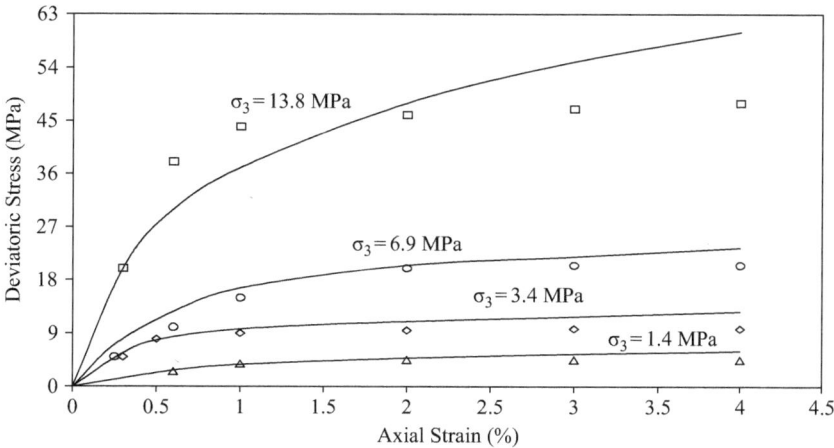

Figure 5 Stress–strain plot for different confining pressures (experimental data after Brown & Trollope, 1970).

4.2 Validation of the FLAC model

4.2.1 Equivalent continuum model

Element tests of jointed rock mass with one to four joints with different orientation and subjected to different confining pressures have been simulated using the proposed equivalent continuum model. To validate the results of equivalent continuum model, it is compared with the actual laboratory experiment results and also by explicit modeling of joints in rock. Equivalent continuum analysis has been carried out for two types of rocks namely Jamrani sandstone and Agra sandstone for both intact and jointed specimens (Arora, 1987, Yaji, 1984). The properties of the intact rocks used for the simulation are presented in Table 4. The jointed rock properties are expressed using Equations 11, 12 and 13. The stress-strain curves for each load increment has been studied for different values of confining pressure for both intact and jointed specimens with different orientations of joints.

The experimental values (Arora, 1987) were then compared with the numerical results. However, testing was performed in three different confining pressures namely 1, 2.5 and 5 MPa. Various Plot for deviatoric stress vs axial strain, at respective confining pressure, for both experimental as well as numerical testing results are shown in Figure 7. It can be seen from the results that the equivalent continuum model developed for the jointed rockmass matches well with the actual experimental test results.

4.2.2 Explicit modeling using FLAC

Explicit modeling of jointed rock sample has also been done to know the efficiency of Numerical modeling with *FLAC-3D*. Introduction of interfaces in cylindrical sample

FLAC-3D is a complicated job and creating more than 4–5 interfaces is not recommended. Here, up to four joints have been tried. Figure 6 shows the cylindrical sample with one and three joints as interfaces. The material properties for explicit modeling, which is assumed to be elastic, are given in Table 4.

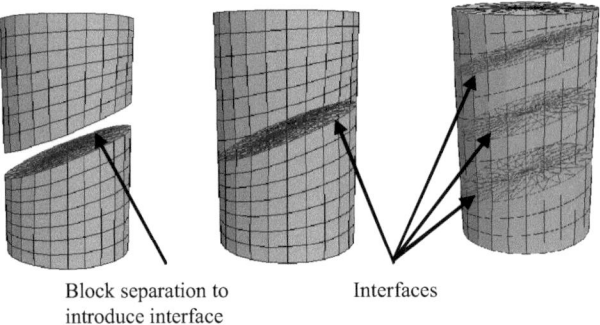

Block separation to   Interfaces
introduce interface

Figure 6 Cylindrical samples with interfaces.

Table 4 Properties of rocks for explicit modeling.

Properties	Sandstone	Agra Sandstone
Density (kN/m^3)	22.50	22.17
UCS (MPa)	70	110
Cohesion (MPa)	13.0	19.22
Friction (°)	44.0	51.0
Elastic Modulus (GPa)	5.1	20
Classification	Hard rock	Very Hard rock

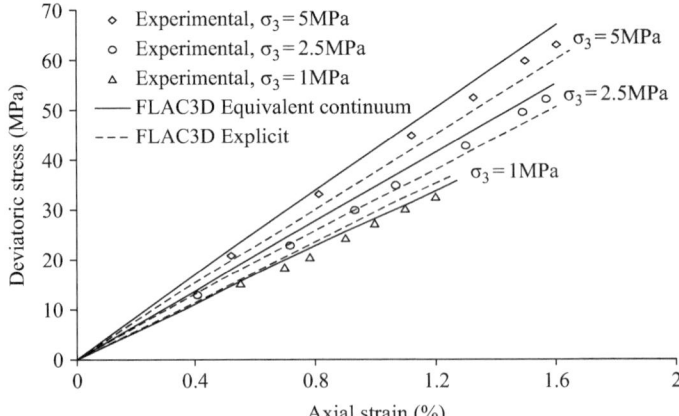

Figure 7 Comparison of numerical and experimental results of sandstone with single joint at 60 degree inclination for 1, 2.5 and 5 MPa confining pressure.

5 FIELD CASE STUDIES

5.1 Finite element analysis of Shiobara power house cavern

The equivalent continuum model developed has been applied for the analysis of a large-scale cavern in jointed rock mass for the Shiobara power station in Japan (Horii *et al.*, 1999). This case study was selected for modeling using the equivalent continuum model because the complete field information, including displacement data, was available for possible analysis and comparison with the proposed model. The Shiobara power station cavern constructed by the Tokyo Electric Company is a large cavern for a pumped storage power station with a maximum output of 900MW. The rock mass surrounding the cavern was characterized mainly as rhyolite consisting of platy and columnar joints. The cavern was located at a depth of 200m below the ground level. The cavern (Figure 8) measured 28m in width, 51m in height and 161m in length. The amount of jointed rock mass excavated due to opening of the cavern was estimated to about 190,000m³. The three in situ principal stresses were recorded as 5.0, 3.9 and 2.8MPa. The reported average intact rock compressive strength and elastic modulus were 83.3MPa and 42.1 GPa, respectively. The elastic modulus of the jointed rock mass was in the range of 3–5 GPa and the strength parameters measured for the jointed rock mass were c=1MPa and f=45°. Multi-point bore hole extensometer (MPBX) data were available at several locations along different measurement lines BI10 to BI19 (Figure 8) around the cavern for a possible comparison with our model. The cross section of the cavern along with the location of the MPBXs is as shown in Figure 8. The measurement lines along which the field displacements are available are also shown in Figure 8. The jointed rock mass surrounding the cavern has been analyzed by the finite element method using the proposed equivalent continuum approach. Equivalent material properties for jointed rock were modeled using the statistical relations given in Equations 4

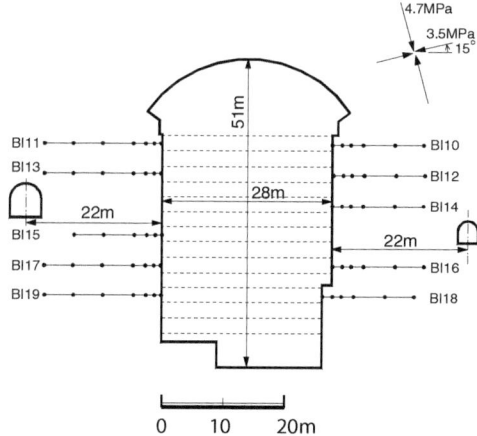

Figure 8 Cross section of cavern and location of displacement transducers.

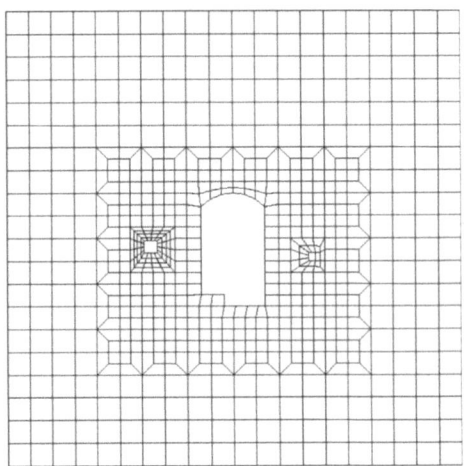

Figure 9 Finite element mesh for the cavern and the surrounding rock.

Table 5 Properties of joint sets for Shiobara power house cavern.

Joint set	Dip angle	Spacing	Joint frequency (J_n)	Joint inclination parameter(n)	Joint strength parameter(r)	Joint factor (J_f) $J_n/(n \times r)$
I	60R	30cm	17	0.46	0.9	41
II	60L	100cm	5	0.46	0.9	12
III	30L	100cm	5	0.05	0.9	111

and 5. The values for a, b, and c constants used here are the same as the ones listed in Table 1. Figure 9 shows the finite element mesh (200m × 200 m) used for the above problem (Sitharam, 2001). The problem was analyzed with the initial stresses existing in the surrounding rock. The joint properties, the dip angle and the average spacing of the three dominant joints present in the surrounding rock are given in Table 5. The joint factor for the above problem has been estimated as follows. The number of joints per meter depth in the surrounding rock which have an influence on the cavern determine the joint frequency value J_n as 5. Since the most critical joint has a joint orientation of 30°, the value of the joint inclination parameter $(n) = 0.05$ and the joint strength parameter (r) is based on the uniaxial compressive strength of the rock and is taken as r = 0.90. The cavern, along with the surrounding rock, has been analyzed using the above approach at the completion of the whole excavation. The results of the finite element analysis have been plotted in terms of relative displacement at the completion of the whole excavation along each measurement line. The measured values of relative displacement given by Horii *et al.* (1999, 2000) are also plotted for comparison. The relative displacement versus distance from the cavern wall along the measurement lines BI10, BI11, BI17 and BI18 are as shown in Figures 10–13 along with the measured values.

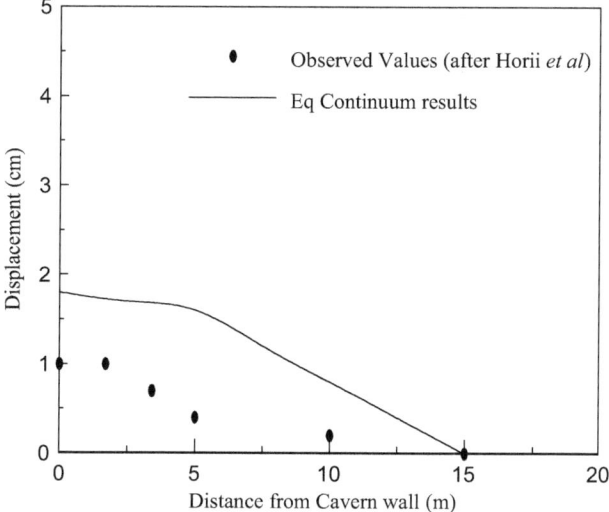

Figure 10 Measured (after Horii *et al.* [1999, 2000]) and calculated relative displacements along the measurement line BI10.

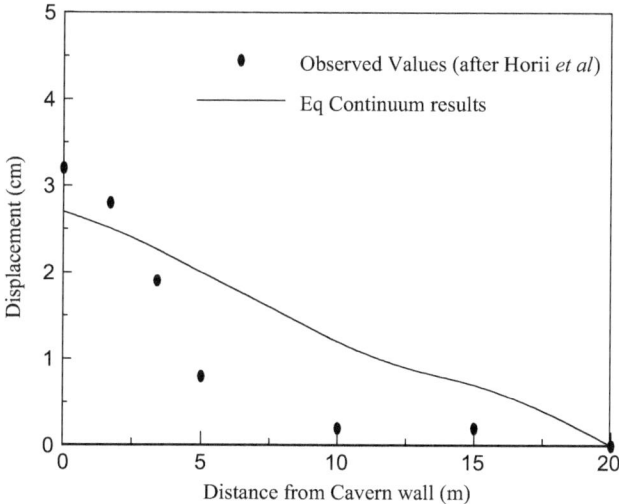

Figure 11 Measured (after Horii *et al.* [1999, 2000]) and calculated relative displacements along the measurement line BI11.

5.2 FLAC analysis of Kiirunavaara mine, Sweden

The Kiirunavaara mine, which is 4000 m long, with an average width of 90 m, is located 144 kilometers north of the Arctic Circle in the city of Kiruna in North Sweden and currently produces 20 Mt of magnetic iron ore annually. The mine was initially

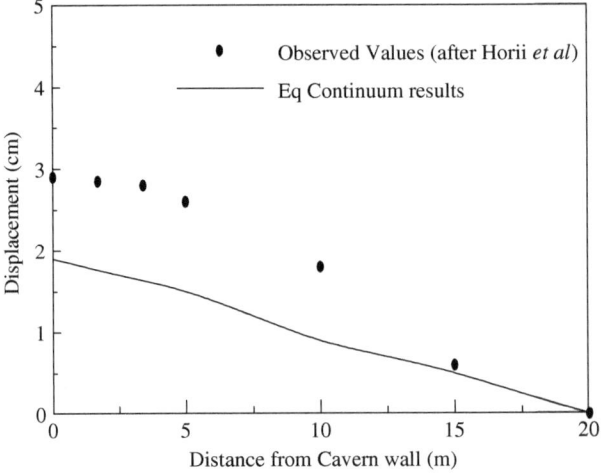

Figure 12 Measured (after Horii *et al.* [1999, 2000]) and calculated relative displacements along the measurement line BI17.

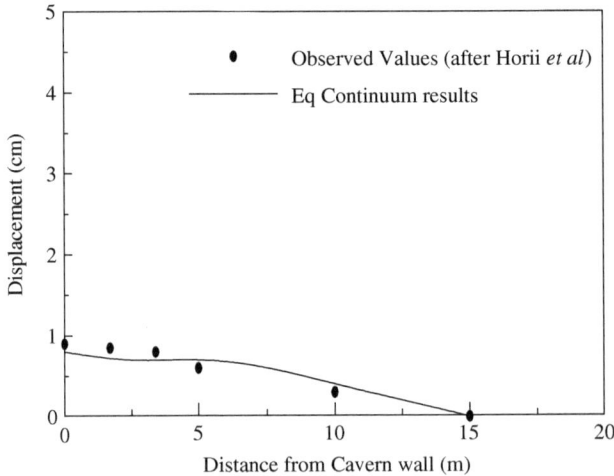

Figure 13 Measured (after Horii *et al.* [1999, 2000]) and calculated relative displacements along the measurement line BI18.

mined as an open pit, starting at the top of Kiirunavaara mountain. Underground mining using sublevel caving started in the 1950s. The mine strikes nearly north-south and dips, on average 60° to the east. The ore body is relatively strong surrounded by competent quartz porphyry on the hanging wall and syenite porphyry on the footwall. The rock mass is relatively jointed, with two to three joint sets occurring in 8 different structural domains of the mine. One joint set is oriented roughly parallel to the ore

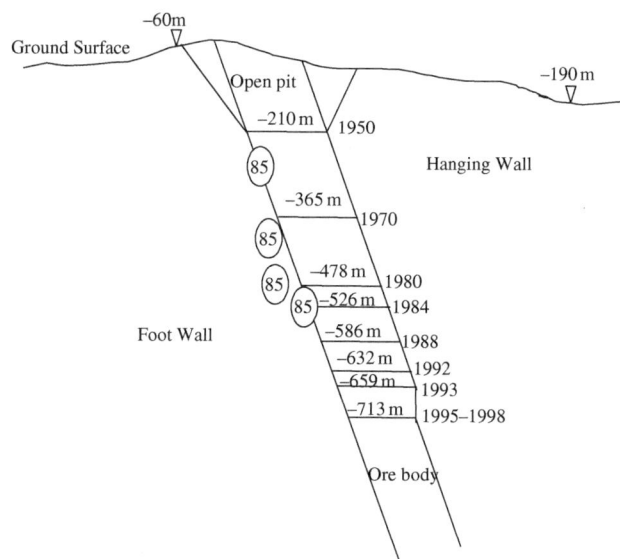

Figure 14 FLAC model for Kiirunavaara mine showing different mining levels.

body as the other two strike obliquely to the ore body and all dip fairly steeply (60° to 90°). A representative section of the mine, namely section Y2300, is selected for numerical modeling. The geometry of the mine at this section showing different mining levels is given in Figure 14. The mine has experienced large-scale stability problems. Because of sub level caving, the hanging wall continuously caves in as the ore is mined. Due to this, large subsidence of the ground surface is observed. Instabilities and large-scale failures were observed both in hanging wall and footwall of the mine. Signs of failure were first observed in the footwall in the year 1985, with more widespread cracking underground in year 1989. The locations where first cracks were observed in 1985 are mapped by Sjoberg (1999) and are presented in Figure 15. Failure surface can be drawn connecting these failure observations, which can be compared using numerical simulations from present study. Qualitative and quantitative estimation of this instability is very important as the city of the Kiruna and the railroads are located on the hanging wall side of the mine. Several researchers attempted to predict the instabilities using different approaches. Sjoberg (1999) simulated the failure using numerical model incorporating Hoek and Brown failure criterion, which involves exact evaluation of various parameters for the equivalent continuum model. The input parameters are selected for the numerical analysis from the observations reported by Lupo (1996) as given in Table 6. The input for the virgin stress distribution is taken from the average stress components presented by Sjoberg (1999) as:

$$\sigma_v = -0.027y_m - 1.62 \quad \sigma_H = -0.036y_m - 2.16 \quad \sigma_h = -0.04y_m - 2.4$$

Where σ_v is the vertical virgin stress, σ_H is the major horizontal virgin stress (Parallel to the ore body), σ_h is the minor horizontal virgin stress (Perpendicular to the ore body) in MPa and z_m is the mine level in meters (z-axis positive upwards).

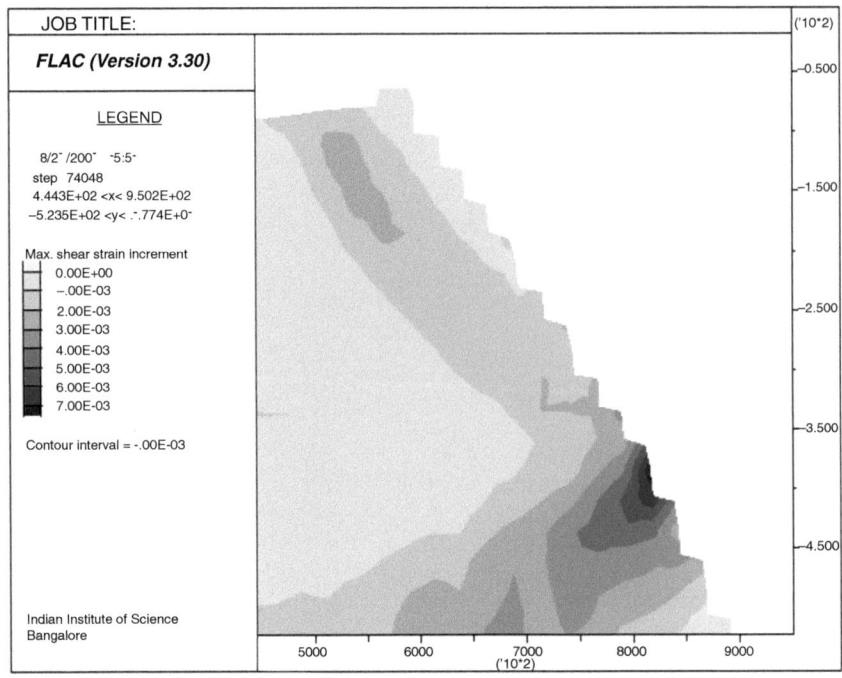

Figure 15 Shear strains for *FLAC* model for −586 m mining level.

Table 6 Input parameters for numerical modeling of Kiirunavaara mine (after Lupo, 1996).

Parameter	Hanging wall	Foot wall	Orebody
Uniaxial Compressive strength (MPa)	100	140	115
Elastic Modulus (MPa)	19000	18000	17000
Poisson's ratio	0.22	0.27	0.25
Density (kg/m^3)	2700	2800	4700
Cohesion (MPa)	1.0	0.6	1.26
Angle of internal friction	35°	35°	36°

Results from the numerical analysis showed that the type of failure occurred in this model are primarily shear failure. In continuum type of model, since the displacements are continuous, we cannot see the development of cracks as observed in the field. Failure can only be observed from the concentration of shear strains in the model. The path of concentrated shear strains represents the failure surface in the model. The failure thus simulated by the numerical model using practical equivalent continuum approach can be compared with the failure observations in the field. Shear failure was observed in the footwall of the model while excavating for the mining level of −586m, agreeing with the field observations as reported by Sjoberg (1999). Typical failure surface for the mining step of −586 m is presented in Figure 15. It can be observed from the figure that failure develops along a curved surface within the foot wall. The figure is zoomed between the excavation steps −365 m to −586 m so as to clearly visualize the failure surface. The shear strains are concentrated near the slope, at a depth of 550 m, which is comparable with the field observations presented in Figure 15. The predicted failure surface matched fairly well with the field observations.

5.3 FLAC Analysis: Nathpa-Jhakri power house cavern

Nathpa-Jhakri hydropower project in the state of Himachal Pradesh in India involved a major opening for powerhouse cavern of dimensions 216 m × 20 m × 49 m (length × width × height) at a depth of 262.5 m below the ground level (Figure 16). The opening is located in the left bank at about 500 m from the Sutlej river. The rock mass around the cavern is mainly quartz mica schist with joints and discontinuities (Bhasin, 1996). The measured in-situ stresses are 4.73MPa and 6.34 MPa in E-W and N-S and directions respectively and 5.89 MPa in vertical direction. The properties of intact rock and joints are given in Tables 7 and 8 respectively (Varadarajan, 2001).

The cavern is excavated in 12 stages. The sequence of excavation with the elevation of each excavation step and the locations of installation of multi point borehole extensometers (MPBX) for displacement measurements are shown in Figure 17. Since the cavern is symmetric, only half of the cavern is analyzed (Sitharam & Latha, 2002). The finite difference grid used for the analysis is of size 210 m × 450 m with 1320 rectangular zones. The excavation steps are simulated in the numerical analysis and the locations of the installation of extensometers are identified for obtaining the displacements for comparison with the measured displacements from instrumentation of the cavern in field. The variation of displacements with time is also obtained from

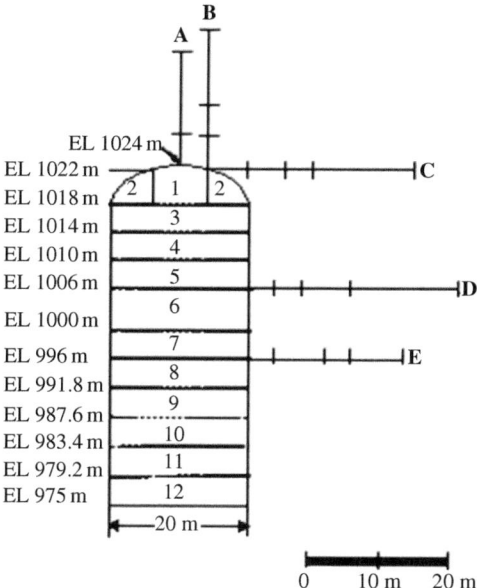

Figure 16 Excavation sequence and locations of extensometers for Nathpa-Jhakri powerhouse cavern.

Table 7 Properties of intact rock for Nathpa-Jhakri powerhouse
cavern.

Modulus number (K)	24500
Modulus exponent (n)	0.24
Cohesion (MPa)	6.38
Poisson ratio	0.29
Angle of internal friction (degrees)	22

Table 8 Properties of joints for Nathpa-Jhakri powerhouse cavern.

Frequency of joints (J_n)	5
Orientation of critical joint (β)	55°
Inclination parameter (n)	0.38
Joint strength parameter (r)	0.6
Joint factor (J_f)	21.9

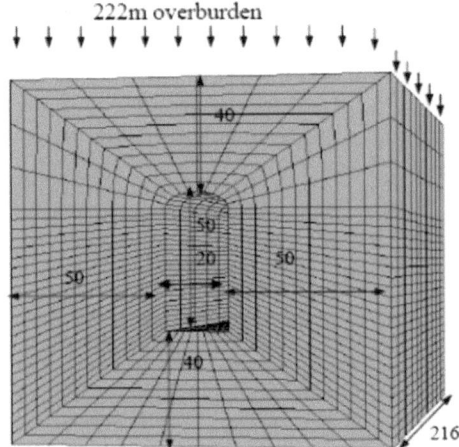

Figure 17 3D Grid of the power house cavern with FLAC-3D (All dimensions are in meters).

numerical analysis by solving for equilibrium after each excavation step. Comparison of the observed and predicted deformations along the measurement line at different locations for various excavation levels after the completion of excavation is presented in Table 9. In all the cases, the minimum value of deformation corresponds to the measuring point of the MPBX positioned close to the cavern wall and the maximum value corresponds to the farthest measuring point from the wall. It can be observed from Table 9 that the numerical model is efficient in predicting the deformations around the cavern wall. The variation of displacement with time for a particular excavation step and with progress of excavation steps is well represented in the analysis.

Table 9 Measured and predicted deformations at for Nathpa-Jhakri cavern.

Stage	Excavation level		Location of MPBX at EL (m)	Deformation along the line (mm)	
	From EL (m)	To EL (m)		Observed	Predicted
1	1024	1018	1024 (A)	13–18	10–14.0
2	1024	1018	1022 (B)	6–12	8.2–13.5
3	1018	1006	1022 (B)	−1.3 to 2.5	1–2.3
4	1006	1000	1018 (C)	1–4	1.4–3.7
5	1000	975	1006 (D)	10–45	13–42.2
6	983	975	996 (E)	1–3	1.3–4.2

Analysis of the cavern has also been extended to *FLAC-3D* for better understanding of stress and deformational behavior of the cavern. For the 3-D numerical simulation, the grid selected is shown in Figure17. As shown in Figure 17, only 40m overburden is taken into model, rest 222m overburden is taken into consideration by applying equivalent amount of pressure at the top. Analysis procedure was similar to *FLAC-2D* as mentioned above, except that here the 3-dimensinal geometry has been considered and the displacements values measured were consistent with the actual experimented values.

5.4 Seismic stability analysis of a river abutment slope

The Chenab river forms about 360m deep gorge in a V-shaped valley in the area between Bakkal and Kauri villages (Figure 18). The railway link between Katra and Laole section in Jammu and Kashmir has been planned to have a steel arch bridge on Chenab river with around 950m span at a height of 360m. The rocks present at the bridge site are heavily jointed. The stability analysis of the right abutment slope at Kauri side of Chenab river between Katra and Laole, in Jammu and Kashmir, India was simulated using FLAC as a plane strain problem. Seismic stability analyses have been

Figure 18 Slope at the Chenab river site.

Table 10 The ratings for surface weathered and fractured dolomitic limestone (NIRM Report, 2004).

Parameter	Surface weathered dolomitic limestone	Fractured dolomitic limestone
RQD	49	10
Q	6.12	1.25
RMR	48	25
GSI	43	27
J_v	20	60

Table 11 Intact rock properties for the slope at right abutment (NIRM Report, 2004).

Property	Average value
Density	2762 Kg/m³
Young's modulus	65 GPa
Poisson's ratio	0.15
UCS	115 MPa
Cohesion	44.44 MPa
Friction	22.76°
Hoek & Brown parameter 'm'	23.5

done using pseudostatic approach and also by applying complete time history of a real earthquake.

The assessment of slope stability of large slopes in jointed rocks requires the use of non-linear material models. Here, the concept of joint factor (J_f) along with hyperbolic model has been used for the study. The subsurface essentially consists of dolomitic limestone with different degrees of weathering and fracturing. The rock mass ratings for surface weathered and fractured dolomitic limestone are given in Table 10. The intact rock properties of the right abutment of the slope are presented in Table 11. Calculated equivalent rock mass properties were, cohesion 1.785 MPa, friction angle 23°, Young's modulus 4.34GPa, and Poisson's ratio 0.15. The Hoek and Brown Parameter 'm' for broken rock is 0.59 whereas the 's' value calculated for RMR of 40 is 0.00127.

The slope is simulated here, with *FLAC* (Figure 19) in plane strain using small strain mode. Relatively high discretization with approximately 2136 zones for a slope height of 359m is considered in the analyses. Horizontal displacements are fixed for nodes along the left and right boundaries while both horizontal and vertical displacements are fixed along the bottom boundary.

Slope stability analysis was done with the shear strength reduction technique, where the simulations are run for a series of trial factor of safety F_{trial} with c and φ adjusted according to the equations:

$$c^{trial} = \frac{1}{F^{trial}} c \tag{14}$$

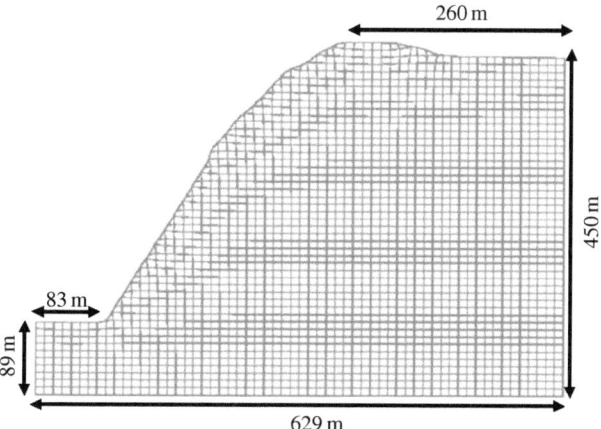

Figure 19 Numerical mesh of the slope using FLAC.

$$\phi^{trial} = \tan^{-1}\left[\frac{1}{F^{trial}}\tan\phi\right] \qquad (15)$$

The value F^{trial} at which slope fails is found efficiently in FLAC using bracketing and bisection method. The static factor of safety for the slope with rock mass properties of cohesion 1.785MPa and friction 23°has been found to be 1.86. Analyses represent the effects of earthquake shaking by pseudostatic accelerations that produce inertial forces, which act through the centroid of the failure mass. The magnitude of the horizontal pseudo static force is

$$F_h = \frac{a_h W}{g} = k_h W \qquad (16)$$

Where, a_h is horizontal pseudostatic acceleration, k_h is dimensionless horizontal pseudostatic coefficient and W, the weight of the failure mass. The horizontal pseudostatic forces are assumed to act in directions that produce positive driving moments. With the values of cohesion, 1.785 MPa and friction 23°, the factor of safety for the peak a_h = 0.31 was found to be 1.02.

Figure 20 shows the maximum shear strain rate for the slope along with velocity vectors. Failure surface is almost circular from top of the slope to the toe. Figure 21 shows the plasticity indicator plot of the slope. They reveal those zones in which the stresses satisfy the yield criterion. A failure is indicated by the contiguous line of active plastic zones that join two surfaces. The yielded zone due to shear can be seen in the Figure 21 and as observed, it also follows a circular/non-circular failure surface.

To carry out the full dynamic analysis with the real earthquake records, the corrected transverse component of acceleration time history was applied

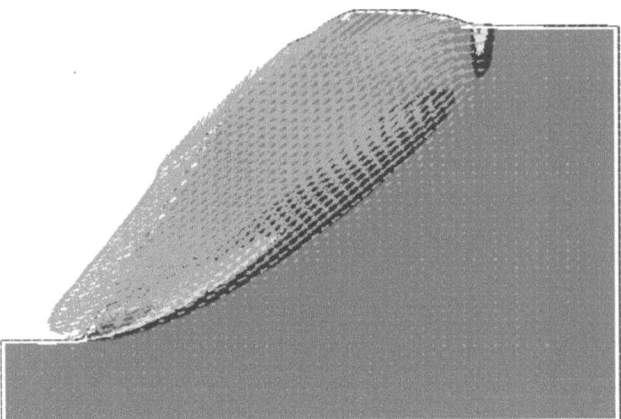

Figure 20 Critical failure surface.

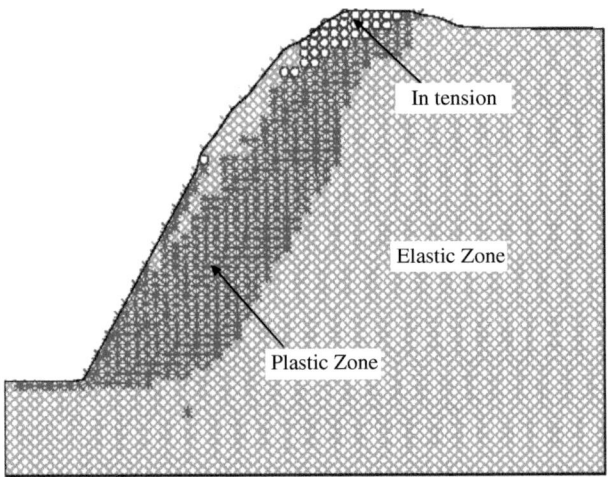

Figure 21 Plasticity indicator plot depicting the yielded zones.

(Figure 22) at the base of the slope. Free field boundary has been used to minimize the wave reflection.

It has been observed that the overall effect of continuous inertial forces may lead to accumulation of the displacement of a particular section of the slope. Once the applied ground motions generated due to inertia forces have ceased, no further deformation has occurred, as there is no marked loss in the strength of rock. The variation in displacements along the slope values were recorded and studied for different Rayleigh damping values.

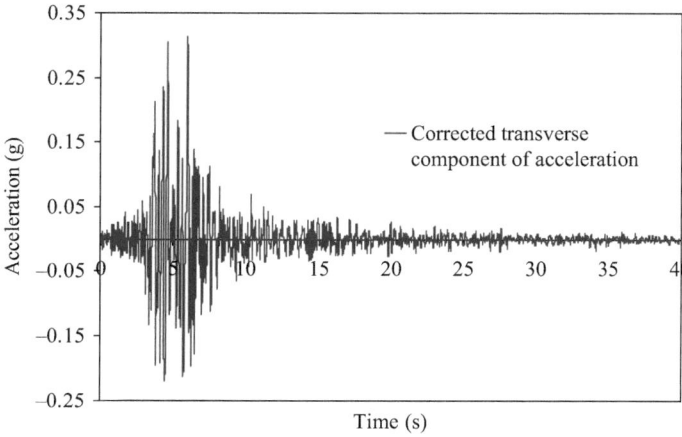

Figure 22 The corrected transverse component of acceleration time history of Uttarkashi earthquake, Oct 20,1991 02:53 IST, used in the study (Data from Dept. of Earthquake Engg., IIT Roorkee).

6 CONCLUDING REMARKS

This work presents the summary of the work on jointed rock mass along with development of practical equivalent continuum model, validation and field case studies. The statistical equations for representing the jointed rock mass properties are very simple and give a fair estimate of jointed rock properties in the absence of reliable experimental data. The accuracy of the estimation of the jointed rock properties depends upon the estimation of joint factor. Joint factor is very important for the description of jointed rock and to find the correlation between the parameters and the geological data. Since the database covers a wide range of rock properties the statistical relationships arrived more or less give a good estimate of uniaxial compressive strength and elastic modulus for all rock types.

Jointed rock has been successfully modeled as an equivalent continuum whose properties represent the properties of the jointed rock. It can be seen from the results that the equivalent continuum model developed works well for single-, multiple- and block-jointed rocks with different joint fabric and joint orientation under a wide range of confining pressures. In the equivalent continuum model, the jointed rock properties are expressed as a simple function of intact rock properties and joint properties. As the joints are not modeled separately, the analysis becomes simpler as highly complex joint fabrics can be modeled using a simple finite element mesh / finite difference *(FLAC)* unlike explicit modeling . The major advantage of explicit modeling of joints is that the mode of failure in the jointed rock mass can be reasonably estimated and the zones most susceptible to failure can be identified which is not possible in the equivalent continuum modeling. Explicit modeling of joints is efficient only when the jointed rock mass has few major joints. Equivalent continuum analysis can be applied to a single-jointed rock mass through to heavily jointed rock mass effectively without compromising the accuracy of the results. Moreover, the input data required for the equivalent continuum model are minimal. It can be inferred from the results that the equivalent continuum model developed provides a reasonable estimate of rock mass behavior in the absence of detailed experimental data. The only input data required for the analysis are the properties of intact rock and the joint properties for estimating the joint factor. It can be concluded from the analysis of various case studies that the equivalent continuum approach simplifies the field problem analysis to a large extent and gives a fair estimate of the behavior of the surrounding jointed rock with minimal input data.

Attempt was made to study the earthquake induced slope instabilities using pseudo static approach and also applying acceleration history of a real earthquake at the base of the slope. The sudden ground displacements due to the earthquake motion induce large inertia forces in the slope, which alter their direction during the earthquake.

ACKNOWLEDGMENTS

I thank Dr. Vidya Bhushan Maji, my doctoral student, for helping me to prepare this paper. Some of the results and content of this paper has appeared in my publication Sitharam (2007) "Equivalent continuum analyses of jointed Rock mass: A Practical Approach", Chapter 22 in the book Engineering in Rocks for slopes, foundations and tunnels, edited by Prof. T. Ramamurthy, Prentice Hall of India, New Delhi (ISBN 978-81-203-3275-1), pp. 518–544. Appropriate acknowledgments have been made to the reference in the text. I thank all my co-workers and doctoral students who are responsible to develop this model and for their efforts.

REFERENCES

Amadei, B. & Goodman, R. E. (1981) A 3-D constitutive relation for fractured rock masses. *Proceedings of the International Symposium on the Mechanical Behaviour of Structured Media*, Ottawa, Canada, Part B, pp. 267–286.

Arora, V. K. (1987) *Strength and deformation behavior of jointed rock*. Ph.D. thesis, Indian Institute of Technology, Delhi, India.

Barton, N. & Bandis, S. (1990) Review of predictive capabilities of JRC-JCS model in engineering practice. *Proceedings of the International Conference on Rock Joints*, Rotterdam, Balkema, pp. 603–610.

Bhasin, R. K., Barton, N., Grimstad, E., Chryssanthakis, P. & Shende, F. P. (1996) Comparison of predicted and measured performance of a large cavern in the Himalayas. *International Journal of Rock Mechanics and Mining Sciences & Geomechanics Abstracts*, 33(6), 607–626.

Brown, E. T. & Trollope, D. H. (1970) Strength of model of jointed rock. *Journal of Soil Mechanics and Foundation Division*, ASCE 96(SM2), 685–704.

Cai, M. & Horii, H. (1992) A constitutive model of highly jointed rock masses. *Mechanics of Materials*, 13, 217–246.

Chen, E. P. (1989) A constitutive model for jointed rock mass with orthogonal sets of joints. *Journal of Applied Mechanics*, Transactions of ASME, 56, 25–32.

Chenab Bridge Project Undertaking (CBPU) AFCONS Infrastructure Ltd (2005) *Geotechnical Investigation Report*, July 2005.

Cundall, P. A. (1976) Explicit finite difference methods in geomechanics. *Proceedings of the Second International Conference on Numerical Methods in Geomechanics*, Blacksburg, Virginia, Vol. I, pp. 132–150.

Duncan, J. M. & Chang, C. Y. (1970) Non-linear analysis of stress and strain in soil. *Journal of Soil Mechanics and Foundation Engineering*, ASCE, 5, 1629–1652.

Duncan, J. M. & Goodman, R. E. (1968) *Finite element analysis of slopes in jointed rocks*. U.S. Army corps of Engineers Report TR, No. 1–68.

Einstein, H. H. & Hirschfeld, R. C. (1973) Model studies in mechanics of jointed rocks. *Journal of Soil Mechanics and Foundation Division*, ASCE, 99, 229–248.

Fosum, A. F. (1985) Effective elastic properties for a randomly jointed rock mass. *International Journal of Rock Mechanics and Mining Sciences & Geomechanics Abstracts*, 22, 467–470.

Gerrard, C. M. (1982) Elastic models of rock masses having one, two and three sets of joints. *International Journal of Rock Mechanics and Mining Sciences & Geomechanics Abstracts*, 19, 15–23.

Horii, H., Yoshida, H., Uno, H., Akutagawa, S., Uchida, Y., Morikawa, S., Yambe, T., Tada, H., Kyoya, T. & Fumio, I. (1999) Comparison of computational models for jointed rock mass through analysis of large scale cavern excavation. *Proceedings of the Ninth International Congress on Rock Mechanics*, Paris, France, ISRM, Vol. 1, pp. 389–393.

Itasca.FLAC Version 3.3 (1995) *User's manuals*, Vol. 1–4. Itasca Consulting Grap Inc., Minneapolis, MN.

Kawamoto, T., Ichikawa, Y. & Kyoya, T. (1988) Deformation and fracturing behavior of discontinuous rock mass and damage mechanics theory. *International Journal of Numerical and Analytical Methods in Geomechanics*, 12, 1–30.

Lupo, J. F. (1996) *Evaluation of deformations resulting from to mass mining of an inclined ore body*. Ph. D thesis, Colorado School of Mines.

NIRM (2004) *Geotechnical mapping at Chenab bridge site near Bakkal and Kauri villages in Udhampur district, J&K, India*. Technical Report, March.

Oda, M., Yamabe, T., Ishizuka, Y., Kumasaka, H. & Tada, H. (1993) Elastic stress and strain in jointed rock masses by means of crack tensor analysis. *Rock Mechanics and Rock Engineering*, 26, 89–112.

Ramamurthy, T. (1986) Stability of rock mass. *8th Indian Geotech Society Annual Lecture, Indian Geotechnical Journal*, 16(1), 1–74.

Ramamurthy, T. (1993) Strength and modulus responses of anisotropic rocks. In: Hudson J.A. (ed.) *Comprehensive Rock Engineering*. Pergamon Press, UK, Vol. 1, no. 13, pp. 313–329.

Ramamurthy, T. (2001) Shear strength response of some geological materials in triaxial compression. *International Journal of Rock Mechanics and Mining Sciences*, 38, 683–697.

Ramamurthy, T. (2004) A geo-engineering classification for rocks and rock masses. *International Journal of Rock Mechanics and Mining Sciences*, 41, 89–101.

Ramamurthy, T. & Arora, V. K. (1993) A classification for intact and jointed rocks. *Proceedings of International Symposium on Geotechnical Engineering of Hard Soils – Soft Rocks*, Athens, Balkema, 1, 235–242.

Ramamurthy, T. & Arora, V. K. (1994) Strength predictions for jointed rocks in confined and unconfined states. *International Journal of Rock Mechanics and Mining Sciences*, 13(1), 9–22.

Roy, N. (1993) *Engineering behavior of rock masses through study of jointed models*. Ph.D. thesis, Indian Institute of Technology, Delhi, India.

Singh, B. (1973a) Continuum characterization of jointed rock masses Part I – the constitutive equations. *International Journal of Rock Mechanics and Mining Sciences & Geomechanics Abstracts*, 10, 311–335.

Singh, B. (1973b) Continuum characterization of jointed rock masses Part II – significance of low shear modulus. *International Journal of Rock Mechanics and Mining Sciences & Geomechanics Abstracts*, 10, 337–349.

Sitharam, T. G. (2007) Equivalent continuum analyses of jointed Rock mass: A Practical Approach. In: T. Ramamurthy (ed.) *Engineering in Rocks for Slopes, Foundations and Tunnels*. Prentice Hall of India, New Delhi, pp. 518–544.

Sitharam, T. G. & Latha, G. M. (2002) Simulation of excavation in jointed rock masses using practical equivalent continuum model. *International Journal of Rock Mechanics and Mining Sciences*, 39, 517–525.

Sitharam, T. G., Sridevi, J. & Shimizu, N. (2001) Practical equivalent continuum characterization of jointed rock masses. *International Journal of Rock Mechanics and Mining Sciences*, 38, 437–448.

Sjoberg, S. (1999) *Analysis of large scale rock slopes*. Ph. D. thesis, Lulea University of Technology, Sweden.

Sridevi, J. & Sitharam, T. G. (2000) Analysis of strength and moduli of jointed rocks. *Geotechnical and Geological Engineering*, 18(1), 3–21.

Sridevi, J. & Sitharam, T. G. (2003) Characterization of strength and deformation of jointed rock mass based on statistical analysis. *International Journal of Geomechanics*, 3(1), 43–54.

Varadarajan, A., Sharma, K. G., Desai, C. S. & Hashemi, M. (2001) Analysis of a powerhouse cavern in the Himalayas. *International Journal of Geomechanics*, 1(1), 109–127.

Vidya Bhushan Maji (2007) *Strength and deformation behaviour of jointed rocks: An equivalent continuum model*. Ph.D. thesis, Indian Institute of Science, Bangalore, India.

Wei, Z. Q. & Hudson, J. A. (1986) Moduli of jointed rock masses. *Proceedings of the International Symposium on Large Rock Caverns*, Helsinki, pp. 1073–1086.

Yaji, R. K. (1984) *Shear strength and deformation of joined rocks*. Ph.D. thesis, Indian Institute of Technology, Delhi, India.

Yoshida, H. & Horii, H. (2000) Micro-mechanics based continuum analysis for the excavation of large-scale underground cavern. *Proceedings of SPE/ISRM Rock Mechanics in Petroleum Engineering*, Vol. 1, pp. 209–218, 1998.

Zienkiewicz, O. C., Kelly, D. W. & Bettess, P. (1977) The coupling of the finite element method and boundary solution procedures. *International Journal of Numerical Methods in Engineering*, 11, 355–375.

Back Analysis

Chapter 7

Back analysis in rock engineering practice

S. Sakurai
Kobe University, Kobe, Japan

Abstract: This paper deals with the use of back analyses for interpreting the results of field measurements carried out during the excavation of rock structures such as tunnels and slopes. Field measurements are important to the rational design and construction of these structures. However, the measurement data are only numbers unless they are properly interpreted. Back analyses play a vital role in interpreting the measurement data properly. It should be noted that back analyses are not simply backward calculations. They must be developed on the basis of an entirely different concept in an analysis in which the mechanical modeling of the materials should not be assumed, but should be back calculated considering the results of field measurements. In this paper, an anisotropic parameter is introduced to simulate the non-linear behavior of geomaterials. By using this anisotropic parameter, a new constitutive equation is proposed, which is applicable to back analyses for interpreting the field measurement data. In addition, since the failure of structures cannot be predicted only through numerical analyses, the critical strain and the critical shear strain of geomaterials are proposed. They can be easily applied to assess the stability of tunnels and slopes. In slope engineering, the factor of safety is popularly used for assessing stability, while in tunneling practice the factor of safety is hardly ever used. This is because the factor of safety always becomes Fs = 1.0 if a plastic zone appears around the tunnels. To overcome this problem, hazard warning levels are proposed on the basis of the critical strain and the critical shear strain to assess the failure of tunnels. Some example problems are given to show just how to employ the hazard warning levels to assess the failure of tunnels.

1 INTRODUCTION

The mechanical properties of rock masses, such as the modulus of deformability and the strength parameters, should be given as input data for the design of rock structures, like tunnels, underground caverns, slopes, etc. In addition, the initial state of stress of the rock masses is also important input data for the design of these structures. In order to determine both the mechanical properties and the initial stress, various laboratory and field tests have been developed. Moreover, many sophisticated computer programs have been created to achieve rational designs for rock structures. However, practicing engineers are aware that the real behavior of structures often differs from the one predicted with sophisticated computer programs. This difference may be due to the fact that modeling rock masses and determining their input data are extremely difficult tasks because of the various uncertainties involved in the complex geological and geomechanical characteristics of rock masses.

To overcome these difficulties, the deformational behavior of structures is measured during the construction stage, and the design parameters used in the original design can be re-evaluated and, if necessary, the original design and the construction method can be modified based on the field measurement results. This construction procedure is called the "observational method"; it is a powerful technique for the construction of rock structures. This method has been especially popular for the excavation of tunnels and slopes, while it has not been popular for the construction of bridges or buildings because all the input data for their design are well documented at the design stage.

It should be noted that field measurement data are only numbers unless they are properly interpreted. In order to properly interpret the measurement results, numerical analyses are a powerful tool. The input data are measured values, such as displacement, strain, stress and pressure, while the output results are the mechanical constants of rocks, the initial state of stress, permeability, etc. This is exactly a reverse calculation (called a "back analysis") as compared to a forward calculation (called a "forward analysis" in this paper and adopted for analyses done at the design stage).

In a forward analysis, the constitutive equations play an extremely important role because the accuracy and the reliability of the calculation results depend entirely on what constitutive equations are used. Therefore, in forward analyses, the modeling of the mechanical characteristics of the materials is a crucial issue, and the constitutive equations are discussed on the basis of the theory of solid mechanics, particularly the theory of plasticity. However, it should be noted that the theory of plasticity was originally developed for metal, for which the normality rule holds, and that the associated flow rules, which are related to a yielding criterion, can be derived by considering Drucker's definition for stable materials. On the other hand, for frictional materials, like soils and rocks, the normality rule does not hold so that flow rules are not associated with a yielding criterion. Hence, they are called non-associated flow rules.

Non-associated flow rules require both a plastic potential function and a yielding criterion, which contain many mechanical parameters. In forward analyses, all these mechanical parameters should be evaluated prior to the design of the structures. However, it is not an easy task to determine all these parameters because of the complexity of the geological and geomechanical characteristics of rock masses. Hence, it is obvious that the accuracy and the reliability of forward analyses, applied to predict the behavior of rock masses, are questionable. Nevertheless, in forward analyses, there is no other way than to adopt the constitutive equation with the non-associated flow rules. In back analyses, on the other hand, we have measured displacements in which valuable information on the mechanical characteristics of *in situ* rock masses must be included implicitly. Therefore, if we consider them in back analyses in a proper manner, we may have the chance to develop constitutive equations for rock masses only for use with back analyses.

One of the objectives of back analyses is to re-evaluate the mechanical parameters of rock masses used at the design stage. In a back analysis, however, the mechanical model (constitutive equation) of a rock mass is usually assumed to be the same as that used in the forward analysis performed at the design stage. For instance, if an elastic model is assumed, the modulus of elasticity and Poisson's ratio will be obtained by a back analysis. However, if an elasto-plastic model is assumed, then the mechanical parameters of cohesion and the internal friction angle will be obtained in addition to the modulus of elasticity and Poisson's ratio, even though the identical input data (measurement results) are used. It is clear that back analyses can derive the correct mechanical parameters of rock masses only if the assumed mechanical model is "true". However, it is almost impossible to assume a true mechanical model for a rock mass. This implies that the mechanical model should not be assumed, but should be back-analyzed from field measurement results. Consequently, we must not use the same constitutive equations for back analyses as those used for forward analyses at the design stage.

Another important objective of back analyses is to predict the possible occurrence of the unpredictable catastrophic failures of structures from field measurement results. Since various types of unpredictable catastrophic failures may occur during construction, due to the various uncertainties involved in rock masses, back analyses should be able to predict all types of failure before the failure occurs. In addition, back analyses must be performed immediately after taking the field measurement data, so as to assess the stability of the structures without delay. To meet this requirement, the number of parameters in the mechanical model for the back analyses should be small, and the calculation algorithm should be simple enough to enable the quick performance of the back analyses, so that the adequacy of the countermeasures can be evaluated in order to avoid any serious failure of the rock structures.

It is not an easy task, however, to predict the failure of rock structures only by numerical analyses, because in predicting the failure of structures, the failure criteria of rock masses should be given as input data for the analyses. The failure criteria, in general, require the strength of rock masses in terms of stress, which can hardly ever be determined by a small specimen in the laboratory because of the scale effect. To overcome this difficulty, Sakurai (1981) and Sakurai *et al.* (1993) proposed "critical strain" and "critical shear strain" for the uniaxial and the triaxial states of stress, respectively. The failure criteria are based on both the critical strain and the critical shear strain; they are greatly advantageous in that there is no need to consider the scale effect. Therefore, the failure criteria determined by a small specimen in the laboratory can be used as the failure criteria for *in situ* rock masses.

2 BACK ANALYSIS REVIEW

In the early days of back analysis, various terms, such as identification, characterization, inverse analysis, etc., were used. The name "back analysis" firstly appeared in a paper published by Kirsten (1976). Ever since that time, various names have been used by different authors. Nevertheless, the term "back analysis" gradually became popular, and it is now commonly used in the rock engineering community. Various types of back analysis procedures have been extensively developed in Geomechanics, ranging from simple elastic problems to far more complex nonlinear problems, and many papers related to back analysis have been published with particular reference to the interpretation of field measurement results (Gioda & Sakurai, 1985).

In the early stages, it was thought that back analyses were mathematical problems. Even if a simple elastic model is assumed, however, back analyses are highly nonlinear problems. Therefore, the main interest of researchers has been to solve such nonlinear problems and to obtain stable solutions with high accuracy. Back analysis procedures are generally classified into two categories: the inverse approach and the direct approach (Cividini *et al.*, 1981). In the inverse approach, the mathematical formulation is just the reverse of that in an ordinary analysis, although the governing equations are identical. It must be noted that the number of measured values should be greater than the number of unknown parameters, so that optimization techniques can be employed to back calculate the unknown parameters. In the case of a ground represented by a simple mechanical model with simple geomechanical configurations, the closed-form solutions in the theory of elasticity and plasticity may be used. On the other hand, for a ground with an arbitrary shape under a more complex geological and geomechanical environment, numerical methods, such as FEM, BEM, FDM, etc., seem to be more promising. For example, Kavanagh (1973) proposed a back analysis formulation based on FEM which may make it possible to obtain the material constants not only for isotropic materials, but also for inhomogeneous and anisotropic materials from both measured displacements and strains.

Gioda (1980) modified Kavanagh's algorithm to back calculate both the bulk and shear moduli by applying static condensation and the least squares method. In order to obtain the material constants, the displacements alone are sufficient. However, to identify the load conditions in addition to the material constants, the measurements of not only the displacements, but also the values for the loads and pressures, are necessary. For this, a unique and general approach to back calculating the load parameters was formulated by Gioda and Jurina (9). Sakurai & Takeuchi (1983) formulated a back analysis algorithm based on the inverse method. By assuming homogeneous and isotropic linear elastic materials, it can determine the strain distribution around a tunnel based on the data from a limited number of measured displacements. This algorithm was extended by Feng & Lewis (1987) in such a way that it could take into account a linearly changing ground stress field, material nonlinearity and the effect of the advancing tunnel face with respect to the installation time of the monitoring devices. It should be pointed out that the inverse approach to back analysis has a great advantage in engineering practice because, in general, no iteration is required. Thus, the computation time can be reduced. However, trouble is often encountered with the inverse approach to obtaining a numerically stable solution for widely scattered values of measurement data which are commonly found in geotechnical engineering problems. Difficulty is also encountered when applying this method to nonlinear problems.

On the other hand, the direct approach is based on an iterative optimization procedure which corrects the trial values of unknown parameters in such a way that the discrepancy between the measured and the computed quantities is minimized. The advantages of this approach are: (1) It may be applied to nonlinear problems without having to rely on a complex mathematical background; and (2) Standard algorithms of mathematical programming, such as the Simplex and Rosenbrock methods, can be used. Gioda & Maier (1980) demonstrated the applicability of the direct method to back calculate the nonlinear material parameters and the load conditions using a numerical example of a pressure tunnel test. Cividini *et al.* (1985) stated that the direct approach could be employed to determine the time-dependent material constants by applying convergence displacement measurement data taken at various stages of the tunnel construction. Shimizu & Sakurai (1983) proposed a back analysis method using the boundary element method (BEM) to determine both Young's modulus and the *in situ* stress from measured displacements. Since the back analysis was conducted with the displacements measured in an underground powerhouse cavern during its excavation, the back-calculated values are those for a large extent of rock masses. Yang & Sterling (1989) also proposed a unique back analysis method for determining both the *in situ* stress and the elastic properties of rock masses using the fictitious stress boundary element method. The back analysis of the *in situ* stress in a nonlinear material was formulated by Zhang *et al.* (1988) using an iterative back analysis algorithm based on the boundary element method. As mentioned above, the direct approach has a great advantage in that it can be easily applied to nonlinear, elasto-plastic material problems. However, this method requires rather time-consuming computations since a large amount of iteration is usually involved.

Both the inverse and direct methods are based on a deterministic concept and provide precise values for material constants and load parameters. However, it is often difficult to determine these values precisely because of the various uncertainties which are usually involved in rock engineering problems. To overcome this difficulty, a stochastic approach is preferable as it is capable of taking these uncertainties into account. The most advantageous feature of this approach is that the final results are expressed in statistical terms, such as mean and variance.

In any case of the back analyses described above, either mechanical models are assumed or the same mechanical models as those used at the design stage are also adopted in the back analyses. However, Sakurai (1997) warned that in back analyses a mechanical model should not be assumed, but should be identified by back analyses. This is due to the fact that, in forward analyses, a one-to-one relationship between the input data and the output results is guaranteed, while in back analyses, it is questionable. This is because the input data, *e.g.*, measured displacements, are obtained at a particular site, so that the mechanical model should be a real one representing the particular site. However, it is almost impossible to constitute the real mechanical model of *in-situ* geomaterials because of their geological and geomechanical complexity. In order to overcome this problem, Sakurai & Shinji (2005) proposed the anisotropic parameter (anisotropic damage parameter) which is easily applicable to the back analyses of tunnels and slopes constructed in grounds consisting of any non-homogeneous and/or anisotropic nonlinear geomaterials. Moreover, it is not necessary to assume the type of mechanical modes, *i.e.*, strain hardening, perfectly plastic or strain softening, which can be automatically determined by back analyses. Sakurai *et al.* (2009) also proposed a back analysis method for slopes, which can determine the factor of safety from measured displacements. It should be emphasized that the critical shear strain proposed by Sakurai *et al.* (1993) makes it possible to back calculate the factor of safety from the measured displacements.

It is obvious that rock masses are extremely complex nonlinear systems that include many parameters. In order to solve these complex systems, Feng *et al.* (2000) proposed a new displacement back analysis approach which is based on a combination of a neural network, an evolutionary technique and numerical analysis methods to identify the mechanical parameters. The method has been successfully applied to the Three Gorge Project permanent shiplock to estimate the mechanical parameters of rock masses. Since rock masses contain various uncertainties and complexities, it is hard to create true mechanical models. Therefore, the neural networks must have a strong capability for self-learning and nonlinear relations between the mechanical parameters and the displacements of rock masses. Feng *et al.* (2004) also proposed another displacement back analysis method to identify the mechanical parameters based on hybrid intelligent methodology, such as the integration of evolutionary support vector machines (SVMs), numerical analysis and a genetic algorithm. Feng *et al.* (2006) also proposed a new identification method based on hybrid genetic programming with the improved particle swarm optimization (PSO) algorithm, which was used to identify the visco-elastic models for rocks.

Feng & Hudson (2010) discussed how to establish the necessary quality and quantity of information required for modeling and designing in rock engineering. In rock engineering projects, observational methods are the most suitable methodology where design, construction and monitoring must be closely related to each other. It is obvious that back analysis plays an extremely important role in these observational methods. It is also important to feed the information obtained *in situ* back into the design and construction without delay.

3 CONSTITUTIVE EQUATION

3.1 Constitutive equations in back analysis

As has already been mentioned, constitutive equations for frictional materials, such as rocks and soils, have been developed by considering the non-associated flow rules in which both the yielding criterion and the plastic potential function play important roles. They require various mechanical parameters of the materials; this causes a complexity to the constitutive equations. (In forward analyses, there is no other way than to use constitutive equations based on the non-associated flow rules.)

In back analyses, however, constitutive equations based on the non-associated flow rules are hardly ever applied to engineering practice, because the number of mechanical parameters is too large to determine all of them from field measurement results. To overcome this difficulty, we have to develop simple constitutive equations. In addition, all the mechanical parameters should be easy to back calculate by the measurement data immediately after the measurements have been taken during the construction.

Field measurement data, such as measured displacements, can be used as input data in back analyses, while the displacements are just one type of output results in forward analyses. Since the measured displacements contain useful information for understanding what is really going on in *in situ* rock masses, we may have a chance to develop a simple constitutive equation by employing the measured displacements. It should be noted that the new constitutive equation does not necessarily follow the conventional theory of plasticity, but was developed on the basis of a completely different idea for the mechanism of the nonlinear mechanical behavior of the materials.

3.2 Anisotropic parameter and anisotropic damage parameter

In order to develop a simple constitutive equation for use in back analyses, we have carried out simple shear tests on sand in a laboratory, and have carefully investigated the mechanism of the mechanical behavior of the frictional materials. The test results reveal that the nonlinear behavior of the materials is mainly caused by the reduction in shear modulus G along the mobilized sliding planes, while both Young's modulus E and Poisson's ratio v never change even after the yielding point and no matter how large of a maximum shear strain occurs (Sakurai & Akayuri, 1998).

Considering the results of the laboratory tests, we have defined an anisotropic parameter m as the ratio of the shear modulus to Young's modulus, as shown in Equations 1 and 2.

$$m = G/E = 1/2(1 + v) \qquad \gamma_S \leq \gamma_L \tag{1}$$

$$m = \frac{1}{2(1 + v)} \exp\{-\alpha(\gamma_S - \gamma_L)\} \quad \gamma_S > \gamma_L \tag{2}$$

where

G: shear modulus, E: Young's modulus, v: Poisson's ratio, γ_S: maximum shear strain on the potential sliding plane, γ_L: maximum shear strain at the elastic limit and α: a material constant.

The laboratory experiments demonstrate that anisotropic parameter m, for soils and soft rock, is given as a function of the maximum shear strain, as shown in Figure 1.

In order to give a more precise physical meaning to the anisotropic parameter, "anisotropic damage parameter d" is defined, as shown in Equation 3.

$$d = 1/2(1 + v) - m \tag{3}$$

where v: Poisson's ratio and m: anisotropic parameter.

It is obvious that the anisotropic parameter becomes $m = 1/2(1 + v)$ in an elastic state, resulting in the anisotropic damage parameter becoming $d = 0$ in the elastic state. In other words, no damage occurs until the materials reach their yielding point. When the materials reach the yielding point, anisotropic damage parameter d starts

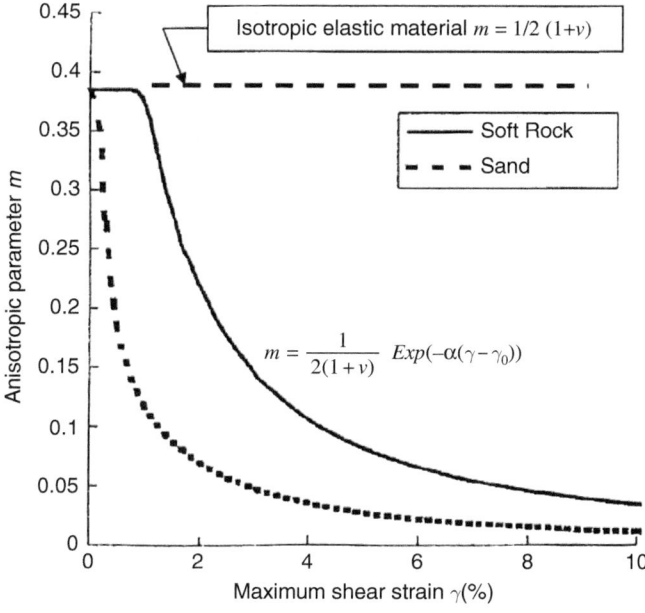

Figure 1 Relationship between anisotropic parameter and maximum shear strain.

to mobilize along a slip plane and increases with an increase in the maximum shear strain, while Young's modulus and Poisson's ratio always remain constant. Thus, anisotropic damage parameter d implies the indication of strain-induced damage and is expressed by a function of the maximum shear strain $d = f(\gamma)$. It should be noted that anisotropic damage parameter d increases monotonically with an increase in the maximum shear strain. Thus, function $f(\gamma)$ is a monotonically increasing function with the maximum shear strain, as shown in Equation 4.

$$\left.\begin{array}{ll} f(\gamma) = 0 & \text{for } \gamma \leq \gamma_0 \\ f(\gamma) = \left(\dfrac{1}{2(1+v)} - \beta\right)[1 - Exp\{-\alpha(\gamma - \gamma_0)\}] & \text{for } \gamma > \gamma_0 \end{array}\right\} \qquad (4)$$

where α, β: material constants
  γ_0: critical shear strain (Sakurai, 1990).

3.3 Stress–strain relationship in plane strain condition

Considering anisotropic parameter m, a stress-strain relationship (constitutive equation) can be expressed, as shown in Equation 5, in the local coordinate system shown in Figure (2), which is for the plane strain condition. It should be noted that this equation is only valid for local coordinates. The x'-axis should be along the mobilized sliding plane. Angle θ, indicating the sliding plane, should be determined by a back analysis of the measured displacements. The stress-strain relationship for the global coordinates should be transformed with the well-known transformation matrix, as shown in Equation 7.

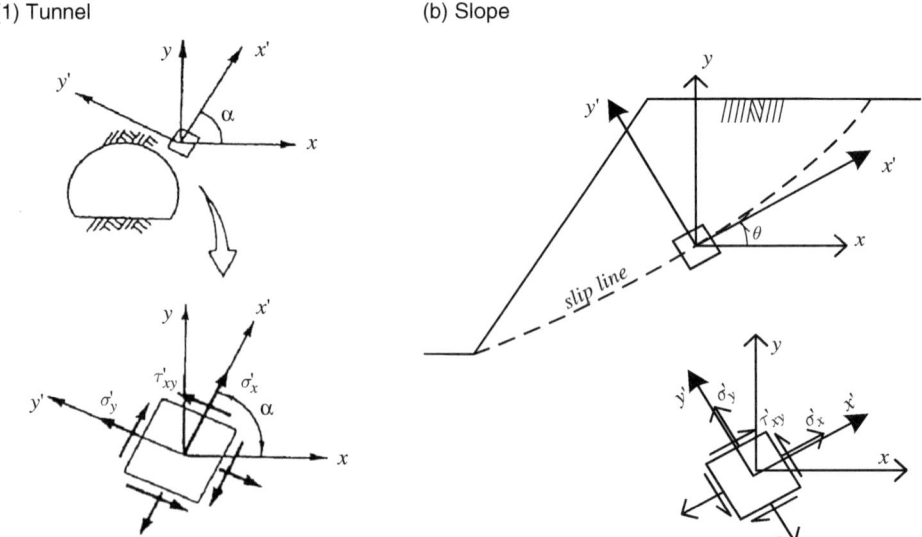

Figure 2 Local and global coordinates.

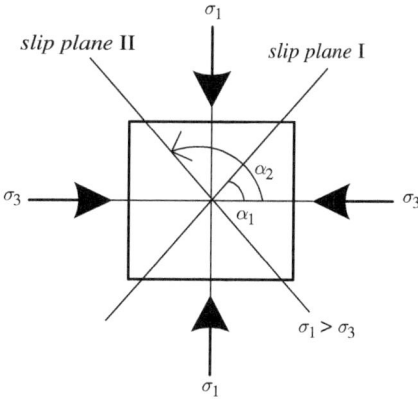

Figure 3 Stress state under plane strain condition ($\sigma_3 = 0$).

$$\{\sigma'\} = [D']\{\varepsilon'\} \tag{5}$$

where

$$[D'] = \frac{E}{1 - v - 2v^2} \begin{bmatrix} 1 - v & v & 0 \\ v & 1 - v & 0 \\ 0 & 0 & m(1 - v - 2v^2) \end{bmatrix} \tag{6}$$

where

E: Young's modulus (always constant), v: Poisson's ratio (always constant) and *m*: *a*nisotropic parameter.

$m = 1/2(1 + v) - d$ In an elastic state $m = G / E$

No dilatancy is assumed.

In the global coordinate, $[D']$ can be transformed as follows:

$$[D] = [T][D'][T]^T \tag{7}$$

where $[T]$: transformation matrix.

3.4 Numerical simulation of uniaxial compressive loading

It is demonstrated by a numerical simulation that both anisotropic parameter *m* and anisotropic damage parameter *d* can be used to simulate the nonlinear behavior of an elasto-plastic material (Sakurai & Shinji, 2005).

The input data for the numerical simulation are as follows:

Young's modulus $E = 1000$ MPa, Poisson's ratio $v = 0.3$, the yielding point is 1.0% in axial strain and the values for *m* and *d* are given to represent the four different cases, *i.e.*, (1) strain hardening, (2) perfectly plastic, (3) strain softening with residual strength and (4) strain softening with no residual strength, as shown in Figure 4(b).

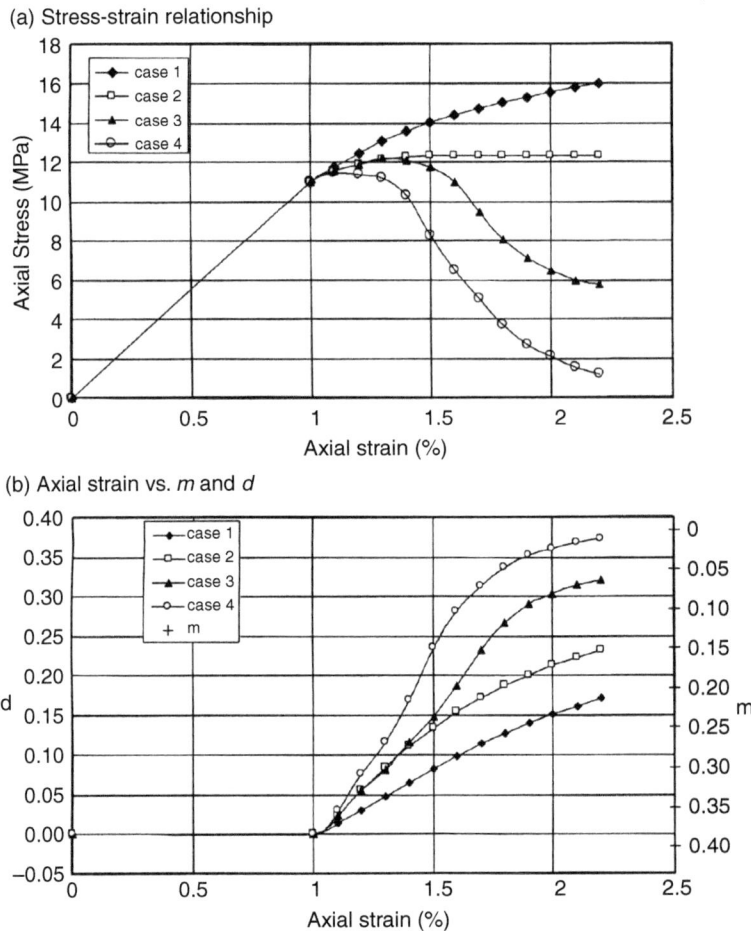

(a) Stress-strain relationship

(b) Axial strain vs. *m* and *d*

Figure 4 Stress-strain relationship under uniaxial compressive state (a) Stress-strain relationship corresponding to different values for *m* and *d*. (b) Values for *m* and *d* vary with increase in axial stress.

It is obvious from the figure that the anisotropic damage parameter is $d = 0$ in an
elastic state, but it gradually increases with an increase in axial strain when the axial
strain is beyond the yielding point of 1.0%

The results of the numerical simulation for the stress-strain relationship under
a plane strain condition are shown in Figure 4(a).

It is surprising to see the results of this simulation, namely, that the three types of
elasto-plastic mechanical behavior, strain-hardening, perfectly plastic and strain-
softening, can be easily simulated only by changing the values for *m* and *d*, both of
which are simple monotonic functions varying with the increase in axial strain, as
shown in Figure 4(b).

The simulation results demonstrate that if the values for *m* and *d* are back calculated
from the measured displacements (strains), the deformational mechanism of the materials can be identified without assuming any mechanical model, such as strain

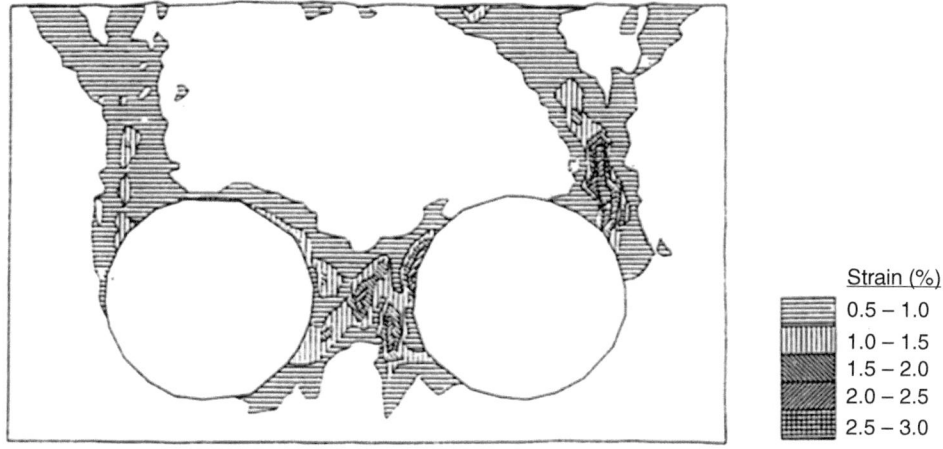

Strain (%)
0.5 – 1.0
1.0 – 1.5
1.5 – 2.0
2.0 – 2.5
2.5 – 3.0

Figure 5 Maximum shear strain distribution around two tunnels (Experiments).

hardening, perfectly plastic, strain softening, etc. It is a great advantage that the values for m and d are simple monotonic functions with displacements (strains), resulting in back analyses that can be easily performed.

3.5 Application of anisotropic parameter for back analysis

3.5.1 Tunnels

A physical model test was carried out in the laboratory for two circular parallel tunnels, which are modeled by granular materials consisting of aluminum bars with different diameters. The maximum shear strain distribution around the tunnels was calculated from the measured displacements using the kinematic relationship between the displacements and strain, as shown in Figure 5 (Sakurai & Akayuri, 1998).

The back analysis was performed in such a way that the error function, given in Equation 8, is minimized for determining anisotropic parameter m and direction of the mobilized sliding plane θ.

$$\delta = \frac{\sum_{i=1}^{M} (u_i^c - u_i^m)^2}{\sum_{i=1}^{M} u_i^m} \rightarrow \min \qquad (8)$$

where

u_i^c: computed displacements
u_i^m: measured displacements
M: amount of measurement data.

Once parameters m and θ have been determined, the strain distribution can be easily calculated with the stress-strain relationship given in Equations 5, 6 and 7. The results

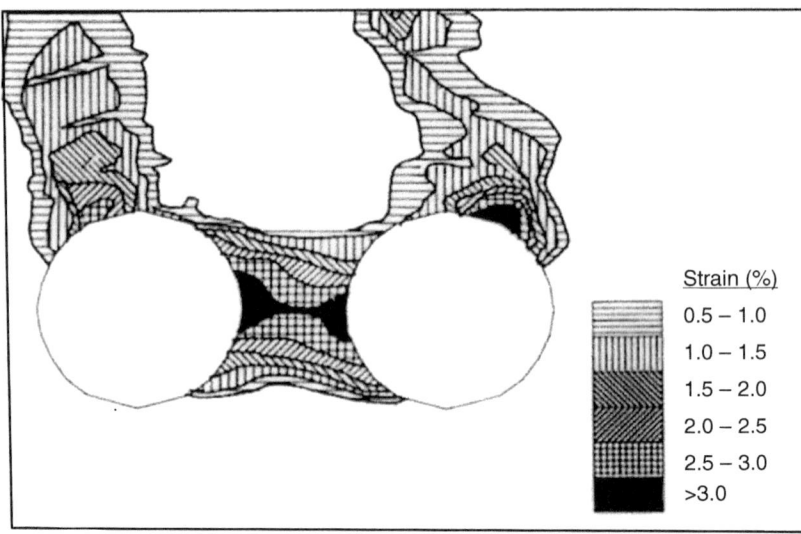

Strain (%)

0.5 – 1.0

1.0 – 1.5

1.5 – 2.0

2.0 – 2.5

2.5 – 3.0

>3.0

Figure 6 Results of back analysis.

of the back analysis for determining the maximum shear strain distribution around the tunnels are shown in Figure 6. It is clear that the back-analyzed maximum shear strain distribution shown in this figure is almost the same as the distribution obtained by the physical model test shown in Figure 5.

It should be emphasized that in this back analysis, we did not assume any mechanical model. Instead, the values for the anisotropic parameter along the mobilizing sliding plane were back calculated from the measurement data, resulting in a yielding zone that can be predicted from the maximum shear strain distribution around the tunnels.

3.5.2 Slopes

In the forward analysis of slopes, we must assume a mechanical model for deformational modes, such as (1) elastic, (2) sliding or (3) toppling, considering the results of geological and geomechanical field explorations. Among these three deformational modes of slope behaviors, the mechanical modeling of the toppling behavior is surely the most difficult. In a back analysis, however, if we use the anisotropic parameter, toppling can be treated just the same as the other two types of behavior without assuming any mechanical model, as shown in the following example.

In order to investigate the toppling mechanism of cut slopes, we performed a physical model test in the laboratory. Since toppling is basically a two-dimensional phenomenon, two-dimensional block models, consisting of a large number of aluminum bars with rectangular cross sections (1 cm × 1 cm), were used for the experiments (See Figure 7) (Sakurai, *et al.*, 1986).

In the experiments, the displacements are measured during the removal of some of the blocks, step by step, to simulate the cutting of the slopes, as shown in Figure 8. In a forward analysis, if we assume a linear elastic material, the displacement

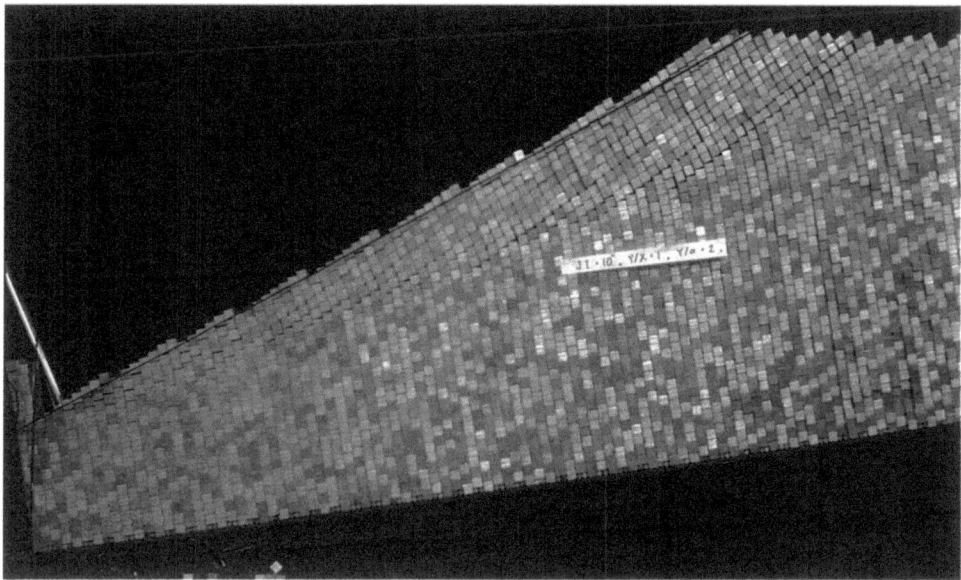

Figure 7 Physical model consisting of large number of aluminum blocks to simulate cut slope (Toppling behavior).

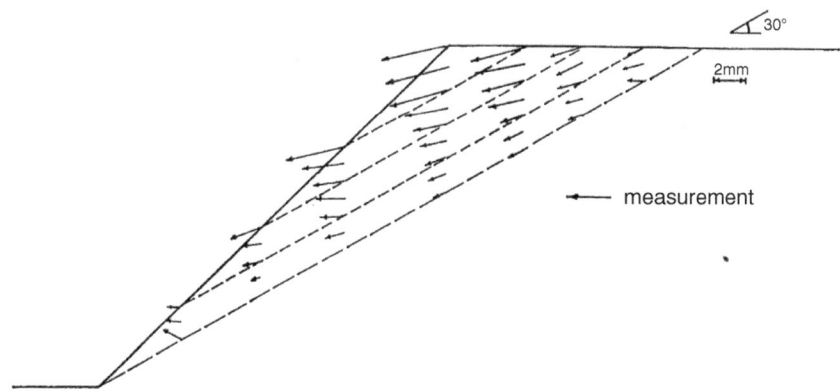

Figure 8 Displacement distribution of slope due to excavation (Results of laboratory experiment).

distribution becomes as shown in Figure 9. It is obvious from this figure that all the displacement vectors are facing upward due to the excavation, *i.e.*, due to rebound. The displacement distribution calculated by a linear elastic forward analysis indicates the rebound due to the excavation, while in reality the laboratory experimental results show the toppling behavior.

The measured displacements shown in Figure 8 have been used as the input data in the back analysis. As a result, the anisotropic parameters have been determined so as to achieve a good agreement between the measured and the back-calculated displacements.

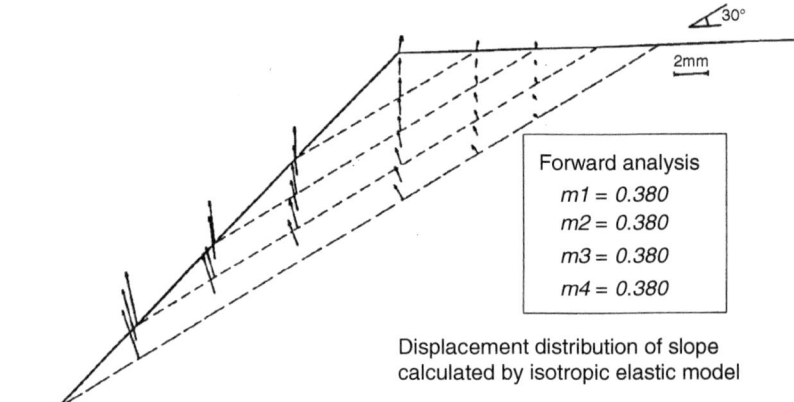

Figure 9 Displacement distribution of slope calculated by isotropic elastic model.

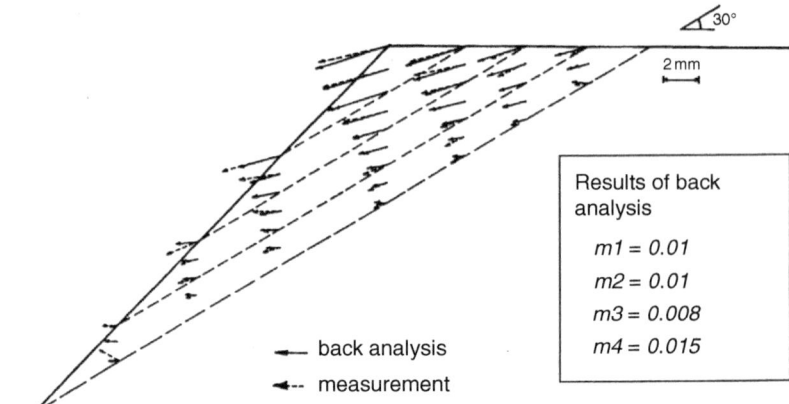

Figure 10 Comparison between measured and back-calculated displacements using anisotropic parameters.

In this calculation, the error function, shown in Equation 8, should be as small as possible. The results of the back analysis are shown in Figure 10, along with the measured displacements, in order that a comparison can be made between them.

Figure 10 indicates a good agreement between the measured and the back-calculated displacements. This good agreement indicates that even for such a discontinuous medium as a block model, a continuum mechanics approach may be applied to analyze the deformational behavior of toppling. Moreover, it is surprising that only four parameters of the anisotropic parameters can work well to simulate a toppling deformation. It is also interesting to see that if all four anisotropic parameters become 0.385, a rebound deformation is indicated, while if the four anisotropic parameters have the values shown in Figure 10, the toppling deformation is simulated.

This implies that anisotropic parameters must be good parameters to use in back analyses. It should be emphasized once again that we did not assume any mechanical model in these back analyses, but tried to find a good agreement between the computed and the measured displacements by changing the values of the anisotropic parameters until we obtained the best match between the two.

4 ASSESSMENT OF FAILURE OF GEOMATERIALS

4.1 Failure criteria used in numerical analyses

In order to assess the stability of geostructures by numerical analyses, the failure criteria of the geomaterials should be carefully investigated. However, the failure criteria of rock masses are hardly ever determined in laboratory experiments because of the scale effect of rocks. In general, the experimental results for the strength and deformability of rock specimens determined in the laboratory are greater than those of *in situ* rock masses. To determine the strength and deformability of rock masses, *in situ* tests, such as plate bearing tests and direct shear tests, can be performed, but they are costly and time-consuming.

In elasto-plastic analyses for assessing the stability of geostructures, yielding criteria are ordinarily used instead of failure criteria, resulting in computed stress levels that never extend beyond the yielding point. Therefore, if the stress reaches the yielding point, it cannot increase any further. The stress remains constant in the case of perfectly plastic materials, while the strain increases monotonically, even beyond the yielding point, until failure. This implies that the failure criteria should be defined in terms of strain.

4.2 Critical strain of rocks and soils

Critical strain in compression is defined as (Sakurai, 1981)

$$\varepsilon_0 = \sigma_c/E \tag{9}$$

where ε_0 is the critical strain, σ_c is the uniaxial compressive strength and E is Young's modulus.

The physical meaning of critical strain is indicated by the stress-strain relationship, as shown in Figure 11. It is obvious that the critical strain is always smaller than the strain at the failure of the materials. The critical strain only becomes the same as the failure strain for brittle materials. This fact shows the linear relationship between stress and strain until failure, and it implies that if the strain occurring around a tunnel is smaller than the critical strain, the tunnel is surely stable. For non-brittle geomaterials, further spontaneous increases in tunnel safety are seen.

Various kinds of rock and soil specimens were tested in the laboratory to determine their critical strain, and the results are plotted in relation to their uniaxial compressive strength in Figure 12. It is interesting to see that critical strain is continuously distributed from soils to hard rocks. This means that it is not necessary to classify the geomaterials surrounding a tunnel into soils or rocks. If we know the uniaxial strength of the materials, then the upper and lower bounds of the critical strain can be easily evaluated from the figure, no matter whether the materials are soils or rocks.

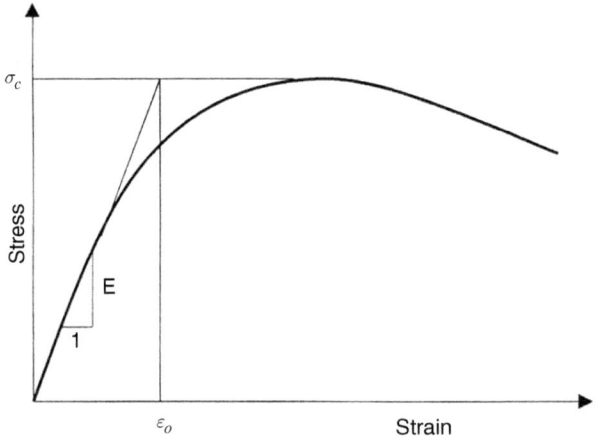

Figure 11 Physical meaning of critical strain.

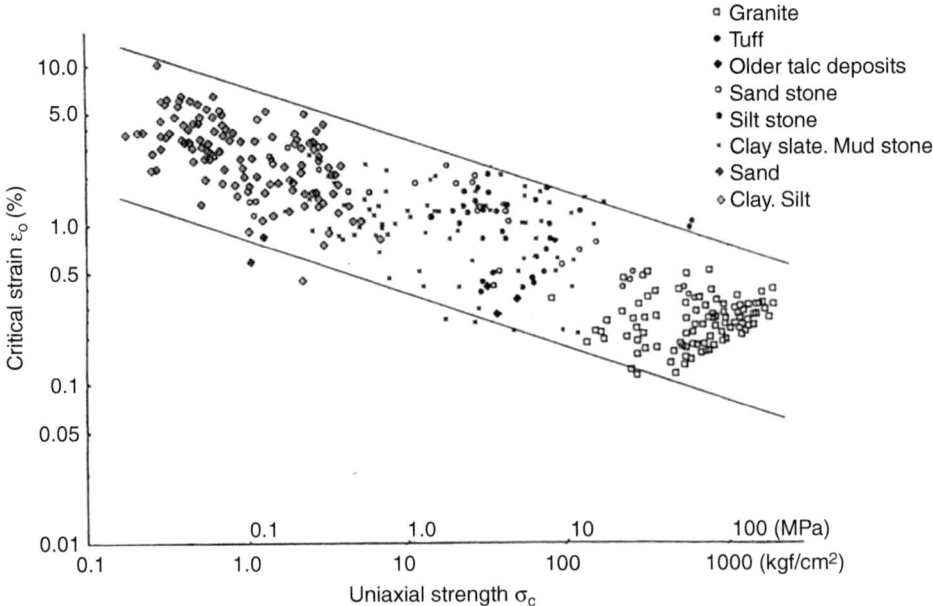

Figure 12 Relationship between critical strain and uniaxial strength of intact rocks and soils.

4.3 Critical strain of rock masses

The values for critical strain shown in Figure 12 were determined from small specimens of soils and intact rocks tested in the laboratory. In engineering practice, however, we need the critical strain of *in situ* materials, not of small-sized specimens. Thus, *in-situ* tests, such as plate bearing tests and direct shear tests, were conducted at a tunnel construction site consisting of highly jointed granite and sandstone. The results of plate bearing tests and direct shear tests can provide Young's modulus E_R and the strength parameters, such as cohesion C and internal friction angle ϕ, respectively, assuming linear Mohr-Coulomb's failure criterion.

Once C and ϕ have been determined, the uniaxial compressive strength of rock masses, σ_{cR}, can be calculated by the following equation:

$$\sigma_{cR} = \frac{C(1 - \sin \phi)}{\cos \phi} \tag{10}$$

The critical strain of rock masses, ε_{OR}, can then be determined by Equation 11. The relationship between the critical strain of rock masses and that of intact rocks is also shown here.

$$\varepsilon_{0R} = \frac{\sigma_{cR}}{E_R} = \frac{m\sigma_c}{nE} = \left(\frac{m}{n}\right)\varepsilon_0 \tag{11}$$

where σ_c: uniaxial compressive strength of intact rocks

$\quad$ E: Young's modulus of intact rocks

$\quad$ m: reduction coefficient for uniaxial compressive strength $(0 < m \leq 1.0)$

$\quad$ n: reduction coefficient for Young's modulus $(0 < n \leq 1.0)$

In order to determine the reduction coefficients, m and n, the intact rocks were drilled from a rock mass at the site where plate bearing tests and direct shear tests had been conducted. The rocks were then tested in the laboratory to determine the critical strain. The reduction coefficients were determined using the test results obtained for both intact rocks and *in situ* rock masses, as shown in Table 1 (Sakurai, 1983).

It is clear from this table that the ratio m/n is always greater than 1.0, ranging approximately from 1.0 to 3.0. This means that the critical strain of jointed rock

Table I Uniaxial compressive strength and Young's modulus of rocks and rock masses.

Rock Type & Class		Intact rock		Rock mass			Reduction coefficients			
		σ_c MPa	E_c GPa	ε_0 %	σ_{CR} MPa	E_m GPa	ε_{0R} %	m	n	m/n
granite	CH	227.4	62.7	0.362	16.70	2.65	0.631	0.0734	0.0422	1.74
granite	CM	211.5	59.8	0.534	11.60	1.96	0.592	0.0548	0.0328	1.67
granite	CL	141.8	47.1	0.301	6.39	1.37	0.466	0.0451	0.0292	1.54
diorite	CH	145.2	37.3	0.389	16.70	2.65	0.631	0.1150	0.0711	1.62
diorite	CM	153.5	38.2	0.402	11.60	1.96	0.592	0.0755	0.0513	1.47
granite	B	130.2	47.5	0.274	20.97	8.60	0.244	0.161	0.181	0.89
granite	CH	117.6	44.3	0.265	14.31	2.98	0.480	0.198	0.067	1.81
granite	CM	44.7	23.7	0.189	6.99	1.21	0.578	0.156	0.051	3.06
sandstone	—	137.2	28.4	0.483	6.27	0.381	1.642	0.046	0.0134	3.43
shale	——	78.4	21.6	0.364	3.82	0.411	0.930	0.049	0.0191	2.57
sandstone		137.8	54.1	0.255	7.89	1.96	0.403	0.057	0.0362	1.57
sandstone		22.9	10.1	0.227	2.49	1.08	0.231	0.109	0.107	1.01
shale		61.1	57.2	0.107	7.81	2.25	0.347	0.128	0.0393	3.26

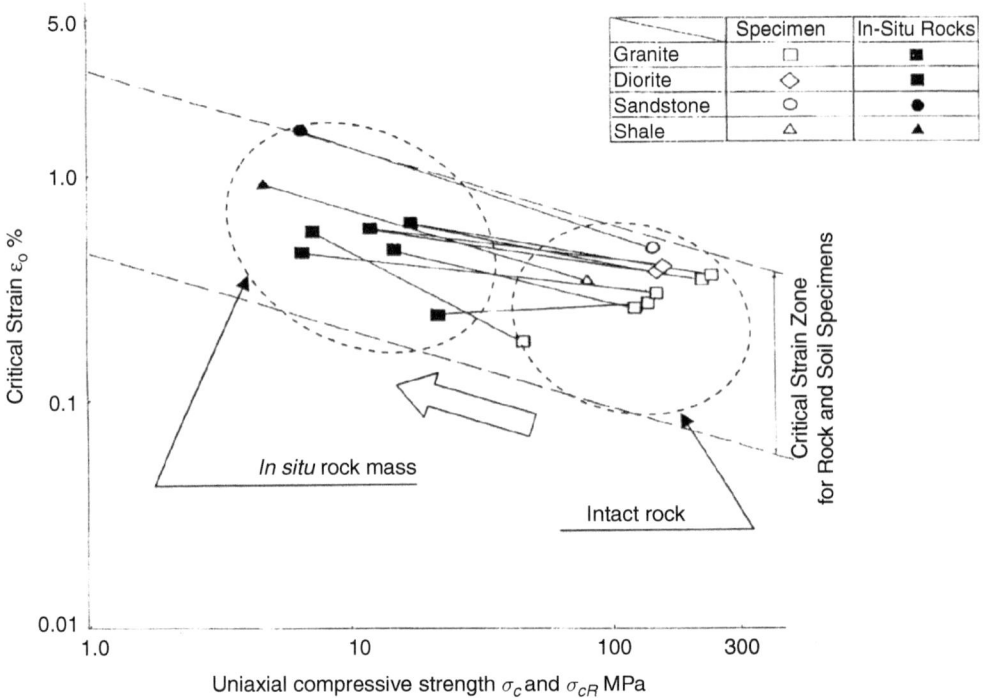

Figure 13 Relationship between critical strain of intact rocks and that of *in situ* rock masses.

masses is approximately 1 to 3 times greater than that of intact rocks. This is the scale effect of the critical strain, but it is an entirely opposite trend to the uniaxial compressive strength and Young's modulus. That is, both the strength and deformability of rock masses are always smaller than those of intact rocks. This must be the reason why laboratory tests on intact rocks are unpopular in rock engineering practice, while laboratory tests on small specimens are quite popular in soil engineering practice. The results obtained in the laboratory tests can be used directly for the design of soil structures.

The critical strain of rock masses, ε_{0R}, is plotted in relation to the uniaxial compressive strength of rock masses, σ_{cR}, in Figure 13. In this figure, the critical strain of the intact rocks is also plotted. The two dotted lines given in the figure are the upper and lower bounds of the critical strain shown in Figure 12. It is of interest to know that the critical strain of rock masses is scattered between the upper and lower bounds of the critical strain of intact rocks, and that the data for intact rocks seem to move parallel to the dotted lines toward those for rock masses with a decrease in uniaxial compressive strength.

It is clear from Figure 13 that the critical strain of rock masses is always greater than that of intact rocks. Thus, if the critical strain of intact rocks is used as the allowable value for strain, a certain amount of safety allowance can always be expected in tunneling practice. In other words, the critical strain of intact rocks can be used to assess the stability of tunnels, while the uniaxial strength and Young's modulus of intact rocks cannot be used directly to assess tunnel stability without considering the scale effect. This is a great advantage in rock engineering practice; that is, the critical strain of jointed rock masses can be estimated from laboratory tests on small specimens.

4.4 Critical shear strain of rocks and soils

The critical shear strain is defined as follows (Sakurai *et al.*, 1993):

$$\gamma_0 = \tau_c/G \tag{12}$$

where γ_0 is the critical shear strain, τ_c is the maximum shear strength and G is the shear modulus.

The critical shear strain of soils can be determined by torsion tests conducted on a hollow cylindrical specimen in the laboratory. For rocks, however, torsion tests can rarely be adopted because of the difficulty of the experiments. To avoid this shortcoming, the critical shear strain, obtained from the critical strain with Equation 13, is derived on the assumption that the geomaterials are elastic until the strain reaches the critical strain of the materials.

$$\gamma_0 = (1 + v)\varepsilon_0 \tag{13}$$

where v is Poisson's ratio and ε_0 is the critical strain.

The critical shear strain of geomaterials, converted from the critical strain using Equation 13, is plotted in relation to the shear modulus, as shown in Figure 14.

The results of torsion tests performed on cohesive soils are also plotted in this figure. It is seen from the figure that all the data obtained in the two different ways, *i.e.*, (1) conversion from the critical strain and (2) torsion tests, fall into the same scattering

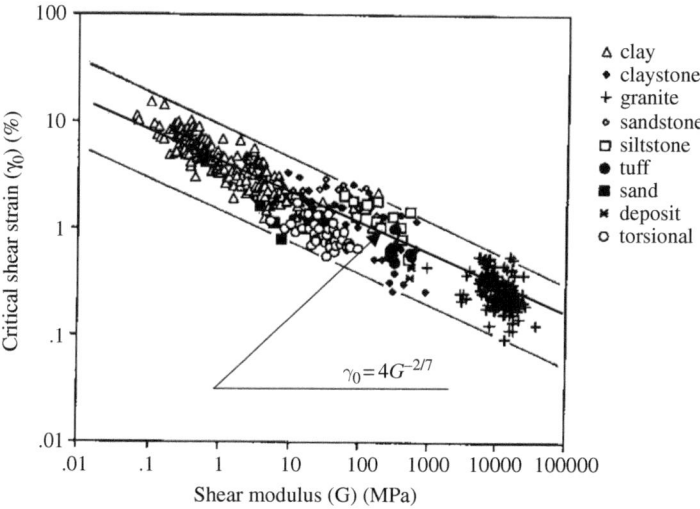

Figure 14 Critical shear strain for various rocks and soils in relation to shear modulus of materials.

zone. This means that the conversion from the critical strain is very reliable for determining the critical shear strain. In this figure, the shear modulus is used as the horizontal axis instead of the shear strength. This is simply because the shear modulus can be determined by performing a back analysis from the measured displacements, as will be described later.

5 HAZARD WARNING LEVELS FOR ASSESSING THE STABILITY OF TUNNLELS

5.1 Factor of safety of tunnels

In assessing the stability of tunnels, the factor of safety is not usually adopted, although it is popularly used for assessing the stability of slopes. This may be due to the fact that the factor of safety is defined in terms of stress. Therefore, if a plastic zone exists around tunnels, the state of stress in the plastic zone cannot go beyond the yielding criterion, resulting in the apparent factor of safety always becoming Fs = 1.0. Even in the case of a ground consisting of strain-hardening materials, the state of stress in the failure zone cannot exceed the failure criterion. In reality, however, tunnels must still be sufficiently stable even though the apparent factor of safety becomes F = 1.0. This is because the plastic zone or the failure zone is always surrounded by an elastic zone which is stable. This means that the factor of safety of tunnels, in terms of stress, cannot be used for assessing the stability of tunnels.

To overcome this shortcoming for the factor of safety of tunnels, Sakurai (1997) proposed "hazard warning levels", expressed in terms of strain, to assess the stability of tunnels. These warning levels can be used for the qualitative assessment of tunnel stability instead of the factor of safety.

5.2 Simple approach to assessing tunnel stability

An approach to assessing tunnel stability should be simple enough to apply to tunneling practice. In order to develop a simple approach, the tunnels are assumed to be approximately circular in shape and to have been excavated in a ground consisting of isotropic and homogeneous materials with a hydrostatic initial state of stress. According to the two-dimensional theory of elasticity, the circumferential strain ε_θ around a circular tunnel is expressed by the following equation:

$$\varepsilon_\theta = \frac{u}{r} \tag{14}$$

where u: radial displacement of the tunnel
r: radial coordinate

In monitoring the stability of tunnels, the crown settlements and the convergences (variation in tunnel diameter) are commonly measured during the excavation. Considering Equation 14, therefore, we can evaluate the circumferential strain at the inner surface of the tunnel either from the crown settlements or the convergences, as follows:

$$\varepsilon_{\theta,r=a} = \frac{\delta}{a} \tag{15}$$

where δ: crown settlement and/or half the convergence
       a: tunnel radius

Now, let us show a case study. A double-lane freeway tunnel was excavated in a ground consisting of heavily jointed granite and welded tuff with many fractures and shear zones. The crown settlements were measured during the excavation, and the circumferential strain at the inner surface of the tunnels was calculated with Equation 15.

The results were plotted in relation to the uniaxial compressive strength of intact rocks, as shown in Figure 15 (Sakurai, 1997). In this figure, the data indicated by the dark circles were obtained at the tunnel sections where serious trouble occurred during the excavation. The numbers with them indicate the kind of trouble, given in Table 2. On the other hand, the white circles were obtained at the tunnel sections where the excavations were completed without any trouble. The two dotted lines shown in the figure indicate the upper and lower bounds of the critical strain of intact rocks, which are shown in Figure 12.

It is obvious from Figure 15 that if the measured strain remained less than the lower bound of the critical strain, the tunnel was excavated without any problems, while if the measured strain surpassed the upper line, a lot of serious trouble surely occurred.

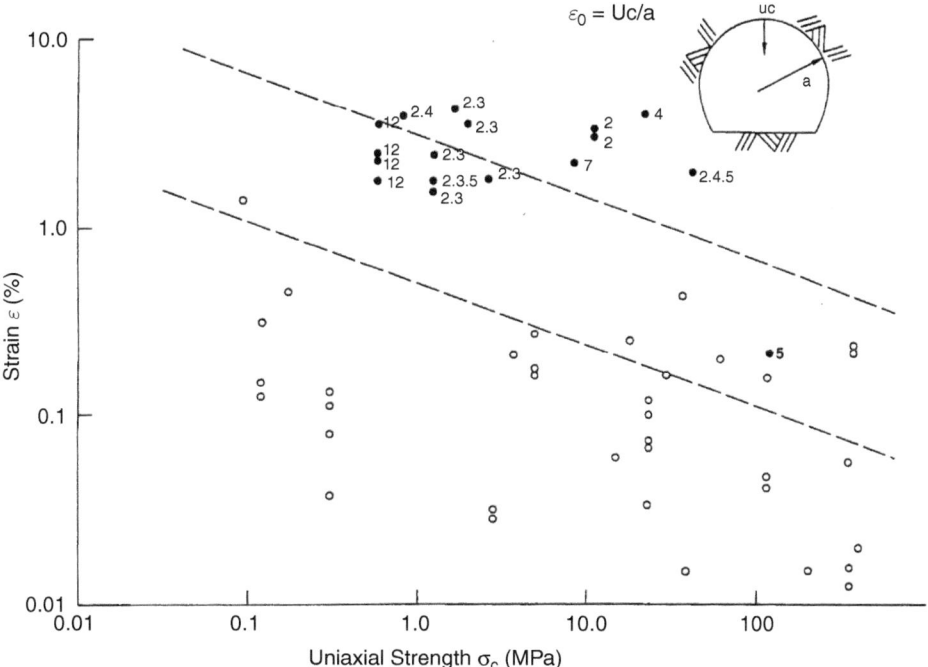

Figure 15 Relationship between measured strain (obtained from crown settlements) and uniaxial compressive strength of intact rocks.

Table 2 Troubles encountered during excavation.

No.	Remarks.
1	Difficulty in maintaining tunnel face
2	Failure and/or cracking in shotcrete
3	Buckling of steel ribs
4	Breakage of rock bolts
5	Fall-in of roof
6	Swelling at invert
7	Miscellaneous (unidentified)

Looking at the figure, the boundary of "trouble" and "no trouble" is located between the upper and lower bounds of the critical strain. In other words, the centerline between the two may be interpreted as a criterion for assessing the stability of tunnels. That is, if the measured strain becomes greater than the middle line, some trouble must have started occurring during the excavation.

5.3 Hazard warning levels in terms of strain

Considering the results shown in Figure 15, hazard warning levels in terms of strain have been proposed for assessing tunnel stability, as shown in Figure 16(a) (Sakurai, 1997). It is seen from the figure that the hazard warning levels are classified into three stages depending on the degree of stability. The geomaterials around tunnels are also classified into three groups (A, B and C) depending on their uniaxial compressive strength.

In order to assess the stability of tunnels during the excavation, it is common to measure the crown settlements and the convergences of the tunnels, and to compare them with their allowable values. At tunnel construction sites, the stability of tunnels must be quickly assessed immediately after the excavation. To fulfill this requirement, the hazard warning levels for strain can be converted into those for displacements, such as the crown settlements and/or half of the convergences, by Equation 16 considering Equation 15.

$$d = \varepsilon_0 a \tag{16}$$

where d: hazard warning level for radial displacements at tunnel surface

   ε_0: critical strain of geomaterials

   a: tunnel radius

The hazard warning levels for the crown settlements can be derived by Equation 16.

As an example, the hazard warning levels for the crown settlements of a tunnel, with a radius of 5 m, are derived. They are shown in Figure 16(b). According to the figure, the tunnel stability is assessed directly from the measured crown settlements by comparing them with the hazard warning levels.

It should be noted, however, that the measured crown settlements must be part of the absolute settlements, because a certain amount of settlements has already taken place when the measurements are begun. This means that a certain amount of crown settlements is missing in the measured data. In engineering practice, however, it is not necessary to evaluate the absolute crown settlements, because the critical strain of

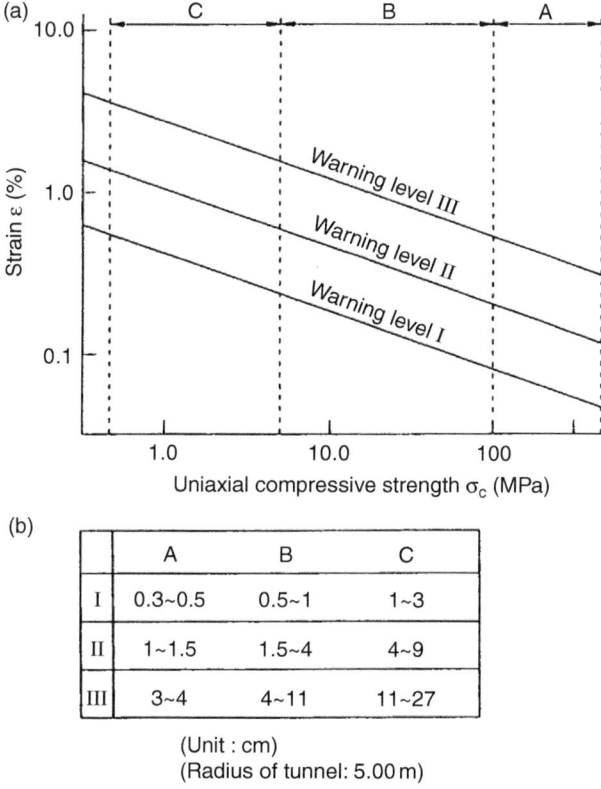

Figure 16 Hazard warning levels for assessing stability of tunnels (a) Relation between σ_c (uniaxial compressive strength) and ε (measured strains) (b) Crown settlements corresponding to Warning Levels I, II and III.

rock masses is always larger than that of intact rocks, as shown in Figure 13. Thus, if we use the critical strain of intact rocks as an allowable value for strain, a certain amount of safety allowance is always guaranteed in assessing tunnel stability. This allowance of safety may compensate for the crown settlements which have already taken place when the measurements are started. In fact, all the data shown in Figure 15 represent the strain measured after the excavation, not the absolute strain.

5.4 Precise assessment of stability of difficult tunnels

When a tunnel is excavated in a geologically and geomechanically difficult ground, various instruments are commonly required for measuring the displacements around the tunnel. Since the failure of geomaterials around tunnels may be caused by a critical value of the maximum shear strain, the critical value must be key for assessing the stability of tunnels excavated in difficult grounds. Thus, the question of how to determine the maximum shear strain distribution around tunnels from measured displacements may arise. The answer is that there are two methods available.

(1) One of the simple ways is to use the kinematic relationship between strains and displacements. The merit of this method is that the maximum shear strain distribution can be determined directly from the measured displacements without the constitutive equations of the materials, while the demerit is that the method requires a great deal of measured displacement data (Sakurai, 1982).

(2) When the amount of measurement data is not large enough to determine the maximum shear strain distribution by the kinematic relationship between strains and displacements, back analysis methods can be adopted to determine the maximum shear strain distribution from the measured displacements. In such back analyses, the anisotropic parameter can be effectively used to obtain a good agreement between the measured and the computed displacements. It should be noted that the back analysis method can be used for any tunnels and underground caverns with complex configurations.

Let us show an example for determining the maximum shear strain distribution from the measured displacements. The measurement data were taken from a subway tunnel in Washington D.C., as shown in Figure 17 (Hansmire et al., 1985).

Once the maximum shear strain distribution around the tunnels is determined from the measured displacements, the stability of the tunnels can be assessed by comparing the maximum shear strain with the critical shear strain given in Figure 14.

For this purpose, hazard warning levels are proposed for assessing the maximum shear strain occurring around tunnels, as shown in Figure 18, in the same manner as for the crown settlements shown in Figure 16(a). In this figure, the centerline, i.e., "Warning Level II", is supposed to be the boundary for tunnels between "trouble" and "no trouble", as shown in Figure 15, which must correspond to the apparent factor of safety Fs = 1.0. Beyond the centerline, some trouble starts to occur in the tunnels, but it does not collapse because the yielding zones are surrounded by an elastic zone which is still stable.

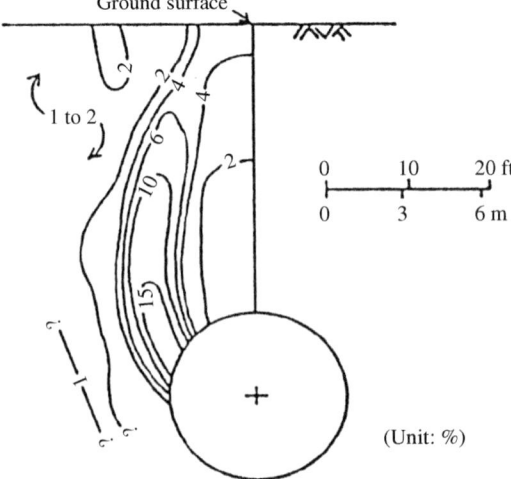

Figure 17 Maximum shear strain distribution around shallow tunnel (after Hansmire et al. 1985).

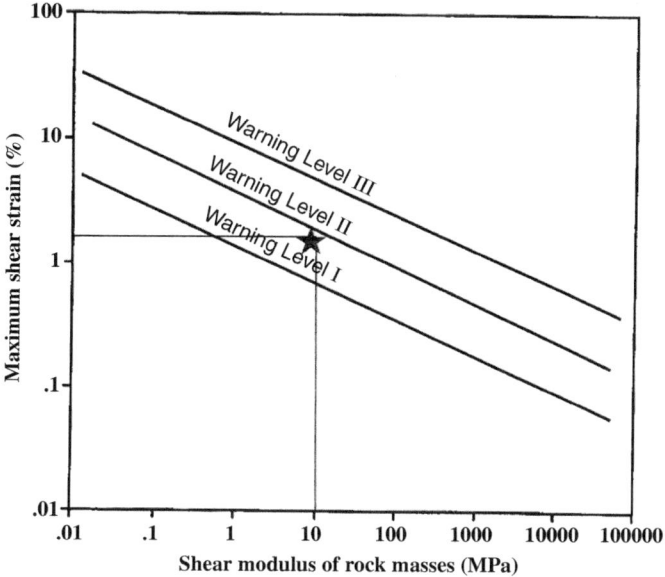

Figure 18 Hazard warning levels for assessing stability of difficult tunnels.

In Figure 18, the horizontal axis is the shear modulus, while the vertical axis is the maximum shear strain. Both the shear modulus and the maximum shear strain can be determined by a back analysis of the measured displacements.

Now, let us show an example of how to use this graph for assessing the stability of tunnels. We assume the maximum value for the maximum shear strain around a tunnel to be back calculated to a value of 1.5%, and the shear modulus of the surrounding rock mass to be back calculated to a value of 10 MPa. As seen in Figure 18, the point of intersection of the maximum shear strain of 1.5% and the shear modulus of the rock of 10 MPa lies near the center line (Warning Level II), but a little under it. This indicates that the risk for tunnel instability approaches Warning Level-II, but the tunnel is still stable at present because the centerline empirically corresponds to Fs ≤ = 1.0, as described above.

This approach to assessing the stability of tunnels was used in the excavation of Sirkeci Station in the Marmaray Project in Turkey (Otsuka *et al.*, 2011).

6 BACK ANALYSIS OF SLOPES

6.1 Paradox in displacement monitoring for slopes

In assessing the stability of slopes, the factor of safety is commonly used when the strength parameters, such as cohesion and the internal friction angle of the geomaterials, are required. Therefore, at the design stage of slopes, various experiments are carried out to obtain data on the strength parameters of the geomaterials. However, the accuracy and the reliability of the data are questionable because it is not easy to determine the strength parameters of *in situ* geomaterials. To overcome this difficulty, monitoring is conducted during the construction stage, and the strength parameters used at the design stage are re-evaluated.

In the monitoring of slopes, the measuring of displacements is usually performed with extensometers, inclinometers, total stations, GPS, etc., although the displacements are not considered at the design stage. This is a paradox between design and monitoring. If we want to solve this paradox, we need a way to determine the strength parameters from the measured displacements.

6.2 Proposed back analysis method for predicting a sliding plane

6.2.1 Procedure of the method

A back analysis method for determining a sliding plane from the displacements measured on a slope surface was proposed by Sakurai & Hamada (1996). Even though the method may be extended to a three-dimensional state, a two-dimensional case is illustrated here.

Let us consider the displacement vectors measured along the slope surface, as shown in Figure 19. It is assumed that the edge of the sliding plane, either point A or point D, is detected by considering unusual deformational behavior, such as cracks, swelling, etc., appearing on the slope surface. If point A is detected as the edge of the sliding plane, a potential sliding plane starts from point A, parallel to displacement vector U_1, until hitting point B on a straight line perpendicular to the slope surface and passing through point E which is located at the center of the two measuring points, (1) and (2). From point B, the sliding plane stretches parallel to displacement vector U_2 until hitting point C. After that, we repeat the same procedure as before until reaching the last point, D. If point D shows some unusual deformational behavior, the predicted sliding plane is correct. If not, some compensation is needed in terms of reconsidering the location of points A and D.

6.2.2 Accuracy of the method

The accuracy of the proposed method for predicting a sliding plane is demonstrated by showing the results of a numerical simulation by FEM in which the anisotropic parameters are considered. In the simulation, a slope with a potential curved sliding plane is

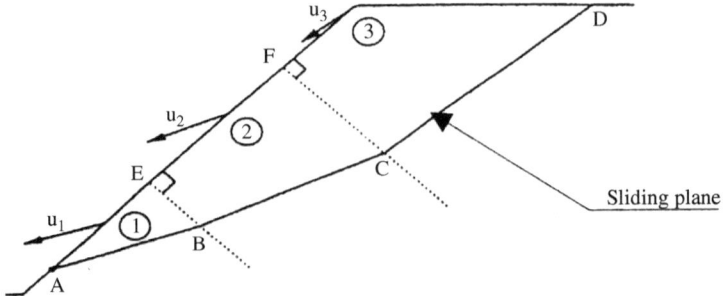

Figure 19 Schematic diagram for procedure of method.

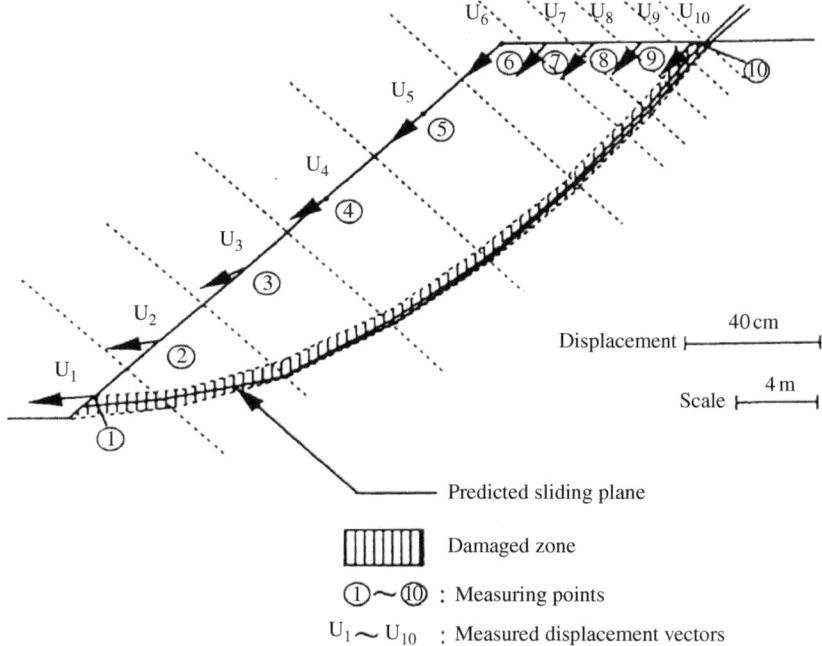

Figure 20 Application of proposed method for predicting sliding plane from displacements measured on slope surface.

taken as an example, as shown in Figure 20. The sliding plane in this figure is assumed as a shaded zone where the shear rigidity decreases with an increase in the maximum shear strain due to strain localization. In the numerical simulation, the anisotropic parameters in the shaded zone decrease with an increase in the maximum shear strain. This results in it being possible for displacements $U_1 - U_{10}$ on the slope surface to be calculated by FEM, as shown in Figure 20. They are used as virtual data for the measured displacements of the simulation.

Let us now suppose these computed displacement vectors along the slope surface as the virtual measured displacements. By using the virtual measured displacements, a sliding plane can be determined by the proposed method. The results are shown in Figure 20. It is obvious that a predicted sliding plane falls exactly within the assumed shaded zone for the strain localization. This means that the proposed method is applicable for predicting the sliding plane from the surface displacements with high accuracy.

6.3 Back analysis procedure for determining the factor of safety from measured displacements

To solve the paradoxes of slope monitoring, Sakurai *et al.* (2009) proposed a back analysis procedure for determining the strength parameters, such as cohesion and the internal friction angle, from the measured displacements. The critical shear strain plays a major role in the procedure.

The back analysis procedure is given as follows:

(1) Young's modulus E and anisotropic parameter m are back calculated from the measured displacements, while Poisson's ratio v is assumed to be constant. However, there is some difficulty in back calculating both Young's modulus and the anisotropic parameter at the same time. Thus, we can assume Young's modulus from one of the intact materials, because it is insensitive to the results of back analyses, as shown in a case study that will be described later.

(2) Shear modulus G is determined by the following equation which is based on the definition for the anisotropic parameter:

$$G = mE \tag{a}$$

(3) Once the shear modulus has been determined by Equation a, critical shear strain γ_0 is obtained from Figure 14, which shows the relationship between shear modulus G and critical shear strain γ_0. Since the data indicate large scattering, we propose using the center line of the scattered data on the basis of the experience obtained for tunnels, *i.e.*,

$$\gamma_0 = 4G^{-2/7} \tag{b}$$

(4) Maximum shear strength τ_c can then be calculated by the following equation, which is based on the definition for the critical shear strain:

$$\tau_c = G\gamma_0 \tag{c}$$

(5) Since the direction of the local coordinates, which is the direction of the mobilized plane, has been determined by the back analyses, internal friction angle ϕ can be determined considering the Mohr-Coulomb failure criterion.

(6) Cohesion C is then calculated by the following equation:

$$C = \frac{1 - \sin \phi}{\cos \phi} \tau_c \tag{d}$$

(7) Using the values for C and ϕ, factor of safety Fs can be calculated by a conventional limit equilibrium method, such as the Fellenius Method, as follows;

$$\boldsymbol{Fs} = \frac{\sum S_i \ell_i}{\sum \tau_i \ell_i} = \frac{\sum (c_i + \sigma_i \tan\phi_i)\ell_i}{\sum \tau_i \ell_i} \tag{e}$$

where ℓ_i is the length of the sliding plane of the "i"th slice, and σ_i and τ_i are the normal and shear stress on the sliding plane of the "i"th slice, respectively.

6.4 Case study (Open pit coal mine)

A case study is shown here, which is an open pit coal mine where large-scale sliding failure occurred. For monitoring purposes, the surface displacements were measured by GPS until the failure occurred. In order to verify the applicability of the back analysis method proposed by Sakurai *et al.* (2009), a back analysis has been carried out with the data obtained from the open pit mine which were gathered one day before the failure occurred.

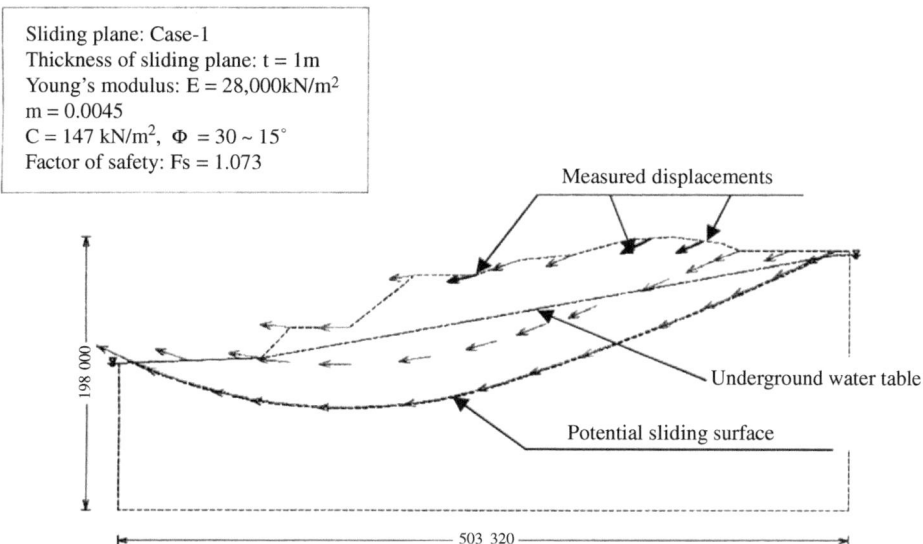

Sliding plane: Case-1
Thickness of sliding plane: t = 1m
Young's modulus: E = 28,000kN/m²
m = 0.0045
C = 147 kN/m², Φ = 30 ~ 15°
Factor of safety: Fs = 1.073

Measured displacements

198 000

Underground water table

Potential sliding surface

503 320

Figure 21 Comparison of computed and measured displacements vectors.

6.4.1 Input data for the case study

The following data are used:

Thickness of sliding surface; t = 1m
Underground water table: Shown in Figure 21 and determined by the field observation
  on the ground surface.
Failure criterion; Mohr-Coulomb type
Young's modulus: Three different values, *i.e.*, E = 28,000, 56,000 and 112,000 kN/m²
  are used, because no information was available.
Poisson's ratio: $v = 0.3$
Displacement data: The data used in this case study were taken one day before the
  failure occurred (GPS measurements). The measured displacement data (vectors)
  were obtained at three measuring points along the cross section of the slope shown in
  Figure 21.

6.4.2 Back analysis and its results

Firstly, the potential sliding surface was back calculated by the proposed method
described in Section 6.2 using the measured displacements shown in Table 3.

Back analyses were then carried out, using the given input data for minimizing
the error function provided in Equation 8, to obtain the optimal value for anisotropic parameter *m* which is key to back analyses. Once the optimal anisotropic
parameter *m* is determined, the strength parameters (C and ϕ) of the materials
can be determined by the proposed back analysis procedure shown in Section 6.3.
Factor of safety Fs can then be calculated by Equation e, and the horizontal and vertical
components of the back-calculated displacements, together with the measured

Table 3 Results of back analyses.

Thickness of sliding plane (m)	E (kN/m²)	m	Fs	Δu²(cm²)	Displacements at measuring points (cm)					
					① δ_x 168.0	① δ_z 46.0	② δ_x 183.3	② δ_z 96.3	③ δ_x 164.4	③ δ_z 69.8
1.0	28000	0.0045	1.073	83	170.0	53.0	172.1	82.8	173.2	77.4
	56000	0.0023	1.004	95	169.5	52.7	173.3	79.9	174.1	77.4
	112000	0.0012	0.959	107	169.2	52.3	174.5	77.8	175.2	77.7

displacements at the three measuring points, are shown in Table 3. The back-calculated displacement vectors are compared with the measured displacement vectors, as shown in Figure 21.

6.4.3 Discussion of the results

Looking at the results of the back analyses for the case study, it is amazing that the factors of safety derived by the back analyses using different values for Young's modulus are almost identical to each other. This demonstrates that Young's modulus is insensitive to the results of back analyses. This is a great advantage of the method, because the Young's modulus of the geomaterials is difficult to evaluate due to the scale effect. This advantage is caused by the fact that if we assume a small value for Young's modulus, then a large value can be derived for the anisotropic parameter so as to minimize the error function given in Equation 8. It is clear that the displacements are greatly influenced by the anisotropic parameter, not by Young's modulus. It is seen from Table 2 that the factor of safety is approximately Fs = 1.0 for all cases no matter what values are used for Young's modulus. This indicates that the slope shown in the case study seemed to be very critical in terms of stability. In fact, the slope failed the very next day after the acquisition of the data which were used for the back analyses.

It should be emphasized that the centerline of the scattering data for the critical shear strain must be a criterion for assessing the failure of both tunnels and slopes. If the strains around tunnels go beyond the centerline, *i.e.*, Warning Level II, some trouble may occur during the excavation. However, the tunnels are still stable because the excavation damage zone (EDZ) is always surrounded by a stable elastic zone.

In slopes, on the other hand, it is assumed that the centerline of the scattering data of the critical shear strain corresponds to the factor of safety Fs = 1.0, considering the results shown in Figure 15. This implies that if the maximum shear strain reaches the centerline, the slopes fail immediately along the sliding plane, because there are no supporting materials to protect the sliding failure. In fact, the case study proves that this assumption is correct, *i.e.*, the centerline corresponds to Fs = 1.0. This may be the reason why the factor of safety is popularly used for assessing the stability of slopes, while it has never been used for tunnels.

7 CONCLUSIONS

(1)   Back analyses can give a quantitative interpretation to field measurement data during the construction of rock structures. It should be noted that back analyses are not simply backward calculations of forward analyses, but the basic concept of back analyses must be different from that of forward analyses. This is because a mechanical model (constitutive equation, deformational modes, etc.) should not be assumed in the analyses, but should be identified by back analyses. Consequently, in back analyses, we should not use the same mechanical models at the design stage as those used for forward analyses.

(2)   Forward analyses for the elasto-plastic behavior of geostructures can be performed on the basis of the conventional theory of plasticity in which non-associated flow rules play a major role. In back analyses, however, the conventional theory of plasticity is hardly ever applicable in geotechnical engineering practice because it contains many mechanical parameters as input data, all of which cannot be determined by back analyses from the limited amount of measurement data.

(3)   In order to overcome the difficulty of using the conventional theory of plasticity for back analyses, the anisotropic parameter (the anisotropic damage parameter) has been proposed, and a new approach to back analysis methodologies has been constructed on the basis of this anisotropic parameter. It is a great advantage that the approach is so simple that it can be easily applied to both tunnel and slope engineering practice. In addition, the new approach to back analyses can simulate any deformational modes of geomaterials, *i.e.*, strain hardening, strain softening or perfectly plastic, which are not necessarily assumed beforehand, but can be identified automatically by the back analyses of measured displacements.

(4)   Since the failure criterion of geomaterials is usually given in terms of stress, the strength parameters, such as cohesion and the internal friction angle of the geomaterials, must be determined. However, it is not an easy task to determine the strength parameters of *in situ* geomaterials because of their scale effect. To overcome this difficulty, the critical strain and the critical shear strain of geomaterials have been proposed as the failure criteria of rock masses. They can be successfully applied during both design and construction stages.

(5)   In slopes, the factor of safety is commonly evaluated for assessing the stability of the slopes at the design stage. To evaluate the factor of safety, the strength parameters, such as cohesion and the internal friction angle must be determined. In monitoring during the construction stage, however, the displacements are measured, but the strength parameters cannot be directly measured. Therefore, the strength parameters should be back calculated from the measured displacement data. In the back analyses, both the anisotropic parameter and the critical shear strain are successfully used to construct a new back analysis procedure. According to the proposed back analysis procedure, the strength parameters can be determined from the measured displacements, resulting in the possibility of re-evaluating the factor of safety during the construction stage.

(6)   In tunnels, on the other hand, the factor of safety can hardly ever be evaluated if a plastic zone exists around the tunnels, because the state of stress in the plastic zone cannot go beyond the yielding criterion, so that the apparent factor of safety always becomes Fs = 1.0. Even in the case of a ground consisting of

strain-hardening materials, the stress distribution in the failure zone should not exceed the failure criterion. Therefore, the apparent factor of safety always becomes Fs = 1.0 as long as the failure criterion is expressed in terms of stress. In reality, however, tunnels must still be sufficiently stable, because the plastic zone or the failure zone is always surrounded by an elastic zone which is stable. In order to avoid this problem, hazard warning levels are defined in terms of either the critical strain or the critical shear strain, and they are used to assess the crown settlements and the maximum shear strains occurring around the tunnels, respectively, resulting in it being possible to assess the stability of tunnels by the measured displacements in a probabilistic way, instead of a deterministic way, with the factor of safety.

(7) In order to demonstrate the applicability of the proposed back analysis procedure, a case study for assessing the slope stability of an open pit coal mine has been introduced, where the displacements were measured by GPS. The back analysis was carried out to determine the strength parameters of the materials from the displacement data obtained one day before the failure occurred. The factor of safety was then calculated, and the results show that the factor of safety is nearly Fs =1.0. It is demonstrated from the case study that the proposed back analysis procedure is highly accurate and reliable for predicting the failure of slopes. It is noted that the back analysis procedure can assess the stability of slopes with the use of only the surface displacements measured by GPS.

REFERENCES

Feng, X.-T., Zhang, Z. & Sheng, Q. (2000) Estimating mechanical rock mass parameters relating to the Three Gorges Project permanent shiplock using an intelligent displacement back analysis method, *International Journal of Rock Mechanics & Mining Sciences*, 37, 1039–1054.

Feng, X.-T. & Yang, C. (2004) Coupling recognition of the structure and parameters of non-linear constitutive material models using hybrid evolutionary algorithms, *International Journal for Numerical Methods in Engineering*, 59, 1227–1250.

Feng, X.-T., Zhao, H. & Li, S. (2004) A new displacement back analysis to identify mechanical geo-material parameters based on hybrid intelligent methodology, *International Journal for Numerical and Analytical Methods in Geomechanics*, 28, 1141–1165.

Feng, X.-T., Chen, B.-R., Yang, C., Zhou, H. & Ding, X. (2006) Identification of visco-elastic models for rocks using genetic programming coupled with the modified particle swarm optimization algorithm, *International Journal of Rock Mechanics & Mining Sciences*, 43, 789–801.

Feng, X.-T. & Hudson, J. (2010) Specifying the information required for rock mechanics modelling and rock engineering design, *International Journal of Rock Mechanics & Mining Sciences*, 47, 179–194.

Feng, Z.L. & Lewis, R.W. (1987) Optimal estimation of *in situ* ground stresses from displacement measurement, *International Journal for Numerical and Analytical Methods in Geomechanics*, 11, 391–408.

Cividini, A., Jurina, L. & Gioda, G. (1981) Some aspects of 'characterization' problems in geomechanics, *International Journal of Rock Mechanics & Mining Sciences & Geomechanics abstracts*, 18, 487–503.

Cividini, A., Gioda, G. & Barla G. (1985) Calibration of a rheological material model on the basis of field measurements. *Proceedings of 5th International Conference on Numerical Methods in Geomechanics*, Nagoya, 1621–1628.

Cividini, A. & Gioda, G. (1994) Deterministic and probabilistic back analysis in rock mechanics, *Proceedings of International Workshop on Inverse Analysis in Geomechanics '94*, Nagoya, Japan, 51–77.

Cividini, A. & Gioda, G. (2003) Back analysis of geotechnical problems, In: Bull, J.W. (ed) *Numerical Analysis & Modelling in Geomechanics*, Spon Press, Taylor & Francis Group, 165–196.

Gioda, G. (1980) Indirect identification of the average elastic characteristics of rock masses, *Proceedings of International Conference on Structural Foundations on Rock*, Sydney, 65–73.

Gioda, G. & Maier, G. (1980) Direct search solution of an inverse problem in elastoplasticity: Identification of cohesion, friction angle and *in-situ* stress by pressure tunnel tests, *International Journal for Numerical Methods in Engineering*, 15, 11823–11848.

Gioda, G. & Sakurai, S. (1987) Back analysis procedures for the interpretation of field measurements in geomechanics, *International Journal for Numerical and Analytical Methods in Geomechanics*, 11, 555–583.

Hansmire, W.H. & Cording, E.J. (1985) Soil tunnel test section: Case history summary, *Journal of Geotechnical Engineering Division by American Society of Civil Engineering*, 111, 1301–1320.

Kavanagh, K.T. (1973) Experiment *versus* Analysis: Computational techniques for the description of static material response, *International Journal for Numerical Methods in Engineering*, 5, 503–515.

Kirsten, H.A.D. (1976) Determination rock mass elastic moduli by back analysis of deformation measurements, *Proceedings of Symposium on Exploration for Rock Engineering*, Johannesburg, 165–172.

Otsuka, I., Sakurai, S., Taki, H., Aoki, T., Shimo, M., Kaneko, T. & Iwano, M. (2011) Observational construction management by field measurement of large scale underground railway station by urban NATM – Railway Bosphorus tube crossing, tunnels and stations, *Proceedings of the 12th ISRM Congress*, Beijing, China, 1769–1772.

Sakurai, S. (1981) Direct strain evaluation technique in construction of underground openings, *Proceedings of the 22nd U.S. Symposium on Rock Mechanics*, Cambridge, MIT, 298–301.

Sakurai, S. (1982) Monitoring of caverns during construction period, *Proceedings of the ISRM Symposium on Rock Mechanics: Caverns and Pressure Shafts*, Aachen, 1, 433–441.

Sakurai, S. (1983) Displacement measurements associated with the design of underground openings, *Proceedings of International Symposium on Field Measurement in Geomechanics*, 2, 1163–1178.

Sakurai, S. & Takeuchi, K. (1983) Back analysis of measured displacements of tunnels, *Rock Mechanics & Rock Engineering*, 16, 173–180.

Sakurai, S., Deeswasmongkol, N. & Shinji, M. (1986) Back analysis for determining material characteristics in cut slopes, *Proceedings of International Symposium on Engineering in Complex Rock Formations*, Beijing, 770–776.

Sakurai, S. (1990) Monitoring the stability of cut slopes, *Proceedings of Mine Planning and Equipment Selection*, Calgary, 269–274.

Sakurai, S., Kawashima, I. & Otani, T. (1993) A criterion for assessing the stability of tunnels, *Proceedings of ISRM International Symposium (EUROCK '93)*, Lisboa, 969–973.

Sakurai, S. & Hamada, K. (1996) Monitoring of slope stability by means of GPS, *Proceedings of the 8th FIG International Symposium on Deformation Measurements*, Hong Kong, 55–60.

Sakurai, S. (1997) Lessons learned from field measurements in tunneling, *Tunnelling and Underground Space Technology*, 12, n 4, 453–460.

Sakurai, S. & Akayuri, C.F.J. (1998.) Deformational analysis of geomaterials considering strain-induced damage, *Proceedings of the 4th European Conference on Numerical Methods in Geotechnical Engineering – NUMGE 98*, 728–738.

Sakurai, S. & Nakayama, T. (1999) A back analysis in assessing the stability of slopes by means of surface measurements, *Proceedings of the International Symposium on Slope Stability Engineering – IS-Shikoku'99*, 1, 339–343.

Sakurai, S. & Shinji, M. (2005) Back analysis of non-linear behavior of soils and rocks considering strain-induced damage of materials, *11th IACMAG*, Torino, Italy.

Sakurai, S., Farazmand, A. & Adachi, K. (2009) Assessment of the stability of slopes from surface displacements measured by GPS in an open pit mine, *Sustainable Exploitation of Natural Resources, Proc. 3rd International Seminar ECOMINING-Europe 21st Century*, Milos Island, Greece, Sept 4–5, 239–248.

Shimizu, N. & Sakurai, S. (1983) Application of boundary element method for back analysis associated with tunneling problems, *Proceedings of the International Conference on Boundary Elements*, Hiroshima, 645–654.

Yang, L. & Sterling, R. (1989) Back analysis of rock tunnel using boundary element method, *Journal of Geotechnical Engineering Division by American Society of Civil Engineering*, 115, 1163–1169.

Zhang, D.-C., Gao X.-W. & Zheng, Y. (1988) Back analysis method of elastoplastic BEM in strain space, *Proceedings of the 6th International Conference on Numerical Methods in Geomechanics*, Innsbruck, 981–986.

Chapter 8

Back-analysis of rock landslides to infer rheological parameters

S. Martino, M. Della Seta & C. Esposito
Department of Earth Sciences, Sapienza University of Rome, Rome, Italy

Abstract: This chapter focuses on the description of a multidisciplinary methodology aimed at a comprehensive modeling of large rock slope instabilities, by considering the rheological behavior of the involved rock masses. Such a methodology is characterized by an integrated approach addressed to take into account the complex set of variables that features rock slope instabilities occurring on a wide range of spatial/temporal scales. Large rock slope failures can be regarded as the paroxysmal phase of time-dependent gravitational processes influenced by several factors that can be grouped in predisposing and conditioning ones. The predisposing factors encompass geological-structural setting and geomechanical properties, the conditioning factors act at different scales and are related to morpho-evolution of slope-to-valley floor systems, due to both surface effects of tectonic deformations and erosional/depositional processes, as well as geodynamic stress regime variations. In addition, triggering factors can be referred to intense and short-duration actions that, among the others, can be related to earthquakes, water pressure changes and anthropic activities. The here illustrated methodology consists in a multi-modeling approach that includes the contributions of a morpho-evolutionary modeling of the slope-to-valley floor system; a detailed engineering-geology modeling, that transposes geomechanical parameters to geological-structural features of; a time-dependent stress-strain numerical modeling, performed through a sequential approach which takes into account the main morph-evolutionary stages of the slope. Two case studies are here presented from the Italian Apennines which experienced such a multi-modeling approach to infer suitable rheological parameters to rock masses involved in creep process on natural slopes.

1 INTRODUCTION

Gravitational deformations widely affect natural slopes where jointed rock masses crop out. These phenomena are generally driven by time-dependent rheology, which is related to stiffness and viscosity of the rock mass including matrix and joint properties. The role of viscosity in gravitational slope deformations consists on regulating the strain rate of the process, thus influencing the final time for slope failures or general collapse. Starting from the 1980s (Savage & Varnes, 1987; Chigira, 1992), such a rheology-constrained evolution of slopes was defined Rock Mass Creep (RMC) while evolution to failure was mainly threated by mechanical approaches referred to discontinuous media as well as to equivalent continuum ones (Sitharam & Sridevi, 2000; Hoek *et al.*, 2002).

Nevertheless, it is actually not realistic to distinguish the time-dependent deformations affecting slopes from the strength-driven failure effects, as the deformations reached within a rock mass result from a combination of stress conditions and strain rates, both of them depending on shape and dimensions of slopes as well as on time available for creep evolution. In this regard, general collapse of slopes occurs after a time interval characterized by stationary creep conditions, *i.e.* when the increased strain rate leads to progressive failure associated to strength reduction (Eberhardt *et al.*, 2004; Stead *et al.*, 2006). Nevertheless, if transitory effects occur (*i.e.* earthquakes, teleseismic events, intense rainfalls, ground water changes, anthropic actions) time for failure could be anticipated or delayed depending on what it happens, *i.e.* on the changes sharply induced in the slope shape and/or in the stress-field (Martino *et al.*, 2004; Maffei *et al.*, 2005; Lenti *et al.*, 2015; Bozzano *et al.*, 2016). As a consequence, it is not obvious to assume that slope stability conditions become more critical with time as transitory effects could induce a re-start of the creep process (*i.e.* zeroing the creep time) or, on the contrary, they could be responsible for an increase in strain rate. Moreover, contrarily to the widely diffused idea that RMC processes can only affect high-dimensioned slopes (*i.e.* km-scale), it is worth stressing that theoretically they can be considered also in case of slopes having more reduced dimensions (*i.e.* hm-scale) if the involved rock masses are characterized by low viscosity and stiffness values (Petley *et al.*, 2005).

As a consequence, to infer a more suitable evaluation of the elapsing time for failure in creep evolving slopes, a multi-modeling approach should be applied to infer rheological parameter values. Such an approach should include the three contributions described in the following list.

1) A morpho-evolutionary modeling of the rock slope, based on geo-morphometric and geochronological techniques addressed to point out geomorphic markers, useful to reconstruct a temporal scanning of the main episodes and rates of morphological variations (*i.e.* due to the combination of tectonic offsets and erosional processes) that affected the slope-to-valley floor system.

2) A detailed engineering-geology modeling, that transposes geomechanical parameters to geological-structural features on significant geological sections adopting an equivalent continuum approach for attributing mechanical parameters to lithotechnical units. Such an approach is particularly suitable for slope-scale processes especially in addition to a geomechanical zoning that takes into account the statistical variability of parameters values.

3) A time-dependent stress-strain numerical modeling, performed by a sequential approach according to the main morpho-evolutionary stages of the slope. The time-dependent stress-strain analysis requires the assignment of rheological parameters that should result from a numerical calibration by means of well constrained back-analyses of already occurred failure events.

Experiencing such a multi-modeling approach allows (Fig. 1) to include in a unique methodology the back-analysis of occurred landslide events and the forecast of slope evolution by considering possible scenarios of suitable destabilizing actions (forced

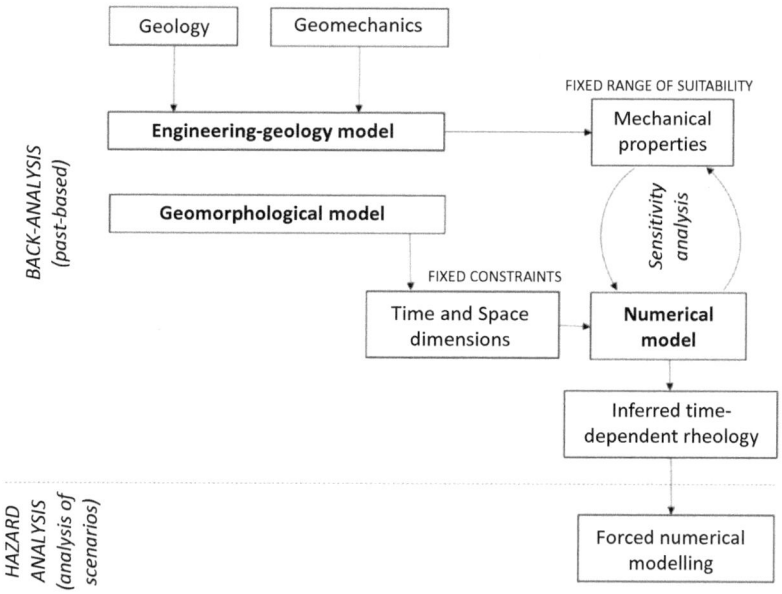

Figure I Flow diagram showing the multi-modeling approach.

modeling). Moreover, the computed strain rates can be considered to establish the best solutions for slope monitoring by terrestrial or remote sensing devices. The multi-modeling methodology to analyze the creep evolution of slopes toward failure or general collapse is here illustrated by two case histories both from the Apennine chain, Italy.

2 METHODOLOGY

The here presented multi-modeling approach allows to infer rheological parameter values for evaluating the strain evolution due to RMC processes that lead natural slopes toward failures or general collapse. Geological and geotechnical contributions are required to exhaustively perform an engineering-geology modeling while time-space evolution of landforms, where slope instabilities occur, is derived from a geomorphological model (Fig. 1). Geomechanical characterization by field surveys and laboratory testing as well as engineering geology modeling through specifically derived cross sections lead to a numerical modeling that makes the link among the collected field evidences, the lithotechnical zoning and the obtained stress-strain outputs. The final goal is to best tune a rheological behavior that reproduces the already occurred strain effects recorded by field evidence, by back-analyzing the past slope vs. valley evolution through a sensitivity analysis to parameter values (Fig. 1). The so tuned solution is suitable for estimating the time elapsing to future slope failures or general collapse as a consequence of creep process as well as of possible occurrence of transient destabilizing actions.

2.1 Description of the multi-modeling approach

The here presented multi-modeling approach extensively adopts the concept of "model" as it means a suitable and reliable transposition of conceptual features and/or elements to a pragmatic tool, that can consist in a drawing reproduction (*i.e.* cross sections, scheme or 3D block-diagrams) as well as in a numerical or physical product (*i.e.* analytical computation, discrete numerical domain, physic-analogue system). Such an approach implies a coupling of quantitative multi-disciplinary models (Fig. 1) to constrain:

a) the morphological slope evolution by a stage-to-stage sequential scheme describing the valley evolution in terms of time for valley deepening and widening phases. The so reconstructed sequence leads to identify a suitable "time zero" for the creep process (t_0) and to discretize the time-step evolution of the slope-to-valley floor system;

b) the geomechanical rock mass properties that are different for rock matrix and joints (according to both discontinuous or equivalent continuum approaches);

c) the rheological parameters (*i.e.* viscosity and stiffness values) that are generally requested to define a general visco-plastic behavior.

These three features are fundamental to perform a back-analysis of slope deformations, also including natural and/or man-induced re-shaping as well as time of their occurrence, in order to constrain the reliable variability of mechanical rock mass parameters and to verify the suitability of both slope deformations and mechanical parameters in terms of rheological behavior.

More in particular, this last goal can be reached by using analytical models (*i.e.* physically-based constitutive laws) to describe the most generalized visco-plastic behavior (Karato, 2008) of a rock mass (*i.e.* according to several literature analytical models like the Kelvin-Voigt, the Maxwell, the Burger or the Zener ones among the others) and by attributing it to numerical codes. The latter, that generally adopt finite difference or finite elements solutions, are requested for computing time-dependent stress-strain effects (Apuani *et al.*, 2007; Discenza *et al.*, 2011; Baroni *et al.*, 2014; Xu *et al.*, 2014).

As an alternative to the numerical modeling, scale-reduced or gravity-forced analogue models could be performed to simulate under laboratory-controlled conditions the stage-to-stage evolutionary scheme of the natural systems using artificial material and/or conditioned stress fields that are down-scaled respect to the real prototype according to physically-based solutions (Chemenda *et al.*, 2005; Bozzano *et al.*, 2013; Bretschneider *et al.*, 2013). These solutions need the geomechanical characterization of the real prototype (*i.e.* the natural jointed rock mass) as well as the dimensioning of the strain-involved slope volume to be previously performed.

2.2 Selected case studies

In order to exemplify the above described multi-disciplinary modeling we selected two experienced case studies (Fig. 2). Both of them are located in the Apennine chain (Italy) but they are characterized by different features as:

Figure 2 Panoramic view of the Scanno (left) and Santa Trada (right) landslides.

– the first one deals with a rock slope instability affecting a whole slope-to-valley floor system of Mt. Rava which is part of the Genzana ridge close to the lake of Scanno (Abruzzo, central Italy) where man-induced stresses are negligible, while geodynamic stress regime variations as well as tectonic offsets and erosional/depositional processes are more prominent: rock mass volumes are on the order of 10^7 m^3 and the considered time interval spans from 1Ma up to present;

– the second one deals with a rock slope instability occurred in the Santa Trada valley (Calabria, southern Italy) on a more reduced spatial and temporal scale where anthropic activities played a relevant triggering role: the volume is on the order of 10^5 m^3 and the elapsed time to failure on the order of 10^5 years.

For both the here presented case histories the geological-structural setting, expressed by a detailed engineering-geology zoning, represented a fundamental predisposing factor to the gravitational slope instability. At the same time, well-constrained failure conditions allowed a suitable back-analysis for calibrating the rheological behavior of the involved rock masses.

3 CASE STUDY 1: SCANNO LANDSLIDE (ABRUZZO, CENTRAL ITALY)

The Scanno landslide area is located in the Central Apennines (Fig. 3), along the Tasso-Sagittario river valley. The study area is part of a complex geological-structural frame, where different geological-structural units belonging to different paleo-geographic domains (the carbonate platforms and the pelagic basin), are put in contact by important fault systems predominantly N–S and NNW–SSE oriented. More in particular, it is characterized by a NE-verging imbricate, fold-thrust belt developed under a Neogene, ENE verging, compressive thrust system. The phase of thrusting was followed by

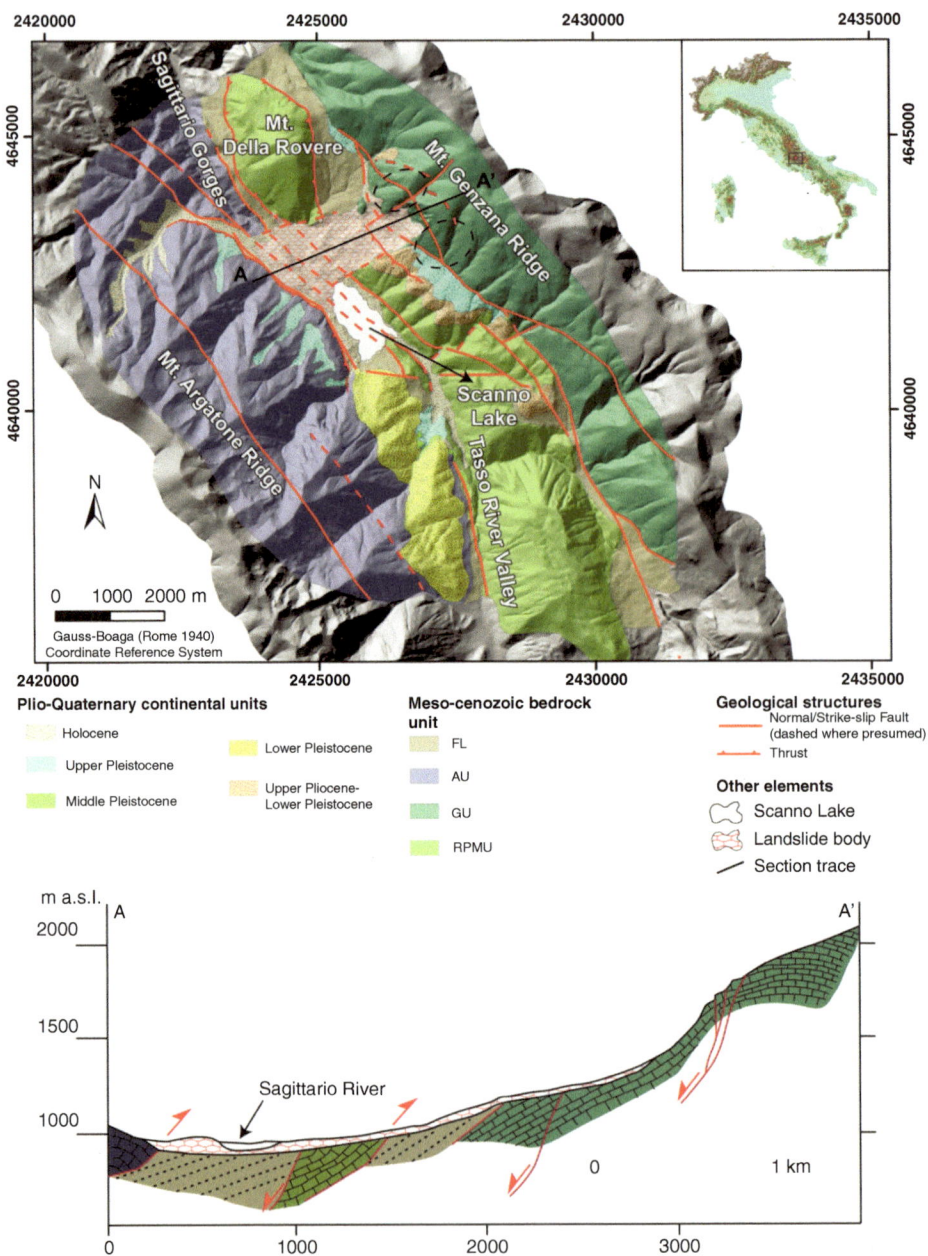

Figure 3 Geological map of the Scanno landslide area and geological cross section.

extensional tectonic activity that began in the Pliocene–Lower Pleistocene (Esposito *et al.*, 2013 and references therein). Regional uplift processes affected the Central Apennines since Lower Pleistocene up to Present. At least three main geological-structural units are present astride the Tasso-Sagittario river valley (Esposito *et al.*, 2013):

1. Mt. Argatone Unit (AU), a homoclinal sequence of Meso-Cenozoic dolomitic limestones and limestones ascribable to the Latium-Abruzzi carbonate platform paleo-geographic domain.
2. Mt. Genzana Unit (GU), which crops out in the Mt. Genzana ridge as a homoclinal sequence of rocks belonging to the transition zone between the Apulian carbonate platform and an ancient basin then erased during the orogenesis. These deposits consist of Meso-Cenozoic limestone and marly limestone.
3. Mt. della Rovere-Piana Malvascione Unit (RPMU), which crops out between Mt. Genzana and Mt. Argatone ridges: it is composed of micritic and detritic limestones with interlayered clayey beds.

The contact between the AU eastern slope and the Tasso-Sagittario river valley is marked by a tectonic element, corresponding to a NNW–SSE oriented and eastwards dipping thrust surface. GU is bordered on its western slope by an important tectonic lineament (Genzana Fault) NNW–SSE oriented and dipping 60–80° westwards. The kinematic indicators point out at least a first strike-slip activity and a subsequent normal dip-slip one. This major tectonic element is part of a fault system NNW-SSE which reaches the valley bottom and which is locally interrupted by normal faults roughly E-W oriented, thus delineating a rombohedral depression. Clayey and sandy-clayey, synorogenic flysch deposits (FL) crop out within the valley bottom: their contacts with the local pre-Miocene sequences are exclusively represented by tectonic elements. The RPMU has tectonic contacts with the surrounding geological units marked by NNW–SSE oriented faults. The local stratigraphic framework is completed by Pleistocene up to recent continental, clastic deposits such as alluvial fans, debris cones, slope talus, alluvial and colluvial deposits.

Geomorphic evidence of a gravity-induced slope deformation (DSGSD, sensu *Auct.*) affecting the western slope of the Mt. Genzana ridge are present over an area of ~ 4 km^2 corresponding to Mt. Rava (Esposito *et al.*, 2013) and involve an area bordered at the bottom by the GF trace and at the top by a sharp morphologic lineation (Fig. 4a, b). Within this sector, the sharpest landform is represented by the Scanno rockslide-avalanche and by related deposit and rockslide-dammed Scanno lacustrine basin (Bianchi-Fasani *et al.*, 2004; Bianchi-Fasani *et al.*, 2011; Nicoletti *et al.*, 1993). According to previous studies (Nicoletti *et al.*, 1993), the landslide involved a rock mass volume of ~ 80 Mm^3 that slid onto the valley bottom and partially ran up on the opposite slope, with a 3.6 km total run-out. After the collapse onto the valley bottom the debris dammed the Palaeo-Tasso River and formed the Scanno Lake (Fig. 4c, d).

The landslide debris covers an area of ~2.6 km^2, mainly developed on the flysch deposits of the valley bottom, and has a typical hummocky topography with a marked frontal ridge. Some road cuts locally show an inverse sorting of the debris, that is composed by a sandy-gravel matrix with interspersed blocks and boulders with volumes up to tens of cubic meters. These morphological and sedimentological features together with the large volume involved and the evidence of a partial run-up on the opposite slope, indicate a rock-avalanche mechanism of emplacement (Bianchi-Fasani *et al.*, 2004; Bianchi-Fasani *et al.*, 2011). The presence of typical landforms such as trenches and counter-slope scarps in the slope section featuring the right flank of the landslide scar, strongly suggest that the occurred rock-avalanche represents a localized

Figure 4 View of the Scanno landslide. a) scar area, right flank; b) scar area, left flank; c) landslide debris filling the Tasso River valley and damming the lake of Scanno; d) Lake of Scanno.

failure within a wider slope deformation. The sharpness of these landforms as well as the characteristics of the infill of the main trench suggest an ongoing activity of the deformation process.

3.1 Morpho-evolutionary modeling

A detailed study of relict intra-vallive erosional surfaces starting from their remnants hanging at different heights upon the present-day valley-floor was performed aimed at: i) recognizing former base-levels; ii) estimating the timing of their deactivations; and iii) reconstructing the stages of landscape evolution. Remnants of the relict surfaces were firstly selected starting from the 5 × 5 m Digital Terrain Model (DTM) as the areas with a slope angle <15°, which are clustered to break the steep slopes along the valley-sides. The final selection was made through a detailed field-based geomorphological analysis that allowed us to discard litho-structural sub-planar surfaces and man-made surfaces. Field surveys were useful to confirm the presence of strath terraces (*sensu* Bull, 1990). Strath terraces were identified as fluvial erosional surface of the Tasso-Sagittario fluvial system (when it was connected), carved directly on bedrock and/or on alluvial-to-slope deposits (*i.e.* fill-cut terraces of Bull, 1990 as reinterpreted in

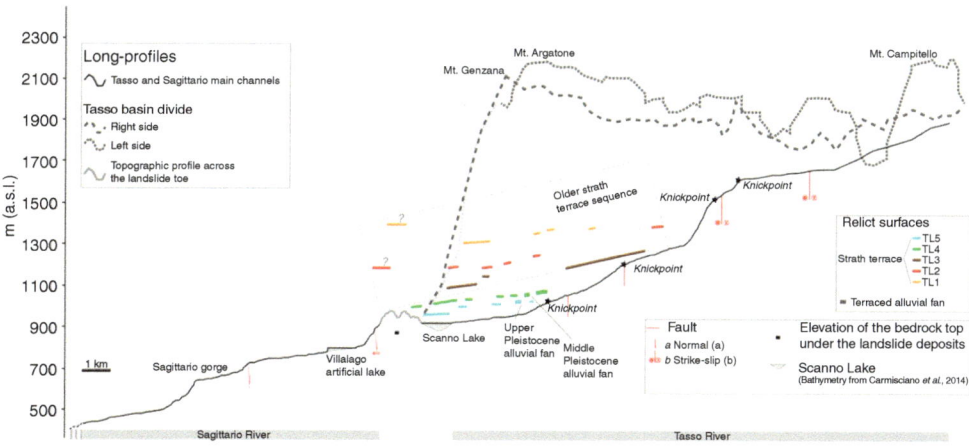

Figure 5 Longitudinal profile along the Tasso river, showing the remnants of relict surfaces correlated with main knickpoints.

Nesci *et al.*, 2012). Whatever their typology, along the fluvial system draining the Apennine chain was demonstrated that strath terraces staircases formed in response to period of prevalently fluvial lateral erosion separated by tectonically-induced downcutting (Wegmann & Pazzaglia, 2009; Gioia *et al.*, 2011; Nesci *et al.*, 2012), the latter being often related to highrate of regional uplift, coseismic uplift, and local faulting. Therefore, the intra-vallive location of strath terraces and their along-valley correlation is useful for reconstructing the past base-levels of erosion and, thus, the former levels of valley-floor. Correlations among the remnants of relict surfaces was performed after having projected them on the stream long-profiles of the Tasso-Sagittario fluvial system, as shown in Figure 5, where a good spatial continuity can be observed among surface remnants and knickpoints along the river profile, both testifying for ancient base levels (Della Seta *et al.*, 2016). Age constraint of the relict surfaces allowing their use as geomorphic markers was inferred indirectly, by taking information from the most recent official geological map available for the area and from associated notes (Carta Geologica d'Italia at the scale of 1:50.000, Sheet 378 "Scanno", available at the website: http://www.isprambiente.gov.it/Media/carg/). In particular, the age of each relict surface was inferred based on its relative height upon the present valley-floor and on its along-valley distribution, on the correlation with Quaternary geological units of known age and on the age of the deposits they are carved into in the case of fill-cut terrace typology.

Relict surfaces in the study area are clustered in five distinct levels hanging at different heights, up to an altitude of ~ 1230 m a.s.l. The higher and older surfaces are characterized by strath terraces carved into early Pleistocene fluvial deposits, slope breccias and bedrock. This older strath terrace sequence is to some extent quite continuous along the Tasso River and headwater sector of the Sagittario River and is often represented by wide sub-planar surfaces. Three terrace levels can be clearly distinguished, named TL1, TL2, and TL3 from the highest to the lowest one, respectively. The early Pleistocene age of the deposits they are carved into, constrains the middle Pleistocene age of the TL1-to-TL3 relict surfaces. The successive relict surfaces (TL4 and TL5) are characterized by sub-planar remnants carved prevalently on bedrock that represent the basal strath terraces, upon which were locally deposited late-middle Pleistocene gravels and sands. The height distribution of the strath staircase in the study area allows constraining the base level lowering since approximatively the last 10^3 ka, *i.e.* since the formation of the first relict landscape after the fluvial deposits were accumulated.

Based on the above discussed geomorphological evidences, it is possible to suggest a conceptual model for the gravitational morpho-genesis of the study area (Fig. 5) that refers to the following 6 steps as reported in the following list.:

1) ~925 ka the erosion of the first strath surface (TL1) was completed, which is carved into the Lower Pleistocene deposits (end of deposition at $t_0 = \sim 1$ Ma).

2) ~925 ka stream entrenchment began reshaping the TL1 surface and a shift to river lateral erosion caused the formation of the second erosional surface (TL2), which was completely shaped ~675 ka. TL2 was carved again into the Lower Pleistocene deposits but at lower elevation than TL1.

3) ~675 ka stream entrenchment began reshaping the TL2 surface and a shift to river lateral erosion caused the formation of the third erosional surface (TL3), which was completely shaped ~450 ka.TL3 is the last surface carved into the Lower Pleistocene deposits and lies at lower elevation than TL1 and TL2.

4) ~450 ka stream entrenchment again deactivated the TL3 strath terrace and was followed by the formation of the fourth erosional surface (TL4) carved prevalently in bedrock and representing the basal surface of Middle Pleistocene alluvial fan deposits. TL4 was completely shaped ~330 ka.

5) Between ~330 ka and ~125 ka the Scanno fault system in the valley bottom likely activated and thick alluvial fan deposition occurred at the tributary junction. The fifth strath surface (TL5) formed carving prevalently in bedrock and representing the basal surface of Upper Pleistocene alluvial fan deposits. A second faulting phase occurred, likely responsible for the fluvial disconnection of the Tasso-Sagittario system, which is testified by a topographic bedrock threshold at the Sagittario headwaters. An endoreic basin developed, as suggested also by a southward draining paleo-channel in the northern sector of the area, and was accompanied by thick alluvial fan deposition at the tributary junction upon the TL5 erosional surface.

6) The present day valley profile is reached after the occurrence of two distinct landside events (Esposito *et al.*, 2013): the Malvascione and the Scanno landslide respectively. As resulted by C^{14} dating (Bianchi Fasani *et al.*, 2004), the Scanno landslide occurred ~10 ka.

3.2 Engineering-geology modeling

Beyond the bedding attitude, the main joint sets identified are basically referable to two conjugate pairs roughly perpendicular to each other. The first couple is formed by the set with dip direction 225 ± 15 and the set 40 ± 10; in both cases the dip angle is quite variable but usually greater than 50°. The second conjugated couple is composed by the

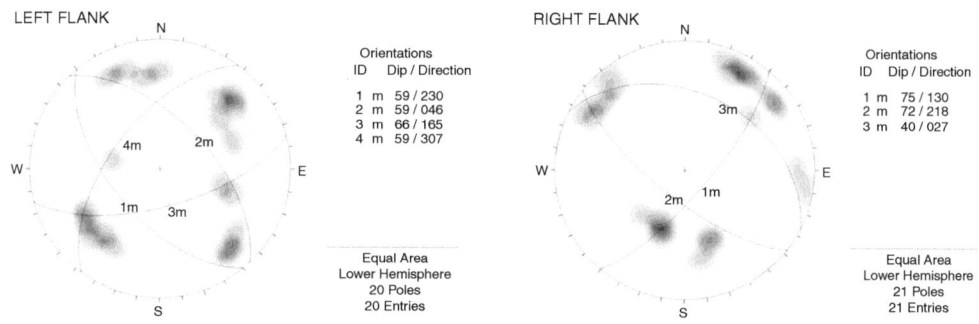

Figure 6 Stereo plots obtained at the left and right flank of the Scanno landslide scar area.

set 300 ± 20 and the set 140 ± 40; also in this case the dip angle is quite high even if locally values of ~ 20° have been measured.

Small scale anticlinal folds NNE–SSW oriented have been also recognized. In this regard, the observed trenches can be considered as the morphological expression of passive opening of the conjugated joint sets with dip directions 225 ± 15 and 40 ± 10, as a response to the downslope movement of the rock-mass. The so reconstructed structural frame allows to point out the main structural constraints that guided the gravity-driven processes (Fig. 6).The detachment area of the rockslide is strongly constrained by structural elements: i) the main scarp is roughly coincident with a NW–SE oriented high angle discontinuity set; ii) the right flank roughly coincides with a high angle, NE–SW oriented plane (Fig. 6); iii) the left flank follows the bedding attitude which is locally represented by bedding planes NW dipping with an angle of 50–60°, just in correspondence of the limb of a meso-scale fold structure (Fig. 6).

Finally, the chair-shaped geometry of the rupture surface is constrained at its base by the bedding attitude, which is NW–SE oriented and dips toward SW with an angle of ~ 30°. Following this reconstruction it is possible to describe the failure mechanism of the Scanno rock avalanche as a (strongly asymmetric) wedge slide (Hungr and Evans, 2004). Such a hypothesis is well constrained by the presence of clayey layers within the stratigraphic sequence, as they represent weaker planes with respect to the shear strength that can be mobilized along the bedding planes.

In order to attribute geomechanical parameter values to the rock masses involved in the landslide, a spatial distribution of geomechanical indices suggested by ISRM (2007), such as Ib (block size index) and Jv (the volumetric joint count), was derived by direct field measurements (Fig. 8); statistical Gauss normal distributions were derived to distinguish rock mass classes in terms of jointing conditions.

For the RPMU the Jv varies from 16 up to 58 joint/m^3 for an increasing distance from main fault lines up to 700 m while for the GU the Jv varies from 6 up to 55 joint/m^3 for an increasing distance from main fault lines up to 350 m.

Based on these indices a continuum equivalent approach was adopted (Sridevi & Sitharam, 2000; Hoek *et al.*, 2002; Esposito *et al.*, 2007) to attribute mechanical parameters to the jointed rock mass; moreover specific triaxial tests were

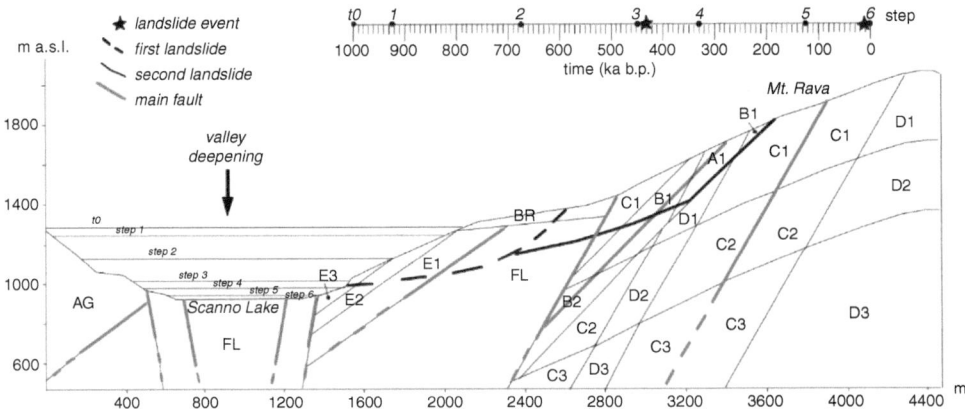

Figure 7 Engineering-geology section along trace AA' of Fig. 3 showing the space-time evolution of the Tasso river valley in correspondence to the Scanno landslide.

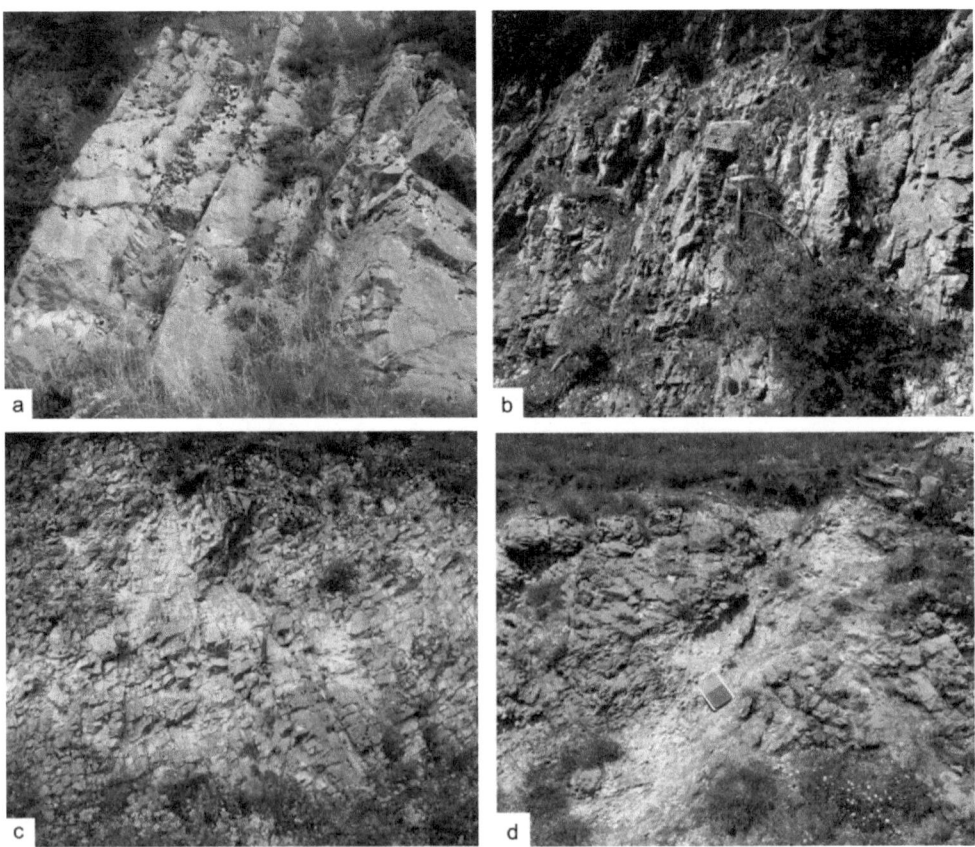

Figure 8 Limestone outcropping on the Scanno landslide slope: the rock mass jointing increases from a) to d), this last one corresponding to a cataclastic fault zone.

performed to measure intact rock mass mechanical parameters (Esposito *et al.*, 2015). According to such an approach, 15 lithotechnical units were distinguished at all (Figure 7 and Table 1), taking into account both jointing conditions and depth. The resulting Young modulus values for the jointed rock mass range from 3 to 40 GPa.

3.3 Stress-strain numerical modeling

A stress strain numerical modeling was carried out through the finite difference code FLAC 7.0 (Itasca, 2011) reproducing the morpho-structural evolution of the SW slope of Mt. Rava in order to calibrate the rheological properties of the rock masses involved in the Malvascione and Scanno landslides (Della Seta *et al.*, 2016). To this aim, a back-analysis was performed of the two landslide events of Malvascione and Scanno respectively, by a sensitivity analysis regarding the attributed viscosity values (Fig. 9). The numerical domain consists in a 550×210 rectangular grid with a square mesh resolution of 10 m covering the 5 km length geological cross-section of Fig. 3 and a range of elevation between 0 and 2100 m a.s.l.

Starting from an initial gravitational equilibrium, a sequential modeling was performed through the 6 steps, referred to the evolutionary model reported in the paragraph 3.2, *i.e.* reproducing the morpho-structural evolution of the SW slope of Mt. Rava strictly related to the one of the Tasso river valley (Fig. 7). At each step, a first elasto-plastic equilibrium was reached, followed by a second equilibrium obtained for an anisotropic stress-strain behavior (*i.e.* attributing a ubiquitous joint constitutive law). After this last equilibrium has been reached, the time-dependent simulation was performed in creep-mode configuration by assuming a visco-plastic rheological model, according to a Burger constitutive law that joint in series a Maxwell visco-plastic element with a Kelvin-Voigt visco-elastic one (Karato, 2008).

The stress field of the numerical domain was constrained to the regional one. Such an effect was simulated by correcting the stress values along the vertical and horizontal directions according to a K-ratio (=horizontal/vertical stress) decreasing from 0.60 to 0.35, *i.e.* related to a stress field due to a normal NW-SE fault activity occurred during the last 260 ka (Bianchi Fasani *et al.*, 2011).

To perform this correction a stress decrease proportional to the vertical stress was applied along the horizontal direction to reproduce the assumed K-ratio variation. The time-dependent numerical solutions were obtained by assuming a $15 \cdot 10^6$ s (almost equal to 6 month) for the time step duration and by attributing viscosity values varying in the range $10^{19} - 10^{22}$ Pa·s, to carry out a numerical sensitivity analysis of the rock mass visco-plastic behavior (Fig. 9).

As it regards the strength properties, they were reduced according to a strain-softening behavior and residual condition were attributed when the strain threshold of 100% was exceeded (*i.e.* the displacement is equal to the mesh resolution). The outputs of the sensitivity analysis performed to calibrate the rock mass viscosity show that the viscosity values of 10^{19} and 10^{22} Pa·s can be regarded as end-members since, in the first case, a generalized failure of the Mt. Rava slope should have occurred at 700 ka while, in the second case, no slope failures should have occurred at all. On the other

Table 1 Mechanical properties of the 15 lithotechnical units reported in Figure 7.

lithotechnical unit	rheology	den (kg/m³)	ϕ (°)	coh (Pa)	ten (Pa)	B_1 (Pa)	G_1 (Pa)	G_2 (Pa)	η_1 (Pa*s)	η_2 (Pa*s)
AG	elastic	2.6E+03	–	–	–	3.7E+09	2.2E+09	–	–	–
FL	elasto-plastic	2.5E+03	32	9.2E+06	1.5E+07	1.5E+08	9.2E+07	–	–	–
BR	elasto-plastic	2.1E+03	30	4.9E+03	5.0E+06	1.2E+07	9.2E+06	–	–	–
A1	visco-plastic	2.5E+03	57	1.6E+07	9.2E+06	6.6E+09	4.0E+09	2.40E+09	1.0E+20	1.0E+21
B1	visco-plastic	2.5E+03	57	1.6E+07	9.2E+06	2.0E+09	1.2E+09	2.40E+09	1.0E+20	1.0E+21
B2	visco-plastic	2.5E+03	57	1.6E+07	9.2E+06	2.4E+09	1.4E+09	2.40E+09	1.0E+20	1.0E+21
C1	visco-plastic	2.5E+03	57	1.6E+07	9.2E+06	1.9E+10	1.2E+10	2.40E+09	1.0E+20	1.0E+21
C2	visco-plastic	2.5E+03	57	1.6E+07	9.2E+06	2.3E+10	1.4E+10	2.40E+09	1.0E+20	1.0E+21
C3	visco-plastic	2.5E+03	57	1.6E+07	9.2E+06	2.5E+10	1.5E+10	2.40E+09	1.0E+20	1.0E+21
D1	visco-plastic	2.5E+03	57	1.6E+07	9.2E+06	3.1E+10	1.9E+10	2.40E+09	1.0E+20	1.0E+21
D2	visco-plastic	2.5E+03	57	1.6E+07	9.2E+06	3.7E+10	2.2E+10	2.40E+09	1.0E+20	1.0E+21
D3	visco-plastic	2.5E+03	57	1.6E+07	9.2E+06	4.1E+10	2.5E+10	2.40E+09	1.0E+20	1.0E+21
E1	visco-plastic	2.3E+03	66	1.1E+07	9.9E+06	2.9E+10	1.7E+10	2.40E+09	1.0E+20	1.0E+21
E2	visco-plastic	2.3E+03	66	1.1E+07	9.9E+06	4.7E+10	2.8E+10	2.40E+09	1.0E+20	1.0E+21
E3	visco-plastic	2.3E+03	66	1.1E+07	9.9E+06	4.2E+10	2.5E+10	2.40E+09	1.0E+20	1.0E+21

Note: den = density; ϕ = friction angle; coh = cohesion; ten = tensile strength; B_1 = bulk modulus of Maxwell rheological element; G_1 = shear modulus of the Maxwell rheological element; G_2 = shear modulus of the Kelvin-Voigt rheological element; η_1 = viscosity of the Maxwell rheological element; η_2 = viscosity of the Kelvin-Voigt rheological element.

hand, if the viscosity values range from 10^{20} to 10^{21}Pa·s a first slope collapse should have occurred at ~330 ka corresponding to the Malvascione landslide event. Nevertheless, only in the case of a viscosity value of 10^{20}Pa·s (Fig. 10), a second slope failure occurs at ~10 ka, *i.e.* in very good agreement with the Scanno landslide C^{14} dating reported by Bianchi Fasani *et al.* (2004). If a 10^{21}Pa·s viscosity value is assumed the second slope failure, ascribable to the Scanno landslide, would occur within the next 2 ka from the present in the same climatic conditions and considering that no external actions (*e.g.* earthquakes) forced the slope.

If a viscosity value of 10^{20}Pa·s is attributed to the Mt. Rava rock masses, the time dependent sequential modeling points out that relevant deformations occurred since 675 ka related to the MRC, evolving toward a shear-zone-driven rock slide within ~300 ka (Fig. 9). More in particular, as it results from the numerical modeling, at ~450 ka a biplanar compound (*sensu* Hutchinson, 1988) enveloped slope deformation

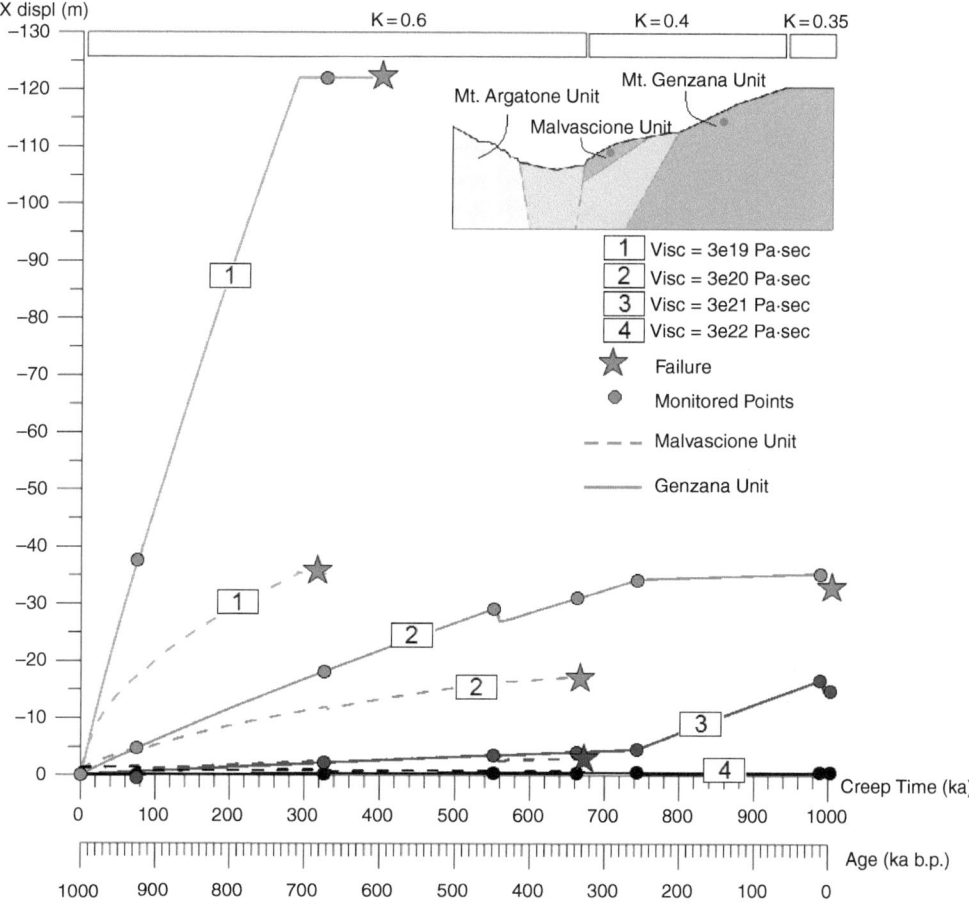

Figure 9 Sensitivity analysis performed for the Scanno landslide back-analysis. The η_1 value was varied to obtain the last slope collapse event at the C^{14} age dated by Bianchi Fasani *et al.* (2004).

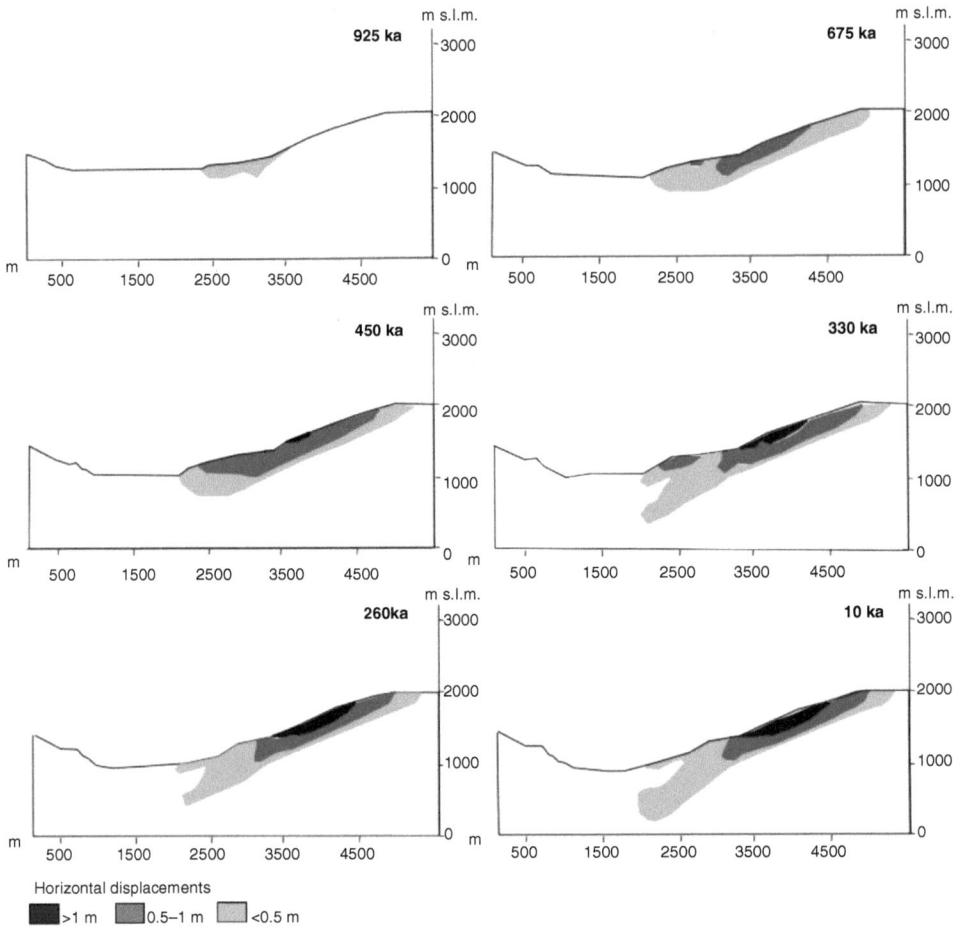

Figure 10 Outputs of the Scanno landslide stress-strain numerical modeling, obtained by zoning the model according to the engineering-geology section of Fig. 7 and by assuming the rheology of Table 1.

clearly outcomes and characterizes the Mt. Rava slope deformation with a tip zone at ~1750 m a.s.l. and a break-out zone at ~1250 m a.sl., *i.e.* hanging ~500 m upon the thalweg of the Tasso river at that time.

At ~330 ka a slope failure occurs involving a rock mass volume of ~40 Mm^3 that develops in correspondence to the lower portion of the deforming mass with respect to the biplanar compound enveloped slope deformation. According to the morpho-structural evolutionary model of the SW slope of Mt. Rava, this failure event should correspond to the Malvascione landslide event (first landslide event in Fig. 7). After this landslide occurrence, the ongoing MRC envelopes a residual deforming zone located uphill, *i.e.* up slope the top of the Mt. Genzana, with a planar enveloped strain-concentration that evolved toward the Scanno landslide event at ~10 ka (second landslide event in Fig. 7).

Figure 11 View of the Santa Trada landslide. a) panoramic view of the Santa Trada creek (the landslide is clearly visible close to the highway) and of the marine terraces along the Tyrrhenian Sea coastline; b) main scarp of the Santa Trada landslide; c) landslide debris visible on February 2009 in the quarry area.

4 CASE STUDY 2: SANTA TRADA LANDSLIDE (CALABRIA, SOUTHERN ITALY)

On 30 January 2009, a rock slide involving approximately 0.23 Mm^3 of highly jointed gneiss was recorded in the Santa Trada creek valley (Calabria, Southern Italy). The landslide occurred after exceptional autumnal rainfalls and involved a quarry whose activity has been documented over the past 10 years by aerial images (Bozzano *et al.*, 2012). The 30 January 2009 landslide involved the Hercynian gneiss outcropping on the left side of the Santa Trada creek valley (Fig. 11a).

Data recorded by the meteorological station of Villa S. Giovanni (Bozzano *et al.*, 2012 and references therein) show that the cumulative rainfall between November 2008 and January 2009 reached approximately 900 mm, which exceeded the expected cumulative rainfall of approximately 280 mm for that period. The landslide contributed to a general re-activation of a pre-existing gravitational deformation due to RMC. Until 9 February 2009, displacements of up to tens of centimeters were recorded by a GBInSARLiSA device that was installed by the Italian National Department of Civil Protection (Bozzano *et al.*, 2012 and references therein). Moreover, vertical displacements up to 2 m were observed at a macroscopic scale along the main scarp located at approximately 190 m a.s.l. (Fig. 11b), corresponding to the crown of a pre-existing gravitational slope deformation.

The landslide mechanism can be classified as a rock slide (Cruden & Varnes, 1996). The crown area has a width of ~110 m, is located at ~190 m a.s.l. and corresponds to a 2 m high scarp (Fig. 11b). The landslide debris is composed of irregular blocks of gneiss with volumes up to several cubic meters (Fig. 11c). Secondary landslides were observed close to the main scar area on both sides. The rock slide mainly involved the quarry area of the slope and it was strongly controlled by a 115/70 (dip direction/dip) joint set; this set pre-existed within the gneiss, as demonstrated by geomechanical measurements obtained from the gneiss outcrops located very close to the landslide area.

The landforms of the slope observed before 30 January 2009 clearly show a pre-existing gravitational slope deformation due to an ongoing RMC process. This deformation was responsible for a convex slope profile and for a wide, multiple scarp-trench system that bounded the landslide mass on the upslope side. The observed pattern of pre-existing scarps and ground failure can be explained by the high-angle translational movement of the jointed rock mass favored by the deepening of the Santa Trada creek. Evidence for this deepening can be derived from the linear erosion of slope debris and alluvial deposits all along the valley. Moreover, a large rock slide event can be clearly observed just in front of the landslide area on the right side of the valley.

The features of this landslide are particularly useful for constraining a time-dependent process at both long (*i.e.*, 10^5-year) and short (*i.e.*, 10^1-year) time scales. The Santa Trada landslide area (Fig. 12) is characterised by outcrops of highly jointed gneiss that can be ascribed to the Hercynian bedrock of the Calabrian arc and part of the Aspromonte metamorphic unit (Bozzano *et al.*, 2012 and references therein). Cataclasitic to mylonitic rocks generally correspond to the main fault lines. The area is also characterized by wide outcrops of deposits corresponding to many orders of marine terraces over marine abrasion platforms. These deposits are composed of reddish-brown gravels and sands with a dip angle of less than 10°. They are transgressive on the Hercynian metamorphic bedrock, as demonstrated by a ~1-meter thick pebbly level that can be observed at their base. Where visible, this level has an attitude of approximately 320/40 (dip direction/dip). The deposits upslope of the Santa Trada landslide slope (263m a.s.l. terrace) that lay over the abrasion platform are dated 243ka (Miyauchi *et al.*, 1994; Dumas *et al.*, 2005). The study area reflects the Campo Piale horst structure, which is bounded to the north by the border fault-system of the Gioia Tauro-Mesima basin and to the south by the Mortille border fault-system of the Reggio Calabria Basin (Ghisetti, 1984; Guarnieri *et al.*, 2004). Two main fault systems occur in the Santa Trada area: the first has a NW-SE orientation, and the second runs approximately parallel to the coast line. Within this structural context, the Santa Trada creek is superimposed on a fault line of the first fault-system, whereas the morphological scarp between 263m a.s.l. and 110m a.s.l. marine terraces corresponds to the main normal fault, which is approximately parallel to the present coast line.

4.1 Morpho-evolutionary modeling

As for the entire Thyrrenian coastal area of southwestern Calabria, the Santa Trada creek valley experienced polycyclic geomorphic processes in response to the sum of climate-driven eustatic oscillations and tectonic crustal deformations (Bozzano *et al.*, 2016). Such a combined input was responsible for the genesis of the spectacular

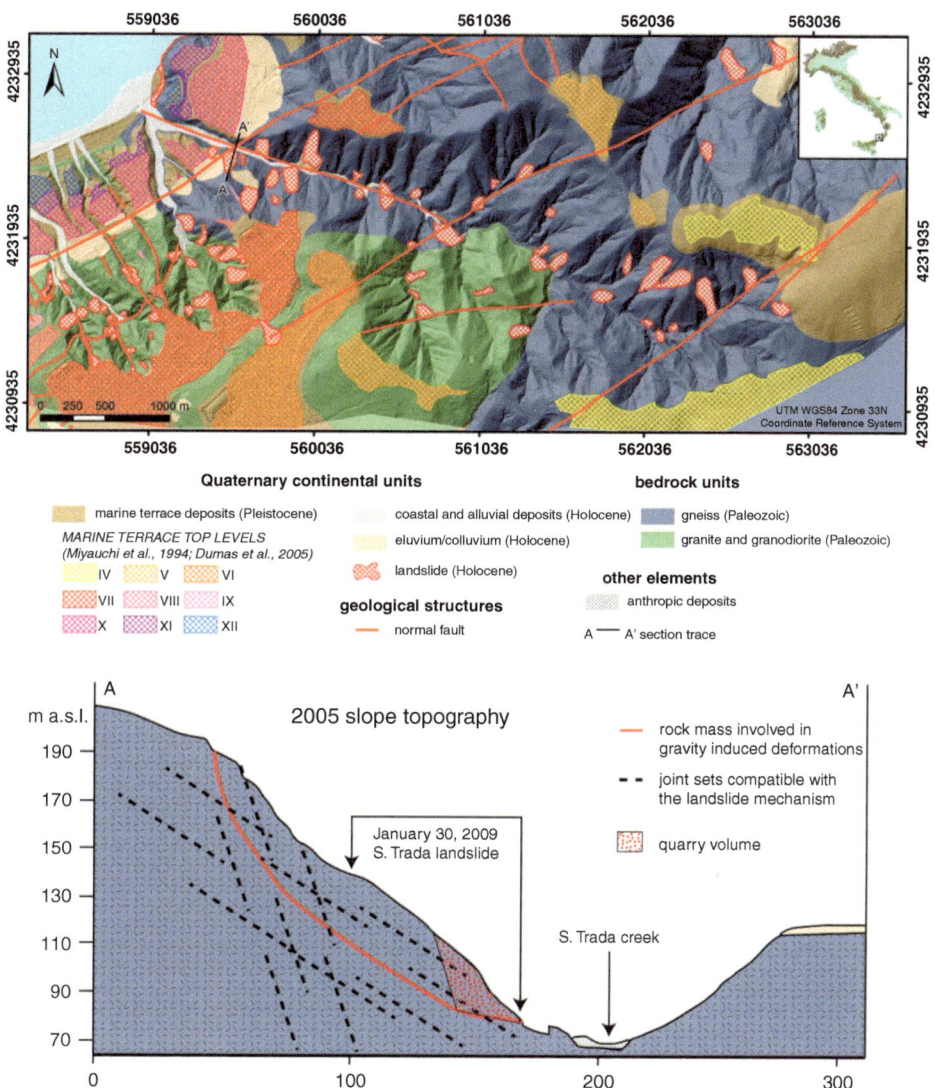

Figure 12 Geological map of the Santa Trada landslide area and geological cross section.

staircase of Quaternary marine terraces, which were progressively formed, uplifted and, once emerged, deeply dissected by river incision, which attempted to re-establish equilibrium conditions along the river profiles with respect to changed base level. All these processes led to a transient landscape, in which relict landforms are still preserved, such as intra-vallive strath terraces.

River incision likely occurred during sea-level fall phases amplified by the combination of regional uplift and cooling climate. During interglacials, instead, the stream power, lowered by sea-level rise, although always fed by regional uplift, was more likely converted by rivers in lateral erosion and formation of strath terraces (Burbank & Anderson, 2012).

Based on the collected geomorphological evidences, seven strath terrace levels were recognized, which are topographically well correlated with the elevation of the inner margins of marine terraces (Fig. 13). Geomorphic evidence show that both the remnants of the marine terrace IV on the left and right upper portion of the drainage divide of Santa Trada creek are at different elevation, thus suggesting that the NW-SE fault was active after their formation. Nonetheless, from the following marine terrace level no evidence of displaced geomorphic markers was observed along this fault. The first (older) strath terrace recognized in the Santa Trada valley correlates well to the marine terrace V and none of the younger levels, which are well correlated with progressively younger marine terraces are displaced by the NW-SE fault. Nonetheless, more recent fault activity was recognized along the fault running parallel to the coastline. In fact, the last three strath terrace levels recognized in the Santa Trada valley apparently do not correlate to the elevation of the three last marine terraces (namely terrace X, at 110 m a. s.l. and 124 ka, terrace XI, at 80 m a.s.l. and 101 ka, terrace XII, at 70 m a.s.l. and 83 ka). However, they are arranged with the same relative height difference, among each other, than the one among the three last marine terraces (Fig. 14). Moreover, if referring to the elevation of such marine terraces in the footwall of the fault (terrace X at 170 m a.s.l., terrace XI at 140 m a.s.l. and terrace XII at 130 m a.s.l.) as reported by Miyauchi *et al.* (1994) few kilometers to the west, the last three strath terraces are very well correlated to them. Therefore, the activity of the WSW-ENE fault is likely to have post-dated the formation of terrace XII.

To summarize, 4 main erosional phases can be roughly related to the deepening of the Santa Trada creek valley (Dumas *et al.*, 2005; Ferranti *et al.*, 2008) in correspondence of the slope affected by the 2009 landslide that is in the footwall of the WSW-ENE fault:

1) a first erosional phase (240 ka – 124 ka) in the period between the formation of the marine terraces at 263 m a.s.l. (t_0) and 170 m a.s.l., respectively;
2) a second erosional phase (124 ka – 101 ka) in the period between the deposition of the marine terrace deposits at 170 m a.s.l. and 140 m;
3) a third erosional phase (101 ka – 83 ka) in the period between the deposition of the marine terrace deposits at 140 m and 130 m a.s.l.;
4) a fourth erosional phase (83 ka – present) in the period between the deposition of the marine terrace deposits at 130 m a.s.l. and the present valley bottom. Due to the lack of geological and geomorphological constraints, in the time interval between two successive marine terrace orders only the main erosional processes were considered for reproducing the sequential evolution of the Santa Trada valley, even if depositional phases or minor erosional ones could have occurred.

Evidence for quarry activity in aerial photographs and satellite images first appears in 2002, while no quarry was present until November 2001. The aerial images acquired between 2001 and 2009 show the existence of a quarry at the bottom of the left-side slope of the Santa Trada creek valley. All of the collected images allowed us to contour the boundaries of the quarry and to evaluate the removed rock volumes over time. Because no DTM was available before 2005 with sufficient resolution, the evidence from the aerial-photo interpretation analyses were reported based on the 2005

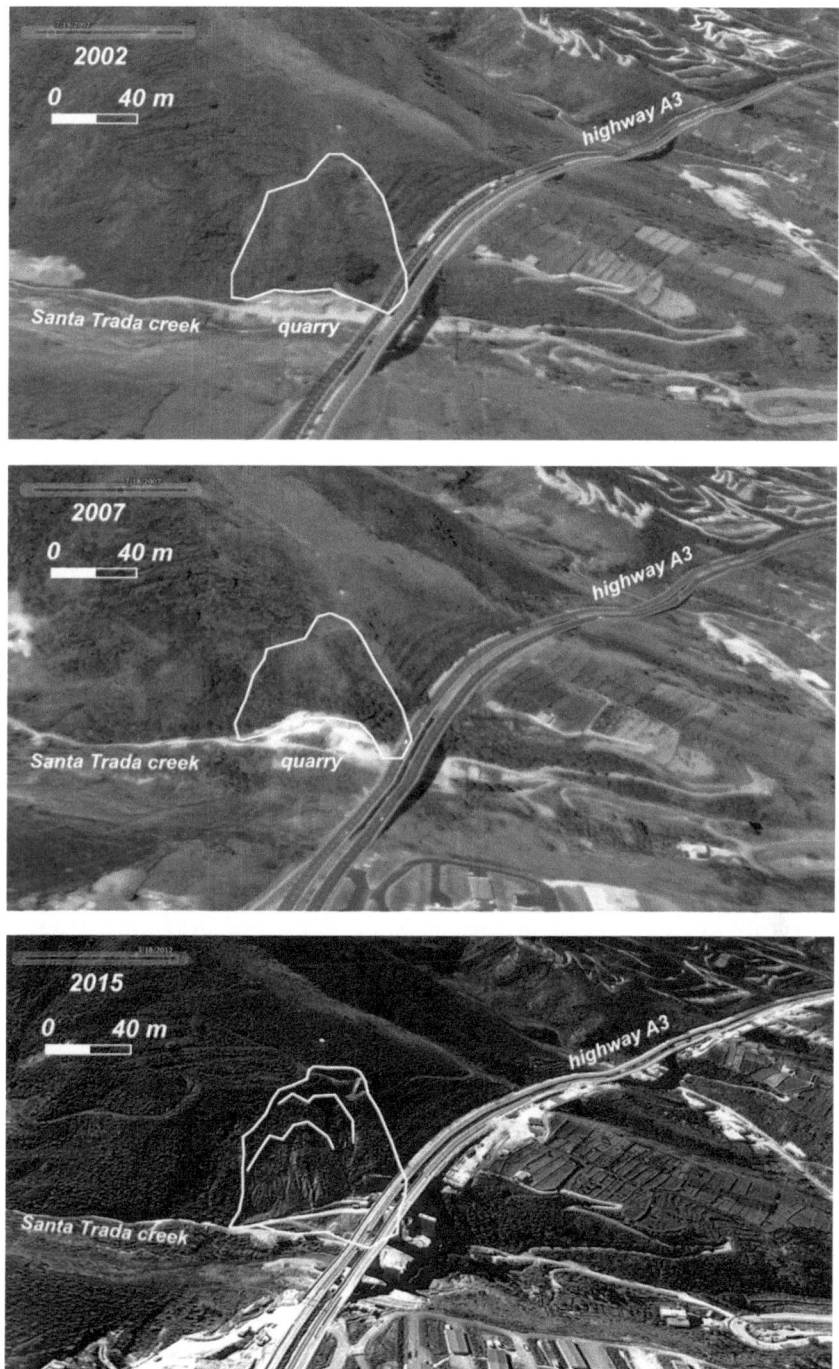

Figure 13 Satellite 3D Google Earth image of the Santa Trada landslide since 2002 to the present. The enlargement of the quarry area at the toe of the already deforming zone is clearly visible from 2002 and 2007, before the slope failure occurred on January 2009.

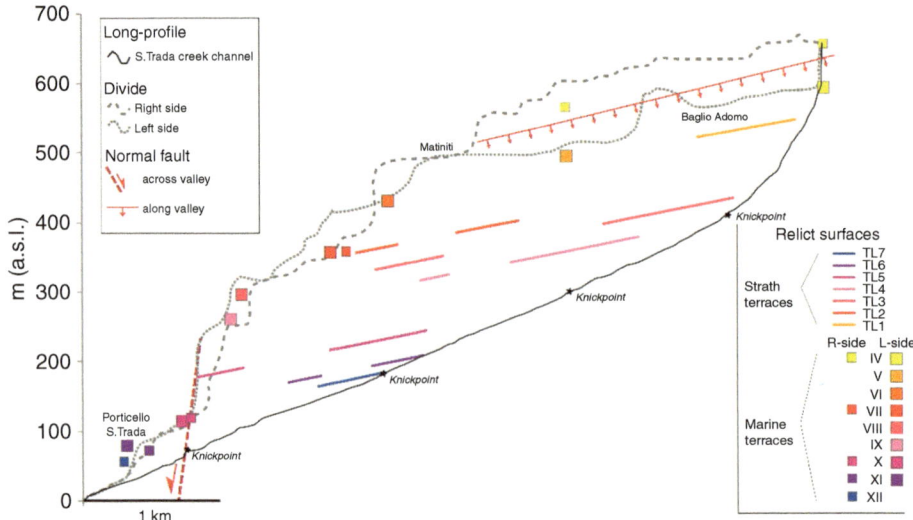

Figure 14 Longitudinal profile along the Santa Trada creek showing the remnants of relict strath surfaces correlated with marine terraces and main knickpoints.

topographic map. The same map was used to obtain a pre-landslide shape of the slope by smoothing the contour lines corresponding to the quarry area, *i.e.*, at the bottom of the considered slope (Fig. 12). The resulting removed volume in the total time interval from 2002 to 2009 is up to 55,000 m^3; more than 90% of the total volume was removed between 2005 and 2009.

4.2 Engineering-geology modeling

The engineering geology model of the landslide slope was developed from on-site geomechanical surveys. These data were collected with the aim of evaluating the indices Ib and Jv, already introduced in the paragraph 3.2, which have been measured through geomechanical scanlines on the outcropping rock masses (Fig. 15) as well as on 140 scanlines collected within a ~ 100 km^2 extended area.

The Young's modulus values measured for the intact rock from triaxial laboratory tests, vary in the range 50–60 GPa and the UCS resulted ~ 94MPa; on the other hand, the rock mass modulus, derived for the jointed rock mass according to the equivalent continuum approach by Sridevi & Sitharam (2000), vary in the range 30–40 GPa and the UCS in the range 70–77 MPa.

The stiffness and strength parameters (Table 2) related to the rock mass classes and to the increasing confining stress on the landslide slope were derived according to the equivalent continuum approaches by Sridevi & Sitharam (2000) and Hoek *et al.* (2002), respectively.

An engineering geology section (Figure 14) was obtained along the trace AA' of Fig. 12, *i.e.*, along the sliding direction of the mass movement showing that the landslide slope is characterised by a band of highly jointed rock mass from approximately 150 m a.s.l. up to 200 m a.s.l. This band has an orientation of approximately

Figure 15 Gneiss outcropping on the Santa Trada landslide slope: the rock mass jointing increases from a) to d).

40/75 (dip direction/dip), according to the highly persistent joint set that exists within all of the gneiss outcrops of the landslide slope.

4.3 Stress-strain numerical modeling

Stress-strain 2D numerical modeling was performed by the FLAC 7.0 (Itasca, 2011) finite difference code to back-analyze the Santa Trada landslide event. For the numerical modeling, a 268 × 160 mesh with a 1.5 m squared resolution was used, assuming the aforementioned engineering-geology section of Fig. 16. The considered topographic profile is between 75 m a.s.l. and 210 m a.s.l., which correspond to the bottom of the Santa Trada creek valley and to the crown area of the time-dependent rock mass deformation, respectively. The sequential modeling was performed by simulating the main erosional stages of the Pleistocene-Holocene evolution of the Santa Trada creek valley cross-section in conjunction with the landslide slope. More in particular, 4 steps due to the river deepening of the Santa Trada creek valley have been reproduced, according to the geomorphological model of Fig. 14, plus 3 other steps (from 5 to 7 of Fig. 16) due to the human quarry activity started on 2002 at the toe of the slope.

The numerical modeling was carried out in creep configuration, *i.e.*, under time-dependent conditions, with a visco-plastic constitutive law according to the Burger

Table 2 Mechanical properties of the 6 lithotechnical units reported in Figure 9.

lithotechnical unit	rheology	den (kg/m^3)	ϕ (°)	coh(Pa)	ten (Pa)	B_1 (Pa)	G_1(Pa)	G_2 (Pa)	η_1 (Pa*s)	η_2 (Pa*s)
A1	visco-plastic	2.7E+03	55	4.1E+05	3.4E+04	1.5E+10	2.5E+10	2.0E+10	4.5E+21	4.5E+22
A2	visco-plastic	2.7E+03	55	4.1E+05	3.4E+04	1.9E+10	3.1E+10	2.0E+10	4.5E+22	4.5E+23
B1	visco-plastic	2.7E+03	52	3.3E+05	2.4E+04	1.2E+10	2.0E+10	2.0E+10	8.3E+20	8.3E+21
B2	visco-plastic	2.7E+03	52	3.3E+05	2.4E+04	1.6E+10	2.7E+10	2.0E+10	8.3E+21	8.3E+22
C1	visco-plastic	2.7E+03	39	2.9E+05	1.1E+04	1.1E+10	1.8E+10	2.0E+10	8.3E+19	8.3E+20
C2	visco-plastic	2.7E+03	39	2.9E+05	1.1E+04	1.7E+10	2.8E+10	2.0E+10	8.3E+20	8.3E+21

Note: den = density; ϕ = friction angle; coh = cohesion; ten = tensile strength; B_1 = bulk modulus of Maxwell rheological element; G_1 = shear modulus of the Maxwell rheological element; G_2 = shear modulus of the Kelvin-Voigt rheological element; η_1 = viscosity of the Maxwell rheological element; η_2 = viscosity of the Kelvin-Voigt rheological element.

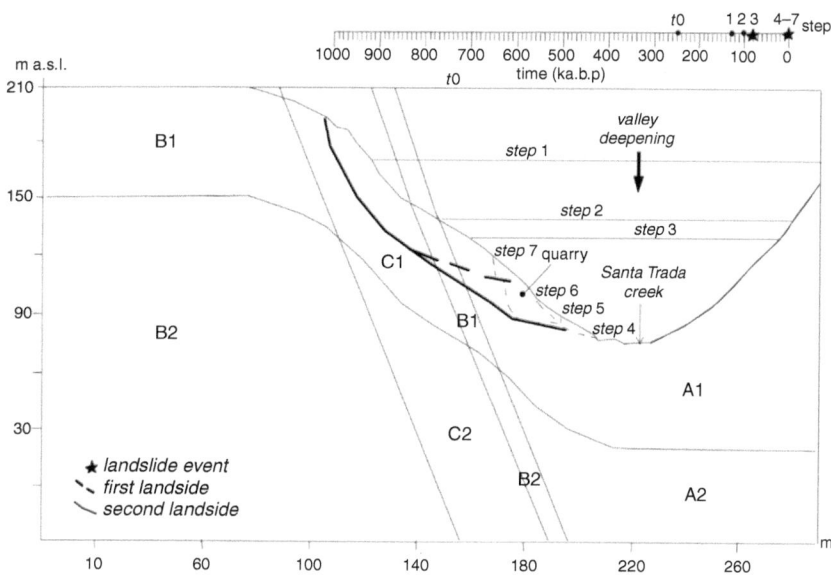

Figure 16 Engineering-geology section along trace AA' of Fig. 7 showing the space-time evolution of the Santa Trada creek valley in correspondence to the landslide.

rheological model (Karato, 2008). Table 2 summarizes the parameter values used in the numerical modeling.

The time-dependent numerical solutions were obtained by assuming a time step duration varying from $30 \cdot 10^6$ s (almost 1 year) to 120 s (almost 2 hours), depending on the time duration of each modeled stage.

The viscosity values for both the visco-elastic and the visco-plastic elements of the Burger model were selected in the range $10^{19} - 10^{23}$ Pa·s and calibrated to match morphological evidences and the displacements recorded as a consequence of the landslide.

The obtained results, demonstrate that the gravitational deformations arising from the time-dependent RMC process that affected the slope need to be considered for justifying both the observed landforms and failures. More in particular, the simulated creep deformations (Fig. 17) are responsible for a plastic-state zone which corresponds to the scarps that can be observed in the aerial images before the beginning of the quarry activity; the main one at 200 m a.s.l. and the secondary ones at 150 m a.s.l.; these scarps coincide with at-yielding zones of the numerical model. The cumulative x-displacement modeled until October 2008 reached 3 m at the main scarp, which is in agreement with the geomorphological elements observed in the pre-failure slope. The simulated failure mechanism occurs on January 2009 close to the cut-wall of the quarry; based on the resulting displacement field, the involved total volume is approximately 0.23 Mm^3.

The x-displacements resulting just behind the cut-wall of the quarry (point B of Fig. 18), clearly show an increasing tendency for tertiary creep displacements during the latest modeled months. In contrast, the deformation involving the crown area of the

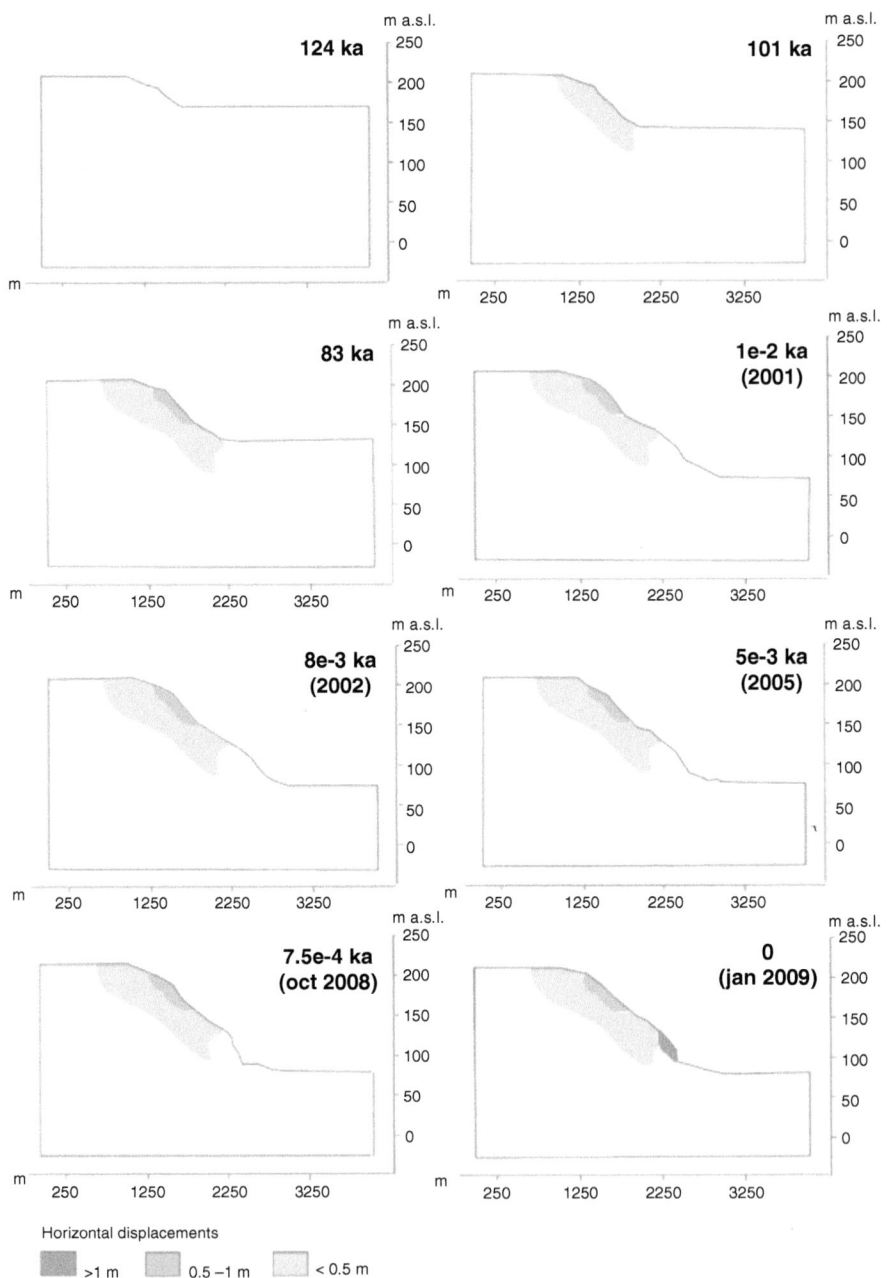

Figure 17 Outputs of the Santa Trada landslide stress-strain numerical modeling, obtained by zoning the model according to the engineering-geology section of Fig. 16 and by assuming the rheology of Table 2.

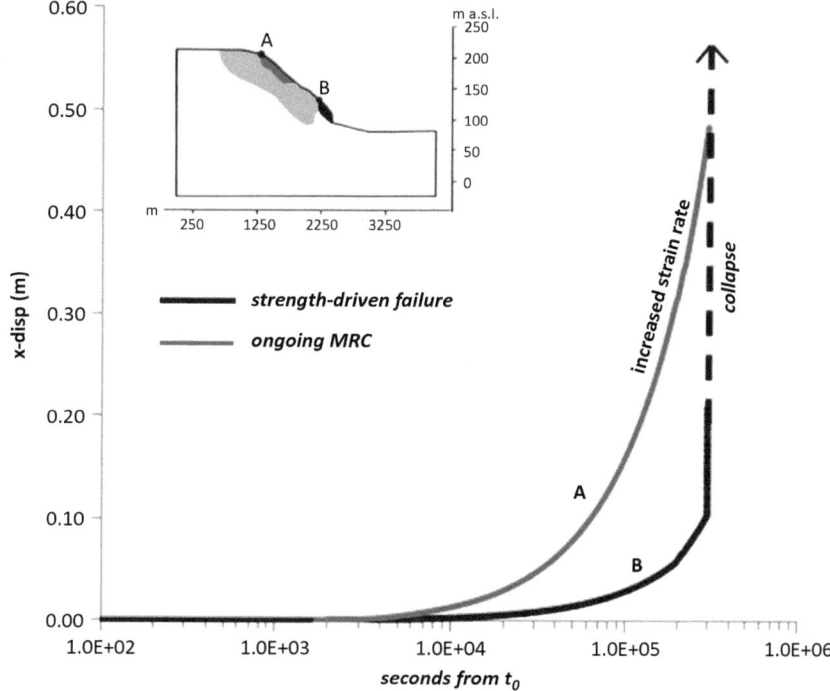

Figure 18 Horizontal displacement vs. seconds from creep t0 resulting from the Santa Trada landslide stress-strain numerical modeling at point A and B respectively. On January 2009 the MRC evolves to failure only at point B while at point A an increased strain rate can be measured (modified from Bozzano *et al.*, 2012).

landslide (point A of Fig. 18) can be attributed to a constant-rate creep (*i.e.*, steady creep). The superimposition of the anthropogenic stress release on the time-dependent deformational evolution is more intense in the low portion of the slope where the 30 January 2009 landslide occurred. Nevertheless, close to the main scarp the deformations are mainly due to the long lasting creep processes. The average strain rate due to the creep deformation obtained via numerical modeling is approximately 1.6 μm/year for the main scarp area and 0.4 μm/year in the portion of the slope where the 30 January 2009 landslide occurred.

The adopted visco-plastic Burger model made it possible to back-analyze the landslide before and after the quarry activity. A sensitivity analysis was performed by using different values for both the stiffness and the strength parameters: the reliable values were selected according to a best fit of the occurred events and the observed morphological elements. The same sequential numerical modeling was performed using a conventional elasto-plastic constitutive law (*i.e.*, the Mohr-Coulomb model). In this case, the resulting displacements did not exceed 10^{-3} m, and no zones were characterised by at-yielding conditions during the Pleistocene-Holocene evolution of the slope and after the beginning of the quarry activity.

5 REMARKS

Creep processes affecting rock masses (RMC) generally evolve toward collapses which represent catastrophic events as they generally involve km- to hm-scale slopes; times for failure strictly depend on the strain-rate due to both viscosity and stiffness parameters as well as to the stress field generated inside the deforming rock-mass volume. To provide a reliable estimation of failure times, a multi-disciplinary modeling approach is needed, including: i) a morpho-evolutionary stage-to-stage model of the slope in the valley system, ii) an engineering-geology model of the deforming slope by attributing specific geomechanical properties to both rock matrix and joints, iii) a stress-strain numerical modeling under time-dependent configuration that uses a rheological behavior to derive strain effects expected along the time.

The here presented case-studies demonstrate the reliability of such a multi-modeling approach to back-analyze the strain evolution of the slope as well as to provide an estimation for expected failure time. The relevance of this results consists on their contribution for a landslide risk management, especially in case of slope interaction with infrastructures having a usage-time that is comparable with the one for slope failures or collapse. Moreover, a best-tuned rheological behavior also permits to estimate strain effects due to destabilizing actions that can involve natural slopes, these last including man-induced interventions, both in case of permanent or transient ones.

In this regard, the MRC process should be considered as responsible for a "ground" strain effect over which transient actions can cause an increasing of the strain rate or a slope failure trigger. Nonetheless it is worth stressing that same magnitude of transient forcing events could lead to different effects depending on their time of occurrence, *i.e.* on the evolution stage of the MRC-induced strains.

The nowadays available numerical solutions, performed by finite differences or finite elements commercial codes, as well as the high reliability of software for performing terrain quantitative analyses make it possible to further improve multi-disciplinary modeling approaches addressed to manage the risk associated to RMC process and to support the design of monitoring systems for early warning, that necessarily request a strain-rate target to be correctly planned. In case of RMC processes, multi-modeling quantitative approaches should be encouraged by natural risk managers as the continuously ongoing strain effects implies that the landslide hazard evaluation is senseless while, on the contràry, the failure time estimation represents the only suitable parameter to estimate the vulnerability of exposed persons or structures to induced damages, *i.e.* to express a judgment on the associated sustainable risk respect to site frequentation or structure usage-time. As time to RMC failure significantly increases the vulnerability reduces since the corresponding strain-rate is negligible, but in case of a fastening in strain evolution, the induced effects or damages could be no more sustainable. To this aim, a major effort should be performed to enrich the knowledge about rheological parameters at a natural-slope scale as several order of magnitude generally distinguish them respect to the laboratory-derived ones (Bretschneider *et al.*, 2013); in this regard, a good challenge for scientific community to be faced for future researches and studies is to collect case studies widespread as well as to long-time monitoring dataset coupled with numerical modeling approaches.

REFERENCES

Apuani, T., Masetti, M. & Rossi, M. (2007) Stress–strain-time numerical modeling of a deep-seated gravitational slope deformation: Preliminary results. *Quaternary International*, 171–172, 80–89.

Baroni, C., Martino, S., Salvatore, M.C., Schilirò, L. & Scarascia-Mugnozza, G. (2014) Thermomechanical stress–strain numerical modeling of deglaciation since the Last Glacial Maximum in the Adamello Group (Rhaetian Alps, Italy). *Geomorphology*, 226, 278–299.

Bianchi-Fasani, G., Esposito, C., Maffei, A., Scarascia-Mugnozza, G. & Evans, S.G. (2004) Geological controls on slope failure style of rock avalanches in Central Apennines (Italy). In: Lacerda, Ehrlich, Fontoura, Sayao (Eds.), Landslides: Evaluation and stabilization. ISBN: 04 1535 665 2, pp. 501–507.

Bianchi-Fasani, G., Esposito, C., Petitta, M., Scarascia-Mugnozza, G., Barbieri, M., Cardarelli, E., Cercato, M. & Di Filippo, G. (2011) The importance of geological models in understanding and predicting the life span of Rockslide Dams: The case of Scanno Lake, Central Italy. In: Evans, S.G., Scarascia-Mugnozza, G., Strom, A. & Hermanns, R. (Eds.), Natural and Artificial Rockslide Dams. Lecture Notes in Earth Sciences, 133. Springer-Verlag, Berlin Heidelberg, pp. 323–345.

Bozzano, F., Bretschneider, A., Esposito, C., Martino, S., Prestininzi, A. & Scarascia Mugnozza, G. (2013) Lateral spreading processes in mountain ranges: Insights from an analogue modeling experiment. *Tectonophysics*, 605, 88–95.

Bozzano, F., Della Seta, M. & Martino, S. (2016) Time-dependent evolution of rock slopes by a multi-modelling approach. *Geomorphology*, 263, 113–131.

Bozzano, F., Martino, S., Montagna, A. & Prestininzi, A. (2012) Back-analysis of a rock landslide to infer rheological parameters. *Engineering Geology*, 131–132, 45–56.

Bretschneider, A., Genevois, R., Martino, S., Prestininzi, A. & Verbena, G. (2013) A physically-based scale approach to the analysis of the creep process involving Mt. Granieri (Southern Italy). In: R. Genevois & A. Prestininzi. International Conference on Vajont 1963–2013, Thoughts and Analyses after 50 Years Since the Catastrophic Landslide, 123–131. Sapienza Università Editrice, Roma (Italy), ISBN: 978-88-95814-96-4

Bull, W.B. (1990) Stream-terrace genesis: Implications for soil development. *Geomorphology*, 3, 351–367.

Chemenda, A., Bouissou, S. & Bachmann, D. (2005) Three-dimensional physical modeling of deep-seated landslides: New technique and first results. *Journal of Geophysical Research*, 110, F04004.

Chigira, M. (1992) Long-term gravitational deformation of rocks by mass rock creep. *Engineering Geology*, 32, 157–184.

Cruden, D.M. & Varnes, D.J. (1996) Landslide types and processes. In: Turner, A.K. & Schuster, R.L. (Eds.), Landslides: Investigation and Mitigation: Transportation Research Board, Spec. Report 247. National Research Council, National Academy Press, Washington, DC, 36–75.

Della Seta, M., Esposito, C., Marmoni, G.M., Martino, S. & Scarascia Mugnozza, G. (2016) Morpho-structural evolution of the valley-slope systems and related implications on slope-scale gravitational processes: New results from the Mt. Genzana case history (Central Apennines, Italy). *Geomorphology*, doi.org/10.1016/j.geomorph.2016.07.003.

Discenza, M.E., Esposito, C., Martino, S., Petitta, M., Prestininzi, A. & Scarascia Mugnozza, G. (2011) The gravitational slope deformation of Mt. Rocchetta ridge (central Apennines, Italy): Geological-evolutionary model and numerical analysis. *Bulletin of Engineering Geology and the Environment*, 70, 559–575.

Dumas, B., Gueremy, P. & Raffy, J. (2005) Evidence for sea-level oscillation by the "characteristic thickness" of marine deposits from raised terraces of Southern Calabria (Italy). *Quaternary Science Reviews*, 24, 2120–2136.

Eberhardt, E., Stead, D. & Coggan, J.S. (2004) Numerical analysis of initiation and progressive failure in natural rock slopes – the 1991 Randa rockslide. *International Journal of Rock Mechanics & Mining Sciences*, 41, 69–87.

Esposito, C., Bianchi Fasani, G., Martino, S. & Scarascia Mugnozza, G. (2013) Quaternary gravitational morpho-genesis of Central Apennines (Italy): Insights from the Mt. Genzana case history. *Tectonophysics*, 605, 96–103.

Esposito, C., Martino, S. & Scarascia Mugnozza, G. (2007) Mountain slope deformations along thrust fronts in jointed limestone: An equivalent continuum modeling approach. *Geomorphology*, 90, 55–72.

Ghisetti, F. (1984) Recent deformations and the seismogenic source in the Messina Strait (Southern Italy). *Tectonophysics*, 210, 117–133.

Gioia, D., Martino, C. & Schiattarella, M. (2011) Long- to short-term denudation rates in the southern Apennines: Geomorphological markers and chronological constraints. *Geologica Carpathica*, 6(2/1), 27–41.

Guarnieri, P., Di Stefano, A., Carbone, S., Lentini, F. & Del Ben, A. (2004) A multidisciplinary approach to the reconstruction of the Quaternary evolution of the Messina Strait with Geological map of the Messina Strait, scala 1:25.000. In Mapping Geology in Italy, Ed. APAT, 45–50.

Hoek, E., Carranza-Torres, C. & Corkum, B. (2002) Hoek-Brown failure criterion – 2002 edition. Proc. NARMS-TAC Conference, Toronto, 1, 267–273.

Hungr, O. & Evans, S.G. (2004) The occurrence and classification of massive rock slope failure. *Felsbau*, 22, 16–23.

ISRM (2007) The complete ISRM suggested Methods for rock characterization, testing and monitoring: 1974–2006. In: Ulusay, R. & Hudson, J.A. (Eds.), Suggested Methods Prepared by the Commission on Testing Methods. International Society for Rock Mechanics, Compilation Arranged by the ISRM Turkish National Group, Ankara, Turkey, 628pp.

Itasca (2011) FLAC 7.0: User Manual, Licence number 213–039–0127–18973. Sapienza-University of Rome, Earth Science Department.

Karato, S. (2008) An Introduction to the Rheology of Solid Earth. Cambridge University Press, 463pp.

Lenti, L., Martino, S., Paciello, A., Prestininzi, A. & Rivellino, S. (2015) Recorded displacements in a landslide slope due to regional and teleseismic earthquakes. *Geophysical Journal International*, 201, 1335–1345.

Maffei, A., Martino, S. & Prestininzi, A. (2005) From the geological to the numerical model in the analysis of the gravity-induced slope deformations: An example from the Central Apennines (Italy). *Engineering Geology*, 78, 215–236.

Martino, S., Prestininzi, A. & Scarascia Mugnozza, G. (2004) Geological-evolutionary model of a gravity-induced slope deformation in the carbonate central Apennines (Italy). *Quarterly Journal of Engineering Geology and Hydrogeology*, 37(1), 31–47.

Miyauchi, T., Dai Pra, G. & Sylos Labini, S. (1994) Geochronology of Pleistocene marine terraces and regional tectonics in the tyrrhenian coast of south Calabria, Italy. *Il Quaternario Italian Journal of Quaternary Sciences*, 7 (1), 17–34.

Nesci, O., Savelli, D. & Troiani, F. (2012) Types and development of stream terraces in the Marche Apennines (central Italy): A review and remarks on recent appraisals. *Géomorphologie: Relief, Processus, Environnement*, 18(2), 215–238.

Nicoletti, P.G., Parise, M. & Miccadei, E. (1993) The Scanno rock-avalanche (Abruzzi south – centralItaly). *Bollettino della Societa Geologica Italiana*, 112, 523–535.

Petley, D.N., Higuchi, T., Petley, D.J., Bulmer, M.H. & Carey, J. (2005) The development of progressive landslide failure in cohesive materials. *Geology*, 33(3), 201–204.

Savage, W.Z. & Varnes, D.J. (1987) Mechanics of gravitational spreading of steep-sides ridges (sackung). *Bulletin of the International Association of Engineering Geology*, 35, 31–36.

Sridevi, J. & Sitharam, T.G. (2000) Analysis of strength and moduli of jointed rocks. *Geotechnical and Geological Engineering*, 18, 3–21.

Stead, D., Eberhardt, E. & Coggan, J.S. (2006) Developments in the characterization of complex rock slope deformation and failure using numerical modeling techniques. *Engineering Geology*, 83, 217–235.

Wegmann, K.W. & Pazzaglia, F.J. (2009) Late Quaternary fluvial terraces of the Romagna and Marche Apennines, Italy: Climatic, lithologic, and tectonic controls on terrace genesis in an active orogen. *Quaternary Science Reviews*, 28, 137–165.

Xu, T., Xu, Q., Deng, M., Ma, T., Yang, T. & Tang, C. (2014) A numerical analysis of rock creep-induced slide: A case study from Jiweishan Mountain, China. *Environmental Earth Sciences*, 72, 2111–2128.

Chapter 9

Multi-approach back analysis in mining

Leandro R. Alejano
Department of Natural Resources & Environmental Engineering, University of Vigo, Vigo, Spain

Abstract: The author describes in this chapter the multi-approach back-analyses of two instability phenomena associated to underground and open pit mining excavations as clear examples of an appropriate control and design strategy in mining. In particular, the collapse of a roof of a mining room excavated in a bedded rock mass and the complex failure phenomenon of a slope in an open pit wall are back-analyzed by means of a multi-approach technique. These approaches have been extremely helpful to understand the mechanisms behind the observed behaviors as well as to identify the most relevant features affecting stability for every case. The acquired knowledge has been a key point to control stability in further mine developments in all cases.

1 INTRODUCTION

Even if a number of relevant breakthroughs have been achieved during the past decades in the rock engineering field, the fact of working with complex natural materials such as rock masses still needs the use of design approaches more empirical and heuristic than those typically encountered in other engineering fields. It is always difficult to have available an accurate description of a rock mass in terms of knowledge and understanding and to be able to obtain significant characteristic values at the scale of the mine or the work. In these circumstances, the use in parallel of different techniques to carry out back-analyses of excavation behavior and instability phenomena can be one of the most reasonable strategies in order to have an appreciation of rock mass behavior at the real scale and to identify key issues and parameters that control the resulting behavior. With this knowledge, future excavations can be designed and controlled in a safer way and operational mine problems can be fixed in a more straightforward manner.

In order to solve most of the design or stability analysis problems posed in mining practice within the rock mass engineering field, one usually has got a short number of limited quality data and very often the initial level of understanding is not particularly high (Starfield & Cundall, 1988). In this way, traditional design approaches commonly utilized in other branches of engineering—where materials to be used can be selected—are typically not suitable, so the fundamental aim of design in rock engineering regards the adequate perception of the consequences of the scarcity, variability and uncertainty of available data on design (Fairhurst, 1993). Therefore, it is convenient not to forget some of the principles which should guide the decision-making process leading to design. According to Bieniawski *et al.* (1993), these principles can be synthesized in the following ones: independence—minimum independent functional requirements—, minimum uncertainty, knowledge —maximum available technology—, optimization and simplicity in the execution of works. Together with these basic principles, one should also account for the particular issues of the mine project including the available budget, initial limitations, local and miner's idiosyncrasies, legal issues and others. In this way, the most relevant key-points in mining rock engineering design and control can be summed up in a proper perception of the information level needed to carry out a design, a reasonable estimate of the acceptable risk and a proper management of uncertainty all along the design process.

At the core of any design methodology in rock engineering a model is needed, such as limit equilibrium and other analytical methods, numerical models or empirical approaches (*i.e.* classification systems), all of which can be complemented with the observational approach. This model regards the mathematical process able to describe the phenomena to be avoided. Such a model should be contemplated as an intellectual tool designed according to the questions to be answered and in regard of the available data, and not as a machine able to produce figures. Certainly, scientific understanding proceeds by way of constructing and analyzing models of the segments or aspects of reality under study. The main aim of these models is neither to give a mirror image of reality, nor to contemplate all its elements in their exact proportions, but rather to single out and make available for intensive investigation those elements which are decisive. In this way, for every case study, there are usually a small number of key parameters (Figure 1). One should abstract from non-essentials, one has to blot out the unimportant to get an unobstructed view of the relevant. 'A model is, and must be,

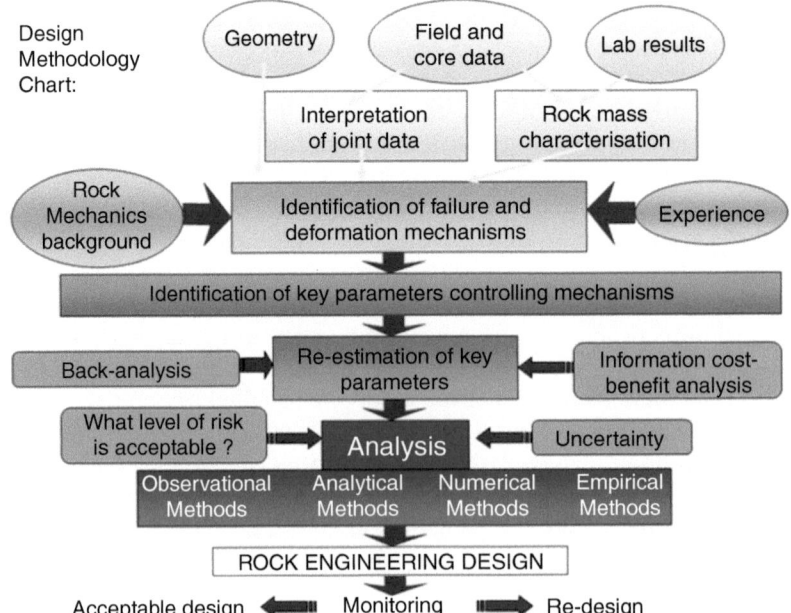

Figure I Design methodology chart proposed by the author for rock engineering applications.

unrealistic in the sense in which the word is most commonly used. Nevertheless, and in a sense, paradoxically, if it is a good model it provides the key to understanding reality' (Baran & Sweezy, 1968).

In this text, the author would like to show that every modelling approach (analytical, numerical, empirical, observational...) has a number of advantages and disadvantages, so the combined used of some of them could led to acquire a good understanding of the occurring phenomena. Two case studies are presented, where various approaches are applied to actual problems in mining rock engineering. These include the collapse of a roof of a mining room excavated in a bedded rock mass and the complex failure phenomenon of a slope in an open pit wall. They are both back-analyzed by means of a multi-approach technique; which has contributed to a better understanding of the occurred phenomena. In the above mentioned cases, analytical and numerical results are compared, and a number of observations are made. Finally, some general conclusions are derived, which could be of interest for mining rock engineering practitioners in order to decide what kind of methods are suitable for a particular case study.

2 CASE STUDY #1: BACK-ANALYSIS OF THE COLLAPSE OF A ROOM IN STRATIFIED ROCK

Roof bed deformation mechanics is a complex topic in mining practice. There is a number of empirical, analytical and numerical techniques that can be used to analyze and design underground excavations in stratified rock masses. Each of them has particular drawbacks, so the use of just one approach in order to obtain consistent and economic designs is not only difficult but also it could lead to inappropriate designs. To deepen the understanding of this topic, it is useful to study real failures in mines. In this chapter, the author describes a roof bed failure in an underground mine excavated in a well-studied stratified rock mass. Then, the failure mechanism is back-analyzed by means of different approaches, namely, the empirical stability graph method, an analytical voussoir technique, and a simulation (with two degrees of accuracy) using the numerical discrete element method. The study enabled ground control engineers to better understand the conditions of failure, to define the most significant parameters affecting the failure and, ultimately, to implement a more suitable reinforcement policy at the mine site.

The roof failure to be studied took place in an underground magnesite room-and-pillar mine, from which some 150,000 tons of carbonate are extracted annually. The deposit consists of a magnesite bed, approximately 14 meters thick, running in the E-W direction (N-92°-E), and with a mean dip of 18°. The mining area is located between two normal faults, practically coinciding with the direction of the maximum bedding dip, running in the NNW-SSE direction.

2.1 Basic mine facts and geological overview

In the underground mining of inexpensive minerals, such as magnesite, mining cavities are conveniently excavated in the mineral itself; thus, all the material goes to the plant for processing and subsequent sale, while the need for environmentally aggressive mining dumps is avoided and haulage and transportation waste costs are saved. Due to environmental and economic issues, the room-and-pillar method is revealed as the most suitable approach and it is applied in the mine.

Since standard mining machines can only move in an efficient manner along roads with slopes of up to 10%, all the mining drifts and rooms are oriented in the mineral bed to form an angle of 19° to the direction of the strata (directions 110° and 250°). Diamond-shaped mining panels have been designed for 11-meter wide rooms and 8-meter wide pillars even if these panels have been conservatively designed to allow 12-meter wide rooms and 7-meter wide pillars, so the geometry control in the operation is not necessary. No problem related to pillar stability has ever been reported, so pillars designed by means of the Hedley & Grant (1972) approach for a factor of safety 1.6 were always stable. Rooms need reinforcement. A general plan of the mine is shown in Figure 2, together with a block diagram showing the bed, ramps and some of the panels.

The mined deposit, a 14-meter thick magnesite bed, is overlying a very thick bed of sandy shale and is overlaid by a 1-meter wide bed of poor quality marly shale, overlaid by a thick bed of sandy shale of a better geotechnical quality. Since this top layer was very weak, it was decided to leave a 1-meter thick bed of magnesite on top of the rooms in order to improve stability (see Figure 3). The initial mining design also recommended reinforcement, and so, 240-kN bolts were placed at roof intervals of 3.5 square meters. Fully grouted 15.2-mm diameter and 6-meter long steel rods were used in this mine.

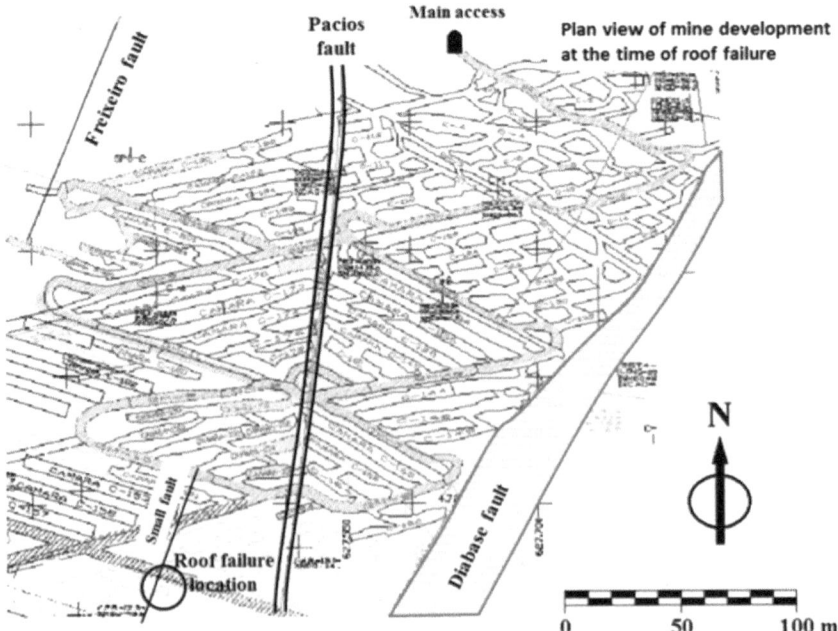

Figure 2 Plan view of the present development of the mine, and block diagram illustrating the geometrical disposition of mine panels, rooms and pillars.

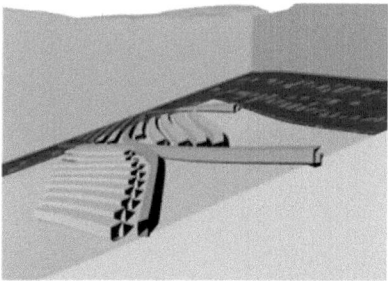

a) Block diagram showing the geometrical disposition of mine panels, rooms and pillars.

b) View of room hanging wall showing installed reinforcement.

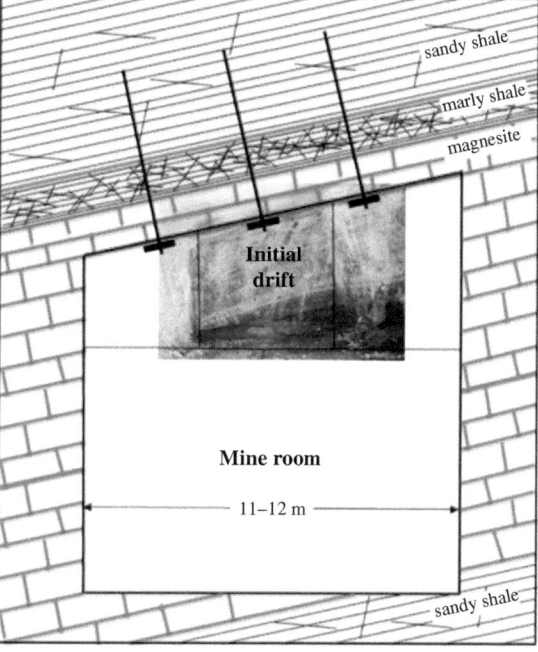

c) Sketch of a room showing the local geology, geometry of the room, location of the initial drift and typical reinforcement.

Figure 3 Standard dimensions of rooms: diagram, cross-cut including reinforcement, and an image of a reinforced room.

2.2 Roof bed failure description

Roof bed failure did not occur in the first 15 years of mining, following the described geometry. In March 2004, however, one of the standard rooms—direction N-110°-E and located at a depth of 200 meters—crossed a significant normal fault with a perpendicular 0.5-meter step, and advanced around 15 meters without support (since room failure was never observed, miners tend to advance some rounds without support and install the support afterwards). This reduced the width of magnesite stratum left in the roof from 1 meter to around half a meter. The room span was measured as between 11.3 and 11.9 meters.

Shortly after, a roof bed failure occurred that affected the whole room, from the fault to the other end of the advancing room. By this time, the contiguous left-hand side room had been finished and supported, whereas the contiguous right-hand room was in the process of being excavated. The roof failure affected the magnesite stratum in the roof, the upper marly shale (around 1 meter thick), and some lower areas of the sandy shale. As Figure 4 illustrates, a bell-shaped zone formed in the roof.

In the following sub-sections, this roof failure is back-analyzed following different approaches in order to show the advantages of this multi-approach strategy in mining. Alejano *et al.* (2008) provided a more detailed description of this case study, including

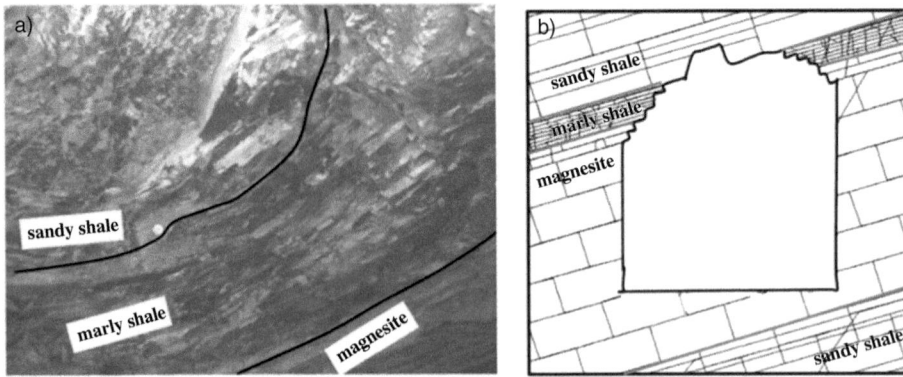

Figure 4 Bell-shaped appearance of an area of the roof bed failure: a) picture of the roof and b) sketch of the geometry.

all the relevant data; Arzúa *et al.* (2015) present further design developments concerning the new support design approaches in the mine.

2.3 Empirical back-analysis

The so-called stability graph method, which dates to the 1980s (Potvin *et al.*, 1988), is based on data compiled from around 100 case histories of Canadian mines. Hoek *et al.* (1995) and others have widely used the method for stope design in mining the rock mechanics field. The method characterizes a rock mass, in which a mining stope wall is to be excavated, according to a stability value N'. This is based on the primary value of Barton's Q (Barton *et al.*, 1974)—called Q'—in which the stress reduction factor (SRF) and the water factor are set to 1. Additional factors, named A, B, and C, are incorporated to account for rock stress, joint orientation and gravity effects, respectively. The so-called stability number is thus estimated as:

$$N' = \frac{RQD}{J_n} \times \frac{J_r}{J_a} \times A \times B \times C = Q' \times A \times B \times C \tag{1}$$

The type of room to be excavated is characterized by the shape factor or hydraulic radius (HR), which is calculated by dividing the cross-sectional area of the analyzed surface by the surface perimeter. Based on these parameters, some authors, for instance Hutchinson & Diederichs (1996), devised stability charts based on a database that became increasingly larger. This chart classified excavation walls and roofs into stable, transition and caved-in zones. Guidelines were also provided for cable bolting design under different conditions.

Application of the stability graph method for the room's roof yielded the data reproduced in Table 1. This regards roofs of rooms 11 to 12-meters wide and the HR ranged between 4.5 and 5.5, depending on the length of the rooms (up to 100 meters). We plot the values obtained for N' and HR on the Diederichs (1999) stability chart (Figure 5). The position of the analyzed rooms is marked by a black circle for the standard mine zones and by a gray circle for the zone surrounding the faults. The

Table 1 Stability numbers for the different mine zones, calculated using stability graph method.

	Q' Modified Q	A Low stresses	B Bedding parallel to wall	C Sub-horizontal wall	N'
Standard mine zones	21	1	0.3	2.25	14.17
Faulted mine zones	3	1	0.3	2.25	2.02

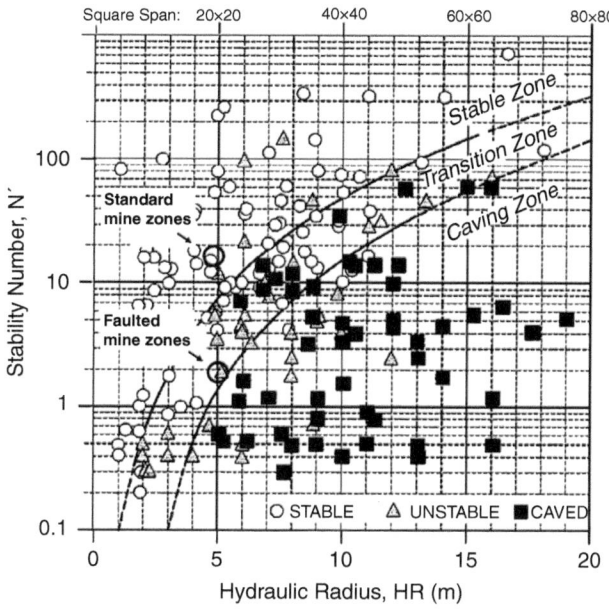

Figure 5 Stability chart for the application of the stability graph method (Diederichs, 1999). The rooms in the analyzed mine are represented by a black circle (standard zones) and a gray circle (faulted zones).

stability graph illustrated in Figure 5 shows that, rooms in the standard mine zones are likely to be stable and so, do not initially need support; but rooms in faulted areas are located in the transition (near to caved) zone, so the use of support is recommended.

As far as the heterogeneous nature of rock masses is concerned, these empirical results should be considered as indicative. However, they seem to concur with observations made in the mine. No failure was observed in the upper part of the mine, even if reinforcement was installed after various rounds (20–25 m advance).

2.4 Analytical back analysis based on voussoir analogue approaches

The main assumption of beam theory regards the fact that the immediate roof can be represented by a beam with a length equal to the room span and abutted by pillars at

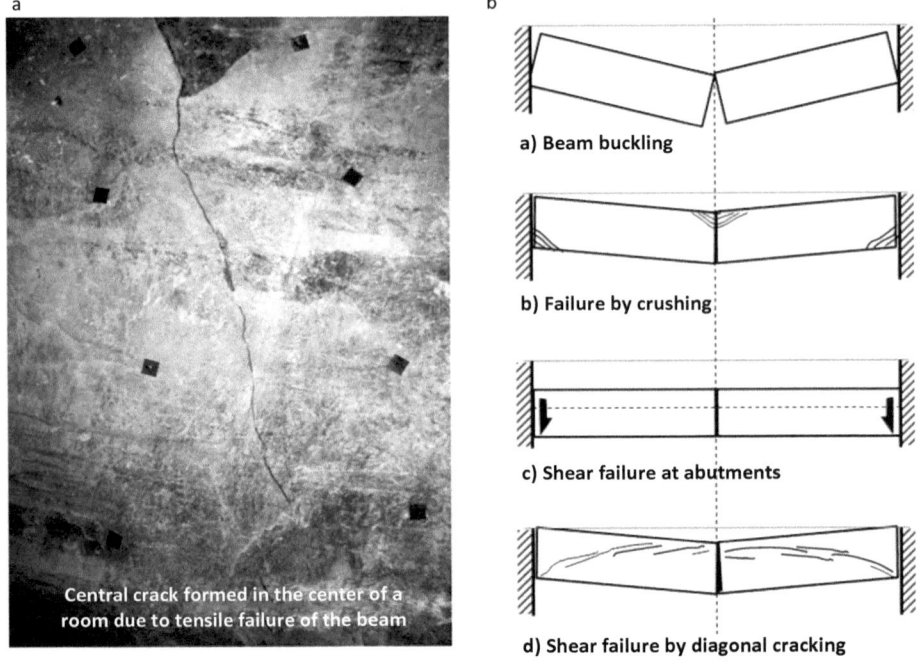

Figure 6 a) Photograph of central crack opened in the center of a roof in a room and b) failure mechanisms or instability processes associated to voussoir beam behavior (modified form Diederichs & Kaiser, 1999a).

each side. This theory is applicable only when the thickness of immediate roof is at least 8 times less than the room span and when plane-strain conditions can be assumed, *i.e.*, the room is not close to an intersection. The factors of safety calculated using this approach indicate the proximity of the beam to the point at which tensile failure of the beam occurs. In practice, such a failure produces three vertical joints—typically 3, *i.e.*, 2 in the abutments and 1 in the beam center (Figure 6.a).

Although these joints do not necessarily produce roof failure, they do make the beam behave as a voussoir beam. Based on this and in many observations in mines, most of the analytical techniques used to design roofs in bedded media are based on the voussoir beam analogue. This analogue is based on the notion of a compressive arch acting along the voussoir, capable of withstanding the weight of the materials and of transmitting it to the abutments. The principles underlying this approach have long been applied to bridge construction techniques—hence the fact that single voussoir non-cemented bridges have stood for centuries.

Since the voussoir beam problem is statically undetermined, at least one reasonable assumption is needed in order to solve it reliably. Depending on the proposed assumption, different calculation techniques have been proposed in recent decades. For the sake of briefness and due to its accurateness, the Diederichs & Kaiser (1999a) approach is used in this analysis. However, other methods are available (*e.g.* Sofianos, 1996).

Voussoir beam methods allow for 3 types of instability processes (Diederichs & Kaiser, 1999a) illustrated in Figure 6.b:

1. Beam buckling, with no significant crushing or spalling;
2. Failure by crushing or spalling of central or abutment voussoirs; and
3. Shear failure at the abutments, which may produce diagonal cracking in weak rock.

Shear failure is typically observed in thick beams with low span/thickness ratios, whereas buckling and crushing are more likely to occur in thinner beams with high span/thickness ratios.

The application of the Diederichs & Kaiser (1999a) analytical method to our back-analysis of the failed room (12-meters wide) follows. In order to obtain information on the behavior of rooms in different areas of the mine, we analyzed both standard and faulted mine zones. Bearing in mind the important role played by the thickness of the initial rock stratum in the roof, and in order to analyze impact on stability, we varied stratum thickness between 0.5 and 1.5 meters. The obtained information is summarized in Table 2, which presents the most significant factor of safety and stability indicators. Some comments are in order.

An elastic beam analysis, following classical approaches (Obert & Duvall, 1967), was based on calculating the factor of safety against tensile fracture of the roof bed. The fact that this factor of safety is less than 1 does not mean that the roof is unstable, but indicates, rather, the occurrence of joints. In this case beam behavior would be better represented by the voussoir analogue.

The voussoir analogue model calculates two factors of safety: one against crushing and one against shearing. As far as buckling is concerned, Diederichs & Kaiser (1999a) preferred to calculate what they called a buckling limit (BL), given that buckling is an evolving phenomenon. Even if a theoretical BL of 100% is required for stability, instability processes—which are very difficult to control—tend to commence from a BL of around 35%, which is why these authors indicated a BL value of less than 35%. They also indicated that this BL is usually achieved when deflection attains a value of around 0.1 times the thickness of the roof bed. For the purpose of our study we defined

Table 2 Factors of safety and stability conditions calculated for a 12–meter room, with a top magnesite stratum (0.5 to 1.5-meters thick) and an overlaid marly shale bed (1-meter thick).

| | | Standard mine zones | | | Faulted mine zones | | |
| | | First bed thickness | | | First bed thickness | | |
		0.5 m	1 m	1.5 m	0.5 m	1 m	1.5 m
Elastic beam model	FoS	1.27	2.34	3.92	1.13	2.30	3.90
Voussoir analogue	$FoS_{crushing}$	3.26	10.46	18.3	2.29	10.25	18.24
Diederichs & Kaiser (1999a)	$FoS_{shear\ (vertical\ joints)}$	8.55	4.35	2.91	8.59	4.32	2.89
	Buckling limit	21	1	1	58	1	1
	$FoS_{deflection}$	1.9	16.4	67	0.52	6.3	24.5
	Deflection (m)	0.032	0.006	0.0022	0.096	0.015	0.006

what we call a deflection factor of safety, calculated as the relationship between a value of 10% of roof bed thickness and deflection calculated.

Table 2 summarizes the results for Factors of Safety and stability conditions calculated for a 12–meter room, with a top magnesite stratum (0.5 to 1.5-meters thick) and an overlaid marly shale bed (1-meter thick). Note that, according to the adopted approach, there is only one unstable case, corresponding to the buckling of a 0.5-meter bed in the faulted area of the mine, with a BL of 58%—substantially greater that the indicated BL of 35%. A nominal deflection of 0.096 meters was calculated, well over 10% of the beam thickness (0.05 meters). Note that an 11-meter wide room produces a BL of 40%, which is the limit equilibrium situation. We are of the opinion that this buckling mechanism was the ultimate cause of roof bed failure although, as explained below, relaxation may also have contributed in a relevant manner. Remark that this voussoir approaches assume fixed abutments and they do not account for the role of horizontal stresses.

2.5 Numerical analysis of roof failure. Scope and fundamentals

An analytical or numerical model lies at the core of any excavation design methodology. This model is a mathematical abstraction that describes the most significant aspects of a phenomenon. Although such models are more a theoretical tool designed to answer specific questions on the basis of the available input data, their use in rock mechanics can enhance our understanding of how failure mechanisms occur in rock masses (Starfield & Cundall, 1988). For the purpose of modeling roof failure, we used UDEC (Itasca, 2004), a code based on the discrete element method. With an explicit time integration scheme, it consists of an assemblage of discrete blocks representing the discontinuous medium.

Initially, a 0.5-meter thick roof bed with cross-joint spacing—equivalent in length to the voussoir—of 1.8 meters was simulated (a value taken from the average spacing between joints of the first joint set (J$_1$) measured in the mine). These voussoirs were modeled as deformable blocks. Properties—those of laboratory magnesite specimens—were as follows: density = 2950 kg/m^3; Young's elastic modulus = 36.5 GPa; Poisson's ratio = 0.25; cohesion = 9.85 MPa; friction angle = 45°; and tensile strength = 7.1 MPa. Cohesion—related to the compressive strength—was reduced to half its value (as in the analytical approach and for the same reason).

The parameters for the cross-joints between voussoirs were particularly important in these models. The value of the peak friction angle was calculated and adjusted according to the normal stress. An initial conservative value of 36° was used. The joint cohesion and tensile strength parameters were set as 0. The deformability parameters —normal and shear stiffness—were also important in the case of buckling failure (they control the flexural behavior of the rock bed). For the faulted area under analysis, normal stiffness could be indirectly estimated, given that the elastic modulus of the rock and of the rock mass were known.

To estimate normal stiffness k_n in the faulted mine area, some prior data was needed. Rock mass elastic modulus (E_{rm}) was calculated as 7.2 GPa (from the rock mass classification systems), and rock elastic modulus (E_r) was averagely measured as 35.6 GPa (from laboratory data). The average cross-joint spacing (s_j) was taken to be 1.8 m. For this rock mass:

$$\frac{1}{E_{rm}} = \frac{1}{E_r} + \frac{1}{k_n \cdot s_j} \tag{2}$$

Therefore, using the data above, a normal stiffness value (k_n) of 5 GPa/m could be obtained—the key value for the calculations. Shear stiffness was not significant, even if it could affect bed deflection and so, a typical smaller value than that for normal stiffness was taken as $k_s = 1$ GPa/m.

2.6 Numerical analysis of roof failure. Simple beam models

A small number of voussoirs or bricks between two abutments were firstly simulated and gradually increased in number (one by one) up to the point of roof failure. Abutments were simulated by means of deformable although very stiff blocks. They were fixed at the external borders in order to represent the conditions of the problem, following the approach of Diederichs & Kaiser (1999a), (D&K, in what follows). The models were run in two stages. The first stage was an initial, elastic one, in which the beam started to deflect without tensile failure—representing the face effect. This was run in order to avoid voussoir free-fall in these simple tests. In the second stage, the actual parameters were used, with the model intended to represent actual behavior.

Simple planar beams were modeled with increasing number of voussoirs. For the planar simple (unloaded) case, up to eight 1.8-meter long voussoirs remained stable (that is a 14.4-meter long beam). See the case of six voussoirs in Figure 7.a. In every case, a good agreement is found between the deflection obtained by UDEC and that derived from the D&K approach.

Then, the loading of the 1-meter thick marly shale bed by means of rigid blocks was included. For this loaded planar case, up to six 1.8-meter long voussoirs remain stable (that is a 10,8-meter long beam). The seven beam voussoir collapses according to UDEC. See the case of six voussoir beam in Figure 7.b, where again, a moderately good agreement is observed between the deflections obtained by UDEC and D&K approach.

Finally, the inclination of the bed was taken into account (17.7°). The 10.8-meter span (six voussoirs) room achieved stable equilibrium with an 82.2-mm deflection

Table 3 Results for the 17.7° inclined loaded beam model obtained using the D&K approach (1999a) and UDEC (Itasca, 2004).

Inclined beam models with load – Faulted mine areas			
Voussoir number		6	7
Room span (m)		10.8	12.6
Diederichs & Kaiser (1999a)	Max. deflection (mm)	58.3	133.3
	Max. comp. stress (MPa)	13.5	29.2
	Buckling limit (%)	37	72
UDEC (Itasca, 2004)	Max. deflection (mm)	82.2	Caves
	Max. comp. stress (MPa)	12	Caves

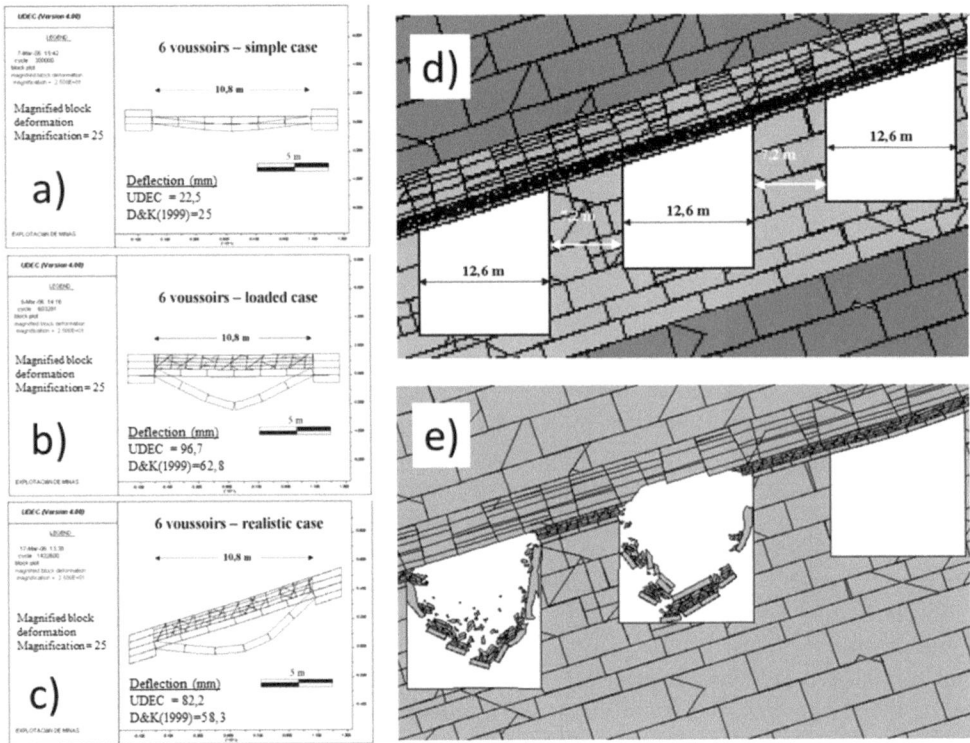

Figure 7 UDEC (Itasca, 2004) results for beam behavior in the left hand side: a) Six voussoir 0.5 m wide horizontal beam, b) Six voussoir 0.5 m wide loaded beam and c) Six voussoir 0.5 m wide inclined and loaded beam, representing actual conditions in the analyzed roof failure. UDEC (Itasca, 2004) results for the complete model representing all the materials: d) different materials and geometry of studied area and e) response of the room showing the caving of roof as an outcome of simultaneous excavation of the left- and right-hand side rooms.

according to UDEC (Figure 7.c). The 12.6-meter span room was unstable, but since it numerically took a long time to cave in, it was not far from equilibrium and so, a 9-meter span would provide the design solution. Results are presented in Table 3. The D&K method yielded a BL of 37% for the 10.8-meter span, slightly above the recommended guidelines. It can be concluded, from both the UDEC and the D&K results, that the slight dip in the bed did not significantly affect results, although it produced a slight reduction in maximum deflection in comparison to the horizontal case.

Given that results and stability levels compare well, these simple models would confirm that the procedure for estimating realistic parameters and the analytical design tools in general, are consistent with the numerical approaches. The next section describes a more realistic model.

2.7 Numerical analysis of roof failure. Complete model

In this model, all the materials are introduced, if only the magnesite beam blocks are considered as deformable ones. The initially stable central room was first excavated

(Figure 8.d), for which a maximum deflection of 137.1 mm was recorded. In the upper part of the model, a vertical displacement of between 80 and 90 mm was observed. The deflection equivalent, therefore, would have been 50 to 60 mm. Comparing this deflection with that obtained in the simple numerical and analytical models this case would, a priori, be more stable, undoubtedly due to the stress effect—which could also be counterproductive, in that it could produce a higher stress level and lead to voussoir compressive failure. When both the rooms on the right and left-hand sides were excavated, movement occurred in the roof strata and the central and left-hand side rooms eventually caved, according to UDEC (Figure 8.e). It can also be observed how the bell-shape of the remaining material was mirrored in this failure, affecting not only the marly shale, but also the first 4-meter layer of sandy shale and a little higher. This seems to accurately represent the failure observed in situ, and would confirm that relaxation may have significantly affected roof stability.

Diederichs & Kaiser (1999a) used a voussoir beam analogue to illustrate the importance of internal boundary normal tensile strength and of abutment relaxation in controlling the stability of spans in laminated rock masses. They also indicated that a few millimeters of hanging wall or back abutment relaxation could significantly shift the no-support limit, inducing failure in previously stable spans. The numerical model described above, which seems to reliably represent the actual phenomenon that occurred in the mine, would confirm these statements by D&K. A comment on the role of relaxation is in order, so it is presented in the next subsection.

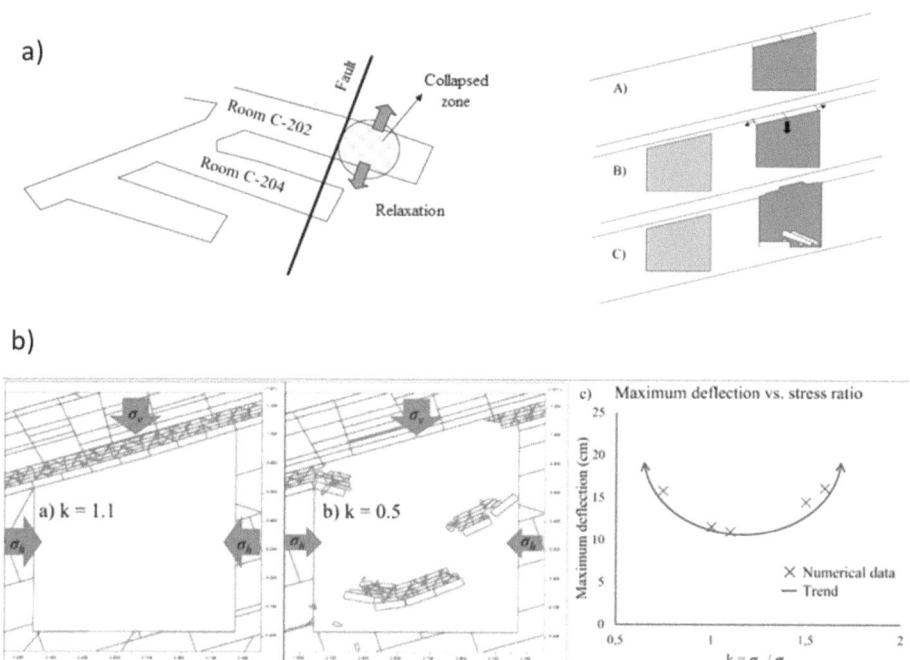

Figure 8 a) Plan view of the caved zone and cross-section in three stages (A, B & C) illustrating roof failure as a result of relaxation, b) Numerical effect of the variation of the stress ratio on roof stability, explanation in the text.

Above all, it highlights the significant role played by relaxation in underground mines. Remark that this issue cannot be handled by analytical approaches, which makes the proposed multi-approach back analysis highly convenient to study and design mining environments.

2.8 The role of relaxation

The role of relaxation in the behavior of underground structures has been reviewed by various authors (Diederichs, 1999), indicating that in complex-geometry underground environments, excavations may affect each other so that relaxation may become a key parameter in controlling stability. Thus, abutment relaxation in rooms excavated in stratified media—for instance due to the opening of a parallel room—may affect roof behavior in a similar way to outer displacement of these abutments. The voussoir analogue approaches developed so far cannot account for this fact, even if it translates into an increase both in deflection and buckling trends in the roof beam. This relaxation could be more significant in destressed areas, as is the case of cross-zones in rooms or faulted zones. A certain degree of relaxation may produce a decrease in the compressive force of the beam and, consequently, an increase in the deflection necessary to attain equilibrium. In stable cases, the beam will be capable of supporting this increase in deflection, but for a beam already close to equilibrium, this relaxation could ultimately cause roof failure.

This analysis leads to speculate that this relaxation, together with the fact that there was no room reinforcement, ultimately caused the roof failure analyzed in this study. This is illustrated in Figure 8.a, which shows a plan view and cross-sections are depicted in 3 stages. Accordingly, roof caving was eventually triggered by the excavation of the parallel room. The excavation work relaxed the abutment of the contiguous room (which was not reinforced) leading to the roof failure.

A numerical model was set up to study the effects of relaxation in the form of stress ratio in the mine. As expected, when the stress ratio is reduced from 1.1 (regular situation) to 0.5 (relaxed zones near faults or excavation of contiguous rooms), the roof collapses (Figures 8.b). The results of considering different stress ratios show that both low and high stress ratios make the roof unstable. In the case of low stress ratio, the instability occurs because the horizontal forces cannot support the beam. In the case of the high stress ratio, the instability occurs because the large horizontal forces produce large bending moments that favor the collapse and ultimate caving (Figure 8.b).

2.9 Epilogue

Ten years after the first collapse observed in the magnesite mine, production has continued at a steady state rhythm of 150,000 tn/year. A new support strategy was proposed, which solve some of the problems encountered, but at least two collapses occurred associated to the occurrence of sub-horizontal joints. This type of discontinuities appears crossing the stratification and if they occur in a roof, the collapse is for sure (Figure 9a and b).

In fact, in these cases the support should be able to load the whole weight of the magnesite beam and that of the loose marly shale if existing. These discontinuities were identified in the joint characterization stage (even if they are more common close to the top of the magnesite seam), but they appear in rather scarce number and with

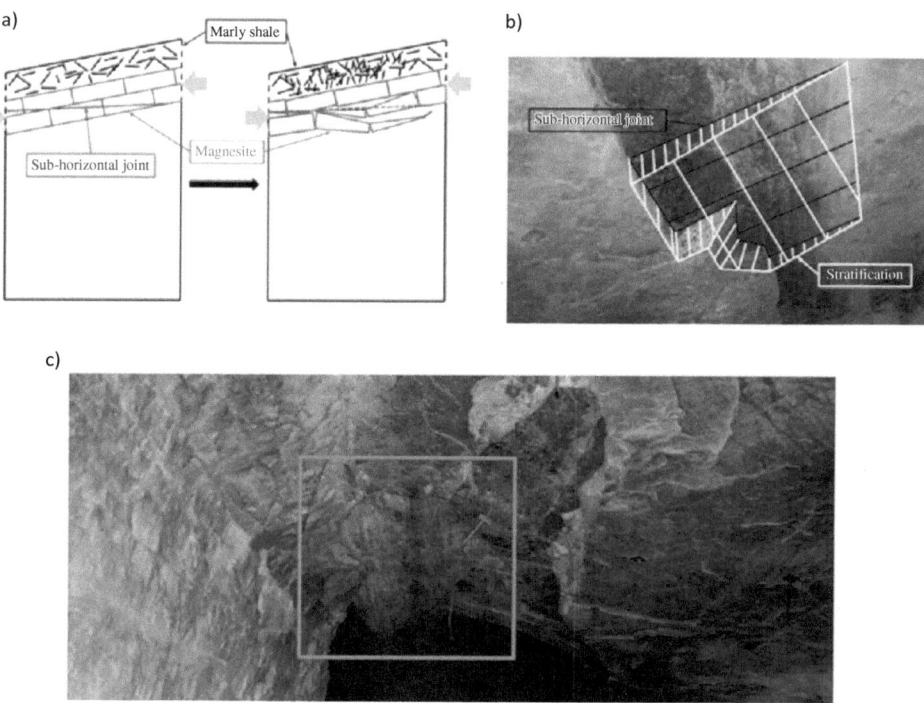

Figure 9 a) Sketch of a room where a sub-horizontal discontinuity crosses the magnesite beam producing collapse, b) an image of such a discontinuity in a room where it does not affect stability due to enhanced support and c) Collapsed room associated to the occurrence of one sub-horizontal joint, magnified in the left-hand size of the picture.

variable direction, so they were not identified as producing a joint set. Additionally, they cannot be readily observed underground, so it is difficult for the miners to identify the problem in advance. Figure 9.c show one of this collapses in an area with enhanced support, where the marly shale is more competent, so the collapsed material corresponds only to the roughly 1-meter thick magnesite bed.

Rock engineering applied in the underground mining field is not a simple matter so, as in the case study presented in this section, many issues tend to affect stability and the normal operational life of a mine. Therefore, the rock engineer should resort to different approaches to study, analyze and understand the mechanisms leading to undesirable instability effects in order to provide design and daily operational solutions, which are not only safe but also economic.

3 CASE STUDY #2: BACK-ANALYSISYS OF A SLOPE FAILURE IN A QUARRY

3.1 Introduction

Because the rock mass behind each slope is unique, there are no standard recipes or routine solutions which are guaranteed to produce the right answer each time they are applied. This sentence taken form the foreword of the seminal book 'Rock Slope Engineering' (Hoek & Bray, 1974), indicated the difficulty of a priori identifying the failure mechanisms producing instabilities in rock slopes. In the last forty years, relevant advances have significantly increased our knowledge in the different mechanisms and associated circumstances leading to instability of slopes, yet the above presented statement still holds true.

The complexity of rock masses and the depth of present open pits and quarries produce a wide range of possible instability mechanisms in slopes. These mechanisms range from simple planar, wedge or circular failure (Wyllie & Mah, 2004), to more complex toppling phenomena or footwall instability mechanisms (Alejano *et al.*, 2011; Stead & Eberhardt, 1997) to different combinations of all these mechanisms (*e.g.* kink band slumping, Preh & Poisel, 2004) which are often very difficult to identify. Interestingly, Sjöberg (1999), in his review of slope stability problems in open pits, remarked that deep-seated toppling can be one of the most common problems associated to large walls.

Again in this case of slopes, analytical and numerical techniques exist, that can be used to study and design slopes. Combined use of these techniques is recommended to have a good understanding of the phenomena. To deepen into the understanding of this topic, it is useful to study real failures in mines.

The main aim of this chapter is to describe, briefly explain and back-analyze the complex failure of an open-pit mine slope in which two mechanisms, each associated with a type of material, combined: block toppling in the upper part and circular failure in the lower part. The toppling affected a limestone rock mass with a persistent discontinuity set dipping toward the slope, separated by a normal fault from a roughly horizontal sedimentary formation of weak rocks in the lower part of the slope and containing mainly claystone and sandstone beds. Although complex failure mechanisms associated with toppling in mining slopes are described in the literature (Hutchison *et al.*, 2000; Coulthard *et al.*, 2001; Alejano, 2014), none reflect the mechanism observed in this mine, even if a similar problem was reported by Ramírez-Oyanguren *et al.* (2014).

Below, the author describes the geological setting, geomechanical characterization and the observed failure. This unusual type of failure was back-analyzed by means of linked standard analytical approaches and using numerical modelling that combined rigid blocks and deformable materials. Further details on this case are found in Alejano *et al.* (2010).

3.2 Geological and mining setting

The open-pit where the failure took place mines a sedimentary formation of claystone and sandstone. The clay is sold to the ceramics industry and the quartz sandstone is used to produce engineered kitchen top slabs and tiles. Mine production is 0.5 Mt/year (roughly half claystone and half quartz sandstone). Around half of extracted material is mineral and the rest goes to waste dumps. The original mine design contemplated 100-meter high and 40° dipping slopes. As a rockfall protection method, 10-meter high benches were included in the initial slope design.

The sedimentary formation is made up of alternating red claystone and white or yellow sandstone beds. Discordantly overlying (direct faulting) these materials, it is a limestone rock mass that outcrops in the mine zone. The average behavior of different beds of the same material is geomechanically similar. The sedimentary formation,

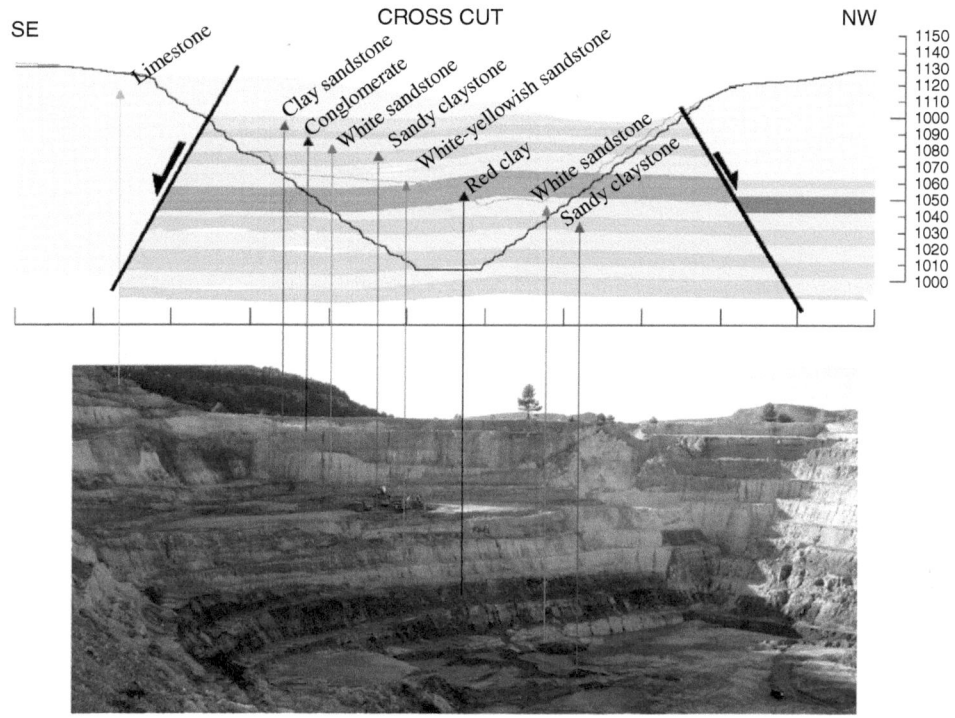

Figure 10 South-easterly—north-westerly geological crosscut of the open-pit mine correlated with a photograph of the south-westerly slope.

which is presented in the field with a minimum thickness of 100 m, has five alternating sandstone and claystone beds, 4 to 12-meter thick. At the top of this formation, there are thin beds of conglomerates and marls. Figure 10 shows a SE-NW crosscut of the sedimentary formations and mine geology correlated with a photograph of the south-western wall of the open-pit mine.

The geological structure of the mine zone is such that the sedimentary package to be mined is located in a block (between two sunken blocks) limited by two sub-vertical normal faults, which also mark the mine limits. Noteworthy is the point of contact between the limestone and the sedimentary formation in the north-western mine wall, as this is key to understanding the slope failure, which occurred due to pushing of the upper limestone slabs onto the sedimentary series.

A brief geomechanical characterization of the materials involved in the phenomenon is presented. A discontinuity survey of the limestone outcrops revealed the rock to be a typical limestone rock mass with three discontinuity sets of highly continuous bedding, with two normal cross-joint sets. Pole distribution in the equal-area lower-hemisphere stereographic projection and the average orientation of these joint sets are illustrated in Figure 11. The slope had a highly continuous well observed joint set J2 (330/73), with spacing ranging between 0.6 and 2 m and with an average value of 1 m. The planes were very continuous and generally not very rough (JRC = 6–8). Tilt tests performed in situ

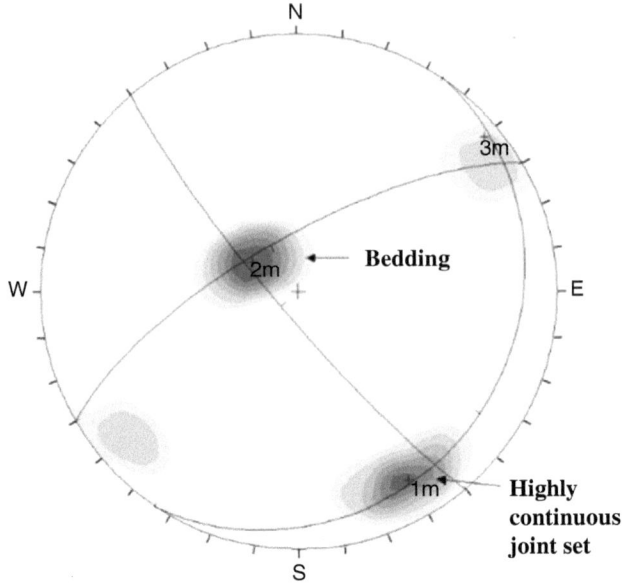

Figure 11 Stereographic distribution of joint poles in limestone and orientations of the 3 joint sets.

Table 4 Strengths and specific weights for the weak materials found in the mine.

	Cohesion (kPa)	Friction (°)	Specific weight (kN/m³)
Sandstone	60	35	20.5
Claystone	50	30	18.6

on these planes revealed a fairly constant friction angle of 40°. Some tests revealed an average UCS value of around 115 MPa for the limestone. The specific weight, as measured in the laboratory, was 25.8 kN/m³. Finally, the combination of laboratory and in situ data enabled the estimation of an RMR in the range 55–65.

To characterize the weak materials, we had to account for the variability observed since the first visit to the mine for the purposes of this study. We used an empirical approach consisting of performing a good number of vane and pocket penetrometer tests. Average values and the material characterization as soils—SW for the sandstone and CL-SC for the claystone—were used to obtain a raw estimate of cohesion and friction that was guided by a specific formulation used in previous applications (Alejano & Carranza-Torres, 2011). Shear tests were also performed. Table 4 illustrates average representative results.

3.3 Failure description

The failure occurred in the north-western wall of the open-pit mine in a 90-meter high slope that dipped 40°. The fault contact between the limestone and the sedimentary

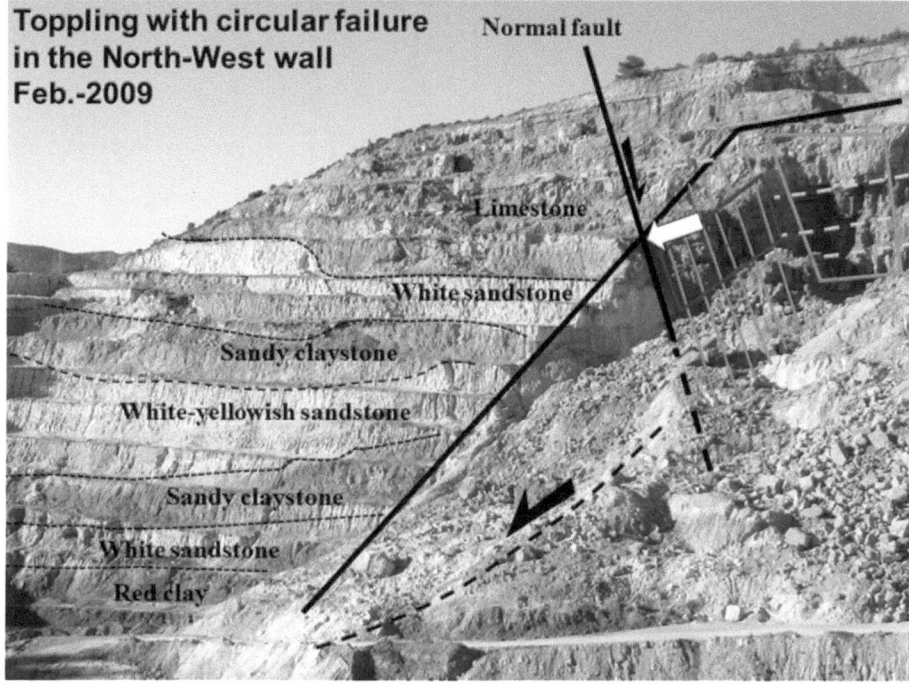

Figure 12 Toppling with circular failure.

formation outcropped in the slope face 20 m below the crest. Apparently, the upper limestone strata had pushed onto the weaker part of the slope, which was suffering circular failure. In a visit to the mine in November 2008, we observed, listened to and recorded raveling. Heave and cracking of the sandstone zone immediately below the normal fault contact was also observed. By the time of a subsequent visit in February 2009, the failure had already occurred. According to the mine crew, it took place following rains in January 2009. The superimposed lines in Figure 12, which depicts the zone after failure, explain the failure mechanisms. Observed in an inspection of the ground behind the slope crest were tensile cracks with stepped counter-slopes up to 40 m behind the crest (Figure 13).

3.4 Analytical back-analysis

Block toppling can be analyzed by means of the Goodman & Bray (1976) method. For this particular case, the method has the advantage that the forces would be transmitted from every block to the contiguous block downwards in the slope. Therefore, for a contact that is parallel to foliation, as in this case, the force transmitted from the upper blocks to this contact can be computed.

The Goodman & Bray (1976) method is applied to the 40°, 90-meter high slope. The continuous joints have the strike parallel to that of the slope, dipping against it in average 73° (dip of J2). For the sake of simplicity and conservativeness, and bearing in

Figure 13 Features indicative of the occurring toppling: a) a step against the lower part of the slope, b) opened triangular cracks and c) cracking and heaving in the sandstone below the contact due to push of the limestone blocks.

mind that some measurements indicate values up to 65°, a value of 70° are selected for all the calculations. The rock slabs have an average thickness of 1 m, corresponding to the average spacing measured in the field. Although the dip of the stepped surface across which toppling occurred is unknown, this value, called β, usually varies in practice from a few up to 20° over the normal to the foliation. Varying this parameter, we solved a number of cases to show that, in our case, the value producing the lowest factor of safety was $\beta = 5°$. This value of β is selected to minimize the factor of safety and so reflects a conveniently conservative approach.

If the slope is entirely composed of limestone slabs, the Goodman & Bray (1976) method will be directly applied using calculation sheets to obtain a final factor of safety—of 1.06 for the case of $\beta = 5°$. In this case, the method is applied in partial manner, that is, to the point where a slab made contact with the sedimentary series. For this slope with 213 slabs, we counted from the upper part to slab number 103, which outcrops at a height of 70 m over the slope foot at the point where the contact fault occurs (Figure 14). For $\beta = 10°$ and 15°, higher factor of safety values—of 1.22 and 1.52, respectively—are obtained.

To analyze this mixed failure mode, the Goodman & Bray (1976) approach was used to estimate the force transmitted from the upper limestone part of the slope to the contact with the sedimentary formation. This force, P102, calculated as 4750 kN, was applied to the lower face of block 103 (28.5-meter high and outcropping at a height of 70 m over the slope face) to stabilize the upper part of the slope. In practice, this force is

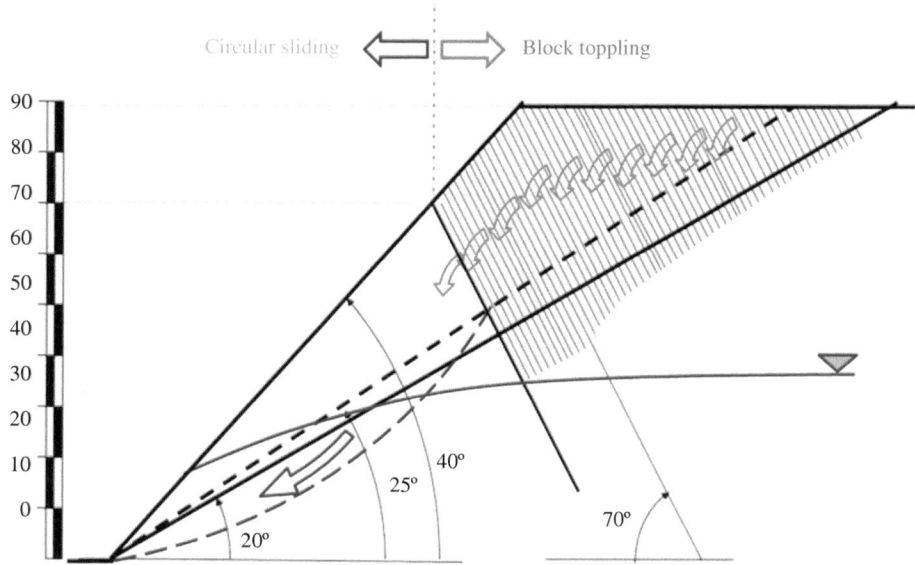

Figure 14 Depiction of the failure mechanism analysis.

not acting at a single point, but it is distributed over the surface of the contact. A conveniently simple but adequately accurate approach is to assume a triangular pressure distribution, with the maximum pressure p (computed from the previous force and the height of the contact area) applied at the upper part of the contact and zero at its lower part:

$$P_{102} = \frac{1}{2} \cdot h_{103} \cdot p \tag{3}$$

This yielded a value of p = 333.9 kPa.

This triangular pressure distribution—representing the push of the upper toppling part of the slope—is applied to the lower and more deformable wedge of the slope, which, due to its low strength, experiences cracking and heaving. To analyze stability, the circular failure stability of the lower part is analyzed, including the distributed pressure transmitted from the toppling blocks of the upper part.

Using the circular failure approach (Bishop, 1955) as implemented in the SLIDE code (Rocscience, 2004) the factor of safety can be computed. We represented the weak sedimentary formation in the slope and included an estimated water level to represent a very rainy period in the area (note that sandstone is much more permeable than claystone). As Figure 15 illustrates, a Bishop's factor of safety of 0.99 is obtained, adequately representing the failure observed in the north-western wall of the mine (see Figure 3).

3.5 Numerical back-analysis

The author used the UDEC code (Itasca, 2004) in previous rock slope stability studies of wall slopes (Alejano *et al.*, 2001) and toppling (Alejano *et al.*, 2006). Its

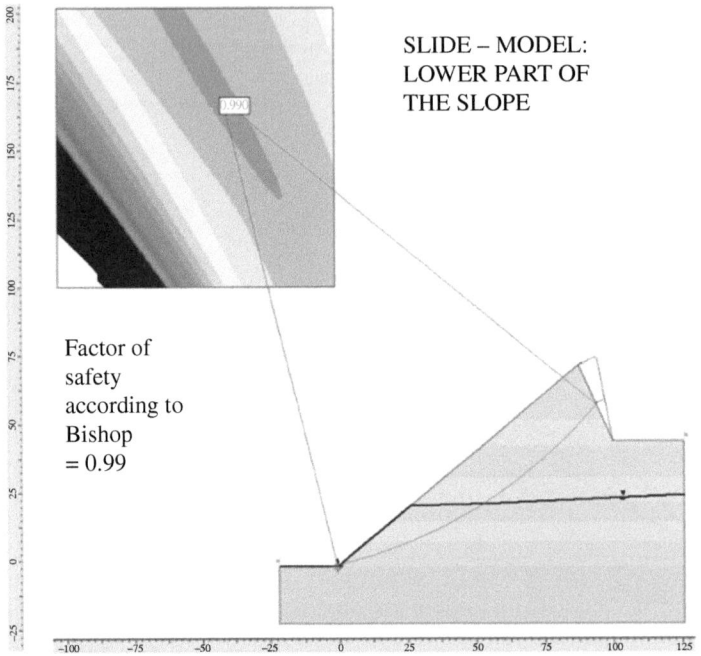

Figure 15 SLIDE model of the failure mechanism in the lower part of the slope.

application to toppling failures has also been validated (Barla *et al.*, 1995). The code has the advantage that it combines rigid and deformable blocks, which is of particular interest in this analysis. In this way, a model is set up, in which the upper part of the slope is made up of rigid limestone blocks or slabs—as considered in the Goodman & Bray (1976) approach. The lower part of the slope, representing weak sedimentary formation, is modelled as deformable materials (grid) with the properties described in Table 2, together with approximate elastic parameters.

Applying the Dawson *et al.* (1999) Shear Strength Reduction Technique (SSRT) to this model representing slope failure in the open-pit mine, a factor of safety of 1.05 is obtained. This is indicative of the failure observed in the mine. Moreover, UDEC adequately represent the failure mechanism identified as a complex toppling-circular failure mechanism (Figure 12). Figure 16 shows the results of plasticity indicating the circular geometry of the failure in the lower part of the slope and the tensile failure cracks or opening of continuous joints in the upper part of the slope and around 40 m behind the slope crest, as observed and recorded in situ. Zones at yield—some meters below the contact zone in the slope face—also correlate well with observed heave and cracking in place.

Remark that the author has not fit the factors of safety obtained by means of the combined analytical (FoS = 0.99) and the numerical (FoS = 1.05) approaches. Obviously, adjusting some parameters this could be an easy task (for instance, slightly increasing the rounding length of the rigid blocks in the numerical model, a slight

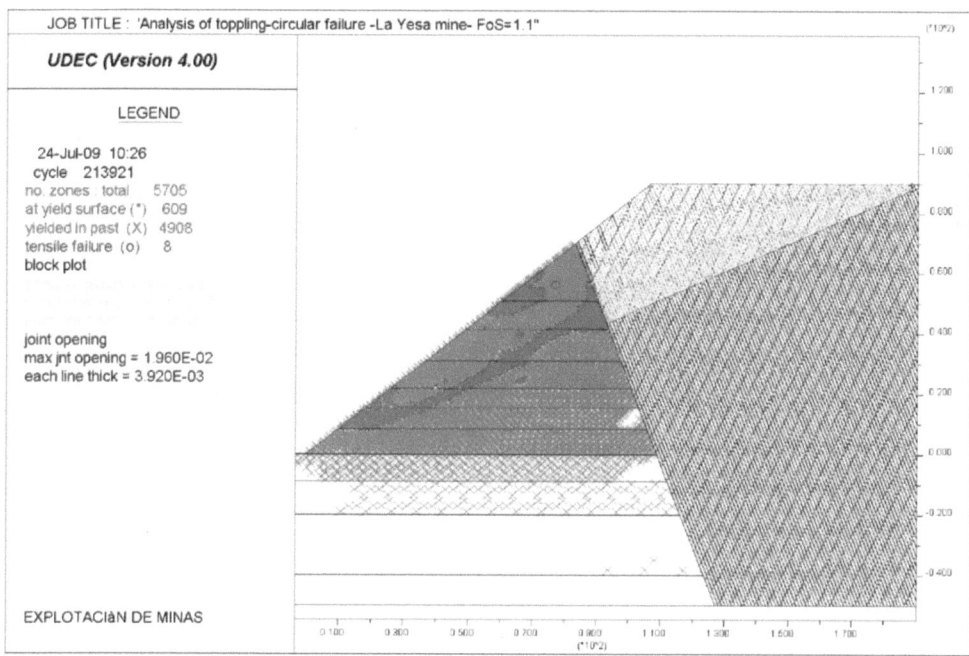

Figure 16 UDEC modelling of the failure mechanism.

decrease of the FoS is easily achieved), but the main aim of the approach was identifying the mechanism and roughly compare the approaches, more than obtaining accurate data for design. In fact, the mine decides to live with this evolving failure mechanism, before the open pit was extended and the extension of this slope was redesigned with an orientation, cinematically avoiding this failure mechanism.

3.6 Conclusions

In this section, a description of a complex rock slope failure mechanism is presented, where toppling and circular failure occurred simultaneously in different parts of a slope. This failure mechanism should be taken into account in slope design that involves a geological structure of the kind described. Although this type of failure is cinematically possible, as far as the author is aware, it has seldom been reflected in the literature (Ramírez-Oyanguren *et al.*, 2014)

The author has shown that this problem can be analyzed by combining classical approaches to toppling and circular failure and it can also be studied by means of distinct element codes, and particularly UDEC (Itasca, 2004). Following the approach described by Dawson *et al.* (1999) and drawing on Stead *et al.* (2006), the use of the SSRT, in combination with an appropriate numerical model, has shown to be a good approach to obtain a reliable factor of safety against this complex failure mechanism to aid in correctly designing rock slopes in this mine.

4 CONCLUSIONS

Unlike in the civil engineering field, in mine designs and operations, the serviceability limit tends to be quite close to the so-called ultimate limit state leading to failure. Very conservative designs are usually non-economic, so an effort is needed to accurately quantify the ultimate limit state. Additionally, the complex nature of rock masses and the difficulties in adequately characterize them, result in the fact that proposing a sufficiently safe and reasonably economic mine design is still a challenging task.

It is relevant to remark that unexpected behaviors are often found in mines. These 'surprising' behaviors are typically associated to poor characterization, rock variability and data uncertainty, particularly complex rock mass structures or over prediction of rock, joint or rock mass strength. For instance, as reported in section 2.9, the occurrence of sub-horizontal joints (misrepresented in the joint characterization) produced a collapse in the mine. Other unexpected behaviors regard: spalling in not so deep zones of a mine, squeezing in drifts in particularly poor quality zones of a rock mass, yielding of pillars coinciding with low quality rock, badly oriented joints or where excavation was not as accurate as expected, uncommon failure mechanisms in slopes (as that presented in section 3) and so on.

In this framework, the observational method should look for preparing plans in advance, which make sure that necessary changes to the preliminary design are put into operation, if the observed behavior during execution of works is unacceptable. This approach is particularly suitable in mining, where we deal with low safety margins. Moreover, a continuous observation of the response of excavation can be the key to early identification of unexpected behavior, and a back-analysis of this observed failure mechanism can be of paramount relevance for future success in design and operation of the mine.

We currently have available a number of sufficiently developed rock mechanic tools in order to reasonably design underground or open pit mines. These techniques can foresee a number of instability mechanisms, which are typical or may occur in each type of exploitation and environment. A proper rock mechanics approach should start with careful geological studies and field data recovery and follow with painstaking laboratory testing. Empirical methods (when available and suitable; remark that for the second case study there is not a suitable empirical method to account for the observed failure mechanism) are convenient to propose a basic design, but analytical and numerical techniques are needed to fine-tune these designs. Analytical solutions are the basis for most of the stability and design analysis procedures in rock engineering. We think that this approach is the best way to solve simple geometry and failure mechanism problems. For more complex problems, where either the failure mechanism or the critical surface, are unknown or where complex behavior appears, numerical methods are preferred. Moreover, different types of control and monitoring (*i.e.* the observational method) are needed, to check the viability of the design implementation, particularly in the early stages of mine operation, or when entering new mine zones or trying new excavation approaches.

In practice, even following a good standard practice of design, a number of instability phenomena may arise, some of which have been illustrated in this text. They should be particularly controlled and monitored by means of observational techniques. The recovered data should be the key to understanding the phenomena and avoid its reproduction in future developments. The application of a multi-approach back analysis technique to the observed undesired phenomena can shed light to understand the problem and to identify the key parameters affecting the resulting behavior. Remark, for instance, how in the first case, the study of the empirical and analytical approaches did not identify horizontal stress as relevant parameter; however, the so-called 'complete' numerical approach and further observation in the mine have put forward the relevant role of this stress.

For demonstrative purposes, the author has described in this chapter the multi-approach back-analysis of two instability phenomena associated to mining excavations as clear examples of the above mentioned strategy. In particular, the collapse of a roof of a mining room excavated in a bedded rock mass and the complex failure phenomenon of a slope in an open pit wall. The multi-approach technique has been extremely helpful to understand the mechanisms behind the observed behaviors as well as to identify the most relevant features affecting stability for every case. The acquired knowledge has been a key point to control the stability in further mine developments in both cases.

Rock masses are complex, but it is not easy to grasp their complexity. As pointed out by Professor Fairhurst: 'Mere increase in the sophistication of mathematical formulation of a problem is of little value if the correspondingly required physical details are not available. Conversely, acquisition of experimental data without guidance from theoretical hypothesis is at best expensive and wasteful, and often misleading'. In words of Professors Harrison and Hudson (2000): 'Currently, our ability to compute has far outstripped our ability to measure the required input parameters and indeed to know whether the computer model is realistic'. The presented approach has the advantage of knowing that the analysis is realistic, and this is indeed an extremely important issue when dealing with such a complex behaving materials we usually call rock masses.

REFERENCES

Alejano, L.R. (2014). Failure mechanisms on rock slopes and slope stability analysis of simple and complex failures in open pit mines and quarries. Chapter 1 of the book: Surface Mining Methods, Technology and Systems. pp. 1–35. Widepublishing, Kolkota, India.

Alejano, L.R. & Carranza–Torres, C. (2011). An empirical approach for estimating shear strength of decomposed granites in Galicia, Spain. *Eng. Geol.* 120: 91–102

Alejano, L.R., Ferrero, A.M., Ramírez-Oyanguren, P. & Álvarez-Fernández, M.I. (2011). Comparison of limit equilibrium, numerical, and physical models of wall slope stability. Int. J. Rock Mech. Min. Sci. 48: 16–26.

Alejano, L.R., García Bastante, F., Alonso, E. & Gómez-Márquez, I. (2001). Stability analysis and design of two quarry slopes with the help of numerical modelling. EUROCK 2001. ISRM Symposium. Rock Mechanics a challenge for society, Espoo, Finland. pp. 801–806. Balkema, Rotterdam.

Alejano, L.R., Gómez-Márquez, I. & Martínez-Alegría, R. (2010).Analysis of a complex toppling-circular slope failure. Eng. Geol. 114: 93–104.

Alejano, L.R, Gómez Márquez, I., Pons, B., Bastante, F.G. & Alonso, E. (2006). Stability analysis of a potentially toppling over-tilted slope in granite. 4th Asian Rock Mechanics Symposium. Asian Society of Rock Mechanics. Singapore, 8–12, November 2006. Ed. Univ. of Singapore.

Alejano, L.R., Taboada, J., Bastante, F.G. & Rodríguez, P. (2008). Multi-approach back-analysis of a roof collapse in a mining room excavated in stratified rock. Int. J. Rock Mech. Min. Sci. 45: 899–913.

Arzúa, J., Alejano, L.R. & Rodriguez, P. (2015). Back-analyses of failures and support re-design of a magnesite room-and-pillar mine excavated in stratified rock. Paper 151. International Congress of the ISRM 2015. Montreal, Canada.

Baran, P.A. & Sweezy, P.M. (1966). Monopoly capital: An essay on the American economic and social order. Monthly Review Press, New York.

Barla, G., Borri-Brunetto, M, Devin, P. & Zaninetti, A. (1995). Validation of a distinct element model for toppling rock slopes. In Proceedings of the International 7th Congress of the ISRM. Tokyo. Japan. Vol. I. pp. 417–421. Balkema, Rotterdam.

Barton, N., Lien, R. & Lunde, J. (1974). Engineering classification of rock Masses for the design of tunnel support. Rock Mech. 6:189–236.

Bieniawski, Z.T., Bauer, S.J. & Costin, L.S. (1993). Geotechnical design methodology workshop. ISRM News. Vol. 4. pp. 42–45.

Bishop, A.W. (1955). The use of the slip circle in the stability analysis of slopes. Geotechnique. 5: 7–17.

Coulthard, M.A., Dugan, K.J. & Hutchison, B.J. (2001). Numerical modelling of complex slope movements at Savage River Mine, Tasmania. pp. 1673–1678, Proceedings of the 10th International Conference on Computer Methods and Advances in Geomechanics, Tucson, USA. Desai, C.S. et al. (eds.)

Dawson, E.M., Roth, W.H. & Drescher, A. (1999). Slope stability analysis by Strength reduction. Geotechnique. 49: 835–840.

Diederichs, M.S. (1999). Instability of Hard Rock masses: The Role of Tensile Damage and Relaxation. Ph.D. Thesis, Department of Civil Engineering, University of Waterloo.

Diederichs, M.S. & Kaiser, P.K. (1999a). Stability of large excavations in laminated hard rock masses: the voussoir analogue revisited. Int. J. Rock Mech. Min. Sci. 36: 97–117.

Diederichs, M.S. & Kaiser, P.K. (1999b).Tensile strength and abutment relaxation as failure control mechanisms in underground excavations. Int. J. Rock Mech. Min. Sci. 36: 69–96.

Fairhurst, C. (1993). Analysis and design in rock mechanics – the general context. Comprehensive Rock Engineering, Vol II, J. Hudson (Ed), Pergamon Press. pp. 1–30.

Goodman, R.E. & Bray, J.W., (1976). Toppling of rock slopes in Rock Engineering for foundation and slopes, Special Conference A.S.C.E., Vol. 2, pp. 201–234. Boulder, Colorado.

Harrison, J.P. & Hudson, J.A. (2000). Engineering Rock Mechanics: Part 2 Illustrative Worked Examples. Pergamon Press, London, UK.

Hedley, D.G.F. & Grant, F. (1972). Stope-and-pillar design for the Elliot Lake Uranium Mines. CIM Bull. 63: 37–44

Hoek, E. & Bray, J.W. (1974). Rock Slope Engineering. Chapman & Hall, London.

Hoek, E.T., Kaiser, P. & Bawden, W.F. (1995). Support of Underground Excavations in Hard Rock. Balkema, Rotterdam.

Hutchinson, D.J. Diederichs, M.S. (1996). Cable-bolting in Underground Mines. Bitech Publishers, Vancouver.

Hutchison, B., Dugan, K. & Coulthard, M.A., (2000). Analysis of flexural toppling at Australian bulk minerals Savage River Mine. GeoEng2000, CDROM Paper. 6. Int. Conf. Geotech. Geol. Eng., Melbourne

Itasca (2004). UDEC v. 4.0. User's Manual. Minnesota, USA.

Obert, L. & Duvall, W.I. (1967). Rock Mechanics and the Design of Structures in Rock. Wiley, New York.

Potvin, Y., Hudyma, M.R. & Miller, H.D.S. (1989). Design guidelines for open stope support. CIM Bull. 82: 53–62.

Preh, A. & Poisel, R. (2004). A UDEC model for "Kink Band Slumping" type failures of rock slopes. In: Numerical Modeling of Discrete Materials in Geotechnical Engineering, Civil Engineering and Earth Sciences. Proceedings of the 1st UDEC/3DEC Symposium, Bochum, Germany, pp. 243–247. Balkema, Rotterdam.

Ramírez-Oyanguren, P, Manera-Bassa, C., González-Philippon, R, & Fernández-Pello, M. (2014). The toppling of large blocks on the northeast slope of the Meirama mine. In: Sympoium EUROCK 2014 – Rock mechanics & Rock Engineering: Structures on and in rock masses. Vigo, Spain. CRC Press.

Rocscience, (2004). Rocscience software products – DIPS, SLIDE. Rocscience Inc., Toronto.

Sjöberg, J. (1999). Analysis of failure mechanisms in high rock slopes. In: Vouille, G., Berest, P. (Eds.), 9 Congrès International de Mécanique des Roches, Paris, France. Vol. 1, pp. 127–130. Balkema, Rotterdam.

Sofianos, A.I. (1996). Analysis and design of an underground hard rock voussoir beam roof. Int. J. Rock Mech. Min. Sci. Geomech. Abstr. 33:153–66.

Starfield, A.M. & Cundall, P.A. (1988). Towards a methodology for rock mechanics modelling. Int. J. Rock Mech. Min. Sci. Geomech. Abstr. 25: 96–106.

Stead, D. & Eberhardt, E. (1997). Developments in the analysis of footwall slopes in surface coal mining. Eng. Geol. 46: 41–61.

Stead, D., Eberhardt, E. & Coggan, J.S. (2006). Developments in the characterization of complex rock slope deformation and failure using numerical modelling techniques. Eng. Geol. 83: 217–235.

Wyllie, D.C. & Mah, C.W. (2004). Rock Slope Engineering. Civil and Mining. Spon Press and Taylor & Francis Group, London.

Risk Analysis

Chapter 10

Groundwater and underground excavations: From theory to practice

M. Sharifzadeh[1] & M. Javadi[2]

[1]*Department of Mining Engineering, Western Australian School of Mine (WASM), Curtin University, Perth, Australia*

[2]*Department of Mining & Metallurgical Engineering, Amirkabir University of Technology, Tehran, Iran*

1 INTRODUCTION

The hydraulic behavior and associated mechanical, physical, and chemical processes of geological formations and rock masses are one of the most important aspects of rock engineering applications, especially for underground excavations. Although the existence of groundwater around underground excavations presents advantageous engineering applications such as hydrocarbon or compressed air storages in unlined rock caverns (Javadi *et al.*, 2014a, 2016b; Froise, 1987; Lee & Song, 2003; Yoshida *et al.*, 2013), most underground excavations usually face different kinds of direct and indirect groundwater challenges (ITA, 1991; Tseng *et al.*, 2001; Yang *et al.*, 2009; Yoo *et al.*, 2012; Gattinoni *et al.*, 2013; Zarei *et al.*, 2012). From the engineering point of view, these groundwater challenges can be categorized into four major groups including in-tunnel (mainly due to the groundwater inflow and construction problems), near zone around tunnel (*i.e.* instability, collapse, swelling), far zone around tunnel (mainly groundwater alteration and drawdown), and support system (*i.e.* deterioration, erosion) challenges that result in increasing time, cost, risk, and environmental hazards and decrease the safety and work efficiency. Successful and appropriate treatment of these challenges requires a problem statement and thorough understanding of the effective features.

The groundwater reaction linked to underground excavations is a result of complex interactions between effective features such as regional hydrology, geological structure, and flow paths that vary from mega to micro-scales. The regional hydrology incorporation with geological conditions mainly control the local hydrogeological conditions of host ground of underground excavations. There is a very close interrelationship between the local hydrogeological conditions and geological structures (Milanovic, 1985; Elhag & Elzien, 2013; Bense *et al.*, 2003; Saul Caine *et al.*, 1996) where some geological structures such as faults may change the groundwater state and anisotropy of rock mass permeability. On the other hand, the groundwater itself can influence the mechanical, physical, and chemical properties of rock mass (Sharifzadeh *et al.*, 2002; Masuda, 2001; Vásárhelyi & Ván, 2006; Brantley *et al.*, 2008), which is called water-rock interaction. Therefore, the realistic analysis of groundwater processes linked with underground excavations should be performed through a framework that briefly reflects the effect of geological structures, rock mass hydraulic behavior, and water-rock interactions. In this case, the conceptual models and classification systems, which present an overview of interactive phenomena, can be utilized to gain an improved understanding and identify the different types of dominant groundwater processes linked to underground excavations.

From the modeling point of view, the rock mass condition and its hydraulic behavior can be considered as the main controlling features of groundwater processes around underground excavations. In many geological formations, the matrix permeability is negligible compared to permeability of fractures and the bulk of fluid flow takes place along preferred flow paths or channels within the fractures (Javadi & Sharifzadeh, 2011; Hitchmough et al., 2007; Odling et al., 1999) as field experiments provide indirect evidence of these preferential pathways (Hsieh & Neuman, 1985; Neuman & Depner, 1988; Novakowski & Bickerton, 1997; Nativ et al., 1999; Wang et al., 1999). In such situations, the hydraulic behavior of rock mass is mainly governed by fractures and their interconnectivity. Generally, the fluid flow processes through rock masses can be investigated by application of continuum and discontinuum modeling methods. The main difference between these methods is mainly referred to as the implementation of fractures effect that is considered implicitly and explicitly in the discontinuum modeling methods, respectively (Javadi et al., 2016a). Proper application of these methods depends on different factors including the heterogeneity intensity of rock mass, geometrical characteristics of fractures, study objects, and the scale of dominant processes relative to underground excavation. It is worth noting that all geological formations, even the ones that are homogeneous, show random variations (spatial nonuniformity) in the values of the hydrogeological parameters (Freeze, 1975) and consequently, there is a considerable amount of uncertainty in the characterizing of the properties of the subsurface rock masses (Renard, 2007). Therefore, it is becoming essential to be able to characterize the uncertainty of groundwater processes and rock mass hydraulic behavior. This can be achieved through implementation of stochastic hydrogeology (Dagan, 2002, 2004; Neuman, 2004; Renard, 2007; Hu et al., 2004; Winter, 2004; Javadi et al., 2016a) that deals with stochastic methods to describe and analyze groundwater processes in both continuum and discontinuum models.

This chapter deals with the link between groundwater processes in geological formation around underground excavations. The main purpose of this chapter is to show the present state of knowledge on the theoretical to practical issues related to hydraulic behavior of geological formation and its effects on the underground excavations. To reach this goal, an overview of the governing equations of groundwater flow in geological formation and its modeling methods is firstly described based on the theoretical framework of rock engineering. Then, the interaction between groundwater and underground excavations is presented mainly in terms of water-rock interaction and challenges that are practically experienced in different projects. Finally, the prediction of groundwater inflow to underground excavations is discussed by focusing on effective features, the role of geological structures, and suitable analysis method.

2 GOVERNING EQUATIONS OF GROUNDWATER FLOW

By applying Newton's second law in the form of conservation of momentum and energy for an arbitrary finite volume of fluid, called a control volume, the equations are derived that describe the motion of fluid. The equations are known as the Navier–Stokes (NS) equations and are also of great interest in many scientific and engineering applications because they describe the physical aspects of fluid motion in different media. For example, the general description of fluid flow in rock fractures or in void space of porous rocks is given by the NS equations. Considering the steady laminar flow of a Newtonian fluid with constant density and viscosity through the void spaces of media, the NS equations can be written in vector form (Javadi et al., 2014b) as

$$\rho\left[\frac{\partial u}{\partial t} + (u.\nabla)u\right] = \mu\nabla^2 u - \nabla p, \tag{1}$$

where $u = (u_x, u_y, u_z)$ is the flow velocity vector, ρ is the fluid density, μ is the fluid dynamic viscosity, t is time, and p is the hydrodynamic pressure. The assumption of fluid incompressibility is acceptable for most of the liquids, e.g. water, under typical subsurface conditions. In order to have a closed system of equations, they must be supplemented by the continuity equation, which represents conservation of mass. For an incompressible fluid, conservation of mass is equivalent to conservation of volume, and the equation takes the form (Brush & Thomson, 2003):

$$\nabla.\mathbf{u} = 0. \tag{2}$$

Equation 1 is composed of a set of coupled nonlinear partial derivatives of varying orders, where the assumptions of incompressible fluid and steady-state condition are not enough to find a general solution of these equations. The main difficulty with solution of Equation 1 arises from the nonlinearity caused by presence of the advection term, (u. ∇)u. Under certain subsurface conditions with very small flow velocity or for small Reynolds number, also named creeping flow, the advection term or inertial forces are small compared with viscous forces. By neglecting the advection term in Equation 1, the much simpler form of governing equation of steady state fluid flow, called Stokes equations, can be obtained as (Kitandis & Dykaar, 1997):

$$\nabla p = \mu\nabla^2\mathbf{u}. \tag{3}$$

By using some vector calculus identities, the Stokes equations can be shown to result in Laplace's equation for the pressure as:

$$\nabla^2 p = 0. \tag{4}$$

for steady-state groundwater flow, Laplace's equation can also be derived by a combination of Darcy's law and continuity equation. Darcy's law is a simple equation that describes linear laminar flow through a porous media or highly fractured rocks. This equation relies on the principle that the amount of flow between two points is directly proportional to the difference in pressure between the points and the ability of the media for fluid flowing. The general form of this equation for anisotropic media can be described as:

$$q_i = -K_{ij}\partial_j g\phi, \tag{5}$$

where q_i [LT^{-1}] represents the components of specific discharge vectors (q_x, q_y, and q_z), K_{ij} [LT^{-1}] are the components of hydraulic conductivity tensor of media, and ϕ is hydraulic head (the sum of elevation and pressure heads). Equation 5 can be described as:

$$q_x = -\left(K_{xx}\frac{\partial\phi}{\partial x} + K_{xy}\frac{\partial\phi}{\partial y} + K_{xz}\frac{\partial\phi}{\partial z}\right), \tag{6a}$$

$$q_y = -\left(K_{yy}\frac{\partial\phi}{\partial y} + K_{yx}\frac{\partial\phi}{\partial x} + K_{yz}\frac{\partial\phi}{\partial z}\right), \tag{6b}$$

$$q_z = -\left(K_{zz}\frac{\partial\phi}{\partial z} + K_{zx}\frac{\partial\phi}{\partial x} + K_{zy}\frac{\partial\phi}{\partial y}\right). \tag{6c}$$

Since the hydraulic conductivity tensor is symmetric, the non-diagonal terms of the hydraulic conductivity tensor components become zero by rotating the coordinate system. In this case, the specific discharges of Equation 6 in the principal directions $(\bar{x}, \bar{y}, \bar{z})$ can be described as:

$$q_{\bar{x}} = -\left(K_{\bar{x}\bar{x}}\frac{\partial\phi}{\partial\bar{x}}\right), \tag{7a}$$

$$q_{\bar{y}} = -\left(k_{\bar{y}\bar{y}}\frac{\partial\phi}{\partial\bar{y}}\right), \tag{7b}$$

$$q_{\bar{z}} = -\left(K_{\bar{z}\bar{z}}\frac{\partial\phi}{\partial\bar{z}}\right). \tag{7c}$$

Back to Equation 2, the conservation of mass of principal flow for incompressible fluids can be rewritten as:

$$\frac{\partial q_{\bar{x}}}{\partial\bar{x}} + \frac{\partial q_{\bar{y}}}{\partial\bar{y}} + \frac{\partial q_{\bar{z}}}{\partial\bar{z}} = 0. \tag{8}$$

By combination of conservation of mass equation and the specific discharges in the principal directions, Equation 8 and Equation 7, the general form of governing equation of steady state incompressible fluid flow can be described as:

$$\frac{\partial}{\partial\bar{x}}\left[(-K_{\bar{x}\bar{x}}\frac{\partial\phi}{\partial\bar{x}}\right] + \frac{\partial}{\partial\bar{y}}\left[-K_{\bar{y}\bar{y}}\frac{\partial\phi}{\partial\bar{y}}\right] + \frac{\partial}{\partial\bar{z}}\left[-K_{\bar{z}\bar{z}}\frac{\partial\phi}{\partial\bar{z}}\right] = 0. \tag{9}$$

For the case of isotropic media, Equation 9 can be reduced to the form of Laplace's equation. Similarly, the general form of groundwater flow governing equation for transient condition can be described as:

$$S_s\frac{\partial\phi}{\partial t} = \frac{\partial}{\partial\bar{x}}\left[(-K_{\bar{x}\bar{x}}\frac{\partial\phi}{\partial\bar{x}}\right] + \frac{\partial}{\partial\bar{y}}\left[-K_{\bar{y}\bar{y}}\frac{\partial\phi}{\partial\bar{y}}\right] + \frac{\partial}{\partial\bar{z}}\left[-K_{\bar{z}\bar{z}}\frac{\partial\phi}{\partial\bar{z}}\right]. \tag{10}$$

where S_s [L^{-1}] is the specific storage coefficient of media that can be interpreted physically as the volume of water released by a unit volume of saturated media due to a unit decline of the groundwater potential.

3 PRINCIPLES OF FLUID FLOW MODELING IN ROCK MASS

3.1 Modeling methods of hydraulic behavior

All geological formations can be represented as a stochastic set of macroscopic and microscopic elements with random spatial variations or nonuniformity in the hydro-geological characteristics and parameters (*i.e.* hydraulic conductivity and porosity), even those that are homogeneous. A proper criterion for nonuniformity is provided by the standard deviation of the frequency distributions. A homogeneous formation under this representation is one in which the frequency distributions do not change through space and reversely, the heterogeneous formations show spatial variable frequency distributions (Freeze, 1975). The natural complication of geological media causes variable and different hydraulic behaviors and consequently complexity in the system description or mathematical model. Therefore, many assumptions and simplifications are necessary to reach a representative mathematical model. Successful predictive mathematical models require a thorough understanding of the physical processes that govern fluid flow in rock mass.

For the purpose of hydraulic modeling, the rock masses can be considered as an assemblage of intact blocks (as a continuous porous medium) that are separated by discontinuities such as joints, fissures, and fractures. Here, the term "fracture" is used as a general term for describing discontinuities (Wilson & Witherspoon, 1974). Each of these portions has distinctive hydraulic characteristics and consequently the hydraulic behavior of rock masses depends on the relative dominance of these portions. By considering such representation of geological media, four different hydraulic processes (Figure 1) can be anticipated as (Adler & Thovert, 1999):

– Flow and transport in single rock fractures
– Flow and transport in fracture networks
– Flow and transport in porous media
– Interactions between porous media and fractures

Based on these hydraulic processes, the geological media can be classified as fractured rock, porous media, or fractured porous rock. The fractured rock is characterized as the media if the blocks of rock surrounded by fractures contain no void space and the major part of the flow occurs through discrete interconnected fractures. This is in contrast to the porous medium, in which it is not practical to distinguish between individual pores, and flow can be assumed to occur in a continuum. The term fractured porous medium is usually used for geological media that contain porous blocks separated by fractures. For such media, the void space consists of two parts including a network of fractures and blocks of porous medium (Sagar & Runchal, 1982; Bear *et al.*, 1993). Considering this classification, the fluid flow processes through rock masses can be investigated by application of continuum and discontinuum modeling methods (Figure 1).

The main difference between continuum and discontinuum modeling methods of hydraulic behavior of rock masses refers to the implementation of fractures effect. Generally, the effect of fractures is considered implicitly and explicitly in the continuum and discontinuum modeling methods, respectively. In the continuum

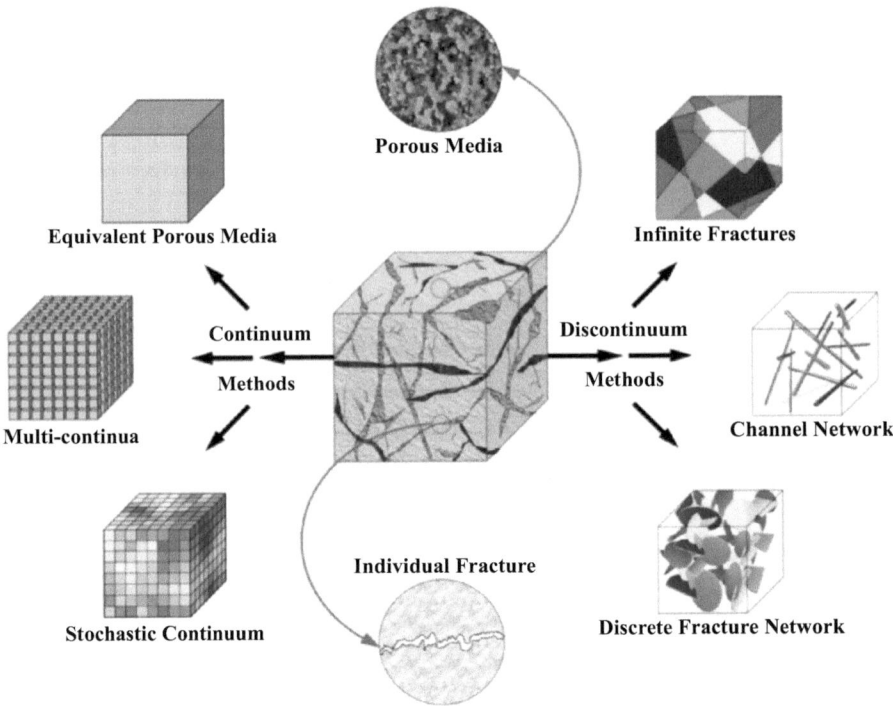

Figure I Different hydraulic processes in rock mass with modeling methods classification.

methods, the fractured rock mass is treated as an equivalent porous medium characterized by "averaged equivalent" hydrogeological properties. In this method, the
modeling of rock mass hydraulic behavior can be performed by utilization of deterministic or stochastic continuum or multi-continua representations. The continuum
representation of fractured rock mass relies on the assumption of homogeneous
porous media that is more likely to occur in extremely long, dense, well-connected
fracture networks with mixed fracture orientations than in sparsely fractured, poorly
connected, and/or strongly anisotropic systems (Niemi *et al.*, 2000; Lin & Lee, 2009);
or where dimensions of the interested domain or scale of application are large
compared to the heterogeneities of the medium (Hsieh & Neuman,1985; Schwartz
& Smith, 1988). For the cases in which these conditions do not hold, the equivalent
continuum representation of fractured rock mass is not appropriate and it is necessary to use the discontinuum methods. In this method, the rock mass is explicitly
represented by individual discrete fractures as geometrically well-defined hydraulic
features. Appropriate application of these modeling methods poses a key question;
when can the hydraulic behavior of a rock mass be represented by an equivalent
porous medium? The answer to this question requires a fundamental comprehension
related to the mathematical description of governing equations, study proposals, and
relative scales of problems. More details about this issue are described in the next
section.

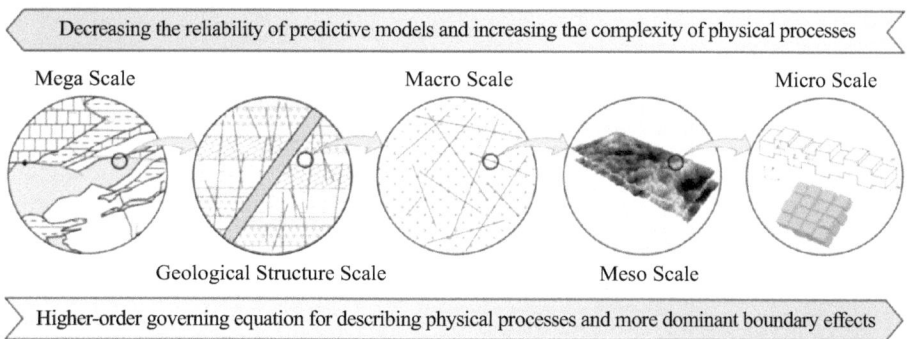

Figure 2   Transforming the size of hydrogeological problem from mega scale to microscopic scale.

3.2 Continuum method

3.2.1 Representative Elementary Volume (REV)

The behavior at all points of a phase in an interstice (a portion of space separated from other portions by well-defined interfaces or interphase boundaries) is characterized by the same set of state variables. The mathematical descriptions including conservation of momentum, energy, and mass, constitutive relations, and boundary conditions can be stated for every point within the considered phase. This is a general "microscopic level" description of physical processes happening at a point within the interstice present in the domain (Figure 2). However, the solving of governing equations under the boundary conditions is very unbearable at microscopic level due to the prodigious complexity in governing equations and detailed geometrical description of interstice and phase boundaries. Here, the major difficulty arises due to the fact that it is rarely possible to measure the physical variables at arbitrary points through the phase (or in microscopic level) for the purpose of model validation and characterizing the media.

In order to overcome the above difficulties, the fluid flow problems can be converted from a microscopic level to a macroscopic one with reformulation in terms of measurable averaged microscopic properties. This procedure is usually referred to as the continuum approach in which the heterogeneous spatial distribution of microscopic entities is replaced by fictitious homogeneous continua at every point within the entire domain. In this case, the effects of the microscopic configuration of interstice and entities are reflected by some parameters, coefficients, or average variables can be taken over an "elementary volume" centered at every point within the media. Since the microscopic interstices are irregularly distributed throughout the media, a set of fluctuating averaged values will be obtained by traversing the domain with a moving elementary volume. The dispersion of these averaged values depends on size of elementary volume and degree of heterogeneity of microscopic interstice. By increasing the size of elementary volume, a volume may be found that sufficiently represents what happens at center point and at its close neighborhood. For this sampling size, the averaged values of domain show negligible spatial dispersion (Figure 3). Such sample size is usually called Representative Elementary Volume (REV). The REV can be characterized in more practical terms:

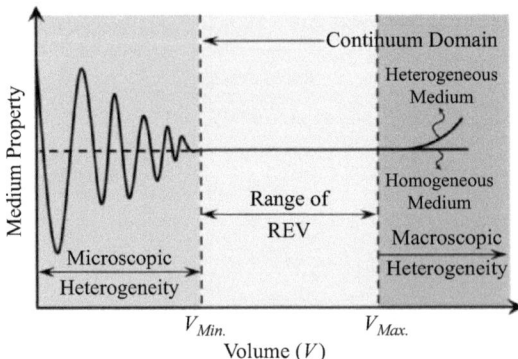

Figure 3 The scale dependency of media properties and concept of REV (After Bear *et al.*, 1993).

- The distributions of all the interstice and entities in the selected REV shall be statistically meaningful.
- The size, shape, and orientation of both REV and averaged values shall be uniform throughout the domain.
- The size of REV shall be much larger than the characteristic length of the interstice (such as fractures spacing or fracture length) and much smaller than macroscopic heterogeneities (such as lithological units).
- The size of the REV must be much smaller than the scale of interest domain (*i.e.* the hydrogeological disturbed zone around the underground excavation).

3.2.2 Equivalent continuum representation of fractured rocks

In continuum approaches, a fractured rock mass is treated as an equivalent porous medium characterized by averaged hydrogeological properties. The advantage of continuum approaches, especially for large scale problems, is that the overall behavior of fractured rocks can be described by implicitly considering the effects of fractures where the detailed geometry and flow processes through each fracture are not required. However, natural fractures have a complicated geometry with different orientation, size, and spatial distribution at all scales and as a consequence, fractured rocks encounter a high level of heterogeneity and anisotropy with respect to the rock mass permeability. Therefore, some indefeasible prerequisites should be established for equivalent porous medium representation of fractured rocks. The prerequisites and possibility of equivalent porous medium representing the hydraulic behavior of fractured rocks has been extensively discussed during the last decades. In most cases, the theory of flow through fractured rocks and homogeneous anisotropic porous media is used to determine when a fractured rock behaves as a continuum. Based on this issue, two criteria should be simultaneously satisfied for a rock mass to be considered as equivalent porous medium:

1. The equivalent permeability of a fractured rock shows insignificant variations in the value with a small addition or subtraction to the sampling volume.
2. An equivalent hydraulic conductivity tensor should exist for rock mass that appropriately predicts the flow vector when the direction of a constant gradient is changed.

The first criterion implies that an REV should exist for a given rock mass. The second criterion refers to the fact that the rock mass hydraulic behavior shall be expressed by a symmetric hydraulic conductivity tensor. To reach this condition, the measured or calculated directional hydraulic conductivities in different directions (for samples larger than REV) should be plotted as an ellipsoid. Where any of these criteria are not met, the hydraulic behavior of the fractured rock cannot be represented by equivalent porous medium and the discontinuum methods should be utilized for hydraulic modeling of media.

3.3 Discontinuum method

In the discontinuum method, the role of fractures in the rock mass hydraulic behavior is explicitly considered in the calculations. In this method, the rock mass is represented by a network of individual district and geometrically well-defined fractures. In this case, the rock mass hydraulic behavior depends rigorously on the geometry of fracture and their interconnection as well as the hydraulic characteristics of fractures. In order to apply the discontinuum method, four main issues should be considered (Javadi *et al.*, 2016a):

1. The geometrical properties of fractures
2. The techniques for representation and generation of fractures
3. The hydraulic characteristics of fractures
4. The techniques for solving the flow equation

3.3.1 Geometrical properties and generation of fractures

The first step in discontinuum representation of rock mass is to define the geometrical properties of fractures in terms of location, orientation, size, shape, spacing, population density, roughness, and aperture. The fractures in rock mass are usually found in different sets where each set has been generally initiated at the same time by the same tectonic process. Therefore, it is generally anticipated that the same fracture set will have similar properties. Some geometrical properties of fractures can be obtained from 1D or 2D observations and site investigation through the surveys in boreholes or outcrops. Generally, 2D observations provide more useful information than 1D ones. However, some geometrical properties of fractures such as shape and size in 3D can be indirectly determined from the observations. The variable nature of rock fractures makes it necessary to describe the geometrical properties using statistical or geostatistical techniques.

After the identification of geometrical properties of fractures, the discontinuum models can be produced by application of generation techniques. To reach this goal, the discontinuum models are usually represented by generated fracture networks. The purpose of fracture network generation is to produce 2D or 3D graphical or numerical realizations of fracture geometry that reflect the geometrical characteristics of real fractures sampled in the site investigations. The fractures in the network are conventionally represented as simplified linear or planner features in 2D or 3D spaces, respectively, as shown in Figure 1. In 3D spaces, the fractures in the network may have different shapes including circular or elliptical disks, polygons, rectangles, and infinite planes. Based on the natural condition of observed fractures, different network models such as hierarchical (short fractures ended when encountering the long fractures), cluster (based on point process), density regionalization (based on the geostatistical methods), conditional (predefined geometry) and correlated models can be utilized for generation of fracture networks.

The fracture network may be generated by use of deterministic, stochastic, or hybrid stochastic–deterministic methods. In deterministic method, the geometrical properties of fractures are explicitly specified in the generation domain. The usual practice in stochastic modeling is to assume that the geometrical properties of each fracture set are statistically distributed and apply the Monte Carlo approach to allocate the properties to fractures in the network realizations. In hybrid methods, a combination of predefined or deterministic geometrical properties (especially for location) in combination with statistical properties (such as size and orientation) is used for fracture network generation. Since in stochastic methods the geometrical properties are randomly allocated to the fractures, each realization will be unique in terms of the local arrangement and interconnection of the fractures. Therefore, large number of fracture network realizations shall be generated to ensure that the results are not dependent on specific fractures geometry arrangements, and to produce more representative uncertain behavior of the fractured media.

3.3.2 Fluid flow through fractures

The hydraulic behavior of rock fractures is significantly dependent upon the geometry of void space (Javadi et al., 2009, 2010; Sharifzadeh et al., 2009; Javadi & Sharifzadeh, 2013). Natural fractures have complicated geometry (at a microscopic scale) where their attributes such as roughness, contact area, aperture field, and matedness play a key role in hydraulic processes (Javadi et al., 2014b; Sharifzadeh et al., 2006). On the other hand, the existence and uniqueness of closed form solution of the full NS equations (the general description of Newtonian fluid motion in a rough-walled fracture) in three dimensions is not yet proven. Therefore, many simplifications in geometry and governing equations are necessary for mathematical description of fluid flow through rock fractures (Javadi et al., 2010). Traditionally, individual fractures have been idealized as a set of parallel plates and NS equations reduce to linear Stokes equation in order to obtain a tractable mathematical description of fluid flow, namely, the cubic law as (Witherspoon et al., 1980):

$$Q = -\frac{wa_h^3}{12\mu}\nabla p = -\frac{kA_f}{\mu}\nabla p, \tag{11}$$

where Q is the volumetric flow rate, w is the fracture width perpendicular to the pressure gradient (∇p), a_h is the hydraulic aperture, k is the permeability, and A_f is the cross-section area of fracture. Since the cubic law is just valid for laminar flow and idealized parallel smooth fracture with constant mechanical aperture, the mechanical aperture in Equation 11 is replaced by the term "hydraulic aperture" to consider the effect of flow tortuosity and channeling, and roughness for natural fracture.

The fluid flow through the fracture network can be calculated if the active (interconnected) fractures in the network are replaced by a network of channels with corresponding hydraulic parameters. In fact, the fracture network is converted to a computational domain of nodes and elements that are fracture intersections and fracture segments between nodes, respectively. In this case, the hydraulic head at each internal node and flow rate in each fracture can be calculated by implementing the cubic law into mass continuity equations. The final matrix form of steady state laminar flow through the fracture network can be expressed as (Sharifzadeh & Javadi, 2011):

$$[D_{kf}]\{H_k\} + [D_{ff}]\{H_f\} = 0, \tag{12}$$

where the subscripts of k and f refer to constrained and free nodes (or boundary and internal nodes), with predefined and unknown hydraulic head values, respectively. H_f and H_k are the vectors contain hydraulic head of internal and boundary nodes, respectively, D_{kf} is the asymmetric matrix of normalized conductance of fracture segments between internal and boundary nodes, and D_{ff} is the asymmetric matrix of normalized conductance of fracture segments between internal nodes. The conductance ratio of fractures is defined by dividing the coefficient term of ∇p in right-hand of Equation 11 by length of fracture. The conductance ratio of i-th fracture in the network is also normalized by the summation of conductance of all the active fractures that are directly connected to i-th fracture. The matrix D_{kf} consists of n rows and m columns that correspond to the number of internal and boundary nodes, respectively, where the component in position i,j is d_{ij} of fracture segment joining i-th boundary node and j-th internal node or is zero if they are not connected. The D_{ff} is an asymmetric square matrix contains n rows and columns, where all the diagonal component of this matrix have the value of –1 and the off-diagonal component in position i,j is d_{ij} of fracture segment (starting from j-th internal node) meeting the i-th internal node or is zero if they are not connected.

4 GROUNDWATER AND UNDERGROUND EXCAVATIONS

4.1 Challenges of groundwater linked with underground excavations

Groundwater through the face or sides of underground excavations and water inflow have always been a major reason for technical problems, construction difficulties, and hazardous conditions that are here referred to as water challenges. Different types of such challenges have been severely experienced in many underground excavations. Some of these challenges in different case studies are presented in Table 1. These challenges result in increased time, cost, risk, and environmental hazards and decreased safety and work efficiency. From an engineering point of view, the water challenges related to underground excavations can be categorized into four major groups:

In-tunnel challenges: these kinds of challenges are mainly expected inside and during the construction of underground excavations. The main resultant problems related to these kinds of groundwater challenges are adverse working conditions, discomfort and inefficiency or worse, endangerment of the labor force or life loss of workers (casualties), damage of construction equipment, inability to bond shotcrete to wet rock, difficult blasting, slow rate of advance, and lost construction time in spite of various attempts to stem the flow. In some cases, the groundwater surrounding the

Table 1a Groundwater challenges linked to tunnel, case study: Alborz Service Tunnel.

Case: Alborz Service Tunnel	Main Refs.: Wenner & Wannenmacher, 2009; Urmiei et al., 2011

Main Challenges: Water burst, H2S gas releasing, Raveling, Damage on transportation system

Description:

Alborz Service tunnel (for the purpose of service, site investigation, and providing access to main tunnels), 6400 m in length and 5.2 m wide, is one of the parts of Tehran Shomal Freeway in the north of Iran. The greater part of this tunnel (about 6000 m) was excavated using an open TBM through the Alborz Mountains. This tunnel passes through very complicated geological formations including Triassic and Jurassic (argillites, sandstones, thin coal layers, and limestone), a 300 m thick fault zone (Kandovan fault), Oligocene sediments (massive anhydrite and gypsum with karstic features at surface level), and Eocene tuffs. Different groundwater challenges such as water burst, collapse, raveling, and harmful gas releasing were experienced during the tunnel advancing. The main challenges were: water burst at Ch. 2+582 km (90 l/s increasing to 125 l/s), water burst (115 l/s) with H2S gas releasing (25 ppm) at Ch. 2+967 km, water burst (290 l/s) with H2S gas releasing (60 ppm) and raveling (100 m3) at Ch. 3+015 km through a fault zone with karstic void fillings, water burst (690 l/s) with high H2S gas releasing (>100 ppm) at Ch. 4+524 km through a water bearing fault zone, and water burst (initially 220 l/s increasing to 600 l/s) through karstic fault zone.

Detrimental effects on electrical installations and transportation system, sedimentation of fine materials in the invert, blocking rails, adverse working conditions with cold water squirting, necessary high speed ventilation, unsuccessful grouting in probe boreholes due to high pressures and flow rates, and at least 12 months' work delay are the main consequences of groundwater challenges in this tunnel.

Water flow with high H2S gas releasing in the TBM backup at Ch. 4+524 km	Adverse working conditions in high H2S concentration environment	Raveling behind the cutter head through a fault zone with karstic void fillings

underground excavation acted as a medium for the entrance of dissolved gases (methane, hydrogen sulfide, sulfur dioxide) that cause dangerous working conditions. Moreover, unexpected and uncontrolled groundwater inrush into an advancing tunnel can stop all forward progress because there is no alternate focus of construction activity while water problems are resolved.

Near zone around tunnel: the groundwater challenges of this group mainly occur due to the alteration of ground conditions around the tunnel or face. Rock mass weathering or weakening and consequent stability problems, instability caused by water pressure or seepage forces acting toward the tunnel face, collapse caused by sudden groundwater ingression, running ground, squeezing of clay-rich-rocks, and swelling of rocks containing clay and anhydrite minerals are the main resultant groundwater challenges around the near zone of tunnel. In permeable soft rocks, typically of silts and sands,

Table 1b Groundwater challenges linked to tunnel, case study: Kuhrang (III) Tunnel.

Case: Kuhrang (III) Tunnel	Main Refs.: Foladgar, 2003; Zarei et al., 2012

Main Challenges: Several water bursts, swelling, collapses, TBM jamming, and work delay

Description:

The Semnan tunnel is located 30 km northeast of the city of Semnan, the capital of Semnan Province of Iran. This tunnel with a length of 3.2 km and 3.2 m wide was executed by drill and blasting and road-header machine. The geological setting of the tunnel consists of a series of Cambrian and Tertiary sedimentary rocks (limestone, dolomites, siltstone, shale, and sandstone) passing through two structural zones of Alborz and Central Iran.

The Kuhrang (III) tunnel with a length of 23.4 km and 4.5 m in diameter was mainly constructed using open TBM and roadheader. This tunnel and four access tunnels (total length 3.7 km) passed through limestone and marlstone layers of the High Zagros Mountains in Central Iran. The high overburden (more than 1200 m), high groundwater level above tunnel (up to 500 m), karstic formations, and several fault zones make this tunnel a very challenging and time-consuming (more than 20 years) project in Iran. Unexpected local groundwater inrushes and consequent tunnel flooding, swelling and consequent large deformations, TBM jamming, and collapses in weak materials are the most important groundwater challenges linked to this tunnel.

The karstic cave exposed in the tunnel floor at Ch. 5+608 km caused water to rush (1200 l/s) into the tunnel. Well-developed karst channels in Cretaceous carbonate rocks around the tunnel with an over-burden of 1100 m and groundwater pressure of 30 bar are the main features of this part of tunnel. The water discharge raised as much as 2.5 m and flooded about 18 km of the tunnel accompanied by 1000 m^3 of clayey material and rock fragments. The karstic zone developed at Ch. 17+705 km to 17+814 km caused water inrushes and tunnel flooding. First, a water inrush at a 70 l/s rate and 16 bar pressure inundated the main tunnel and halted the TBM performance. Therefore the excavation had to be continued from another access point by blasting method. Again, water rushed into the tunnel at Ch. 17+705 at a 250 l/s rate that caused the complete flooding of the access tunnel and water flow out from the portal.

Large deformations were experienced through the green marlstone layers from Ch. 0+500 to Ch. 0+750. The pressure from the ground caused breaking in the steel frames and shotcrete of the initial lining. Extra steel beams were also installed to control the deformations. However, the lateral deformations and floor uplift exceeded 40 cm and 50 cm, respectively. TBM jamming in sticky moist material occurred during the construction of the tunnel at Ch. 8+063 km. The sticky materials flowed from a large karstic cave into the tunnel that caused TBM trap in filling material. The excavation duration of 20 m of this zone exceeded 50 days.

Schematic view and photograph of large deformations (swelling) in marlstone	Schematic view and photograph of sticky materials flow from karstic cave	Schematic view and photograph of water burst from karstic cave

Table 1c Groundwater challenges linked to tunnel, case study: Long Zagros Tunnel.

Case: Long Zagros Tunnel	Main Refs.: Grandori et al., 2011; Zarei et al., 2011a

Main Challenges: Several water bursts, H2S gas releasing, raveling, and lining corrosion

Description:

The Long Zagros tunnel with a total length of 49 km in three lots (lot 1A: 14 km, lot 1B: 9 km, lot 2: 26 km) is under construction with three D.S. TBMs in the west of Iran. This tunnel with diameter of 6.1 to 6.7 m passes along different folded formations through High and Folded Zagros Mountains. The main formations of the tunnel route are EP or Pabdeh (Eocene to Oligocene age containing shale, marl, limestone, limy shale), Kg or Gurpi (Late Cretaceous to Paleocene age containing marl, shale, limy shale, shaly limestone, and thin layers of argillaceous limestone), Ki or Ilam (Late Cretaceous age containing well-bedded argillaceous limestone, shale, limy shale, and shaly limestone), Kga or Garu (Early Cretaceous age containing altered argillaceous limestone, pyritic and bituminous shale, radiolaria and chert), Jk or Khami (Jurassic age containing thin bedded to massive limestone and argillaceous limestone), and J or Surmeh (Jurassic age containing dolomitic limestone, marl/marlstone). The excavation of this tunnel along high groundwater level (up to 800 m) and overburden (1200 m), folded and faulted zones, and hydrocarbon source rocks is associated with different challenges related to groundwater such as water burst, collapse, raveling, and harmful gas releasing.

Several problems related to sticky moist materials (shale and marl) such as cutter blockage, muck clogging of scraper vent, and big lumps on the handling conveyer have been experienced during the tunnel (lot 2) construction through Gurpi formation that resulted in the very slow advance rate. Raveling of mud and debris from the tunnel face into the TBM shield and back up has been experienced in water bearing fault zones in tunnel interval between Ch 1+050 km to 1+101 km (lot 1B). This phenomenon occurred in association with chimney collapses in front of the face and water inrush. Several water bursts in different sections have been experienced during the construction of tunnel. In addition, the water inflow was associated with releasing of dissolved harmful Hydrogen Sulfide, H2s gas, in many cases, where a notable relation was found between measured concentration of H2s and groundwater flow rate. In lot 1B, water burst of 70 l/s with releasing H2s gas at Ch. 0+317 km in bituminous shale and limestone (Garu formation), water burst of 320 l/s at Ch.0+585 km in limestone of Garu formation, and water burst of 120 l/s at Ch. 1+101km have been faced during the tunnel construction. The first indicative challenge of H2s gas releasing (130 ppm) was experienced at Ch. 3+720 km (lot 2) after entrance of tunnel from Gurpi (shale) into the Ilam (limestone) formations that has an anticlinal structure. The gas releasing with water burst caused serious damage to the TBM, a 6 month work interruption, and the death of 4 workers. Different variations of groundwater head (100 to 350 m above tunnel), H2s gas releasing (30 to 130 ppm), and water burst (100 to 350 l/s) have been experienced through the anticlinal structure of the Ilam formation from Chs. 3+720 to 4+940 km (lot 2). In addition, the water burst of 110 l/s with H2s gas releasing up to 300 ppm occurred from Chs. 8+240 to 8+260 km (lot 2) through the Gurpi formation. The worst conditions of water burst of 800 to 1050 l/s and H2s gas releasing up to 700 ppm (maximum measurable with sensors) have been experienced in the tunnel (lot 2) interval between Chs. 13+600 to 14+000 km of the Ilam formation and anticlinal structure).

Water burst at Ch.0+585 km (Lot 1B)	Water burst and raveling at Ch.1+065 km	Chemical corrosion of concrete segment

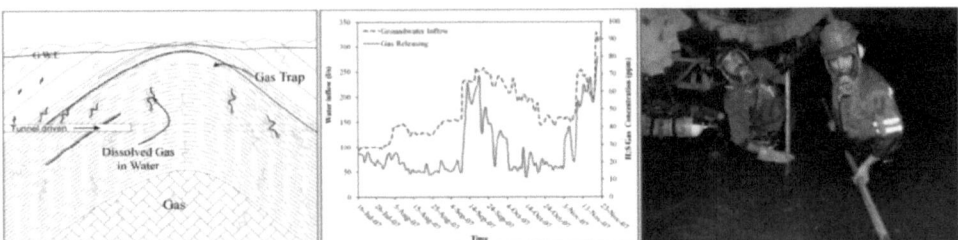

Gas releasing from anticlinal structure	Water inflow-gas releasing relationship	Work conditions in harmful H2S gas

Table 1d Groundwater challenges linked to tunnel, case study: Semnan Tunnel.

Case: Semnan Tunnel	Main Refs.: Zarei et al., 2011b; Zarei et al., 2013

Main Challenges: Several water bursts, collapses, and work delay

Description:

The Semnan tunnel is located 30 km northeast of the city of Semnan, the capital of Semnan Province of Iran. This tunnel with length of 3.2 km an 3.2 m width was executed by drill and blasting and road-header machine. The geological setting of the tunnel consists of a series of sedimentary rocks (limestone, dolomites, siltstone, shale, and sandstone) of Cambrian and Tertiary passing through two structural zones of Alborz and Central Iran.

Unexpected local groundwater bursts are the most important challenges of this tunnel that are mainly related to geological features such as fractures, faults, and dikes. An open fracture with a maximum aperture of 100 mm and dip of 70° was encountered at Ch. 2+119 km through carbonate rocks. There is an intrusive dike close and almost parallel to this open fracture that acts as a barrier and stores of groundwater. A sudden and violent water burst with over 750 l/s and over 3 months delay in operation process were the results of this open fracture.

The Bashm thrust fault as a conduit-barrier system with 10 m non-cohesive core (barrier component) intersects the tunnel at Ch. 0+947 km to 1+056 km. The extension of highly fractured zone along the footwall (81 m) is wider than that along the hanging wall (18 m). The 236 m initial groundwater level above this fault released increasing inrush from 7 l/s in footwall fractured zone to about 20 l/s immediately behind the fault core. The tunnel face collapse in the fault core accompanied with sudden groundwater release caused several construction difficulties and 9 months delay in operation.

Dike core fault (DCF) as a conduit-barrier system interrupts the tunnel at Ch. 1+180 to 1+240 km. A dike emplaced in the fault acts as a water barrier in fault zone. This normal fault with dip of 70° and 283 m groundwater level above released 72 l/s water inflow into the tunnel. The groundwater burst and collapse occurred after the excavation of the dike that delayed the excavation for over 7 months.

F2 strike-slip fault with dip of 60° intersects perpendicularly the tunnel at Ch. 2+211 km. The groundwater head of 262 m above the fault caused water inrush of 110 l/s and about 500 m3 mud and debris collapsed materials into the tunnel. F1 strike-slip fault with dip of 70° and intersection at Ch. 3+056 km presented a conduit flow system. The groundwater head of 77 m released rainy form water inflow with continuous recharging system.

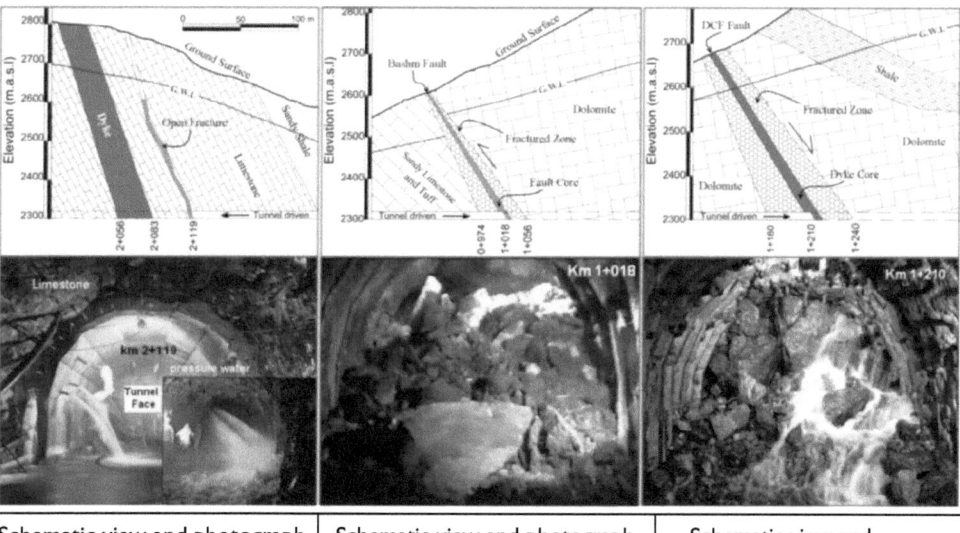

Schematic view and photograph of large deformations (swelling) in marlstone	Schematic view and photograph of sticky materials flow from karstic cave	Schematic view and photograph of water burst from karstic cave

transport of fines from the surrounding ground into the tunnel may result in higher groundwater inflows and more settlement or plastic zone around the tunnel.

Far zone around tunnel: the problems of this group are mainly related to the groundwater drawdown and consequent environmental impacts. Altering the groundwater regime, lowering of the water table, decrease in the discharge (or even a drying up) of wells, short-term impact and long-term depletion of the natural water resources in the surrounding area, and altering the groundwater chemistry are the direct environmental consequences of groundwater drawdown in the far zone around the underground excavations. The related large-scale consolidation and subsidence occurring as a result of the lowering of the water pressure in the rock mass are the other aspect of environmental consequences of groundwater drawdown. In addition, different types of physicochemical and biological processes such as mineralogical composition, alteration in rock mass mineralogy and geochemistry, groundwater degassing and bubble trapping, hydrogeochemical reactions, and anaerobic microbial activities may occur in the effective zone around underground excavations.

Support system: the groundwater originating from the surroundings of the underground excavation can have adverse effects on support system, particularly during their working life. The most important groundwater challenges related to the support system are deterioration of mortar internally, degradation/reduction in strength of concrete, erosion of mortar (masonry-lining), corrosion and chemical reaction of reinforcement, internal fittings, and cast iron lining, lack of tightness, clogging drainage due to fines transport, and swelling soil-lifting or damage to invert.

4.2 Water-rock interaction

Groundwater (water content or moisture) is one of the most important agents influencing the different rock mass behaviors. The effect of water on rock mass behaviors can be categorized into different classes including hydraulic, mechanical, physical, and chemical or combinations of interactions such as physicochemical processes. Porosity, texture, strength, petrology, mineralogy, chemical composition, structure, and fracturing degree are the main properties of rock media that control the water-rock interactions. Swelling, dissolution, hydrolysis, physical and chemical weathering, corrosion, oxidation, softening, deterioration, slaking and weakening are the results of water-rock interaction that cause severe challenging conditions related to the underground excavations.

The influence domain of water-rock interactions varies wildly from the negligible effects to quiddity gradation depending on the rock mass properties, hydrogeological conditions and time. However, quantification or mathematical description of water-rock interactions will be practically inexecutable due to the complexity of interactions and variety of effective features. In such situations, a classification system can be utilized to gain an improved understanding and identify the different types of interactions. Such classifications present an overview of phenomena that can be considered as a common communication language between engineers. Based on the physical, chemical and mechanical effects of water on intact and fractured rock, a classification system is proposed here. In this classification system, the effect of water on rocks can be classified into four groups (Figure 4) as (Sharifzadeh *et al.*, 2002):

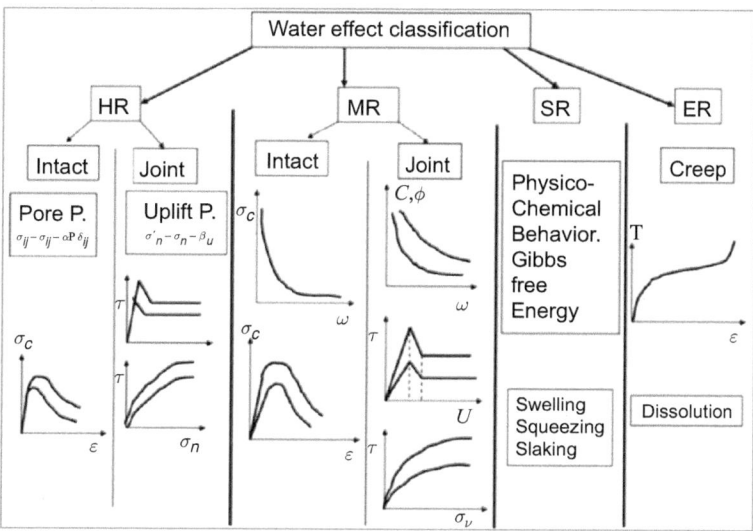

Figure 4   Classification system for effect of water on rocks.

1. *Hard and competent Rocks (HR):* The high compressive strength, high specific gravity, compact texture, and very low porosity are the main properties of this group. The water-rock interaction in this group of rocks is mainly related to the mechanical behavior. The chemical effect of water on these rocks is negligible; hence the strength parameters, cohesion and friction angle (c, Ø) or m and s in Hoek-Brown criterion, are not significantly changed due to the water absorption. The strength reduction of intact and fractured rocks of this group is mainly related to the reduction of effective stress due to the pore pressure, where only the stresses in the failure criteria are replaced by effective ones. It should be noted that the strength reduction is proportional to the saturation level, which is applied as pore pressure.

2. *Medium and structurally weak Rocks (MR):* This group of rocks is characterized by inter-structural weakness, which can be contacts between crystals or minerals, interfaces between formation times, and cracks resulting from weathering or unloading. The medium compressive strength, coarse grain texture, medium to high porosity, and weathered cracks or interfaces are the main properties of this group. Water absorption can cause decrease of cohesion and deterioration of inter-structural weakness. Consequently, deformability module and strength parameters of both intact rock and joints (*i.e.* JCS) decrease due to the water absorption that should be applied in the failure criteria. The reduction strength of both intact rock and joints can be idealized with negative exponential function of moisture content, where the reduction rate of strength is higher at lower values of water content and approaches constant (or zero) at higher water content values. The main difference between HR and MR groups appears in the varied water content. The main strength reduction in HR is observed in saturation state, while this issue happens at low water content for MR.

3. *Sensitive Rocks (SR)*: This group of rocks contains water sensitive grains such as clay minerals, montmorillonite, shale, pyrite, talc, and so on. Water absorption in these rocks causes physicochemical processes such as polarization of mineral planes, and electrostatic and magnetostatic fields. These processes are usually accompanied by phenomena such as swelling, squeezing, and slaking. For this group of rocks, the stress-strain relations are not sufficient to describe the behavior; nevertheless the physicochemical equations or thermodynamic potential such as Gibbs free energy may lead to better prediction of behavior.

4. *Evaporite Rocks (ER)*: water absorption by evaporite rocks (such as potash, gypsum, salt, and anhydrite) causes weakness in interatomic and molecules bonds (chemical bonds). These types of rocks are characterized by viscoelastic and time-dependent behavior that cause creep. At low moisture, when chemical bonds have not yet completely failed, the creep is usually expected, where the creep intensity increases by moisture. Increasing the moisture causes failing in all chemical bonds and the rock will dissolve in water.

5 GROUNDWATER INFLOW INTO UNDERGROUND EXCAVATIONS

5.1 The aim of groundwater inflow prediction

Prediction of groundwater inflow to underground excavations is one of the essential tasks and critical aspects of the design faced by engineers and hydrogeologists. The main purposes of prediction of groundwater inflow into underground excavations are:

1. Determining intensity of water-related problems linked to underground excavations and amount of groundwater inflow for determining the extent of controlling measures (*i.e.* grouting) and appropriate pumping equipment
2. Decreasing the time, cost, risk, and hazardous environmental consequences of construction activities
3. Increasing the safety, efficiency, performance, and utility of a project
4. Optimization of design for tunnel route and depth, construction techniques, drainage systems, support system, and necessary measures for long-term objectives

For these reasons, the possibility and the probable water inflow into excavations must be somehow predicted in advance. Successful predictions of groundwater inflow to underground excavations require a thorough understanding of the physical processes that govern flow in the surrounding rock mass and the role of effective features. However, the prediction of groundwater inflow into underground excavations, particularly deep ones, can result in a very difficult and time-consuming operation with uncertain outputs due to the complicated nature of effective features controlling the groundwater inflow mechanisms. Therefore, before one can predict the intensity of groundwater inflow, it is first necessary to identify and comprehensively explain the effective features.

5.2 Effective features in groundwater inflow

The development of realistic and robust predictive tools of groundwater inflow into underground excavations requires a thorough understanding of the inflow mechanism and physical processes that govern flow in host rock mass. Groundwater inflow into underground excavations is a complicated event resulting from a multitude of factors and physical processes. The governing physical processes and effective features are very complicated due to the different scales from mega to micro-scales with complex interdependency. Some of the most important effective features that control the mechanism of inflow into underground excavations with their interdependency are shown in Figure 5.

As shown in Figure 5, the regional hydrology incorporation with geological conditions mainly controls the local hydrogeological conditions of host ground of underground excavations. Similarly, the incorporation of local hydrogeological and geological features specifies the groundwater and rock mass conditions. There is a very close interrelationship between the groundwater, rock mass condition, and geological features where some geological structures such as faults may change the groundwater state and anisotropy of rock mass permeability, or development of karst as a geological feature is a function of changes in groundwater table and water solution phenomenon. The attributes of underground excavation such as shape, size, depth and construction method can disturb the rock mass and groundwater conditions due to the redistribution of stress and consequent hydromechanical processes.

Regardless of the complex interaction between effective features, the groundwater and rock mass conditions can be considered as the main controlling features of groundwater inflow into underground excavations from a modeling point of view. In fact, the groundwater and rock mass conditions provide the two main factors of reservoir (aquifer) of groundwater and flow paths, respectively, that mainly control the groundwater inflow

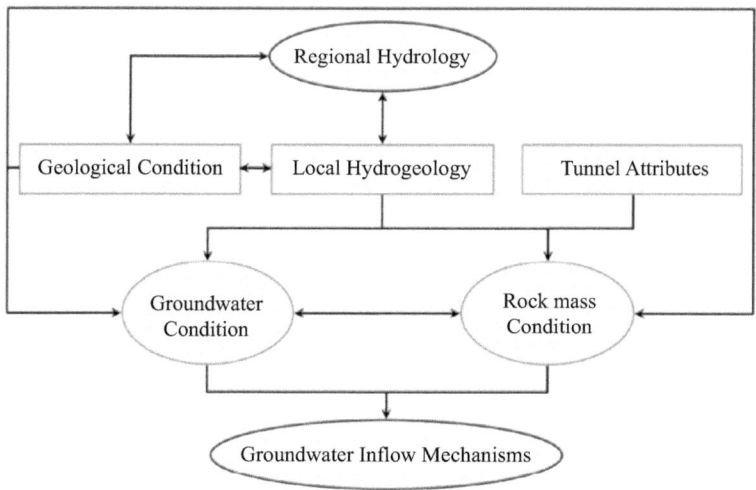

Figure 5 Effective features associated with mechanism of groundwater inflow into underground excavations.

mechanisms. The interconnected vacuity spaces through geological media such as pores, holes, dolines, karst cavities, and fractures provide the essential storage capacity of groundwater. The main flow paths are the interconnected void pores (primary porosity), and fractures (secondary porosity) that provide the essential permeability of rock mass. Therefore, a detailed local hydrogeological setting (hydraulic behavior of rock mass and groundwater conditions) of the host area will be necessary to evaluate the groundwater inflow mechanisms (Sharifzadeh *et al.*, 2012).

5.3 Effects of geological structures

5.3.1 Conceptual model

The rock mass hydraulic behavior and main flow paths through it are mainly dependent on the tectonic regime and geological structures. Most of the geological structures are complicated with multiple-scale effects due to the origin and tectonic regime. Although it is practically difficult to acquire appropriate geological data individually for each geological feature, the effect of geological structures on physical fluid flow processes shall be taken into account for more realistic and robust modeling and predicting the groundwater inflow into underground excavations. A conceptual model to analyze and explain the role of geological structure on groundwater inflow mechanisms and consequence problems is shown in Figure 6. The most important geological structures with respect to groundwater inflow into underground excavations containing porosity, bedding, fracture, dike, fold, fault, and karst were considered in this conceptual model.

The anisotropy, uniformity, and heterogeneity of geological structures and consequently the uncertainty of physical processes increase by rotating in anti-clockwise direction from porosity to karst structures (Figure 6). Therefore, the locality occurrence of groundwater inflow into underground excavations and construction risks increase by

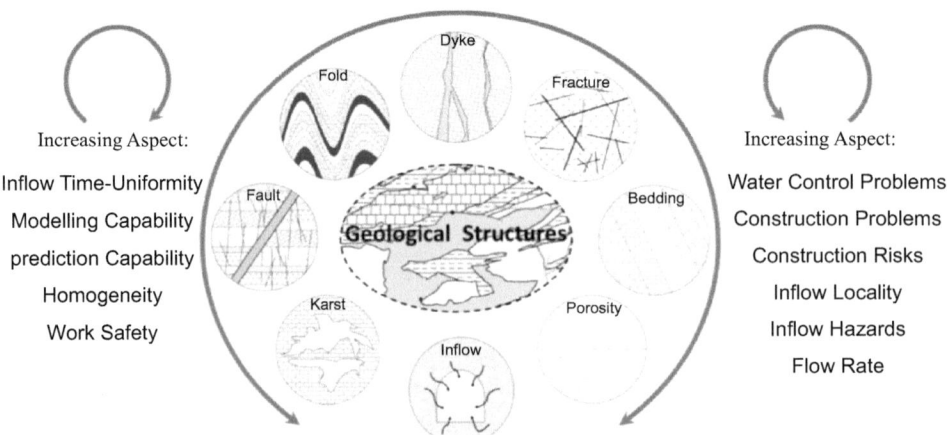

Figure 6   A conceptual model to analyze the groundwater inflow into underground excavations through the geological structures (Sharifzadeh et al., 2012).

Table 2 Classification of groundwater inflow rate (Lit/min per meter length of 6 m diameter tunnel) with consequence, dominate geological structures and flow mechanisms.

Class	Inflow Rate	Description	Inflow Mechanism	Dominate Geological Structure	Construction Consequence
I	<12.5	Very Low	Dripping	Porosity	Insignificant
II	12.5–35	Low	leakage	Porosity, Bedding	Decreasing construction rate
III	35–150	Medium	Seepage	Fracture, Dike	1–7 days work delay
IV	150–350	High	High Flow	Fold, Fault	1–30 days work delay, simultaneous construction and ground improvement
V	350–1000	Very High	Inrush	Fault, Karst	More than one month work delay, ground improvement necessary before construction starts
VI	>1000	Extremely High	Water Burst	Fault, Karst	Stops the construction, complex ground improvements necessary before construction starts

rotating in an anti-clockwise direction. In addition, the water storage capacity of geological structures increases by rotating in anti-clockwise direction from porosity to karst. Therefore, by rotating in the anti-clockwise direction, the intensity of groundwater inflow into underground excavations and related challenges (construction risks and problems, inflow hazards, safety challenges) will be increased in the case of existence of flow paths between the free groundwater storage and underground excavation. In fact, by rotating in anti-clockwise direction from porosity to karst the mechanism of groundwater inflow into underground excavations changed from dripping to water burst, where more inflow rate and locality imply more hazard potential, risks, and water control problems. Moreover, the computational modeling capability decreases due to the increase in the heterogeneity and complexity of geological structures by rotating in anti-clockwise direction from porosity to karst. Table 2 gives a classification of water inflow rate into a tunnel with respect to dominant geological structures and flow mechanism.

5.3.2 Modeling strategies and characterization of geological structures

The geological structures present different and individual hydraulic behaviors in terms of groundwater reservoir and flow paths that are mainly controlled by the origin and tectonic regime. The variety of hydraulic behaviors calls for different modeling approaches and strategies. Therefore, reliable characterization and hydrogeological models are necessary for evaluating the effect of geological structures on groundwater inflow into underground excavations. A framework of geological structures respect to prediction of water inflow into tunnels throughout practical models is shown in Figure 7. The practical models are categorized into four groups including analytical solution, continuum numerical modeling, discontinuous numerical modeling, and empirical models based on engineering geology.

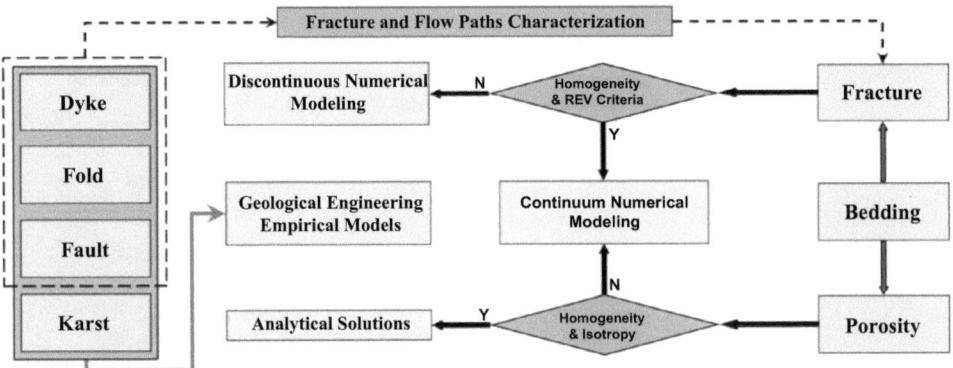

Figure 7 A framework of modeling method selection for water inflow prediction into underground excavations through the geological structures (Sharifzadeh et al., 2012).

The selection of each practical model (Figure 7) strongly depends on the purpose of groundwater inflow predictions and characteristics of geological structures. For isotropic and homogeneous media (such as soils and unfractured rocks), the groundwater inflow into circular tunnels can be predicted through analytical solutions (El Tani, 2003; Park et al., 2008). The homogeneity assumption linked with continuum numerical modeling approaches based on the equivalent porous medium assumption can be held for some cases such as uniform bedding with sparse fracturing. The validity of continuum approaches for fractured rocks depends on several parameters, the most important being the existence of a REV. If these criteria do not hold for the tunnel-surrounding rock mass, discontinuous numerical modeling is performed for more accurate predictions. In such situations, the distinct fracture network (DFN) concept is an alternative to the discontinuous representations of fractured rock and may appear much more adapted for fluid flow simulation. In addition, the effect of pseudo-scale geological structures such as fold, fault zone, and dike on groundwater inflow can be implicitly studied via numerical models and reliable hydrogeological characterization of fractures.

Each geological structures has individual hydraulic behavior dependent on the geological characterization related to the groundwater reservoir and flow paths. In most of the geological structures, the matrix permeability is negligible in comparison to permeability of fractures and rock mass hydraulic behavior is controlled by fractures. Natural fractures in rock mass have a complicated geometry with different orientation, size, and spatial distribution at all scales. These heterogeneous situations are encountered with a high level of uncertainty with respect to the rock mass permeability due to the highly inhomogeneous spatial distribution of hydraulic conductivity. Most of the geometrical characteristics of fractures are mainly dependent on the tectonic regime and main geological structures such as fault and fold. Therefore, the fracture is the most important feature in characterization of geological structures with respect to groundwater inflow into underground excavations.

Soils and unfractured rocks can be considered local homogeneous media. Porosity is primarily controlled by the shape, size and arrangement of the media grains and decreases with depth due to the compaction and confining stress. The porosity of unfractured rocks changes from less than 2% for igneous rocks, 1–5% for most of the sandstones, and more than 20% for soft limestones. For granular soils depending on the grain size, the porosity varies from 13% to 50%. In such media, the most important parameter of soils and unfractured rocks is the effective porosity (interconnected pores) that controls the permeability and contributes to groundwater inflow into underground excavations.

Bedding planes serve as source-sink boundaries (discontinuity) of adjacent porous sediments and have a major influence on the hydraulic properties of the rock mass, particularly the anisotropy of permeability. In most cases, the fracture characteristics control the permeability and hydraulic behavior of rock bedding formations. Based on the thickness of bedding layers, the general structure of rock mass can be categorized as massive structure (layer thickness more than 30cm) or bedded structure (layer thickness less than 30cm). Fracture spacing in layered sedimentary rocks is proportional to the thickness of the fractured layers (Bai & Pollard 2000). Opening-mode fractures in layered sedimentary rocks are often observed to propagate across, terminate at, or step-over at a layer contact or bedding contacts. In interbedded brittle and ductile rocks, fractures initiate within stiffer beds and terminate at the contact with more ductile beds. Strength of bedding contacts controls the type of fracture intersection where fracture termination is favored at very weak bedding contacts, whereas fractures propagate straight through strong contacts, and moderate-strength contacts may develop step-over fractures. Moreover, thicker beds may produce greater amounts of step-over than thinner beds (Cooke & Underwood, 2001). On the other hand, where layering plays an important role in restricting joint growth, a lognormal distribution reflects the true length population of joints whereas in more massive rocks, a power law distribution was more appropriate (Odling et al., 1999).

Macro and pseudo-scale geological structures such as fold, fault zone, dike, and karst have different hydraulic behaviors and are almost the main source of high local groundwater inflow into underground excavations. In such situations, proper evaluation of groundwater inflow into underground excavations requires a reliable hydrogeological model that briefly describes the characteristics of surrounding environment (Javadi et al., 2015). Setting up this model one encounters the lack of accurate geological input data due to the complexity and unknown conditions of rock mass (in most of the cases). Moreover, numerical modeling of these geological structures is practically tedious, time-consuming and expensive due to the complexity of input data and governing conditions. In such situations, an empirical model based on engineering geology has a key role in the inflow predictions. Such models linked with fracture characterization can be adapted for more efficient predictions and understanding the physical processes.

Dikes are long and thin imposition of mostly fine-grained igneous rock with steep or vertical and approximately parallel sides. Groundwater inflow through dikes shows duality in behavior that strongly depends on the tectonic history (weathering and fracturing characteristics of intrusive) after dike emplacement. The basic igneous dike rocks such as dolerite can weather montmorillonite clays that are noted for their swelling characteristics and for acting as water barrier. On the other hand, fractured dikes could act as conduit for groundwater inflow into underground excavations. By contrast, non-fractured dikes act as a barrier against groundwater flow.

Fold-related fractures can be classified in three classes including an axial extensional set successively parallel to the fold axis, a cross-axial extensional set oriented perpendicular to the fold axis and two sets of conjugate shear fractures oblique to the fold axis with their obtuse angle intersecting the trend of the fold axis. However, it is necessary to draw attention to the occurrence within folded strata of fracture sets having originated before folding and being unrelated to either fold geometry or kinematics. Occurrence of such pre-folding fracture sets within folded layers clearly changes the common view of fracture–fold relationships.

Faults with complex characteristics and duality behaviors in hydraulic processes have different impacts on groundwater inflow into underground excavations. They may act as conduits, barriers, or combined conduit-barrier systems that enhance or impede groundwater inflow. The hydraulic behavior of normal, strike-slip and reverse faults highly is highly related to fault zone architecture, origin mechanisms, weathering, and fracture characteristics. The primary components of fault zones are core, damage zone, and protolith, where the scalar relationship cannot be implied between these components. The stage of fault evolution specifies the hydraulic behavior of faults, where the fault core may act as a conduit during deformation or as a barrier when open pore space is filled by mineral precipitation or grain-size reduction following the deformation. Unlike the core, the damage zone permeability is dominated by the hydraulic properties of the fractures. The damage zone is characterized by complex fracture networks and minor faults, the intensity of which is higher than that of the adjacent host rock. The fracture network characterization related to groundwater inflow into underground excavation can be performed based on the fault types and their origins. In fact, fault type represents the stress regime during the last deformation that can determine the fracture characteristics in the fault zone as shown in Figure 8.

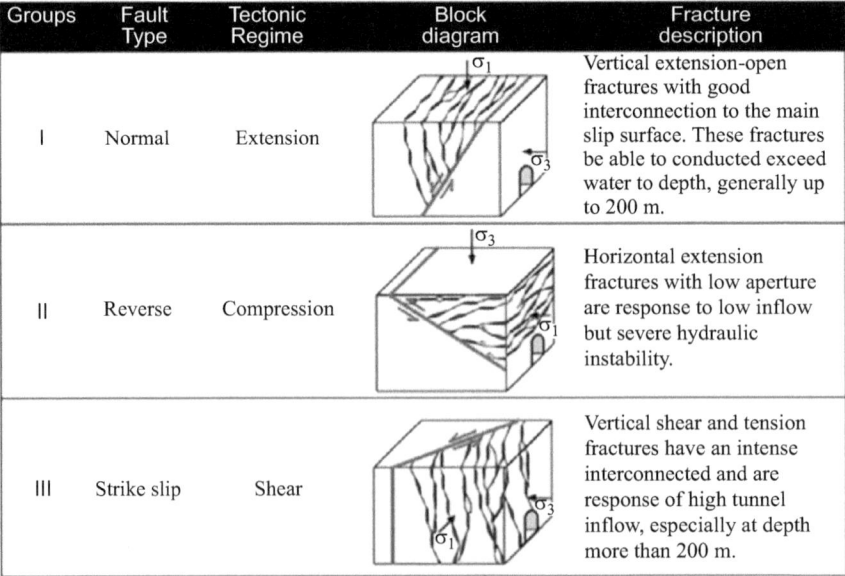

Groups	Fault Type	Tectonic Regime	Block diagram	Fracture description
I	Normal	Extension		Vertical extension-open fractures with good interconnection to the main slip surface. These fractures be able to conducted exceed water to depth, generally up to 200 m.
II	Reverse	Compression		Horizontal extension fractures with low aperture are response to low inflow but severe hydraulic instability.
III	Strike slip	Shear		Vertical shear and tension fractures have an intense interconnected and are response of high tunnel inflow, especially at depth more than 200 m.

Figure 8 Classification of fracture characteristics in fault zone (Zarei et al., 2013).

Strike-slip faults have sub-vertically conjugate fractures oriented obliquely to the strike of the fault; hence, significant percolation of water to depth is expected. Normal faults form in extensional tectonic regime with vertical extension fractures parallel to the strike of the fault; therefore remarkable transmissibility is expected. On the contrary, thrust faults are created in compressive tectonic regime and extension fractures oriented sub-horizontally, often not interconnected; therefore, minor transmissibility is expected. In general, vertical fractures show high transmissibility and inflow compared to horizontal fractures (Zarei *et al.*, 2013). On the other hand, occurrence of pre-faulting fracture sets within the rock layers obviously changes the general hydraulic behavior of faults.

5.4 Suitable analysis method based on the scales of problems

One of the most important issues with the selection of hydraulic modeling methods is the dependency of dominant physical processes on the scale of hydrogeological problems. In fact, the engineering physical-based approach to modeling of hydrogeological problems requires the deep understanding of scale dependency of dominant processes. The suitable analysis method for different hydrogeological problems and relevant modeling methods will depend on the size of the excavation relative to the geometrical attributes of fractures and the object of study. Figure 9 illustrates the transition from very large domain to a very local zone around an underground excavation (UE) with increasing detail of hydrogeological attributes of hypothetical geological medium. Based on this scale-dependency, the different hydrogeological problems related to UEs can be classified into four categories:

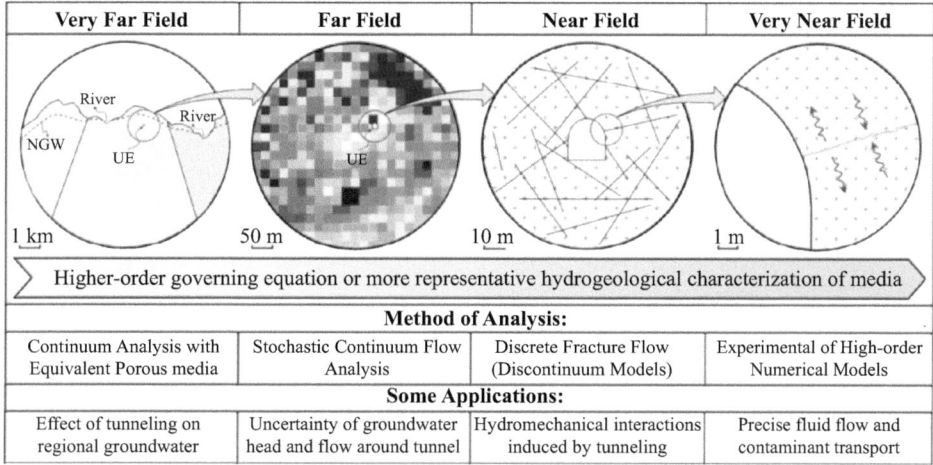

Very Far Field	Far Field	Near Field	Very Near Field
1 km	50 m	10 m	1 m

Higher-order governing equation or more representative hydrogeological characterization of media

Method of Analysis:			
Continuum Analysis with Equivalent Porous media	Stochastic Continuum Flow Analysis	Discrete Fracture Flow (Discontinuum Models)	Experimental of High-order Numerical Models
Some Applications:			
Effect of tunneling on regional groundwater	Uncertainty of groundwater head and flow around tunnel	Hydromechanical interactions induced by tunneling	Precise fluid flow and contaminant transport

Figure 9 Schematic representation of scale transition of hydrogeological problems related to the underground excavations (UE) and suitable analysis method.

1. *Very far field*: The hydrogeological problems involve very large distances around the underground excavation. For such problems, the size of dominant hydrogeological processes is sufficiently large to consider the entire domain as individual continuum units with deterministic equivalent porous media properties. The application of continuum model for such problems assumes the existence of an REV for each hydrogeological unit in the domain. For these kinds of problems, the size of the flow domain is so large that it leads to some difficulties in hydrogeological characterization and computational constraints. An example of application of very far field problem is the effect of tunneling on changes in the regional groundwater state.

2. *Far field*: The dominant hydrogeological processes occur at relatively large distances around the underground excavation. The size of flow domain is sufficiently large for existence of REV and to utilize the continuum models. However, the macroscopic heterogeneity in comparison with the dominant physical processes calls for more stringent hydrogeological characterization than very far field models (individual deterministic continuum units). In such situations, there is a considerable amount of uncertainty in characterizing the properties of the subsurface and practical application of hydrogeological models. This issue can be resolved through implementation of stochastic continuum methods to describe and analyze groundwater processes. Evaluation of the uncertainty of water inflow, groundwater drawdown, hydraulic confinement, and groundwater head redistribution around the storage cavern can be considered as examples of far field problems.

3. *Near field*: The hydrogeological processes mainly occur in a relatively small domain around the underground excavation, where the flow domain contains distinctive fractures. The geometrical definition of the fractures may be deterministic or statistical (in most cases) in terms of location, shape, size, and orientation. For these problems, the discontinuum representation of fractured rock appears much more adapted for fluid flow simulation due to the size of dominant physical processes in comparison to complexity of fractures geometry. In such a situation, the discrete fracture network (DFN) models are introduced as an irreplaceable tool for modeling fluid flow and transport phenomena around underground excavation because the dominance of the fracture geometry at small and intermediate scales can be approximated explicitly and in detail. Due to the statistical definition of fractures geometry (in most cases), a large number of DFN realizations shall be constructed and analyzed to describe the statistical behavior of media.

4. *Very near field*: The physical processes of interest are focused on fluid flow and contaminant transport through a well-defined single fracture and possibly in the bounded porous blocks. The geometry and location of the fracture intersecting the underground excavation shall be defined as well as high-order governing equations. The hydrogeological processes in the very near field domain involve the various classes of problems that occur locally at very small distance from the underground excavation. For example, the local problems corresponding to flow and contaminant transport in the very close vicinity of a contaminant source (such as underground noxious waste repositories) are represented as the very near field.

6 CONCLUSION

This chapter deals with the combination of issues related to groundwater, hydrogeology, geological structures and underground excavations. Different kinds of issues such as challenges of groundwater linked to underground excavations, rock-water interactions, and the role of geological structures on groundwater inflow are described to improve the reliability of analysis and fluid flow modeling in rock mass. Considering the present state-of-the-art of reciprocal interactions between groundwater, rock mass, and underground excavations, it is felt that more reliable predictive models, investigations, rock mass and geological structure characterization, and hydrogeological models are still needed for successful predictions. Through them, implementation of stochastic hydrogeology that deals with stochastic methods to analyze groundwater processes around underground excavations in both continuum and discontinuum numerical methods in combination with reliable characterization of geological structures will be necessary in future studies.

REFERENCES

Adler, P. & Thovert, J. (1999) *Fractures and fracture network*. Dordrecht, Kluwer academic Publisher, p. 428.

Bai, T., & Pollard, D.D. (2000) *Fracture spacing in layered rocks: a new explanation based on the stress transition. Journal of Structural Geology*, 22, 43–57.

Bear, J., Tsang, C.F. & De Marsily, G. (1993) *Flow and contaminant transport in fractured rock*. San Diego, Academic Press, p. 432.

Bense, V.F., Van Balen, R.T. & De Vries, J.J. (2003) The impact of faults on the hydrogeological conditions in the Roer Valley Rift System: an overview. *Netherlands Journal of Geosciences / Geologieen Mijnbouw*, 82 (1), 41–54.

Brantley, S.L., Kubicki, J.D. & White, A.F. (2008) *Kinetics of water-rock interaction*. New York, Springer.

Brush, D. & Thomson, N.R. (2003) Fluid flow in synthetic rough-walled fractures: Navier–Stokes, Stokes, and local cubic law simulations, *Water Resources Research*, 39 (4), 1085–1099. DOI: 10.1029/2002WR001346.

Cooke, M.L. & Underwood, C.A. (2001) Fracture termination and step-over at bedding interfaces due to frictional slip and interface opening. *Journal of Structural Geology*, 23, 223–238.

Dagan, G. (2002) An overview of stochastic modeling of groundwater flow and transport: from theory to applications. *Eos, Transactions American Geophysical Union*, 83 (53), 621–625. DOI: 10.1029/2002EO000421.

Dagan, G. (2004) On application of stochastic modeling of groundwater flow and transport. *Stochastic Environmental Research and Risk Assessment*, 18, 266–267. DOI: 10.1007/ s00477-004-0191-7

Elhag, A.B. & Elzien, S.M. (2013) Structures controls on groundwater occurrence and flow in crystalline bedrocks: A case study of the El Obeid area, Western Sudan. *Global Advanced Research Journal of Environmental Science and Toxicology*, 2 (2), 037–046.

El Tani, M. (2003) Circular tunnel in a semi-infinite aquifer. *Tunnelling and Underground Space Technology*, 18, 49–55.

Foladgar, A. (2003) Introduction of Kuhrang tunnel project and excavation methods. In: *Proceedings of the 6th Iranian Tunnelling Conference*, 11–13 November 2003, *Tehran, Iran*. Iranian Tunnelling Association (IRTA) Publication.

Freeze, R.A. (1975) A stochastic-conceptual analysis of one dimensional groundwater flow in non uniform homogeneous media. *Water Resources Research*, 11 (5), 725–741.

Froise, S. (1987) Hydrocarbon storage in unlined rock caverns: Norway's use and experience. *Tunnelling and Underground Space Technology*, 2 (3), 265–268. DOI: 10.1016/0886-7798(87)90033-2

Gattinoni, P., Scesi, L., Adbin, E.C. & Cremonesi, D. (2013) Hydrogeological risk and mining tunnels: The Fontane-Rodoretto Mine Turin (Italy). *International Journal of Environmental, Ecological, Geological and Geophysical Engineering*, 7 (1), 8–12.

Grandori, R., De Biase, A. & Nicoletti, I. (2011) General overview and adverse geological conditions related to the presence of high water inflows and harmful gases (H_2S). In: Sharifzadeh, M. (ed): Underground Spaces for Sustainable Development: *Proceedings of the First Asian and 9th Iranian Tunneling Symposium, 31 October–3 November 2011, Tehran, Iran.* Tehran, Iranian Tunnelling Association (IRTA) Publication. Paper Code: ATS11-02122.

Hitchmough, A.M., Riley, M.S., Herbert, A.W. & Tellam, J.H. (2007) Estimating the hydraulic properties of the fracture network in a sandstone aquifer. *Journal of Contaminant Hydrology*, 93, 38–57.

Hsieh, P.A. & Neuman, S.P. (1985) Field determination of the three-dimensional hydraulic conductivity tensor of anisotropic media, 2. Methodology and application to fractured rocks. *Water Resources Research*, 21 (11), 1667–1676.

Hu, B.X., Wu, J. & He, C. (2004) On stochastic modeling of groundwater flow and solute transport in multi-scale heterogeneous formations. *Computational and Applied Mathematics*, 23 (2–3), 121–151.

ITA. (1991) Report on the damaging effects of water on tunnels during their working life. *Tunnelling and Underground Space Technology*, 6 (11), 11–76.

Javadi, M. & Sharifzadeh, M. (2011) Assessment of inflow possibility into underground excavation using DFN and percolation concepts. In: Eskikaya, Ş. (ed): *Proceedings of 22th World Mining Congress & Expo, 11–16 September 2011, Istanbul, Turkey.* Ankara, AydoğduOfset. Vol. IV, pp. 3–10.

Javadi, M. & Sharifzadeh, M. (2013) Microscopic evaluation of non-linear fluid flow through rough-walled fractures using 'T Model'. In: Pyrak-Nolte, L.J., Morris, J., Chan, A., Rostami, J. & Dershowitz, W. (eds): *Proceedings of the 47th US Rock Mechanics / Geomechanics Symposium, ARMA 2013, 23–26 June 2013, San Francisco, California, USA.* New York, Curran Associates Publisher. Paper Code: ARMA 13–392.

Javadi, M., Sharifzadeh, M. & Shahriar, K. (2010) A new geometrical model for non-linear fluid flow through rough fractures. *Journal of Hydrology*, 389, 18–30.

Javadi, M., Sharifzadeh, M., Shahriar, K. & Mehrjooii, M. (2009) Non-linear fluid flow through rough-walled fractures. In: Vrkljan, I. (ed.): *Rock Engineering in Difficult Ground Conditions – Soft Rocks and Karst: Proceedings of the International Symposium of the International Society for Rock Mechanics, EUROCK 2009, 29–31 October 2009, Cavtat, Croatia.* London, © 2010 Taylor & Francis Group. pp. 261–266.

Javadi, M., Sharifzadeh, M., Shahriar, K. & Sayadi, S. (2014a) Numerical modeling of hydraulic confinement around crude oil storage cavern in Fractured Rocks: Direct Application of DFN Concept. In: Negro, A., Cecílio, M.O. & Bilfinger, W. (eds): *Tunnels for a better Life: Proceedings of the World Tunnel Congress, WTC 2014, 9–15 May 2014, Iguassu Falls, Brazil.* São Paulo, CBT/ABMS. p. 253.

Javadi, M., Sharifzadeh, M., Shahriar, K. & Mitani, Y. (2014b) Critical reynolds number for non-linear flow through rough-walled fractures: The role of shear processes. *Water Resources Research*, 50, 1789–1804. DOI: 10.1002/2013WR014610.

Javadi, M., Sharifzadeh, M., Shahriar, K. & Sayadi, S. (2015) Effective features on groundwater Inflow to underground excavation with emphasis on discontinuum numerical method. In:

Proceedings of the 33th National Geosciences Symposium, 22–24 February 2015, *Tehran, Iran*. pp. 535–536.

Javadi, M., Sharifzadeh, M. & Shahriar, K. (2016a) Uncertainty analysis of groundwater inflow into underground excavations by stochastic discontinuum method: Case study of Siah Bisheh pumped storage project, Iran. *Tunnelling and Underground Space Technology*, 51, 424–438.

Javadi, M., Sharifzadeh, M., Shahriar, K. & Sayadi, S. (2016b) Migration tracing and kinematic state concept embedded in discrete fracture network for modeling hydrocarbon migration around unlined rock caverns. *Computers & Geosciences*, 91, 105–118.

Kitandis, P.K. & Dykaar, B.B. (1997) Stokes slow in a slowly varying two-dimensional periodic pore. *Transport in Porous Media*, 26, 89–98.

Lee, C.-I. & Song, J.-J. (2003) Rock engineering in underground energy storage in Korea. *Tunnelling and Underground Space Technology*, 18 (5), 467–483. DOI: 10.1016/S0886-7798(03)00046-4

Lin, H.-I. & Lee, C.-H. (2009) An approach to assessing the hydraulic conductivity disturbance in fractured rocks around the Syueshan tunnel, Taiwan. *Tunnelling and Underground Space Technology*, 24, 222–230.

Masuda, K. (2001) Effects of water on rock strength in a brittle regime. *Journal of Structural Geology*, 23, 1653–1657.

Milanovic, P. (1985) Hydrogeological and engineering geological problems of hydrotechnical construction in karst. In: Günay, G. & Johnson, A.I. (eds): *Proceedings of the International Symposium on Karst Water Resources*, 7–19 July 1985, *Ankara, Turkey*. IAHS Publication No. 161.pp. 151–177.

Nativ, R., Adar, E.M. & Becker, A. (1999) Designing a monitoring network for contaminated ground water in fractured chalk. *Ground Water*, 37 (1), 38–47.

Neuman, S.P. (2004) Stochastic groundwater models in practice. *Stochastic Environmental Research and Risk Assessment*, 18, 268–270.

Neuman, S.P. & Depner, J.S. (1988) Use of variable-scale pressure test data to estimate the log hydraulic conductivity and dispersivity of fractured granites near Oracle, Arizona. *Journal of Hydrology*, 102, 475–501.

Niemi, A., Kontio, K., Kuusela-Lahtinen, A. & Poteri, A. (2000) Hydraulic characterization and upscaling of fracture networks based on multiple-scale well test data. *Water Resources Research*, 36 (12), 3481–3479.

Novakowski, K.S. & Bickerton, G.S. (1997) Borehole measurement of the hydraulic properties of low-permeability rock. *Water Resources Research*, 33 (11), 2509–2517.

Odling, N.E., Gillespie, P., Bourgine, B., Castaing, C., Chiles, J.P., Christensen, N.P., Fillion, E., Genter, A., Olsen, C., Thrane, L., Trice, R., Aarseth, E., Walsh, J.J. & Watterson, J. (1999) Variations in fracture system geometry and their implications for fluid flow in fractured hydrocarbon reservoirs. *Petroleum Geoscience*, 5, 373–384.

Park, K.-H., Owatsiriwong, A. & Lee, J.-G. (2008) Analytical solution for steady-state groundwater inflow into a drained circular tunnel in a semi-infinite aquifer: A revisit. *Tunnelling and Underground Space Technology*, 23, 206–209.

Renard, P. (2007) Stochastic Hydrogeology: What Professionals Really Need? *Ground Water*, 45 (5), 531–541.

Sagar, B. & Runchal, A. (1982) Permeability of Fractured rock: Effect of fracture size and data uncertainties. *Water Resources Research*, 18 (2), 266–274.

Saul Caine, J., Evans, J.P. & Forster, C.B. (1996) Fault zone architecture and permeability structure. *Geology*, 24 (11), 1025–1028.

Schwartz, F.W. & Smith, L. (1988) A continuum approach for modelling mass transport in fractured media. *Water Resources Research*, 24 (8), 1360–1372.

Sharifzadeh, M., Fahimifar, A. & Esaki, T. (2002) Classification and modeling of water effect on mechanical behavior of intact rock and rock joint. In: *Proceedings of the Fourth International Summer Symposium*, 3–4 August 2002, *Kyoto, Japan*. pp. 263–266.

Sharifzadeh, M. & Javadi, M. (2011) Near-Field Application of Aperture Back Calibrated Distinct Fracture Network. Published at 12th International Congress on Rock Mechanics. In: Qian, Q. & Zhou, Y. (eds): *Harmonising Rock Engineering and the Environment: Proceedings of the 12th International Congress on Rock Mechanics, 18–21 October 2011, Beijing, China*. London, © 2010 Taylor & Francis Group. pp. 1361–1365.

Sharifzadeh, M., Javadi, M., Shahriar, K. & Mehrjooii, M. (2009) Effect of surface roughness on velocity fields through rock fractures. In: Vrkljan, I. (ed): *Rock Engineering in Difficult Ground Conditions – Soft Rocks and Karst: Proceedings of the International Symposium of the International Society for Rock Mechanics, EUROCK 2009, 29–31 October 2009, Cavtat, Croatia*. London, © 2010 Taylor & Francis Group. pp. 351–356.

Sharifzadeh, M., Javadi, M. & Zarei, H.R. (2012) The Role of Geological Structures to Tunnel Inflow, Modelling Strategies and Predictions. In: Phienwej, N. & Boonyatee, T. (eds): *Tunnelling and Underground Space for a Global Society: Proceedings of the ITA-AITES World Tunnel Congress, WTC 2012, 21–23 May 2012, Bangkok, Thailand*. Engineering Institute of Thailand (EIT) (THA). pp 483–484.

Sharifzadeh, M., Mitani, Y. & Esaki, T. (2006) Rock Joint surfaces measurement and analysis of aperture distribution under different normal and shear loading using GIS. *Rock Mechanics and Rock Engineering*, 41 (2), 299–323.

Tseng, D.-J., Tsai, B.-R. & Chang, L.-C. (2001) A case study on ground treatment for a rock tunnel with high groundwater ingression in Taiwan. *Tunnelling and Underground Space Technology*, 16, 175–183.

Urmiei, A., Mozaffari, A., Sharifzadeh, M. & Zarei, H.R. (2011) Engineering geological challenges for rock tunnels in Iran. In: Sharifzadeh, M. (ed): Underground Spaces for Sustainable Development: *Proceedings of the First Asian and 9th Iranian Tunneling Symposium, 31 October–3 November 2011, Tehran, Iran*. Tehran, Iranian Tunnelling Association (IRTA) Publication. Paper Code: ATS11-033.

Vásárhelyi, B. & Ván, P. (2006) Influence of water content on the strength of rock. *Engineering Geology*, 84, 70–74.

Wang, J.S.Y., Trautz, R.C., Cook, P.J., Fineterle, S., James, A.L. & Birkholzer, J. (1999) Field tests and model analyses of seepage into drift. *Journal of Contaminant Hydrology*, 38, 323–348.

Wenner, D. & Wannenmacher, H., 2009. Alborz Service Tunnel in Iran: TBM Tunnelling in Difficult Ground Conditions and its Solutions. In: *Underground Spaces for Safety, Better Envirnoment & Energy: Proceedings of the 8th Iranian Tunnelling Conference, 18–20 May 2009, Tehran, Iran*. Iranian Tunnelling Association (IRTA) Publication. pp. 343–353.

Wilson, C.R. & Witherspoon, P.A. (1974) Steady state flow in rigid networks of fractures. *Water Resources Research*, 10 (2), 328–335.

Winter, C.L. (2004) Stochastic hydrology: Practical alternatives exist. *Stochastic Environmental Research and Risk Assessment*, 18, 271–273. DOI: 10.1007/s00477-004-0198-0

Witherspoon, P.A., Wang, J.S.Y., Iwai, K. & Gale, J.E. (1980) Validity of cubic law for fluid flow in a deformable rock fracture. *Water Resources Research*, 16 (6), 1016–1024.

Yang, F.-R., Lee, C.-H., Kung, W.-J. & Yeh, H.-F. (2009) The impact of tunneling construction on the hydrogeological environment of "Tseng-Wen Reservoir Transbasin Diversion Project" in Taiwan. *Engineering Geology*, 103, 39–58. Doi: 10.1016/j.enggeo.2008.07.012.

Yoshida, H., Maejima, T., Nakajima, S., Nakamura, Y. & Yoshida, S. (2013) Features of fractures forming flow paths in granitic rock at an LPG storage site in the orogenic field of Japan. *Engineering Geology*, 152 (1), 77–86. DOI: 10.1016/j.enggeo.2012.10.007

Yoo, C., Lee, Y., Kim, S.-H. & Kim, H.-T. (2012) Tunnelling-induced ground settlements in a groundwater drawdown environment – A case history. *Tunnelling and Underground Space Technology*, 29, 69–77. DOI: 10.1016/j.tust.2012.01.002

Zarei, H.R., Uromeihy, A. & Sharifzadeh, M. (2012) Identifying geological hazards related to tunneling in carbonate karstic rocks – Zagros, Iran. *Arabian Journal of Geosciences*, 5 (3), 457–464. DOI: 10.1007/s12517-010-0218-y

Zarei, H.R., Uromeihy, A. & Sharifzadeh, M. (2013) A new tunnel inflow classification (TIC) system through sedimentary rock masses. *Tunnelling and Underground Space Technology*, 34, 1–12.

Zarei, H.R., Sharifzadeh, M. & Uromeihy, A. (2011a). Gas ground risks and geological investigations for TBM tunneling in Iran. In: Sharifzadeh, M., (ed): Underground Spaces for Sustainable Development: *Proceedings of the First Asian and 9th Iranian Tunneling Symposium, 31 October–3 November 2011, Tehran, Iran.* Tehran, Iranian Tunnelling Association (IRTA) Publication. Paper Code: ATS11-02115

Zarei, H.R., Uromeihy, A. & Sharifzadeh, M. (2011b) Evaluation of high local groundwater inflow to a rock tunnel by characterization of geological features. *Tunnelling and Underground Space Technology*, 26, 364–373.

Chapter 11

Reliability assessment of ultimate and serviceability limit states of underground rock caverns

Anthony T.C. Goh[1] & Wengang Zhang[2]
[1] *School of Civil and Environmental Engineering, Nanyang Technological University, Singapore*
[2] *School of Civil Engineering, Chongqing University, P.R. China*

Abstract: In the design of underground rock caverns, failure of both the ultimate and serviceability limit states need to be considered. Ultimate limit state refers to collapse, in which the stresses exceed the strength of the rock masses. Failure of the serviceability limit state refers to excessive deformations resulting in difficulties during excavation such as lining placement and reinforcement installation. In addition, it is widely recognized that a deterministic analysis of either the factor of safety or the calculated displacement gives only a partial representation of the true margin of safety/reliability, since the uncertainties in the design parameters affect the probability of failure. This chapter uses numerical modeling to assess both the ultimate and the serviceability limit states of underground single and twin rock caverns. The global factor of safety is used as the criterion for the ultimate limit state and the calculated percent strain around the cavern opening is adopted as the serviceability limit state criterion. Based on the numerical results, simple logarithmic regression models were developed for estimating the global factor of safety and the induced percent strain of the single and twin caverns, respectively. Simplified procedures are proposed for evaluating the failure probabilities for deep underground rock caverns for preliminary design applications.

I INTRODUCTION

One of the major considerations in design of an underground rock cavern is the evaluation of its stability since the excavation of the rock causes a redistribution of the stresses in the proximity of the underground opening, which can lead to failure. Empirical and numerical approaches are commonly adopted tools in the design of underground rock caverns. Many empirical methods are essentially based on quantification of the rock mass through a classification system such as the tunneling quality index (Q) and the rock mass rating (RMR). As empirical methods are simpler to use, they are generally applied at the early stages of a site investigation and the design phase. A shortcoming of empirical methods is that they are mainly developed to assess structural resistance, which is only one of the design issues to be accounted for in the design. In addition, they are based on past case studies and assumed rock behavior. Furthermore, these empirical methods fail to take into account other important design parameters such as the in situ stress fields and the cavern shapes. Many researchers have used numerical methods to study the stability of rock caverns, and a number of approaches have been developed and applied (*e.g.*, Chryssanthakis *et al.*, 1995;

Zhu & Li, 2000; Sitharam & Latha, 2002; Fan *et al.*, 2004; Zhu & Zhao, 2004; Hao & Azzam, 2005; Cai *et al.*, 2007; Jia & Tang, 2008; Hoek, 2011; Goh & Zhang, 2012; Mohanty & Vandergrift, 2012; Tsesarsky *et al.*, 2013; Zhang *et al.*, 2014; Zhang & Goh, 2014, 2015a, 2015b).

For multiple caverns, the construction of a new cavern close to an existing cavern modifies the state of stresses and movements around the existing cavern in an area called the "influence zone". Usually the size of this influence zone depends on the ground type, in situ stress, cavern span, width of the pillar separating the caverns, and excavation sequence. The subject of interaction between parallel caverns/tunnels has been studied by several authors who have reported the results of field measurements or analytical studies of the problem *(e.g.,* Barla & Ottoviani, 1974; Ghaboussi & Ranken, 1977; Gercek, 2005; Zhao & Ma, 2009; Mortazavi *et al.*, 2009; Karademir, 2010; Esterhuizen *et al.*, 2011; Zhang & Goh, 2015a).

Conventional evaluation of stability of geotechnical structures and underground openings involves the use of a factor of safety *(FS)* which considers the relationship between the resistance and the load or the calculated displacements/strains. The former is usually used as the criterion for assessing the ultimate limit state while the latter is adopted as the serviceability limit state criterion. For underground caverns, however, neither the *FS* nor the induced strain is known explicitly. Instead, it may be estimated only through repeated point-by-point numerical analyses with different input values. Generally, the performance function, expressing the dependent responses as a function of input design variables, is constructed artificially using polynomial or logarithmic regression (LR) methods *(e.g.,* Basarir, 2008; Zhu *et al.*, 2008; Goh & Zhang, 2012; Zhang & Goh, 2012; Siahmansouri *et al.*, 2012; Zhang & Goh, 2015b). Alternatively, the Multivariate adaptive regression splines (MARS) algorithm and the Artificial Neural Network approach (ANN) are also used to develop surrogate response surface models (Goh & Zhang, 2012; Lü *et al.*, 2012; Mahdevari & Torabi, 2012; Rafiai & Moosavi, 2012; Adoko *et al.*, 2013; Zhang & Goh, 2013; Zhang & Goh, 2014). Though slightly inferior to the MARS and ANN methods in terms of predictive capacity, the regression models remain popular due to its simplicity and model interpretability.

Even after obtaining the limit state surfaces, this single *FS* or displacements/strain value cannot be used to quantify reliability since the stochastic nature of the design parameters is usually not considered. The alternative is to use probabilistic design approaches which take uncertainty and complexity into account to assess the probability of failure P_f. The calculation of P_f involves the determination of the joint probability distribution of the resistance R and the load (stress) S and the integration of the probability density function (PDF) over the failure domain. For a problem with multiple n random variables, the calculation of P_f involves the determination of a multi-dimensional joint PDF of the random variables and the integration of the PDF over the failure domain. A Monte Carlo simulation (MCS) is a procedure, which seeks to simulate stochastic processes by random selection of input values to an analytical model in proportion to their joint PDF. An alternative to assess P_f is known as the First-Order Reliability Method (FORM) (Hasofer & Lind, 1974), which is an approximate method. The FORM approach involves the transformation of the limit state surface into a space of standard normal uncorrelated variables, wherein the shortest distance from the transformed limit state surface to the origin of the reduced variables is the reliability index β (Cornell, 1969).

In this chapter, both the ultimate and the serviceability limit states of underground single and twin rock caverns are numerically investigated. The global *FS* obtained using the shear strength reduction technique is used as the criterion for the ultimate limit state and the calculated percent strain around the opening is adopted as the serviceability limit state criterion. Based on the numerical results, LR models were developed for estimating the global *FS* and the induced percent strain of the single and twin caverns, respectively. Charts were developed for preliminary design use, for checking the cavern stability. Subsequently, probabilistic assessment on ultimate and serviceability limit states of underground rock caverns was performed using MCS and FORM based on the developed regression models. It is found that these models can be easily implemented into MCS and FORM to calculate P_f for cavern reliability.

2 METHODOLOGIES

2.1 Rock mass classifications and correlations

When performing numerical analysis, particular attention must be given to the selection of appropriate input parameters, especially in the preliminary stage of an engineering design. Various indirect empirical relations have been proposed to calculate the rock mass properties such as the deformation modulus E_m, the shear strength parameters cohesion c and friction angle ϕ, the rock mass uniaxial compressive strength σ_{cm} and the tensile strength σ_t. For the numerical analyses that were carried out, the following equations (Equations 1–7) were adopted for determining the rock mass properties. The empirical equation from Tugrul (1998) was used to estimate *RMR* from *Q* instead of Barton (1995) and Bieniawski (1984) as it gives more conservative estimations of *RMR*.

$$RMR = 7\ln Q + 36 \qquad \text{(Tugrul, 1998)} \tag{1}$$

$$E_m(GPa) = 10^{(RMR-10)/40} \ (RMR \leq 50) \quad \text{(Serafim \& Pereira, 1983)} \tag{2}$$

$$E_m(GPa) = 2RMR - 100 \quad \text{(Bieniawski, 1978)} \tag{3}$$

$$c(MPa) = 0.005(RMR - 1.0) \quad \text{(Bieniawski, 1989)} \tag{4}$$

$$\phi(°) = 0.5RMR + 4.5 \quad \text{(Bieniawski, 1989)} \tag{5}$$

$$\sigma_{cm}(MPa) = RMR \quad \text{(Palmstrom, 2000)} \tag{6}$$

$$\sigma_t(MPa) = \sigma_{cm}/15 \tag{7}$$

Adopting the above empirical equations, the *Q* value of each category and its corresponding rock properties are shown in Table 1. In Table 1, the Poisson's ratio v values are assumed based on commonly used values. For simplicity, density of 2670 kg / m^3 is assumed for rock mass for all the ranges of *Q*. It should be noted that herein the use of *Q* and the empirical relationships may introduce errors. These errors include measuring errors in determining *Q* values and uncertainties in transforming *Q* to engineering properties through empirical equations. Hence these relationships are intended to provide the initial estimates of the rock mass properties for preliminary design and should be used with great caution in engineering design.

Table 1 Rock mass properties with different Q values.

Q	c (MPa)	$\phi(°)$	E_m (GPa)	μ	σ_t (MPa)
0.4	0.14	19.3	3.1	0.35	1.97
1	0.18	22.5	4.5	0.35	2.40
4	0.22	27.4	7.8	0.20	3.05
10	0.26	30.6	11.3	0.20	3.47
40	0.30	35.4	19.7	0.16	4.12
100	0.34	38.6	28.6	0.16	4.55

2.2 Shear strength reduction technique

In this study, the global stability *FS* are assessed using the shear strength reduction technique (SSR). This technique has been used by various authors including Matsui & San (1992), Dawson *et al.* (1999) and Dawson *et al.* (2000), and is now available in many commercial finite element (FEM) and finite difference (FDM) programs. This procedure essentially involves repeated analyses by progressively reducing the shear strength properties until collapse occurs. For a Mohr-Coulomb material, by reducing the shear strength by a factor *F* the shear strength equation becomes:

$$\frac{\tau_f}{F} = \frac{c}{F} + \sigma_n \frac{\tan\phi}{F} \tag{8}$$

$$F = \frac{\tau_f}{c^* + \sigma_n \tan\phi^*} \tag{9}$$

where τ_f is the shear strength, σ_n is the normal stress, and $c^* = c/F$ and $\phi^* = $ arctan$(\tan\phi/F)$ are the new Mohr-Coulomb shear strength parameters. Systematic increments of *F* are performed until the finite element or finite difference model does not converge to a solution (*i.e.* failure occurs). The critical strength reduction value which corresponds to non-convergence is taken to be the global factor of safety *FS*. It should be noted that this *FS* indicates the global stability. Consequently, global *FS* < 1 implies that total collapse has occurred. The technique has been applied to a number of underground excavation problems including rock caverns (Hammah *et al.*, 2007; Zhang & Goh, 2012; Goh & Zhang, 2012) and tunnels (Vermeer *et al.*, 2002; Xia *et al.*, 2014; Wang *et al.*, 2014).

2.3 Logarithmic regression

In this chapter, the logarithmic regression (LR) model is adopted to model the non-linear relationship between the dependent and independent variables since the LR model is more accurate in predictive capacity. The basic form of the logarithmic regression models is as follows:

$$y = a(X_1)^b(X_2)^c(X_3)^d(X_4)^e... \tag{10}$$

in which *y* is the dependent response, X_i are the input parameters (independent variables) and the coefficients *a*, *b*, *c*, *d*, *e*, etc are determined through least-squares estimation.

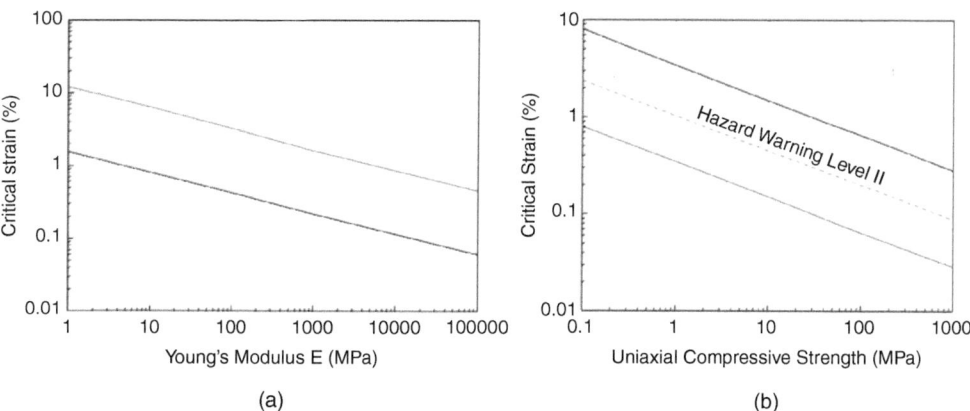

Figure 1 Sakurai's relationship between: (a) ε_c and E_m and (b) ε_c and σ_{cm} and Hazard warning level II (adapted from Sakurai (1986, 1997)).

2.4 Concept of critical strain

The threshold percent strain values are required to assess the serviceability limit state. An appropriate value can be assigned to the percent strain threshold through the concept of critical strain, defined as the strain value above which construction problems are likely to occur. Sakurai (1986, 1997) proposed a relationship between critical strain and Young's modulus E_m and uniaxial compressive strength σ_{cm}, and suggested Figs. 1*a* and 1*b* as representing the limiting bounds for these relationships. These relationships are based on various rock tests and field measurements. Sakurai (1997) also showed that critical strain results for in situ rock masses also fall within these bounding lines. Hazard warning levels for stability assessment of tunnels have been established by Sakurai on the basis of the relationship between critical strain ε_c and σ_{cm} in laboratory specimens and then subsequently confirmed through in situ displacement monitoring. Hazard warning level II (Fig. 1*b*) is coincident with the central line of the critical strain versus σ_{cm} limit zone, and represents the transition from a stable to unstable tunnel/cavern subjected to shear failure.

Table 2 shows the range of critical strain values obtained from the upper and lower bound curves in Figs. 1*a* and 1*b*, and from the hazard warning level II in Fig. 1*b*. The method proposed by Sakurai relates the critical strain values to the strength and stiffness parameters of the rock mass.

2.5 MCS and FORM

A Monte Carlo Simulation (MCS) is a procedure, which seeks to simulate stochastic processes by random selection of input values to an analytical model in proportion to their joint probability density function. The basis of MCS is the use of random numbers which were originally either manually or mechanically generated by using such techniques as spinning wheels, dice rolling or card shuffling. It is a powerful technique that is applicable to both linear and non-linear problems, but can require a large number of

Table 2 Critical strains derived from Sakurai's relationships.

Q	E_m (GPa)	σ_{cm} (MPa)	Critical strain ε_c (%)		
			From Fig. 1a	From Fig. 1b	Hazard warning level II (From Fig. 1b)
0.4	3.1	29.6	0.20–1.60	0.11–1.10	0.43
1	4.5	36.0	0.16–1.30	0.10–1.00	0.41
4	7.8	45.7	0.14–1.10	0.09–0.90	0.38
10	11.3	52.1	0.12–1.00	0.08–0.80	0.36
40	19.7	61.8	0.10–0.80	0.07–0.70	0.33
100	28.6	68.3	0.09–0.70	0.06–0.60	0.26

simulations to provide a reliable distribution of the response. In general, if the maximum error of P_f is e at confidence $1-\alpha$, then the required number of realizations n is

$$n = \hat{p}_f \hat{q}_f \left(\frac{z_{\alpha/2}}{e}\right)^2 \tag{11}$$

in which $\hat{p}_f$ is the estimated value of P_f, $\hat{q}_f = 1 - \hat{p}_f$, and $z_{\alpha/2}$ is the point on the standard normal distribution satisfying $P[\, Z > z_{\alpha/2}] = \alpha/2$. For example, if the estimated P_f is 0.1% and the desired maximum error on P_f is 0.0001 at confidence 90% ($\alpha = 0.10$), then the required n is

$$n = \hat{p}_f \hat{q}_f \left(\frac{z_{\alpha/2}}{e}\right)^2 = 0.001(0.999)\left(\frac{1.645}{0.0001}\right)^2 \approx 270000 \tag{12}$$

Hasofer and Lind (1974) proposed an invariant definition for the reliability index. The approach is referred to as the First-order reliability method (FORM). In classical FORM, the original correlated basic random vector **x** in X-space are transformed into independent standard normal vector **u** in U-space. The reliability index β is defined as:

$$\beta = \min_{u \in F} \sqrt{\mathbf{u}^T \mathbf{u}} \tag{13}$$

where F denotes the failure domain in which the performance function g(**u**) ≤ 0. The point denoted by **u*** which minimizes Equation 13 and satisfies **u** ∈ F is the design point (DP), and is also the point on the limit state surface (LSS) (g(**u**) = 0) that is closest to the origin of the U-space. An iterative algorithm may be used to search for **u***, and has been well-documented (Ang & Tang 1984; Ditlvsen & Madsen 1996; Melchers 1999; Baecher & Christian 2003; Choi *et al.*, 2006). A new algorithm for FORM was proposed by Low & Tang (2007), which interpreted the reliability index based on the perspective of an expanding ellipsoid in the original X-space of the basic random variables, and the reliability index β can be computed as:

$$\beta = \min_{x \in F} \sqrt{\left(\frac{x_i - \mu_i}{\sigma_i}\right)^T [\mathbf{R}]^{-1} \left(\frac{x_i - \mu_i}{\sigma_i}\right)} \tag{14}$$

where x_i is the set of n random variables, μ_i is the set of mean values, $\mathbf{R}$ is the correlation matrix and F is the failure region. The minimization in Equation 14 is performed over F corresponding to the region $g(X) = 0$. Low and Tang (2007) had shown that an EXCEL (Microsoft) spreadsheet environment can be used to perform the minimization and determine β. If the random variables have probability distributions close to normal, then P_f can be obtained from the expression:

$$P_f \approx 1 - \Phi\,(\beta) \tag{15}$$

in which $\Phi\,(\beta)$ is the value of the cumulative probability. This value can be obtained from tables of the standard cumulative normal distribution function found in many textbooks or from built-in functions in most spreadsheets.

3 NUMERICAL MODELS AND MODELING RESULTS

3.1 Numerical model of a single cavern case

The basic assumptions of the numerical analyses are:

- The study was a two-dimensional plane-strain problem;
- The rock material obeyed the Mohr-Coulomb failure criterion that follows the elastic perfectly-plastic stress-strain relationship, and has been commonly adopted for modeling rock mass behavior (Hatzor *et al.*, 2002; Sitharam & Latha 2002; Cai 2008, 2011);
- Only stress-induced failure was considered, the discontinuous nature of the rock mass is incorporated implicitly in the Mohr-Coulomb constitutive relationship used to represent the mass as an equivalent continuum;
- Caverns are unsupported, and creep is not considered;
- Full-face excavation.

The FDM FLAC^{3D} code (Itasca, 2005) was utilized for the numerical experiments. The cross-section of the cavern, the side view, boundary conditions and the details of cavern periphery are shown in Fig. 2. The cavern roof arc is semi-circular and the overburden height D from the ground surface to the top of the side wall is fixed at 100 m. The plane strain conditions are enforced by including a thin 1 m slice of material in the long-itudinal direction and imposing boundary conditions on the two off-plane surfaces that allow movement vertically but are restrained against displacements normal to these planes. Outer boundaries are located far from the cavern to minimize the boundary effects. No surface loading above ground surface is considered. The initial vertical in situ stress σ_v is induced by self-weight of the rock. The horizontal stress σ_h is calculated using $K_0 \times \sigma_v$.

The four variables for single cavern analysis and their respective ranges are listed in Table 3. The two dependent responses are the global factor of safety FS_{g_s} and the percent strain ε_s. The former is calculated by the shear strength reduction technique while the latter is derived through the displacements of key points including crown C, springline S, middle sidewall M, and invert I as illustrated in Fig. 2. Strain ε_{Xx} is calculated as follows (Sakurai, 1997):

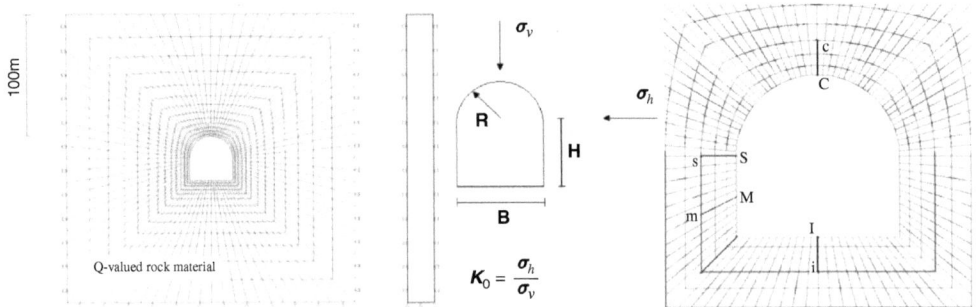

Figure 2 Cavern configuration, boundary fixity, side view and details of cavern periphery.

Table 3 Factors used in parametric studies for single cavern case.

Parameter	Description	Values
K_0	In situ stress ratio	0.5, 1, 2
B	Cavern span (m)	5, 10, 15, 20, 25, 30
B/H	Ratio of cavern width to side wall height	0.25, 0.5, 1, 2
Q	Tunneling quality index	0.4, 1, 4, 10, 40

$$\varepsilon_{Xx}(\%) = \frac{u_X - u_x}{l_{Xx}} \times 100 \qquad (16)$$

and the percent strain is:

$$\varepsilon_s = \max(\varepsilon_{Xx}) \qquad (17)$$

where X is C, S, M or I and x is c, s, m or i; u_x is the displacement of the corresponding inner node x and l_{Xx} is the length between nodes X and x (Fig. 2). It should be noted that the selected length of l_{Xx} is typical of many extensometers and the locations of inner nodes such as x are chosen at the grid of the third layer elements around the opening. Numerical trials carried out on different locations of inner nodes and different mesh densities indicate that the choice of the reference length and the density of the mesh at the opening have minimal influence on the magnitude of the maximum strain. For brevity, these results have been omitted.

3.2 Modeling results of a single cavern case

The computed global factor of safety values for single caverns (FS_{g_s}) are summarized in Table 4.

The percent strain values are calculated except for the cases with $FS_{g\_s} < 1$ since FS_{g_s} less than unity denotes collapse for the purposes of this study and consequently the nodal displacements are large. A total of 219 cases are analyzed. It should be noted that based on the numerical modeling results, K_0 has minimum influence on FS_{g_s} derived via SSR. In SSR, the global factor of safety FS is mainly influenced by the shear strength properties and the geometrical dimensions. However, K_0 has significant influence on the cavern deformation and consequently the cavern serviceability limit state.

Table 4 Results from numerical experiments for FSg_s.

B (m)	Q	FS_{g_s}			
		B/H = 0.25	B/H = 0.5	B/H = 1	B/H = 2
5	0.4	0.93	1.12	1.29	1.38
	1	1.13	1.35	1.56	1.64
	4	1.42	1.69	1.95	2.02
	10	1.65	1.97	2.27	2.35
	40	1.95	2.32	2.67	2.76
10	0.4	0.69	0.86	1.00	1.08
	1	0.84	1.04	1.21	1.32
	4	1.05	1.31	1.51	1.65
	10	1.23	1.52	1.77	1.93
	40	1.46	1.79	2.08	2.27
15	0.4	0.58	0.74	0.85	0.92
	1	0.71	0.90	1.03	1.13
	4	0.88	1.12	1.29	1.42
	10	1.03	1.31	1.51	1.64
	40	1.22	1.55	1.78	1.94
20	0.4	0.49	0.65	0.76	0.82
	1	0.60	0.79	0.92	1.01
	4	0.75	0.99	1.14	1.26
	10	0.87	1.15	1.34	1.47
	40	1.04	1.36	1.58	1.73
25	0.4	0.44	0.58	0.69	0.75
	1	0.53	0.70	0.84	0.92
	4	0.66	0.88	1.04	1.15
	10	0.78	1.03	1.12	1.33
	40	0.94	1.22	1.44	1.58
30	0.4	0.39	0.52	0.64	0.69
	1	0.48	0.64	0.77	0.85
	4	0.60	0.80	0.97	1.06
	10	0.70	0.93	1.13	1.23
	40	0.85	1.12	1.33	1.46

3.3 Numerical model of twin cavern case

The geometric model including the twin caverns and the design variables considered are shown in Fig. 3. In addition to the assumptions for the single cavern case outlined earlier, two additional assumptions are:

− The twin caverns are of equal size, both horse-shoe shaped, with semi-circular roof, the span-to-side wall height ratio (B/H) is 2, and horizontally aligned;
− The excavation involves six stages: heading, first benching, second benching of the right cavern, followed by heading, first benching, second benching of the left cavern, as depicted in Fig. 3.

The variables and the ranges considered are shown in Table 5. The two dependent responses are the global factor of safety of twin caverns (FS_{g_t}) and the percent strain ε_t.

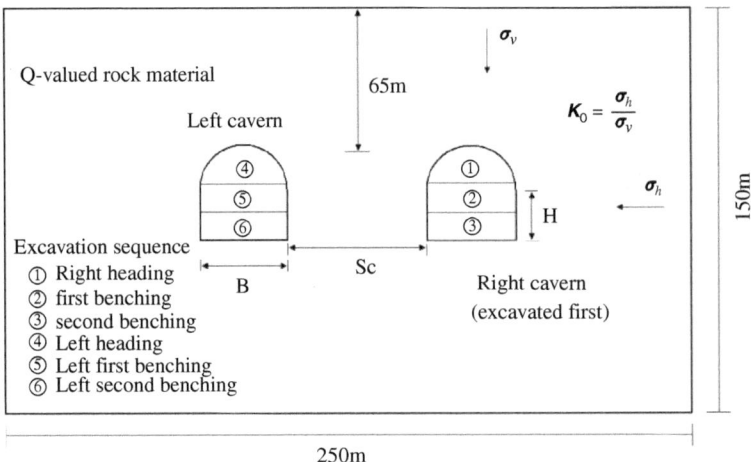

Figure 3 Geometrical model and basic design parameters.

Table 5 Factors used in parametric studies for twin cavern case.

Parameter	Description	Values
K_0	In situ stress ratio	0.5, 1, 2
B	Cavern span (m)	10, 20, 30
S_c/B	Ratio of pillar width to cavern span	1, 1.5, 2, 2.5
Q	Tunneling quality index	1, 4, 10, 40, 100

ε_t is derived the same way as the single cavern case. The only difference is that more peripheral nodes around the twin caverns are involved, as shown in Fig. 4. The strain in each of the key points, such as ε_{XR} (R denotes the right cavern) for example, is:

$$\varepsilon_{XR} = \max\left(\varepsilon_{XR\_xjR}\right) \tag{18}$$

where X is C, S, M or I and x is c, s, m and i; ε_{XR_xjR} (j=1, 2, 3) is defined as

$$\varepsilon_{XR\_xjR}(\%) = \frac{u_{XR} - u_{xjR}}{l_{XR\_xjR}} \times 100 \tag{19}$$

The percent strain of the twin caverns ε_t is the maximum of the strains of the eight key points:

$$\varepsilon_t = \max\left(\varepsilon_{XR}, \varepsilon_{XL}\right) \tag{20}$$

3.4 Modeling results of twin cavern case

The computed FS_{g_t} values are summarized in Table 6.

The percent strain values are calculated except for the cases with $FS_{g\_t} < 1$. A total of 147 cases are analyzed.

Table 6 Results from numerical experiments for *FSg_t*.

B (m)	Q	FS_{g_t}			
		$S_c/B = 1$	$S_c/B = 1.5$	$S_c/B = 2$	$S_c/B = 2.5$
30	100	1.38	1.47	1.59	1.66
	40	1.24	1.31	1.42	1.48
	10	1.04	1.10	1.19	1.25
	4	0.90	0.96	1.03	1.08
	1	0.72	0.77	0.83	0.87
20	100	1.55	1.64	1.72	1.79
	40	1.38	1.46	1.53	1.61
	10	1.17	1.23	1.29	1.35
	4	1.01	1.06	1.12	1.17
	1	0.81	0.85	0.90	0.94
10	100	2.13	2.15	2.20	2.24
	40	1.90	1.92	1.96	2.00
	10	1.61	1.62	1.65	1.68
	4	1.39	1.40	1.43	1.46
	1	1.11	1.12	1.15	1.17

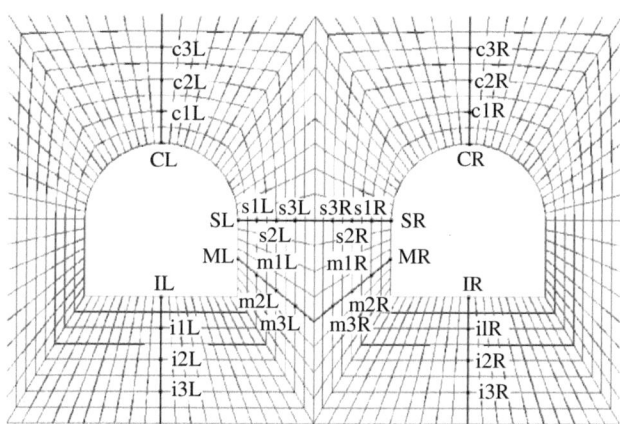

Figure 4 Details of cavern peripheral nodes between twin caverns.

4 PREDICTIVE MODELS AND DESIGN CHARTS

4.1 Predictive model for single cavern FS_{g_s} and design charts

Based on the above results, the LR model was developed for predicting FS_{g_s} in terms of Q, B and B/H, as shown in Equation 21

$$FS_{g\_s\_LR} = 2.9562Q^{0.1605}B^{-0.4127}(B/H)^{0.2314} \tag{21}$$

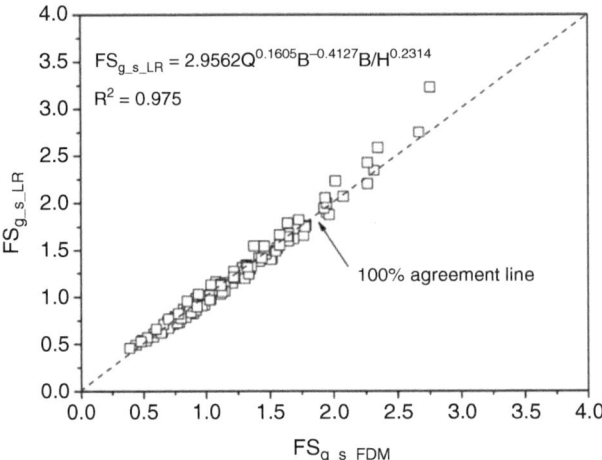

Figure 5 Comparison between $FS_{g\_s\_FDM}$ and $FS_{g\_s\_LR}$.

A comparison between $FS_{g\_s\_LR}$ and $FS_{g\_s\_FDM}$ (the global factor of safety obtained from FLAC^{3D}) is shown in Fig. 5. The high coefficient of determination R^2 of 0.975 indicates that the LR predictions are in good agreement with the target FDM FS_{g_s} values.

Based on the mathematical equation Equation 21, a series of charts relating FS_{g_s} to Q, B and B/H have been developed as shown in Fig. 6. The proposed charts are potentially useful for preliminary design and checking of single cavern stability. Houghton & Stacey (1980) proposed similar plots between the unsupported cavern span and Q value for prediction of stability. Their curves for FS of 1.0 and 1.2 are incorporated into Fig. 6 for comparison with the numerical results. It is obvious that curves from Houghton & Stacey (1980) show similar and consistent trends with the design charts proposed in this study with the empirically derived curves from Houghton & Stacey (1980) being more conservative.

4.2 Predictive model for single cavern ε_s

The LR model developed for predicting ε_s in terms of Q, B, B/H and K_0 is shown below

$$\varepsilon_{s\_LR}(\%) = 0.749 Q^{-0.8088} B^{0.2842} (B/H)^{-0.3476} K_0^{0.5421} \tag{22}$$

A plot of ε_{s_LR} versus ε_{s_FDM} shown in Fig. 7 indicates that the LR predictions are generally in agreement with the target FDM ε_s, particularly for ε_s less than 2%. It should be noted that for case with $\varepsilon_s > 2\%$, the equation does not follow a 1:1 relationship, indicating that this LR model underestimates the induced percent strain value. Thus, for large strain problems, Equation 22 should be used with caution.

4.3 Predictive model for twin caverns FS_{g_t} and design charts

The LR model was developed for predicting FS_{g_t} in terms of Q, B and S_c/B, as shown in Equation 23

(a)
FS=1.0 (Houghton & Stacey 1980)
FS=1.2 (Houghton & Stacey 1980)
B (m)
Q
FS=1.0
FS=1.2
FS=1.5
FS=2.0
FS=2.5
FS=3.0
(b)
(c)
(d)

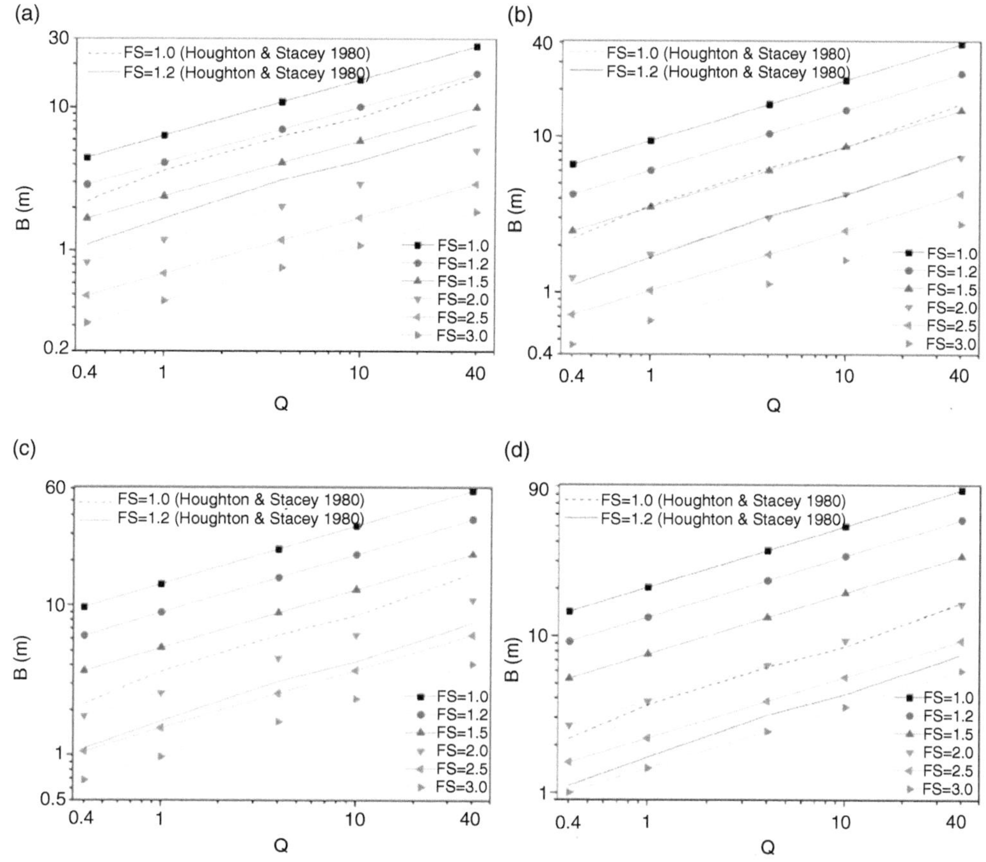

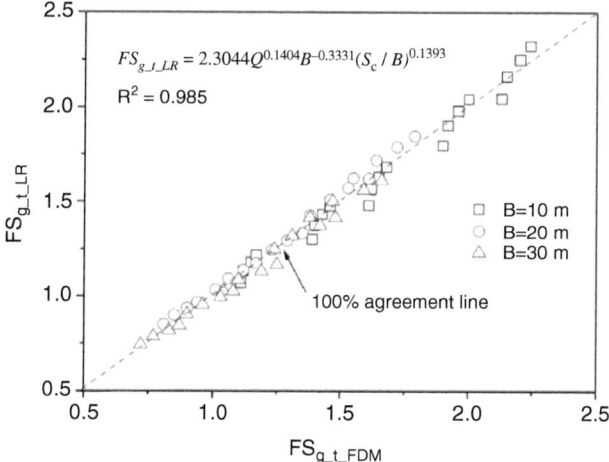

Figure 6 Design curves of FS_{g_s}: (a) B/H=0.25, (b) B/H=0.5, (c) B/H=1, and (d) B/H=2.

$FS_{g_t_LR} = 2.3044Q^{0.1404}B^{-0.3331}(S_c / B)^{0.1393}$
$R^2 = 0.985$
$FS_{g_t_LR}$
$FS_{g_t_FDM}$
B=10 m
B=20 m
B=30 m
100% agreement line

Figure 7 Comparison between ε_{s_FDM} and ε_{s_LR}.

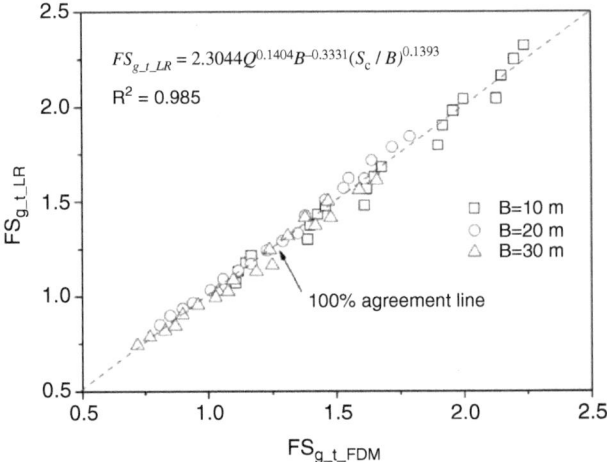

Figure 8 Comparison between $FS_{g\_t\_FDM}$ and $FS_{g\_t\_LR}$.

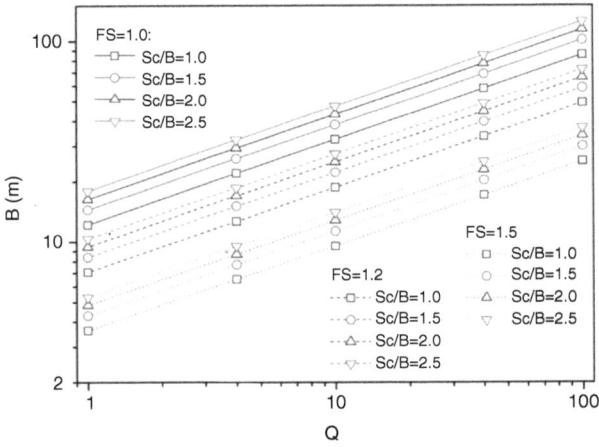

Figure 9 Design curves for twin caverns FS_{g_t}.

$$FS_{g\_t\_LR} = 2.3044Q^{0.1404}B^{-0.3331}(S_c/B)^{0.1393} \tag{23}$$

A comparison of FS_{g_t} between the predictions from Equation 23 and the $FS_{g\_t\_FDM}$ as shown in Fig. 8 indicates that the LR predictions are in good agreement with the target FDM FS_{g_t} values.

Based on Equation 23, charts relating FS_{g_t} to Q, B and S_c have been developed as shown in Fig. 9. The proposed charts are potentially useful for preliminary design and checking of twin cavern stability. By specifying the pillar geometry S_c/B for a given Q as well as the cavern span B, the expected global factor of safety can be estimated. Alternatively, by assuming the required global factor of safety for a given Q and cavern span, the appropriate pillar width can be determined.

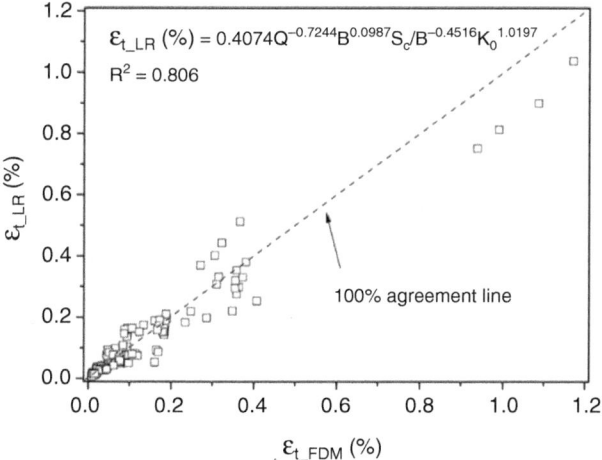

Figure 10 Comparison between ε_{t_FDM} and ε_{t_LR}.

4.4 Predictive model for twin caverns ε_t

The LR model developed for predicting ε_t in terms of Q, B, S_c/B and K_0 is shown below

$$\varepsilon_{t\_LR}(\%) = 0.4074Q^{-0.7244}B^{0.0987}(S_c/B)^{-0.4516}K_0^{1.0197} \tag{24}$$

A plot of ε_{t_LR} versus ε_{t_FDM} shown in Fig. 10 indicates that the LR predictions are generally in agreement with the target FDM ε_t, particularly for ε_t less than 0.2%. It should be noted that the strain values of the twin caverns in Fig. 10 are generally smaller than those for a single cavern (Fig. 7) since the lower bound Q value is 1 for the analysis of the twin caverns, compared with a lower bound Q value of 0.4 for the analysis of the single cavern cases. In addition, as with the single cavern case, for large strain problems, Equation 24 should be used with caution.

4.5 Parameter relative importance for built predictive models

Parameter relative importance is carried out based on method of Crick *et al.* (1987). Fig. 11*a* plots the parameter relative importance comparison for FS_{g_s} and ε_s models of the single cavern case. It is obvious that B is the most important variable in determining FS_{g_s}, followed by B/H and Q, indicating the significances of the cavern size and shape. The plot also indicates that ε_s is primarily influenced by Q, followed by K_0 and B/H, implying that serviceability problems are mainly caused by incompetent rock mass and unfavorable in situ stress fields. Fig. 11*b* presents the parameter relative importance comparison for FS_{g_t} and ε_t models of the twin cavern case. It can be noted that B is most important in determining FS_{g_t}, and that S_c/B and Q are equally significant. ε_t is mostly influenced by K_0, followed by Q and S_c/B.

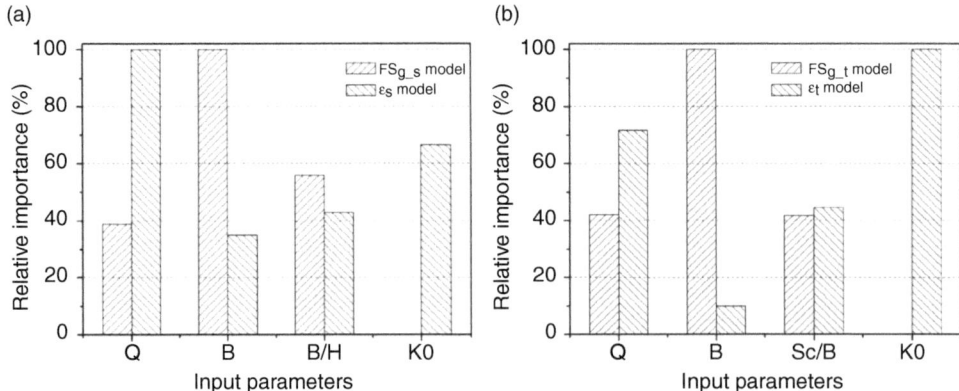

Figure 11 Parameter relative importance for: (a) single cavern case, (b) twin cavern case.

5 RELIABILITY ASSESSMENTS

5.1 Reliability assessment of ultimate limit state for a single cavern

Before describing the probabilistic assessments on ULS, it should be mentioned that for the FORM and MCS analyses in the present and the following sections, the parameters are assumed to be uncorrelated. With the determination of the ultimate limit state function for single caverns in Equation21, reliability assessment of the FS_{g_s} can be carried out using either MCS or FORM which is computationally less intensive. Failure in terms of global stability occurs if the predicted FS_{g_s} is smaller than a threshold factor of safety $FS_{g\_s\_cr}$ ($FS_{g\_s\_cr}$ is assumed as unity in this chapter). Probabilistic analysis is performed by calculating the reliability index β, to which the probability of failure is directly related. The reliability index β can be used to quantify the expected performance level. Typical values for β and probability of failure P_f for representative geotechnical components and systems and their expected performance levels have been proposed (US Army Corps of Engineers, 1997), as shown in Table 7.

Table 7 Target reliability indices (US Army Corps of Engineers, 1997).

β	$P_f \approx 1 - \Phi(\beta)$	Expected performance level
1.0	0.16	Hazardous
1.5	0.07	Unsatisfactory
2.0	0.023	Poor
2.5	0.006	Below average
3.0	0.001	Above average
4.0	0.00003	Good
5.0	0.0000003	High

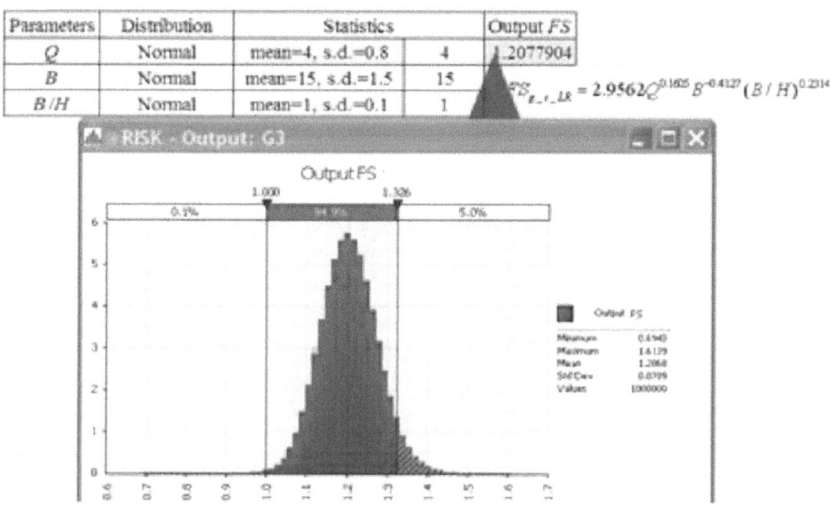

Parameters	Distribution	Statistics		Output FS
Q	Normal	mean=4, s.d.=0.8	4	1.2077904
B	Normal	mean=15, s.d.=1.5	15	
B/H	Normal	mean=1, s.d.=0.1	1	

$$FS_{g\_r\_LR} = 2.9562 Q^{0.1605} B^{-0.4127} (B/H)^{0.2314}$$

Figure 12 MCS with @RISK based on the developed LR model.

EN 1990 proposed a table relating the target reliability index and probability of failure to different limit state and length of reference periods as in Table 8 below:

Table 8 Target reliability index, β, and target probability of failure, P_f, values.

Limit state	Target Reliability Index, β		Target Probability of failure, P_f	
	I year	50 years	I year	50 years
ULS	4.7	3.8	1×10^{-6}	7.2×10^{-5}
SLS	2.9	1.5	2×10^{-3}	6.7×10^{-2}

One of the most useful software packages for MCS probability analysis is a Microsoft Excel add-in program called @RISK (http://www.palisade.com) which can be used for risk evaluations using sampling techniques including Latin Hypercube and Monte Carlo samplings. In this section, the derived LR model is incorporated into @RISK cell for FS_{g_s} estimation. For each MCS simulation, the number of iterations is 1 000 000 and Latin Hypercube sampling is adopted. The MCS procedures and the calculation of P_f are illustrated in Fig. 12.

In the FORM procedure, the performance function derived through LR model was incorporated into an EXCEL spreadsheet environment based on the approach by Low and Tang (2007) from which β and P_f can be determined. The typical FORM procedures and the calculation of P_f are illustrated in Fig. 13. Fig. 13 shows a sample spreadsheet for computing β where Q, B and B/H are assumed to be normally distributed. The spreadsheet cells A2:A4 allows the selection of distribution types for the input variables, including Normal, Lognormal, Triangular etc as explained in Low & Tang (2007). For nonnormals, the nonnormal distribution types are replaced by an equivalent Normal ellipsoid, centered at the equivalent Normal mean. Cells C2:F4 are parameters which are set corresponding to the variable distribution types. For Normal

Table 9 Statistical information of the input variables for reliability assessment on ULS.

Case No.	Variables	Distribution	Statistics
1	Q	Normal	Mean = 4, COV = 0.1, 0.2, 0.3, 0.4
	B	Normal	Mean = 10, COV = 0.1
	B/H	Normal	Mean = 0.5, COV = 0.1
2	Q	Normal	Mean = 4, COV = 0.1, 0.2, 0.3, 0.4
	B	Normal	Mean = 15, COV = 0.1
	B/H	Normal	Mean = 1, COV = 0.1
3	Q	Normal	Mean = 10, COV = 0.1, 0.2, 0.3, 0.4
	B	Normal	Mean = 10, COV = 0.1
	B/H	Normal	Mean = 0.25, COV = 0.1
4	Q	Normal	Mean = 10, COV = 0.1, 0.2, 0.3, 0.4
	B	Normal	Mean = 15, COV = 0.1
	B/H	Normal	Mean = 0.5, COV = 0.1
5	Q	Normal	Mean = 10, COV = 0.1, 0.2, 0.3, 0.4
	B	Normal	Mean = 25, COV = 0.1
	B/H	Normal	Mean = 1, COV = 0.1
6	Q	Normal	Mean = 10, COV = 0.1, 0.2, 0.3, 0.4
	B	Normal	Mean = 30, COV = 0.1
	B/H	Normal	Mean = 2, COV = 0.1

	A	B	C	D	E	F	G	H	I	J	K	L	M	N
1	Distribution	Variables	Para1	Para2	Para3	Para4	x_i^*	Correlation matrix [R]			n_i	$g(x)$	β	$P_f(\%)$
2	Normal	Q	4	0.8			2.535	1	0	0	-1.831			
3	Normal	B (m)	15	1.5			18.56	0	1	0	2.372	-5E-03	3.278	0.052
4	Normal	B/H	1	0.1			0.867	0	0	1	-1.33			
5							$g(\mathbf{x}) = FS_{g\_s\_LR} - FS_{g\_s\_cr} = FS_{g\_s\_LR} - 1 = 2.9562Q^{0.1605}B^{-0.4127}(B/H)^{0.2314} - 1$							

Figure 13 FORM incorporated with the developed LR model.

distributions, cells C2:C4 correspond to the mean values and cells D2:D4 correspond to the standard deviations. The correlation matrix [**R**] cells H2:J4 are used to define the correlations between Q, B and B/H. The n_i vector in cells K2:K4 contains equations for $(x_i - \mu_i^N)/\sigma_i^N$. Cell L3 contains the ULS function g(**x**) = $FS_{g\_s\_LR}$–$FS_{g\_s\_cr}$. The DP (x* values) was obtained by using the spreadsheet's built-in optimization routine SOLVER to minimize the cell, by changing the x* values, under the constraint that the ULS function g(x*) = 0. Prior to invoking the SOLVER search algorithm, the x* values in cells G2:G4 were set equal to the mean values (4, 15, 1) of the original random variables. Iterative numerical derivatives and directional search for DP x* were automatically carried out in the spreadsheet.

P_f values of eight cases are calculated using the two reliability assessment methods for comparison. The assumed statistical information of the input variables for the eight cases is listed in Table 9.

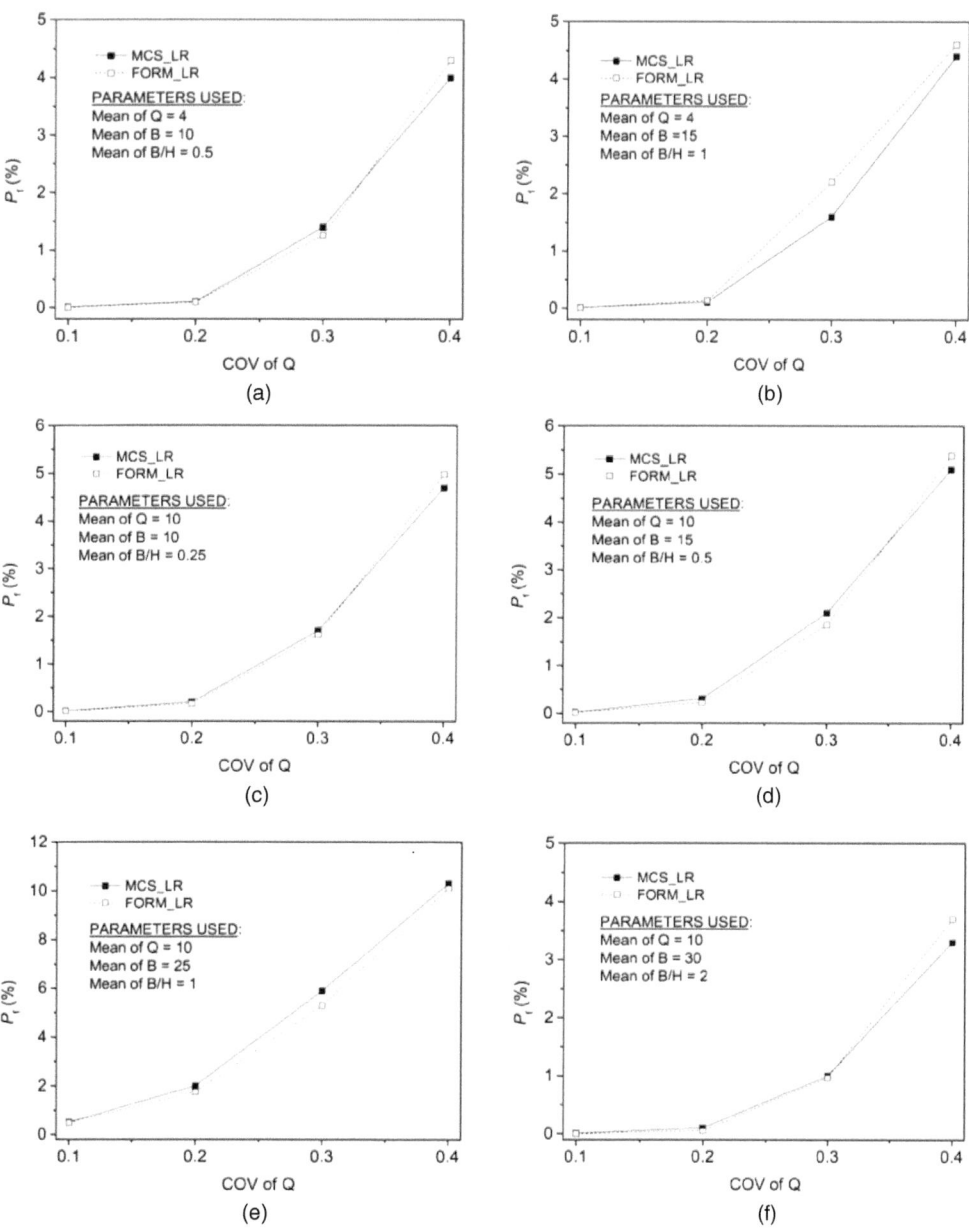

Figure 14 ULS P_f derived for case: (a) case 1, (b) case 2, (c) case 3, (d) case 4, (e) case 5, and (f) case 6.

The values of P_f derived for the eight cases are shown in Fig. 14 for comparison. The relationship between FS_{g_s} and P_f depends on the uncertainties and variability of the system under consideration. It is clear that P_f increases with the increase of COV of Q, demonstrating that a single FS_{g_s} value which is determined by a combination of mean

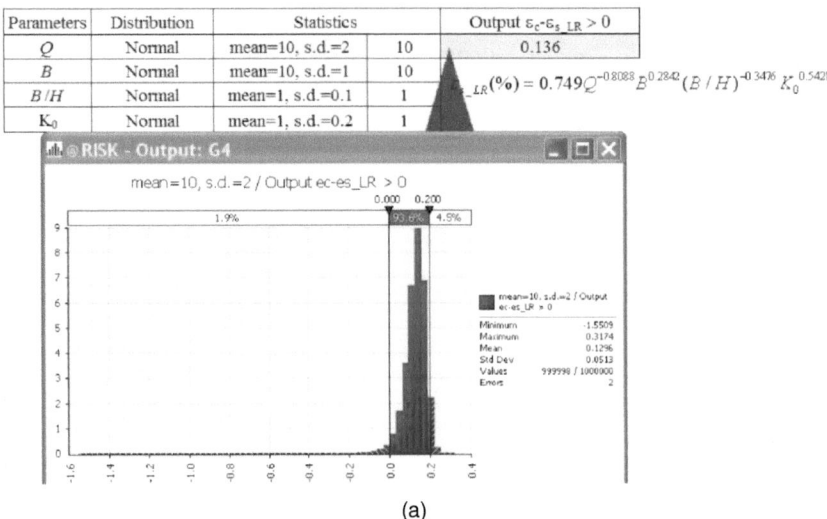

Parameters	Distribution	Statistics		Output ε_c-$\varepsilon_{s\_LR} > 0$
Q	Normal	mean=10, s.d.=2	10	0.136
B	Normal	mean=10, s.d.=1	10	
B/H	Normal	mean=1, s.d.=0.1	1	
K_0	Normal	mean=1, s.d.=0.2	1	

$$\varepsilon_{LR}(\%) = 0.749 Q^{-0.8088} B^{0.2842} (B/H)^{-0.3476} K_0^{0.5421}$$

(a)

Distribution	Variables	Para1	Para2	Para3	Para4	x_i^*	Correlation matrix [R]				n_i	$g(\underline{x})$	β	$P_f(\%)$
Normal	Q	10	2			5.664	1	0	0	0	-2.168		2.195	1.407
Normal	B (m)	10	1			10.21	0	1	0	0	0.211	0.000		
Normal	B/H	1	0.1			0.973	0	0	1	0	-0.271			
Normal	K_0	1	0.2			1	0	0	0	1	0			

$$g(\underline{x}) - \varepsilon_c - \varepsilon_{s\_LR} - \varepsilon_c - 0.749 Q^{-0.8088} B^{0.2842} (B/H)^{-0.3476} K_0^{0.5421}$$

(b)

Figure 15 Typical procedures for calculation of P_f: (a) MCS of @RISK and, (b) FORM.

values of Q, B and B/H may not always provide an adequate quantification of safety. It is obvious that for most of the cases considered, the differences between MCS and FORM are marginal.

5.2 Reliability assessment of serviceability limit state for a single cavern

The typical MCS and FORM procedures and the calculation of P_f are illustrated in Fig. 15 a and b, respectively. For each MCS simulation, the number of iterations is 1 000 000 and the sampling technique is Latin Hypercube. For the FORM procedures and the developed spreadsheet environment, the explanations are the same as those for ULS.

P_f values of six cases are calculated using the reliability assessment methods for comparison. The assumed statistical information of the input variables for the six cases is listed in Table 10.

The values of P_f derived using the two probabilistic methods for the six cases are shown in Fig. 16 for comparison. It is clear that P_f increases with the increase of COV

Table 10 Statistical information of the input variables for reliability assessment on SLS.

Case No.	Variables	Distribution	Statistics
1	Q	Normal	Mean = 10, COV = 0.1, 0.2, 0.3, 0.4
	B	Normal	Mean = 10, COV = 0.1
	B/H	Normal	Mean = 2, COV = 0.1
	K_0	Normal	Mean = 1, COV = 0.2
2	Q	Normal	Mean = 10, COV = 0.1, 0.2, 0.3, 0.4
	B	Normal	Mean = 10, COV = 0.1
	B/H	Normal	Mean = 1, COV = 0.1
	K_0	Normal	Mean = 1, COV = 0.2
3	Q	Normal	Mean = 10, COV = 0.1, 0.2, 0.3, 0.4
	B	Normal	Mean = 15, COV = 0.1
	B/H	Normal	Mean = 2, COV = 0.1
	K_0	Normal	Mean = 1, COV = 0.2
4	Q	Normal	Mean = 10, COV = 0.1, 0.2, 0.3, 0.4
	B	Normal	Mean = 10, COV = 0.1
	B/H	Normal	Mean = 1, COV = 0.1
	K_0	Normal	Mean = 0.5, COV = 0.2
5	Q	Normal	Mean = 10, COV = 0.1, 0.2, 0.3, 0.4
	B	Normal	Mean = 15, COV = 0.1
	B/H	Normal	Mean = 1, COV = 0.1
	K_0	Normal	Mean = 0.5, COV = 0.2
6	Q	Normal	Mean = 40, COV = 0.1, 0.2, 0.3, 0.4
	B	Normal	Mean = 15, COV = 0.1
	B/H	Normal	Mean = 0.5, COV = 0.1
	K_0	Normal	Mean = 1, COV = 0.2

of Q, demonstrating that a single ε_s value which is deterministically determined may not always provide an adequate quantification of serviceability. The differences between MCS and FORM are marginal for the six cases. For the same Q and B/H values, increase of the cavern span B results in significant increase of P_f.

5.3 Reliability assessment of ultimate limit state for twin caverns

Assuming the critical FS_{g_t} value is $FS_{g\_t\_cr}$, the ultimate limit state function based on LR model is then given by

$$g(x) = FS_{g\_t\_LR} - FS_{g\_t\_cr} = 2.3044Q^{0.1404}B^{-0.3331}(S_c/B)^{0.1393} - FS_{g\_t\_cr} \qquad (25)$$

For brevity, the reliability assessment of the FS_{g_t} was carried out using only MCS ($FS_{g\_t\_cr}$ is assumed as 1). The implementation of the developed FS_{g_t} model into MCS is not illustrated here since it is very similar to Fig. 12.

Fig. 17 plots the influences of the choice of $FS_{g\_t\_cr}$ on the calculated P_f using the LR model. It is obvious that $FS_{g\_t\_cr}$ substantially influences P_f.

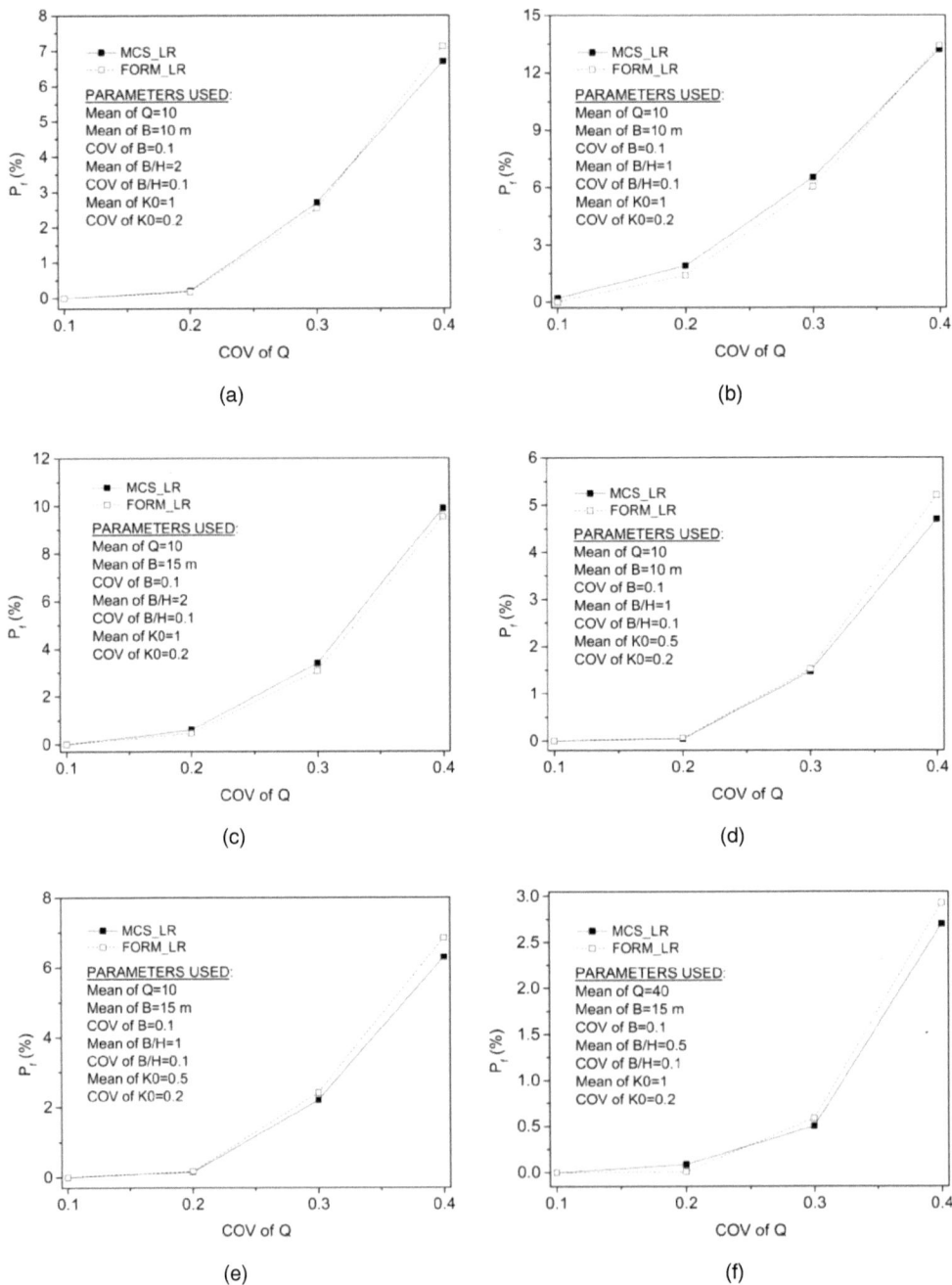

Figure 16 SLS P_f derived for case: (a) case 1, (b) case 2, (c) case 3, (d) case 4, (e) case 5, and (f) case 6.

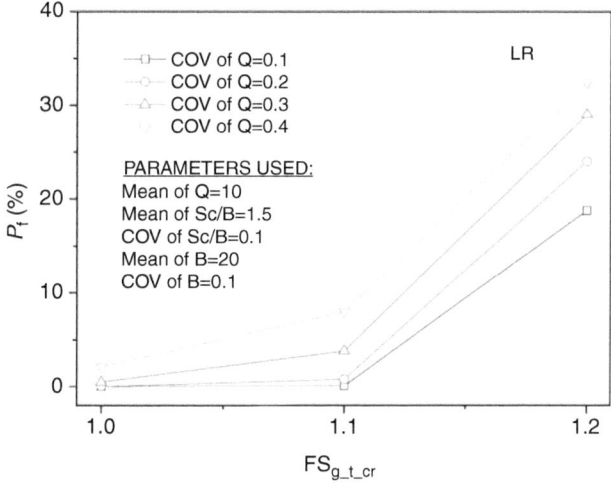

Figure 17 Influence of choice of $FS_{g\_t\_cr}$ on P_f.

5.4 Reliability assessment of serviceability limit state for twin caverns

The serviceability limit state function for the twin cavern case is given by

$$g(\pmb{x}) = \varepsilon_{t\_cr} - \varepsilon_{t\_LR} = \varepsilon_{t\_cr} - 0.4074Q^{-0.724}B^{0.0987}(S_c/B)^{-0.4516}K_0^{1.0197} \qquad (26)$$

The reliability assessment on Equation 26 was carried out using only FORM. The implementation of the developed ε_{t_LR} model into FORM spreadsheet is not illustrated since it is similar to Fig. 13.

Fig. 18 plots the influence of the various design parameters on P_f of SLS. It is clear that both the COV and the average value of Q significantly influence the P_f. However, the influence of B on P_f is not as significant as that for S_c/B. In addition, from Fig. 18c it can be observed that the influence of K_0 on P_f is also significant when Q=40 compared with Q=100. The P_f under $K_0 = 1$ is generally higher than that under $K_0 = 0.5$ as the ε_t for $K_0 = 1$ is larger than for $K_0 = 0.5$ but ε_c is the same.

6 SUMMARY AND CONCLUSIONS

This chapter numerically investigated both the ultimate and the serviceability limit states of underground single and twin rock caverns. Predictive models and design charts for global factor of safety of both the single and twin rock caverns were developed, for preliminary design use and checking of cavern(s) stability. Models of induced percent strain prediction were developed for estimating the possible deformation during construction. The concept of the critical strain was used to limit the induced strains. Simplified probabilistic assessment procedures are proposed for evaluating the probabilities that either ultimate or serviceability limit state was exceeded for deep underground rock caverns for preliminary design applications.

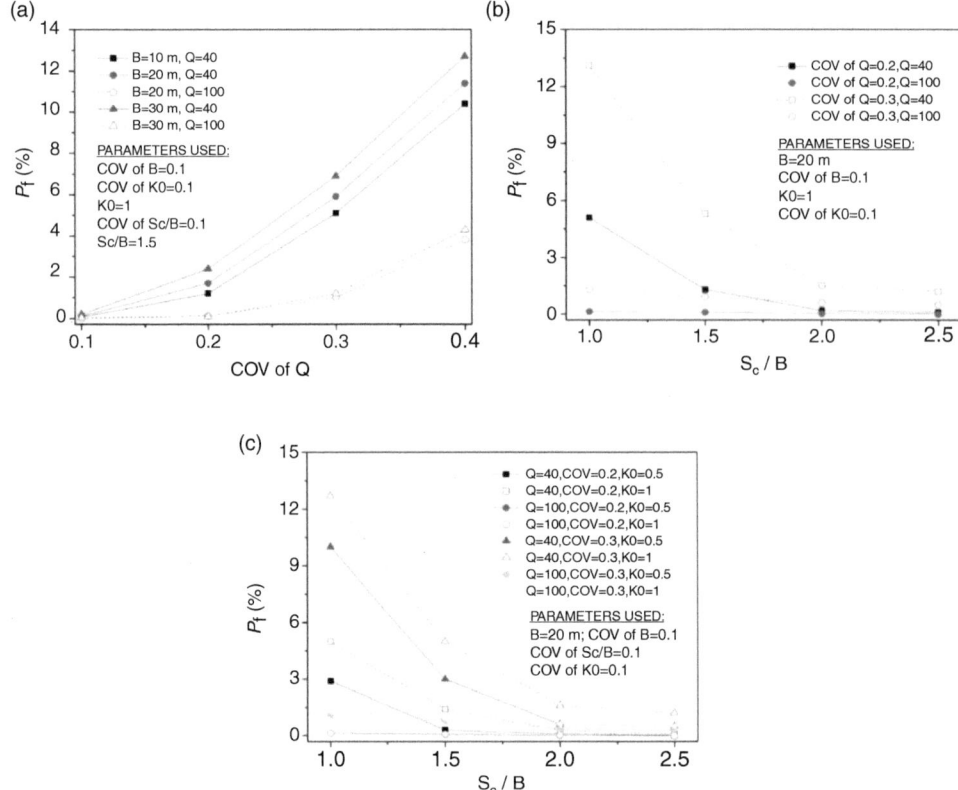

Figure 18 Influence of Q, S_c/B, B, and K_0 on P_f of SLS: (a) *COV* of Q, B and mean Q, (b) S_c/B, *COV* and mean Q, and (c) S_c/B, K_0, *COV* and mean Q.

REFERENCES

Adoko, A.C. Jiao, Y.Y. Wu, L. Wang, H. & Wang, Z.H. (2013) Predicting tunnel convergence using Multivariate Adaptive Regression Spline and Artificial Neural Network. *Tunneling and Underground Space Technology*, 38, 368–376.

Ang, A.H.S. & Tang, W.H. (1984) *Probability concepts in engineering planning and design, Vol. II – Decision, risk and reliability*, New York, John Wiley and Sons.

Baecher, G.B. & Christian, J.T. (2003) *Reliability and Statistics in Geotechnical Engineering*, John Wiley, San Francisco.

Barla, G. & Ottoviani, M. (1974) Stresses and displacements around two adjacent circular openings near to the ground surface. *Proceedings of the 3rd International Congress on Rock Mechanics*, 1–7 September Denver, Colorado: National Academy of Sciences. 2, 975–980.

Barton, N. (1995) *The influence of joint properties in modeling jointed rock masses.* Keynote Lecture, In: 8th ISRM Congress, Balkema, Rotterdam. 3, 1023–1032.

Basarir, H. (2008) Analysis of rock-support interaction using numerical and multiple regression modeling. *Canadian Geotechnical Journal*, 45, 1–13.

Bieniawski, Z.T. (1978) Determining rock mass deformability: Experience from case histories. *International Journal of Rock Mechanics and Mining Sciences & Geomechanics Abstracts*, 15, 237–247.

Bieniawski, Z.T. (1984) *Rock Mechanics Design in Mining and Tunneling*, the Netherland, A.A. Balkema, Rotterdam. 97–133.

Bieniawski, Z.T. (1989) *Engineering Rock Mass Classifications*, New York, John Wiley and Sons.

Cai, M. Kaiser, P.K. Morioka, H. Minami, M. Maejima, T. Tasaka, Y. & Kurose, H. (2007) FLAC/PFC coupled numerical simulation of AE in large-scale underground excavations. *International Journal of Rock Mechanics and Mining Sciences*, 44(4), 550–564.

Cai, M. (2008) Influence of stress path on tunnel excavation response – Numerical tool selection and modeling strategy, *Tunneling and Underground Space Technology*, 23, 618–628.

Cai, M. (2011) Rock mass characterization and rock property variability considerations for tunnel and cavern design, *Rock Mechanics and Rock Engineering*, 44, 379–399.

Choi, S.K. Grandhi, R.V. & Canfield, R.A. (2006) *Reliability-Based Structural Design*, Springer, London.

Chryssanthakis, P. Barton, N. & Monsen, K. (1995) Dynamic loading of physical and numerical models of very large underground openings. *Proceedings of the 8th International Conference on Rock Mechanics*, 25–30 September, Tokyo, Japan, Balkema. 3, 1313–1316.

Cornell, C.A. (1969) A probability-based structural code. *ACI Structural Journal* 66(12), 974–985.

Crick, M.J. Hill, M.D. & Charles, D. (1987) The Role of Sensitivity Analysis in Assessing Uncertainty. *In: Proc. of an NEA Workshop on Uncertainty Analysis for Performance Assessments of Radioactive Waste Disposal Systems*, 24–26 February, Seattle, United States, Paris, OECD, 116–129.

Dawson, E.M. Motamed, F. Nesarajah, S. & Roth, W.H. (2000) Geotechnical stability analysis by strength reduction. *Proceedings of Sessions of Geo-Denver 2000-Slope Stability*. GSP 101, 289, 99–113.

Dawson, E.M. Roth, W.H. & Drescher, A. (1999) Slope stability analysis by strength reduction. *Géotechnique*, 49(6), 835–840.

Ditlvsen, O. & Madsen, H.O. (1996) *Structural Reliability Methods*, John Wiley & Sons, Chichester.

EN 1990. (2002), Eurocode 0: *Basis of structural design*, CEN (European Committee for Standardization), Brussels, Belgium.

Esterhuizen, G.S. Dolinar, D.R. & Ellenberger, J.L. (2011) Pillar strength in underground stone mines in the United States. *International Journal of Rock Mechanics and Mining Sciences*, 48, 42–50.

Fan, S.C. Jiao, Y.Y. & Zhao, J. (2004) On modeling of incident boundary for wave propagation in jointed rock masses using discrete element method. *Computers and Geotechnics*, 31(1), 57–66.

Gercek, H. (2005) Interaction between parallel underground openings. *Proceedings of the 19th International Mining Congress and Fair of Turkey*, 9–12 June, Izmir Turkey, IMCEV2005, 73–81.

Ghaboussi, J. & Ranken, R.E. (1977) Interaction between two parallel tunnels. *International Journal of Rock Mechanics and Mining Sciences & Geomechanics Abstracts*, 1, 75–103.

Goh, A.T.C. & Zhang, W.G. (2012) Reliability assessment of stability of underground rock caverns. *International Journal of Rock Mechanics and Mining Sciences*, 55, 157–163.

Hammah, R.E. Yacoub, T. & Curran, J.H. (2007) Serviceability-based slope factor of safety using the shear strength reduction (SSR) method. *The Second Half Century of Rock Mechanics 11th Congress of the International Society for Rock Mechanics*, Taylor & Francis, 1137–1140.

Hao, Y.H. & Azzam, R. (2005) The plastic zones and displacements around underground openings in rock masses containing a fault. *Tunneling and Underground Space Technology*, 20(1), 49–61.

Hasofer, A.M. & Lind, N. (1974) An exact & invariant first-order reliability format. *Journal of Engineering Mechanics ASCE* 100(1), 111–121.

Hatzor, Y.H. Talesnick, M. & Tsesarsky, M. (2002) Continuous and discontinuous stability analysis of the bell-shaped caverns at Bet Guvrin, Israel, *International Journal of Rock Mechanics and Mining Sciences*, 39, 867–886.

Hoek, E., 2011. *Cavern reinforcement and lining design* (A note prepared for RocNews). [Online] Available from: http://www.rocscience.com/assets/files/uploads/8500.pdf.

Houghton, D.A. & Stacey, T.R. (1980) Application of probability techniques to underground situations. *Proceedings of the 7th Regional Conference for Africa on Soil Mechanics and Foundation Engineering*, A. Balkema. 2, 879–883.

Jia, P. & Tang, C.A. (2008) Numerical study of failure mechanism of tunnel in jointed rock mass. *Tunneling and Underground Space Technology*, 23, 500–507.

Karademir, S.M. (2010) A parametric study on three dimensional modeling of parallel tunnel interactions. *Ph.D. Thesis*, Middle East Technical University, Ankara, Turkey.

Low, B.K. & Tang, W.H. (2007) Efficient spreadsheet algorithm for first-order reliability method, *Journal of Engineering Mechanics*, 133(12), 1378–1387.

Lü, Q. Chan, C.L. & Low, B.K. (2012) Probabilistic evaluation of ground-support interaction for deep rock excavation using artificial neural network and uniform design. *Tunneling and Underground Space Technology*, 32, 1–18.

Mahdevari, S. & Torabi, S.R. (2012) Prediction of tunnel convergence using Artificial Neural Networks. *Tunneling and Underground Space Technology*, 28, 218–228.

Matsui, T. & San, K.C. (1992) Finite element slope stability analysis by shear strength reduction technique. *Soils and Foundations*, 32(1), 59–70.

Melchers, R.E. (1999). *Structure Reliability Analysis and Prediction*, 2nd ed. Wiley, New York.

Mohanty, S. & Vandergrift, T. (2012) Long term stability evaluation of an old underground gas storage cavern using unique numerical methods. *Tunneling and Underground Space Technology*, 30, 145–154.

Mortazavi, A. Hassani, F.P. & Shabani, M. (2009) A numerical investigation of rock pillar failure mechanism in underground openings. *Computers and Geotechnics*, 36, 691–697.

Palmstrom, A. (2000) On classification systems. *Proceedings GeoEng2000*, Melbourne, Australia.

Rafiai, H. & Moosavi, M. (2012) An approximate ANN-based solution for convergence of lined circular tunnels in elasto-plastic rock masses with anisotropic stresses. *Tunneling and Underground Space Technology*, 27, 52–59.

Sakurai, S. (1986) Field measurement and hazard warning levels in NATM. *Soils and Foundations*, 34(2), 5–10.

Sakurai, S. (1997) Lessons learned from field measurements in tunneling. *Tunneling and Underground Space Technology*, 12(4), 453–460.

Serafim, J.L. & Pereira, J.P. (1983) Considerations of the geomechanics classification of Bieniawski. *Proceedings of the International Symposium on Engineering Geology and Underground Construction*, the Netherlands, A.A. Balkema, Rotterdam. 1, 1133–1142.

Siahmansouri, A. Gholamnejad, J. & Marji, M.F. (2012) A new method to predict ratio of width to height rock pillar in twin circular tunnels. *Journal of Geology & Geosciences*, 1: 103. doi: 10.4172/2329-6755.1000103

Sitharam, T.G. & Latha, G.M. (2002) Simulation of excavations in jointed rock mass using a practical equivalent continuum approach. *International Journal of Rock Mechanics and Mining Sciences*, 39, 517–525.

Tsesarsky, M. Gal, E. & Machlav, E. (2013) 3-D global-local finite element analysis of shallow underground caverns in soft sedimentary rock. *International Journal of Rock Mechanics and Mining Sciences*, 57, 89–99.

Tugrul, A. (1998) The application of rock mass classification systems to underground excavation in weak lime stone, Ataturk dam. *Turkey Engineering Geology*, 50, 337–345.

U.S. Army Corps of Engineers. (1997) Engineering and design introduction to probability and reliability methods for use in geotechnical engineering, *Engineering Technical Letter No. 1110-2-547*, Department of the Army, Washington, DC.

Vermeer, P.A. Ruse, N. & Marcher, T. (2002) Tunneling heading stability in drained ground. *Felsbau*, 20(6), 8–18.

Wang, Y.F. Zhu, H.H. & Zheng, Y.R. (2014) Stability analysis and failure mechanism of jointed rock tunnel. *Tunneling and Underground Construction GSP 242, ASCE*, Shanghai, 106–115.

Xia, C.C. Zhao, X. Zhang, G.Z. & Wang, C.B. (2014) Stability analysis by strength reduction method in shallow buried tunnels. *Tunneling and Underground Construction GSP 242, ASCE*, Shanghai, 321–331.

Zhang, W.G. & Goh, A.T.C. (2012) Reliability assessment on ultimate and serviceability limit states and determination of critical factor of safety for underground rock caverns. *Tunneling and Underground Space Technology*, 32, 221–230.

Zhang, W.G. & Goh, A.T.C. (2013) Multivariate adaptive regression splines for analysis of geotechnical engineering systems. *Computers and Geotechnics*, 48, 82–95.

Zhang, W.G. & Goh, A.T.C. (2014) Multivariate adaptive regression splines model for reliability assessment of serviceability limit state of twin caverns. *Geomechanics and Engineering*, 7(4), 431–458.

Zhang, W.G. & Goh, A.T.C. (2015a) Numerical study of pillar stresses and interaction effects for twin rock caverns. *International Journal for Numerical and Analytical Methods in Geomechanics*, 39(2), 193–206.

Zhang, W.G. & Goh, A.T.C. (2015b) Regression models for estimating ultimate and serviceability limit states of underground rock caverns. *Engineering Geology* 188, 68–76.

Zhang, Y.H. Fu, X.D. & Sheng, Q. (2014) Modification of the discontinuous deformation analysis method and its application to seismic response analysis of large underground caverns. *Tunneling and Underground Space Technology*, 40, 241–250.

Zhao, B.Y. & Ma, Z.Y. (2009) Influence of cavern spacing on the stability of large cavern groups in a hydraulic power station. *International Journal of Rock Mechanics and Mining Sciences*, 46, 506–513.

Zhu, W. & Li, S.C. (2000) Optimizing the construction sequence of underground openings using dynamic construction mechanics and a rock mass fracture damage model. *International Journal of Rock Mechanics and Mining Sciences & Geomechanics Abstracts*, 37(5), 517–523.

Zhu, W. & Zhao, J. (2004) *Stability analysis and modeling of underground excavations in fractured rocks* (Elsevier Geo-Engineering Book Series), Elsevier.

Zhu, W.S. Sui, B. Li, X.J. Li, S.C. & Wang, W.T. (2008) A methodology for studying the high wall displacement of large scale underground cavern complexes and its applications. *Tunneling and Underground Space Technology*, 23, 651–664.

Chapter 12

Risk assessment of CO$_2$ injection processes and storage

L. Ribeiro e Sousa[1,2], R. Leal e Sousa[3], Eurípedes Vargas Jr.[4], Raquel Velloso[4] & Karim Karam[3]

[1]*China University of Mining & Technology, Beijing, China*
[2]*University of Porto, Portugal*
[3]*Masdar Institute of Science and Technology, Abu Dhabi, UAE*
[4]*Catholic University of Rio de Janeiro, Brazil*

Abstract: Different options for carbon capture, utilization and storage (CCUS) technologies exist and have been or are in the process of being implemented worldwide. With recent technological advancements, large scale CCUS schemes have become economically viable particularly when the captured CO$_2$ is used in the enhancement of oil (EOR) or gas (EGR) recovery, or when government incentives such as carbon credits are in place. CO$_2$ can also be stored in coal beds, in unminable coal seams and well-sealed abandoned coal mines. In this chapter, the risks associated with CO$_2$ storage are described by identifying and examining different types of hazards. The importance of modeling and monitoring is emphasized and different methodologies to assess risk are discussed. There are a number of models available for data analysis, representation and risk analysis. Bayesian networks (BN) are described in detail, and some applications of BNs in CO$_2$ injection and storage are presented for different hazard scenarios. Finally, several conclusions are drawn.

1 INTRODUCTION

Various technologies have been developed over recent years to address the increasingly urgent demand for the protection of the Earth's atmosphere from depleting emissions, and ensure sustainability. This is true for different gaseous emissions, most significantly for carbon dioxide (CO$_2$), where significant advances have been made in the processes of emissions reduction. Furthermore, it is no longer acceptable to simply reduce (or eliminate) emissions, but rather capture and store atmospheric CO$_2$ in processes that lead to an overall reduction of the atmospheric CO$_2$ resulting in carbon-negative facilities, cities and counties.

Different options for carbon capture, utilization and storage (CCUS) technologies exist and have been or are in the process of being implemented worldwide. European Energy Policies and Initiatives such as "Energy 2020" and "Energy Roadmap 2050" (EASAC, 2013a, b; Rodrigues *et al.*, 2015) take into account not only energy demand but also environmental requirements. The Energy 2020 Initiative was established by the European Union Council in 2007 taking into account the long term prospects of the energy industry. The Energy Roadmap 2050 Initiative was established in March 2011 to ensure security of energy supply and competitiveness in a low-carbon economy.

In general, it is now accepted that CCUS is a viable technological solution to reduce GGH emissions and can be introduced as a strategy within the larger context of climate change policies (Qin, 2013; Langfeld & Agasty, 2013; Shi *et al.*, 2013; Rodrigues *et al.*, 2015).

With these technological advancements, the economics of large, commercial-scale CCUS schemes has become more viable. This is particularly true when governmental entities have policies in place that promote CCUS such as carbon credit incentives (as opposed to but possibly coupled with taxation on carbon emissions). More recently, much research is being conducted on innovative techniques to increase the financial viability of CCUS such as the use of the captured carbon in enhancing the recovery of oil and gas fields where conditions are favorable. Continued research in every stage of the CCUS process is only likely to decrease costs and make CCUS more economically viable.

Despite these advancements, CCUS, as with any operation in the underground, is uncertain. These uncertainties come from various sources, and there are hazards and risks associated with them. This chapter addresses the technical risks associated with CCUS from various sources. A detailed economic feasibility study and study of the economic risks involved is beyond the scope of this chapter.

The storage of CO_2 in deep onshore and offshore geological formations uses many of the technologies developed by oil and gas industry, and can be coupled with enhanced oil and gas recovery schemes (EOR and EGR) and in saline aquifers (IPCC, 2005; EASAC, 2013a; Vercelli *et al.*, 2014). CO_2 can also be stored in coal beds, particularly in unminable deep coal seams and well-sealed abandoned coal mines (Piessens, 2012). CO_2 should be safely injected and stored at well characterized and properly managed sites in order to assure the long-term safety of a geological CO_2 storage project (Total, 2007; He *et al.*, 2012; Rodrigues *et al.*, 2015).

CO_2 is injected in deep geological formations where conditions of pressure and temperatures are favorable for CO_2 to exist in the so-called supercritical or liquid form, requiring less volume than in its gaseous form (EASAC, 2013a). Adequate planning of the injection and storage processes is essential to ensure safety with no CO_2 leakage during the entire storage period. At depths below about 800-1,000m, CO_2 has a liquid-like density, which makes the potential use of porous sedimentary rocks as underground reservoirs possible.

There exist two broad disciplines for the reduction of atmospheric CO_2, namely biological fixation and geological sequestration. This chapter is concerned with geological sequestration which is described in more detail in the following. There are different possibilities for geological sequestration as illustrated in Figure 1 (IPPC, 2005; Sousa & Rodrigues, 2008). Several geological solutions (closed systems) can be considered as feasible, such as depleted oils and gas reservoirs, deep saline aquifers, shale gas, coal seams, abandoned coal mines, among others. Coal seams and abandoned coal mines are currently being used to produce enhanced coal bed methane (ECBM), as well shale as gas technologies which can be viable in terms of permanent CO_2 disposal (He, 2012; Rodrigues *et al.*, 2015).

Different approaches are available (Kuhn *et al.*, 2010; EASAC, 2013a; Köhler *et al.*, 2013; Vercelli *et al.*, 2014) to determine the suitability of a site for CO_2 storage. For studying the feasibility of the CCUS projects, it is also necessary to perform pilot tests as was the case at the Lacq Project Site in France (Total, 2007; EASAC, 2013a, b).

Geological storage requires constructing facilities to capture the significant atmospheric CO_2 emissions from facilities such as power plants, and factories such as those

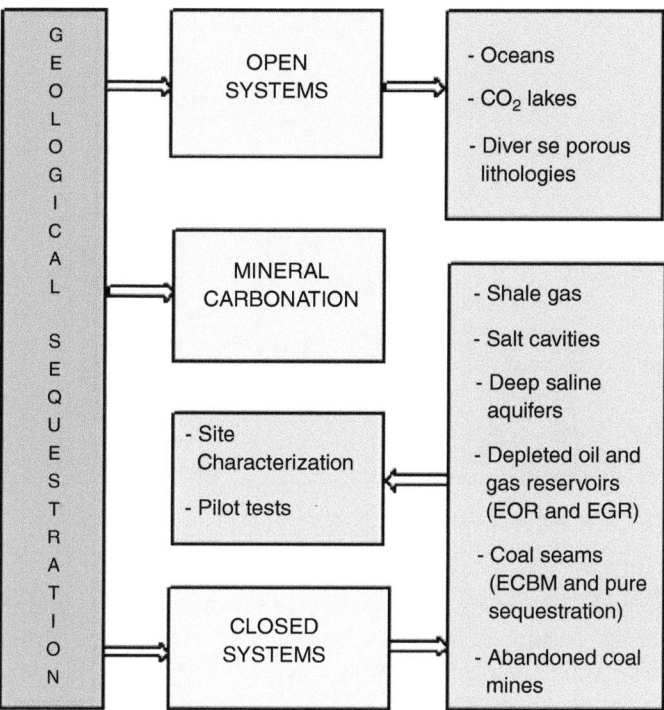

Figure 1 Different situations for geological sequestration.

for the production of cement, steel, ethanol, among others. It is also possible to act on fuel, that is, instead of proceeding to the capture of CO$_2$ after combustion of coal or natural gas can be drawn primarily engaged in industrial units, the carbon present in them (IEA-UNIDO, 2011; Sousa, 2012a). The captured CO$_2$ is then transported by pipelines or in ships, to underground storage sites. Most of the mechanisms related to this technology are not new, since they are already employed by the oil and gas industries.

There is a pressing need to develop advanced clean energy technology since emissions of CO$_2$ will more than double by 2050. To solve this challenge, the International Energy Agency (IEA) prepared roadmaps to advance innovative energy technology (IEA-UNIDO, 2011). The main issue in regard to CCUS technologies is to match sources and reservoirs, and so it is fundamental to create large infrastructure to transport CO$_2$. Figure 2 illustrates the 2050 projected industrial emissions and the geological storage suitability.

As with any operation, particularly underground, there are uncertainties associated with CCUS schemes. One of the main concerns of CCUS is the escape of the injected CO$_2$ from geological formations. The risks associated with this and other uncertainties need to be systematically assessed and managed.

Risk assessment should be an integral element of the entire lifecycle of CCUS schemes from project conception, planning to delivery and long term operation and

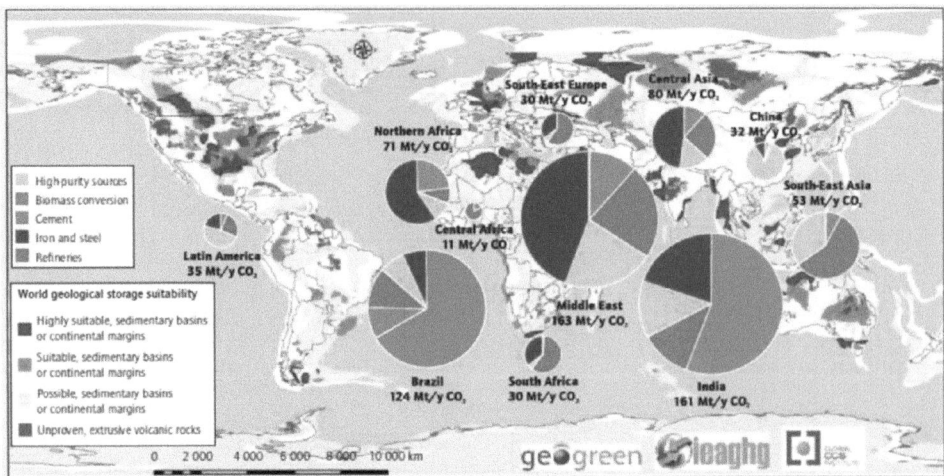

Figure 2 Matching sources and reservoirs (IEA-UNIDO, 2011).

termination. The risk assessment should include for example site selection, site characterization, storage system, design, monitoring and if necessary remediation. There are already several well-functioning CO_2 storage operations in depleted oil and gas reservoirs and in saline aquifers (IPCC, 2005; He *et al.*, 2011; Sousa, 2012a; EASAC, 2013a). The major issues to be considered are related with risks of CO_2 leakage that has consequences for safety and the environment. Leaks can occur as a result of pressure build-up caused by the injection of CO_2, and the possibility of induced seismicity and long-range impacts.

This chapter addresses different risks particularly those related to the storage in reservoirs and through wellbores (Sousa & Sousa, 2012; Grossmann & Dahmke, 2013). Different methodologies for risk assessment are described, and emphasis is placed on the use of Bayesian Networks (BN) for risk assessment. BNs are a graphical representation of knowledge for reasoning under uncertainty and they have become a popular representation for encoding uncertain expert knowledge in an expert system. BNs can be used at any stage of a risk analysis, and provide a good tool for decision analysis. Furthermore, they can be extended to influence diagrams, including decision and utility nodes in order to explicitly model a decision problem (Sousa, 2012b).

The chapter is organized as follows. A brief introduction to CCUS projects is presented, followed by technical aspects of injection and safety storage in different underground geological formations, including carboniferous formations, are referred to and analyzed in detail in section 2. In section 3, risk assessment and risk concepts are referred to followed by the reviewing and discussing known hazards for carbon sequestration. Also strategies to be followed will be presented to be followed in order to reduce risk and deal with disruptions in normal operations. In section 4, available analysis procedures, carried out mainly through numerical modeling at several scales are described, comprising the evaluation and inclusion of coupling processes. The development of methodologies for risk assessment is outlined for different hazard scenarios in section 5, with special emphasis in the use of BNs. An application of a BN related to activate faults is presented in section 6. In section 7, the particular problem of wellbore integrity is analyzed based on a methodology using dynamic BNs. Section 8 deals with updating risk by monitoring of reservoirs and wellbores. Finally, some conclusions are drawn in the last section.

2 CO₂ INJECTION AND STORAGE

The basic principle behind carbon capture, utilization and storage (CCUS) is to trap CO_2 generated by different sources and transport it to suitable underground geological sites as illustrated in Figure 3, thereby reducing atmospheric CO_2. The captured CO_2 must be stored in appropriate sites or geological reservoirs or using mineral carbonation (IPCC, 2005; Sousa, 2012a; He *et al.*, 2012; EASAC, 2013a, Xie *et al.*, 2015b). There are several reservoirs in nature where the CO_2 can be trapped, particularly in the pores of sedimentary rocks. It is necessary to consider several aspects when determining the suitability of a site for the underground storage of CO_2 (Gomes, 2010; Sousa &

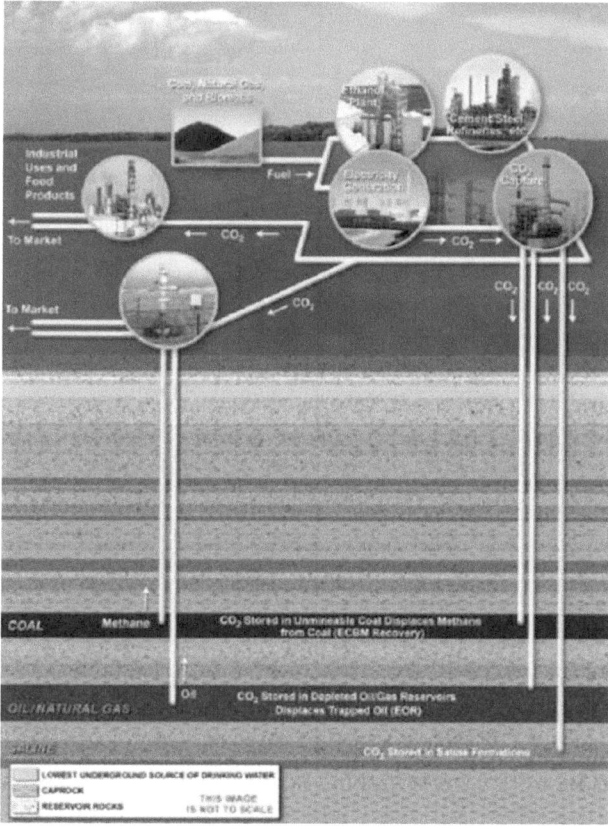

Figure 3 Illustration of a CCUS unit (cited in EASAC, 2013a).

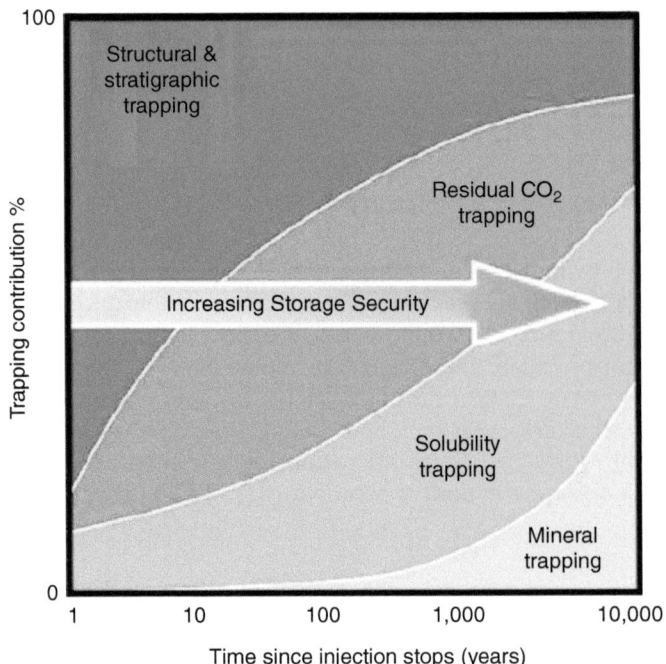

Figure 4 Evolution of the trapping mechanism with time (IPCC, 2005).

Sousa, 2012): the storage period should be long, preferably in the hundreds or thousands of years; the methodologies used cannot violate any national or international laws or regulations; formal environmental impact assessments should be made to ensure minimal impact; and risks should be systematically assessed and mitigated, for example to ensure the likelihood of accidents is very low. Once CO_2 is injected, it is trapped in the geological formations by mechanisms that change over time as illustrated in Figure 4 (IPCC, 2005; EASAC, 2013a).

The CO_2 injection and sequestration in carboniferous reservoirs can be performed either in deep unminable coal seams or in abandoned coal mines (Piessens, 2012; Ramos & Falcone, 2013; Sousa *et al.*, 2015). Coal formations contain cleats that impart some permeability to the system. Between cleats, coal has a large number of micropores into which gas molecules can diffuse and be absorbed. Gaseous CO_2 injected through wells will flow through the cleat system, diffuse into the coal matrix and be absorbed onto the micropore surfaces (Figure 5). The cleat spacing is usually uniform and ranges from the order of millimeters to centimeters (Shi & Durucan, 2005). If CO_2 is injected into coal seams it can displace methane gas, thereby enhancing coal bed methane recovery (Sousa, 2012a; Langfeld & Agasty, 2013; Xie *et al.*, 2015b).

CO_2 can also be stored in abandoned coal mines. The use of abandoned coal mines as reservoirs can be complex as illustrated in Figure 6. The problems associated with using these mines as reservoirs are discussed in Piessens (2012).

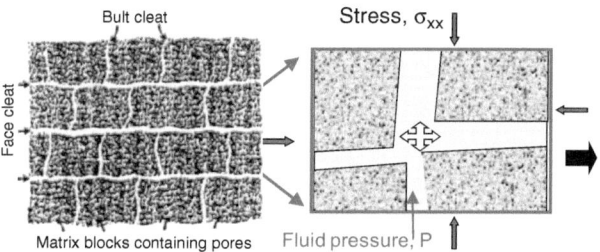

Figure 5 Micro structure of coal (Qu *et al.*, 2012).

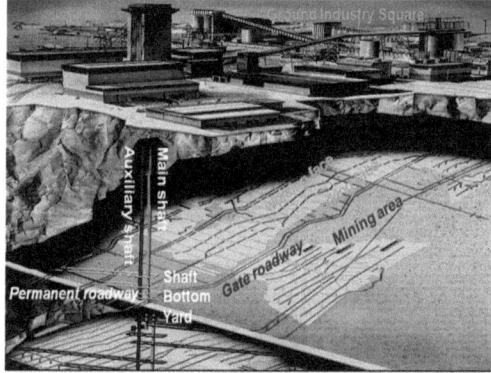

Figure 6 Coal mine in China (He, 2012).

3 RISK AND RISK ASSESSMENT

3.1 General

The systematic risk assessment and management of the lifecycle of CCUS projects is essential to ensure the safe delivery, operation and termination of such systems. The risk assessment begins with the site selection. During the design and planning stages, it is necessary to evaluate risks of the geological storage site itself, and information on site characterization of provides the input necessary for the risk assessment calculations. During the operation phase, many sources of risk exist, and it is necessary to evaluate the risks associated with the different sources. In particular, the risks associated with the possibility of CO$_2$ leakage, and reaching the surface and near-surface environment poses a significant risk (Figure 7). In order to evaluate these risks it is necessary to identify possibly pathways along which CO$_2$ can leak, to estimate both the likelihood of this happening, as well as the consequences if it did.

The risks associated with early stages of CO$_2$ storage in coal formations with or without methane production are discussed in Myer (2003); IPPC (2005); Sousa & Sousa (2012) and Sousa *et al.* (2015). In abandoned coal mines, risks are also associated sealing wells and shafts and these mines require major rehabilitation work before being

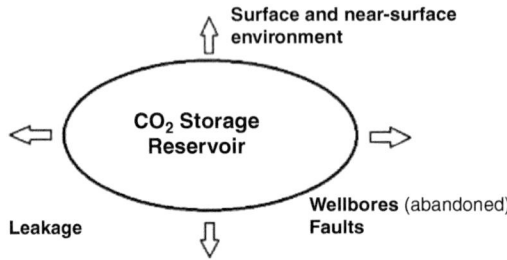

Figure 7 Geological CO_2 storage system.

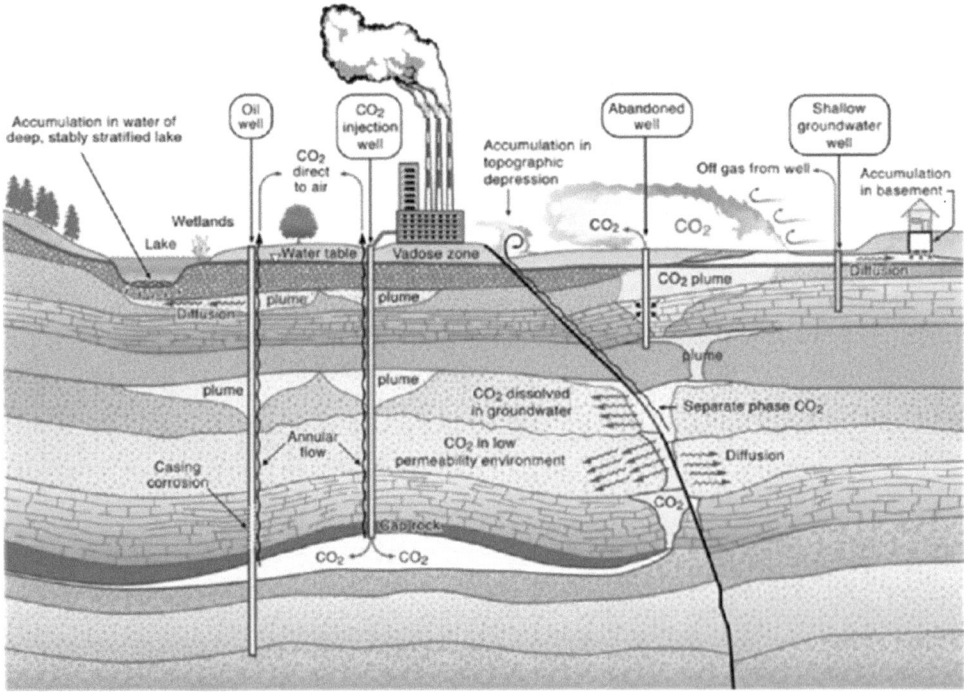

Figure 8 Potential escape mechanisms (Benson, 2005).

able to be used safely. Figure 8 provides a summary of some possible leakage mechanisms of CO_2. The causes of geomechanical problems, the risks and their factors are discussed in Sousa & Sousa (2012).

In general, the geological storage of CO_2 can have impacts on human health and safety; biodiversity; atmospheric environment (including GGH emissions); water (groundwater and surface water); geology (soils and underground space); waste (including construction debris); socio-economic impact on populations (Barros *et al.*, 2012). Because of these potentially detrimental impacts, it is necessary to systematically assess and manage the risks associated with the geological storage of CO_2.

Risk can be expressed in many different ways (see *e.g.* Einstein, 1988); and the definitions used in this paper are based on the ISSMGE recommended Glossary of risk assessment terms (TC32). Risk (or expected risk) can be expressed as:

$$E[R] = P[T] \times \sum_i P[C_i|T] \times C_i \tag{1}$$

where $E[R]$ is the expected risk, $P[T]$ the probability of threat 'T', or the hazard defined as the probability that a particular threat occurs within a given period of time; $P[C_i|T]$ the conditional probability that consequences C_i occur given the threat, or the vulnerability, and $u(C_i)$ the utility of consequence 'C_i'.

Consequences can be described with different attributes such as effect on humans in form of injuries or fatalities, destruction of infrastructure, impairment of operations, etc. A decision maker can define utility functions that describe his/her relative preference between attributes (Keeney & Raiffa, 1976; Howard & Abbas, 2015).

Risk assessment and risk management for CCUS systems requires, as an initial step, the identification of the threats that are possible which can lead to detrimental consequences. Evaluating the likelihood or probability of each threat forms the hazard assessment stage of the risk assessment and management. A comprehensive risk assessment should include the detailed assessment of each hazard.

3.2 Threat identification

The major threats associated with CCUS systems are shown in Table 1. These are based on several studies which include Myer (2003), IPCC (2005), Price *et al.* (2008), Sousa & Sousa (2012), and Grossmann & Dahmke (2013). For the storage of CO$_2$ in geological formation in coal eight distinct threats are identified as shown in Table 1.

The major threats are related to the safety of the selected site namely in terms of possible leakage which can occur as a result of the increase of pressure caused by the injection of CO$_2$, an eventual induced seismicity and/or by the long range impacts (EASAC, 2013a).

Existing pumping or injection wells increase the threat of potential escape problems of CO$_2$. Wells make the connection between the surface and subsurface reservoirs,

Table 1 Threat identification scenarios for CO$_2$ storage in coal.

Hazard	Description
R_1	Escape of CO$_2$ from pumping wells or injection of fluids
R_2	Slow and steady leakage of CO$_2$ from geological storage
R_3	Fast and large discharge of CO$_2$ from geological storage
R_4	Leakage from geological storage to groundwater
R_5	Leakage of CO$_2$ from geological storage to fossil fuel assets
R_6	Leakage of CO$_2$ that eliminates the benefits of geological storage
R_7	Induced fracturing or seismicity
R_8	Leakage from abandoned coal mines

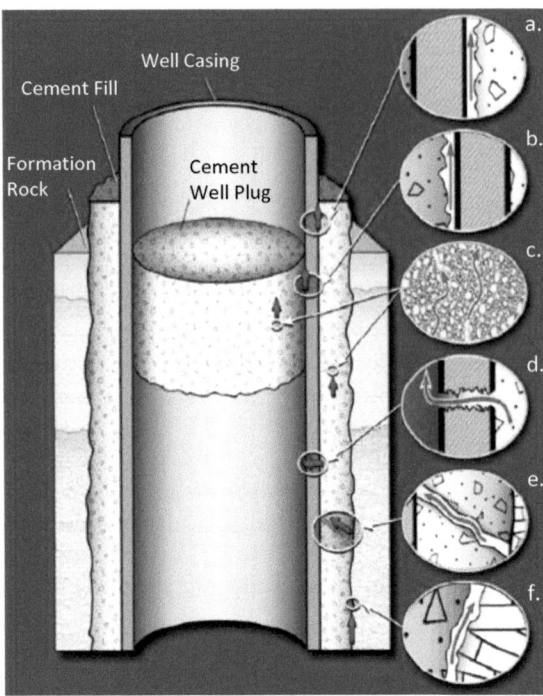

Possible escape pathways along wells:

a) Flow through the interface of the well casing and cement layer on the inside face of the coating.

b) There is an escape mechanism similar to the previous case but the flow occurs between the case and the cement.

c) Shows the mechanism of percolation of CO_2 through the cement seal.

d) and e) Flow crossing the final layer of concrete and masonry.

f) Represents another way of leakage, between the cement and strata surrounding the well.

Figure 9 Potential escape pathways along wells (Cited in Sousa *et al.*, 2015).

consequently crossing all rock strata and creating an eventual path for the leakage of CO_2 due to imperfections. All the elements of the pumping or injection system need to be studied to minimize the threat of CO_2 leakage. These include: the sealing caprock, the walls of the well and the annular of interface with walls, the first layer of cement case and the rock mass surrounding the wells (Sousa & Sousa, 2012; Bai *et al.*, 2015b). This situation corresponds to first scenario R_1 represented in Table 1. Figure 9e shows potential escape paths of CO_2 along wells.

Leakage of CO_2 from the reservoir may be slow and steady (threat R_2) or fast with large discharge of CO_2 (threat R_3). For threat R_2 the releases are too small to cause significant injuries or potential loss of life, but can have significant economic consequences. For threat R_3, on the other hand, a large amount of CO_2 can be released in a short period of time causing injuries and possible fatalities. A large number of fatalities is rare but has occurred namely in volcanically active areas such as at Lake Nyos in Cameroon (EASAC, 2013a). This is an example of a case where the probability of the threat, *i.e.* the hazard is low, but the consequences are very high, and therefore the risk is high. Mitigation measures can and should be taken in order to reduce the probability of such threats occurring (hazard). In the site selection phase, areas that are known to have faults, and areas around lakes should be avoided (Price *et al.*, 2008). In addition, sites should avoid locations near areas with high population density. If locating the reservoir near a large and deep lake cannot be avoided, a continuous monitoring system to observe levels of CO_2 concentration should be installed.

Threat R_4 relates to the migration of CO_2 from the reservoir to the surface potentially affecting shallow groundwater which is used for potable water and industrial and agricultural needs. Dissolved CO_2 forms carbonic acid which can have indirect effects such as the mobilization of (toxic) metals or sulfates, and possibly giving the water an odd odor, color or taste. In extreme circumstances, contamination might reach dangerous levels, prohibiting the use of groundwater for drinking or irrigation (IPCC, 2005). In such cases, it is necessary to develop inspection and monitoring stations for data collection and analysis. Analysis methodologies can include for example dynamic Bayesian Networks (Sousa & Sousa, 2012), among other techniques.

The threat R_5 corresponds to leakage to fossil fuel assets. The injection of CO_2 at high pressures can lead to the seepage of fossil deposits through faults and other discontinuities. The major risk is associated with the contamination of reservoirs which can as a consequence cause severe economic losses. Actions to reduce these risks include appropriate site selection away from fossil fuel assets and (or) that are likely to retain their CO_2 for a long period.

The threat R_6 corresponds to the leakage of CO_2 to the atmosphere that eliminates the benefits of geological storage and as a result the CCUS system is a redundant scheme with additional costs. In the case of the threat materializing it would be necessary to perform remedial measures that only act to increase costs further. A continuous monitoring system of the wells to ensure proper and time-invariant sealing is necessary.

The threat R_7 corresponds to induced fracturing or seismicity. Such events are, in general, very small, causing no damage. According to EASAC (2013a), induced seismicity has not been experienced in the field, although research on potential induced seismicity is underway. However, there is a potential for induced fracturing and seismicity in large-scale CCUS (Zoback & Gorelink, 2012). Actions suggested to reduce the risks of induced fracturing are discussed in Price et al. (2008).

Finally, threat R_8 is associated with leakage from abandoned coal mines which is a possibility for example in China due to the great number of such mines as discussed by He (2012). In coal mines, high concentrations of CO_2 can be reached by a sudden release of CO_2. CO_2 is denser than air and high concentrations can occur in confined areas near the surface and cause risks to humans, as has happened in volcanic lakes (EASAC, 2013a). The effect of active faults is an important issue to be considered (Beck & Franz, 2010; Sousa & Sousa, 2012). An abandoned coal mine when used as a reservoir can be considered as a long gallery conceptual model (Figure 10). The so-called Jar Pot Model provides a good image of the constraints and is an important tool to understand the pressure evolution in the mine once injection has stopped (Piessens, 2012).

Evaluating the likelihood or probability of each of the threats described in the previous section materializing forms the hazard assessment. There are various techniques to assess the hazards that range in complexity from being qualitative to semi-quantitative to quantitative. Probabilities can be assessed using:

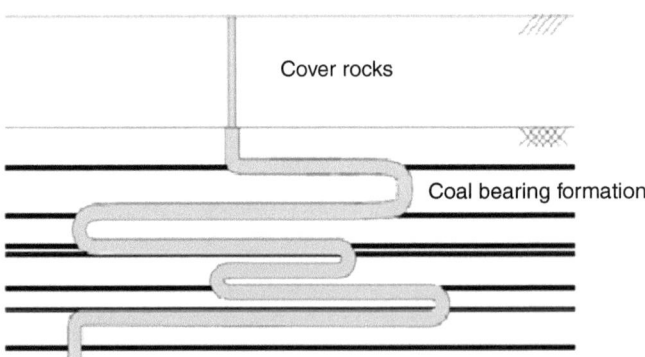

Figure 10 Simplified representation of an abandoned coal mine (Piessens & Dusar, 2003).

(a)   Statistical methods: these methods are based on a statistical analysis of historical data to evaluate hazards. It may be difficult to use such methods when evaluating the hazards of CCUS schemes given the limited amount of historical data available.

(b)   Heuristic methods: these are typically qualitative or semi-quantitative methods that rely on engineering experience and judgment. Absolute or relative values can be assigned to the hazards of different threats.

(c)   Physical and/or mechanical models: these engineering models attempt to represent the physical processes that occur in reality. Given the complexity of the phenomena these models are most frequently based on numerical methods and techniques as described in the following section.

4 NUMERICAL MODELING

Storage of large volumes in CO_2 in carboniferous reservoirs implies large-scale pressures that may cause fractures in the cap-rock and may drive CO_2 or brine leakage through localized pathways. Therefore predictions are required and a detailed analysis of the behavior of reservoirs is necessary (Vargas *et al.*, 2012; EASAC, 2013a). Analytical tools should be able to simulate the short and long processes for the CO_2 reservoirs in a similar way as for the oil and gas reservoirs. Numerical models are commonly used today for predicting the behavior of these reservoirs. Hazard and risk analyses may be carried out at the wellbore scale and at the reservoir scale and should be performed simulating different scenarios with parametric and sensitivity analyses and using numerical commercial codes or in-house developed codes.

It is fundamental to identify and to describe the involved processes that comprise physical, chemical and biological processes as well as their coupling as well (Elsworth *et al.*, 2012; Qu *et al.*, 2012). A detailed analysis of these problems is presented in Vargas *et al.* (2012), as well as the formulations for the numerical modeling on the rock mass continuum and pore scales.

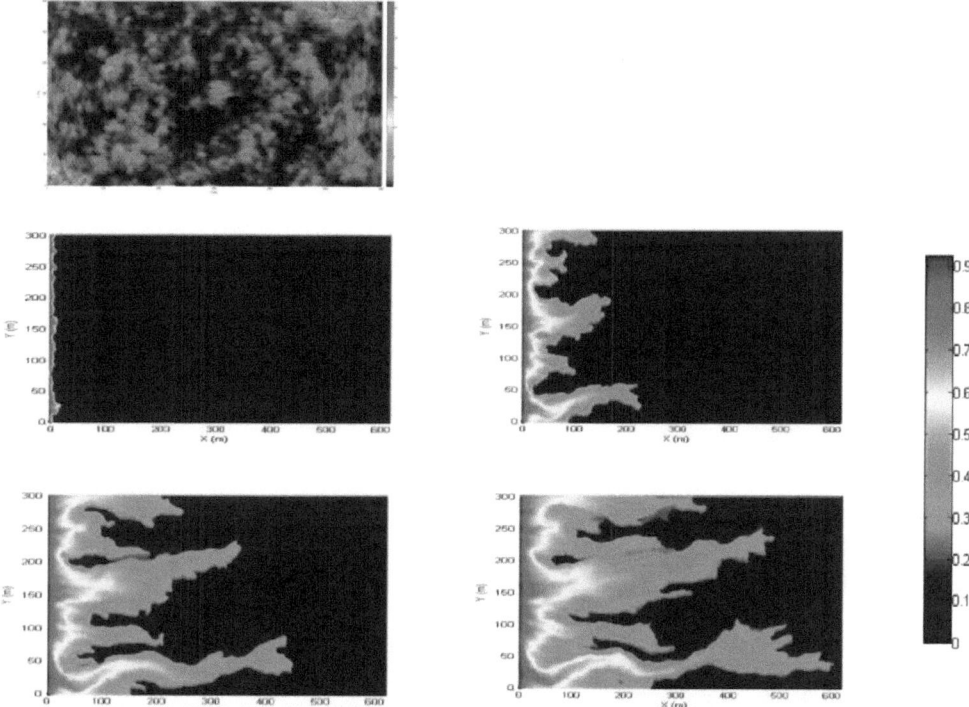

Figure 11 Geometry of a porous reservoir (properties given in Table 2) with space varying permeabilities where a non-wetting fluid is displaced by a wetting fluid. Images represent 100, 200
and 300 days of simulation time. Injection occurs from the left-hand side. Colors represent
permeabilities (upper left-hand side) and saturations of the injected fluid.

In the following, examples are presented that illustrate some of the difficulties
encountered in the numerical modeling of the above mentioned processes at
different scales. Figure 11 presents the injection of wetting phase into a porous
reservoir containing a non-wetting phase. The simulation in this case is carried out
using finite elements for two-phase flow incorporating a Raviart-Thomas (RT)
post-processing for velocities and Discontinuum-Galerkin (DG) techniques for the
wetting phase saturation (Passos *et al.*, 2014). These procedures guarantee mass
conservation in the system. The porous medium displays space varying permeabilities given by a stationary stochastic field, log-normal distribution (Murad
et al., 2013). The simulation properties are given in Table 2. It is clear from
Figure 11 that heterogeneities can give rise to the growth of fingers that will affect
the geometry of the regions in the reservoir to be reached by the injected fluid.
Figure 12 shows that this pattern of flow concentration appears to be more
flagrant when the medium contains fractures/faults occur in the medium as
shown in the medium. The properties are given in Table 3.

At the pore structure level, an example is presented related to the injection of a non-
wetting fluid into a fissured geometry representative of a coal micro-structure

Table 2 Properties of porous medium.

Domain dimensions	$600\,\text{m} \times 300\,\text{m}$
Rock properties	$\phi = 0,3$
	$k^m = 1\,\text{a}\,30 \times 10^{-13}\,m^2$
Fluid properties	Viscosities $\mu_w = 1\text{cp}, \mu_n = 10\text{cp}$
	$\rho_w = \rho_{nw} = 1000\text{kg/m}^3$
Relative permeabilities	Brooks Corey model
	$K_{rw} = Se^m, K_{rnw} = (1 - Se)^m, m = 2$
Residual saturation	$S_{rw} = 0, S_{rn} = 0,0$
Injection rate	30 m³/day
Simulation time	300 days
Mesh geometry	9995 triangles

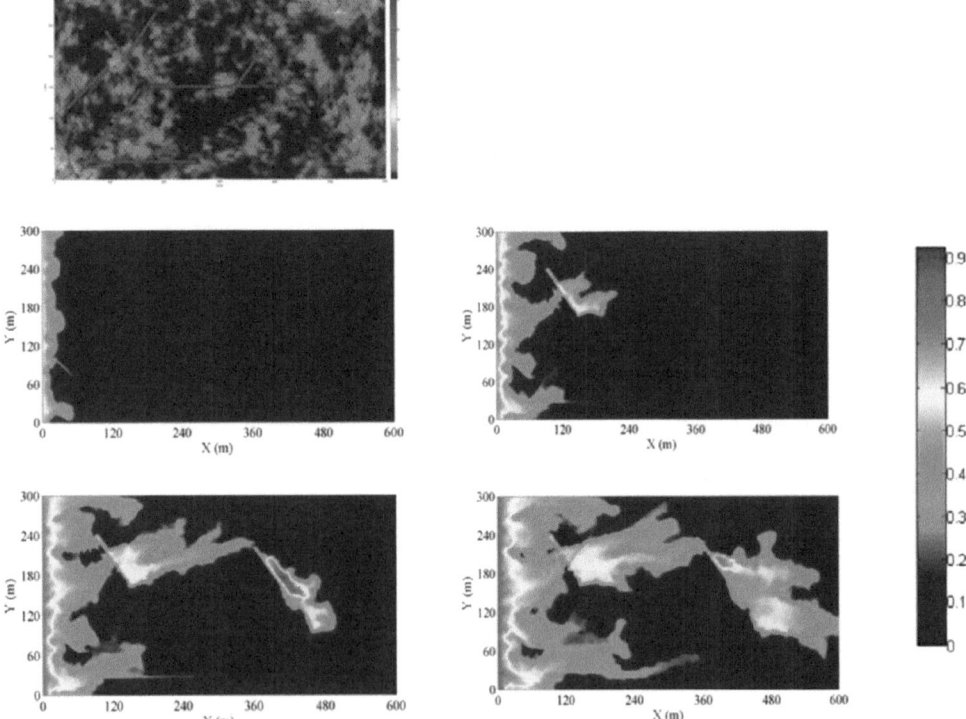

Figure 12 Geometry of a porous reservoir (properties given in Table 3) containing fractures with space varying permeabilities where a non-wetting fluid is displaced by a wetting fluid Injection from the left-hand side. Images represent 100, 200 and 300 days of simulation time. Injection occurs from the left-hand side. Colors represent permeabilities (upper left-hand side) and saturations of the injected fluid.

considering Lattice-Boltzmann methods (Vargas *et al.*, 2012). Figure 13 shows the geometry used in the analysis of the injection process of CO_2 which is representative of cleats of coal formations. The diffusion of CO_2 is not considered but these processes can also be modeled.

Table 3 Properties of porous medium and fractures.

Domain dimensions	$600\,\text{m} \times 300\,\text{m}$
Rock properties	$\phi = 0,3$
	$k^m = $ variable between $1\,\text{a}\,30 \times 10^{-13}\,m^2$
Fracture properties	$\phi^f = 1,0$, $K^f = 8 \times 10^{-7} m^2$; $e = 1\,mm$
Fluid properties	$\mu_w = 1\text{cp}$, $\mu_n = 10\text{cp}$
	$\rho_w = \rho_{nw} = 1000\text{kg/m}^3$
Relative permeabilities	Brooks-Corey model
	$K_{rw} = Se^m$, $K_{rnw} = (1 - Se)^m$, $m = 2$
Residual saturation	$S_{rw} = 0$, $S_{rn} = 0,0$
Injection rate	$30\,\text{m}^3/\text{day}$
Simulation time	300 days
Mesh geometry	9995 triangles and 105 fracture elements

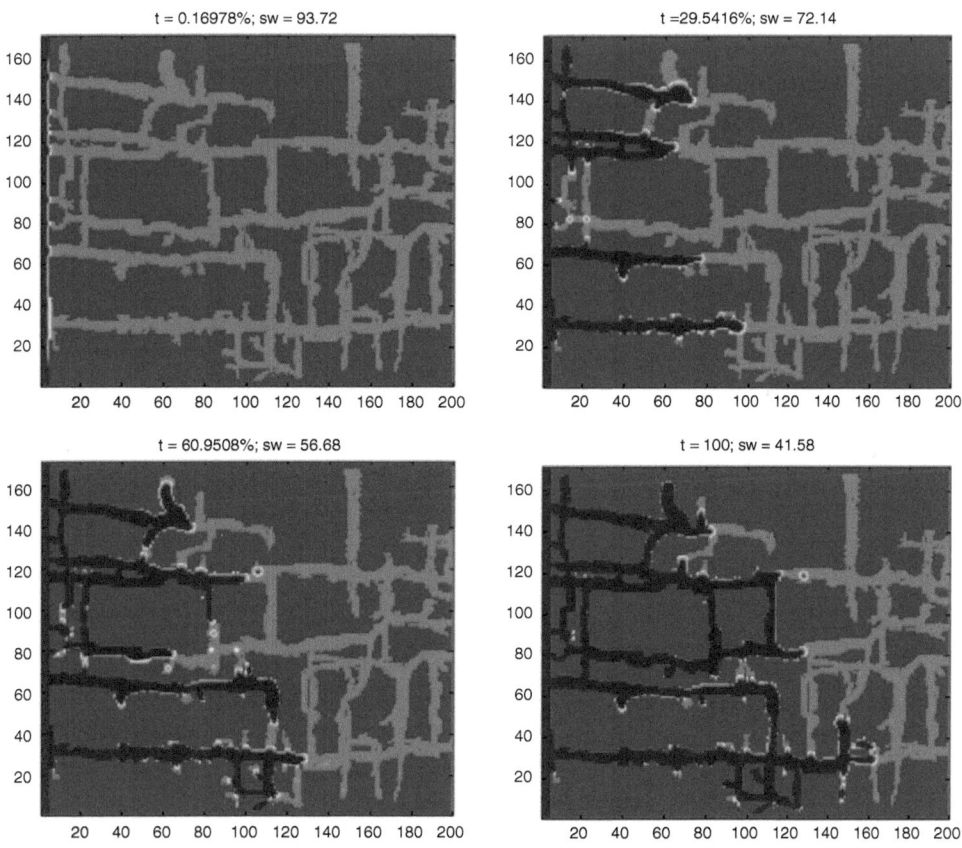

Figure 13 Injection process of CO_2 in coal formation cleats (Vargas *et al.*, 2012).

The development of techniques and strategies is necessary for a better site characterization before injecting the CO_2. In addition, monitoring plans are needed to validate the conceptual and numerical models (IPCC, 2005; EASAC, 2013a).

5 DEVELOPMENT OF FORMAL RISK ASSESSMENT METHODOLOGIES

5.1 Introduction

The systematic risk assessment and management of the lifecycle of CCUS projects is essential to ensure the safe delivery, operation and termination of such systems, and various methodologies exist to do this.

Decision theory and decision analysis provides a framework that can, and has been used in engineering problems to improve the decision making process. Decision analysis uses decision theory principles supplemented by judgment psychology (Henrion *et al.*, 1991). The application of decision analysis to problems of underground engineering has been described in the context of landslide risk assessment in Einstein & Karam (2001), and tunnel exploration planning in Karam *et al.* (2007a,b). It also provides a framework by which complex decision problems can be simplified, allowing one to more clearly make decisions based on alternatives (Howard & Abbas, 2015). Decision analysis incorporates uncertainty, values, and preferences in a structure that models a decision (Howard & Matheson, 1984). The process is actually also the standard decision making process used in engineering in which one determines parameters, includes them in models and makes decisions based on the model results. The updating cycle can represent the observational method in geotechnical engineering (Einstein, 1988).

In more complex problems with an incomplete understanding of the many factors and processes involved and the interaction between them, and with the recent advancements in computer processing power, other techniques have been used in risk assessment. These include methodologies for data analysis and representation, such as rule-based systems, fuzzy-rule-based systems, artificial neural networks, and Bayesian networks (BN). BN has been used in risk analysis in Sousa (2012b) and Sousa *et al.* (2015). Knowledge representation systems (or knowledge based systems) use various computational techniques of AI (Artificial Intelligence) for representation of human knowledge and inference. More recently there has been a resurgence of interest among many AI researchers in the application of probability theory, decision theory and analysis to several problems, resulting in the development of BN and influence diagrams, an extension of BN designed to include decision variables and utilities. A description of the main methodologies is presented in the following (Solomon, 2007; Sousa, 2012b).

Rule-Based Systems or KBS (Knowledge-Based Systems) are computer models of experts in a certain domain. The building blocks for modeling the experts are called production rules (Sousa, 2012b). A production rule is of the form: If A then B, where A (premise) is an assertion, and B (conclusion) can be either an action or another assertion. A rule-based system consists of a library of such rules. These rules reflect essential relationships within the domain, or they reflect ways to reason about the domain. When specific information about the domain comes in, the rules are used to draw conclusions and to point out appropriate actions. One of the major problems of rule-based systems is how to treat uncertainty. There are many schemes for treating uncertainty in rule-based systems and the most common are fuzzy logic, certainty factors and Dempster – Shafer belief functions (Russell & Norvig, 2003). However, it is not easy to capture reasoning under uncertainty with inference rules for production rules. The reason for this is that in all the schemes for treating uncertainty, mentioned above, the uncertainty is treated locally. More specifically, it is difficult to combine uncertainties from different rules. Despite their shortcomings, KBS have been used in many applications in different domains, such as Medicine, Banking, Aerospace Engineering for scheduling operations and in Civil Engineering for the recommendation system MATUF in the maintenance and repairing of tunnels (Sousa *et al.*, 2009) and as recommendation systems for repairing bridges (Sousa, 2000), among others.

There are many schemes for treating uncertainty in rule based systems. Fuzzy logic is one way of introducing uncertainty into rule based systems. It is a superset of conventional logic that has been extended to handle the concept of "partial truth", *i.e.* a value between (completely) true and (completely) false (Zadeh, 1999). Based on fuzzy logic, fuzzy rule expert systems were created. They use a collection of fuzzy membership functions and rules drawn-out from the experts (Sousa, 2012b; Sousa *et al.*, 2015). The inference mechanisms of these rules have some weaknesses; they have a weak theoretical foundation, inconsistency and sometimes oversimplification of the real world. Despite their shortcomings, fuzzy logic has been applied to several domains. In geotechnical engineering an application of fuzzy logic is the use of Fuzzy set rules in rock mass characterization (Sonmez *et al.*, 2003; Sousa, 2012b).

Another methodology is based on Artificial Neural Networks (ANN) (Russell & Norvig, 2003; Suwansawat & Einstein, 2006). ANN are interconnected groups of artificial neurons, similar to the network of neurons in the human brain, that use a computational model for information processing based on a connectionist approach to computation. An ANN consists of multiple layers of single processing elements called neurons and of their connections. Each neuron is linked to some of its neighbors with a varying coefficient of connectivity (weight) that represents the strength of these connections. This is stored as a weight value on each connection. The ANN learns new knowledge by adjusting these weights and the connections between neurons. The ANN rely on data to be trained, adjusting their weights and connections to optimize their behavior as pattern recognizers, decision makers, system controllers, predictors, etc. The strength of these models is their adaptability, without requiring a deep knowledge about the complex relationships of the domain of application. This adaptability allows the system to perform well even when the system that is being modeled, or controlled, changes over time. The objective of using an ANN is to make predictions in the future. Although an ANN network could provide almost perfect answers to the set of data with which it was trained, it may fail to produce an adequate answer when "new" data surfaces. This is a result of 'overfitting' (Suwansawat & Einstein, 2006). In order to perform adequately and produce good results, these systems require a large number of sample data in order to be trained. Also, since there is not a complete understanding of the learning process, the analysis of the results may be difficult. Thus, this is not the right approach in cases in which one needs to have a complete understanding of the problem domain and relationship among variables of the domain.

Classic tools of decision theory include influence diagrams, fault trees, event trees and decision trees, and these can be used to help visualize the decision problem graphically. Fault trees and event trees can be used to model on one hand the different ways an event can occur (fault tree) and on the other hand, systematically identify the possible sequence of events and their consequences (event tree) (Eskesen *et al.*, 2004). Fault trees and event trees (or decision trees) can be combined. Note that the same fault tree can be combined with a decision where one can assess whether or not it would be worth taking measures to avoid or mitigate damage. A decision tree is a formal representation of the various components of a decision problem. It consists of a sequence of decisions, namely a list of possible alternatives; the possible outcomes associated with each alternative; the corresponding probability assignments; monetary consequences and utilities (Ang & Tang, 1975). The valuation of an outcome can be done using utility theory. Utility is an arbitrarily set dimensionless transformation function that describes the relative preference of the decision maker toward different outcomes, as well as the decision maker's attitude toward risk. The utility function for a risk neutral decision maker is linear whereas that for a risk adverse decision maker is exponential.

Utility functions are commonly based on monetary values for simplicity, but can also be based on other values such as those for environmental effects. Multi-attribute theory provides a means to combine different measures of preference into a single scalar utility value that represents the relative preference of the outcomes while considering the decision maker's attitude toward risk.

While decision trees have the advantage of being a graphical representation of the decision problem, and hence are more intuitive to the decision maker, they can quite quickly become very cumbersome, and difficult to represent.

5.2 Bayesian Networks (BN)

A BN, also known as belief network, is a graphical representation of knowledge for reasoning under uncertainty. Over the last decade, BN has become a popular model for encoding uncertain expert knowledge in expert systems (Heckerman, 1997). BN can be used at any stage of a risk analysis, and may substitute both fault trees and event trees in logical tree analysis. While common cause or more general dependency phenomena pose significant complications in classical fault tree analysis, this is not the case with BN. They are in fact designed to facilitate the modeling of such dependencies. Because of what has been stated, BN provide a good tool for decision analysis. Furthermore, they can be extended to influence diagrams, including decision and utility nodes in order to explicitly model a decision problem.

The concepts of Bayes' theorem are essential for BN (Ang & Tang, 1975). A BN is a concise graphical representation of the joint probability of the domain that is being represented by the random variables, consisting of (Russell & Norvig, 2003): i) a set of random variables that make up the nodes of the network; ii) a set of directed links between nodes (the links reflect cause-effect relations within the domain; each variable has a finite set of mutually exclusive states; iii) the variables together with the directed links form a directed acyclic graph; and iv) attached to each random variable A with parents $B_1, \ldots, B_n$ there is a conditional probability table $P(A = a|B_1 = b_1, \ldots, B_n = b_n)$, except for the variables in the root nodes (root nodes have prior probabilities).

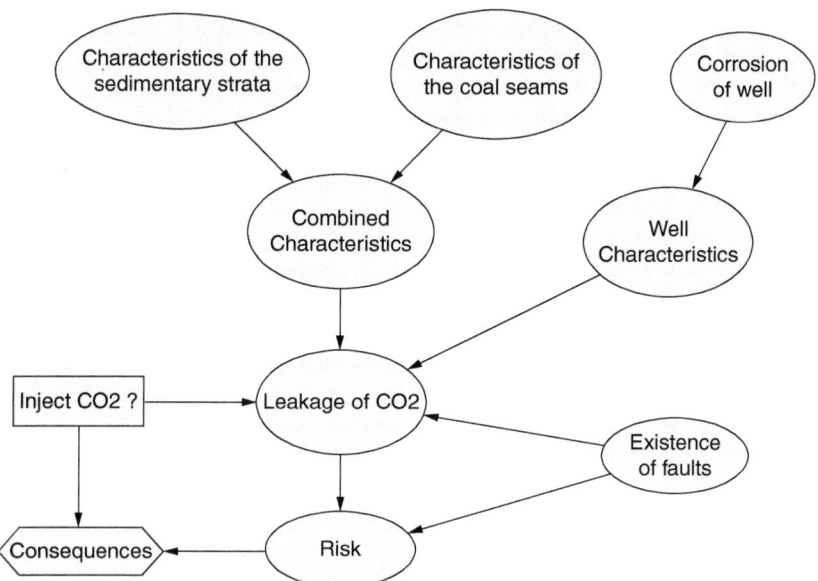

Figure 14 BN for risk analysis of storage of CO$_2$ (Sousa & Sousa, 2012).

Figure 14 is an illustration of a BN for risk analysis due to CO$_2$ injection in carboniferous formations. The arrows going from one variable to another reflect the relations between variables.

A BN is a joint probability distribution of all the variables, taking into account that some variables are conditionally independent. The simplest conditional independence relationship encoded in BN is that a node is independent of any ancestor nodes given its parents, *i.e.* that a node only depends on its direct parents. Thus, the joint probability of a BN over the variables U = {A$_1$,..., A$_n$}, can be represented by the chain rule:

$$P(U) = \prod_{i}^{n} P(A_i = a_i | parents\ (A_i))$$ (2)

where "parents (A$_i$)" is the parent set of A$_i$.

Since a BN defines a model for variables in a domain and their relationships, it can be used to answer probabilistic queries about them. This is called inference. The most common types of queries are the following:

(a)   A *priori probability* distribution of a variable.

$$P(A = a) = \sum_{X_1} ... \sum_{X_k} P(x_1, ..., x_k, A = a)$$ (3)

where A is the query-*variable* and X$_1$ to X$_k$ are the remaining variables of the network. This type of query can be used during the design phase of a tunnel for example to assess its probability of failure for the design conditions (geology, hydrology, etc.).

(b)   *Posterior* distribution of variables given evidence (observations). This query consists of updating the state of a variable (or subset of variables) given the observations (new information).

$$P(A = a|e) = \frac{P(A = a, e)}{\sum_{X_1} \cdots \sum_{X_k} \sum_{A} P(x_1, \ldots, x_k, A = a, e)} \qquad (4)$$

where e is the vector of all the evidence, and A is the query variable and X_1 to X_k are the remaining variables of the network. This type of query is used to update the knowledge of the state of a variable (or variables) when other variables (the evidence variables) are observed. It could be used, for example, to update the probability of failure of a tunnel, after construction has started and new information regarding the geology crossed becomes known (Sousa & Einstein, 2012).

The most straightforward way to make inference in a BN, if efficiency were not an issue, would be to use the equations above to compute the probability of every combination of values and then marginalize out the ones one needed to get a result. There are several algorithms for efficient inference in BN, and they can be grouped as follows: exact inference methods and approximate inference methods. The most common exact inference method is the Variable Elimination algorithm that consists of eliminating (by integration or summation) the non-query, non-observed variables one by one by summing over their product. This approach takes into account and exploits the independence relationships between variables of the network. Approximate inference algorithms are used when exact inference may be computationally expensive, such as in temporal models, where the structure of the network is very repetitive, or in highly connected networks.

There are several types of BN. Table 4 summarizes different types of BNs, indicating the more significant characteristics. An application of a DBN was presented in Sousa &

Table 4 Different types of BNs.

Type of BN	Characteristics
Discrete	a. Each node is an event described by a random variable that may have several states. b. All nodes coupled with parent ones are defined by the table of conditional probabilities or by the function of conditional probabilities. c. For the nodes without parents, the probabilities of states are unconditional.
Continuous	a. In many cases events are described by a continuous random variable. b. Are used to simulate stochastic processes in the state space with continuous time.
Hybrid	a. Combines discrete random variables with continuous random variables. b. Discrete variables cannot have continuous parents. c. Continuous variables should have a normal law of distribution based on conditional probabilities. d. Continuous variable X with discrete parents Y and continuous parents Z is normally
Dynamic (DBN)	a. Combines in general two Bayesian models, one for risk assessment and other for decision making. b. Are used to simulate processes with time.

Sousa (2012) for the hazard R4 (Leakage from geological storage to groundwater) and also was applied with success to the Porto Metro where three collapses occurred during construction (Sousa, 2010; Sousa & Einstein, 2012).

6 SIMULATING AN ACTIVATE FAULT USING BAYESIAN NETWORKS

An example of a BN is presented to illustrate their potential use for risk analysis in CO$_2$ injection processes. The first example was developed for a situation where one is looking at different mitigation measures, for reducing the leakage of CO$_2$, due to induced fracturing (R$_4$, R$_5$ and R$_7$), assessing the risk of each option and choosing the one that minimizes it.

Figure 15 presents the BN model for the initial state, where no information is gathered, and consists of 10 change/deterministic nodes, 3 utility nodes and one decision node. The probability tables associated with the chance nodes of the BN are presented in an Annex. The decision choices are: stop injection; reduce injection; and no action.

The Chance Nodes are defined as follows:

i)   *Sedimentary strata conditions*, where three possible states were considered: *Good*, *Bad* and Very *Bad*.
ii) *Coal seam*, with three possible states: *Good*, *Bad* and *Very Bad*.
iii) *Combined characteristics*, which combines the properties of the sedimentary strata and the coal seams. The values were attributed in function of the

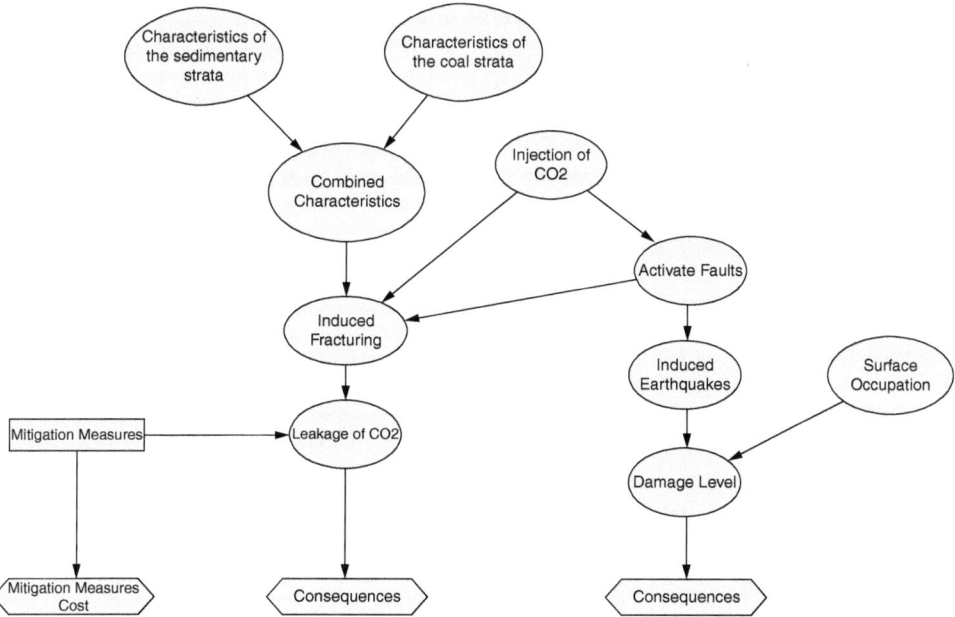

Figure 15 BN for risk analysis of storage of CO$_2$ with an activate fault (Cited in Sousa & Sousa, 2012).

properties defined to both formations, with the following possible states: *Good*, *Bad* and *Very Bad*.

iv)   *Activate faults* which represent the possibility of faults, existing within the reservoir, being activated by the injection of CO_2. The possible states are: *Yes* and *No*

v)   *Induced fracturing* was considered to represent the activation of faults during the injection process. This variable can take on two distinct states: *Yes* or *No*.

vi)   *Leakage of* CO_2, which considers the possibility of CO_2 leaking from theinjection site. It depends on the characteristics of both formations involved (coal seams and sedimentary strata), the existence of faults, and of course whether or not CO_2 is injected. Its possible states are: Yes and No.

vii) *Induced earthquakes*. This node represents the possibility of induced earthquakes due to the process of CO_2 injection and the activation of faults. Two distinct states (*Yes* or *No*) were considered.

viii) *Damage Level* represents a random variable assigned to the damage caused if an earthquake occurs, and can take on the possible states "No Damage", "Level 1", "Level 2", each with an associated probability.

The Deterministic Nodes are:

i)   *Injection of* CO_2, represents the process of injection of CO_2 into the reservoir/well. It can take on the values: *Yes* or *No*.

ii)   *Surface Occupation*, represents the type of occupation at the surface. It can take on values: *remote* or *near a population*.

The Decision Node is:

i)   *Mitigation Measure*: represents the different decision options, in this case the options are to "Stop Injection", "Reduce Injection" and "No Action".

Utility Nodes are:

i)   The nodes *Consequence* represent the different utilities associated with the consequences of each level of damage, "No Damage", "Level 1" and "Level 2", or the consequences associated with Leakage of CO_2.

ii)   The node *Mitigation measures Cost* represents the costs associated with the three decision options, "Stop Injection", "Reduce Injection" and "No Action".

Numerical simulations were performed using the software Genie (http://genie.sis.pitt.edu/-downloads.html). Two cases of different Geomechanical conditions were considered as illustrated in Table 5. In case 1 the geomechanical characteristics of the sedimentary strata are considered to be *Bad* and those of the coal strata are considered *Good*. Case 2, represents a situation where both the geomechanical characteristics of the sedimentary and the coal strata are *Very Bad*.

Figures 16 and 17 illustrate the results of BN calculations for cases 1 and 2, presented in Table 5. The results of the show that: i) for case 1, the decision that minimizes the risk is No action, ii) for case 2, the optimal mitigation measure is to reduce the injection of CO_2.

Table 5 Different hypothesis considered in the BN.

Case	Characteristics sedimentary strata	Characteristics of coal strata	Injection of CO$_2$	Surface occupation	No Action	Reduce Injection	Stop Injection	Optimal mitigation measure
1	Bad	Good	yes	near population	−71.125	−71.25	−84.25	No action
2	Very bad	Very bad	yes	near population	−90.25	−82.5	−86.5	Reduce Injection

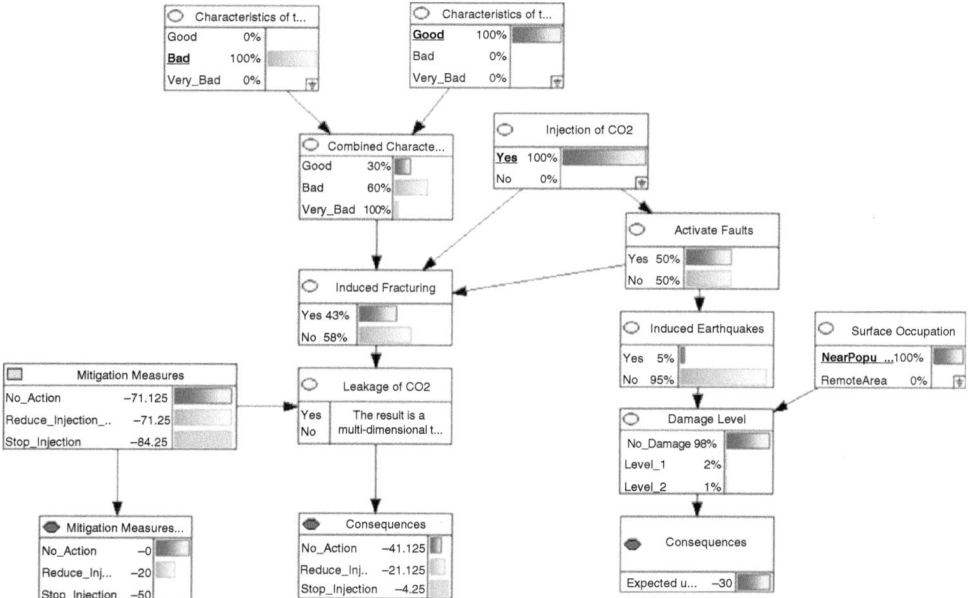

Figure 16 Case 1 BN results.

Sensitivity analysis

Sensitivity analysis were performed to investigate the effect of the mitigation measures, P (induced earthquakes| activate faults = yes) and the P (induced fracturing |injection CO$_2$=yes, Activate Faults=yes). The results are presented in Figures 18 to 21.

7 WELLBORE INTEGRITY ANALYSIS USING DYNAMIC BAYESIAN NETWORKS

7.1 Methodology for wellbore integrity

The existence of pumping wells or injection of fluids is a major source of potential escape problems of CO$_2$. This situation corresponds to the hazard R$_1$ (Figure 22). It is

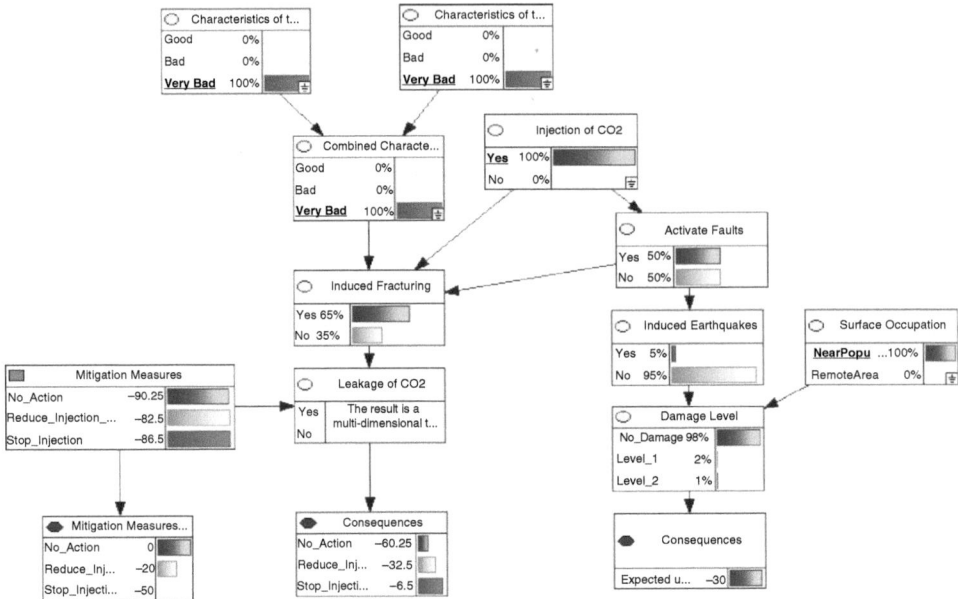

Figure 17 Case 2 BN results.

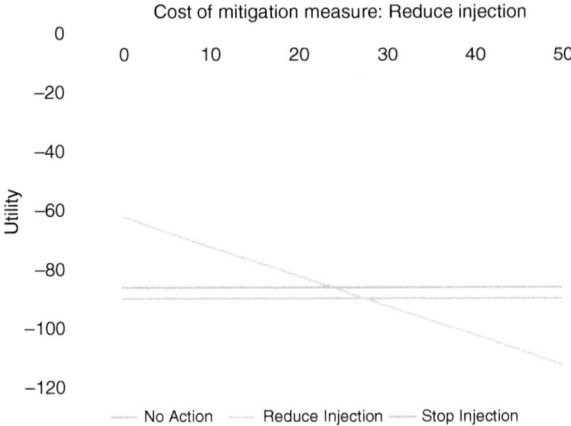

Figure 18 Varying cost of mitigation measure: Reduce injection.

one of the most important risks because the wellbore can form a direct liaison between the reservoir and the surface and CO_2 can react with the cement of the different barriers, primary and secondary (Figure 23). Wellbore integrity analysis involves measuring and assessing the quality of the barriers using pressure tests and cement

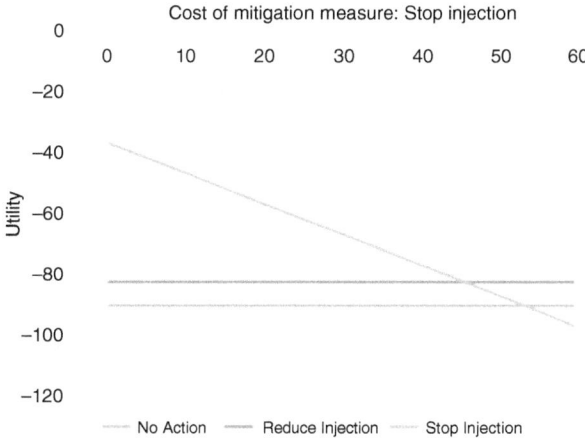

Figure 19 Varying the cost of mitigation measure: Stop injection.

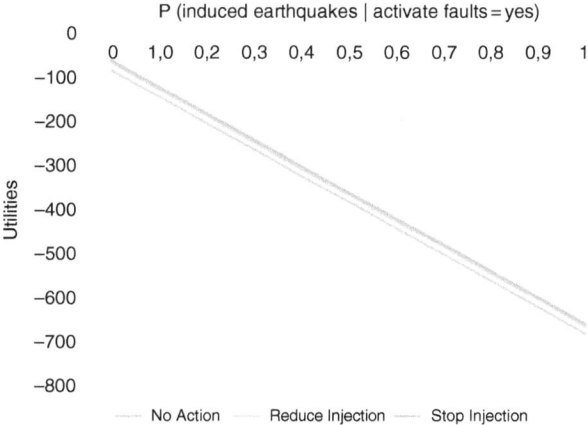

Figure 20 Varying P (induced earthquakes| activate faults = yes).

evaluation documentation (Reinicke & Fichter, 2010; Brecht *et al.*, 2013; Hou *et al.*, 2013; Vercelli *et al.*, 2014; Bai *et al.*, 2015a). The abandoned wellbores that are closed at the end of their working time deserve a special attention due to the fact they were closed in different periods in the past using different criteria. In consequence complex and iterative studies have to be studied using numerous measuring techniques (Reinicke & Fitcher, 2010; Bai & Reinickes, 2013).

To assess this situation a wellbore integrity methodology was established using DBN (Sousa & Sousa, 2013; Sousa *et al.*, 2015). In the schematic example presented in this section the wellbore integrity problem is modeled with time coupled with monitoring of the wellbore.

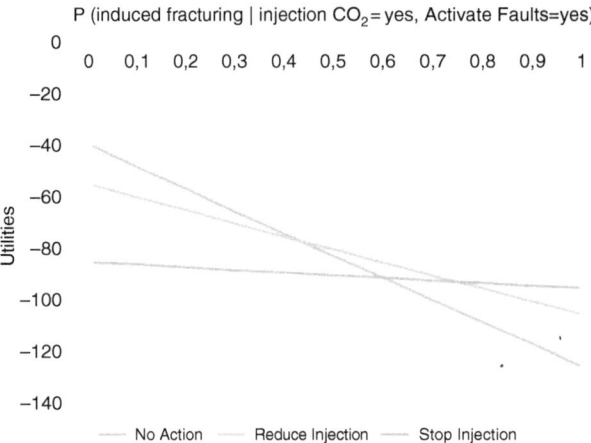

P (induced fracturing | injection CO$_2$= yes, Activate Faults=yes)

No Action — Reduce Injection — Stop Injection

Figure 21 Varying P (induced fracturing |injection CO$_2$=yes, Activate Faults=yes).

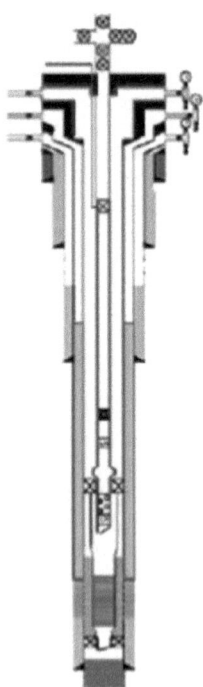

Figure 22 Wellbore highlighting different barriers (Vercelli *et al.*, 2014).

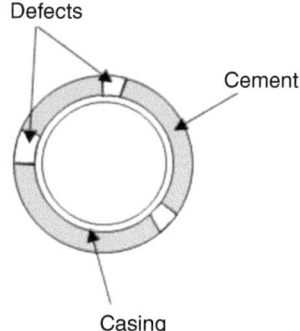

Figure 23 Interactions between steel casings, cement with defects and the formation.

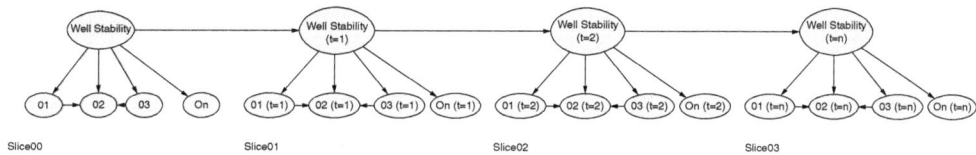

Figure 24 Modeling the integrity of a wellbore.

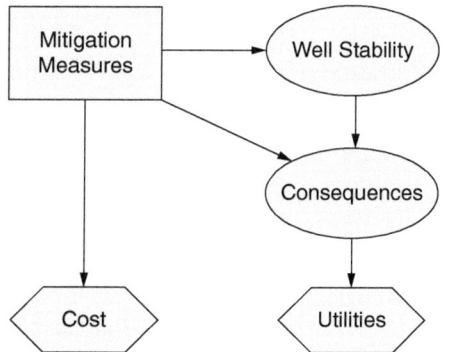

Figure 25 Bayesian model based on the measurements of the wellbore.

Two models can be built. One for the modeling the integrity of the wellbore takes into accounts the monitored measurements (Figure 24). The variables O represent the observations or results of tests performed in the well (integrity tests) that will allow one to infer regarding the wellbore integrity. The other model, the Bayesian decision model, is indicated in Figure 25. The decision will be made on the optimal remedial solution for

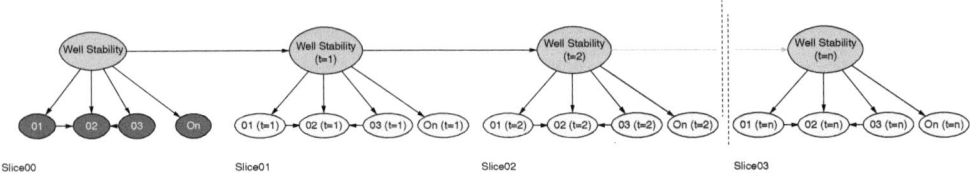

Figure 26 Modeling the integrity for slice 00.

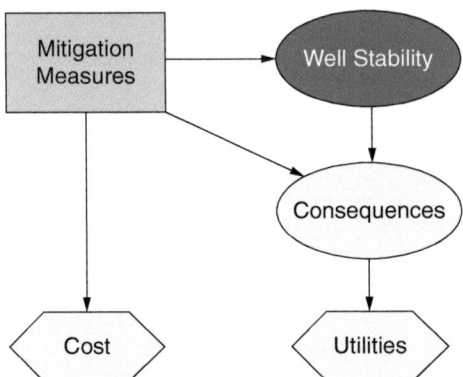

Figure 27 Bayesian model for the wellbore stability based on the measurements.

an eventual problem that passes through a decision about mitigation measures of wellbore.

The way the integrity methodology works is as follows (Figure 26):

i)   Monitoring and observation is made at time slice00 and entered in the network.
ii)   The evidence is propagated through the network and the probability of occurrence of an integrity problem is determined.
iii) The evidence is propagated through into the future and the probability is evaluated for the next slices.

Once the prediction model is executed, these results can be input into a decision model to determine the optimal remedial measures. An optimal remedial solution can be also doing nothing, *i.e.* no remedial solution. Figure 27 shows a decision model which evidence permits to evaluate wellbore stability based on the measurements. As time progresses the same steps are repeated.

7.2 Example illustrating the methodology

An example, designed to illustrate this technique is presented below. For illustrative purposes a simplified DBN was developed, containing only one observation node, *Pressure measurements*, which represents a situation where annular pressures are being monitor in order to assess wellbore integrity. Figure 28 shows the first three slices of the DBN, and associated probability tables. The DBN contains two chance

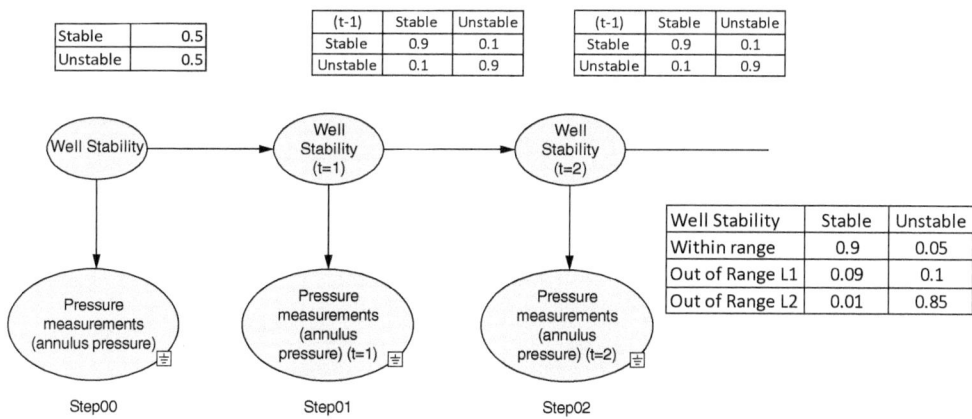

Figure 28 Pressure monitoring DBN.

nodes: i) *Well stability* which represents the integrity of the well. It can take on the values *Stable* and *Unstable*; ii) *Pressure measurements*, which represents the monitoring of annular pressures in the well. It can take on the possible values *within range (i.e.* within expected range), *Out of range L1* (out of expected range: alarm level 1), out of range L2 (out of expected range: alarm level 2, which represents a larger deviation from the expected pressure range than level 1).

For this example three observations were made at time t1, t2 and t3. The pressure readings were: t1: within range; t2: out of range L1; t3: out of range L2. Figure 29

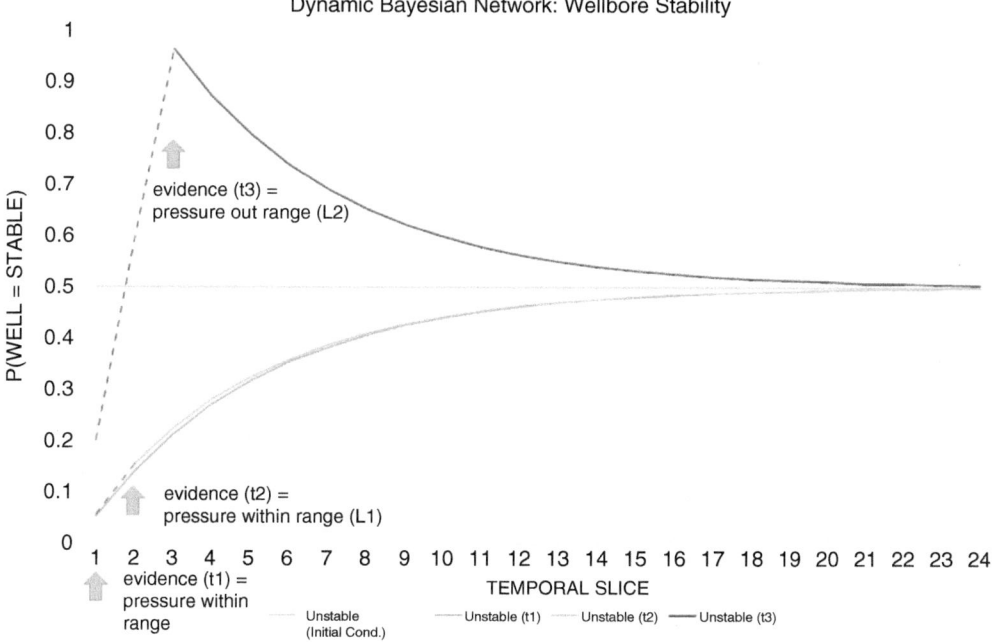

Figure 29 DBN wellbore stability. Numerical simulations.

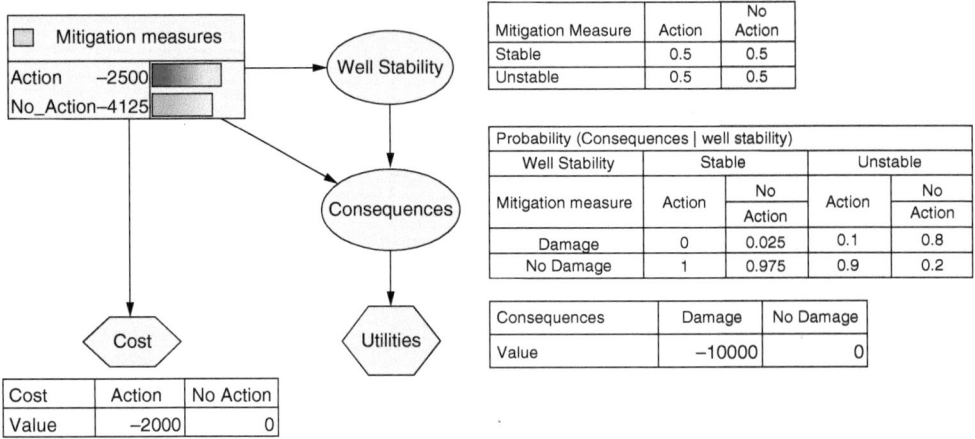

Figure 30 Well stability decision model: Initial conditions.

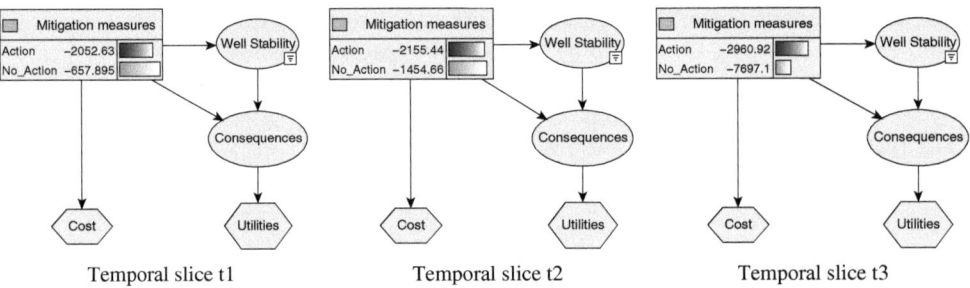

| Temporal slice t1 | Temporal slice t2 | Temporal slice t3 |

Figure 31 Well Stability Decision Model: temporal slices t1, t2, t3.

shows the results of the propagation of evidence (observation) through the network (*i.e.* the probability of well instability with time, given the observations).

The DBN in Figure 28 can be coupled with a decision model to determine the best mitigation at a given time. Once the prediction model is executed, the results are input into a decision model which is used to determine the best remedial measure. Figure 30 shows the Decision model, initial conditions and the best remedial measure at time 0.

Figure 31 shows the results of the decision model for the temporal slices t1, t2 and t3. The results show that, given the monitoring results (*i.e.* propagation of evidence) the optimal remedial measures are *No action* at temporal slice t1 and t2 and *Action* at temporal slice t3.

8 UPDATING RISK: MONITORING OF RESERVOIRS AND WELLBORES

Risk assessment and management is a dynamic process. Once the initial risk assessment is done, risk can and should be continuous updated during the operational phase of

CCUS systems. To do this, certain parameters should be made continuously measured in order to assess the behavior of the CO_2 reservoir system. The results of the monitoring must compared with ones predicted by modeling and risk analysis. The models can be updated after careful interpretation of a set of observed results (Solomon, 2007; Sousa & Sousa, 2012).

Monitoring is performed for various purposes, such as (IPCC, 2005):

a) to ensure and to document the volume injected into wells, specifically to monitor the conditions of the injection well and measuring the rates of injection, as well as the pressures on the top of the well and in the formation;

b) to verify the amount of injected CO_2 that was stored by different mechanisms;

c) to optimize the efficiency of the storage project, through the knowledge of the volume storage, the most appropriate injection pressures and the need for drilling new wells;

d) to demonstrate, with appropriate monitoring techniques, that CO_2 is still contained in the intended storage formations;

e) to detect leaks and to provide an early warning of any occurrence, so that the situation can be remedied by appropriate mitigation measures;

f) to know the integrity of wells that are being used or are abandoned;

g) to calibrate and verify models for determining the performance;

h) to detect micro-seismicity associated to storage projects.

Details of the methodologies followed in monitoring are described in the publication of IPCC (2005). The actual state of technology is sufficient to meet the needs of monitoring, namely about natural tracers, water composition, displacements at surface,

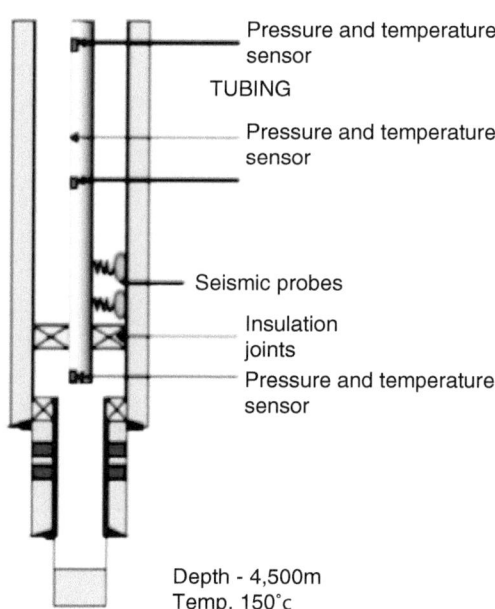

Figure 32 Monitoring the storage pilot test (Total, 2007).

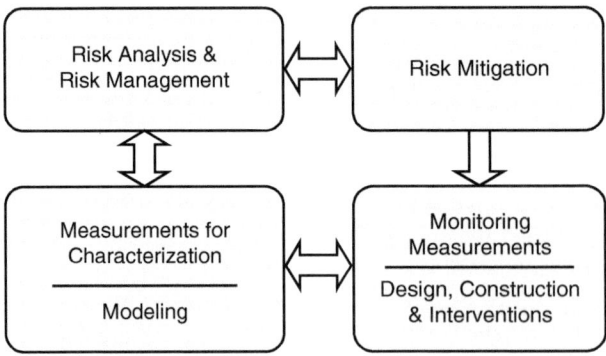

Figure 33 Methodology for integrity analysis of a wellbore (Cited in Sousa & Sousa, 2012).

subsurface pressure and temperature, well logs, seismic imaging, soil gas sampling, among other measuring techniques. Figure 32 shows technologies that were being implemented at wellbores at the storage reservoir Rousse (Total, 2007). These measurements will be done during and after the injections of CO_2 and compared with predictions by numerical modeling. Since the injection stopped, the wellbore will be accessible for in-site measurements (fluid, rock and cement). Figure 33 presents a methodology that can be used for long term integrity analysis of a wellbore in terms of risk evaluation (Sousa & Sousa, 2012).

Information of detecting surface wellbores sensing techniques related with CO_2 injection for Salah project and for the pilot storage site Ketzin can be seen in the following publications (Onuma & Ohkawa, 2009; Busch, 2010; Kohler *et al.*, 2013). Isotope measurements that have been applied for decades in oil and gas industry can also being applied for CCUS projects with further refinements (Myrttinen *et al.*, 2010; Nowak *et al.*, 2013). Finally it is important to develop strategies for evaluating the wellbore integrity using different measuring techniques, particularly the verification 'pressure tight' and 'loss free' (Reinicke & Fitcher, 2010).

9 CONCLUSIONS

Geological carbon sequestration presents a considerable potential for mitigating global climate changes. CO_2 can be safely injected and stored at well characterized and properly managed sites for a long term. Depleted oil and gas reservoirs, saline aquifers, shale gas, coal seams, abandoned coal mines, among others, can be used for storage of CO_2.

Geological carbon sequestration entails risk that may be large and significant. However risks can be limited or reduced. Development of methodologies for risk evaluation based on BN was made and some relevant applications were performed regarding storage of CO_2, the influence of active faults and for wellbore integrity analysis.

The applications of BNs permitted to conclude the following: in risk management, BNs are a powerful tool in the decision analysis; BNs allowed the extension of influence diagrams including the use of decision nodes and also utilities nodes; BNs combine the knowledge of experts and available data through statistical methods; the beneficial use of DBN in decision processes involving time is very relevant and the application can be also very important for the wellbore integrity analysis.

ACKNOWLEDGMENTS

The authors would like to thank CRC Press, Taylor & Francis Group for including information of Chapters published in the book entitled 'CO$_2$ Storage in Carboniferous Formations and Abandoned Coal Mines', in 2012. Also they would like to express to Professor Lemos de Sousa from Fernando Pessoa University their gratitude for his help.

ANNEX

Table 1 P (Characteristics of the sedimentary strata).

Good	0.5
Bad	0.3
Very Bad	0.2

Table 2 P (Characteristics of the coal strata).

Good	0.5
Bad	0.3
Very Bad	0.2

Table 3 P (Combined Characteristics| char. sedimentary strata, char. coal strata).

Characteristics of the coal strata	Good			Bad			Very Bad		
Characteristics of the sedimentary strata	Good	Bad	Very Bad	Good	Bad	Very Bad	Good	Bad	Very Bad
Good	0.9	0.5	0.2	0.3	0	0	0.1	0	0
Bad	0.1	0.4	0.3	0.6	0.6	0.3	0.3	0.3	0
Very Bad	0	0.1	0.5	0.1	0.4	0.7	0.6	0.7	1

Table 4 P(Activate Faults| Injection of CO$_2$).

Injection of CO$_2$	Yes	No
Yes	0.5	0
No	0.5	1

Table 5 P (Induced fracturing| Combined Characteristics, Injection of CO_2, Activate Faults).

Combined Characteristics	Good				Bad				Very Bad			
Injection CO_2	Yes		No		Yes		No		Yes		No	
Activate faults	Yes	No	Yes	No	Yes	No	Yes	No	Yes	No	Yes	No
Yes	0.5	0.1	0	0	0.7	0.2	0	0	0.9	0.4	0	0
No	0.5	0.9	1	1	0.3	0.8	1	1	0.1	0.6	1	1

Table 6 P (Leakage of CO_2| Induced Fracturing, Mitigation Measures).

Induced Fracturing	Yes			No		
Mitigation Measures	No Action	Reduce Injection pressure	Stop Injection	No Action	Reduce Injection Pressure	Stop Injection
Yes	0.9	0.5	0.1	0.05	0	0
No	0.1	0.5	0.9	0.95	1	1

Table 7 P (Induced Earthquakes| Activate Faults).

Activate Faults	Yes	No
Yes	0.1	0
No	0.9	1

Table 8 P (Damage level| Induced Earthquakes, Surface Occupation).

Induced Earthquakes	Yes		No	
Surface Occupation	Near a population	Remote Area	Near a population	Remote Area
No Damage	0.5	0.9	1	1
Level 1	0.4	0.1	0	0
Level 2	0.1	0	0	0

Table 9 Mitigation Measures (decision options).

Mitigation Measures:
No Action
Reduce Injection Pressure
Stop Injection

Table 10 Surface Occupation Options.

Surface Occupation:
Near a population
Remote Area

Table 11 Consequences (due to leakage of CO_2).

Leakage of CO_2	Yes	No
Value	−100	0

Table 12 Consequences (due to Damage).

Damage Level	No Damage	Level 1	Level 2
Value	0	−1000	−2000

Table 13 Mitigation Measure cost.

Mitigation Measure	No Action	Reduce Injection Pressure	Stop Injection
Value	0	−20	−50

REFERENCES

Ang, A.; Tang, W.H. (1975). Probability concepts in engineering planning and design. Vol. 1, Basic Principles, New York, John Wiley & Sons.

Bai, M.; Reinicke, K. (2013). Numerical simulation of CO_2 leakage through abandoned wells during CO_2 underground storage. Conf. Clean Energy Systems in the Subsurface, pp. 197–210.

Bai, M.; Sun, J.; Song, K.; Reinicke, K.; Teoduriu, C. (2015a). Evaluation of mechanical well integrity during CO_2 underground storage. Journal of Earth Sciences, Special Issue on Subsurface Energy Systems in China: Production, Storage and Conversion, 11p.

Bai, M.; Sun, J.; Song, K.; Reinicke, K.; Teoduriu, C. (2015b). Risk assessment of abandoned wells affected by CO_2. Journal of Earth Sciences, Special Issue on Subsurface Energy Systems in China: Production, Storage and Conversion, 11p.

Barros, N.; Oliveira, G.; Sousa, M.L. (2012). Environmental impact assessment of carbon capture and sequestration: General overview. Conf. on Energy Future of the Role of Impact Assessment, 6.

Beck, H.P.; Franz, O.T. (2010). Energy storage in abandoned mines – A method to stabilize the German power grid. 2nd Sino-German Energy Conference, pp. 261–269.

Benson, S. (2005). Carbon dioxide capture and storage: Overview with an emphasis on geological storage. Tutorial Presented at the AGU Annual Meeting, 5, San Francisco.

Brecht, A.; Edler, D.; Rehmer, K.P. (2013). SEW – A new software application that supports the safety evaluation of underground storage wells. Conf. Clean Energy Systems in the Subsurface, pp. 413–420.

Busch, W. (2010). Experiences with satellite radar monitoring of ground movements over underground gas and CO_2 reservoirs. 2nd Sino-German Energy Conference, pp. 33–37.

EASAC (2013a). Carbon capture and storage in Europe. European Academies Science Advisory Council, EASAC Report 20, 86p.

EASAC (2013b). Capturing carbon to tackle climate change. A nontechnical summary of carbon capture and storage in Europe. European Academies Science Advisory Council, EASAC Report 20, 16p.

Einstein, H.H. (1988). Landslide risk assessment procedure. Proc. 5th Intl. Symp. on Landslides, Lausanne.

Einstein, H.H.; Karam, K.S. (2001). Risk assessment and uncertainties. Conf. on Landslides; Causes, Impacts and Countermeasures, Davos, Switzerland.

Elsworth, D.; Wang, S.; Izadi, G.; Kumar, H.; Mathews, J.; Liu, J.S.; Lee, D.-S.; Pone, D. (2012). Complex process coupling in systems pushed far-from-equilibrium: Applications to CO_2 sequestration in carboniferous formations. Workshop CO_2 Storage in Carboniferous Formations and Abandoned Coal Mines, Beijing, pp. 55–67.

Eskesen, S.; Tengborg, P.J.; Veicherts, T. (2004). Guidelines for tunnelling risk management: International Tunnelling Association, Working Group No. 2. Tunnelling and Underground Space Technology, 19 (3), pp. 217–237.

Gomes, A. (2010). CO_2 injection processes in carboniferous formations (in Portuguese). MSc Thesis, University of Porto, 116p.

Grossmann, J.; Dahmke, A. (2013). Chances and risks of geologic CO_2 storage. Conf. Clean Energy Systems in the Subsurface, pp. 29–38.

He, M. (2012). Considerations on CO_2 storage in abandoned coal mines in China. Workshop CO_2 Storage in Carboniferous Formations and Abandoned Coal Mines, Beijing, pp. 25–36.

He, M.; Sousa, L.R.; Eslworth, D.; Vargas, Jr., E. (editors) (2012). CO_2 storage in carboniferous formations and abandoned coal mines. Workshop CO_2 Storage in Carboniferous Formations and Abandoned Coal Mines, Beijing, 201p.

He, M.; Sousa, L.R.; Sousa, R.L.; Gomes, A.; Vargas, E.; Zhang, N. (2011). Risk assessment of CO_2 injection processes and storage in carboniferous formations: A review. Journal of Rock Mechanics and Geotechnical Engineering, 3 (1), pp. 39–56.

Heckerman, D. (1997). A tutorial on Learning with Bayesian networks. Data Mining and Knowledge Discovery, 1, pp. 79–119.

Henrion, M.; Breese, J.; Horvit, E. (1991). Decision analysis and expert systems. AI Magazine, 12 (4), pp. 64–91.

Hou, M.Z.; Qiao, Z.; Luo, X.; Wang, Q. (2013). Numerical study of CO_2-injection borehole integrity with consideration of thermo-mechanical effects. Conf. Clean Energy Systems in the Subsurface, pp. 197–210.

Howard, R.A.; Abbas, A.E. (2015). Foundations of decision analysis. First Edition. Prentice Hall, Harlow, UK.

Howard, R.A.; Matheson, J.E. (editors) (1984). Readings on the principles and applications of decision analysis. 2 vol., Menlo Park, CA, Strategic Decisions Group.

IEA-UNIDO (2011). Technology roadmap carbon capture and storage in industrial applications. IEA Publications, Paris, 46p.

IPCC (2005). Carbone dioxide and storage. IPCC Special Report, Cambridge University Press, 431p.

Karam, K.S.; Karam J.S.; Einstein, H.H. (2007a). Decision analysis applied to tunnel exploration planning. II: Consideration of uncertainty. ASCE Journal of Construction Management, 133 (5), pp. 354–363.

Karam, K.S.; Karam J.S.; Einstein, H.H. (2007b). Decision analysis applied to tunnel exploration planning. I: Principles and case study. ASCE Journal of Construction Management, 133 (5), pp. 344–353.

Keeney, R.L.; Raiffa, H. (1976). Decision analysis with multiple conflicting objectives. New York, John Wiley and Sons.

Köhler, S.; Zemke, J.; Becker, W.; Wiebach, J.; Liebscher, A.; Moller, F.; Bannach, A. (2013). Operational reservoir monitoring at the CO_2 pilot storage site Ketzin, Germany. Conf. on Clean Energy Systems in the Subsurface, pp. 53–63.

Kuhn, M.; Martens, S.; Liebscher, A.; Moller, F.; Ketzin Team (2010). Progress report on the first European on-shore CO_2 storage site at Ketzin, Germany. 2nd Sino-German Energy Conf., pp. 29–31.

Langfeld, O.; Agasty, A. (2013). Production of coal bed methane in Germany. Conf. Clean Energy Systems in the Subsurface, pp. 177–185.

Murad, M.; Borges, M.; Obregon, J.; Correa, M. (2013). A new locally conservative numerical method for two-phase flow in heterogeneous poroelastic media. Computers and Geotechnics, 48, pp. 192–207.

Myer, L. (2003). Geomechanical risks in coal bed carbon dioxide sequestration. Lawrence Berkeley National Laboratory, Earth Sciences Division, Berkeley, 24p.

Myrttinen, A.; Becker, V.; Geldern, R.; Barth, J.A.; Nowak, M. (2010). Stable isotope and dissolved inorganic carbon sampling, monitoring and analytical methods in CO_2 injection and EGR projects. 2nd Sino-German Energy Conf., pp. 75–78.

Nowak, M.; Myrttinen, A.; Geldern, R. Becker, V.; Mayer, B.; Barth, J. (2013). A brief overview of isotope measurements carried out at various CCS pilot sites worldwide. Conf. Clean Energy Systems in the Subsurface, pp. 75–87.

Onuma, T.; Ohkawa, S. (2009). Detection of surface deformation related with CO_2 injection by DInSAR at In Salah, Algeria. Energy Procedia, pp. 2177–2184.

Passos, N.C.; Juvinao, A.; Vargas Jr., E.A.; Vaz, L.E.; Muller, A. (2014) Analysis of alternative numerical procedures for simulation of two-phase flow in a heterogeneous porous-fractured media. International Journal of Modeling and Simulation for the Petroleum Industry, 8 (2), pp. 43–54.

Piessens, K. (2012). The conceptual model for an abandoned coal mine reservoir. Work. CO_2 Storage in Carboniferous Formations and Abandoned Coal Mines, Beijing, pp. 179–200.

Piessens, K.; Dusar, M. (2003). CO_2 sequestration in abandoned coal mines. Royal Belgian Institute for Natural Sciences, Geological Survey of Belgium.

Price, P.; McKone, T.; Sohn, M. (2008). Carbon sequestration risks and risk management. Lawrence Berkeley National Laboratory, Environment Energy Technologies Division, Berkeley, 19p.

Qin, T. (2013). Regulation of carbon capture and storage in China: Lessons from the EU CCS directive. Conf. Clean Energy Systems in the Subsurface, pp. 1–19.

Qu, H.; Liu, J.S.; Chen, Z.; Pan, Z.; Connell, L. (2012). A fully coupled gas flow, coal deformation, and thermal transport model for the injection of carbon dioxide into coal seams. Workshop CO_2 Storage in Carboniferous Formations and Abandoned Coal Mines, Beijing, pp. 69–93.

Ramos, E.; Falcone, G. (2013). Recovery of the geothermal energy stored in abandoned mines. Conf. Clean Energy Systems in the Subsurface, pp. 143–155.

Reinicke, K.M.; Fitcher, C. (2010). Measurement strategies to evaluate the integrity of deep wells for CO_2 applications. 2nd Sino-German Energy Conf., pp. 67–74.

Rodrigues, C.; Dinis, M.A.; Sousa, M.L. (2015). Review of European energy policies regarding the recent "carbon capture, utilization and storage" technologies scenario and the role of coal seams. International Journal of Earth and Environmental Sciences, Springer-Verlagno, 1, 9p.

Russell, S.; Norvig, P. (2003). Artificial Intelligence. A modern approach. Second Edition. Prentice Hall Series in Artificial Intelligence.

Shi, J.Q.; Durucan, S. (2005). CO_2 storage in caverns and mines. Oil & Gas Science and Technology – Rev. IFP, 60 (3), pp. 569–571.

Shi, Y.; Wang, F.; Yang, Y.; Lei, H.; Jin, N.; Xu, T. (2013). Use of a CO_2 geological storage system to develop geothermal resources: A case study of a sandstone reservoir in the Songliao basin of northeast China. Clean Energy Systems in the Subsurface, pp. 89–103.

Solomon, S. (2007). Carbon dioxide storage: Geological security and environmental issues – Case study on the Spleiner gas field in Norway. Bellona Foundation Report, May, 28p.

Sonmez, H.; Gokceoglu, C.; Ulusay, R. (2003). An application of fuzzy sets to the Geological Strength Index (GSI) system used in rock engineering. Engineering Applications of Artificial Intelligence, 16 (3), pp. 251–269.

Sousa, L.R. (2012a). Present day conditions in the world of CCS projects. Workshop CO_2 Storage in Carboniferous Formations and Abandoned Coal Mines, Beijing, pp. 1–23.

Sousa, L.R.; Sousa, R.L. (2012). Risk associated to storage of CO_2 in carboniferous formations. Application of Bayesian Networks. Workshop CO_2 Storage in Carboniferous Formations and Abandoned Coal Mines, Beijing, pp. 153–178.

Sousa, L.R.; Sousa, R.L. (2013). Considerations about deep oil and gas reservoirs in pre-salt formations. Cases in Brazil and China. Euro-Asia Regional Round Table, Forum 2, Xian.

Sousa, L.R.; Sousa, R.L.; Silva, C.; Freitas, V. (2009). Maintenance methodologies of old railway tunnels. New Developments in Rock Mechanics and Engineering, Sanya, pp. 401–406.

Sousa, L.R.; Sousa, R.L.; Vargas Jr., E. (2015). Risk assessment: Application to CO_2 geological storage in coal. Chapter for Coal Utilization, Porto, 23p.

Sousa, M.L.; Rodrigues, C.F. (2008). CO_2 abatement: State of engineering. PDID&D Unit, Fernando Pessoa University, Porto.

Sousa, R.L. (2000). Knowledge based system for distress identification and diagnosis in concrete structures (in French). MSc Thesis, Ecole Nationale des Ponts et Chaussées, Paris.

Sousa, R.L. (2010). Risk analysis for tunneling projects. PhD. Thesis. MIT, Cambridge, 589p.

Sousa, R.L. (2012b). Methodologies for risk analysis and decision making. Work. CO_2 Storage in Carboniferous Formations and Abandoned Coal Mines, Beijing, pp. 125–152.

Sousa, R.L; Einstein, H. (2012). Risk analysis during tunnel construction using Bayesian Networks: Porto Metro case study. Tunneling and Underground Space Technology,. 27, pp. 86–100.

Suwansawat, S.; Einstein, H. (2006). Artificial neural networks for predicting the maximum surface settlement caused by EPB shield tunneling. Tunnelling and Underground. Space Technology, 21 (2), pp. 133–150.

Total (2007). Lacq CO_2 capture and geological storage pilot project. Project Information Dossier, Combevoie, 52p.

Vargas Jr., E.; Velloso, R.Q.; Ribeiro, W.N.; Ribeiro, W.N.; Muller, A.L.; Vaz, L.E. (2012). Considerations on the numerical modelling of injection processes of CO_2 in geological formations with emphasis on carboniferous formations and abandoned coal mines. Workshop CO_2 Storage in Carboniferous Formations and Abandoned Coal Mines, Beijing, pp. 107–123.

Vercelli, S.; Beaubien, S.E.; Lombardi, S.; Modesti, F.; Bigi, S. (2014). Selection and characterization of CO_2 storage sites: research highlights for the SiteChart project. SiteChart Report D9.2, 28p.

Xie, H.; Liu, T.; Zhengmeng, H.; Wang, Y.; Wang, J.; Liang Tang, L.; Jiang, W.; He Y. (2015a). Using electrochemical process to mineralize CO_2 and separate Ca^{2+}/Mg^{2+} ions from hard water to produce high value-added carbonates. Environmental Earth Sciences, DOI 10.1007/s12665-015-4401-z.

Xie, H.; Xie, J.; Gao, M.; Zhang, R.; Zhou, H.; Gao, F.; Zhang Z. (2015b). Theoretical and experimental validation of mining-enhanced permeability for simultaneous exploration of coal and gas. Environmental Earth Sciences, Special Issue on 'Unconventional Gas Resources in China, 12p.

Zadeh, L. (1999). Fuzzy sets as the basis for possibility. Fuzzy Sets and Systems, 1, pp. 3–28, 1978. (Reprinted in Fuzzy Sets and Systems 100 (Supplement): 9–34).

Zhang, N.; Sousa, L.R. (2012). Carbon Capture and Storage (CCS) activities in China. Workshop CO$_2$ Storage in Carboniferous Formations and Abandoned Coal Mines, Beijing, pp. 37–54.

Zoback, M.; Gorelick, S. (2012). Earthquake triggering and large-scale geologic storage of carbon dioxide. PNAS Early Edition, 5p.

Design and Stability Analysis:

Overviews

Chapter 13

Empirical design methods in underground mining

P.B. Hughes[1], D. Milne[1], R.C. Pakalnis[2] & Antonio Samaniego[3]

[1]Department of Civil and Geological Engineering, University of Saskatchewan, Saskatoon, Canada
[2]Pakalnis & Associates, Vancouver, British Columbia, Canada
[3]SVS Ingenieros SAC, Lima, Peru

1 INTRODUCTION

Empirical mine design methods are used extensively by industry, partly because of their ease of use. Another advantage is they allow the engineer to easily compare their mining conditions with conditions at other mines. Some rock mechanics practitioners suggest that empirical design methods are ideally suited to initial design assessments, but more 'advanced' numerical approaches should be taken when underground development has provided better data for modeling. Both empirical design methods and numerical modeling methods rely on good input data and both methods should be applied throughout the life of a mine.

1.1 Definition of empirical design in underground mining

Empirical design methods are based on four fundamental attributes:

- experience
- documented case histories
- a quantified, repeatable assessment of rock mass conditions and
- a basic understanding of rock mechanics theory.

Without all four of these attributes we have 'rules of thumb' or precedence based design. Arguably the most important of these four requirements are documented case histories. Well documented case histories, with accurate assessments of mining and rock mass conditions, are key to empirical design methods. It is of critical importance that empirical design practitioners understand the importance, and limitations, of the case histories used in the development of the design method (Pakalnis, 2014; Pakalnis et al., 2006).

Some of the following discussion is taken from Milne & Pakalnis (1997). There are some inherent differences between numerical modeling based design and empirical design approaches in rock mechanics. Numerical modeling design methods are based on an estimate of the constitutive behavior of a rock mass. Constitutive behavior includes an estimate of rock mass failure criteria which may be the Hoek-Brown m & s criteria, Mohr-Coulomb criteria or some measure of the interaction of joint surfaces and blocks of intact rock. Once a constitutive rock mass behavior has been arrived at, it becomes a part of the design method and is not varied. Empirical design methods are primarily based on an estimate of the constitutive properties of a rock mass. Based on the collection and analysis of case histories, the relative influence on stability of factors such as joint orientation, strength and spacing are estimated. Once an estimate of these values are made they become part of the design method and are not varied. Failure conditions are then empirically derived based on weighted constitutive properties of the rock mass, some assessment of stress conditions and a measure of the underground geometry.

Differences between numerical modeling based design and empirical design are best seen in the approach to back analysis. With empirical design approaches, back analysis consists of adjusting conditions which represent the boundary between stable and unstable conditions. The empirical approach evolves with the collection of more case histories, and / or with more detailed case histories. With numerical modeling approaches, back analysis or model calibration usually consists of adjusting the model input representing the rock mass properties to agree with observed cases of stability and instability. Since the numerical modeling process is not easily shared with other practitioners, it tends to evolve with the experience and the knowledge of individual practitioners.

Empirical design methods in underground mining started with rock mass classification systems. Prior to the development of modern rock mass classification, which started with Terzaghi (1946), a quantified, repeatable assessment method of rock mass conditions did not exist. Classification systems were developed as methods to quantify rock mass conditions, as well as an approach to design underground openings. Early methods concentrated on civil engineering applications, but included both coal mining and hard rock mining cases. The following section discusses the development of rock mass classification and early empirical design methods that were associated with them.

2 ROCK MASS CHARACTERIZATION

Rock mass characterization consists of quantifying the parameters governing rock mass behavior. These properties can be expressed as intact rock characteristics, discontinuity (joint) characteristics and the density and pattern of discontinuities and groundwater. It should be noted that groundwater conditions are sometimes omitted from rock mass characterization. Due to the importance of groundwater on rock mass behavior, and the tendency to ignore water pressure in many empirical and numerical design methods, water must be included in rock mass characterization. When values of rock mass characterization include groundwater conditions, care must be taken to ensure that groundwater conditions are not also used as a factor in the design loading conditions. It should also be noted that the characterization of rock mass conditions should be done for the conditions expected for the excavation under design. Groundwater problems may be present in a stope undercut being mapped, but dewatering or other changes may be expected prior to stope mining and these should be taken into account in the rock mass characterization for stope design.

As mentioned, the Terzaghi classification (Terzaghi, 1946) is the first classification system, combining both loading conditions and a numerical, though fairly subjective, assessment of the rock mass conditions. The Terzaghi method, and most early classification systems, were developed as design methods with loading, or excavation conditions built into the classification. The following sections describe only the commonly used classification systems that are applied to hard rock mining. The early classification systems were all developed as design tools for underground linear development (tunnels and drifts).

2.1 Rock Quality Designation (RQD)

RQD is defined as the total length of intact core greater than 100mm in length, divided by the total length of the core run. When a core run includes two distinct, significantly different units, RQD may be assessed over a shorter interval than the core run. Intact lengths of core only consider core broken by joints or other naturally occurring discontinuities so drill breaks must be ignored.

Care must be taken to ensure that the core intervals considered intact are also competent. This can be ensured with the use of a term called 'handled' RQD (HRQD) where only core that cannot be broken by handling can be considered intact (Robertson, 1988). This ensures that very weak rock or clay gouge, which may be cored in long sections, cannot be assigned a high RQD.

RQD may also be assessed without core. If line mapping data is available with joint spacing, Equation 1, by Priest and Hudson (1976) may be used to estimate RQD.

$$RQD = 100 \, e^{-0.1\lambda}(0.1\lambda + 1) \tag{1}$$

Where: λ = the joint frequency per meter

If a good exposure of rock is available, RQD can be estimated based on the number of joints observed per cubic meter (Equation 2) (Palmstrom, 1982).

$$RQD = 115 - 3.3Jv \tag{2}$$

Where: Jv = the number of joints estimated in a cubic meter of rock

The RQD classification system is a key parameter in many of the more recently developed classification systems. A span design approach was developed based with RQD and span (Merritt, 1972), however more detailed classification systems have developed and the RQD span design approach is not commonly used.

2.2 RMR classification

The RMR classification system was developed by Bieniawski (1973) for estimating support requirements, primarily for civil engineering tunneling work. It is based on five parameters describing the properties of the rock mass. A sixth factor is added to assess the influence of the interaction between the orientation of engineering structure and the orientation of the discontinuity set controlling stability. The weighting of the first 5 parameters totals 100 points with the 6th factor, joint orientation interaction, subtracting from this total. The weightings for the classification parameters have evolved since the technique was developed with new weightings occurring in 1973, 1974, 1975, 1976, 1979 and 1989. Most empirical design methods use the weightings assigned to the 1976 where the following parameter weightings were used:

Rock strength   15 points
RQD   20 points
Discontinuity Spacing   30 points
Groundwater   10 points
Joint condition   25 points
Joint orientation correction of up to -12 points.

The initial empirical design method associated with this classification system is the RMR Stand-up Time Graph.

2.2.1 Data base

The initial database determined the stand-up time of an opening based on the recorded RMR system which includes the data points used to develop the empirical design method. The data consists of 351 case histories of which 123 were of failures and 228 were stable cases (Bieniawski, 1989). Data came from coal mining, civil tunnelling and hard rock mining case histories.

2.2.2 Design application

The graph is supported with data from both civil and mining applications. The method is designed for tunneling applications where the geometry of the opening is defined by the tunnel span or width. The empirical design method provides an estimate of the length of time an unsupported opening back will remain stable. The stand-up time method is no widely used in industry, rather the Grimstad & Barton (1993) method described below is preferred.

2.2.3 Limitations

It should be noted that no direct assessment of stress conditions has been made in this study, suggesting that stress was not a significant problem for the case histories analyzed. In all case histories opening span was used to assess the geometry, suggesting that complex opening geometries were not considered. Also, the data is presented in a fashion that does not allow the reliability of the design method to be assessed. The raw data used in the development of this design method is available in Bieniawski (1989). Only limited application of this design method exists for underground mining. Drift backs are seldom left unsupported for extended lengths of time prior to supporting, though some application may be found for limiting drift advance rate in very poor ground conditions.

2.2.4 RMR for rock mass characterization

For mining applications the RMR rock mass classification system has evolved to a rock mass characterization system. Unlike tunneling, which is usually done in a single direction, mine drifting is done under changing directions and elevations. Incorporating a factor relating opening geometry to joint orientation would make rock classification values highly variable and confusing. For this reason the joint orientation correction is not applied for rock mass characterization in mining and the RMR_{76} system is used, denoted as RMR_{76}' when the joint orientation correction is omitted.

2.3 Q classification system

The Q classification system was developed by Barton *et al.* (1974). The factors considered in this classification system are similar to those assessed in the RMR classification system, but take a more rigorous approach. The classification value is based on six measures of the rock mass which are:

RQD = Rock Quality Designation (Section 2.1)
  Jn = Number of joint sets present
  Jr = Joint surface roughness
  Ja = Joint surface alteration
  Jw = Groundwater assessment
 SRF = Stress Reduction Factor (Assessment of stress conditions)

The Q classification value is calculated from the following equation:

$$Q = \frac{RQD}{Jn} x \frac{Jr}{Ja} x \frac{Jw}{SRF} \tag{3}$$

The Q classification system has not gone through as many changes as the RMR system. The only change of note is with the SRF term where the maximum SRF value increased from 20 to 400, as reported by Barton (2002). Q values range from 0.001 to 1000 representing extremely poor to excellent rock mass conditions. Initial design methods associated with the Q classification system focused on the design of back support.

2.3.1 Data base

The Q Support Graph is based on 1050 case histories collected since 1993 (Barton, 2002), shown in Figure 1. This support graph has evolved with rock support improvements and case histories. Data was collected from walls and backs from both civil and mining tunnels. Openings used for different applications were compared with a term called the excavation support ratio (ESR) which acted as a factor of safety term to modify the span to account for different excavation types.

2.3.2 Design application

The graph is supported with data from civil and mining applications. It is designed for tunnels or drifts and cannot account for two-way spanning. The design approach has been updated to follow improvements in support practices, with emphasis on tunnel linings such as shotcrete.

2.3.3 Limitations

This empirical design technique accounts for stress conditions, unlike the RMR Stand-up time graph, however, unfavorable joint orientations are not considered in the analysis. Cases where jointing orientation is a concern, or wedge instability a possibility, should not be assessed with this design approach. Adjustments to the SRF value have been made so it can be used in bursting conditions, however, a more extensive

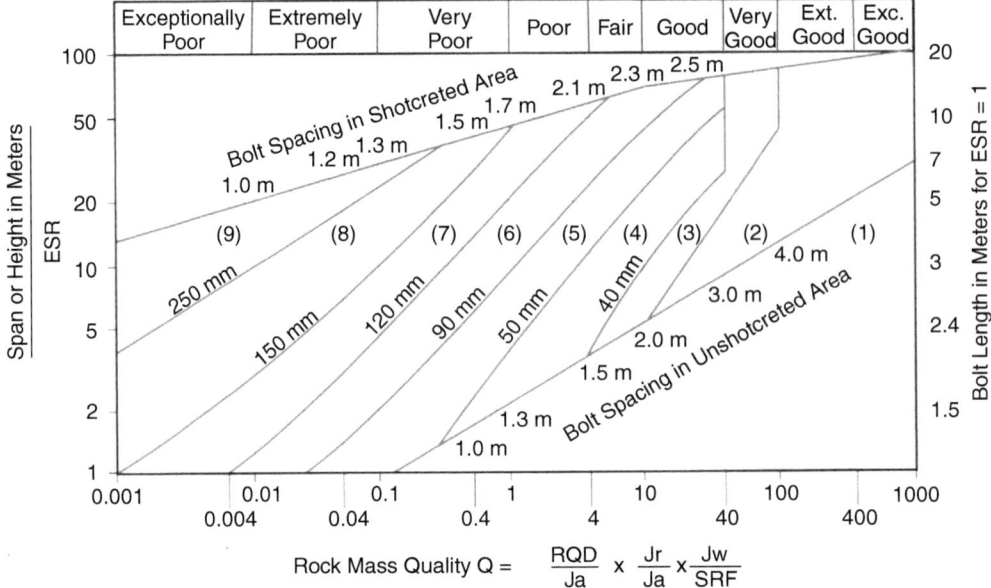

Reinforcement Categories
(1) Unsupported
(2) Spot Bolting
(3) Systematic Bolting
(4) Systematic Bolting with 40–100 mm Unreinforced Shotcrete
(5) Fibre Reinforced Shotcrete, 50–90 mm and Bolting
(6) Fibre Reinforced Shotcrete, 90–120 mm, and Bolting
(7) Fibre Reinforced Shotcrete, 120 –150 mm, and Bolting
(8) Fibre Reinforced Shotcrete, > 150 mm, with Reinforced Ribs of Shotcrete and Bolting
(9) Cast Concrete Lining

Figure 1 Q support graph indicating support type required for varied opening spans and Q classification. (After Grimstad & Barton, 1993) (From Potvin & Nedin, 2003).

database of mining case histories in bursting conditions would be needed for reliable application in mining.

2.3.4 Q for rock mass characterization

For most applications of the Q system for empirical mine design the SRF factor, related to stress conditions, is set to 1.0. Stress conditions in underground mines often vary significantly due to changes in mining depth so the assessment of stress in the classification system would make the classification values highly variable and difficult to use. Stress conditions are not included in rock characterization in mining to avoid this problem, however, the effect of stress is applied later in the empirical design process. The Q' term represents rock mass characterization with the SRF stress term set to 1.0. It must be noted that the Jw term for groundwater is included in the Q' term. Most empirical design methods in mining do not encounter groundwater frequently so when water does occur, it is necessary to rely on the assessment of water incorporated in the Q system (Pakalnis & Milne, 2012).

2.4 Geological Strength Index (GSI)

The GSI value was developed as a tool for estimating rock mass strength criteria m and s, used in a failure criteria applied to numerical modeling. When initially developed, the GSI value was equated to RMR_{76} for values above 25 with groundwater set to dry conditions and joint orientation set to favorable conditions (Hoek *et al.*, 1995). GSI can also be estimated based on a subjective description of the pattern of jointing and an assessment of joint surface conditions. More recently it was suggested that the GSI could be estimated by a combination of half the RQD value plus 1.5 times the joint condition assessment used with the RMR89 classification system (for a maximum value of 50 + 45) (Carter & Marinos, 2014). The GSI system is not considered in this study because it is not applied to empirical design methods developed with documented case histories.

3 MINE STOPE STABILITY GRAPH

Large temporary openings that are unique to mining environments are not well suited to the methods described in the previous section. To account for the unique openings required in mining, Mathews *et al.* (1981) proposed a method to empirically determine the stability of a stope based on stope dimensions, rock mass rating (Q'), and three factors accounting for stress, structure, and relative structural orientation. Subsequent modifications provided by Potvin (1988); Nickson (1992); Milne (1997); Clark & Pakalnis (1997); Mawdesley (2002); and Capes (2009) have added valuable knowledge to the original database.

The stope stability graph requires the quantification of the Modified Stability Number and the Hydraulic Radius of the mine opening. The Modified Stability Number (N') was proposed by Potvin (1988) and refines the original definition as proposed by Mathews *et al.* (1981). The Modified Stability Number is defined as follows:

$$N' = Q' \times A \times B \times C \tag{4}$$

Where:

$$Q' = \frac{RQD}{Jn} \times \frac{Jr}{Ja} \times J_w \tag{5}$$

A   Stress Parameter that is a measure of the ratio of intact rock strength to induced stress along the stope boundary under analysis (Figure 2). The A parameter reduces to reflect the potential of the stope wall to yield under comparatively high stresses

B   Structural orientation parameter that is a measure of the relative orientation of the dominant structural discontinuity with respect to the stope wall under analysis (Figure 3). Structures sub-parallel to parallel to the stope wall under consideration are considered worse from a stability standpoint; perpendicular structural features have a low influence on stability.

C   The geometric factor is a measure of the influence of gravity on the stability of the stope face under consideration (Figure 4). Overhanging surfaces or structural features which are oriented unfavorably with respect to gravity sliding influence the geometric factory.

The Hydraulic Radius (HR) is defined as the ratio of the area to the perimeter, and is sometimes referred to as the shape factor. HR allows for the comparison of various excavation shapes as it accounts for the distance to the load carrying abutments from the center of the opening (Milne, 1997).

It should be noted that the Q' value used in the N' calculation is sometimes reported without the Jw term. The original database consisted of stope with little or no evidence of water, so the Jw term was not considered. It is, however, widely recognized that the presence of water in a rock mass is detrimental to stability (Milne & Pakalnis, 2012). If

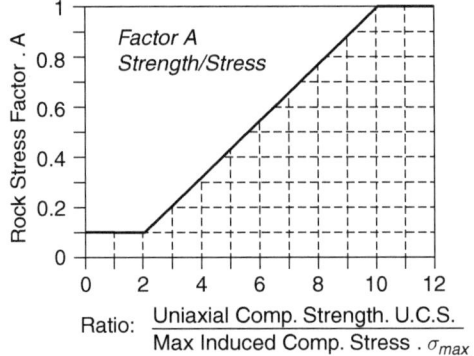

Figure 2 A Parameter Chart (From Potvin, 1988).

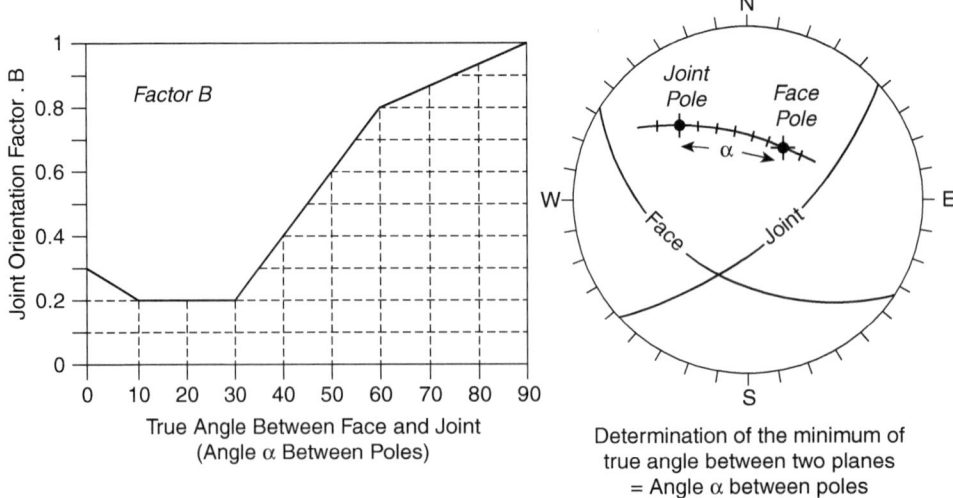

Figure 3 B Factor Chart (From Potvin, 1988).

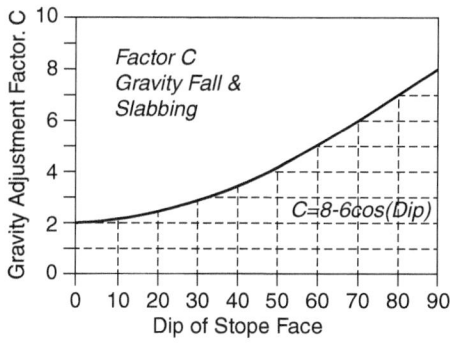
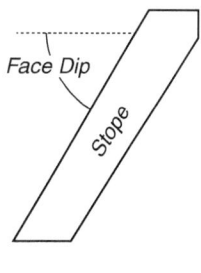

Figure 4 C Factor Chart (From Potvin, 1988).

water is present it must be included in the assessment of Q', since the influence of water is not included elsewhere in the design method. Another common area is in the assessment of Jr and Ja for Q'. Once the joint critical for stability is determined (as assessed for Factor C), the Jr and Ja values for this critical joint must be used to calculated Q'.

Once the N' and HR are determined for the individual stope surface, the value is plotted on the stability graph in zones that are described as stable, potentially stable and potentially unstable (Mathews *et al.*, 1981) or, stable, unsupported transition, stable with support, supported transition and caved (Nickson, 1992).

3.1 Databases

3.1.1 Mathews

The original stability graph (Mathews *et al.*, 1981) evaluates the shape factor of the stope based on the hydraulic radius to a rock stability index called the stability number N and determines the performance of the stope based on case studies. The database for the original study is based on 26 case studies from 3 mines complemented with 29 case studies from literature. The performance of the stopes tends to be qualitative, with refinements on the definition of stope performance presented by Stewart and Forsyth (1993). The general rule of thumb is that unplanned dilution increase as stope designs plot closer to the caving zone.

3.1.2 Potvin

Potvin (1988) provided an additional 176 case studies from 32 mines were added to the original Mathews databased (Mathews *et al.*, 1981). Some minor modification for Factors A, B and C were proposed for the stability number and referred to as N'. However, the major contribution from Potvin (1988) came from refining the stability zones on the graph: the transition zone between stable and "caving" was refined such that it could be used as design tool in stope planning.

3.1.3 Nickson

Nickson (1992) further refined the stope stability chart with the incorporation of cable bolt stability support zones determine from 15 unsupported and 46 cable bolt supported stope case studies (Figure 5). The design zones per Nickson (1992) are defined as follows:

- *Stable: low dilution*
- *Unsupported Transition Zone: case histories of stopes experienced dilution and ground falls causing operational problems; sensitive to blast vibrations and effect of time.*
- *Stable with support: require some form of ground support. Reasonable confidence in maintaining stability is achieved by installing pattern support. Unsupported stopes in the zone will have sever ground control problems*
- *Supported Transition Zone: reduced confidence in maintaining stability through installation of pattern support. Unsupported stopes in the zone will have sever ground control problems*
- *Caved Zone: Stope surface is likely unsupportable. Severe ground control problems can be expected.*

Potvin (2014) states that the Nickson (1992) graph has become a de-facto industry standard.

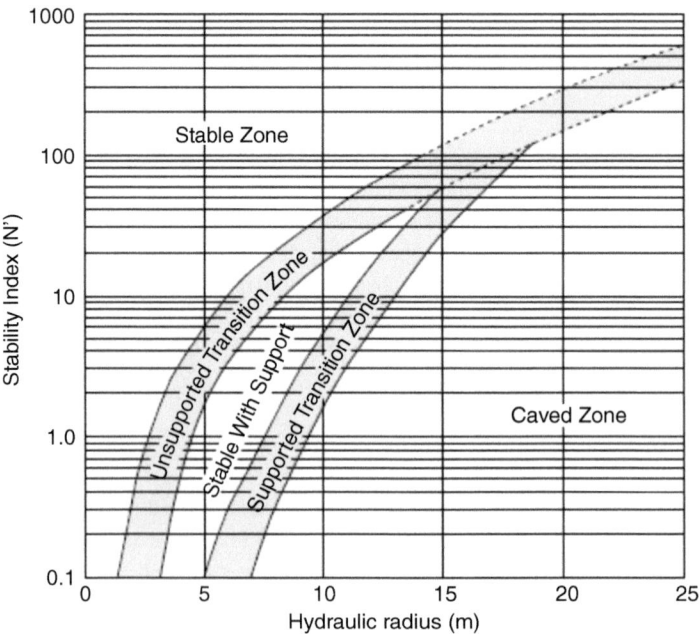

Figure 5 Nickson Stope Stability Chart.

3.1.4 Equivalent Overbreak Linear Slough (ELOS) and the modified stability graph

Clark & Pakalnis (1997) quantified the amount of dilution that can be expected for a stope based on the Hydraulic Radius and Modified Stability number. A case study of 88 stopes reconciliations (CMS surveys after mining was completed) determined the dilution that occurred based on the size, geometry and ground conditions present (Figure 6). The dilution was normalized to linear length of the stope (ELOS) and can be used by the designer in estimating the amount of dilution that can be expected given a HR and Stability Index for a Mine. Additional case histories were added to the dilution graph to account for poorer ground conditions (Capes *et al.*, 2005; Brady *et al.*, 2003).

3.2 Design applications

The modified stability graph method can be applied at different stages of a mining project with different objectives. It remains a design method that provides an initial

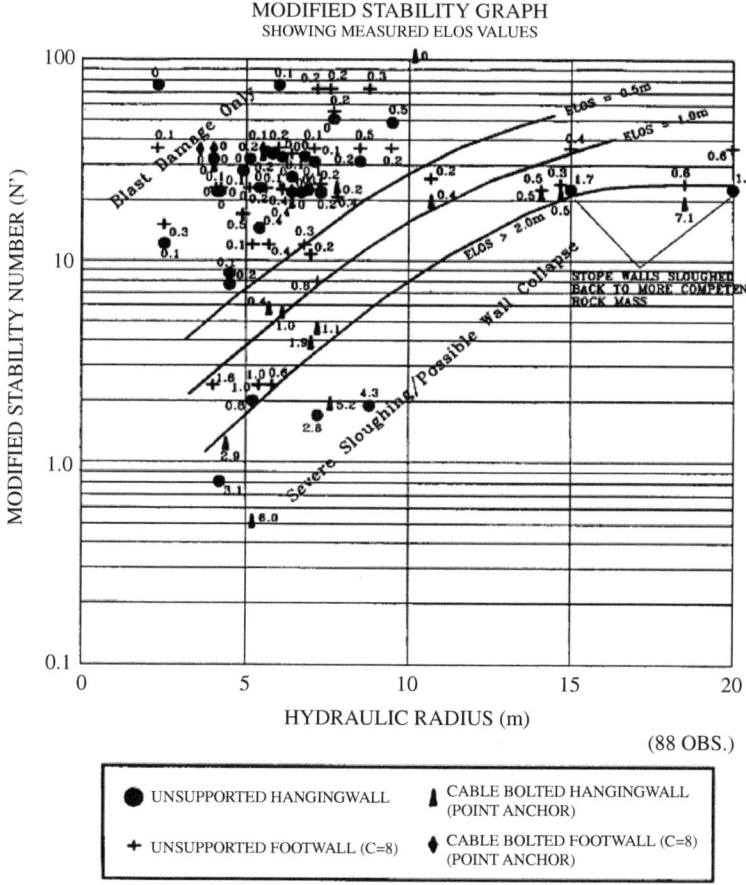

Figure 6 Modified Stability Graph with ELOS values (after Clark & Pakalnis, 1997).

concept of design sizes; it is wells suited for feasibility level of designs. The use of the Modified stability curve at the feasibility stage allows mine designers to determine critical aspects of mine plans including sub-level interval, stope dimension and the need for cable bolt support stabilization. (Potvin, 2014).

Continuing, the stope stability analysis is a useful tool in short term planning for individual stopes. With an increase understanding of the rock mass behavior through empirical study of previously mined stopes, a refinement of the original design can be achieved. At this stage, alterations to the stope dimensions, modification of cable bolt support and a better understanding of dilution volumes can be achieved.

3.3 Limitations

Potvin (2014) and Pakalnis *et al.* (2006) discuss the limitations, common mistakes and misconceptions in using the modified stability graph method for open stope design.

*   A different Q' must be used for stope back and walls.
*   A critical joint with strong influence on stability should have a Factor B approaching 0.2 and a joint with little influence on stability (toppling situation, or joints sub-perpendicular to the stope surface) should have a Factor B approaching 1.0.
*   In Q' values should not be assigned to rock mass on a domain basis, without considerations to the stopes or other mine structure being designed.
*   The joint roughness (Jr) and joint alteration (Ja) parameters must be assessed on the joint set that potentially will most critically affect the stability of the stope surface considered for design.
*   Factor C, for the sliding cases should not be applied to vertical walls; a value of 8.0 should be applied to all vertical walls.
*   Employing Factor A other than 1.0 indicates the mechanism of failure is stress induced whereby the HW fails other than sloughing into the stope. If the observed mechanism of failure is that of crushing/overstressed, shear or tensile failure, then Factor A can be considered less than 1.0 (Pakalnis, 2002)

3.4 Surface assessment for irregular geometry

As mentioned in section 3, the hydraulic radius term assesses the geometry of the excavation surface being designed. The hydraulic radius term is ideally suited for quantifying stope dimensions where the span cannot capture the two way spanning that is influencing stability. Equation 6 can be used to calculate hydraulic radius (HR) of a rectangular surface.

$$H.R = \frac{Area}{Perimeter} = \frac{axb}{2(a+b)} = \frac{2}{\frac{1}{4}\left(\frac{1}{.5a} + \frac{1}{.5b} + \frac{1}{.5a} + \frac{1}{.5b}\right)} \tag{6}$$

Where terms 'a' and 'b' are the sides of a rectangle and .5a and .5b are the distance to the abutments (rectangle boundary), measured from the centre.

The hydraulic radius term does not adequately assess more complex geometries, such as found with room and pillar mining. A term called the radius factor (RF) was developed for assessing geometries that are not well described by hydraulic radius.

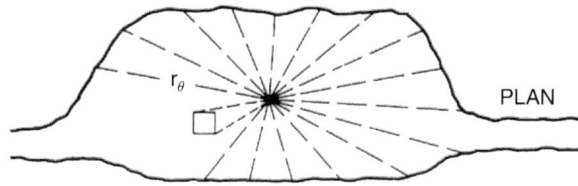

Radius Factor

$$RF = \frac{.5}{\frac{1}{n}\Sigma_{\theta=1}^{n}\frac{1}{r_\theta}}$$

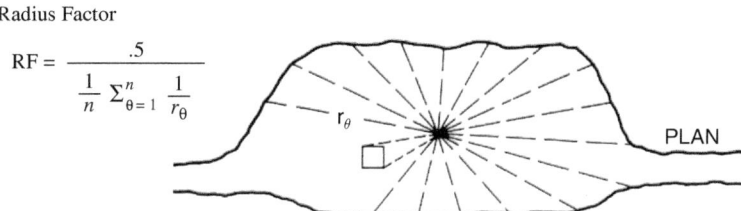

Figure 7 two approaches for assessing the influence of a pillar for assessing the geometry of a back (DRAFT).

The RF assesses complex geometries by measuring the distance from the centre of a surface to the surface abutments (and pillars) at small angular increments θ and calculating a value based on the following equation:

$$RF = \frac{0.5}{\frac{1}{n}\sum_{\theta=1}^{n}\frac{1}{r_\theta}} \tag{7}$$

Figure 7 demonstrates the application of the RF value for assessing the geometry of an opening surface.

4 MINER-ENTRY SPAN STABILITY

The Critical Span Curve (Lang, 1994) was originally a site-specific database for the Detour Mine that investigated the maximum stable back span for a given RMR. The original case study has been augmented by case histories of stable and unstable entry-type stope backs from six (6) different mines (Wang *et al.*, 2000). Subsequent updates of the database by Ouchi (2008) expands on the original database to incorporate lower rock mass ratings (RMR<45%).

   The critical span is defined as the diameter of the largest circle that can be inscribed within the boundaries of the exposed back as viewed in plan (Figure 8). The measured exposed span is then related to the rock mass rating in the immediate back to determine the stability of the span. The design span refers to spans which have used no support and/or spans which include surficial pattern bolting (1.8m long bolts on 1.2m x 1.2m) for local support; no cable support is considered in the design method. The rock mass rating as proposed by Bieniawski (1976) is employed to assess the rock mass in the immediate back with the following corrections: reducing RMR rating by "10" if shallow joints (dip under 30°) are present; reducing the RMR by 10 if there are signs

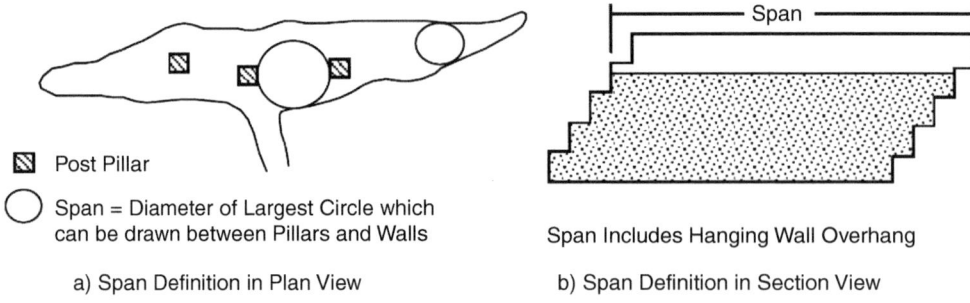

a) Span Definition in Plan View

☒ Post Pillar

◯ Span = Diameter of Largest Circle which
can be drawn between Pillars and Walls

b) Span Definition in Section View

Span Includes Hanging Wall Overhang

Figure 8 definition of inscribed span (after Pakalnis & Vongpaisal 1998).

of high stress such as corner spalling and by "20" if bursting conditions are present (Kumar *et al.*, 2002).

Once the maximum span and the RMR (with corrections) is obtained, the stability of the excavation is classified into three categories:

a) Stable Excavation
   ◦ No uncontrolled falls of ground.
   ◦ No movement of back observed
   ◦ No extraordinary support measures have been implemented.
b) Potentially Unstable Excavation
   ◦ Extra ground support may have been in-stalled to prevent potential falls of ground
   ◦ Movement within back
   ◦ Increased frequency of ground support rehabilitation
c) Unstable Excavation
   ◦ The area has collapsed
   ◦ Failure above the back is approximately 0.5 x span in the absence of major structure
   ◦ Support was not effective to maintain stability.

4.1 Database

The database for the miner-entry span stability was originally a site-specific database from the Detour Lake mine owned by Placer Dome Inc. and included 172 unique points that had RMR76 values ranging between 60 and 80 (Lang, 1994). The pillar data used to develop this methodology is a compilation of seven individual pillar stability databases that have been published worldwide.

The database was expanded to 292 observations in 2000 with case histories from six underground mines (Wang *et al.*, 2000) and is shown in Figure 9. The expanded database includes RMR76 values from 24 to 87, with 63% of the cases having an RMR76 between 60 and 80 (Brady *et al.*, 2003). The updated span curve utilized artificial neural networks to determine the transitions between the unstable, potentially unstable and stable boundaries.

(Brady *et al.*, 2003) augmented the existing Wang *et al.* (2000) database to incorporate weak rock masses that were under represented in the original database. An

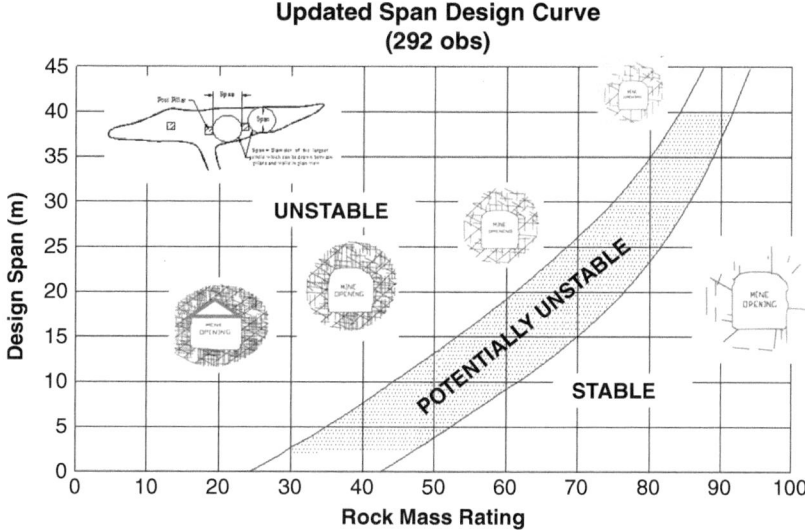

Figure 9 Updated Span Design Curve. Based on 292 observations (Wang *et al.*, 2000).

additional 463 points were added to the database with RMR76 ranging between 15 and 60. Further classification of the span design curve was performed to account for type of surficial ground support installed in the span. Ouchi (2007) built on work by Brady *et al.* (2003) and investigated the effect the type of ground support installed had on the maximum allowable span. An example of the curve for Swellex support is shown in Figure 10.

4.2 Design applications

The Design Span Miner-entry curve is an easy to use method for determining the maximum span possible for a given RMR. However, the curve must be used with proper engineering judgment. The effect of high stresses and adverse structure must be included in the holistic design of the opening, in addition to the use of the Design Span Curve in assessing the maximum span width from a rock mass rating perspective.

The use of the design span curve involves the mapping of all available headings and faces with the RMR$_{76}$ system. Upon determining the rock mass rating in and around a potential opening, the designer can then determine the maximum span that is possible for miner-entry openings. The Design Span Curve is also invaluable in assessing the stability of existing openings of old headings. Once old headings are assessed, the engineer can then independently assess the required ground support for a stable opening.

4.3 Limitations

The design span curve is not indicative of the minimum required rock mass rating to develop unsupported (bald) headings. The design span curve considers that surficial ground support is installed to support the immediate back.

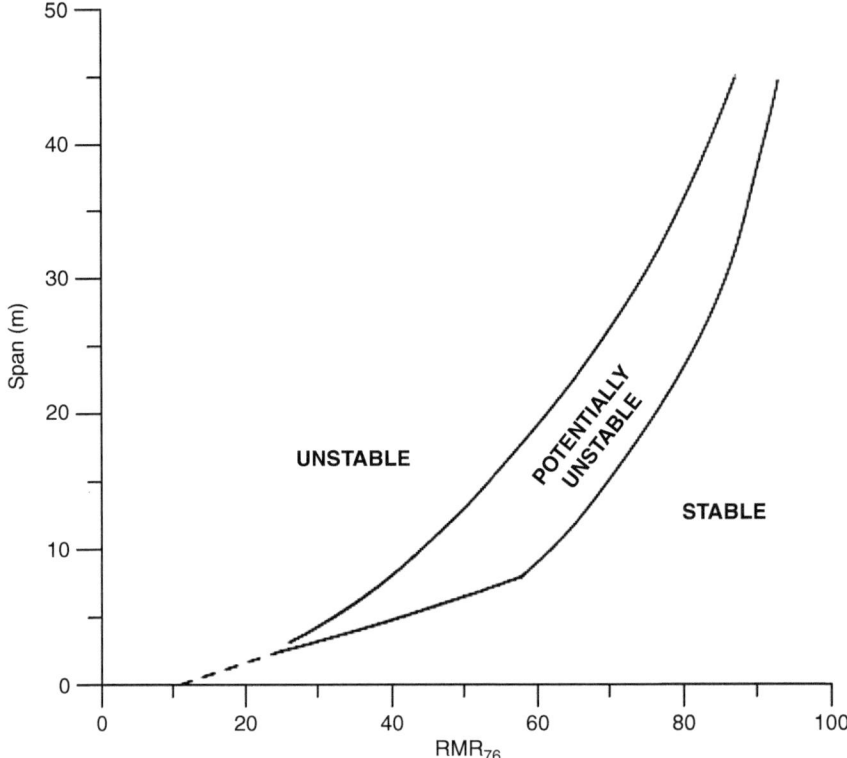

Figure 10 Weak Rock Mass Design Span Curve for 1.8 m long split set bolts on 1.2 m x 1.2 m pattern (Ouchi, 2008).

The design span curve is calibrated to RMR_{76} values. The use of other rock mass rating systems is not compatible with the design span curves.

The use of the miner-entry span curve is suitable in areas where there is no adverse structure, discrete wedges need identified and supported prior to assessment. The design span curve is applicable to short term stability (less than 3 months) (Lang, 1994). The stability curve should only be used in cases where a flat back exists and is not suited for 'shanty-back' style openings.

5 PILLAR STABILITY

The empirical stability of pillars consists of two databases: pillars for underground mining (square, rib, and sill) and crown pillars required for long-term mine stability. A comprehensive pillar database that relates geometry, loading conditions, in situ rock strength and stability condition has been developed by Lunder (1994) built on work by Hudyma (1988).

For the specific case of crown pillars, Carter (2014) presents a summary of the empirical study of crown pillar stability. As seen in Figure 11, the rock mass present

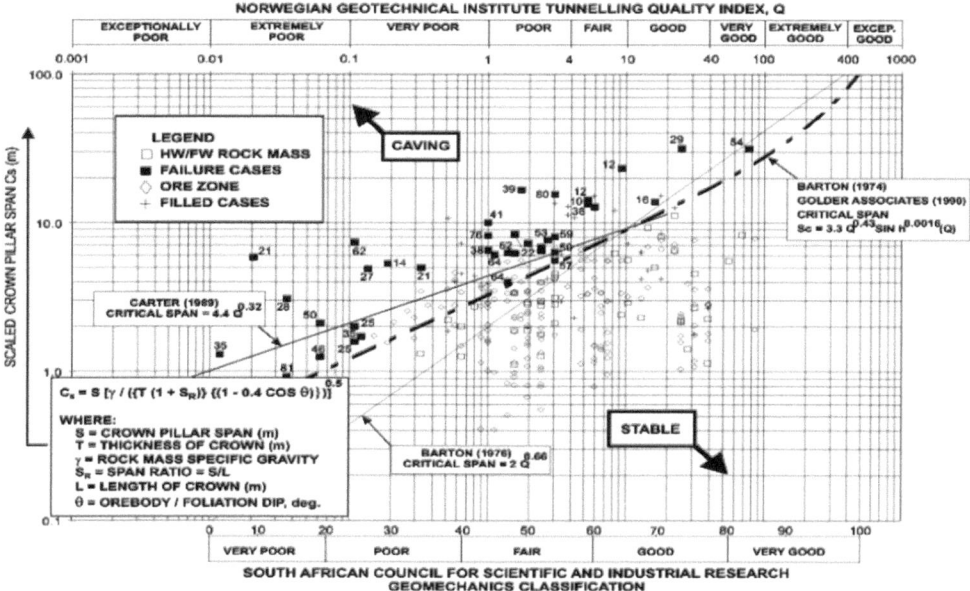

Figure 11 Scaled Span Chart, showing stable and failed case records plotted as Scaled Spans versus Rock Quality (after Carter, 2014).

in the crown pillar is compared to the calculated scaled crown pillar span (Cs) to determine the stability of the crown pillar under consideration.

5.1 Database

A total of 178 stability cases have been included where each case example represents a failed pillar, an unstable pillar or a stable pillar as compiled by Lunder (1994). The pillar data used to develop this methodology is a compilation of seven (7) individual pillar stability databases that have been published worldwide. Five of the seven databases originate from massive sulfide deposits and all of the databases have reported rock mass ratings in excess of 65% representing good to very good quality rock mass conditions (Pakalnis, 2014).

5.2 Design applications

Two primary factors are used in this design methodology, a geometric term that represents pillar shape and a strength term that includes the intact rock strength and the predicted pillar load. Rock mass strength is dependent upon the amount of confining stress applied to a sample. In the case of mine pillars, the more slender a pillar, the less confining stress will be available resulting in a lower strength for a given rock type. The pillar load is to be considered the core stress in center of the pillar. The Pillar Strength Graph was developed by plotting the ratio of pillar load/UCS (unconfined compressive strength) as shown in Figure 12 to the Wp/Hp. The pillar width (Wp) is defined as the dimension normal to the direction of the induced stress whereas.

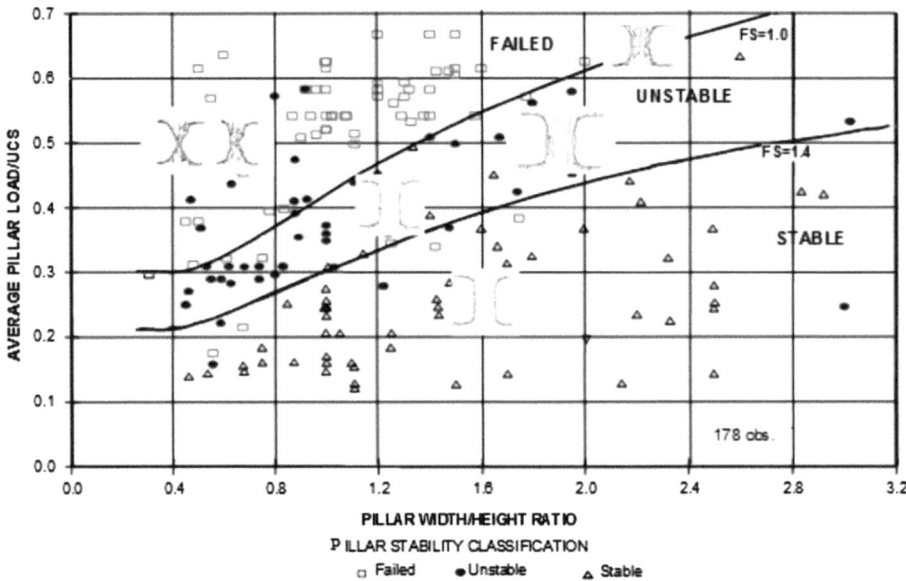

Figure 12 Underground Pillar Stability Graph. 178 Observations. (Lunder, 1994).

5.3 Limitations

The pillar stability graph is best suited to the following conditions per Hudyma (1988):

- 70 MPa < UCS < 316 MPa
- 9 meters < Width of pillar < 45 meters
- 60 < RMR of Pillar < 78
- (Average Pillar Load/ UCS) < 0.5

The type of pillar failure is not considered in this method. Yielding and sloughing pillar failure are all possible failure modes for pillars, however failure associated with bursting conditions is not considered. Judgment has to be applied by the engineer in determining the appropriate Factor of Safety for the designed pillars.

6 UNDERHAND CUT AND FILL SILL BEAM STABILITY

The cemented backfill forms the back in UCF mining, replacing weak or highly stressed rock where back stability is a concern. Large spans under cemented backfill are common in sill pillar mining, common with long-hole open stoping and cut and fill mining. Typically the sill beam stability is assessed with analytical equations proposed by Mitchell & Roettger (1989) to determine the minimum required strength against possible modes of failure. Work by Pakalnis *et al.* (2006) and more recently Hughes (2014) provide a database of design strength of backfills for a given span width.

6.1 Database

Work by Pakalnis *et al.* (2006) and more recently Hughes (2014) provide a database of design strength of backfills for a given span width.

The Pakalnis *et al.* chart (2006) has been modified and updated by Hughes (2014) provides the maximum mining span and design strengths for sixteen operating mines (Figure 13). The curves shown in Figure 13 is based on flexural instability employing fixed beam analysis with surcharge loading and a Factor of Safety of 2.0 for various sill beam geometries.

Basing research off of Pakalnis *et al.* (2006), Hughes (2014) performed a historical study of Stillwater mine to determine the maximum span vs. lab strength of fill for the Stillwater Mine complex (Figure 14). The results of the study calibrate the design curves for the Stillwater Mine to include updated tensile strength, stope closure and refinement on crushing failure equations based on the historical study.

6.2 Design applications

The use of the underhand cut and fill design database is suited for preliminary planning on minimum required backfill strength for a given span width. When using the design curves in Figure 13, the designer will obtain the minimum required backfill strength to withstand flexural failure within the stope; a Factor of Safety of

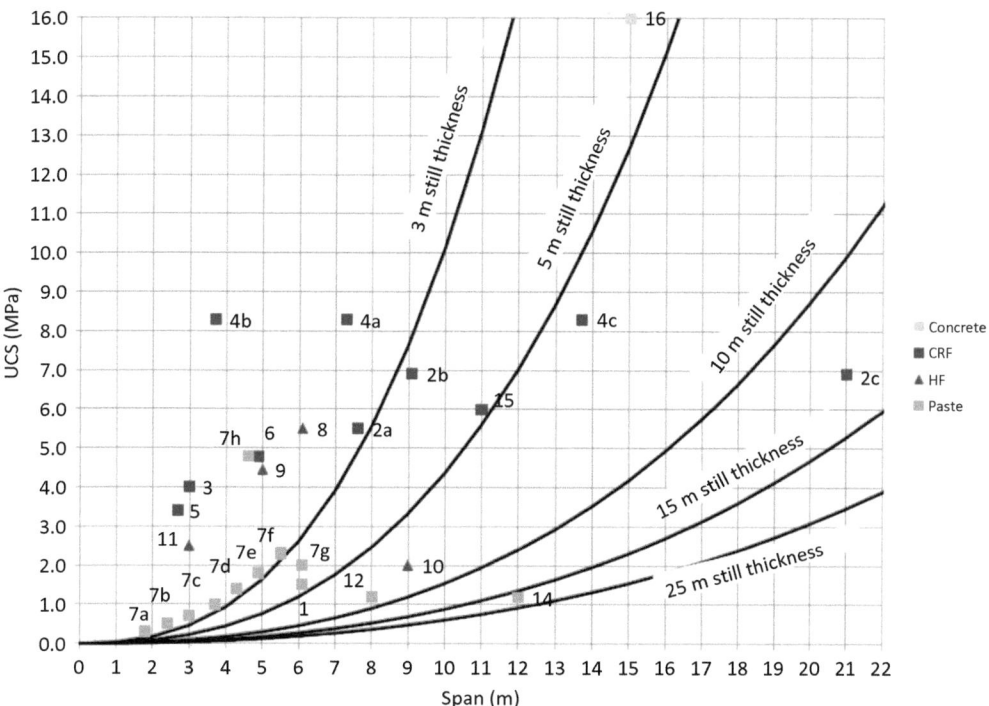

Figure 13 Empirical database of fill strength vs. span width after Pakalnis *et al.* (2006).

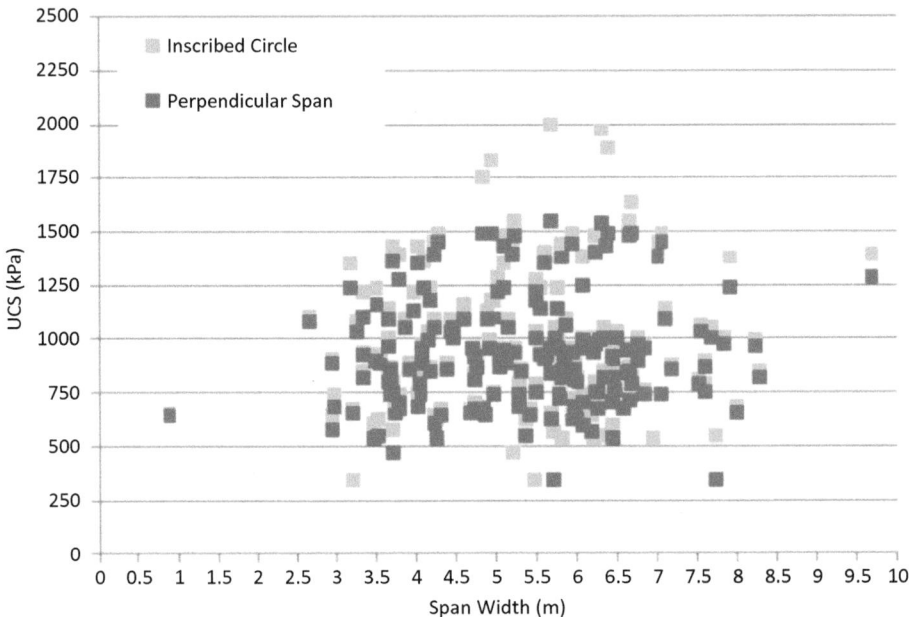

Figure 14 Stillwater Mine historic database of mining under cemented fill. Number of observations: 406 (Hughes, 2014).

2.0 has been applied to the design curves. Once the type of minimum fill strength is known, design of the backfill recipe can progress with a known minimum backfill design strength.

6.3 Limitations

The underhand cut and fill design should only be used as an initial design estimate for minimum backfill strength for a given maximum span.

The design curves consider that flexural failure is the mode of failure and the method is insensitive to stope-dip. Caving, rotational, shear, and crushing failure are not considered.

The design strength of fill is based on the design strength of the 28-day UCS tests. The strength of the fill within the stope may be less for CRF cases (Yu, 1995; Stone, 1993) and greater strength for cases where paste backfill is placed in the stope (Johnson *et al.*, 2015).

The database does not consider the ground stresses at the mine and does not consider the effect of closure on the sill beam. Research by Hughes (2014) finds that, in narrow stopes, the minimum required backfill strength for a stable sill is sensitive to the amount of closure within the stopes.

The database considers that the tensile strength of the backfill is 1/10th of the UCS. Studies by Hughes (2014) and Johnson *et al.* (2015) indicate that this is a conservative value for the tensile strength of fill. The tensile strength of the fill is closer to 15-20% of the UCS of the cemented paste backfill.

7 DISCUSSION

Three of the most significant advances in mine design in the last 50 years are:

1. Modern rock mass characterization systems that provide a repeatable numerical assessment of rock mass conditions.
2. Empirical design methods based on a comprehensive database of case histories
3. Advances in numerical modelling methods to assess induced stresses around mine excavations and associated rock mass failure criteria.

Rock mass characterization coupled with empirical design methods are easily applied in a mining environment and allow mining engineers to compare their mining situation to other operations. This paper summarized the more commonly used empirical design methods and highlights some of common errors in application. Some general guidelines are of value.

7.1 Rock mass characterization

As discussed in the preceding sections, rock mass characterization is a key component of all empirical design methods. The detail at which rock mass characterization data has been collected will correspond directly with the detail of the empirical design results. For example, stope dilution analysis (Section 3.1.4) may be conducted on mine stope hanging walls based on detailed rock characterization in limited, representative areas of the mine. In this case, the dilution estimates should represent average stope behavior, but may give very inaccurate estimates of individual stope hanging wall behavior. As with all design methods, the analysis is only as good as the input data.

7.2 Empirical design database

An understanding of the database used in the development of empirical design methods is critical to their application. As mentioned in section 5.1, the Lunder pillar design method (Figure 12) is based upon pillar case histories with RMR values in excess of 65. Using this empirical design method in a rock mass with lower RMR values would not be appropriate and would likely give unconservative results.

A less obvious factor that highlights the importance of the database for empirical design concerns mining practices. The Stability Graph method, as modified by Nickson (1992) (Section 3.1.3) is based on case histories collected in the late 1980s and early 1990s. Incorporated in this database is some degree of blast damage and borehole deviation commonly associated with mining at that time. Improved blasting and drilling practices will reduce blast damage and will tend to make the results from the Stability Graph more conservative. It should be noted that changes to mining practice will have the same effect on failure criteria for numerical modelling methods, such as the Hoek-Brown criteria, which has evolved to become less conservative.

7.3 Failure mode

The expected failure mode must be taken into account when applying empirical design methods, which is why a basic understanding of rock mechanics is one of the attributes required for empirical design methods. For example, as mentioned in Section 4.3, the Span Design method does not assess the presence or stability of discrete wedges; analytical design methods are required. Also, discrete shears, which may not contribute to the overall rock mass characterization, may influence stability, in which case the Span Design method may not be reliable.

In summary, variables influencing stability must be assessed in the empirical design method for the approach to be valid. Since empirical design methods predominantly assess the rock mass with rock mass characterization methods, failure mechanisms that are not rock mass controlled, such as bursting, require different failure analysis methods.

7.4 Advances in empirical design

Empirical design advances require more documented case histories. To continue to improve existing empirical design methods, detailed case histories with stope and drift specific rock mass classification values, reasonable stress condition estimates and recorded mining practice are needed. With more data and reliable data, the influence of changing mining practice and ground conditions will be more closely related to excavation performance. Shared databases, such as those collected by graduate students at the University or British Columbia, University of Queensland and many other institutions, are key to advances in empirical design.

REFERENCES

Barton, N. (2002). Some new Q-value correlations to assist in site characterization and tunnel design. International Journal of Rock Mechanics & Mining Sciences, Vol 39, pp. 185–216.

Bieniawski, Z.T. (1976). Rock Mass Classifications in Rock Engineering. Proceedings: Symposium on Exploration for Rock Engineering, Johannesburg, South Africa, pp. 97–106.

Brady, T., Martin, L. and Pakalnis, R. (2003). Empirical Approaches for Weak Rock Masses. 98th Annual AGM-CIM Conference, Montreal, QC.

Capes, G.W. (2009). Open stope hangingwall design based on general and detailed data collection in rock masses with unfavorable hangingwall conditions. Ph.D. Thesis, University of Saskatchewan, 248p.

Capes, G., Milne, D. and Grant, D. (2005). Stope Hangingwall Design Approaches at the Xstrata Zinc George Fisher Mine, North Queensland, Australia. American Rock Mechanics Symposium, Fairbanks, USA.

Carter, T.A. (2014). Guidelines for Use of the Scaled Span Method for Surface Crown Pillar Stability Assessment. Presented at 1st International Conference on Applied Empirical Design Methods in Mining – Lima, Peru, June.

Carter, T. and Marinos, V. (2014). Use of GSI for Rock Engineering Design. Presented at 1st International Conference on Applied Empirical Design Methods in Mining – Lima, Peru, June.

Clark, L.M. and Pakalnis, R.C. (1997). An Empirical Design Approach for Estimating Unplanned Dilution from Open Stope Hangingwalls and Footwalls. 99th CIM-AGM, Vancouver, CD-Rom.

Grimstad, E. and Barton, N. (1993). Updating the Q-system for NMT. Proceedings of the International Symposium on Sprayed Concrete, Fagernes (eds. Kompen, Opshal and Berg), Norwegian Concrete Association, Oslo.

Hoek, E., Kaiser, P.K. and Bawden, W.F. (1995). Support of Underground Excavations in Hard Rock. Rotterdam: Balkema.

Hughes, P. (2014). Underhand cut and fill cemented paste backfill sill beams. Ph.D. Thesis, University of British Columbia, Vancouver, Canada.

Johnson, J.C., Seymour, J.B., Martin, L.A., Stepan, M., Arkoosh, A. and Emery, T. (2015). Strength and Elastic Properties of Paste Backfill at the Lucky Friday Mine, Mullan, Idaho. 49th US Rock Mechanics/Geomechanics Symposium, San Francicso, CA.

Kumar, P., Pakalnis, R., Roque, P. and Corey, G. (2002). Development of Empirical and Numerical Design Techniques in Burst Prone Ground at Goldcorp Red Lake Mine. Presented at AGM-CIM, Vancouver, 8pp.

Lang, B. (1994). Span design for entry type excavations. M.Sc. Thesis, University of British Columbia, Vancouver, Canada.

Lunder, P. (1994). Hard rock pillar strength estimation: an applied empirical approach. M.A.Sc. Thesis, University of British Columbia, p. 166.

Mathews, K.E., Hoek, E., Wyllie, D.C. and Stewart, S.B.V. (1981). Prediction of stable excavation spans for mining at depths below 1000 m in hard rock mines. Canmet Report DSS Serial No. OSQ80–00081.

Mawdesley, C.A. (2002). Predicting rock mass cavability in block caving mines. Ph.D. Thesis, University of Queensland, 410p.

Milne, D. (1997). Underground design and deformation based on surface geometry. Ph.D. Thesis, University of British Columbia, Vancouver, Canada.

Milne, D. and Pakalnis, R. (2012). Advances in Methods of Empirical Stope Design. Proceedings of the 46th US Rock Mechanics/Geomechanics Symposium held in Chicago, IL, USA, 24–27 June.

Mitchell, R.J. and Roettger, J.J. (1989). Analysis and modelling of sill pillars. Innovations in Mining Backfill Technology. Balkema: Rotterdam, pp. 53–62.

Nickson, S.D. 1992. Cable support guidelines for underground hard rock mine operations. M.A. Sc. Thesis, The University of British Columbia, 223p.

Ouchi, A. 2002. Empirical design of span openings in weak rock. M.A.Sc. Thesis, University of British Columbia, Vancouver, Canada.

Palmström, A. (1982). The volumetric joint count-a useful and simple measure of the degree of rock mass jointing. In Proceedings of the Fourth International Congress of the International Association of Engineering Geology, New Delhi, India, val. V., theme 2, V.221–8.

Pakalnis, R. (2014). Empirical Design Methods – Update (2014). Presented at 1st International Conference on Applied Empirical Design Methods in Mining – Lima, Peru, June.

Pakalnis, R., Caceres, C., Clapp, K, Brady, T., Williams, T., Blake, W. and MacLaughlin, M. (2006). Design spans – underhand cut-and-fill mining. CIMM Bulletin, Vol. 99, No 7pp.

Pakalnis, R. and Vongpaisal, S. (1998). Empirical design methods. 100th AGM-CIMM, Montreal.

Potvin, Y. (1988). Empirical open stope design in Canada. Ph.D. Thesis, The University of British Columbia, 350p.

Potvin, Y. (2014). The modified stability graph method; more than 30 years later. Presented at 1st International Conference on Applied Empirical Design Methods in Mining – Lima, Peru, June.

Potvin, Y. and Nedin, P. (2003). Management of Rockfall Risks in Underground Metalliferous Mines. Minerals Council of Australia, Canbera.

Priest, S.D. and Hudson, J.A. (1976). Discontinuity spacings in rock. International Journal of Rock Mechanics and Mining Sciences & Geomechanics Abstracts, Vol. 13, pp. 135–148. Pergamon Press 1976. Printed in Great Britain.

Robertson, A.M. (1988). Estimating weak rock strength. AZMESME Annual Meeting, Phoenix, Arizona, USA, Preprint #88–145.

Stone, D.M.R. (1993). The Optimization of Mix Designs for Cemented Rockfill. In Proceedings of Minefill 93, Johannesburg, South Africa.

Terzaghi, K. (1946). Rock Defects and Loads on Tunnel Support. Rock Tunneling with Steel Supports, eds. Proctor, R.V. and , T. White. Commercial Shearing Co., Youngstown, Ohio.

Wang, J., Pakalnis, R., Milne, D. and Lang, B. (2000). Empirical Underground Entry-type Excavation Span Design Modification. 53rd Annual Conference of the Canadian Geotechnical Society, p. 8.

Yu, T.R. (1995). Consolidated rockfill. Course Notes Presented at Cheng-Kung University, 257pp.

Chapter 14

20 years of intelligent rock mechanics

X.T. Feng[1] & C.X. Yang[2]

[1]Institute of Rock and Soil Mechanics, Chinese Academy of Sciences, Wuhan, China
[2]School of Resources & Civil Engineering, Northeastern University, Shenyang, China

Abstract: Recently, the methodology of intelligent rock mechanics (IRM) has been becoming an important aspect in rock engineering design, as one of the most popular non 1:1 mapping techniques. It has been 20 years since the proposal of IRM in 1993, aiming to provide an alternative research scheme for complicated rock mechanical problems. With extensive successful applications in different rock engineering fields, the past 20 years have seen enormous progress in the new direction, both in rock mechanics modeling and rock engineering design. This work reviews the 20 years of development of IRM. It covers some of the most important intelligent modeling methods and integrated intelligent design concepts that have been developed, with a focus on the evolution of the integrated intelligent design concepts with applications to some of the most complicated rock engineering constructed in China during the last decade.

1 INTRODUCTION

With the steadily incensement of the scale of rock engineering projects, more and more complicated problems have been met in the field of rock mechanics. One of the most important aspects is the concomitant increase in the modeling complexity. The deterministic method inherited from the conventional sciences like material strength and elasto-plastic-viscosity are not always satisfactory. Sometimes, the mathematical descriptions of some rock mechanics mechanisms are weak or incomplete and there may even be no generally accepted conceptual model. Moreover, even if the mechanisms are understood, it may not be practical to obtain the related parameter values for a given rock mass.

To overcome such complexities, approaches from artificial intelligence were introduced and trialed in the early 1980s. A more systematic research of intelligent methods in rock mechanics was started by Feng and his coworkers in 1986. Based on the encouraging achievements in investigation of artificial intelligence and expert systems and their application to rock mechanics and engineering problems during the following several years, a new term "intelligent rock mechanics (IRM)" was named to define a new research scheme in 1993. The publication of "A new direction-intelligent rock mechanics and rock engineering" (Feng et al., 1997) thereafter outlines the research framework of IRM. According to Feng et al. (1997), IRM uses non 1:1 mapping methods, based on new ideas of self-learning, nonlinear dynamic possessing, evolutionary identification, global optimization, distributed representation and the comprehensive integration of different techniques, to model the actual behavior of jointed rock mass. It is an open and more interdisciplinary science concerning artificial intelligence, nonlinear science, system science, mechanics, geosciences and engineering science. Useful knowledge can be learned by data mining from experienced cases and improved through nonlinear dynamic process in the context of successive applications to approximate actual behavior. Evolutionary identification methods are employed to find the reasonable models under the conditions that the exact relationship cannot be assumed beforehand. And the distributed representation techniques are capable of finding the non 1:1 relationship.

Since the suggestion of the research scheme of IRM, great efforts have been made to advance forward along the new direction, both in basic theories, fundamental techniques, algorithms and tools and applications. Moreover, some new design concepts are progressively programmed for rock slopes, underground openings and underground mines. A more sophisticated framework of IRM was presented in 2000 (Feng, 2000). During the 2000s, the methodology of IRM was extensively applied to several large engineering conducted in China. Significant developments and supplements were made both in modeling methods and design concepts. In the sequential publications about rock mechanics modeling and rock engineering design approaches in Feng & Hudson (2004), Hudson & Feng (2007), and Feng & Hudson (2010), IRM has been considered and discussed as a standard non 1:1 methodology, as seen in the summary chart (see Figure 1) about rock mechanics modeling methods suggested by Feng & Hudson (2011).

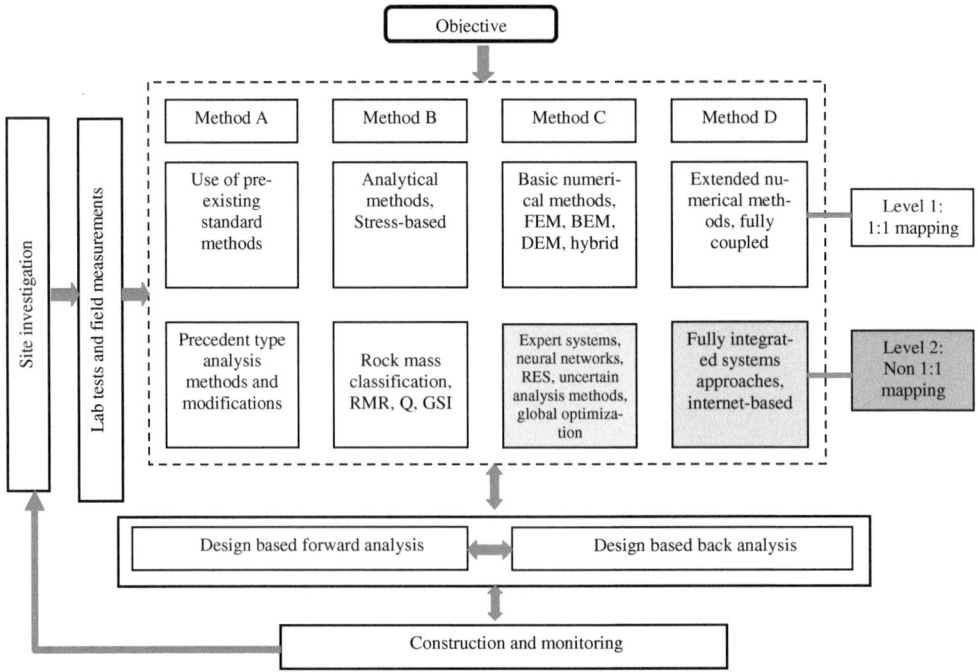

Figure 1 Rock mechanics modeling methods (Feng & Hudson, 2011).

The past 20 years have seen enormous progress in the development of IRM, not only in the techniques, algorithms and tools, but also, definitely more important, in the application aspects and design concepts. In the early publications (Feng *et al.*, 1997; Feng, 2000; Wang & Feng, 1997; Feng & Wang, 2000), the basic ideas and some methods and applications were surveyed to present an overview of IRM. In this review, we focus on the most significant developments with respect to the applications to some of the most complicated rock engineering problems constructed in China during the last decade. This paper is organized in four sections. Section 2 briefly lists the important landmarks in the development history of IRM. Section 3 summarizes some important intelligent modeling methods for rock mechanics that have been developed and validated in different rock engineering. Several integrated intelligent design concepts for rock engineering with complicated geological conditions are presented in Section 4. Finally, some concluding remarks and perspectives are given in Section 5.

2 DEVELOPMENT HISTORY OF IRM

Some important landmarks in development history of IRM can be listed as following:

– Approaches from artificial intelligence were introduced and trialed in the early 1980s.
– A more systematic research of intelligent methods in rock mechanics was started by Feng and his coworkers from 1986.
– A new term "intelligent rock mechanics (IRM)" was named to define a new research scheme of intelligence of rock mechanics in 1993.
– The publication of "A new direction-intelligent rock mechanics and rock engineering" thereafter in 1997 (Feng *et al.*, 1997) draws the research framework of IRM.
– With some new design concepts are progressively programmed for rock slopes, underground openings and underground mines, a more sophisticated framework of IRM was then presented in 2000 (Feng, 2000).
– During the 2000s, the methodology or IRM was extensively applied to several large engineering conducted in China. Significant developments and supplements were made both in modeling methods and design concepts.
– In the sequential publications about rock mechanics modeling and rock engineering design approaches in Feng & Hudson (2004), Hudson & Feng (2007), Feng & Hudson (2010) and Feng & Hudson (2011), IRM has been considered and discussed as a standard non 1:1 methodology.
– More and more scholars have been closely watching the direction and starting relative study.
– Recently, IRM has been listed as a special topic in series conferences sponsored by ISRM.

Intelligent methods developed/extended for different rock mechanics modeling tasks:

– Rock mechanics expert systems: Rock mass classification, support design, recognition of failure mode, rockburst risk estimation
– Neural networks (NN)/support vector machine (SVM): Learning nonlinear relationship from instance data, stress-strain response information, deformation time series

- Genetic programming (GP): Identification of the mathematical expression
- Genetic algorithm (GA)/particle swarm optimization (PSO): Parameter estimation with known model expression
- Evolutionary neural network-genetic algorithm-numerical computation considering economical aspects: Optimization of excavation and support scheme
- Evolutionary neural networks-genetic algorithm-numerical computation, evolutionary neural network-particle swarm optimization-numerical computation, multi-information back analysis with sensitive analysis and correlation analysis: Back analysis of rock mass mechanical models and parameters
- Comprehensive integration of different modeling tasks: Back analysis of mechanical parameters⇒prediction analysis⇒dynamically recognition of control method

3 INTELLIGENT METHODS FOR ROCK MECHANICS AND ROCK ENGINEERING

In addition to the increases in project scale and complexity, there has been a concomitant increase in the modeling complexity. For those engineering projects constructed in difficult and unexpected geological conditions such as large scale, complicated strata, high stress, great overburden depth, etc. there is little prior experience and limit data available for engineers to make decisions. A series of intelligent feedback analysis techniques have been developed to recognize different kinds of rock mechanics models under such complicated conditions. This section reviews these methods with an emphasis on the application aspects for rock engineering design.

3.1 Rock mechanics expert system

Expert systems are knowledge-orientated computer programs that try to simulate how a human expert solves a specific problem. They attempt to identify, formulize, encode and use the knowledge of human experts for a high-performance program. They can be used as a guide for inexperienced people or a decision support tool for experts themselves to make a decision. Generally, an expert system consists of a knowledge base, data base, inference engine, explanation system, knowledge acquisition and man-machine interface. Construction of an expert system includes the procedure of knowledge acquisition, knowledge representation and encoding and inference engine design. The knowledge is usually represented as rules with an IF-THEN format. Once the user introduces a fact, the inference engine is triggered to reach an adequate answer.

Expert system methods were firstly used for rock mass classification analysis in the latter 1980s (Zhang et al., 1988; Juang et al., 1989). To improve the construction efficiency, Feng (1991) presented an integrated rock mechanics expert system model. A neural network based knowledge learning procedure was added to find hidden knowledge and a new "uncertain reasoning" method to deal with fuzzy, uncertain and random information. Figure 3 illustrates a typical reasoning process for rock mass classification, where the result shows that the probability of the rock mass belonging to rank 1 is 0.875. The model has the comprehensive representation of

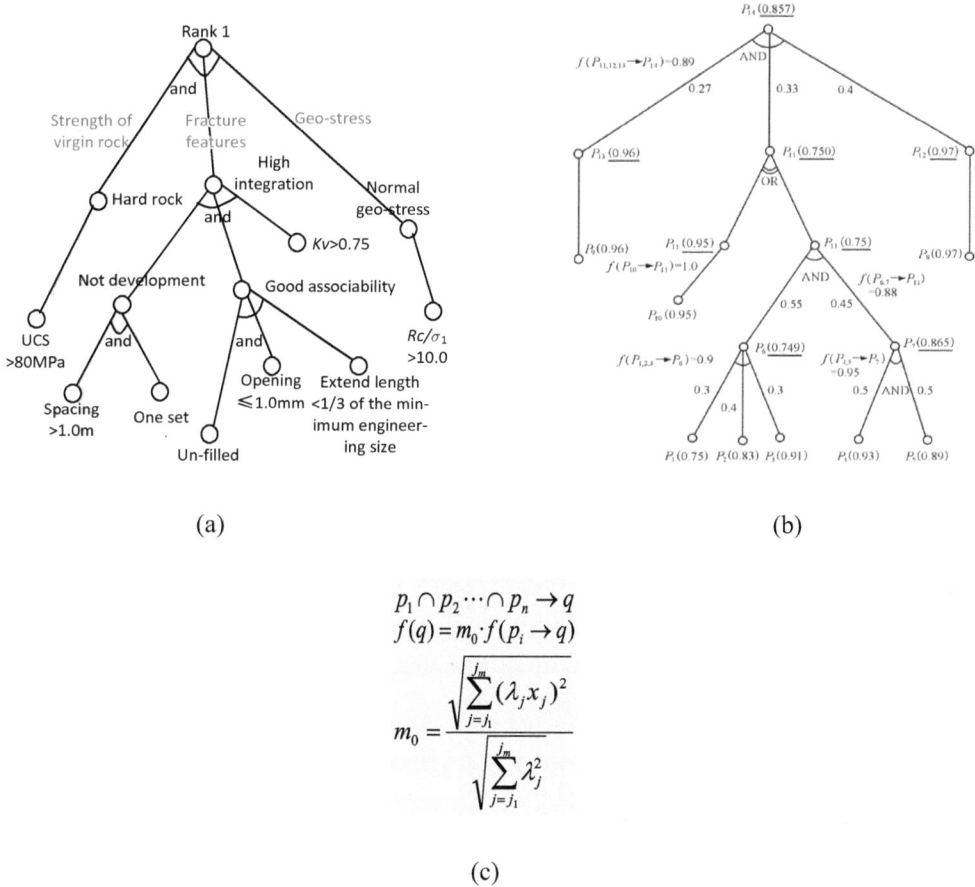

Figure 2 Uncertain reasoning method for rock mass classification: (a) Source information; (b) Reasoning process; (c) Reasoning rules.

rules, semantic network, frames, predicates and mathematical models. Various expertise, numerical models and experience criteria can be input and encode automatically, providing an environment of integrated decision-making. This made expert systems more applicable for complicated rock mechanics problems. By using the model as developing environment, several expert systems were developed for different problems (Feng, 1991; Lin & Feng, 1991; Feng & Lin, 1992a, 1992b, 1993), including discontinuity property evaluation, comprehensive rock classification, optimal zone design of underground opening support, slope stability analysis and selection of underground mining methods. They have been widely used in different large open and underground mines. Inspired by their works, expert system has been applied extensively for different tasks, especially as an additional support tool which can be combined with other methods and techniques for rock engineering design (Zhang *et al.*, 1995; Tapia, 1998; Cai *et al.*, 1998; Fuenkajorn & Kamutchat, 2001; Yu & Chern, 2007; Schubert & Grossauer, 2007; Khandelwal, 2012; Zhang *et al.*, 2012).

3.2 Intelligent recognition of rock mechanical models

Constitutive modeling is one of the main topics in IRM. Generally, constitutive models can be described as nonlinear input-output mapping relationships between stress vector and strain vector, as well as some affecting factors. Namely

$$\boldsymbol{\sigma} = f(\boldsymbol{\varepsilon}, \ldots)$$
$$f : \mathbf{R}^n \to \mathbf{R}^m \tag{1}$$

Thus, the modeling task is to find the reasonable formation of f, $i.e.$ the model structure and its parameters. Apparently, the model structure defines the fundamental behavior and should be firstly determined. According to the sufficiency of the information available, there are three kinds of methods developed in IRM for rock constitutive modeling.

3.2.1 Neural network constitutive modeling methods

Besides the mechanical factors, there are now new design demands in projects where the interactions between rock stress, water flow, heat and chemistry over long time periods need to be understood and modeled in order that the necessary predictive process required for design can be generated. The mechanisms of most of the factors are far from well-known. Neural network (NN) modeling methods are capable of learning the fundamental relationship from laboratory test results or field measurements without resorting to a physical description of each factor.

NN-based constitutive modeling method was firstly presented by Ghaboussi *et al.* (1991, 1994) for modeling the constitutive behavior of composites and sands. A comprehensive conceptual model was extended by Feng *et al.* (1999) for modeling of coupled T(thermal)-H(hydraulic)-M(mechanical)-C(chemical) processes of rocks, as seen in Figure 3. By adding the time-related parameters, they established a NN model to simulate the creep behavior of some soft rock (Yang *et al.*, 2008). To model the effect of chemical environment, by introducing the chemical parameters, pH value, a neural network model is successfully established to model the mechanical behavior of limestone under different water-chemical environment (Chen *et al.*, 2010). Some of the results are shown in Figure 4.

To learn directly from the stress-strain response, different laboratory experiments have to be designed to investigate each constitutive factor under consideration. On the other hand, the macro response of rock structure, such as the load-deformation and excavation-deformation responses, may contain abundant constitutive information. This information can also be extracted through numerical computation and be used for learning, as the works in Millar & Calderbank (1995) and Wu *et al.* (2000).

This kind of method has also been used in modeling the failure criteria of rocks (Rafiai & Jafari, 2011; Rafiai *et al.*, 2013).

3.2.2 Modeling based on global optimal search of nonlinear items

By adding some nonlinear items to simple models (*e.g.*, linear elastic), Feng & Yang (2001) presented another way for constitutive modeling. Usually, the nonlinear levels are dependent on a series of parameters, for example, the coefficients of polynomial, the exponential parameters of the time dependent item, *et al.* The purpose of the modeling

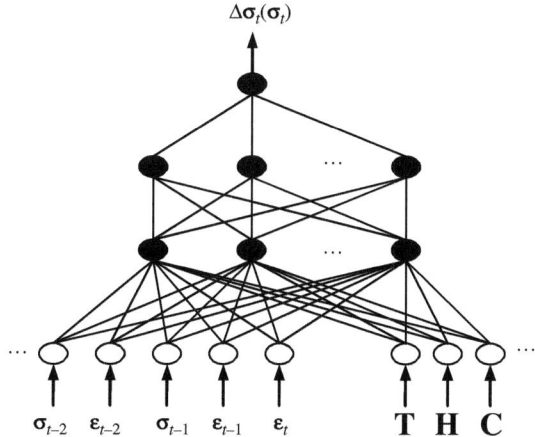

Figure 3 A comprehensive neural network model for modeling of coupled T-H-M-C processes of rock material.

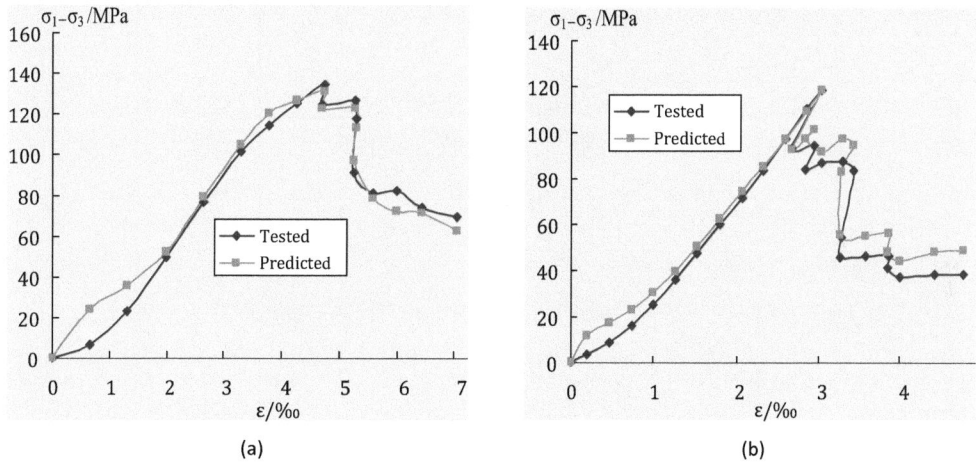

Figure 4 Comparison between the tested results and the simulated results based on neural network constitutive model of limestone under different water-chemical environment: (a) $CaCl_2$, pH=2, confining stress 5MPa; (b) Na_2SO_4, pH=12, confining stress 2.5MPa. (After Chen *et al.*, 2010).

problem is to find these parameters by minimizing the discrepancy between the quantities $\mathbf{u}_i^*$ obtained in lab or field tests and the corresponding data $\mathbf{u}_i$ obtained by the model output analysis.

$$\text{Minimize} \quad F(\mathbf{p}) = \left\{ \frac{1}{n} \sum_{i=1}^{n} [u_i(\mathbf{p}) - u_i^*]^2 \right\}^{1/2} \tag{2}$$

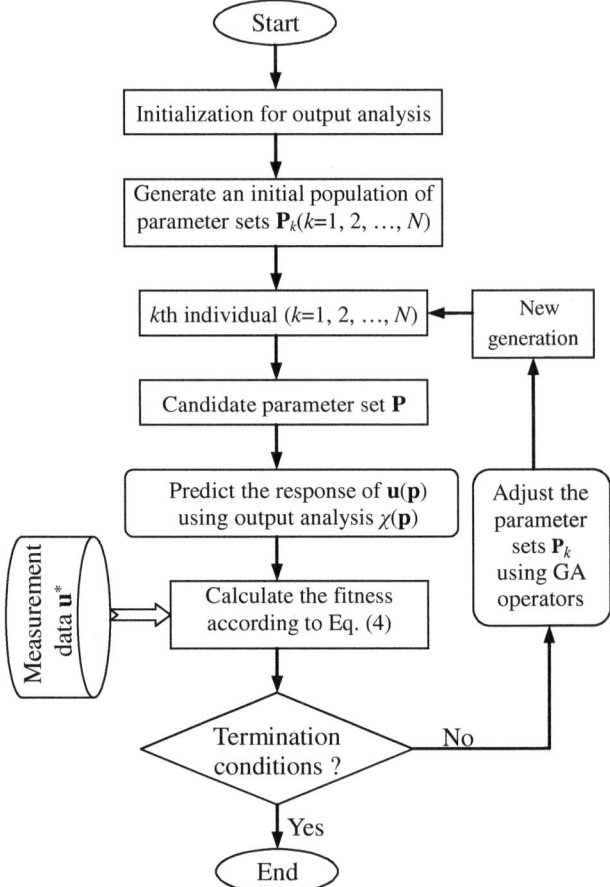

Figure 5 Flow chart for global optimal search of nonlinear parameters using genetic algorithm.

where n is the number of testing points, $\mathbf{P} = (p_1, p_2, \ldots)^\mathrm{T}$ is the vector of unknown parameters.

To solve such nonlinear optimization problems, global optimal search techniques, such as genetic algorithms (GAs), particle swarm optimization (PSO), are employed. The flow chart of a solution procedure using GA is shown in Figure 5. The solution procedure starts from a population of candidate solutions $\mathbf{P}_k$ (k=1, 2, ..., N), where N is the population size. Each possible solution is used for forward analysis to predict the response of $\mathbf{u}_i$. The predicted $\mathbf{u}_i$ is then compared with $\mathbf{u}_i^*$ to evaluate the fitness of the population of the candidate solutions. The estimated fitness values are then used by the GA operations to obtain a new population of candidate parameters and thus evolve into a new generation. The population of candidate parameter sets is updated until the prescribed stopping criterion is met.

Using these kinds of modeling methods, they established the nonlinear elastic constitutive model of laminated composites by adding some polynomial items associated with the stress history to linear model, and a nonlinear finite element procedure is

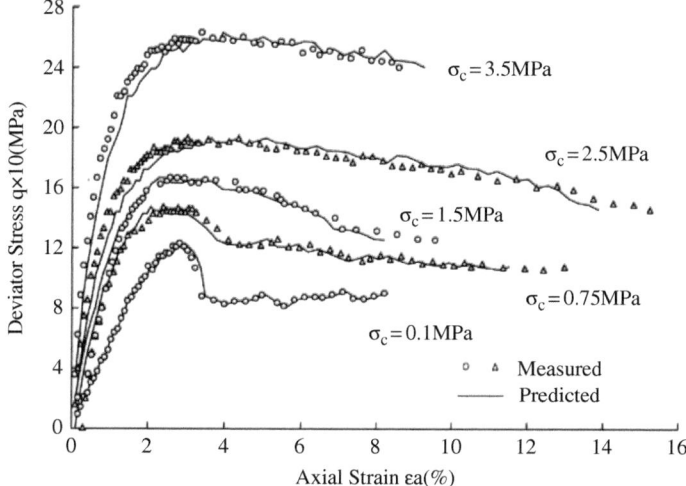

Figure 6 Simulation results based on the obtained stress-strain-time model for the diatom soft rock under different consolidation stress, also the tested results are presented for comparison. (After Feng *et al.*, 2002).

employed to obtain the stress-strain response from the tested load-deformation response (Feng & Yang, 2001). Based on the Cambridge model, the time dependent behavior of diatom soft rock was investigated and the time-associated parameters are globally searched to simulate the nonlinear stress-strain-time relationship (Feng *et al.*, 2002). Some of the results are shown in Figure 6.

3.2.3 Modeling based on coupled recognition of model structure and its parameters

In each modeling process, the model structure defines the essential behavior and has to be determined before the estimation of model parameters. These kinds of methods view the constitutive model as a nonlinear combination of a set of input variables, such as the stress, strain, stress history, strain history, *et al.*, or a set of constitutive units, such as the elastic unit, viscous unit and plastic unit. Feng & Yang (2004) proposed using genetic programming (GP) method to evolve the best model structure through so-called symbolic regression. And an extra parameter estimation procedure is used to determine the model parameters.

By coupling the GP and GA, a hybrid evolutionary algorithm was proposed for simultaneously recognition of the structure of nonlinear constitutive model and its coefficients, as shown in Figure 7. Starting from a set of building blocks (constitutive variables and constitutive units, and mathematic functions), GP uses a layered computer program coding to combine those building blocks to model structures randomly. These model structures are translated to a mechanic model thereafter. Then, a GA procedure is used to determine the model parameters to construct the definite models. Other parameter estimation procedure (*e.g.* PSO) can be used instead. The

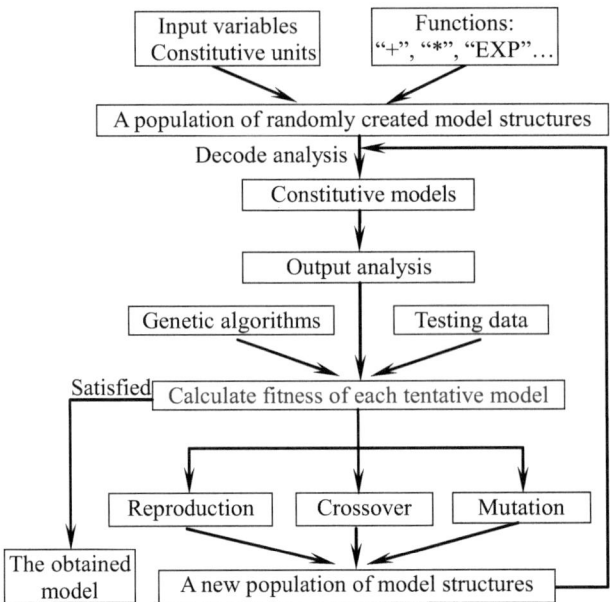

Figure 7 Flowchart showing genetic programming and the genetic algorithm modeling progress for simultaneously recognition of model structure and its parameters.

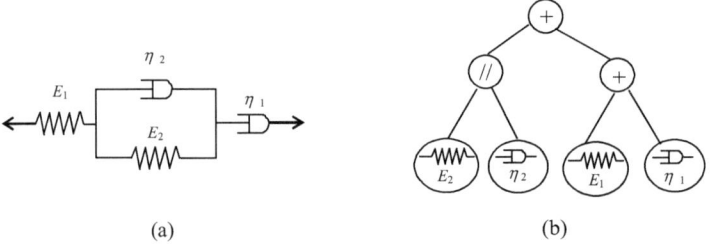

(a)   (b)

Figure 8 Genetic programming representation of Burgers body.

obtained models are evaluated and selected into further evolution through genetic operations.

The proposed recognition method has been successfully used in recognition of the nonlinear stress-strain relationship of lamina plate (Feng & Yang, 2004). By introducing the rheological elements as variables and parallel and serial connection mechanism as functions, an evolutionary modeling scheme based on the PSO-GP algorithm was proposed to recognize the visco-elastic models of rock materials, celadon argillaceous rock and the fuchsia argillaceous rock related to the Goupitan hydropower station engineering constructed in China (Feng *et al.*, 2006). The typical representation of the visco-elastic model and the evolution operations are shown in Figure 8 and Figure 9 respectively. The recognized models are well consistent with the testing results (see Figure 10).

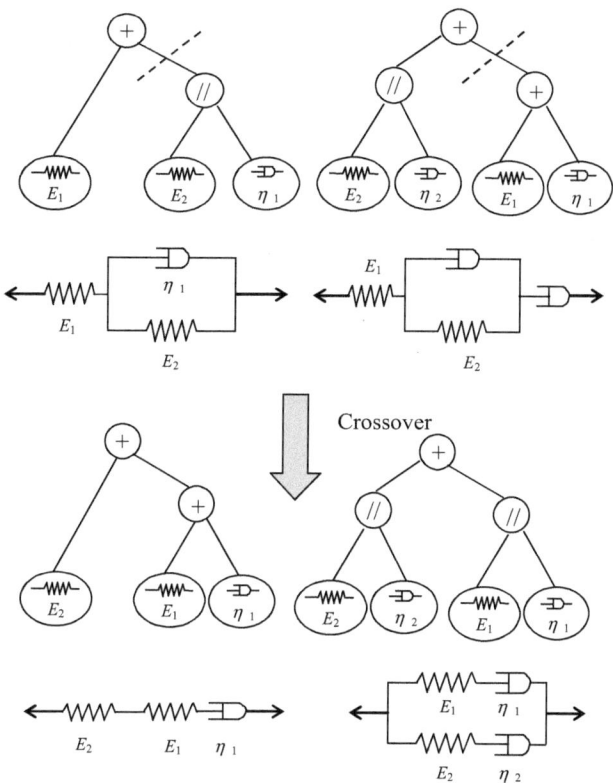

Figure 9 Genetic evolution operations on visco-elastic models.

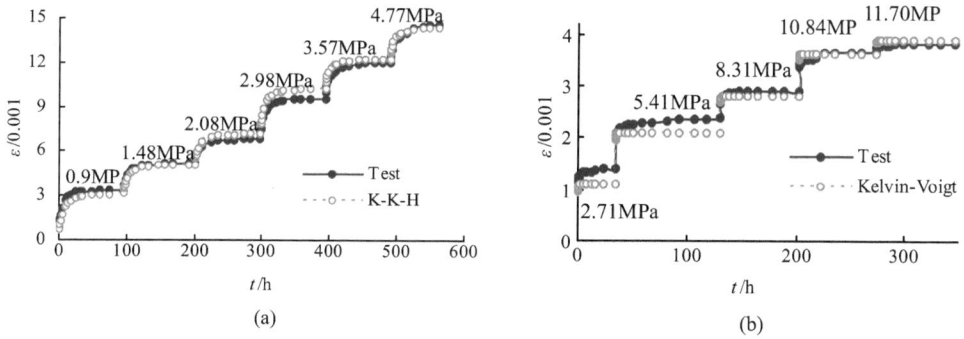

Figure 10 Comparison of predictions of the recognized models with creep testing results, the model is obtained using genetic programming and an improved PSO algorithm: (a) celadon argillaceous rock; (b) fuchsia argillaceous rock. (After Feng *et al.*, 2006).

These intelligent recognition methods are summarized in the context of constitutive modeling, which is one of the main topics of IRM. Due to their universal applicability, they can be used extensively to other rock mechanics modeling tasks, as done in recognition of the mathematic model for calculating the safety factor of rock slopes (Yang *et al.*, 2004) and the displacement time series forecasting models (Yang & Feng, 2004). Furthermore, they can be used combining with numerical simulating techniques to perform intelligent numerical analysis with progressively improved models and parameters.

3.3 Intelligent estimation of rock mechanical parameters

3.3.1 Intelligent modeling of nonlinear relationship between simple testing indices and mechanical parameters

Rock mechanical parameters can be experimentally determined using direct or indirect methods. Limited laboratory facilities and difficulties in obtaining high-quality core samples make it costly and time-consuming for direct testing, especially for rock types with complicated structures, such as intercalated clay layers and carbonate rocks. The indirect methods use simple index parameters and/or mineralogical analyses as well as basic mechanical tests, *e.g.* physical tests including weight and porosity, elastic wave velocity, Schmidt hammer and point-load index, to establish predictive models through regression analysis. However, the relations between rock mechanical parameters (*e.g.* uniaxial compressive strength and elasticity modulus) and those simple parameters are usually nonlinear. In any cases, it is difficult to estimate exactly the relevant mathematical equations by traditional regressive analysis. Feng & Wang (1994) presented a NN-based method with several improved training algorithms to model the nonlinear relation between residual strength and grain-size indices of intercalated clay layers. It is concluded that the relative error of the learned relations for extrapolating analysis was less than 7% on the average. This method has been used extensively by others in estimation of uniaxial compressive strength (Meulenkamp & Grima,1999; Singh *et al.*, 2001; Tiryaki, 2008; Cevik *et al.*, 2011; Yagiz *et al.*, 2012; Rabbani *et al.*, 2012; Kumar *et al.*, 2013; Yesiloglu-Gultekin *et al.*, 2013; Yurdakul & Akdas, 2013), elasticity modulus (Sonmez *et al.*, 2006; Yilmaz & Yuksek, 2008, 2009; Yagiz *et al.*, 2012; Ocak & Seker, 2012; Kumar *et al.*, 2013) and tensile strength (Ceryan *et al.*, 2013) of intact rocks, and deformation modulus (Majdi & Beiki, 2010) of rock mass.

3.3.2 Intelligent back analysis of the rock mechanical parameters

Back-analysis procedure can be defined as an optimization problem for finding the mechanical parameters of rock masses by minimizing the difference between the observed and calculated values, *e.g.*, displacement, stress, stain and excavation damage zone (EDZ) thickness. Thus, the solving procedure shown in Figure 5 can be easily extended to solve such optimization problem by using numerical simulation for forward analysis instead. This method has been validated that can give reasonable balance between the uniqueness and precision of the estimation results (Yang *et al.*, 2010). To solve such nonlinear inversion problem for large scale rock engineering with complicated conditions, Feng *et al.* (2000, 2004) proposed a more comprehensive method. The object function can be expressed as

$$F(\mathbf{p}) = \left\{ \frac{1}{N} \sum_{i=1}^{q} \sum_{j=1}^{w} \left[SVM_{ij}(\mathbf{p}) - u_{ij} \right]^2 \right\}^{1/2} \tag{3}$$

where $\mathbf{p}$ is a set of parameters to be recognized, $SVM_{ij}(\mathbf{p})$ is the calculated displacement component j at the ith monitoring point using a support vector machine (SVM) model trained to simulate the forward numerical analysis process (a NN model can be used as well), u_{ij} is the measured displacement, q is the number of monitoring points, w is the number of displacement component, w × q = N. To construct the SVM/NN model for forward analysis, an extra training process has to be conducted. The learning and testing cases are generated by inputting sample parameter sets obtained through orthogonal design method (uniform design method) into the numerical model to calculated the sample measurements. The obtained SVM/NN model, combined with a GA/PSO based global search procedure, is then used for problem solving. A flow chart of the intelligent back analysis method is illustrated in Figure 11. These kinds of methods have been successfully used in the highwall stability analysis of the permanent shiplock at the Three Gorges Project in China (Feng *et al.*, 2000), and in estimation of

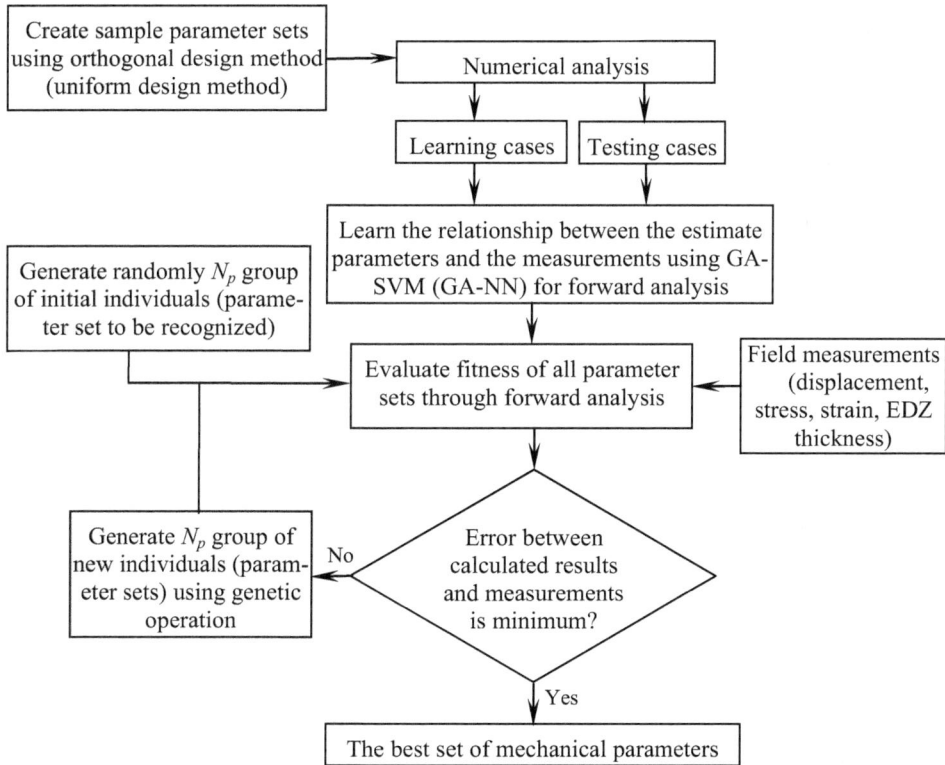

Figure 11 Flow chart for intelligent back-analysis using GA-SVM, GA and FEM (N_p is the population size).

the mechanical parameters of subclay, strongly weathered tuff and weakly weathered tuff of Bachimen slope, Funing expressway, Fujian, China (Feng *et al.*, 2004b). For the later application, an improved reinforcement scheme was suggested and applied based on the back-analyzed results, in which large amount of anchor cables were eliminated while some drainage measures were added. The in situ monitoring results indicated that the slope reinforced using the new scheme is stable. Compared to the original budget, more than 11 million RMB was saved in such a remedial engineering. Because of the great performance, their work was quickly followed by many others (Deng & Lee, 2001; Chen & Zhou, 2002; Zhao, 2006; Yu *et al.*, 2007; Wan *et al.*, 2011; Qu *et al.*, 2012; Jiang & Wang, 2012; Zhang *et al.*, 2013).

For some extremely complicated sites, *e.g.* high ground stress, a single type of measurements at different points may not be sufficient for back analysis of different kinds of rock mechanical parameters (deformation parameters and strength parameters). A multi-information back analysis method is presented for understanding the particular mechanical activity of rock mass under high earth stress (Jiang *et al.*, 2007). EDZ thickness and increment displacement of rock mass were selected, through sensitivity analysis of key rock mechanical parameters, to define a compound object function, namely

$$
\begin{aligned}
F(p) = \alpha \sum_{i=1}^{q} [NN(n; h_1, ..., h_k; m)_{disp,i}(\mathbf{p}) - u_{disp,i}]^2 \\
+ (1-\alpha) \sum_{j=1}^{w} [NN(n; h_1, ..., h_k; m)_{EDZ,j}(\mathbf{p}) - u_{EDZ,j}]^2
\end{aligned}
\tag{4}
$$

where $NN(n; h_1, ... , h_k; m)_{disp,i}(\mathbf{p})$ and $NN(n; h_1, ... , h_k; m)_{EDZ,j}(\mathbf{p})$ are the calculated displacement increment at point i and EDZ thickness at point j, respectively, $u_{disp,i}$ and $u_{EDZ,i}$ are the measured displacement increment and EDZ thickness, respectively. q is the number of displacement, w is the number of EDZ thickness monitoring points, α is the weight coefficient determined through sensitivity analysis. The method has been applied to stability analysis for the large caverns of Laxiwa hydropower station in China, constructed in the site with high tectonic stress (Jiang *et al.*, 2007). The five key rock mechanical parameters of CWFS model are recognized.

3.4 Integrated intelligent modeling of excavated deformation dynamics of large engineering

For large rock engineering, it is usually excavated in steps. Deformation of the excavated rock mass in each step is one of the most important information for assessing the stability status. By combining the neural dynamic modeling methods proposed by Feng *et al.* (1996, 1999, 2004), Feng & Seto (1998, 1999) and the intelligent numerical analysis with progressively improved models, an integrated intelligent analysis method for estimation of the deformation dynamics was presented. As seen in Figure 12, the method was integrated with NN (or SVM)-based time series modeling and numerical modeling based on intelligent back analysis method. For the NN (or SVM)-based time series modeling method, the relation between displacement data monitored at previous steps t-p-1, ... , t−1 and current step t is firstly learned by NN (or SVM) to construct a predicting model. The model is then used to extrapolate the time series to predict

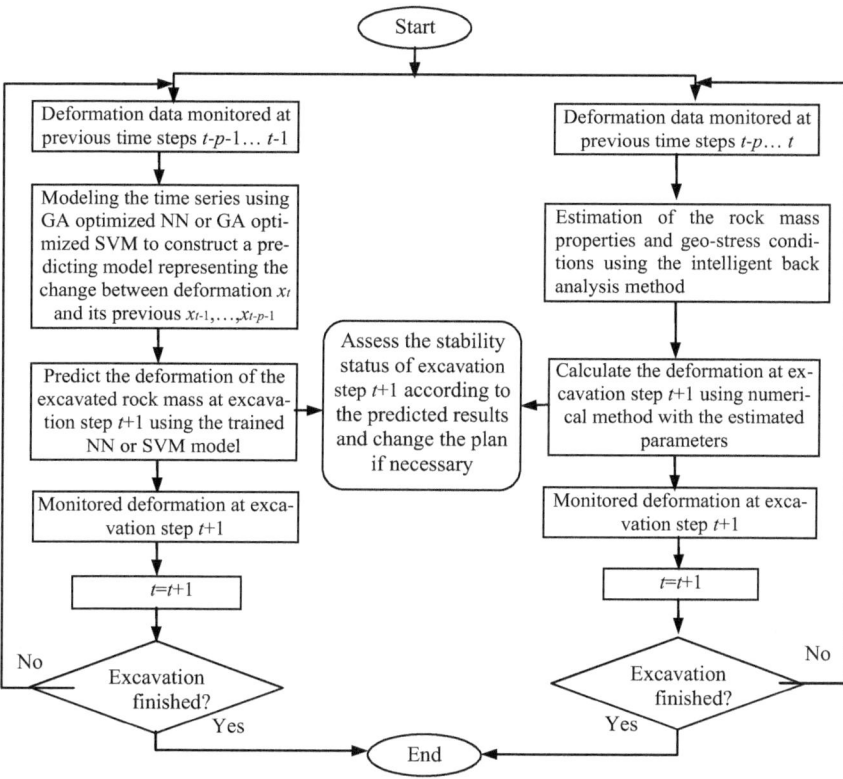

Figure 12 Integrated intelligent estimation of excavated deformation for rock slope engineering.

deformation at step $t+1$ with input of the monitored displacement data at time step $t-p, \ldots, t$, and so on. The new monitoring data can be fed back continuously as input for model re-learning. In the intelligent numerical method, the monitored displacement data at previous excavation steps can also be used for back analysis of the rock mechanical parameters. The estimated results can then be input to perform a forward numerical analysis to predict the deformation of the rock mass at the next steps. The back-analysis and prediction can be continuously performed in this way. The prediction results of these two methods can be used to assess the stability status and to help making decisions. The efficiency of this integrated intelligent method has been validated in the analysis of the high rock slope of the permanent shiplock of the Three Gorges Project (Feng *et al.*, 1999) and the horizontal deformation at depth in the Bachimen landslide in Fujian Province, China (Feng *et al.*, 2004a).

3.5 Intelligent optimal design of excavation sequences and support parameters

For high slope and large underground openings/cavern group excavated in steps, the excavation sequences and support parameters may significantly affect the stability of the engineering, as well as the construction cost. An intelligent optimal design method

(a)

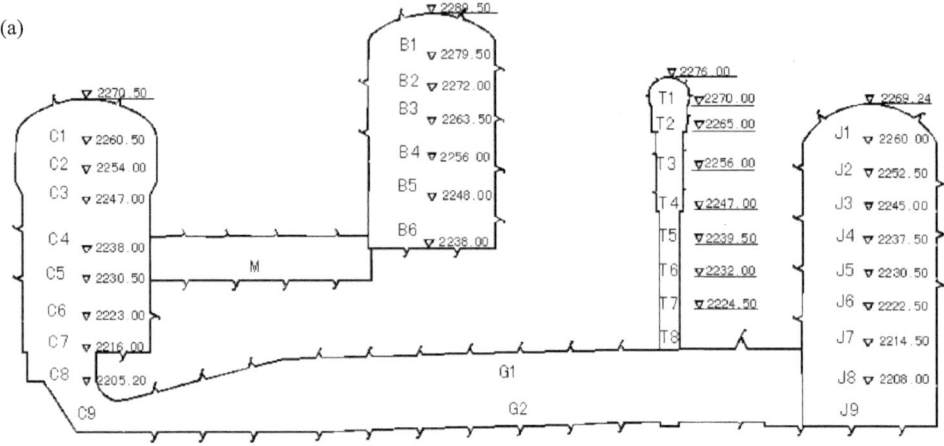

(b)

Stage	Main power house	Main transformer cave	Pressure balance chamber	Draft tube gate chamber	Others
1	C1	B1	J1	T1	
2	C2	B2	J2	T2	
3	C3	B3	J3	T3	
4	C4	B4	J4	T4	M
5	C5	B5	J5	T5	
6	C6	B6	J6	T6	
7	C7		J7	T7	
8	C8		J8	T8	G1
9	C9		J9	T9	G2

(c)

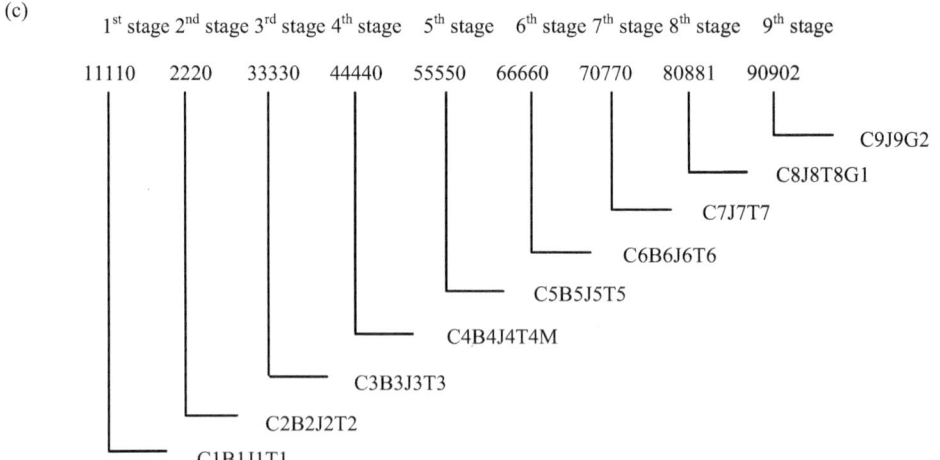

Figure 13 A typical coding of the excavation sequences of a cavern group: (a) Excavation blocks of the cavern group; (b) Example of some excavation scheme; (c) Coding of the excavation sequences.

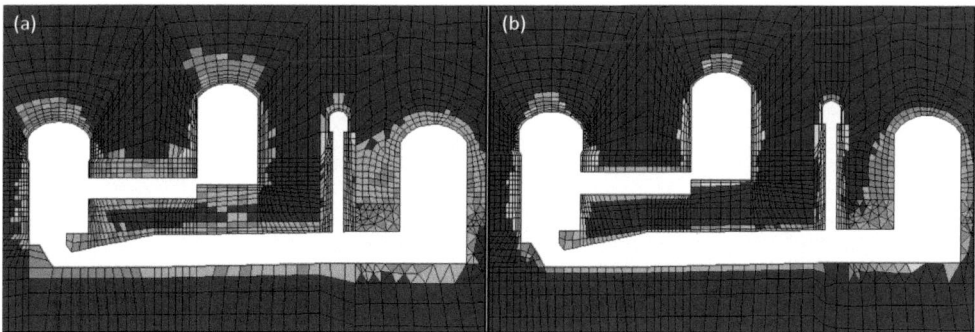

Figure 14 Comparison of excavation induced plasticity area between (a) Original design, (b) Optimized design using the methodology of IRM.

has been presented by extending the intelligent back analysis method to such a multi-object optimization problem (Su *et al.*, 2007). The object function can be expressed as

$$f(x) = \sum_{i}^{n} w_i p_i \tag{5}$$

where $\mathbf{x}$ is a coding of excavation sequences and support parameters to be optimized. The coding of excavation sequences is illustrated in Figure 10. p_i is the standardized indices calculated as

$$p_i = x_i/s, s = \left(\frac{1}{m-1}\sum_{i=1}^{m}(x_i - \overline{x})^2\right)^{1/2}, \quad \overline{x} = \frac{1}{m}\sum_{1}^{m}x_i$$

w_i is the weight coefficient determined using group decision method according to the specific engineering. n is the number of optimization indices. The optimization indices can include the displacement in key points, plastic area, energy release rate, and also economic aspects estimated based on the support parameters.

The intelligent optimal design method was used to optimize the excavation sequence and support parameters of the large caverns of Laxiwa hydropower station located on the Yellow River in China. The results are shown in Figure 14, where the excavation induced plasticity area reduced significantly by using the optimized design.

3.6 Intelligent modeling of rockbursts

Rockbursts have been becoming one of the most serious geological disasters in rock engineering. However, due to the numerous affecting factors with complicated relationship, the mechanism is still far from well-know. Theoretic models based on single criterion (strain energy index, stress-strength ratio, brittleness index and so on) cannot satisfy the need of accurate prediction. Indirect methods, such as experience analogy analysis, drilling bits, underground sound monitoring, water content rate determination and so on, have difficulties in deriving deterministic expressions or in dealing with massive data with many uncertainties. Feng and his coworkers (Feng & Wang,

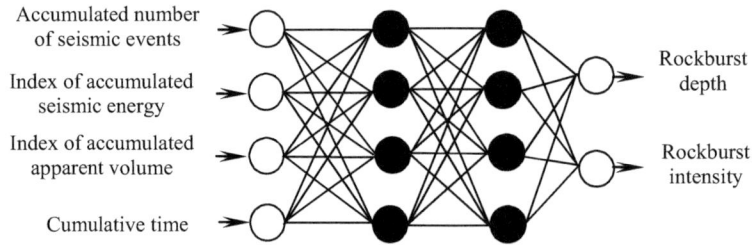

Accumulated number
of seismic events

Index of accumulated
seismic energy

Index of accumulated
apparent volume

Cumulative time

Rockburst
depth

Rockburst
intensity

Figure 15 Neural network model structure for rockburst risk prediction based on monitored microseismic information.

1994; Feng *et al.*, 1998) developed a NN model by considering different kinds of index parameters including stress-strength ratio, brittle index of rock, intersection angle between main joint set and maximum principal stress, stress descent index and elastic energy index. These models were successfully applied in the rockburst risk assessment in deep gold mines in South Africa. This method inherited by others for different engineering (Bai *et al.*, 2002; Chen *et al.*, 2003; Li & Xiao, 2008; Sun *et al.*, 2009). Recently, with application to excavation of deep and long diversion tunnels of Jinping II hydropower power station in China, a new NN model for data mining from micro-seismic monitoring data was developed (Chen *et al.*, 2013). Figure 15 shows the NN model structure.

4 INTEGRATED INTELLIGENT DESIGN CONCEPTS FOR COMPLICATED ROCK ENGINEERING

The above mentioned IRM methods have been applied extensively for different modeling tasks in large rock engineering constructed in China during the past decade. They have been used as complementary approaches to standard approach to deal with such complicated systems with multi-information and multi-task. Great performances were obtained through benefiting from the advantages of each. Thus, some more sophisticated design concepts have come into being for different kinds of large rock engineering. Figure 16 depicts the general diagram of comprehensively integrated design concepts based on collaborative problem solving. In this section, the integrated intelligent design concepts for rock slope engineering, landslide treatment engineering and underground caverns are typically discussed.

4.1 Integrated intelligent design for rock slopes

The integrated system includes integration of multiple tasks and integration of analysis methods (Figure 17). The former includes intelligent recognition of rock mass parameters and rock mechanical models, rock mass classification, determination of slope angle with NN model trained with experienced cases, recognition of potential failure modes suing integrated rock mechanics expert system model, estimation of the safety factor, and analysis of slope performance. The latter includes integration of potential methods used to solve one task or multiple tasks. There are four methods used to analyze stability of the slope: limit equilibrium analysis, numerical analysis,

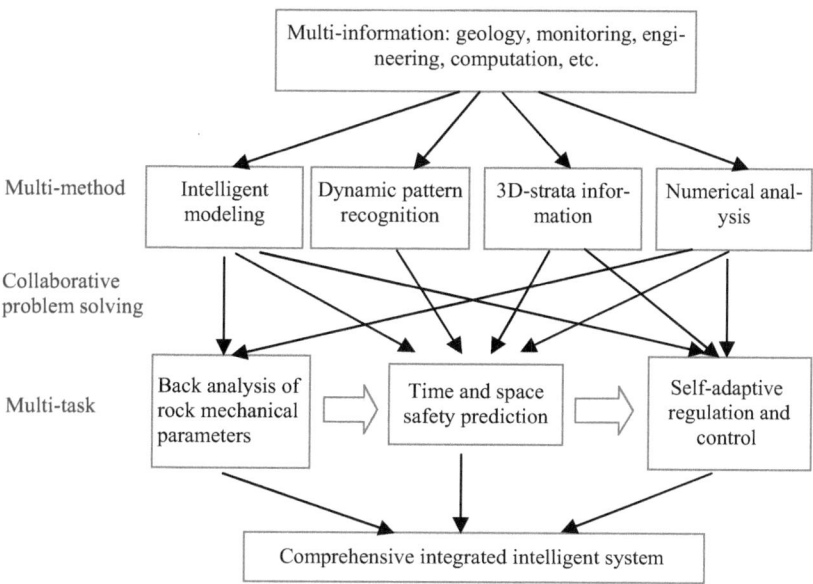

Figure 16 Diagram of integrated intelligent design concepts for rock engineering based on collaborative problem solving with multi-information, multi-method and multi-task.

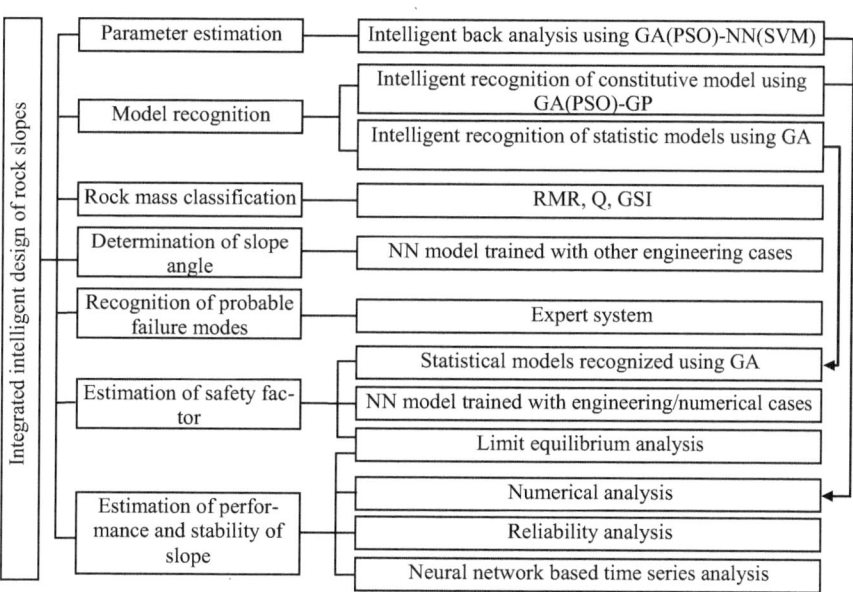

Figure 17 Hierarchical chart of integrated intelligent design of rock slope.

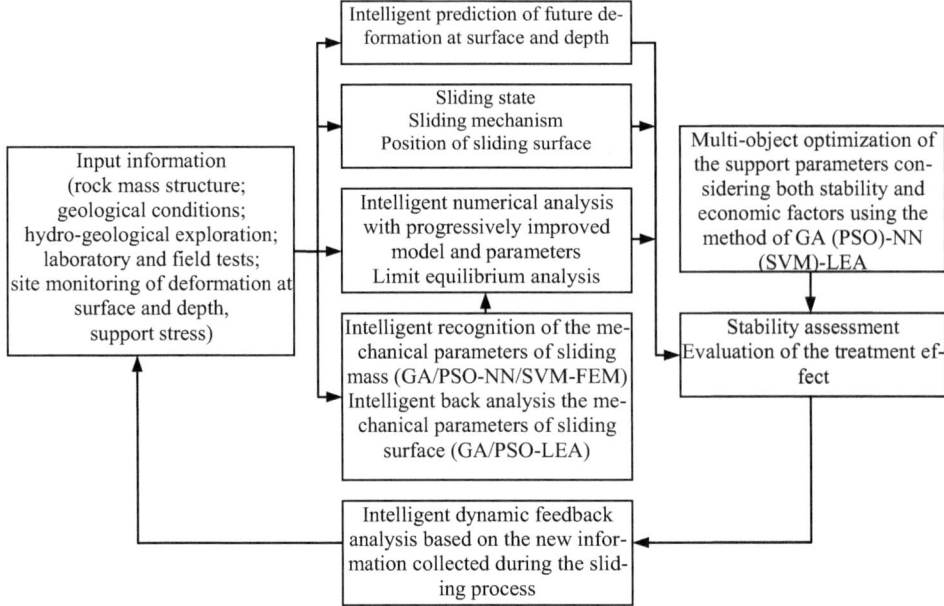

Figure 18 Integrated intelligent design for landslide treatment engineering.

reliability analysis and neural network-based time series analysis. There are three methods used to estimate the safety factor in the system: statistical models recognized using GA, NN models trained with engineering cases or numerical cases, and limit equilibrium analysis. The data and cases dealt with in the analysis process can be saved into a database and casebase. If the slope performance is not as expected, it is fed back to check the determination of slope angle and recognition of potential failure modes.

4.2 Integrated intelligent design for landslide engineering

Figure 18 shows an integrated intelligent design concept for landslide treatment engineering. A systematic exploration of the rock mass structure, geological conditions, hydrogeological conditions, as well as laboratory test of rock mass properties, are firstly conducted to determine the sliding state, sliding mechanism and position of sliding surface. Based on this information, a numerical model (FEM, FLAC^{3d}, etc.) and a limit equilibrium analysis (LEA) model can be established, respectively. The results from intelligent time series forecasting method, numerical simulation with mechanical parameters of the sliding mass obtained using intelligent back analysis method, and LEA performed with the mechanical parameters of sliding surface recognized through a GA(PSO)-LEA back analysis procedure, are comprehensively analyzed to access the stability status and evaluate the treatment effect. The analysis results can be used either for optimization of the support parameter accounting for both stability and economic factors or to making adjustment decisions. With the successively monitored information, the whole analysis process is dynamically fed back to improve the treatment scheme progressively.

Treatment analysis for several landslides in Three Gorges Project reservoir area and Bachimen landslide in Fujian Province, China was performed with the integrated intelligent method. The optimized treatment measures not only satisfy the stability well but also bring great economic savings.

4.3 Intelligent design for underground caverns

During the excavation of large scale engineering, the input information, such as the geological conditions, rock mechanic properties, rock stress, etc., may vary dramatically. Usually it is improbable to obtain a full understanding beforehand and it has to be revealed progressively during the excavation. Based on these views, an intelligent and dynamic design scheme is proposed for the stability analysis, design and construction of a large cavern group. The method aims at the safe and economic construction for the large cavern group under complicated geological conditions. There are two ways to access the stability status. One uses the expert system to model the existed experiences and using NN (or SVM) to gain new experience knowledge from other rock engineering that has been constructed. The second way is through conventional rock mass stability ranking analysis and numerical analysis based on intelligent recognition of rock mechanic models and intelligent back analysis of rock mass parameters. The stability assessment results are presented for intelligent optimization of excavation sequences and support parameter to making adjustment decisions for the next excavation. The whole analysis process receives dynamic feedback from the new information revealed by successive excavations.

A typical operating process of the integrated intelligent design concepts is shown in Figure 19. It has been in good practice of design and construction of underground

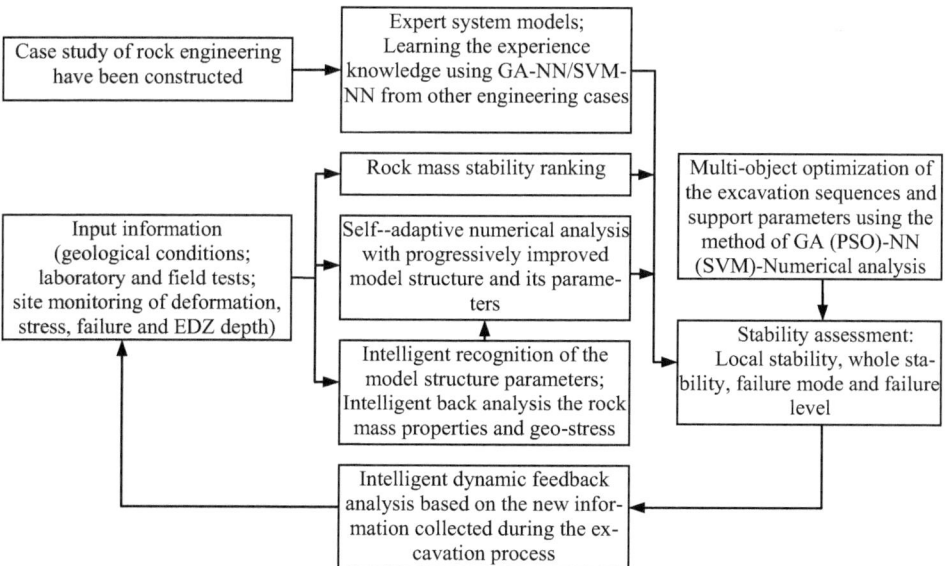

Figure 19 Integrated intelligent design scheme for large underground caverns.

powerhouse at Shuibuya hydropower station (Hubei, China) associated with complicated strata consisting of alternating soft and hard rocks, Laxiwa hydropower station with high tectonic stress and the Jinping II hydropower station with great overburden depth.

5 CONCLUSIONS AND FUTURE PERSPECTIVES

This work reviews the progress in research into IRM since it was proposed 20 years ago. The evolution of an integrated intelligent dynamic design concept for rock engineering is presented, showing the trends and progress accomplished during the past two decades. Some of the most important methods and applications with respect to the complicated rock engineering conducted in China in the recent decade are also presented. From what was discussed in this review it can be observed that the IRM has been a powerful tool for complicated rock engineering problems. It provides a powerful selection to the questions, "How do we select one or several analysis methods for the given geological conditions and engineering required? How do we decide what kind of results are good or more suitable?" as asked for IRM in its origin site [2].

The details of the techniques, algorithms and tools developed for IRM are not exhaustively presented since there are too many candidates, such as neural networks, support vector machine, genetic algorithms, genetic programming, particle swarm optimization, expert systems, fuzzy mathematics, rough sets, chaos, fractal and their various hybrid systems. Most of them work in a data-based manner. The data must be carefully collected, prepared and transmitted to represent the distinctive characters of the rock engineering under analysis as has been specified in [8].

It can be noted that design and management aspects of rock engineering has been concerned more and more closely in IRM. This offers IRM a new life in application to rock engineering. It also provides us new and inspiring development fields.

IRM is an open research direction consisting of very deep and rich contents. For example, collaborative research of the IRM on the internet is one of the most interesting aspects that will provide a powerful information sharing platform for domain scientists from various fields. A conceptual model has been suggested in [4]. Significant efforts are still required to bring the new branch into worldwide application.

REFERENCES

Bai, M.Z., Wang, L.J., Xu, Z.Y. (2002) Study on a neutral network model and its application in predicting the risk of rock burst. *China Safety Science Journal*, 12, 65–69.

Cai, J.G., Zhao, J., Hudson, J.A. (1998) Computerization of rock engineering systems using neural networks with an expert system. *Rock Mechanics and Rock Engineering*, 31(3), 135–152.

Ceryan, N., Okkan, U., Samui, P., *et al.* (2013) Modeling of tensile strength of rocks materials based on support vector machines approaches. *International Journal for Numerical and Analytical Methods in Geomechanics*, 37, 2655–2670.

Cevik, A., Sezer, E.A., Cabalar, A.F., *et al.* (2011) Modeling of the uniaxial compressive strength of some clay-bearing rocks using neural network. *Applied Soft Computing*, 11(2), 2587–2594.

Chen, B.R., Feng, X.T., Yao, H.Y., *et al.* (2010) Study on mechanical behavior of limestone and simulation using neural network model under different water-chemical environment. *Rock and Soil Mechanics*, 31(4), 1173–1180.

Chen, H.J., Li, N.H., Ni, D.X., *et al.* (2003) Prediction of rockburst by artificial neural network. *Chinese Journal of Rock Mechanics and Engineering*, 22, 762–768.

Chen, Y.F., Zhou, C.B. (2002) Back analysis of elastoplastic mechanical parameters of complex dam foundation on the basis of orthogonal experiments. *Rock and Soil Mechanics*, 23(4), 450–454.

Deng, J.H., Lee, C.F. (2001) Displacement back analysis for a steep slope at the Three Gorges Project site. *International Journal of Rock Mechanics and Mining Sciences*, 38(2), 259–268.

Feng, X.T. (1991) *Study on intelligent system of optimal design of open support*. PhD Dissertation, Northeastern University, 1991.

Feng, X.T. (2000) *An Introduction to Intelligent Rock Mechanics*. Beijing, Science Press (in Chinese).

Feng, X.T., Chen, B.R., Yang, C.X., *et al.* (2006) Identification of visco-elastic constitutive models of rock materials using coupling genetic programming with particle swarm optimization algorithm. *International Journal of Rock Mechanics and Mining Sciences*, 43(5), 789–801.

Feng, X.T., Hudson, J.A. (2004) The ways ahead for rock engineering design methodologies. *International Journal of Rock Mechanics and Mining Sciences*, 41(2), 255–73.

Feng, X.T., Hudson, J.A. (2010) Specifying the information required for rock mechanics modeling and rock engineering design. *International Journal of Rock Mechanics and Mining Sciences*, 47(2), 179–194.

Feng, X.T., Hudson, J.A. (2011) *Rock Engineering Design*. London, CRC Press.

Feng, X.T., Katsuyama, K., Wang, Y.J., *et al.* (1997) A new direction-Intelligent rock mechanics and rock engineering. *International Journal of Rock Mechanics and Mining Sciences*, 34(1), 135–141.

Feng, X.T., Lin, Y.M. (1992a) An expert system for rock mass classification of underground engineering. *29th International Geological Congress*, Japan.

Feng, X.T., Lin, Y.M. (1992b) Research of zoning engineering stability of rock mass. *Proc. of International ISRM Symposium on Rock Characterization*, Thomas Telford, UK.

Feng, X.T., Lin, Y.M. (1993) *Expert Systems in Rock Mechanics*. China, Liaoning Science and Technology Press. (in Chinese)

Feng, X.T., Li, S.J., Liao, H.J., *et al.* (2002) Identification of nonlinear rock-like material constitutive models using genetic algorithm. *International Journal for Numerical and Analytical Methods in Geomechanics*, 26(8), 815–830.

Feng, X.T., Seto, M. (1998) Neural network dynamic modelling of rock microfracturing sequences under triaxial compressive stress conditions. *Tectonophysics*, 292(3–4), 293–309.

Feng, X.T., Seto, M. (1999) A new method of modelling the rock microfracturing process in double-torsion experiments using neural networks. *International Journal for Numerical and Analytical Methods in Geomechanics*, 23(9), 905–923.

Feng, X.T., Wang, Y.J. (1994) A study on relationship of grain-size indices and strength of intercalated clay layers using neural network tenchniques. *Proc. of International Conference on Computationsl Methods in Structural and Geotechnical Engineering*, Hong Kong.

Feng, X.T., Wang, Y.J. (2000) Some thoughts concerning the development of intelligent rock mechanics. *ISRM News Journal*, 6(2),14–17.

Feng, X.T., Wang, L.N. (1994) Rock-burst prediction based on neural networks. *Transactions of Nonferrous Metals Society of China*, 4(1), 7–14.

Feng, X.T., Wang, Y.J., Yao, J.G. (1996) Neural network model for real-time roof pressure prediction in coal mines. *International Journal of Rock Mechanics and Mining Sciences & Geomechanics Abstracts*, 33(6), 647–653.

Feng, X.T., Webber, S., Ozbay, M.U. (1998) Neural network assessment of rockburst risks for deep gold mines in South Africa. *Transactions of Nonferrous Metals Society of China*, 8(2), 335–341.

Feng, X.T., Yang, C.X. (1999) Intelligent rock mechanics (2) – Intelligent recognition of input parameters and constitutive models. *Chinese Journal of Rock Mechanics and Engineering*, 18(3), 350–353.

Feng, X.T., Yang, C.X. (2001) Genetic evolution of nonlinear material constitutive models. *Computer Methods in Applied Mechanics and Engineering*, 190(45), 5957–5973.

Feng, X.T., Yang, C.X. (2004) Coupling recognition of the structure and parameters of nonlinear constitutive material models using hybrid evolutionary algorithms. *International Journal for Numerical Methods in Engineering*, 59, 1227–1250.

Feng, X.T., Zhang, Z.Q., Xu, P. (1999) Adaptive and intelligent prediction of deformation time series of high rock excavation slope. *Transactions of Nonferrous Metals Society of China*, 9(4), 842–846.

Feng, X.T., Zhang, Z.Q., Sheng, Q. (2000) Estimating mechanical rock mass parameters relating to the Three Gorges Project permanent shiplock using an intelligent displacement back analysis method. *International Journal of Rock Mechanics and Mining Sciences*, 37(7), 1039–1054.

Feng, X.T., Zhao, H.B., Li, S.J. (2004a) Modeling non-linear displacement time series of geo-materials using evolutionary support vector machines. *International Journal of Rock Mechanics and Mining Sciences*, 41, 1087–1107.

Feng, X.T., Zhao, H.B., Li, S.J. (2004b) A new displacement back analysis to identify mechanical geo-material parameters based on hybrid intelligent methodology. *International Journal for Numerical and Analytical Methods in Geomechanics*, 28(11), 1141–1165.

Fuenkajorn, K., Kamutchat, S. (2001) Rock slope design using expert system: ROSES program. The 6th Mining, Metallurgical, and Petroleum Engineering Conference, October 24–26, 2001, Bangkok, Thailand.

Ghaboussi, J., Garrett, J.H. Jr., Wu, X. (1991) Knowledge-based modeling of material behavior with neural networks. *Journal of Engineering Mechanics*, 117(1), 132–153.

Ghaboussi, J., Sidarta, D.E., Lade, P.V. (1994) Neural network based modeling in geomechanics. In: Siriwardane and Zaman (eds.) *Computer Methods and Advances in Geomechanics*, Balkema.

Hudson, J.A., Feng, X.T. (2007) Updated flowcharts for rock mechanics modelling and rock engineering design. *International Journal of Rock Mechanics and Mining Sciences*, 44(2), 174–195.

Jiang, A.N., Wang, J.X. (2012) Research of intelligent displacement back analysis of dalian subway tunnel excavation disaster prediction. *Disaster Advances*, 5(4), 1504–1509.

Jiang, Q., Feng, X.T., Su, G.S., *et al.* (2007) Intelligent back analysis of rock mass parameters for large underground caverns under high earth stress based on EDZ and increment displacement. *Chinese Journal of Rock Mechanics and Engineering*, 26(s1): 2654–2662.

Juang, C.H., Lee, D.H. (1989) Development of an expert system for rock mass classification. *Civil Engineering Systems*, 6(4), 147–156.

Khandelwal, M. (2012) Application of an expert system to predict thermal conductivity of rocks. *Neural Computing & Applications*, 21(6), 1341–1347.

Kumar, B.R., Vardhan, H., Govindaraj, M., & Vijay, G.S. (2013) Regression analysis and ANN models to predict rock properties from sound levels produced during drilling. *International Journal of Rock Mechanics and Mining Sciences*, 58, 61–72.

Li, T.-B., Xiao, X.P. (2008) Comprehensively integrated methods of rockburst prediction in underground engineering. *Advanced Earth Science*, 23, 533–540.

Lin, Y.M., Feng, X.T. (1991) Expert system of discontinuity properties evaluation. In: Roegiers (ed.) *Rock Mechanics as a Multidisciplinary Science*, Balkema.

Majdi, A., Beiki, M. (2010) Evolving neural network using a genetic algorithm for predicting the deformation modulus of rock masses. *International Journal of Rock Mechanics and Mining Sciences*, 47, 246–253.

Meulenkamp, F., Grima, M.A. (1999) Application of neural networks for the prediction of the unconfined compressive strength (UCS) from Equotip Hardness. *International Journal of Rock Mechanics and Mining Sciences*, 36, 29–39.

Millar, D.L., Calderbank, P.A. (1995) On the investigation of a multilayer feedforward neural network model of rock deformability behaviour. In: Fujii, T. (Ed.), *Proceedings of the 8th ISRM Congress on Rock Mechanics*, Balkema, Rotterdam

Ocak, I., Seker, S.E. (2012) Estimation of elastic modulus of intact rocks by artificial neural network. *Rock Mechanics and Rock Engineering*, 45, 1047–1054.

Qu, C., Qiao, L., Cheng, F. (2012) Optimization of design parameters and intelligent back-analysis of high bench dump slope in open pit. *Journal of Theoretical and Applied Information Technology*, 46(1), 420–425.

Rabbani, E., Sharif, F., KoolivandSalooki, M., *et al.* (2012) Application of neural network technique for prediction of uniaxial compressive strength using reservoir formation properties. *International Journal of Rock Mechanics and Mining Sciences*, 56, 100–111.

Rafiai, H., Jafari, A. (2011) Artificial neural networks as a basis for new generation of rock failure criteria. *International Journal of Rock Mechanics and Mining Sciences*, 48,1153–1159.

Rafiai, H., Jafari, A., Mahmoudi, A. (2013) Application of ANN-based failure criteria to rocks under polyaxial stress conditions. *International Journal of Rock Mechanics and Mining Sciences*, 59, 42–49.

Schubert, W., Grossauer, K. (2007) *The application of an expert system for the basic design of tunnels*. In: Eberhardsteiner, J. *et al.* (eds.) *ECCOMAS Thematic Conference on Computational Methods in Tunnelling*, August 27–29, 2007, Vienna, Austria.

Singh, V.K., Singh, D., Singh, T.N. (2001) Prediction of strength properties of some schistose rocks from petrographic properties using artificial neural networks. *International Journal of Rock Mechanics and Mining Sciences*, 38(2), 269–284.

Sonmez, H., Gokceoglu, C., Nefeslioglu, H.A., *et al.* (2006) Estimation of rock modulus: For intact rocks with an artificial neural network and for rock masses with a new empirical equation. *International Journal of Rock Mechanics and Mining Sciences*, 43, 224–235.

Su, G.S., Feng, X.T., Jiang, Q., *et al.* (2007) Intelligent method of combinatorial optimization of excavation sequence and support parameters for large underground caverns under condition of high geostress. *Chinese Journal of Rock Mechanics and Engineering*, 26(s1), 2800–2808.

Sun, J., Wang, L.G., Zhang, H.L., *et al.* (2009) Application of fuzzy neural network in predicting the risk of rockburst. *Procedia Earth Planetary Science*, 1, 536–543.

Tapia, M.A., Valverde, M.A., Amadei, B., *et al.* (1998) The REX expert system: A new alternative for rock excavation design. *International Journal of Rock Mechanics and Mining Sciences*, 35(4/5), 675–676.

Tiryaki, B. (2008) Predicting intact rock strength for mechanical excavation using multivariate statistics, artificial neural networks, and regression trees. *Engineering Geology*, 99, 51–60.

Wan, L.Y., Zhang, X.F., Liu, K.Y. (2011) Intelligent displacement back analysis method of three-dimension applied in unsymmetrical pressure tunnel with shallow depth. *Applied Mechanics and Materials*, 90, 2286–2291.

Wang, Y.J., Feng, X.T. (1997) Intelligent rock mechanics: Development and application. *Computer Methods and Advances in Geomechanics*, 1, 645–650.

Wu, H.C., Zhang, X.B., Bao, T., *et al.* (2000) Modeling the stress-strain relation for granite using finite element-neural network hybrid algorithms. *Computer Methods and Advances in Geomechanics: Proceedings of the 10th International Conference on Computer Methods and Advances in Geomechanics*, 7–12 January, 2001, Tucson, Arizona, USA, CRC Press, pp. 200–241.

Yagiz, S., Sezer, E.A. and Gokceoglu, C. (2012) Artificial neural networks and nonlinear regression techniques to assess the influence of slake durability cycles on the prediction of uniaxial compressive strength and modulus of elasticity for carbonate rocks. *International Journal for Numerical and Analytical Methods in Geomechanics*, 36, 1636–1650.

Yang, C.X., Feng, X.T. (2004) Evolutionary identification of deformation dynamics of geotechnical structures. *Key Engineering Materials*, 274–276, 949–954.

Yang, C.X., Feng, X.T., Chen, B.R. (2008) *Modeling creep behavior of soft rock using an evolutionary neural computing method*. In: Fan, J.H. & Chen, H.B. (eds.) *Advances in Heterogeneous Material Mechanics*, Lancaster, DEStech Publications, pp 456–459.

Yang, C.X., Tham, L.G., Feng, X.T., *et al.* (2004) Two-Stepped evolutionary algorithm and its application to stability analysis of slopes. *Journal of Computing in Civil Engineering*, 18(2), 145–153

Yang, C.X., Wu, Y.H., Hon, T. (2010) A no-tension elastic-plastic model and optimized back-analysis technique for modeling nonlinear mechanical behavior of rock mass in tunneling. *Tunneling and Underground Space Technology*. 25(3), 279–289.

Yesiloglu-Gultekin, N., Gokceoglu, C., Sezer, E.A. (2013) Prediction of uniaxial compressive strength of granitic rocks by various nonlinear tools and comparison of their performances. *International Journal of Rock Mechanics and Mining Sciences*, 62,113–122

Yilmaz, I., Yuksek, A.G. (2008) An example of artificial neural network (ANN) application for indirect estimation of rock parameters. *Rock Mechanics and Rock Engineering*, 41(5), 781–795.

Yilmaza, I., Yuksek, G. (2009) Prediction of the strength and elasticity modulus of gypsum using multiple regression, ANN, and ANFIS models. *International Journal of Rock Mechanics and Mining Sciences*, 46, 803–810.

Yurdakul, M., Akdas, H. (2013) Modeling uniaxial compressive strength of building stones using non-destructive test results as neural networks input parameters. *Construction and Building Materials*, 47, 1010–1019.

Yu, C.W., & Chern, J.C. (2007) Expert system for D&B tunnel construction. *Underground Space The 4th Dimension of Metropolises*, London, England, pp 799–803.

Yu, Y., Zhang, B., Yuan, H. (2007) An intelligent displacement back-analysis method for earth-rockfill dams. *Computers and Geotechnics*, 34(6), 423–434.

Zhao, H. (2006) Back analysis of intelligent displacement based on particle swarm optimization. *Chinese Journal of Rock & Soil Engineering*, 28(11), 2035–2038.

Zhang, L., Ding, X., Budhu, M. (2012) A rock expert system for the evaluation of rock properties. *International Journal of Rock Mechanics and Mining Sciences*, 50, 124–132.

Zhang, Q., Mo, Y., and Tian, S. (1988) An expert system for classification of rock masses. In: Cundall, P. A., Starfield, A. M. and Sterling, R. L. (Eds.), *Proceedings of the 29th US Symposium on Rock Mechanics*, Balkema, Rotterdam

Zhang, Q., Nie, X. Y., Tian, W. T., et al. (1995). A case-based reasoning system for tunnel support design. In: Fujii, T. (Ed.), *Proceedings of the 8th ISRM Congress on Rock Mechanics*, Balkema, Rotterdam.

Zhang, X., Yang, S., Wu, S., *et al.* (2013) Intelligent Back-analysis of Mechanical Parameters for Landslide Based on Sensitive Grey Correlation Analysis. *EJGE The Electronic Journal of Geotechnical Engineering*, 18, 1205–1214.

Chapter 15

Rock Engineering Systems (RES): An update[1]

John A. Hudson

Emeritus Professor, Earth Science and Engineering, Imperial College London, UK
E-mail: john.a.hudson@gmail.com

Abstract: The Rock Engineering Systems (RES) approach was developed by the author to provide a method of identifying the key factors and their relative importance in a rock engineering project in order to assist in design and in risk reduction. The method enables structured elicitation of the information known by the various parties involved, together with the method for establishing the interaction and dominance of the key factors. The approach is explained and case examples of RES use in the 1992–2015 literature are highlighted.

1 THE PURPOSE OF THE ROCK ENGINEERING SYSTEMS (RES) APPROACH

When embarking on a rock mechanics and engineering project, it is necessary to establish the key information for design and the risks inherent in the project. Given the many factors involved in a rock engineering project, such as geological setting, rock stress, the presence of fractures, hydrogeological conditions, project location, types of anticipated excavation and support methods, etc., it is vital to have a method of establishing the interactions between these factors and their relative interaction intensities and dominance or subordinacy in the rock engineering system. Additionally, it is helpful to have a methodology for eliciting all the relevant information from the key personnel involved in the project. The RES approach enables all this to be done in a structured way, thus assisting in design and in the reduction of risk.

As its basis, the RES approach uses an interaction matrix in which the main parameters governing a particular circumstance (*e.g.*, slope failure, use of blasting or tunnel boring machines, underground stability and support) are selected and the interactions between them are considered. This involves a comprehensive assessment of the factors and interactions, the advantage being that all potential influencing factors can be included initially. The reason why the RES approach reduces epistemic (knowledge) uncertainty is because study of the interactions between the factors indicates which of these are most interactive in the system being considered, which are dominant, and which have a lesser or insignificant contribution.

Also, once the structure of the problem is developed in this way, the acceptability of simplifying assumptions in the project or model is clarified, again reducing the

---

1 This Chapter is an abridged version of Chapter 3 in the 2015 book "Rock Engineering Risk" by J.A. Hudson and X.T. Feng, also published by CRC Press / Balkema, Taylor & Francis Group.

epistemic uncertainty. Moreover, the likelihood of a major hazard being overlooked is reduced. In addition, the subjectivity introduced into a project or model when it is studied by a single person is reduced through a group approach. Although epistemic uncertainty is emphasized here, the reduction of aleatory (chance) uncertainty can also be incorporated into the RES approach through the use of probability distributions and other strategies.

The author developed the Rock Engineering Systems (RES) approach and authored the earlier book "Rock Engineering Systems: Theory and Practice" (Hudson, 1992). Since 1992, *i.e.*, in the 20+ years since then, the approach has been used for a variety of rock engineering problems, including, *inter alia*, the stability of natural and artificial slopes, excavation methods, the stability of underground openings, and organizing the many factors (features, events and processes) in radioactive waste disposal. Moreover, there are now many case studies available from the application of RES to engineering problems in a variety of countries, *i.e.*, Bangladesh, China, Greece, Iran, Italy, Korea, Spain, Sweden, Turkey, UK, and the USA. Firstly, the RES structured approach will be explained, together with the consequential developments that follow naturally from the use of an interaction matrix. Then, a series of RES applications described in the literature will be reviewed.

2 A REVIEW AND EXPLANATION OF THE ROCK ENGINEERING SYSTEMS (RES) METHODOLOGY

The RES methodology is an analytic approach, rather than a synthetic approach: in other words the rock mechanics/engineering model is not built up by assuming certain variables should automatically be included (synthesis), but by studying the problem, breaking the problem down into its constituent variables (analysis), and assessing their significance so that their relative importance can be established—and an appropriate model then constructed. One of the key aspects is firstly to establish the objective of the rock engineering project—because the importance of the respective variables depends on the project objective. A helpful output from the analysis process is the ability to assess the risk of inappropriate modeling and/or design.

2.1 The interaction matrix

The basic tool of the RES methodology is the interaction matrix, as illustrated in Figure 1. This enables a compilation of the manner in which each particular system factor affects all other system factors, together with the complementary manner in which all system factors affect that particular factor. When these effects are established for all the factors being considered, many operations can be conducted on the interaction matrix for a variety of purposes.

A simple 2×2 matrix is illustrated in Figure 1(a) having the two factors/variables Subject A and Subject B. These are placed in the diagonal boxes from top left to bottom right, this being known as the 'leading diagonal' of the matrix. The influence of Subject A on Subject B is contained in the top right box, and the influence of Subject B on Subject A is contained in the bottom left box. In this way, a clockwise influence convention is used. If the influence of Subject A on Subject B is the same as the influence

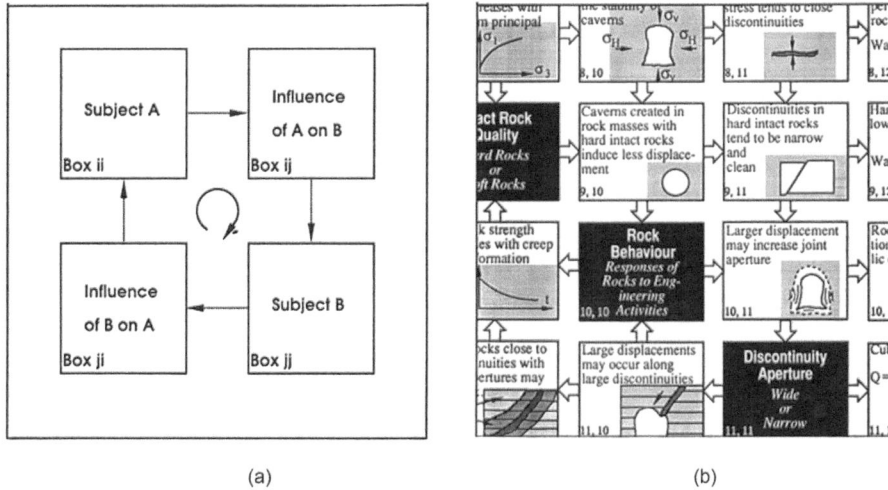

(a)   (b)

Figure 1 The interaction matrix, the basic tool of the RES methodology. (a) a 2x2 interaction matrix with leading diagonal terms Subject A and Subject B, from Hudson (1992); (b) a portion of a larger illustrated interaction matrix with 12 leading diagonal terms.

of Subject B on Subject A, *i.e.*, the two off-diagonal terms in Figure 1(a) are the same, the matrix is termed 'symmetrical'. In the problems we will discuss, the off-diagonal terms will be different: *e.g.*, in the case of Subject A being the rock stress and Subject B being a rock fracture, the influence of the rock stress on a fracture is not the same as the influence of a fracture on the rock stress. The boxes in the square interaction matrix are indexed with the rows being i (i = 1 to N) and the columns being j (j = 1 to N). In Figure 1(a), we have just Boxes 1,1; 1,2; 2,1; 2,2 but, in Figure 1(b)—which is a portion of a 12 × 12 matrix—there are higher numbered boxes.

Let us say that we are interested in the interactions within a rock mass having the three main variables of rock structure, rock stress and water flow, all critically important in many rock engineering problems. The six separate binary interactions are shown in Figure 2. So, including the extra variable, construction, the resulting 4 × 4 interaction matrix is shown in Figure 3.

Note that each row in the interaction matrix contains the influences that the leading diagonal variable in that row has on all the other variables. For example, **Row 2** in Figure 3 contains the three binary influences of the *in situ* stress on rock mass structure, water flow and construction. Conversely, each column in the matrix indicates how the other variables affect the variable in that column. For example, **Column 2** in Figure 3 contains the three binary influences of the three variables rock mass structure, water flow and construction on *in situ* stress. Note especially that, by locating the construction variable in the lower right corner of the interaction matrix, *i.e.*, Box 4,4 in Figure 3, Row 4 indicates the way in which construction affects the other variables and Column 4 indicates the way in which the other variables affect construction. This splits the matrix into those boxes related to rock mechanics and those boxes related to rock engineering, as illustrated in Figure 4. The interaction matrices described and illustrated in the later sub-sections of this Chapter will follow the convention of locating

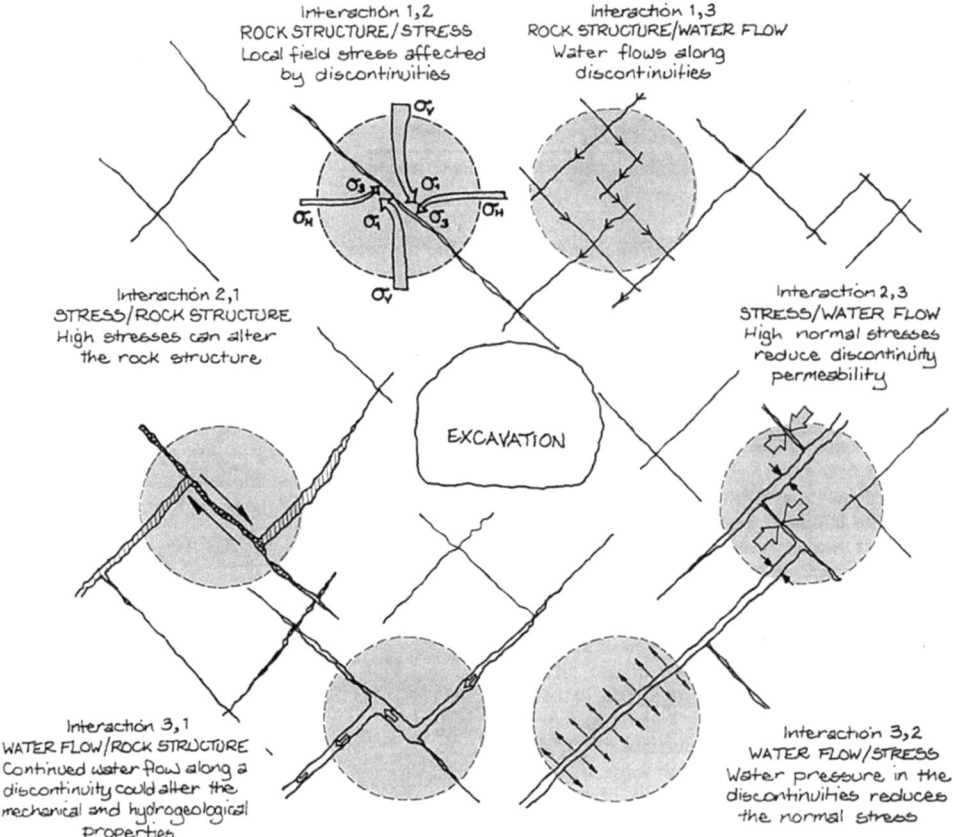

Figure 2 Consideration of the interactions in a rock mass with the three variables: rock structure, rock stress and water flow (from Hudson, 1989).

the key engineering variable in the bottom right hand box of the matrix, *i.e.*, Box N,N. In the case of studying the safety of natural rock slopes, the variable of interest, potential instability, would thus be located in Box N,N.

When any rock engineering problem is being considered, an interaction matrix can be constructed by firstly establishing the primary variables for the leading diagonal terms and then identifying the interactions for the off-diagonal boxes. In Figure 5, a simple example is presented for rock fractures. When studying natural fractures in a rock mass, assume that we are interested in the 'variables'[2] of fracture

---

2 The term 'variable' has been used to describe the leading diagonal terms—because of the potential use of mathematics to analyze the interaction matrices. However, for Figure 5, it may be felt that the term 'parameter' or 'factor' would be more appropriate. The correct use of the terms 'parameter' and 'variable' is as follows: the term 'parameter' refers to the *a*, *b* and *c* in an expression such as $ax + by = c$, *i.e.*, the coefficients making the relation specific; whereas, the term 'variable' refers to the unspecified, unknown *x* and *y* in the equation. Any of the terms variables/parameters/factors will be used in this Chapter for the leading terms of the interaction matrix, depending on the context

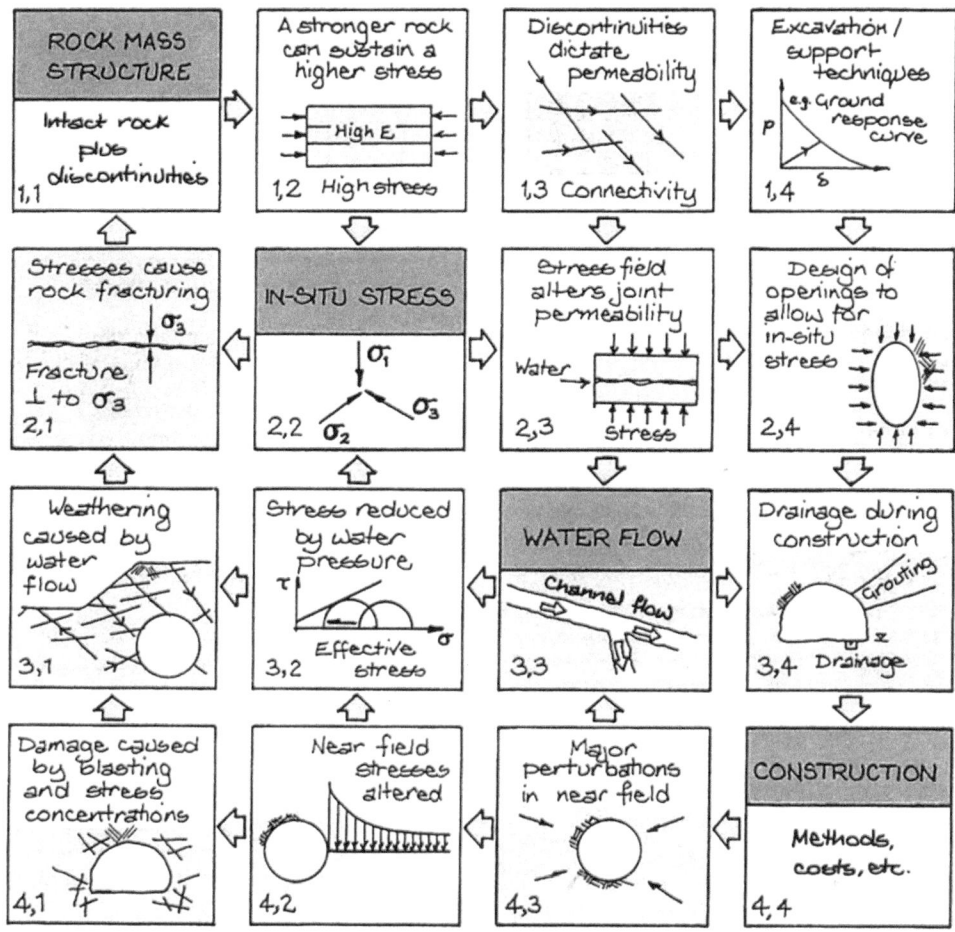

Figure 3 A 4 × 4 interaction matrix with the four main variables rock mass structure, *in situ* stress, water flow and construction along the leading diagonal, and the 12 binary interactions between the pairs of main variables in the off-diagonal boxes (from Hudson, 1989).

orientation, spacing, extent (or persistence) and roughness. Using these four para-meters as the leading diagonal terms of a 4 × 4 interaction matrix, we can identify the content of the 12 off-diagonal boxes and hence show that these variables are likely to be related.

The interactions listed in the 12 off-diagonal boxes of the Figure 5 matrix are of several forms: direct cause and effect, commonly observed correlations, and effects caused by sampling. The existence of the interactions indicates that there will be relations between the orientations, spacings, extents and roughnesses of fractures in a given rock mass.

An example of a generic rock mechanics/rock engineering 12 × 12 interaction matrix can be found in Hudson (1991). This matrix has the 12 leading diagonal terms: excavation dimensions, rock support, depth of excavations, excavation methods,

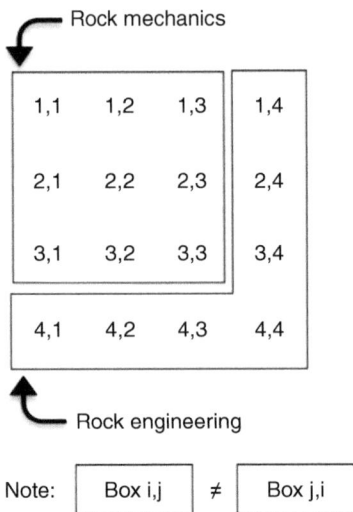

Figure 4 By locating Construction in Box 4,4, the interaction matrix shown in Figure 3 is conveniently split into the interactions relating to rock mechanics and those relating to rock engineering (from Hudson, 1989).

rock mass quality, discontinuity geometry, rock mass structure, *in situ* stress, intact rock quality, rock behavior, discontinuity aperture and hydraulic conditions. Because there are 144 boxes, there are 132 off-diagonal boxes (with brief sketches illustrating the interactions). A similar 12 × 12 interaction matrix has been constructed for rock slopes (Hudson, 1992) and an interesting 15 × 15 interaction matrix also containing sketches was compiled by Cancelli & Crosta (1994).

An interaction matrix of this type can be made for any rock engineering problem by first establishing the leading diagonal variables and then filling in the off-diagonal boxes. The off-diagonal boxes can then be allocated values according to their significance enabling further analyses, as will be explained in the following sections. Even before further analysis, the compilation of an interaction matrix for any rock engineering problem will always clarify the situation. Moreover, it enables the composite knowledge of a group of people to be organized and recorded. In other words, the interaction matrix provides the structure for eliciting and presenting the information relevant to a particular problem—and hence reducing the epistemic uncertainty.

2.2 Coding the interaction matrix, and the cause–effect plot

Having constructed an interaction matrix, the next step is to 'code' the off-diagonal components in order to express their importance or to enable mathematical manipulation of the matrix. There are five main methods to accomplish this coding for the off-diagonal boxes in an interaction matrix (Hudson, 1992):

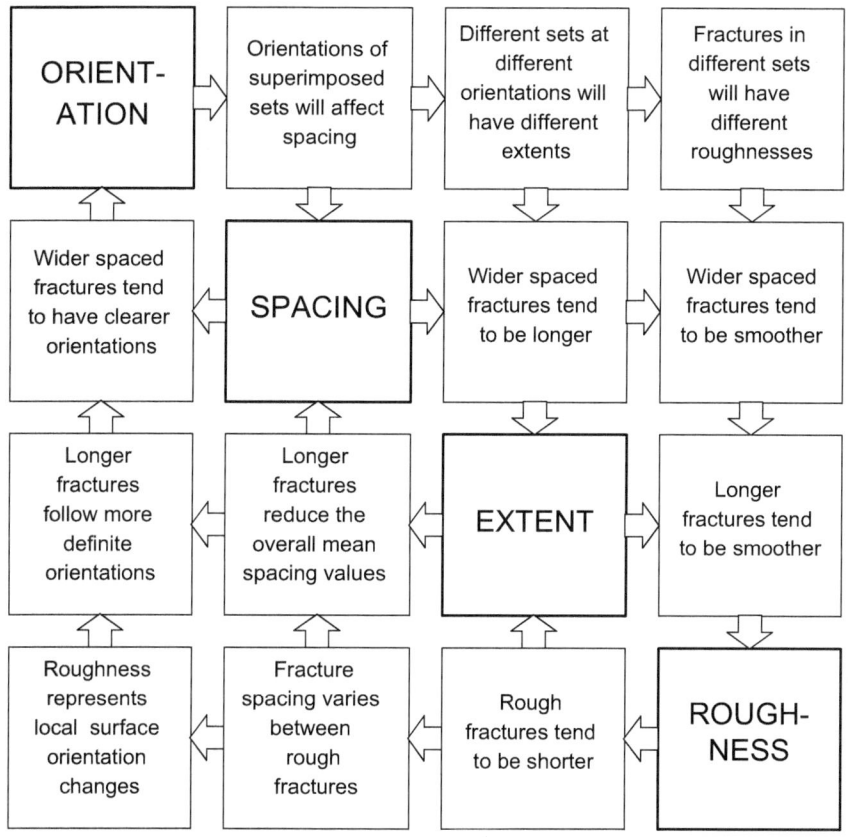

Figure 5 Example interaction matrix for rock fracture characteristics with the four leading diagonal terms Orientation, Spacing, Extent and Roughness (from Harrison & Hudson, 2000).

1. Binary: the mechanisms in the off-diagonal boxes are either switched on or off, so the coding is either as 1 or 0.
2. **Expert Semi-Quantitative: a number from 0 to 4 is allocated as follows:**
    0 – No interaction
    1 – Weak interaction
    2 – Medium interaction $\Big\}$ (or on a 1 – 5 scale)
    3 – Strong interaction
    4 – 'critical' interaction
3. According to the slope of an assumed linear relation
4. More numerically via a partial differential relation
5. Explicitly via complete numerical analysis of the mechanism

By far the most widely used of the five coding methods is Method 2, emboldened above. This is because Method 1 does not provide enough discrimination, and the information for Methods 3–5 is rarely available. Method 2 provides the necessary

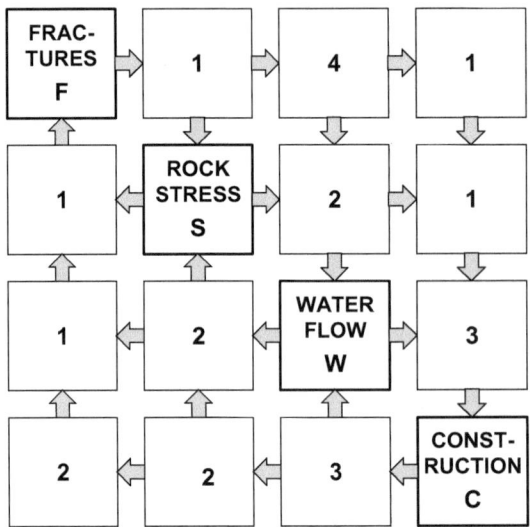

Figure 6 The interaction matrix in Figure 3 coded using the Expert Semi-Quantitative method. Summing the values in a row gives the *C* ordinate for a leading diagonal variable; summing the values in the column through the same variable gives the associated *E* ordinate. For example, the *C–E* co-ordinates for the leading diagonal term Fractures are (6,4) (from Harrison & Hudson, 2000). **Note that this simplified example is only included to illustrate how the C–E co-ordinates are obtained; a larger matrix is required for real applications.**

discrimination and the 0–4 (or 1–5) interaction values can be established by one person, or preferably by discussion within a group of persons familiar with the project being considered.

Consider the coded simple interaction matrix in Figure 6—which is the Figure 3 matrix coded using the Expert Semi-Quantitative method just described (Harrison & Hudson, 2000). In the first row of the matrix, for example it is considered that the ESQ values for the influence of fractures on rock stress, water flow and construction are 1,4,1 respectively. Now the boxes in Row 1 of the Figure 6 matrix contain the influences of rock fractures on all the other leading diagonal variables in the matrix, which we can sum to a value of 6. In a complementary way, we note that Column 1 contains the influences of all the other variables on rock fractures, *i.e.*, 1,1,2 which totals 4. We term the total of Row 1 as the *Cause* (because this is the total way in which the first variable influences the system) and we term the total of Column 1 as the *Effect* (because this is the total way in which the system affects the first variable. So the *Cause–Effect* (*C–E*) co-ordinates for the first variable are (6,4). Performing the same operation on the other three variables gives the *C–E* co-ordinates for the four variables in Figure 6 as (6,4), (4,5), (6,9), (7,5). This ESQ method of matrix coding can be undertaken for any sized interaction matrix. If the matrix is 12x12, then there will be 12 pairs of *C–E* co-ordinates, one for each variable in the leading diagonal. The next step is to prepare a table of the ESQ values, as in Table 1, list the sum and difference and plot the pairs of co-ordinates, as in Figure 7.

Table 1 List of the C–E co-ordinates from Figure 6, together with the sum and difference for each leading diagonal variable.

Leading diagonal variable	C	E	C+E	C-E
			Interactive intensity	Dominance/ Subordinacy
Fractures, F	6	4	10	2
Rock Stress, S	4	5	9	−1
Water Flow, W	6	9	15	−3
Construction, C	7	5	12	2 ·
Sum	23	23		
Mean	5.75	5.75		

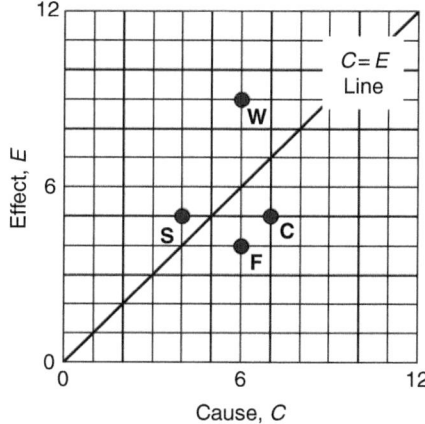

Figure 7 Cause–Effect plot for the leading diagonal variables in Figures 3 and 6: F–Fractures; S–Rock Stress; W–Water Flow; C–Construction (from Harrison & Hudson, 2000).

Note that, in Table 1, the $C + E$ value represents how active that particular variable is within the matrix system: the higher the value, the more active the variable. The C-E value represents how dominant the variable is within the system: a positive value indicates that the variable is affecting the system to a greater extent than the system is affecting the variable. So, positive values of $C - E$ represent a dominant variable, whereas negative values of $C - E$ represent a subordinate variable—the system is affecting the variable more than the variable is affecting the system. Another point with reference to Table 1 is that the sum of the C values (ΣC) for the whole matrix is in fact the sum of all the values in the off-diagonal boxes, but the same applies to the sum of all the E values (ΣE). Thus, the mean of the C values equals the mean of the E values. The next step is to plot the (C,E) co-ordinates on a *Cause–Effect* plot, as in Figure 7.

With reference to Figure 8, more interactive variables, *i.e.*, with larger $C + E$ values, will plot further away from the origin, *e.g.*, W with a $C + E$ value of 15. The more dominant factors, $C > E$, plot to the right of the $C = E$ line—as is the case for F and C. The overall conclusion is that we have a moderately interactive system structure in which Water Flow has the strongest interaction. Fractures and Construction slightly

Four types of Variable Groups in the *Cause–Effect* Plot

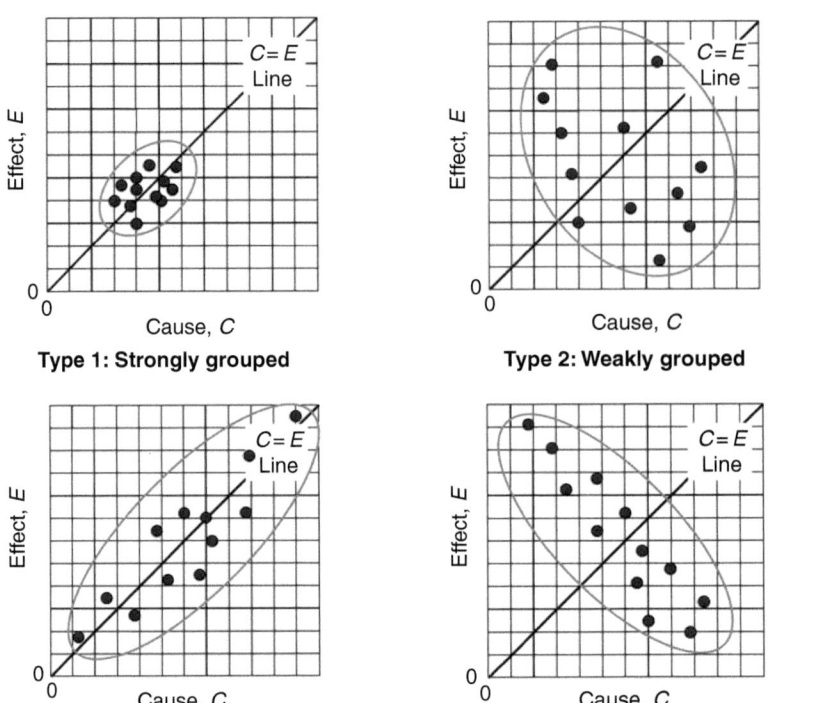

Type 1: Strongly grouped

Type 2: Weakly grouped

Type 3: Variable intensity, similar dominance

Type 4: similar intensity, variable dominance

Note: Variables with more interactive intensity plot further along the $C = E$ line
Dominant variables plot to the night of the $C = E$ line (*Cause > Effect*).
Subordinate variables plot to the left of the $C = E$ line (*Effect > Cause*).
The center of gravity of the points always lies on the $C = E$ line (Sum of C_{ij}/n = Sum of E_{ij}/n)

Figure 8 Four different types of variable clustering in the *Cause–Effect* plot.

dominate the system (being slightly to the right of the $C = E$ line) and Rock Stress and Water Flow are slightly dominated by the system (being slightly to the left of the $C = E$ line). These conclusions depend, of course, on the values assigned to the interactions. Note that the center of gravity of the *C–E* co-ordinates will always lie on the (0,0) to (12,12) diagonal whatever coding values are assigned to the off-diagonal interactions in the matrix.

In this Section, the intention has been to explain the interaction matrix coding, the *Cause–Effect* co-ordinates and the *Cause–Effect* plot together with its implications. For this reason, a highly simplified 4 × 4 matrix has been used in Figure 6 for illustrative purposes. However, the method used for this demonstration applies in exactly the same way regardless of the number of variables, *i.e.*, the dimension of the interaction matrix. Thus, it is emphasized that this example is only to explain how the *C–E* co-ordinates are obtained and utilized; practical examples will be represented by larger interaction matrices with typically between nine and fifteen leading diagonal terms.

In the general case and for the larger interaction matrices included later in this Chapter, the type of constellation formed by the positions of the leading diagonal variables in the *Cause–Effect* plot indicates the type of system being investigated. There are four main types of *Cause–Effect* plot:

Type 1: Clustered around the center of gravity of the points
Type 2: Dispersed around the center of gravity of the points
Type 3: In an elliptical zone around the $C = E$ diagonal
Type 4: In an elliptical zone around the other diagonal

These four types are illustrated in Figure 8 for the case of four different 12×12 interaction matrices.

C–E co-ordinate points plotting further along the $C = E$ diagonal represent system variables with a greater interactive intensity because $C + E$ is greater; whereas those plotting significantly to the right of the $C = E$ diagonal represent dominant variables because $C \gg E$. This is illustrated in Figure 9.

Meanwhile, we can consider a *C–E* plot for the generic underground excavation matrix (Hudson, 1991—this matrix, with sketches as the off-diagonal terms, is too large to include in the Chapter). The 12 variables are as follows:

1. Excavation Dimensions	7. Rock Mass Structure
2. Rock Support	8. *In Situ* Rock Stress
3. Depth of Excavations	9. Intact Rock Quality
4. Excavation Methods	10. Rock Behavior
5. Rock Mass Quality	11. Discontinuity Aperture
6. Discontinuity Geometry	12. Hydraulic Conditions

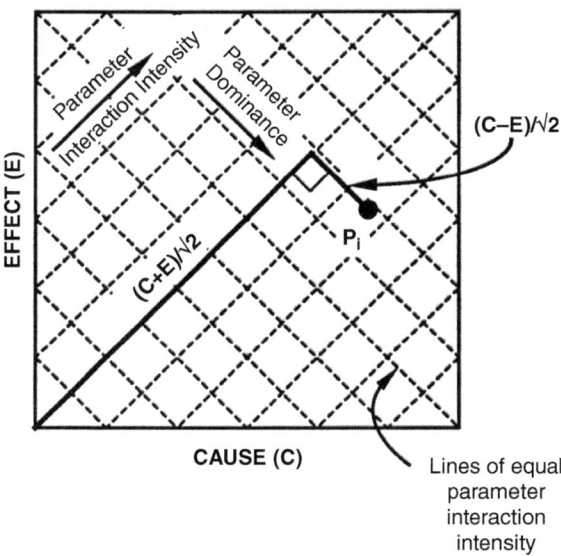

Figure 9 Interpreting the position of a parameter in *Cause–Effect* space (from Hudson, 1992).

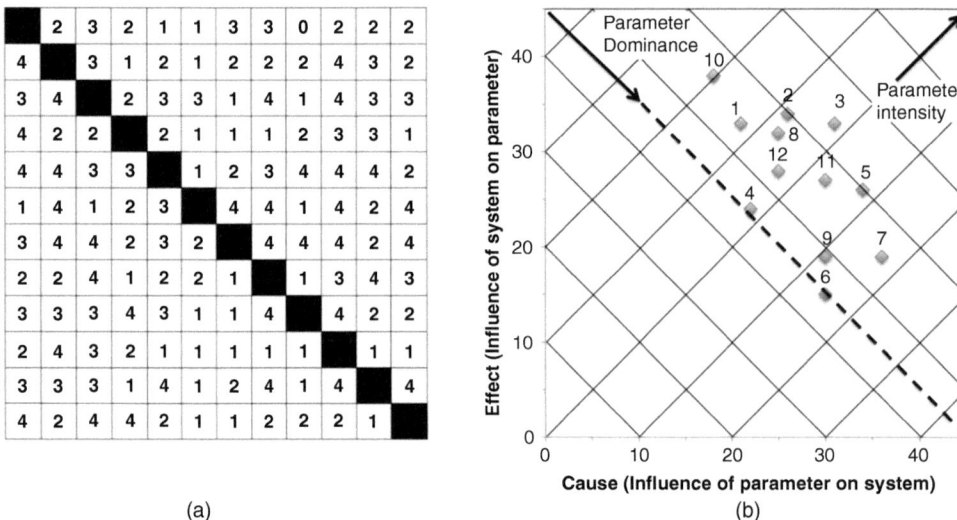

(a)   (b)

Figure 10 Matrix ESQ coding and the resultant *Cause–Effect* plot for a 12 × 12 interaction matrix for underground excavations.

Using the ESQ coding for the significance of the off-diagonal interactions given earlier, one group of engineers coded the matrix with the values given in Figure 10(a). Summing the values in the row and column through each leading diagonal term and plotting the *C–E* co-ordinates for each variable results in the plot shown in Figure 10(b).

Interpreting the constellation of points in Figure 10(b) with reference to the guidance in Figures 8 & 9, we can make the following observations.

− The system is significantly interactive because all the points are on or above the 50% interactivity line ($E = 44$–C). Also, many of the parameters have a similar interactivity.
− Parameter 10, Rock Behavior, is the most dominant, being the furthest to the right of the $C = E$ line, followed by Parameters 7 and 6, Rock Mass Structure and Discontinuity Geometry.
− Parameters 1, 2 and 8 are the most subordinate, these being Excavation Dimensions, Rock Support and *In Situ* Rock Stress.

A similar 12 × 12 matrix has been developed and ESQ coded for rock slopes (Hudson, 1992). In this case, the 12 leading diagonal terms are

1. Overall Environment	7. Hydraulic Conditions
2. Intact Rock Quality	8. Slope Orientation and Location
3. Discontinuity Geometry	9. Slope Dimensions
4. Discontinuity Mechanical Properties	10. Proximate Engineering Disturbance
5. Rock Mass Properties	11. Support/Maintenance
6. *In Situ* Rock Stress	12. Construction

The ESQ coding and the resultant *Cause–Effect* plot are shown in Figure 11.

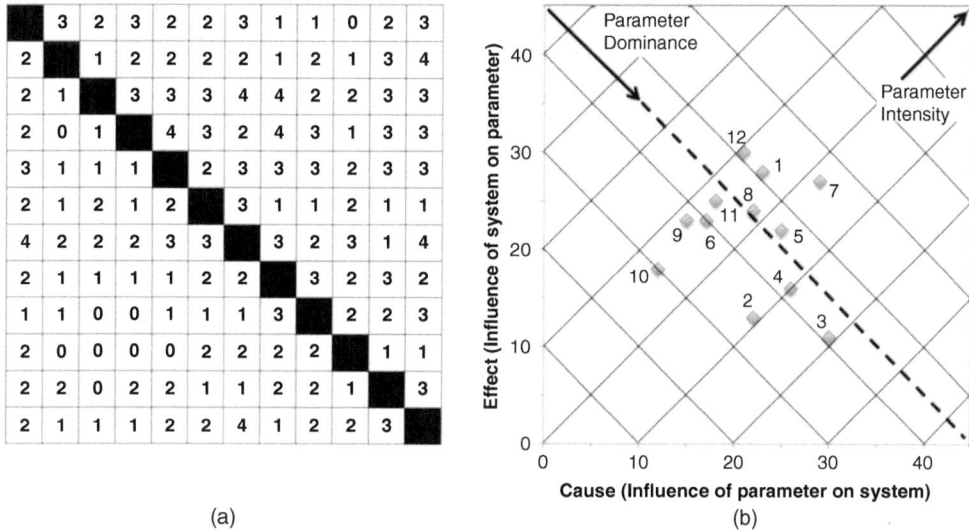

Figure 11 Matrix ESQ coding and the resultant *Cause–Effect* plot for the 12x12 interaction matrix for rock slopes.

We note that in a similar way to the *Cause–Effect* plot in Figure 10(b) for underground excavations, the parameters in Figure 11(b) for rock slopes have similar interactive intensities BUT the interactive intensity is less than for the underground excavations case. This is because the underground rock mass system is more closely linked than the surface rock mass system. In this case, Parameter 7, Hydraulic Conditions, is the most interactive and Parameter 3, Discontinuity Geometry, is the most dominant.

3 EXAMPLES OF ROCK ENGINEERING SYSTEMS (RES) APPLIED TO ROCK MECHANICS AND ROCK ENGINEERING DESIGN

In the 20+ years since the publication of the RES book (Hudson, 1992), the approach has been used for studying a wide variety of systems, not only in rock engineering but also in other subjects: these cover, *inter alia*, surface blasting, natural and artificial slope stability, underground blasting, tunnel boring machines, underground support, siting facilities and radioactive waste disposal.

3.1 Surface blasting, and slope stability

In this sub-section, the applications of RES to surface blasting and slope stability are described.

3.1.1 Surface blasting

Latham and Lu (1999) developed an RES application for the quantitative assessment of the blastability of rock masses for surface mass blasting. The authors point out that,

Table 2 List of parameters and their units used by Latham and Lu (1999) for the RES development of a blastability index.

Parameters and units for RES Blastability Index
1.   Strength (Uniaxial compressive strength, *UCS*, MPa, also via Point Load Index)
2.   Resistance to fracturing (Uniaxial tensile strength, σ_t, MPa)
3.   Sturdiness of the rock (Density, ρ, t/m^3)
4.   Elasticity of rock (Young's modulus, *E*, GPa)
5.   Resistance of rock to dynamic loading (P-wave velocity, V_p, km/s)
6.   Hardness of rock (Schmidt rebound hardness value, *SHV*, rebound height scale)
7.   Deformability (Poisson's ratio, v, dimensionless)
8.   Resistance of rock to breaking (Fracture toughness, K_{Ic}, MPa-m$^{1/2}$)
9.   *In situ* block sizes (Mean of block size distribution, mean)
10.   Fragility of rock mass (Fractal dimension of rock block sizes, *D*)
11.   Integrity of rock mass (Ratio of field: lab P-wave velocities, R_v, dimensionless)
12.   Fracture plane's strength (Cohesion, *c*, MPa and friction angle, ϕ, degrees)

"The failure to promote blast design tools beyond rules of thumb might have resulted from the fact that the influence of *in situ* rock properties, discontinuity structures and their interactions are often too difficult to be quantitatively isolated and identified." They also note that, "The problem of obtaining a satisfactory measure of blastability from an assessment of numerous potentially influential factors has at least three features which have often been neglected in early attempts to investigate blastability. One is the interactions between factors. Another is the degree of influence (or the weighing) to be attributed to each factor or coupled factors. A third is the need to treat subjective data, a situation often encountered in geotechnical engineering with systems of soils, rocks, fluids and discontinuities."

The parameters and associated units that the authors used for their 12 × 12 interaction matrix are given in Table 2. The resultant *Cause–Effect* plot indicated that the blasting system has a low/medium intensity (because the parameters were found to be nearly half way along the *C* = *E* line). Also, all the parameters had essentially equal dominance/subordinacy (because they were clustered around the *C* = *E* line. By analyzing the relative contributions of the parameters, they created a Blastability Index (*BD*).

The greater the value of *BD*, the more difficult the rock is to blast. This blastability assessment system was applied to a case study at a highway improvement cutting site in North Wales, UK. The range in parameter interaction intensity was found to be quite wide, so only those factors contributing to a total of 72.5% of the (*C* + *E*) in the ordered histogram, that is, the eight parameters, P1, P2, P3, P4, P5, P6, P9, P10 in Table 2 were chosen as the main contributory factors to the blastability of the rock masses at the site. Applying the measured values of the parameters on site, the *BD* indicated that the rock masses are, in general, difficult or moderately difficult to blast.

In their article, Latham and Lu point out that their work is related to "uncontrollable factors governed by *in situ* geological conditions and the term 'blastability' has been deliberately restricted to quantify this intrinsic resistance of the rock mass". In their scheme, the emphasis is on the intact rock properties, which would be expected because the purpose of blasting is to reduce the natural block size distribution to the required fragment size distribution. Note that, in Table 2, Parameters 3, 4, 5 and 7 are intact rock properties not directly related to failure, but are included because they characterize the quality of the rock. Parameter 6 is more strongly correlated with the failure properties, and Parameters 1, 2 and 8 are direct measurements of rock strength. In terms of the pre-existing fracturing in the rock mass, Parameters 9, 10 and 11 are indicators of the degree of fracturing present in the rock mass, but the only parameter explicitly representing the fracture failure properties is Parameter 12, the Mohr–Coulomb values. However, the art in applying engineering rock mechanics principles and the RES approach to rock engineering design is to adopt a pragmatic approach by successfully capturing the essence of the problem without introducing unnecessary complications—which the authors have done. Moreover, the use of a blastability index such as the one described in their paper, reduces the risk of adopting an inappropriate blasting scheme. For a full description of this RES application, the reader is referred to the philosophy and full case study presented in Latham and Lu (1999).

In a later publication, (Faramarzi *et al.*, 2012) describe the development of an RES-based model for risk assessment and prediction applied to backbreak in ANFO (Ammonium Nitrate–Fuel Oil) bench blasting. The term 'backbreak' refers here to the extension of damaged rocks beyond the last row of production holes in a surface mine. The authors explain that such "backbreak is an adverse phenomenon in rock blasting operations, which causes safety reduction due to the instability of mine walls, high dilution, increasing loading and hauling costs, poor fragmentation, increasing water inflow due to fractured rock, and uneven burden in subsequent blasts".

Based on the work by Benardos & Kaliampakos (2004a), a vulnerability index (*VI*) was developed for predicting backbreak and to analyze the associated risk encountered during surface blasting. Faramarzi *et al.* (2012) explain that the first step is to identify the parameters that are responsible for the occurrence of risk in the case of backbreak, analyze their behavior, and evaluate the significance (weight) that each one has in the overall risk conditions. They chose the 16 parameters listed in Table 3. In the second step, a Vulnerability Index, *VI*, can be determined, using Equation 1.

$$\text{VI} = 100 - \sum_{i=1} a_i \frac{Q_i}{Q_{\max}} \tag{1}$$

Table 3 The 16 parameters used by Faramarzi *et al.* (2012) for the RES evaluation of blasting breakback at a surface mine. The reader is referred to the authors' paper for a detailed description of these properties.

Parameter			
P_1	Burden	P_9	Time delay
P_2	Maximum instantaneous charge	P_{10}	Discontinuities orientation to face
P_3	Last row powder factor to total powder factor	P_{11}	Velocity of detonation
P_4	Powder factor	P_{12}	Blasthole deviation
P_5	S/B ratio	P_{13}	RMR
P_6	S_T/B ratio	P_{14}	Blasthole inclination
P_7	Number of rows	P_{15}	Hole diameter
P_8	Stiffness ratio (H/B)	P_{16}	B/D ratio

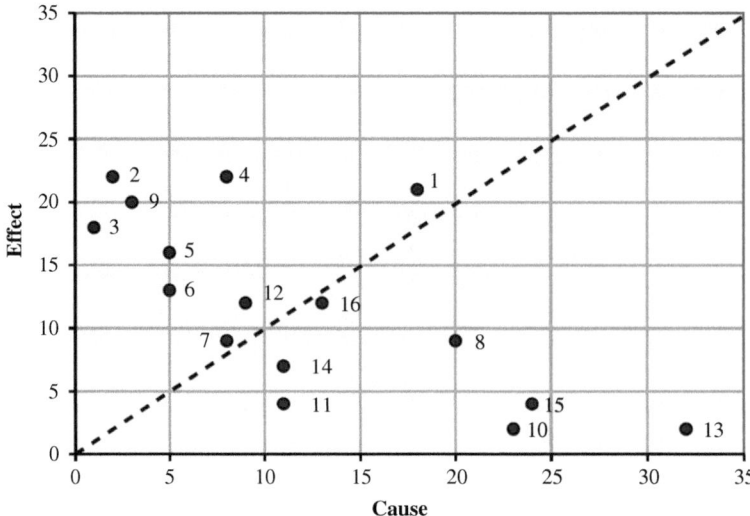

Figure 12 RES *Cause–Effect* plot used for estimating blasting breakback, from Faramarzi *et al.* (2012), see Table 3 for parameter identification.

where a_i is the weighting of the ith parameter, Q_i the value (rating) of the ith parameter, and Q_{max} is the maximum value assigned to the ith parameter (as a normalization factor). Based upon the estimated *VI* (expressed on a 0–100 scale), the level of backbreak risk can be identified. The authors also use a Backbreak Index, $BBI = (1 - VI)$.

Using the parameters in Table 3, Faramarzi *et al.* (2012) constructed a 16×16 interaction matrix and coded the off-diagonal terms using the ESQ method. The resultant *Cause–Effect* plot is shown in Figure 12. Note that, because the mean of all the *Cause* values is the mean of all the coded off-diagonal terms in the interaction matrix and the mean of all the *Effect* values is also the mean of all the coded off-diagonal terms in the interaction matrix, the mean *Cause* equals the mean *Effect*, i.e., the center of gravity of all the parameter points in Figure 12 must lie on the *C=E* diagonal line, which is the dashed line in Figure 12.

Interpreting this *Cause–Effect* plot, we note that the system has a relatively low total interactivity (the mean of the parameter points is less than a third of the way along the dashed diagonal. The three most dominant parameters (*Cause >> Effect*) are 13, 10, 15, i.e., RMR (Rock Mass Rating), Discontinuity Orientation, and Hole Diameter, respectively. The three most subordinate parameters are (*Effect >> Cause*) are 3, 2, 9, i.e., Powder Factor Ratio, Maximum Instantaneous Charge, and Time Delay, respectively. This is a significant conclusion—because it indicates that the breakback phenomenon itself is mainly dominated by the rock conditions, with the blasting parameters having less significance. In other words, in this application, the engineer has less control on the (breakback) outcome.

Faramarzi *et al.* (2012) used the method described to predict backbreak and the level of risk corresponding to each blast for 30 blasts carried out at Sungun copper mine, western Iran. They state that, "the results obtained were compared with the backbreak measured for each blast, which showed that the level of risk achieved is consistent with

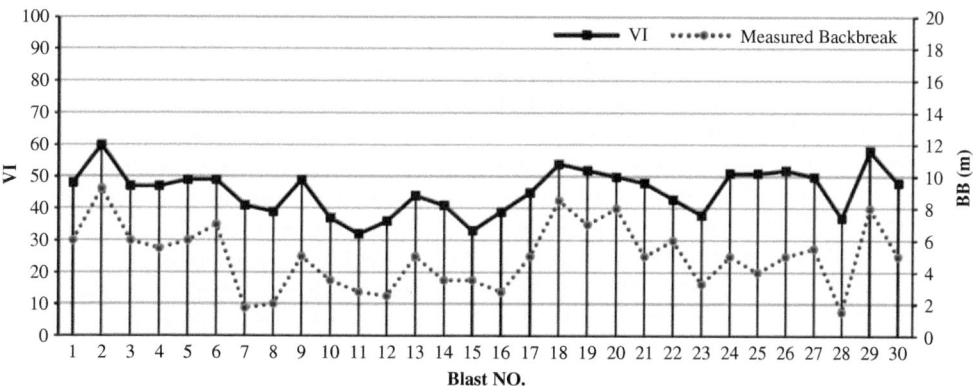

Figure 13 Variation in backbreak measured and vulnerability index (*VI*) for different blasts at Sungun copper mine.

the backbreak measured.", see Figure 13. Thus, the RES methodology and the vulnerability index, *VI*, can be used for blasting breakback risk assessment at surface mines using the technique described in which 16 parameters were used—a mixture of rock conditions and blasting factors.

3.1.2 Natural slopes

One of the earlier RES papers on natural rock slope stability set the scene for a variety of subsequent rock slope stability papers based on the same approach: this paper was "A comprehensive method of rock mass characterization for indicating natural slope instability" by Mazzoccola & Hudson (1996). The purpose of this paper was to use the RES methodology to develop a new rock mass classification sensitive to large-scale instabilities in natural slopes and hence suitable for indicating unstable slopes, the area under study being in the Italian Central Alps. The possibility of the occurrence of large landslides in these regions limits land-use and threatens existing urban centers.

For natural slopes, analysis is often difficult because of a lack of data, geological complexity, the scale of the instability phenomena and the high number of interacting factors. So, in order to be able to have a structured approach to such complexity, the RES approach was adopted. Following this approach, 19 parameters relating to the general environment and to the rock mass characteristics were chosen, see Figure 14. Their causes and effects were analyzed to weight each parameter according to its degree of interactivity in the system. The rock mass instability index developed takes into account the variability of the parameter values for different slopes when assigning ratings to different classes of the parameter values. In parallel, a predictability rating was computed, according to the presence in the field of a number of 'indicators of instability'. Both indices allow discrimination of critical slopes, and were found to be in good agreement with field evidence.

The rock types in the study area are mainly represented by two lithologies: a porphyritic, massive granitic gneiss and different rock types of sedimentary origin.

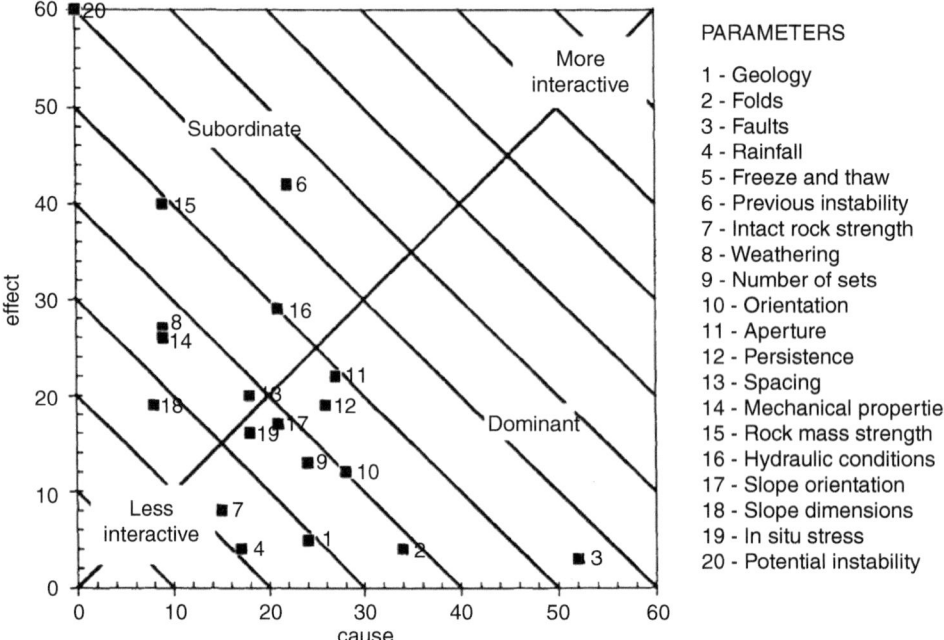

Figure 14 The *Cause-Effect* plot using the co-ordinates established from the ESQ coding method described earlier in the Chapter.

The high degree of landslide hazard in the area results mainly from the interaction between two major factors: the morphogenetic agents and the nature of the structural features. The morphogenetic agents result from the combined action of uplifting and erosion. The structural features are linked to both high tectonic disturbance and post-glacial stress relief which cause the opening of wide and persistent fractures, frequently already sheared, which in turn induce deep-seated gravitational movements.

A 20 × 20 interaction matrix was compiled and coded using the ESQ method described earlier in this Chapter. The 20 leading diagonal factors chosen to characterize this natural slope stability circumstance were: 1–geology; 2–folds; 3–faults; 4–rainfall; 5–freeze and thaw cycles; 6–previous instability; 7–intact rock strength; 8–weathering; 9–number of fracture sets; 10–fracture orientation; 11–fracture aperture; 12–fracture persistence; 13–fracture spacing; 14–fracture mechanical properties; 15–rock mass strength; 16–hydraulic conditions; 17–slope orientation; 18–slope dimensions; 19–*in situ* stress; and 20–potential instability (this last one being the factor being studied). The 20 parameters are plotted in Figure 14 according to their resultant (*Cause, Effect*) co-ordinates using the method described earlier.

At this stage, having defined the relative interactive intensity as a measure of the significance of the parameters, the actual parameter values must come into play and a more detailed data input is needed from the field. The parameter values were chosen from a 'pulldown menu'. The list of the 19 relevant parameters was used in the field to collect data on 20 slopes, located in the Cimaganda rockslide area. Note that the parameter 'potential instability' is, of course, not used and so the number of indicator parameters is 19. Some parameters were described qualitatively; others were described quantitatively. For this reason, it was not possible to utilize the actual parameter values directly to compute an instability index, but a rating was assigned to different classes of parameter descriptions and values. Three classes of parameter values were set, with ratings of 0 for 'low contribution', 1 for 'contributory' and 2 for 'strongly contributing'.

The Rock Mass Instability Index was then defined by Equation 2 as

$$RMII_j = \sum_{i=1}^{19} a_i P_{ij} \qquad (2)$$

where i refers to parameters 1 to 19, j refers to the slopes (from 1 to 20), a_i is the $C+E$ scaled value for each parameter and P_j is the rating assigned to different classes of parameter values and is different for different slopes (*i.e.* the jth slope). A Predictability Rating (PR) was also developed—more detail being given in Mazzoccola & Hudson (1996). This method of establishing a RES 'vulnerability' index, which is explained in detail in Hudson (1992) and Mazzoccola & Hudson (1996), has been used in many of the RES applications described in the literature.

Rozos *et al.* (2008) used RES to rank the instability potential of natural slopes in Karditsa County, Greece, and hence to develop a method of zoning landslide risk in the area. The authors already had general data on 388 case studies and specific data for 224 of these failure sites on the main dimensions, width and length, of the area affected by landslide activity. A predictive index was required because the landslides affect both urban and cultivated areas, as well as engineered structures. By far the greatest frequency of landslides at this location occurs in flysch and molasses formations, with the remainder occurring in schist–cherts and in transition zone beds, with most landslides being of the rotational type.

The thirteen parameters selected for the interaction matrix leading diagonal by Rozos *et al.* (2008) were lithology, rainfall, slope inclination, slope orientation, geometry of discontinuities, tectonic regime, altitude, geological structure, geomechanical action of water, thickness of weathering mantle, human intervention on slope geometry, human intervention on vegetation—plus potential instability of the slope. Each of the first 12 parameters was separated into five categories representing specific conditions and a number ranging from zero to four was assigned for each category, with the 0 category representing the most stable conditions and the 4 category the most potentially unstable conditions. The details of these categories are given in Table 1 of Rozos *et al.* (2008).

The matrix interactions were then coded (Figure 15) using the Expert Semi-Quantitative (ESQ) method with values of 0 to 4. The *Cause + Effect* values were used as weighting coefficients, which express the proportional share of each parameter (as a failure-causing factor) in slope failure. The resultant *Cause–Effect* plot is shown in Figure 16. Note that this plot is similar to the Type 4 case in Figure 8, *i.e.*, having similar parameter intensities but variable parameter dominances.

It can be seen from Figure 16, that the most dominant parameters are 1,2,5,8 and 10, *i.e.*, lithology, rainfall, geometry of discontinuities, tectonic regime and altitude (where 'altitude' here means whether the condition is 'plain', 'semi-hilly', 'hilly', 'semi-mountainous', and 'mountainous'). Based on the *Cause* and *Effect* values,

Interaction matrix													
P1	0	3	3	0	1	3	1	3	2	4	4	3	27
1	P2	2	0	0	3	3	0	4	0	3	4	4	24
0	0	P3	1	0	2	1	0	1	0	2	2	4	13
0	1	2	P4	1	2	2	1	0	0	2	2	3	16
1	0	3	3	P5	0	4	0	3	1	4	0	3	22
2	0	3	1	0	P6	2	0	2	0	0	3	3	16
0	0	1	1	0	1	P7	0	1	0	1	1	3	9
3	0	3	1	2	0	2	P8	2	2	4	2	4	25
1	0	2	0	0	3	2	1	P9	0	4	2	2	17
1	3	2	0	0	1	2	0	2	P10	1	3	1	16
0	0	3	3	3	2	3	1	3	0	P11	3	3	24
0	0	2	2	0	2	3	2	3	2	1	P12	4	19
0	0	0	0	0	0	0	0	0	0	0	0	P13	0
9	4	26	15	6	17	27	6	24	5	26	26	37	228

Cause–C

Effect–E

P1 = Lithology	P2 = Rainfall	P3 = Slope inclination	P4 = Slope orientation
P5 = Geometry of main discontinuities	P6 = Human intervention on vegetation	P7 = Human intervention to slope geometry	P8 = Tectonic regime
P9 = Geomechanical action of water	P10 = Altitude	P11 = Geological structure	P12 = Thickness of weathering mantle
			P13 = Potential instability

Figure 15 Interaction matrix for evaluating natural rock slope instability with Expert Semi-Quantitative (ESQ) coding of the off-diagonal terms, from Rozos *et al.* (2008).

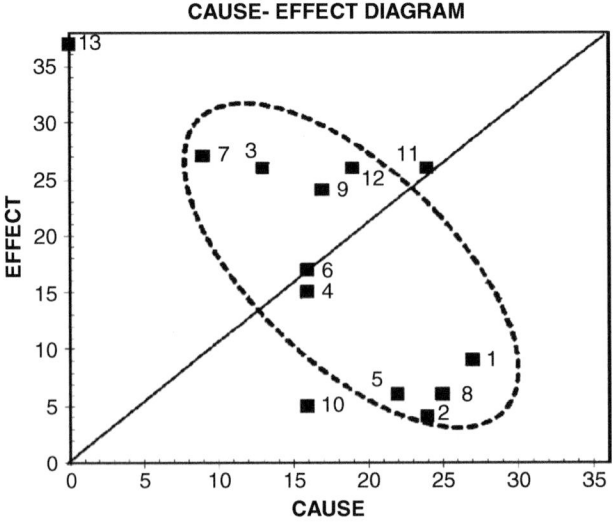

Figure 16 The *Cause–Effect* plot for the coded matrix in Figure 15, from Rozos *et al.* (2008).

Rozos *et al.* (2008) created a slope Instability Index. The authors found that, not only did the Instability Index provide an indication of a slope's instability potential *per se*, but it also indicated the size of the area likely to be affected by the consequential landslide—leading to the related implications for land use and development planning processes in landslide susceptible areas.

In a paper on analyzing earthquake-induced slope instability for the purpose of risk assessment, Castaldini *et al.* (1998) used the RES approach in a case study from the Northern Apennines in Italy. They used a multidisciplinary approach incorporating tectonics, seismology, geology, hydrogeology, geomorphology, and soil/rock mechanics. Their study area was Montese in the Garfagnana region of the Apennines which has been subjected to large earthquakes from historical times to the present. Castaldini *et al.* (1998) provide considerable information on the regional geology, the historical earthquake record, plus the rock mass and landslide characteristics. They used a comprehensive 21 × 21 interaction matrix with the following leading diagonal factors: lithology, active faults, inactive faults, *in situ* stress, seismic magnitude, seismic ground response, slope orientation, slope dimension, previous instability, rainfall, intact rock strength, weathering, number of discontinuity sets, orientation of discontinuities, aperture of discontinuities, persistence of discontinuities, spacing of discontinuities, mechanical properties of discontinuities, rock mass strength, hydraulic conditions, and potential instability. Their coded matrix is shown in Figure 17.

From their *Cause–Effect* plot, they found that the dominant variables for this RES case described by Castaldini *et al.* (1998), *i.e.*, those furthest to the right of the C = E diagonal with C>>E, are active faults, inactive faults, lithology, intact rock strength and the orientation of discontinuities. In order to assess the level of potential instability of the slopes as induced by earthquakes, Castaldini *et al.* (1998) explain that they "defined a Rock Mass Instability Index *(RMII),* in such a manner that the higher the index value is, the more critical the slope will be. The relative importance of each parameter is expressed by the sum of *Cause* and *Effect*, which, in its turn, is expressed in terms of the percentage of the total (C+E) and scaled in a way that, when all ratings are equal to a maximum value of 2, the maximum possible *RMII* is 100. Having scaled both the value of this sum a_j for each parameter and the rating of each parameter for each slope, the *RMII* can be computed according to the formula: $RMII_i = \Sigma a_j \times P_{ij}$ where i refers to the slope number, j refers to the parameter number (1 to 20); a_j is the scaled sum (*Cause* + *Effect*) for each parameter; and P_{ij} is the rating assigned to each parameter."

They also note that, "The distribution of *RMII* values usually seems to show three main portions: an upper part which groups slopes with similar values; a middle part where the values more or less gradually decrease; and a lower portion where the values stabilized around minimum values." They found that, for the non-seismic case, there were the three classes of low (L), intermediate (I) and high (H) relative proneness to instability with the *RMII* values: 0–42; 43–59; > 60, and that, with the consideration of seismicity in the area, the values are similar but some slopes then become classified in the H region.

The authors comment in their paper that the parameters in the interaction matrix need to be easily detectable through field work or simple laboratory tests and that the ESQ coding method may be too subjective, saying that it would be improved by using

A Lithology
B Active faults
C Inactive faults
D *in situ* stress
E Seismic magnitude
F Seismic ground response
G Slope orientation
H Slope dimension

I Previous instability
L Rainfall
M Intact rock strength
N Weathering
O N of discontinuity sets
P Orientation of discontinuities
Q Aperture of discontinuities

R Persistence of discontinuities
S Spacing of discontinuities
T Mechanical properties of discontinuities
U Rock Mass Strength
V Hydraulic conditions
Z POTENTIAL INSTABILITY

CAUSE

A	1	1	1	0	3	0	2	2	0	3	2	2	1	1	2	2	3	2	0	2	30
0	**B**	0	3	2	2	2	1	3	0	0	2	3	3	2	4	4	3	3	3	4	44
0	0	**C**	1	0	1	2	1	2	0	0	2	3	3	2	4	4	2	3	3	2	35
0	0	0	**D**	0	0	0	0	1	0	1	0	1	4	3	2	2	3	2	3	2	24
0	0	0	0	**E**	4	0	0	4	0	0	0	3	0	2	2	3	1	3	3	4	29
0	0	0	0	0	**F**	0	0	3	0	0	0	1	0	1	1	1	1	2	1	4	15
0	0	0	1	0	3	**G**	2	3	3	0	3	0	0	0	1	1	0	1	2	3	23
0	0	0	1	0	3	0	**H**	2	0	0	0	2	0	3	2	1	2	2	1	4	23
1	0	0	4	0	2	2	2	**J**	0	0	3	0	0	2	0	0	2	3	3	2	26
0	0	0	0	0	0	0	0	3	**L**	0	3	0	0	0	0	0	2	1	4	3	16
0	0	0	0	0	2	0	3	2	0	**M**	2	2	0	0	1	1	2	3	0	2	20
1	0	0	0	0	2	0	1	3	0	2	**N**	0	0	2	0	0	3	2	1	2	19
0	0	0	1	0	3	2	2	3	0	0	2	**O**	0	0	0	0	0	3	3	4	23
0	0	0	1	0	1	4	1	3	0	0	1	0	**P**	1	1	1	1	0	2	4	21
0	0	0	2	0	3	0	1	3	0	0	2	0	0	**Q**	1	0	3	3	4	3	25
0	0	0	1	0	1	3	2	3	0	0	2	0	0	0	**R**	1	2	3	3	4	25
0	0	0	1	0	3	1	1	3	0	0	2	0	0	1	1	**S**	0	3	3	3	22
0	0	0	1	0	3	2	2	3	0	0	0	0	0	0	0	0	**T**	3	0	4	18
0	0	0	2	0	3	0	3	3	0	0	0	0	0	0	0	1	0	**U**	0	4	16
0	0	0	3	0	3	0	2	3	0	0	3	0	0	2	1	0	3	2	**V**	4	26
0	0	0	0	0	0	0	0	0	0	0	0	0	0	0	0	0	0	0	0	**Z**	0
2	1	1	23	2	42	18	26	52	3	6	29	17	11	22	23	22	33	44	39	64	480

EFFECT

Figure 17 The interaction matrix of Castaldini *et al.* (1998) coded using the Expert Semi-Quantitative (ESQ) method with the range 0–4, and indicating the *Cause* and *Effect* co-ordinates.

mathematical relations in the off-diagonal boxes of the matrix. Indeed, this latter suggestion has been considered by the current author, possibly by including the necessary relations and then taking the Laplace transform of these in order to eliminate time from all the off-diagonal relations. Another alternative is to use a neural network approach.

3.2 Underground rock engineering

The three subjects described in this sub-section are underground blasting, tunnel boring machines and underground support.

3.2.1 Underground blasting

Andrieux & Hadjigeorgiou (2008) discuss a Destressability Index methodology for the assessment of the likelihood of success of a large-scale confined destress blast in an underground mine pillar. The work was aimed at reducing the rock stress in the particular context of large-scale choked destress blasts in mine pillars. The use of the RES approach has led to a destressability index which was applied to back-analyze a fully instrumented large-scale confined destress blast at Brunswick Mine, in Canada. The index indicates whether a given situation is conducive to being destressed by means of a large-scale confined destress blast, and, if so, whether the design of the blast is appropriate to achieve this goal.

The authors explain that, "Destress blasting can be defined as any attempt involving the usage of confined explosive charges (*i.e.*, without free faces) to reduce the ground stresses in a particular region, and in which the blasted material is left in place...it is the process of using confined explosive charges in order to damage the rock mass, for the purpose of softening its behavior, reducing its capacity to carry high stresses and, hence, reducing the potential for it to undergo violent failure." For this RES approach to large-scale choked panel destress blasting and based on much case study information, the authors used and explained in detail the following nine parameters for their interaction matrix:P1, the stiffness of the rock; P2, the brittleness of the rock; P3, the degree of fracturing of the rock mass; P4, the proximity of the rock mass to (static) stress-induced failure; P5, the orientation of the destress blast; P6, the width of the destress blast; P7, the unit explosive energy; P8, the confinement of the explosive charges, and P9, the result of the destress blast. The ESQ-coded interaction matrix is shown in Figure 18 and the associated *Cause–Effect* plot is shown in Figure 19.

									Causes
P_1	2	2	2	1	1	2	1	2	13
2	P_2	1	1	1	2	2	2	2	13
1	1	P_3	2	1	2	2	2	3	14
1	1	2	P_4	1	1	3	1	4	14
2	2	2	1	P_5	1	1	1	3	13
3	2	4	2	2	P_6	1	1	3	18
4	3	4	4	1	1	P_7	3	4	24
2	2	3	2	1	2	2	P_8	3	17
3	3	4	4	2	1	2	1	P_9	20
Effects 18	16	22	18	10	11	15	12	24	146

Figure 18 The ESQ-coded (0–4) interaction matrix for large-scale choked pillar destress blasts, from Andrieux & Hadjigeorgiou (2008).

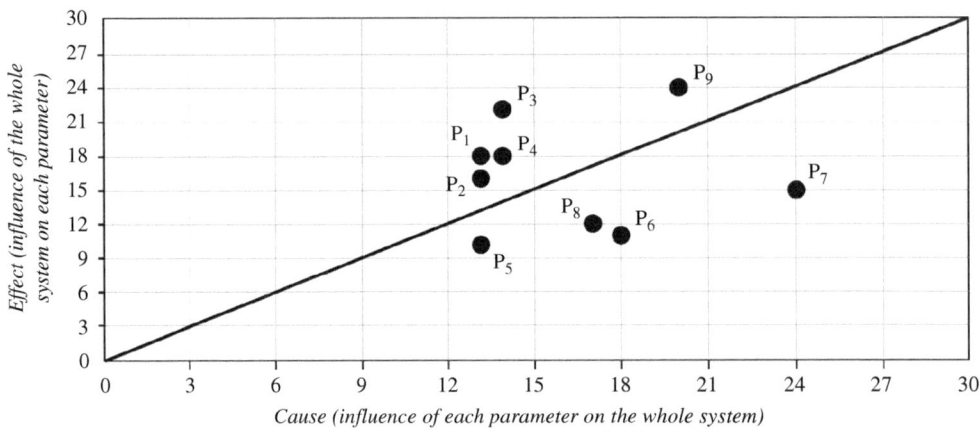

Figure 19 Cause–Effect plot for large-scale choked panel destress blasting, from Andrieux & Hadjigeorgiou (2008).

Given that the dominant parameters $(C > E)$ are always below the $C = E$ line, it can be seen from Figure 19 that the dominant parameters are P7, P6, P8 and P5, *i.e.*, the unit explosive energy, the width of the destress blast, the confinement of the explosive charges, andthe orientation of the destress blast. The authors then developed a 'destressability index' rating, emphasizing that this is, "…not as a direct design procedure, but, rather, an assessment of the likelihood of success of a proposed design in a given situation." The methodology was applied to a case study at the Brunswick Mine near Bathurst, NB, in Atlantic Canada by Andrieux *et al.* (2003). A cross-section through the case study instrumented pillar is included as Figure 20.

The authors conclude that a particular appeal of the RES approach, "is that it provides a series of easily implemented steps that result in a rational assessment of the likelihood of success of a given destress blast design in a given situation of rock mass conditions and stress regime… considering that (1) large-scale confined pillar destress blasts are usually a last resort endeavor with no possible second attempt, (2) the cost associated with their implementation is typically substantial and (3) the consequences of failure generally lead to significant ore losses and lost production (these blasts are only considered in the first place when large amounts of valuable ore are at risk, either directly or indirectly), this type of blast deserves sound engineering in order to maximize the likelihood of success." A detailed description of the approach and application is given in their paper.

3.2.2 Tunnel Boring Machines (TBMs)

The use of TBMs in rock has run the full spectrum: from high advance rates with no problems to becoming irretrievably stuck and having to be removed or abandoned. The first RES example illustrated here in the tunneling context has been reported in the paper, "A methodology for assessing geotechnical hazards for TBM tunneling— illustrated by the Athens Metro, Greece" by Benardos & Kaliampakos (2004a). The authors state that, "The methodology presented in this paper aims at the identification

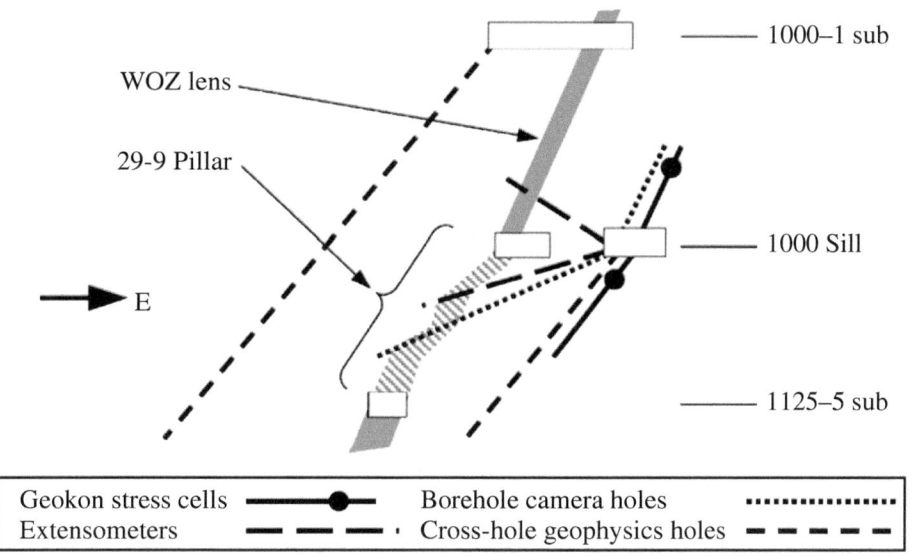

| Geokon stress cells | ●━━ | Borehole camera holes | ●●●●●●●●●●●●●● |
| Extensometers | ━ ━ ━ · | Cross-hole geophysics holes | ━ ━ ━ ━ ━ |

Figure 20 Cross-section through the 29-9 Pillar at the Brunswick Mine, looking north and showing the projected location of the various instrumentation, from Andrieux *et al.* (2003).

of risk-prone areas, incorporating, at the same time, the uncertainty of ground conditions...the methodology assesses the hazards by introducing the concept of a vulnerability index, to identify the weighting of the parameters, and with probabilistic modeling to address the uncertainty in the parameters' values." The authors' flowchart relating to the proposed vulnerability assessment methodology is shown in Figure 21.

The proposed model is illustrated via the Athens Metro case study, used also for validating its performance under actual construction conditions. Their risk analysis addressed: face instabilities/collapses and overbreaks; surface settlements; and water inflows. They used the RES approach together with probability assessments for the data values and a GIS application.

They used eight parameters: rock mass fracture degree as represented by RQD—(P1), weathering degree of the rock mass—(P2), overload factor—stability factor (N)—(P3), rock mass quality represented by RMR classification—(P4), uniaxial compressive strength of the rock—(P5), overburden, construction depth—(P6), hydrogeological conditions represented by the watertable surface relative to the tunnel depth—(P7), rock mass permeability—(P8). These parameters were rated on a 0–3 scale for each of the 11 locations along the route being studied.

Each parameter's weighting factor, a_i, was than calculated from Equation 3

$$a_i = \frac{(C_i + E_i)}{(\Sigma_i C_i + \Sigma_i E_i)}(\%) \tag{3}$$

and the final weighting of the principal parameters indicated the dominant parameters (highest a_i values) as rock mass rating (RMR), fracture degree (RQD), hydrogeological conditions, and rock mass weathering. Finally, a vulnerability index (*VI*) is estimated via Equation 4

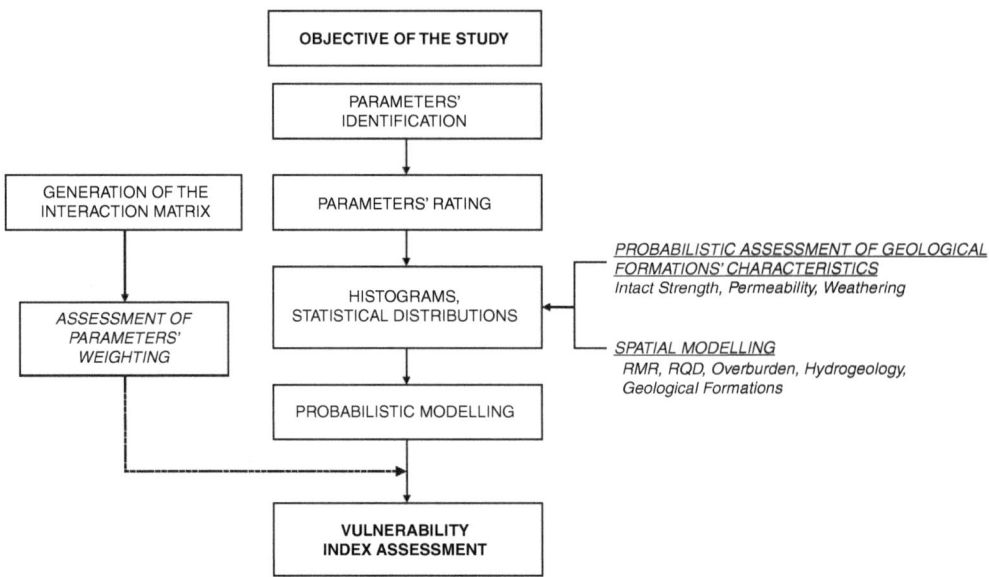

Figure 21 Flowchart of the proposed vulnerability assessment methodology, from Benardos & Kaliampakos (2004a).

$$VI = \left(100 - \sum_{i=1}^{n} a_i \frac{P_i}{P_{\max}} \right) \qquad (4)$$

where *VI* (range 0–100) is in the form of a probability distribution and a_i is the weighting of the ith parameter in the system, P_i the value (rating) of the ith parameter, and $P_{\max}$ the maximum value a parameter can take (a normalization factor). In terms of general categories, *VI* values 0–33 represent low vulnerability, 33–66 medium vulnerability, and 66–100 high vulnerability. So, for each of the 11 tunnel study lengths, the *VI* is expressed as a discrete probability distribution.

The authors note that, "the mean values of *VI* and *AR* for the 11 examined tunnel segments have a high negative correlation coefficient, about 0.92, signifying the coherent behavior of the proposed vulnerability index methodology." A companion paper, Benardos & Kaliampakos (2004b), uses an alternative method of predicting tunnel progress for the Athens Metro, but making use of artificial neural networks (ANNs).

3.2.3 Tunnel stability

Kim *et al.* (2008) explain that quantitatively identifying rock behavior expected in excavating tunnels can assist engineers in selecting the best tunneling method and support system and in evaluating tunnel stability through numerical analysis adjusted to rock behavior. They used RES to develop a Rock Behavior Index (RBI) for assessing plastic deformation and ground failure and illustrated the proposed model via a case study on the Seoul Metro Line 9. Based on work by Cai *et al.* (2004) who provided a list of parameters that should be considered when describing a rock mass and using the results for design purposes, the authors settled on seven parameters influencing the rock behavior: unconfined compressive strength, Rock Quality Designation (RQD), joint surface condition, stress, groundwater, earthquake, and tunnel span. They coded the interaction matrix by the ESQ method (0–4) and, using a similar method to the previous case examples, created three RBI indices: one for rock fall, one for cave-in and one for plastic deformation. These RBIs then indicated the potential for these cases according to the scale: 0–20, very low probability; 20–40, low probability; 40–60, moderate probability; 60–80,high probability; 80–100, very high probability.

Shin *et al.* (2009) developed a methodology for quantitative hazard assessment for tunnel collapses based on case histories in Korea. They proposed a Tunnel Collapse Hazard Index (KTH-Index), an index system for assessing the hazard level of collapse at a tunnel face based on a sensitivity analysis of a database containing past collapse cases; 56 sets of such data were utilized. For the sensitivity analysis, the authors used a neural network based technique and RES. The assessment system was applied to the section of the SYK tunnel where large-scale collapses had already occurred and it was found that the predicted hazard levels were in good agreement with the field data already known.

3.3 Underground radioactive waste disposal

The previous sections have covered a wide range of RES applications with a significant number of papers relating to natural rock slope stability which emphasize the large number of factors involved. Another subject with a large number of factors is underground radioactive waste disposal—because of the large number of features, events and processes (FEPs) that have to be studied during the compilation of a disposal license application, a process which typically takes 20+ years. Structuring the FEPs in a coherent way is most helpful and RES can provide the necessary capability.

Skagius *et al.* (1995, 1997) describe the process of structuring the FEPs in their paper, "Performance assessment of the geosphere barrier of a deep geological repository for spent fuel: the use of interaction matrices for identification, structuring and ranking of features, events and processes." The authors explain that, "the main purpose of the assessment is to identify the important issues affecting the long-term behavior of, and the radionuclide migration within, the far-field rock of an underground repository for spent fuel." They used a 13x13 interaction matrix. Skagius *et al.* (1997) highlight an example content of one off-diagonal box as "Natural fracture system—transport of radionuclides, molecular diffusion of radionuclides in the natural fracture system. This will affect important transport parameters such as the surface area available for sorption and matrix diffusion. The sorption capacity of the rock is affected by the fracture minerals."Each such off-diagonal box has a reference to the Svensk Kärnbränslehantering AB (SKB) Swedish FEP database. The authors conclude by noting that the compiled information is valuable because it presents the fundamental background material in a structured and consistent manner. They also provide advice on the assembling of interaction matrices in the radioactive waste context. However, in this application it should be noted that the interaction matrix is being used only for structuring information.

4 CHAPTER SUMMARY

The Rock Engineering Systems (RES) approach has been outlined through the explanation of the interaction matrix—with its leading diagonal of the key factors and the off-diagonal boxes containing the interactions between the key factors. The procedure involves coding the off-diagonal boxes according to their significance—the most popular coding method being the Expert Semi-Quantitative (ESQ) method, *i.e.*, an integer from 0 to 4. Other potential coding methods are via the slope of an assumed linear relation, more numerically via a partial differential relation, and explicitly via complete numerical analysis of the mechanism. Once the interaction matrix has been coded, the *Cause–Effect* diagram can be constructed which indicates directly by eye the relative interactivity and dominance/subordinacy of the component factors. A series of examples was highlighted based on published papers and describing applications in the rock engineering context

The essence of the procedure in using RES is as follows.

- Consider the project objective,
- Choose the leading diagonal terms for the interaction matrix,
- Code the off-diagonal terms in the interaction matrix,
- Create the *Cause–Effect* plot,
- Determine the parameters with the most interactivity and dominance,
- Create an assessment index, *e.g.*, an instability or vulnerability index (as in the case examples described),
- Calibrate the index via test cases, and then
- Use the index for the project being considered.

This procedure provides considerably more knowledge about the project, as evidenced by the suite of RES applications described in this Chapter. In fact, the knowledge may already have been present in the minds or computers of a project team; it is the RES procedures that enable structured elicitation of that knowledge for the project in hand—which, in turn, increases understanding concerning the project, reduces the epistemic uncertainty, and enables reduction of the risk.

REFERENCES

Andrieux, P., Brummer, R., Liu, Q., Mortazavi, A. & Simser, B. (2003) Large-scale panel destress blast at Brunswick Mine. *Bull Can Inst Min Metall Pet*, 96, 78–87.

Andrieux, P. & Hadjigeorgiou, J. (2008) The destressability index methodology for the assessment of the likelihood of success of a large-scale confined destress blast in an underground mine pillar. *Int J Rock Mech Min*, 45, 407–421.

Benardos, A.G. & Kaliampakos, D.C. (2004a) A methodology for assessing geotechnical hazards for TBM tunneling—Illustrated by the Athens Metro, Greece. *Int J Rock Mech Min*, 41, 987–999.

Benardos, A.G. & Kaliampakos, D.C. (2004b) Modelling TBM performance with artificial neural networks. *Tunn Undergr Sp Tech*, 19, 597–605.

Cai, M., Kaiser, P.K., Uno, H., Tasaka, Y. & Minami, M. (2004) Estimation of rock mass deformation modulus and strength of jointed hard rock masses using the GSI system. *Int J Rock Mech Min*, 41, 3–19.

Cancelli, A. & Crosta, G. (1994) Hazard and risk assessment in rockfall prone areas. In: Skipp, B.O. (ed.), *Risk and Reliability in Ground Engineering*. Springfield: Thomas Telford. pp. 177–190.

Castaldini, D., Genevois, R., Panizza, M., Puccinelli, A., Berti, M. & Simoni, A. (1998) An integrated approach for analysing earthquake-induced surface effects: a case study from the Northern Apennines. Italy. *J Geodyn*, 26 (2–4), 413–441.

Faramarzi, F., Ebrahimi Farsangi, M. & Mansouri, H. (2012) An RES-based model for risk assessment and prediction of backbreak in bench blasting. *Rock Mech Rock Eng*, Online August, 1–11.

Harrison, J.P. & Hudson, J.A. (2000) *Engineering Rock Mechanics: Illustrative Worked Examples*. Oxford: Elsevier. (Also published in Chinese by Science Press of Beijing, 2009).

Hudson, J.A. (1989) *Rock Mechanics Principles in Engineering Practice*. London: Butterworths.

Hudson, J.A. (1991) Atlas of rock engineering mechanisms. Part 1—Underground excavations. *Int J Rock Mech Min*, 28, 523–526.

Hudson, J.A. (1992) Atlas of rock engineering mechanisms. Part 2—Slopes. *Int J Rock Mech Min*, 29, 157–159.

Hudson, J.A. (1992) *Rock Engineering Systems: Theory & Practice*. New York: Ellis Horwood.

Hudson, J.A. & Feng, X.-T. (2015) *Rock Engineering Risk*. CRC Press/Balkema, Taylor & Francis Group, London, p. 572.

Kim, M.-K., Yoo, Y.-I. & Jae-Joon Song, J.-J. (2008). Methodology to quantify rock behavior around shallow tunnels by Rock Engineering Systems. *Geosyst. Eng.*, 11(2), 37–42.

Latham J-P. & Lu Ping (1999) Development of an assessment system for the blastability of rock masses. *Int J Rock Mech Min*, 36, 41–55.

Mazzoccola, D.F. & Hudson, J.A. (1996) A comprehensive method of rock mass characterization for indicating natural slope instability. *Q JEng Geol*, 29, 37–56.

Rozos, D., Pyrgiotis, L., Skias, S. & Tsagaratos, P. (2008) An implementation of rock engineering system for ranking the instability potential of natural slopes in Greek territory: an application in Karditsa County. *Landslides*, 5, 261–270.

Shin, H.S., Kwon, Y.C., Jung, Y.S., Bae, G.J. & Kim, Y.G. (2009) Methodology for quantitative hazard assessment for tunnel collapses based on case histories in Korea. *Int J Rock Mech Min*, 46, 1072–1087.

Skagius, K., Ström, A. & Wiborgh, M. (1995) *The use of interaction matrices for identification, structuring and ranking of FEPs in a repository system: application on the far-field of a deep geological repository for spent fuel*. Svensk Kärnbränslehantering AB, Stockholm, SKB Technical Report 95–22.

Skagius, K., Wiborgh, M., Ström, A. & Morén, L. (1997) Performance assessment of the geosphere barrier of a deep geological repository for spent fuel—the use of interaction matrices for identification, structuring and ranking of features, events and processes. *Nucl Eng Des* 176, 155–162.

Design and Stability Analysis:

Coupling Process Analysis

Chapter 16

Multiphysics of coal-gas interactions

*M.Y. Wei[1], Z.W. Chen[2], J.S. Liu[3], G.L. Cui[1], Y.L. Tan[1] &
S.W. Zhang[1]*

[1] *State Key Laboratory of Geomechanics and Geotechnical Engineering, Institute of Rock and Soil Mechanics, Chinese Academy of Sciences, Wuhan, China*
[2] *School of Mechanical and Mining Engineering, The University of Queensland, Brisbane, Queensland, Australia*
[3] *School of Mechanical and Chemical Engineering, The University of Western Australia, Perth, Western Australia, Australia*

Abstract: Advances in our understanding of the geomechanics of coal-gas interactions have changed the manner in which we treat coal seam gas: from mitigating its dangers as a mining hazard to developing its potential as an unconventional gas resource recovered as a useful by-product of CO_2 sequestration. When coal seam gas is recovered, complex coal-gas interactions have strong controlling effects on the extraction efficiency of coal or gas. These include influences on gas sorption and flow, coal deformation, porosity change and permeability modification. This chain of reactions can be defined as "coupled processes" implying that one physical process affects the initiation and progress of another. Therefore, the inclusion of cross couplings is the key to rigorously formulate the geomechanics of coal-gas interactions.

A wide variety of coal permeability models incorporating the correct physics have been proposed to simulate the interactions of multiple processes triggered by the CO_2 injection or production of methane. Coal permeability models, representing the effects of sorption, swelling and effective stresses on the dynamic evolution of permeability, are normally required to define the mechanics of coal-gas interactions. The inclusion of the effective stress concept led to a generalized coal permeability model being developed. This contribution transformed the pore pressure-based models into a new generation of effective stress-based model. Consistent efforts have done to include that the impact of coal matrix-fracture compartment interactions into the mechanics of coal-gas interactions. A series of permeability models have advanced this subject from the pore pressure-based theory to the effective stress-based theory, to the effective stress transfer-based theory. These models couple the transport and sorption of a compressible fluid within a deformable medium where the effects of deformation are rigorously accommodated. This paper reports this novel framework on geomechanics of coal-gas interactions and its applications primarily in the field of coal seam gas extraction.

1 INTRODUCTION

1.1 Coalbed methane extraction and coal mine safety

Coalbed methane (CBM) is naturally occurring methane gas (CH_4) in coal seams. Methane was long considered a major problem in underground coal mining but now CBM is recognized as a valuable resource. The world's coalbed methane resource

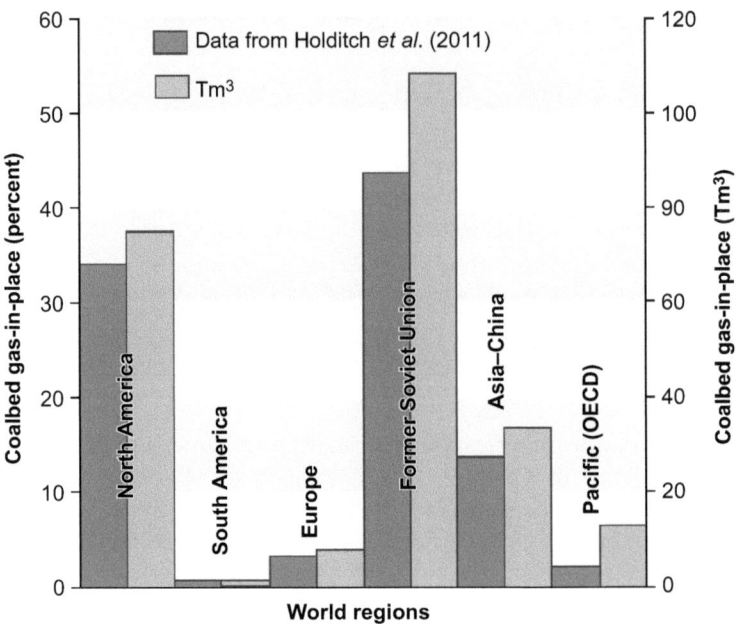

Figure 1 Variability of the coalbed gas-in-place in percent and volume of various regions in the world. Tm^3, trillion cubic meter (Makogon *et al.*, 2007).

estimates range from 2980 to 9260 trillion cubic feet (TCF) or $8.433 \times 10^{13} m^3$ to $2.621 \times 10^{14} m^3$ (Palmer, 2008). Advances in our understanding of coal–gas interactions have changed the manner in which we treat coalbed methane: from mitigating its dangers as a mining hazard to developing its potential as an unconventional gas resource (Liu *et al.*, 2011).

CBM has been considered a major mine hazard since the first documented coal mine gas explosions occurred in the United States in 1810 until the mid-19th century. Since the late 20th century CBM has received increased emphasis as a potential energy resource. For the past 100 years, efforts have focused on improving technology and safety of coal mines worldwide to prevent gas explosions associated with underground coal mine operations. Coal seam degasification and its efficiency are directly related to mitigate this hazard, and results in the beneficial recovery of a clean-burning and low-carbon fuel resource (Noack, 1998; Zhu *et al.*, 2011). The number of coal mine outbursts and related incidents decreased significantly from 1926 through 1975, which is related to the widespread use of gas drainage systems and application of control prevention management of coal mines in America (Dubaniewicz Jr., 2009; Msha, 2011).

With the growing international concern over the issue of global warming, geological sequestration of CO_2 is a significant contender in the mix of greenhouse mitigation options. Since deep coal seam is one of geological media to potentially sequestrate huge amounts of CO_2, CO_2 sequestration in deep coal seams has attracted attention as a method of reducing the output of greenhouse gases to the atmosphere (Gale & Freund, 2001), where coal serves as a receptor for the injected CO_2 which is sequestered in the naturally fractured medium. The micro-pores and pores in the coal matrix provide the main storage space for gas and the micro-fractures through macro-fractures comprise rapid pathways for gas seepage and migration.

CO$_2$-enhanced coalbed methane (CO$_2$-ECBM) production involves the injection of CO$_2$ into a coal seam to promote the desorption of chemically-bound methane while simultaneously sequestering CO$_2$ into coal media. This process exploits the greater affinity of carbon dioxide (CO$_2$) to adsorb onto coal relative to methane (CH$_4$), resulting in the net desorption of methane and its potential recovery as a low-carbon fuel (Connell *et al.*, 2011). Laboratory isotherm measurements for pure gases have demonstrated that coal can adsorb approximately twice the volume of CO$_2$ (on a molar basis) as CH$_4$ (White *et al.*, 2005). Other laboratory experiments show that this ratio could be even larger at reservoir pressure higher than 9.6 MPa, where the gaseous CO$_2$ changes to supercritical CO$_2$ (Chen *et al.*, 2008; Hall *et al.*, 1994; Krooss *et al.*, 2002).

1.2 Multiphysics of CBM extraction

When coal reservoir is disturbed due to dewatering, gas extraction or gas injection, complex interactions of stress and chemistry exist and have strong influence on the transport and adsorptive properties of coal seams. These include influences on gas sorption and flow, coal deformation, porosity change and permeability modification as shown in Figure 2.

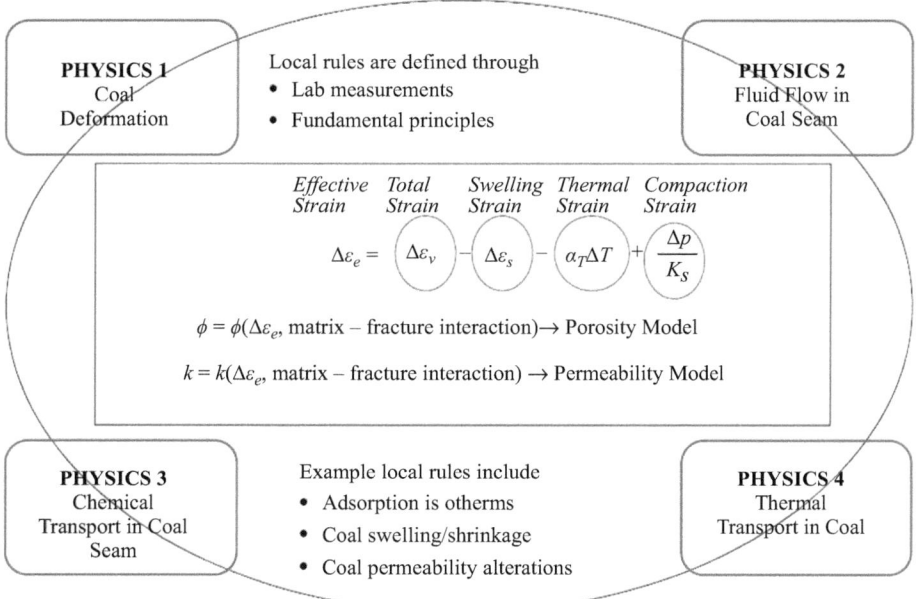

Figure 2 Interactions of multiple coupled processes through a stress-controlled coal porosity model and coal permeability model defined as a function of the effective strain, $\Delta\varepsilon_e$, during CBM extraction.

The causes of instantaneous gas outbursts are directly associated with the complex interaction of multiphysics. Various models and mechanisms have been proposed to explain the complex processes involved in bursts and bumps (Beamish & Crosdale, 1998; Lama & Bodziony, 1998). These models have a common feature that a spatial variation of stresses, gas pressures, damage, permeability, and desorption rate exists ahead of the mining face in underground coal seams. This is due mainly to the sudden stress redistribution induced by mining (Harpalani, 1985). Changes in one zone influence adjacent zones and are of great consequence in controlling the stability of coal seams. But also coupling of the effects of stress, permeability and desorption provide a potential positive feedback to the liberation of gas.

1.3 Cross-couplings

Gas flow within coal seams is quite different from that of conventional reservoirs. Detailed studies have examined the storage and transport mechanisms of gas in coal seams. In-situ and laboratory data indicate that the storage and flow of gas in coal seams is associated with the matrix structure of coal and the absorption or desorption of gas. CBM extraction triggers a series of complicated gas–coal interactions including gas sorption and flow, coal deformation, porosity change and permeability modification. In addition, sorption-induced strain of coal can change the porosity, the permeability and the storage capacity of coal seam via feedback to in situ stresses via displacement constraints. This chain of reactions can be defined as "coupled processes" implying that one physical process affects the initiation and progress of another (Zhang et al., 2008).

The relative roles of stress level, gas pressure, and fracture distribution are intimately connected to the processes of gas desorption, diffusion, transport, and coal shrinkage. As gas pressure reduces below the desorption point, methane is released from coal matrix to the fracture network and coal matrix shrinks. For gas production, the reduction of gas pressure increases effective stress which in turn closes fracture aperture and reduces the permeability (McKee et al., 1988; Palmer & Mansoori, 1996; Seidle & Huitt, 1995). As a direct consequence of this matrix shrinkage the fractures dilate and fracture permeability correspondingly increases (Harpalani & Schraufnagel, 1990). Thus a rapid initial reduction in fracture permeability (due to change in effective stress) is supplanted by a slow increase in permeability (with matrix shrinkage). Whether the ultimate, long-term, permeability is greater or less than the initial permeability depends on the net influence of these dual competing mechanisms (Chen et al., 2008; Connell, 2009; Shi & Durucan, 2004). Furthermore, other factors, like heterogeneity of coal, gas composition and water content, also contribute to the complexity of gas–coal interactions (Han et al., 2010; Kang et al., 2011; Wu et al., 2010b). All of these lead to permeability hardly be predicted and change dramatically: up to 100 times in the San Juan basin (Palmer, 2009). Moreover, permeability of a reservoir has a close relationship with productivity of CBM. Information on permeability is in favor of long-term production design.

A wide variety of coal permeability models incorporating the correct physics have been proposed to simulate the interactions of multiple processes triggered by the CO_2 injection or production of methane. However, these models have partially succeeded in explaining in situ data. Compared with experimental data, most of these models have so far failed to explain experimental results from conditions of the controlled stresses (Peng et al., 2014). The primary goal of this work is to review our previous studies that advanced CBM extraction from the pore pressure-based theory to the effective stress-based theory, to the effective stress transfer-based theory.

2 FORMULATION OF COAL-GAS MULTIPHYSICS

Coal permeability is probably the most important factor determining coalbed methane production performance. The influence of sorption-induced coal matrix deformation on the evolution of porosity and permeability of fractured coal seams has been widely studied.

2.1 General permeability models

The porosity-based model considers factors such as the volume occupied by the free-phase gas, the volume occupied by the adsorbed phase gas, the deformation-induced pore volume change, and the sorption-induced coal pore volume change. More importantly, these factors are quantified under in situ stress conditions. In the following, we briefly explain each of the models currently available in the literature, and refer the reader to the source papers for further details.

2.1.1 Seidle-Huitt model

This model does not include the elastic strain of the coal and assumes that all permeability changes are caused by the sorption-induced strain only. Under these assumptions, the porosity and permeability are defined as (Seidle & Huitt, 1995):

$$\frac{\phi}{\phi_0} = 1 + \frac{\varepsilon_L}{3}\left(1 + \frac{2}{\phi_0}\right)\left(\frac{\phi p_0}{P_L + p_0} - \frac{p}{P_L + p}\right) \tag{1}$$

$$\frac{k}{k_0} = \left[1 + \frac{\varepsilon_L}{3}\left(1 + \frac{2}{\phi_0}\right)\left(\frac{p_0}{P_L + p_0} - \frac{p}{P_L + p}\right)\right]^3 \tag{2}$$

2.1.2 Palmer-Mansoori model

Unlike the Seidle-Huitt model, the Palmer-Mansoori model considers the elastic deformation of coal under uniaxial stress conditions. The porosity model is defined as (Palmer, 2009):

$$\frac{\phi}{\phi_0} = 1 + \frac{c_m}{\phi_0}(p - p_0) + \frac{\varepsilon_L}{\phi_0}\left(\frac{K}{M} - 1\right)\left(\frac{p}{P_L + p} - \frac{p_0}{P_L + p_0}\right) \tag{3}$$

$$\frac{k}{k_0} = \left[1 - \frac{c_m}{\phi_0}(p - p_0) + \frac{\varepsilon_L}{\phi_0}\left(\frac{K}{M} - 1\right)\left(\frac{p}{P_L + p} - \frac{p_0}{P_L + p_0}\right)\right]^3 \tag{4}$$

where $M = \frac{(1-\nu)}{(1+\nu)(1-2\nu)}$, $c_m = \frac{1}{M} - (\frac{K}{M} + f - 1)\gamma$, f is a fraction between 0 and 1, and γ is given compressibility.

2.1.3 Shi-Durucan model

The assumptions are same as the Palmer-Mansoori model (Shi & Durucan, 2004).

$$\frac{k}{k_0} = \exp\left\{3c_f\left[\frac{v}{1-v}(p-p_0) + \frac{\varepsilon_L}{3}\left(\frac{E}{1-v}\right)\left(\frac{p_0}{P_L+p_0} - \frac{p}{P_L+p}\right)\right]\right\} \tag{5}$$

where c_f is cleat volume compressibility.

2.1.4 Cui-Bustin model

This model has a general form but it is only applied to the same assumed situation as the Palmer-Mansoori model (Cui & Bustin, 2005).

$$\frac{\phi}{\phi_0} = \exp\left\{\left(\frac{1}{K} - \frac{1}{K_p}\right)[(\sigma - \sigma_0) - (p - p_0)]\right\} \tag{6}$$

$$\frac{k}{k_0} = \exp\left\{-\frac{3}{K_p}[(\sigma - \sigma_0) - (p - p_0)]\right\} \tag{7}$$

where K_p is the bulk modulus of pores.

2.1.5 Robertson-Christiansen model

In this model, the deformation of coal grains is neglected and equal axial and radial stresses are assumed (Robertson & Christiansen, 2005b).

$$\frac{k}{k_0} = \exp\left\{-3c_0\frac{1-\exp[\alpha_c(p-p_0)]}{\alpha}\right.$$

$$\left. + \frac{9}{\phi_0}\left[\frac{1-2v}{E}(p-p_0) - \frac{\varepsilon_L}{3}\left(\frac{P_L}{P_L+p_0}\right)\ln\left(\frac{P_L+p}{P_L+p_0}\right)\right]\right\} \tag{8}$$

where c_0 is the initial fracture compressibility and α_c is the change rate in fracture compressibility.

2.2 General formulation of effective strain based model

Considering a porous medium containing solid volume of V_s and pore volume of V_p, we assume the bulk volume $V=V_p+V_s$ and the porosity $\Phi = V_p/V$. The volumetric evolution of the porous medium with the load of $\bar{\sigma}$ and p can be described in terms of $\Delta V/V$ and $\Delta V_p/V_p$, the volumetric strain of coal and volumetric strain of pore space, respectively (Detournay & Cheng, 1993). The relations are

$$\frac{\Delta V}{V} = -\frac{1}{K}(\bar{\sigma} - \alpha p) + \varepsilon_s \tag{9}$$

$$\frac{\Delta V_p}{V_p} = -\frac{1}{K_p}(\bar{\sigma} - \beta p) + \varepsilon_s \tag{10}$$

Where $\beta = 1 - K_p/K_s$. K represents the bulk modulus of coal, K_s represents the bulk modulus of coal grains and K_p represents Bulk modulus of pore volumetric strain.

Without the gas sorption effect, the volumetric variation of the porous medium satisfies the Betti-Maxwell reciprocal theory (Detournay & Cheng, 1993), $\frac{\partial V}{\partial p}|\bar{\sigma} = -\frac{\partial V_p}{\partial \bar{\sigma}}|\bar{p}$, and we obtain

$$K_p = \frac{\phi}{\alpha} K \tag{11}$$

Using the definition of porosity, the following expressions can be deduced as

$$\frac{\Delta V}{V} = \frac{\Delta V_s}{V_s} + \frac{\Delta \phi}{1 - \phi} \tag{12}$$

$$\frac{\Delta V_p}{V_p} = \frac{\Delta V_s}{V_s} + \frac{\Delta \phi}{\phi(1 - \phi)} \tag{13}$$

Solving Equations 9, 10, 12 and 13, we obtain the relationship as

$$\Delta \phi = \phi \left(\frac{1}{K} - \frac{1}{K_p} \right) (\bar{\sigma} - p) \tag{14}$$

Substituting Equation 11 into the above equation yields

$$\Delta \phi = \phi \left(1 - \frac{\alpha}{\phi} \right) \frac{\bar{\sigma} - p}{K} \tag{15}$$

Because generally $(\bar{\sigma} - p)/K \ll 1$, the above equation can be simplified into

$$\frac{\phi}{\phi_0} = 1 + \frac{\alpha}{\phi_0} \Delta \varepsilon_e \tag{16}$$

Where $\Delta \varepsilon_e$ is defined as the total effective volumetric strain.

It is clear that there is a relationship between porosity, permeability and the grain-size distribution in porous media. This relationship is expressed as

$$k = \frac{d_e^2 \phi^3}{72(1 - \phi)^2} \tag{17}$$

where d_e is the effective diameter of grains. Based on the Equation 17, one obtains

$$\frac{k}{k_0} = \left(\frac{\phi}{\phi_0} \right)^3 \left(\frac{1 - \phi_0}{1 - \phi} \right)^2 \tag{18}$$

When the porosity is much smaller than 1.0 (normally less than 10%), the second factor of the right-hand side asymptotes to unity. This yields the cubic relationship between permeability and porosity for the coal matrix

$$\frac{k}{k_0} = \left(\frac{\phi}{\phi_0}\right)^3 \tag{19}$$

Therefore, the permeability ratio is evolved by

$$\frac{k}{k_0} = \left[1 + \frac{\alpha}{\phi_0}\Delta\varepsilon_e\right]^3 \tag{20}$$

$$\Delta\varepsilon_e = \Delta\varepsilon_v + \frac{\Delta p}{K_s} - \Delta\varepsilon_s \tag{20a}$$

or

$$\Delta\varepsilon_e = -\frac{\Delta\sigma - \Delta p}{K} \tag{20b}$$

where $\Delta\varepsilon_e$ is defined as the total effective volumetric strain, $\Delta\varepsilon_v$ is total volumetric strain increment, $\Delta p/K_s$ is coal compressive strain change, $\Delta\varepsilon_s$ is gas sorption-induced strain increment.

Equations 16 and 20 are models for coal porosity and permeability that are derived based on the fundamental principles of poroelasticity.

2.3 Extended formulations effective strain based model

2.3.1 A dual poro-elastic model

Although the influence of gas sorption-induced coal deformation on porosity and permeability has been widely recognized, most studies are under conditions of no change in overburden stress and effective stress-absent where effective stresses scale inversely with applied pore pressures. Zhang *et al.* (2008) introduced a new permeability model which relaxed the above assumptions and couples the transport and sorption of a compressible fluid within a dual porosity medium where the effects of deformation are rigorously accommodated.

Permeability models for single-porosity medium ignore the mass transfer between matrix blocks and fractures. In reality, coal is a typical dual-porosity medium consisting of matrix blocks and fracture networks. The gas flows primarily in the fracture network and stores mainly in the matrix blocks principally in the adsorbed form. When gas is injected into coal, it first flows along fractures. At this moment, gas pressure difference between fractures and matrix blocks is created so gas in fracture diffused into matrix blocks. This diffusion process takes much longer time than gas flow in fractures (Larsen, 2004). In other words, assumption that mass transfer finishes immediately is not true. This is a major problem of permeability models for single-porosity medium. In order to understand well the impact of coal matrix-fracture compartment interactions, a permeability model for dual-porosity medium was proposed (Wu *et al.*, 2010a).

Coal is conceptualized as in Figure 3 (Wu *et al.*, 2010a). It consists of coal matrix and fractures. In this model, matrix permeability model is same as Equation 20 and fracture permeability is determined by both total volumetric deformation of coal and matrix swelling/shrinkage due to effective stress decrease/increase and adsorption/desorption.

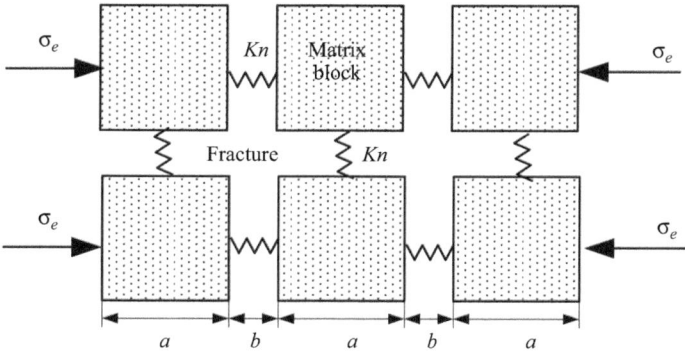

Figure 3 Dual-porosity fractured medium.

The cubic matrix length and the fracture aperture are represented by a and b, respectively. K_n is the fracture stiffness, and σ_e is the effective stress.

For the fracture system comprising continuous orthogonal fractures, the cubic law for fracture permeability can be expressed as (Wu *et al.*, 2010a):

$$k_f = \frac{b^3}{12a} \tag{21}$$

The dynamic permeability of the fracture system can be expressed as

$$\Delta k_f = \frac{3b^2 \Delta b}{12a} - \frac{b^3 \Delta a}{12a^2} = k_f\left(\frac{3\Delta b}{b} - \frac{\Delta a}{a}\right) \tag{22}$$

or
$$\Delta k_f = k_f(3\Delta\varepsilon_f - \Delta\varepsilon_v) \tag{22a}$$

where, $\varepsilon_v = \sigma_e^m / K + \varepsilon_s$. So

$$\frac{\Delta k_f}{k_f} = 3\frac{\Delta\sigma_e^f}{K_n} - \frac{\Delta\sigma_e^m}{K_s} - \Delta\varepsilon_s \tag{23}$$

If the initial permeability of the fracture system is k_{f0} at the initial effective stress σ_e^0, the permeability can be expressed as

$$k_f = k_{f0}\exp\left[3\frac{\sigma_e^f - \sigma_e^{f0}}{K_n} - \frac{\sigma_e^m - \sigma_e^m}{K_s} - (\varepsilon_s - \varepsilon_s^0)\right] \tag{24}$$

where σ_e is effective stress, f represents fractures and m represents matrix blocks, 0 denotes the initial value of variables, K_n is the normal stiffness of individual fractures and K_s is the grain elastic modulus. Therefore, Equations 16–24 define a model for coupled gas flow and coal deformation in dual-porosity medium. Cross-couplings between the field equations of coal deformation and gas flow are illustrated in Fig. 4.

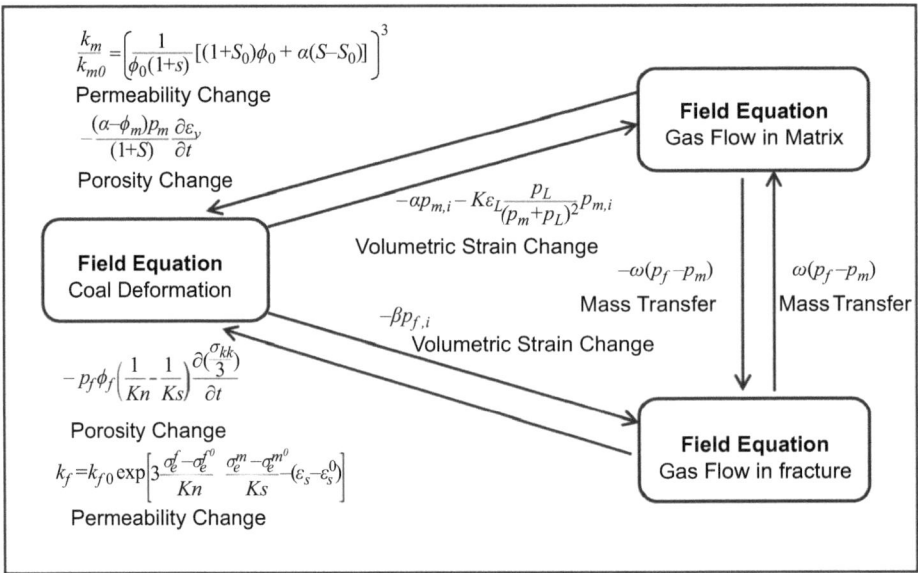

Figure 4 Cross coupling of the field equations. (Wu *et al.*, 2010a).

2.3.2 Concept of transition from local swelling to macro swelling

Although considering characteristics of microscopic structure of coal and mass transfer between matrix blocks and fractures, experimental observations could not be predicted by above dual-permeability models. From experiments conducted by Siriwardane *et al.* (2009) and Mazumder & Wolf (2008), it clearly presented that permeability relates to time when the pore pressure is constant. Other laboratory observations (Robertson & Christiansen, 2005b; Wang *et al.*, 2011) have shown that coal permeability under the influence of gas adsorption can change instantaneously from reduction to enhancement. It is commonly believed that this instantaneous switching of permeability is due to the fact that the matrix swelling ultimately ceases at higher pressures and the influence of effective stresses takes over. In this study, our previously-developed poroelastic model is used to uncover the true reason why coal permeability switches from reduction to enhancement. This goal is achieved through explicit simulations of the dynamic interactions between coal matrix swelling/shrinking and fracture aperture alteration, and translations of these interactions to permeability evolution under unconstrained swellings.

In the study of Liu *et al.* (2011), our results have revealed the transition of coal matrix swelling from local swelling to global swelling as a novel mechanism for this switching. The proposed a concept of the switch from local swelling to global swelling is shown in Figure 5. When gas is injected into coal, matrix swelling has two stages: local swelling and global swelling. Initially a coal is in the initial equilibrium state. When gas is injected, the fracture pressure reaches the injection pressure much faster than the matrix pressure and as a consequence the maximum imbalance between matrix pressure and fracture pressure is achieved. This imbalance diminishes as the gas penetrates

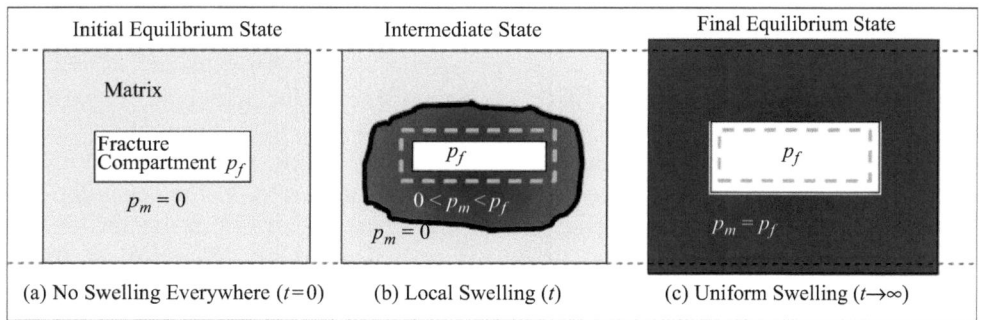

Initial Equilibrium State	Intermediate State	Final Equilibrium State
Matrix; Fracture Compartment p_f; $p_m = 0$	p_f; $0 < p_m < p_f$; $p_m = 0$	p_f; $p_m = p_f$
(a) No Swelling Everywhere $(t=0)$	(b) Local Swelling (t)	(c) Uniform Swelling $(t\to\infty)$

Figure 5 Illustration of the difference between local swelling and global swelling (Liu *et al.*, 2011).

into the coal matrix which makes the pore pressure increase. At this stage, the coal matrix swells but this swelling is confined in the vicinity of the fracture voids because area of gas penetration limits around fracture voids. This localized swelling reduces the fracture aperture thus the fracture permeability drops immediately. As the gas penetration progresses, the swelling zone extends further into the coal matrix. When the swelling zone front moves away from the fracture void, the impact of localized swelling on the fracture aperture starts to decline. At this stage, the local swelling becomes the global swelling and the fracture permeability recovers. When matrix pressure equals fracture pressure again, the final equilibrium state is achieved (Peng *et al.*, 2004).

Specific observations include: (1) at the initial stage of CH_4 injection, matrix swelling is localized within the vicinity of the fracture compartment. As the injection continues, the swelling zone is extending further into the matrix and becomes macro-swelling. Matrix properties control the swelling transition from local swelling to macro swelling; (2) matrix swelling processes control the evolution of coal permeability. When the swelling is localized, coal permeability is controlled by the internal fracture boundary condition and behaves volumetrically; when the swelling becomes global swelling, coal permeability is controlled by the external boundary condition and behaves non-volumetrically; and (3) matrix properties control the switch from local swelling to global swelling and the associated switch in permeability behavior from reduction to recovery.

This conceptual model is useful to understand the impact of coal matrix-fracture interactions on the evolution of coal permeability. Although it could explain and match experimental observations, it cannot reflect how gas flows in fractures. Therefore, we need to transform our conceptual model into a continuous mechanics based approach.

2.3.3 Effective stress transfer model

Coal permeability models are derived normally under three common assumptions: (1) uniaxial strain; (2) invariant total stress; and (3) local equilibrium. Experimental measurements are normally conducted under constant effective stress or free swelling conditions. The inconsistency between model assumptions and the experimental conditions determines that these coal permeability models cannot be used to analyze the experimental data. Based on the theory of poroelasticity, coal permeability is determined by the effective stress only. Therefore, there would be no permeability change when the effective stress remains constant. This theoretical conclusion contradicts with the "V" shape profile of coal permeability as widely observed through experiments. This "abnormal" behavior has been explained through a novel dual-permeability model. The model is then formulated based on our previous concepts of local swelling, global swelling and their evolutions from the initial equilibrium state to the final equilibrium state. In the formulation, four strains were defined: coal global strain, fracture local strain, matrix global strain, and pore local stain. Coal permeability is defined as a function of these strains. Their evolutions are determined by the effective stress transfer between the matrix system and the fracture system, and regulated by the gas diffusion process from the fracture system to the matrix system. The strain evolutions were used to define how coal permeability changes with time or gas pressure in the matrix system

The porosity of fracture is related to the effective strain of fractures as (Liu *et al.*, 2011).

$$\frac{\phi_f}{\phi_{f0}} = 1 + \frac{1}{\phi_{f0}} \Delta \varepsilon_{fe} \tag{25}$$

where ϕ_{f0} is the initial fracture porosity and ϕ_f is the current fracture porosity.

As mentioned in conceptual model, the effective strain of fractures is the resultant of the coal global strain and the fracture local strain. The change of effective volumetric strain is:

$$\Delta \varepsilon_{fe} = \Delta \varepsilon_v + \Delta \varepsilon_{fl} \tag{26}$$

The fracture local strain comprises adsorption-induced strain around fracture walls and compressive strain of matrix blocks. The adsorption-induced strain narrows fracture apertures while compressive strain of matrix blocks increases fracture apertures. The fracture local strain is defined as:

$$\Delta \varepsilon_{fl} = \frac{(p_f - p_m)}{K_m} - \Delta \varepsilon_{fs} \tag{27}$$

So the effective strain of fractures is:

$$\Delta \varepsilon_{fe} = \Delta \varepsilon_v + \frac{(p_f - p_m)}{K_m} - \Delta \varepsilon_{fs} \tag{28}$$

where the first item represents global strain which is obtained from governing equation of mechanical deformation of coal; the second item represents the compressive strain of matrix blocks and the last item is the adsorption-induced strain around fracture walls. They are functions of fracture and matrix pressure which could be solved from equation of gas flow in fracture networks and equations of matrix-fracture mass transport.

The typical relationship between porosity and permeability is cubic law (Liu *et al.*, 2011):

$$\frac{k_f}{k_{f0}} = \left(\frac{\phi_f}{\phi_{f0}}\right)^3 \tag{29}$$

Substituting Equation 25 into Equation 29 to obtain fracture permeability model:

$$\frac{k_f}{k_{f0}} = \left(1 + \frac{1}{\phi_{f0}}\Delta\varepsilon_{fe}\right)^3 \tag{30}$$

Where k_{f0} is the initial fracture permeability and k_f is the current fracture permeability.

Similar to models for fracture networks, the porosity of matrix is related to the effective volumetric strain as (Liu et al., 2010):

$$\frac{\phi_m}{\phi_{m0}} = 1 + \frac{\alpha_m}{\phi_{m0}}\Delta\varepsilon_{me} \tag{31}$$

Similar with fractures, the effective strain for pores inside of matrix blocks is:

$$\Delta\varepsilon_{me} = \Delta\varepsilon_{mg} + \Delta\varepsilon_{gl} \tag{32}$$

As mentioned in conceptual model, the matrix global strain is less than the coal global strain because fracture pressure will compress the matrix blocks and the matrix global strain is:

$$\Delta\varepsilon_{mg} = \Delta\varepsilon_v - \frac{(p_f - p_m)}{K_m} \tag{33}$$

The adsorption-induced strain decreases the pore volume while matrix pressure compresses coal grains and increases the pore volume:

$$\Delta\varepsilon_{gl} = \frac{p_m}{K_s} - \Delta\varepsilon_{ms} \tag{34}$$

Substituting Equations 33 and 34 into Equation 32, the effective strain for pores inside matrix blocks is:

$$\Delta\varepsilon_{gl} = \frac{p_m}{K_s} - \Delta\varepsilon_{ms} \tag{35}$$

where the first item represents global strain which is obtained from governing equation of mechanical deformation of coal; the second item represents the compressive strain of matrix blocks; the third item is the adsorption-induced strain around fracture walls and the last item demonstrates the compressive strain of coal grains. They are functions of fracture and matrix pressure which could be solved from equation of gas flow in fracture networks and equations of matrix-fracture mass transport.

According to cubic law, the matrix permeability is:

$$\frac{k_m}{k_{m0}} = \left(1 + \frac{\alpha_m}{\phi_{m0}}\Delta\varepsilon_{me}\right)^3 \tag{36}$$

where ϕ_{m0} is the initial matrix porosity, ϕ_m is the current matrix porosity, k_{m0} is the initial matrix permeability and k_m is the current matrix permeability.

Finally, a novel model is then formulated based on our concepts of local swelling, global swelling and their evolutions from the initial equilibrium state to the final equilibrium state. This model solved the mystery of this "abnormal" behavior that traditional theoretical conclusion contradicts with the "V" shape profile of coal permeability as widely observed through experiments

2.4 Finite element solutions

Numerical modeling methods have been widely used to solve the complex coal-gas interaction due to the capacity to deal with issues that analytical analysis cannot normally handle. Commonly assumptions of traditional numerical models: (a) Coal is a homogeneous, isotropic and elastic continuum. (b) Strains are much smaller than the length scale. (c) Gas contained within the pores is ideal, and its viscosity is constant under isothermal conditions. (d) The rate of gas flow through the coal is defined by Darcy's law. (e) Conditions are isothermal. (f) Coal is saturated by gas. (g) Compositions of the gas are not competitive, *i.e.*, one gas component at a time.

2.4.1 Governing equations for coal seam deformation

The strain-displacement relation is expressed as

$$\varepsilon_{ij} = \frac{1}{2}\left(u_{i,j} - u_{j,i}\right) \tag{37}$$

where ε_{ij} is the component of the total strain tensor and u_i is the component of the displacement. The equilibrium equation is defined as

$$\sigma_{i,j,j} + f_i = 0 \tag{38}$$

where σ_{ij} denotes the component of the total stress tensor and f_i denotes the component of the body force. The constitutive relation for the deformed coal seam becomes

$$\varepsilon_{ij} = \frac{1}{2G}\sigma_{ij} - \left(\frac{1}{6G} - \frac{1}{9K}\right)\sigma_{kk}\delta_{ij} + \frac{\alpha}{3K}p\delta_{ij} + \frac{\varepsilon_s}{3}\delta_{ij} \tag{39}$$

where K represents the bulk modulus of coal and K_s represents the bulk modulus of coal grains. G is the shear modulus of coal, E is Young's modulus of coal and v is Poisson's ratio of coal. α is defined as the Biot coefficient. δ_{ij} is the Kronecker delta. p is the gas pressure within the pores.

2.4.2 Governing equations for gas flow

The equation for mass balance of the gas is defined as

$$\frac{\partial m}{\partial t} + \nabla \cdot (\rho_g q_g) = Q_s \tag{40}$$

where ρ_g is the gas density, q_g is the vector of Darcy velocities, and Q_s is the gas source or sink. m is the gas content including free-phase gas and absorbed gas and is defined as (Saghafi *et al.*, 2007).

$$m = \rho_g \phi + \rho_{ga}\rho_c \frac{V_L p}{p + P_L} \tag{41}$$

The Darcy velocity is defined as

$$q_g = -\frac{k}{\mu}\nabla p \tag{42}$$

Substituting Equations 41 and 42 into Equation 40, we obtain

$$\left[\phi + \frac{\rho_c \rho_a V_L P_L}{(p + P_L)^2}\right]\frac{\partial p}{\partial t} + p\frac{\partial \phi}{\partial t} - \nabla \cdot \left(\frac{k}{\mu}p\nabla p\right) = Q_s \tag{43}$$

Equations 38 and 43 are coupled through the porosity-permeability relation.

All equations define the coupled model including coal deformation and gas flow in a dual-porosity medium. The governing equations, particularly the gas flow equation, are nonlinear second-order partial differential equations (PDEs) in space and first-order PDEs in time. These equations cannot be theoretically solved because of the nonlinearity in both the space and time domains. Therefore, they are implemented into, and solved using a certain suitable software.

General comments: my understanding is that book chapter is more like a review paper, where you summarize other's work and highlight the differences among them. This work primarily focused on the work has been done by our group, but I feel probably it's better if we also include other model as well, such as Palmer's model, Shi-Duracan model *et al.* Just to show the comprehension of the review.

3 EVOLUTION OF COAL PERMEABILITY THROUGH NUMERICAL MODELING

3.1 Coal permeability models

Example is presented to illustrate effects of interactions between fractures and matrix on the evolution of coal permeability and to reveal the reason why coal permeability changes under the condition of free swelling. The common coal sample in laboratory is a cylinder confined by constant stress and the gas is injected from bottom as shown as Figure 6(a). Because of the symmetrical geometry of this sample and its symmetrical load, the 3-D coal sample shown in Figure 6(a) could be represented by the 2-D model as shown in Figure 6(b). The injection pressure, P_{in}, is applied at the bottom boundary and no flow boundary is applied on other boundaries. The initial pressure for both fracture networks and matrix blocks is P_0. The confining pressure, P_{con}, is applied at all boundaries. This boundary condition allows the domain to free swell. Input parameters for simulations are listed in Table 1. The data are collected from Zhang *et al.* (2008).

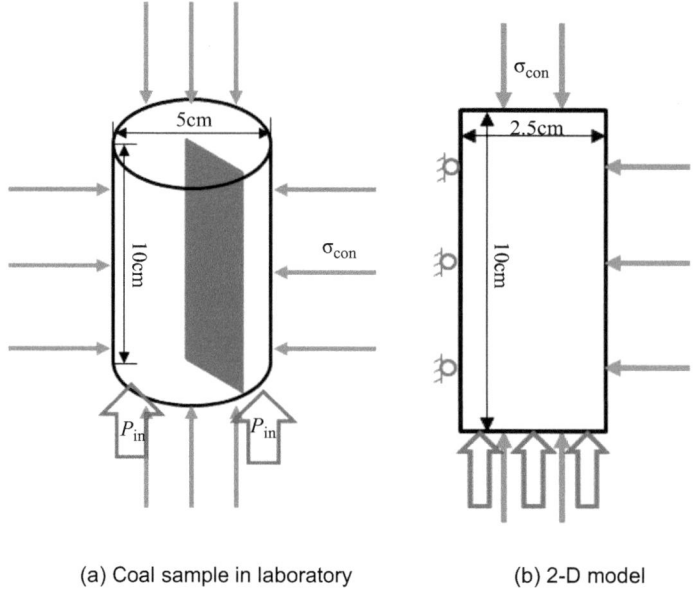

(a) Coal sample in laboratory   (b) 2-D model

Figure 6 Simulation model of coal sample.

Table 1 Input parameters for simulations.

Symbol	Value	Description	Unit
E	2.713	Young's modulus of Coal	GPa
v	0.339	Poisson's ratio of Coal	–
α_f	0.1	Biot coefficient of Fracture	–
α_m	2/3	Biot coefficient of Matrix	–
μ	1.84×10^{-5}	Viscosity	Pa·s
ϕ_{m0}	2	Initial matrix porosity	%
ϕ_{f0}	2	Initial fracture porosity	%
k_{m0}	1×10^{-18}	Initial matrix permeability	m²
k_{f0}	1×10^{-17}	Initial fracture permeability	m²
P_0	1	Initial pressure	MPa
P_{in}	3	Injection pressure	MPa
P_L	6.109	Langmuir pressure constant of Methane	MPa
P_{con}	0	Confining pressure	MPa
V_L	0.015	Langmuir sorption capacity of Methane	m³/kg
ε_L	0.02295	Langmuir strain constant of Methane	–
ρ_c	1250	Coal density	kg/m³
P_a	0.1	Atmosphere pressure	MPa
a	1	Shape factor	m⁻²

3.2 Comparisons of permeability models

Numerical results were compared with the experimental data to valid the coupling model. To verify our new model, six sets of numerical simulation were conducted and comparison results to permeability ratio between our new model and experimental data from Robertson and Christiansen (2005) were presented (Peng *et al.*, 2014). In Robertson and Christiansen's work (2005), the experimental data were collected after 24 hours or longer. Therefore, the numerical simulation results of fracture permeability ratio of this model were also collected after 24 hours. This result is called as the "Result I of new model" in the following. In reality, the injection of absorbing gases such as CO_2 may induce heterogeneous sorption-associated swelling and volumetric strains depending on the lithotypes present and gas pressure (Karacan, 2007; Karacan & Mitchell, 2003). However, this model does not deal with these factors so it is hard to exactly match the experimental data at 24 hours. If the numerical simulation results around 24 hours were collected this result as the "Result II of new model" in the following. Another two typical permeability models were also used: Palmer-Mansoori model (PM model) and Shi-Durucan model (SD model), as counterparts of this model. The Figure 7 shows the comparison results between this model, PM model and SD model.

PM and SD models are derived for the case of uniaxial strains based on the theory of poroelasticity. These model conditions are not consistent with the experimental ones where constant stresses were applied. More importantly, these models can predict one permeability value for a specific injection pressure because they cannot consider the evolution of matrix–fracture interactions. These inconsistencies between model assumptions and experimental conditions may be the reason for the mismatches. However, the new model can simulate how experiments are performed under the exact same conditions. This is why this model can match the experimental data reasonably well (Peng *et al.*, 2014).

3.3 Summary

The main characteristic of this model is that it implements the switch of the fracture local strain to the coal global strain due to effective stress transfer between fractures and matrix blocks. The model could accurately obtain the coal permeability profiles. The "V" shape profile represents that the transition of coal permeability from the initial equilibrium state to final equilibrium state takes a long period of time from a few days to months. The "Langmuir" shape profile represents that the transition takes a shorter period of time from a few minutes to hours. In reality, the shape profile of coal permeability should be somewhere in between.

4 APPLICATION EXAMPLES

In the following, we present two simulation examples to illustrate the resultant effects of coupled gas sorption and coal deformation in the process of gas extraction and CO_2 injection based on the new permeability model.

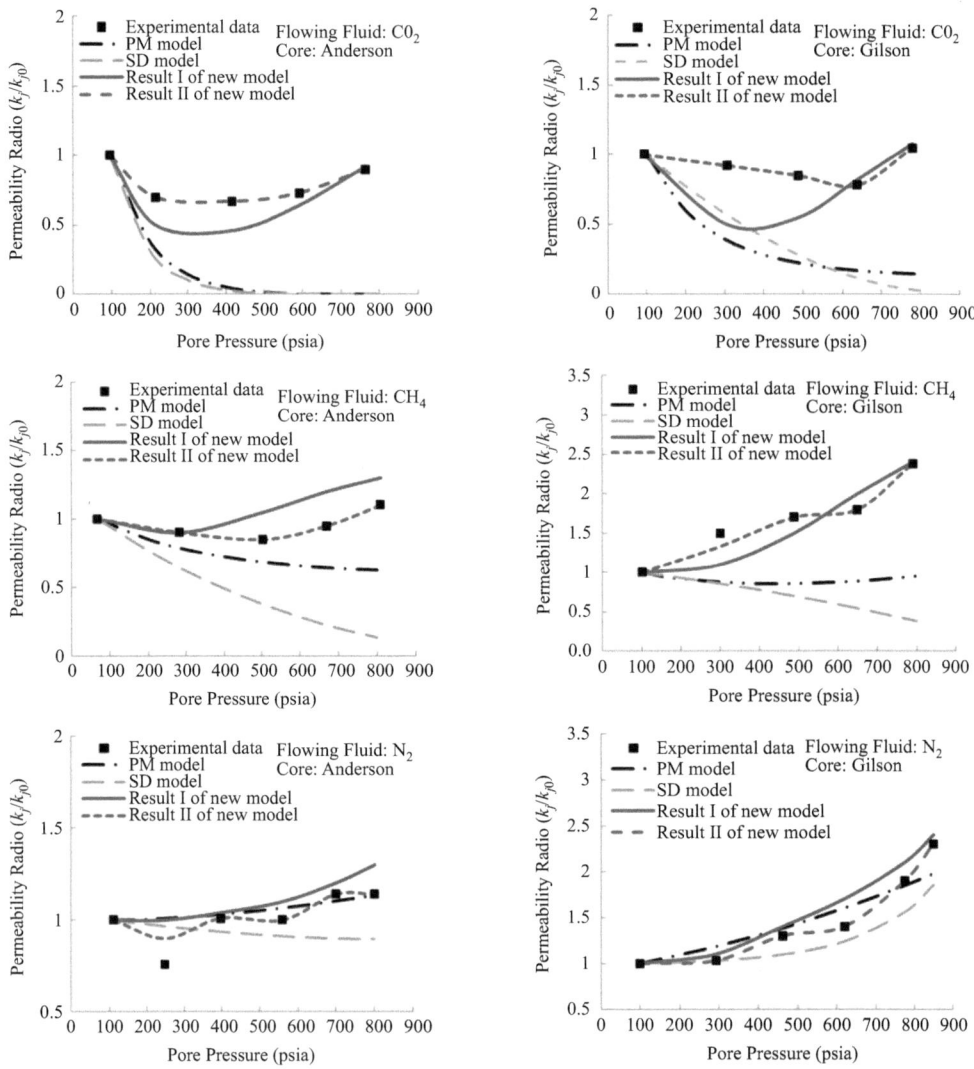

Figure 7 Comparison between different models (Peng *et al.*, 2014).

4.1 Evolution of coal permeability around the production well

A field scale model is used to simulate the performance of coalbed methane production under in-situ conditions (Liu *et al.*, 2010). Input parameters for this simulation are identical to the parameters used in Table 1. The simulation model geometry is 200 m by 200 m with a methane production well located at the lower left corner. For the coal deformation model, all four sides are constrained in the normal direction. For the gas transport model, the coal is saturated initially with CH_4 and the initial pressure is 6.12 MPa. A condition of atmospheric pressure is applied at the boundary representing the production well. Simulation results are presented in Figures 8–10.

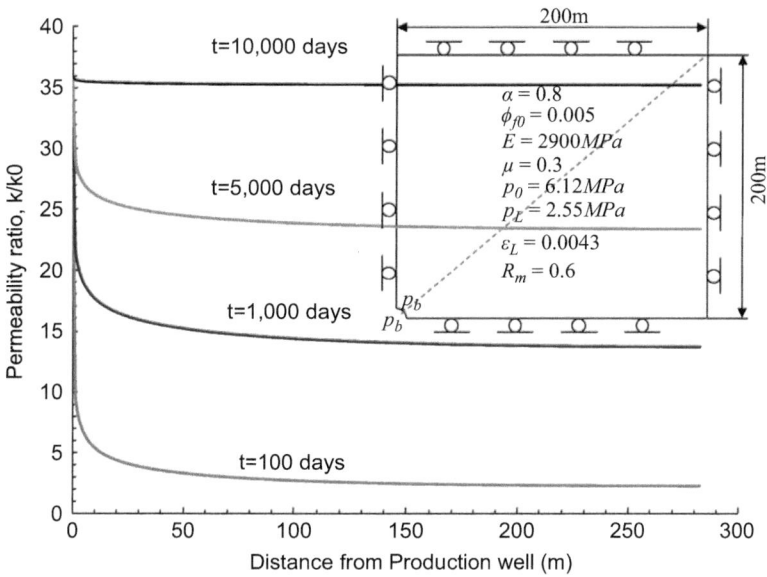

Figure 8 Spatial and temporal evolution of coal permeability ratios on a diagonal radial traverse from the production well.

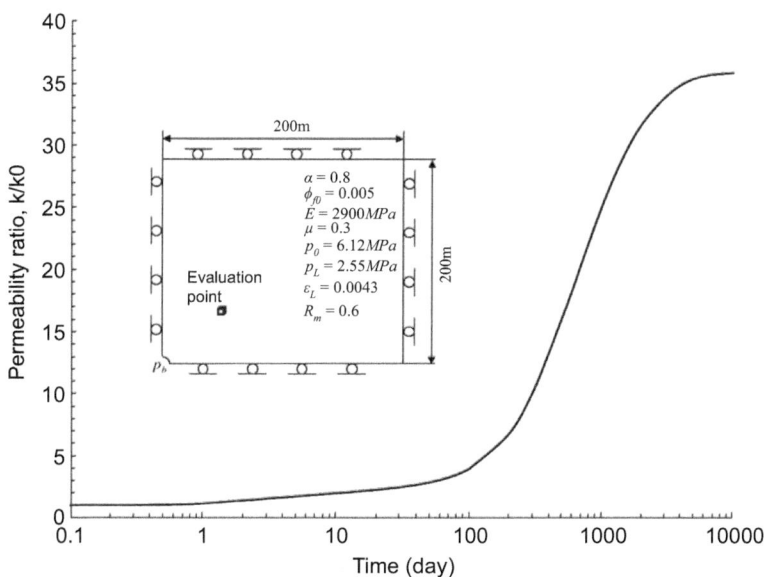

Figure 9 Evolution of coal permeability at a specific evaluation point.

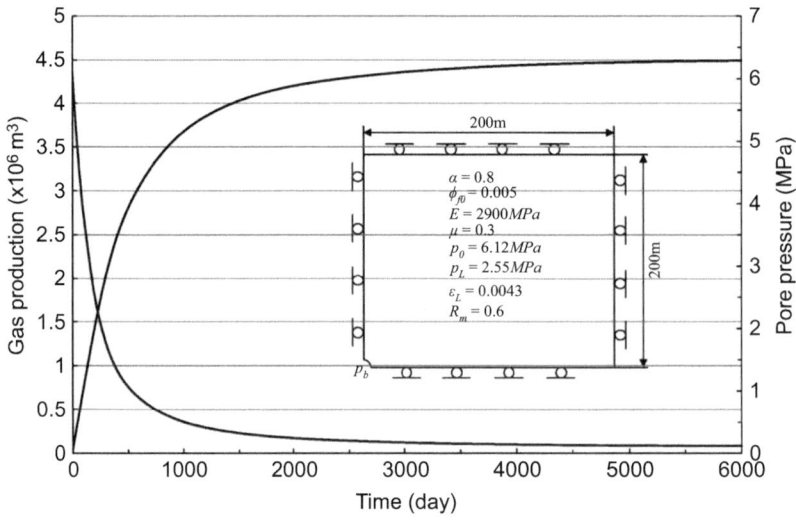

Figure 10 Evolution of the cumulative gas production and pore pressure.

In this simulation, the reservoir volume remains unchanged throughout the production. This assumption requires that global strains within the coal seam are zero. However, this constraint does not preclude sorption-induced shrinkage (or swelling) of individual coal blocks and complementary opening (or closing) of fractures. The direct consequence of these internal transformations is the isotropic change in permeability.

As shown in that equation, the change in coal permeability is defined only by the swelling strain. This represents the ideal case, *i.e.*, 100% of the swelling strain contributes to the effective stress-induced coal deformation. However, only a portion of the effective stress-induced coal deformation contributes to the permeability change as this is modulated through the parameter Rm (elastic modulus reduction ratio). The parameter 1–Rm represents the ratio of the partitioned strain for the fracture system to the total equivalent strain. In this case, 1–Rm is equal to 0.4. This means that only ~40% of the total effective stress-induced coal strain (equal to the swelling strain) is directly responsible for the permeability growth, as shown in Figs. 8 and 9. Figure 10 shows the relation of the cumulative gas production and the pore pressure with time. The gas production was calculated based on 5 m height of coal seam, and the nonlinear change trend between cumulative gas production and pore pressure indicates the influence of gas desorption.

4.2 Impact of CO_2 injection and differential deformation on CO_2 injectivity under in-situ stress conditions

The five-spot well pattern model geometry of 100 m by 100 m is shown in Figure 11 with the CO_2 injection well centered within the block and with the four CH_4 recovery wells located at the vertices. For the coal deformation model, all four sides are confined in the normal direction while the production and injection wells are unconfined. For the binary gas transport model, the coal is saturated initially with CH_4 and the initial pressure is 4.3

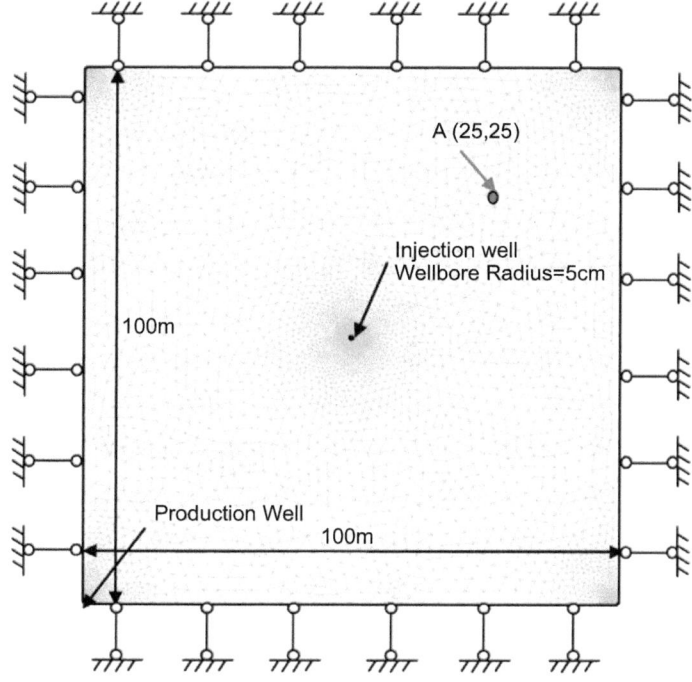

Figure 11 Field-scale model: five-spot well pattern model with the CO_2 injection well centered and the four CH_4 recovery wells located at the vertices.

MPa. The Neumann boundary conditions are specified at the four production wells, and the constant injection pressure condition is specified at the injection well.

In order to investigate the mechanical response of the coal seam to CO_2 injection, four sets of simulations were conducted to investigate the effects of coal mechanical properties on the gas injection performance under different coal matrix Young moduli and Poisson ratios; to calculate the influence of initial permeability on the implementation of CO_2-ECBM technology under different values; and to investigate the influence of gas components on CO_2 injection efficiency and CH_4 production under three different gas swelling strain constants. The detailed simulation strategies are shown in Table 2.

Table 2 Simulation strategies.

Case 1	Impact of Young moduli on the resulting response $E = 2.71$; 4.07; 5.42 MPa
Case 2	Impact of Poisson ratios on the resulting response $\mu = 0.20$; 0.34; 0.40
Case 3	Impact of initial permeabilities on the resulting response $k0 = 3 \times 10^{-16}$; 3×10^{-17}; $3 \times 10^{-18} m^2$
Case 4	Impact of gas swelling strain constants on the resulting response $\varepsilon_{\infty 2} = 0.0119$; 0.0237; 0.0474

4.2.1 Impact of coal mechanical properties

Figure 12(a) shows that a smaller Young modulus results in a less reduction in coal permeability as pore pressure increases. When E = 2.71 GPa, initially, the permeability decreases with gas injection until permeability ratio reduces to 0.88 (pore pressure is about 11 MPa), followed a rebound in permeability with CO_2 injection. The turning point marks the transition from the dominance of the gas sorption-induced permeability change to the dominance of the effective stress change induced permeability change. However, the turning points were not observed for moduli of E = 4.07 GPa and 5.42 GPa where the permeability decreases monotonically with CO_2 injection for the whole injection process. This illustrates that sorption induced permeability change plays a more significant role than the influence of effective stress within this range of pore pressure. As shown in

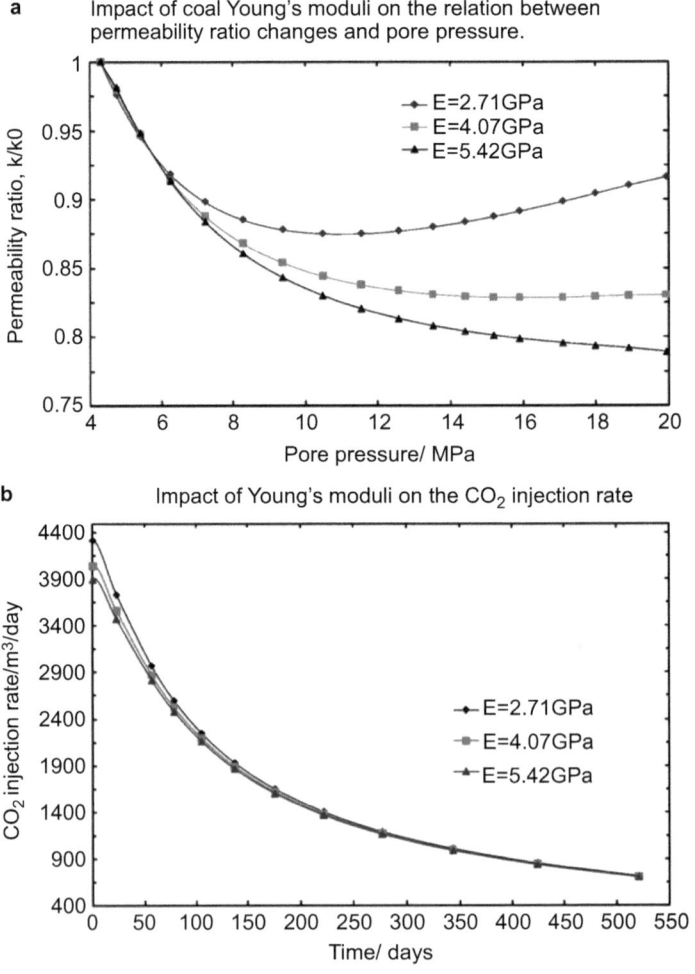

Figure 12 Sensitivity of the model to Young's modulus.

Figure 12(b), the peak values of the gas injection rate are similarly regulated by the magnitudes of coal moduli: a lower coal modulus results in a larger CO_2 injection rate since permeability remains elevated even as gas pressures increase throughout the reservoir.

Figure 13(a) shows that the magnitude of the Poisson ratio also has a significant impact on the permeability. When $\mu = 0.20$, the permeability decreases from the initial 1.0 to 0.92, and then rebounds to 1.04 at the end of gas injection. A similar pattern is observed for the case of $\mu = 0.34$. The turning point from decrease to increase marks the transition from the dominance of the gas sorption-induced permeability change to the dominance of the effective stress induced permeability change. However, the turning point was not observed for the case of $\mu = 0.40$. In this case, the permeability decreases monotonically with CO_2 injection for the whole injection process. Figure 13(b) shows

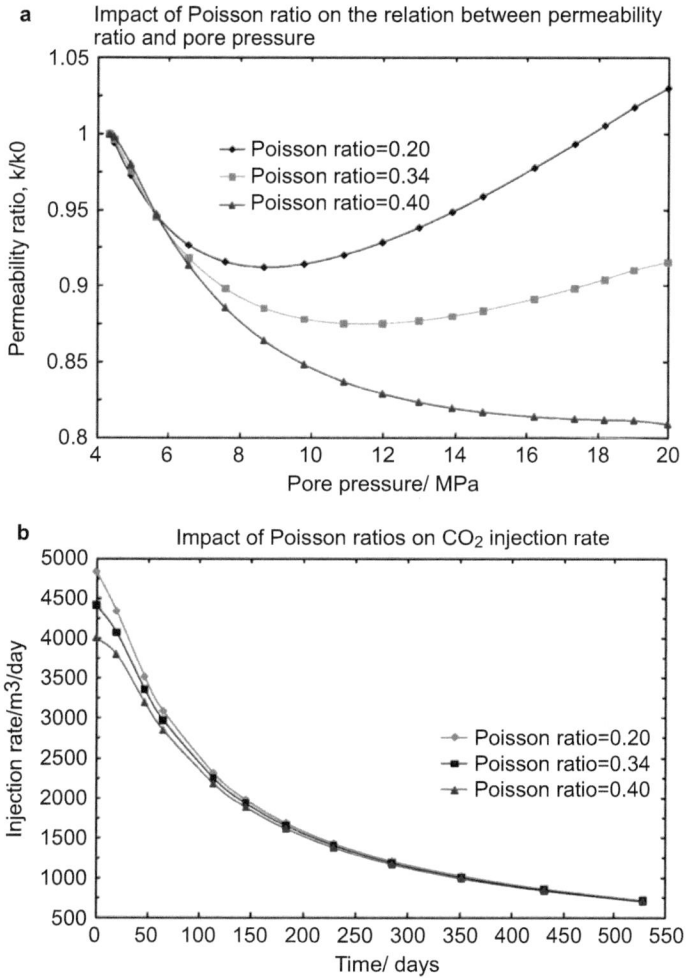

Figure 13 Sensitivity of the model to Poisson ratio.

the impact of Poisson ratios on the injection rate, again reflecting the influence of permeability change modulated by Poisson ratio.

4.2.2 Impact of initial permeability

Figure 14 shows the impact of the initial permeability on the evolutions of permeability ratio, gas concentration, injection rate, and pore pressure. When the gas pressure increases, CO_2 and CH_4 advection changes the composition of gas components in the coal matrix. The increase of pore pressure results in a decrease of effective stress. The reduction in effective stress enhances the coal permeability. In contrast, the binary gas diffusion induced differential swelling of the coal matrix reduces the cleat apertures and decreases the permeability. The net change in permeability accompanying gas diffusion is controlled competitively by the influence of effective stresses and differential matrix swelling. Therefore the impact of the initial permeability is minimal as shown in Figure 14(a). The transport of gases is controlled both by the dispersion and the convection processes. Because the coal permeability is controlled both by the initial permeability and by the change ratio, the initial permeability has a significant impact on the movement of the gas diffusion front. As shown in Figure 14(b), this movement is inversely proportional to the initial permeability. Under constant injection pressure, the injection rate is primarily controlled by the coal permeability. The coal permeability is a summation of

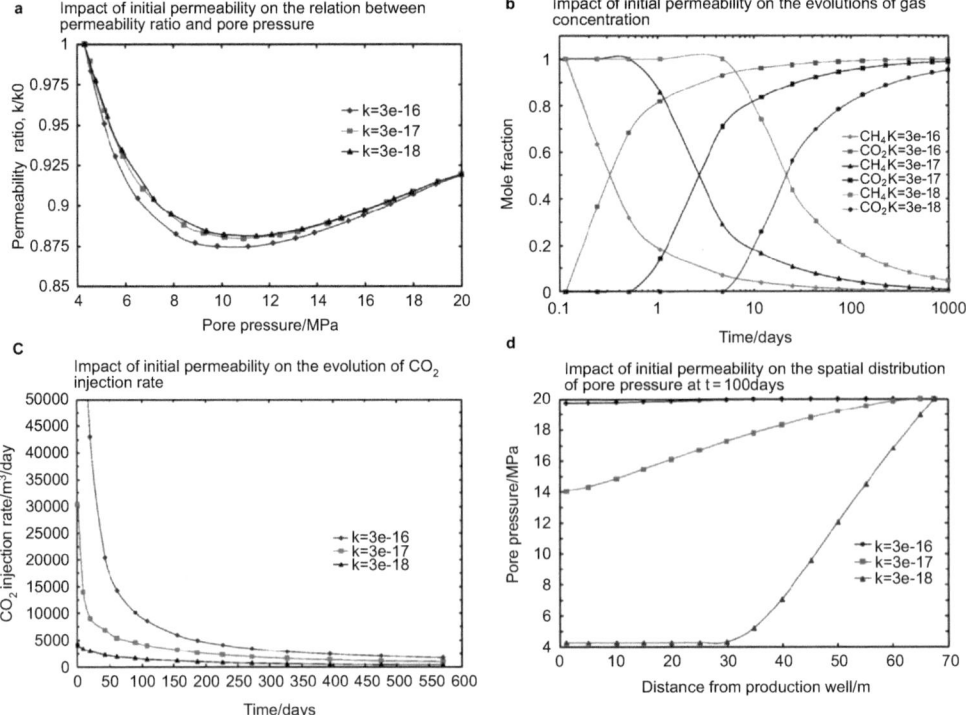

Figure 14 Sensitivity of the model to initial permeability.

the initial permeability and the permeability change regulated by the influence of effective stresses and differential matrix swelling. As shown in Figure 14(a), the permeability change is from 0 to $0.88k_0$ for all three cases. This means that the gas injection rate is primarily controlled by the initial permeability, as shown in Figure 14(c). Under the constant injection pressure, the CO_2–CH_4 displacement front is primarily by convection together with a small effect from CO_2 and CH_4 dispersion. Therefore, the front displaces at a velocity proportional to the initial permeability as shown in Figure 14(b). The pore pressure in the coal is close to the injection pressure behind the front. As shown in Figure 14(d), the pore pressure is almost constant when the initial permeability is equal to $3 \times 10^{-16} m^2$. This is because the displacement front has reached the outside boundary. For the other two cases with smaller permeability, the pore pressures are significant less that the injection pressure because the displacement front has not reached the outside boundary.

4.2.3 Impact of gas swelling strain constants

The net change in permeability accompanying binary gas diffusion is controlled competitively by the influence of effective stresses and differential matrix swelling. In this simulation, three different magnitudes of the gas swelling strain constant were used to alter the balance between the sorption-induced permeability change and the effective stress-induced permeability change. Modeling results are shown in Figure 15.

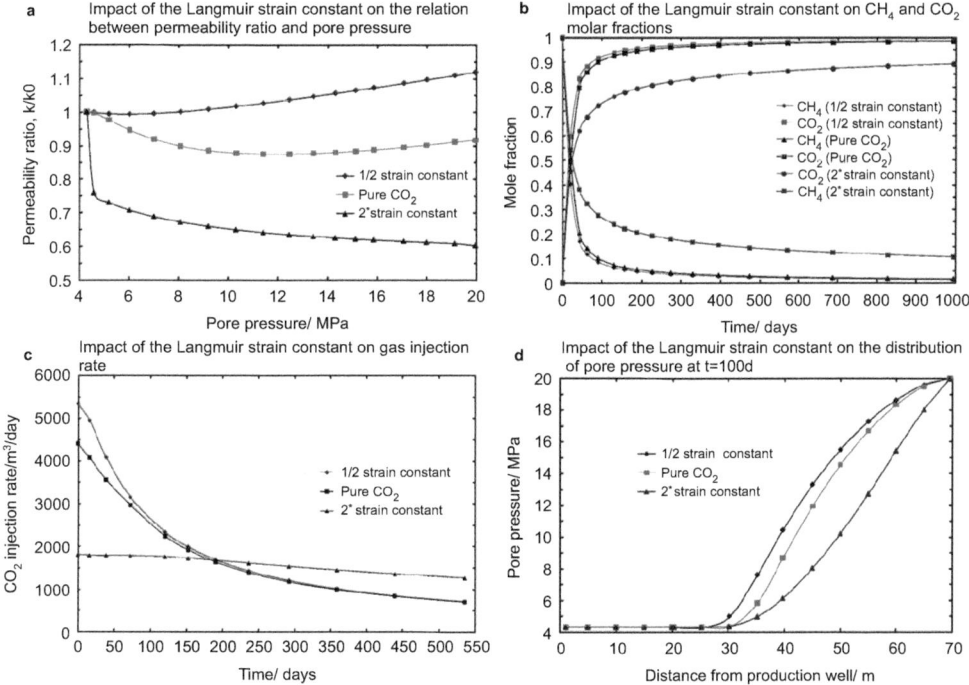

Figure 15 Sensitivity of the model to the injection gas Langmuir strain constants.

The model of with the strain constant reduced by one half (case 1) is to simulate the binary gas injection of an N_2/CO_2 mixture, maybe representing direct injection of flue gas, as a method to dramatically increase gas injectivity but with a lower CO_2 concentration per unit mass of gas injected (*e.g.* Shi *et al.*, 2008). Conversely the model with twice strain constant (case 3) is to represent the affinity to other coals of different rank with much larger gas swelling strain constants (Mazumder *et al.*, 2008; Robertson & Christiansen, 2005a; Shi *et al.*, 2008).

The impact of the gas swelling strain constant on the permeability change is shown in Figure 15(a). For case 1, the influence of effective stress change on permeability is dominant, while the impact of the sorption induced permeability change is secondary; for the pure CO_2 model, the gas sorption-induced permeability change is dominant over the effective stress change induced permeability in the initial stages; as injection proceeds, the mechanical influence takes over the dominant role. However, for case 3, the gas sorption-induced permeability change always plays a dominant role during CO_2 injection within this range of pore pressure. Figure 15(b) shows the evolution of gas mole fractions at the reference location under different gas swelling strain constants. For case 1 and the pure CO_2 models, gas mole fractions reach steady values within about 200 days, but for case 3, this percentage is much higher (about 20%) and it still needs long time to complete CH_4 displacement. These observations illustrate that the gas swelling strain constant affects both the gas dispersion and the gas convection. Figure 15(c) shows the impact of the gas swelling strain constant on the gas injection rate. For case 1 and for pure CO_2 models, the permeability changes gradually from the initial value to decrease to increase. However, an almost instant 25% reduction takes place for case 3 model. These changes in permeability are consistent with the changes in injection rate. Figure 15(d) shows the impact of the gas swelling strain constant on the distribution of pore pressure. For all three cases, the displacement front has not reached the outside boundary. This is why the pore pressures are much smaller than the outside boundary pressure (or the injection pressure).

5 CONCLUSIONS

Understanding of coal-gas interaction is very important for the full interpretation of gas production behavior. A wide variety of coal permeability models are derived normally under three common assumptions: (1) uniaxial strain; (2) invariant total stress; and (3) local equilibrium. This paper documents the evolution of our research efforts on the development of coal permeability models, their implementation into the framework of coal-gas interaction geomechanics, and their applications. Research concepts are advanced through the following four milestones: (1) Effective Stress Concept. The inclusion of the effective stress concept led to a generalized coal permeability model was developed. This contribution transformed the pore pressure-based models into a new generation of effective stress-based ones. In fact, these pore pressure-based coal permeability models can be considered as special cases of our effective stress-based ones. (2) Dual Porosity Concept. The inclusion of the dual porosity concept led to a generalized dual porosity and dual permeability model for coals. This contribution transformed the single poroelasticity-based coal-gas interaction geomechanics into the dual poroelasticity-based one. (3) Local and Global Swelling Concept. The inclusion of the local and global swelling concept led to a series of generalized and coal matrix-fracture interaction-based models. This contribution provided novel mechanisms for the explanation of the observed "abnormal" behaviors. (4) Effective Stress Transfer Concept. The inclusion of this concept led to a continuous mechanics-based approach that can accommodate the local and global swelling concept. This approach provided the basis for the development of novel CBM production modeling tools. Strain evolution is used to define how coal permeability changes with time or gas pressure in the matrix system.

This proposed study is a logical extension and integration of long-term persistent efforts. Future work is focused on a novel numerical simulation tool of permeability generator. A simulator will be developed to transform our monitorings of gas sorption and coal local deformation into coal permeability profiles. It is expected that these profiles are regulated by the coal matrix properties.

Nomenclature.

V_s	Solid volume
V_p	Pore volume
V	Bulk volume
Φ	Porosity
α	Biot coefficient
p	Gas pressure
K	Bulk modulus of coal
K_s	Bulk modulus of coal grains
K_p	Bulk modulus of pore volumetric strain
d_e	Effective diameter of grains
ε_e	Total effective volumetric strain
ε_v	Total volumetric strain
ε_s	Gas sorption-induced strain
Δ	Increment
K_n	Fracture stiffness
a	Cubic matrix length
b	Fracture aperture
k	Permeability
k_f	Permeability of the fracture
k_{f0}	Initial fracture permeability
Φ_{f0}	Initial fracture porosity
Φ_f	Fracture porosity
ε_{fe}	Effective strain of fractures
ε_{fl}	Fracture local strain
ε_{fs}	Fracture adsorption-induced strain
Φ_{m0}	Initial matrix porosity
Φ_m	Matrix porosity
α_m	Biot coefficient for matrix
ε_{me}	Effective strain for pores of matrix blocks
ε_{mg}	Matrix global strain
ε_{gl}	Matrix local strain
ε_{ij}	Component of the total strain tensor

(continued)

u_i	Component of the displacement
G	Shear modulus of coal
E	Young's modulus of coal
v	Poisson's ratio of coal
σ_{ij}	Component of the total stress tensor
f_i	Component of the body force
δ_{ij}	Kronecker delta
ρ_g	Gas density
q_g	Vector of Darcy velocities
Q_s	Gas source or sink

Subscript

f	Fractures
m	matrix blocks
0	initial value of variables

REFERENCES

Beamish, B. B., and Crosdale, P. J. (1998) Instantaneous outbursts in underground coal mines: An overview and association with coal type. International Journal of Coal Geology, 35, no. 1–4, 27–55.

Chen, Z., Liu, J., Connell, L., Pan, Z., and Zhou, L. (2008) Impact of effective stress and CH_4–CO_2 counter-diffusion on CO_2 enhanced coalbed methane recovery, in Proceedings SPE Asia Pacific Oil and Gas Conference and Exhibition, Perth, Australia.

Connell, L. (2009) Coupled flow and geomechanical processes during gas production from coal seams. International Journal of Coal Geology, 79, no. 1, 18–28.

Connell, L.D, Sander, R., Pan, Z., et al. (2011) History matching of enhanced coal bed methane laboratory core flood tests[J]. International Journal of Coal Geology, 87, no. 2, 128–138.

Cui, X., and Bustin, R. M. (2005) Volumetric strain associated with methane desorption and its impact on coalbed gas production from deep coal seams. The American Association of Petroleum Geologists, 1181–1202.

Detournay, E., and Cheng, A. H. D. (1993) Fundamentals of poroelasticity, Chapter 5, in Fairhurst, C., ed., Analysis and Design Methods: Oxford, Pergamon, pp. 113–171.

Dubaniewicz Jr., T. H. (2009) From Scotia to Brookwood, fatal US underground coalmine explosions ignited in intake air courses. National Institute for Occupational Safety and Health, Pittsburgh Research Laboratory, Pittsburgh, PA. Available from: http://dx.doi.org/10.1016/j.jlp.2008.08.010, 12 pages references, 37 figures and 4 tables. Accessed 16.12.11.

Gale, J., and Freund, P. (2001) Coal-bed methane enhancement with CO_2 sequestration world-wide potential. Environmental Geosciences, 8, no. 3, 210–217.

Hall, F. E., Zhou, C., Gasem, K. A. M., Robinson, R. L. J., and Yee, D. (1994) Adsorption of pure methane, nitrogen, and carbon dioxide and their binary mixtures on wet Fruitland coal. Society of Petroleum Engineers, Eastern Regional Conference and Exhibition, Charleston, WV, pp. 329–344.

Han, F., Busch, A., van Wageningen, N., Yang, J., Liu, Z., and Krooss, B. M. (2010) Experimental study of gas and water transport processes in the inter-cleat (matrix) system of coal: Anthracite from Qinshui Basin, China. International Journal of Coal Geology, 81, no. 2, 128–138.

Harpalani, S. (1985) Gas Flow through Stressed Coal: California University, Berkeley (USA).

Harpalani, S., and Schraufnagel, R. A. (1990) Shrinkage of coal matrix with release of gas and its impact on permeability of coal. Fuel, 69, no. 5, 551–556.

Kang, S. M., Fathi, E., Ambrose, R. J., Akkutlu, I. Y., and Sigal, R. F. (2011) Carbon dioxide storage capacity of organic-rich shales. SPE Journal, 16, no. 4, 842–855.

Karacan, C. Ö. (2007) Swelling-induced volumetric strains internal to a stressed coal associated with CO_2 sorption. International Journal of Coal Geology, 72, no. 3–4, 209–220.

Karacan, C. Ö., and Mitchell, G. D. (2003) Behavior and effect of different coal microlithotypes during gas transport for carbon dioxide sequestration into coal seams. International Journal of Coal Geology, 53, no. 4, 201–217.

Krooss, B. M., Van Bergen, F., Gensterblum, Y., Siemons, N., Pagnier, H. J. M., and David, P. (2002) High-pressure methane and carbon dioxide adsorption on dry and moisture equilibrated Pennsylvanian coals. International Journal of Coal Geology, 51, no. 2, 69–92.

Lama, R. D., and Bodziony, J. (1998) Management of outburst in underground coal mines. International Journal of Coal Geology, 35, no. 1–4, 83–115.

Larsen, J. W. (2004) The effects of dissolved CO_2 on coal structure and properties. International Journal of Coal Geology, 57, no. 1, 63–70.

Liu, J., Chen, Z., Elsworth, D., Miao, X., and Mao, X. (2010) Linking gas-sorption induced changes in coal permeability to directional strains through a modulus reduction ratio. International Journal of Coal Geology, 83, no. 1, 21–30.

Liu, J., Chen, Z., Elsworth, D., Qu, H., and Chen, D. (2011) Interactions of multiple processes during CBM extraction: A critical review. International Journal of Coal Geology, 87, no. 3–4, 175–189.

Makogon, Y. F., Holditch, S. A., and Makogon, T. Y. (2007) Natural gas-hydrates – A potential energy source for the 21st Century. Journal of Petroleum Science and Engineering, 56, no. 1–3, 14–31.

Mazumder, S., Wolf, K., Van Hemert, P., and Busch, A. (2008) Laboratory experiments on environmental friendly means to improve coalbed methane production by carbon dioxide/flue gas injection. Transport in Porous Media, 75, no. 1, 63–92.

Mazumder, S., and Wolf, K. H. (2008) Differential swelling and permeability change of coal in response to CO_2 injection for ECBM. International Journal of Coal Geology, 74, no. 2, 123–138.

McKee, C. R., Bumb, A. C., and Koenig, R. A. (1988) Stress-Dependent Permeability and Porosity of Coal and Other Geologic Formations. SPE Formation Evaluation, 3, no. 1, 81–91.

Msha (2011) Historical Data on Mine Disasters in the United States: U.S. Mine Safety and Health Administration.

Noack, K. (1998) Control of gas emissions in underground coal mines. International Journal of Coal Geology, 35, 57–82.

Palmer, I. (2008) Coalbed methane wells are cheap, but permeability can be expensive. Energy Tribune, 10–13.

Palmer, I. (2009) Permeability changes in coal: Analytical modeling. International Journal of Coal Geology, 77, no. 1–2, 119–126.

Palmer, I., and Mansoori, J. (1996) How permeability depends on stress and pore pressure in coalbeds: A new model. Society of Petroleum Engineers, Denver, Colorado.

Peng, Y., Liu, J., Zhu, W., Pan, Z., and Connell, L. (2014) Benchmark assessment of coal permeability models on the accuracy of permeability prediction. Fuel, 132, 194–203.

Robertson, E. P., and Christiansen, R. L. (2005a) Measurement of sorption-induced strain. International Coalbed Methane Symposium: University of Alabama, Tuscaloosa, p. 0532.

Robertson, E. P., and Christiansen, R. L. (2005b) Modeling permeability in coal using sorption-induced strain data, in Proceedings SPE annual technical conference and exhibition, Society of Petroleum Engineers Dallas, Texas.

Saghafi, A., Faiz, M., and Roberts, D. (2007) CO_2 storage and gas diffusivity properties of coals from Sydney Basin, Australia. International Journal of Coal Geology, 70, no. 1–3, 240–254.

Seidle, J. R., and Huitt, L. G. (1995) Experimental measurement of coal matrix shrinkage due to gas desorption and implications for cleat permeability increases. Society of Petroleum Engineers.

Shi, J., and Durucan, S. (2004) Drawdown induced changes in permeability of coalbeds: A new interpretation of the reservoir response to primary recovery. Transport in Porous Media, 56, no. 1, 1–16.

Shi, J.-Q., Durucan, S., and Fujioka, M. (2008) A reservoir simulation study of CO_2 injection and N_2 flooding at the Ishikari coalfield CO_2 storage pilot project. Japan. International Journal of Greenhouse Gas Control, 2, no. 1, 47–57.

Siriwardane, H., Haljasmaa, I., McLendon, R., Irdi, G., Soong, Y., and Bromhal, G. (2009) Influence of carbon dioxide on coal permeability determined by pressure transient methods. International Journal of Coal Geology, 77, no. 1–2, 109–118.

Wang, S., Elsworth, D., and Liu, J. (2011) Permeability evolution in fractured coal: The roles of fracture geometry and water-content. International Journal of Coal Geology, 87, no. 1, 13–25.

White, C. M., Smith, D. H., Jones, K. L., Goodman, A. L., Jikich, S. A., LaCount, R. B., DuBose, S. B., Ozdemir, E., Morsi, B. I., and Schroeder, K. T. (2005) Sequestration of carbon dioxide in coal with enhanced coalbed methane recovery, a review. Energy and Fuels, 19, no. 3, 659–724.

Wu, Y., Liu, J., Elsworth, D., Chen, Z., Connell, L., and Pan, Z. (2010a) Dual poroelastic response of a coal seam to CO_2 injection. International Journal of Greenhouse Gas Control, 4, no. 4, 668–678.

Wu, Y., Liu, J., Elsworth, D., Miao, X., and Mao, X. (2010b) Development of anisotropic permeability during coalbed methane production. Journal of Natural Gas Science and Engineering, 2, no. 4, 197–210.

Zhang, H., Liu, J., and Elsworth, D. (2008) How sorption-induced matrix deformation affects gas flow in coal seams: A new FE model. International Journal of Rock Mechanics and Mining Sciences, 45, no. 8, 1226–1236.

Zhu, W. C., Wei, C. H., Liu, J., Qu, H. Y., and Elsworth, D. (2011) A model of coal–gas interaction under variable temperatures. International Journal of Coal Geology, 86, no. 2–3, 213–221.

Chapter 17

Geomechanical stability and integrity of nuclear waste disposal mines in salt structures

S. Heusermann, S. Fahland & R. Eickemeier
Federal Institute for Geosciences and Natural Resources (BGR), Hannover, Germany

Abstract The use of salt structures for the disposal of low, medium, and high level radioactive waste makes it necessary to perform safety analyses, in particular geomechanical model calculations, to demonstrate the stability of the repository and the integrity of the salt barrier. The analysis should comprise all necessary steps used for geoscientific investigations, from geological exploration to numerical modeling. The different steps like geological-geophysical exploration and geological modeling, classification of salt layers, constitutive modeling and parameter determination, definition of mechanical and thermal loading, formulation of stability and integrity criteria as well as evaluation of numerical results are described. Two examples of geomechanical safety analyses of radioactive waste repositories in salt structures are discussed. The first example describes the Morsleben repository used for the disposal of LLW and MLW and covers the evaluation of the stability of the mine and the integrity of the salt barrier affected mechanically by large old mining rooms. The second example considers a fictive HLW repository in the Gorleben salt dome and illustrates the thermomechanical response and integrity of the salt barrier under thermal loading caused by the heat-generating waste.

1 INTRODUCTION

For a number of decades now rock salt structures have been considered in several countries as a potential host rock for the final disposal of radioactive waste from nuclear research facilities and nuclear power plants. In view of its excellent physical properties, especially its pronounced ductile behavior, rock salt has been favored by numerous international experts, in particular from the USA and Germany. Up to now, two regular radioactive waste repositories in rock salt have been operated: the WIPP site in New Mexico, USA, for transuranic (TRU) waste (Mora, 1999) as well as the Morsleben repository, Germany, a former salt and potash mine which was used until 1998 for the disposal of medium level waste (MLW) and low level waste (LLW), see Preuss *et al.* (2002) and Brennecke (2011). Furthermore, the Gorleben salt dome, Germany, has been extensively investigated over several decades to demonstrate the suitability of the site for the disposal of high level waste (HLW), see BMWI (2008), Klinge *et al.* (2007), Köthe *et al.* (2007), Bornemann *et al.* (2008) and Bräuer *et al.* (2011).

The fundamental requirement for a radioactive waste repository, *i.e.* the prevention of radioactive substances from entering the biosphere, is met through the use of a system of barriers. The public acceptance of a waste repository depends on the assurance that these barriers are sufficient to provide the necessary protection. Safety analysis, therefore, is central to the planning and approval of a repository. The geological barriers are the most important part of the multi-barrier system of repositories in salt rocks. Thus, the load-bearing capacity and geomechanical integrity of the rock, its geological and tectonic stability as well as its geochemical and hydrogeological development are important foci for the safety analysis. Therefore, not only the engineering but also the geological aspects have to be considered. The safety analysis must review the basic safety concept which in turn must take into account the possibilities for failures that could occur during the excavation stage, the operational stage, and the post-closure stage as well as look at the measures to prevent failure (Hudson & Feng, 2015). As the most important item, geomechanical model calculations have to be carried out to analyse the stability of the components of the repository such as shafts, drifts and rooms and demonstrate the mechanical integrity of the geological salt barrier (Langer, 1999).

2 METHODOLOGY OF A GEOMECHANICAL SAFETY ASSESSMENT

Especially for the disposal of radioactive waste in deep geological formations like salt structures, a geotechnical safety analysis must be performed to guarantee that the repository and the associated geological barriers are stable and do not represent a risk to the environment as a result of unacceptable deformation, stress or leakage. Thus, the safety analysis is a fundamental part of the planning, excavation and subsequent sealing and closure of the repository. In particular for repositories in ductile rock types like salt rock, the analysis must include an evaluation that considers very long time frames and cannot be based on measurement results or observations usually carried out over short time frames. Consequently, realistic theoretical models are required to predict the physical processes in the rock salt mass expected over the long term.

In detail, the geomechanical safety analysis encompasses the following steps (Figure 1):

- Geological exploration and subsequent construction of a model of the geological structure of the salt formation and the overburden as well as of the geometry of the mine or repository,
- Description of material behavior, *i.e.* constitutive models and parameters for the time, temperature and stress dependent deformation and strength behavior of the host rock, in particular rock salt,
- Definition of the loading of the structure, *i.e.* initial stress state, thermal effects caused by heat-generating waste, additional mechanical and/or thermal effects caused by backfilling drifts and rooms,
- Geomechanical modeling of the structure including the simplification of the geological and geometrical situation, the classification of homogeneous parts with similar material behavior as well as the definition of the thermal and mechanical loadings of the structure,
- Discretization of near-field and far-field finite element models based on the geomechanical model,

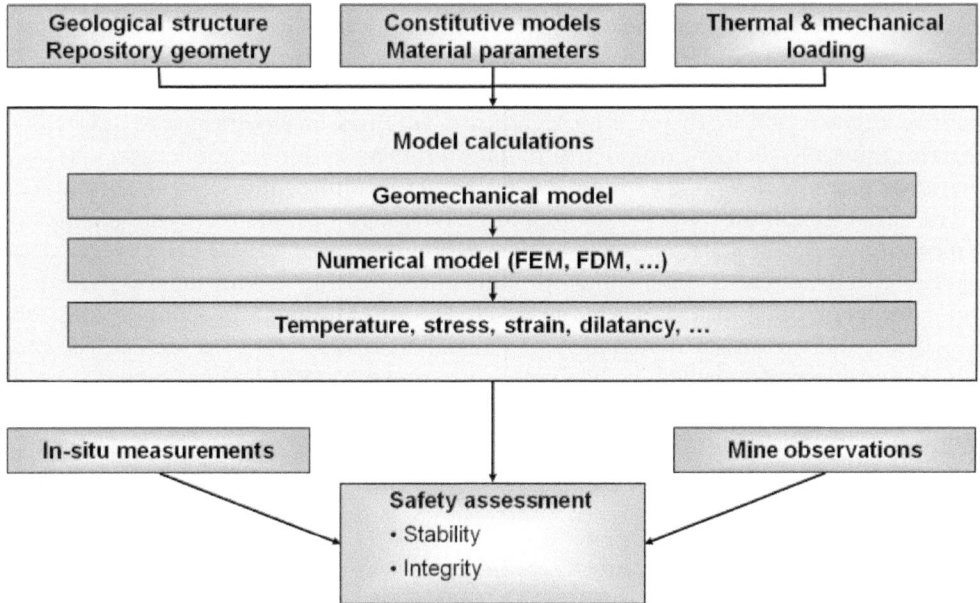

Figure 1 Necessary steps for a geomechanical safety analysis of radioactive waste repositories in rock salt.

– Calculation of temperatures, thermal and mechanical strains, stresses, displacements, dilatancy, and related factors,
– Analysis of numerical results to evaluate the stability of mine components like shafts, drifts, and rooms as well as the integrity of the salt barrier considering the results of in-situ measurements and mine observations.

3 GEOLOGICAL INVESTIGATIONS

Geoscientific exploration and geological modeling of a candidate site for radioactive waste disposal are fundamental and essential steps to produce the necessary data for geomechanical modeling. The exploration of a salt structure usually includes surface and underground investigations.

3.1 Geological in-situ exploration

The objectives of geological exploration from the surface are summarized as follows: Determination of the structure, composition, and hydrogeology of the overburden, determination of the subrosion behavior of the salt structure, creation of a geological map of the salt table, designation of drilling locations for shaft excavation, determination of the stratigraphy of the salt formation, determination of the mineralogical and geochemical composition of the salt layers, and creating a preliminary geological map of the designated exploration and repository level (Hammer *et al.*, 2010).

The main objective of the underground exploration is to develop a detailed geological model of the salt structure as well as to demonstrate that the suitable salt layers are of an adequate scale in terms of area, volume and distance to the flanks of the salt structure to serve as a repository. Furthermore, it should be demonstrated that the planned repository contains neither significant volumes of gas, brines or thermally unstable minerals, such as carnallite, and that there is no hydraulic connection via fluid migration paths between the exploration level and the aquifer above the salt structure.

The surface exploration program should include methods like the hydrogeological exploration and geological mapping of the overburden, geophysical surveys like seismic reflection and refraction, drilling salt table boreholes and sinking deep exploration and pilot shaft boreholes (Bornemann *et al.*, 2008). The underground exploration program usually consists of detailed geological mapping of the shafts, drifts and exploration boreholes drilled in the exploration mine. Typical logging methods used in selected exploration boreholes include borehole deviation logs, temperature logs, permeability tests and electromagnetic reflection surveys (EMR).

3.2 Geological modeling

To evaluate the suitability of sites for radioactive waste repositories geological 3-D models are essential basic tools during the planning, construction and operating stages as well as for performing long-term safety analysis. Moreover, the geological model data can be converted to generate geomechanical numerical models. Geological modeling can be performed using suitable tools, *e.g.* the openGEO^{TM} code developed especially for modeling rock salt structures (Wilke *et al.*, 2004; Hammer *et al.*, 2010).

Different basic exploration data, *e.g.* borehole, mine and geophysical data, can be taken into account when developing and constructing a geological 3-D model. Borehole data are usually in the form of interpreted geophysical log measurements, *e.g.* from gamma-ray and density logs. The stratigraphic correlation derived from drill cores is used to validate these interpretations. Mine data include the geological maps of shafts, drifts and rooms, geological sections (based on mine and borehole data), maps of depth and thickness of the relevant layers as well as geometric models of the position, size and geometry of the underground openings. Mineralogical-geochemical data from rock samples may help refine the stratigraphic classification and correlation of strata.

Geological 3-D models can be used for further interpretation (Figure 2), generating profiles and intersections in any direction, visualizing virtual borehole drilling to optimize the position and orientation of new exploration boreholes, working out the distance between selected points and the volumes of the geological bodies, as well as for creating maps at different depths and contour lines. The discrete 3-D bodies of the model can be transferred to other tools and applications, *e.g.* geomechanical modeling tools.

3.3 Geological-geotechnical classification of homogeneous parts in the rock salt mass

Since the ductility of the layers may vary considerably, numerical safety analyses of repositories in salt formations require detailed knowledge of the material behavior, especially the creep behavior of the diverse salt layers. Therefore, the rock salt has to be

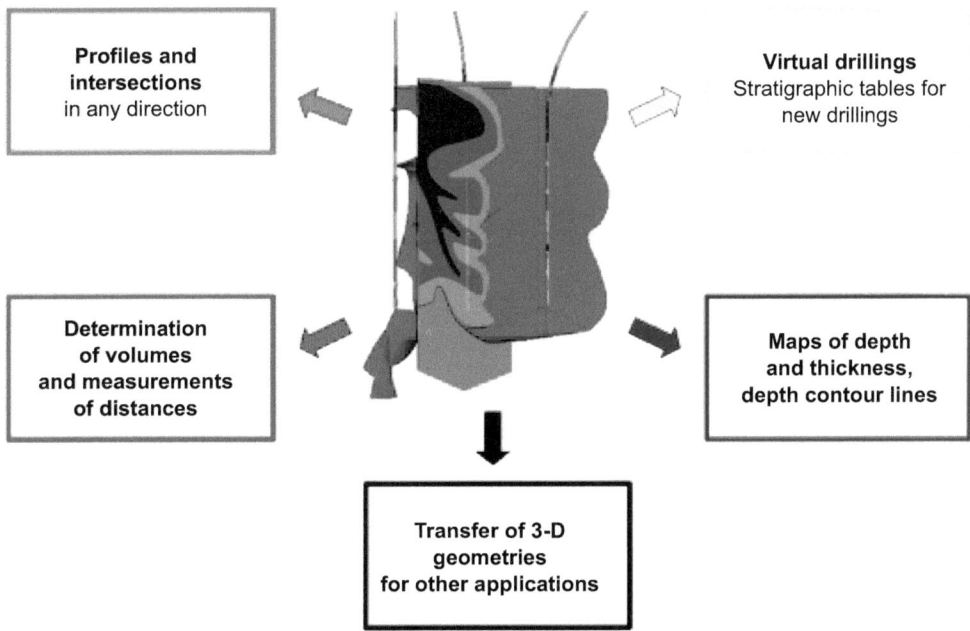

Figure 2 Information derived from 3-D geological modeling (Hammer *et al.*, 2010).

characterized appropriately with respect to steady-state creep behavior and the homogeneous parts have to be classified on the basis of creep test results. These homogeneous parts have to be taken into proper account when developing the geomechanical and numerical models (Plischke, 2007).

The correlation of mapped homogeneous areas with geological and petrographic attributes makes it possible to derive statements on areas in the rock salt mass about which petrographic and stratigraphic descriptions have been made, but cannot be investigated in detail by means of drilling and laboratory testing (see Figure 3).

4 CONSTITUTIVE MODELS

4.1 Overview

Over recent decades numerous and diverse constitutive models have been developed and continuously improved to provide theoretical descriptions of the complex mechanical behavior of rock salt. These models consider the relevant physical effects which must be taken into account for the numerical safety analysis of waste repositories in salt structures, *e.g.* transient and steady-state creep, dilatancy, damage, creep rupture and healing.

Several overviews of the status of constitutive models for rock salt have been compiled and published, *e.g.* by Hunsche & Schulze (1994), Schulze *et al.* (2007), Hampel *et al.* (2012), Hampel *et al.* (2015). The following comprehensive (but not necessarily complete) list comprises exemplary different constitutive models in approximately chronological order: M/D model (Munson & Dawson, 1979), modified

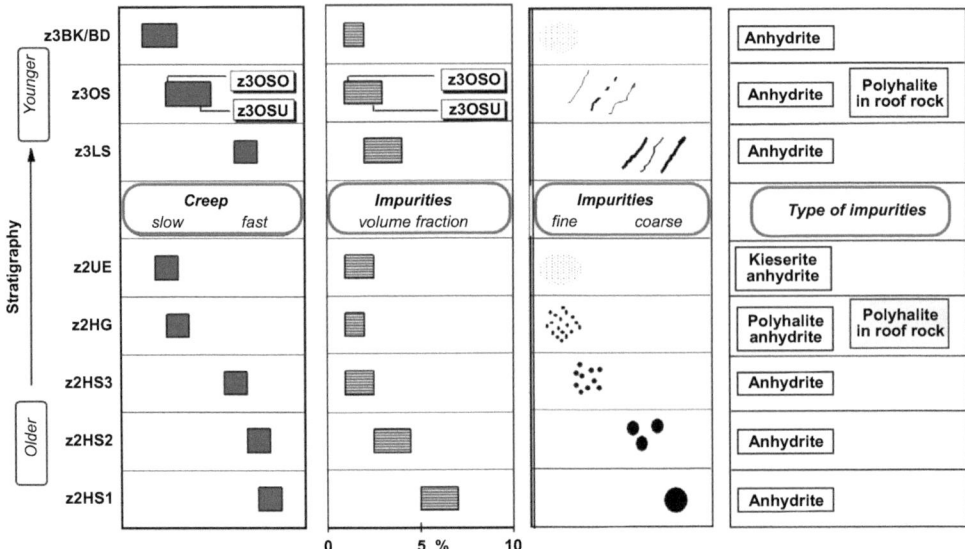

Figure 3 Correlation between creep behavior and petrographic attributes (Plischke, 2007).

Burgers model/LUBBY2 (Heusermann, 1982), BGRa model (Hunsche, 1984), BGRc model (Wallner, 1984), BGRb model (Hunsche & Schulze, 1994), TUB salt model (Döring *et al.*, 1998), Hou/Lux model (Hou & Lux, 1999), LUBBY-MDCF model (Rokahr *et al.*, 2004), CDM model (Hampel & Schulze, 2007; Hampel, 2012; Hampel, 2015), Minkley model (Minkley & Mühlbauer, 2007), Günther/Salzer model (Günther & Salzer, 2012) and the Lux/Wolters model (Wolters *et al.*, 2012).

Some typical constitutive models for rock salt are considered in detail below. These examples focus especially on creep and dilatancy, are well established and have been used over many years for the numerical safety analysis of several radioactive waste disposal sites in salt formations located in the USA and Germany.

4.2 Creep

4.2.1 The BGRa creep model

The BGRa creep model includes steady-state creep considering one creep deformation mechanism depending on temperature and stress (Hunsche, 1984). This is the standard creep model for rock salt and has been widely used for the long-term safety analysis of nuclear waste repositories in Germany.

$$\dot{\varepsilon}^{cr}_{dev} = A \cdot e^{-\frac{Q}{R \cdot T}} \cdot \left(\frac{\sigma_{dev}}{\sigma^*}\right)^n \tag{1}$$

Here, typical parameters are: A= 0.18 [d^{-1}] (structural factor), Q = 54 [kJ·mol^{-1}] (activation energy), R = 8.3143·10^{-3} [kJ·mol^{-1}·K^{-1}] (universal gas constant), σ*=1.0 [MPa] (reference stress), n = 5 [-] (stress exponent).

4.2.2 The BGRb creep model

The BGRb creep model is used to determine the creep more precisely by considering two mutually independent deformation mechanisms (Hunsche & Schulze, 1994). This model is valid for the higher temperatures (about 100 °C up to 200 °C) in the rock salt found in the near-field around the high-level waste canisters and casks.

$$\dot{\varepsilon}_{dev}^{cr} = \left[A_1 \cdot e^{-\frac{Q_1}{R \cdot T}} + A_2 \cdot e^{-\frac{Q_2}{R \cdot T}} \right] \cdot \left(\frac{\sigma_{dev}}{\sigma^*} \right)^n \tag{2}$$

Here, typical parameters are: A_1 = 2.3·10^{-4} [d^{-1}], A_2 = 2.1·10^6 [d^{-1}], Q_1 = 42.0 [kJ·mol^{-1}], Q_2 = 113.4 [kJ·mol^{-1}], σ^* = 1.0 [MPa], n = 5 [–].

4.2.3 The BGRc creep model

The BGRc creep model includes strain-hardening transient creep and two different steady-state creep deformation mechanisms which depend on temperature and stress.

$$\dot{\varepsilon}_{dev}^{cr} = A_1 \cdot e^{-\frac{Q_1}{R \cdot T}} \cdot \left(\frac{\sigma_{dev} - \sigma_{RN}}{\sigma^*(1-z)} \right)^{n_1} + A_2 \cdot e^{-\frac{Q_2}{R \cdot T}} \left(\frac{\sigma_{dev} - \sigma_{RN}}{\sigma^*(1-z)} \right)^{n_2} \tag{3}$$

If the strain hardening parameters z and σ_{RN} are neglected, this formulation can be reduced to solely steady-state creep considering two different deformation mechanisms (Wallner, 1984).

$$\dot{\varepsilon}_{dev}^{cr} = A_1 \cdot e^{-\frac{Q_1}{R \cdot T}} \cdot \left(\frac{\sigma_{dev}}{\sigma^*} \right)^{n_1} + A_2 \cdot e^{-\frac{Q_2}{R \cdot T}} \left(\frac{\sigma_{dev}}{\sigma^*} \right)^{n_2} \tag{4}$$

Here, typical parameters are: A_1 = 1.30·10^5 [d^{-1}], A_2 = 0.18 [d^{-1}], Q_1 = 113.0 [kJ·mol^{-1}], Q_2 = 54.0 [kJ·mol^{-1}], σ^* = 1.0 [MPa], n_1 = 5.5 [–], n_2 = 5.0 [–].

4.2.4 The LUBBY2 creep model

The LUBBY2 creep model was originally developed for the stability analysis of salt caverns, see Heusermann (1982), Heusermann *et al.* (1983). Initially, it was referred to as a modified nonlinear Burgers model, but was renamed later to LUBBY2 (nickname based on the acronym for the "Lehrgebiet für Unterirdisches Bauen (LUB)", University of Hannover). This model includes a transient (primary) creep part as well as a steady-state creep part. A strain-hardening formulation was developed for the transient creep stage.

$$\dot{\varepsilon}_{dev}^{cr} \left(\varepsilon_{pr}^{cr} \right) = \left[\frac{1}{\overline{\eta}_K(\sigma_{dev})} \cdot \left(1 - \frac{\varepsilon_{pr}^{cr}}{\sigma_{dev}} \cdot \overline{G}_K(\sigma_{dev}) \right) + \frac{1}{\overline{\eta}_M(\sigma_{dev})} \right] \cdot \sigma_{dev} \tag{5}$$

If transient creep is neglected, the following formulation for steady-state creep can be used:

$$\dot{\varepsilon}_{dev}^{cr} = \frac{1}{\overline{\eta}_M(\sigma_{dev})} \cdot \sigma_{dev} \tag{6}$$

with

$\dot{\varepsilon}_{dev}^{cr}(\varepsilon_{pr}^{cr}) = $ (strain dependent) deviatoric creep strain rate (1/d),
$\sigma_{dev} = $ deviatoric stress (MPa),
$\varepsilon_{pr}^{cr} = $ deviatoric primary creep strain (–),
$\overline{G}_K(\sigma_{dev}) = $ (stress dependent) Kelvin module (MPa),
$\overline{\eta}_K(\sigma_{dev}) = $ (stress dependent) Kelvin viscosity (MPa·d),
$\overline{\eta}_M(\sigma_{dev}) = $ (stress dependent) Maxwell viscosity (MPa·d).

The nonlinear stress dependency of the creep rate is taken into account by using stress-dependent creep parameters $\overline{G}_K$, $\overline{\eta}_K$ and $\overline{\eta}_M$. This dependency can be expressed as an exponential correlation.

$$\overline{G}_K = \overline{G}_K^* \cdot e^{k_1 \cdot \sigma} \tag{7}$$

$$\overline{\eta}_K = \overline{\eta}_K^* \cdot e^{k_2 \cdot \sigma} \tag{8}$$

$$\overline{\eta}_M = \overline{\eta}_M^* \cdot e^{m \cdot \sigma} \tag{9}$$

The material parameters can be determined from multi-stage creep tests, including the transient and steady-state creep stage, for details see Lux & Heusermann (1983). As an example, uniaxial creep tests on rock salt samples from the Asse salt mine located in the north-west of Germany yielded the following results for the parameters in Equations 7–9: $\overline{G}_K^* = 1.88 \cdot 10^5$ MPa, $k_1 = -2.54 \cdot 10^{-1}$ 1/MPa, $\overline{\eta}_K^* = 4.98 \cdot 10^5$ MPa · d, $k_2 = -2.67 \cdot 10^{-1}$ 1/MPa, $\overline{\eta}_M^* = 1.21 \cdot 10^8$ MPa · d, $m = -3.27 \cdot 10^{-1}$ 1/MPa.

If the steady-state creep model requires additional temperature dependency, the following extended formulation can be used including temperature T (in °C) and parameter l:

$$\overline{\eta}_M = \overline{\eta}_M^* \cdot e^{m \cdot \sigma} \cdot e^{l \cdot T} \tag{10}$$

Figure 4 illustrates a comparison of theoretical creep rates calculated for different stresses and temperatures using the BGRc creep model (only steady-state creep, see Equation 4) and the LUBBY2 creep model (only steady-state creep, see Equation 10). Here, both models result in very similar creep rates if the relevant stress domain from 1 to 20 MPa is considered and appropriate creep parameters are determined (Figure 4). In Figure 4, the following parameters are used for the BGRc model: $A_1 = 7.173 \cdot 10^{-7}$ [d^{-1}], $A_2 = 7.109 \times 10^{-3}$ [d^{-1}], $Q_1 = 35.54$ [kJ·mol^{-1}], $Q_2 = 35.97$ [kJ·mol^{-1}], $\sigma^* = 1.0$ [MPa], $n_1 = 6.17$ [–], $n_2 = 1.92$ [–]. The parameters for the LUBBY2 model are $\overline{\eta}_M^* = 4.16 \cdot 10^{14}$ MPa · d, $m = -0.339$ 1/MPa, and $l = -0.047$ [1/K].

4.2.5 The Munson/Dawson creep model

The Munson/Dawson creep model includes steady-state creep and considers a number of different creep deformation mechanisms as well as transient creep effects (Munson & Dawson, 1979). This is the standard creep model for rock salt for

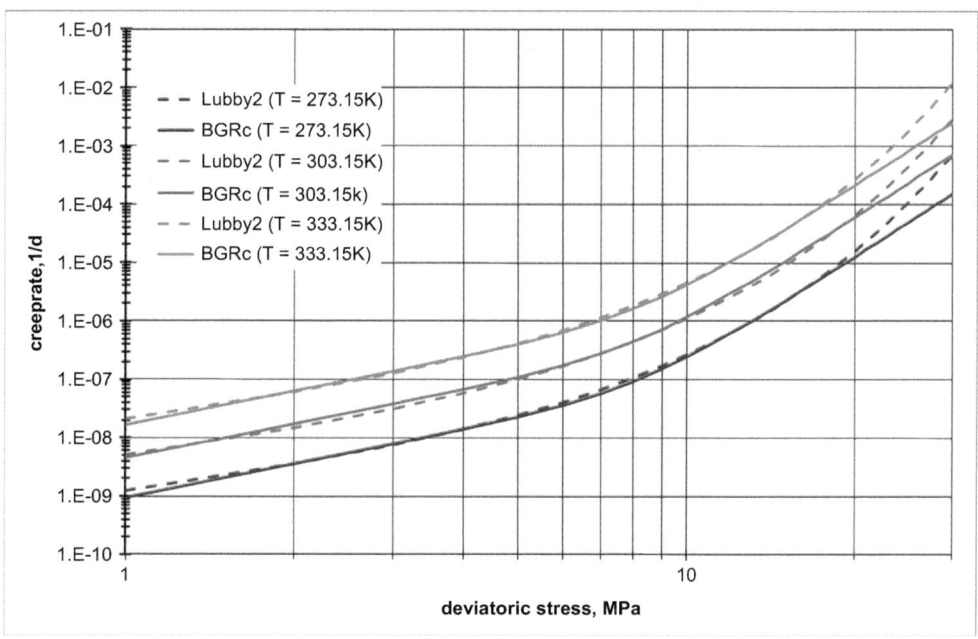

Figure 4 Comparison of creep rates determined using the BGRc and the LUBBY2 models.

numerical analysis of the WIPP site, USA. The three dominant steady-state creep mechanisms are

a)   Dislocation climb

$$\dot{\varepsilon}^{cr}_{dev,1} = A_1 \cdot e^{-\frac{Q_1}{R \cdot T}} \cdot \left(\frac{\sigma_{dev}}{\mu} \right)^{n_1} \tag{11}$$

b)   Undefined mechanism

$$\dot{\varepsilon}^{cr}_{dev,2} = A_2 \cdot e^{-\frac{Q_2}{R \cdot T}} \cdot \left(\frac{\sigma_{dev}}{\mu} \right)^{n_2} \tag{12}$$

c)   Dislocation glide

$$\dot{\varepsilon}^{cr}_{dev,3} = 2 \left(B_1 \cdot e^{-\frac{Q_1}{R \cdot T}} + B_2 \cdot e^{-\frac{Q_2}{R \cdot T}} \right) \cdot \sinh \left[q \left(\frac{\sigma_{dev} - \sigma_0}{\mu} \right) \right], \quad \sigma_{dev} > \sigma_0 \tag{13}$$

where the A's and B's are structure factors, Q's are activation energies, n's are stress exponents, q is an activation volume, σ_0 is a cut-off stress, and μ is the shear modulus. The reduced BGRc formulation without strain hardening (according to Equation 4) is very similar to the Munson/Dawson formulation for dislocation climb and the undefined mechanism (Equations 11 and 12). Dislocation glide (Equation 13) is only present at high deviatoric stresses of $\sigma_0 > 20$ MPa and in view of the usual depth of repositories in salt structures of between 800 – 900 m and the related stress states is only of minor significance.

4.3 Dilatancy

To evaluate the mechanical integrity of the salt barrier, first of all it is important to study the dilatant behavior of the rock salt mass. Dilatancy is the starting point for possible subsequent processes, like damage and creep rupture of rock salt. Therefore, an adequate long-term safety analysis must primarily include the formulation of a dilatancy boundary and the calculation of potentially dilatant areas in the salt barrier.

To this end a dilatancy concept based on extensive lab test series on rock salt has been developed according to Hunsche & Schulze (2003). Here, the volumetric rate of dilatancy is correlated to the deviatoric creep rate via an empirical relation using r_v (depending on σ_{dev} and the minimum compressive stress σ_{min}):

$$\dot{\varepsilon}_{dil,vol} = r_v \cdot \dot{\varepsilon}_{dev}^{cr} \tag{14}$$

If the deviatoric stress σ_{dev} exceeds a boundary stress $\sigma_{dev,dil}$, (see Equation 16) the factor r_v is positive and dilatancy will occur:

$$r_v = a \cdot \left| \frac{\sigma_{dev} - \sigma_{dev,dil}}{|\sigma_{min} - 1/3 * \sigma_{dev}|} \right|^2 \quad if \ \sigma_{dev} > \sigma_{dev,dil} \tag{15}$$

$$\sigma_{dev,dil} = b \cdot (\sigma_{min} - 1/3 * \sigma_{dev})^c \tag{16}$$

with empirical material parameters a = 0.8165, b = 3.2, and c = 0.78.

5 GEOMECHANICAL MODELING

The geomechanical modeling of a radioactive waste repository in rock salt involves the following steps:

- The detailed geological model is simplified by combining geological units with similar mechanical behavior in the rock salt mass, in the overburden and the adjoining rock mass. Homogeneous salt layers are defined considering their steady--state creep behavior with respect to structural factor A_{cr} (Plischke, 2007). Thus, the creep capability of the layers can be taken into account using factor A^*, related to the reference value A = 0.18 1/d in the BGRa creep model (see Equation 1):

$$A_{cr} = A^* \cdot A \tag{17}$$

- The different values of A^* have to be determined in laboratory creep tests on samples taken from several salt layers in the particular repository. Furthermore, the geological model should include the geometry of the planned repository or the existing mine. If possible, the geological model should be automatically converted to a basic numerical model.
- The mechanical material parameters of all geological units of the salt structure, the overburden and the adjacent rock must be determined experimentally in

accordance with different constitutive models, *e.g.* elasticity, creep and dilatancy. Furthermore, the thermal parameters of the geological layers, in particular for the salt layers, have to be identified if thermomechanical calculations are needed.

- The initial temperature and stress conditions of the entire rock mass have to be defined. Usually, appropriate in-situ measurements are carried out to obtain the necessary experimental data.
- The different types of loading of the entire structure must be properly defined. If the disposal of HLW is being considered, the thermal load of the waste must be defined. The disposal concept, *e.g.* disposal in drifts or in boreholes, plays a big role here. Preparatory thermal calculations are needed to determine important values like size, geometry and depth of the repository as well as distance or length of the disposal drifts and boreholes. Furthermore, thermal and mechanical loading may result from the backfilling of drifts and boreholes, *e.g.* if saltcrete is used. Additionally, hydraulic aspects must be taken into account if fluids are present in the overburden.
- Finally, the numerical model must be discretized (for use with finite element codes or finite difference codes), taking the geometry of the geological structure and of the repository into account. The analysis data have to include all the thermal, mechanical and hydraulic aspects described above.

6 STABILITY AND INTEGRITY CRITERIA

6.1 Stability of repositories in rock salt

The stability of repositories in rock salt and their components (shafts, rooms and drifts) has to be analyzed at the following scales (Heusermann *et al.*, 2009):

- Local stability of the salt rock, considering especially the risk of local failure, *e.g.* occurrence of roof fall or fractures in the roof, wall or floor of drifts,
- Stability of medium-size areas of the repository, considering especially the risk of the collapse of pillars and roofs as well as major parts of the mine structure,
- Overall stability and integrity of the entire structure, including the impact of potential damage zones in and around the repository.

To verify the stability of the rock salt several criteria based on laboratory tests on rock salt samples can be considered, *e.g.* the short-term triaxial compression strength, the uniaxial tensile strength, the dilatancy boundary as well as permissible strains in the rock salt.

To analyze the stability of the mine at small and medium scales, at least two criteria should be taken into account. First, critical areas of tensile stress in the rock salt mass should be determined with respect to the very low tensile strength of rock salt. Second, the deformation state of roofs and pillars in terms of the deviatoric strain should be calculated and compared with the maximum allowable deviatoric strain measured in the laboratory. Since rock salt must be considered to be a very strain-rate sensitive material, the values obtained in the laboratory must be extrapolated to the in-situ strain-rate conditions, *e.g.* see example in Section 7.1.

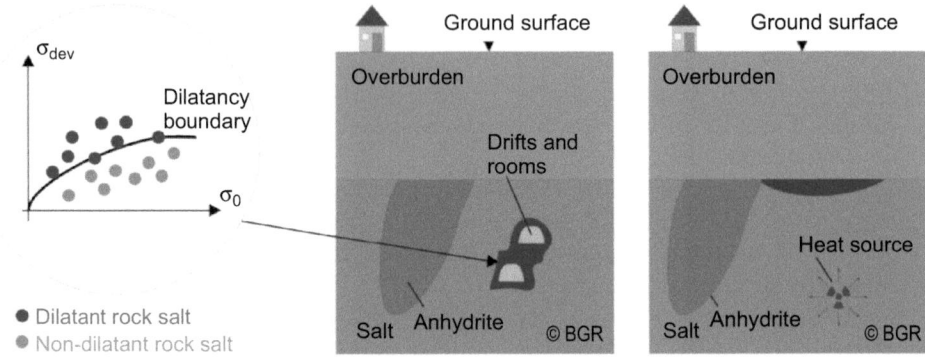

Figure 5   Illustration of the dilatancy criterion (according to Heusermann *et al.*, 2007).

6.2 Integrity of the salt barrier

To analyse the integrity of the salt rock barrier from a geomechanical point of view, the following criteria, which both have to be satisfied, can be used (Langer & Heusermann, 2001):

–   Dilatancy criterion: The geomechanical integrity of the salt barrier is not guaranteed if the deviatoric stresses exceed the dilatancy boundary (see Equation 16). Then microcracks will form and cause progressive damage as well as the appearance of and subsequent increase in the permeability of the salt rock (Figure 5, potential dilatant zones in the rock salt are colored red).

–   Fluid pressure criterion: The geomechanical integrity of the barrier is not guaranteed if the hydrostatic pressure of an assumed fluid column extending to the ground surface exceeds the minimum principal stress at the considered location on the salt body contour, *e.g.* top of the salt structure, contact area between salt and anhydrite layers connected hydraulically to the overburden (Figure 6, potential fracture-risk zones in the rock salt are colored red):

$$p_{fl} - \sigma_{min} < 0 \tag{18}$$

–   Advantageous tensile strength of rock salt is neglected. Usually, the density of the fluid is assumed to be 1.2 kg/dm^3. If Equation 18 is not satisfied, potential migration of fluids from the overburden or the adjacent rock into the salt barrier cannot be ruled out.

7 EXAMPLES

7.1 Geomechanical analysis of an actual LLW/MLW repository in rock salt

The Morsleben repository for radioactive waste (ERAM) was constructed in a former salt and potash mine, consisting of several mining sections. The repository was used for

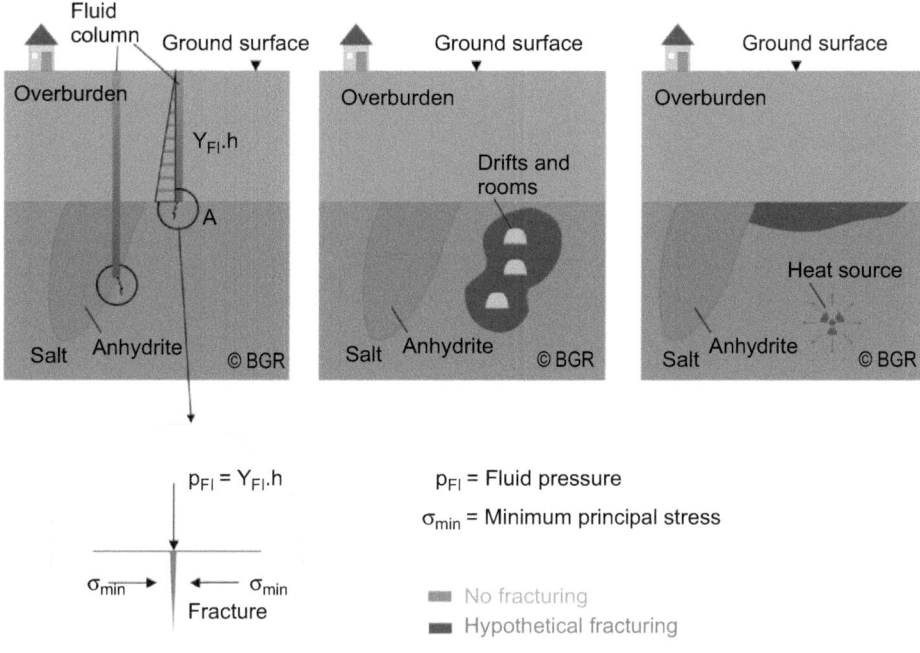

Figure 6   Illustration of the fluid pressure criterion (according to Heusermann *et al.*, 2007).

the disposal of low radioactive waste (LLW) and medium radioactive waste (MLW) from 1972 to 1998. Especially the southern, the western, and the eastern parts located at the periphery of the mine were used for waste disposal (Fahland & Heusermann, 2013).

The Morsleben repository is located in the Allertal fault zone. The top of the salt structure lies at approximately 140 m below mean sea level, or about 270 m below ground level. The thickness of the salt structure varies between 380 m and 500 m. The exploration and modeling of the geological structure of the salt rock and the over-burden in several characteristic cross sections of the different parts in the mine are based on the geological mapping of drifts, rooms, and numerous drill cores from the site, as well as ground-penetrating radar measurements (Behlau & Mingerzahn, 2001).

The safety analysis, here, focuses on the disposal area in the southern part of the Morsleben repository. This part is characterized by an unfavorable, steep configuration of old mining rooms, forming several roofs between the rooms (see Figure 7). The repository closure concept specifies that most of the old mining rooms are backfilled with saltcrete in order to stabilize the mine and to improve the long-term integrity of the salt barrier.

The salt structure in the southern part of the Morsleben repository is characterized by a distinct folding of the salt layers and a large amount of main anhydrite layers (z3HA) of the Leine-sequence. The main units of the Zechstein strata (salt layers z2HSO, z2HSB,

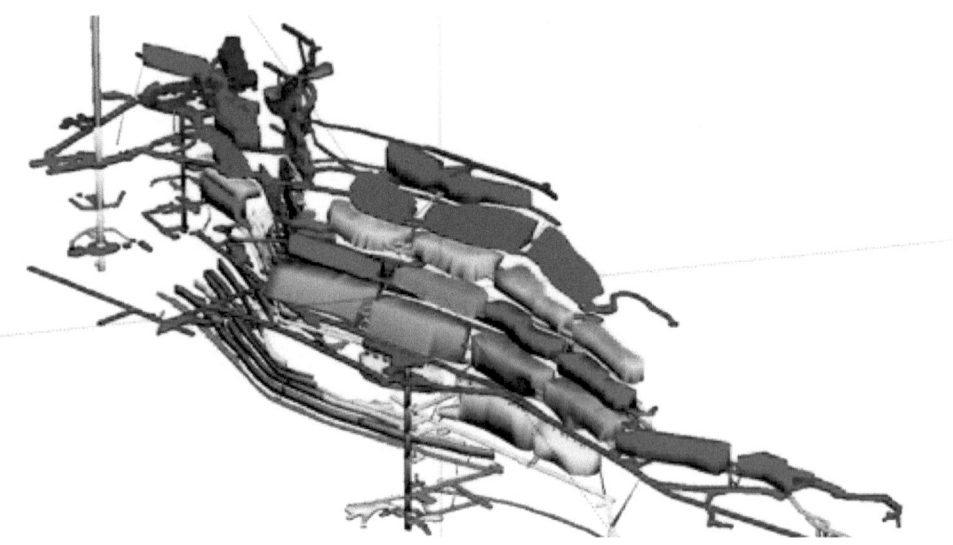

Figure 7   3-D view of old mining rooms in the southern part of the Morsleben repository (after Heller *et al.*, 2004).

z2HSW, z2W, z2SF, z3O, z3LS, z3AM, and anhydrite layers z1WA, z3HA) and composites of the main units (z3OS-BK/BD, z3-z4) are considered. The Hauptsalz z2HS was separated into several parts (z2HSW, z2HSO, and Z2HSB), on the basis of different creep behavior. The structure of the overburden was idealized with respect to the main layers: caprock (cr), Keuper-Jurassic (k-j), and Quaternary (q) (Figure 8).

A simplified geomechanical 3-D model was created taking into account a characteristic geological cross section oriented orthogonally to the axis of the geological structure and the mining rooms. For reasons of symmetry the model includes half of the length of the rooms and of the pillar at the head of the rooms (Figure 9). The entire 3-D model is 800 m in height, 1,000 m wide and 65 m in length. The main objective of the model calculations was to analyse the stability of the old mining rooms and the long-term integrity of the salt barrier.

For the analysis, it was assumed that the rooms were instantaneously excavated in the year 1940. For the operational stage, a period of 100 years up to the year 2040 was considered. For the post-operational stage, a period of 10,000 years was simulated. Additionally, the planned partial backfilling of the old mining rooms during the repository closure was taken into account. Then, two cases were considered: (a) without backfilling, and (b) with backfilling of the mining rooms. It was assumed that backfilling would be carried out 80 years after excavation of the rooms up to different degrees of filling using saltcrete or lignite fly-ash. In the model calculations, the steady-state creep behavior (according to Equation 1) and the dilatant behavior (according to Equations 14–16) of the salt layers were considered. Calculations were performed using the special purpose JIFE code (Faust *et al.*, 2011).

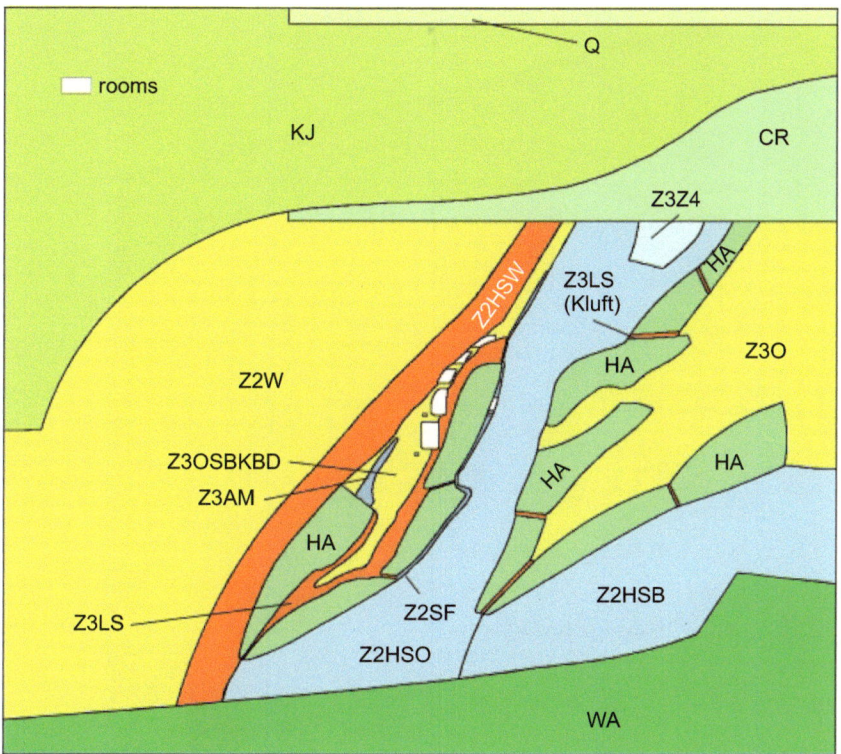

Figure 8 Idealized geological model of the southern part of the Morsleben repository (Fahland *et al.*, 2015).

7.1.1 Stability analysis of rooms in the southern part of the Morsleben repository

To analyse the stability of old mining rooms in the southern part of the Morsleben repository, the deviatoric strains in the roofs were calculated and compared with the maximum allowable deviatoric strain measured in the laboratory.

Figure 10 shows the results of short-term deformation and strength tests on rock salt samples from the southern part of the Morsleben repository. The laboratory tests were performed on Liniensalz (z3LS) and Orangesalz (z3OS) samples at different strain rates $d\varepsilon/dt = 5.0\cdot10^{-8} - 1.0\cdot10^{-5}$. The regression lines calculated from the measured ultimate deviatoric strain values and plotted in the semi-logarithmic diagram yield maximum allowable deviatoric strains ε of approximately 8.5 % (z3OS) and 17.5 % (z3LS) at the estimated in-situ strain rate of $d\varepsilon/dt = 1.0\cdot10^{-12}$.

Figure 11 depicts the calculated deviatoric strain in the rock salt layers of the southern part for a period of 80 years after room excavation corresponding to the present situation. As to be expected, significant strain values of more than 5% are determined in the roofs between the rooms in the z3OS-BK/BD layers. The highest value of about 10% is calculated in the roof between the two upper rooms. Compared with the

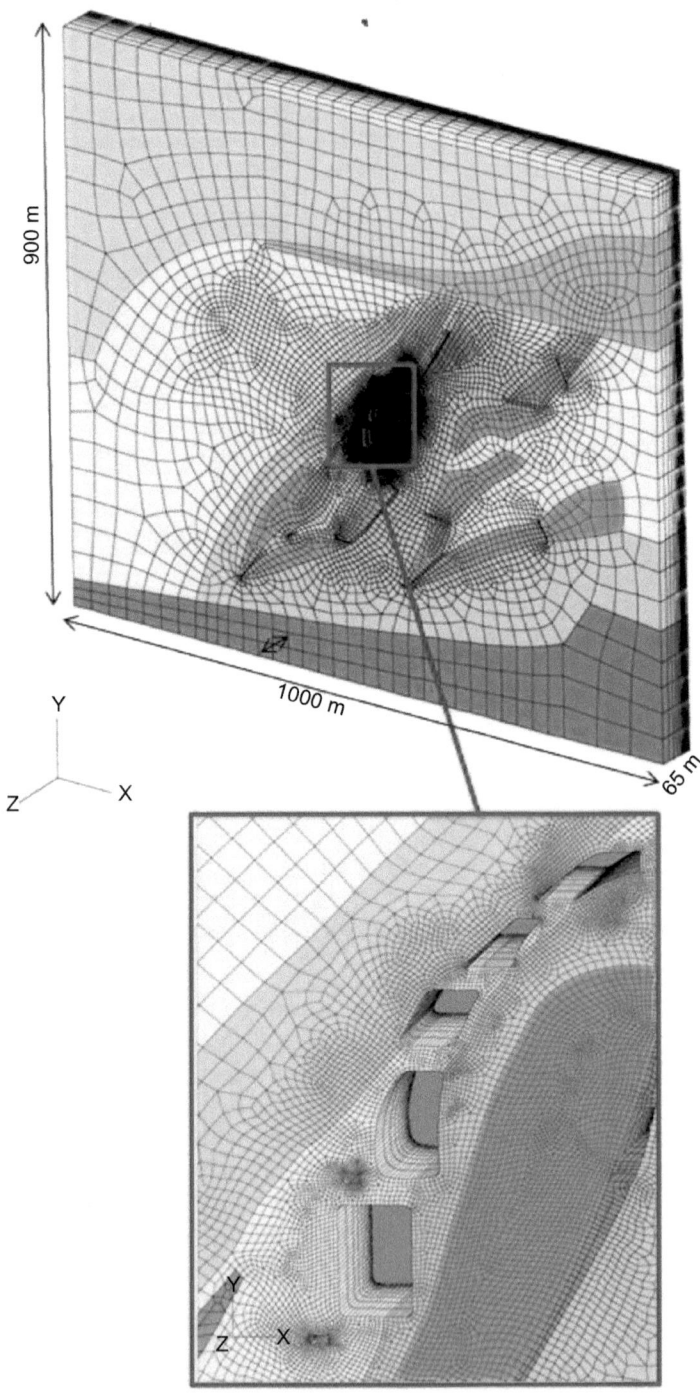

Figure 9 Simplified 3-D finite-element model of the southern part – entire model and detail showing the pillar at the head of the rooms.

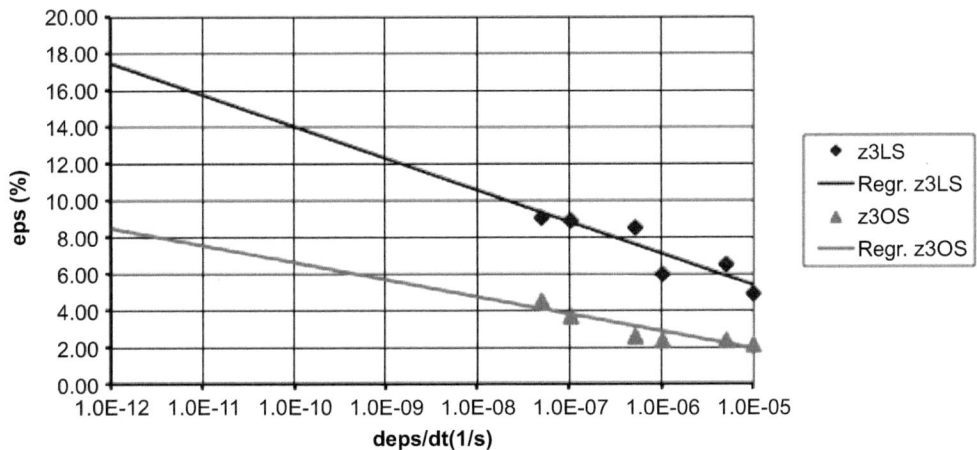

Figure 10 Maximum allowable deviatoric strain at different strain rates for salt layers z3LS and z3OS (Heusermann & Fahland, 2005).

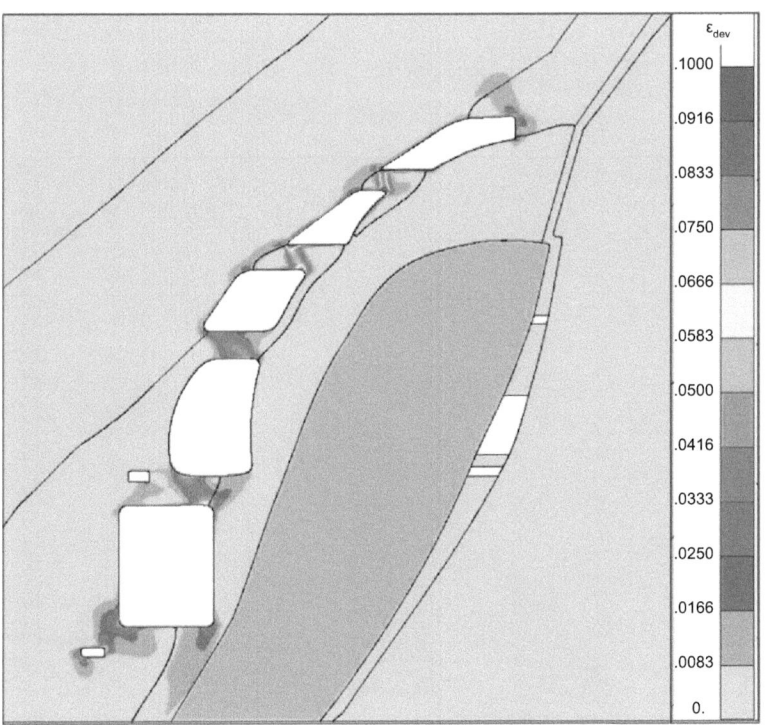

Figure 11 Calculated deviatoric strain in the rock salt layers of the southern part.

extrapolated laboratory results (maximum allowed strain 8.5%, see Figure 10), this roof must be classified as strongly deformed. This theoretical finding is in sound qualitative agreement with in-situ observations of several pronounced fractures in the roofs of the rooms in the southern part.

7.1.2 Integrity analysis of the salt barrier in the southern part of the Morsleben repository

According to the first integrity criterion (see section 6), the dilatant rock zones of the southern part were calculated for a period of 10,000 years without considering the backfilling of the mining rooms. Typical results are plotted in Figure 12, where dilatant areas are colored blue to red. As to be expected, dilatancy, caused by the excavation of the rooms, occurs in the rock salt around all rooms. But extensive parts of the salt barrier between the rooms and the top of the salt dome are free of dilatancy.

According to the second integrity criterion (see section 6), the frac-risk zones in the salt barrier were calculated for a period of 10,000 years. Since the minimum principal stress is significantly reduced around the rooms, extensive frac-risk areas occur in the salt barrier (Figure 13, yellow to red colored areas). From a very hypothetical point of view, the stress conditions appear to be unfavorable with respect to a theoretical fluid pressure which exceeds the minimum principal stress in the salt rock around the rooms, as well as between the anhydrite layers and the rooms. Thus, the model calculations indicate a significant area in the salt barrier in which the fluid pressure criterion is not satisfied.

Figure 12 Dilatant zones in the rock salt after 10,000 years without backfilling the rooms.

Figure 13 Hypothetical frac-risk zones after 10,000 years without backfilling the rooms.

To consider the favorable mechanical effect of backfilled rooms (see gray-colored parts of the rooms in figures 14 and 15), additional calculations were carried out for a period of 10,000 years. The dilatant rock zones were determined (Figure 14), according to the dilatancy criterion. The results are very similar to the case without backfilling, because no healing of the rock salt is considered.

According to the fluid pressure criterion, a zone around the rooms in the southern part with a distinct reduction in the minimum principal stress is revealed (Figure 15, yellow to red colored area). The frac-risk zone is considerably smaller compared with the model for the case without backfilling the rooms. Because large parts of the salt barrier show neither dilatancy nor hypothetical exposure to fluid-induced fracturing and no hydraulic connection exists between the anhydrite and the overburden, the long-term integrity of the salt barrier in the southern part will be guaranteed if back-filling of the rooms is realized.

7.2 Thermomechanical analysis of a generic HLW repository in rock salt

Geoscientific surface exploration to study the suitability of the Gorleben salt dome as a potential site for a geologic repository for heat-generating radioactive waste began in 1977. This work was followed up by extensive underground exploration of the site beginning in 1983. As part of a moratorium the exploration was interrupted for ten years from 2000 to 2010 and was restarted in November 2010. Since autumn 2012 the exploration of the Gorleben site has been put on hold again. Parallel to the exploration

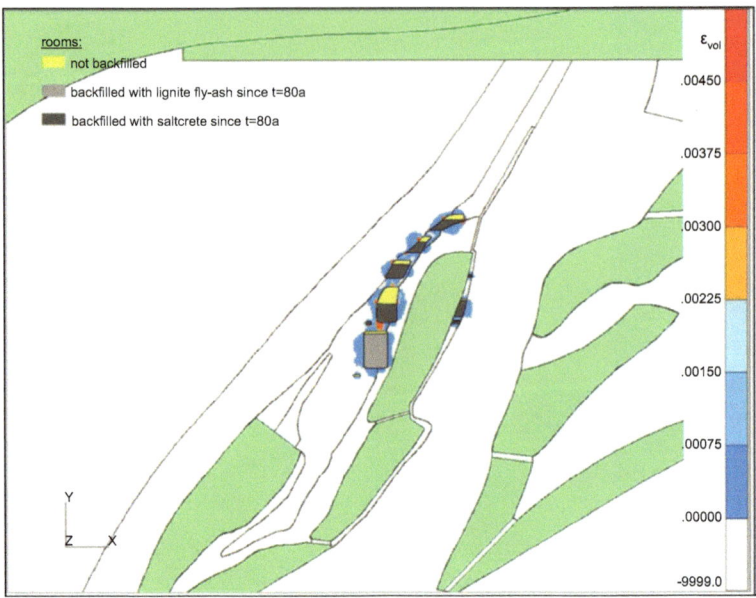

Figure 14 Dilatant zones in the rock salt after 10,000 years with backfilling the rooms.

Figure 15 Hypothetical frac-risk zones after 10,000 years with backfilling the rooms.

of the site, numerous model calculations were undertaken during recent decades to analyse and predict the thermomechanical response of the salt barrier to the planned disposal of heat-generating waste.

According to the accepted methodology for a geomechanical safety assessment (see section 2), the first important step comprises geological investigations. Cross section QS1West through the EB1 exploration level along the 1W cross cut was used (Figure 16, top) to describe the geological structure and the different geological layers of the Gorleben salt dome, composed of Zechstein (nearly 250 Ma old) rock salt. The cross section is based on the results from deep exploration boreholes drilled from the surface and shows an interpretation of the structure from the overburden down to the base of the salt dome. The following homogeneous layers of the overburden, characterized by their distinct elastic material properties, were taken into consideration: Quaternary (q), Tertiary (t), cap rock (cr), Upper Cretaceous (kro), Lower Cretaceous (kru), Jurassic – Keuper (j – k), Bunter Sand-stone (so – su), Rotliegendes (r). The main stratigraphic horizons of the salt dome are the Aller Series rock salt (z4), Leine Series rock salt (z3), Staßfurt Series rock salt (z2), see Bornemann *et al.* (2008) and Hammer *et al.* (2010).

In addition to the brittle Hauptanhydrit (z3HA) with reference to the creep behavior the following ductile homogeneous zones in the vicinity of the EB1 exploration level were differentiated: Liniensalz (z3LS), Kaliflöz Staßfurt (z2SF), Hauptsalz/Kristallbrockensalz (z2HS3), Hauptsalz/Streifensalz (z2HS2), Hauptsalz/Knäuelsalz (z2HS1), see Figure 16 (bottom).

Two different disposal concepts were considered: The drift emplacement (DE) concept and the borehole emplacement (BE) concept (Bollingerfehr *et al.*, 2008). The former concept emplaces POLLUX casks in horizontal drifts located at the 860 m level, while the latter concept emplaces BSK3 fuel element canisters in vertical boreholes at a depth of 870 to 1170 m.

Regarding the drift emplacement concept, Figure 17 depicts a simplified arrangement of the disposal drifts. The emplacement area covers a total length of 3400 m and a width of 324 m, and is located in the main salt layers z2HS, confined by anhydrite layers z3HA. A total of 4173 POLLUX casks can be placed in this area. The initial heat, taken into account for the numerical calculation, amounts to 769 W per meter drift. For reasons of symmetry only half of the structure was considered.

Figure 18 shows the layout and simplified arrangement of the disposal boreholes in the borehole emplacement concept. The vertically oriented boreholes are 300 m in length. The emplacement area including the boreholes covers a total length of 1,150 m and a width of 303.6 m, and is also located in the main salt layers z2HS. A total of 9148 canisters can be placed in this area. The initial heat power taken into account for the numerical calculation amounts to 324 W per meter borehole. For reasons of symmetry only half of the structure was considered.

The numerical modeling employed two different 3-D finite-element models: Model 1 for considering the drift disposal concept, Model 2 for considering the borehole emplacement concept. Both models were developed on the basis of the geological 2-D model of cross section QS1West of the Gorleben salt dome (Figure 16). Figure 19 depicts a plot of Model 1, 4,000 m in height, 9,000 m wide and 6,000 m in length, including about 551,000 isoparametric 8-node elements and 1,680,000 degrees of freedom. Since the analysis addresses the far-field area of the salt structure, underground openings like shafts, disposal drifts, and disposal boreholes are not considered

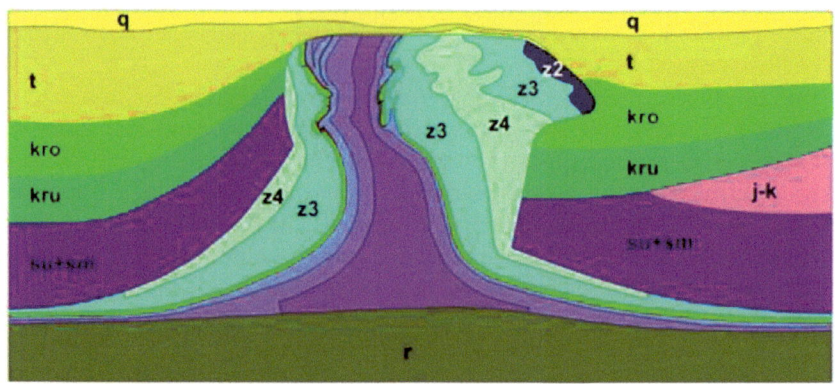

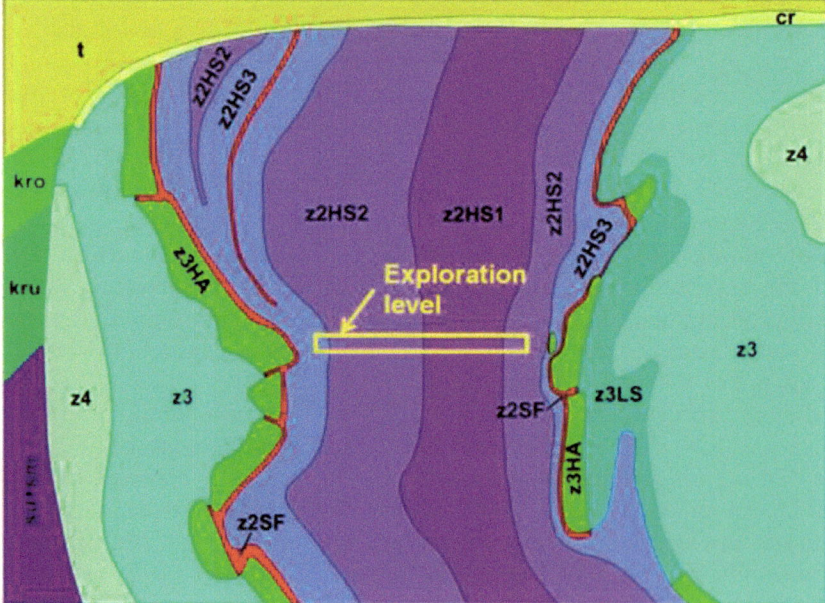

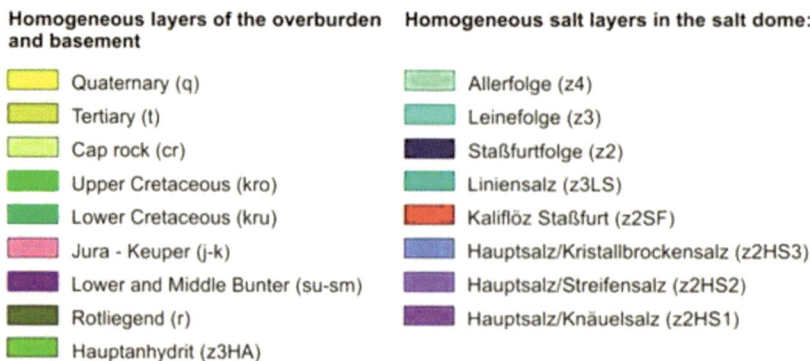

Homogeneous layers of the overburden and basement

- Quaternary (q)
- Tertiary (t)
- Cap rock (cr)
- Upper Cretaceous (kro)
- Lower Cretaceous (kru)
- Jura - Keuper (j-k)
- Lower and Middle Bunter (su-sm)
- Rotliegend (r)
- Hauptanhydrit (z3HA)

Homogeneous salt layers in the salt dome:

- Allerfolge (z4)
- Leinefolge (z3)
- Staßfurtfolge (z2)
- Liniensalz (z3LS)
- Kaliflöz Staßfurt (z2SF)
- Hauptsalz/Kristallbrockensalz (z2HS3)
- Hauptsalz/Streifensalz (z2HS2)
- Hauptsalz/Knäuelsalz (z2HS1)

Figure 16 Simplified geological cross section of the Gorleben salt dome along the QS1West drift (according to Bornemann *et al.*, 2008), top: complete salt formation and overburden, bottom: detail of the salt formation.

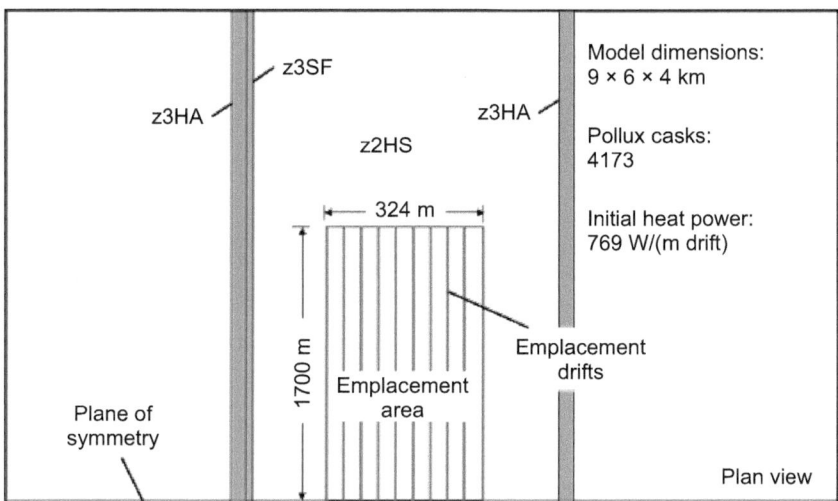

Figure 17 Plan view and simplified arrangement of the disposal drifts in the DE concept.

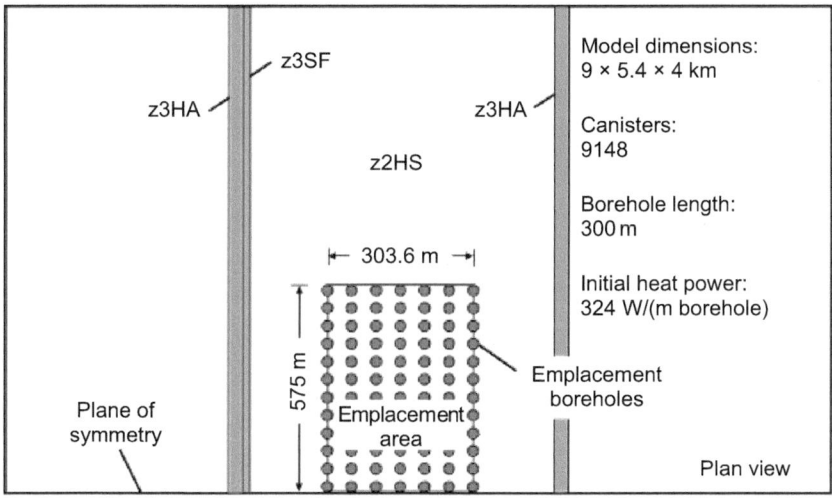

Figure 18 Plan view and simplified arrangement of the disposal boreholes in the BE concept.

in the models. The initial stress and temperature conditions assumed in the models are based on experimental data obtained in geotechnical in-situ measurements at the Gorleben site (Bräuer *et al.*, 2011). An instantaneous emplacement of the entire waste was assumed for both models *i.e.* ignoring the temporal and spatial disposal sequence. The calculations were performed using the special purpose JIFE code (Faust *et al.*, 2011) for a period of 10,000 years. Steady-state creep behavior (according to

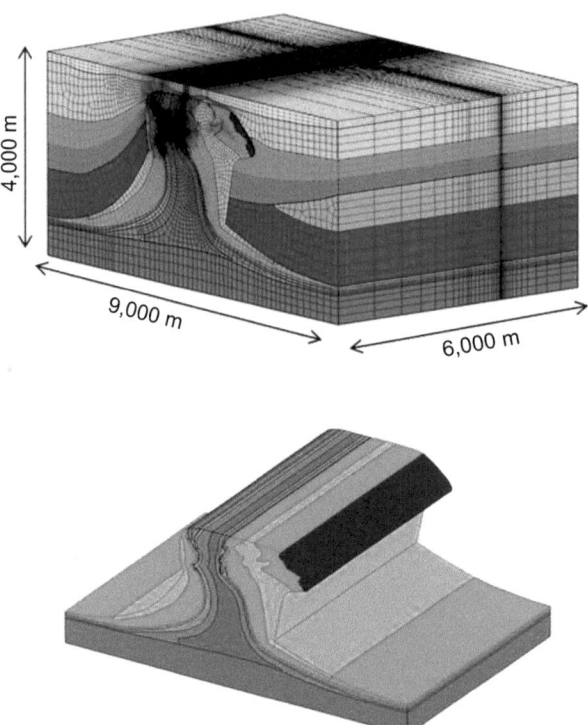

Figure 19 Three-dimensional modeling of the Gorleben salt dome. Top: complete finite element model, bottom: detail of the salt dome.

Equations 1 and 2) and the dilatant behavior (according to Equations 14–16) of the salt layers were considered.

Figure 20 depicts a time history of the temperature values calculated for several selected points: ground surface, top salt, in the salt rock mass between top salt and the disposal level as well as in the center of the disposal area. Maximum temperature values of up to 150 °C arise around the central emplacement drift about 80 years after waste emplacement. At the top salt point maximum values of 35 °C are predicted for a period of about 3,000 years.

Figure 21 shows the calculated temperature distribution in the salt dome for the symmetry level of the model for a period of 82.1 years after the emplacement of waste. Over this time, the maximum temperature is reached in the center of the disposal area. The heat sources in the disposal drifts appear as individual small concentric isotherms in the proximal zone of the drifts. At increasing distances from the drifts, the discrete heat sources act like a single homogeneous heat source, and the isotherms are shown with ellipsoidal shapes in the zones with higher temperature increases.

Figure 22 depicts a time history of the temperature values calculated for the top salt point, for the borehole disposal area as well as for certain points in the salt rock mass between the top salt and the disposal level. Maximum temperature values of up to 198 °C arise in the center of the borehole cluster about 345 years after waste emplacement.

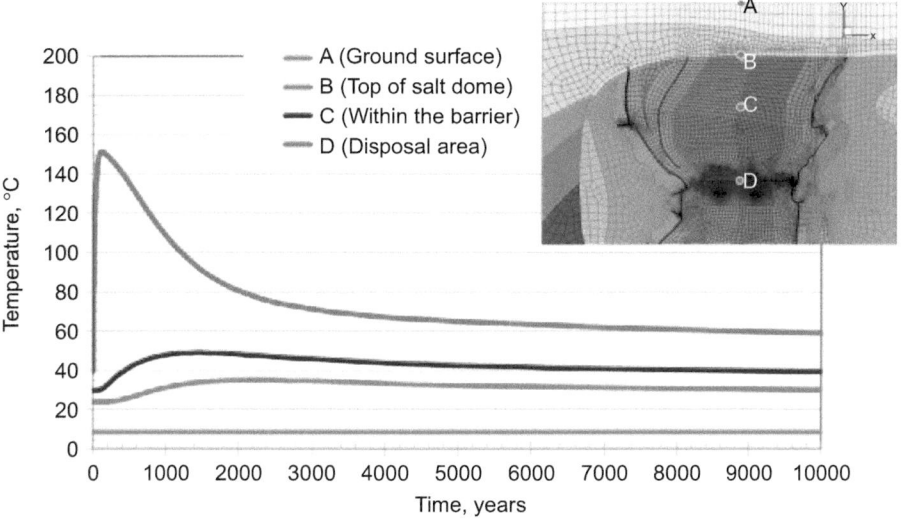

Figure 20 Time history of calculated temperatures (Model 1, drift emplacement).

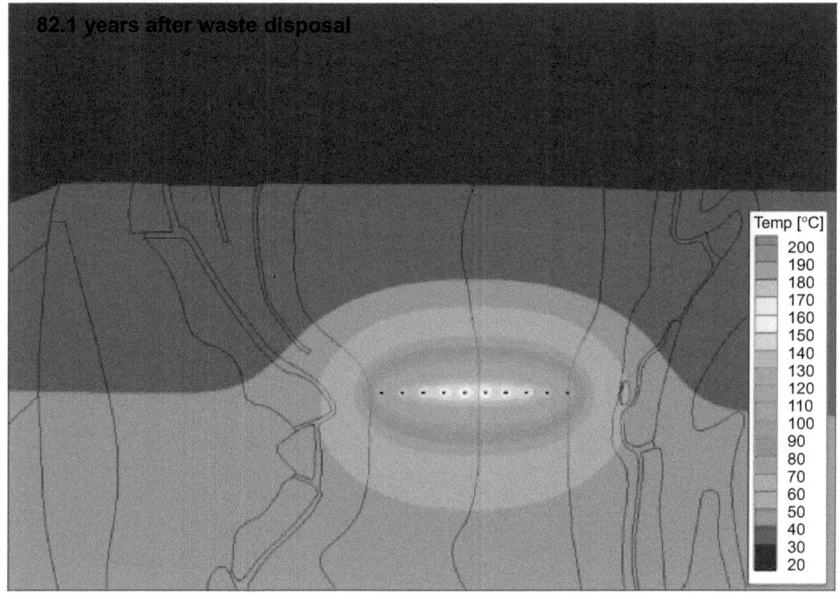

Figure 21 Maximum temperature in the rock salt 82.1 years after waste disposal (Model 1, drift emplacement).

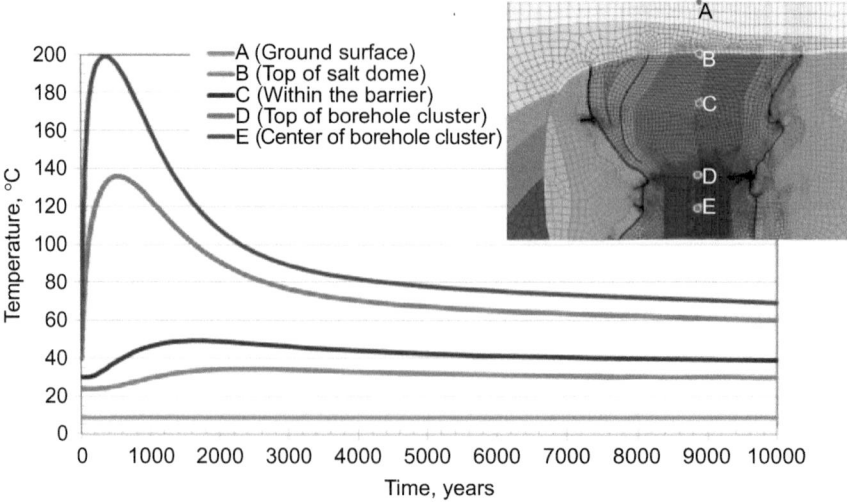

Figure 22 Time history of calculated temperatures (Model 2, borehole emplacement).

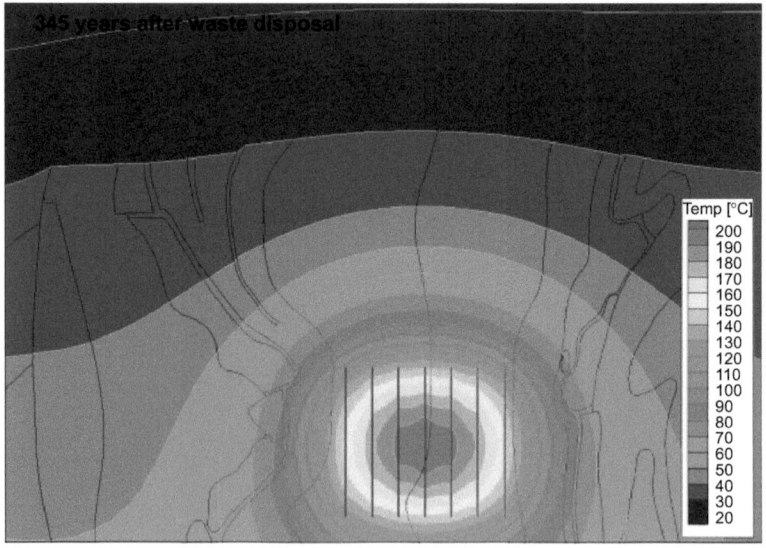

Figure 23 Maximum temperature in the rock salt 345 years after waste disposal (Model 2, borehole emplacement).

After 10,000 years temperatures of 65 to 75 °C are predicted for this point. At the top salt point, maximum values of about 35 °C are predicted for a period of about 2,500 years.

Figure 23 shows the calculated temperature distribution in the salt dome for a period of about 345 years after the emplacement of the waste. At this point in time,

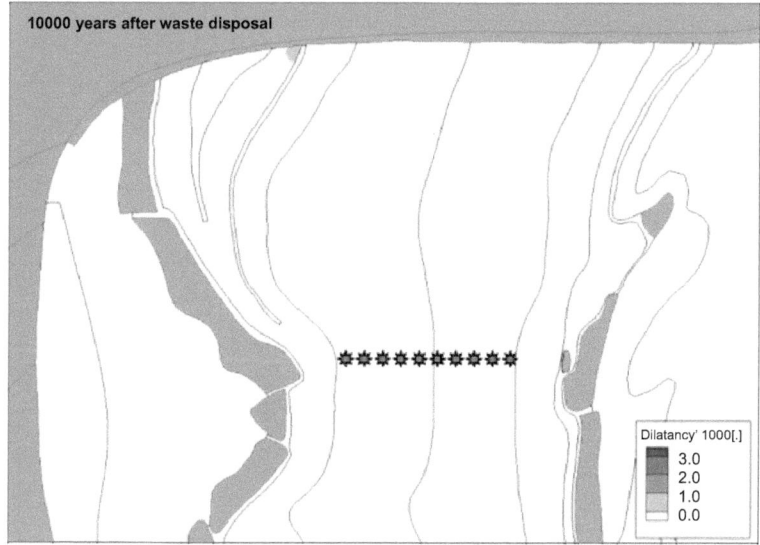

Figure 24 Dilatant zones in the rock salt after 10,000 years (Model I, drift emplacement).

the maximum temperature of about 198 °C is reached in the center of the borehole cluster.

For both models the maximum extent of the dilatant zones is reached within 10,000 years. Dilatancy remains very tightly restricted to the zones in the Kristallbrockensalz lying directly above the emplacement zone. The dilatant zone does not impinge on rocks with potential migration paths, which means that no new paths for the infiltration of fluids are created as a result of this effect (figures 24 and 25).

Concerning the drift disposal concept in Model 1, the fluid pressure criterion in the symmetry level within the Hauptsalz horizons is violated after a period of 29.4 years down to a maximum depth of 90 m below the top of the salt (Figure 26). The spatial extent of the affected zones reduces over time and is no longer detectable after ca. 747 years except for a small limited area at the top salt level. The spatial extent of the zones in which the fluid pressure criterion is violated remains restricted to the zones at the top of the salt lying above the disposal area (Figure 27).

The thermomechanical modeling of the borehole emplacement concept provides very similar results to the drift emplacement concept (figures 28 and 29). With respect to the thickness of the salt barrier of more than 500 m, the predicted values are not critical.

The temporary violation of the fluid pressure criterion at the top of the salt (see figures 26 and 28) is caused by the far-field thermally induced stress state and stress redistribution. This is illustrated in Figure 30, where the horizontal stress component orthogonal to the strike of the geological structure is depicted along the vertical axis of the repository at different points of time. Curve 1 represents the initial stress state in the virgin salt dome before waste disposal. Curve 2 illustrates the horizontal stress state 29.4 years after disposal of the heat generating waste corresponding to the maximum temperature in the emplacement area. Curve 3 describes the horizontal stress state approximately

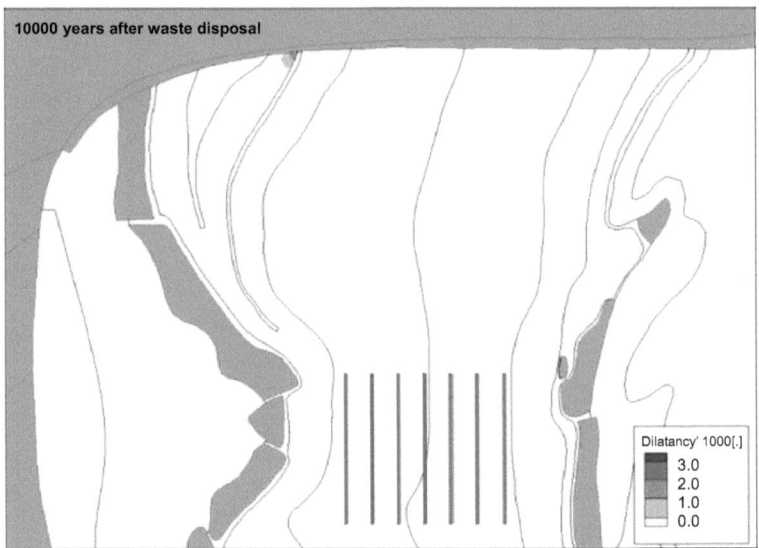

Figure 25 Dilatant zones in the rock salt after 10,000 years (Model 2, borehole emplacement).

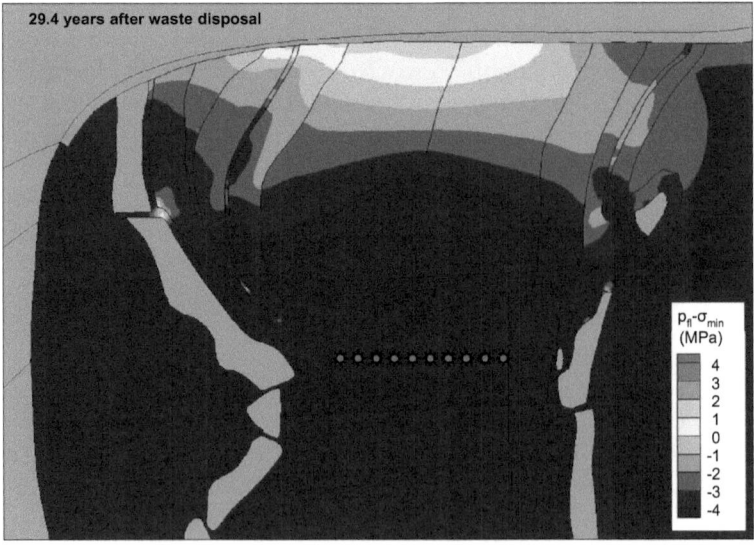

Figure 26 Frac-risk zones in the salt barrier after 29.4 years (Model 1, drift emplacement).

100 years after waste disposal. Curve 4 shows the distribution of horizontal stress after a period of 10,000 years. Due to the heat input of the waste, the horizontal stress component increases significantly in the salt rock at greater depth. This effect is compensated by a distinct reduction of the horizontal stress component in the upper

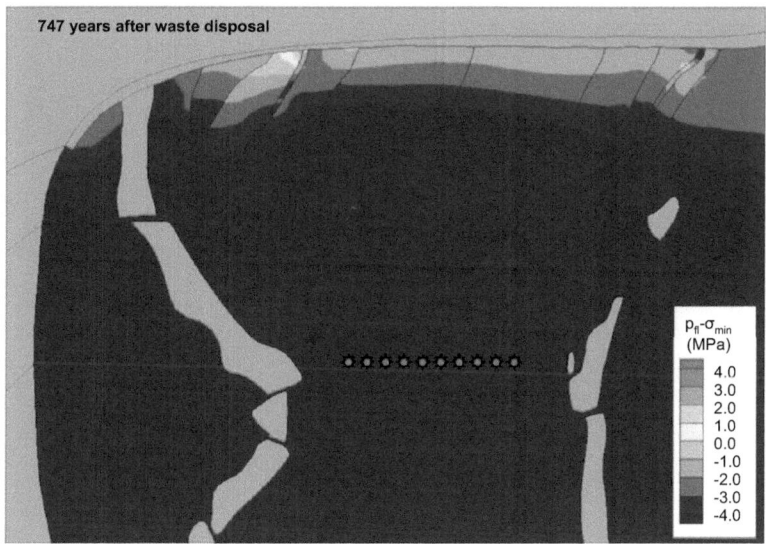

Figure 27 Frac-risk zones in the salt barrier after 747 years (Model I, drift emplacement).

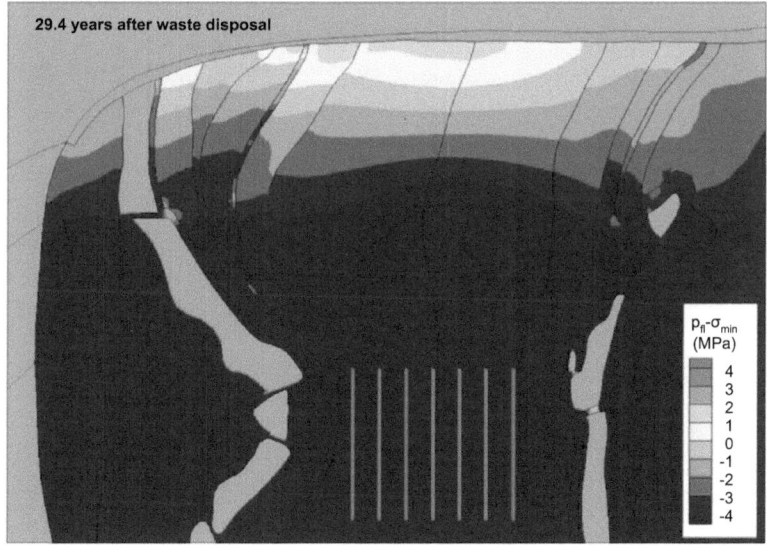

Figure 28 Frac-risk zones in the salt barrier after 29.4 years (Model 2, borehole emplacement).

zone of the salt dome where, according to Equation 18, the violation of the fluid pressure criterion occurs for a limited period.

Considering the fluid pressure criterion, Figure 31 shows the time history of the minimum principal stress at the most critical point T at the top of the salt structure above the center of the disposal areas. The red line depicts the constant value of the hypothetical

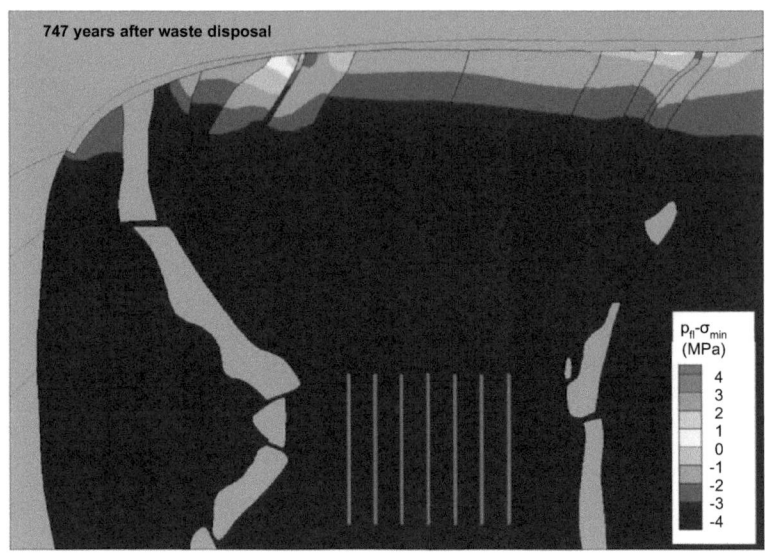

Figure 29 Frac-risk zones in the salt barrier after 747 years (Model 2, borehole emplacement).

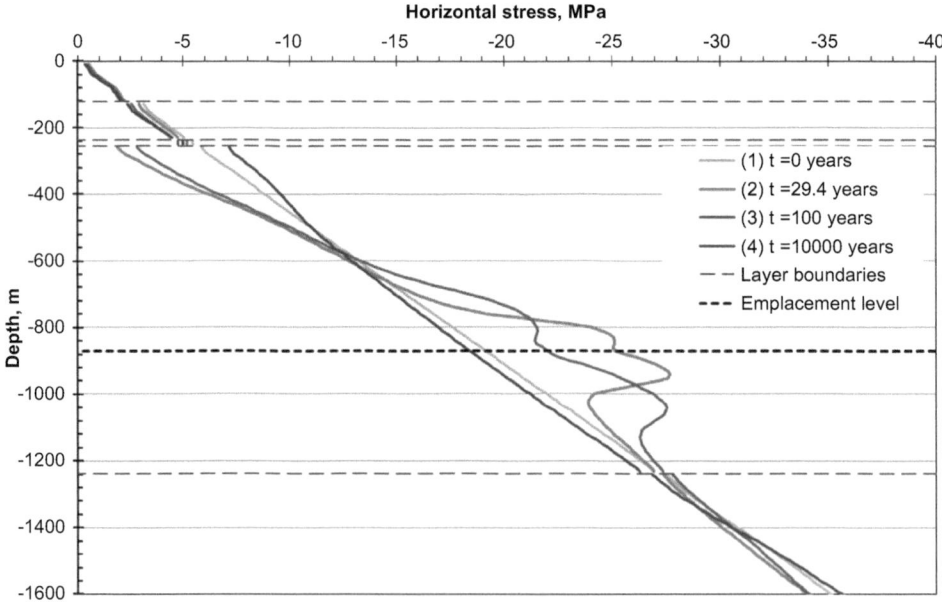

Figure 30 Horizontal stress distribution in the central vertical axis of the repository.

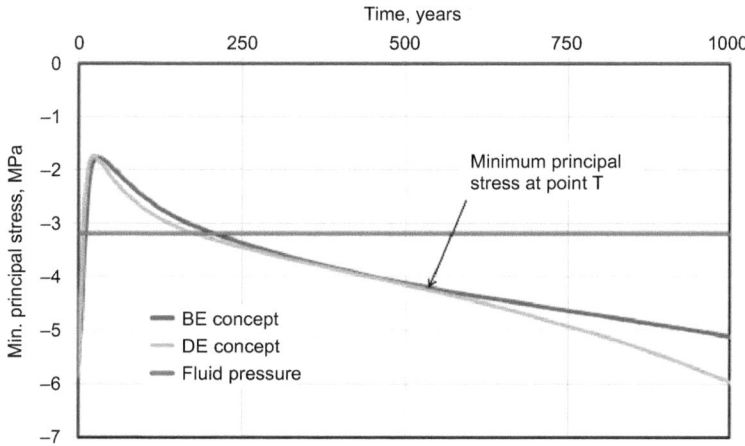

Figure 31 Time history of the minimum principal stress at the top of the salt dome (Heusermann *et al.*, 2015).

fluid pressure acting on the surface at the top of the salt dome. The green and blue curves represent the calculated thermally induced variation of the minimum principal stress over time.

For both disposal concepts it is evident that the fluid pressure is higher than the minimum principal stress for a time frame of about 200 years after waste emplacement. For this short period, the fluid pressure criterion is theoretically violated at the top of the salt structure. After about 200 years, favorable stress conditions will be re-established by a gradual decrease of the heat input, by the creep of the rock salt and by stress redistribution. Within the time frame of 10,000 years considered in the calculations the amount of the minimum principal stress at point T increases continuously. Thus, for periods of more than 200 years the fluid pressure criterion is not violated. Since both criteria are satisfied, the long-term integrity of the salt barrier is established.

8 SUMMARY

The use of salt structures for the disposal of low, medium, and high level radioactive waste makes it necessary to perform quantitative safety analyses, in particular geomechanical model calculations, to demonstrate the stability of the repository and the long-term integrity of the salt barrier. The analysis comprises the following main steps:

– Geological-geophysical exploration
– Geological modeling
– Classification of salt layers
– Constitutive modeling and parameter determination
– Definition of mechanical and, where required, thermal loading
– Formulation of stability and integrity criteria
– Numerical calculations
– Evaluation of numerical results

The methodology of the geomechanical safety analysis is shown in two case studies: the first study comprises the analysis of the stability and integrity of the Morsleben repository used for the disposal of LLW and MLW. The second study considers a prediction of the thermomechanical response of the salt barrier under HLW conditions assuming a fictive HLW repository in the Gorleben salt dome. It can be seen that the long-term integrity of the salt barrier is well established in both case studies.

REFERENCES

Behlau, J. & Mingerzahn, G. (2001) Geological and tectonic investigations in the former Morsleben salt mine (Germany) as a basis for the safety assessment of a radioactive waste repository. In: Langer, M. & Talbot, C. J. (eds.) *Geosciences and Nuclear Waste Disposal. Engineering Geol.*, Spec. Issue, Vol. 61, No. 2–3, 83–97, Amsterdam, Elsevier.

BMWI (2008) *Final disposal of high-level radioactive waste in Germany – The Gorleben repository project*. Federal Ministry of Economics and Technology (BMWI), Public Relations/IA8: Berlin, Germany.

Bollingerfehr, W., Filbert, W., Wehrmann, J. & Bosgiraud, J.-M. (2008) New transport and emplacement technologies for vitrified waste and spent fuel canisters. In: Davies, C. (ed.) *Proc. Seventh European Commission Conference on the Management and Disposal of Radioactive Waste – Euradwaste '08*, 259–267, EUR 24040, Brussels, Belgium.

Bornemann, O., Behlau, J., Fischbeck, R., Hammer, J., Jaritz, W., Keller, S., Mingerzahn, G. & Schramm, M. (2008) *Description of the Gorleben Site – Part 3: Results of the geological surface and underground exploration of the salt formation*. Federal Institute for Geosciences and Natural Resources (BGR), 223p., Hannover, Germany, ISBN 978-3-9813373-6-5.

Bräuer, V., Eickemeier, R., Eisenburger, D., Grissemann, C., Hesser, J., Heusermann, S., Kaiser, D., Nipp, H.-K., Nowak, T., Plischke, I., Schnier, H., Schulze, O., Sönnke, J. & Weber, J.R. (2011) *Description of the Gorleben Site – Part 4: Geotechnical exploration*. Federal Institute for Geosciences and Natural Resources (BGR), 176p., Hannover, Germany, ISBN 978-3-9814108-0-8.

Brennecke (2011) Radioactive waste disposal challenges in Germany – 11442. *WM2011 Conference*, February 27–March 3, 2011, Phoenix, AZ.

Döring, T., Kiehl, J. & Erichsen, C. (1998) Ein räumliches Stoffgesetz für Steinsalz unter Berücksichtigung von primärem, sekundärem und tertiärem Kriechen, Dilatanz, Kriech-und Zugbruch sowie Nachbruchverhalten. *Geotechnik* 21(3), 254–258.

Fahland, S. & Heusermann, S. (2013) Geomechanical analysis of the integrity of waste disposal areas in the Morsleben repository. In: Feng, X.-T., Hudson, J.A. & Tan, F. (eds.) *Rock Characterisation, Modelling and Engineering Design Methods*, 345–350, London, Taylor & Francis, ISBN 978-1-138-00057-5.

Fahland, S., Heusermann, S. & Schäfers, A. (2015) Geomechanical analysis and assessment of the integrity of the southern part in the Morsleben repository. In: Roberts, L., Mellegard, K. & Hansen, F. (eds.) *The Mechanical Behavior of Salt VIII*, 373–380, London, Taylor & Francis, ISBN 978-1-138-02840-1.

Faust, B., Krüger, R., Lucke, A. & Tertel, S. (2011) *JIFE – Java application for interactive nonlinear finite-element analysis in multi-physics*. User's manual, IFF, Berlin, Germany.

Günther, R.-M. & Salzer, K. (2012) Advanced strain-hardening approach: A powerful creep model for rock salt with dilatancy, strength and healing. In: Berest, P., Ghoreychi, M., Hadj-Hassen, F. & Tijani, M. (eds.) *Mechanical Behavior of Salt VII*, 13–22, London, Taylor & Francis, ISBN 978-0-415-62122-9.

Hammer, J., Mingerzahn, G., Behlau, J., Fleig, S., Kühnlenz, T. & Schramm, M. (2010) *Geological exploration and 3D-modelling of a saliferous host rock formation – Gorleben salt dome.* KIT Scientific Reports 7569, Projektträger Karlsruhe (PTKA-WTE), KIT Scientific Publ., Karlsruhe, Germany, 669–716, ISSN 1430-6751, ISSN 1869-9669.

Hampel, A. (2012) The CDM constitutive model for the mechanical behaviour of rock salt: Recent developments and extensions. In: Berest, P., Ghoreychi, M., Hadj-Hassen, F. & Tijani, M. (eds.) *Mechanical Behavior of Salt VII*, 45–55, London, Taylor & Francis, ISBN 978-0-415-62122-9.

Hampel, A. (2015) Description of damage reduction and healing with the CDM constitutive model for the thermo-mechanical behaviour of rock salt. In: Roberts, L., Mellegard, K. & Hansen, F. (eds.) *The Mechanical Behavior of Salt VIII*, 301–310, London, Taylor & Francis, ISBN 978-1-138-02840-1.

Hampel, A., Günther, R.-M., Salzer, K., Minkley, W., Pudewills, A., Yildirim, S., Rokahr, R.B., Gährken, A., Missal, C., Stahlmann, J., Herchen, K. & Lux, K.-H. (2015) Joint project III on the comparison of constitutive models for the thermo-mechanical behavior of rock salt – I. Overview and results from model calculations of healing of rock salt. In: Roberts, L., Mellegard, K. & Hansen, F. (eds.) *The Mechanical Behavior of Salt VIII*, 349–359, London, Taylor & Francis, ISBN 978-1-138-02840-1.

Hampel, A., Salzer, K., Günther, R.-M., Minkley, W., Pudewills, A., Leuger, B., Zapf, D., Staudtmeister, K., Rokahr, R.B., Herchen, K., Wolters, R. & Lux, K.-H. (2012) Joint projects on the comparison of constitutive models for the mechanical behavior of rock salt II. Overview of the models and results of 3-D benchmark calculations. In: Berest, P., Ghoreychi, M., Hadj-Hassen, F. & Tijani, M. (eds.) *Mechanical Behavior of Salt VII*, 231–240, London, Taylor & Francis, ISBN 978-0-415-62122-9.

Hampel, A. & Schulze, O. (2007) The Composite Dilatancy Model – A constitutive model for the mechanical behavior of rock salt. In: Wallner, M., Lux, K.-H., Minkley, W. & Hardy Jr., H.R. (eds.) *The Mechanical Behavior of Salt – Understanding of THMC Processes in Salt*, 99–107, London, Taylor & Francis, ISBN 978-0-415-44398-2.

Heller, M., Mauke, R., Mohlfeld, M. & Skrzyppek, J. (2004) ERAM-SIS – A spatial information system for visualization and management of geotechnical data. *Proc. Int. Conf. on Radioactive Waste Disposal (DisTec2004)*, April 26–28, 2004, 130–136, Berlin, Germany.

Heusermann, S. (1982) *Kritische Gegenüberstellung und Bewertung von Stoffgesetzen zur Beschreibung des Kriechverhaltens von Steinsalz auf der Grundlage von Laboruntersuchungen und In-situ-Messungen.* Forschungsergebnisse aus dem Tunnel-und Kavernenbau, Heft 6, Institut für Unterirdisches Bauen, Universität Hannover.

Heusermann, S., Eickemeier, R. & Fahland, S. (2015) Thermomechanical analysis of a fictive HLW repository in the Gorleben salt dome. In: Roberts, L., Mellegard, K. & Hansen, F. (eds.) *The Mechanical Behavior of Salt VIII*, 401–409, London, Taylor & Francis, ISBN 978-1-138-02840-1.

Heusermann, S. & Fahland, S. (2005) Long-term geomechanical stability and integrity of the salt barrier in the central part of the Bartensleben salt mine. *Proc. Post-Mining 2005*, November 16–18, 2005, 12p., Nancy, France.

Heusermann, S., Fahland, S. & Eickemeier, R. (2009) Geomechanical stability and integrity of radioactive waste repositories in salt rock. *SINOROCK 2009 – Proc. International Symposium on Rock Mechanics: Rock Characterisation, Modelling and Engineering Design Methods*, May 19–22, 2009, Hongkong, China.

Heusermann, S., Lux, K.-H. & Rokahr, R. (1983) *Entwicklung mathematisch-mechanischer Modelle zur Beschreibung des Stoffverhaltens von Salzgestein.* Schlussbericht zum Forschungsauftrag ET2011A der PLE Jülich, Institut für Unterirdisches Bauen, Universität Hannover.

Heusermann, S., Nipp, H.-K., Eickemeier, R., Fahland, S. & Preuss, J. (2007) Geomechanical integrity of waste disposal areas in the Morsleben repository. *Int. Conf. on Radioactive Waste*

Disposal in Geological Formations (REPOSAFE), November 6–9, 2007, Braunschweig, Germany.

Heusermann, S., Vogel, P., Eickemeier, R. & Nipp, H.-K. (2012) Thermomechanical modelling of the Gorleben exploration site to analyse the integrity of the salt barrier. In: Qian, Q. & Zhou, Y. (eds.) *Harmonising Rock Engineering and the Environment*, 413–418, London, Taylor & Francis, ISBN 978-0-415-80444-8.

Hou, Z. & Lux, K.-H. (1999) A material model for rock salt including structural damages as well as practice-oriented applications. In: Cristescu, N.D., Hardy Jr., H.R. & Simionescu, R.O. (eds.) *Basic and Applied Salt Mechanics*, 55–59, Swets & Zeitlinger, Lisse, ISBN 9058093832.

Hudson, J.A. & Feng, X.-T. (2015) *Rock engineering risk*. Taylor & Francis, London, ISBN 9781138027015.

Hunsche, U. (1984) Results and interpretation of creep experiments on rock salt. In: Hardy Jr., H.R. & Langer, M. (eds.) *Proc. First Conf. on the Mechanical Behavior of Salt, November 9–11, 1981*, 159–167, Trans Tech Publications, Clausthal-Zellerfeld, Germany.

Hunsche, U. & Schulze, O. (1994) Das Kriechverhalten von Steinsalz. *Kali & Steinsalz*, Band 11, Heft 8/9, 238–255.

Hunsche, U. & Schulze, O. (2003) The dilatancy concept – A basis for the modelling of coupled TMH processes in rock salt. *Proc. European Commission CLUSTER Conference on the Impact of EDZ on the Performance of Radioactive Waste Geological Repositories, November 3–5, 2003*, Luxembourg.

Klinge, H., Boehme, J., Grissemann, C., Houben, G., Ludwig, R.-R., Rübel, A., Schelkes, K., Schildknecht, F. & Suckow, A. (2007) *Description of the Gorleben site – Part 1: Hydrogeology of the overburden of the Gorleben salt dome*. Federal Institute for Geosciences and Natural Resources (BGR), 145p., Hannover, Germany, ISBN 978-3-9813373-4-1.

Köthe, A., Hoffmann, N., Krull, P., Zirngast, M. & Zwirner, R. (2007) *Description of the Gorleben site – Part 2: Geology of the overburden and adjoining rock of the Gorleben salt dome*. Federal Institute for Geosciences and Natural Resources (BGR), 220p., Hannover, Germany, ISBN 978-3-9813373-5-8.

Langer, M. (1999) Principles of geomechanical safety assessment for radioactive waste disposal in salt structures. *Eng. Geol. 52*, 257–269, Amsterdam, Elsevier.

Langer, M. & Heusermann, S. (2001) Geomechanical stability and integrity of waste disposal mines in salt structures. *Eng. Geol. 61*, 155–161, Amsterdam, Elsevier.

Lux, K.-H. & Heusermann, S. (1983) Creep tests on rock salt with changing load as a basis for the verification of theoretical material laws. *Proc. 6th Symp. on Salt*, Vol. I, 417–435, Toronto, Canada.

Minkley, W. & Mühlbauer, J. (2007) Constitutive models to describe the mechanical behavior of salt rocks and the imbedded weakness planes. In: Wallner, M., Lux, K.-H., Minkley, W. & Hardy Jr., H.R. (eds.) *The Mechanical Behavior of Salt – Understanding of THMC Processes in Salt*, 119–127, London, Taylor & Francis, ISBN 978-0-415-44398-2.

Mora, C.J. (1999) *Sandia and the Waste Isolation Pilot Plant 1974 – 1999*. SAND99-1482, Sandia Laboratories, Albuquerque, NM, USA.

Munson, D.E. & Dawson, P.R. (1979) *Constitutive Model for the low temperature creep of salt (with application to WIPP)*. SAND-79-1853, Sandia Laboratories, Albuquerque, NM, USA.

Plischke, I. (2007) Determination of mechanical homogeneous areas in the rock salt mass using creep properties for a classification scheme. In: Wallner, M., Lux, K.-H., Minkley, W. & Hardy Jr., H.R. (eds.) *The Mechanical Behavior of Salt – Understanding of THMC Processes in Salt*, 321–325, London, Taylor & Francis, ISBN 978-0-415-44398-2.

Preuss, J., Eilers, G., Mauke, R., Müller-Hoeppe, N., Engelhardt, H.-J., Kreienmeyer, M., Lerch, C. & Schrimpf, C. (2002) Post closure safety of the Morsleben repository. *WM2002 Conference, February 24–28, 2002*, Tucson, AZ, USA.

Rokahr, R., Staudtmeister, K. & Zander-Schiebenhöfer, D. (2004) Application of a continuum damage model for cavern design – Case study: Atmospheric pressure. *Solution Mining Research Institute, SMRI-Meeting, April 18–21, 2004*, Wichita, KS, USA.

Schulze, O., Heemann, U., Zetsche, F., Hampel, A., Pudewills, A., Günther, R.-M., Minkley, W., Salzer, K., Hou, Z., Wolters, R., Rokahr, R. & Zapf, D. (2007) Comparison of advanced constitutive models for the mechanical behavior of rock salt – results from a joint research project – I. Modeling of deformation processes and benchmark calculations. In: Wallner, M., Lux, K.-H., Minkley, W. & Hardy Jr., H.R. (eds.) *The Mechanical Behavior of Salt – Understanding of THMC Processes in Salt*, 77–88, London, Taylor & Francis, ISBN 978-0-415-44398-2.

Wallner, M. (1984) Analysis of thermomechanical problems related to the storage of heat producing radioactive waste in rock salt. In: Hardy Jr., H.R. & Langer, M. (eds.) *Proc. First Conf. on the Mechanical Behavior of Salt*, November 9–11, 1981, 739–763, Trans Tech Publications, Clausthal-Zellerfeld, Germany.

Wilke, F., Schweinsberg, J., Behlau, J. & Bornemann, O. (2004) Geological 3-D model for predicting new cavern locations. *Solution Mining Research Institute, SMRI-Meeting*, April 18–21, 2004, *Wichita, KS, USA*.

Wolters, R., Lux, K.-H. & Düsterloh, U. (2012) Evaluation of rock salt barriers with respect to tightness: Influence of thermomechanical damage, fluid infiltration and sealing/healing. In: Berest, P., Ghoreychi, M., Hadj-Hassen, F. & Tijani, M. (eds.) *Mechanical Behavior of Salt VII*, 425–434, London, Taylor & Francis, ISBN 978-0-415-62122-9.

Chapter 18

Rock engineering in underground compressed air energy storage

D. Ryu[1], H. Kim[2] & D. Park[1]
[1]*Underground Space Department, Korea Institute of Geoscience and Mineral Resources, Daejeon, Korea*
[2]*Department of Energy and Mineral Resources Engineering, Sejong University, Seoul, Korea*

Abstract: The concern about climate change and global warming has triggered global paradigm shift and different energy industrial environment. Energy storage system (ESS) comes into the spotlight as an emerging industry. Especially, compressed air energy storage (CAES), which has a long history of commercialization, gets reappraisal as a competitive ESS technology at a utility scale. However, it has been mentioned that CAES has a weakness of site limitation. Rock engineering is expected to make its key roles for CAES to overcome the limitation and penetration into ESS market successfully. We here consider the CAES types with underground storage cavern in a rock salt and hard rock, and key issues of rock engineering in implementing these types of CAES are presented. In a rock salt, the shape and pillar width of multiple storage caverns are important for bulk storage implementation. In a hard rock, a field experiment of air tightness, structural stability, energy balance and efficiency analysis during operation in the storage system should be interesting topics. We introduce a sophisticated apparatus and evaluation procedure for the characterization of air tightness in the components of storage system. We also demonstrate the results of reliability-based analysis and sensitivity analysis of the parameters for the stability of the structure at a shallow depth. And the results of energy balance and efficiency analysis through the coupling analysis are presented under the various operation modes of the system.

1 INTRODUCTION

The recent need to reduce greenhouse gas emissions and to introduce mixed energy sources, which integrate renewable powers with conventional sources, has dramatically changed industrial environment of electrical power generation. Under the jolt, ESS is recognized as one of the underpinning technologies to have great potential in meeting characteristic demand of unpredictable daily and seasonal variations. A number of ESS technologies, that are economical over each time scale, have been being developed and proposed, but only two technologies-CAES (compressed air energy storage) and PHS (pumped hydroelectric storage) are cost-effective at utility scales (; with a duration of from several hours to days and at a scale of hundreds of MW power). Rock engineering has played an important role in the design and construction of utility caverns for PHS. Recent increasing interest in CAES as an ESS will bring a great chance to rock engineering as well. Both the existing commercial CAES plants and those under planning have used rock salts as a storage space because of their low permeability and air tightness. It naturally resulted in a fixed idea that CAES has the critical weakness of a site limitation in a comparison with other electro-chemical ESS, since the rock salt formations are not everywhere. However, the state of the art rock engineering technologies will overcome this stereotype through the technical and economic feasibility studies in a hard rock using the concept of LRC (lined rock cavern).

In this chapter, the function of rock engineering and its key roles for CAES to penetrate into ESS market is introduced. We here consider the CAES with underground storage cavern types in a rock salt and hard rock, and key issues are presented. In a rock salt, the shape and pillar width of multiple storage caverns are important for bulk storage implementation. In a hard rock, a field experiment of air tightness, structural stability, energy balance and efficiency analysis during operation in the storage system should be interesting topics. For the characterization of air tightness in the components of storage system, a sophisticated apparatus and evaluation procedure are introduced. Reliability-based analysis and sensitivity analysis of the parameters for the stability of the structure at a shallow depth are explained. And energy balance and efficiency are considered through the coupling analysis at the various operation modes.

These basic concepts for CAES are expected to be expanded and utilized for the bulk storage of compressed natural gas and gaseous hydrogen.

2 CAES (COMPRESSED AIR ENERGY STORAGE)

2.1 Background

CAES is a type of commercialized ESS technology which can provide electric power of over scores megawatt with a single unit for arbitrage or black-start of utility power plant. During the periods of low power demand, the surplus electricity drives a reversible motor/generator (M/G) unit in turn to run a multi-train of compressors for injecting air into a storage facility, which is either an underground cavern or above-ground tanks. When the power generation cannot meet the load demand in a grid, the compressed air stored in the storage facility is released and heated by a heat source to provide electricity. Generally, CAES can be divided into a diabatic and adiabatic system, which are based on whether the heat in a compression process is cooled into outer system or stored in a thermal storage facility for a heat source. The diabatic CAES as conventional system heats compressed air by a heat source from the combustion of fossil fuel and the adiabatic system uses the heat recovered from the compression process. Finally, the compressed air energy is captured by the expanders. The waste heat from the exhaust can be recycled by a recuperator unit (Figure 1).

2.2 Past, present and future of CAES

The first utility-scale CAES plant, the Huntorf power plant, was installed in Germany in 1978 (Succar & Williams, 2008; Madlener & Latz, 2013; Raju & Kummar, 2012). It uses two salt domes as the storage caverns and it runs on a daily cycle with 8 hours of compressed air charging and 2 hours of operation at a rated power of 290 MW (Succar & Williams, 2008). This plant provides black-start power to nuclear units, back-up to local power systems and extra electrical power to fill the gap between the electricity generation and demand. The second utility-scale CAES plant started operation in

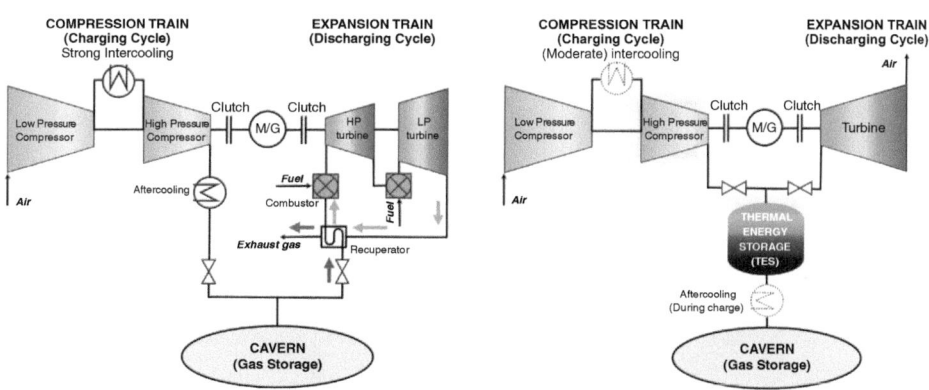

Figure 1 Schematic diagram of diabatic (left) and adiabatic (right) CAES (revised from Finkenrath *et al.*, 2009).

McIntosh, U.S., in 1991 (Succar & Williams, 2008; Madlener & Latz, 2013; Raju & Kummar, 2012). The 110 MW McIntosh plant can operator for up to 26 hours at full power, which also utilizes a salt cavern as a bulk storage facility. A recuperator is operated to reuse the exhaust heat energy, which reduces the fuel consumption by 22~25% and improves the cycle efficiency from ~42% to ~54% in comparison with the Huntorf plant (Chen *et al.*, 2009; Finkenrath *et al.*, 2009). Both the commercial plants have the common characteristic of the usage of rock salt for bulk storage of compressed air. Recently, there have been innovations of CAES system, which including top-side equipments with higher efficiency and the fuel-free concept of isothermal process and adiabatic process. Nevertheless, CAES has unavoidable problem of bulk storage, which should be cost-efficient and site-flexible to blow away the competition in ESS market.

Heretofore the salt dome seems to be a highly cost-efficient storage facility because CAES has provided service for arbitrage like PHS and can use the existing transmission infra-structure. When considering the renewables integration and ancillary service, the limitation of the site selection due to a specific geologic structure like a dormal salt is a critical problem to be solved. In the U.S., most of the salt domes region has very poor wind resources that are not economically exploitable (Figure 2). If the aim of storage is to provide backup for large quantities of wind power, salt domes will not play a large role in the United States (Succar & Williams, 2008).

While bulk storage in hard rock might be used, their development will likely be more challenging and costly than the salt dome CAES systems that have been deployed with considering transmission cost. Therefore, rock engineering should contribute to the cost-efficient storage and the enhancement of site-flexibility in order that CAES has a promising future in ESS market (Figure 3).

2.3 Key issues of rock engineering in CAES

Main components of CAES like multi-train compressors and turbines specify maximum storage pressure during charging and operation pressure range during discharging for overall system efficiency. Approximately, the maximum storage pressure is from 5 MPa to 7.5 MPa and the minimum storage pressure is from 2 MPa to 4 MPa with considering

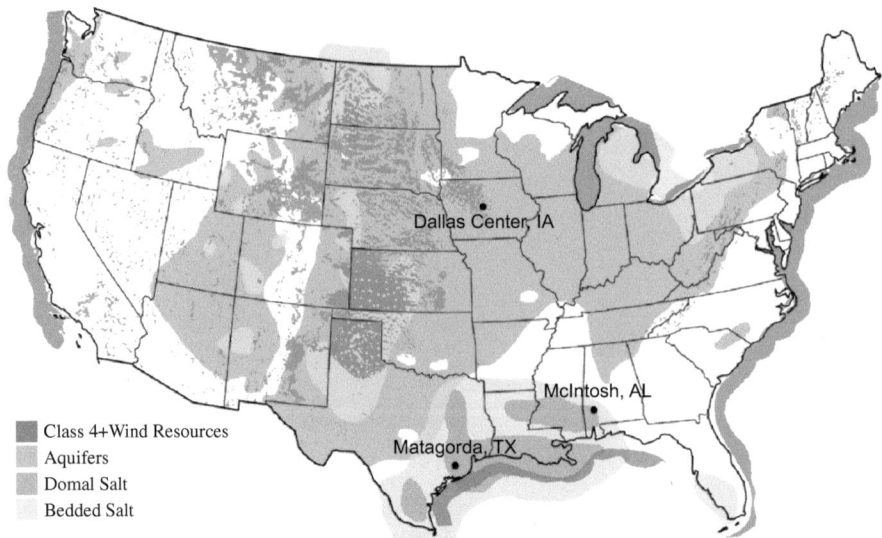

Figure 2 Map of geological formations suitable for CAES plants in the U.S. (Succar & Williams, 2008).

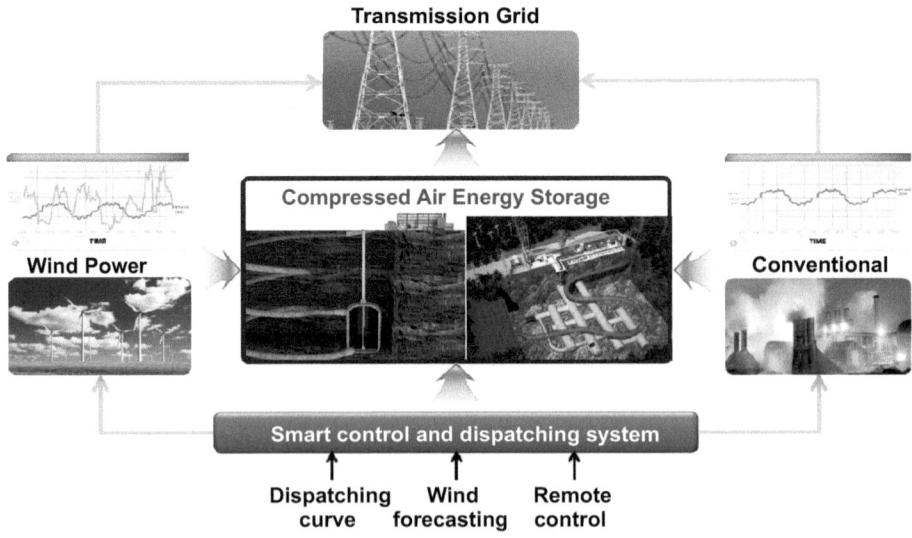

Figure 3 A key role of CASE in renewable-integration: how to integrate wind power with conventional sources.

two exiting commercial plants, which depend on the specification of main components and system engineering. Isothermal CAES, which is reciprocal type and fuel-free, may require the operation pressure range from 4.5 MPa to 20 MPa. In the issue, under the severe operation conditions, rock engineering should be able to provide solutions to a reliable and economic storage facility.

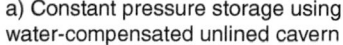

a) Constant pressure storage using   (b) Variable pressure storage using LRC
water-compensated unlined cavern

Figure 4 Two types of underground compressed air bulk storage in a hard rock.

Underground compressed air storage has two types of storage according to the condition of storage pressure (Figure 4). One is to keep the storage pressure constant through pressure compensation by surface water or groundwater and compensating water, which has advantages of system efficiency and minimization of volume of storage cavern. Because of the high excavation costs associated with the larger volumes required by other storage schemes, a CAES reservoir in hard rock would probably be a constant pressure, water-compensated cavern. The constant pressure storage methods like groundwater reservoir and unlined hard rock with water curtain and water shaft have still critical engineering problems about a leakage, fatigue failure, a hysteresis of fluid flow in porous medium during charging and discharging and a champagne effect (Nakagawa *et al.*, 2003).

The other is to permit storage pressure variable during charging and discharging cycles, which doesn't need compensating water. Generally, the variable pressure storage method requires the relatively huge volume in comparison with constant pressure storage method. Nevertheless, the storage method using LRC is still attractive because of a site-flexibility. Salt domes by solution mining and LRC (hard rock cavern with inner containment system) have been considered and adopted for underground compressed air bulk storage.

Their relevance to compressed air bulk storage, and the identification of required research areas, are identified throughout this chapter. Based on recent researches, key issues are established and described from the viewpoint of 1) mechanical considerations in rock salt like cavern shape and pillar width, 2) mechanical and hydraulic considerations in unlined rock cavern, 3) technologies for performance and economics enhancement like air tightness evaluation, optimal depth and reliability-based design scheme in LRC, and 4) coupled behaviors in compressed air storage system.

3 MECHANICAL CONSIDERATIONS IN ROCK SALT

Underground storage facilities like solution-mined rock salt caverns, excavated unlined caverns with water curtain system and lined caverns in hard rock are much safer in terms of safety and environmental protection than steel and concrete tank farms at the ground surface. Particularly, a rock salt comparing to other rock formation has the following advantages for a storage cavern.

1)   The permeability of rock salt is very low (ranging from 10^{-21} to 10^{-24} m²) enough to ensure the sealing of cavern.
2)   Rock salt, when damaged potentially by frequent changes of inner storage pressure, has a self-recovery ability to ensure its safety.
3)   Rock salt is usually dissolved into water, so that solution mining, which facilitates the construction and shape control of caverns, is applicable.
4)   Rock salt is relatively widely distributed and abundant mineral resources with large reserves.

In this section, two mechanical issues of CAES within a rock salt will be reviewed; one is to optimize the shape of the storage cavity built by a solution mining, and the other is the span length (; pillar width) between multiple storage caverns.

3.1 Optimal shape of cavity

Underground salt caverns have been widely used mainly for hydrocarbon bulk storage, since caverns can be comparatively easily leached out from salt formation. It is reported that there are even thousands of such caverns implemented throughout the world (Thoms & Gehle, 2000), but their shape and dimension are significantly different. Sobolik & Ehgartner (2006) divided these various cavern shapes into four groups (; cylinder, enlarged top, enlarged middle, and enlarged bottom, Figure 5) and evaluated relative performance of each shape so as to derive an optimum cavern dimension. They considered, in their evaluation, four factors of dilatant damage safety factor in salt, cavern volume closure, axial well strain in the caprock, and surface subsidence, and concluded from the performance comparison that the enlarged bottom shape provides the worst overall performance. Wang et al. (2011) also reported that ellipsoidal cavern performs much better than pear-shaped cavern by studying the maximum displacement, plastic volume, and pillar width of salt cavern gas storages, and proposed that the optimum ratio of long and short axes of ellipsoidal cavern as 7/3.

Due to the frequent change of storage pressure during repetitive compression and decompression of cavern, dilatant damage can occur at the surrounding salt rock. The

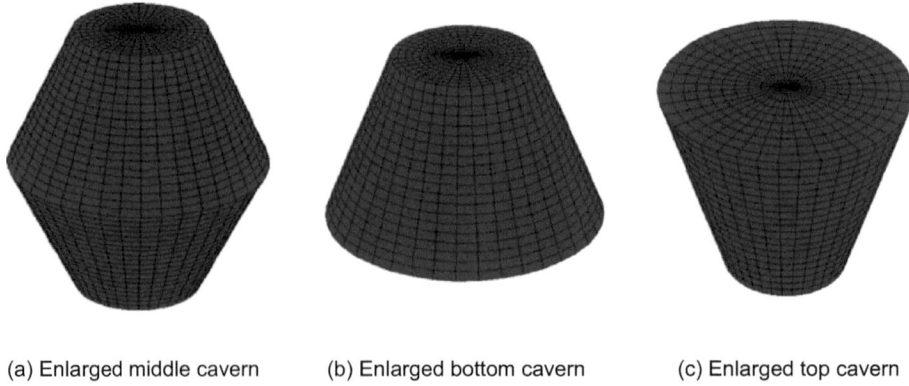

(a) Enlarged middle cavern   (b) Enlarged bottom cavern   (c) Enlarged top cavern

Figure 5 Salt cavern gas storages with different shapes (Sobolik & Ehgartner, 2006).

dilatancy accompanies microfracturing in the rock and results in the significant increment in permeability. Typical factor of safety (FS) as a dilatants criteria, that has been used in evaluating cavern shape performance, is expressed in terms of two invariants of stress tensor; one is the mean stress invariant (I_1) and the other is the square root of the shear stress invariant (J_2).

$$FS = \frac{b.I_1}{\sqrt{J_2}} \tag{1}$$

where, b is the material constant, and has a typical value of 0.27, $I_1 = \sigma_1 + \sigma_2 + \sigma_3$, $J_2 = [(\sigma_1 - \sigma_2)^2 + (\sigma_2 - \sigma_3)^2 + (\sigma_3 - \sigma_1)^2]$, and σ_1, σ_2, and σ_3 are the maximum, intermediate and minimum principal stress, respectively.

When the FS is less than 1.0, the shear stress is greater than the average stress and dilatant damage occurs in salt rock.

Wang *et al.* (2013) recently introduced the concepts of slope instability and pressure arch into the shape design of cavern, and proposed the mathematical models of cavern lower structure and cavern upper structure separately under the assumption that the failure mechanism of two parts are different from each other (Figure 6).

For the cavern upper structure design, they assumed that the collapse of overlying strata, which results from the loss of support afforded by rock salt at the location of cavern, may stop after a certain extent since the structure becomes stable with the help of so-called pressure arch effect. The shape of cavern upper structure should be designed such that pressure arch can improve its stability.

Once salt cavern is constructed, the initial loads are redistributed and transferred to the surrounding salt rock. Due to stress redistribution and concentration, the instability of salt cavern can be induced, which is represented by a slope sliding in the cavern lower structure. The shape design of the cavern lower structured, therefor, is considered to a slope instability problem to increase the factor of safety against the slope formed by the cavern shape (Figure 7).

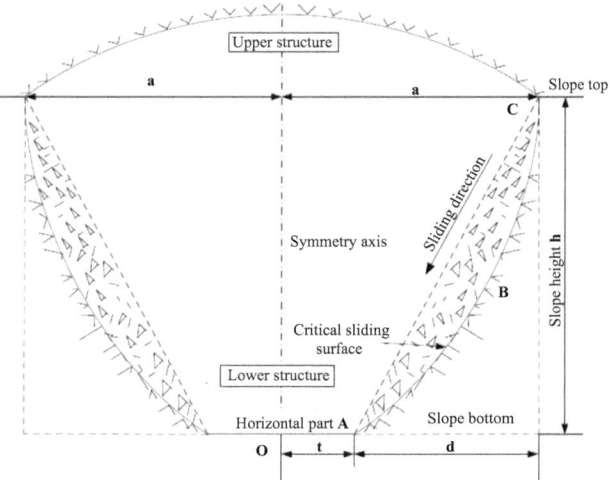

Figure 6 Schematics of cavern structure for shape design analysis (Wang *et al.*, 2013).

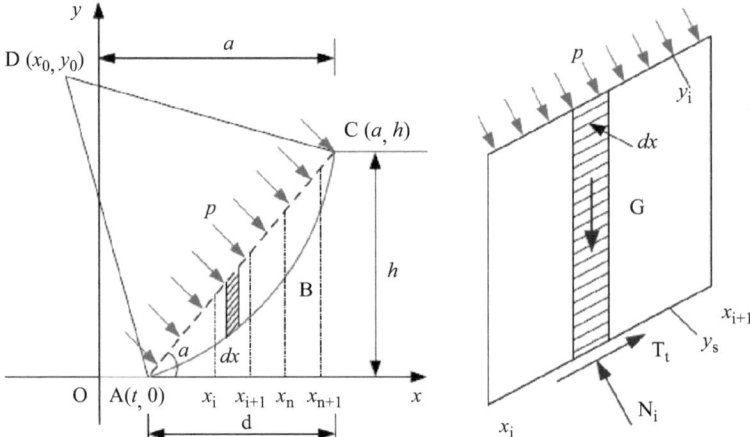

Figure 7 Model element for calculating the factor of safety against slope sliding at cavern lower structure (Wang *et al.*, 2013).

It was shown that the maximum compressed-air pressure determines the shape and dimension of cavern lower structure, while the minimum compressed-air pressure decides that of cavern upper structure.

3.2 Pillar width of multiple caverns

For a commercial scale operation, multiple caverns may be required for storing a sufficiently large amount of compressed air. One of the influential factors in designing the multiple caverns is appropriate pillar width, which corresponding to the unmined rocks between the neighboring excavated storage caverns (Figure 8). If the pillar thickness is designed to be too large, the design would be unacceptable from an economical viewpoint, since it results in wasting of rock salt resources. On the other hand, when the pillar is too thin, it may lead to the failure of the pillar and overall instability of storage caverns.

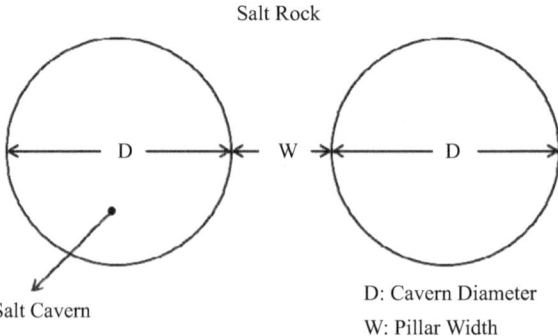

Figure 8 Pillar width between two adjacent caverns in salt rock (the pillar width, W is expressed as the times of the diameter, D).

In principle, the stress induced within pillars on the excavation of storage caverns is dependent on cavern dimensions. The cavern dimensions should be appropriately determined so that the stress concentration is limited to be less than the strength of the pillar rock. High internal pressure of the storage caverns and potential seepage of the compressed air into the pillar and ambient rock of the caverns may also affect the pillar design. Furthermore, rock salt is well known to show a time-dependent deformation increment so that a typical creep behavior should be included for a long-term operational design of the multiple caverns in salt rocks.

Wang *et al.* (2015) carried out a series of numerical analyses and proposed an allowable pillar thickness between two adjacent caverns in a salt mine, China. The pillar width of 2.0 to 2.5 times the cavern diameters was proposed from the investigations of various designing factors such as injection-production mode, pillar width, air pressure, depth, running time and cavern pressure difference on the vertical stress, displacement, plastic deformation zone, and safety factors of pillars. The vertical stress on the pillar naturally increased with decreasing pillar width, and asynchronous injection-production mode of two storage caverns may increase the pillar instability due to the larger pressure distance, which was more critically dominant in the narrower pillar conditions. Figure 9 shows how the plastic zones around the cavern are created,

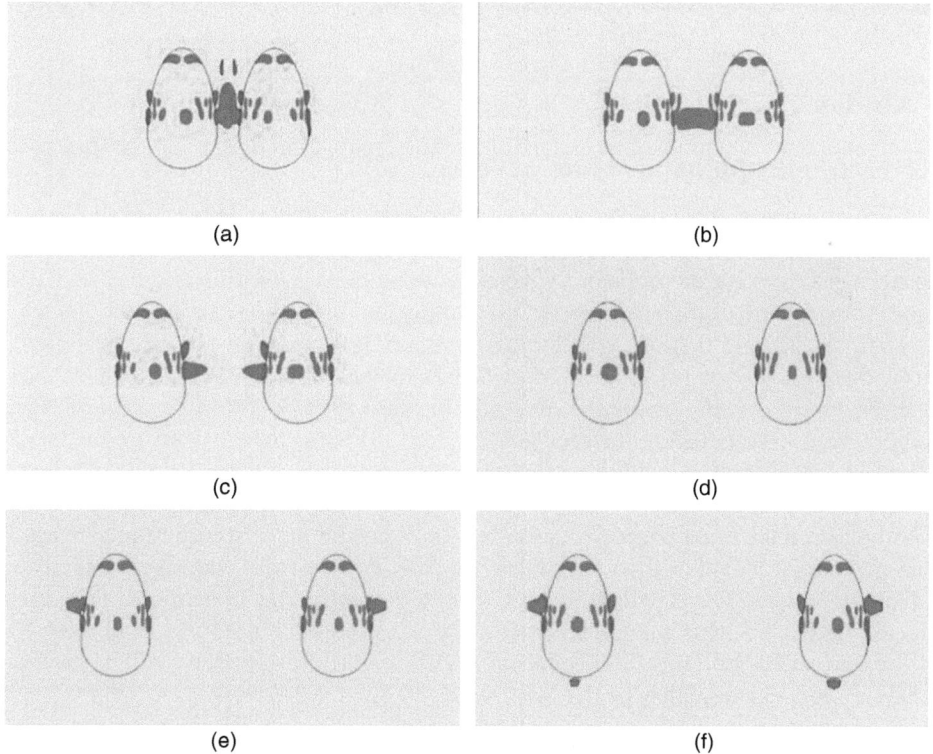

(a)

(b)

(c)

(d)

(e)

(f)

Figure 9 Plastic zones around two rock salt caverns for pillar width (W) as a function of the diameter (D) (Wang *et al.*, 2015).

depending on the pillar width as a function of the diameter (D). The less pillar width results in the larger connected plastic zones of higher pillar instability.

It is noted that the proposed pillar width in this study is somewhat smaller than that of typical design values of 1.5 to 3.0 times of the maximum diameter, since the creep deformation of the rock salt may lead to more uniform stress concentration within the pillar and reduce the pillar instability. However, the creep parameters are site-specific and these results may be different depending on the site specific geological formations of rock salts. For example, the results will be changing between at a rock salt dome and bedded salt rock with the interfaces of different deformation characteristics from the salt.

4 MECHANICAL AND HYDRAULIC CONSIDERATIONS IN A HARD ROCK

In this section, the mechanical and hydraulic issues and their solutions of CAES in a hard rock are described. There are two types of CAES in a hard rock, which can be classified into unlined and lined rock caverns. Unlined rock caverns provide a constant pressure condition in charging and discharging modes whereas lined rock caverns are operated under a variable pressure condition. The topic of unlined rock caverns will be handled with a focus on hydraulics and a case study in Japan. Lined rock caverns has key issues about site characterization, stability, air tightness evaluation, effects of EDZ (excavation damaged zone) and so on.

4.1 Unlined rock cavern

4.1.1 Hydraulics for unlined rock caverns

Unlined rock caverns provide a constant pressure condition in charging and discharging mode so that one fundamental requirement for preventing air leakage for the unlined rock storage cavern is that ambient hydrostatic water pressure within country rock must balance the pressure of injected and stored compressed air (Allen *et al.*, 1982).

A water pressure in balance with storage pressure is a common practice in designing hydrocarbon gas storage in unlined caverns and sometimes a so-called water curtain is created by injecting water above the cavern to increase allowable storage pressure (Liang & Lindblom, 1994). In the unlined rock cavern for CAES, the cavern pressure can be kept unchanged even during the withdrawal of the compressed air with the help of compensating water shaft that provides a constant back pressure of the stored air (Figure 10).

The balanced ambient pressure, however, may not be an absolute requirement for CAES if the rock mass is sufficiently impermeable. For example, the storage facility at the Norton Mine in Ohio will be located at 670 m depth, at a hydrostatic pressure of about 6.7 MPa, whereas the CAES operating pressure is planned to range from 6 to 11 MPa (Succar & Williams, 2008). In such a case, the sealing capacity will be provided by the very low permeability of the host rock.

Most igneous and metamorphic intact rocks such as granite, limestone and dolomite have the favorably low permeability. However, inter-connected fractures and joints embedded in these rocks can cause severe air leakage when they are drained of groundwater. To prevent the leakage, these discontinuities should be either saturated or filled

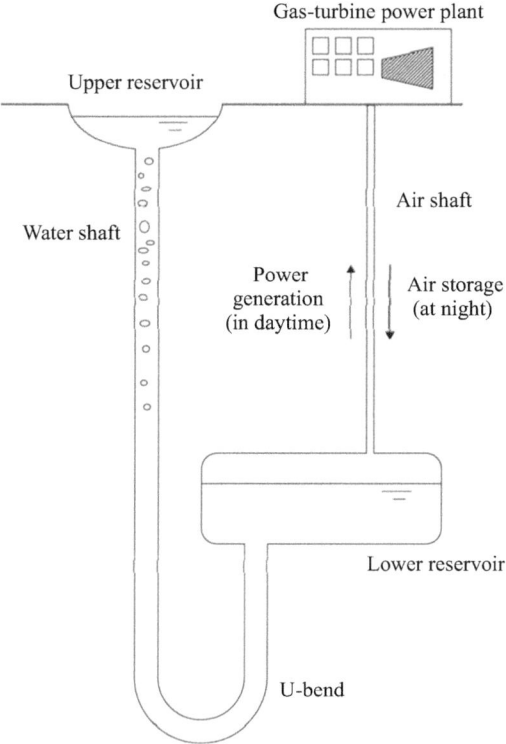

Figure 10 Conceptual design of unlined rock cavern for CAES with water compensation shaft (Nakagawa *et al.*, 2003).

with water with sufficient hydrostatic pressure or appropriately grouted to reduce the permeability. Komada *et al.* (1980) reported that overall hydraulic conductivity of rock mass must be characterized by less than 10^{-8} m/s for less than 2% of the total stored air volume per day. Kim *et al.* (2012) have performed numerical studies of the influence of in-situ rock permeability on the storage caverns at 100 m depth, and an acceptable air loss of less than 1% can be obtained when the rock mass permeability is lower than 10^{-17} m^2.

It is noted that in-situ measurement of rock mass permeability is not always possible and precise in practice. Thus, in principle, unlined rock caverns for CAES should be excavated at a depth where the hydrostatic pressure of ground water equals or slightly exceeds the storage pressure. In addition, the cavern should be located at a rock with low permeability and less fracture occurrence. Fractures can even be created in rock mass around the storage caverns during excavation. These fracture within the excavation disturbed or damaged zone should be either grouted to reduce permeability or filled with water of sufficiently high pressure to prevent substantial air leakage. And artificial water supply (*e.g.*, water curtain), if necessary, should be implemented for higher pressure compensation.

One peculiar phenomenon in unlined rock caverns, especially in pressure compensated CAES system (Figure 10), is a champagne effect. As air saturated water rises in the compensating water shaft, the solubility decreases causing the blowout of air as bubbles. To prevent this kind of hydraulic instability problem, a U-bend water seal between the reservoir floor and the water shaft to balance the buoyant head and overcome the inertia of the column can be constructed (Allen *et al.*, 1998).

4.1.2 Case study in Japan

The only field study of unlined rock cavern for CAES is a pilot-scale project in Japan, which was carried out in the early 2,000s at a former mine in Kamioka (Nakagawa *et al.*, 2003). The air was stored in a 51 m long, 3 m × 3 m cavern in a hard rock at 450 m depth but groundwater level was located at 180 m above the cavern, which corresponds to the depth of the cavern entrance. The cavern was excavated and reinforced mainly by rock bolts so that the stored compressed air is directly faced with surrounding rock mass. The storage cavern was inclined at 1:5, and a concrete plug was installed to prevent air leakage to the access cavern (Figure 11).

The results of a pressurization test reported in Nakagawa *et al.* (2003) indicated that subsequent air leakage occurs only after the cavern pressure becomes greater than initial groundwater pressure of the surrounding rock mass, which distinctly shows that a hydraulic confining method do work for CAES in unlined rock caverns. Moreover, cyclic compression and decompression of the cavern air resulted in different air leakage behaviors, and the air leakage rate during the decompression mode was greater than that of the compression even at the same pressure level, which can be explained by the lower degree of water saturation around the rock mass surrounding

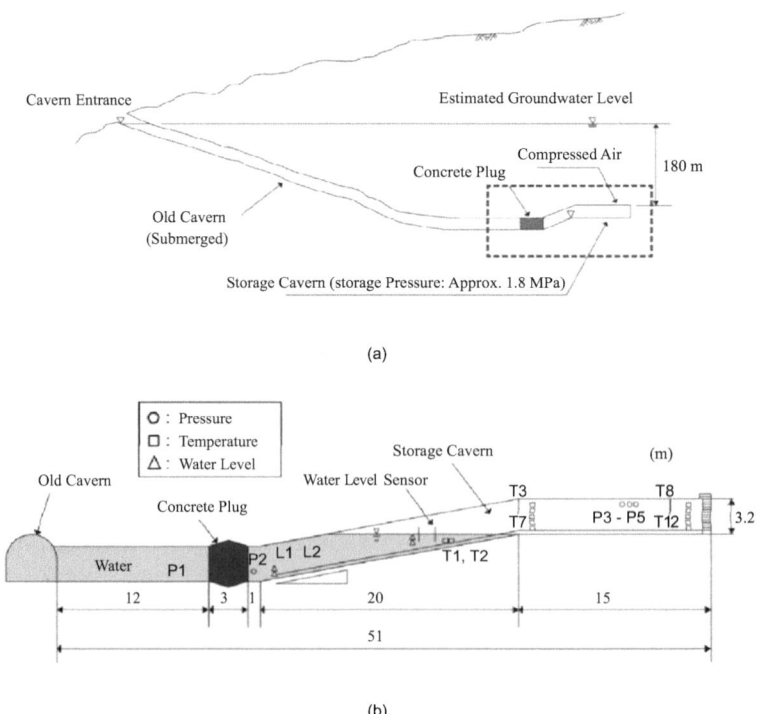

(a)

(b)

Figure 11 Schematics of the pilot testing for CAES in unlined rock cavern in Japan (Nakagawa *et al.*, 2003).

the storage cavern. It was also noted that the critical pressure, which air leakage is subsequently increasing and was equal to the initial ground water pressure, a bit decreases during the cyclic operation, but recovers back to the initial water pressure level when the compression is followed by the decompression after a sufficiently long time.

4.2 Lined rock cavern

4.2.1 Evaluation of air tightness

One of the key challenges in underground storage of compressed air in LRC is the risk of air leakage from the storage caverns. Pressurized air can leak through an initial defect in the inner impermeable sealing liners (*e.g.*, imperfect welds of steel liners) that may be created at the time of construction or later by induced failures of the liners due to repetitive air compression and decompression. The air leakage from storage caverns may cause a reduction in CAES efficiency and also increase the risk of the structural instability of the facility as a result of its highly concentrated and localized occurrence. Thus, detection of air leakage at underground LRCs for CAES is important for the successful implementation of CAES and an improvement of the overall safety of CAES facilities.

Kim *et al.* (2012) conducted a parametric sensitivity study to investigate impact of permeabilities of LRC components on the performance of the CAES system through numerical simulations. These components consist of inner linings and surrounding rocks. Aforementioned defects and fractures, if any, within the linings and rocks are considered as so-called equivalent permeability concept in their study, which is based on a continuum model approach.

Assuming an ideal gas, the total mass stored in a storage cavern at a certain pressure and temperature can be estimated from the ideal gas law as

$$m = \frac{P V_{cavern}}{R_{air} T} \tag{2}$$

where m is mass of gas (kg), P is absolute pressure (Pa) within the cavern, V_{cavern} is cavern volume (m^3), R_{air} is the specific gas constant for air (= 286.9 J/kg K), and T is absolute temperature (K). The total volume of the open cavern inside the concrete lining is considered constant during the compression and decompression cycles.

During the compression and decompression cycles, pressure and temperature within the cavern, that is varying nonlinearly with time, are calculated from the numerical simulations and then a leakage rate can be determined according to

$$\Delta m = m_1 - m_2 = \frac{V_{cavern}}{R_{air}} \left(\frac{P_1}{T_1} - \frac{P_2}{T_2} \right) \tag{3}$$

where subscripts 1 and 2 indicate initial and later state, respectively.

Figure 12 presents the evolution of leakage rate and daily leakage percentage for various combinations of lining and rock permeability. The results show that leakage of less than 1% would be achieved if permeability of the concrete lining were less than 1×10^{-18} m^2, even if the permeability of the rock were as high as 1×10^{-15} m^2. A less than 1% leakage rate is also achieved if the rock mass permeability were less than 1×10^{-17} m^2, even with a comparatively permeable concrete lining.

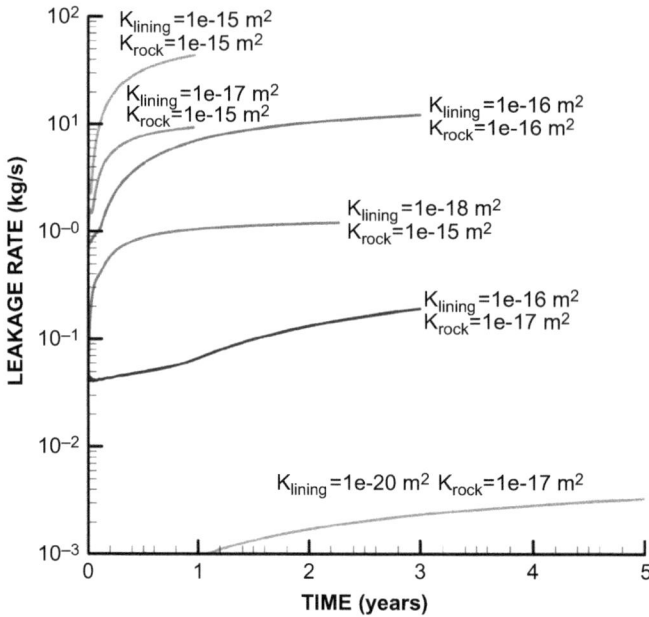

Figure 12 Calculated evolution of daily leakage percentage for different combinations of concrete lining and rock permeability (Kim et al., 2012).

Leakage rate is also influenced by the water saturation as well as the permeability in the concrete lining. The impact of the water saturation in the lining was investigated from the same numerical simulation, and the leakage rate increases with decreasing water saturation (; increasing gas saturation) in the concrete lining, which can be explained by a relatively permeability concept.

The air leakage from storage caverns may not only cause a reduction of CAES efficiency but can also increase the risk of a structural instability of the facility as a result of its highly concentrated and localized occurrence. Thus, detection of air leakage at underground LRCs for CAES is important for the overall safety of CAES facilities.

Rutqvist et al. (2012) proposed that a shut-in test could be an effective method to detect air leakage and to calculate the leakage rate. Figure 13 demonstrates one simulation of the shut-in method after cracking occurred at the 3rd day cycle and leakage could be suspected. By reading P_1, P_2, T_1 and T_2 accurately, the leakage rate can be calculated from Equation 3. In this particular case of $T_1 = 290.32$ K, $T_2 = 288.37$ K, $P_1 = 7.62$ MPa, $P_2 = 7.55$ MPa, $V_{cavern} = 19.63$ m^3 and $R_{air} = 286.9$ J/(kg K), we can determine that an air mass of 4.47 kg has been lost from the CAES system over 100 h, leading to a leakage rate of 1.24×10^{-5} kg/s (1.07 kg/day). The shut-in test was useful for detecting air leakage and for estimating the air leakage rate, but it may have limitation in identifying the location of air leakage. Most importantly, regular CAES operation should be stopped to apply the method and check leakage.

As an alternative of a shut-in test, Kim et al. (2015) demonstrated by numerical simulations that a pressure monitoring at the interface between lining and surrounding rock mass can be useful in detecting a leakage and identifying the leakage location. They

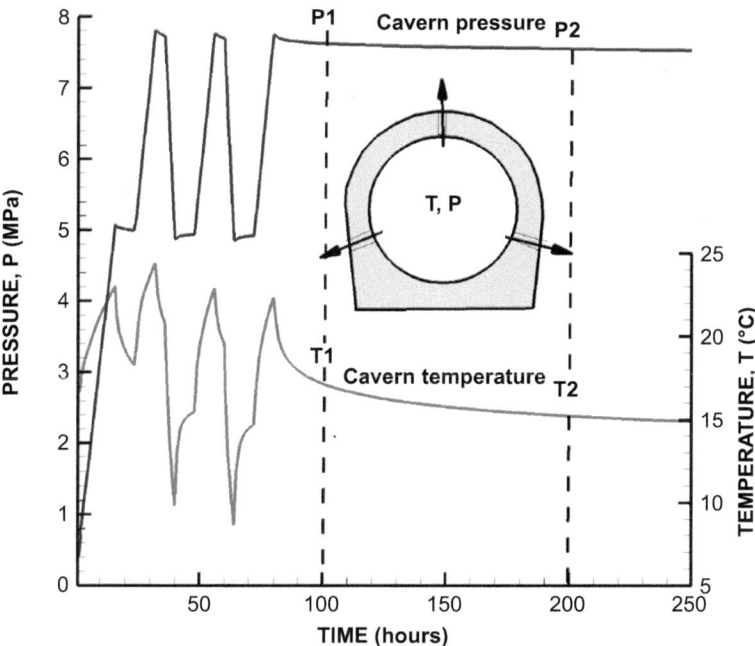

Figure 13 Calculated evolution of cavern pressure and temperature during simulated shut-in test for the case of air leakage through the cracks in the concrete lining (Rutqvist *et al.*, 2012).

noted that the contact between the concrete lining and the rock could be imperfect, providing a permeable interface and dominant leakage pathway, which was observed in their previous experiments (Kim *et al.*, 2014a). Once the leakage occurs through the lining, the pressure inside the cavern quickly penetrates and spread along the permeable interfaces.

When leakage occurs, the pressure data at the monitoring locations along the interface display a synchronized behavior with the cavern pressure, but with a certain delay time. The delay time should be proportional to the distance between the monitoring locations. By calculating the cross-correlations between the cavern pressure and each pressure monitoring history along the interface and by quantifying the delay time, the leakage can be characterized. For two pressure monitoring histories, $P_1(t)$ and $P_2(t)$, the cross correlation coefficient $R_{P_1P_2}(\tau)$ is defined as

$$R_{P_1P_2}(\tau) = \frac{1}{T}\int_0^T P_1(t)P_2(t+\tau)dt \tag{4}$$

where τ and T are the time lag and integration time period, respectively.

In this case, $P_1(t)$ is the cavern pressure history, and $P_2(t)$ is one of the pressure measurements along the interface. Then, calculated $R_{P_1P_2}(\tau)$ shows a peak value at $\tau = t_{peak}$, which represents that the pressure measurements are the most correlated to the cavern pressure with this amount of time delay.

Figure 14 illustrates one simulation of leakage characterization through pressure monitoring at the interface. The locations of sensors for pressure monitoring and the monitored

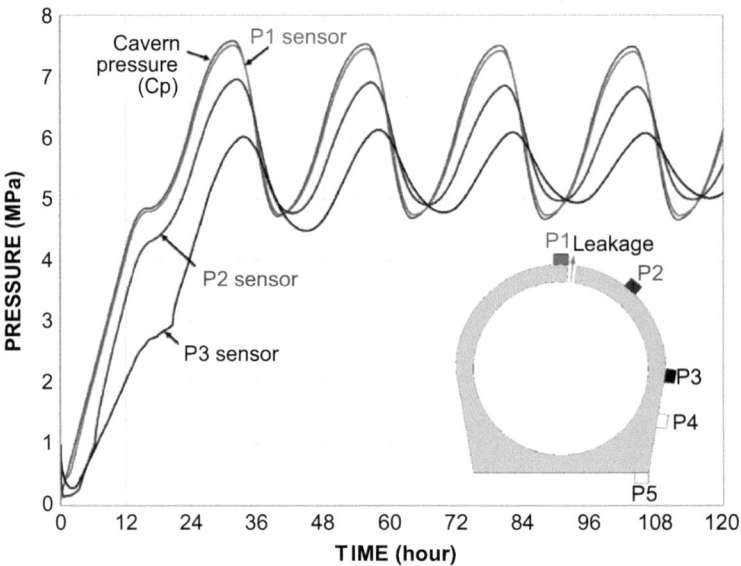

(a) Cavern pressure and pressure evolution at the monitoring sensor locations

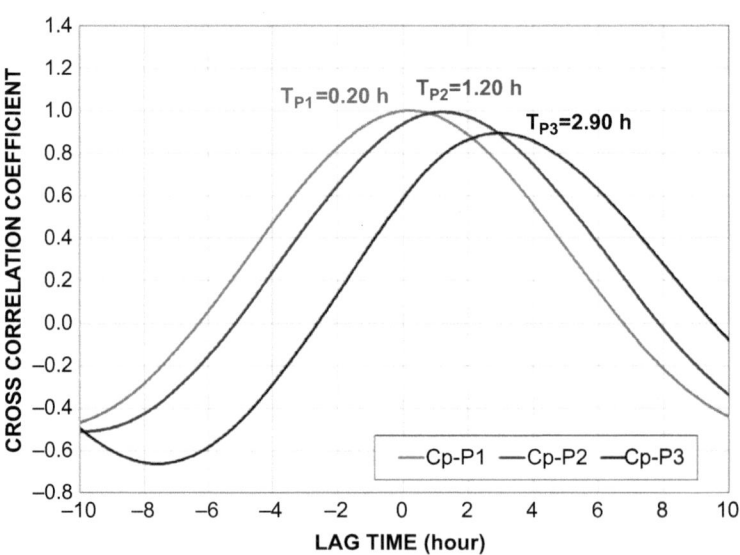

(b) Cross-correlation coefficients and the peak delay times

Figure 14 Simulation of leakage characterization through pressure monitoring (Kim *et al.*, 2015).

pressure responses at each location was shown in the Figure 14(a). The pressure monitoring sensors for leakage detection were located at the top (P1), at 45 degrees upwards from the side wall (P2), at the side wall (P3), at 45 degrees downward from the side wall (P4) and at the bottom (P5) of the LRC at the interface between the lining and surrounding rock. In the simulation, it was assumed that leakage occurs on the top of the cavern. When the cavern pressure fluctuates between approximately 5 and 8 MPa, the pressure histories at different monitoring locations show a cyclic evolution with both delayed peak time and smaller pressure amplitude, depending on the measurement location. The results of cross correlation (Figure 14(b)) indicate that the delay time of individual pressure responses provides a clear quantitative parameter for analyzing leakage. The shortest delay time of 0.2 hour can be obtained from the sensor at P1 that is located closest to the leaking point. In addition, the ratio of distances between the three closest sensor locations (P1:P2:P3) from the leakage point are 1:5:14.3 (=0.3 m:1.5 m:4.3 m), and the calculated ratio between the delay times are 1:6:14.5 (=0.2 h:1.2 h:2.9 h). Thus, there exists a strong proportional relationship between the peak delay time and the distance of the measurement from the leakage point, which allows one to localize the air leakage.

From these examples of leakage monitoring, a leakage warning system can be designed and earlier localization of the leakage point in the LRC can be accomplished in terms of pressure monitoring at the interface without stopping operation of CAES, and quantitative leakage rate may also be estimated by a subsequent shut-in test.

4.2.2 Sensitivity analysis in LRC at a shallow depth

The principle idea behind the LRC storage concept is to rely on surrounding rock mass to absorb all the pressure in caverns transferred through concrete linings, and to install an additional air-tight liner (usually steel liners) inside the concrete linings to enclose the air in the cavern and to prevent leakage. The LRC technology thus can allow a significant increase in maximum operating pressure over unlined storage cavern concept, since the air in storage is completely contained due to an impervious liner. However, the increase of maximum operating pressure in caverns may generate unfavorable instability of storage caverns. One of the important design aspects of underground pressurized caverns is the safety against ground uplift.

The important design parameters to be taken into consideration in estimating ground uplift above the LRCs, are the depth of the cavern, the cavern geometry including shape, diameter, and length, the internal pressure level, in-situ stress conditions, rock mass strength and the layout of caverns and so on. For example, in case that overburden rock masses above the cavern are not sufficiently strong enough or located at too shallow depth to resist the upward lifting pressure due to the internal high pressurized compressed air in a cavern, crack may generate at the cavern periphery and result in overall instability of the cavern.

Kim *et al.* (2013a) investigated the effect of these design parameters of underground storage caverns on the safety against ground uplift in terms of a limit equilibrium analysis using a simple mathematical solution within which failure plane is assumed to be straight upward to ground surface (Figure 15).

Ground uplift occurs when total upward lifting force (F_u) exceeds the sum of weight (W_r) of overburden rock mass above the cavern and shear resistant force (τ_r) at the failure plane. Shear resistant force on the failure plane was considered to be the summation of the resisting force by cohesion (τ_c) and friction angle (τ_f). Total upward lifting force consists of the lifting force caused by the pressure (P) in cavern, buoyant force against storage cavern (F_{bc}) and the overburden rock mass (F_{br}).

Figure 16 shows that the calculated factor of safety against ground uplift is increasing with the increment of storage depth as well as with the decrement of the storage pressure.

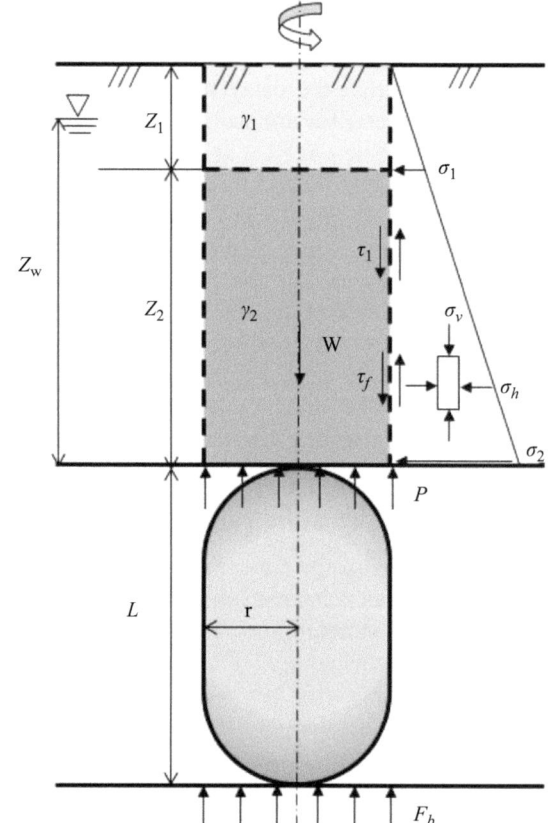

Figure 15 A straight upward failure model of a silo-type storage cavern for ground uplift evaluation (Kim et al., 2013a). (P: Max. storage pressure in MPa, r: Radius of storage cavern in m, L: Length of storage cavern in m, Z_1: Thickness of upper weathered rock in m, Z_2: Thickness of lower fresh rock in m, Υ_1: Unit weight of weathered rock in kN/m^3, γ_2: Unit weight of fresh rock in kN/m^3, γ_w: Unit weight of groundwater in kN/m^3, Z_w: Groundwater table above cavern in m.

In this specific example, factor of safety greater than 2.0 would not be obtained, unless the storage depth is approximately 4 times larger than the cavern radius and the pressure in the cavern is less than 4 times of the vertical stress by overburden. Cavern shape also impacts the stability, and the storage geometry of the silo type has relatively higher factor of safety rather than the horizontal cavern type in these cases. Linear relationship is observed between in-situ stress ratio and factor of safety against ground uplift. The less stress ratio indicates that the normal force applied to failure surface (; horizontal stress) is decreasing and consequent reduction of shear resistant force on failure plane, and finally resulted in the decrease of factor of safety. However, it should be noted that higher horizontal stress and lower in-situ stress ratio may not be preferable in the viewpoint of cavern stability of which shape is the silo type with vertically longer length.

Figure 17 shows the results of relative sensitivity of all the parameters, which are obtained by perturbing the parameter values, including storage pressure, cavern size, overburden depth, in-situ stress state, and rock physical properties such as cohesion and friction angle. Nearly horizontal line indicates that small change of the design parameter

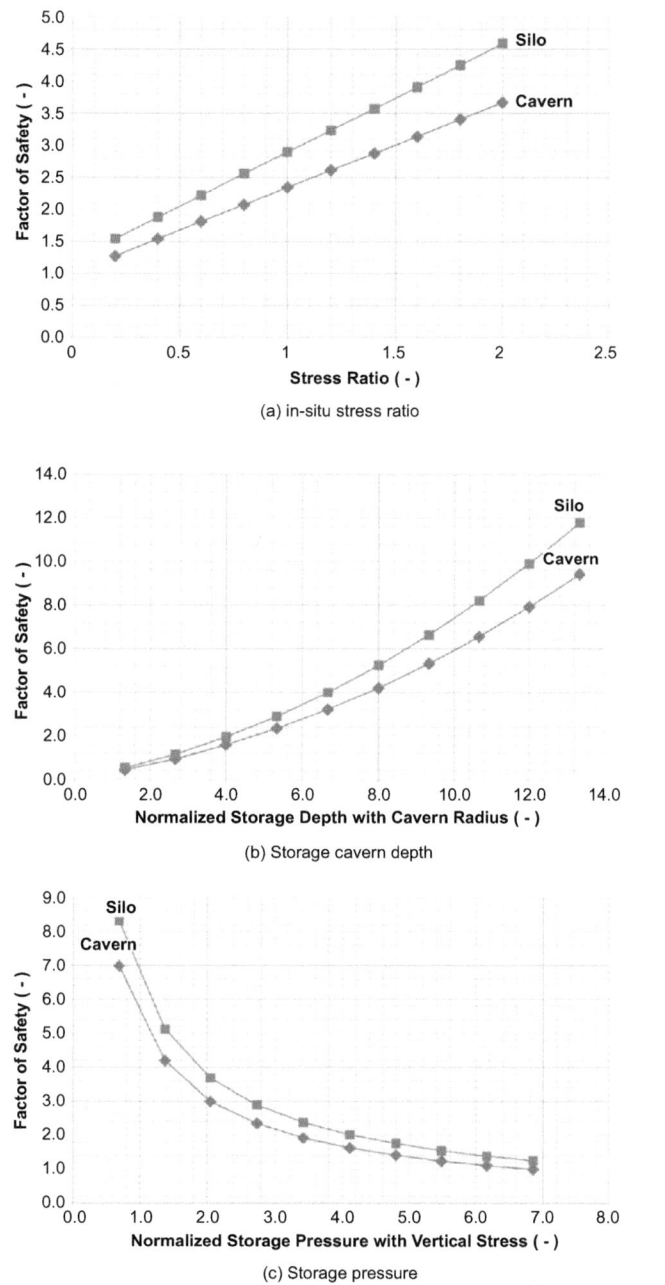

Figure 16 Effect of parameters on factor of safety against ground uplift (Kim *et al.*, 2013a).

do not have much impact on the factor of safety. Groundwater table above storage cavern showed sub-horizontal line and the least influence on ground uplift among these parameters. The thickness of fresh rock, maximum storage pressure and radius of storage cavern showed relatively steeply dipping lines, indicating higher correlation with ground

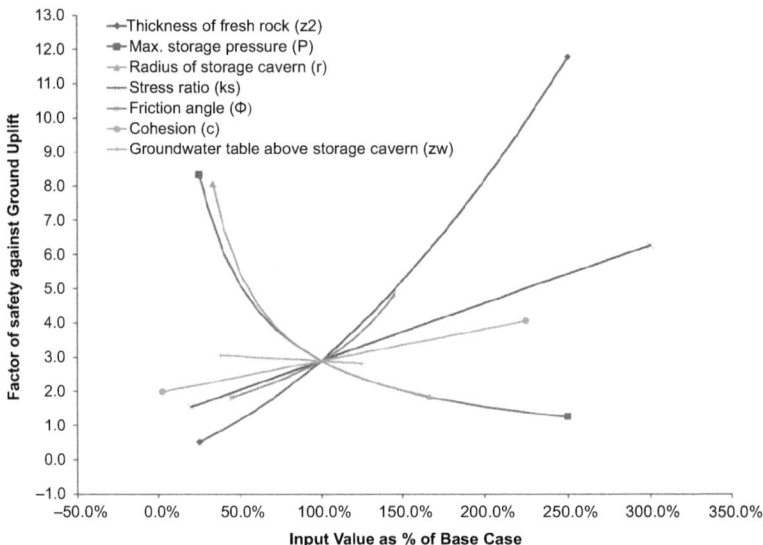

Figure 17 Factor of safety of each parameter against ground uplift (Kim *et al.*, 2013a).

uplift. And maximum storage pressure and radius of storage cavern showed nonlinear negative relationship.

Acknowledging that the simplified model provides rather a conservative results and is applicable for a single independent caverns where the failure plane above the cavern becomes close to a straight upward, these results may provide useful information in selecting a candidate site and preliminary design screening stage for pressurized underground caverns before detail design based on more complex numerical models.

4.2.3 Reliability-based stability analysis for LRC

As an alternative of a conventional deterministic analysis, where uncertain physical and mechanical properties of materials are conservatively assigned as a single fixed value determined by a designer's engineering judgment, probabilistic approaches can be useful for reliability-based design optimization because geotechnical uncertainties are quantified by statistical probability distributions and thus the reliability of the structures under investigation is assessed in terms of the probability of failure in a limit-state. In this section, we present an example of a probability-based design for the inner containment structures (*i.e.*, steel and concrete liners) used for CAES. For computational efficiency, a point estimation method (PEM) combined with numerical analysis is employed rather than simulated-based Monte Carlo methods which usually demands a large number of numerical simulations and significant calculation times (Park *et al.*, 2013).

Figure 18 demonstrates the specifications of LRC considered in the analysis. The storage cavern is planned to be 5 m in diameter and located at a depth of approximately 100 m. The liner system of the storage cavern is designed with support from steel and concrete liners. The role of the steel liner, as an impermeable barrier to air storage, is to ensure absolute air-tightness at a high internal pressure; therefore, cracks in the steel

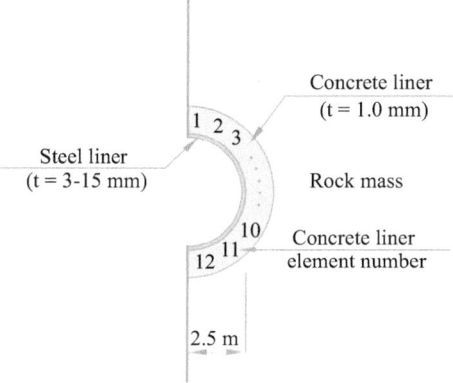

Figure 18 Liner system for CAES at a pressure of 15 MPa (Park *et al.*, 2013).

liner are absolutely not permissible. However, the objective of the concrete liner, which is placed between the steel liner and the surrounding rock mass, is to uniformly transfer the compressed-air pressure to the surrounding rock mass and act as a smooth base for the steel liner. Thus, some minor cracks in concrete are permissible, if the cracks are smaller than an allowable tolerance. In this feasibility study, the concrete liner was designed to be 1 m thick.

In the analysis, the stress ratio, the properties of the rock mass, and the friction angle of inter-face between the concrete and steel linings are considered to be random variables, and their statistical distributions are determined through the 3σ rule (Park *et al.*, 2013).

To examine what types of steel liners could be applicable at a high air pressure, we assessed the reliability of the steel liner in terms of the probability of failure by varying the tensile strength and thickness of the steel liner. The applied tensile strengths for the steel liner were 350, 400, 450, and 500 MPa, and its thicknesses were 3, 6, 9, 12, and 15 mm. Figure 19 shows the relationships between the steel liner's thickness and probability of failure at different tensile strengths of the steel liner. From this figure, it was observed that the probability of failure for the steel liner tends to decrease as the thickness of the steel liner increases. The colored regions in this figure indicate areas where structural design information is available. As indicated in Figure 19, the required minimum thicknesses of steel liners with tensile strengths of 400 and 450 MPa were 11.68 and 4.12 mm, respectively, for a target probability of failure of 10^{-3}, and 13.66 and 6.03 mm, respectively, for a target probability of failure of 10^{-4}. In the present LRC problem, the target probability of failure for the steel liner was set as low as 10^{-3} (0.1%), accounting for structural risks due to cavern pressurization; the steel type with an ultimate tensile strength of 400 MPa, was applied, which is one of the most commonly used steel strengths. In the end, the steel liner was designed to be 14 mm in thickness, with consideration of corrosion of the steel liner due to the compressed-air pressure in the cavern and its abrasion due to interaction with the concrete liner; the margin for both corrosion and abrasion was 1 mm.

Figure 20 illustrates the distribution of the probability of exceeding the 1.5 mm crack width tolerance in the concrete liner, and X and Y in this figure denote the horizontal and vertical axes, respectively. The probabilities of exceeding the 1.5 mm crack width tolerance

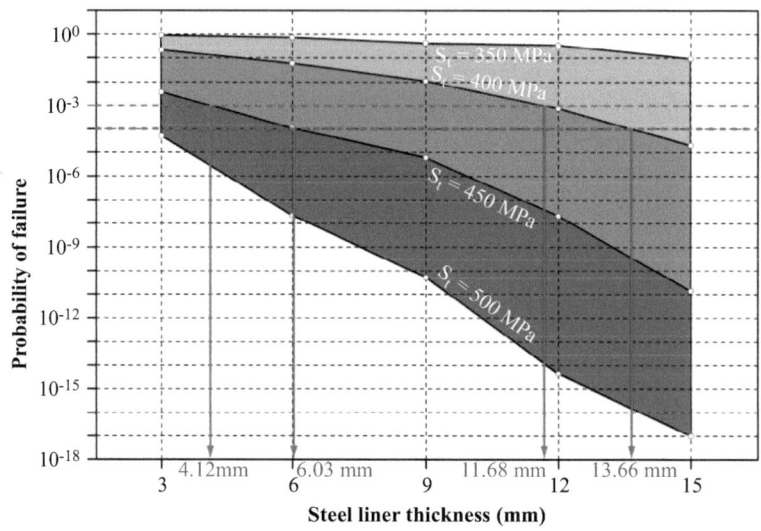

Figure 19 Design chart for determining strength and thickness of the steel liner for LRC. S_t is the ultimate tensile strength of steel (Park *et al.*, 2013).

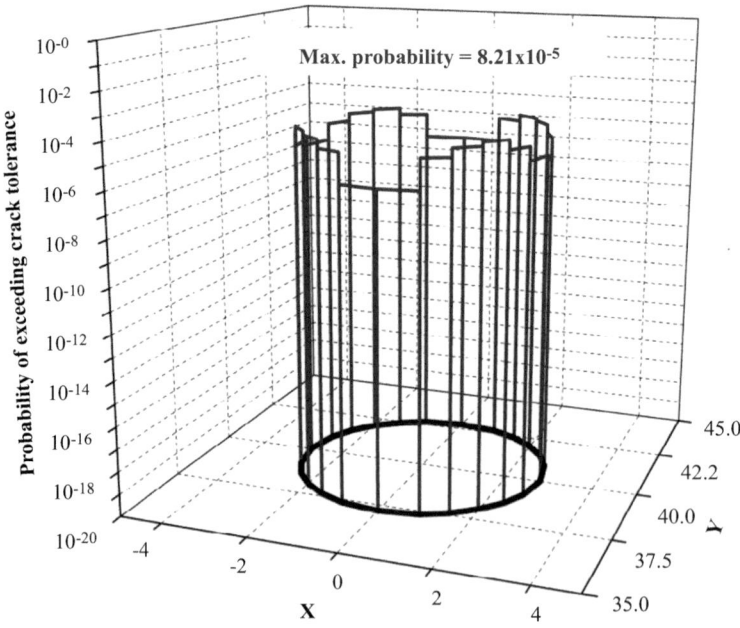

Figure 20 Probabilities of exceeding the 1.5 mm crack width tolerance of the concrete liner elements (Park *et al.*, 2013).

were less than 1.0×10^{-4} (the maximum probability of exceeding 1.5 mm = 8.21×10^{-5}), which suggests that the cracks in the concrete liner very rarely exceeds a 1.5 mm width. From these probabilistic results, we may conclude that the concrete liner is very likely able to function as an intermediate layer for uniform load transfer to the surrounding rock mass.

Table 1 Effects of different properties of EDZ and concrete on the displacement and stress in the concrete linings of LRCs for underground CAES.

Cases	C1	C2	C3	C4
Young's modulus of EDZ (GPa)	7.0	35.0	7.0	35.0
Young's modulus of concrete lining (GPa)	23.0	23.0	35.0	35.0
Compliance of EDZ (GPa/mm)$^{-1}$	0.086	0.017	0.017	0.086
Compliance of concrete lining (GPa/mm)$^{-1}$	0.022	0.022	0.014	0.014
Total compliance (GPa/mm)$^{-1}$	0.107	0.039	0.031	0.1
Radial displacement at the inner surface of concrete lining (mm)	0.73	0.5	0.46	0.66
Tangential effective tensile stress in the concrete lining (MPa)	4.07	2.09	2.4	5.26

Note: Thickness of EDZ and concrete lining were 0.6 and 0.5 m, respectively

From this example, we demonstrated that how we may evaluate the reliability of LRC with steel and concrete liners for CAES and derive probability-based structural design information such as the required minimum thickness of the steel liner at different strength and target probabilities of failure and the probability of exceeding a crack width tolerance of the concrete liner.

4.2.4 Characterization of EDZ for LRC

Construction of underground excavations in geological formations usually results in an EDZ, where significant changes in geomechanical, hydrogeological, and thermal properties may occur as a result of in situ stress redistribution and damage from the excavation process. Within the EDZ, where both microcracking of intact rock and failure of rock mass are induced, the permeability and deformation modulus may increase at potentially several orders of magnitude higher than that of the undamaged ambient rock. Consequently, occurrence of EDZ may affect the performance of CAES, particularly the increment of deformation modu-lus can result in geomechanical instability of storage caverns.

Kim *et al.* (2013b) demonstrated that the impact of the EDZ on the stability of CAES storage caverns by simulating and investigating radial displacement and tangential stress in concrete linings of LRC. Table 1 presents the calculated evolution of stress with the concrete lining and the radial displacement of the inner surface of the concrete lining under different compliance conditions of the EDZ and concrete linings. The compliance is inversely proportional to the elastic Young's modulus, which is an indicator of stiffness. In the less compliant conditions of the EDZ (C2 and C3), this was 0.6 m/35 GPa = 0.017 m/GPa, whereas in the more compliant conditions (C1 and C4), it was 0.6 m/7 GPa = 0.086 m/GPa. The total compliance, which is experienced from the inner surface of the lining, could be estimated by adding the compliance of concrete lining to that of the EDZ. The increase in radial displacement is related primarily to the increased compliance of the total system. It is noted that the maximum radial displacement is not so significant from an engineering viewpoint, ranging from 0.46 to 0.73 mm. The magnitude of tangential tensile stress in the concrete lining ranging from 2.4 to 5.26 MPa could be more important for potential tensile fracturing and consequent air leakage through it. It is observed that that tensile stress increased with a more compliant EDZ, whereas it decreased with a more compliant concrete lining. Consequently, the highest tangential stress occurred for the case of a relatively compliant EDZ with a stiff concrete lining.

A parameter study showed that the radial displacement and tangential stress in the concrete lining could be effectively reduced if the compliance of the EDZ, defined by EDZ thickness multiplied by EDZ compressibility, could be minimized. It was also noted that a reduction of compliance (increase in stiffness) of the concrete lining resulted in the negligible decrease of radial displacement, but a significant increase of tangential stress that could potentially result in tensile fracturing through the lining.

This finding shows that the most favorable design for reducing tensile tangential stress in the lining would be a relatively compliant concrete lining and relatively stiff (uncompliant) rock that is not significantly softened in the EDZ. Because the EDZ compliance depends on its compressibility (or modulus) and thickness, care should be taken during drill-and-blast operations to minimize the damage from the blasting. The previous analysis (Rutqvist *et al.*, 2012) showed that the benefit of maintaining an inner impermeable seal that would reduce the effective tensile stress in the concrete and thereby tend to prevent tensile fracturing. Thus, for CAES in LRCs, we can offer some general recommendations to minimize damage induced by excavation, to reinforce EDZ right after excavation, and to employ an inner impermeable seal. All these actions would help to minimize the risk for pressure-induced failure in the concrete lining.

5 COUPLED BEHAVIOR IN COMPRESSED AIR STORAGE

Thermodynamic and geomechanical behavior of CAES in LRC can be effectively simulated by Thermal-Hydraulic-Mechanically (THM) coupled model analysis. Permeability of concrete linings and rock masses, and air and/or water flow through these medium are important in assessing air leakage. And pressure and temperature in the cavern, as well as heat transfer through the lining and rocks should be considered for energy efficiency analysis. In addition, strength and deformation characteristics of EDZ, rock mass and concrete lining are important for geomechanical stability analysis. Most importantly, these processes are not independent with each other, but highly correlated with each other.

Another advantage of employing the THM coupled analysis is that both inside the cavern and surrounding lining and rock mass system can be examined using a single numerical simulation. Of course, we may obtain more accurate pressure and temperature evolution using heat transfer analysis including only inside the cavern, and may have more detail information on the geomechanical deformation from stress analysis modeling only outside of the cavern. However, the current coupled analysis can effectively simulate thermodynamic as well as geomechanical behaviors of both inside and outside of the LRC from a single simulation.

In modeling coupled thermodynamic and geomechanical behavior, it is important to capture the entire construction and operation sequence of LRCs. The CAES system in LRCs can be simulated by the following steps:

1)   Initial simulation to achieve steady-state vertical gradients of pressure, temperature and stress in the rock mass as initial conditions before excavation.
2)   Excavate the cavern and keep it open for a certain period (*e.g.*, 1 week) at atmospheric pressure within the open cavern, to allow the cavern to converge

mechanically and to achieve new distributions of pressure, temperature and stress in the rock mass.

3) Install the concrete linings at a specific initial saturation (*e.g.*, 70%), atmospheric air pressure, and near-zero effective stress, and to keep atmospheric air pressure in the cavern.

In step 1 initial pressure, temperature, and stress gradients are set for a cavern depth (*e.g.*, 100 m), with the identified water table information (*e.g.*, close to the ground surface). Initial temperature was set using a vertical temperature gradient (*e.g.*, 0.03 °C/m) and with a constant temperature at the ground surface. After step 2, a pressure sink develops around the excavation, with atmospheric pressure within the excavation and a small inflow of water into the cavern. Finally, after step 3, the initial conditions before start of the CAES operation is achieved. This includes atmospheric cavern pressure, a concrete lining an initial saturation of *e.g.*, 70% and zero stress, a fully or partially saturated, depending on the groundwater level, surrounding rock mass with excavation induced gradients of pressure, and stress concentrations around the cavern.

After this initialization, the simulated CAES operation is started by injecting and withdrawing air from the cavern for various modeling cases. Daily compression and decompression cycles for a typical CAES operation is simulated with a pre-designed cavern pressure range (*e.g.*, from 5 to 8 MPa), injection rate and temperature of compressed air, which are subject the specification of air compressor in hand. The daily cycles of air compression and withdrawal is also simulated by a pre-defined operation scenarios (*e.g.*, first injecting air at a constant rate for 8 hours, storing it for an additional 4 hours, then producing at a constant rate for 4 hours, and finally waiting for another 8 hours till the start of a new compression cycle).

In the following, we present an energy-balance analysis of the CAES operation, which can be obtained from the current THM coupled analysis, in order to understand the energy loss and the relative contributions from air leakage and heat loss to the overall energy balance.

From the first law of thermodynamics, the change in total energy stored in the CAES underground cavern (ΔE_s) can be expressed as the summation of the change in internal energy (ΔE), the work done (ΔW) by injected compressed air, and the sum of outflows by production, air leakage, and heat transfer (ΔQ):

$$\Delta E_s = \Delta E + \Delta W + \Delta Q \tag{5}$$

Since the volume change in the underground cavern is restricted and very small, the internal energy is determined by air-mass flow, specific heat, and air temperature. The work done by the air movement and leakage of the stored air is the product of pressure (P) and volume (V) changes in the air, which results in the function of air mass flow and temperature by the ideal gas law. Heat exchange between the air filled cavern and the concrete lining occurs by two distinct mechanisms: heat conduction (Q_c) and advection (Q_a) with air flow.

$$\Delta E = C_{air} \cdot T \cdot m \tag{6}$$
$$\Delta W = P \cdot V = R_{air} \cdot T \cdot m \tag{7}$$
$$\Delta Q = \Delta Q_c + \Delta Q_a \tag{8}$$

where m (kg/s) is the air mass flow, C_{air} (J/kg K) is the specific heat at the constant volume, T(K) is the temperature, P (Pa) is pressure, V (m³) is volume, ΔQ_c (J/s) is heat conduction rate, and ΔQ_a (J/s) is heat advection rate.

Then, Equation 5 can be written as

$$C_{air}T_s\Delta m_s = \frac{C_{air}}{R}V_s\frac{\partial P_s}{\partial t} = (C_{air} + R_{air})T_m m_m - (C_{air} + R_{air})T_l m_l + (\Delta Q_c + \Delta Q_a)$$

(9)

where Δm_s is the rate of change of stored air mass, m_m and m_l are the mass flows (kg/s) by air movement due to injection and production as well as air leakage, and T_s, T_m, and T_l are the temperatures (K) of stored air, injected/produced, and leaking air, respectively. Here, V_s is the volume of stored air, equal to cavern volume. T_m is equal to the injection temperature (21° = 294.15 K) during the compression, and equal to cavern temperature, i.e., temperature of stored air T_s during decompression. The temperature of leaking air (T_l) is taken to be equal to the temperature of stored air (T_s).

In the simulation, TOUGH-FLAC simulator (Rutqvist, 2011) is used to determine the values of each term in Equation 5.5, including the rate of mass change (Δm_s), the storage pressure (P_s) and temperature (T_s) of air, mass flow rate of air movement (m_m), and leakage (m_l).

Figure 21 presents the case of a leaky concrete lining of which permeability was prescribed intentionally higher than practical values in order to show the results of the present energy balance analysis more distinctly. Figure 21(a) is the rate of energy change in storage cavern, which corresponds to the left-hand side of the Equation 9. Figure 21(b) is the temperature evolution during a daily cycle, which shows that heat transfer may occur from the cavern to the lining during the compression, and vice-versa during the decompression due to the temperature gradient between the cavern and the lining. The hatched area in Figure 21(c) is the rate of energy change during injection and production subtracts the rate of energy change in storage cavern, which corresponds to the energy-loss rate including both heat conduction and advection by air leakage. The energy loss by the air leakage only is shown in the hatched area in Figure 21(d).

More quantitative energy balance analysis can be obtained by time integrating each term in Equation 9. In this example, the energy loss by heat conduction is much greater than that of air leakage, but both are within the same order of magnitude. Due to the energy loss by air leakage, total energy loss during the compression phase is as much as 36% of total injected energy, and approximately 26% is recovered through heat conduction during the decompression phase. Thus, ultimately 10% of the total injected energy is lost during a single cycle. It is noted that heat conduction between the stored air and the concrete lining plays a major role in the CAES energy balance and is even more significant than air leakage in this example. However, it should be emphasized that most heat loss to the liner during compression is regained during subsequent decompression.

The influence of injection temperature and thermal insulation condition of the concrete lining on the overall energy balance was also investigated in the same way. If the injection temperature is reduced to similar to the ambient temperature of surrounding rocks, the heat loss through the lining as well as total energy can be reduced to 1 % of the total energy injected into the cavern during the compression phase. As the injection temperature increases, the heat loss becomes more significant, and the total energy loss increases to 11 % through the cycle at the injection temperature of 46 °C in this specific

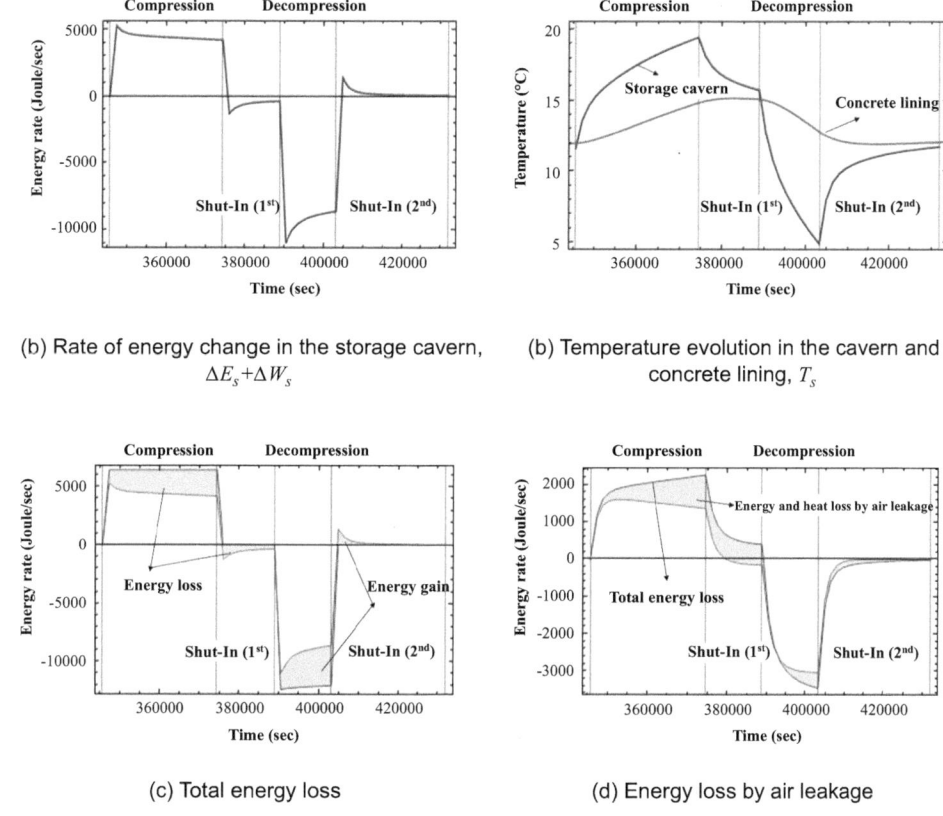

(b) Rate of energy change in the storage cavern, $\Delta E_s + \Delta W_s$

(b) Temperature evolution in the cavern and concrete lining, T_s

(c) Total energy loss

(d) Energy loss by air leakage

Figure 21 Results of energy balance analysis of the LRC for CAES (Kim *et al.*, 2012).

case. This is because the maximum and minimum tempera-ture achieved during compression depends on the temperature of the injected air, and the rate of heat exchange is proportional to the temperature difference between the cavern and the lining.

The effect of insulation condition was comparatively insignificant to storage efficiency. However, the CAES operational scheme (; the duration of compression-decompression phase during a daily operation) affects the efficiency such that relatively longer injection period (C4 case in Figure 22) proved to be more preferable, since the temperature increment inside the LRC can be mitigated under this condition due to longer heat transfer period.

From these results, it can be recommended that air injection temperature should be cooled down to similar to ambient temperature of storage cavern and the compression and decompression period during a daily operation should be designed as longer as possible for an optimum energy efficiency of CAES in LRCs.

6 SUMMARY

CAES is a viable option to meet the growing demand for large scale ESS which is driven by non-demand oriented power generation. Despite this advantage, it has been

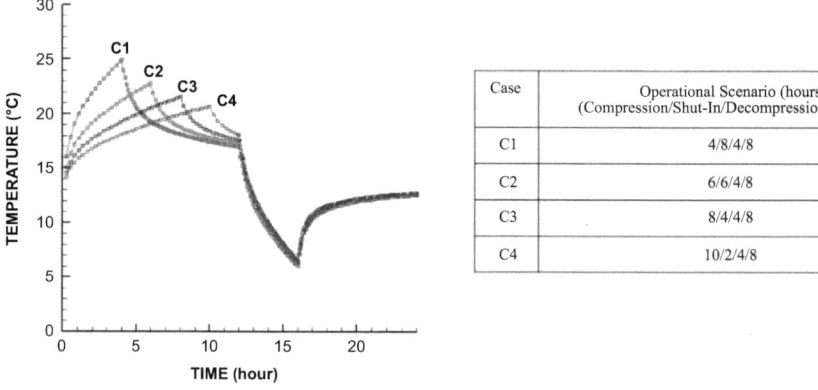

Case	Operational Scenario (hours) (Compression/Shut-In/Decompression/Shut-In)
C1	4/8/4/8
C2	6/6/4/8
C3	8/4/4/8
C4	10/2/4/8

(a) Temperature and operational scenario

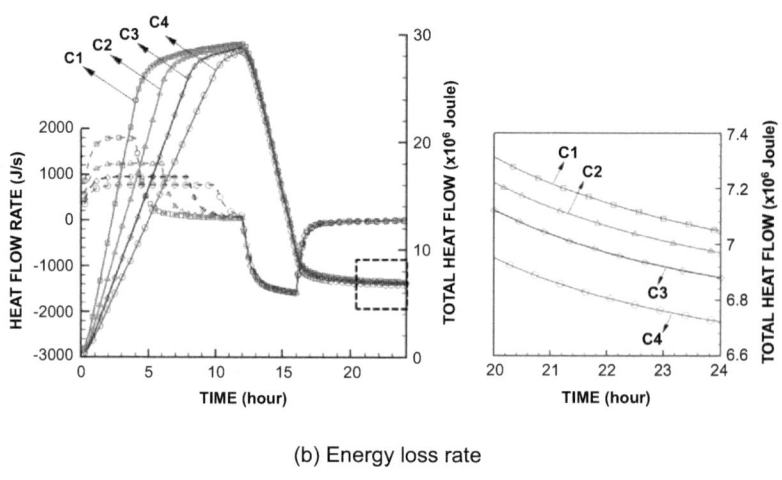

(b) Energy loss rate

Figure 22 Effect of operational scenario on energy efficiency of LRC (Kim *et al.*, 2014b).

recognized that the siting of CAES facilities may be limited by specific geologic conditions such as rock salt and their construction involves considerably high investment costs. Thus, the flexibility in site-selection and the cost-effective way of CAES are prerequisites for ESS market penetration.

In this chapter, an overview of CAES was presented and key issues of rock engineering for underground compressed-air bulk storage, which include air tightness, site selection, structural design and operational strategy, were reviewed.

Air tightness is a critical factor influencing both storage performance and structural stability of CAES caverns. The air leakage from storage caverns may cause a reduction of CAES efficiency but also increase the risk of a structural instability of the facility as a result of its highly concentrated and localized occurrence. Thus, the strategy and method of air leakage detection at underground LRCs for CAES were introduced through the parametric sensitivity analysis of LRC components and the simulation of a shut-in test was proposed to detect air leakage and to calculate the leakage rate.

The sensitivity analysis of LRC at a shallow depth demonstrated the effect of the design parameters of underground storage caverns on cavern safety and stability. These results were obtained from rather simplified analysis. Nevertheless, they can be effectively used at a preliminary design screening stage for selecting the candidate site of an underground CAES facility.

The LRC used for CAES should be equipped with an inner containment system for the tightness of highly compressed air. The inner containment system is typically composed of steel liner, separator and back-filled concrete. The design optimization of these structural components is crucial to economics, performance and structural stability of LRC. CAES is a form of energy storage that lacks established design criteria, so that it may be difficult to depend on experience from existing projects and there are no existing specific codes and standards available for CAES design. Because rock mass is a stochastic material with significant variations in its properties, it can be recommended to use a probabilistic approach for CAES design rather than a conventional deterministic approach. In addition, probabilistic approaches enable engineers to design the structural components of CAES caverns in a reliability-based manner. Furthermore, there is an international trend toward probability-based design of underground structures.

Drill and blast are still considered as the most efficient and economical excavation method especially in a hard rock. However, blast-induced damaged zones, *i.e.*, EDZs (Excavation Disturbed or Damaged Zone) around the caverns may cause a change in both mechanical and hydraulic properties, such as a decrease in deformation modulus and strength and an increase in permeability. This can lead to potential defects of the structural components of LRC. Through characterization of EDZ, some general recommendations to minimize the effect of blast-induced damage was provided, which include the reinforcement (*e.g.*, rock grouting) of EDZ after excavation and the installation of impermeable membranes inside the caverns. All these actions can help to minimize the risk of the pressure-induced failure of CAES caverns.

Finally, the coupled thermodynamic and geomechanical analysis including the entire construction and operation sequence of LRCs was introduced to investigate an operational strategy of CAES. The results of the energy-balance analysis during a daily operation, which was obtained from the THM coupled modeling, showed that air injection temperature should be cooled down to similar to ambient temperature of storage cavern and the compression and decompression period during a daily operation should be designed as longer as possible for an optimum energy efficiency of CAES in LRCs.

The recent achievements in rock engineering for CAES have contributed to confirm that CAES can be installed safely and economically in a hard rock. The paradigm shift in energy business will increase substantial R&D needs in rock engineering and will be new challenges to rock engineers.

REFERENCES

Allen, R.D., Doherty, T.J., Fossum A.F. (1982) Geotechnical Issues and Guidelines for Storage of Compressed Air in Excavated Hard Rock Caverns. Pacific Northwest Laboratory, PNL-4180, April, 163pp.

Chen, H., Cong, TN., Yang, W., Tan, C., Li, Y., Ding, Y. (2009) Progress in electrical energy storage system: a critical review. Progress in Natural Science, 19(3), 291–312.

Finkenrath, M., Pazzi, S., D'Ercole, M. (2009) Status and technical challenges of advanced Compressed Air Energy Storage (CAES) technology. International Workshop on Environment and Alternative Energy, Munich, Germany. http://www.neuralenergy.info/2009/06/caes.html

Kim, H.M., Rutqvist, J., Ryu, D.W., Choi, B.H., Sunwoo, C., Song, W.K. (2012) Exploring the concept of compressed air energy storage (CAES) in lined rock caverns at shallow depth: A modeling study of air tightness and energy balance. Applied Energy, 92, 653–667.

Kim, H.M., Park, D., Ryu, D.W., Song, W.K. (2013a) Parametric sensitivity analysis of ground uplift above pressurized underground rock caverns. Engineering Geology, 135–136, 60–65.

Kim, H.M., Rutqvist, J., Jeong, J.H., Choi, B.H., Ryu, D.W., Song, W.K. (2013b) Characterizing excavation damaged zone and stability of pressurized lined rock caverns for underground compressed air energy storage. Rock Mechanics and Rock Engineering, 46(5), 1113–1124.

Kim, H.M., Lettry, Y., Ryu, D.W., Song, W.K. (2014a) Mock-up experiments on permeability measurement of concrete and construction joints for air tightness assessment. Materials and Structures, 47(1–2), 127–140.

Kim, H.M., Kim, H., Ryu, D.W., Park, E.S. (2014b) Influence of insulation characteristics and operational modes on the storage efficiency in underground caverns for compressed air energy storage (CAES). Journal of Korean Society of Mineral and Energy Resources Engineers, 51(6), 820–828 (in Korean).

Kim, H.M., Rutqvist, J., Kim, H., Park, D., Ryu, D.W., Park, E.S. (2015) Failure monitoring and leakage detection for underground storage of compressed air energy in lined rock caverns. Rock Mechanics and Rock Engineering, 49, 574–584.

Liang, J., Lindblom, U.E. (1994) Critical pressure for gas storage in unlined rock caverns. International Journal of Rock Mechanics and Mining Sciences, 31, 377–381.

Madlener, R., Latz, J. (2013) Economics of centralized and decentralized compressed air energy storage for enhanced grid integration of wind power. Applied Energy, 101, 299–309.

Nakagawa, K., Shidahara, T., Ikegawa, Y., Suenaga, H., Miyamoto, Y. (2003) Experimental study on water confining of high-pressure air in an unlined underground cave – Air leakage test with a lateral cavern. Central Research Institute of Electric Power Industry, Abiko Research Laboratory Report No. U02050 (in Japanese).

Park, D., Kim, H.M., Ryu, D.W., Choi, B.H., Han, K.C. (2013) Probability-based structural design of lined rock caverns to resist high internal gas pressure. Engineering Geology, 153, 144–151.

Raju, M., Kumar Khaitan, S. (2012) Modeling and simulation of compressed air storage in caverns: a case study of the Huntorf plant. Applied Energy, 89, 474–481.

Rutqvist, J. (2011) Status of the TOUGH-FLAC simulator and recent applications related to coupled fluid flow and crustal deformations. Computers & Geosciences, 37, 739–750.

Rutqvist, J., Kim, H.M., Ryu, D.W., Synn, J.H., Song, W.K. (2012) Modeling of coupled thermodynamic and geomechanical performance of underground compressed air energy storage in lined rock caverns. International Journal of Rock Mechanics and Mining Sciences, 52, 71–81.

Sobolik, S., Ehgartner, B. (2006) Analysis of shapes for the strategic petroleum reserve. Albuquerue, NM: Sandia National Laboratories.

Succar, S., Williams, R.H. (2008) Compressed air energy storage: theory, resources, and applications for wind power. Energy Systems Analysis Group, Princeton Environmental Institute, 81.

Thoms, R.L., Gehle, R.M. (2000) A brief history of salt cavern use (keynote paper). Proceedings of the 8th World Salt Symposium, R.M. Geertman, ed., Elsevier. Vol. I, 207–214.

Wang, T., Yan, X., Yang, H., Yang, X., Jiang, T., Zhao, S. (2013) A new shape design method of salt cavern used as underground gas storage. Applied Energy, 104, 50–61.

Wang, T., Yang, C., Yan, X., Daemen, J. (2015) Allowable pillar width for bedded rock salt caverns gas storage. Journal of Petroleum Science and Engineering, 127, 433–444.

Design and Stability Analysis:

Blast Analysis and Design

Chapter 19

Blast-induced damage

F.G. Bastante

Department of Natural Resources and Environmental Engineering, University of Vigo,
Vigo, Pontevedra, Spain

Abstract: The chemical energy released by blasting shatters and moves rock but also damages the remaining rock mass, affecting its strength properties. The notion of trying to limit blast-induced damage — which emerged in the middle of the last century, motivated by practical financial and safety reasons — has led to the development of controlled blasting techniques. A growing interest in this issue in recent decades has resulted in the emergence of several proposals for predicting the extent of blast-induced rock mass damage. From the engineering point of view, the most useful predictive models are theoretically grounded and supported by experimental results. Nonetheless, since the theories are rather simplistic and experimental data — despite the complexity of the topic — are scarce, modeling results tend to be crude. Despite their limitations, however, such models do assist engineers in their task of limiting blast-induced damage. This chapter reviews theoretical-empirical models of practical use for blasting operations.

I INTRODUCTION

Concerns arising in the last century regarding the damage caused to the rock mass by blasting have motivated the development of controlled or contour blasting techniques aimed at limiting damage. These techniques, applied to the row of perimetral boreholes that form the excavation contour, include line drilling, presplitting, cushion blasting, trim or smooth blasting and air-deck blasting; where appropriate, the buffer blasting technique is also applied to one or more rows adjacent to the perimeter row.

Most of these methods, developed in the field mainly by trial and error (Persson *et al.*, 1993), are applied on the basis of a priori monitoring of test blasts and subsequent adjustment of the initial design based on the results obtained.

The need for greater control over the extent of blast-induced damage (BID) has led to the development of various predictive models which, although based on extremely simplistic and questionable assumptions, are of undeniable interest in practical blasting operations.

This chapter describes some of the more useful predictive models. The main assumptions underpinning these models are described objectively. However, before describing these models, I briefly review the complex mechanisms that are triggered in the rock mass when explosive charges are detonated and also review the concept of BID and the motives for this area of research.

1.1 Rock mass fracturing mechanisms triggered by explosive charges

As an explosive charge contained within a borehole progresses, gases with very high internal energy are generated that rapidly collide against the adjacent rock. This impact leads to a train of stress and strain (P and S) waves that interact with the joints, fractures, lithology changes, free faces, etc., in turn generating new waves. The dynamic strain field becomes extremely complex, given that each point of the rock mass will be affected not only by the direct waves generated by each blasthole, but also by the transmitted, reflected and refracted waves that develop in response to each change in rock mass acoustic impedance.

The wave train becomes attenuated as the distance from the borehole grows due to geometric divergence and also because the waves lead to non-elastic phenomena which typically break up the rock in the vicinity of the borehole.

As for the residual energy in the gases, after impact and before loss of confinement, this energy may continue to be sufficiently strong to also contribute to rock mass fracturing.

Various mechanisms are proposed to explain rock mass fracturing by the combined action of the waves and gases. By way of a brief explanation, as the waves move away from the borehole, the rock is crushed if dynamic compressive strength is overcome in the initially compressive, dynamic stress state. Shear cracks may occur adjacent to this area. Subsequently, due to the rapid drop in tangential stress, radial cracks may form, some of which will extend more than the rest depending on factors like rock mass structure and micro-structure or the prior stress state. At this point the waves cannot crack the rock but they may continue to develop or extend from pre-existing pores or cracks. When the waves reach free faces almost all the energy is reflected due to low acoustic air impedance, reversing their characteristics: compressive radial waves that reach free faces become tensile waves, with the opposite occurring with tangential waves. These waves may cause rock spalling (if they exceed the dynamic tensile strength of the rock in the strain-rate regime) or may extend pre-existing cracks.

As for the gases, after impact against the rock mass, mean pressure decreases sharply due to the fact that their rapid expansion causes rarefaction waves. The stress field produced by this pressure, acting over a relatively lengthy time period, can cause or lead to cracks extending along the main stress trajectories. The gases also tend to enter and enlarge cracks and fissures and cause the rock mass to move. When the borehole has insufficient relief, the vertical component of this movement tends to tear the rock mass in the direction of the surface.

In conclusion, therefore, the waves generated in the initial impact of the gases against the rock mass and the residual energy of the gases both contribute to rock mass damage.

1.2 The concept of damage

The term blast-induced damage, in a general sense, refers to a weakening in the mechanical properties of the rock mass due to the complex interaction of the different mechanisms described in the previous section. From a practical standpoint, the rock mass is damaged when the change in its strength properties is measurable or perceptible, affecting, to varying degrees, its stability and, in turn, the safety of excavation works, as corroborated by Krauland (1994) and Hustrulid (1994). A practical implication of this view is that interpretation of BID will depend on the blasting context.

For example, blasting to arrive to the final slope of a bank measuring a few meters high may not jeopardize stability. Even if backbreak occurs and rock mass fracturing increases in the vicinity of the borehole, the impact on costs may be insignificant and, consequently, the damage may not be rated as such, at least not in the short or medium term. In contrast, blasting in tunnels may affect safety more; consequently, the cost of operations to address blast damage will be much greater and the impact of the damage will therefore be rated as much greater.

This would indicate that, in interpreting damage, there is always a component deriving from the actual context in which blasting takes place.

1.3 Reasons for controlling rock mass damage

The interest in controlling rock mass damage has very practical grounds: the remaining rock mass is usually required to remain stable over long periods of time. Such control reduces the risk of injury to persons and machinery and may also lead to savings in costs for the overall mining operation or increased operating income.

Ore dilution, increased water flow and subsequent rock weathering are just some of the harmful effects of damage documented in reviews by different authors from over last fifty years.

Langefors & Kihlström (1963) note the following: *"In the case of water power tunnels, the effect of badly blasted contours will be felt continuously in greater resistance in the flow of water [...] Inconvenience and expense are generally incurred [...] if blocks of stone have to be removed after completion of the work and when dressing of the roof is needed and further safety measures have to be taken [...] The demands for accurate blasting apply especially where cavities caused for surplus rock have to be filled up with concrete at very high cost [...], and in rock of low strength where it is essential for an uninterrupted rhythm of the job to get a contour with the best curvature and to avoid or delay slides in the rock"*.

Bauer (1982) points out that large-diameter drilling that requires large amounts of explosive per blasthole *"...can result in severe backbreak problems for final pit walls. If backbreak is not controlled, a decrease in the overall pit slope angle will ultimately be necessary, and consequences, such as decreased recoverable ore reserves and increased waste to ore ratios, will result"*. Bauer also indicates that *"...safety berms will be less effective or nonexistent. Hazardous working conditions can also result. Remedial measures, such as scaling large areas and using wire mesh or other artificial support, are very expensive and difficult to implement."*

Persson *et al.* (1993) corroborated Bauer regarding the need, in large mining operations, *"for high steep slopes which must remain stable to avoid failures and costly scaling."*

More recently Iverson *et al.* (2013) noted as follows: *"The field investigations into drifting practices were important to outline safety problems linked to ground control. The obvious safety problems are: overbreak results in wider spans that require additional ground support and an increased likelihood of failure if not properly assessed; rough and undulating back and wall surfaces occur due to aggressive blasting and likely increase the hazards associated with scaling and the installation of bolts and support accessories; ..."*.

In short, optimal damage control aims to achieve an excavation contour without affecting the stability of the remaining rock mass, while devising the most efficient design in terms of safety and cost.

2 DAMAGE PREDICTION MODELS

Rock mass damage is the outcome of a series of complex interactions between different kinds of waves transmitted through the rock mass, with residual gas energy remaining after impact against borehole walls before release to the atmosphere. Damage will therefore depend on the explosive used, the rock mass and the geometry and initiation sequencing of blasting.

Numerical analysis, which can, in theory, model the blasting process, requires knowledge of both the explosive and the rock constitutive model and also the parameters required for modeling purposes. In practice, however, such models are not used, among other reasons, because most parameters — often numerous — are unknown.

Moreover, there are a number of specific parameters that potentially influence BID, including blasthole diameter, coupling ratio, blasthole pattern, bottom and column explosive charges and their respective lengths, type of initiation and water in boreholes (Ouchterlony *et al.*, 2002).

In contrast with numerical models, most predictive models — as described in this chapter — are extremely simple to use as they require very few parameters that can be measured and calculated. Given the large number of parameters eliminated from these models that may have a significant influence on BID, the interpretation of results in terms of both absolute values and dependence relationships between variables and parameters can only be approximate.

The interest in implementing predictive models lies, therefore, in making an initial estimate of the magnitude of the BID and, once blasting impact has been analyzed in situ, using this information to consider possible lines of action aimed at controlling the extent of damage.

Numerical models can help understand the mechanisms causing rock mass damage, their interrelationships and relevance and, ultimately, are an indispensable tool that will establish the relationship between damage and the (rock, explosive, operational variables) vector.

2.1 The Holmberg-Persson model

In their book *The Modern Technique of Rock Blasting*, Langefors & Kihlström (1963) included an exercise to design the row adjacent to the contour row in a carefully conducted blasting operation. Used as the damage control parameter was the charge level, defined as $w = Q_R/D_R^{1.5}$, where D_R represents the radial distance to the borehole and Q_R is the explosive mass corresponding to a charge height of $2D_R$. The authors related the charge level parameter with the law of propagation of ground vibrations, which, for Swedish hard rock, was around $PPV \sim w^{0.5}$, where PPV is the peak particle velocity. Langefors, establishing the threshold for rock fracturing at $w \geq 1$, analyzed the value of the linear charge concentration l for which buffer row damage would not affect the rock mass more than contour row damage.

Years later, Holmberg & Persson (1978) developed a model (H-P model) to predict the BID when the PPV was directly used as a parameter to represent damage. This model, supported by some experimental results (Homberg & Krauland, 1977), was well received in operational practice. Fifteen years after its proposal, Persson *et al.* (1993) noted: *"This technique, although crude, has been used successfully to predict damage to the remaining rock for tunnel and slope stability purposes."*

2.1.1 Derivation of the Holmberg-Persson model

According to Persson *et al.* (1993): *"Damage is a result of the induced strain ε which, for an elastic medium in the sine-wave approximation, is given by the equation ε = v/c"*, where *v* is vibration velocity and *c* is the velocity of seismic wave propagation. Subsequently the authors added: *"The extent of rock damage can be approximately correlated with the peak particle velocity [...] as a measure of the damage potential of the wave motion."* According to the above, different PPV thresholds associated with different levels of damage can be defined. Consequently, the first issue is how to estimate PPV for the vicinity of the borehole.

The problem of determining the law of propagation of ground vibrations arises regularly in engineering blasting operations since maximum vibration velocity together with the dominant frequency are key parameters in controlling blast damage to buildings. The most widely used predictive model establishes the following relationship between *PPV*, the explosive charge (*Q*) and the distance from the blast to the measurement point (*D*) (Duvall *et al.*, 1963):

$$PPV = KQ^{\alpha}D^{-\beta} \tag{1}$$

where *K, a* and *β* are constants to be determined by means of a statistical analysis of blast data (*PPV, Q* and *D*) recorded in situ using a seismograph.

Although the relationship described in Equation 1 is statistical in nature, using the dimensional analysis technique, Ambraseys and Hendron (1968) and Hendron (1973) derived what is known as cube-root scaling — $PPV \sim (D/Q^{1/3})^{-\beta}$ — which equips the predictive model with a certain theoretical grounding. However, the most widely used model for elongated explosive charges is what is called square-root scaling, according to which $PPV \sim (D/Q^{1/2})^{-\beta}$.

Given that, in order to determine the constants of Equation 1, the vibrations are usually measured at a considerable distance from the blastholes, Holmberg & Persson (1978, 1979) extended this model for use in the vicinity of the blastholes, by considering the differences in wave arrival time when PPV is achieved to be insignificant.

From Equation 1 these authors defined a variable called vibration intensity (*w*) — equivalent to Langefors' charge level — and expressed *Q* in terms of charge concentration *l* and charge length *H*:

$$w = (PPV/K)^{1/\alpha} = D^{-\beta/\alpha}Q = D^{-\beta/\alpha}lH \tag{2}$$

Next, assuming that the contribution of each charge differential to vibration intensity would be represented by the differential expression $dw=D^{-\beta/\alpha}dQ$, they derived their model. Taking the charge axis as the X coordinate axis and the axis corresponding to the radial coordinate from the X axis as the Y axis, for the coordinates (*xo, ro*) we have:

$$w = \int D^{-\alpha/\beta} l dx \rightarrow PPV = K\left[l\int D^{-\beta/\alpha} dx\right]^{\alpha} \quad with \quad D = \left[ro^2 + (x - xo)^2\right]^{1/2} \quad (3)$$

In these expressions, since D is the distance between each differential of the charge dx and the point of measurement, then D is a function of charge length H. Note that the variable that is summed is not the PPV, but vibration intensity, which is indicative of the energy released by the explosive and transmitted to the rock mass. For example, for square-root scaling, dw is a damage contribution indicator for the energy released by an elementary charge dQ, with this energy attenuating due to geometric divergence. Vibration intensity therefore establishes a relationship between the explosive energy transmitted to each point and kinetic energy (characterized by the PPV obtained in the model), equivalent to the total for this scaled energy. Use of the PPV as a damage criterion is, perhaps, more motivated by the need for a quantitative parameter to assist blast design than any attempt to physically represent the phenomenon.

It now remains to establish the correlation between the PPV and the damage caused in the rock mass. Persson *et al.* (1993) indicate the most direct method for this is: *"by a comparison of the crack frequency before and after the blast by using core logging or with a borehole periscope. In addition, [...] the incipient blast damage can be evaluated by extensometers fastened in drillholes."* These authors distinguished five decreasing-damage zones as distance from the blasthole increased, whose limits — based on experiments in hard Scandinavian granitic or gneissy bedrock — were 15 m/s (crushing), 5 m/s (good fragmentation), 2.5 m/s (marginal fragmentation), 1 m/s (incipient damage) and 0.7 m/s (incipient swelling). For these rock masses the authors established the following values for Equation 1: $K = 0.7$ m/s, $\alpha = 0.7$, $\beta = 1.5$ and critical PPV for incipient fracture, 0.7–1 m/s.

Applying Equation 3, we can predict BID for different charge concentrations and lengths and use this information to design blastholes for the contour row and adjacent rows.

2.1.2 Other PPV-based models

Alternative formulations of the H-P model for predicting PPV in the vicinity of boreholes have been proposed, briefly discussed below.

Harries (1983) used the analytical expression of the radial velocity of the vibration produced by a spherical charge in an infinite and isotropic elastic medium (Favreau, 1969). To extend the model to the case of an elongated charge, a model of superimposed waves (Plewman & Starfield, 1965; Starfield & Pugliese, 1968) was applied based on the assumption that this could be represented in a series of stacked elementary spherical charges that would be successively initiated in line with the detonation velocity of the explosive. Since the Favreau solution considered an elastic medium, Harries introduced a component of inelastic attenuation according to distance in his model. This model yields estimates of the velocity time histories for each point.

Hustrulid *et al.* (1992) and Hustrulid (1994, 1999a) proposed what is called the Colorado School of Mines (CSM) approach, a simplification of Harries' model, which can be applied when there is little or no temporal overlap between the waves generated by the elementary charges that model the cylindrical charge. This model is expressed as follows:

$$PPV_R = 0.61(P_b/\rho c_p)(\phi/D_R)e^{C_I(D_R-0.61\phi)} \qquad (4)$$

where PPV_R is the peak value of the radial component of vibration velocity, ρ and c_p are the density and the propagation velocity of the P waves in the rock, respectively, ϕ is the borehole diameter, D_R is the radial distance to the blasthole axis, C_I is the inelastic attenuation coefficient (determined experimentally) and P_b is the pressure in the gases as they expand to completely fill the blasthole.

Ouchterlony *et al.* (1993) proposed calculating the PPV directly from Equation 1 but including a correction factor $fc = atan(H/2D_R)/(H/2D_R)$ that affects charge length H and, therefore, the total charge contributing to damage.

Hustrulid (1999b) used Neiman's hydrodynamic model (1979) to predict the vibration velocity along an axis perpendicular to the charge and passing through its center. The expression, only valid for points close to the charge, is as follows:

$$PPV_R = I\psi^{-1} \ with \ I = \sqrt{\rho_e Q_e/8\rho} \ and \ \psi = (r/L)\sqrt{2\left(r^2 + \frac{L^2}{4}\right)sinh^{-1}L} \qquad (5)$$

where ρ_e and Q are the density and explosion heat of the explosive, respectively, r is D_R/ϕ, L is H/ϕ and $sinh^{-1}$ is the inverse hyperbolic sine function. Since the assumptions underpinning the Neiman model (instantaneous detonation of the entire charge, an elastic and uncompressible surrounding medium, explosive energy entirely converted into kinetic energy, etc.) do not correspond to reality, Hustrulid proposed calibrating the model with experimental data. If from the experimental data we obtain $PPV_e = I_e \ \Psi^{-\beta_h}$, then the scaling factor is defined by the relationship between PPV_e and PPV_R given in Equation 5. In practice, vibration measurements are often taken at distances that fulfill $r >> L$; hence, from Equation 5 we deduce that $\Psi \sim r^2$. Comparing this relationship with that reflected in Equation 1 we have $\beta_h \approx \beta/2$

In the previous model Tesarik *et al.* (2011) accounted for the effect of decoupling the charge by introducing the coupling factor (f) in the calculation of I, that is: $I_f = f \ I$.

Iverson *et al.* (2008) proposed directly using Equation 1 to determine near-field PPV, with D representing the mean distance from the measurement point to the charge (the National Institute for Occupational Safety and Health (NIOSH)-modified Holmberg-Persson approach). Iverson *et al.* (2009) later suggested that the distance should instead be calculated using a weighted approach, with the inverse of the distance raised to the power β used to weight the distance for each element of the charge.

Lu & Hustrulid (2003), using Heelan's (1953) radiation pattern for a cylindrical charge, obtained a similar solution to that of the CSM model represented in Equation 4, replacing the terms for PPV attenuation with distance with the attenuation terms in Equation 1. We thus have: $PPV \sim P_b(\phi/D)^\beta/(\rho \ c_p)$.

2.2 The Ouchterlony model

In the 1990s, the Swedish Rock Engineering Research Foundation (SveBeFo) conducted blast-damage research at Vånga dimension-stone quarry in Sweden. To measure damage to the rock mass after each blast — initially with a single hole (Olsson & Bergqvist, 1993) and later with three or four holes in a row (Olsson & Bergqvist, 1996) — the researchers extracted blocks from the remaining rock, sawed them along a plane perpendicular to the blasthole axis and carried out a dye-penetrant inspection to map fractures.

Ouchterlony (1997), on the basis of a thorough analysis of the results of Olsson & Bergqvist (1996), developed a predictive BID model, condensing the results of these authors to produce seven values corresponding, mostly, to mean values for the longer radial cracks resulting from different blasts. Tests were performed in small blastholes (22–64 mm) with linear charge concentrations of up to about 0.5 kg/m, in the absence of water and with simultaneous initiation.

2.2.1 Derivation of the Ouchterlony model

Ouchterlony used the Grady & Kipp model (1987) to estimate the critical pressure of crack initiation (P_{BIDc}) in a cylindrical borehole of radius equal to the BID and with an elastic medium. Imposed was the condition of the formation of at least two fractures, with a linear relationship assumed between dynamic pressure on the blasthole walls ($P_{\phi/2}$) and radial vibration velocity. The following expression was obtained:

$$P_{BIDc} = 1.08 K_{IC} BID^{-0.5} \tag{6}$$

where K_{IC} is fracture toughness. Thus, if the value of the peak dynamic tension in the rock (P_{DR}) as a function of radial distance from the blasthole (D_R) is known, we can estimate the BID applying Equation 6. Ouchterlony used expressions of Liu & Katsabanis (1993) to calculate this function.

These authors applied the one-dimensional hydrodynamic theory of detonation to estimate peak dynamic pressure on the blasthole walls that exerted a fully coupled charge. It was assumed that the gases expanded isentropically and behaved polytropially and that rock behavior was defined by a shock equation of state. Using regression, they then related the dynamic pressure to a set of parameters that roughly represented the detonation pressure and the impedance contrast Z (the rock mass/explosive impedance ratio). Ouchterlony included the effect of charge decoupling in the results of Liu & Katsabanis, obtaining an equation that can be expressed in general form as:

$$P_{\Phi/2} = P_b Z^a \tag{7}$$

where a is a constant. Applying detonation theory and the hypothesis of isentropic gas expansion, the following relationships are obtained between P_b, explosion pressure (P_e) and detonation pressure (P_{cj}):

$$P_{cj} = \frac{\rho_e D_e^2}{\gamma + 1}; \quad P_e = P_{cj}\left(\frac{\gamma}{\gamma + 1}\right)^{\gamma}; \quad P_b = P_e f^{c\gamma^1} \tag{8}$$

where D_e is detonation velocity, γ is the isentropic expansion factor of the gaseous products in the state of detonation (assumed to remain constant until the gases return to the initial explosive density ρ_e), γ^1 is the average isentropic expansion factor of the gases from ρ_e to their final expanded state in the borehole, f is the coupling factor and c is a constant usually set to 2 for elongated charges.

In subsequent calculations, Ouchterlony used $\gamma^1 = \gamma = (1 + D_e^2/2Q_e)^{0.5}$, which implies the assumption of perfect gas behavior.

To estimate stress wave attenuation as a function of distance, Liu & Katsabanis assumed that rock density between the shock wave front and the blasthole — which expands by the action of gases — is held constant. Using numerical method, and maintaining the hypotheses for calculating $P_{\phi/2}$, they finally obtained the function P_{DR}. Analyzing various rock/explosive combinations and using regression, they obtained an expression of the type:

$$P_{DR}/P_{\Phi/2} = [D_R/(\Phi/2)]^n \quad with \quad n \approx 1.5(D_e/c_p)^{0.25} \tag{9}$$

Setting P_{DR} as equal to P_{BIDc}, considering the multiplicative constants in Equation 6 and Equation 9 as parameters to be determined by regression of the experimental data obtained by Olsson & Bergqvist (1996) and rearranging the terms we have:

$$(2BID/\Phi)^{D_e/c_p} = \lambda \left(\frac{P_{\Phi/2}\sqrt{BID}}{K_{IC}}\right)^\delta \tag{10}$$

In his statistical analysis, Ouchterlony considered different cases that include or exclude some of the factors in Equations 7 to 10, such as Z, (D_e/c_p) and γ^1. Finally, after a review of the parameter values obtained from each regression and of the respective coefficients of determination, Ouchterlony obtained the values $c = 2.2$, $\lambda = 0.566$ and $\delta = 2/3$ and also extracted both the Z and γ^1 factors from the model (in this case it can be taken as assumed that $c = 2$ for elongated charges and, hence, that $\gamma^1 = 1.1$). The following expression resulted:

$$\left(\frac{2BID}{\Phi}\right) = \left(\frac{P_b}{P_{bcr}}\right)^e \quad with \quad P_{bcr} = \frac{3.3K_{IC}}{\sqrt{\Phi}} \quad and \quad e = 2/\left[3(D_e/c_p)^{0.25} - 1\right] \tag{11}$$

Subsequent SveBeFo research led to inclusion in the model of four multiplicative factors affecting BID calculated so as to incorporate the effect of the relationship between blasthole spacing (S) and the burden (B) and also the initiation system, the presence of water in blastholes and the rock mass structure.

Thus, Olsson & Ouchterlony (2003) established, for $S/B < 1$ relationships with simultaneous initiation (dispersion in firing time under 1 ms) and small charge concentrations, a value of 1 for the factor corresponding to the initiation system. However, this factor was 2 for simultaneous initiation and $S/B > 2$ and was applied to pyrotechnic initiation independently of the S/B relationship. As for the presence of water in the borehole, this can sharply increase BID. For example, for the case of the Olsson & Bergqvist tests conducted in dry boreholes, the presence of water would have increased BID by 2.5 to 4.5 times, depending on the coupling factor (f) and the explosive type, that is, on the relationship between detonation velocities confined by water and unconfined.

Finally, these same authors later modified the model given by Equation 11. First, they reintroduced impedance contrast Z in the model and, second, they eliminated the term D_e/c_p. This model (Olsson et al., 2008) can be expressed as:

$$\left(\frac{2BID}{\Phi}\right) = \left(\frac{P_{\Phi/2}}{P_{bcr'}}\right)^{1.525} \quad with \quad P_{bcr'} = \frac{10.8K_{IC}}{\sqrt{\Phi}} \tag{12}$$

$P_{\Phi/2}$ is obtained from Equation 7 and Equation 8, with $a = 1/3$ and $c = 1.835$. The authors note that in the statistical fit, they assumed a γ value of 3.4 for Kimulux 42 — an emulsion explosive with ρ_e and Q equal to 1100 kg/m^3 and 3.35 MJ/kg, respectively — rather than use the value obtained using the perfect gas approach.

2.3 The NIOSH model

The NIOSH has developed other BID prediction models (Hustrulid & Johnson, 2008; Hustrulid & Iverson, 2010). One of these models uses gas pressure (P_b) — calculated from Cook's (1947) isotherm — and rock density as parameters to estimate BID, and, as reference, ammonium nitrate/fuel oil (ANFO) and rock with density 2650 kg/m^3. Denoting as P_{eANFO} the ANFO explosion pressure, we have:

$$\left(\frac{2BID}{\Phi}\right) = 25\sqrt{P_b/P_{eANFO}} \ \sqrt{2650/\rho} \tag{13}$$

Hustrulid & Iverson (2010) used Equation 13 to evaluate drifting data from the Kiruna mine during an LKAB Group drift-driving program involving measurement of overbreak from several blasts. The authors started from the assumption that overbreak is largely independent of the perimeter charge if buffer row design is correct. In comparing the overbreak measurements with results obtained by applying Equation 13 to the charge in the buffer holes, Hustrulid and Iverson found that the pressure-based BID was a good predictor of overbreak.

Another NIOSH model, also referred to in the above-mentioned articles and described in detail in Iverson *et al.* (2013), uses explosive energy (Q_e), rather than pressure as a parameter to determine BID. This model is developed below.

2.3.1 NIOSH explosive energy model

Ash (1963) documented, on the basis of data compiled from numerous mines, the relationship between the different geometric operating variables used in blasting, establishing, in particular, $B = K_B \phi$ as the relationship between the burden (B) and the drilling diameter (ϕ), where K_B is a constant for each explosive-rock combination, for a value normally lying between 20 and 40.

On the basis of the hypothesis that degree of fragmentation is a function of explosive consumption per unit of blasted rock, Hustrulid (1999a) expressed a simplified relationship between B and ϕ as follows:

$$K_B = 25\sqrt{\rho_e Q_e/\rho_{ANFO} Q_{ANFO}} \ \sqrt{2650/\rho} = 25\sqrt{RBS} \ \sqrt{2650/\rho} \tag{14}$$

where RBS is the relative bulk strength for ANFO. Again, ANFO and a rock density of 2650 kg/m^3 were used as reference. On the simple geometrical basis of just-touching damage circles, Hustrulid suggested — as a practical way to estimate BID — the following expression:

$$\left(\frac{2BID}{\Phi}\right) = K_B \tag{15}$$

Iverson *et al.* (2013) compared the results of applying this model with the results obtained for test blasting of concrete blocks. Three blasts were implemented, each in a concrete block, using a single 38-mm blasthole and a coupled charge. Two different explosives were used but with very similar RBS characteristics. Various techniques were used to estimate the extent of damage: wire sawing to expose the radial cracks and micro-velocity probe and crosshole velocity probes to measure variations in S and P wave velocity at a distance from the blasthole. Brazilian tests were also performed — to analyze variations in rock strength away from the blasthole — and data was recorded on strain waves in nearby blastholes (Johnson, 2010).

Analysis of the results indicates that BID calculated using Equation 15, referred to by the authors as practical radius damage, is in agreement with the damaged area value obtained using seismic methods, namely, 56–65 cm. According to the law of attenuation of peak strain (calculated from recorded data), at these distances the range of critical strain ranges between 1000 μm/m and 1330 μm/m. Moreover, if critical strain is computed from the mechanical properties of each concrete block — assumed to have elastic behavior — values between 1400 μm/m and 1600 μm/m are obtained, with critical particle speeds of 3.8 m/s to 5 m/s. BID predicted from Equation 15 would seem to be located at the limits of "*crushing*" and "*good fragmentation*" defined by Persson *et al.* (1993). These authors themselves indicate that the radial cracks could be expected to extend beyond the limit defined by this equation.

Hustrulid (2010) introduced the coupling factor (*f*) into the model, which he called the modified Ash energy (MAE) approach:

$$\left(\frac{2BID}{\Phi}\right) = 25f\sqrt{RBS}\ \sqrt{2650/\rho} \tag{16}$$

For tunnel blasting design, the authors recommend locating holes adjacent to the contour row at a distance equal to BID with spacing (S) of (1.3–1.5) BID. The contour blastholes should be located in the area of low energy coverage by the buffer row, that is, staggered with respect to the buffer row holes and, therefore, with the same between-hole spacing (S).

To calculate the charge for the contour row they used the methodology described by Hustrulid and Johnson (2008), who, in turn, used the expression of Sanden (1974), which determines the maximum spacing (S) between two blastholes experiencing static pressure P_b that would achieve tensile breakage of the rock, assuming elastic behavior. Denoting σ_t as its tensile strength, we have:

$$S = \Phi\left(\frac{P_b + \sigma_t}{\sigma_t}\right) \tag{17}$$

Since S is a known design variable [(1.3–1.5)BID], via Equation 17 we obtain the P_b required to achieve a cut between the boreholes in the contour row, there being some leeway in the design to ensure that P_b does not exceed the compressive strength of the rock.

To calculate P_b these authors used Cook's (1947) isothermal expansion, taking the state of explosion (P_e) of the explosive as the origin. The co-volume value of the gases for each state was calculated with Hustrulid's (2007) expression. Thus:

$$P_b = P_e f^2 \left(\frac{1000 - 1.1\rho_e e^{-473 \times 10^{-6}\rho_e}}{1000 - 1.1\rho_b e^{-473 \times 10^{-6}\rho_b}} \right) \quad with \quad \rho_b = \rho_e f^2 \tag{18}$$

Hence, from Equation 17 we obtain P_b and from Equation 18 we obtain the coupling factor (f) and, therefore, the required diameter of the charge.

2.4 Models based on Langefors' theory

Hustrulid (2010), on the basis of Langefors' theory for calculating burden in blasting, applied an approach similar to the MAE, expressed as:

$$\left(\frac{2BID}{\Phi} \right) = 23 f_f \sqrt{\frac{RBS}{C_e f_f}} \tag{19}$$

where f_f is the rock fixation factor at the bottom of the blasthole and C_e is the specific charge (kg/m^3). We understand that C_e refers to the specific charge required to achieve a certain degree of fragmentation.

Bastante et al. (2012), assuming that BID could be represented as a function of the maximum burden (B_m) that can be broken with bottom charge concentration l in a rock mass, estimated this function by solving for B_m and compared the results with experimental measurements. This model is described in the next section.

2.4.1 Development of the Bastante model

According to Langefors' theory of rock blasting (Langefors & Kihlstrom, 1963), the bottom charge concentration required to tear burden B_m by blasting a row of holes containing elongated charges is given by:

$$l = 0.9\bar{c}B_m^2/s \quad with \quad \bar{c} \approx 0.07/B_m + 1.2c_0 \tag{20}$$

where s is the relative weight strength of the explosive with respect to Langefors' standard dynamite and $\bar{c}$ is the rock factor, with c_0 determined by the minimum specific charge needed to break the bottom of a borehole with a concentrated charge. A factor of 1.2 is introduced as a margin to account for c_0 variations in the rock mass. According to Langefors, c_0 is usually between 0.28 kg/m^3 and 0.35 kg/m^3.

The hypothesis is that BID is a function of B_m, in turn dependent on the energy placed in the blasthole. Polytropic behavior is assumed in order to account for the effect of charge decoupling (f) in internal energy variation during gas expansion. The rock factor term $1/B_m$ is excluded, leaving the following expression:

$$BID = F\left(f^{(\gamma^1 - 1)} \sqrt{ls/c_0} \right) = F(\Omega) \tag{21}$$

The function F(Ω) is obtained by least squares regression of the experimental measurements made by Olsson & Bergqvist (1993, 1996), and also by Kilebrant et al. (2010), who, for six tests in hard granite gneiss, measured BID in core samples using various observation techniques (core mapping and ultrasonic velocity, porosity and density measurements).

The difficulty implied by Equation 21 is that γ^1 depends on the type of explosive and the degree of gas expansion. Using data obtained from different sources, trying to avoid bias in the regression and seeking a simple model, the authors assumed a value of γ^1 dependent solely on f. Thus, γ^1 is assumed to be constant and equal to 1.7 when f is greater than 1/3. For lower values of f they consider that the gases continue expanding with a constant slope equal to 1.1 and, therefore γ^1 gradually decreases $(\gamma^1 \approx 1.1-0.66 Ln^{-1} f)$. Moreover, the authors used as a reference explosive a gelatin with $Q_e = 4.6$ MJ/kg and a rock mass value of $c_o = 0.35$ kg/m^3.

As a result of the statistical fit, they obtained a BID = 0.8Ω and, considering $s \approx 672$ RBS/ρ_e and $l = f^2 \rho_e \Pi \Phi^2/4$, the model is expressed as:

$$\left(\frac{2BID}{\Phi}\right) = 36 f^{\gamma^1} \sqrt{\frac{RBS}{c_o}} \tag{22}$$

According to Bastante *et al.* (2012), for coupled charges ($f = 1$) the values obtained from Equation 22 lie within the range for burden in routine production blasting operations. The difference between this equation and Equation 14 lies in the fact that the latter uses rock density rather than the rock constant as the parameter to determine the burden.

2.5 The seed waveform model

Plewman & Starfield (1965) and Starfield & Pugliese (1968) described a model to predict the strain wave resulting from blasting at each point of the rock mass. Although the model today is used frequently to analyse vibrations, it is little used for strain analyses.

2.5.1 Development of the seed waveform model

In this model the explosive charge column is discretized into a set of N concentrated charges — small and equal in length — that will initiate in line with the detonation velocity. Each of these charges will then produce an elementary strain wave $\xi_s(t)$ equal in magnitude and form and attenuated with distance according to a scaling law ($\sim D^{-\beta}$). For each point of the rock mass reached by the elementary strain wave train, the model assumes that the resulting strain $\xi(t)$ can be obtained by linearly superimposing the waves of the train. In their model the authors only take into account the strain component in the radial direction $\xi_R(t)$ with regard to the axis of the charge, hence, according to the above, for each rock mass point we have:

$$\xi_R(t) = \sum_{i=1}^{N} KD_i^{-\beta} \xi_s(t - \tau_i)cos^2\theta_i \tag{23}$$

where K and β are constants estimated from experimental data, just like the function $\xi_s(t)$; τ_i is the time to arrival of the elementary wave corresponding to the concentrated charge i at the rock mass point being analyzed, and θ_i is the angle formed by the vector obtained from joining that point with charge i and the horizontal axis.

Starfield & Pugliese (1968) applied the model to experimental data obtained in the 1960s via US Bureau of Mines tests on confined charges, some of which were published in Duvall *et al.* (1966). Measured for different parts of the Lithonia granite rock mass was the radial strain produced for a short charge (0.76 m) and a long charge (8.2 m) of high-pressure gelatin in a 76-mm diameter blasthole.

Starfield and Pugliese, using parameters for the model (K, β and $\xi_s(t)$) obtained from information provided by tests with gelatin short-charges, simulated the strain resulting from both short and long charges of gelatin at different points of the rock mass. Tests and simulations were also performed for ANFO. Despite the simplicity of the model the results were surprisingly consistent with radial strain wave measurements. This model is a good predictor of the strain wave to peak values of at least some 200–300 μm/m in Lithonia granite.

3 PREDICTIVE MODEL APPLICATIONS

Of the different models described above, the H-P is the most general and most widely cited (and also the most criticized) model, as it allows different damage thresholds to be defined and can be used for both small- and large-diameter drilling. The Ouchterlony & Bastante models refer to a specific type of damage and are calibrated using data obtained in small-diameter blasting operations. The NIOSH model — Equations 16–18 — also seems especially oriented towards underground drifting applications, and, as the authors point out, is focused on *practical damage radius*, by which *"it is meant that if the rock mass lying outside of this ring were removed, the rock remaining within the ring would easily break apart (Iverson et al., 2013)"*.The Ouchterlony model is strongly supported experimentally and, therefore, it seems reasonable to use it — within its range of application — to predict mean maximum radial crack extent values. The MAE approach, underpinned by the pragmatism of Hustrulid, establishes a relationship between damage level and the practical value for blasting burden, which would seem to be solidly grounded. More questionable is the relative arbitrariness of the definition of *practical damage radius*.

The Bastante model can be considered a variation of the MAE approach in which practical burden and rock density are replaced by maximum burden and the rock constant, respectively. Calibrated with experimental data, this model also defines BID as the maximum length of the radial fractures. In this regard it should be noted that, although the regression fit was statistically satisfactory, the fact that the analyzed data did not constitute a homogeneous population undermines the fit.

As for the seed waveform model, this methodology can accommodate particle velocities as usually recorded in civil engineering operations. Of the described models it is the only one that represents the phenomenon, even if very simplistically. However, implementation is complex, although this problem is partly offset by the high-level programming languages available nowadays. It is, in any case, a useful analytical tool.

Figure 1 depicts the response of some of the predictive models to variations in the linear charge concentration *l*, and where appropriate, the coupling factor *f*. Experimental data for hard Scandinavian bedrock (Olsson & Bergqvist, 1993, 1996; Kilebrant *et al.*, 2010) are also included. For these data, the coupling factors used for testing — low (blasting with detonating cord), medium and high — are shown. Linear charge concentration *l* is expressed in Kimulux 42-equivalent kilograms (kg_{Kx}/m).

The representation of the H-P model, shown in Figure 1a, is as follows. Rock properties, from Johnson (2010) and Olsson & Bergqvist (1993), were ρ = 2620 kg/m^3,

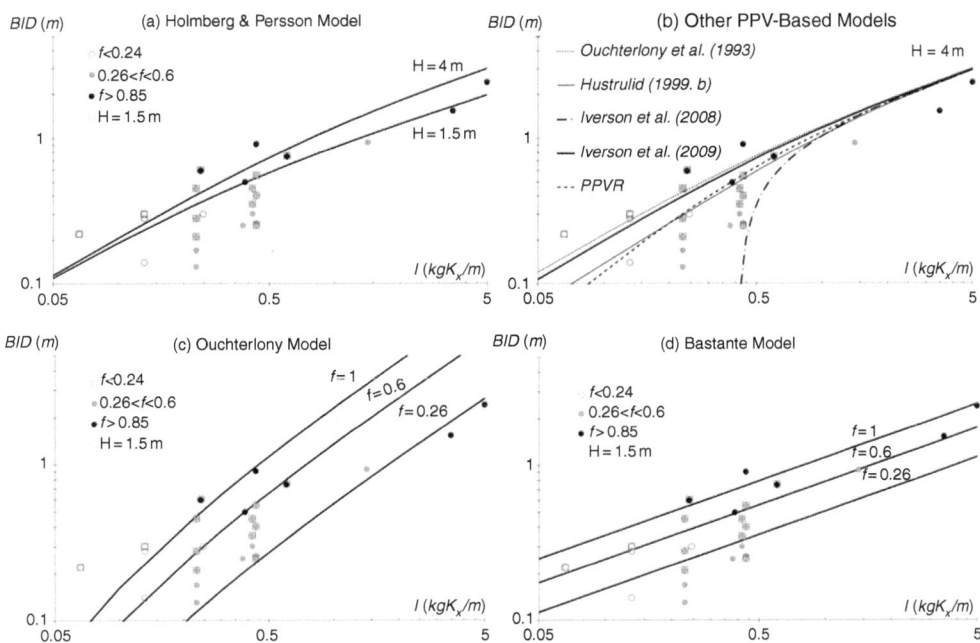

Figure 1 Extent of blast-induced damage (BID) in contrast with the linear charge concentration (*l*). Results for predictive models using, as applicable, Holmberg & Persson parameters for hard rock (*K* = 700, *α* = 0.7, *β* = 1.5) and a peak particle velocity (PPV) threshold of 1 m/s. Some experimental results are included (see references in the text). *H* and *f* indicate charge length and coupling factor, respectively.

c_p = 4500 m/s and σ_t = 12 MPa, respectively. The PPV threshold for tensile strength was calculated, using the H-P model hypothesis, as 1 m/s (220 μm/m) and, using the H-P values for *K*, *α* and *β* in Equation 3, the radial distance corresponding to said PPV was determined for an axis perpendicular to the charge center. Figure 1 also shows the model results for the charge lengths used in the tests: *H* = 1.5 m, denoted by squares, and *H* ≈ 4 m, for the other tests.

It cannot be inferred from the experimental data that greater charge lengths result in greater BID, although this may be due to different testing conditions. The H-P model orders of magnitude and trends with respect to the data seem quite reasonable. However, the statistical nature of Equation 1, which supports this model, should be noted. The value of the *K* parameter depends, as well as on the rock mass, on the type of explosive and on the *f* factor. Another issue is that the greatly simplified calculation of the damage parameter does not take into account strain-rate effects. In operational practice the model needs to be calibrated with BID measurements taken in situ in order to establish the relationship between BID and the threshold PPV.

Figure 1b shows other PPV-based models, calculated using the same parameters as for the H-P model. The results for the Ouchterlony *et al.* (1993) and Iverson *et al.* (2009) models are virtually the same as those for the H-P model. Included for illustrative purposes are the results for a model labeled PPVR, based on using PPV rather than (as in the H-P model) vibration intensity.

The Ouchterlony model (Figure 1c) represents Kimulux 42 behavior for various values of f, for which the Olsson $et\ al.$ (2008) equation derived from Equations 7, 8 and 12 was used, expressed as follows:

$$BID = K_{er}f^{1.05}l^{0.88}D_e^{2.55} \tag{24}$$

where K_{er} is a constant that depends on both the explosive and the rock. Note that a different equation is required for each explosive even though l is represented in kg_{Kx}/m in the X-axis. Therefore, it is recommended not to judge the reliability of the method according to the information in Figure 1, which only aims to show model orders of magnitude and trends.

Equation 24 predicts the BID value for other conditions of f and l. For example, the authors take into account variation in D_e with charge diameter; hence, BID $\sim f^{1.05}l^{1.1}$ is satisfied for Kimulux. Great BID sensitivity to l results in higher BID values than obtained with the H-P model, for higher l than those used for calibration (up to about 0.5 kg/m).

Ouchterlony recommends caution in using this model when $l > 0.6$ kg/m, recommending instead the H-P model or the Ouchterlony $et\ al.$ (1993) model when $l > 1.6$ kg/m. Also evident in Figure 1c is how greatly the coupling factor affects the extent of damage.

Figure 1d shows the Bastante model, also based on Kimulux 42 (RBS = 1.18) and obtained using Equation 22 and the relationship between him l and ϕ. BID is less sensitive to the l and f design variables than in the previous models: BID $\sim f^{[0.5-0.7]}l^{0.5}$.

Figure 2 shows some Starfield & Pugliese (1968) results from applying the seed waveform model to radial strains. The $\xi_s(t)$ function used is a damped sine wave:

$$\xi_s = K_1 e^{-K_2 t}\sin(K_3 t) \tag{25}$$

The constants K_1, K_2 and K_3 — as also those of Equation 23, K and β — are determined from blasting measurements made for short-charge blasts. K and K_1 are scaling factors, β is the slope of the law of attenuation of peak strain with distance in a log-log plot and K_2 and K_3 define wave attenuation over time and wave frequency, respectively.

The inset in Figure 2a shows ξ_s, obtained by the authors from records for blasting in Lithonia granite based on a 0.76-m gelatin charge and 63.5-mm diameter. Figure 2a also shows both experimental results and calculated results, $\xi_R(t)$, at a radial distance

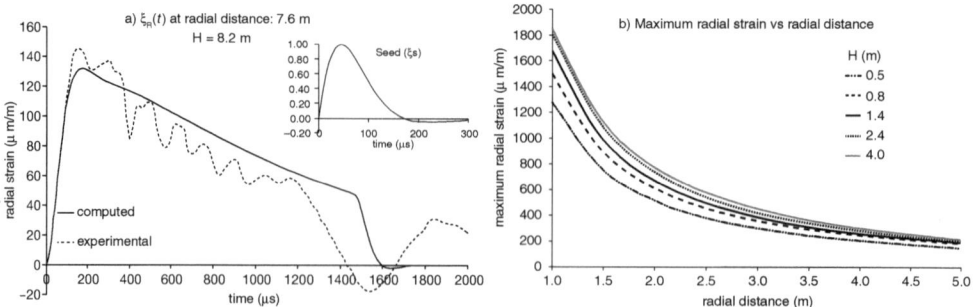

Figure 2 Examples of seed waveform model applications: (a) predicted radial strain wave for a rock mass point from the indicated seed (Starfield & Pugliese, 1968); (b) using the same seed, predicted maximum radial strain in response to radial distance for different charge lengths (H).

from the charge center of 7.6 m for an elongated charge (H = 8.2 m). Despite being a good prediction, the fact that ξ_S varies with distance must be taken into account, so values for this parameter cannot be extrapolated. The same is true of the attenuation parameter β, when rock breakage occurs in the vicinity of the blasthole. This same disadvantage applies to all predictive models that use this parameter.

Figure 2b shows how the maximum value of $\xi_R(t)$ varies for different charge lengths according to radial distance. It can be observed how the model represents the law of diminishing returns in response to variations in charge height. While the increase in charge length is held constant at about 70%, the average peak strain increase by 19% when length goes from 0.5 m to 0.8 m, and by 6% when length goes from 2.4 m to 4 m. Greater charge lengths would not generate more than a marginal increase in peak strain.

4 CONCLUSIONS

It is undeniable that, nearly 40 years after it was developed, the H-P model continues to predominate for the purpose of predicting BID. The strong and rather controversial assumptions underlying the model and the development of new PPV-based variants have not managed to render obsolescent a model that ultimately requires, like all other such models, experimentation and good judgment in its application. Non-PPV-based models calibrated with experimental data still require further research to confirm their usefulness outside their range of application, that is, in different experimental conditions to those for which they were calibrated. In this regard, Finn Ouchterlony's (1997) advice regarding his own model seems particularly relevant to using any of the models, namely, that any crack length predictions to be used in real blasting operations must be interpreted with great care.

The relatively high number of publications concerning blast damage control in rock masses in recent decades indicates the high level of interest in this field of study. Their reading, however, reflects some pitfalls that need to be overcome. First, there is no clear terminology regarding the definition of damage, due in part to differences in blasting contexts. Second, the lack of a standard methodology for estimating the extent of damage makes it difficult to thoroughly analyse the results of different authors. Finally, little experimental work has been published, especially taking into account how SveBeFo research has demonstrated the great difficulty in predicting damage, even when clearly defined, given the complexity of the phenomenon and the large number of variables.

No methodology exists that, combining technical, financial and safety criteria, enables a specific damage threshold to be established in blasting design. The challenge continues to be to establish a relationship between the parameters used to characterize rock mass damage from blasting and its extent and to determine what changes blasting causes in the geomechanical behavior of the rock mass.

REFERENCES

Ambraseys, N.R. & Hendron, A.J. (1968) Dynamic behavior of rock masses. In: Stagg, K.G. & Zienkiewicz, O.C. (eds.) *Rock Mechanics in Engineering Practices*. London, Wiley. pp. 203–227.

Ash, R.L. (1963) The mechanics of rock breakage (Parts I–IV). *Pit and Quarry*, 56 (2), 98–100 56 (3), 118–123; 56 (4), 126–131; 56 (5), 109–118.

Bastante, F.G., Alejano, L. & González-Cao, J. (2012) Predicting the extent of blast-induced damage in rock masses. *International Journal of Rock Mechanics & Mining Sciences*, 56, 44–53.

Bauer, A. (1982) Wall control blasting in open pits. In: Baumgartner, P. (ed.) *Rock Breaking and Mechanical Excavation, 14th Canadian rock mechanics symposium*, 13–14 May 1982, *Vancouver*. The Canadian Institute of Mining and Metallurgy, CIM special volume 30. pp. 3–10.

Cook, M.A. (1947) An equation of state for gases at extremely high pressures and temperatures from the hydrodynamic theory of detonations. *Journal of Chemical Physics*, 15 (7), 518–524.

Duvall, W.I., Atchison, T.C. & Fogelson, D.E. (1966) Empirical approach to problems in blasting research. In: American Rock Mechanics Association (publisher) *Failure and Breakage of rock, the 8th U.S. Symposium on Rock Mechanics, 15–17 September 1966, Minneapolis, Minnesota*. pp. 500–523.

Duvall, W.I., Johnson, Ch.F., Meyer, A.V.C. & Devine, J.F. (1963) *Vibrations from instantaneous and millisecond-delayed quarry blasts*. U.S. Department of the Interior, Bureau of Mines. Report RI 6151.

Favreau, R.F. (1969) Generation of strain waves in rock by an explosion in a spherical cavity. *Journal of Geophysical Research*, 74, 4267–4280.

Grady, D.E. & Kipp, M.E. (1987) Dynamic rock fragmentation. In: Atkinson, B.K. (ed.) *Fracture Mechanics of Rock*. London, Academic Press. pp. 429–475.

Harries, G. (1983) The modelling of long cylindrical charges of explosive. In: Holmberg, R. & Rustan, A. (eds.) *Proceedings of the First International Symposium on Rock Fragmentation by Blasting, 23–26 August 1983, Lulea, Sweden*. pp. 419–431.

Heelan, P.A. (1953) Radiation from a cylindrical source of finite length. *Geophysics*, 18 (3), 685–696.

Hendron, A.J. (1973) *Scaling of ground motions from contained explosions in rock for estimating direct ground shock from surface bursts on rock*. Army Engineer. Omaha District. Technical Report number: 15.

Holmberg, R. & Krauland, N. (1977) *Evaluation of the fracture frequency before and after blasting with 250 mm holes at Aitik*. Swedish Detonic Research Foundation Report DS 1977:12 (in Swedish).

Holmberg, R. & Persson, P-A. (1978) The Swedish approach to contour blasting. In: Konya, C.J. (ed.) *Proceedings of the 4th Conference on Explosives and Blasting Technique, 1–3 February 1978*. New Orleans. Society of Explosives Engineers. pp. 113–127.

Holmberg, R., & Persson, P-A. (1979) Design of tunnel perimeter blasthole patterns to prevent rock damage. In: Jones, M.J. (ed.) *Proceedings Tunneling '79, 12–16 March 1979*. London. Institution of Mining and Metallurgy. pp. 280–283.

Hustrulid, W. (1994) *The practical blast damage zone in drift driving at the Kiruna Mine*. In: Proceedings of the seminar Skadezon vid tunneldrivning. Swedish Rock Engineering Research, Stockholm. SveBefo Report number: 8, appendix 3.

Hustrulid, W. (1999a) *Blasting Principles for Open Pit Mining. Volume 1: General Design Concepts*. Rotterdam, Netherlands, A.A. Balkema.

Hustrulid, W. (1999b) *Blasting Principles for Open Pit Mining. Volume 2: Theoretical Foundations*. Rotterdam, Netherlands, A.A. Balkema.

Hustrulid, W. (2007) A Practical, yet technically sound, design procedure for pre-split blasts. In: *Proceedings of the 33rd Annual ISEE Conference on Explosives and Blasting Technique, Volume 1, 28–31 January 2007, Tennessee. Cleveland. Ohio*. International Society of Explosives Engineers.

Hustrulid, W. (2010) Some comments regarding development drifting practices with special emphasis on caving applications. *Mining Technology*, 119 (3), 113–131.

Hustrulid, W., Bennet, R., Ashland, F. & Lenjani, M. (1992) A new method for predicting the extension of the blast damaged zone. In: *Proceedings of the Sprängteknisk Konferens, 15–16 January 1992, Göteberg-Kiel*, Nitro Nobel.

Hustrulid, W. & Iverson, S.R. (2010) Evaluation of Kiruna mine drifting data using the NIOSH design approach. In: Sanchidrián, J.A. (ed.) *Proceedings of the 9th International symposium on rock fragmentation by blasting-FRAGBLAST 9, 13–15 September 2009, Granada*. London. CRC Press, Taylor & Francis Group. pp. 497–506.

Hustrulid, W. & Johnson, J.C. (2008) A Gas Pressure-Based Drift Round Blast Design Methodology. In: Schunnesson, H. & Nordlund, E., (eds.) *Proceedings of the 5th International Conference & Exhibition on Mass Mining, MASSMIN 2008, 9–11 June 2008, Lulea*. Lulea University of Technology. pp. 657–669.

Iverson, S.R., Hustrulid, W. & Johnson, J.C. (2013) *A New Perimeter Control Blast Design Concept for Underground Metal/nonmetal Drifting Applications*. National Institute for Occupational Safety and Health (NIOSH) Publication number: 2013-129, Report of Investigations 9691.

Iverson S.R., Hustrulid, W., Johnson, J.C., Tesarik, D. & Akbarzadeh, Y. (2009) The extent of blast damage from a fully coupled explosive charge. In: Sanchidrián, J.A. (ed.) *Proceedings of the 9th International symposium on rock fragmentation by blasting-FRAGBLAST 9, 13–15 September 2009, Granada*. London. CRC Press, Taylor & Francis Group. pp. 459–468.

Iverson, S.R., Kerkering J.C. & Hustrulid, W. (2008) Application of the NIOSH-Modified Holmberg-Persson Approach to Perimeter Blast Design. In: *Proceedings of the 34th Annual Conference on Explosives and Blasting Technique, 27–30 January 2008. New Orleans, Louisiana*. Cleveland. International Society of Explosives Engineers. 34pp.

Johnson, J.C. (2010) *The Hustrulid Bar–A Dynamic Strength Test and Its Application to the Cautious Blasting of Rock*. [Online] PhD dissertation, University of Utah, Utah. ProQuest, UMI Dissertations Publishing. Available from: http://gradworks.umi.com/34/12/3412454. html [Accessed 18th March 2015].

Kilebrant, M., Norrgård, T. & Jern, M. (2010) The size of the damage zone in relation to the linear charge concentration. In: Sanchidrián, J.A. (ed.) *Proceedings of the 9th International symposium on rock fragmentation by blasting-FRAGBLAST 9, 13–15 September 2009, Granada*. London. CRC Press, Taylor & Francis Group. pp. 449–457.

Krauland, N. (1994) *Experiences from damage zones in the mining business*. In: Proceedings of the seminar Skadezon vid tunneldrivning. Swedish Rock Engineering Research, Stockholm. SveBefo Report number: 8, appendix 2 (in Swedish).

Langefors, U. & Kihlström B. (1963) *The modern technique of rock blasting*. New York, Wiley.

Liu, Q. & Katsabanis P.D. (1993) A theoretical approach to the stress waves around a borehole and their effect on rock crushing. In: Rossmanith, H-P. (ed.) *Proceedings of the Fourth International Symposium on Rock Fragmentation by Blasting-FRAGBLAST 4, 5–8 July 1993, Vienna*. Rotterdam. A.A. Balkema. pp. 9–16.

Lu, W. & Hustrulid, W. (2003) The Lu-Hustrulid approach for calculating the peak particle velocity caused by blasting Explosives and Blasting Technique. In: Holmberg, R. (ed.) *Proceedings of the EFEE 2nd World Conference, 10–12 September 2003, Prague*. The Netherlands. A.A. Balkema a member of Swets & Zeitlinger. pp. 291–300.

Neiman, I.B. (1979) Determination of dimensions of the zone of crushing of rock in place by blasting. *Soviet Mining*, 15 (5), 480–485.

Olsson, M. & Bergqvist, I. (1993) *Crack growth in rock during cautions blasting*. Swedish Rock Engineering Research, Stockholm. SveBeFo Report number: 3 (in Swedish).

Olsson, M. & Bergqvist, I. (1996) *Crack growth during multiple hole blasting*. Swedish Rock Engineering Research, Stockholm. SveBeFo Report number: 18 (in Swedish).

Olsson, M. & Ouchterlony, F. (2003) *New formula for blast-induced damage in the remaining rock*. Swedish Rock Engineering Research, Stockholm. SveBeFo Report number: 65 (in Swedish).

Olsson, M., Svärd, J. & Ouchterlony, F. (2008) *Blast damage from string emulsion, field tests and a damage zone table including simultaneous initiation.* Swedish Blasting Research Centre och Luleå tekniska universitet, Stockholm och Luleå. Swebrec Report number: 2008:1 (in Swedish).

Ouchterlony, F. (1997) Prediction of crack lengths in rock after cautious blasting with zero interhole delay. *International Journal for Blasting and Fragmentation*, 1, 417–444.

Ouchterlony, F., Olsson, M. & Bergqvist, I. (2002) Towards New Swedish Recommendations for Cautious Perimeter Blasting. *International Journal for Blasting and Fragmentation*, 6 (2), 235–261.

Ouchterlony, F., Sjöberg, C. & Jonsson, B.A. (1993) Blast damage predictions from vibration measurements at the SKB underground laboratories at Äspö in Sweden. In: *Proceedings of the Ninth Annual Symposium on Explosives and Blasting Research, January 31 – February 4, 1993, San Diego, California.* Cleveland, Ohio. ISEE. pp. 189–197.

Persson, P-A., Holmberg, R. & Lee, J. (1993) *Rock blasting and explosives engineering.* Boca Raton, FL, CRC Press.

Plewman, R.P. & Starfield, A.M. (1965) The effects of finite velocities of detonation and propagation on the strain pulses induced in rock by linear charges. *Journal of the South African Institute of Mining and Metallurgy*, 66 (3), 77–96.

Sanden, B.H. (1974) *Pre-Split Blasting.* MSc Thesis, Mining Engineering Department, Queen's University, 125pp.

Starfield, A.M. & Pugliese, J.M. (1968) Compression waves generated in rock by cylindrical explosive charges. *International Journal of Rock Mechanics and Mining Sciences & Geomechanics Abstracts*, 5, 65–77.

Tesarik, D.R., Hustrulid, W. & Nyberg, U. (2011) Assessment and application of a single-charge blast test at the Kiruna mine, Sweden. *Blasting and Fragmentation*, 5 (1), 47–72.

Chapter 20

Blasting propagation velocity

C. González-Nicieza & M.I. Álvarez-Fernández
Department of Mining Engineering, University of Oviedo, Oviedo, Spain

Abstract: Blast induced ground vibration is an impact from the use of explosives that has historically been an extremely difficult problem to effectively mitigate. When an explosive charge detonates, the rapid reaction of the explosive components produces an intense dynamic wave (shock wave) set around the blast hole. The energy carried by these waves crushes the rock to a fine powder. Beyond the shock zone, although the energy of the waves gets attenuated through processes of geometric spreading and energy dissipation is enough to cause the radial cracking of the rock mass. Finally, the shock wave degenerates into seismic waves.

There are many variables involved in the transmission of ground vibration that when combined, result in the formation of a complex vibration waveform. A more profound knowledge of how stress waves propagate in fractured rock masses is needed to control and manage vibration impacts: traditional methods of prediction such as empirical equations must be complemented with artificial intelligence methods (AIMs) or numerical models based on the waveform superposition method. On the other hand, advanced signal analysis confirms the convenience of the revision of vibration and seismic safety standards.

I INTRODUCTION

During the detonation of an explosive charge, the small mass of explosives is converted into a large volume of gas immediately releasing a lot of energy, that is linked to the high speed detonation (VOD) of the explosive components (Ngo *et al.*, 2007). These VODs can reach 8000 m/s in commercial explosives, although speeds of more than 11000 m/s have been documented.

As the amount and speed of input of energy is greater that the transmission capacity of the ground, there is a state of "shock wave". That is a disturbance that passes through a medium faster than the speed of propagation of the waves in such a medium. It is a very high peak of dynamic pressure of over 10^4 MPa. The shape of the pressure-time history curve is characterized by a steep increase in pressure followed by an exponential decay (Stewart *et al.*, 2014), whose shape is often determined by the Friedlander equation.

Shortly after the shock wave passes, a partial vacuum is created since the surrounding rock cannot respond at the speed the wave tries to impose, causing its disintegration or 'break', an abrupt change in its properties.

In terms of stresses, when the shock wave reaches the wall of the hole, the pressure pulse generates a stress state σ_r–σ_θ. Initially, while the wave passes σ_r is compression and σ_θ of tensile, then the sign reversal is performed. In this triaxial state the rock mass is fragmented due to different breakage mechanisms such as crushing.

The pressure peak decreases greatly, however, σ_θ is enough to produce radial cracking (may extend outside the blasting hole to distances 4 to 6 times longer than the load radius) and reflection breakage in the presence of a free face.

As the energy consumption goes on in the fracture, the intensity of the stress wave diminishes and decreases the speed at which the wave is transmitted, until at last it equals the speed of transmission in the ground, time in which shockwave becomes an elastic seismic wave (which means that no permanent deformation occurs in the rock mass). These waves are transmitted into the ground in the form of seismic waves whose front moves radially (or cylindrically) from the source of detonation. However, the transmission is not uniform due to pre-existing fractures.

The dynamic of the explosion shows that the energy released by a blast produces different effects. Only a fraction of the explosive energy is efficiently consumed in the actual breakage and displacement of the rock mass (over 30 %), and the remaining energy is spent on undesirable effects, such as ground vibrations. Usually, the productive effects are (Ipromin, 2010):

- The movement of a pre-established rock volume;
- The rock fragmentation into well defined elements of regulated dimensions;
- The projection and resettlement of rocks on and at a certain distance against their initial position.

Unwanted consequences are:

- Excessive breaking of some of the rock.
- Projection (throw) in an excessive manner ("flying rocks").
- Permanent fractures and deformations of the rock mass.
- Ground and air vibrations.

For all this it is necessary to estimate the factors on which the energy transferred to rocks depends. The energy developed by the explosive reactions is an intrinsic thermal-dynamic feature of the explosive. Its value may be calculated and is expressed in thermal and mechanical units. The energy transfer is influenced both by the explosive and the rock mass and depends on the acoustic impedance of the two.

- The explosive impedance I_e is defined as the product of explosive density (ρ_e) and detonation velocity (VOD).
- The rock mass impedance (I_r) is defined as the product between its density (ρ_r) and the propagation velocity of the elastic waves (C).

So, the energy that is to be transferred is impacted by the impedance factor η_1, represented by the following equation:

$$\eta_1 = 1 - \frac{(I_e - I_r)^2}{(I_e + I_r)^2} \tag{1}$$

Another factor that influences the energy transfer onto the rock mass is the coupling ratio, expressed through the ratio between the blasthole diameter and the load diameter

(φ_b/ φ_e). So, the energy transfer onto the rock during a non-ideal situation is developed at a coupling factor of η_2 (Berta, 1990):

$$\eta_2 = \frac{1}{e^{\frac{\phi_b}{\phi_e}} - (e-1)}$$ (2)

2 GROUND VIBRATIONS

Ground motion is the most important environmental side effect of rock blasting. According to Gutowski (1976) the basis of almost all of the analytical work on sources and transmission paths in ground is contained within the pioneering work of Lamb, who investigated the response of isotropic, homogeneous and elastic half-spaces to various harmonic and impulsive loads. If an oscillating point load is applied to an unloaded elastic half-space, two types of waves will emanate from the loading point.

One type is termed *body waves*, one of which propagates at the longitudinal wave speed of an elastic solid, and whose vibrational motion is in the direction of propagation. The characteristic of the longitudinal waves is that they propagate through successive compressions and decompressions along the propagation direction.

There is another *body wave* type, called transverse wave, that travels at the shear wave speed and which is slower that the longitudinal wave speed. The material particle is travelling within a plane that is tangent to the wave front. The motion vector may change its direction within this plane. This makes the trajectory of the particle within the tangent plane a curve line, like an ellipse, circle etc.

In addition to the *body waves*, there are *surface waves* (mainly Rayleigh & Love). The formation of the Rayleigh waves is the result of the non-plane nature of the volume wave's front that impacts on the land free surface.

It is also of interest to note that, for a small rigid disk vibrating on a half-space surface, 67% of the input energy goes into Rayleigh waves, 26% into shear waves, and only 7% goes into the longitudinal waves (Gutowski, 1976).

The vibration experienced at some distance from the blast normally comprises a complex combination of wave types that are difficult to separate in practice (Brinkmann, 1987). The character of ground vibration is fully described by measurements, in three mutually perpendicular directions, of the time history of the acceleration, the velocity, or the displacement at a point (particle) in the ground. Blast vibration is described using the following terms (Bender, 2007):

- Displacement: the distance the particles move.
- Particle Velocity: how fast the particles move.
- Acceleration: the rate at which the particle velocity is changing, measured in mm/s^2 or in g's.
- Frequency: The number of oscillations per second that a particle makes under the influence of seismic waves, measured in Hertz (cycles per second).
- Propagation Velocity: The speed at which a seismic wave travels away from the blast, measured in feet/second or m/s. Propagation velocity is several orders of magnitude greater than particle velocity.

In order to estimate and analyze the blast vibration effect and consequences, different indicators have been proposed such as: peak particle velocity (PPV, defined by Dehghan, 2010) as the highest speed at which an individual earth particle moves or vibrates as the waves pass through a particular site), peak particle acceleration (PPA), peak particle displacement (PPD), etc.

The peak ground motion (velocity, acceleration or displacement) is the maximum vector sum of the three components (and is not the maximum vector sum, calculated with the maxima for each component). This *false* maximum vector sum may be up to 40% greater than the other one. A true vector summation can provide a more precise measurement of particle velocity at each instant in time by summing the three components of motion into an amplitude vector (Farhad *et al.*, 2014). On the other hand, the *true vector sum* could be as much as 73% higher than the highest individual channel although, in practice, it is usually only 15 to 20% higher (Bender, 2007).

The wave motion spreads concentrically from the blast site in all directions and its amplitude necessarily decays with distance from the source. This decay is usually called attenuation. Mathematically, attenuation may be represented as (Triviño, 2012):

$$\frac{d\dot{u}_{max}}{dr} \quad \text{or} \quad \frac{d(ln\ \dot{u}_{max})}{dr} \tag{3}$$

where $\dot{u}_{max}$ is the peak particle velocity at a distance r from the source.

The primary reason for attenuation in rock blasting, common to any material, is the geometric spreading of energy. At any time after the blast, the wave front defines a sphere centered on the source, so the wave energy is distributed on the surface of this sphere. As the sphere surface increases with the square of the distance and the total wave energy remains constant, the energy density at any given point decays by a factor r^2. Since wave energy is proportional to the square particle velocity amplitude, the geometric attenuation of body waves is proportional to r. In the case of surface waves, the energy is distributed in a cylindrical way, rather than a spherical one, so the energy density decays by r instead of r^2, and geometric attenuation is proportional to $\sqrt{r}$.

In addition to geometric spreading, seismic waves experience loss of energy caused by friction and other forms of energy dissipation. Since this energy loss per cycle of deformation is a material property, it is called *material damping*. The classical expression for wave attenuation by *material damping* is:

$$\dot{u}_{max2} = \dot{u}_{max1} \cdot e^{-\alpha \cdot (r_2 - r_1)} \tag{4}$$

where α is the attenuation coefficient, $\dot{u}_{max1}$ and $\dot{u}_{max2}$ are the peak particle velocity at a distances r_1 and r_2 from the source.

Geometric spreading and material damping attenuation can be combined in a single expression as follows:

$$\dot{u}_{max2} = \dot{u}_{max1} \cdot \left(\frac{r_1}{r_2}\right)^n \cdot e^{-\alpha \cdot (r_2 - r_1)} \tag{5}$$

with n = 1 for body waves and n = 0.5 for surface waves.

Another parameter that characterizes the seismic wave is its frequency. As a result of an explosion a wide frequency range appears, but some frequencies (or frequency ranges) are predominant. Since wave decay caused by material damping increases proportionally with the number of deformation cycles, and a higher frequency wave passes through more deformation cycles than a lower frequency wave for the same travelled distance, the attenuation coefficient α increases with frequency. So, predominant frequencies decrease from the source.

Rock mass anisotropy (discontinuities, faults, heterogeneities, etc.) and topography also influence the transmission of vibrations, generating refractions and reflections in the waves, so actual attenuations are different from theoretical ones.

2.1 Factors affecting ground vibrations

Factors influencing ground vibration can be classified in two main groups. One group defines the characteristics of the surroundings, affecting transmission:

- The type of rock and rock mass (rock mass rating, RMR), which has a natural transmission velocity and frequency that favors wave propagation for that frequency. The presence of families of discontinuities and their characteristics (as separation, filling or water) also affect transmission. This factor affects the wave energy too.
- The distance to be covered by the wave between the source and the control point.
- The presence of significant discontinuities (faults), special geological structures, etc.

The second group of factors defines wave energy and is related to the source, in this case, blasting:

- The type of explosive, as high-velocity explosives will generate a higher intensity shock wave.
- The weight of explosive detonated per delay, which defines the amount of energy used in the generation of the shock wave.
- Blast geometry, defined by parameters such as perforation diameter, blasthole length, spacing, burden, number of blastholes.
- The blasting sequence, since the waves from successive blastholes can behave in one of three ways: accumulate to amplify the overall wave; counteract each other to minimize the overall vibration; or act independently.

At short distances, factors which define wave energy are more significant, but at high distances transmission parameters are determining.

2.2 Prediction models

The effect of vibration caused by blasting in the vicinity has become an urgent problem to be solved. The prediction of ground vibration components has a great importance in the minimization of the environmental complaints. While any of the three kinemetric descriptors (displacement, velocity and acceleration) could be employed to describe ground motion, the great correlation between cracks induced by blasting and the particle velocity (described by three perpendicular components: longitudinal, transverse and vertical) makes this parameter the most preferable to be used.

The problem of predicting the transmission of vibration through the ground is complex (Gutowski, 1976). The reasons for this complexity include the lack of a comprehensive understanding of ground behavior, the difficulty of determining accurate values of its properties, and the difficulty of modeling precisely the sources of vibration and the resulting near- and far-field behavior. However, in spite of these and other obstacles, it is possible to make reasonable assessments of ground-transmitted vibration through a judicious use of the empirical and theoretical results that are available.

Empirical research shows that the relation between the ground velocity may be estimated through the following equation:

$$PPV = K \cdot \frac{Q^\alpha}{D^\beta} \tag{6}$$

where:

- PPV is the maximum velocity of the particle (mm/s);
- Q is the maximum explosive charge per delay (kg);
- D is the distance to explosion (m);
- k, α, β are coefficients depending on the properties of explosives, the geological environment or the blasting technology and those are established experimentally for each site. The term K reflects the source energy and the coupling efficiency of the explosive to the blasthole wall (Yilmaz & Unlu, 2014). Higher values of K indicate high energy and well coupled explosives. The term β represents the loss of vibrational energy with distance. Higher values of β represent less competent rock mass which attenuates vibrational energy more quickly while lower values represent a competent rock mass, which transmits the vibrational energy with little attenuation (Scott, 2009).

The concept of scaled distance was put forward in order to calculate the attenuation of particle velocity in the ground. The scaled distance is derived by combining the distance between source and measurement points, and the maximum charge per delay. This scaled distance is defined by the equation below:

$$SD = \frac{D}{\sqrt[a]{Q}} \tag{8}$$

where:

- SD is the scaled distance, and
- a is 2 (USBM, 1959) or 3 (Ambraseys & Hendron, 1968). For a fully confined spherical explosion in a elastic infinite semi-space, the theory suggests that a = 3. By means of a dimensional analysis it can reach a value of a = 2.

Several empirical formulae have been established (Table 1).

For the calculation of the acceleration, an empirical formula proposed by Hudson (1961) may be used:

$$a = 877,95 \cdot \frac{Q^{0,75}}{D^2} \tag{7}$$

By assuming a log-normal occurrence of the PPV at any particular scaled distance, Dowding (1985, 1996) presented an approximate statistical method to derive the 84

Table 1 Empirical formulae to predict ground motion.

Formula	Author
$PPV = K \cdot \left[\dfrac{D}{\sqrt{Q}} \right]^{-\beta}$	USBM (Duvall & Petkof, 1959)
$PPV = K \cdot \left[\dfrac{D}{\sqrt[3]{Q}} \right]^{-\beta}$	(Ambraseys & Hendron, 1968)
$PPV = K \cdot Q^{0.55} \cdot \left(\dfrac{h}{D} \right)^{0.1}$	(Awojobi et al., 1974)
*h is the depth where the explosive load is detonated	
$PPV = K \cdot \left[\dfrac{\sqrt{Q}}{\sqrt[3]{D}} \right]^{\beta}$	(Langefors & Kihlstrom, 1963)
$PPV = K \cdot \left(\dfrac{D}{\sqrt{Q}} \right)^{-\beta} \cdot e^{-\alpha D}$	(Ghosh-Daemen, 1983)
$e^{-\alpha D}$ represents the inelastic attenuation factor	
$PPV = n + K \cdot \left(\dfrac{D}{\sqrt{Q}} \right)^{-1}$	CMRI (Pal Roy, 1993)

n = site constants which is influenced by rock properties and geometrical discontinuities,
K = site constants which related to design parameters

and 95% confidence lines of the PPV using the standard error of the prediction. The 95% confidence line refers to the PPV value that will not be exceeded in 95 out of 100 blasts. In other words, the 95% confidence line can be viewed as the 95th percentile line. However, these current approaches do not incorporate the posterior uncertainty of the fitting coefficients. In order to resolve this problem, Yan & Yuen (2015) propose a Bayesian method to derive the site-specific fitting coefficients based on a small amount of data collected at an early stage of a blasting project.

The standard scaling law itself has been questioned for not considering the wave nature of the radiating signals from a blast (Blair, 1990), and has also been qualified as inadequate to predict blast-induced damage (Fleetwood *et al.*, 2009).

In González-Nicieza *et al.* (2014) a new equation is proposed, established relating the velocity of vibration produced per unit charge of explosive (PPV/Q) with distance via the following expression:

$$\frac{\text{PPV}}{\text{Q}} = K \cdot D^{\beta} \tag{9}$$

where K y β are two parameters that will depend on the geological conditions and are obtained by the statistical fitting of the equation of the obtained curve (Figure 1).

Due to the fact that the number of influencing parameters is high, artificial neural networks (ANN) and several artificial intelligence methods (AIM) were developed to predict rock blasting vibrations (Azizabadi *et al.*, 2014).

A neural network can be considered as an intelligent cube with the capacity to predict output behavior when it recognizes input behavior. It is trained by processing a large number of input behavior patterns and shows the output behavior patterns. Following suitable training, when presented with a new input model the neural network is capable of recognizing similar behaviors; the result is a prediction of outputs.

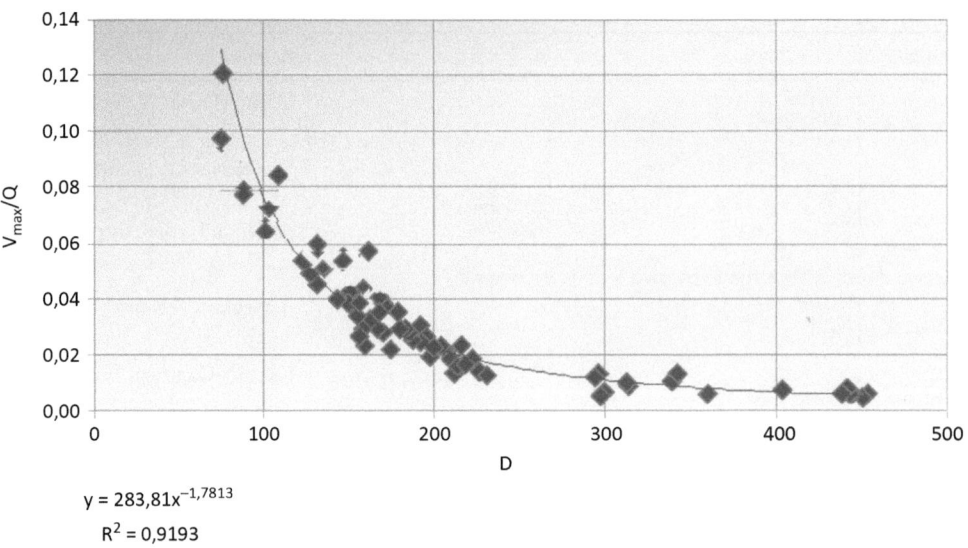

$$y = 283{,}81x^{-1{,}7813}$$
$$R^2 = 0{,}9193$$

Figure 1 Law of transmission per unit charge.

Neural networks are capable of learning from previous events. Thus, a neural network trained with a suitable number of input datasets can make predictions regarding outputs associated with new input data, provided this is of the same kind that the network used during training. Figure 2, left, depicts the ANN model. The values x_i are the input data and w_{ij} are the corresponding weights. The sum block builds the linear combination of the data with the weights as coefficients adding the optional term b_i, known as the bias. The neuron output is a function y_i, obtained in the activation block through one of the many kinds of activation or transfer functions, although the most widely used is the sigmoidal activation function.

Some experiences indicate a satisfactory prediction for peak particle velocities and frequency using the ANN methodology (Khandelwal & Singh, 2007; Mohamed, 2011; Kamali & Ataei, 2010; Saadat *et al.*, 2014). In Álvarez-Vigil *et al.* (2012) the predictions obtained were compared with those obtained using conventional statistical methods. The correlation coefficients obtained with ANN was 0.98 for peak particle velocity (PPV) and 0.95 for frequency (f), as can be seen in Figure 3, compared to 0.50

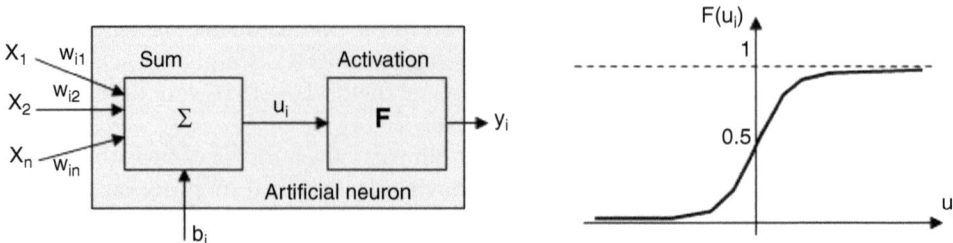

Figure 2 Artificial neural network model and sigmoidal activation function.

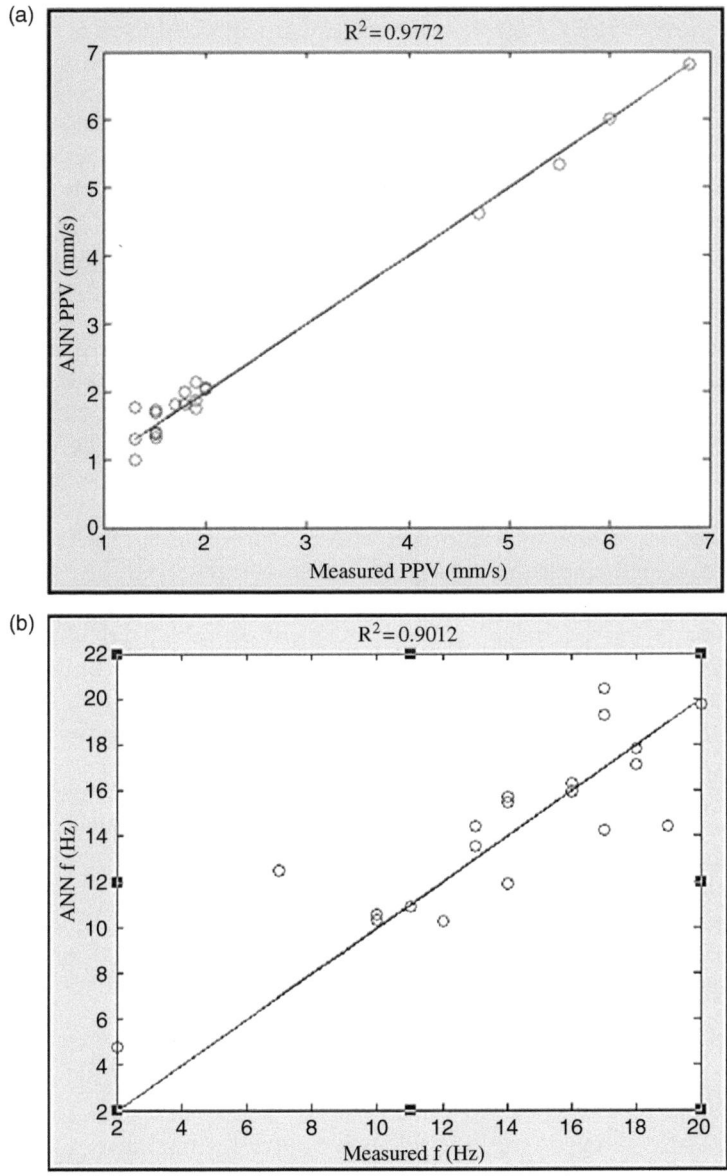

Figure 3 (a) Correlation between ANN-predicted and measured PPV values (b) Correlation between ANN-predicted and measured f values.

and 0.15, respectively, for conventional statistical methods (Multiple Linear Regression).

The network used in Álvarez-Vigil *et al.* (2012) was a three layer network (input, hidden and output), with 11 neurons in the input layer, 5 neurons in the hidden layer and 2 neurons in the output layer. To train the network, a series of 60 data was used corresponding to an initial measurement campaign implemented over 10 months, by varying the number of holes (between 5 and 40), their diameters (between 105 and 140 mm) and their charges. Vibration measurement was conducted at four sites using accelerometers. To test and validate the ANN methodology, 20 sets of new input-output data were selected. These data were not used in training the network.

Other artificial intelligence methods (AIM) such as fuzzy interface systems (FIS), adaptive neuro-fuzzy inference system (ANFIS) or RS-FNN (Rough Set – Fuzzy Neural Network) comprehension method have been adopted to predict the peak particle velocity and principal frequency.

Empirical methods and AIM only provide an estimation of the maximum amplitude of particle velocity and give no information about the complete seismic waveform. Waveform superposition modeling that includes all parameters, *i.e.* modeling of the complete seismic waveform produced by a blast, would overcome these weaknesses.

The waveform superposition method is based on the principle that the measured time history at a given point is the result of linear superposition in the time domain of the time histories emitted by each of the single-hole charges. Due to the linearity of the problem and the superposition principle in which any distributed source can be described as the sum of multiple point sources, there is no additional difficulty in modeling a complex source. In addition to spectral amplitudes, all phase effects from the superposition are included in the synthetic seismogram.

The procedure begins by drilling a single hole and loading it by a charge similar to the holes of the actual blast pattern. The displacement or velocity time history is then measured at locations where the ground vibrations are to be predicated or reduced. The next step is to simulate the complete blast seismogram at specific locations by superposition method. The single assumption for the medium is that the paths of waves for the single-hole shot and the production blast are the same.

Blair (2004, 2011) has comprehensively reviewed the waveform superposition technique and the charge weight scaling technique. The study concluded that the charge weight scaling technique remains the most widely used approach to predict vibrations due to rock blasting.

A dimensional analysis (DA) technique has been performed by Khandelwal & Saadat (2014) on various blast design parameters to propose a new formula for the prediction of the peak particle velocity (PPV) obtaining a CoD value of 0.830 between PPV predicted and measured.

2.3 Influence of geology and depth

Dogan *et al.* (2013) showed that PPV values for underground blasts are 47% to 95% smaller than surface blasts during evaluation of blast induced effects. In a more recent investigation carried out by Caylak *et al.* (2014), it is shown that ground vibrations show different spreading properties in different directions, depending on the geological structure of the region.

González-Nicieza *et al.* (2014) published a study to discuss and test the existence of damping in ground vibrations transmission laws between quarry blasts and an old railroad tunnel which is located under one of the quarry benches (Figure 4).

The study collects data from two year-recordings. Information gathered from two borehole triaxial geophones (GEO1, GEO 2) has been used to analyze the vibration caused by the blastings carried out in the old railway tunnel. These geophones were

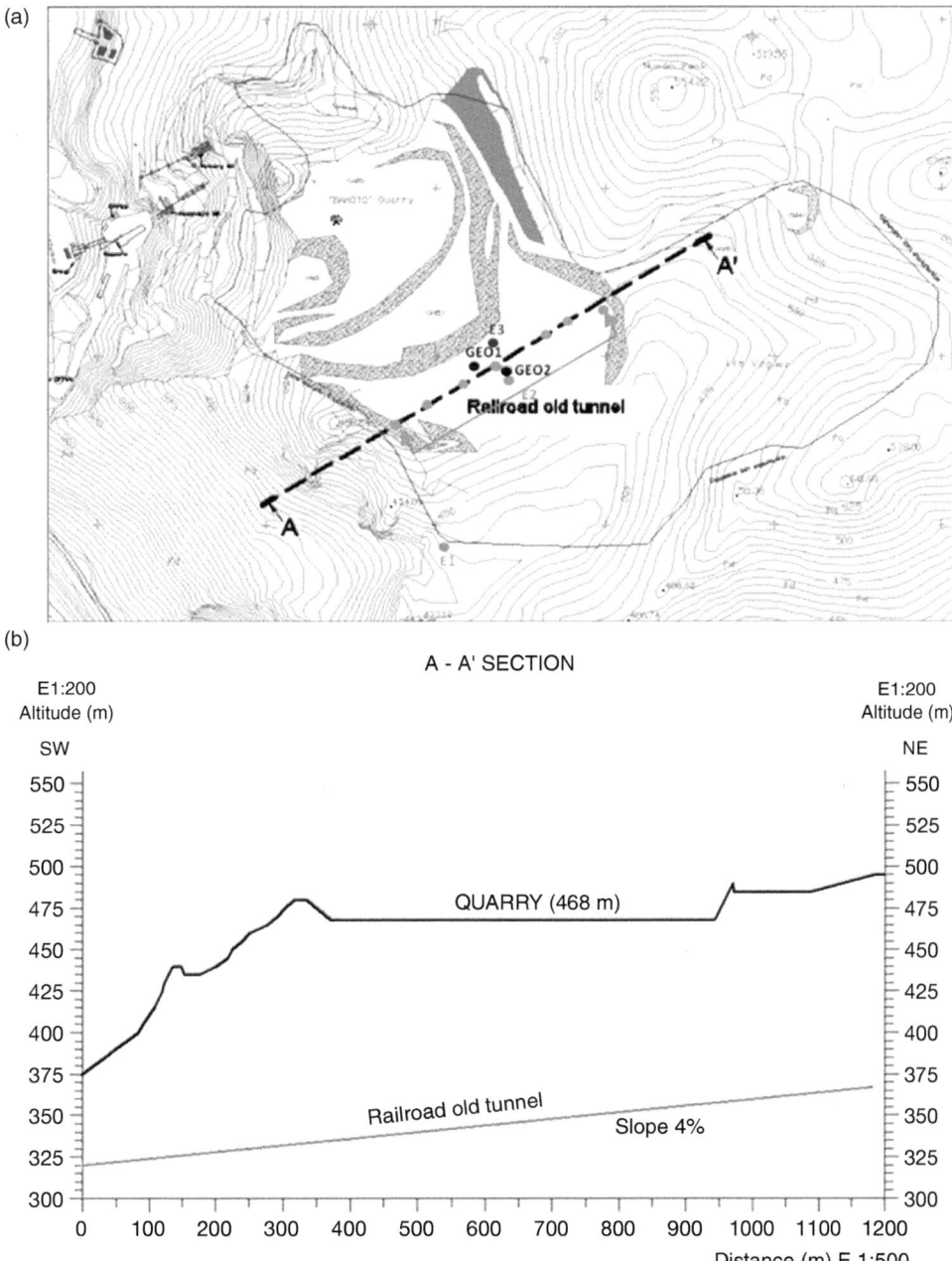

(a)

(b)

A - A' SECTION

Figure 4 Location of the old railroad tunnel, measure stations and topographical profile.

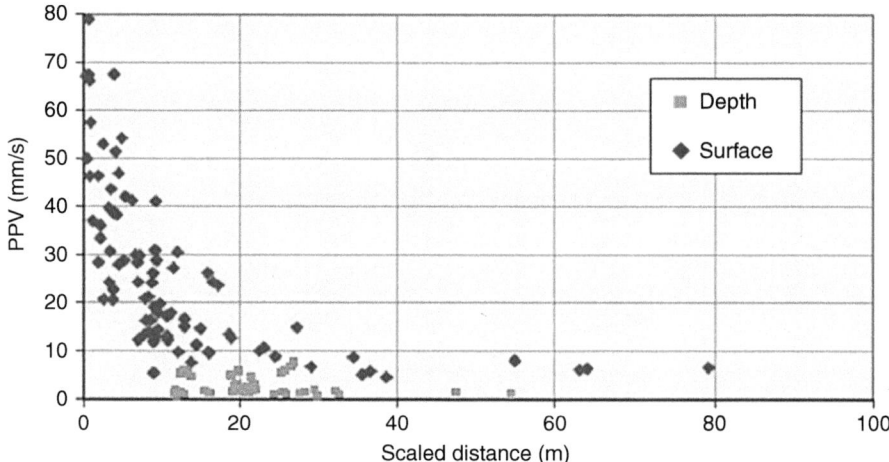

Figure 5 PPV versus the scaled distance from surface accelerometers and underground geophones.

located next to the tunnel, one on each side, and at same level in depth (Figure 4). To measure vibrations on surface three tri-axial high dynamic range stations, model ETNA, have been used. One of the accelerometers was settled on a permanent basis, on a station called E1 that was located on the right bank of the tunnel. The other equipment, E2, was moving along the trace of the tunnel surface, across the square from the quarry,

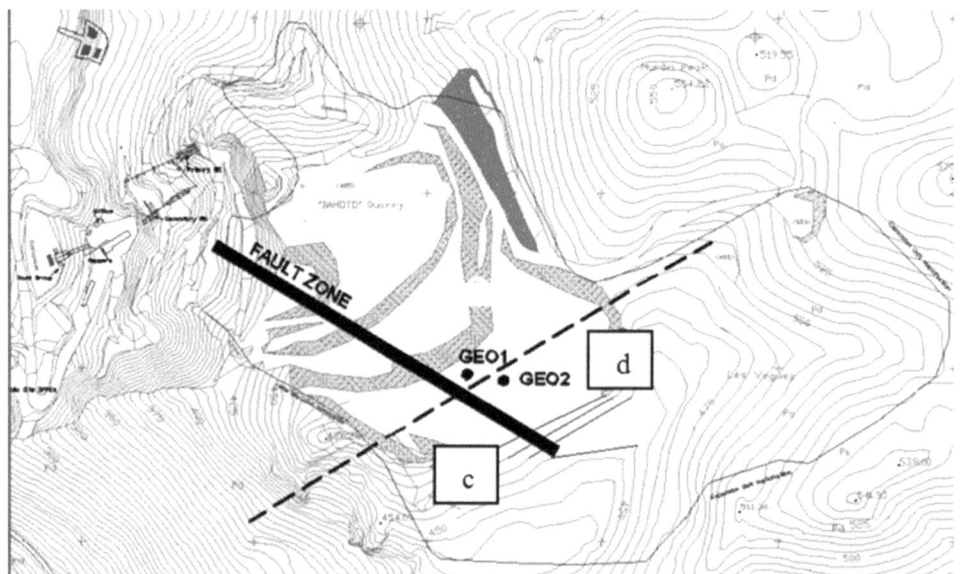

Figure 6 Location of benches *c* and *d* with respect to the fault running through the quarry.

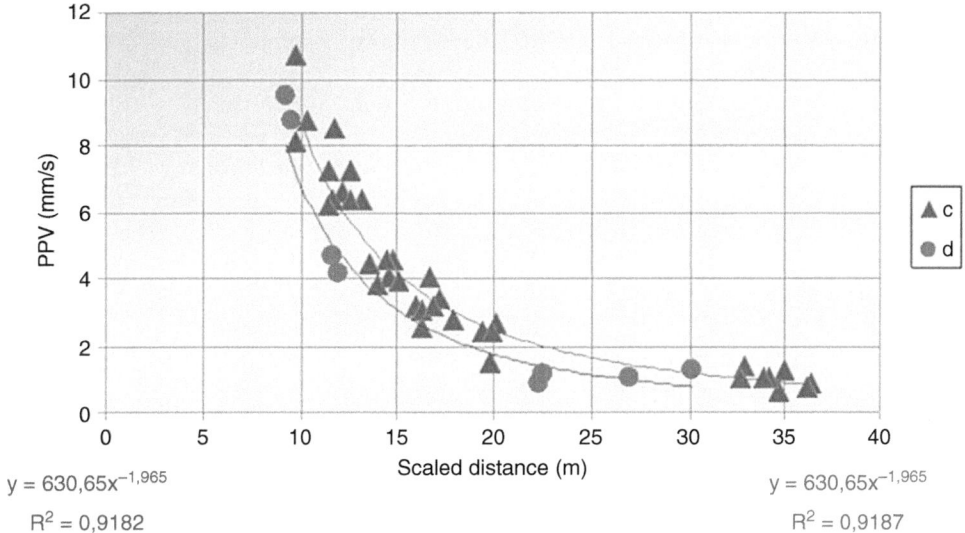

$y = 630{,}65x^{-1{,}965}$

$R^2 = 0{,}9182$

$y = 630{,}65x^{-1{,}965}$

$R^2 = 0{,}9187$

Figure 7 Law of transmission for benches *c* and *d*.

for records at different distances from the controlled blasting. Then, five blasting records were obtained with other equipment located in the square of the quarry (E3).

Verification of the existence of damping of vibrations with depth was carried out by jointly analyzing the data obtained on the surface and underground. Figure 5 shows the values of the PPV versus the scaled distance of the measuring device from the blast, for both the measurements on the surface and those carried out underground.

If the PPV measures obtained at the same time on the surface accelerometers and depth geophones are checked, it is noticed that the recordings obtained on the surface are between 70% and 80% higher than those registered underground.

With the fittings obtained for the calculated laws of transmission of vibrations, it may be stated that the analyzed rock mass presents a fairly uniform structure. However, the law of vibrations was analyzed for benches *c* and *d*, between which there is a fault whose alignment is shown in Figure 6.

The slight difference that exists between the laws of the two benches can be observed in Figure 7, PPV of bench *c* is lower than PPV of bench *d*. As bench *d* is situated on the same side of the fault as the two geophones, the readings of the blasts in this bench are higher than the readings of the blasts in bench *c*, which is located on the opposite side of the fault.

3 MEASURE INSTRUMENTS

The sensors are the first elements of the measuring system (Ipromin, 2010). The seismograph is a special seismic instrument used to measure seismic waves. A seismograph has at least three channels allowing the record of three orthogonal directions: longitudinal, transversal and vertical. Because of the importance of frequency, the full waveform should be record.

The seismograph is usually consisting of three blocks:

- Seismic transducers block: There are three types of seismic transducers: motion transducers, velocity transducers (geophones) and acceleration transducers. The two last ones are the most commonly used. Accelerometers are good at recording high frequency vibrations, whereas geophones are well suited for low frequency vibrations. The seismic records need to be made for three directions: one vertical and two horizontal. One of the horizontal directions needs to be radial and the other one transversal onto the direction blast center-measuring point.
- Amplifier block.
- Command and record block: Usually, the amplifier and the command-record blocks are built together and are known as the central unit. In some seismographs these three blocks are placed under the same casing.

The main characteristics of transducer are sensitivity and frequency response (Dowding, 1993). Sensitivity is the ratio of its electrical output to its kinemetric measure (acceleration, velocity or displacement). Frequency response is the frequency range over which the relation between electrical output and mechanical motion is constant (lineal). The state of the art in hardware design is far ahead of the state of the art in user technology. It is the responsibility of users to develop an adequate level of understanding of the sensors that they select.

Planning a monitoring program using seismic instruments is similar to other engineering design efforts. A typical engineering design effort begins with a definition of an objective and proceeds through a series of logical steps to preparation of plans and specifications. Similarly, the task of planning a monitoring program should be a logical and comprehensive engineering process that begins with defining the objective and ends with planning how the measurement data will be implemented (Dunnicliff, 1993).

A critical aspect is the mounting of transducers in the field. The results shown by (Segarra *et al.*, 2015) have important implications in vibration control studies where measurements on a hard surface are done. The worse the transfer of vibrations from rock to sensor is the higher is the error in vibration data. Depending on the frequency and the coupling method used, the rock motion in the measurement can be modified by a factor from 0.16 to 1.25 for the range of frequencies (17–200 Hz) and the vibration levels studied (5 and 20 mm/s):

- Free placed sensors tested at 5 mm/s amplify, in general, rock motion.
- Sandbag mounts encompass a complex transmission of vibrations.
- Anchoring performance is independent of the mount characteristics and of the input velocity. It appears as the only method that leads to a stiff rock-to-mount coupling, with resonant frequency much higher than the usual frequency from blasting, ensuring consistent measurements for the frequencies commonly found in blasting.

4 SIGNAL ANALYSIS

To analyze vibration signals is necessary to take into consideration the objective of the analysis: damage control, cut design in underground excavations, blast design, optimization of delays, etc.). Some of the more commonly used methods for analysis are:

1. Time Analysis: The time history of wave (velocity or acceleration) must be analyzed. It is fundamental to evaluate which are the more important component of the movement (L, T or V), the true and the false maximum vector sum, the direction of true maximum vector sum and the evolution with distance. If it is possible, is interesting to characterize the delays between boreholes.

2. Frequency Analysis: This kind of analysis is more difficult and its objective is to obtain the dominant frequency of the vibration. There are some methodologies with different principles and results.

3. Energy spectrum Analysis: The energy transferred to the rock in the form of seismic waves is calculated as the integral of the energy flow past a control surface at a given distance from the blast. The energy flux (the power or rate of work, per unit area) is the scalar product of the stress at the surface and the particle velocity (Achenbach, 1975). Ram Chandar & Sastry (2015) have proposed another concept in estimating energy by analyzing the complete wave forms obtained, with geophones, from ground vibration monitoring. Wave forms produced from blasting operations are complex and, are analyzed with the help of a signal processing software, on the base of the square of the amplitude can be integrated to get the area of the curve and the energy. Summation of energy in all three directions is taken for estimating the total wave energy from each event.

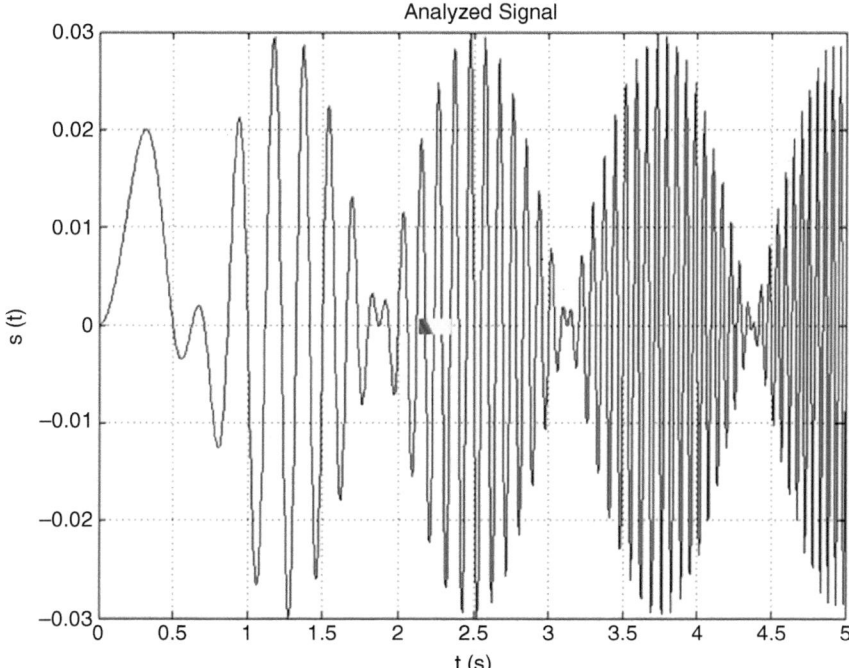

Figure 8 Analyzed synthetic signal.

Table 2 Synthetic signal details.

Amplitud máxima	0.03 cm/s
Duración	5 s
Frecuencia de muestreo (fs)	200 Hz
Frecuencia mínima	0 Hz (t = 0 s)
Frecuencia máxima	10 Hz (t = 5 s)

Some of these methods are exposed as follows. It will use a synthetic speed signal, assuming that it represents the record of speed using implementations gathered in MathLab software (MathLab 2015-b) (longitudinal, transversal or vertical) from a sensor. This signal (shown in Figure 8), of variable frequency, is constructed with the following function:

$$V(t) = 0.03 \cdot \cos\left(\frac{4}{5}\pi t\right) \cdot \sin(4\pi t^2) \, \text{cm/s} \tag{10}$$

Its characteristics are summarized in Table 2.

4.1 Fourier transform calculation

The Fourier Transform is the old and most used spectral analysis method, mathematically defined as:

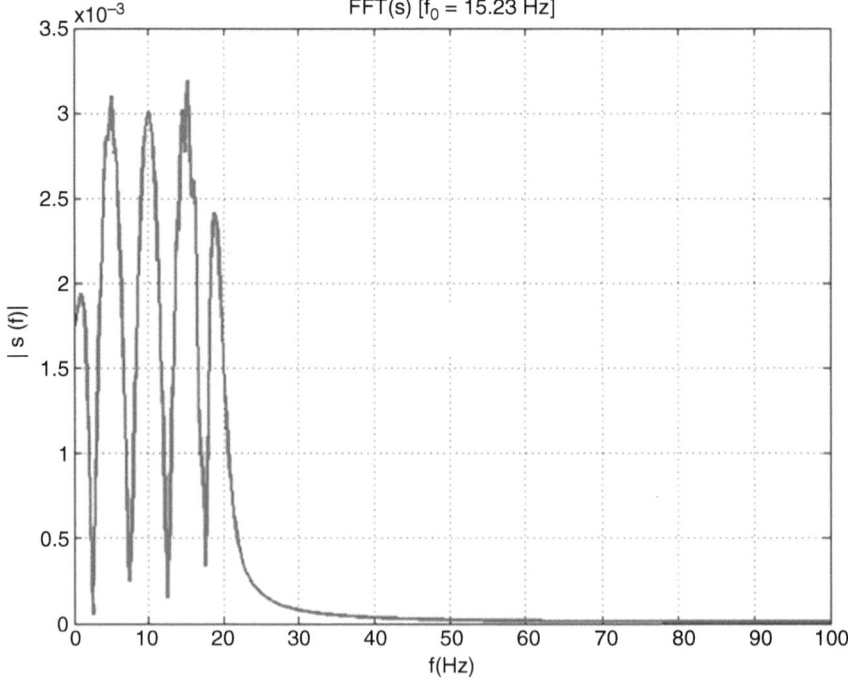

Figure 9 FFT of analyzed signal.

$$F(f) = \int_{-\infty}^{\infty} V(t) \cdot e^{-j2\pi ft} dt \tag{11}$$

From a numerical point of view, Fourier transform is calculated with the FFT (Fast Fourier Transform) and in the analyzed signal's case, the resulting FFT can be seen in Figure 9. There are five local maxima that roughly correspond to frequencies of 1, 5, 10, 15 and 18 Hz. There is no clear dominant frequency, while the maximum corresponds to 15.23 Hz.

4.2 Power Spectral Density (PSD)

The spectral analysis aim is to describe the power frequency distribution contained in a discrete signal using a finite set of data. The power spectral density (PSD) of a discrete random signal V(n), is defined as:

$$P_d(f) = \frac{1}{Lf_s} \left| \sum_{n=0}^{L} V(n)e^{-\frac{j2\pi fn}{f_s}} \right|^2 \tag{12}$$

where L is the number of the signal samples and f_s the sampling frequency.

There are many methods to improve PSD calculation. Out of all of these, Welch's algorithm will be used (Welch, 1967). In this case, DSP is shown in Figure 10. Three

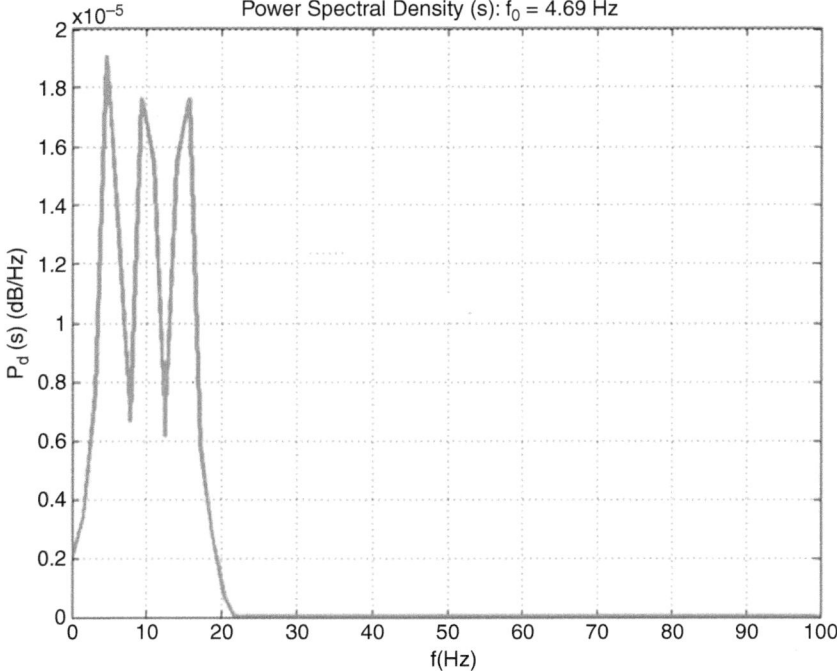

Figure 10 PSD of the analyzed signal (Welch algorithm).

peaks are roughly located at the frequency values of 5, 10 and 15 Hz. PSD corresponds to the highest density of values of the signal in time, which are the three central lobes of the signal in Figure 8.

4.3 Short Time Fourier Transform (STFT)

For a non-standing signal, as those registered from seismic events or blasting analysis, the above mentioned techniques indicate the presence of different frequencies but don't show the distribution of these frequencies in time. One way to improve the signal analysis is assuming as stationary through a small window in time. Then, its Fourier transform will provide its content in frequency in a short period of time. Adjusting the window's width properly and shifting in time, the content of frequencies in the signal can be removed to a frequency-time graphical representation: the STFT (Nawab & Quatieri, 1988). Mathematically, the STFT for an instant τ and a frequency f, is defined as:

$$STFT(\tau, f) = \int_{-\infty}^{\infty} V(t) \cdot g(t - \tau) e^{-j2\pi ft} dt \tag{13}$$

where $V(t)$ is the analyzed signal, g (t) is the window function and $e^{-j2\pi ft}$ is the Fourier kernel. In this case, the STFT is shown in Figure 11. The linearity of the frequency over time can be seen. Although the three maximums are almost similar, the STFT finds the maximum at the instant 3.715 s for a frequency of 14.85 Hz. Also other two peaks can

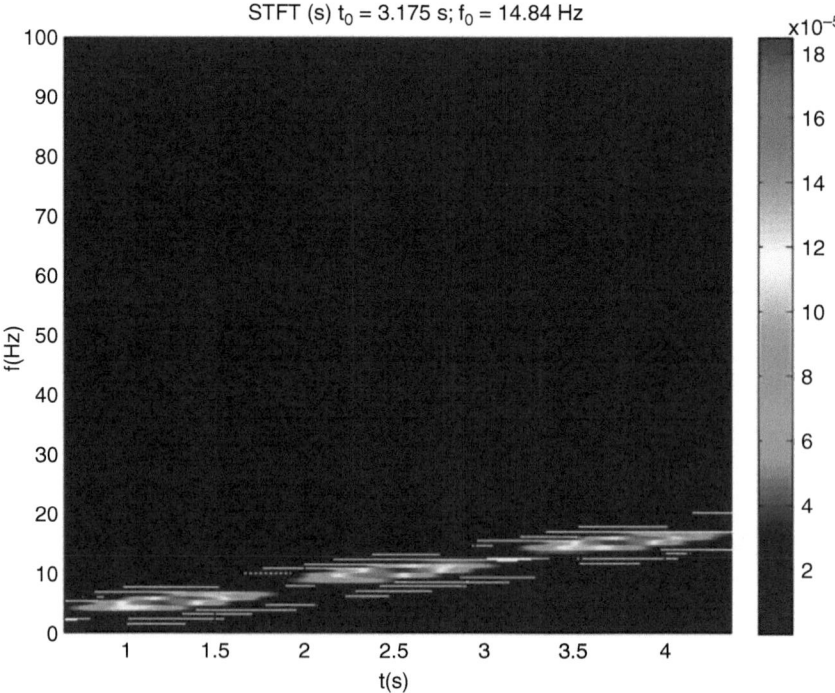

Figure 11 STFT of the analyzed signal.

be noticed at the instants 1.25 s and 2.5 s corresponding to frequencies of 5 and 10 Hz, respectively.

4.4 Continuous Wavelet Transform (CWT) and Wavelet Packed Analysis (WPA)

A continuous wavelet transform (CWT) is used to separate a continuous-time function into wavelets. Unlike Fourier Transform, the continuous wavelet transform possesses the ability to construct a time-frequency representation of a signal that offers very good time and frequency localization.

A wavelet is a wave limited to a short time length which means an average value of 0. Comparing Wavelets to sinusoidal waves, which are the basis of Fourier analysis, there are two main differences: sinusoidal waves have unlimited length in time (extend from $-\infty$ to $+\infty$) are smooth and can be predicted. However, wavelets tend to be irregular and asymmetrical.

Mathematically, the continuous wavelet transform (CWT) is defined as a sum of the signal $V(t)$ over the time, multiplied by and scaled-time-shifted versions of the mother wavelet function, φ, i.e.:

$$CWT(a, b) = \frac{1}{\sqrt{a}} \int_{-\infty}^{\infty} V(t) \cdot \varphi\left(\frac{t - b}{a}\right) dt \qquad (14)$$

The parameters a and b are named the scale and the shift time, respectively. The scale parameter stretches or compresses the wavelet (lower scale higher compression). In Figure 12 one of the most used wavelet, known as *db5*, is depicted. Its central frequency to a scale and a sampling period equal to 1 is of 0.6667 Hz.

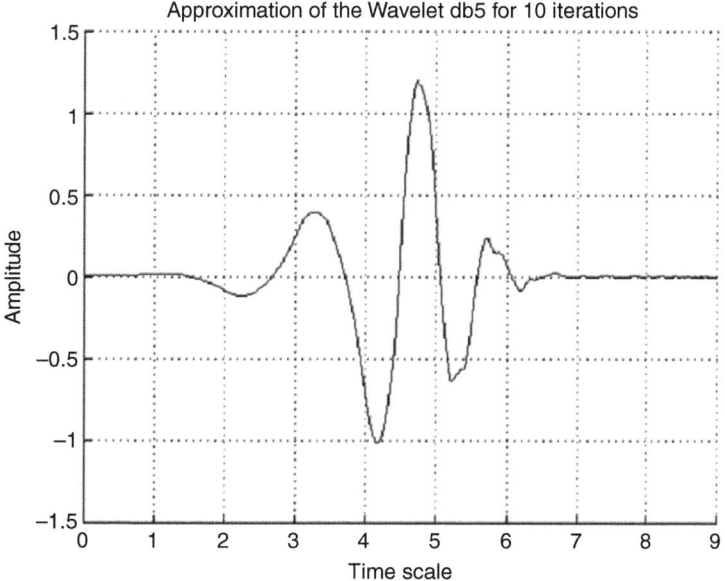

Figure 12 Wavelet db5.

In Figure 13 can be seen the CWT for the signal using the wavelet *db5*, with 64 scales. It has been represented the time on the x-axis, frequency on the y-axis and the value of the coefficients on the z-axis. The range of colors on the right side represents the iso-value wavelet transformation coefficients. The highest coefficient corresponds to a frequency of 3.81 Hz found at the instant t = 1.130 s. All the local maxima of the coefficients are found at very low frequencies and the lobes that presents the signal in time can be seen clearly (Figure 8).

The CWT is very useful for detecting very fast transient events which are not easy to say from the time-signal representation. Moreover, the wavelet analysis method is better to distinguish initiation time of millisecond delay detonator, and phenomena of super-position and cut of seismic wave by different delay detonators can be found. Wavelet transform analysis can provide a method to evaluate the blast-induced damage in the rock mass and to detect singularities (such as faults, discontinuities, caves) hidden in conventional analysis.

For many signals (as in this case), the low frequency content is the most important part. It is what gives to signal their identity. On the other hand, the high frequency content provides details of the signal. Applied to the human voice, if the high frequency content is removed, the voice sounds different but understandable. However, if the low frequency content is removed, it becomes unintelligible.

In analysis wavelet *approximation* and *details* are the words mentioned. *Approximations* are the high scales (low frequencies) and *details* are the lower scales (high frequencies). So, it proceeds to the decomposition of the signal in successive approximations and details. In the case of the analysis wavelet packaging (WPA) the

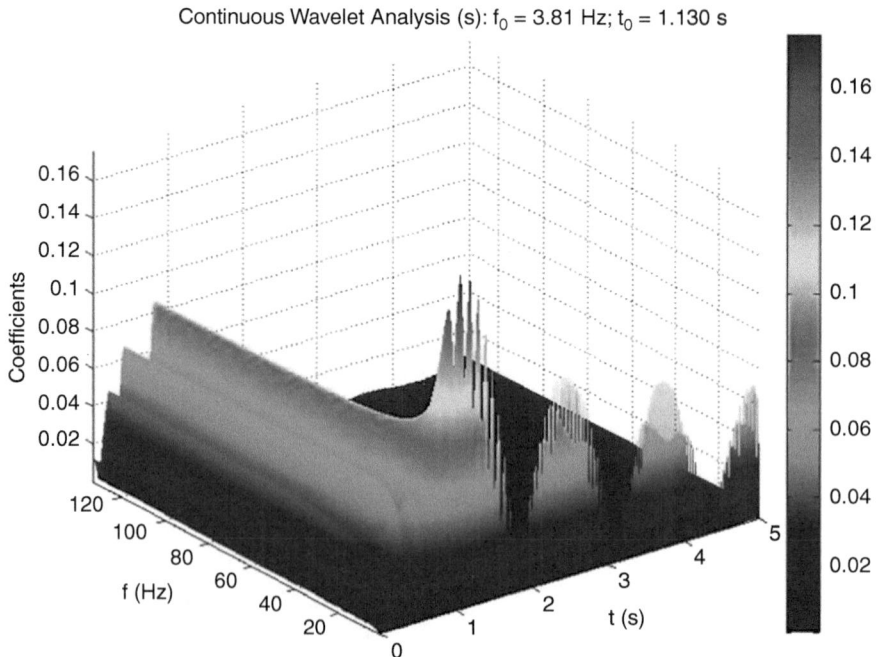

Continuous Wavelet Analysis (s): $f_0 = 3.81$ Hz; $t_0 = 1.130$ s

Figure 13 CWT of the analyzed signal db5, 64 scales.

original signal (S) decomposes, at the first level, in the part of low-frequency (A1) and the part of high frequency (D1). In a second level of decomposition, the A1 signal is decomposed into AA2 and DA2 and the high-frequency signal level 1 (D1) is decomposed into two. The decomposition can continue up to the desired level.

4.5 Energy spectrum analysis

When a signal $V(t)$ is decomposed using WPA to level i, $V(t)$ can be expressed in the form (Zhong, 2015):

$$s(t) = \sum_{j=0}^{2^i-1} \varphi_{i,j}(t_j) = \varphi_{i,0}(t_0) + \varphi_{i,1}(t_1) + \dots + \varphi_{i,2^i-1}(t_{2^i-1}) \tag{15}$$

where $\varphi_{i,j}(t_j)$ is the reconstructed signal at node (i,j) (level i, j sub-band). If the lowest and maximum frequency of $s(t)$ are 0 and f_m, then the frequency width at the ith level is $f_m/2^i$.

According to spectral analysis Parseval's theorem, the energy spectrum WPA of the signal, at (i, j) node, can be obtained in the form:

$$E_{i,j}(t_j) = \int_{-\infty}^{\infty} |\varphi_{i,j}(t_j)|^2 \, dt = \sum_{k=1}^{m} |x_{j,k}|^2 \tag{16}$$

where $x_{j,k}$ are the coefficients of the reconstructed signal at node (i,j) and m is the number of samples of the signal $V(t)$. Consequently, the total energy of $V(t)$ is obtained from the expression:

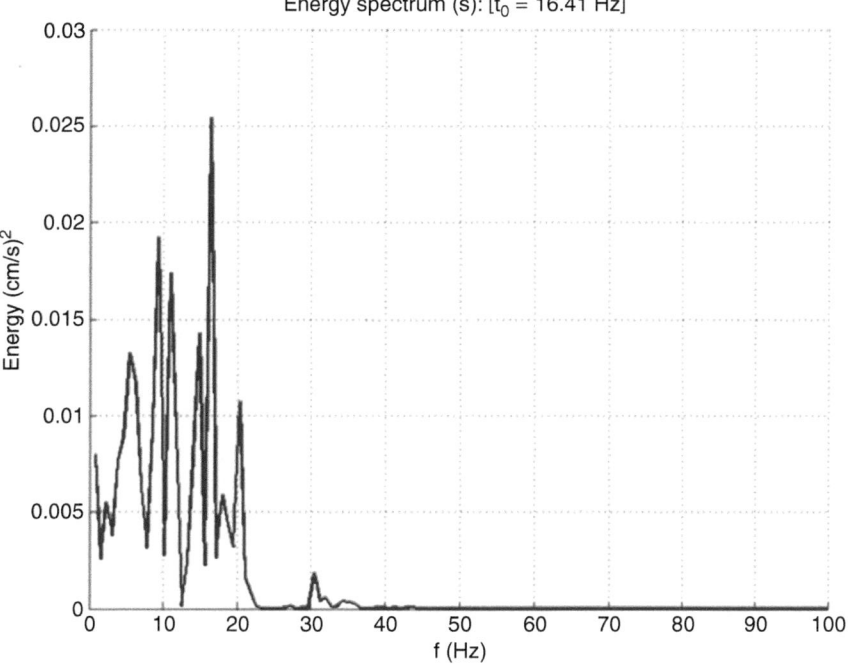

Figure 14 Spectrum of energy from the analyzed signal.

$$E = \sum_{j=0}^{2^i-1} E_{i,j}(t_j) \tag{17}$$

Equations 15 and 17 show that the signal, decomposed by the WPA method in different frequency components, reflect the effect of the vibration frequency as well as energy band and, simultaneously, intensity and length of vibration. As a result, the energy obtained from the WPA of the signal reflects the three elements of vibration (intensity, frequency and length).

Figure 14 shows the spectrum of energy obtained with wavelet db5. The dominant frequency in the energy spectrum is 16.41 Hz. There are some frequencies in the spectrum, near the 10 Hz and also around 5 Hz and 20 Hz (values close to those obtained with the FFT).

5 DAMAGE CRITERIA AND PREVENTION

When it comes to establishing structural safety against the effects of a seismic event (*e.g.* a blast or an earthquake), criteria which reflect factors such as the intensity, frequency and length of vibration, as well as other factors inherent to the structure which suffers vibration, such as its natural frequency and the damping coefficient should be taken into account.

5.1 Structure response

Understanding the dynamic response and damage characteristics of structures to blast ground motions is essential for safe and economic design of mining and construction blasting, as well as for protecting structures. The response of structures to blast vibration is conditioned on:

1. Vibration parameters as duration, frequency, energy or amplitude.
2. Parameters of the ground structure foundation.
3. Behavior of the structure and its elements.

In accordance with fundamental material properties, the relatively high frequency noted is less damaging to structures than waves of lower frequencies. This is exemplified in blasting regulations with regards to higher maximum PPVs at higher frequencies (ISEE, 1998).

Structures consist of many components, and two of most important are walls and superstructural skeletons. Superstructure response, measured at a corner, is associated with the shearing and torsional distortion of the frame, while the wall response, which measured in the middle of the wall, is associated with bending of that particular wall. The wall and superstructure continue to vibration freely after the passage of the ground motion, according to Dowding (1985). He also indicated that the wall motion tends to be larger in amplitude than the superstructure motions and tends to occur at higher frequencies during free vibration than those of the superstructure. Detailed studies (Dowding *et al.*, 1980; Medearis, 1976) have shown that the natural frequencies of walls range from 12 to 20 Hz and those of superstructures from 5 to 10 Hz.

A simplified model is to assume that the structure consists of a single degree of freedom. In this case, with a vibration force of $F_i sin \omega_i t$, the equation that governs its dynamic behavior is:

$$\ddot{x} + 2\xi_0 \omega_0 \dot{x} + \omega_0^2 x = \frac{F_i}{m} sin \omega_i t \qquad (18)$$

being:

ω_0, the system natural pulsation:

$\omega_0 = \sqrt{\frac{k}{m}}$, where m is the mass and k rigidity.

ξ_0, the damping coefficient.

Damping is a function of building construction and to some extent the intensity of vibration. Measurement reveals a wide range of damping for residential structure with an average of 5% (Dowding et al., 1981).

Structural dynamics theory indicates that the factor of amplification β_i, which corresponds to the movement of the structure at a frequency $f_i = \frac{\omega_i}{2\pi}$, is given as:

$$\beta_i = \frac{1}{\sqrt{(1 - \omega_i^2/\omega_0^2)^2 + 4\xi_0^2(\omega_i^2/\omega_0)^2}} \qquad (19)$$

The amplification factor (that is, the response to the vibration of a structure) is a dimensionless physical quantity, clearly related to its natural frequency and the damping coefficient.

The probability of damage in structures depends on the relationship between dominant frequency of the ground vibration and natural frequency of the structure. If, in addition, the damping coefficient is very small, the response of the structure (its displacement), will be maximum. Amplification is defined as the increase in the amplitude measured in the structure with respect to ground amplitude due to the transfer of the exciting wave on the ground to the structure.

Based on this idea, a way of analyzing the response of a structure under a vibration is to build the response spectrum of the signal. A response spectrum is a plot of maximum responses of different SDF (Single Degree of Freedom) systems to the same blast vibration. Different authors have shown that the measured structural response correlates better with calculated SDF response than with peak ground motion. In the calculated relative displacement time history there will be a maximum. If that maximum is multiplied by the structural natural frequency, the resulting product is called pseudovelocity. Response spectrum is a plot of the pseudovelocities calculated from different natural frequencies.

On the other hand, the blast vibration response of free structures (including ground structures as slopes) is very different to that of restrained ones (tunnels, pipelines, basement walls, underground mines). Movements, on free structures, continue after the ground motion has passed and can amplify or reduce selectively incoming ground motions. Usually, strains in a restrained structure will be the same as for the surrounding ground.

Besides, during underground excavation, blast induced vibrations can cause micro-cracks in the fresh concrete lining and decrease its strength. Some researchers (Zhang *et al.*, 2005) reported that concrete can be successfully vibrated up to 4 hours from the time of mixing and the strength can be increased by up to 14%. If revibration takes place 5 or 6 hours later, it can damage the concrete and decrease its strength. That time window is often too short to finish one round of excavation in the field. Others (Ahmed & Ansell, 2014) give recommendations emphasizing that blasting should be avoided during the first 12 hours after shotcreting and that distance and shotcrete thickness are important factors for how much additional time of waiting is possibly needed.

Chen *et al.* (2007) introduced the dynamic response characteristics of a rock-anchored beam (RAB) under blasting vibration. It is found that the key to ensure the safety of RAB is to prevent the split of the bonding interface between the RAB and a vertical rock wall, and to control the horizontal peak particle velocity (PPV). A PPV of 70 mm/s for a RAB older than 28 days would be of some safety. For a RAB in the age range of 3 to 7 days the safe PPV is 30 to 70 mm/s and for an age range of 0 to 3 days, 20 to 30 mm/s.

5.2 Damage criteria

Both the operators and the general public need adequate safeguards based upon factual data to protect their specific interests. Industry needs a reliable basis on which to plan and conduct blasting operations to minimize or abolish legitimate damage claims and eliminate the nuisance variety of complaint. The public would benefit by the absence of conditions which would create damage.

An exhaustive review of historic efforts concerning damage criteria is exposed by Mostafa (2010). Recently, much effort has been expended on developing numerical models to estimate dynamic response and damage of structures to blast ground motions. However, there is a plethora of standards available the world over based on various aspects of ground vibrations and no attempt has made to reconcile regulations between countries. Different countries have set their own standards on the basis of their extensive field investigations carried out in their mines for several years.

Most of the existing standards currently use PPV and the fundamental frequency of vibration to assess the effects of vibrations. Thus it occurs, for example, with the Spanish standard UNE 22-381-93, Standard German DIN 4150, Australian standard AS 2187, English standard BS 7385-2, American standard USBM RI 8507, Switzerland Sn 640 312A, CMRI Indian standard standards or Chinese standard GB6722. It stands out with the Swedish SN640312 (1978) regulation which took into account the same criteria and has been replaced by the SS 460 48 66 (1991), which does not take account frequency directly.

On the other hand, there is another group of criteria for damage to structures based on energy criteria. An early energy criterion was proposed by Crandell (1949). It is based on considerations of kinetic energy and the assumptions of simple harmonic motion, defined as:

$$ER = 10.8 \cdot \frac{A^2}{f^2} \tag{20}$$

where ER is the energy ratio, A is the peak acceleration (in m/s^2), and f is the frequency of peak amplitude (in Hz). Crandell's damage criteria were based on pre- and post-blast investigations of over 1000 residential structures. He recommended that the threshold level at which minor damage occurs is about 3 while above 6 there is more danger.

Nicholls (1971) points out that, although not used by Crandell, this equation is readily rewritten to express energy ratio as a function of velocity as follows:

$$ER = \frac{V^2}{2353} \tag{21}$$

where the units for particle velocity (V) are in mm/s.

More recently Guosheng *et al.* (2011) assessed the energy carried by the wave on the frequency characteristic of a particular structure through the energy spectrum obtained from the WPA of the signal. Assuming that f is the frequency of the vibration, ξ_0 the coefficient of damping and f_0 the natural frequency of the structure, according to Equation 19, the response coefficient of a structure, for a given band of frequencies, obtained by WPA, at node (i, j), is defined as:

$$\varepsilon_{i,j} = \frac{2^i}{f_m} \int_{f_{i,j-1}}^{f_i} \frac{1}{\sqrt{(1 - f^2/f_0^2)^2 + 4\xi_0^2(f/f_0)^2}} \, df \tag{22}$$

where $f_{i,j-1}$ y f_i are the upper and lower frequencies of the corresponding sub-band frequency at node (i, j) of the wavelet decomposition.

Figure 15 shows the response coefficient of the frequency band for the case of the analyzed signal, decomposed by WPA to level 7, for a sampling frequency of 100 Hz,

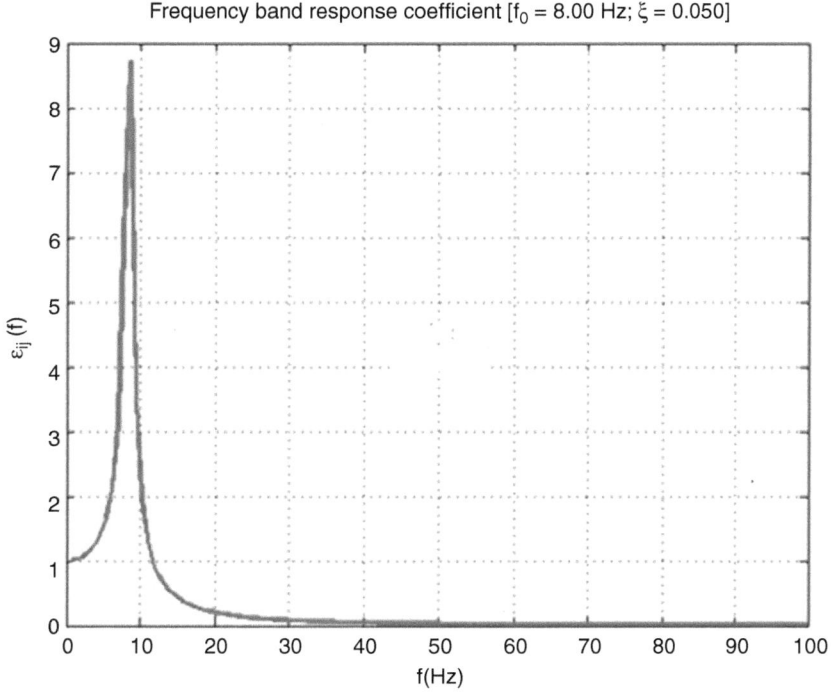

Figure 15 Response coefficient of band frequencies.

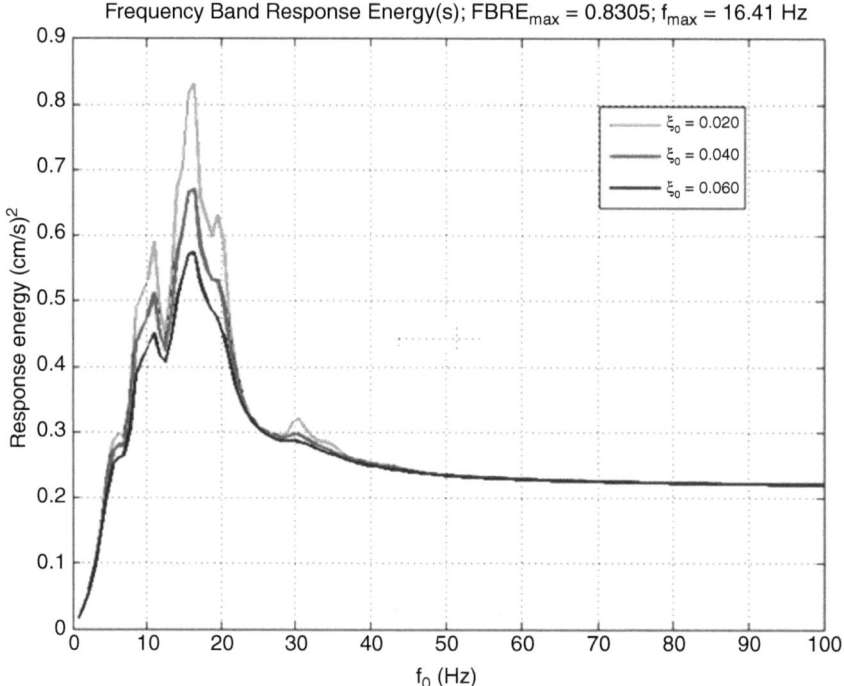

Figure 16 FRBE of the analyzed signal.

assuming that the structure suffering vibration has a natural frequency of 8 Hz and a 5% damping coefficient.

However, Guosheng *et al.* (2011) do not take into account that there are different frequencies of impact on the structure. To improve their criteria, energy WPA must be multiplied by the corresponding coefficients of band response. In this way the frequency band response energy (FBRE) is obtained. In the analyzed case, the FBRE is shown in Figure 16, for three usual damping coefficients in structures: 2%, 4% and 6%. As can be seen, the energy response presents a maximum of 0.8305 (cm/s) at frequency 16.41 2 Hz, for a damping coefficient of 2%.

To check this methodology a comparative analysis with actual data is carried out in four housing structures. Four seismic monitoring stations type ETNA were installed in some other buildings referred to as ED-E, ED-H, ED-M and ED-R. Their locations and the analyzed blasting are shown in Figure 17.

Figure 18 shows the velocity record, in cm/s, for the most unfavorable component in each control station. The recorded PPV values were very low: 0.0971; 0.0531; 0.0951 and 0.0517 cm/s, respectively, while the dominant frequencies determined by the FFT were 6.64, 5.66, 8.79 and 6.64 Hz respectively. In accordance with standard UNE 22-381-93, none of the recorded signals would cause damage to buildings (Group II of the Table 3).

Figure 19 shows the FBRE, for each of the signals. For each sub-band three damping coefficients are considered: 2%, 4% and 6%.

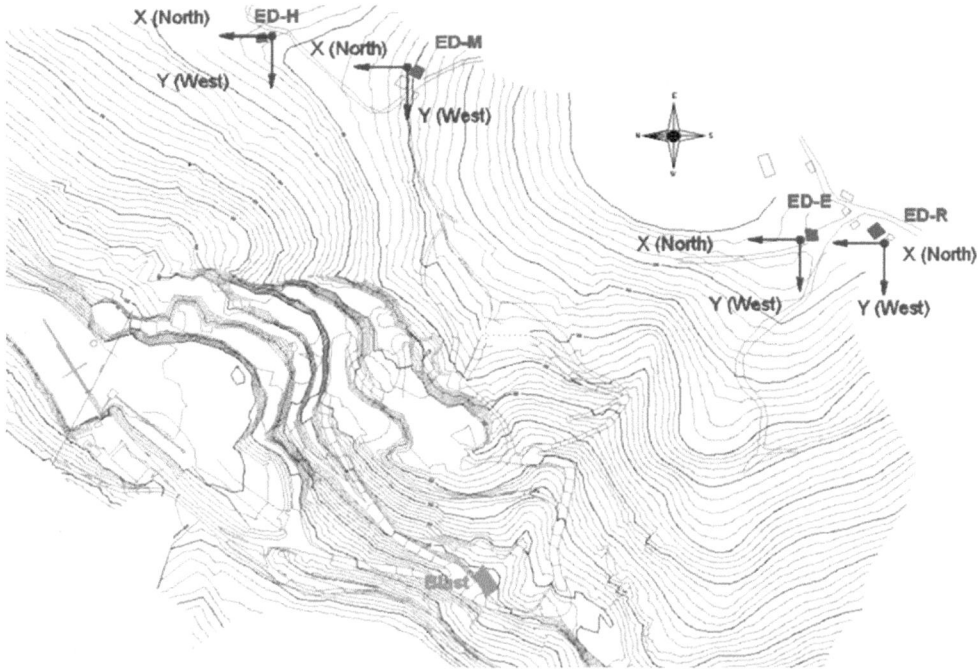

Figure 17 Location of blasting and control stations.

The maximum response energy is obtained, obviously, to a damping of 2% in the four buildings (Table 4). It should be noted that the differences between the response energy in the four buildings are much more remarkable than in the velocity field.

Figure 20 shows the comparison of FRBE of each signal with standard UNE 22-381-93, adapted to units of energy (using square limit velocity values from the standard).

The analysis of these figures allows establishing that:

- The signal recorded in the ED-H building does not exceed safety limits in any case.
- In building ED-M, the maximum FRBE exceeds the regulation for buildings of Group III in a narrow frequency band between 7.5 and 12 Hz, but not for Group II.
- ED-E and ED-R buildings exceed safety limits for buildings of Group II marked by adapted standard, even for sound insulation of 6%, in the frequency range of 5 to 8 Hz. This means that if the natural frequency of the buildings were within that band, safety limits would be exceeded and the blasting would be classified as dangerous. Remember that the superstructure responds to lower frequency (5 to 10 Hz).

The convenience of a revision of the methodology of analysis of vibration and seismic safety (blasts and earthquakes) standards is clear, as the wavelet decomposition-based analysis (WPA) proves to be more precise and restrictive than the methodologies currently used.

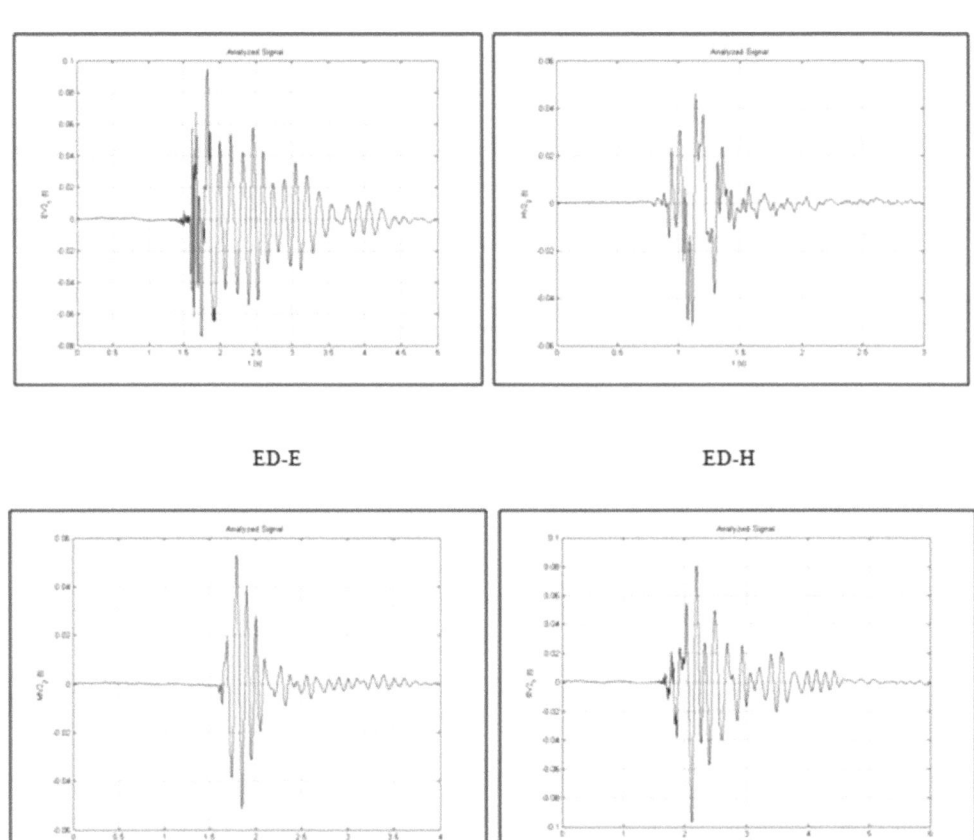

ED-E   ED-H

ED-M   ED-R

Figure 18 The most unfavorable velocity signal in each control station.

Table 3 Maximum permissible velocities (UNE 22-381-93).

Group	< 15 Hz	15 – 75 Hz	> 75 Hz
III	0.4	0.4 – 2.0	2.0
II	0.9	0.9 – 4.5	4.5
I	2.0	2.0 – 10.0	10.0

Note: Group I: buildings and industrial buildings. Group II: housing buildings, shopping malls or recreation. Group III: Structures of archaeological or historical value.

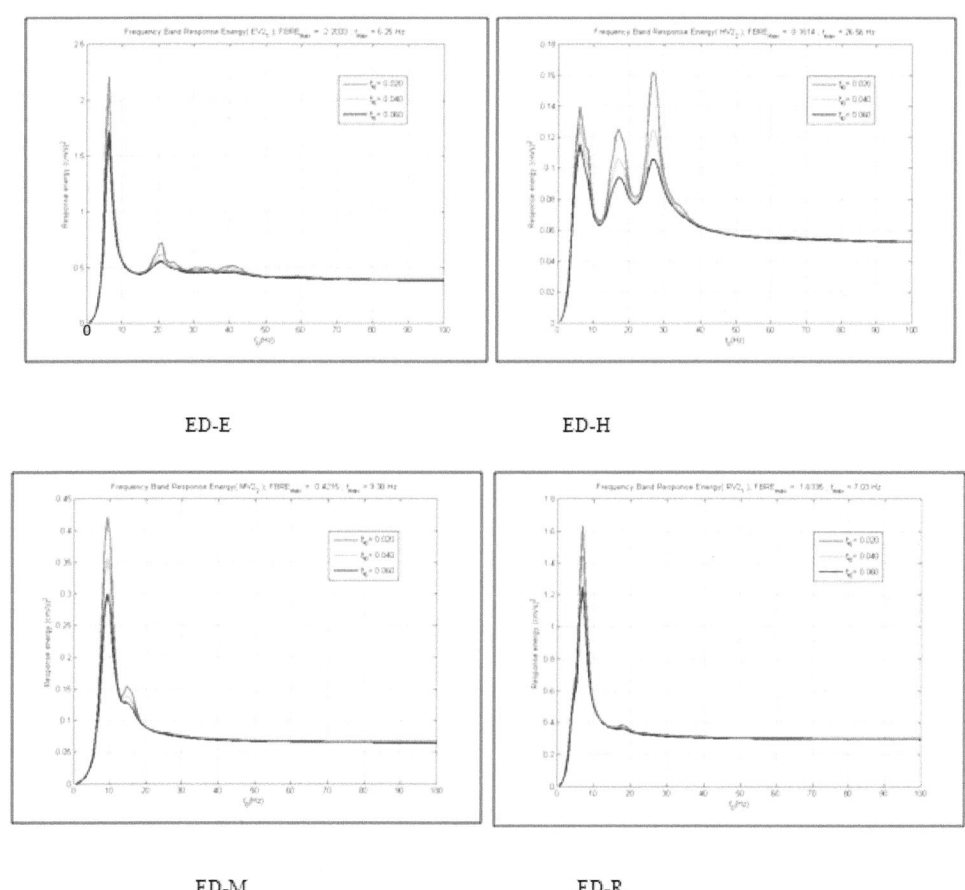

ED-E   ED-H

ED-M   ED-R

Figure 19 Response energy to frequency band.

Table 4 Maximum response energy and related frequency.

Building	Maximum velocity (cm/s)	Maximum response energy (cm/s)^2	Frequency for maximum energy (Hz)
ED-E	0.0951	2.2033	6.25
ED-H	0.0517	0.1614	26.56
ED-M	0.0531	0.4215	9.38
ED-R	0.0971	1.6335	7.03

5.3 Rock mass damages

Blast-induced damage in rock is a significant yet poorly understood area in the rock excavation industries (Triviño, 2012). The prediction and control of blast damage has been traditionally done by approximate methods mostly based on experience rather than on understanding of the physical phenomenon. Perhaps the difficulties of experimentation and modeling in blasting, added to the significant imperfections of natural

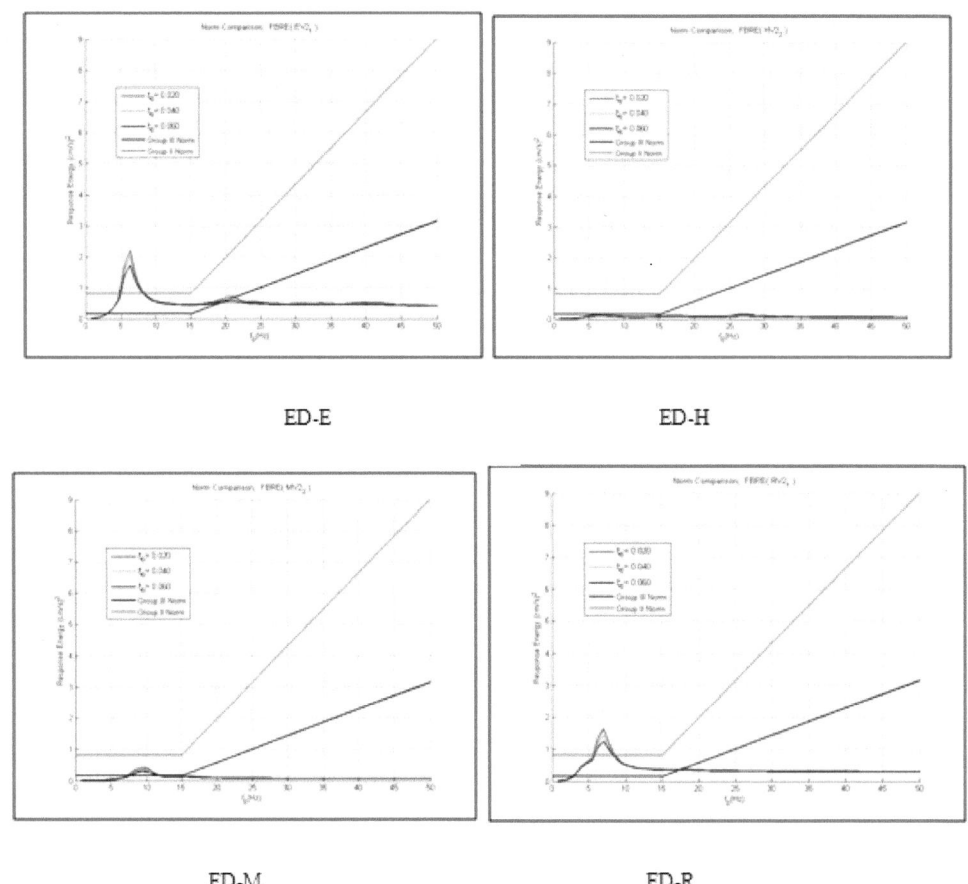

ED-E   ED-H

ED-M   ED-R

Figure 20 Comparison with the standard UNE 22-381-93, adapted to energy units.

rock masses at every scale, plus the limited knowledge on material behavior at very large stresses and loading rates, has significantly limited the research in this area and therefore its understanding.

Damage is a change in the rock mass properties which degrades its performance and behavior. From the mining point of view, it is the structural performance of the rock, which is of importance because the remaining rock has to support load in the form of back, wall or a pillar. That is why blast damage determines an important link between the excavation process and the structural stability of the rock mass. During the excavation process, the redistribution of in situ stresses and releases of seismic energy also induce rock mass damage, which can sometimes overshadow that caused by blasting.

Some failure criteria consider this damage as a disturbance factor reducing strength and deformation moduli. All of these suppose an increased cost in the installation and maintenance of supports in underground excavations, open pits and slopes.

Damages over rock masses go since breakdown of the inherent interlocking of the weakness planes (changes on rugosity, separation, fatigue), affecting the stability of wedges and blocks, to permeability changes. In underground excavations, with horizontal stratified roofs, vibrations could cause the beam resonance and the reduction in the maximum unsupported span and stand up time.

Langefors & Kihlstrom (1973) have proposed for tunnels that PPV's of 305 mm/s and 610 mm / s results in fall of rock in unlined tunnels and formation of new cracks respectively. Bauer & Calder (1970) observed that:

– No fracturing of intact rock will occur for a PPV of 254 mm/s.
– PPV of 254 – 635 mm/s results in minor tensile slabbing.
– PPV of 635 – 2540 mm/s would cause strong tensile and some radial cracking.
– Break up of rockmass will occur at a PPV of 2540 mm/s.

Holmberg & Persson (1979) stated that damage is a result of induced strain (ε) which is given by:

$$\varepsilon = \frac{PPV}{C} \tag{23}$$

where C is characteristic propagation velocity of P wave. It was also observed by them that the proposed generalized PPV equation is valid only for the distance that are long in comparison to charge length, so that charge can be considered as concentrated. In order to estimate blast-induced damage based on this model, Holmberg (1984) proposed some *PPV* threshold values for different rock conditions: 1000 mm/s for hard rock with strong joints; 700 to 800 mm/s for medium hard rock with no weak joints; and 400 mm/s for soft rock with weak joints.

Assuming a brittle failure mode of rock, a threshold of critical peak particle velocity PPV_{crit} which can be withstood by the rock before tensile failure occurs can be computed from the equation given by Persson *et al.* (2001):

$$PPV_{crit} = \frac{\sigma_t \cdot C}{E} \tag{24}$$

where σ_t is the tensile strength. In addition, the vibration level above which some minor damage may be expected can be estimated from the following equation given by McKenzie *et al.* (1995):

$$PPV_{crit} = 1.2 \cdot \frac{\sigma_t}{C \cdot \rho_r} \tag{25}$$

where ρ_r is the density of the rock (kg/m^3).

The Holmberg-Persson model has, however, been shown to have several shortcomings, including physical inconsistencies (Blair & Minchinton 1996, 2006), and even to be mathematically erroneous (Hustrulid & Lu, 2002)

Bogdanhoff (1995) monitored near field blast acceleration of an access tunnel in Stockholm. Vibration measurements were done at distances between 0.25 and 1.0 m. outside tunnel perimeter holes with accelerometers. Altogether eight blasts were monitored and the vibrations were filtered and PPV in the assumed damage range was found to be between 2000 and 2500 mm/s. These results are similar with Iverson *et al.* (2008), who have reported that the limiting PPV for new crack formation is 1850 mm/s for the Montana magmatic rocks (Norite).

5.4 Techniques for prevention

The adverse effects of vibrations caused by explosives used in mining and civil works can be controlled by suitably planning blasting operations, appropriately managing and optimizing the equipment used, and monitoring the processes that cause these vibrations.

An important objective is to define possible interventions to be considered in the blasting process, as actions able to prevent the environmental problems instead of mitigating them. The acceptable techniques for reduction and control of vibrations are:

1. The use of delay blasting techniques establishes internal free faces from which compressional waves produced later in the blast can delay patterns. A change in delay from 5 ms to 9 ms reduces vibration amplitude by a factor of 2 or 3.
2. Delays can generate destructive interferences, preventing the superposition of peak amplitudes.
3. In general, when blasting multiple row patterns, greater relief can be obtained by using a longer delay between rows than between the holes within a single row. A delay of at least 2 or 3 ms/m of burden between the holes within a row is recommended for the necessary relief and best fragmentation (Mostafa, 2010).
4. Use a spacing/burden ratio greater than one. With this, it is possible a lower confinement level during detonation. The presence of weak seams or irregular back break may dictate the local use of a spacing/burden ratio close to one.
5. Control drilling of blast holes as closely as possible. Restrict the amount of sub drilling to the level required to maintain good floor conditions. Excessive sub drilling can increase vibration because of the lack of a nearby free face to create reflection waves.
6. Use techniques to reduce charge weight per delay: reduce hole depths with lower bench heights and increase specific drilling, use smaller diameter holes, subdivide explosive charges in holes by using inert decks and fire each explosive deck with initiators using different delays, use electronic or mechanical timers to increase the available number of periods of delay electric blasting caps and to increase timing flexibility. Non electric delays coupled with surface delay connectors can provide similar flexibility.
7. Use explosive, such as water gels, which are much less sensitive than dynamite to hole to hole propagation. Hole to hole propagation occurs when the explosive charges or blast holes are only a little distance apart, as in trenching, decked holes, or underwater excavations, or at greater distances when blasting interbedded soft and hard layer rock, such as coral or mud-seamed rock, that is saturated with water.
8. Other recommendations are to reduce the number of holes and its diameter, use decoupled loads, increase the dip of the borehole (this decreases the confinement), decrease the burden (although this increases the acoustic wave).
9. In underground works, when the perimeter holes are detonated first (pre-shearing or pre-splitting), a boundary is created around the excavation reflecting the stress waves back into the heading being blasted and minimizing damages.

Besides these blast control techniques there are other possible actuations. One of these is the use of trenches as vibration mitigation (Zoccali *et al.*, 2015; Garinei *et al.*,

2014; Karlström & Boström, 2007). The complexity of the intervention design process depends on the high number of parameters that influence its efficacy, such as trench geometry size (width, depth, length), section shape, used in-filled material, distance among source-trench-receiver and frequency content of the source.

Some researchers have shown, through the analysis of numerical simulation results, that with a model of double millisecond holes in bench blasting, it can be obtained that peak particle velocity and energy attenuation are occurred near the blasting source. This provides basic theory for protection of structures near the blasting source.

To effectively mitigate blast waves, a novel protection mechanism has been proposed recently (Xu *et al.*, 2014), based on nanofluidic energy capture (NEC). When an intensive stress wave is acted on a non-wetting liquid-nanoporous solid composite material system, a significant part of the incident energy can be quickly converted to the potential energy of infiltrated water molecules and is temporarily captured (until the nanopores are filled up), leading to a prominent reduction in both the transmitted force and energy. The system has the advantage of high energy mitigation density and low working pressure, fast response time (in the order of ns) and it can be used repeatedly.

REFERENCES

Achenbach, J.D. (1975) *Wave Propagation in Elastic Solids*. Elsevier, Amsterdam, p. 166.

Ahmed, L. & Ansell, A. (2014) Vibration vulnerability of shotcrete on tunnel walls during construction blasting. *Tunnelling and Underground Space Technology*, 42, 105–111.

Álvarez-Vigil, A.E., González-Nicieza, C., López Gayarre, F. & Alvarez-Fernández, M.I. (2012) Predicting blasting propagation velocity and vibration frequency using artificial neural network. *International Journal of Rock Mechanics and Mining Sciences*, 55, 108–116.

Ambraseys, N.R. & Hendron, A.J. (1986) *Dynamic behaviour of rock masses: rock mechanics in engineering practices*. London: Wiley.

Ataei, M. & Kamali, M. (2012) Prediction of blast-induced vibration by adaptive neuro-fuzzy inference system in Karoun 3 power plant and dam. *Journal Vibration Control*, 22, 1–9.

Awojobi, O. & Sobayo, O.A. (1974) Ground vibrations due to seismic detonation in oil exploration. *Earthquake Engineering & Structural Dynamics*, 3 (2), 171–181.

Azizabadi, H.R.M., Mansouri, H. & Fouché, O. (2014) Coupling of two methods, waveform superposition and numerical, to model blast vibration effect on slope stability in jointed rock masses. *Computers and Geotechnics*, 61, 42–49.

Bauer, A. & Calder, P.N. (1970). *Open Pit and Blasting Seminar* Mining Engineering Deptartment Publication, Queen's University, Kingston, Ontario, p. 3.

Bender, W.L. (2007) *Understanding Blast Vibration and Airblast, their Causes, and their Damage Potential. Spring 2006 and Fall 2007 workshops of the Golden West*. Chapter of the International Society of Explosives Engineers.

Berta, G (1990) Explosives: An Engineering Tool. Milano: Italesplosivi.

Blair, D.P. (1990) Some problems associated with standard charge weight vibration scaling laws. *In Proceedings of the 3rd International Symposium on Rock Fragmentation by Blasting – Fragblast 3*, Brisbane, Australia. pp. 149–158.

Blair, D.P. (2004) Charge weight scaling laws and the superposition of blast vibration waves. *Fragblast*, 8 (4), 221–239.

Blair, D.P. (2011) A probabilistic analysis of vibration based on measured data and charge weight scaling. *In: Proceedings of the 6th EFEE World Conference on Explosives and Blasting Technique, September 2011*, Lisbon, Portugal. pp. 319–337.

Bogdanoff, I. (1995) Vibration measurements in damage zone in tunnel blasting. *Proceedings of Rock fragmentation by blasting, FRAGBLAST-5*, 23–24 August, Montreal, Quebec, Canada. (Ed) Mohanty, pp. 177–185.

Brinkmann, J.R. (1987) The control of ground vibration from colliery blasting during the undermining of residential areas. *Journal of the South African Institute of Mining and Metallurgy*, 87 (2), 53–61.

Caylak, C., Kocaslan, A., Gorgulu, K., Buyuksarac, A. & Arpaz, E. (2014) Importance of ground properties in the relationship of ground vibration – structural hazard and land application. *Journal of Applied Geophysics*, 104, 6–16.

Chen, M., Lu, W. & Yi, C. (2007) Blasting vibration criterion for a rock-anchored beam in an underground powerhouse. *Tunnelling and Underground Space Technology*, 22, 69–79.

CMRI (1993) Vibration standards, Central Mining Research Institute, Dhanbad.

Crandell, F.J. (1949) Ground vibration due to blasting and its effect upon structures. *Journal of the Boston Society of Civil Engineers*, April, 222–245.

Dehghan Banadaki, M.M. (2010) *Stress-wave induced Fracture in Rock due to Explosive Action*. [Ph.D. Thesis] Department of Civil Engineering, University of Toronto.

Dogan, O., Anil, Ö., Akbas, S.O., Kantar, E. & Erdem, R.T. (2013) Evaluation of blast-induced ground vibration effects in a new residential zone. *Soil Dynamic Earthquake Engineering*, 50, 168–181.

Dowding, C. H., Beck, W.K. & Atmatzidis, D.K. (1980) Blast Vibration Implications of Cyclic Shear Behavior of Model Plaster Panels. *Geotechnical Testing Journal*, 3 (2), pp. 80–88.

Dowding, C. H., Murray, P.D. & Atmatzidis, D.K. (1981) Dynamic Response Properties of Residential Structures Subjected to Blasting Vibrations. *Journal of Structural Engineering*, 107 (2), 1233–1249.

Dowding, C.H. (1985) *Blast Vibration Monitoring and Control*. Prentice-Hall, Englewood Cliffs, p. 297.

Dowding, C.H. (1993) Blast vibration monitoring for rock engineering. In: *Comprehensive rock engineering. Principles, Practice & Projects*. Hudson ed. *Excavation, support and monitoring, Volume 4*. Oxford: Pergamon press. pp. 111–135.

Dowding, C.H. (1996) *Construction Vibrations*. Prentice-Hall, Englewood Cliffs.

Dunnicliff, J. (1988) *Geotechnical Instrumentation for Monitoring Field Performance*. New York: John Wiley & Sons. ISBN 0-471-00546-0.

Duvall, W.I. & Petkof, B. (1959) Spherical propagation of explosion generated strain pulses in rock. *USBM Report of Investigation 5483*, p. 21.

Faramarzi, F., Ebrahimi, F., Mohammad, A. & Mansouri, H. (2014) Simultaneous investigation of blast induced ground vibration and airblast effects on safety level of structures and human in surface blasting. *International Journal of Mining Science and Technology*, 24, 663–669.

Fleetwood, K.G., Villaescusa, E. & Li, J. (2009) Limitations of using PPV damage models to predict rock mass damage. *In Proceedings of 35th Annual Conference on Explosives and Blasting Technique*, Denver, USA. pp. 349–363.

Garinei, A., Risitano, G. & Scappaticci, L. (2014) Experimental evaluation of the efficiency of trenches for the mitigation of train-induced vibrations. Transportation Research Part D: Transport and Environment, 32, 303–315.

Ghosh, A. & Daemen, J.K. (1983) A simple new blast vibration predictor. *Proceedings of the 24th US symposium on rock mechanics*, College Station, Texas. pp. 151–161.

González-Nicieza, C., Álvarez-Fernandez, M.I.; Alvarez-Vigil, A.E., Arias-Prieto D., López-Gayarre, F. & Ramos-Lopez, F. L. (2014) Influence of depth and geological structure on the transmission of blast vibrations. *Bulletin of Engineering Geology and Environmental*, 73 (4), 1211–1223.

Gutowski, T. G. & Dym, C.L. (1976) Propagation of ground vibration a review. *Journal of Sound and Vibration*, 49(2), 179–193.

Guosheng, Z., Jiang, L. & Kui, Z. (2011). Structural safety criteria for blasting vibration based on wavelet packet energy spectra. *Mining Science and Technology (China)* 21, 35–40.

Holmberg, R. & Persson, P.A. (1979) Swedish approach to contour blasting, *Proceedings of Fourth Conference on Explosive and Blasting Techniques.* New Orleans. pp. 113–127.

Holmberg, R. (1984) Improved stability through optimized rock blasting. *Proceedings of 10th Conference on Explosives and Blasting Technique.* Orlando, Florida. pp. 234–248.

Hudson, D.E., Alford, J.L. & Iwan, W.D. (1961) Ground accelerations caused by large quarry blasts. *Bulletin of the Seismological Society of America,* 51, 191–202.

IPROMIN *Assessment Study on the Seismic Effect of Blasting Explosions on the Protected Sites and Methods used to mitigate this effect of explosions – control and monitoring procedures – Rosia Montana.* S.C. IPROMIN S.A. Bucharest.

Iverson, S., Kerkering, C. & Hustrulid, W. (2008) Application of the NIOSH-modified Holmberg-Persson approach to perimeter blast design. In: *Proceedings of the 34th Annual Conference on Explosives and Blasting Technique.* Cleveland, OH: International Society of Explosives Engineers 2. pp. 1–33.

Kamali, M. & Ataei, M. (2010) Prediction of blast induced ground vibrations in Karoun III power plant and dam: a neural network. *The Journal of The Southern African Institute of Mining and Metallurgy,* 110, 481–490.

Karlström, A. & Boström, A. (2007) Efficiency of trenches along railways for trains moving at sub- or supersonic speeds. *Soil Dynamics and Earthquake Engineering,* 27 (7), 625–641.

Khandelwal, M. & Singh, T.N. (2007) Evaluation of blasting induced ground vibration predictors. *Soil Dynamic Earthquake Engineering,* 27, 116–125.

Khandelwal, M. & Saadat, M.A. (2014) Dimensional Analysis Approach to Study Blast-Induced Ground Vibration. *Rock Mechanic and Rock Engineering.* DOI 10.1007/s00603-014-0604-y.

Krehl, P. (2001). Chapter 1 – History of Shock Waves. *Handbook of Shock Waves,* Volume *1.* pp. 1–142.

Langefors, U. & Kihlstrom, B. (1963) *The Modern Technique of Rock Blasting.* New York, Ed. Wiley.

McKenzie, C.K., Scherpenisse, C.R., Arriagada, J. & Jones, J.P. (1995). Application of computer assisted modeling to final wall blast design. *EXPLO '95.* 285–292, Brisbane, Australia.

Medearis, K. (1976) The Development of Rational Damage Criteria for Low-Rise Structures Subjected to Blasting Vibrations. *Kenneth Medearis Associates, Final Rept. to National Crushed Stone Assn., August 1976, Washington, DC.* p. 93.

Mohamed, M.T (2011) Performance of fuzzy logic and artificial neural network in prediction of ground and air vibrations. *International Journal of Rock Mechanic and Mining Science,* 48, 845–851.

Mostafa, M. (2010) *Vibration Control.* Michael Lallart (Ed.), In Tech: Sciyo. pp. 355–380. ISBN: 978-953-307-117-6. Available from: http://www.intechopen.com/books/vibration-control/vibration-control.

Nawab, S. & Quatieri, T. (1988). Short-time Fourier transform, in Lim J., Oppenheim A., Eds. *Advanced Topics in Signal Processing.* Prentice Hall Signal Processing Series, pp. 289–337.

Ngo, T., Mendis, P., Gupta, A, & Ramsay, J. (2007) Blast loading and blast effects on structures – an overview. *Electronic Journal of Structural Engineering,* Special Issue, 7, 79–91.

Nicholls, H.R., Johnson, C.F. & Duvall, W. (1971) Blasting vibrations and their effects on structures. V.S. Bureau of Mines, *Bulletin 656.*

Ozgur, Y. & Tugrul, U. (2014) An application of the modified Holmberg–Persson approach for tunnel blasting design. *Tunnelling and Underground Space Technology,* 43, 113–122.

Persson, P.A., Holmberg, R. & Lee, J. (2001). *Rock Blasting and Explosives Engineering.* 6th printing. USA: CRC Press, pp. 540.

Ram Chandar, K. & Sastry, V.R. (2015) A New Method of Estimating Wave Energy from Ground Vibrations. *Geomaterials*, 5, 45–55 Available from: April 2015 in SciRes. http://file.scirp.org/pdf/GM_2015030215430081.pdf.

Saadat, M., Khandelwal, M. & Monjezi, M. (2014). An ANN-based approach to predict blast-induced ground vibration of Gol-E-Gohar iron ore mine, Iran. *Journal of Rock Mechanics and Geotechnical Engineering*, 6, 67–76.

Scott, A. (2009). *Appendix 16 – Review of the Impact of Blasting on Rock Mass Permeability*, Bickham Coal WRA & Draft WMP, p. 43.

Segarra, P., Sanchidrián, J.A., Castedo, R., López, L.M. & del Castillo, I. (2015). Performance of some coupling methods for blast vibration monitoring. *Journal of Applied Geophysics*, 112, 129–135.

Stewart, L.K., Freidenberg, A., Rodriguez-Nikl, T., Oesterle, M., Wolfson, J., Durant, B., Arnett, K., Asaro, R.J. & Hegemier, G.A. (2014) Methodology and validation for blast and shock testing of structures using high-speed hydraulic actuators. *Engineering Structures*, 70, 168–180.

Triviño Parra, L.F. (2012) *Study of Blast-Induced Damage in Rock with Potential Application to Open Pit and Underground Mines.* [Ph.D. Thesis]. Department of Civil Engineering, University of Toronto.

Welch, P.D. (1967) The Use of Fast Fourier Transform for the Estimation Power Spectra: A Method Based on Time Averaging Over Short Modified Periodograms. *IEEE Trans. Audio Electroacustics*, AU-15, 70–73.

Xu, B., Qiao, Y. & Chen, X. (2014) Mitigating impact/blast energy via a novel nanofluidic energy capture mechanism. *Journal of the Mechanics and Physics of Solids*, 62, 194–208.

Yan, W.M. & Yuen, K. (2015) On the Proper Estimation of the Confidence Interval for the Design Formula of Blast-Induced Vibrations with Site Records. *Rock Mechanic and Rock Engineering*, 48, 361–374.

Zhang, C., Hu, F. & Zou, S. (2005) Effects of blast induced vibrations on the fresh concrete lining of a shaft. *Tunnelling and Underground Space Technology*, 20, 356–361.

Zhong, G., Li J., Zhao K., (2015) Structural safety criteria for blasting vibration based on wavelet packet energy spectra. Mining Science and Technology (China), 21, 35–40.

Zoccali, P., Cantisani, G. & Loprencipe, G. (2015) Ground-vibrations induced by trains: Filled trenches mitigation capacity and length influence. *Construction and Building Materials*, 74, 1–8.

Chapter 21

Statistical estimation of blast fragmentation

Jeong-Hun Han & Jae-Joon Song
Department of Energy Resources Engineering, Seoul National University, Korea

Fragmentation of blasted rock blocks is one of the indispensable indices to evaluate the efficiency of the blast design. This is why new algorithms and software have been continuously developed and suggested for fragmentation analysis. Popular software available in this field such as Split Desktop, WipFrag, FragScan, etc. is mainly based on two-dimensional image processing technology. Stereophotogrammetry is one of the promising technologies for the fragmentation analysis, which can provide the 3D coordinates of an object from a stereo-image constructed by two or more photos.

This chapter introduces the statistical analysis of block size, which is subsequent to the fundamentals of blast fragmentation and the image acquisition of fragmented blocks. The stereophotogrammetry was adopted as a main tool for the image acquisition of fragmented blocks. 3D models of rock blocks scattered on the ground and surfaced on a pile were used for the statistical estimation of fragmentation. In the laboratory experiments, results from stereophotogrammetry and two-dimensional image processing were compared with the physical values measured by a water tank. Finally, a new algorithm for automatic block delineation has been introduced as well as its application. This chapter consists of the following topics.

I FUNDAMENTALS OF BLAST FRAGMENTATION

As a fundamental way to evaluate the efficiency of the blast design and productivity, the fragmentation of the blasted rock blocks in the field is generally assessed by analyzing the size distribution of the blasted rock. The result from the fragmentation analysis can be used to evaluate the achievement of blast design, and can also be used as feedback data for the design of the subsequent blast, allowing the optimal blast to be achieved. Especially in quarry blast where blasting causes huge blasted rock blocks, it is crucial to measure and evaluate the accurate sizes of the blasted rock blocks because the size of these huge rock blocks becomes a very important element in designing the subsequent blast, transportation system, and its capacity.

The quantitative evaluation of fragmentation can be most accurately executed through sieve test. The sieve test, however, is more suitable for materials with relatively smaller sizes and thus is difficult to be applied to those with bigger sizes or volumes. As a result, an indirect evaluation method is used in rock engineering to execute the quantitative evaluation of the blasted rock blocks. One of the most commonly used methods is photogrammetry. Photogrammetry is the survey technology to extract and measure the information of an object from the image. In rock engineering, since Hagan (1980) conducted a research on joint survey using photogrammetry, a number of researchers have used photogrammetry to conduct rock engineering research (Tsoutrelis *et al.*, 1990; Reid & Harrison, 2000; Chen *et al.*, 2004; Wang *et al.*, 2005; Ohnishi *et al.*, 2006; Firpo *et al.*, 2011; Valenca *et al.*, 2013). With regard to the fragmentation analysis, the basic idea of photogrammetry is to obtain images of the blasted rocks, and analyze the sizes of the blasted rocks in the images using a scale bar in the same image (Van Aswegen & Cunningham, 1986). This method is based on 2D image processing, which first converts an image into the binary image, and sorts out data for blasted rocks by removing the background information from the image based on threshold. Finally the size distribution can be obtained using the size information based on a scale bar. With the development of computer technologies, engineers have researched ways to automatically analyze the fragmentation with 2D image processing. These methods automatically detect the boundaries between objects using brightness information and analyze information on each object. Split Desktop, WipFrag, FragScan, and GoldSize are the most popular software packages, and have its definite advantage of the automatic fragmentation analysis (Girdner *et al.*, 1996; Kleine & Cameron, 1996; Maerz *et al.*, 1996; Siddiqui *et al.*, 2009; Kemeny *et al.*, 2001).

The interpretation using the 2D image processing was reported to overestimate the median fragment size which represents the diameter at 50 % of passing by about 50~100% depending on the method (Liu & Tran, 1996). This type of 2D image processing merely uses the boundary and area information of an object shown in the image, and thus does not utilize the information of an object in a three-dimensional way. In addition, the result may vary depending on images with different camera angles. As a result, the blasted rock blocks are examined on a plane as a conveyor belt or bucket to obtain as much information as possible. In order to improve the estimation, researchers focused on estimating the size distribution in a three-dimensional way. Its basic principle is to obtain the area information of an object within the image using 2D image processing, and calculate the equivalent diameter of a circle that has the same area with the area information above, and then calculate the volume using this equivalent diameter (Maerz *et al.*, 1987; Sudhakar *et al.*, 2006). These researches, however, have the restriction in the way that, as discussed above, the origin of the material is two-dimensional and therefore ultimately cannot use three-dimensional information. In order to overcome such a restriction, researchers have focused on restoring three-dimensional information and analyzing the fragmentation based on this information (Noy, 2006; Müller *et al.*, 2009; Thurley, 2011).

This paper presents the approach of statistical fragmentation analysis using the information restored from stereophotogrammetry. Stereophotogrammetry is a survey technology to restore three-dimensional information of an object using two or more photographic images with different camera angles, and extract the various information such as size, shape, area, volume, etc. (Paul & Bon, 2000; Karl, 2007). It has the advantage to utilize more geometrical features of an object such as 3D point clouds, and thus provide more accurate information compared to 2D image processing.

2 PHOTOGRAMMETRIC APPROACHES FOR SIZE ANALYSIS OF FRAGMENTED BLOCKS

2.1 2D image processing method

In this study, Split Desktop software was used for two-dimensional fragmentation analysis. Split Desktop is software specialized for fragmentation analysis based on 2D image processing. An object which will be used as a scaling standard is placed around the target object when the image is taken. After scaling, the boundaries of objects are primarily delineated using auto-delineation function. In the next step, the manual editing is implemented on the first outcome. Then, the size distribution of the blasted rock blocks can be obtained by analyzing the outcome from the manual editing step. 2D image processing may result in the certain circumstance where one large block is divided into several small blocks in automatic delineation, or in contrast, several small blocks can be recognized as one large block. As this problem is known to occur often in 2D image processing technology (Maerz *et al.*, 1996; Liu & Tran, 2006; Ozkahraman, 2006), an extra care is required in the step of manual editing process. Figure 1 shows the processing flow in Split Desktop.

2.2 Stereophotogrammetric method

In this study, PhotoModeler Scanner was used to restore the three-dimensional information of rock blocks. As software specialized in three-dimensional modeling, PhotoModeler Scanner has its advantage in utilizing the three-dimensional information of an object. When acquiring stereo images of an object for the purpose of three-dimensional modeling, the scaling and referencing standard is placed around the object. After camera and lens distortion are correct, the relative location between images is determined among the input images in a referencing step, and the scaling processing comes after. Following the modeling the focus area, the three-dimensional result can be obtained about the object. This study extracted the object's volume and three-dimensional point clouds. Figure 2 shows the processing flow in PhotoModeler Scanner.

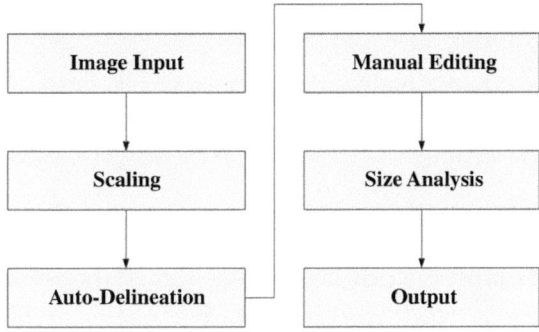

Figure 1 Flowchart in Split Desktop (after Han & Song, 2014).

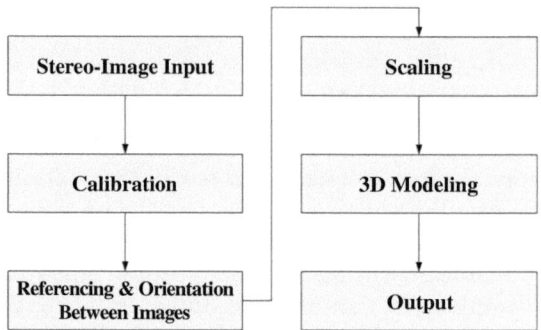

Figure 2 Flowchart in PhotoModeler Scanner (after Han & Song, 2014).

3 ANALYSIS PROCEDURE

3.1 Size distribution with 2D image processing

A few assumptions are applied to implement the fragmentation analysis. First, blasted rock blocks are assumed to be resulted from the same blast surface, and therefore, all blasted rock blocks are assumed to have the same unit weight (γ). Based on this assumption, the passing can be calculated through the volume ratio, as shown in Eq. (1).

$$m = \gamma V, \; Passing \; (\%) = \frac{m_x}{m_{total}} \times 100 = \frac{\gamma V_x}{\gamma V_{total}} \times 100 = \frac{V_x}{V_{total}} \times 100 \qquad (1)$$

(m_x and V_x are the cumulative mass and volume till x, respectively)

In order to compare the result from Split Desktop's analysis with the result measured by water tank, following steps are implemented. Split Desktop provides the area information of an object. With the assumption that the shape of a rock block is a perfect circle in two dimensions, the equivalent diameter can be calculated from the area formula of a circle so that it has the same area with the area information above, as shown in Eq. (2). Here, D_{equi} and A represent the equivalent diameter and the area information, respectively.

$$D_{equi} = \sqrt{\frac{4A}{\pi}} \qquad (2)$$

Then, based on the assumption that the shape of a rock block is a perfect sphere in three dimensions, the volume can be calculated from the volume formula of a sphere using D_{equi}. Finally, the passing can be estimated using Eq (1).

3.2 Size distribution with stereophotogrammetry

First, for the purpose of three-dimensional analysis, each rock block is modeled using PhotoModeler Scanner. PhotoModeler Scanner provides the volume information of an object. With the assumption that the shape of a rock block is a perfect sphere in three dimensions, the equivalent diameter can be calculated from the volume formula of a sphere so that it has the same volume with the volume information above, as shown in Eq. (3). Here, D_{equi} and V represent the equivalent diameter and the volume information, respectively. The passing can be estimated using Eq (1).

$$D_{equi} = 2 \times \sqrt[3]{\frac{3V}{4\pi}} \tag{3}$$

3.3 Size distribution with stereophotogrammetry and statistical analysis

The survey of the entire blasted rock blocks is the best way to analyze the accurate size distribution. Considering the field situation, however, it is difficult to analyze hundreds and/or thousands of blasted rock blocks resulted from the blast. Moreover, it is difficult to obtain the information of blasted rock blocks lying inside a muckpile, while the information of surface blocks in a muckpile can be most easily and intuitively obtained. This study presents the Surface-Based Estimation (SBE), the statistical estimation that uses such information of surface blocks in a muckpile.

Following cases resulted in a distorted estimation, as shown in Figure 3. Therefore, modeling results from the following cases were excluded in the fragmentation analysis

– In case where PhotoModeler Scanner models rock blocks using images taken from less than two directions

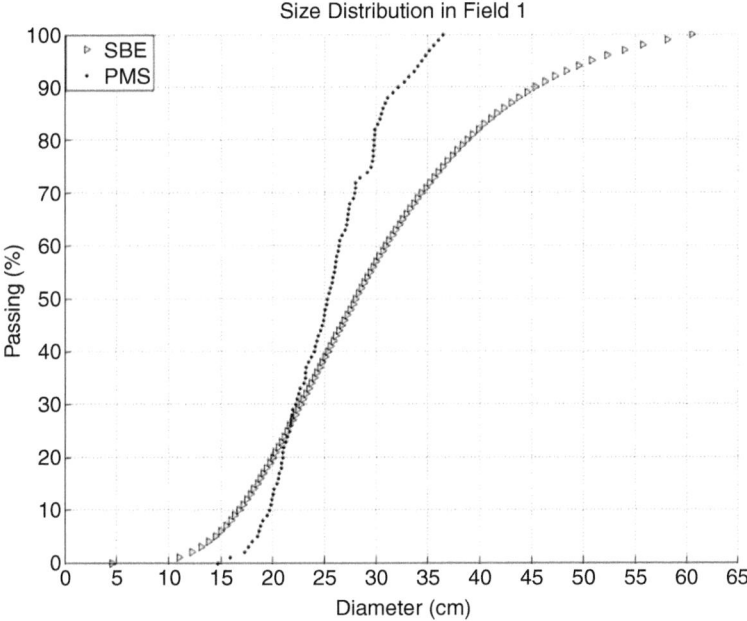

Figure 3 Distorted estimation result using improper input parameters (after Han & Song, 2014).

- In case where PhotoModeler Scanner cannot identify the overall shape of a rock block because it is lying between other rock blocks
- In case where the upper and the lower limits for the statistical estimation are not established

In this study, the maximum and minimum diameters of surface blocks are used as the upper and the lower limit, respectively, to prevent the case where the diameters are estimated senselessly low and large.

In heaped blocks, a part of rock blocks is invisible from outside because rock blocks exist in between other rock blocks, as shown in Figure 4. Therefore, errors can occur in the amount of the invisible parts when using stereophotogrammetry (Han & Song, 2014). Maerz and Zhou (1998) reported that the analysis of heaped blocks based on 2D image processing underestimated the result, and compensated for the result using Rosin-Rammler model. Rosin-Rammler model, shown in Eq. (4), predicts the passing F(x) using two parameters, xc that represents the diameter at 63.2% of passing and n that describes the material uniformity. This model, however, is known to be better applicable to object with fine size (Van Breugel, 1995), while it is weak in analyzing the size distribution of large sized diameter (Rizk & Lefebvre, 1985). In addition, according to the research of Sanchidrián et al. (2014) and Sanchidrián (2015), this model provides its best performance for the relatively small range of rock fragments.

$$F(x) = 1 - exp\left(-\left(\frac{x}{x_c}\right)^n\right) \tag{4}$$

In order to measure errors of information between the entire muckpile and surface blocks, the size information from the laboratory and field experiments were analyzed. First, the mean diameter of 39 rocks collected for the laboratory

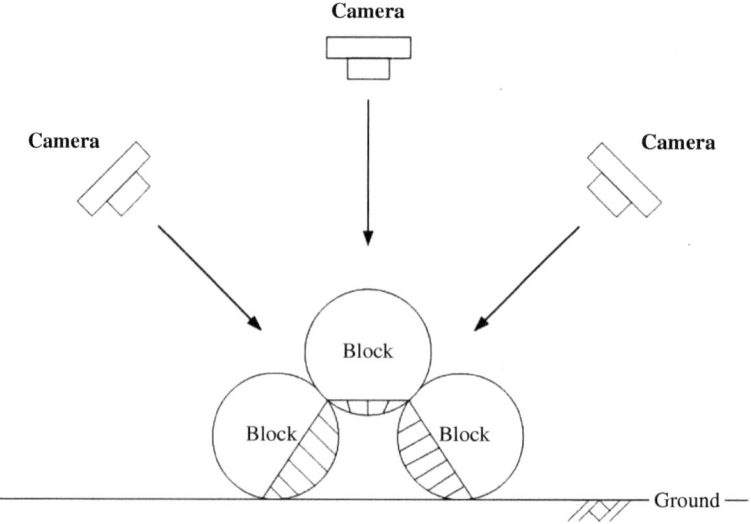

Figure 4 Possible source of error of heaped blocks (Han & Song, 2014).

Table 1 Errors of surface blocks in heaped blocks (after Han & Song, 2014).

			Mean diameter (cm)	Compensation ratio (%)
Laboratory experiments	Entire		18.83	
	Surface in Case 1		16.26	13.65
	Surface in Case 2		16.79	10.83
Field experiments	Field 1	Entire	24.65	-
		Surface	21.06	14.56
	Field 2	Entire	26.40	-
		Surface	23.11	12.46
	Field 3	Entire	22.95	-
		Surface	17.60	23.31
Mean compensation ratio (%)				14.96

experiments was 18.83 cm. Two cases of laboratory experiments for heaped blocks were carried out. The mean diameter of surface blocks was 16.26 cm for Case 1, while it was 16.79cm for Case 2. For the field experiments, blasted rock blocks were collected in mines at the Donghae and Samchuck branches of the Ssangyong Resource Development Corporation in Gangwon, Korea. A total of 100, 101 and 103 rock blocks were collected at Donghae (Field 1 and Field 2) and Samchuck (Field 3) branches. Three cases of field experiments were carried out. In Field 1, the mean diameters of the entire rock blocks and surface blocks were 24.65 cm and 21.06 cm, respectively. In Field 2, the mean diameters of the entire rock blocks and surface blocks were 26.40 cm and 23.11 cm, respectively. In Field 3, the mean diameters of the entire rock blocks and surface blocks were 22.95 cm and 17.60 cm, respectively.

The compensation ratio was estimated using the ratio of the mean diameter of the surface blocks on that of the entire muckpile and, and it is indicated in Table 1. In this study, 15 % of error ratio was compensated for in analyzing diameters. This amount of compensation is expected to be improved through more experiments for the blasted rock blocks with various patterns of arrangement.

Following is the fragmentation analysis process based on the statistical estimation. First, every individual surface block is modeled using stereophotogrammetry, and its volume and diameter information is analyzed. Then, through the frequency count analysis for the surface blocks, a probability density function (PDF) is determined so that it is most comparable to the size distribution of the surface blocks. In the next step, the Cumulative Distribution Function (CDF), which has the horizontal axis of diameter and the vertical axis of cumulative probability, can be determined for the PDF. The SBE of this study generated the random probability and estimated its corresponding diameter using the inverse function of the CDF. Figure 5 shows the processing flow of the SBE.

4 APPLICATION

In blast fragmentation analysis, the economically significant sizes of blasted fragments can be usually classified into three categories as follows (Cunningham, 1996).

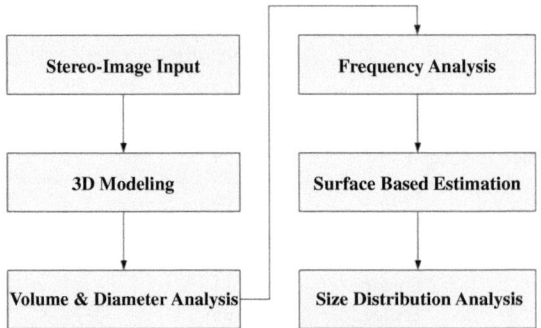

Figure 5 Procedure for the Surface Based Estimation, SBE (after Han & Song, 2014).

–   Oversize: The boulder size above which secondary breaking is necessary before further handling. In underground mines this can be as little as 300 mm, while in opencast mines it is seldom defined as greater than 1,000 mm.
–   Mid-range: Those fragment sizes that have significant, but not terminal importance for handling and the ability to achieve premium pricing
–   Fines: The particle size below which the product can either not be sold, or becomes difficult to handle, due to flow or other properties. A common minimum size of fines for coal or dolomite is 6 mm, but in gold ores, this size may be as small as 1 mm.

The subsequent blast is necessary in a case where the sizes of blasted rock blocks are bigger than 'oversize'. The rock fragments of 'oversize', therefore, become an essential element for the design of loading and transportation system (Sirotjuk, 1970; Maclachlan & Singh, 1989). In addition, it was slightly difficult to model and analyze fine sized rock blocks in this study. Therefore, this study targeted blocks larger than mid-range in diameter.

4.1 Laboratory experiment

4.1.1 Scattered blocks

A total of 39 rock blocks with different shapes and sizes were collected and scattered on a plane for the laboratory experiments, as shown in Figure 6. The volume of each rock block was measured using a water tank. The volume was in the range of 600 cm^3 ~ 13,000cm^3, and the total volume was approximately 170,000cm^3. Figure 7 shows the result analyzed from Split Desktop. In Case 1, the diameters at 20 %, 50 %, 80 % and 100 % of passings were 167.24 mm, 245.31 mm, 328.80 mm and 487.65 mm, respectively. In Case 2, the diameters at 20 %, 50 %, 80 % and 100 % of passings were 174.02 mm, 242.44 mm, 322.83 mm and 488.64 mm, respectively. Figure 8 shows the size distributions analyzed from the water tank measurement and estimated through the procedure discussed in section 3.1.

The modeling outcome of rock blocks using PhotoModeler Scanner is shown in Figure 9, and Figure 10 shows the size distributions analyzed from the water tank measurements and estimated through the procedure discussed in section 3.2. As shown

(a) Case 1   (b) Case 2

Figure 6 Scattered blocks of different arrangements (after Han & Song, 2014).

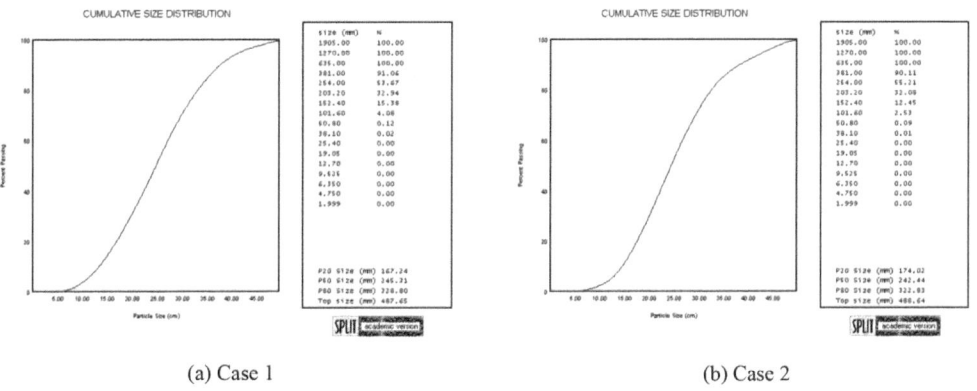

(a) Case 1   (b) Case 2

Figure 7 Size distributions using Split Desktop for scattered blocks.

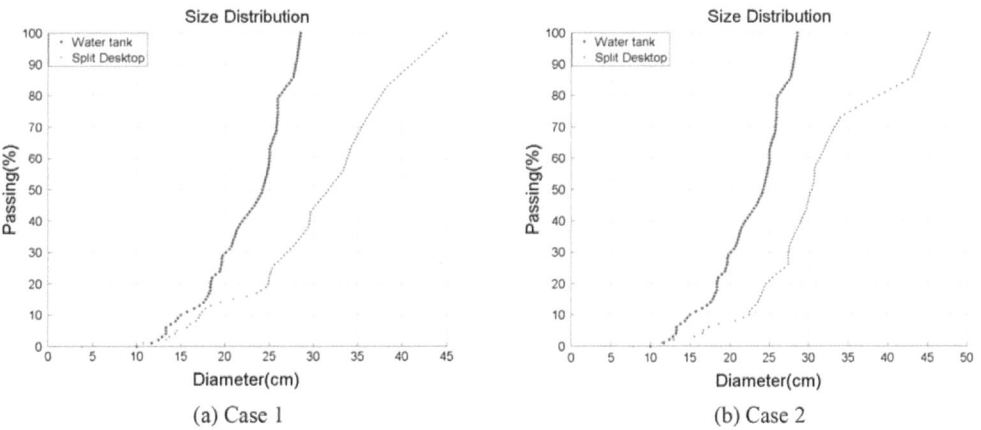

(a) Case 1   (b) Case 2

Figure 8 Size distributions from the water tank measurement and the Split Desktop analysis for scattered blocks (after Han & Song, 2014).

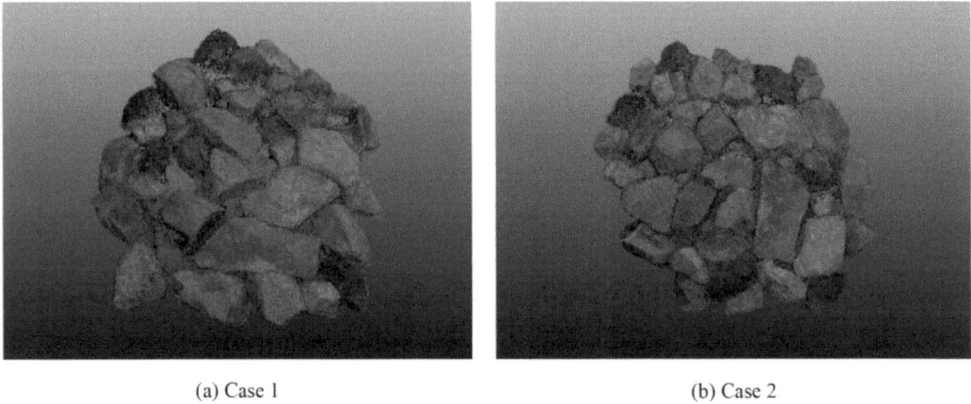

(a) Case 1
(b) Case 2

Figure 9 Modeling images using a PhotoModeler Scanner for scattered blocks.

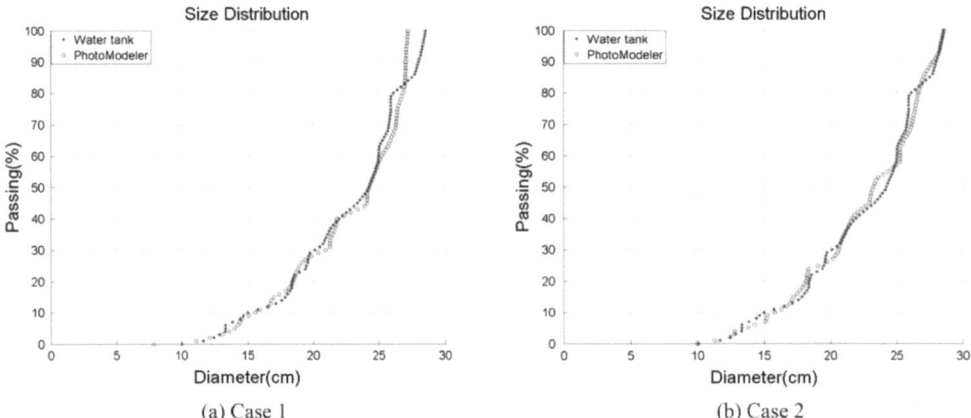

(a) Case 1
(b) Case 2

Figure 10 Size distributions from the water tank measurement and the PhotoModeler Scanner analysis for scattered blocks (after Han & Song, 2014).

in the result, the fragmentation analysis based on the information from three-dimensional modeling enabled the result to be more accurate.

4.1.2 Heaped blocks

The same rock blocks were heaped with different arrangements, as shown in Figure 11. Figure 12 shows the size distributions analyzed from Split Desktop. In Case 1, the diameters at 20 %, 50 %, 80 % and 100 % of passings were 142.89 mm, 206.42 mm, 294.46 mm and 419.04 mm, respectively. In Case 2, the diameters at 20 %, 50 %, 80 % and 100 % of passings were 141.75 mm, 208.63 mm, 289.05 mm and 437.63 mm, respectively. Figure 13 shows the size distributions analyzed from the water tank measurement and estimated through the procedure discussed in section 3.1.

(a) Case 1　　　　　　　　　　　　　(b) Case 2

Figure 11 Heaped blocks of different arrangements (after Han & Song, 2014).

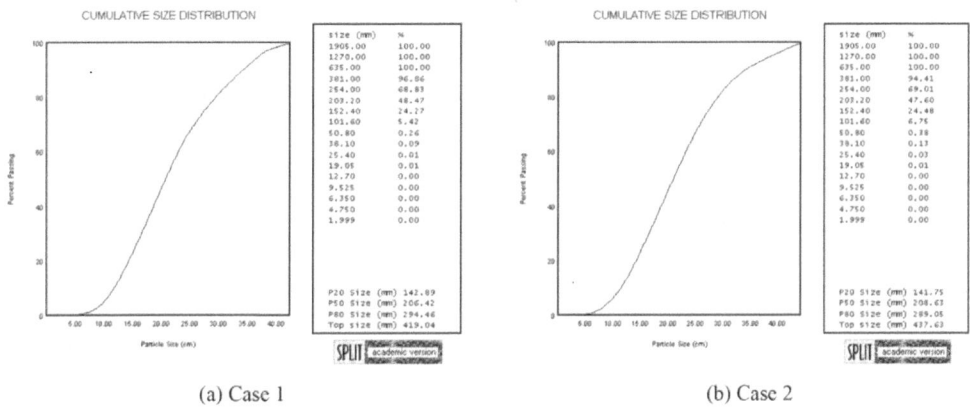

(a) Case 1　　　　　　　　　　　　　(b) Case 2

Figure 12 Size distributions using Split Desktop for heaped blocks.

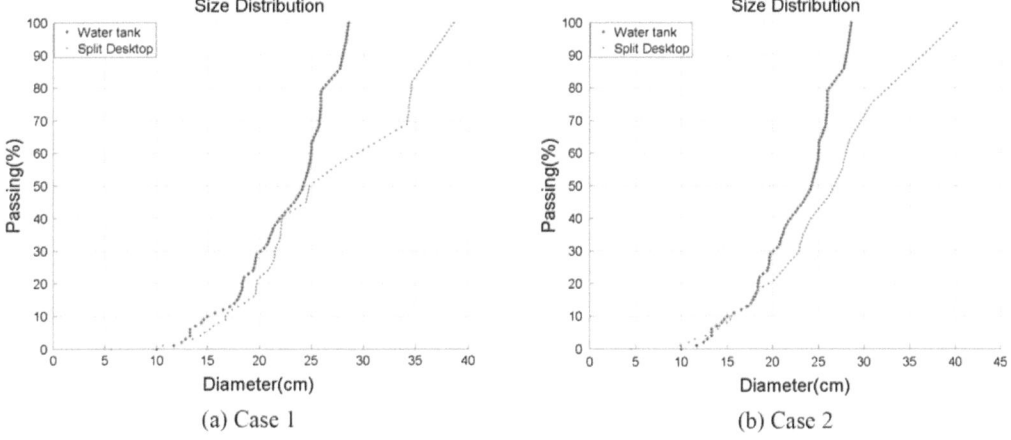

(a) Case 1　　　　　　　　　　　　　(b) Case 2

Figure 13 Size distributions from the water tank measurement and the Split Desktop analysis for heaped blocks (after Han & Song, 2014).

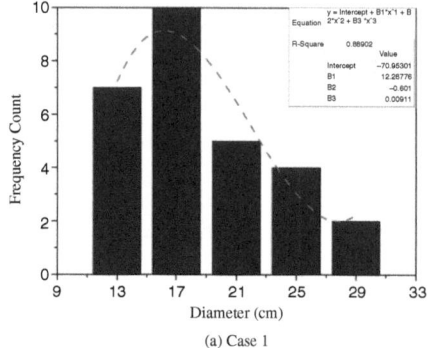
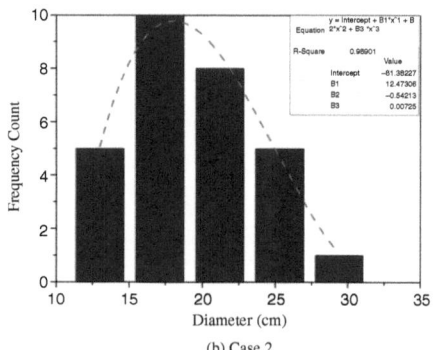

| (a) Case 1 | (b) Case 2 |

Figure 14 Frequency count of surface blocks analyzed with a PhotoModeler Scanner for heaped blocks (after Han & Song, 2014).

Next, the fragmentation for heaped blocks way analyzed based on the procedure discussed in section 3.3. As a result of the frequency count analysis of the surface blocks, both cases obeyed a lognormal distribution, as shown in Figure 14. Eq. (5) represents the Probability Density Function (PDF) of a lognormal distribution with the mean of μ and the standard deviation of σ. Eq. (6) represents the Cumulative Distribution Function (CDF), and Eq. (7) represents the inverse function for the CDF. As shown in Eqs. (5), (6) and (7), the lognormal distribution is a distribution with the mean of μ and the standard deviation of σ as input variables. Accordingly, the size information was estimated using the mean and the standard deviation for the surface blocks as input data, as shown in Eq. (7). Additionally, the upper and the lower limits were established, as discussed in section 3.3. The statistical estimation with the upper and the lower limits is as follows. Eq. (8) represents the Truncated Probability Density

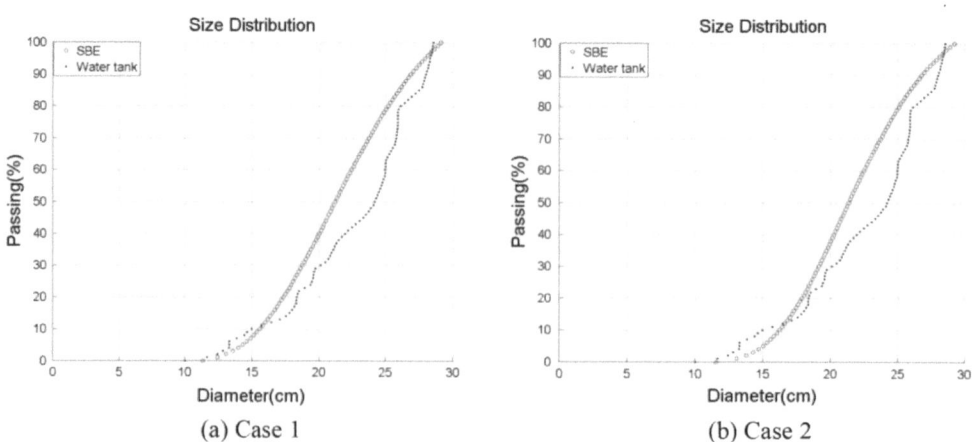

| (a) Case 1 | (b) Case 2 |

Figure 15 Size distributions from the SBE and the water tank measurement for heaped blocks (after Han & Song, 2014).

Function (TPDF) of a lognormal distribution that has the upper limit of a and the lower limit of b in diameter x. Eq. (9) represents the Truncated Cumulative Distribution Function (TCDF) for the TPDF. 15 % of error ration was compensated for in the result analyzed from PhotoModeler Scanner, as discussed in section 3.3. Figure 15 shows the size distributions analyzed from the water tank measurement and estimated through the SBE. In Case 1, the maximum difference between the water tank measurement and the SBE was 3.01 cm, while the difference at 50 % of passing was 2.99 cm. The average difference was 1.55 cm. In Case 2, the maximum difference between the water tank measurement and the SBE was 2.85 cm, while the difference at 50 % of passing was 2.85 cm. The average difference was 1.52 cm.

$$f(x|\mu,\sigma) = \frac{1}{x\sigma\sqrt{2\pi}} e^{-\frac{(\ln x - \mu)^2}{2\sigma^2}}, \quad x > 0 \tag{5}$$

$$F(x|\mu,\sigma) = \frac{1}{2} + \frac{1}{2} erf\left[\frac{\ln x - \mu}{\sigma\sqrt{2}}\right] = \frac{1}{\sigma\sqrt{2\pi}} \int_0^x \frac{e^{-\frac{(\ln x - \mu)^2}{2\sigma^2}}}{t} dt \tag{6}$$

$$x = F^{-1}(p|\mu,\sigma) \tag{7}$$

$$f_T(x) = \frac{f(x)}{\int_a^b f(x)dx}, a \le x \le b \tag{8}$$

$$F_T(x) = \int_a^x f_T(t)dt = \frac{\int_a^x f(t)dt}{\int_a^b f(t)dt} = \frac{F(x) - F(a)}{F(b) - F(a)} \tag{9}$$

4.2 Field experiment

The fragmentation for the blasted rock blocks in the blast field was analyzed using stereophotogrammetry. As shown in Figures 16(a), 16(b), and 16(c), a of total 100, 101, and 103 blasted rock blocks from Field 1, 2 and 3 were collected and analyzed for the field experiment. It is improbable to measure the sizes of blasted rock blocks in the field using a water tank. In addition, as shown in the result from the laboratory experiments, the analysis based on the modeling for each rock block yielded the result almost equivalent to the result estimated using a water tank. Accordingly, PhotoModeler Scanner was used as a substitute for a water tank measurement. First, each rock block was modeled using several images taken with different angles, and the size information was analyzed. Figure 16(d) shows an example of the modeling. The volume was in the range of about 1,500 cm^3 ~ 70,400 cm^3, and the total volume was approximately 1,020,000 cm^3 in Field 1. The volume was in the range of about 1,500 cm^3 ~ 42,700 cm^3, and the total volume was approximately 1,100,400 cm^3 in Field 2. The volume was in the range of about 1,700 cm^3 ~ 25,300 cm^3, and the total volume was approximately 730,000 cm^3 in Field 3. The diameter distributions of the surface blocks in every case obeyed a lognormal distribution, as shown in Figure 17.

(a) Field 1

(b) Field 2

(c) Field 3

(d) Modelling of individual block

Figure 16 Muckpile in the field (after Han & Song, 2014).

Figure 18 shows the size distributions for the field experiments. In Field 1, the maximum difference between the water tank measurement and the SBE was 6.42 cm, while the difference at 50 % of passing was 0.81 cm. The average difference was 1.88 cm. In Field 2, the maximum difference between the water tank measurement and the SBE was 5.07 cm, which represented the result of a linear extrapolation at 0 % of passing. The maximum difference excluding this data was 1.67 cm. The difference at 50 % of passing was 0.90 cm, and the average difference was 0.58 cm. In Field 3, the maximum difference between the water tank measurement and the SBE was 3.10 cm, while the difference at 50 % of passing was 1.45 cm. The average difference was 1.32 cm.

Compared to the analysis based on 2D image processing which only uses two-dimensional boundary and area information, the analysis based on the three-dimensional modeling yielded the result with the improved accuracy because it uses information from every side of an object shown in images for calculating three-dimensional modeling and volume calculation.

However, bias may exist as the sizes of the rock blocks collected in the blast field were limited to the possible size to be transported by manpower. In order to reduce influence of this bias, rock blocks with various shapes and sizes were collected from different places. Considering the analysis result, however, an approach for the fragmentation analysis using stereophotogrammetry and the statistical estimation presented in this

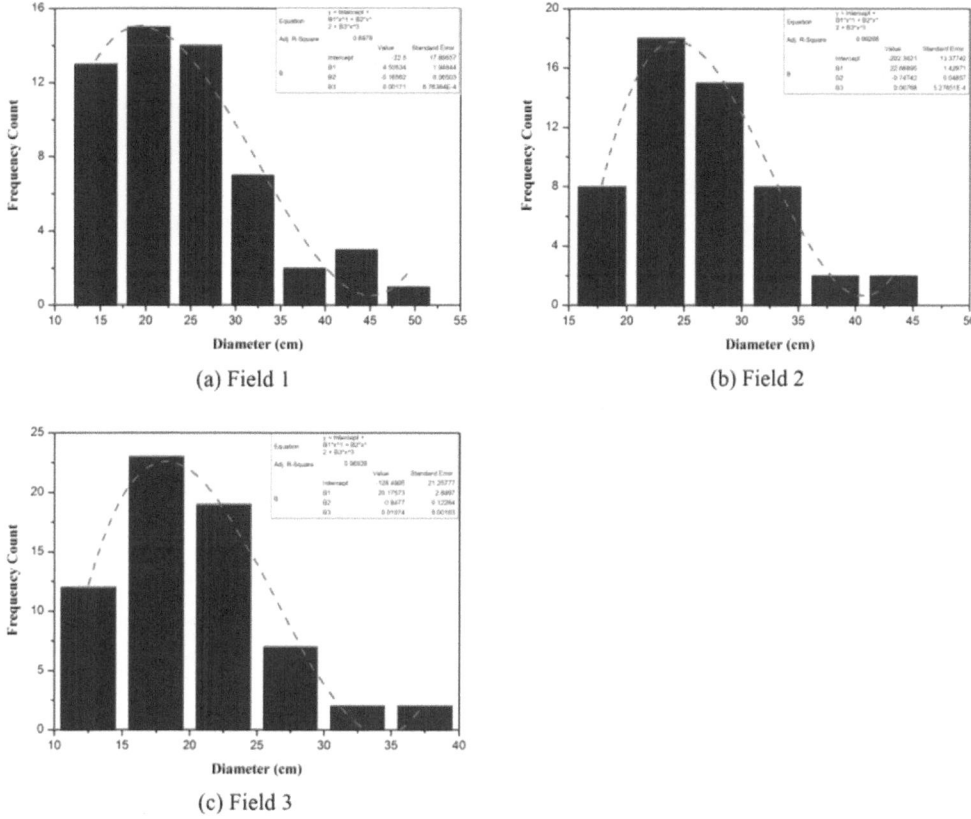

Figure 17 Frequency count of surface blocks analyzed with a PhotoModeler Scanner for a muckpile in the field.

study is expected to be applicable to the fragmentation analysis for blasted rock blocks with wider range of diameters

5 AUTOMATIC BLOCK DELINEATION

Compared to the result of the fragmentation analysis using stereophotogrammetry, the result of the fragmentation analysis based on 2D image processing was in lack of the accuracy as it only used two-dimensional area information. In addition, as mentioned in Section 2.1, the process of delineating blocks through manual editing was necessary after extracting boundaries by automatic filtering. In contrast, while the fragmentation analysis using stereophotogrammetry provided the improved result, the process of modeling and analyzing each block of the surface blocks was necessary. This process consumed most time in the entire process. Therefore, this study aimed to develop an algorithm which identifies rock blocks from the entire muckpile, and this algorithm

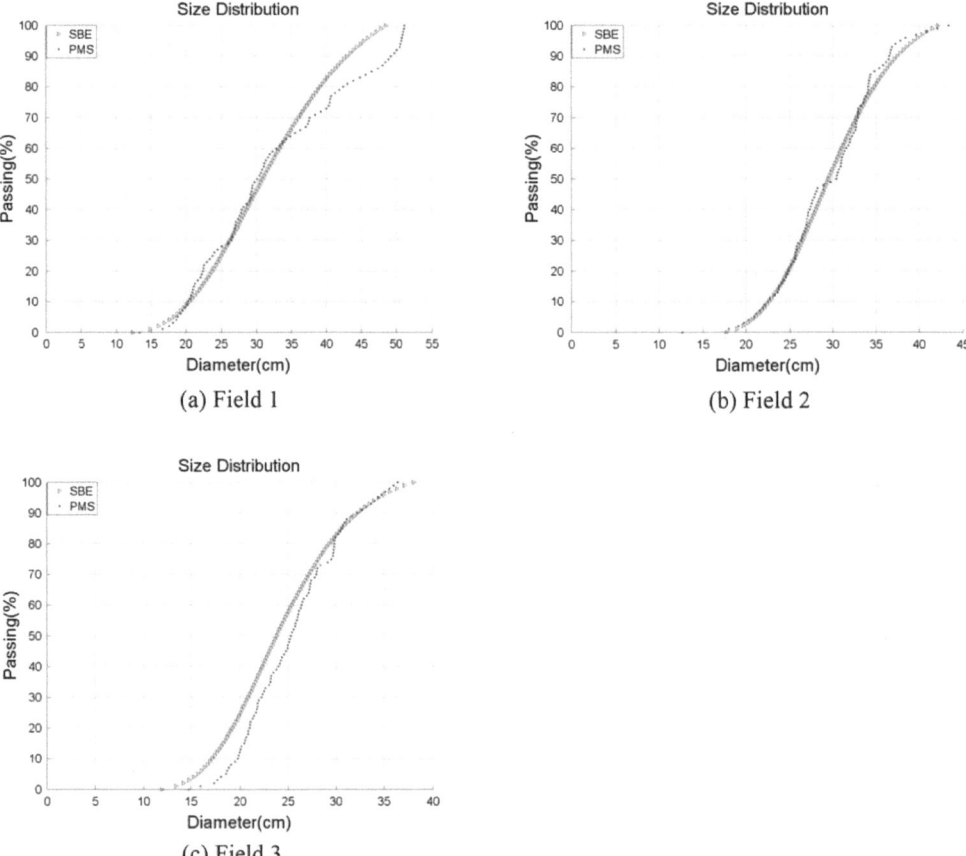

Figure 18 Size distributions from the SBE and the water tank measurement for a muckpile in the field (after Han & Song, 2014).

was integrated into the SBE. Finally, the applicability of this integrated algorithm for the fragmentation analysis was verified.

5.1 Pattern recognition

Pattern recognition is a technology for distinguishing and classifying data based on features that the elements in a certain group have in common. It is used in various fields such as character recognition such as number and letter recognition (Likforman-Sulem & Sigelle, 2008; Perwej & Chaturvedi, 2011), biometric recognition which involves recognizing faces, voices, and fingerprints (Anusuya & Katti, 2009), and weather forecast (Bannayan & Hoogenboom, 2008). The design of pattern recognition system can be categorized into five steps as follows (Richard *et al.*, 2001):

1   Data collection: Collecting as much data as possible is necessary for a stabilized performance of the system. While simple modeling test is plausible in cases where

data characteristics are typical enough to be predicted, sufficient amount of data which can ultimately represent the system should be acquired.

2   Characteristics selection: This step is the most important step in the system design, and it is crucial to obtain adequate prior knowledge by analyzing the target objects. It is important to select the common characteristics that are simple and strong against noises.

3   Model selection: As it is not possible to evaluate the performance of the selected model when there is no information of the actual objects, it is important to choose a proper model through analyzing the target objects.

4   Training: This is a step which develops the most efficient classifier by learning.

5   Evaluation: This is a step that evaluates the system performance and improves the system as necessary.

5.2 Algorithm

The algorithm developed in this study (Han *et al.*, 2016), the Block Delineation Algorithm (BDA), distinguishes each block based on the features that the block shape have in common, and its input material is the 3D point clouds obtained from PhotoModeler Scanner. The basic assumption applied to this algorithm is that the shape of a rock block is convex. The overall procedure of BDA is as follows. First, the algorithm identifies the peak that is the highest point of the entire blocks, and delineates the first block that has this peak. After removing the information of the first block from the entire blocks, the algorithm iterates the same procedure until the final block is delineated. The algorithm suggested in this study is mainly composed of four steps: preprocessing, block delineation, additional operation and block analysis.

5.2.1 Preprocessing

In some cases of modeling using stereophotogrammetry, a few singular values that do not associate with the entire model can be contained in the result, while stereophotogrammetry provides favorable modeling results of target objects in general. This phenomenon occurs when incorrect color information is referenced in the process of finding the corresponding point, or when the conjugate points in the stereo-pair of images have different color information due to the different camera angle. Such values, therefore, is defined as noise in this study, and are discarded ahead of delineating blocks. The details for the noise removal function are as follows:

- Noise removal function
  1) Saving the total number of point clouds for the entire data, n_1
  2) Calculation of the distance between the point (a_i, b_i, c_i) and the rest of the points $(1 \leq i \leq n_1, i$ is an integer.)
  3) Establishment of the equation for a sphere $(x-a_i)^2+(y-b_i)^2+(z-c_i)^2 < r^2$
  4) Counting the number of points within the sphere above
  5) If any number is not counted (*i.e.*, all data exists outside the sphere), the point (a_i, b_i, c_i) is classified into a singular value and removed as this point is in lack of consistency with the rest of the points.
  6) Iteration of the steps of 2 ~ 5 for the entire data

7)   Saving the number of the rest of the points, n_2
8)   If $n_1 \neq n_1$, then iteration of the steps of 1 ~ 7. Otherwise, closing the function as it means there are no more noises defined in this study

The radius of a sphere, r was determined as 2 cm, considering the mesh refinement to be implemented next.

The quality of the modeling outcome may depend on the number of the generated point clouds. However, a large number of points do not mean the best efficiency in the calculation. Rather, it can cause a significant amount of calculation time. For this reason, the entire point clouds are reconstructed so that they are arranged with the interval of 1 cm along the x–y direction, and it contributes to a reduction in calculation time.

• Peak detection function
   In reality, it is highly improbable for rocks to have perfectly flat and horizontal surfaces, which means there is a peak point that has the highest z coordinate in the point clouds of a rock. The block delineation algorithm distinguishes blocks based on the peaks defined in this study. The details for the peak detection function are as follows:

   1)   Starting with the first point, the area of interest is established with a certain range along x-y direction: a semi-infinite range greater than or equal to the z coordinate of a starting point, as shown in Fig. 19.
   2)   Counting the number of points included within the area of interest
   3)   If a single point is counted, the current point is classified into the peak.
   4)   Iteration of the steps of 1 ~ 3 for the entire input data

The number of peaks detected with this function depends on the size of the area of interest. If the large area of interest is established, a small number of peaks are detected. In contrast, if the small area of interest is established, a large number of peaks are detected. Therefore, it is recommended to establish a proper size of the area of interest.

5.2.2 Block delineation

Blocks are delineated based on the peaks detected in the preprocessing step. Block delineation is mainly composed of four functions: parallel transference, set-up of the area of interest, block shape refinement and valley cutting.

• Parallel transference function
   1)   Sorting the peaks detected in the preprocessing step in descending order
   2)   The entire data set is moved so that the first peak information satisfies $(x, y, z) = (0, 0, z)$
   3)   Deletion of the first peak information from the entire data set
   4)   Iteration of the steps of 1 ~ 3 to the last peak data after each block is delineated

• Set-up of the area of interest function
   The algorithm suggested in this study identifies the boundary of a block using the point clouds which are projected on the x–z plane ($y=0$), and restore the projected points to its original 3D point clouds. In the preprocessing step, the entire point

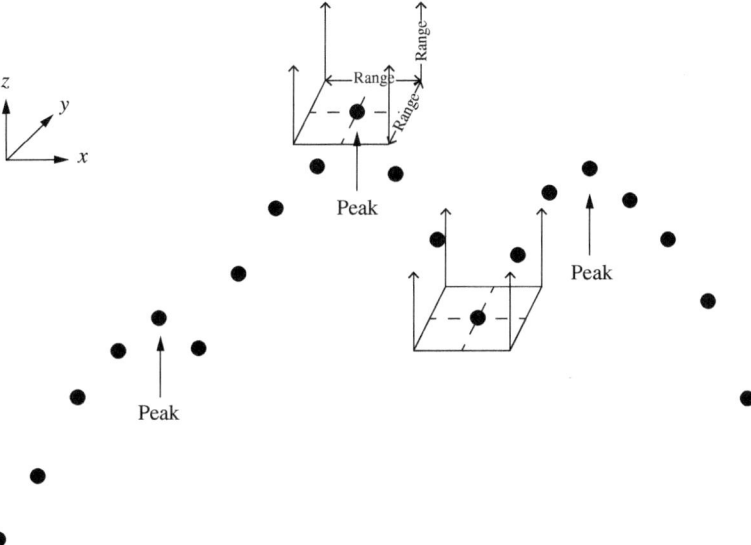

Figure 19 Peak detection (Han & Song, 2016).

clouds are reconstructed so that the data has the interval of 1 cm in x–y direction. Under this condition, if the point clouds are projected on the x–z plane, the projected points are arranged with the same distance, as shown in Figure 20 (a). However, if the point clouds are rotated, it is in lack of connectivity on the x–z plane, as shown in Figure 20(b). The area of interest, therefore, is established ahead of delineating blocks, as shown in Figure 20(b). Then, the data projected on the x–z plane is used in the next step. Considering the data interval reconstructed in the preprocessing step, the size of the area of interest is set as $-1 < y < 1$.

- Block shape refinement function
  If the projected points on x–z direction are used and there is a rapid change at the height, it is hard to identify the boundary of a rock block, as shown in Figure 21(a). Therefore, the projected data needs to be refined so that it is easy to identify the

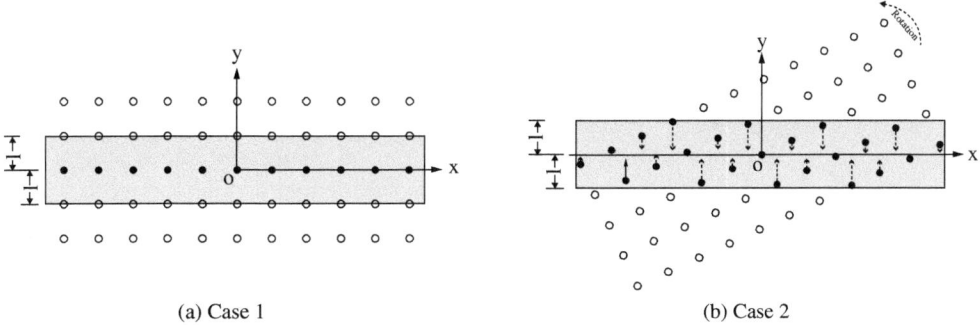

(a) Case 1   (b) Case 2

Figure 20 Data of interest after rotation (after Han & Song, 2016).

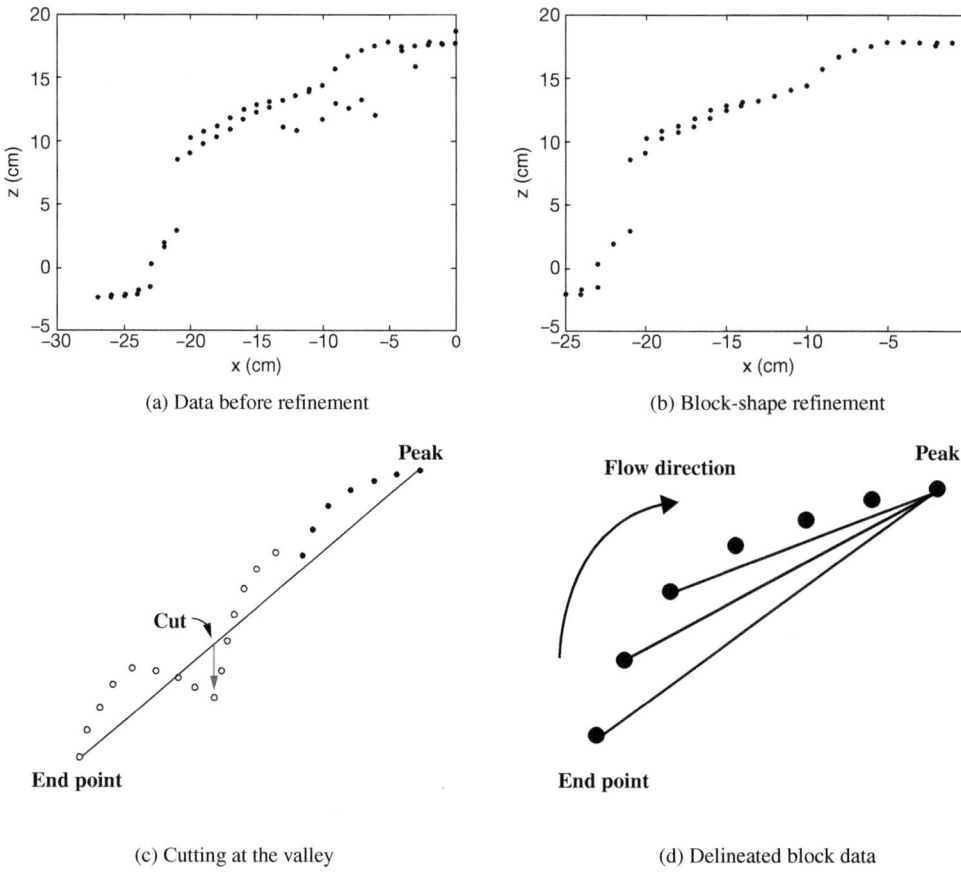

(a) Data before refinement

(b) Block-shape refinement

(c) Cutting at the valley

(d) Delineated block data

Figure 21 Detecting Block (after Han & Song, 2016).

boundary of a block. First, the initial five points are saved into the temporary variable. The next steps are to find the highest z coordinate among those initial points, and replace the z coordinate of the first point with the highest z coordinate. The original coordinate of the first point are not changed. The boundary of a block is refined with the highest boundary through this process, the original coordinate are recovered after delineating the block. These steps are iterated for the rest of the point clouds. Figure 21(b) shows the result of block shape refinement.

- Valley cutting function
  Considering that rock blocks are in the shape of a sphere and these blocks are arranged side by side, it is easy to recognize that there exists a valley between them. In this study, the boundary of a block is delineated based on this valley. The details for the valley cutting function are as follows.

  1) Counting the total number of the entire data, n
  2) Establishment of a line that connects the first and the last points, as shown in Eq. (10)

$$z = \frac{z_{end} - z_1}{x_{end} - x_1} \times (x - x_1) + z_1 \qquad (10)$$

3) For the entire x_i, calculation of the difference between the original z-coordinate, z_i and the z coordinate calculated from Eq. (10)
4) Saving z_{max} which is the largest difference value calculated in the step of 3 and its location
5) If $z_{max} \geq tmp$, then the data from the initial point to the extent of the corresponding location are saved. Otherwise, *count=0*
6) Calculation of z_{max} using Eq. (10)
7) If $z_{max} \geq tmp$, then going to the step of 1. Otherwise, count=count+1 and the iteration of the steps of 6 and 7 for $1 \leq i \leq n - 1$
8) If $n = count$, then the current points are classified into the block data.
   (Considering the uneven shape of a block, *tmp* is empirically set to one.)

Figure 21(c) shows the basic concept of the valley cutting. The first step is to construct a line that connects the first and the last points. The next steps are to find the deepest valley and its location, and remove the data after this location from the data set of interest. Through the iteration of the steps above, the final block can be delineated, as shown in Figure 21(d). After rotating the entire point clouds 2° degrees for every iteration of this valley cutting function, the whole procedure is iterated while the rotated angle, *deg* satisfies *deg*<180°. This method causes the duplicated data in the final blocks. The final step, therefore, is to remove those duplicated data from the final blocks.

5.2.3 Additional operation

In a case where the boundary of a block is ambiguous as shown in Figure 22(a) or the peak cannot be detected due to the influence of an overburden block as shown in

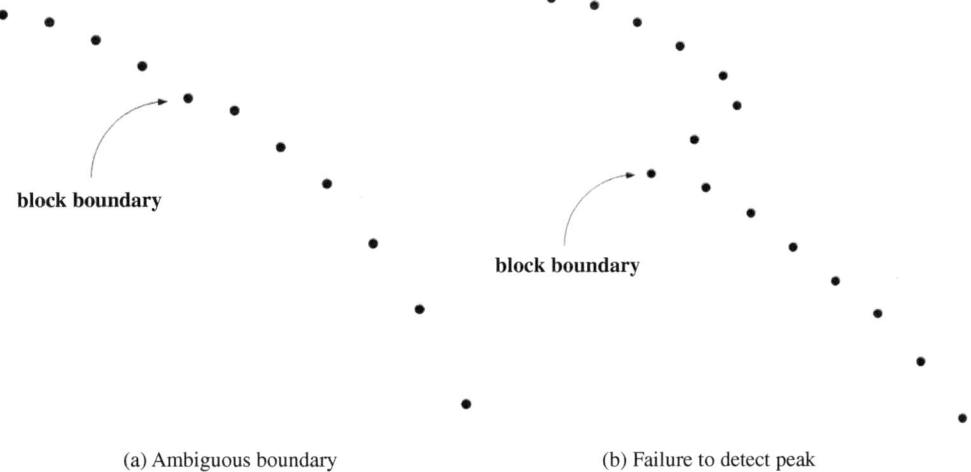

(a) Ambiguous boundary   (b) Failure to detect peak

Figure 22 Cases where the proposed algorithm is not applied.

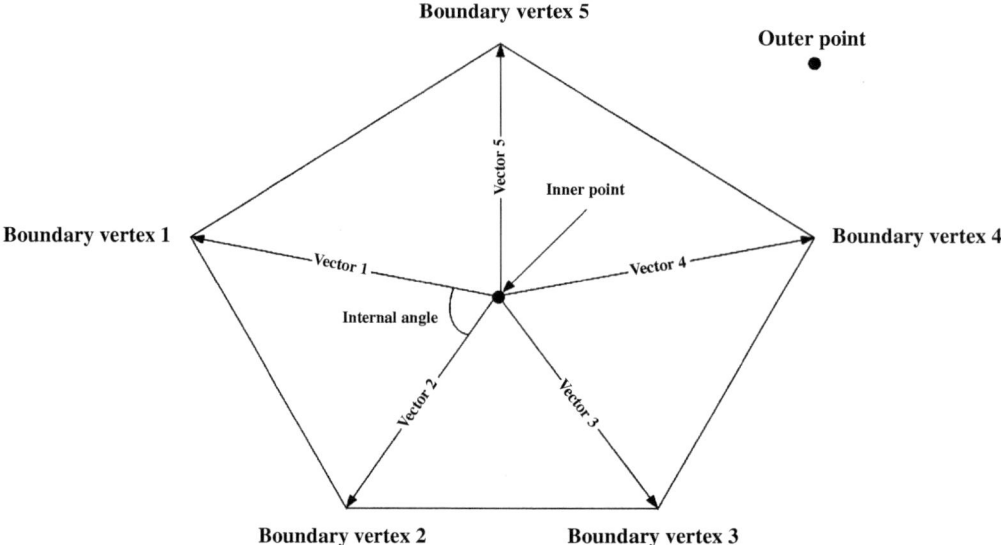

Figure 23 Method for inner point identification (after Han & Song, 2016).

Figure 22(b), the block detection becomes inapplicable. For these cases, blocks can be delineated by a user's operation. The fundamental principle of this function is to find the boundary of a block within the area established by a user, and Figure 23 exemplifies the extraction method. First, the information of vertices entered by a user is saved. Then, for the first point, vectors that connect this first point and every vertex are constructed. If this first point exists inside the area established by a user as shown in Figure 23, the summation of internal angles between vectors becomes 360°. The data within the area established by a user can be extracted through the iteration of the procedure above for the entire data set. A certain range of the end part of the extracted points is inspected to find the boundary, and the point with the lowest z value is identified as the boundary of a block, as shown in Figure 24. The details for the additional operation are as follows:

1) Input of vertices for establishing the area of interest
2) Construction of vectors that connects a point of interest and each boundary vertex
3) Calculation of internal angles between vectors using dot product, as shown in Eq. (11)

$$\theta = cos^{-1}\left(\frac{A \cdot B}{|A||B|}\right) \tag{11}$$

4) If $\Sigma\theta = 360$, then the current point is saved into the data set of interest
5) Iteration of the steps of 2 ~ 4 for the entire data
6) Input of an arbitrary point within the boundary
7) Implementation of the parallel transference so that the information of the input point in the step of 6 satisfies $(x, y, z) = (0, 0, z)$

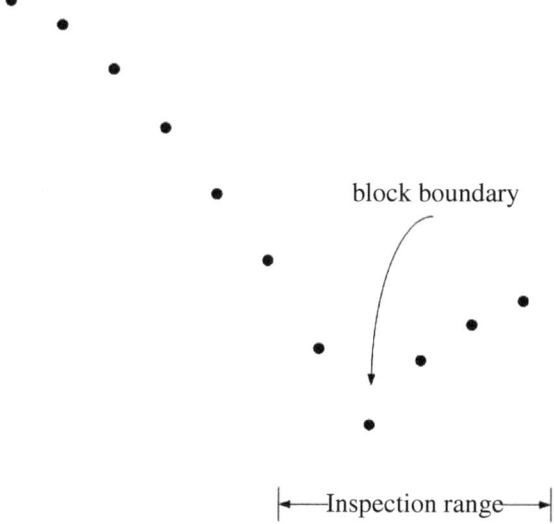

Figure 24 Block boundary decision in the manual analysis process (after Han & Song, 2016).

8)   Boundary determination at the point with the lowest z value by inspecting a certain range of the end part of the area of interest
9)   Iteration of the steps of 1 ~ 8 for the rest blocks

5.2.4 Block analysis and algorithm integration

In the preprocessing step, the point clouds are reconstructed so that the data has the interval of 1 along x–y direction. In this study, therefore, a block is assumed to be composed of cuboids that have 1 cm in width and length and z in height. As a result, the volume of a block can be expressed in the summation of z coordinate information of each cuboid. However, if a top block in heaped blocks is modeled and delineated, it would look like it is floating in the air. For the volume calculation of such a block, therefore, this block needs to be moved onto the ground so that its information becomes $(x, y, z) = (x, y, 0)$. The equivalent diameter can be calculated using the volume formula of a sphere based on the assumption that the shape of a rock block is a perfect sphere, as discussed in section 3.2.

The block delineation algorithm (BDA) suggested in this study was integrated into the SBE. 3D point clouds obtained from PhotoModeler Scanner modeling were used as the input materials for this integrated algorithm. Blocks were delineated using the BDA, and then its size distribution was analyzed based on the SBE. Figure 25 shows the overall processing flow of the integrated algorithm.

5.3 Application to laboratory and field experiments

The BDA was applied to two laboratory experiments for the fragmentation analysis of scattered blocks, while the integrated algorithm (BDA+SBE) was applied to two

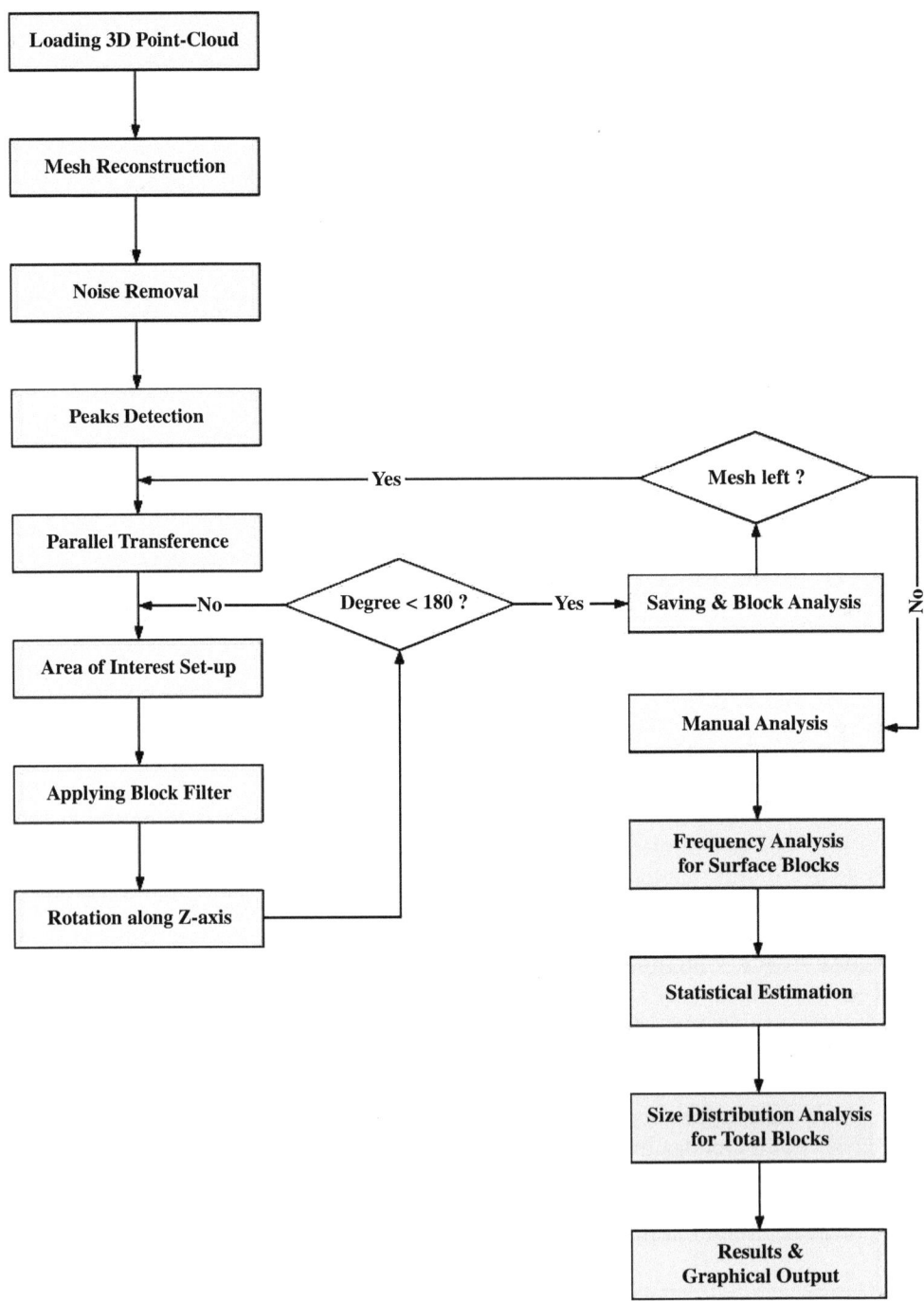

Figure 25 Flowchart of the integrated algorithm (Han & Song, 2016).

laboratory experiments and three field experiments for the analysis of heaped blocks. The same rock blocks collected for the laboratory and field experiments in section 4 were used for this application.

5.3.1 Scattered blocks

Figures 26(a) shows the result window of the BDA. In the left-top corner, the block delineated with the BDA is shown in an accumulative way, while the right-top corner specifies the location of the current block in the entire blocks. Peaks are marked to specify the blocks to be delineated, and the number that identifies the current block is marked in the right-top corner to enable the block to be deleted later based on that number if the algorithm fails to delineate the block. The left-bottom corner displays the current block in space, while the right-bottom corner shows the additional operation.

In the arrangements of scattered blocks, the statistical estimation discussed in section 3.3 is not necessary because the overall shapes of blocks can be detected easily. In Case 1, 40 blocks were delineated using the BDA, while the wrong information for 6 blocks was included in the results. Such wrong information was deleted from the results, and 5 more blocks were delineated through the additional operation. In Case 2, 41 blocks were delineated using the BDA, while the wrong information for 5 blocks was included in the results. The wrong information was deleted from the results, and 3 more blocks were delineated through the additional operation.

Figure 27 shows the distribution curves analyzed from the water tank measurement and analyzed with the BDA. In Case 1, the maximum difference between the water tank measurement and the BDA was 1.55 cm, while the difference at 50 % of passing was 0.33 cm. The average difference was 0.44 cm. In Case 2, the maximum difference between the water tank measurement and the BDA was 2.64 cm, while the difference at 50 % of passing was 0.33 cm. The average difference was 0.87 cm. The average difference was 0.96 cm.

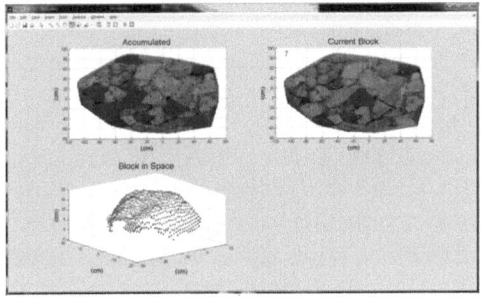
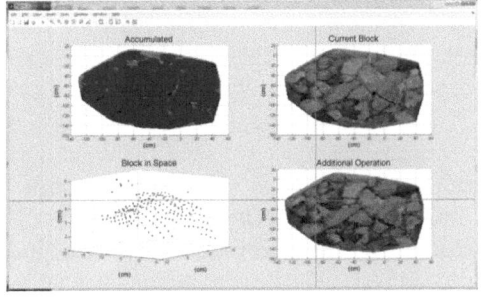

(a) Auto-delineation   (b) Additional operation

Figure 26 Description of the Block Delineation Algorithm results (after Han & Song, 2016).

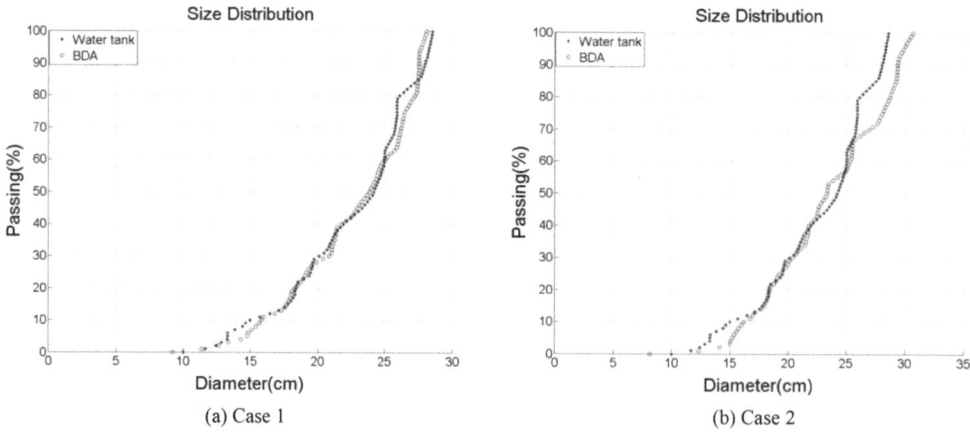

Figure 27 Size distributions from the water tank measurement and the BDA for scattered blocks (Han & Song, 2016).

5.3.2 Heaped blocks

For heaped blocks, the surface blocks were delineated first with the BDA, and its size distribution was analyzed with the SBE, as discussed in section 3.3. In addition, 15 % of error rate was compensated for in the results, as shown in Table 1.

In Case 1 of the laboratory experiments, 25 blocks were delineated using the BDA, while the wrong information for 6 blocks was included in the results. The wrong information was deleted from the results, and 10 more blocks were delineated through the additional operation. The minimum and the maximum diameters of the surface blocks were 9.02 cm and 31.96 cm, respectively, while the mean and the standard deviation of these diameters were 20.19 cm and 6.06 cm, respectively. In Case 2, 25 blocks were delineated using the BDA, while the wrong information for 4 blocks was included in the results. The wrong information was deleted from the results, and 9 more blocks were delineated through the additional operation. The minimum and the maximum diameters of the surface blocks were 11.17 cm and 29.33 cm, respectively, while the mean and the standard deviation of these diameters were 20.57 cm and 4.39 cm, respectively. The frequency count analysis showed the size distributions of the surface blocks for both cases obeyed a lognormal distribution, as shown in Figure 28. Figure 29 shows the distribution curves analyzed from the water tank measurement and analyzed with the BDA+SBE. In Case 1 of the laboratory experiments, the maximum difference between the water tank measurement and the BDA+SBE was 3.45, while the difference at 50 % of passing was 0.81 cm. The average difference was 0.91 cm. In Case 2, the maximum difference between the water tank measurement and the BDA+SBE was 2.93, while the difference at 50 % of passing was 1.86 cm. The average difference was 0.99 cm.

In Field 1 of the field experiments, 38 blocks were delineated using the BDA, while the wrong information for 10 blocks was included in the results. The wrong information was deleted from the results, and 14 more blocks were delineated through the additional operation. The minimum and the maximum diameters of the surface blocks were 12.11 cm and 44.42 cm, respectively, while the mean and the standard deviation of these diameters were 25.17 cm and 7.30 cm, respectively.

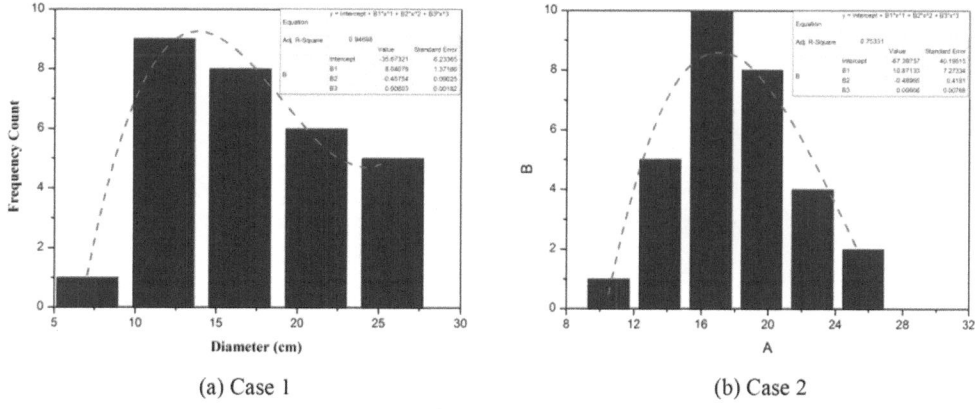

(a) Case 1 (b) Case 2

Figure 28 Frequency count of surface blocks analyzed with the BDA for heaped blocks (after Han & Song, 2016).

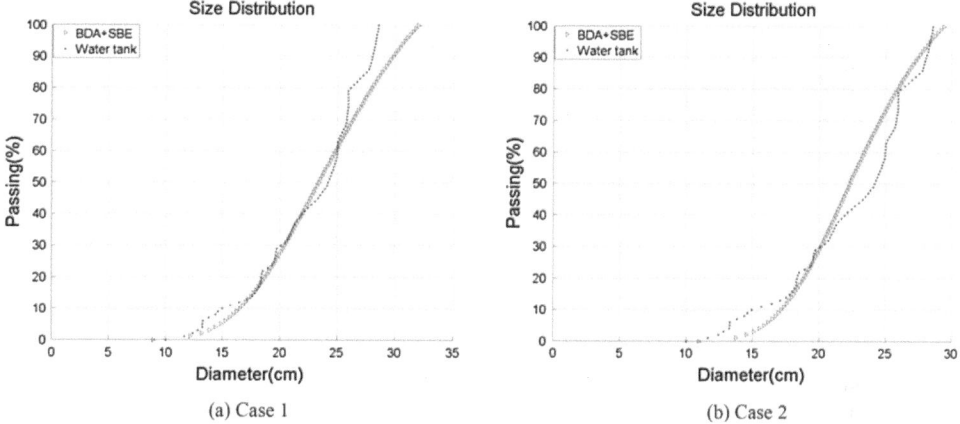

(a) Case 1 (b) Case 2

Figure 29 Size distributions from the integrated algorithm and the water tank measurement for heaped blocks (after Han & Song, 2016).

In Field 2, 42 blocks were delineated using the BDA, while the wrong information for 11 blocks was included in the results. The wrong information was deleted from the results, and 15 more blocks were delineated through the additional operation. The minimum and the maximum diameters of the surface blocks were 15.51 cm and 42.58 cm, respectively, while the mean and the standard deviation of these diameters were 27.68 cm and 7.48 cm, respectively. In Field 3 of the field experiments, 55 blocks were delineated using the BDA, while the wrong information for 19 blocks was included in the results. The wrong information was deleted from the results, and 8 more blocks were delineated through the additional operation. The minimum and the maximum diameters of the surface blocks were 11.15 cm and 38.33 cm, respectively, while the mean and the standard deviation of these diameters were 24.27 cm and 6.63 cm, respectively.

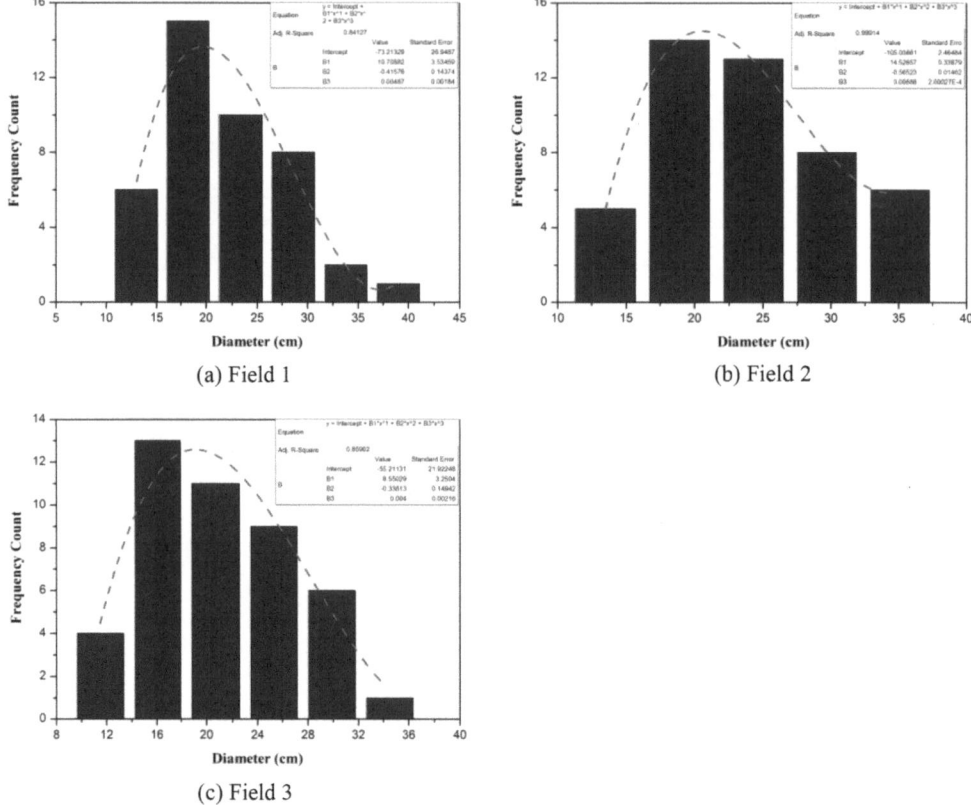

Figure 30 Frequency count of surface blocks analyzed with the BDA for a muckpile in the field (after Han & Song, 2016).

The frequency count analysis showed the size distributions of the surface blocks for every case obeyed a lognormal distribution, as shown in Figure 30. Figure 31 shows the distribution curves analyzed from the water tank measurement and analyzed with the BDA+SBE. In Field 1 of the field experiments, the maximum difference between the water tank measurement and the BDA+SBE was 9.37, while the difference at 50 % of passing was 0.01 cm. The average difference was 2.71 cm. In Field 2, the maximum difference between the water tank measurement and the BDA+SBE was 3.77, while the difference at 50 % of passing was 0.94 cm. The average difference was 1.76 cm. In Field 3, the maximum difference between the water tank measurement and the BDA+SBE was 3.96, while the difference at 50 % of passing was 2.62 cm. The average difference was 2.41 cm.

6 DISCUSSION AND CONCLUSION

In the analysis of the information in images, it is probably the most accurate way to perceive an object with human eyes. For the fragmentation analysis as well, an object in

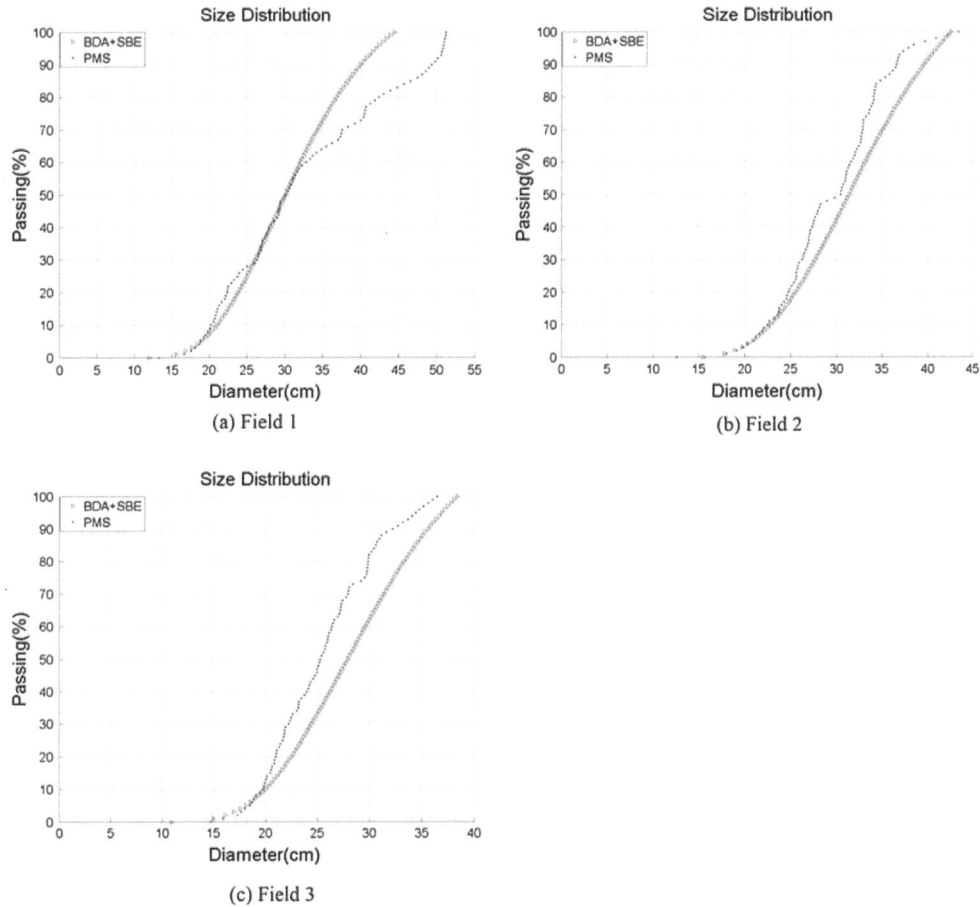

Figure 31 Size distributions from the integrated algorithm and the PhotoModeler Scanner analysis for a muckpile in the field (after Han & Song, 2016).

an image could be perceived much better with human eyes. Considering the efficiency, however, it is helpful to use the automatic process such as Split Desktop. For the purpose of the automatic fragmentation analysis, an algorithm that detects blocks based on shape features of blocks and analyzes them in a three-dimensional way was developed in this study. Considering variety and irregularity of the shapes of rocks, however, it may not be possible to develop an algorithm that can perfectly detect every block. Accordingly, the block delineation algorithm suggested in this study may not be applicable to every shape of rock blocks. As shown in the results, when the algorithm failed to perceive a block, it had a negative influence on the result.

The delineation rate of the BDA is shown in Table 2. For scattered blocks, it had approximately 90 % of the delineation rate, while it had about 70 % for heaped blocks. As the boundaries between blocks in scattered blocks were comparatively clear, most blocks were able to be detected by the BDA. Compared to the analysis for scattered blocks, the delineation rate decreased in the analysis for heaped blocks. This was due to

Table 2 Delineated rate using the Block Delineation Algorithm (Han & Song, 2016).

Experiments		Laboratory		Field 1	Field 2	Field 3
		Case 1	Case 2			
Scattered Blocks	Delineated	34	36	-		
	Additional	5	3			
	Rate (%)	87	92			
Heaped Blocks	Delineated	19	21	28	31	36
	Additional	10	9	14	15	8
	Rate (%)	66	70	67	67	82

the tendency that the algorithm cannot easily detect blocks when the boundaries between neighboring blocks are ambiguous. Therefore, this algorithm is expected to be applicable to the fragmentation analysis of a muckpile on a plane such as a conveyor belt.

The volume calculation method in this study yielded slightly different results from the water tank measurement and the PhotoModeler Scanner analysis. The difference for all experiments was approximately 3 cm on average, and the maximum difference was 10 cm. This maximum difference, however, was not due to the volume calculation method in this study but due to the case that the algorithm did not perceive a block properly. In this study, the equivalent diameter was calculated using the cubic root of the volume as shown in Eq. (3), and used in the fragmentation analysis. For this reason, the volume calculation method in this study had a little effect on the result.

The algorithm suggested in this study delineates blocks based on peaks defined in this study, and the total processing time depends on the number of peaks. Therefore, it is recommended to establish a proper size of the search area for peak detection. When a shadow or sunlight is cast on the boundary, the image processing may not delineate blocks because it cannot detect the change of the color information between blocks. For this reason, this study did not take the color information in images into consideration.

Considering the uneven shapes of rock blocks, the algorithm suggested in this study may not be applicable to the analysis for all shapes of blocks. In addition, the larger and heavier blocks are likely to be placed at the bottom in a muckpile, and this may result in the underestimation of the volume and the size. The performance of the algorithm in this study, therefore, can be improved through the further study on these issues.

REFERENCES

Anusuya M.A., Katti S.K., 2009, Speech Recognition by Machine, International Journal of Computer Science & Information Security, Vol. 6, No. 3, pp. 181–205.

Bannayan M., Hoogenboom G., 2008, A Tool for Real-Time Prediction of Daily Weather Data Realizations Based on a Modified k-nearest Neighbor Approach, Environmental Modelling & Software, Vol. 23, pp. 703–713.

Chen S., Yue Z.Q., Tham L.G., 2004, Digital Image-based Numerical Modeling Method for Prediction of Inhomogeneous Rock Failure, International Journal of Rock Mechanics & Mining Sciences, Vol. 41, pp. 939–957.

Cunningham C.V.B., 1996, Optical Fragmentation Assessment, Proceedings of the Fragblast-5 Workshop on Measurement of Blast Fragmentation, pp. 13–19.

Firpo G., Salvini R., Francioni M., Ranjith P.G., 2011, Use of Digital Terrestrial Photogrammetry in Rocky Slope Stability Analysis by Distinct Elements Numerical Methods, International Journal of Rock Mechanics & Mining Sciences, Vol. 48, pp. 1045–1054.

Girdner K.K., Kemeny J.M., Srikant A., McGill R., 1996, The Split System for Analyzing the Size Distribution of Fragmented Rock, Proceedings of the Fragblast-5 Workshop on Measurement of Blast Fragmentation, pp. 101–108.

Hagan T.O., 1980, A Case for Terrestrial Photogrammetry in Deep-mine Rock Structure Studies, International Journal of Rock Mechanics & Mining Sciences, Vol. 17, No. 4, pp. 191–198.

Han J.H., Song J.J., 2014, Statistical Estimation of Blast Fragmentation by applying Stereophotogrammetry to Block Piles, International Journal of Rock Mechanics & Mining Sciences, Vol. 68, pp. 150–158.

Han J.H., Song J.J., 2016, Block Delineation Algorithm for Rock Fragmentation Analysis, International Journal of Rock Mechanics & Mining Sciences, Vol. 82, pp. 48–60.

Karl K., 2007, Photogrammetry, Walter de Gruyter.

Kemeny J., Mofya E., Kaunda R., Lever P., 2001, Improvements in Blast Fragmentation Models using Digital Image Processing, Proceedings of the 38th U.S. Symposium on Rock Mechanics, pp. 213–216.

Kleine T.H., Cameron A.R., 1996, Blast Fragmentation Measurement using GoldSize, Proceedings of the Fragblast-5 Workshop on Measurement of Blast Fragmentation, pp. 83–89.

Likforman-Sulem L., Sigelle M., 2008, Recognition of Degraded Characters using Dynamic Bayesian Networks, Pattern Recognition, Vol. 41, pp. 3092–3103.

Liu Q., Tran H., 1996, Validation of FragScan, WipFrag, Split, Proceedings of the Fragblast-5 Workshop on Measurement of Blast Fragmentation, pp. 151–155.

Maclachlan R.R., Singh A., 1989, Photographic Determination of Oversize Particles in Heaps of Blasted Rock, Journal of the South African Institute of Mining & Metallurgy, Vol. 89, No. 5, pp. 147–152.

Maerz N.H., Franklin J.A., Rothenburg L., Coursen D.L., 1987, Measurement of Rock Fragmentation by Digital Photoanalysis, Proceedings of the 6th International Society for Rock Mechanics Congress, pp. 687–692.

Maerz N.H., Palangio T.C., Franklin J.A., 1996, WipFrag Image Based Granulometry System, Proceedings of the Fragblast-5 Workshop on Measurement of Blast Fragmentation, pp. 91–99.

Maerz N.H., Zhou W., 1998, Optical Digital Fragmentation Measuring Systems, International Journal for Blasting & Fragmentation, Vol. 2, No. 4, pp. 415–431.

Müller B., Hausmann J., Niedzwiedz H., 2009, Control of Rock Fragmentation and Muck Pile Geometry during Production Blasts, Proceedings of the 9th International Symposium on Rock Fragmentation by Blasting: Fragblast 9, pp. 277–286.

Noy M.J., 2006, The Latest in on-line Fragmentation Measurement – Stereo Imaging over a Conveyor, Proceedings of the 8th International Symposium on Rock Fragmentation by Blasting: Fragblast 8, pp. 61–66.

Ohnishi Y., Nishiyama S., Yano T., Matsuyama H., Amano K., 2006, A Study of the Application of Digital Photogrammetry to Slope Monitoring Systems, International Journal of Rock Mechanics & Mining Sciences, Vol. 43, pp. 756–766.

Ozkahraman H.T., 2006, Fragmentation Assessment and Design of Blast Pattern at Goltas Limestone Quarry, Turkey, International Journal of Rock Mechanics & Mining Sciences, Vol. 43, pp. 628–633.

Paul R.W., Bon A.D., 2000, Elements of Photogrammetry, McGraw-Hill.

Perwej Y., Chaturvedi A., 2011, Neural Networks for Handwritten English Alphabet Recognition, International Journal of Computer Applications, Vol. 20, No. 7, pp. 1–5.

Reid T.R., Harrison J.P., 2000, A Semi-automated Methodology for Discontinuity Trace Detection in Digital Images of Rock Mass Exposures, International Journal of Rock Mechanics & Mining Sciences, Vol. 37, No. 7, pp. 1073–1089.

Richard O.D., Peter E.H., David G.S., 2001, Pattern Classification, John Wiley & Sons.

Rizk N.K., Lefebvre A.H., 1985, Drop-Size Distribution Characteristics of Spill-Return Atomizers, Journal of Propulsion & Power, Vol. 1, No. 1, pp. 16–22.

Sanchidrián J.A., 2015, Ranges of validity of some distribution functions for blast-fragmented rock. In: Spathis AT *et al.*, editors. Proceedings of the 11th International Symposium on Rock Fragmentation by Blasting – FRAGBLAST 11, Sydney, Australia, 24–26 August 2015. Carlton Victoria: The Australasian Institute of Mining and Metallurgy, pp. 741–748.

Sanchidrián J.A., Ouchterlony F., Segarra P., Moser P., 2014, Size distribution functions for rock fragments. International Journal of Rock Mechanics and Mining Sciences, Vol. 71, pp. 381–394.

Siddiqui F.I., Ali Shah S.M., Behan M.Y., 2009, Measurement of Size Distribution of Blasted Rock using Digital Image Processing, Engineering Sciences, Vol. 20, No. 2, pp. 81–93.

Sirotyuk G.N., 1970, A Method of Calculating the Fragment-Size Composition of Blasted Rock from the Given Oversize Fragmentation Dimension, Mining-Metallurgical Institute, Vol. 1, pp. 72–79.

Sudhakar J., Adhikari G.R., Gupta R.N., 2006, Comparison of Fragmentation Measurements by Photographic and Image Analysis Techniques, Rock Mechanics & Rock Engineering, Vol. 39, No. 2, pp. 159–168.

Thurley M.J., 2011, Automated Online Measurement of Limestone Particle Size Distributions using 3D Range Data, Journal of Process Control, Vol. 21, pp. 254–262.

Tsoutrelis C.E., Exadactylos G.E., Kapenis A.P., 1990, Study of the Rock Mass Discontinuity System using Photoanalysis, Proceedings of International Conference on Mechanics of Jointed and Faulted Rock, pp. 103–112.

Valenca J., Dias-da-Costa D., Júlio E., Araújo H., Costa H., 2013, Automatic Crack Monitoring using Photogrammetry and Image Processing, Measurement, Vol. 46, pp. 433–441.

Van Aswegen H., Cunningham C.V.B., 1986, The Estimation of Fragmentation in Blast Muckpiles by means of Standard Photographs, Journal of the South African Institute of Mining & Metallurgy, Vol. 86, No. 12, pp. 469–474.

Van Breugel K., 1995, Numerical Simulation of Hydration and Microstructural Development in Hardening Cement-Based Materials, Cement & Concrete Research, Vol. 25, No. 2, pp. 319–331.

Wang W., Li L., Hakami E., 2005, Image Analysis of Multiple Rock Fractures, Proceedings of the IEEE International Conference on Mechatronics & Automation, pp. 1272–1276.

Chapter 22

Blasting parameters

Ebrahim Ghasemi[1] *& Mehdi Rahmanpour*[2]

[1]*Department of Mining Engineering, Isfahan University of Technology, Isfahan, Iran*
[2]*Department of Mining and Metallurgical Engineering, Amirkabir University of Technology, Tehran, Iran*

Abstract: Many parameters affect the blasting results which can be divided into four categories: parameters related to explosive properties, parameters related to geomechanical properties of rock mass, parameters related to blast geometry, and parameters related to initiation pattern. This chapter focuses on these parameters and their role on blasting operation.

I INTRODUCTION

Despite the recent achievements in mechanical rock cutting technologies, blasting is the first choice for rock fragmentation and loosening in many mines. The statistics shows that a large amount of explosives is used for rock blasting in mines every year. Furthermore, poor rock blasting has been the main cause of many hazards and human injuries and fatalities. It has an important role on mine safety and costs. Optimum design of blasting is a suitable solution to increase the efficiency and safety of this operation. The main goal of optimum blasting is to increase the fragmentation, production and safety and to decrease the drilling and blasting costs, dilution, damage to the walls and negative effects on environment. Many parameters affect the optimum blasting design, which can be classified into four categories (Jimeno *et al.*, 1995; Bhandari, 1997; Roy, 2005):

1. Parameters related to explosive properties,
2. Parameters related to geomechanical properties of rock mass,
3. Parameters related to blast geometry, and
4. Parameters related to initiation pattern.

From another point of view, these parameters can be broadly divided into two categories, namely, controllable parameters and uncontrollable parameters. Controllable parameters can be changed, while uncontrollable parameters are natural and cannot be changed. Based on this definition, all mentioned categories are controllable except for parameters related to geomechanical properties of rock mass.

In the following each of these categories is introduced briefly and its role on blasting results is explained.

2 PARAMETERS RELATED TO EXPLOSIVE PROPERTIES

Blast design is an engineering challenge and it is a kind of a technical and economic decision making problem. Selection of a proper explosive is an important step in any blast design and its optimization. Knowing the properties and performance of explosives will improve the efficiency of blast results. Explosives release gas and heat as a result of a rapid reaction (Gokhale, 2011). There are various types of explosives with different characteristics. For example, ANFO type explosives are loose, free-flowing, granular compositions, whereas emulsion explosives have a consistency that varies from that of syrup to firm putty. Almost 80% of mining blasts are made using ANFO (Gokhale, 2011). There are also various blends of emulsion and ANFO type explosives, notably so-called heavy ANFOs. Water gel (slurry) explosives are also used in some countries (RioTinto, 2007). These properties provide an estimation of the blasting results such as fragmentation, displacement and vibration. The physical properties of the explosive can dictate the handling system used to charge the explosive into blastholes. The proper explosive should be selected considering the specifications of the usage. Although, there are a variety of explosives with different strength and characteristics but it should be noted that the main criterion that guides the blasters to select an explosive is the cost issues (Dutta *et al.*, 1973; Roy, 2005; Fox, 2012). The most important characteristics are strength/energy, velocity of detonation, water resistance, sensitivity, density, desensitization, detonation pressure, critical diameter. In this section the main aspects of explosives are discussed.

2.1 Density

Explosives are supplied in different densities in order to provide more control on the total energy released in a blast (Jimeno *et al.*, 1995; Bhandari, 1997; Mader, 2008; Gokhale, 2011). Therefore explosives should be selected such that they suit the particular blasting conditions of site. Density of explosives varies between 0.6 and 1.7 g/cm^3. But it should be noted that the in-hole density (or the loading density) of explosives has a significant effect on the energy per meter of charge length. Loading density is normally lower than the density of explosive itself; however it depends on the compaction level of charge within a blasthole. Higher-density explosives generate more energy and breakage. There is an optimum density for each explosive as a blasting agent which is also known as death-density. Agents with low density are very sensitive to detonation cord and initiate before the detonation of the primer cartridge. In this respect, high density agents are insensitive and they do not detonate. In hard rock blasting where fine fragmentation is required, a dense explosive is recommended. Contrarily, low-density explosives are recommended for easily broken rock or where fine fragmentation is not a necessity. Also, when conducting an underwater blasting, the density of explosive should exceed water density in order to sink in blastholes. In a controled blasting operation, the charge is normally decoupled. In this condition, an empty space is leaved between the charge and the wall of the blasthole (Jimeno *et al.*, 1995). This space should be considered in calculating the charge density.

The other important factor with regard to density is the compaction of explosives in blastholes. If any explosive is compacted in a blasthole, then its density increases. If the explosive density is increased beyond a certain amount, then its sensitivity is reduced significantly. This density is called 'critical density'. Beyond this level even a good primer may not detonate the blasting material (Bhandari, 1997).

2.2 Velocity Of Detonation (VOD)

Velocity of detonation (VOD) is the speed with which the detonation wave propagates through the explosive and it is given in feet or meter per second. It is measured confined and unconfined (Olofsson, 1988; Fox, 2012; Mahadevan, 2013). Detonation or reaction velocities of low explosives are 600–900 m/s. In high explosives the detonation velocity exceeds 1500 m/s (Konya, 1995). High explosives are classified as primary, secondary and tertiary explosives. Primary explosives are used as initiating explosives and they are very sensitive to heat, shock, and friction. Secondary explosives are relatively insensitive compared with initiating explosives and they could be applied in large quantities. These types of explosives need primary explosives for their detonation. Tertiary explosives are very insensitive to shocks. In order to detonate them, a small quantity of secondary explosives is required. Tertiary explosives are also called blasting agents (Gokhale, 2011). Detonation velocity of commercial explosives varies from 1500 to 9000 m/s. Detonation velocity defines the rhythm of energy release. The confined and unconfined detonation velocities of explosives are very different and they are known as upper and lower detonation velocities respectively. For some special seismic explosives (*e.g.* WGeosit) the upper and lower detonation velocities are the same (Meyer *et al.*, 2002).

The velocity of detonation is higher than the sonic waves. If the velocity is less than the sonic waves then it is called deflagration. Two explosives having the same strength but different VOD may perform quite differently in a blast. As a general rule the higher the VOD, the greater the shock energy and lower the heave energy. However, it is important not to correlate shock energy directly with fragmentation energy. The VOD of many explosives increases with charge diameter and confinement. Because of their high degree of refinement and efficiency, emulsion explosives can maintain very high VOD even with poor confinement and in small diameters.

Velocity of detonation is a measure of C-J plane of detonation wave when it travels through the explosive column (Cooper, 1997; Gokhale, 2011; Mahadevan, 2013). Velocity of detonation is a function of composition, particle size, confinement, loading density, diameter of the blasthole, presence of voids in the rock mass, rock mass temperature, initiating system and aging. Among them density, diameter and confinement of explosives increase the detonation velocity. VOD for two explosives is compared using Equation 1.

$$V_{e1} = V_{e0} + K(d_{e1} - d_{e0}) \tag{1}$$

Where, V_{e0} is the detonation velocity of the explosive with a density of d_{e0}, and V_{e1} is the detonation velocity of the explosive with a density of d_{e1}. In this equation, K is a constant and depending on the explosive type it varies between 3000 and 4000 m/s.

Considering the equation, as the density of an explosive increases it improves the detonation velocity. Thus, as an explosive is charged compactly within a blasthole then the detonation velocity increases. VOD in a not-compacted explosive is 70%–80% of the VOD in a compacted one. This statement is not always true and for each explosive there is a density beyond which the detonation does not occur. The particle size of an explosive is the other factor that controls the density of an explosive and the corresponding detonation velocity. As the particle size shrinks the velocity of detonation increases. Also the type of initiating system causes the VOD to become low or high. The other factor is the charge diameter. As the charge diameter increases the velocity of detonation increases too.

2.3 Detonation pressure

Detonation pressure (P_d) is a function of explosive density and detonation velocity (Cooper, 1997; Jimeno et al., 1995; Bhandari, 1997; Roy, 2005; Gokhale, 2011). There are some theoretical and also empirical models to calculate detonation pressure (Equation 2–4). It can also be calculated by analysis of chemical equations during the detonation process. It is obvious that factors such as density, diameter and particle size of the explosive which affect detonation velocity will also influence detonation pressure (Cameron & Torrance, 1990; Roy, 2005; Mahadevan, 2013).

$$P_d(kbar) = \left(4.18 \times 10^{-7} \times SG_e \times V_e^2 (ft/s)\right) \Big/ (1 + 0.8 \times SG_e) \tag{2}$$

$$P_d(^N/_{m^2}) = \left(V_e^2(m/s) \times d_e(kg/m^3)\right) \Big/ 3.8 \tag{3}$$

$$P_d(GPa) = (1/2) \times d_e(kg/m^3) \times V_e(m/s) \times 10^{-9} \tag{4}$$

Where, SG_e is the specific gravity of the explosive, d_e is the density of the explosive, and V_e is the detonation velocity of the explosive.

In dense and hard rock blasting operation an explosive with a high detonation pressure is recommended. The maximum detonation pressure occurs in the direction of shock waves. Thus, it gains its maximum value in the explosive cartridge at the opposite end from where the initiation occurs. At the side of the cartridge the detonation pressure is nearly zero because detonation waves do not extend to the edge of the cartridge. To apply the maximum detonation pressure on the rock, it is necessary to place the explosive on the rock and initiate it from the opposite end. Also, the contact area between the rock and explosive should be as large as possible. This technique is widely used in mud capping or plaster shooting of boulders.

The pressure exerted on the wall of the blasthole by the expanding gases is called the 'blasthole pressure'. It depends upon the detonation pressure and the confinement of the explosive in the blasthole. It is equal to 50%–60% of the detonation pressure. In jointed rock masses the blasthole pressure tends to be very low (Gokhale, 2011).

2.4 Shock energy, gas energy and heave energy

Following detonation, high-pressure gases compress and crush the rock immediately surrounding the explosives. The energy that is released by the explosive can be partitioned into two main types, the shock energy and the heave energy. The shock energy transmitted to the rock depends on detonation pressure of the explosive (Roy, 2005; Fox, 2012). It is also related to the extent and the rate of the borehole expansion to a so called equilibrium state and includes the effects due to sub-optimal initiation. This energy is termed 'shock energy' which is primarily responsible for conditioning the rock and initiating mechanisms that generate fractures. 'Gas energy', 'heave energy' or 'blasthole pressure' is delivered during the later expansion of the explosive products into the crack network in the rock. Once a fracture network is established the gas is able to expand into the network, both extending the fracture process and causing movement of the rock. As this happens, the gas pressure drops until it vents to the atmosphere. Without gas pressure, the fractured rock remains interlocked and would not move. A combination of shock and heave energy breaks and moves the rock mass into a pile of fragmented rocks that are easy to dig, load and haul (Hustrulid, 1999).

Gas pressure is related to the chemical composition, heat and the volume of gas liberated per unit weight of explosives during the detonation. Gas energy produces less pressure than shock energy does (Roy, 2005). For simplicity it is assumed that gas pressure is approximately half the detonation pressure. But, in case of some explosives such as ANFO, gas pressure exceeds the detonation pressure. The action of gas pressure is loosening, expanding and throwing of the rock mass. As the temperature of explosion increases the gas pressure increases too. Also, in a constant temperature, the explosive that produces more gas has the highest gas energy. High explosives can produce both the shock and gas energy, but low explosives can only produce the gas energy.

After detonation the shock wave propagates as a compressive wave in all directions. Just behind the shock wave is the reaction zone where the chemical decomposition is taking place (Mahadevan, 2013). Then depending on the relative force of the wave and the compressive strength of the rock, it crushes the surrounding rock to some extent. When the force exceeds the elastic limits of the rock, bends the rock and cause it to crack radially (Figure 1). When the shock wave reaches a free face, the wave is reflected back into the rock as a tension wave. The speed and the energy of the wave have been

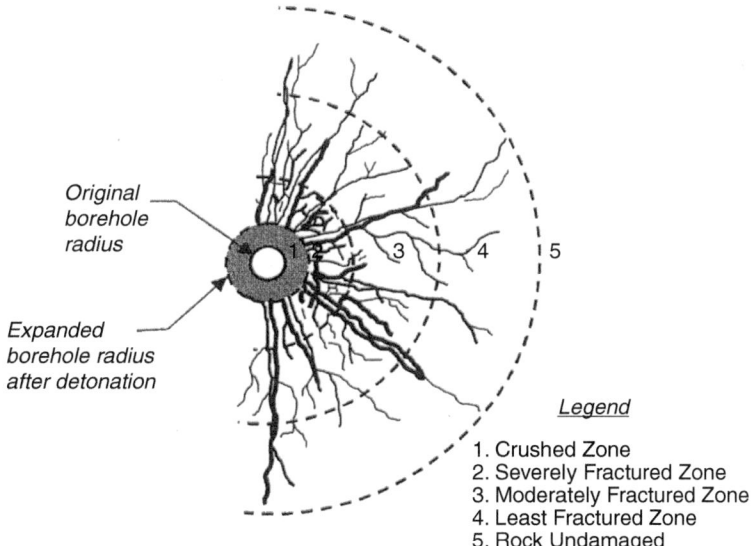

Original borehole radius

Expanded borehole radius after detonation

Legend
1. Crushed Zone
2. Severely Fractured Zone
3. Moderately Fractured Zone
4. Least Fractured Zone
5. Rock Undamaged

Figure 1   The mechanism of blasting (Roy, 2005).

lowered. For optimal burdens, the wave still carries enough energy to overcome the tensile strength of the rock. Too large burdens – such as behind the borehole – do not permit rock to bending and no radial fractures occur. Also, there will be no wave reflection and no lateral cracks will generate. In these conditions the wave energy is simply absorbed by the rock (Jimeno *et al.*, 1995; Roy, 2005).

2.5 Strength and brisance

The energy of an explosive expresses its ability to do work or to produce mechanical effects. Strength of an explosive is determined by factors such as density, detonating velocity as well as the amount of the gas released per unit weight of explosives (Meyer *et al.*, 2002; Moser *et al.*, 2003; Gokhale, 2011; Fox, 2012). An explosive with greater energy will be able to do more work on the surrounding rock, and it is calculated as a percentage of another explosive. Normally, TNT (Three Nitroglycerin) is chosen as standard explosive with a strength of 100 and the other explosives are assigned a relative strength with respect to TNT. Energy produced by an explosive can be calculated using thermodynamic codes and measured using a variety of techniques. These techniques include Traulz test, Ballistic mortar test, seismic strength test, crater charge test, cylinder compression test, plate dent method, double pipe test and under-water test. Apart from that some empirical models are also developed to determine the strength of explosives. For example Equation 5 could be applied to determine the relative weight strength of any explosive compared with a standard explosive.

$$\text{Relative Weight Strength} = \left(\frac{\rho_e \times V_e^2}{\rho_s \times V_s^2} \right)^{1/3} \tag{5}$$

Where, ρ_e is the density and V_e is the velocity of detonation of the explosive in question. In this equation ρ_s and V_s refer to the standard explosive.

Brisance, on the other hand, is the fragmentation effect of an explosive on the rocks. It is the ability of an explosive to break the rock and it depends on charge density, detonation velocity, gas yield and explosion heat (Olofsson, 1988; Meyer *et al.*, 2002). It means that, as the rate of releasing the energy increases, then an amount of gas and heat released in a small step of time causes more force to break the rock. As an explosive is compacted in a hole its density and the resulting detonation velocity increases. This will increase the detonation pressure. Thus, density has a leading effect on strength and brisance.

2.6 Sensitivity

Sensitivity depends upon the type of external factor that affects the explosive which determines application and handling safety of explosives (Jimeno *et al.*, 1995; Meyer *et al.*, 2002). It is a measure of the ease with which an explosive can be initiated by electricity, heat, friction impact, or shock (Gokhale, 2011; Fox, 2012). The trend in commercial explosives is towards lower sensitivity to initiation without detracting from detonation efficiency. In this case material can be classified as primer-sensitive and detonator-sensitive explosives. Primer-sensitive explosives have relatively low sensitivity to shock, friction and impact, resulting in excellent safety and handling characteristics. The reliable detonation of primer-sensitive explosives requires initiation by a primer that is in good contact with the charge. Ammonium nitrate is the major ingredient of most primer sensitive explosives. Detonator-sensitive explosives can be initiated by a single #8 strength detonator or by a strand of 10 g/m detonating cord. In small diameter blast-holes, sensitive explosives such as cap sensitive emulsions or water gels are recommended. The smaller the blastholes, the more sensitive the explosives need to be.

2.7 Critical diameter

The critical diameter of an explosive is the diameter below which a stable detonation does not occur (Olofsson, 1988; Bhandari, 1997; Fox, 2012; Mahadevan, 2013). The lowest diameter at which detonation velocity is maximal and the ideal detonation is achieved is termed as 'ideal diameter'. For diameters below the critical diameter, the detonation wave does not propagate or it travels with a low velocity (Figure 2). Some explosives like ANFO are more sensitive to diameter changes. Generally speaking, this diameter is about 10–35 millimeters (Gokhale, 2011). Critical diameter of explosives are different and it depends on particle size, density, confinement and the reactivity of it components. To ensure reliable initiation under normal conditions of use, explosive suppliers recommend a minimum diameter for each of their products. For better blasting results under most conditions, the recommended minimum diameter is larger than the critical diameter.

As the charge diameter increases the velocity of detonation increases too. When the charge diameter reaches 76–152 millimeters (equals to 3–6 inches) velocity of detonation becomes constant (Figure 3).

In practice, for blastholes with a diameter of less than 50 mm though costly but it is better to use slurries or cartridge type explosives. For bench blasting with blastholes with a diameter of larger than 50 mm ANFO is adequate.

2.8 Detonation transmission

It is a sympathetic behavior of explosives and it is also known as 'flash-over'. It means that when a cartridge of explosive detonates, it impulse another adjacent cartridge or another charged blasthole to explode (Olofsson, 1988; Jimeno et al., 1995; Meyer et al., 2002; Nyberg et al., 2003; Mahadevan, 2013). It is normally tested by detonating a charge and repeatedly increasing the distance between two charges and checking

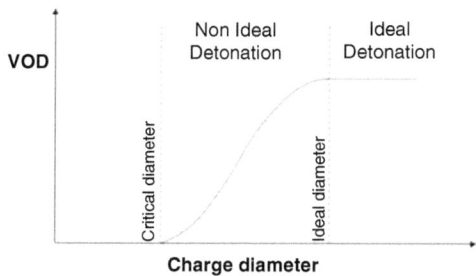

Figure 2 The VOD curve with respect to charge diameter.

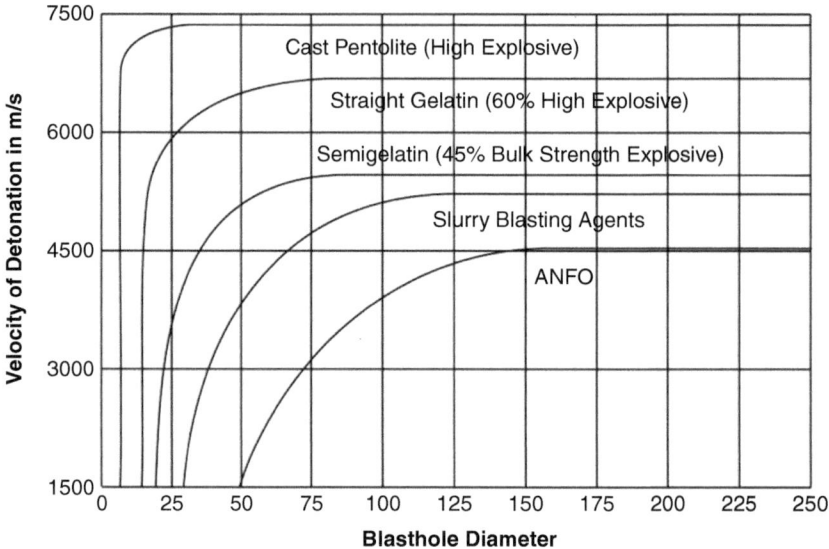

Figure 3   The VOD curve of some explosives with respect to charge diameter (Jimeno *et al.*, 1995; Gokhale, 2011).

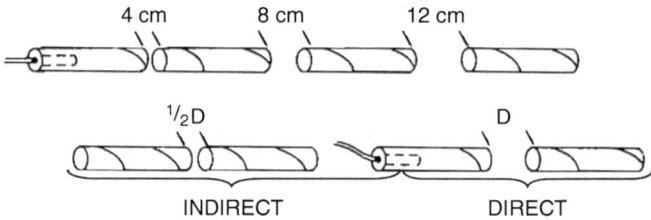

Figure 4   Test of sympathetic behavior of explosives (Jimeno *et al.*, 1995).

whether the other charge is being detonated or not. It is also measured directly or indirectly (Figure 4). It should be noted that only half of the energy is transmitted in the later case. This phenomenon guarantees that all the explosives are completely detonated within a blasting pattern. For the case of industrial explosives the maximum distance at which the detonation is transmitted is between 2 and 8 times their diameter. Aging, charge diameter and the type of material between two consecutive charges affect detonation transmission.

2.9 Initiating systems

Initiating systems are used to safely initiate charges of explosives at predetermined times by carrying a firing signal from one place to another, using chemical or electrical energy (Fox, 2012). Modern initiating explosives incorporate various explosive and inert components, which are partly or wholly consumed in the blast. Small quantities of signal tubing or wire often remain in the muck-pile. Non-electric initiating explosives use pyrotechnic compositions or explosives to store and transmit energy by controlled shock waves, detonation or burning. Electric initiating systems require an exploder to generate an electrical charge, which is transmitted along wires. Blast timing is usually controlled by pyrotechnic (burning) delay elements located inside detonators. Non-electric initiating systems based on a signal tube are currently the most widely used for blasting in surface metal mines. Most mines now use nonelectric detonators inside blastholes, with remote initiation of blasts using a non-electric firing system. Electronic blasting systems are becoming more common, and differ from electric and non-electric delay systems in that the delay time is controlled by a programmable integrated circuit, resulting in very precise timing. The accuracy and programmability of electronic detonators allow for blast timing to be tailored to the geometry, geology and unique requirements of any blasting operation to more effectively use explosives energy.

Also, the type of initiating system lowers or improves VOD. If the initiation is not sufficiently energetic then the detonation starts at a low speed and it decreases the detonation velocity and the strength of explosives.

2.10 Desensitization

Most explosives become less sensitive at higher densities and this phenomenon is known as desensitization. In other words, when density exceeds beyond a certain level the sensitivity of explosive diminishes (Jimeno et al., 1995; Bhandari, 1997; Fox, 2012). This phenomenon is common in those explosives that lack TNT in their composition (such as water gels and ANFO-type explosives). Figure 5 shows the behavior ANFO's detonation velocity with respect to density changes. Beyond a density of 1.1 gr/cm^3 – this is the death density of ANFO – detonation velocity decreases drastically.

Desensitization can occur at deep holes due to the hydrostatic pressure. It is also possible to occur for explosives through the dynamic pressure caused by nearby earlier firing charges. In the latter case, desensitization will be caused by weak initial detonator, placing of a charge in a blasthole with a large diameter or by the pressure of some adjacent charge. In all the cases the shock or gas energy exerted initially will deform and compress the remaining charges. This deformation will increase their density and desensitize them.

2.11 Stability

Explosives should be chemically stable and not decompose under normal conditions for a period of time (Jimeno et al., 1995; Meyer et al., 2002). There is a method to check the stability of explosives called Abel test. In this test a sample of explosive is heated at a specific temperature. The time that takes the explosive to decompose shows the stability level of that explosive. For example, it takes 20 minutes for nitroglycerine to decompose at 80 °c. The stored explosives are tested regularly at certain intervals to check for their condition. This property indicates the maximum storage time of any explosive. The stability of explosives depends on their chemical composition, storing temperature, presence of water and exposure to sun light.

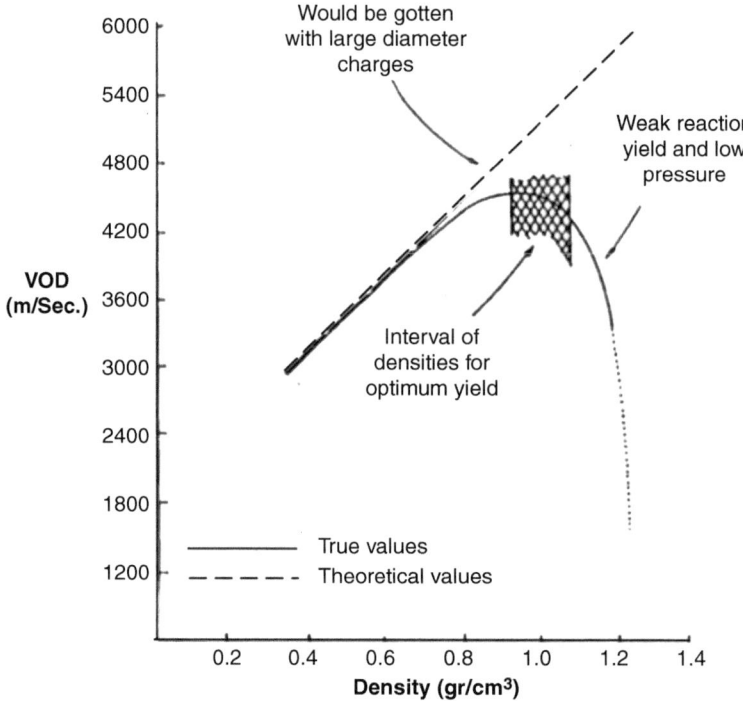

Figure 5 Detonation velocity of ANFO as a function of density (Jimeno *et al.*, 1995).

2.12 Toxicity of fumes

This parameter is very important in underground blasting activities. After blasting, some toxic gases (known as fumes) such as CO, CO_2, SH_2 and SO_2 fill the blasting area that can seriously affect worker's health. Incomplete detonation increases the volume of fumes. Presence of NO2 gives a red or orange colored cloud of fumes. Explosives should be selected with regard to their place of application and regulation (Zimmermann, 2003; Gokhale, 2011). According to the Institute of Makers of Explosives in USA a good detonation is one that produces 4.5 liters of toxic gas per 200 grams of explosives, maximum. In poor blasting operations it reaches to 19 liters per 200 grams of explosives.

2.13 Water resistance

This characteristic is important because there is always a degree of humidity in the air that may affect the stability of explosives (Bhandari, 1997; Fox, 2012). Also, in some circumstances there is a need to conduct a blasting operation in presence of water. In these situations a part of energy is used to evaporate the water which reduces the blasting heat and energy. Therefore, explosives should maintain their characteristics in these situations. Water causes the explosives to decompose and to produce nitric acid. Water reduces the stability, sensitivity, energy and strength of explosives and decreases

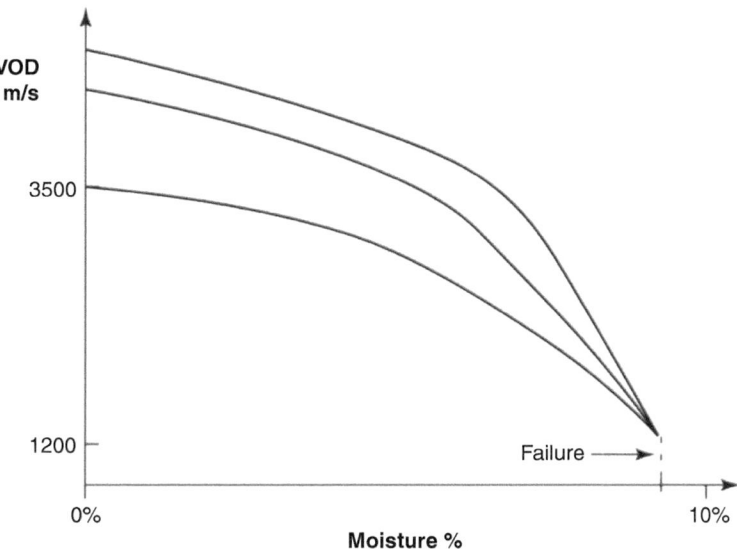

Figure 6 Effect of water on VOD (Mahadevan, 2013).

the detonation velocity. As shown in Figure 6 water could decrease the VOD of ANFO significantly.

The water resistance is the capacity of an explosive to withstand exposure to water without losing its characteristics and varies considerably. There are some methods to determine water resistance of explosives (Meyer *et al.*, 2002). It is related to the chemical composition of explosives and the amount of nitroglycerine or any other additives. Emulsions have excellent water resistance; heavy ANFOs is somehow resistant while ANFO has negligible water resistance (Fox, 2012; Pradhan & Jade, 2013). Explosives with excellent water resistance could withstand water for more than 12 hours.

2.14 Other parameters

There are other parameters such as aging, storage life, flammability and volatility (Jimeno *et al.*, 1995; Bhandari, 1997; Gokhale, 2011; Fox, 2012). Aging diminishes the detonation velocity, because the number and the volume of the air bubbles decreases as the explosive ages. Air bubbles particularly in gelatin type explosives are the generators of hot spots. Flammability is a characteristic of explosives that should be considered during handling and storage of explosives. In order to reduce the flammability of explosives, the amount of liquid oxygen in their composition has been reduced. Volatile explosives are hard to handle and store therefore, explosives should not be volatile.

Resistance to high and low temperature is another characteristic of explosives. Those explosives with nitroglycerine in their composition tend to freeze in temperatures below 8°C. Adding a certain amount of nitroglycol improves this characteristic of explosives to a degree of $-20°C$. In high temperatures the liquid fluid within ANFO tends to evaporate which should be controlled.

3 PARAMETERS RELATED TO GEOMECHANICAL PROPERTIES OF ROCK MASS

When selecting explosives, the first considerations should be the site geology and the end objectives of blasting. The geological and geomechanical characteristics of rock mass have a great influence on blasting results. These parameters affect rock fragmentation, land and air vibration, fly rock and other design variables. In this section the main aspects of rock mass properties are discussed.

3.1 Rock strength

Rock strength determines whether cracks will occur due to blast vibration (Bhandari, 1997; Roy, 2005; Gokhale, 2011). As a rule of thumb, a high-velocity explosive is preferred for hard rock blasting. A low-velocity explosive with a heaving energy may give satisfactory results in soft and highly jointed rock. It is noted that an explosive which has an excellent effect in hard rock grounds may not be effective in soft rocks. When a hole is blasted initially the compressive strain waves strike the free face, and a shear and a tensile reflected wave is generated (Figure 7). As the tensile strength of rock is far less than the compressive strength, the reflected tensile wave may have considerable breaking effect on rock. This is true in theory but high speed cameras indicate that in practice it does not occur (RioTinto, 2007).

When a compression wave reaches a free face it reflects back as a tension wave. These waves generate some radial cracks near the bench face. Since there is no confinement in the free face, they separate from the face and fall down. This phenomenon is called spalling (Gokhale, 2011). The reason behind the absence of spalling may be caused by the existence of internal discontinuities. These internal discontinuities reflect the compressive

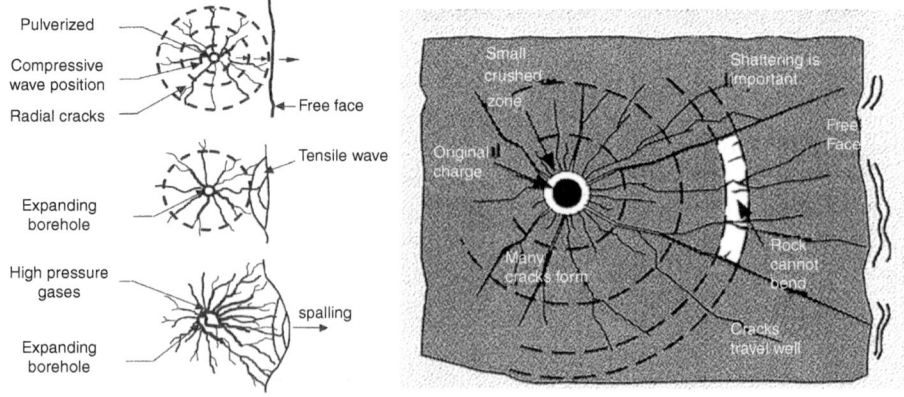

Figure 7 Crack formation in hard rock (RioTinto, 2007).

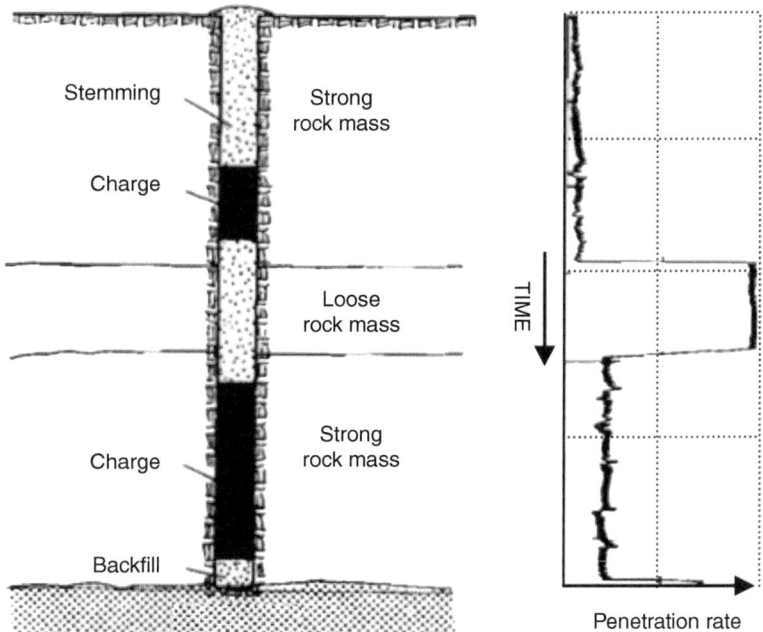

Figure 8 Different rock beds and the resulting penetration rate in a bench (Jimeno *et al.*, 1995).

waves before they reach the free face. The reflected tensile waves are too weak to initiate new fractures but they interact with radial cracks and cause them to extend.

In blasting, dynamic strength of rock is of most interest. Dynamic strength is 5–13 times more than the static strength (Rinehart, 1960). When an explosive produces a shock energy that is greater than dynamic strength, this excessive energy will be used to crush the rocks surrounding the blasthole. This has little effect on fragmentation of rock and it will waste the shock energy. Therefore, it is recommended that an explosive should be chosen such that it produces a shock energy equal or less than the dynamic strength of rock. This point should be considered in smooth and contour blasting operations. Sometimes rock strength differs through the blasthole. In such cases deck charging is used to increase the yield of blasting. There are some good criteria to infer the material strength through the blasthole during the drilling stage. Penetration rate and rotary torque are such criteria that are easy to measure in drilling stage. In those rocks with high compressive strength penetration rate is low and rotary torque are comparatively high. When drilling of a blasthole is finished, the recorded penetration rates will reveal the areas with similar strength. This is a guide for blasthole loading (Figure 8). Correct distribution of explosives through the blastholes will improve the blasting results.

3.2 Density of rock mass

Density and strength of rock are highly correlated. Generally, low density rocks are easy to fragment and displace (Jimeno *et al.*, 1995; Bhandari, 1997; Gokhale,

2011). In case of high density rocks some precautions must be taken. In such a case, explosives with high bubble energy should be used and stemming should be effectively conducted. Bubble energy defines the time interval between the release of shock and gas energy. Also, in order to elevate gas pressure, blasthole diameter should be increased. Fortunately, the density of different rocks varies within a small range (Gokhale, 2011).

3.3 Rock porosity and voids

Porosity affects the shock wave attenuation and reduces the dynamic strength of rock. In porous rocks application of explosives with high bubble energy such as ANFO is recommended (Jimeno *et al.*, 1995; Bhandari, 1997). The voids, cavities and empty spaces are much larger and they differ from porosity. These empty spaces affect drilling and blasting efficiency. Excessive loading of these voids may cause fly rock. These spaces can interrupt wave propagation and gas energy that could descend the yield of blasting. In these situations deck charging and intermediate stemming should be conducted (Figure 9).

3.4 Discontinuities in rock mass

The effect of joints and discontinuities in rock mass is usually more significant than the geomechanical properties of rock mass. Researches show that this issue is more important than explosive properties and blast geometry (Bhandari, 1997). This parameter is normally described in the RMR system as rock quality designation (RQD). Sometimes, presence of joints and discontinuities help the rock breaking operation by decreasing the powder factor (Gokhale, 2011). But, these discontinuities may act as a shortcut for blasting gases to escape. This will waste the gas pressure and decline the blasting efficiency which is not an interested situation. Blasting in jointed rock may be accompanied by toe formation, poor fragmentation, fly rock and vibration (Jimeno *et al.*, 1995; Suthar, 2013). In highly jointed rocks those explosives with a high gas pressure is recommended (*e.g.* ANFO). In the rock mass without any joint and weak plans it is better to use high density explosives (*e.g.* slurries and emulsions). These explosives have a high detonation velocity and shock energy. Thus, they are able to create radial crack easily.

Sometimes the distance between discontinuities is large and they form a rock mass with blocky structure. In these situations, rock fragmentation and excessive back-break

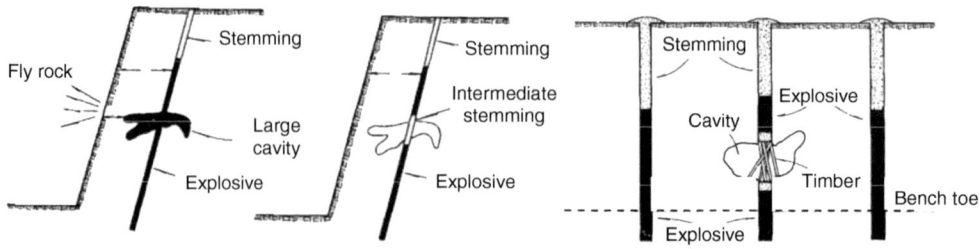

Figure 9   Deck charging to avoid large cavities and fly rock (Jimeno *et al.*, 1995).

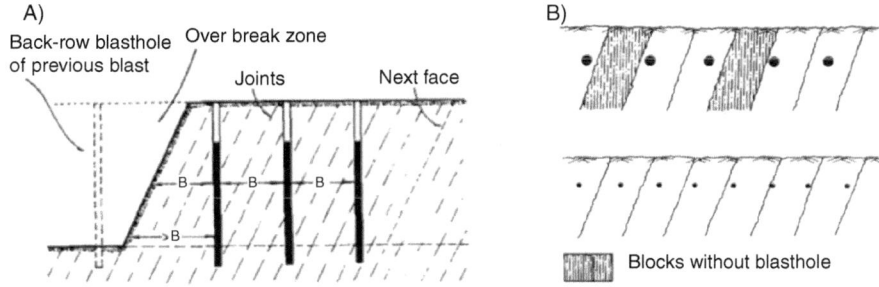

Figure 10 Structural form of discontinuities affect the burden and blasting result (Jimeno *et al.*, 1995).

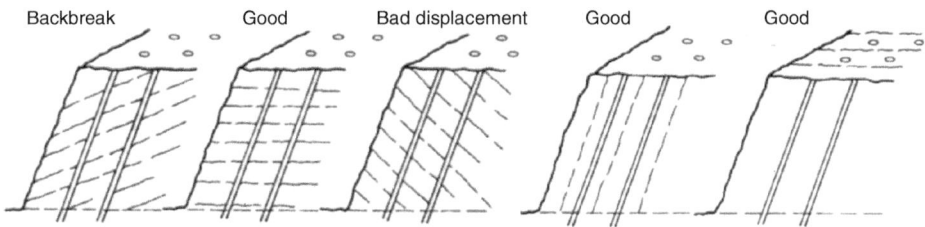

Figure 11 Orientation of discontinuities and preferred blasting patterns.

are controlled by the blasthole patterns. The relative orientation of joint sets and the free face determines the resistance of the rock mass to blast movements and damage (Roy, 2005). The relative orientation of blasting face and blastholes should be placed such that the existing discontinuities help rock breakage (Figure 10 and Figure 11). According to Figure 10-a and Figure 11, when joints and blast orientations are somehow parallel to each other, then the possibility of back-break is high. In order to control this type of back-breaks it is recommended to drill the blastholes parallel to the face. Then it is possible to maintain a constant burden for each blasthole. Air deck charging is recommended in jointed ground formations. This method provides a better utilization of explosive energy and reduces back break, ground vibration and throw (Suthar, 2013; Jhanwar, 2013). Also, the blasting pattern should be selected such that, there exist a blasthole in each block generated by joints (Figure 10-b).

In blocky grounds the blasting pattern should be changed in order to drill a blasthole in each block (Figure 10-b). In order to do that, the spacing should be reduced. This needs blastholes' diameters to decrease. This operation increases the number of holes and the drilling and blasting costs but it increases the blasting efficiency.

The other issue that affects the joint behavior is aperture (Figure 12). As aperture increases, joints absorb the shock energy of detonation. This could reach to 80% in joints with an aperture of about 1 mm. This can occur in joints where the aperture is filled with some loss material, air or water. The point to remember is that discontinuities will act as a free face (RioTinto, 2007).

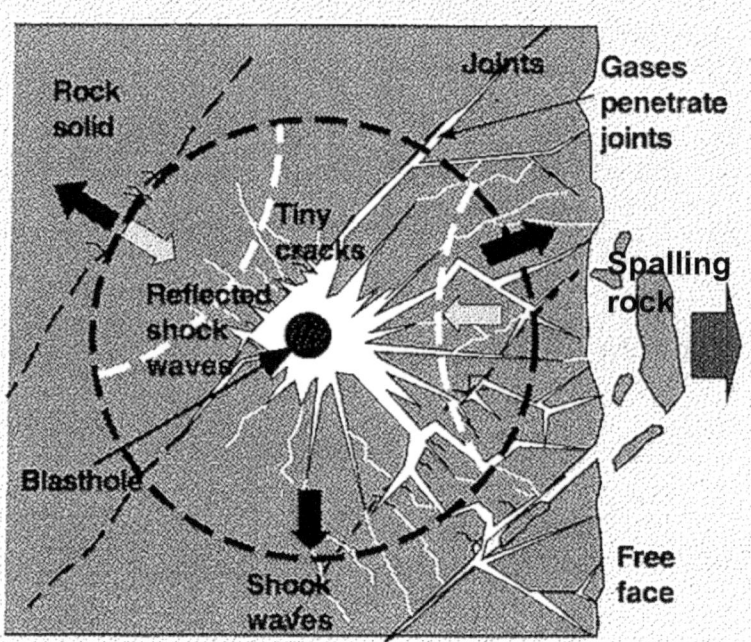

Figure 12   Rock breakage process (RioTinto, 2007).

3.5 Presence of water

Water causes the explosives to decompose and reduces their sensitivity and energy. For example, when humidity is more than 10% in the atmosphere, ANFO is altered and stops detonation. Therefore, before conducting a blasting operation in wet grounds ANFO should be placed inside plastic containers and then loaded to the holes. In cases where water level is high inside the blasthole, it should be dewatered. If dewatering is impossible then water gels and emulsion can be used in blastholes. Contrariwise, water reduces the tensile and compressive strength of rock. It also increases the propagation velocity of shock waves in porous area.

3.6 Wave velocity

There are two types of seismic waves that are produced due to detonation. They are classified as body and surface wave. Body waves travel within a medium but surface waves are propagated on the free face. Body waves include P-wave[1] and S-wave[2]; and two types of surface waves are common in blasting operation including Rayleigh and Love waves (Davis, 1997; Roy, 2005). The longitudinal wave velocity is the velocity at which a rock transmits the shock waves. The higher the longitudinal wave velocity, higher the required detonation velocity (Hemphill, 1981; Bhandari, 1997). P waves

1 Primary or longitudinal waves or simply P-wave
2 Secondary or shear waves or transversal waves or simply S-wave

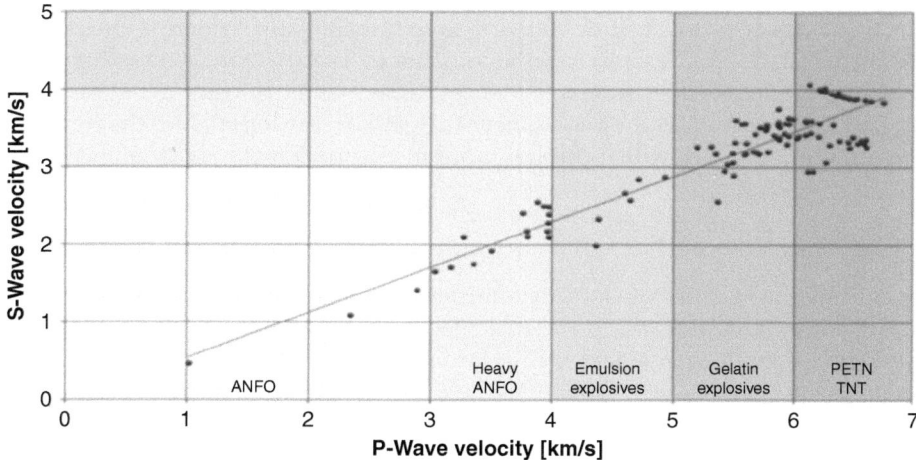

Figure 13   Relation between P and S waves in different rocks for some of the most important explosives (Mueller & Rossmanith, 2013).

travel twice faster than S waves (Figure 13). In homogeneous and isotropic rocks the stress wave propagates in all direction with the same speed. In homogeneous rocks as the density of rock increases the waves are transmitted efficiently. In inhomogeneous layers of rock with different mineral composition the stress wave change, accelerate or retard and hence, the wave will be distorted (Mueller & Rossmanith, 2013). Explosive type and detonation velocity is chosen with respect to the longitudinal velocity of rocks. In this context modulus of elasticity is of importance. It reflects the stiffness of rock and shows the ability of a rock sample to resist deformation. Rocks with high modulus of elasticity are hard to break.

When an explosive detonates it produces a strong initial shock wave which then decays into stress waves in the surrounding rock. The pressure caused by the shock wave exceeds rock strength which leads to crushing of rock around the blasthole. As the waves move away from the blasthole, its pressure decreases and the wave becomes an elastic compressive wave. This wave is called P-wave. Wave strength is approximated using peak particle velocity (ppv). As the distance from blasthole increases, the wave strength and the resulting ppv decrease. When the P-wave reaches the free face it forms the S-wave and Rayleigh wave. Rayleigh wave is a combination of P-wave and S-wave. Its amplitude decays exponentially with distance from the free face. Its velocity is about 90% of the S-wave and it depends on the Poisson's ratio of the rock (Mueller & Rossmanith, 2013). Hamdi *et al.* (2003) studied the effects of rock properties on wave velocity.

3.7 Rock impedance

It is the product of sonic wave velocity and the density of the material. It characterizes material with respect to its transfer properties. It is normally defined in terms of Pa/m, N/m^3 or kg/m^2s. The impedances play an important role in the reflection and transmission of waves across interfaces of rock formations, joints and faults. This characteristic is the product of density and longitudinal wave velocity (Jimeno *et al.*, 1995; Bhandari, 1997). It indicates the amount of energy that is transferred from explosive to the rock mass. If the impedances of rock and explosive are close to each other, then main portion of the explosion energy will be transferred into the rock mass. Therefore, the explosives with a similar impedance to rock mass are recommended.

3.8 Rock hardness

From an engineering point of view rock-hardness indicates the amount of stress that is necessary to cause failure within the rock. As the hardness of rock increases then its compressive strength increase as well. Hardness of a rock depends on the type of minerals inside the rock structure. Silica with a hardness of 7 in Mohr scale is the most common mineral in the earth crust. Rock hardness and its abrasiveness guide the selection of drilling equipments (*i.e.* rotary or percussive) and blasthole diameter. For blastholes with a small diameter, burden and spacing are small and the drilling pattern is dens.

3.9 Engineering rock quality

Rock quality affects the blasting results and the amount of damage in the adjacent blast area. The most common methods available for rock classification are Q and RMR. These two quality criteria are related as given in Equation 6.

$$Q = (RQD/J_n) \times \left({J_r}/{J_a} \right) \times \left({J_w}/{SRF} \right)$$
$$RMR = 9 \operatorname{Ln} Q + 44$$

(6)

Where, RQD is the rock quality designation ($RQD = 115 - 3.3J_v$), RMR is rock mass rating, J_n is the number of joints, J_r is their roughness, J_a is the alteration level, J_w is water reduction, SRF is the stress reduction father, and finally J_v is number of joints in a cubic meter of rock. These classification systems contain the parameters that reflect the resistance of rock mass to seismic loading of blasting (Roy, 2005). Attempts have been made to study the relation between rock mass rating and peak particle velocity and ground vibration (Sauvage & Blanchier, 2003; Sinha, 2013; Sirveiya & Thote, 2013; Sauvage, 2013).

3.10 Blastability of rock

Blastability is defined as a rock mass property to a specified blast design, site requirements, safety regulations and explosive characteristics. It defines how easy it is to blast a rock mass under a specified condition. In that regard the most important rock mass properties are rock impedance, rock structure and properties of discontinuities. There are different empirical models that are developed to determine the blastability of a rock mass (see Jimeno *et al.*, 1995; Roy, 2005; Gokhale, 2011). A simple blastability index is the relationship between compressive (C) and tensile (T) strength of rock mass. As the relation $\left({C}/{T} \right)$ increases it eases the fragmentation of rock.

3.11 Other parameters

There are other parameters such as internal friction and conductivity in rock. Internal friction also known as 'specific dumping capacity' is a measure to assess the ability of rock to attenuate waves. This capacity converts the shock energy into heat. Other parameters such as porosity, joints, permeability and water content increase the internal friction considerably. ANFO is the recommended explosives for these rocks.

Conductivity is the ability of the ground to transmit electrical current. This is more common in salt, magnetite and sulfides. In these grounds the electricity cables and detonators should be used carefully and if possible other blasting system such as Nonel must be installed.

4 PARAMETERS RELATED TO BLAST GEOMETRY

Many effective parameters on blasting are related to the geometry of drilling and blasting pattern such as blasthole diameter, burden, spacing, drilling pattern, and etc. (Figure 14). This category of parameters is controllable and has a great role on fragmentation, displacement, explosive consumption and detrimental effects of blasting. In this section these parameters are introduced and their effects on blasting results are explained.

4.1 Blasthole diameter

The determination of optimum blasthole diameter (D) is governed by several factors, such as rock mass properties, type and size of loading equipment, explosive type and critical diameter, bench height, drilling cost, scale of operation, the required degree of fragmentation, unit cost of production and environmental constraints (Jimeno *et al.*,

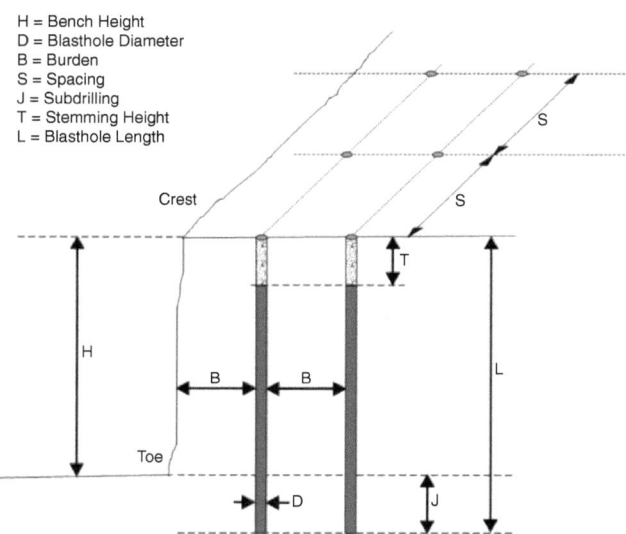

H = Bench Height
D = Blasthole Diameter
B = Burden
S = Spacing
J = Subdrilling
T = Stemming Height
L = Blasthole Length

Figure 14 Parameters related to blast geometry in bench blasting (Gokhale, 2011).

Table 1 Classification of blastholes according to the diameter (Rustan *et al.*, 2010).

Blasthole size	Underground mines (mm)	Surface mines (mm)
Small blastholes	<50	<50
Intermediate blastholes	50–100	50–200
Large blastholes	>100	>200

1995). When the diameter of the blastholes is small, the costs of drilling, priming and initiation are high and charging, stemming and connection take a lot of time and labor. Small diameter blastholes usually need lower powder factor than large diameter blastholes. Thus, they give better distribution of energy in the rock mass and hence will break the rock satisfactorily. It should be noted that the minimum blasthole diameter must be greater than the critical diameter of the explosive especially in open pit mines. Every explosive has a critical diameter which is the minimum diameter at which the detonation process, once initiated, becomes self-supporting in the column of the charge. When the diameter is smaller than the critical diameter, the detonation is not supported and is extinguished.

When the blasthole diameter is large, the following advantages are obtained: lower overall costs of drilling and blasting, easier mechanization of the explosive charge, higher detonation velocity, higher drilling productivity, and increase in loading equipment performance. Also there are some disadvantages with large blasthole diameters such as coarse fragmentation due to increased burden and spacing, higher costs of loading and crushing and environmental problems such as ground vibration, flyrock and air pressure. Furthermore, in this case larger loading and hauling equipment should be applied. A wide range of blasthole diameter from 25 to 508 mm is used in blasting operations. A classification of blasthole diameters is given in Table 1 (Rustan *et al.*, 2010).

For optimum fragmentation in surface mines, the blasthole diameters equal to the bench height divided by 120 and the maximum blasthole diameter should be equal to the bench height divided by 60 (Bhandari, 1997). Most tunneling operations employ the blasthole diameter between 38–51 mm. Only cut holes are larger than 51 mm (Heinio, 1999). In underground mines, the blasthole diameter depends on the size of blasting operation. For example, in sublevel stoping method the blasthole diameter ranges from 76 to 165 mm (small diameter used in ring drilling and large diameter used in parallel drilling) and in vertical crater retreat (VCR) method the diameter of 165 mm is common (Hustrulid & Bullock, 2001; Darling, 2011).

4.2 Bench height

One of the primary factors that control the design of a blast is bench height (H). The effective factors on bench height include statutory regulations (excessively high benches are unsafe and, therefore, not permitted), rock mass properties, the type and size of digging and loading equipment, grade control requirements, and the need to maximize the overall cost efficiency of drilling and blasting. Bench heights vary within wide limits. In surface mines, bench heights of 15–20 m are common, although benches with heights up to 30 m are occasionally encountered. In general, benches with heights of about 10–18 m have been considered the most economic and least hazardous to work (Bhandari, 1997).

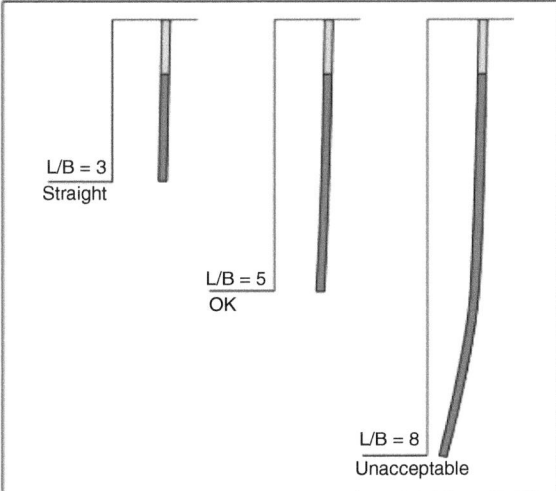

Figure 15 Effect of drill deviation with increasing bench height (Darling, 2011).

When the bench height is large, drill deviation occurs (Figure 15), which means a variation in the distribution of explosive energy at the bottom of the blastholes (Darling, 2011). In areas where blastholes are far from each other, inadequate breakage will occur, which often leads to pockets of high (or hard) toe. When blastholes come too close to the face, excessive flyrock and air blast can occur. Furthermore, when the bench height increases, the drilling speed decreases dramatically especially in high depth.

The stiffness of the parallelepiped rock located in front of the blastholes exerts great influence on the results of blasting. When the bench height to burden ratio is large, it is easy to displace and deform rock, especially at the bench center. The optimum ratio is greater than 3. If ratio equals to 1, the fragments will be large, with ground vibration, flyrock, overbreak and toe problems. With ratio equal to 2, these problems are attenuated and are completely eliminated when ratio is greater than 3 (Darling, 2011).

In surface blasting design, if the bench height is not predetermined then it should be greater than or equal to 3 times the burden. In drilling bench heights greater than 4 times the burden, hole deviation is an issue. Equation 7 can be used for initial estimation of bench height:

$$H \geq {}^{D}\!/_{15} \tag{7}$$

where H is bench height (m) and D is blasthole diameter (mm).

4.3 Burden

Burden (B) is one of the most critical parameters in blasts design. Burden is the minimum distance from the axis of a blasthole to the free face (Figure 14). This

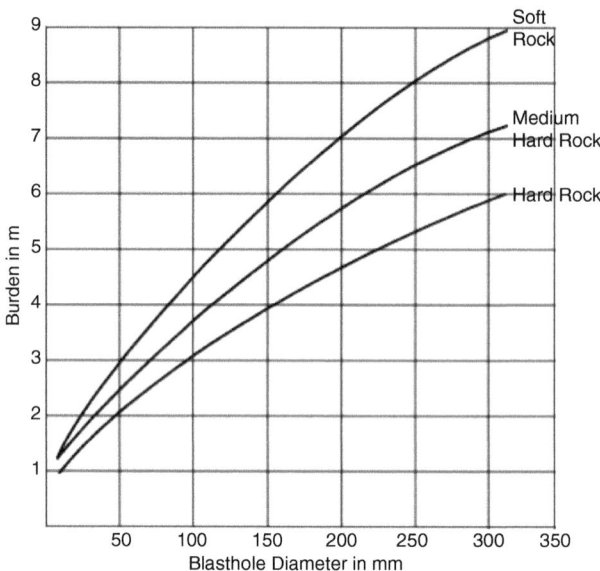

Figure 16 Burden as a function of blasthole diameter (Gokhale, 2011).

parameter depends basically upon drilling diameter, properties of the rocks and of the explosives, bench height and the desired degree of fragmentation and displacement (Jimeno *et al.*, 1995).

Correct selection of burden has a great influence on blasting results. If the burden is too small, detonation gases escape into the atmosphere in the form of noise and air blast therefore less energy is available for the fragmentation. Where the burden value is too large, gases are confined for a time interval longer than desired; which can result in higher ground vibrations, excessive backbreak, toe and uneven floor.

In surface mines, burden is about 3–10 m based on rock and explosive characteristics and operation size. Numerous formulas have been suggested to calculate the burden, most of which utilize either charge volume, charge weight or blasthole diameter as the basic parameter; however, their values fall in the range of 25 to 40 D (Darling, 2011). The appropriate value of burden depends on the hardness of rock mass. Figure 16 shows the variation of burden for different diameters of blastholes and different rocks (Gokhale, 2011).

For underground mines, burdens are usually lower than those for surface mines because of greater confinement and demand for fine fragmented rock.

4.4 Spacing

The distance between adjacent blast holes, measured perpendicular to the burden is defined as spacing (S) which controls the mutual effect between blastholes (see Figure 14). Spacing is calculated as a function of burden, blasthole depth, relative

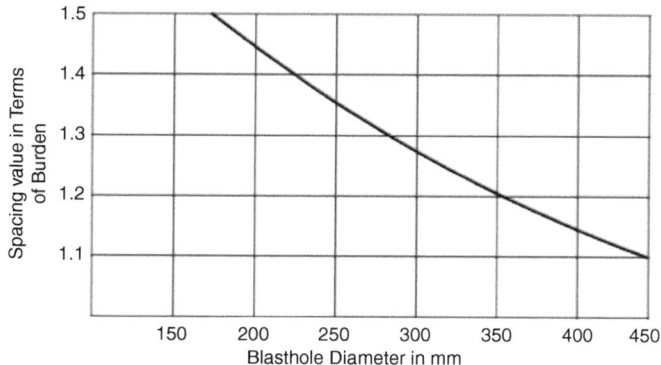

Figure 17 Spacing to burden ratio as a function of blasthole diameter (Gokhale, 2011).

primer location between adjacent charges and initiation time interval (Jimeno *et al.*, 1995; Gokhale, 2011). Very small spacing causes excessive crushing between charges and superficial crater breakage, large blocks in front of the blastholes and toe problem.

Excessive spacing between blastholes causes inadequate fracturing between charges, along with toe problems and an irregular face with overhang in the new bench. Adequate results can normally be obtained when spacing exceeds the burden. Spacing less than the burden tend to cause premature splitting between blastholes and early loosening of the stemming. Both these effects encourage rapid release of gases to the atmosphere, production of large size fragmentation and considerable backbreak. In bench blasting, it is normal to use spacing to burden (S/B) ratio of 1.1:1 to 1.5:1. The value of spacing equal to 1.1 B has been found to be more appropriate for large diameter blastholes and 1.5 B is more suitable for small diameter blastholes (Figure 17). Furthermore, determining the spacing to burden ratio is greatly affected by rock type. This ratio is lower for soft rocks. Another effective parameter on spacing is joints orientation. When the joints are across the face, a close spacing is needed and when the joints are parallel to free face, larger spacing is possible. Spacing less than 1.0 B should never be used except for controlled blasting techniques. Furthermore, when adjacent charges are initiated separately with a sufficient time delay interval to permit each charge to break separately; there is no interaction between the holes. In such cases spacing should approximate the burden.

In tunnel blasting, drilling blastholes in face can be classified into four categories: cut holes, stoping and helper holes, contour holes (side and back holes), and lifter holes (Figure 18). Spacing to burden ratio of blastholes at different parts of a tunnel face should be different because the confinement gets released gradually as the tunnel is enlarged from the cut holes to the contour holes (Singh & Goel, 2006). Spacing to burden ratio for various types of blastholes in a tunnel is provided in Table 2. In underground mines, the spacing is usually equal to 1.3 B (Hustrulid & Bullock, 2001; Darling, 2011).

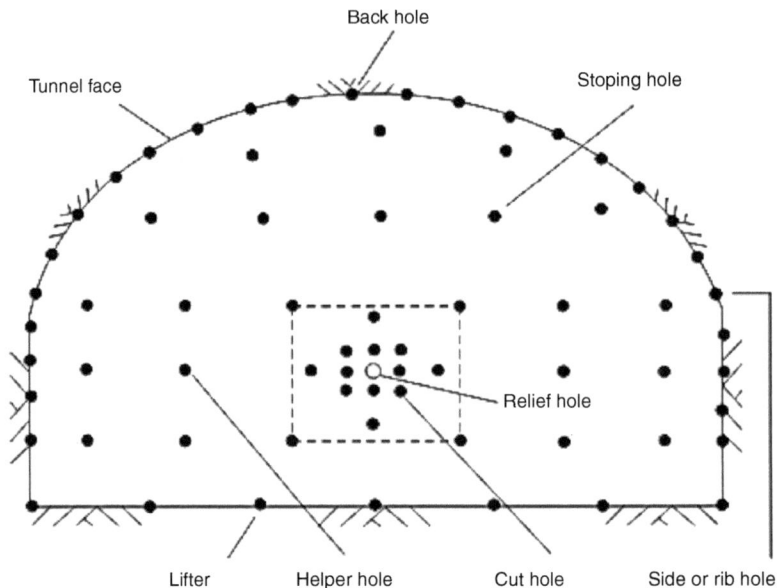

Figure 18 Blastholes in different parts of a tunnel section (Singh & Goel, 2006).

Table 2 Spacing to burden ratio for different types of blastholes in a tunnel (Singh & Goel, 2006).

Type of blasthole	Spacing to burden ratio
Stoping and helper	1.25
Lifter	1
Contour	0.8

4.5 Subdrilling

In bench blasting, poor fragmentation at bench level will form a series of hard toes and an irregular bench floor because of the larger confinement in bench level. To avoid these conditions, it is necessary to drill the blastholes below the floor level. This is termed subdrilling or underdrilling (J). The optimum effective subdrilling depends on the structural formation and rock density, the type of explosive, the blasthole diameter, the blasthole inclination, the effective burden, and the location of initiator in the charge. If subdrilling is small, the rock will not be completely sheared off at floor level, which will result in toe appearance and a considerable increase in loading costs. However, if subdrilling is excessive, the following will occur: an increase in drilling and blasting costs, an increase in vibration level, excessive fragmentation in the top part of the underlying bench, and affecting slop stability (Jimeno *et al.*, 1995; Bhandari, 1997).

The accepted amount of subdrilling is one-third of the burden (Darling, 2011). It ranges from 0.2 to 0.5 times the burden (which is longer for higher benches and harder rocks). Subdrilling can also be calculated as 5 to 8 times the diameter of the hole.

In order to reduce subdrilling, the use of explosives which give a high concentration of energy per unit of length in the bottom part of the charge and the inclined blastholes drilling are recommended. Subdrilling should not be used in horizontally bedded sediments that have well-defined bedding planes such as coal mines because these can be used as shear planes at the bottom of the blast to provide easy relief and a smooth floor.

It should be mentioned that when blastholes are vertical, the length of blasthole is the sum of bench height and subdrilling length (Equation 8) and for inclined blastholes with an angle (β), Equation 9 can be used for calculating the hole length (Jimeno *et al.*, 1995; Gokhale, 2011).

$$L = H + J \tag{8}$$

$$L = \frac{H}{\cos\beta} + \left[\left(1 - \frac{\beta}{100} \right) \times J \right] \tag{9}$$

4.6 Stemming

Stemming (T) is the portion of blasthole which has been packed with inert material above the charge to confine and retain the gases produced by the explosion, thus improving the fragmentation process (Figure 14). Optimum stemming length depends mainly on blasthole diameter, stemming material, and surrounding rock properties (Atlas Copco, 2012). The effects of stemming length on blast results are similar to those of burden. That is, a short stemming length will allow the explosive gases to vent, generating flyrock and airblast problems and reducing the effectiveness of the blast, while a too great stemming length will give poor fragmentation of the rock above the column load (see Figure 19).

Dry granular materials are best for stemming because they have inertial resistance and high frictional resistance to ejection. Materials that behave plastically or that tend to flow are not suitable for stemming, *e.g.* water, mud, and wet clay. Stemming length can be reduced significantly if effective stemming is used, resulting in better explosive distribution and improved overall fragmentation. The optimum size of the stemming material increases with the diameter of the blasthole, and the average size of the stemming particles should be about 0.05 times the diameter of the blasthole (Darling, 2011). The common stemming length is about 0.7 to 1.5 times the burden. When the burden has a high frequency of natural cracks and planes of weakness, relatively long stemming columns can be used. When the rock is hard and massive, the stemming should be short which prevents excessive noise, airblast and backbreak. If unacceptably large blocks are obtained from the top of the bench, even when the minimum stemming column is used, fragmentation can be improved by locating a small pocket charge centrally within the stemming. Furthermore, in underground blasting with long blasthole method, the intermediate stemming between sequenced deck charges must be calculated to avoid simultaneous sympathetic initiation (Jimeno *et al.*, 1995). The

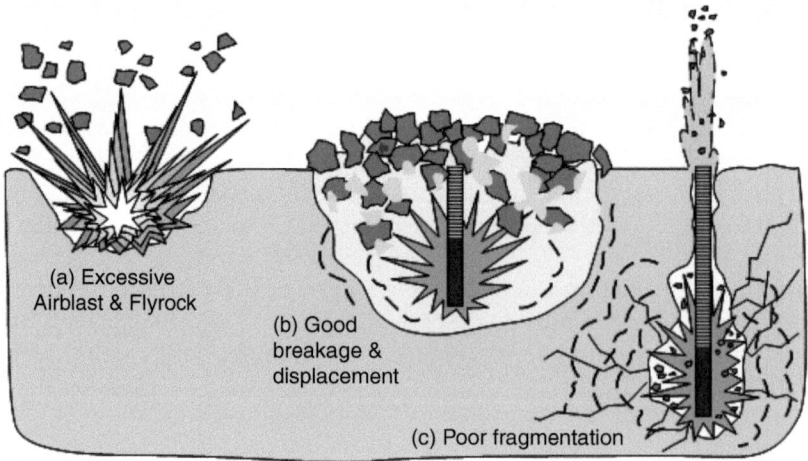

Figure 19 Effect of correct and incorrect stemming (Atlas Copco, 2012).

stemming between decks in dry holes should have a minimum length of 6 times the diameter of the blasthole and in wet holes 12 times the diameter of the blasthole (Darling, 2011).

4.7 Blasthole inclination

In bench blasting, often blastholes are drilled at some angle rather than the usual vertically downward direction. Properties of rock mass, particularly the orientation of joints and other types of discontinuities play an important role in the choice of inclination angle (Gokhale, 2011). The main advantages of inclined blasthole are better fragmentation, displacement and swelling of the muckpile, less probability of misfire, smoother and safer slopes in the newly created benches, less subdrilling, less backbreak, better use of the explosive energy, lower powder factor, reduced over-crushing of the rock mass and higher yield of blast (the volume of blasted rock per meter length of the blasthole) (Jimeno *et al.*, 1995; Gokhale, 2011).

Drilling of blasthole at angels up to 20 degrees from the vertical are recommended. Angles greater than 25 degrees are seldom used because of increased drilling length, difficulty in maintaining blasthole alignments, difficulty in positioning of the drills and in collaring operations, excessive bit wear, higher flyrock hazards, lesser penetration rate, and difficulty in charging blastholes (Jimeno *et al.*, 1995; Gokhale, 2011).

4.8 Drilling (blasthole) pattern

The drilling pattern ensures the distribution of the explosive in the rock and desired blasting result. Blasthole patterns depend on blasthole diameter, rock properties,

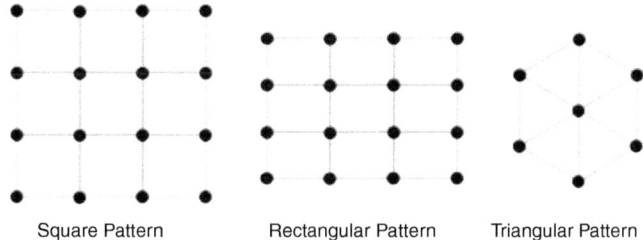

Square Pattern   Rectangular Pattern   Triangular Pattern

Figure 20 Drilling (blasthole) patterns (Gokhale, 2011).

explosive properties, bench height, and the results needed (Gokhale, 2011; Atlas Copco, 2012). In bench blasting, according to the Figure 20, blastholes are drilled in one of the three patterns: square, rectangular, and staggered (triangular). Operating experience and blast modeling results have shown that, in massive rocks, better fragmentation and productivity are obtained with staggered patterns, especially those drilled on an equilateral triangular grid, than with either square or rectangular patterns. Equilateral triangular patterns provide optimum distribution of explosion energy in the rock. This pattern produces the best fragmentation, with a spacing ratio of $S = 1.15B$ for vertical blastholes and $S = 1.15B\cos\beta$, where β is the angle with respect to the vertical in inclined holes.

A square pattern is more appropriate for a confined shot, such as sumps used for dewatering of the next bench level. A square pattern may also be appropriate in sinking cuts when developing a new level. In construction blasting, a square or rectangular pattern helps maintaining straight sidewalls, especially in road cuts (Jimeno *et al.*, 1995). Furthermore, blasting in underground mines such as sublevel stoping with parallel blastholes, upward cut and fill and shrinkage stoping is similar to bench blasting and square or rectangular patterns are used (Hustrulid & Bullock, 2001).

In tunneling, selection of a suitable pattern of blastholes can accomplish the following (Jimeno *et al.*, 1995; Heinio, 1999; Johansen & Mathiesen, 2000; Singh & Goel, 2006): tunnel's desired shape, size, orientation and gradient, reduced over-break/under-break, thereby achieving smooth configuration and compact heaping of the blasted muck after blasting at the working face. Figure 18 shows expression and nomenclature used in tunnel driving. The main aim of cut holes is creation of an opening on tunnel face. This opening acts as an initial free face that reduces considerably the amount of drilling and explosives needed. Thus, creation of a proper cut is a precondition for satisfactory tunnel blast. A number of different types of cuts have been developed over the years concurrent with the development of the drilling equipment. The cuts can basically be divided into two groups:

* Parallel hole cuts, and
* Angled hole cuts.

The first group is most used in operations with mechanized drilling, whereas the second group is generally not used due to the difficulty in drilling and is only applied in small excavation. Parallel cuts are economic when the tunnel area exceeds 25 m² and angled cuts are economic when the tunnel area is less than 25 m². The position of the cut affects the rock projection, fragmentation and also the number of blastholes. Of the three positions, corner, lower center and upper center, the latter is usually chosen as it avoids the free fall of the material, the profile of the broken rock is more extended, less compact and better fragmented.

Stoping and helper blastholes can be geometrically compared to bench blasting although it requires 4 to 10 times higher powder factors. These blastholes are almost drilled parallel to the axis of the tunnel.

Contour (side and back blastholes) and lifter blastholes are those which establish the final shape of the tunnel and are placed with little spacing. If the contour and lifter blastholes are drilled parallel to the tunnel axis, the tunnel face gets smaller and smaller after each round. To ensure that the correct tunnel profile is maintained, these blastholes are drilled at a slight angle into the tunnel wall, which is called the look-out angle.

4.9 Powder factor (specific charge)

The amount of explosive used per unit of volume or weight of rock blasted is called powder factor. The powder factor is obtained by division of the mean charge per blasthole to the blast volume (B×S×H) (Jimeno *et al.*, 1995). As a result, a vast range of powder factors is used. The range of powder factors for different types of mining and construction is given in Figure 21 (Darling, 2011). In general, the average powder factor in underground mines is twice that of surface operations, simply because underground workings are more confined, and as a result, oversize material is quite difficult to deal with and displacement of rock into a loose muckpile may require more energy. Oversize material is usually avoided by using a higher powder factor. When powder factor is high, apart from giving good fragmentation, there are less toe problems and it helps to achieve the optimum cost effectiveness of the operation. It should be noted that excessive powder factor can result in decreasing of operation safety and problems such as flyrock, backbreak and etc.

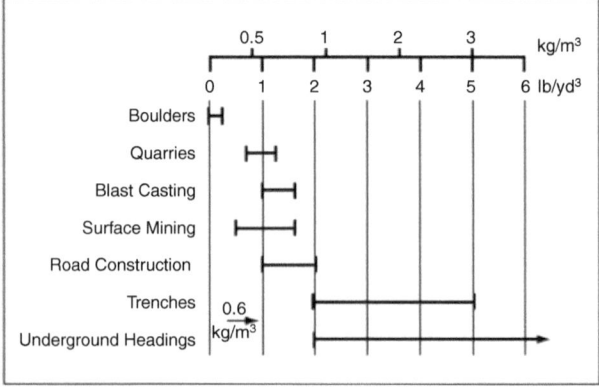

Figure 21 Typical range of powder factor for different types of blasting (Darling, 2011).

5 PARAMETERS RELATED TO INITIATION PATTERN

A typical blast for a mining and construction project may contain as many as 100 blastholes, which in total contain several thousand kilograms of explosive. Simultaneous detonation of this quantity of explosive would not only produce very poorly fragmented rock, but would also damage the rock in the walls of the excavation and create large vibrations in nearby structures. In order to overcome this situation, the blast is broken down into a number of sequential detonations by delays. Thus, firing sequence (delay pattern) and delay time are important for proper fragmentation, proper throw of blasted materials and fewer blasting hazards.

5.1 Firing sequence

The sequence of detonation timing is called firing sequence (delay pattern). There is a large number of factors to be considered when deciding firing sequence: type of desired fragmentation, surface or in-hole delays, firing direction, shape of muckpile and loading equipment, number of delays available, type of trunkline system, and environmental constraints (ground vibration/air blast, etc.). In bench blasting, five commonly adopted firing patterns are used (Gokhale, 2011): "V" (chevron) pattern (Figure 22), echelon delay pattern (Figure 23), flat face pattern (Figure 24), channel pattern (Figure 25), and sinking hole pattern (Figure 26).

In tunnel blasting, the firing sequence should provide progressive enlargement of free face and firing of blastholes towards the maximum free face. Maximum free face should be made before the contour holes are blasted so that the minimum explosive quantity is required to break rock at the contour and the blast-induced damage is minimized (Singh & Goel, 2006).

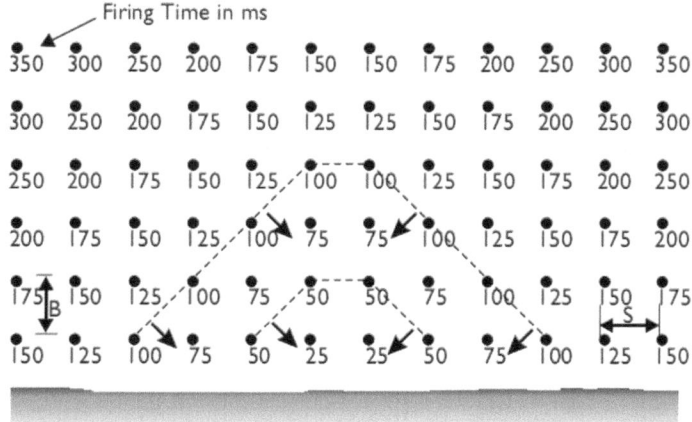

Figure 22 "V" pattern (Gokhale, 2011).

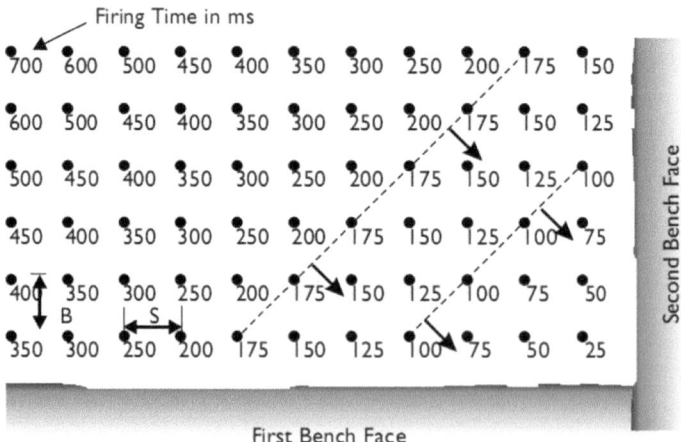

Figure 23 Echelon delay pattern (Gokhale, 2011).

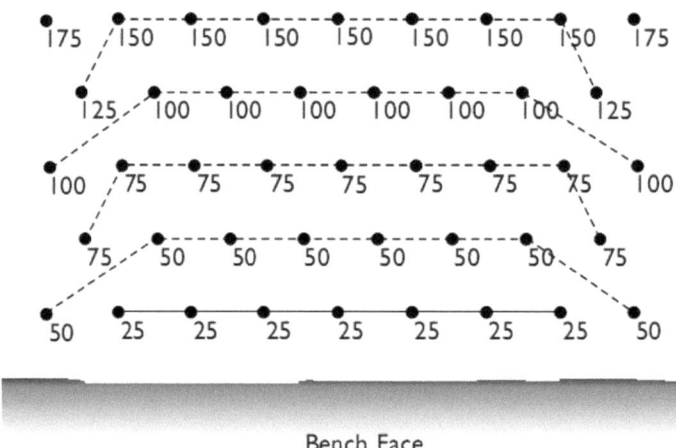

Figure 24 Flat face pattern (Gokhale, 2011).

5.2 Delay time

The purpose of using delay times is to reduce the maximum charge per delay and thereby reduce ground vibrations and to improve the fragmentation. Optimum delay allocation for a blast depends on many factors including (Atlas Copco, 2012):

- Rock mass properties (strength, Young's modulus, density, porosity, structure, etc.);
- Blast geometry (burden, spacing, bench height, free faces, etc.);
- Diameter, inclination and length of blasthole;

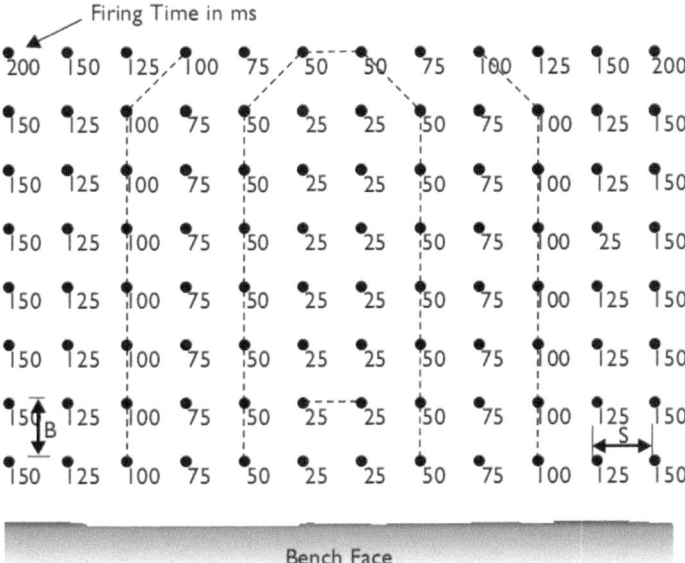

Figure 25 Channel delay pattern (Gokhale, 2011).

- Explosive characteristics, degree of coupling, decking, etc.;
- Initiating system (surface or in-hole delays, type of downline, non-electric or electronic, etc.);
- Type and location of primer;
- Environmental constraints (air and ground vibration levels and frequency); and
- The desired result (fragmentation, muckpile displacement and profile, etc.).

There are two main types of delay in a blast pattern: the hole-to-hole (also known as "inter-hole") delay and the row-to-row (or "inter-row") delay (Wyllie & Mah, 2004; Darling, 2011). The delay time between adjacent blastholes in a row is sometimes called the intra-row delay. The required delay time is related to distance between blastholes by the following two relationships (Wyllie & Mah, 2004).

For row-to-row detonation:

$$t_r = (10 - 13) \times B \tag{10}$$

where t_r is delay time between rows (ms) and B is burden (m).

For hole-to-hole detonation:

$$t_h = k \times S \tag{11}$$

where t_h is delay time between holes (ms), k is delay constant, and S is spacing (m). Delay constant depend on the rock type as shown in Table 3.

The delays in blasting operations are created using delay detonators (non-electric or electronic detonators) which are available with millisecond or long delays, with approximately ½-second timing intervals. Generally, for producing the best blasting results, long-period (½-second) delays are used in underground blasting applications,

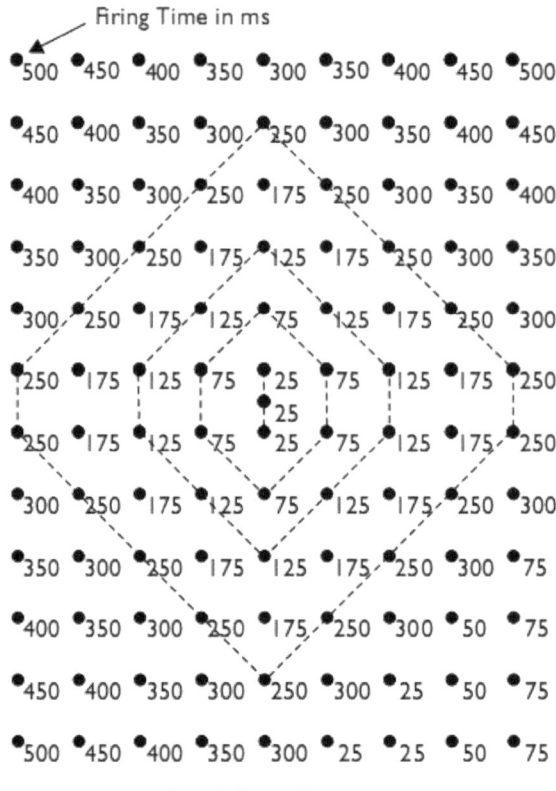

Firing Time in ms

Bench Face Absent

Figure 26 Sinking hole pattern (Gokhale, 2011).

Table 3 Delay constant based on rock type (Wyllie & Mah, 2004).

Rock type	Delay constant (ms/m)
Sand, loam, marl, coal	6–7
Soft limestone, shale	5–6
Compact limestone and marble, granite, basalt, quartzite, gneiss, gabbro	4–6
Diabase, diabase porphyrites, compact gneiss and micaschist, magnetite	3–4

whereas, detonators with relatively short-period timing intervals from 25–50 milliseconds are used in surface blasting.

REFERENCES

Atlas Copco (2012) Blasthole Drilling in Open Pit Mining, 3rd Edition. Garland, Texas, Atlas Copco Drilling Solutions LLC.

Bhandari, S. (1997) *Engineering rock blasting operations*, Balkema, Rotterdam, Brookfield, 375pp., ISBN: 90-5410-658-1.

Cameron, A.R. & Torrance, A.C. (1990) *Underwater evaluation of the performance of bulk commercial explosives*, SEE Annual Conference, Orlando, FL.

Cooper, P.W. (1996) *Explosives engineering*, Wiley-VCH Verlag GmbH, Weinheim.

Cooper, P.W. (1997) Introduction to detonation physics, chapter 4 in Zukas, J.A. & Walters, W. P. (Eds), *Explosive effects and applications*, Springer, 115–136, ISBN: 0-387-98201-9.

Darling, P. (2011) *SME Mining Engineering Handbook*, 3rd Edition. Littleton, Colorado, Society for Mining, Metallurgy, and Exploration, Inc. (SME).

Davis, W.C. (1997) Shock waves, rarefaction waves, equations of state, chapter 3 in Zukas, J.A. & Walters, W.P. (Eds), *Explosive effects and applications*, Springer, 47–112, ISBN: 0-387-98201-9.

Fox, B. (2012) *Blasting in open cut metal mines, in Blasthole drilling in open pit mining*, 3rd Edition, Atlas Copco, 59–68.

Gokhale, B.V. (2011) *Rotary drilling and blasting in large surface mines*, Taylor & Francis Group, London, UK, ISBN 978-0-415-87878-4, 47–463.

Hamdi, E., Audiguier, M., Mouza, J. du, & Fjader, K. (2003) Blast induced micro cracks assessment in muck pile blocks: P-wave velocity and porosity measurements, in Holmberg, R. (Ed.), *Proceedings of EFEE second world conference on explosives and blasting technique*, 10–12 September 2003, Prague, Czech Republic.

Heinio, M. (1999) *Rock excavation handbook for civil engineering*. Sweden, Sandvik, Tamrock.

Hemphill, G.B. (1981) *Blasting operations*, Mc Grow Hill, New York, 258 pages, cited in Bhandari, 1997.

Hustrulid, W.A. (1999) *Blasting principles for open pit mining*, 2 Vol., Balkema, Rotterdam, Netherlands, 1013pp., ISBN: 90-5410-459-7.

Hustrulid, W.A. & Bullock R.L. (2001) Underground Mining Methods: Engineering Fundamentals and International Case Studies. Littleton, Colorado, Society for Mining, Metallurgy, and Exploration, Inc. (SME).

Jhanwar, J.C. (2013) Investigation into the influence of air-decking on blast performance in opencast mines in India: A case study, in Ghose, A.K. & Joshi, A. (Eds), *Blasting in Mines – New Trends*, Taylor & Francis Group, London, 105–110, ISBN 978-0-415-62139-7.

Jimeno, C.L., Jimeno, E.L., & Carcedo, F.J.A. (1995) *Drilling and blasting of rocks*, Ramiro, Y. V.D. (translate by), A. A. Balkema, Rotterdam, Brookfield, ISBN: 90-5410-199-7, 391pp.

Johansen, J. & Mathiesen, C.F. (2000) Modern Trends in Tunnelling and Blast Design. Rotterdam, A.A. Balkema.

Konya, C.J. (1995) *Blast design*, Intercontinental Development Corporation, Montville, OH.

Mader, C.L. (2008) *Numerical modeling of explosives and propellants*, 3rd edition, CRC Press, Taylor & Francis Group, 25–26, ISBN: 978-1-4200-5238-1.

Mahadevan, E.G. (2013) *Ammonium nitrate explosives for civil applications – slurries, emulsions and ammonium nitrate fuel oils*, ISBN: 978-3-527-33028-7, 214pp.

Meyer, R., Kohler, J. & Homburg, A. (2002) *Explosives*, Fifth Edition, Wiley, ISBN 3-527-30267-0.

Moser, P., Grasedieck, A., Mouza, J. du, & Hamdi, E. (2003) Breakage energy in rock blasting, in Holmberg, R. (Ed.), *Proceedings of EFEE second world conference on explosives and blasting technique*, 10–12 September 2003, Prague, Czech Republic.

Mueller, B. & Rossmanith, H.P. (2013) New physical findings revolutionize the drilling and blasting technology as well as the prediction of ground vibrations—Part 1: The new blasting model, in Ghose, A.K. & Joshi, A. (Eds), *Blasting in Mines – New Trends*, Taylor & Francis Group, London, 29–39, ISBN 978-0-415-62139-7.

Nyberg, U., Arvanitidis, I., Olsson, M., & Ouchterlony, F. (2003) Large size cylinder expansion tests on ANFO and gassed bulk emulsion explosives, in Holmberg, R. (Ed.), *Proceedings of EFEE second world conference on explosives and blasting technique*, 10–12 September 2003, Prague, Czech Republic.

Olofsson, S.O. (1988) *Applied explosives technology for construction and mining*, 2nd Edition, Published by APPLEX, Sweden, ISBN 91-7970-634-7, 304pp.

Pradhan, M. & Jade, R.K. (2013) Detonation behavior of bulk emulsion explosive in water filled blast holes, in Mohanty, B. & Singh, V.K. (Eds), *Performance of explosives and new developments, Workshop hosted by Fragblast 10, the 10th international symposium on rock fragmentation by blasting*, New Delhi, India, 24–25 November 2012, Taylor & Francis Group, London, ISBN 978-0-415-62142-7.

Rinehart, J.s. (1960) *On fractures caused by explosions and impacts*, Quart. Colorado school of mines, 56(4), 166, cited in Bhandari, 1997.

RioTinto (2007) *Safe and efficient blasting training manual*, Report No. TCHSMS ML 4.07-1, 219pp.

Roy, P.P. (2005) *Rock blasting effects and operation*, A. A. Balkema Publishers, ISBN: 04-1537-230-5, 345pp.

Rustan, A., Cunningham, C., Fourney, W., Simha, K.R.Y., & Spathis, A.T. (2010) *Mining and Rock Construction Technology Desk Reference: Rock Mechanics, Drilling and Blasting*. Boca Raton, CRC Press.

Sauvage, A.C. (2013) Applied method integrating rock mass in blast design, Singh & Sinha (Eds), Rock fragmentation by Blasting, *Fragblast 10, Proceedings of the 10th international symposium on rock fragmentation by blasting*, New Delhi, India, 26–29 November 2012, Taylor & Francis Group, London, ISBN 978-0-415-62143-4.

Sauvage, A.C. & Blanchier, A. (2003) Determination of most active rock mass heterogeneity on blast: their use in an experimental design based on statistical process control, In Holmberg, R. (Ed.), *Proceedings of EFEE second world conference on explosives and blasting technique*, 10–12 September 2003, Prague, Czech Republic.

Singh, B. & Goel, R.K. (2006) Tunnelling in Weak Rocks. Elsevier.

Sinha, B.P. (2000) *Monitoring of ground vibrations data for regression analysis for estimation of safe charge*, Mining Eng. Jr. Vol. 1, pp. 26–31.

Sinha, B.P. (2013) Linking relationship between parameters of rock mass and ground vibrations, In Ghose, A.K., & Joshi, A. (Eds) *Blasting in Mines – New Trends*, Taylor & Francis Group, London, 29–39, ISBN 978-0-415-62139-7.

Sirveiya, A.K. & Thote, N.R. (2013) Assessing the effect of rock mass properties on rock fragmentation, in Sanchidrian, J.A. & Singh, A.K. (Eds), *Measurement and Analysis of Blast Fragmentation, Workshop hosted by Fragblast 10, the 10th international symposium on rock fragmentation by blasting*, New Delhi, India, 24–25 November 2012, Taylor & Francis Group, London, ISBN 978-0-415-62140-3.

Suthar, S.H. (2013) Performance enhancement by adopting improved blasting techniques in a limestone mine: A case study, in Ghose, A.K. & Joshi, A. (Eds), *Blasting in Mines – New Trends*, Taylor & Francis Group, London, 123–127, ISBN 978-0-415-62139-7.

Winzer, S.R. & Ritter, A.P. (1980) The role of stress waves and discontinuities in rock fragmentation: a study of fragmentation in large limestone block, In *Proceedings of 21st US Symp. Rock Mech.*, Rolla, pp. 362–370.

Wyllie, D.C. & Mah, C.W. (2004) Rock Slope Engineering: Civil and Mining, 4th Edition. Oxon, Spon Press.

Zimmermann, R. (2003) Safety aspects of permitted explosives for use in underground coal mines, In Holmberg, R. (Ed.), *Proceedings of EFEE second world conference on explosives and blasting technique*, 10–12 September 2003, Prague, Czech Republic.

Rock Slope Stability Analysis and Design

Chapter 23

Application of fracture mechanics to rock slopes

D. Elmo[1] & D. Stead[2]
[1]*NBK Institute of Mining Engineering, University of British Columbia, Vancouver, Canada*
[2]*Department of Earth Sciences, Simon Fraser University, Vancouver, Canada*

Abstract: This chapter provides a synthesis of the application of fracture mechanics to rock slopes, both natural and engineered. The characterization of brittle fracture in rock slopes is described with a focus on the types of damage, stress-path dependency and the evolution of damage during slope failure. The state-of-the-art in the numerical modeling of brittle fracture in rock slopes is presented along with methods of characterizing damage both in numerical models and in the field. The importance of scale effects and the challenges facing numerical modeling of brittle fracture in rock slopes are also discussed. The authors emphasize the need to develop improved methodologies for characterising the factors influencing brittle fracture in rock slopes; it is important to ensure that data collection maintain pace with computing developments, as considerable improvements in both parameter and model uncertainty are still required for the analysis of large scale slopes.

1 INTRODUCTION

This chapter provides a synthesis of the application of fracture mechanics to rock slopes, both natural and engineered. The application of fracture mechanics to rock engineering is essentially based on Irwin's modification and extension of the Griffith fracture theory, which addresses crack propagation and provides an explanation of the mechanics of fracture. Rock fracture mechanics refers to the discrete initiation and propagation of cracks/discontinuities that are inherent structural features in most rock masses. The principal concepts in rock fracture mechanics are presented in Irwin & de Wit (1983), Liu (1983), Atkinson (1987) and Whittaker *et al.* (1992). Figure 1 illustrates the three fundamental modes of fracture termed Mode I (tensile), Mode II (in-plane shear) and Mode III (anti-plane shear) respectively:

- Mode I: the crack is subjected to a normal tensile stress and the faces separate symmetrically with respect to the crack front resulting in displacements of the crack surfaces that are perpendicular to the crack plane.
- Mode II: loading subjects the crack to an in-plane shear stress; crack surfaces slide relative to each other and general displacement of the rock surfaces are in the crack plane and perpendicular to the crack front.
- Mode III: loading subjects the crack to an anti-plane shear stress; as for mode II, crack surfaces slide past each other but in general displacements are in the crack plane and parallel to the crack front.

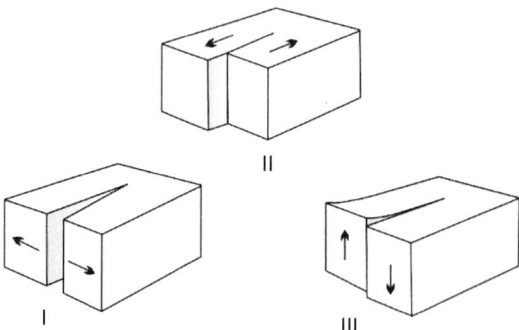

Figure 1 Modes of fracture. Mode I Tensile, Mode II in-plane shear and Mode III Anti-plane shear.

The application of fracture mechanics to rock slope stability was demonstrated by Tharp & Coffin (1985) and Singh & Sun (1989), who analyzed single isolated and edge cracks under varied loading conditions including mode I, mode II, and mixed mode I-II. These authors showed the importance of both fracture length and fracture toughness on rock slope stability. Scavia & Castelli (1996) and Scavia (1990) investigated the mechanical behavior of intact rock bridges in rock slope stability using the displacement discontinuity method. Kemeny (2003) considered sub-critical crack growth and time-dependent degradation of rock joint cohesion and showed how a time dependent reduction in both cohesion and the factor of safety may have importance in the long-term stability of rock slopes. Engineers must carefully consider tectonic, geomorphic and excavation-induced stresses, particularly for high rock slopes or deep open pit mines where stress induced fractures may provide kinematic freedom for a previously stable block and result in slope failure (Havaej, 2015). Brideau *et al.* (2009) discussed the importance of brittle failure and stress induced damage as the controlling factors in the development of failure surfaces, as well as lateral and rear-release mechanisms for the Hope slide (British Columbia, Canada) and Randa rock slide (Switzerland). Brittle fracturing has also been considered as the main controlling factor that led to the catastrophic Vajont failure (Wolter, 2014; Havaej *et al.*, 2015; Paronuzzi & Bolla, 2015).

2 CHARACTERIZATION OF BRITTLE FRACTURE IN ROCK SLOPES

Consideration of damage mechanics in geotechnics has increased markedly in the last decade including applications in blasting, underground excavation, petroleum geomechanics and fluid flow studies. Applications to date in rock slope analysis have however been less common. Tang (1998) describes the use of damage mechanics in the analysis of the Banyan Shan rock slope in China. Eberhardt *et al.* (2004) simulate the influence of damage due to unloading by glacial ice and cohesive strength degradation with time in the stability of the Randa rockslide using finite-element and finite-discrete element modeling. Xu *et al.* (2011, 2014) describe a combined microseismicity damage mechanics approach in stability investigations for the left abutment slope of the Jinping dam in southwestern China.

2.1 Damage processes

The growing importance of damage mechanics in rock engineering is highlighted by an increasing need to address the phenomena of time-dependent progressive failure in high rock slopes. Kemeny (2003) clearly showed that progressive time-dependent damage may play a major role in rock slope instability including both external and internal forms of rock slope damage as described in Stead & Eberhardt (2013). Although damage constitutive criteria have been used extensively within underground rock engineering, particularly within salt rock mechanics, only recently has significant research been devoted to their application to rock slopes. In Table 1 the authors suggest the types of rock slope damage that may influence failure mechanisms in rock slopes. Damage in rock slopes may be defined at several scales. If we consider the damage, D_{fs}, along a failure surface within a representative volume element (RVE) then:

$$D_{fs} = \frac{\partial S_D}{\partial S} \tag{1}$$

Where D_{fs} is the damage on potential failure surface, ∂S_D represents the areas of microcracking on failure surface and ∂S is the total area of failure surface considered.

The damage will vary from 0 (undamaged) to 1 (damage occupies the entire failure surface). This could be related to progressive removal of asperities due to shear induced tensile fracturing, accompanied by a similar reduction in properties such as cohesion, friction angle, joint roughness coefficient (JRC), joint compressive strength (JCS) and normal and shear joint stiffness values. This would consequently also lead to a reduction in resisting forces and the factor of safety of the slope.

Similarly, the rock mass strength might be reduced due to cumulative damage, associated with localized reductions in rock mass rating (e.g. tunneling quality index, Q, and Geological Strength Index, GSI). The importance of disturbance due to the excavation method has been recognized in the 2002 version of the Hoek-Brown criterion through the use of the disturbance factor (D), which is simply one component of the total rock slope mass damage. As Table 1 indicates damage within the rock mass may be associated with varied processes. It is important that field methods be developed to characterize the damage within a rock mass. Algliardi et al. (2013), Stead & Eberhardt (2013) and

Table 1 Types of rock slope mass damage.

Type of damage	Process
Brittle	Crack initiation, comminution
Ductile	Rigid block overlying soft rock
Creep	Sub-critical crack growth over prolonged time; e.g. sackung deformation, joint asperity breakdown
Fatigue	Low/high frequency, loading/unloading (glaciation, isostasy, pore water pressure fluctuation, freeze-thaw, wet/dry, seismic activity)
Tectonic	Preconditioned damage due to faulting, folding, in-situ stress, etc.
Anthropogenic	Stress-induced damage due to excavation
Physico-chemical	Hydrothermal alteration, weathering, corrosion
Geomorphic	Stress-induced damage associated with valley formation, cambering, erosion, thermal cycling etc.

Stead & Wolter (2015) emphasize the importance of an understanding of structural geology when considering rock slope damage processes. Numerous methods to measure damage have been used within continuum mechanics including visual observation of the density of microcracking, variation of elastic modulus, hardness, acoustic and electrical properties Kim & McCarter (1998). Stead *et al.* (1996) and Eberhardt *et al.* (1997) illustrate the use of acoustic emission in laboratory experiments to characterize time dependent damage in rock under uniaxial stress. Kim *et al.* (2015) similarly describe the use of acoustic emission and stress-strain techniques in the quantitative assessment of damage in brittle rock. Future characterization of damage within rock slopes is essential to the success of fracture mechanics simulation and should include the use of both direct and indirect (*e.g.* geophysical) damage characterization within rock slope masses.

2.1.1 Stress path dependency and rock slope mass damage

The potential stress paths that a rock mass experiences may be significantly different to the one determined at the laboratory scale under monotonic loading conditions, as it is likely to involve increase and decrease of principal stresses, as well as a rotation of the applied stresses. Recent studies have demonstrated the applicability of S-shaped failure criteria to large open pit slopes (Wesseloo & Dight, 2009; Stead & Eberhardt, 2013). As shown in Figure 2 below, high gravitational and low confining stresses represent the condition for brittle failure mechanisms. These conditions may be particularly relevant

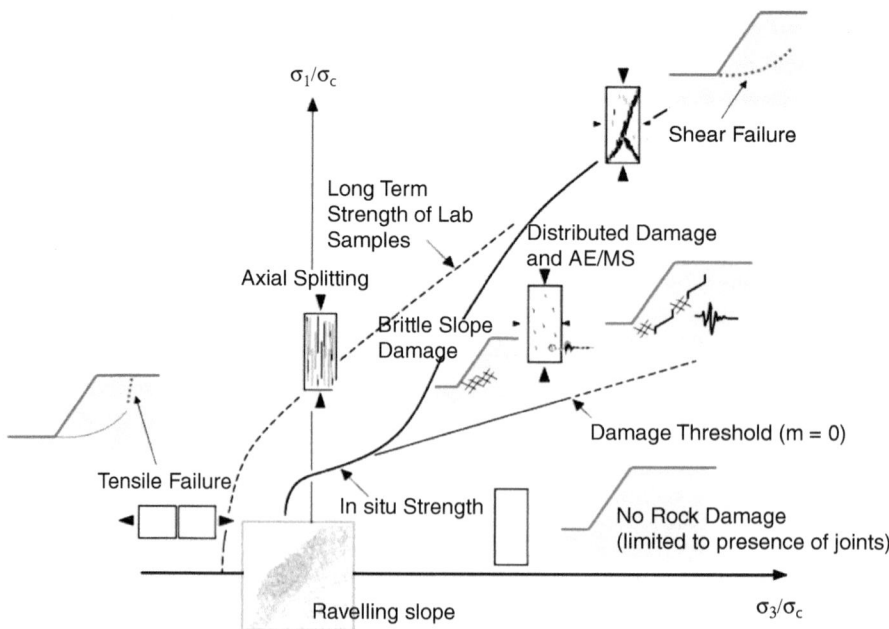

Figure 2 S-shaped failure criterion and its applications to common stress conditions found in rock slope (after Stead & Eberhardt, 2013).

at the toe or at the crest of high rock slopes, where stress-induced cracking may lead to high damage concentration zones. In engineered slopes, the stress path that may develop within a given slope is related to the engineering design of the expansion phases (*i.e.* push-back or cutbacks). In a real open pit scenario the geometry of a pushback is site-specific, and controlled by a variety of parameters, including ore-body geometry, financial and geotechnical considerations, mining equipment, etc. When considering numerical analyses, these will likely not include all push-backs as the practice is to reduce the number of excavation stages to be simulated. Additionally, the majority of the numerical codes used to simulate brittle failure in rock slopes employ an explicit solution scheme, and each excavation stage is defined in terms of a numerical time (generally seconds) or computational steps. Results from stress analysis (Duran *et al.*, 2014) show that stress paths may vary significantly in different zones of the slope. The sudden removal of a large amount of material may thus cause spurious damage zones to occur within the model resulting from unbalanced forces rather than being a true manifestation of slope failure. These effects are amplified in the lower section of a slope, as this would experience a greater loss of confinement. Similar considerations can be drawn for natural slopes, as the retreat of glaciers has been reported (*e.g.* Evans & Clague, 1994; McColl, 2012) as a major cause of slope instability. In order to better capture the correct stress path leading to the occurrence of rock mass damage, numerical models of natural slopes would need to include slope evolution and the simulation of over steepening, debuttressing and unloading processes associated with glacier retreat (Leith, 2013).

2.2 Rock slope brittle fracture and progressive failure (primary, secondary and tertiary brittle fracture processes)

Brittle fracture associated with rock slope instability involves three stages, Stead *et al.* (2007). Primary rock slope brittle fracture includes processes that occur prior to the onset of failure such as propagation of failure surfaces through fracture tip growth, coalescence of fractures and failure of intact rock bridges and shearing along discontinuities involving removal of asperities.

Primary brittle fracture processes may result in rock slope failures through sliding along discrete daylighting planes of weakness, step-path failure surface generation and in even some cases toe-breakout or fracture of key blocks within a slope leading to kinematic release. Stead & Coggan (2006) suggested rock slope instability could be described using a total slope failure analysis terminology modified after Couture *et al.* (1999). Strouth & Eberhardt (2009) used a similar approach to analyze an unstable rock slope in Washington State, U.S. In this framework, primary brittle fracture processes are predominantly associated with the slope initiation or the trigger zone.

Following the onset of primary slope movements within a slope, secondary brittle fracture processes may commence such as the development of rear and lateral release surfaces (allowing global slope failure) and internal slope deformation through fracturing and dilation of the rock mass associated with translational failure, toppling or multiple complex interacting failure mechanisms. Secondary rock slope brittle fracture processes may reflect a transition from the initiation to transportation stages in a slope failure; gradual reduction in rock mass strength and removal of kinematic restraint occurring prior to global rock slope failure and debris transportation. The final stages in the rock slope brittle fracture involved the comminution of the rock mass associated with transport leading up to final debris deposition. These can be considered as tertiary brittle fracture processes and are recognized to be particularly important when characterizing the distance that rock failure debris will travel or the "runout" of slope failure debris. Stead *et al.* (2007) emphasized the need to consider runout and tertiary slope failure processes in large open pits as well as natural slopes; the recent large open pit slope failure at the Kennecott mine clearly confirmed the importance of considering runout in large open pits. Relevant studies on runout in open pit mines are given by Rose & Hungr (2006) and Whittal *et al.* (2015).

2.3 Techniques for characterizing damage in rock slope models

Various techniques have been proposed to characterize damage in rock slopes. For instance a method has been introduced by various authors (Tuckey, 2012; Gao, 2013; Hamdi *et al.*, 2013; Havaej, 2015) that borrows from the DFN community, in which fracture intensity is expressed with reference to a unified system of fracture intensity measures that provide an easy framework to move between differing scales and dimensions. Similarly, damage intensity is referred to as D_{ij} intensity, where the subscript i refers to the dimensions of sample, and subscript j refers to the dimensions of measurement. Gao (2013) used D_{21} (ratio of total crack damage length per unit area) to characterize damage in simulation of roof failure in underground coal mines. Hamdi *et al.* (2013) utilized D_{21} and D_{32} (ratio of total crack area to unit volume) to quantify damage in 2D/3D numerical simulation of unconfined compression and Brazilian tests. Havaej (2015) used D_{21} and D_{32} to characterize the progressive failure of 2D ploughing models and 3D simulation of non-daylighting wedges respectively. Zhang & Stead (2014) describe the use of the D_{32} parameter in PFC3D (Itasca, 2015) to characterize damage in a rock mass.

Readapting a technique used in structural geology, in which a 3D strain ellipsoid (or 2D strain ellipse) is defined as an imaginary deformed sphere (or circle in 2D) defined by three vectors (ε_1, ε_2 and ε_3), Havaej employed 3D ellipsoids of damage and 2D ellipses of damage to inscribe the cracks generated within slope models. The dimensions of each ellipsoid/ellipse are defined based on the extent of crack development away from the center of the damage zone. Geometrical characteristics of the ellipsoid including volume, length and orientation of its axes thus allow for quantification of damage within the model and also an assessment of the importance of anisotropic damage (Havaej, 2015).

Lim *et al.* (2012) used rosette plots to characterize induced microcracks for Lac-du-Bonnet granite. Hamdi (2015) integrated damage intensity measures (D_{21}) and rosette plots to analyze the distribution of crack length and orientation for laboratory scale models. When properly calibrated to account for the larger mesh sizes required to simulate large slope models the same approach could be easily applied to characterize damage developing within much larger slope models. Stead (2015) discussed the importance of damage in the failure of shale rock slopes emphasizing the need to consider damage not only in the plane of the anisotropy but also trans-laminar damage. Morgan & Einstein (2014) describe a particularly relevant investigation on the development of cracking in opalinus shale with varied orientation to loading.

2.4 Rock slope brittle fracture and step-path analysis

The development of deeper open pit mines has given significant impetus to the research on brittle fracture. At the microscopic scale, brittle failure in rock is characterized by the initiation and propagation of new inter- and intra-grain tensile cracks (*e.g.* Martin, 1994; Eberhardt, 1998; Diederichs, 1999; Hajiabdolmajid, 2001). Microcracks are more likely to nucleate in the presence of microdefects or develop at the tip of existing natural fractures. Under loading, flaws or micro-cracks in rock act as stress concentrators and the source of further cracking/fracturing of the rock due to their coalescence. Any material or geometrical discontinuity may promote crack initiation and propagation in compression/tension, therefore the micro-scale problem is analogous to the slope excavation problem, as the interaction between intact rock brittle fracture and natural discontinuities may ultimately lead to the formation of a continuous failure surface (Stead *et al.*, 2006). Because of the difference between the shear strength (defined by the cohesion and frictional terms) of the intact rock and the shear strength of natural fractures, the existence of relatively small intact rock bridges can significantly increase the degree of stability of a rock slope. Even when they occupy only a very small percentage of the fracture-coplanar area, intact rock bridges may provide internal or self-supporting load carrying capacity equivalent to conventional underground support systems (Diederichs, 2003). Jennings (1970) was the first to fully document the importance of intact rock bridge fracturing for open pit mining, and his work led to the development of several probabilistic limit equilibrium methods incorporating step-path analysis into rock slope design (McMahon, 1979; Read & Lye, 1984; Baczynski, 2008).

The question of what constitutes a rock bridge is not a trivial issue. For instance, in Figure 3a the 2D rock bridge can be easily defined as the shortest distance between two existing fractures, Yan *et al.* (2007). However, for a highly fractured rock mass (Figure 3b) the definition of rock bridge would not be unique, varying from a purely geometrical distance to a more complex damage zone within which a critical path is defined by extension and coalescence of existing fractures, Elmo *et al.* (2007). A condition of minimum shear resistance is achieved along the critical path, which

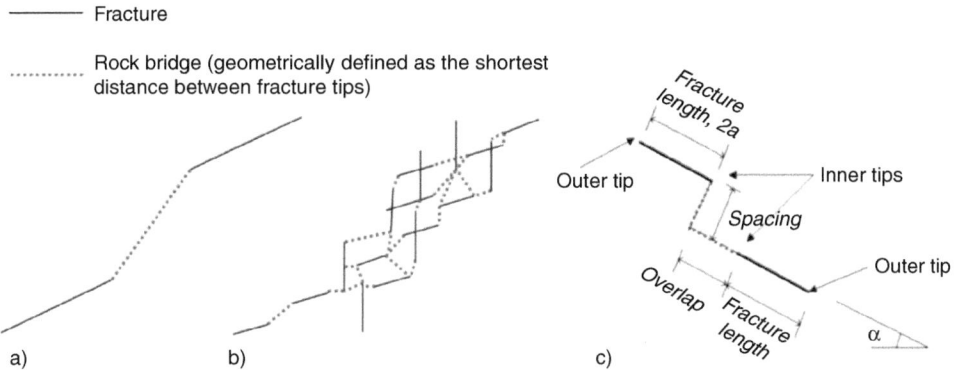

Figure 3 Simple (a) and complex (b) examples of a 2D rock bridge and (c) an attempt to define a 2D step-path geometry (modified from Yan *et al.*, 2007).

ultimately makes the definition of the effective rock bridge a complex interaction between geometry, material properties and applied field stress.

Although considerable attention is being devoted to the development of complex numerical models, comparatively little research has focused on the field description of step-paths and fracturing of intact rock bridges. Furthermore, the need to define rock bridges as 3D entities is essential if brittle fracture simulation of open pits is to be realistic and non-conservative (Elmo *et al.*, 2007). The definition of 3D rock bridges and the incorporation of their effects in numerical models (either explicitly or implicitly) depend directly on the quality and quantity of available field data; however, at the pre-feasibility stage engineers typically have access to limited fracture length data. In particular, quantification of fracture length is a critical component of brittle failure and step-path analysis, requiring both the use of new data collection techniques and rigorous data interpretation through fracture network analysis.

Measurements of fracture length can be obtained by mapping rock exposures, using conventional (scanline or window mapping) or remote sensing techniques (*e.g.* photogrammetry and/or LiDAR). In general remote sensing techniques process fracture length as the diameter of a disc inscribing the mapped feature. Alternatively, assuming a fracture to be a polygon (with n sides and $n \geq 3$), it is possible to define an equivalent fracture radius as the radius of a disk with an equivalent area to the otherwise polygonal fracture. Whatever approach is taken to define the shape of a fracture, the mapped fracture length would be the explicit measure of the 2D trace that a fracture makes with a geological surface or mining exposure; therefore it would not necessarily correspond to the diameter of a circular disk.

Furthermore, the definition of 3D rock bridges requires geotechnical engineers to be familiar with truncation and censoring biases: values below a certain fracture length are omitted (truncation), or relatively larger values cannot be measured because of the limited extent of the rock exposure (censoring) (Mauldon, 1998; Zhang & Einstein, 2000; Sturzenegger *et al.*, 2011). Truncation bias also plays a major role in defining the correct fracture intensity used to build synthetic 3D fracture network models. The importance of fracture connectivity cannot be over-emphasized if realistic brittle fracture analysis of slopes is to be undertaken (Kalenchuk *et al.*, 2006; Kim *et al.*, 2007; Elmo *et al.*, 2014; Alghalandis, 2014).

2.4.1 2D and 3D step-path analysis

Failure of intact rock bridges in the field may include a variety of complex failure mechanisms. Failure surfaces can be a single fracture plane (planar failure), or comprise two or more intersecting fracture planes, (wedge failure). For planar failure to occur lateral release surfaces must be present. Such lateral release surfaces may often be vary from simple planar features to complex three dimensional step-paths as recognized by Jennings (1970) and Stead & Eberhardt (2013) and shown in (Figure 4b). For both planar and wedge failure, a rear release surface at the slope crest may be present in the form of a tension crack. Clearly, the 2D simplification of the problem shown in Figure 3 and Figure 4(a) is a limiting factor and the question arises to whether 2D models can realistically capture step-path failure mechanisms in an effective manner. For 2D step-path failure analysis, engineers should apply a filtering process to the initial 3D fracture model to account for plane strain

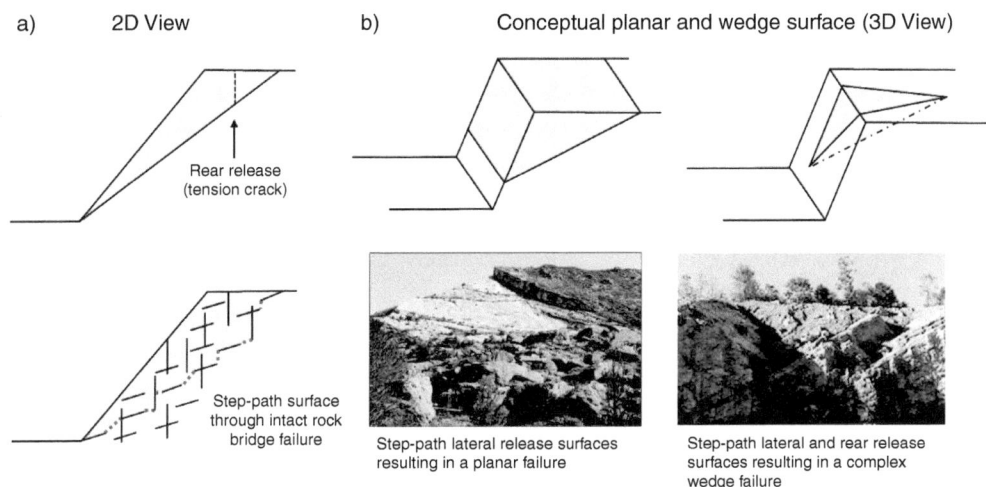

Figure 4 Limitations of a simple 2D step-path failure analysis which fail to consider the effects of complex 3D geometries where for example planar failure and wedge failure involve lateral step-path release joints.

conditions by including only those fracture sets with a dip direction parallel to subparallel to the assumed 2D section. However, the effects of brittle failure mechanism on lateral release surfaces cannot be included in a 2D analysis, which is limited to considering planar failure mechanisms.

Effects of rock bridge failure on a potentially unstable rock wedge can only be considered by using a 3D analysis. Examples are given by Groneng *et al.* (2010), who used FLAC3D to assess the long term stability of the Aknes slope in Norway, Hungr & Amman (2011), who investigated the influence of intact rock on lateral constraint surfaces of rockslides, and Havaej *et al.* (2012), who employed a lattice spring model to simulate a 3D toe-break out problem, Figure 5.

3 PHYSICAL MODELING OF BRITTLE FRACTURE PROCESSES AND ROCK SLOPES

The essentially 3D geometry of rock slides makes modeling of brittle failure a challenging problem, particularly if one attempts to simulate the fracturing process directly in 3D. Various authors have proposed a physical (experimental) modeling approach, which may also incorporate fractures/faults and different lithologies.

Early simple physical models by Muller & Hofmann (1970) were particularly instructive clearly showing the importance of dilation and damage in the failure of biplanar slope geometries. Stacey (1974) provided an early example of the use centrifuge models to test 2D and 3D slope models, which included bedding planes and cross joints. The results of those physical tests showed that failure was controlled entirely by the character the rock mass fabric, expressed in terms of the ratio of the joint spacing to

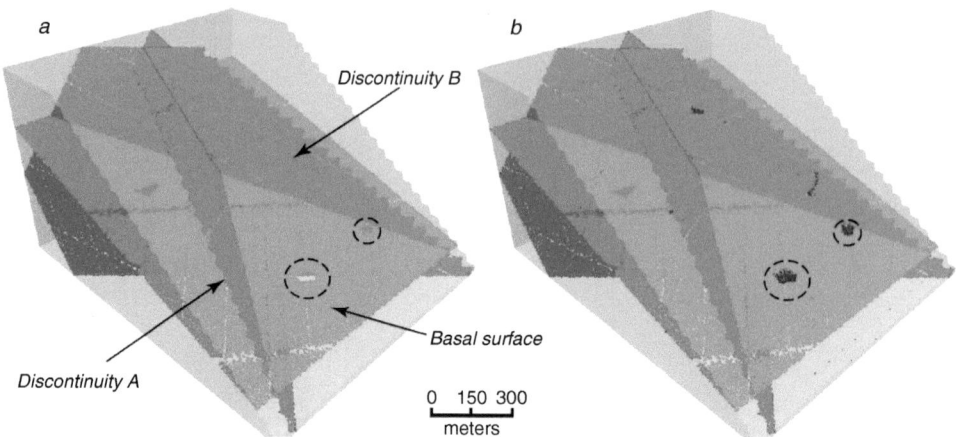

Figure 5 (a) Lattice Spring model of a 3D pentahedral wedge with 1% intact rock bridges located along the basal surface (dashed circles show the patches of rock bridge). b, Failure of the intact rock bridges (black dots within the dashed circles). After Havaej *et al.*, 2012.

the bedding plane spacing. More importantly, the physical tests clearly illustrate that even in the absence of any distinct major failure surfaces, failure can progressively occur on discontinuities throughout the slope. Adhikary & Dyskin (2007) provide an excellent example of the use of centrifuge models to simulate brittle fracture/damage associated with toppling rock slope failure; this work has served as a benchmark in the development of subsequent brittle fracture numerical models, *e.g.* Alzo'ubi (2009). Van Sint Jan & Prudencio (2008) describe the use of physical models in simulating the influence of non-persistent joints (rock bridges) on rock mass strength with results of importance for rock slope stability, Figure 6.

Bachmann *et al.* (2006) used a combined physical modeling-numerical simulation approach to analyze the development of deep seated gravitational slope deformations. The results of their physical model show fracturing at various locations in the model including near the slope crest and in the middle of the model (Havaej, 2015). The authors suggest that the importance of physical modeling in understanding rock slope failure mechanisms and as a constraint for numerical models is liable to increase in the foreseeable future with the development of new technologies allowing for the reproduction of fractured rock masses and the development of new improved monitoring methods applied to physical models including particle image velocity, fiber optics, x-ray tomography and similar techniques.

4 NUMERICAL METHODS FOR MODELING BRITTLE FRACTURE IN ROCK SLOPES

The excavation of large engineered slopes, results in the initiation of complex failure processes, including the deformation and fracturing of intact pieces of rock, the displacement of large individual blocks and the potential for opening/closing and

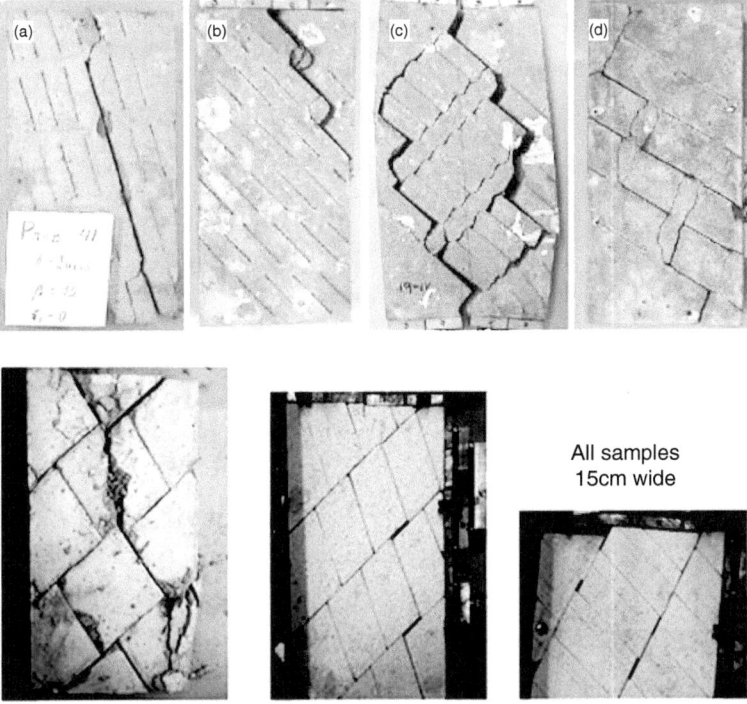

Figure 6 (Top) Observed failure in biaxial tests on physical models with type 1 joints where non-persistent joints end in a rock bridge and (Bottom) with type 2 non-persistent joints terminating against other joints of different orientation (modified from Van Sint Jan & Prudencio, 2008).

sliding of rock joint surfaces relative to each other. A careful analysis of the stresses and associated displacements is required in order to better understand and simulate the behavior of such large excavations. Provided they are used correctly, numerical models are essential for studying the fundamental processes occurring at the rock mass scale when analyzing brittle fracture in rock slopes. The most common types of numerical models that have found application in the modeling of brittle failure in rock slopes can be grouped as follows:

– Continuum methods (*e.g.* Boundary Element Method - BEM, Finite Element Method – FEM and Finite Difference Method - FDM);
– Discontinuum methods (*e.g.* Discrete Element Methods - DEM, and Particle flow models);
– Hybrid models (*e.g.* Hybrid FEM/DEM); and
– Lattice spring models.

Discrete Fracture Network (DFN) models represent a special class of DEM and generally do not include constitutive criteria to analysis stress-displacement problems. Rather, the DFN approach seeks to explicitly represent discrete fractures in 2D and 3D space, which can then be embedded within standard FEM, FDM, DEM, Hybrid,

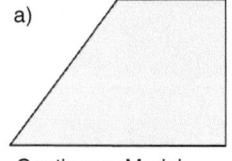

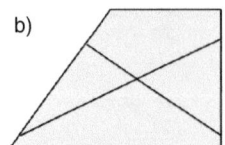

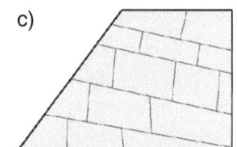

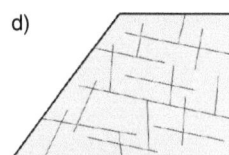

a) Continuum Model: Use of equivalent continuum properties

b) Continuum Model: Use of equivalent continuum properties but also incorporating major structures

c) Discontinuum Model: Accounting for block kinematics but no explicit intact rock failure (deformable or rigid blocks)

d) Hybrid Models: Accounting for both block kinematics and explicit intact rock failure (new fractures, thus new blocks can be formed in the model)

Figure 7 Simplified examples of different numerical methods for modeling a rock slope problem in relation to the existing fracture system geometry: (a) continuum method, (b) continuum with fully persistent fractures, (c) discontinuum method without intact rock fracturing and (d) discontinuum method with explicit intact rock fracturing.

PFC and lattice spring models. An example of different numerical conceptualization of a rock slope problem is given in Figure 7.

The continuum approach may circumvent some of the difficulties associated with the discrete method, in terms of complexity of the model and impracticality of modeling every fracture in a deterministic way. However, an inherent limitation of the equivalent continuum approach is that the stress acting on a specific fracture is usually not the same as that deduced from the overall stress, since it depends on the stiffness of the fracture itself and on the stiffness of the surrounding matrix (Cai & Horii, 1993). In addition, block relative displacements and interlocking, with associated internal moments produced by block rotations, cannot be adequately accounted for in a continuum model. The continuum approach trades material complexity for geometrical simplicity, requiring proper homogenization techniques to identify the material parameters associated with specified constitutive equations for the equivalent continuum; the homogenization process is usually very complex and valid only over a certain representative elementary volume or REV (Jing, 1998).

4.1 Modeling challenges

In ideal context, a slope model should try and reflect the inherent Discontinuous, Inhomogeneous, Anisotropic, and Non-Elastic nature of the rock mass (*i.e.* the "DIANE" concept, after Harrison & Hudson, 2000). However, the objective of a slope model is not to create a perfect imitation of reality, rather the work of the analyst should be to determine which rock mass features should be considered explicitly and which can be represented in an average way (Hoek *et al.*, 1990). In relation to the discontinuous nature of a rock mass, and the fact that brittle processes represent a transition from continuum to discontinuum, numerical models should incorporate the ability to explicitly include pre-existing fractures, as well as simulate stress-induced fractures. The physical processes and the modeling techniques chosen will eventually influence the extent to which these features can be incorporated in the model. Parameter representability associated with sample size, representative elemental volume and homogenization/upscaling represent additional fundamental problems associated with modeling of rock slopes.

Most numerical codes used in geotechnical engineering for rock slope analysis are formulated based on the linear Mohr-Coulomb failure criterion, which considers the internal cohesion and internal friction of the rock material. Tensile strength is also an important parameter and a tension cut-off is generally employed to minimize the limitations of what is basically a shear-based failure criterion. Internal cohesion and friction are strain dependent and research now acknowledges that cohesive strength is mobilized first; once the cohesive strength is depleted and shear displacement can occur, the frictional strength is then mobilized. Hajiabdolmajid & Kaiser (2002) used this criterion in the analysis of the Frank Slide, while Eberhardt *et al.* (2004) applied a similar approach to model the Randa rock slide.

Limited consideration has generally been given to the realism of adopting a continuum "plastic deformation" approach for moderate to strong rock, particularly in relation to the presence of natural discontinuities, rock slope scale and associated yield stress. It is argued that in open pits less than 500m deep (or limited height natural slopes), induced stresses may not be sufficient to cause true plastic failure, unless considering weak and altered rock masses. In moderate to strong rock masses stress concentrations, tensile and/or compressive, are more likely to cause brittle micro-fracturing. Whether or not the macro scale deformation of the rock slope can be considered plastic, in practice it would originate by micro-fracturing occurring at stresses far below the plastic yield stress for the rock mass. Where simulations involve large slopes then the potential for plastic yield becomes more important; however it remains difficult to evaluate the relative roles of compressive induced micro-fracturing, compressive induced plastic yield and the kinematics of the rock slope. Furthermore, in most situations rock slope failures are controlled by the orientation, shear strength and stiffness properties of adversely oriented discontinuities. In the initial description of the Hoek-Brown criterion (Hoek, 1983) and the Geological Strength Index (Hoek *et al.*, 1995) it was clearly emphasized that where discrete structures control failure then modeling using an equivalent continuum approach alone may be inappropriate. Realistic simulations would necessarily require models capable of considering rock mass plastic yield, explicit simulation of tensile fracturing and ability to consider kinematic processes.

The question also arises as to over what scale can such tensile failure processes be simulated in a realistic manner. Brittle fracture analysis of slopes involves mechanical processes which may range from small to medium scale problems to large open pits. However, limited attention has been given to the scaling up of fracture mechanics approaches to the modeling of large scale structures (see also Section 5) While realistic slope models have been developed that agree with observational data it remains a major area for research to provide a smoother transition between laboratory scale fracture mechanics derived properties and large scale rock mass properties.

A tiered approached toward scale of modeling could be introduced that is strongly related to both structural observations and rock mass quality. However, any simplification process should be such that it does not result in the 'smoothing-out' and even non-consideration of critical structurally controlled key blocks or adversely oriented structures *i.e.* the kinematics of the slope failure should be preserved as much as

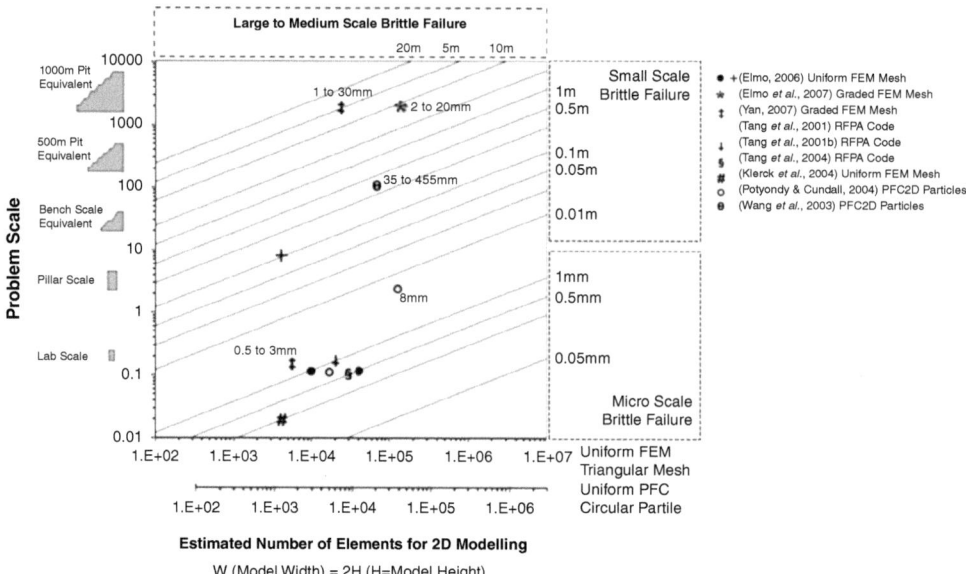

Figure 8 Relationship between model scale, mesh size and number of elements for 2D analysis of brittle failure processes (modified from Stead *et al.*, 2007).

possible. Consideration should be given to model design to highlight critical features so that the mesh may be refined appropriately. Constitutive criteria incorporating fracture mechanics could be used for specific regions within a model as opposed to the whole slope thereby allowing faster computing times. Where large open pits are modeled in two and three dimensions, there is also an inherent requirement to use relatively large mesh dimensions due to increasing computing limitations. Figure 8 illustrates the relationship between model scale, mesh size and number of elements for two dimensional slope engineering problems. It is evident, even for simple two dimensional brittle fracture models, that modeling large engineering structures requires significantly large element sizes. The computing challenge engineer's face is immediately apparent and arguably non-trivial if brittle fracture processes are to be explicitly included in the model and simulated at a realistic scale in three dimensions.

4.1.1   Uncertainty and variability: The importance of collecting structural data

A more representative characterization of discontinuities is essential if brittle facture analysis is to be achieved. At the same time brittle fracture modeling of slopes has highlighted an often extreme shortage of the appropriate data with which to optimize the use of increasingly sophisticated codes. The process of data collection is strongly linked to the concepts of variability, epistemic and aleatory uncertainty (Harrison, 2012; Feng & Hudson, 2015). Variability is defined as the observable manifestation of heterogeneity of physical parameters and/or processes, while uncertainty relates to the

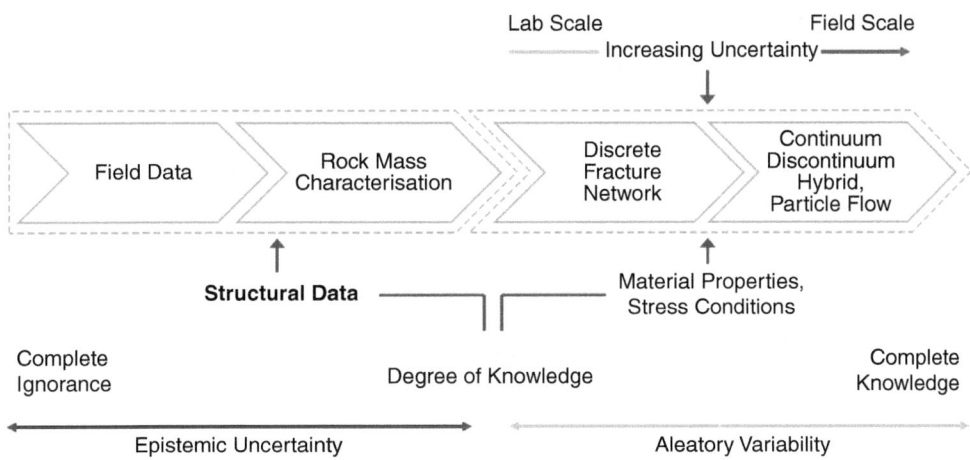

Figure 9 Importance of data collection; qualitative relationship between degree of knowledge, uncertainty, variability and problem scale.

engineer's state of knowledge, and thus reflects the decision to recognize and describe the observed variability in a qualitative or quantitative manner (Uzielli, 2008). Model uncertainty would also increase with problem scale due to an increasing number of unknown parameters, Figure 9.

While the randomness of geological processes is responsible for the manifested variability of the parameters used to describe natural discontinuities, it is important that engineers recognize such variability of the parameters may not necessarily be described by stochastic (aleatoric) models. For instance, mapping of natural discontinuities can be carried out according to either a subjective (biased) or objective sampling approach (ISRM, 1981). Sampling of natural discontinuities should include procedures to capture total uncertainty, defined as the combination of aleatory variability, *i.e.* variability inherent in random processes, as well as epistemic uncertainty, *i.e.* lack of knowledge (Feng & Hudson 2015; Elmouttie & Poropat, 2011). Subjective sampling considers only those discontinuities that are deemed important by the engineer carrying out the sampling. Conversely, objective sampling considers all natural discontinuities intersecting a given line or contained within a given area. Therefore, only the latter is a random process, and accordingly the data being collected could be characterized using statistical analysis; when subjective sampling methods are adopted, any variation in the mapped data would be considered the result of limited knowledge, and accordingly the data should be characterized using epistemic models. Limited knowledge has an inherently larger degree of uncertainty associated to it, which can only be minimized by improving the data collection procedure (quantity and quality of data being collected). Using a model to carry out a sensitivity analysis may improve understanding of mechanical behavior and highlights the impact of specific material properties on the modeling results, but it should not be viewed as an alternative to justify limited data.

Recent developments in digital imaging techniques including laser scanning and ground based photogrammetry could overcome some of the limitations of data accessibility for large scale problems. The availability of these new techniques means we should adjust the scale of our observations to capture large scale features that may be critical for instability. Digital imaging techniques have also the capability to undertake rock discontinuity characterization at the scales that are important and relevant not only to bench or highway cuts but to multiple bench large open pits - leading to the potential recognition of large open pit scale persistence/roughness classifiers (Stead *et al.*, 2007).

4.2 Numerical modeling of fracture processes

Considerable attention has been given to the choice of appropriate models for the simulation of complex rock slope problems. The variety of numerical methods and fracture models developed in an attempt to simulate the fracture process of rock material reflects the complexity of the fracture process itself. However, not all numerical models can realistically simulate the progressive fracture process, including crack initiation, propagation, interaction and coalescence (Liu, 2003), as the explicit simulation of brittle fracture requires robust and efficient methods to account for multiple interacting cracks and rigorous fracture models to reflect the rock mass fabric characteristics. Although conventional continuum and discontinuum models can provide useful analysis for interpretation of failure, neither approach can capture the interaction of existing discontinuities and the creation of new fractures through fracturing of the intact rock material (Stead *et al.*, 2004; Coggan & Stead, 2005).

4.2.1 Continuum models

The majority of published brittle rock fracture modeling remains based on boundary element methods including the displacement discontinuity method. BEM models and in particular the displacement discontinuity method have found application in the modeling of rock fractures. A coupled DDM/splitting model was proposed by Tan *et al.* (1996) to simulate side crack propagation and indentation, while De Bremaecker & Ferris (2004) described a 2D numerical approach based on the DDM method to model shear fractures. This method has been used extensively in the analysis of step-path features at the laboratory scale. Boundary element methods usually allow the consideration of varied fracture modes such as Modes I, II and mixed fracture modes. Basic fracture mechanics research on step-path intact rock fracture in laboratory specimens continues to be a subject of intense research particularly in three-dimensions. This should be borne in mind when extrapolating brittle fracture modeling from the laboratory to the bench to the mine scale. Singh & Sun (1989), Scavia & Castelli (1996) and other researchers have demonstrated the use of displacement discontinuity techniques in fundamental step-path research on rock slopes.

Continuum techniques, where the rock is discretized into domains, may incorporate fracture mechanics algorithms to allow simulation of brittle fracture. FEM models for simulation of rock fractures include SICRAP (Saouma & Kleinoski, 1984), CRACKER (Swenson & Ingraffea, 1998), FRAN2D (Wawryznek & Ingraffea, 1989), DIANA & NUMA (Alehossein & Hood, 1996) and R-T2D (Liu, 2003). Finite element modeling of brittle fracture in rock slopes has been demonstrated using codes such as RFPA

(Tang *et al.*, 2001, 2004). Although continuum codes can be used in simulation of elastic-plastic yielding of rock slopes and successfully indicate zones of tension they have yet to be used routinely to simulate brittle fracture specifically in engineering structures. When used with appropriate failure criteria (*e.g.* Mohr Coulomb with cohesion weakening and frictional hardening - Hajiabdolmajid, 2001), continuum codes have the capability of simulating areas of potential intact rock fracture. Spreafico (2015) describes the use of the FEM code, Phase2 incorporating polygon Voronoi joints and DFN's to simulate the brittle fracture involved in the 2014 San Leo rock slope failure, Italy. Similarly, Stacey *et al.* (2003) adopted an extension strain criterion, 3D elastic analyses and physical models to show the importance of considering extensile strains (potential brittle fracture) in association with large open pit slopes for varying slope curvature and in situ stress ratios.

4.2.2 Discontinuum models

Various authors have indicated the potential for modeling brittle fracture using discontinuum codes. Discontinuum codes enable the simulation of blocky media which may be either rigid or deformable (in the latter case the blocks are meshed and deform according to assumed continuum constitutive criteria). Deformation along joints is controlled by specified contact constitutive criteria. Among others, Christianson *et al.* (2006), Alzo'ubi (2009) and Havaej *et al.* (2014) have demonstrated the application of the Voronoi technique using the 2D discontinuum code UDEC (Itasca, 2015). Tuckey (2012) applied the UDEC Voronoi approach to investigate the effects of intact rock bridges on slope stability. Voronoi tessellation divides up the rock mass into randomly-sized polygonal blocks. This allows for an indirect (*i.e.* not formulated upon principles of fracture mechanics) approach for the simulation of crack propagation, and fracturing occurs if the contact strength between Voronoi blocks is exceeded. Gao (2013) proposed a new Trigon logic designed to overcome some of the limitations of the conventional Voronoi approach. Reduced mesh sensitivity and better prediction of friction angle and failure patterns under varied conditions are among the advantages of the Trigon logic; additionally, the use of a Trigon mesh helps reducing the overlapping potentially generated by the conventional Voronoi tessellation and provide a greater degree of freedom for the blocks to rotate and slide against each other. Vivas (2014) applied the Trigon logic to brittle failure for a 150m high rock slope, with and without added pore water pressure.

Particle flow codes such as PFC (Potyondy & Cundall, 2004) represent an alternative discontinuum approach that has been extensively used to numerically investigate brittle fracturing in rocks. PFC treats the rock mass as an assemblage of circular particles (2D) or spheres (3D), and there exists the possibility to embed specified joints. The sliding mechanisms may be captured using a clumping of particles and appropriately oriented joints or incorporating a sliding joint model (Pierce *et al.*, 2007). Wang *et al.* (2003) demonstrated the application of PFC in the analysis of heavily jointed rock slopes, while Camones *et al.* (2013) used PFC as a tool for modeling step-path failure mechanism in fractured rock masses. Recent developments into the application of 3D bonded particle codes include the YADE DEM code, which Sholtès & Donzé (2015) used to model intact rock bridge failure for the 1991 Randa rockslide, and Jiang *et al.* (2015), who used a DEM bond contact model to simulate the behavior of a jointed rock slope.

Examples of modeling of brittle failure processes using the Voronoi tessellation, Trigon, PFC and DEM bonded contact models are given in Figure 10. Gao & Stead (2014) describe the development of a 3D brittle fracture modeling method using 3DEC; Figure 10(g) shows a preliminary model of a failure of a coastal rock slope using 3DEC Trigon. Both the Voronoi tessellation approach and the particle flow code require a calibration process to define the contact properties for either the Voronoi/Trigon polygons or the bonded circular/spherical particles. The calibration process generally

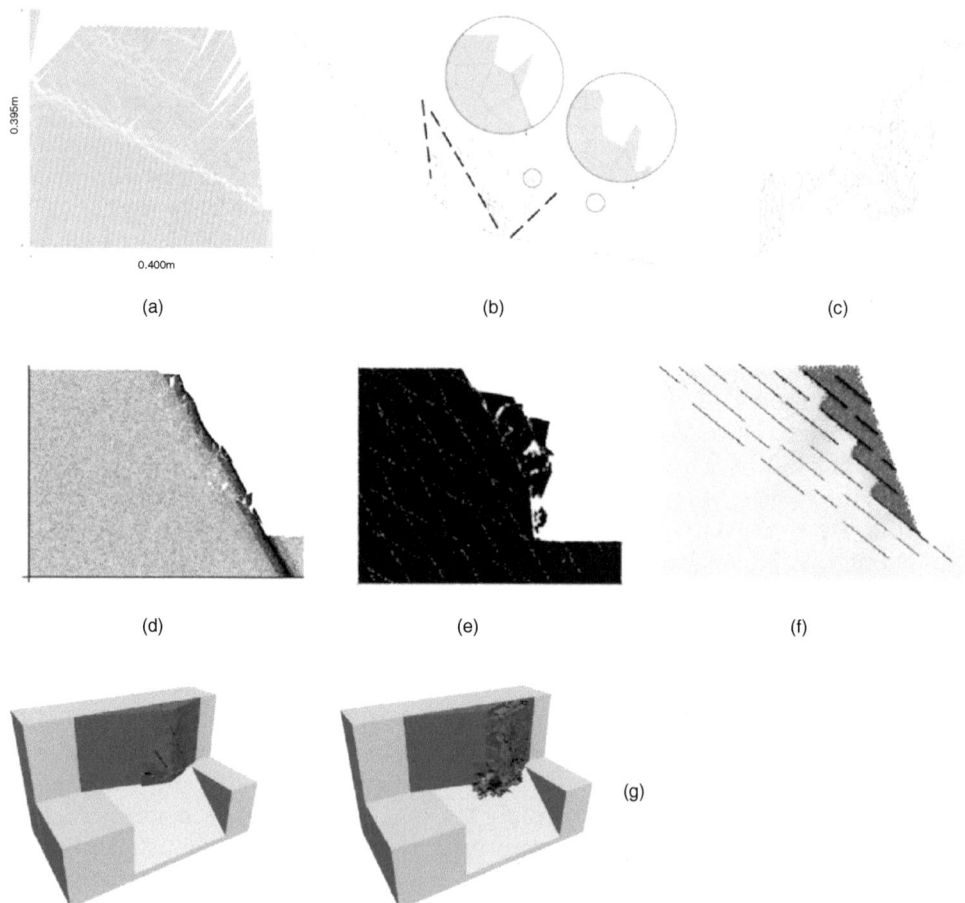

Figure 10 (a) UDEC Voronoi simulations for toppling failure (Alzo'ubi, 2009); (b) UDEC Voronoi simulation of extensive shear and tensile damage (Tuckey, 2012); (c) UDEC Trigon simulations of fracturing within a rock slope (Vivas, 2014); (d) simulation of rock slope failure in PFC code (Wang *et al.*, 2003); (e) jointed slope model during collapse (Jiang *et al.*, 2015); (f) jointed slope model in YADE Open DEM (Sholtès & Donzé, 2015); and (g) preliminary model of a failure of a coastal rock slope using 3DEC Trigon (Gao & Stead, 2014; Gao, 2013 personal communication).

involves using laboratory results for uniaxial/triaxial tests. Careful consideration should be given to upscaling the modeled contact properties to large scale models as the contact properties will be dependent on the assumed discretization size (Voronoi/ Trigon element size or radius of the circular/spherical particle).

Another example of discontinuum approach to model brittle fracture is given by the Discontinuous Deformation Analysis (DDA). Applications of the DDA model to block fracturing have been presented in Amadei *et al.* (1994) and Ke & Goodman (1994).

4.2.3 Hybrid models

Hybrid continuum-discontinuum techniques (FDEM) incorporating fracture mechanics criteria provide an attractive option for modeling discrete brittle fracture in rock slopes. These codes combine a finite element-discrete element approach and allow the transition from a continuum to a discontinuum (as observed during primary to tertiary slope brittle fracture processes). In general terms, the finite element–based analysis of continua is merged with discrete element–based transient dynamics, contact detection, and contact interaction solutions (Munjiza, 2004). The numerical analysis of fracturing processes in rock, in addition to its intrinsic discrete/discontinuous nature, has to consider that such problems are often highly dynamic, with rapidly changing domain configurations, requiring sufficient resolution and allowing for multi-physics phenomena. Additionally, contact behavior also gives rise to a very strong nonlinear system response. For these reasons, such problems are typically simulated employing time-integration schemes of an explicit nature (Owen *et al.*, 2004). An FDEM solution strategy allows for a better description of the physical processes involved with brittle failure mechanisms, accounting for diverse geometric shapes and effective handling of large numbers of contact entities with specific interaction laws. The implementation of specific fracture criteria to simulate progressive fracturing and the ability to account for the full representation of the anisotropic and inhomogeneous effects of natural jointing are additional advantages of the hybrid FDEM approach.

ELFEN (Rockfield, 2015), Y-Geo (Lisjak & Grasselli, 2014) and IRAZU (Geomechanica, 2015) represents hybrid FDEM codes successfully applied to model a variety of brittle failure processes, including rock slope failure. In ELFEN, explicit fracturing can be simulated using either an intra-element or inter-element algorithms (Klerck *et al.*, 2004). The former may require local adaptive remeshing to allow for an acceptable element topology. Y-Geo and IRAZU codes employ an inter-element algorithm with a discrete crack inserted along an existing element edge most favorably oriented with respect to the expected failure plane. The commercial version of ELFEN simulates fracturing according to Mode I, while Y-Geo and IRAZU allow modeling of both Mode I and Mode II fracturing. For example, using Mode I and the inter-element algorithm, a new crack would be inserted along an element edge most favorably oriented (parallel) with respect to the locally induced major principal stress. Following the crack insertion, the contact along the two newly-created surfaces is modeled using a penalty or Lagrangian multiplier method (Lisjak & Grasselli, 2014).

Eberhardt *et al.* (2004) and Stead *et al.* (2006) applied FDEM modeling to simulate brittle failure processes in rock slopes, as in the case of the Randa rock slide in

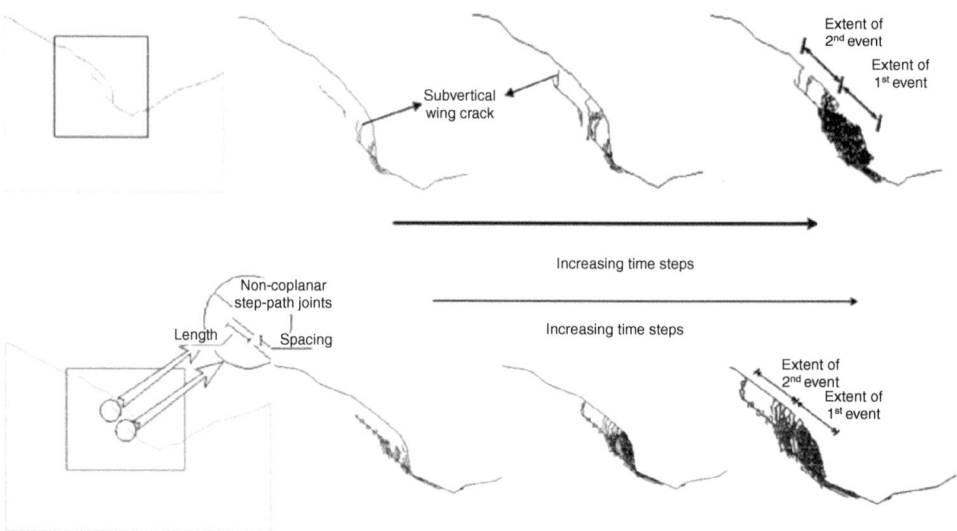

Figure 11 Simulation of the Randa rockslide assuming (top) persistent discontinuities and (bottom) limited persistence discontinuities and step-path failure. Modified from Eberhardt *et al.* (2004).

Switzerland (Figure 11). Earlier models considered simple step-path problems by including discrete joint surfaces within an otherwise continuum FEM mesh. Stead *et al.* (2007), Elmo *et al.* (2008) and Vyazmensky *et al.* (2010) developed advanced FDEM brittle fracture simulations of rock slopes by using an integrated DFN-FDEM approach. These authors were able to simulate more complex joint controlled brittle failure mechanisms including flexural toppling and interaction between underground and surface mining. FDEM modeling results shown in Figure 12 provide a graphical interpretation of the concepts of primary, secondary and tertiary brittle fracture processes.

4.2.4 Lattice Spring Models (LSM)

Lattice Spring Models (LSM) provide an alternative approach to DEM for modeling brittle fracture processes. The rock mass is represented by a lattice whose points are connected by linear and angular springs. Slope Model (Itasca, 2015) uses a 3D lattice-spring configuration where point masses are connected by springs that are assigned nonlinear force-displacement laws (Cundall & Damjanac, 2009). Fracturing is simulated by breaking of one or more springs once the applied stress exceeds the shear or tensile strength of the spring(s); a DFN model can also be superimposed on the lattice springs (Itasca, 2015), Figure 13. In its current version, Slope Model is a small strain code and coordinates are not updated as a result of displacements, thus the model does not simulate secondary and tertiary brittle failure processes. Notwithstanding, Havaej *et al.* (2015), successfully applied Slope Model to simulate brittle fracturing as a controlling factor in the initiation of the Vajont landslide, including the effects of groundwater on the modeled brittle failure processes, Figure 13(b).

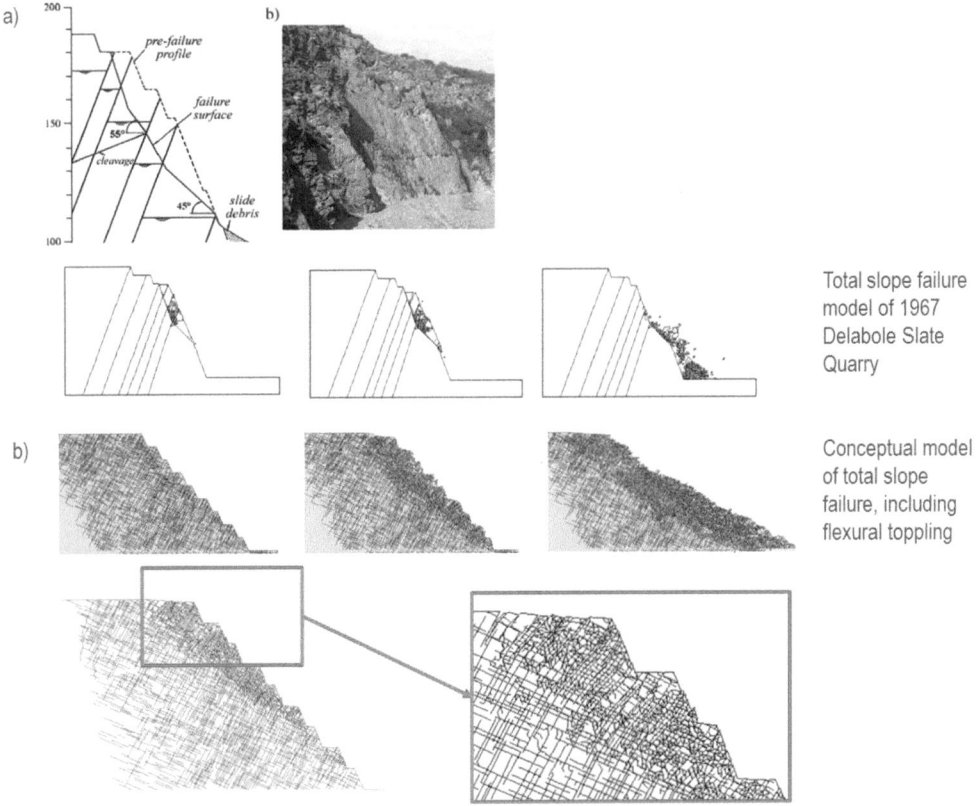

Figure 12 a) FDEM total slope failure model of 1967 Delabole Slate Quarry (after Stead *et al.*, 2006) and FDEM modeling of primary, secondary and tertiary brittle failure processes.

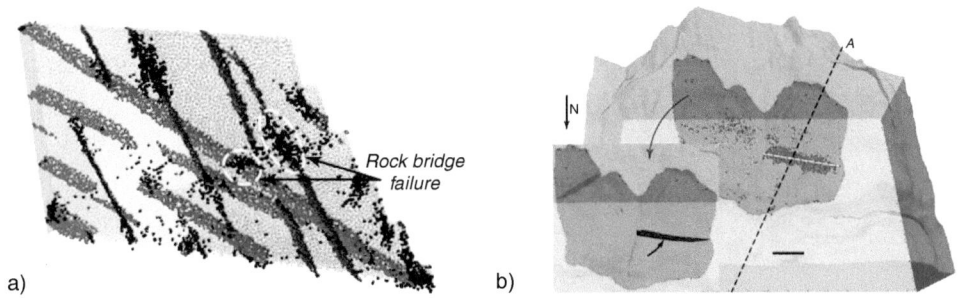

Figure 13 a) slope failure through fracture of intact rock bridges modeled with Slope Model (black dots represent induced cracks) (*Lorig et al.*, 2010); and b) simulation of the Vajont landslide showing development of a stress induced brittle fracture discontinuity zone discontinuity (dashed yellow line) (after Havaej *et al.*, 2015).

5 IMPORTANCE OF SCALE EFFECTS AND CONSIDERATIONS FOR BRITTLE FAILURE MODELING

The importance of considering scale effects is a topic widely discussed in the literature, and it is generally acknowledged that rock mass strength decreases with increasing sample size. The importance of scale effects can be considered in two categories: i) laboratory scale samples and the apparent effects of randomly distributed flaws within otherwise intact rock specimen and ii) larger rock mass volumes containing a number of natural discontinuities related to systematic jointing patterns. Numerical modeling of brittle failure processes in rock slopes arguably spans both categories, as fracture mechanics theory deals with nucleation, growth and coalescence of stress induced cracks at the microscale. As fracturing continues, these micro cracks develop into macroscale features and potentially interact with the rock mass fabric. Within this context, two questions arise that have potentially important implications: at what scale can we model? And, are we modeling brittle fracture at appropriate scales? As previously discussed in Section 4.1, computational requirements often limit the extent to which numerical modeling of large scale slope problems can explicitly simulate fracturing processes at the microscale level, as this would require using a sub-millimeter discretization size (mesh, particle size, length of lattice springs). Similarly, the extent to which the natural rock mass fabric can be embedded within a FEM, DEM or lattice model is a function of fracture spacing and the minimum element size used to discretize the model. The alternative is to scale-up rock mass properties to implicitly represent the degree of natural fracturing not included in the model using a Synthetic Rock Mass (SRM) approach, which combine the effects of the intact and fractured portions of the rock mass into a unique set of equivalent continuum properties; the approach is graphically illustrated in Figure 14.

SRM models can be used to define
"equivalent" rock mass properties for a portion
of the rock mass up to the Representative
Elementary Volume (REV)

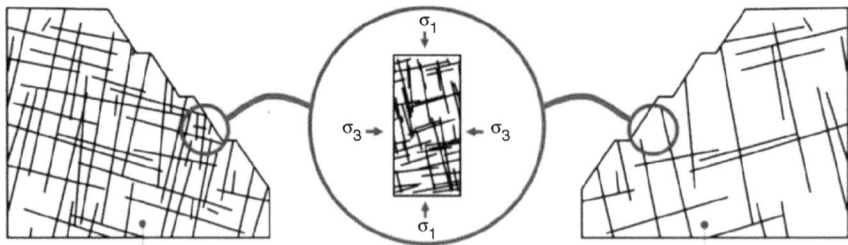

Large scale slope problem with a detailed representation of the natural rock mass fabric: the use of a millimetre/centimetre scale mesh size would make the model computationally expensive and inefficient

Same large scale slope problem: only major structures are explicitly included in the model thus allowing the use of a metre scale mesh size, which drastically reduce computation times

Figure 14 Conceptual definition of the use of equivalent rock mass properties to represent small scale fracturing.

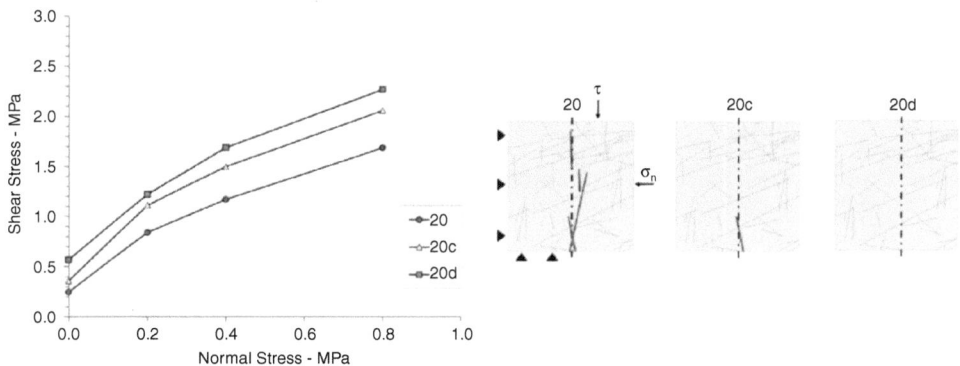

Figure 15 Results for synthetic shearing tests performed on 20m x 20m models with varying fracture length along the assumed shearing direction (modified from Elmo *et al.*, 2011).

The SRM approach was originally proposed using either the PFC2D or PFC3D codes (*e.g.* Cundall, 2008; Mas Ivars, 2008). Alternative formulations using a hybrid FDEM code are given by Elmo & Stead (2010). Independently of the numerical platform used, the SRM approach provides a significant opportunity to improve upon the more traditional method of deriving rock mass properties from empirical rock mass classification schemes. SRM results can be incorporated into a continuum finite element or finite difference models (*e.g.* Sainsbury *et al.*, 2008), and can be used for the modeling of anisotropic rock slopes (Sainsbury & Sainsbury, 2013).

An SRM approach also allows the evaluation of the rock mass strength for an equivalent discontinuity, along which failure occurs partly along existing joints and partly through the intervening intact rock. Synthetic FDEM models of shearing mechanisms clearly show one of the most important variables governing rock mass behavior is fracture length, Figure 15. Similarly, hybrid FDEM models of shearing mechanisms at different scales (2m, 5m, 10m and 20m) demonstrates that an SRM approach can effectively capture the reduction of rock mass strength and equivalent GSI rating as the model size is increased, Figure 16.

However, there are factors that need to be carefully considered when planning a SRM analysis, including:

–   The computational time and cost associated with SRM construction and testing. Because both the intensity and geometrical location of the discontinuities (relative to the proportion of intact rock) determine the strength of the rock mass (Cundall, 2008), it is argued that the strength of the synthetic rock mass will be a function of the underlying DFN fracture pattern embedded in the models. The stochastic nature of the DFN process is such that there are an infinite, but equi-probable, number of possible realizations of the fracture systems based on the specified input parameters. Ideally, SRM modeling should include testing a representative number of DFN realizations to account for uncertainty related to the fracture pattern;

–   The parameterization of joint properties used within the SRM model. As SRM models require the knowledge of joint properties, the problem of establishing a scale dependent relationship for rock mass properties is also directly associated

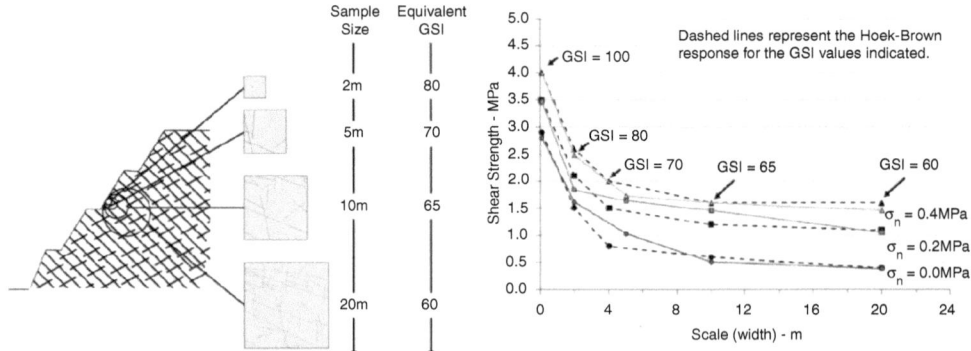

Figure 16 Variation of modeled shear strength as a function of sample width and applied normal stress and correlation between a back-calculated equivalent GSI rating and model scale (modified from Elmo, 2012).

with the extrapolation of mine scale shear strength properties from laboratory tests (*i.e.* mm scale); and

- The inherent variability within a particular geotechnical domain which makes the number of fracture network realizations needed to derive a reliable SRM a mostly undefined problem. With the exception of either a very closely fractured or an almost massive rock mass, the mechanical response is generally non-uniform due to the orientation, spacing and persistence of the discontinuities (Pine & Harrison, 2003).

6 CONCLUSIONS, RECOMMENDATIONS AND FUTURE RESEARCH DIRECTIONS

In this chapter the authors have attempted to show the importance of considering brittle fracture and the principles of fracture mechanics in rock slope investigations. We have emphasized the need to develop improved methodologies for characterizing the factors influencing brittle fracture in rock slopes. We suggest that the following areas of research are important in rock slope characterization:

- Development of improved data collection methods tailored to brittle fracture-damage modeling;
- Further use of video fractography as described in Stead & Eberhardt (2013) to study the growth of brittle fractures during rock slope failure;
- Increased use of UAV's and terrestrial remote sensing to characterize brittle fracture in rock masses;
- Development of robust methods for characterizing discontinuity persistence, rock bridges, connectivity and brittle fracture damage; and
- Better procedures for mapping/delineation of damage domains in slopes, both blasting and excavation/slope movement-induced.

The authors have also presented the state-of-the-art in modeling of brittle fractures in rock slopes using both 2 and 3D methods. We suggest that modeling of brittle fracture in rock slopes can be improved by:

- Increased consideration of geostatistics in brittle fracture modeling;
- Improved knowledge of in-situ stresses in rock slopes in order to constrain brittle fracture modeling;
- More consideration of the stress-path followed in the excavation/formation and evolution of rock slopes;
- Better understanding of transition from the slope failure initiation to the runout stage using energy based brittle fracture models;
- Improved integration of rock slope modeling- characterization and monitoring; and
- More rigorous constraint of simulated model displacements and brittle fracturing against slope monitoring using microseismic systems, ground-based Radar/LiDAR and state-of-the-art continuous borehole inclination sensors.

As engineered rock slopes increase in height to over 1km and population growth increasingly extends into mountainous areas with high rock slopes the need to consider brittle fracture mechanisms and rock slope damage becomes more and more important. Few rock slopes, if any fail without the involvement of brittle facture whether it is at the microscale in the fracture and removal of joint asperities or at the macroscale with the growth of persistent discontinuities to form step-paths or to allow daylighting of failure surfaces. Conventional continuum approaches which do not consider fracture may in many cases predict yielding and indicate slope failure. In some cases however, yielding may be predicted but failure not indicated due the lack of kinematic freedom. The inclusion of fracture mechanics in such slopes may however clearly indicate the change in the kinematic freedom of key blocks within a slope due to stress-induced fracture growth and lead to sometimes unexpected failure. As the stresses that induce such changes in kinematics increase with higher slopes there will be an increasing need to incorporate the principles of fracture mechanics in future rock slope investigations.

REFERENCES

Adhikary, D.P. & Dyskin, A.V. (2007) Modelling of progressive and instantaneous failures of foliated rock slopes. *Rock Mech. and Rock Eng.*, 40 (4), 349–362.

Agliardi, F., Crosta, G.B., Meloni, F., Valle, C. & Rivolta, C. (2013) Structurally-controlled instability, damage and slope failure in a porphyry rock mass. *Tectonphysics*, 605, 34–47.

Alehossein, H. & Hood, M. (1996) State of the art review of rock models for disc roller cutters. In: *Rock Mechanics*. Ed. Aubertin, Hassani, F. and Mitri, H. Balkema / CRC Press, Taylor & Francis Group. 2&4 Park Square, Milton ParkAbingdon OX14 4RN UK. 693–700.

Alghalandis, Y.F. (2014) *Stochastic modelling of fractures in rock masses*. Ph.D. Thesis, University of Adelaide, Australia.

Alzo'ubi, A.M. (2009) *The effect of tensile strength on the stability of rock slopes*. Ph.D. Thesis, University of Alberta, Canada.

Amadei, B., Lin, C.T., Sture, S. & Jung, J. (1994) Modelling fracturing of rock masses with the DDA method. *Proc. of the 1st North American Rock Mechanics Symposium*. Austin, Balkema. 585–590.

Atkinson, B.K. (1987). *Fracture mechanics of rock*. London: Academic Press. 534pp.

Bachmann, D., Bouissou, S. & Chemenda, A. (2006) Influence of large scale topography on gravitational rock mass movements: New insights from physical modeling. *Geophysical Research Letters*, 33, L21406.

Baczynski, N.R.P. (2008) STEPSIM4 Revised: network analysis methodology for critical paths in rock mass slopes. *Proceedings of the 2008 Southern Hemisphere International Rock Mechanics Symposium*, Perth, 405–418.

Brideau, M.A., Yan, M. & Stead, D. (2009) The role of tectonic damage and brittle rock fracture in the development of large rock slope failures. *Geomorphology*, 103.

Cai, M. & Horii, H. (1993) A constitutive model and FEM analysis of jointed rock masses. *Int. J. Rock Mech. Min. Sci. Geomech. Abstr.*, 30, 351–359.

Camones, L.A.M., Vargas Jr., E.A., de Figueiredo, R.P. & Vellos, R.Q. (2013) Application of the discrete element method for modeling of rock crack propagation and coalescence in the step-path failure mechanism, *Eng. Geology*, 153, 80–94.

Christianson, M.C., Board, M.P. & Rigby, D.B. (2006) UDEC simulation of triaxial testing of lithophysal tuff. *Proc. 41st U.S. Symposium on Rock Mechanics*. Paper 968. 41st U.S. Symposium on Rock Mechanics (USRMS): "50 Years of Rock Mechanics – Landmarks and Future Challenges.", Golden, Colorado, June 17–21, 2006.

Coggan, J.S. & Stead, D. (2005) Numerical modelling of the effects of weak mudstone on tunnel roof behaviour. *Proc. 58th Canadian Geotechnical Conference*. Saskatoon. Canada, Paper GS502.

Couture, R., Evans, S.G., Locat, J., Hadjigeorgiou J. & Antoine, P. (1999) A proposed methodology for rock avalanche analysis. *Slope Stability Engineering: Proceedings of the International Symposium*, IS-Shikoku '99. Hardcover. Ed. J.C. Jiang, Yagi Norio, and T. Yamagami. 1369–1378.

Cundall, P.A. (2008) Recent advances in numerical modelling for large-scale mining projects. *ACG News*, 30, (June) 1–7.

Cundall, P.A. & Damjanac, B. (2009) A comprehensive 3D model for rock slopes based on micromechanics. *Proc. Slope Stability Conference*, Santiago, Chile.

De Bremaecker, J.C. & Ferris, M.C. (2004) Numerical models of shear fracture propagation. *Engineering Fracture Mechanics*, 71, 2161–2178.

Diederichs, M.S. (1999) *Instability of hard rock masses: The role of tensile damage and relaxation*. Ph.D. thesis, University of Waterloo.

Diederichs, M.S. (2003) Rock fracture and collapse under low confinement conditions. *Rock Mech. and Rock Eng.*, 36, 339–381.

Duran, F.I., Diederichs, M.S. and Hutchinson, D.J. (2014) A Numerical Analysis of Stress Path and Rock Mass Damage in Open Pit Rock Slopes. *Proc. of the American Rock Mechanics Association conference*. June, Minneapolis, paper 14-7358.

Eberhardt, E. (1998) *Brittle rock fracture and progressive damage in uniaxial compression*. Ph. D. Thesis, University of Saskatchewan.

Eberhardt, E., Stead, D. & Coggan, J.S. (2004) Numerical analysis of initiation and progressive failure in natural rock slopes-the 1991 Randa rockslide. *Int. J. Rock Mech. Min. Sci.*, 41, 69–87.

Eberhardt, E., Stead, D., Stimpson, B. & Read, R.S. (1998) Identifying crack initiation and propagation thresholds in brittle rock. *Canadian Geotechnical Journal*, 35(2), 222–233.

Elmo, D. (2012). FDEM & DFN modelling and applications to rock engineering problems. *MIR 2012 – XIV Ciclo di Conferenze di Meccanica e Ingegneria delle Rocce – Nuovi metodi di indagine, monitoraggio e modellazione degli ammassi rocciosi*. Faculty of Engineering, Turin University, Italy.

Elmo, D., Rogers, S., Stead, D. & Eberhardt, E. (2014) A Discrete fracture network approach to characterise rock mass fragmentation and implications for geomechanical upscaling. *Transactions of the Australian Inst. Mining Metallurgy*, 123(3), 149–161.

Elmo, D., Schlotfeldt, P., Beddoes, R. & Roberts, D. (2011) Numerical simulations of scale effects under varying loading conditions for naturally fractured rock masses and implications for rock for rock mass strength characterization and the design of overhanging rock slopes. *Proc. 45th US Rock Mechanics Symposium*, San Francisco, CA.

Elmo, D. & Stead, D. (2010) An integrated numerical modelling – discrete fracture network approach applied to the characterisation of rock mass strength of naturally fractured pillars. *Rock Mech. and Rock Eng.*, 43(1), 3–19.

Elmo, D., Vyazmensky, A., Stead, D. & Rance, J. (2008) Numerical analysis of pit wall deformation induced by block-caving mining: A combined FEM/DEM-DFN synthetic rock mass approach. *Proc. of the 5th Conference and Exhibition on Mass Mining*, Lulea, Sweden.

Elmo, D., Yan, M., Stead, D. & Rogers, S. (2007) The importance of intact rock bridges in the stability of high rock slopes: Towards a quantitative investigation using an integrated numerical modelling – discrete fracture network approach. *Proc. Int. Symp. on Rock Slope Stability in Open Pit Mining and Civil Engineering*, Perth, Australia.

Elmouttie, M.K. & Poropat, G.V. (2011) Uncertainty propagation in structural modeling. *Proc. Int. Symp. on Rock Slope Stability in Open Pit Mining and Civil Engineering*. Vancouver, Canada.

Evans, S.G. & Clague, J.J. (1994) Recent climatic change and catastrophic geomorphic processes in mountain environments. *Geomorphology*, 10, 107–128.

Feng, X-T. & Hudson, J. (2015) *Rock engineering risk*. ISRM Book Series # 1., CRC Press.

Gao, F. (2013) *Investigation on the failure mechanisms of underground coal mine roadways through discrete element modelling*. Ph.D. thesis, Simon Fraser University.

Gao, F. & Stead, D. (2014) The application of a modified Voronoi logic to brittle fracture modelling at the laboratory and field scale. *Int. J. Rock Mech. & Min. Sci.*, 68, 1–14.

Geomechanica (2015) *IRAZU FEMDEM software*. Toronto, Canada, http://www.geomechanica .com.

Groneng, G., Lu, M., Nilsen, B. & Jenssen, A.K. (2010) Modelling of time dependent behaviour of the basal sliding surface of the Aknes rock slide area in western Norway. *Eng. Geology*, 114, 414–422.

Hajiabdolmajid, V. & Kaiser, P.K. (2002) Slope stability assessment in strain-sensitive rocks. *EUROCK 2002, Proceedings of the ISRM International Symposium on Rock Engineering for Mountainous Regions, Funchal, Madeira*. Sociedade Portuguesa de Geotecnica, Lisboa, 237–244.

Hajiabdolmajid, V.R. (2001) *Mobilization of strength in brittle failure of rock*. Ph.D. Thesis, Queen's University.

Hamdi, P., Stead, D. & Elmo, D. (2013) Numerical simulation of damage during laboratory testing on rock using a 3D-FEM/DEM approach. *Proc. 47th U.S. Rock Mechanics/ Geomechanics Symposium*, San Francisco.

Harrison, J.P. (2012). Rock Engineering, uncertainty and Eurocode 7: implications for rock mass characterisation. *MIR 2012 – XIV Ciclo di Conferenze di Meccanica e Ingegneria delle Rocce – Nuovi metodi di indagine, monitoraggio e modellazione degli ammassi rocciosi*, Faculty of Engineering, Turin University, Italy.

Harrison, J.P. & Hudson, J.A. (2000) *Engineering rock mechanics. Part 2: Illustrative workable examples*. In: Sarkka, P., Eloranta, P. Eds. Oxford: Pergamon.

Havaej, M. (2015) *Characterisation of high rock slopes using an integrated numerical modelling – remote sensing approach*. PhD Thesis, Simon Fraser University, Burnaby, Canada.

Havaej, M., Stead, D., Eberhardt, E. & Fisher, B.R. (2014) Characterization of bi-planar and ploughing failure mechanisms in footwall slopes using numerical modelling. *Eng. Geology*, 178, 109–120.

Havaej, M., Stead, D., Lorig, L. & Vivas, J. (2012) Modelling rock bridge failure and brittle fracturing in large open pit rock slopes. *Proc. 46th US Rock Mechanics/Geomechanics Symposium, Chicago*.

Havaej, M., Wolter, A. & Stead, D. (2015) Exploring the potential role of brittle rock fracture in the 1963 Vajont Slide, Italy, *Int. J. Rock Mech. & Min. Sci.*, 78, September 2015, 319–330.

Hoek, E.T. (1983) Strength of jointed rock masses. Rankine Lecture. *Geotechnique*, 33, 187–223.

Hoek, E.T., Grabinsky, M.W. & Diederichs, M.S. (1990) Numerical modelling for underground excavation design. *Trans. Instn Min. Metall. Sect. A: Min. industry*, 100, A22–A30.

Hoek, E.T., Kaiser, P.K. & Bawden, W.F. (1995) *Support of underground excavations in hard rock*. A.A. Balkena Rotterdam.

Hungr, O. & Amman, F. (2011) Limit equilibrium of asymmetric laterally constrained rockslides. *Int. J. Rock Mech. & Min. Sci.*, 48, 748–758.

International Society of Rock Mechanics. (1981) *ISRM suggested methods for rock characterization testing and monitoring*. In: Brown, E.T. Ed., Imprint. Oxford: Pergamon. 211pp.

Irwin G.R. & de Wit R. (1983) A summary of fracture mechanics concepts. *J. Testing and Evaluation*, 11, 56–65.

Itasca (2015) Itasca Consulting Group inc. UDEC, 3DEC, PFC2D, PFC3D and Slope Model.

Jennings, J.E. (1970) A mathematical theory for the calculation of the stability of slopes in open cast mines. *Proc. Conference on Planning of Open Pit Mines*. Johannesburg, 87–102.

Jiang, M. Jiang, T., Crosta, G.B., Shi, Z., Chen, H. & Zhang, N. (2015) Modeling failure of jointed rock slope with two main joint sets using a novel DEM bond contact model. *Eng. Geology*, 193, 79–96.

Jing, L. (1998) Formulation of discontinuous deformation analysis (DDA) - an implicit discrete element model for block systems. *Eng. Geology*, 49, 371–381.

Kalenchuk, K.S., Diederichs, M.S. & McKinnon, S. (2006) Characterizing block geometry in jointed rock masses. *Int. J. Rock. Mech. & Min. Sci.*, 43, 1212–1225.

Ke, T.C. & Goodman, R.E. (1994) Discontinuous deformation analysis and the artificial joint concept. *Proc. 1st NARMS conference*. Austin, 599–606.

Kemeny, J. (2003) The time-dependent reduction of sliding cohesion due to rock bridges along discontinuities: A fracture mechanics approach. *Rock Mech. & Rock Eng.*, 36(1), 27–38.

Kim, B.H., Cai, M., Kaiser, P.K. & Yang, H.S. (2007) Estimation of block sizes for rock masses with non-persistent joints. *Rock Mech. & Rock Eng.*, 40, 145–168.

Kim, D.S. & McCarter, M.K. (1998) Quantitative assessment of extrinsic damage in rock materials. *Rock Mech. & Rock Eng.*, 31(1), 43–62.

Kim, J.S., Lee, K.S., Cho, W.J., Choi, H.J. & Cho, G.C. (2015). A comparative evaluation of stress–strain and acoustic emission methods for quantitative damage assessments of brittle rock. *Rock Mech. & Rock Eng.*, 48, 495–508.

Klerck, P.A., Sellers, E.J. & Owen, D.R.J. (2004) Discrete fracture in quasi-brittle materials under compressive and tensile stress states. *Comput. Methods Appl. Mech. Engrg.*, 193, 3035–3056.

Leith, K.J. (2012) *Stress development and geomechanical controls on the geomorphic evolution of alpine valleys*. Ph.D. dissertation, ETH Zürich, Switzerland.

Liang, Z., Xu, N., Tang, S. & Chunan, T. (2013) Microseismic Monitoring and Numerical Simulation of Rock Slope Failure. *International Journal of Distributed Sensor Networks*, Article 845191.

Lim, S.S., Martin, C.D. & Åkesson, U. (2012) In-situ stress and microcracking in granite cores with depth. *Eng. Geology*, 1–13. doi:10.1016/j.enggeo.2012.07.006.

Lisjak, A. & Grasselli, G. (2014) A review of discrete modeling techniques for fracturing processes in discontinuous rock masses. *J. Rock Mech. & Geotechn. Eng.*, 6, 301–314.

Liu, H.W. (1983) On the fundamental basis of fracture mechanics. *Eng. Fracture Mech.*, 17, 425–438.

Liu, H.W. (2003) *Numerical modelling of the fracture process under mechanical loading*. Ph.D. Thesis. Dept. of Civil and Mining Engineering. Lulea University of Technology.

Martin, C.D. (1994) *The strength of massive Lac du Bonnet granite around underground openings*. Ph.D. Thesis, University of Manitoba.

Mas Ivars, D., Pierce, M., Darcel, C., Reyes-Montes, J., Potyondy, D., Young, R.P. & Cundall, P.A. (2011) The synthetic rock mass approach for jointed rock mass modelling. *Int. J. Rock Mech. & Min. Sci.*, 48, 219–244.

Mauldon, M. (1998) Estimating mean fracture trace length and density from observations in convex windows. *Rock Mech. & Rock Eng.*, 31, 201–216.

McColl, S.T., Davies, T.R.H. & McSaveney, M.J. (2012) The effect of glaciation on the intensity of seismic ground motion. *Earth Surf. Processes Landforms*, 37, 1290–1301.

McMahon, B.K. (1979). *Report to Bougainville Copper Limited on Slope Design Studies*. Pan Hill. McMahon, Burgess and Yeates, Sydney.

Morgan, S.P. & Einstein, H.H. (2014) The effect of bedding plane orientation on crack propagation and coalescence in shale. *Proc. 48th US Rock Mechanics / Geomechanics Symposium held in Minneapolis*, MN, USA.

Muller, L. & Hofmann, H. (1970) Selection, compilation and assessment of geological data for the slope problem. *Symposium on the Theoretical Background to the Planning of Open Pit Mines with Special Reference to Slope Stability, Johannesburg, South Africa*, Van Rensburg, P. W. J (Editor), 153–170.

Munjiza, A. 2004. *The combined finite-discrete element method*. Chichester, UK, John Wiley & Sons Ltd.

Owen, D.R.J., Feng Y.T., de Souza Neto, E., Cottrell, M., Wang, F., Andrade Pires, F. & Yu J. (2004) The modelling of multi-fracturing solids and particulate media. *Int. Jour. Num. Meth. Eng.*, 60, 317–339.

Paronuzzi, P. & Bolla, A. (2015) Gravity-induced rock mass damage related to large en masse rockslides: Evidence from Vajont. *Geomorphology*, 234, 28–53.

Pierce, M., Cundall, P.A. & Potyondy, D. (2007) A synthetic rock mass model for jointed rock. *Proc. 1st Canada-US Rock Mechanics Symposium*, Vancouver, 341–349.

Potyondy, D.O. & Cundall, P.A. (2004) A bonded-particle model for rock. *Int. J. Rock Mech. & Min. Sci.*, 41, 1329–1364.

Read, J.R.L. & Lye, G.N. (1984) Pit slope design methods: Bougainville Copper open cut. *Proceedings of the 5th International Congress on Rock Mechanics*, Melbourne.

Rockfield (2015) Rockfield Software Ltd. Technium, Kings Road, Prince of Wales Dock, Swansea, SA1 8PH, UK. http://www.rockfield.co.uk/elfen.htm.

Rose, N.D. & Hungr, O. (2006) Forecasting potential rock slope failure in open pit mines using the inverse-velocity method. *Int. J. Rock Mech. & Min. Sci.*, 44, 308–320.

Sainsbury, B., Pierce, M. & Mas Ivars, D. (2008) Analysis of Caving Behaviour Using a Synthetic Rock Mass-Ubiquitous Joint Rock Mass Modelling Technique. *Proc. Of the First Int. FLAC/ DEM Symposium on Numerical Modelling*, Minneapolis, U.S. R. Hart *et al.*, Eds.

Sainsbury, D. & Sainsbury, B. (2013) Three-Dimensional Analysis of Pit Slope Stability in Anisotropic Rock Masses. *Proc. of Slope Stability*, Brisbane, Australia.

Saouma, V.E. & Kleinosky, M. (1984) Finite element simulation of rock cutting: a fracture mechanics approach. *Proc. 25th U.S. Symp. on Rock Mechanics*, 792–799.

Scavia, C. (1990) Fracture mechanics approach to stability analysis of rock slopes. *Eng. Fracture Mech.*, 35 (4/5), 899–910.

Scavia, C. & Castelli, M. (1996) Analysis of the propagation of natural discontinuities in rock bridges. *Eurock' 96*, 445–451.

Sholtès, L. & Donzé, F.V. (2015) A DEM analysis of step-path failure in jointed rock slopes. *C. R.Mecanique*, 343, 155–165.

Singh, R.N. & Sun, G.X. (1989) Fracture mechanics applied to slope stability analysis. *Int. Symp. on Surface Mining – Future Concepts*. University of Nottingham, England, 93–97.

Spreafico, M. (2015) *Gravitational instability in heterogeneous rock slabs in Valmarecchia: long-term evolution and mitigation strategies*. PhD Thesis, University of Bologna, Italy.

Stacey, T.R. (1974) The behaviour of two-and three-dimensional model rock slopes. *Q. J. Engng. Geol.*, 8, 67–72.

Stacey, T.R., Xianbin, Y., Armstrong, R. & Keyter, G.J. (2003) New slope stability considerations for deep open pit mines. *The Journal of the South African Institute of Mining and Metallurgy*, 103, 373–389.

Stead, D. (2015) The influence of shales on slope instability, *Rock Mech. and Rock Eng.*, 49(2), pp. 635–651.

Stead, D. & Coggan, J.S. (2006) Numerical modelling of rock slopes using a total slope failure approach. In: *Landside from Massive Rock Slope Failure*, Ed., Evans, S., Hermans, R., and Strom, A. Springer, Dordrecht, Netherlands. 131–142.

Stead, D., Coggan, J.S. & Eberhardt, E. (2004) Realistic simulation of rock slope failure mechanisms: the need to incorporate principles of fracture mechanics. *Int. J. Rock Mech. & Min. Sci.*, 41.

Stead, D., Coggan, J.S., Elmo, D. & Yan, M. (2007) Modelling brittle fracture in rock slopes: Experience gained and lessons learned. Australian Centre for Geomechanics. *Int. Symp. Rock Slope Stability in Open Pit and Civil Engineering, Slope Stability07*, Perth, Western Australia, September 12–14 2007, 239–252.

Stead, D. & Eberhardt, E. (2013) Understanding the mechanics of large landslides. Invited keynote and paper, *International Conference Vajont 1963–2013. Italian Journal of Engineering Geology and Environment*, 6, 85–112.

Stead, D., Eberhardt, E. & Coggan, J.S. (2006) Developments in the characterisation of complex rock slope deformation and failure using numerical modelling techniques. *Eng. Geology*. 83, 217–235.

Stead, D., Szczepanik, Z. & Gaskin, W. (1996) Acoustic characterisation of potash. *Proc. 4th Conference on the Mechanical Behavior of Salt. June 1996, Ecole Polytechnique Montreal*, 16–21.

Strouth, A. & Eberhardt, E. (2009) Integrated back and forward analysis of rock slope stability and rockslide runout at Afternoon Creek, Washington. *Canadian Geotechnical Journal*, 46, 1116–1132.

Sturzenegger, M., Stead, D. & Elmo, D. (2011) Terrestrial remote sensing-based estimation of discontinuity frequency, mean trace length, block size/shape and dependence on observation scale. *Eng. Geology*, 119(3–4), 96–111.

Swenson D.V. & Ingraffea A.R. (1988) Modelling mixed-mode dynamic crack propagation using finite elements: theory and application. *Computational Mechanics*, 3, 187–192.

Tang, C.A., Lin, P., Wong, R.H.C. & Chau, K.T. (2001) Analysis of crack coalescence in rock-like materials containing three flaws – Part II: numerical approach. *Int. J. Rock Mech. & Min. Sci.* 38, 925–939.

Tang, C.A., Xu, T., Yang, T.H. & Liang, Z.Z. (2004) Numerical investigation of mechanical behaviour of rock under confining pressure and pore pressure. *Proc. of SINOROCK2004 symposium*. Three Gorges Dam Project, Yangtze River, China.

Tang, H. (1998) A study on rock slope stability by the method of damage mechanics. *Proc. 8th IAEG, Conference, Vancouver*, 1293–1298.

Tan, X.C., Kou, S.Q., & Lindqvist, P.A. (1996) Simulation of crack propagation by indenters using DDM and fracture mechanics. *Proc. of 2nd North American Rock Mechanics conference*. Montreal, Canada, 685–692.

Tharp, T.M. & Coffin, D.F. (1985) Field application of fracture mechanics analysis to small rock slopes. *Proc. 26th U.S. Symp. on Rock Mech.*, 667–674.

Tuckey, Z. (2012) *An integrated field mapping-numerical modelling approach to characterising discontinuity persistence and intact rock bridges in large open pit slopes*, M.Sc. Thesis, Simon Fraser University.

Uzielli, M. (2008) *Statistical data of geotechnical data. Geotechnical and Geophysical Site Characterisation*. London. Taylor and Francis. 173–194.

Van Sint Jan, M.L. & Prudencio, M.G. (2008) The influence of non-persistent joints on the failure modes of large rock slopes. *SHIRMS 2008*, Y. Potvin, J. Carter, A. Dyskin, R. Jeffrey (eds), 517–527. 2008 Australian Centre for Geomechanics, Perth.

Vivas, J. (2014) *Groundwater characterization and modelling in natural and open pit rock slopes*. MASc Thesis, Simon Fraser University, Burnaby, Canada.

Vyazmensky, A., Stead, D., Elmo, D. & Moss, A. (2010) Numerical analysis of block caving induced instability in large open pit slopes: A finite element / discrete element approach, *Rock Mech. & Rock Eng.*, 43(1), 21–39.

Wang, C., Tannant, D.D. & Lilly, P.A. (2003) Numerical analysis of the stability of heavily jointed rock slopes using PFC2D, *Int. J. Rock Mech. & Min. Sci.*, 40, 415–424.

Wawrzynek, P. & Ingraffea, A.R. (1989) An interactive approach to local remeshing around a propagation crack. *Finite Element in Analysis & Design*, 5, 87–96.

Wesseloo, J. & Dight, P. (2009) Rock mass damage in hard rock open pit mine slopes. *Proceedings Slope Stability 2009*, Santiago.

Whittaker, B.N., Singh, R. & Sun, G. (1992) *Rock fracture mechanics: principles, design and applications*. Amsterdam. Elsevier. 592pp.

Whittall, J., Eberhardt, E., Hungr, O. & Stead, D. (2015) Runout of open pit slope failures: using and abusing the fahrböshung method. *Proceedings Slope Stability 2015*, Cape Town.

Wolter, A. (2014) *Characterisation of Large Catastrophic Landslides Using an Integrated Field, Remote Sensing and Numerical Modelling Approach*. Ph.D. dissertation, Simon Fraser University, Burnaby, Canada.

Xu, N.W., Dai, F., Liang, Z., Z., Zhou, Z., Sha, C. & Tang, C.A. (2014) The dynamic evaluation of rock Slope stability considering the effects of microseismic damage. *Rock Mech. & Rock Eng.*, 47, 621–642.

Xu, N.W., Tang, C.A., Li, L.C., Zhou, Z., Sha, C., Liang, Z.Z. & Yang, J.Y. (2011) Microseismic monitoring and stability analysis of the left bank slope in Jinping first stage hydropower station in southwestern China. *Int. J. Rock Mech. & Min. Sci.*, 48(6), 950–963.

Yan, M., Stead, D. & Sturzenegger, M. (2007) Step-path characterization in rock slopes: An integrated numerical modeling-digital imaging approach. *11th Congress of the International Society for Rock Mechanics. Lisbon, Portugal 9–13 July 2007*, 693–696.

Zhang, L. & Einstein, H.H. (2000) Estimating the intensity of rock discontinuities. *Int. J. Rock Mech. & Min. Sci.*, 37, 819–837.

Zhang, Y. & Stead, D. (2014) Modelling 3D crack propagation in hard rock pillars using a Synthetic Rock Mass approach. *Int. J. Rock Mech. & Min. Sci.*, 72, 199–213.

Chapter 24

Development and application of numerical limit analysis for geotechnical materials

Zheng Yingren
Department of Civil Engineering, Logistical Engineering University, Chongqing, China

Abstract: Solid material develops from elastic to plastic then to failure after undertaking some load, which means that yield is different from failure. The system of yield criteria for geotechnical material is discussed and two definitions for point failure and surface failure in stress field. Also, the criterion of intact surface-failure is defined through classic limit analysis method. Simultaneously, the deficiencies for classic limit analysis and numerical analysis are pointed out, and based on the combination of these two methods, a newly developed numerical limit analysis method is built, which has enlarged the application range of limit analysis. The significances and features of classic limit analysis method and numerical limit analysis method are studied, as well as their reliabilities. The wide applicabilities of numerical limit analysis in slope (landslide) engineering, tunnel engineering and foundation engineering are illustrated.

I INTRODUCTION

The limit analysis method of geotechnical materials (Hill, 1950; Zheng, 2010) has been widely applied in geotechnical engineering as a design basis for over a century. Classic geotechnical stability problems including slope stability, soil pressure and bearing capacity of the foundation, are based on the limit analysis theory. The limit analysis method of geotechnical material originates from Coulomb's law in 1773, then the limit equilibrium method (in the 1920s), slip line field method (method of characteristic line) (in the 1940s), the upper and lower bound methods of limit analysis (in the 1950s) were established. After a century of development, the limit analysis method has 9 perfect and has been well applied in the design of geotechnical engineering, especially in the stability of rock and soil. But this method is not valid for complex engineering. With the development of numerical methods in geotechnical mechanics, numerical limit method is established and widely application. Zienkiewicz *et al.* (1975) put forward the strength reduction finite element method (strength reduction FEM) and the step-loading finite element method (step-loading FEM). Those methods that made it possible to obtain the safety factor of stability and ultimate load of the geotechnical material have been widely used in slope and landslide stability analyses (Matsui, 1992; Griffiths & Lane, 1999, 2000; Dawson, 1999; Zhao & Zheng *et al.*, 2002).

According to the mechanics means, the strength reduction FEM and the step-loading FEM are also called the numerical limit method or finite element (including finite element, finite difference, discrete element, etc.) limit analysis method (Zheng et al., 2005, 2007, 2011, 2012). They are the development of the classic limit analysis method and the applications of numerical method for solving the problem of limit analysis. So not only the safety factor and the ultimate load, but also the failure location and form of materials can be calculated.

The numerical limit method not only includes the strength reduction FEM and the step-loading FEM, but also can be extended to various failure conditions, for example the permafrost slope instability due to temperature increasing, the reservoir bank slope instability due to water level decline rate increasing, etc. What's more, the numerical limit method is available to all kinds of solid materials, as long as its force process develops from elastic to plastic and then to the failure under the external force.

2 THE FAILURE PROCESS OF MATERIALS

With the stress increasing, the solid materials firstly enter the elastic state. Then some parts of materials reach the elastic limitation, which means these parts have entered the plastic state. Finally, with plastic developing into the plastic limit, the materials reach the failure state. At the beginning, only a few points have reached the yield state. However, due to the inhibitory effect of the surrounding material, failure does not occur. So the yield is not equal to the failure, it means that the materials have entered the plastic state and have been damaged. When the plastic develops into a certain extent, local cracks will appear in the stress concentration areas. This state is called the local failure or point failure of materials.

After continual loading, the local cracks will be cut through. When the failure surface occurs, the intact surface-failure takes place in the materials. The classic limit analysis methods have provided the intact surface-failure condition of the materials and have been applied in engineering. Therefore, the failure criterion can be obtained based on the principle of limit analysis. So the ultimate load and stability safety factor can be calculated by using the two criteria (yield criterion and failure criterion) and the equilibrium equation.

2.1 Yield criterion

Yield means that materials reach the elastic limit and enter the plastic state. The force process of materials changes from the initial yield, the subsequent yield to the plastic limit. Material developing from elastic to plastic is determined by the yield criterion, which is built on certain theory, experimental phenomena or experience. There are a lot of yield criteria including Tresca and Mises criteria for metal material, Mohr-Coulomb criterion (M-C criterion) and Drucker-Prager criterion (D-P criterion) for geotechnical material. All the criteria above are based on the elasticity theory. The M-C criterion does not consider the effect of intermediate principal stress, so yield criteria with triple shear planes have appeared, which are obtained by the true triaxial tests, such as Matsuoka criterion, Lade criterion etc.

Recently, based on the energy theory and three shear stress vectors, Gao et al. derived energy yield criterion with triple shear planes, called triple shear energy criterion (Gao & Zheng, 2007), that is the common yield criterion of both geotechnical material and metal material, and a variety of well-known criteria are its special case (Zheng & Gao, 2008, 2011). When not considering internal friction angle, it is Mises criterion. If we consider the angle of internal friction without considering the intermediate principal stress, it is simplified to a single shear state and is the Mohr-Coulomb criterion; if it is assumed that the Lode angle is constant, it is the Drucker-Prager criterion. So the yield criterion system of geotechnical and metal materials can be listed (see Table 1).

If the ultimate strength curve of material is linear, all criteria above can be derived strictly from theories. The experiment result shows that the ultimate strength curve of soil is generally linear, while ultimate strength curve of rock is nonlinear. In the current project, rock's curve is treated as straight. To gain the reasonable value of cohesion and internal friction angle, the limit curve of rock should be regarded as piecewise linear.

2.2 Failure criterion

2.2.1 Definition of failure

Generally, the definition of failure in plastic mechanics is that materials enter the infinite plastic state. It means that stress is constant while strain is indefinitely increasing. According to plastic mechanics, materials changes from the initial yield into the subsequent yield, and then reach failure. The subsequent yield is related to history parameters and stress path, while the initial yield and failure state are independent of historical parameters and stress path. Therefore, researches of the initial yield and the failure state of material can be carried out in an ideal plastic state, which is independent of history parameters.

As is well-known, the initial yield state and the failure state are same in an ideal plastic state. They are of the same stress, but different strains. Strain of the former corresponds to the strain of material just entering the plastic state, while the latter corresponds to plastic limit strains of the material from the plastic state into failure. The definition of failure in classic plastic mechanics is not applicable to ideal plastic material, because it makes yield and failure the same. So, failure should be defined as that material strain reaches the plastic limit strains, namely ultimate strain of materials, This definition is applicable to hardening materials, softening materials and ideal plastic materials.

2.2.2 Shear failure criteria of geotechnical materials

Geotechnical materials failure including tension failure and shear failure, is highly complex. This paper conducts exploratory research only on shear failure, which can be divided into point-failure and intact surface-failure. There is no mature theory for point failure of geotechnical materials. Some scholars judge point failure by the ultimate strain of material. The typical stress-strain curves of material are shown in Figure 1. In the figure, ε_y is the initial yield strain of material, ε_f is the plastic limit strain (ultimate strain) of the material (Zheng, 2012). Therefore, for defining the criterion of point-failure, the ultimate strain need be found. Intact surface-failure can be judged by the classic limit analysis method. Therefore, the classic limit analysis method has essentially provided the intact surface-failure criterion.

Table 1 System of yield criteria expressed by stress.

Shear State	Simple Shear		Tri-shear	
	Yield Criterion	Formula	Yield Criterion	Formula
Metal Material	Tresca	$\sigma_1 - \sigma_3 = k$	Mises	$J_2 = C$
Geotechnical Material	Mohr-Coulomb	$p\sin\varphi + \dfrac{q}{3}\left(\sqrt{3}\cos\theta_\sigma - \sin\theta_\sigma\sin\varphi\right) = c\cos\varphi$	Gao-Zheng	$p\sin\varphi + \dfrac{q}{3}\left(\sqrt{3}\cos\theta_\sigma - \sin\theta_\sigma\sin\varphi\right) =$ $2c\cos\varphi\sqrt{\dfrac{1-\sqrt{3}\tan\theta_\sigma\sin\varphi}{3+3\tan^2\theta_\sigma - 4\sqrt{3}\tan\theta_\sigma\sin\varphi}}$
	Drucker-Prager θ_σ is constant	$aI_1 + \sqrt{J_2} - k = 0$	Drucker-rager θ_σ is constant	$a_a I_1 + \sqrt{J_2} - k_a = 0$
	$\theta_\sigma = 30°$ (Triaxial Compression)	$\alpha = \dfrac{2\sin\varphi}{\sqrt{3}(3-\sin\varphi)}$ $k = \dfrac{6c\cos\varphi}{\sqrt{3}(3-\sin\varphi)}$	$\theta_\sigma = 30°$ (Triaxial Compression)	$a_a = \dfrac{2\sin\varphi}{\sqrt{3}(3-\sin\varphi)}$ $k_a = \dfrac{6c\cos\varphi}{\sqrt{3}(3-\sin\varphi)}$
	$\theta_\sigma = -30°$ (Triaxial Tensile)	$\alpha = \dfrac{2\sin\varphi}{\sqrt{3}(3+\sin\varphi)}$ $k = \dfrac{6c\cos\varphi}{\sqrt{3}(3+\sin\varphi)}$	$\theta_\sigma = -30°$ (Triaxial Tensile)	$a_a = \dfrac{2\sin\varphi}{\sqrt{3}(3+\sin\varphi)}$ $k_a = \dfrac{6c\cos\varphi}{\sqrt{3}(3+\sin\varphi)}$
	$\theta_\sigma = 0°$ (Non-associated, Plane Strain)	$\alpha = \sin\varphi$ $k = c\cos\varphi$	$\theta_\sigma = 0°$ (Non-associated, Plane Strain)	$a_a = \sin\varphi$ $k_a = \dfrac{2}{\sqrt{3}}c\cos\varphi$

Note: In the Table, σ_1 and σ_3 are the first and the third principal stress. a, κ, a_a, κ_a are coefficient. I_1 and J_2 are the first invariant of stress tensor and the second invariant of deviatoric stress tensor. θ_σ is Lode Angle.

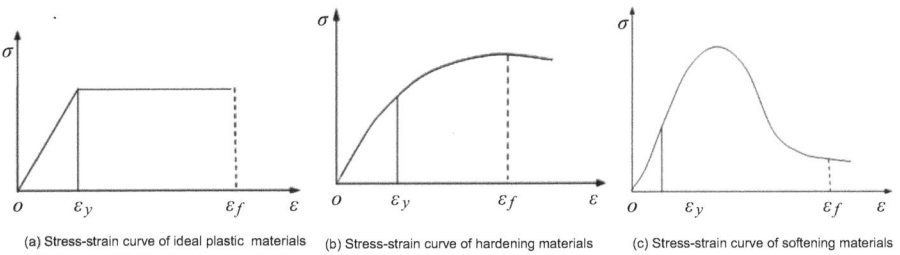

Figure I Stress-strain relation curves of typical materials.

Whether the material reach the limit state is conditional. 1) For reaching the limit state, material must produce a certain displacement, geotechnical strength can be fully played at this time. So the requirement that geotechnical body should have sufficient displacement to reach the limit state is the first condition of intact surface-failure. 2) From the theoretical analysis and engineering practice, geotechnical body is a field failure, which is surface failure in three-dimensional calculation and line failure in two-dimensional calculation. There is a series of yield surfaces in the geotechnical body while just one of the yield surfaces produces the failure firstly. Therefore, the geotechnical failure surface does not mean all of the yielded surfaces, but refers to the most dangerous one (that is the surface in which the sliding force and sliding resistance force are closest, the safety factor is smallest). Thus, yield surface cut-through is the necessary condition to limit state. 3) The connected yield surfaces in geotechnical body can be multiple while only one the failure surface exists. In ideal plastic case, failure occurs when the sliding force produced by the external force is equal to the sliding resistance force of material on the failure surface. Therefore the balance between the sliding force and the sliding resistance force is the third condition and also the most important condition. Forces on the sliding surface can be calculated by the limit equilibrium method or slip line field method. When using the method of upper limit analysis, failure occurs when the external work done by external forces and the internal energy dissipation power satisfies the equation of virtual work. This shows that the third condition of intact surface failure can be described as the external work on failure surface equals to the internal energy dissipation power.

In summary, Zheng states that the condition of the intact surface-failure of geotechnical body can be described as: at the ultimate state and the ideal plastic case, the force on the failure surface meets Equation 1 (Zheng, 2011, 2012):

$$F' = Q \tag{1}$$

Where F' is the external force produced sliding force on the failure surface, Q is the sliding resistance force produced by the strength of material and the external force.

$$W = D \tag{2}$$

Where W is the external power of geotechnical body, D is the internal energy dissipation power along the discontinuous surface. When calculating, pay attention that the direction of displacement speed shall satisfy the requirement of the constitutive relation. Equation 1 and 2 can be regarded as the criteria of intact surface-failure of geotechnical materials.

3 THE FINITE ELEMENT LIMIT ANALYSIS METHOD

3.1 The principal of finite element limit analysis method

In 1975, Zienkiewicz presented an approach that uses the numerical method to solve the limit problem directly, and the step-loading FEM and the strength reduction FEM appeared. These methods are called the numerical limit analysis method or the finite element limit analysis method. Different from the classic limit analysis method, the numerical limit analysis method uses the numerical method with an ideal elastic-plastic model. In the solving process, it is made to reach the failure state eventually in the numerical calculation by constantly reducing the strength of materials (reduce the cohesion c and friction coefficient $\tan\varphi$ of soil by the same proportion) or increase load. The strength reduction factor corresponding to failure state is just the safety factor, while the load at failure is just the ultimate load. This method does not require presupposition of failure surface. When computing reaches the failure state, software will automatically generate failure surface and destruction information is given out. For the widely used M-C material, the safety factor ω can be expressed as:

$$\tau = (c + \sigma \tan\varphi)/w = c' + \sigma \tan\varphi' \tag{3}$$

$$w = c/c', \quad w = (\tan\varphi)/\tan\varphi' \tag{4}$$

Where c' and φ' are the cohesion and internal friction angle at failure state.

The numerical limit analysis method obtains the ultimate load or safety factor directly while it neither requires to know the failure surface in advance nor requires to calculate sliding force and sliding resistance force on the failure surface. Because the method is simple, accurate, widely applicable and practicable, and can be used to determine the location and form of the failure surface too. It has a very broad application prospect.

3.2 Selection of yield criteria

The yield criterion is very important in the finite element limit analysis method because it has great effects on the computational results. The Mohr-Coulomb yield criterion is often adopted in practical geotechnical engineering. The Mohr-Coulomb yield surface on the π-plane is an irregular hexagon (shown in Figure 2) that lead to a difficulty in numerical calculations. Therefore, it is necessary to modify the Mohr-Coulomb yield surface approximately or to adopt the generalized Mises yield criterion that is related to the Mohr-Coulomb yield criterion. The generalized Mises yield criterion based on the Mises yield criterion, and considers the average compressive stress. It can be expressed as follows:

$$\alpha I_1 + \sqrt{J_2} = k \tag{5}$$

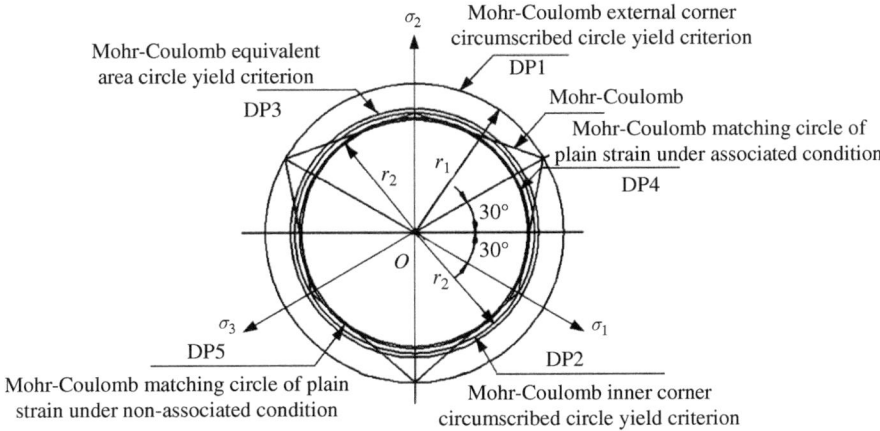

Figure 2 The yield surface on the π-plane.

Where I_1 is the first invariant of stress tensor, J_2 is the second invariant of the deviator stress tensor, and the material constants a, k are coefficient related to the Coulomb's material constants (c,φ), as described in Table 2 (Zhang *et al.*, 2002, 2003; Deng *et al.*, 2006; Zheng & Kong, 2010).

Generalized Mises yield criteria have five different yield conditions (DP1, DP2, DP3, DP4, DP5) associated with the Mohr-Coulomb yield criterion, which represents different circles on the deviator plane (Figure 2). Equation 5 was prepared by Drucker and Prager in 1952. So, the generalized Mises yield criterion is also known as the Drucker-Prager (D-P) yield criterion. This criterion is visualized as a conical surface in principle stress space. The cross-sectional shape of this cone on the π-plane is a circle (Fig. 2), which readily facilitates numerical computation.

In the generalized Mises yield criterion, DP1, DP2 and DP4 are widely used by international customary software, but the yield criteria DP1 and DP2 have no strict theoretical basis, which lead to considerable calculation error. While DP3, DP4, DP5 can be used in numerical calculation directly. The DP3 criterion is

Table 2 The α, k value of different yield criteria.

Yield criteria	α	k
DP1: Mohr-Coulomb external corner circumscribed circle yield criterion	$\dfrac{2\sin\varphi}{\sqrt{3}(3-\sin\varphi)}$	$\dfrac{6c\cos\varphi}{\sqrt{3}(3-\sin\varphi)}$
DP2: Mohr-Coulomb inner corner circumscribed circle yield criterion	$\dfrac{2\sin\varphi}{\sqrt{3}(3+\sin\varphi)}$	$\dfrac{6c\cos\varphi}{\sqrt{3}(3+\sin\varphi)}$
DP3: Mohr-Coulomb equivalent area circle yield criterion	$\dfrac{2\sqrt{3}\sin\varphi}{\sqrt{2}\sqrt{3}\pi(9-\sin^2\varphi)}$	$\dfrac{6\sqrt{3}c\cos\varphi}{\sqrt{2}\sqrt{3}\pi(9-\sin^2\varphi)}$
DP4: Mohr-Coulomb matching circle of plain strain under associated condition	$\dfrac{\sin\varphi}{\sqrt{3}\sqrt{3+\sin^2\varphi}}$	$\dfrac{3c\cos\varphi}{\sqrt{3}\sqrt{3+\sin^2\varphi}}$
DP5: Mohr-Coulomb matching circle of plain strain under non-associated condition	$\dfrac{\sin\varphi}{3}$	$c\cos\varphi$

appropriate for 3D problem, and the DP4 and DP5 are suitable for plane strain. For the plane strain problem, when using DP4 criterion under associated flow rules or DP5 criterion under non-associated flow rules, high precision can be reached.

3.3 Safety factor transformation in different D-P criteria

In D-P criteria, there are many expressions of α and k (see Table 2). Although the safety factors obtained under different yield conditions are different, the yield conditions can be transformed each other. At present, international customary software only provides three D-P criteria including DP1, DP2 and DP4, while the DP3, DP4 and DP5 are normally used. Thus, the transformation of yield conditions is very necessary. Let suppose c_0, φ_0 as the initial strength parameters, the safety factor of the external corner circumscribed circle yield criterion (DP1) is ω_1, the safety factor of the equivalent area circle yield criterion (DP3) is ω_2. The following transformational equation of the safety factor can be obtained (Zhao & Zheng, 2006).

$$w_2 = \left\{ [3\sqrt{3}(3(cos^2\varphi_0 w_1{}^2 + sin^2\varphi_0)^{1/2} - sin\varphi_0)^2 - 8sin\varphi_0]/18\pi cos^2\varphi_0 \right\}^{1/2} \quad (6)$$

It is the transformational relation of the safety factor between the external corner circumscribed circle yield criterion (DP1) and the equivalent area circle yield criterion (DP3). As long as the safety factor ω_1 of the external corner of the circumscribed circle yield criterion is obtained, the safety factor ω_2 of the equivalent area circle yield criterion can be calculated by the Equation 6. The same can obtain the transformational relation of the safety factor between the external corner circumscribed circle yield criterion (DP1) and the Mohr-Coulomb matching circle of plain strain under non-associated rule (DP5) (Zhao & Zheng, 2006).

$$w_2 = \left\{ [3(cos^2\varphi_0 w_1{}^2 + sin^2\varphi_0)^{1/2} - sin\varphi_0)^2 - 12sin\varphi_0]/12cos^2\varphi_0 \right\}^{1/2} \quad (7)$$

Where ω_1 is the safety factor of the external corner circumscribed circle yield criterion under non-associated rule (DP1); ω_2 is the safety factor of Mohr-Coulomb matching circle of plain strain under non-associated rule (DP5).

3.4 Judged failure criteria in the numerical limit analysis method

However, the numerical limit analysis method requires effective criteria to judge the failure of geotechnical mass in the calculation process. Without the judged failure criterion, even if the geotechnical mass has been failure, the solution cannot be obtained. The key point of the numerical limit analysis method is how to judge whether the geotechnical mass has achieved the failure state according to the result of numerical computation. At present, there are three recognized judged failure criteria under the static state (Zhao & Zheng et al., 2005): 1) Connection of plastic strain in the geotechnical mass. The connectional plastic zone from the inside to the ground or free face is considered regarded as a judged failure criterion. The connection of the plastic zone only means the achievement of the yield state, but not necessarily means the intact failure state of geotechnical mass. It shows that the connection of the plastic zone is just the necessary condition instead of the sufficient condition of failure. 2) In the process of numerical computation, geotechnical mass instability and the misconvergence of numerical computation occur at the same time. Currently, the international customary software generally regards the misconvergence of displacement and force in the process of numerical computation as the judged failure criterion, excepting the misconvergence caused by error of finite element calculation. 3) Failure of geotechnical mass means that the strain and displacement on the failure surface mutate, meanwhile the curve of the reduction factor and displacement mutate. Thus, this can also be used as a judged failure criterion.

3.5 Examples

3.5.1 Example 1: Two-dimensional slope

Two-dimensional slope belongs to a plane strain problem. According to plane strain, the model is built and use large-scale finite element software ANSYS to calculate. This is a homogeneous soil slope, its height is $H=20$ m, the gravity $\gamma =20$ kN/m3, the cohesion $c=42$ kPa, the internal friction angle $\varphi=17°$, and slope angle is $\beta= 30°, 35°, 40°, 45°$ and $50°$ respectively. In plane strain calculation, DP4 criterion under associated flow rules or DP5 criterion under non-associated flow rules is adopted. The safety factor under the associated flow rule is listed in Table 3 and non-associated one is listed in Table 4. The classic limit equilibrium method uses the slope stability analysis software SLOPE/W. The calculation results show that, the safety factors obtained by both the DP4 and DP5 are same as the result obtained by the classic limit equilibrium method (Zheng et al., 2011).

Table 3 Safety factors under associated flow rule.

Calculation Criteria	The Safety Factor				
	30°	35°	40°	45°	50°
DP4(Associated flow rule)	1.56	1.42	1.32	1.22	1.13
SPENCER limit equilibrium method	1.55	1.41	1.30	1.20	1.12
Error	0.01	0.01	0.01	0.02	0.01

Table 4 Safety factors under non-associated flow rule.

Calculation Criteria	The Safety Factor				
	30°	35°	40°	45°	50°
DP5(Non-associated flow rule)	1.56	1.42	1.31	1.21	1.12
SPENCER limit equilibrium method	1.55	1.41	1.30	1.20	1.12
Error	0.01	0.01	0.01	0.01	0.00

Note: The 30° etc. is the slope angle.

3.5.2 Example 2: Weightless foundation

Fig 3 is the Prandtl's failure surface, Fig 4 is the finite element solution and the failure surface of weightless foundation.

According to the theory of slip line field, the foundation ultimate bearing capacity can be approximated as,

$$P_{\mathrm{u}} = cN_{\mathrm{c}} + qN_{\mathrm{q}} + \frac{1}{2}B\gamma N_{\gamma} \tag{8}$$

Where P_u is the ultimate bearing capacity, N_c, N_q and N_y are bearing force coefficients.

Bearing factor coefficients N_c and N_q have the theoretical solution while N_y only have the empirical solution. Table 5 shows that the theoretical solution of N_q is same as the numerical one while theoretical solution of N_y is more accurate than the empirical one.

Table 6 shows the sizes of Prandtl failure surface. Table 7 shows the relevant sizes of failure surface be computed by FEM. Therefore the sizes and failure surface obtained by the two methods are same (Zheng, 2012).

The example 1 and 2 show that the calculation results of the classic analysis method and the numerical analysis method are very close. It proves that the classic analysis method and the numerical analysis method have the similar accuracy, and further indicates the reliability of the limit analysis method.

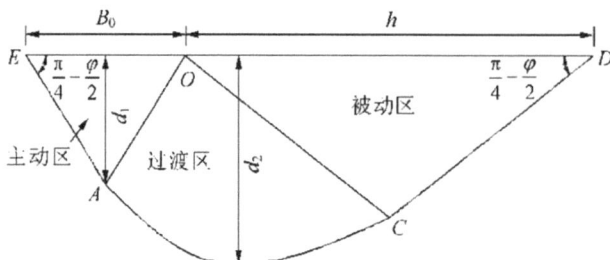

Figure 3 Failure surface of Prandtl.

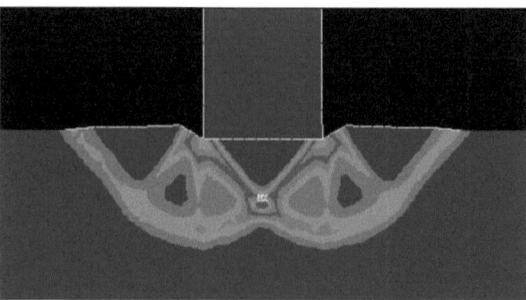

Figure 4 Failure surface of shallow foundation in no weight soils.

Table 5 Comparison of Nc, Nq and Nγ between of FEM solution and other solutions.

Factors	Method	Results						
		0°	5°	10°	15°	20°	25°	30°
N_c	Prandtl	5.14	6.49	8.34	10.98	14.84	20.72	30.14
	FEM (smooth)	5.22	6.60	8.50	11.19	15.18	21.21	31.00
	Reissner	1.00	1.56	2.47	3.93	6.38	10.62	18.32
N_q	FEM (smooth)	1.01	1.60	2.51	4.01	6.63	11.03	18.92
	Terzaghi	0.00	0.09	0.46	1.41	3.52	8.07	17.99
	Meyerhof	0.00	0.07	0.37	1.12	2.86	6.73	15.58
N_γ	Vesic	0.00	0.45	1.22	2.64	5.37	10.83	22.29
	W.F. Chen	0.00	0.46	1.31	2.93	6.17	12.90	27.51
	FEM (smooth)	0.00	0.21	0.83	1.79	3.87	10.55	18.35

Table 6 Sizes of Prandtl failure surface.

Size(m)	Results						
	0°	5°	10°	15°	20°	25°	30°
d1	0.5	0.55	0.6	0.65	0.71	0.79	0.87
d2	0.71	0.79	0.89	1.01	1.16	1.35	1.59
h	1	1.25	1.57	1.99	2.53	3.27	4.29

Table 7 Relevant sizes of failure surface be computed by FEM.

Size (m)	Results						
	0°	5°	10°	15°	20°	25°	30°
d1	0.49	0.53	0.60	0.65	0.70	0.75	0.89
d2	0.70	0.80	0.90	1.05	1.19	1.35	1.62
h	0.98	1.25	1.50	1.92	2.51	3.15	4.20

3.5.3 Example 3: Developing process of landslide

Taking a landslide for an example, the formatting and developing process of sliding surface is researched with the strength of the geotechnical mass reducing. In this example, with the strength reduction factor increasing, the safety factor is reducing corresponding, and yield zone is developing gradually. Fig. 5 shows the shear strain increment nephogram of sliding surface under different safety factor. In the figures, when the safety factor is 1.03, the plastic zone of sliding surface cut through but not failure. When the safety factor is 1.0, the calculation does not converge and landslide becomes instability (Zheng, 2012).

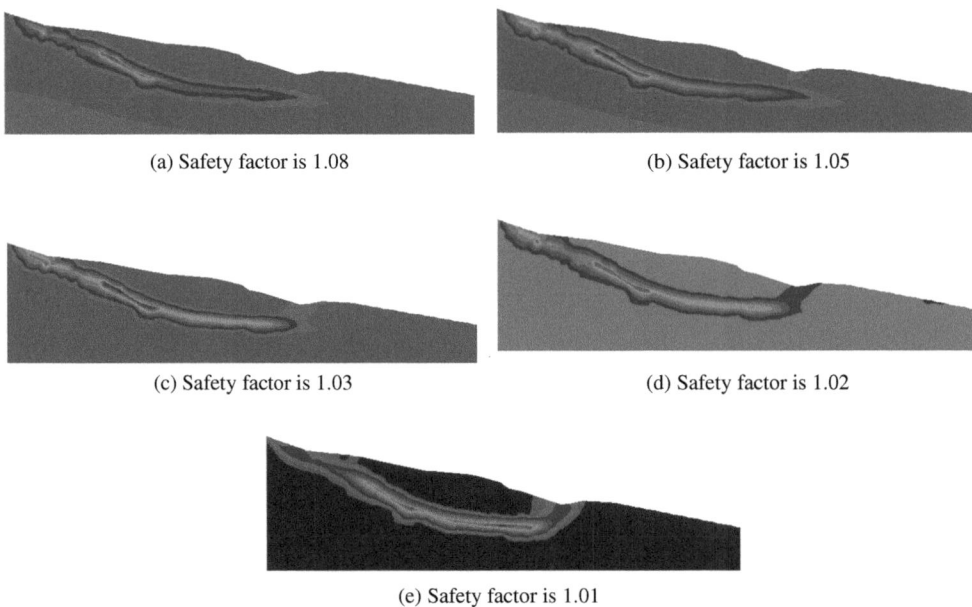

(a) Safety factor is 1.08   (b) Safety factor is 1.05

(c) Safety factor is 1.03   (d) Safety factor is 1.02

(e) Safety factor is 1.01

Figure 5 Development of sliding surface displayed by the shear strain increment nephogram.

4 THE APPLICATION OF THE NUMERICAL LIMIT ANALYSIS METHOD IN GEOTECHNICAL ENGINEERINGS

4.1 Application in slope engineering

4.1.1 Application in soil and rock slopes

Slopes can be classified by the types of materials and the slope structure, The slopes can be classified as soil slope, rock slope and rock-soil composite slope according to the types of materials. They can be classified as similar homogeneous slope, layered slope and cross layer slope of soft and hard rock according to the types of slope structure. Similar homogeneous slopes can be divided into soil slope, cataclastic rock slope and loose rock slope. Layered slopes can be divided into bedding layer slope (bedding slope, buckling slope and bidirectional bedding slope), counter-tilt layered slope (cutting layer slope and toppling slope). Cross layer slope of soft and hard rock can be sorted as upper soft and lower hard (accumulated soil slope and rock-soil composite slope) and upper hard and lower soft (soft rock extrusion slope).

4.1.1.1 Failure model of similar homogeneous slope

Soil slope, cataclastic rock slope and loose rock slope can be seen as similar homogenous slopes. The physical and mechanic parameters of rock and structure plane of slope I are shown in Table 8. Safety factor of this slope computed by strength reduction FEM is 1.016, and the corresponding graph of deformation and failure

Table 8 Material parameters of various slopes.

Materials		ρ/kN·m^{-3}	Shear strength		
			c/kPa	φ/°	Dip angle /°
similar homogeneous					
	Rock-soil	24	145	23	
slope I					
Bedding rock mass slope II	Rock mass	27	1100	39	
(one group structure plane)	Structure plane	25	100	20	30
Bedding rock mass slope III	Rock mass	27	1100	41	
(one group structure plane)	Structure plane	25	100	20	70
Bedding rock mass slope IV	Rock mass	25	1000	38	
(two groups structure planes)	The first group Structure plane	17	120	24	75
	The second group Structure plane	17	120	24	75
Toppling rock slope V	Rock mass	24	1100	41	
	Structure plane	24	100	20	70
Three dimensional wedge rock slope VI	The structure plane of symmetrical wedge	20	20	20	
	The structure plane of asymmetrical wedge	20	50	30	
	Rock mass	26	1×103	45	

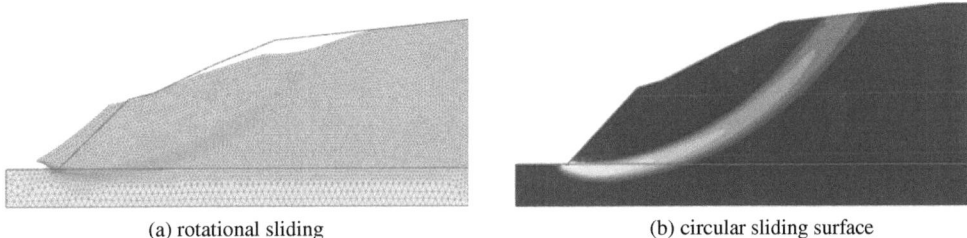

(a) rotational sliding   (b) circular sliding surface

Figure 6 The model of similar homogeneous slope I.

surface is shown in Figure 6. It presents the slope body slides rotationally along the circular sliding surface obviously.

4.1.1.2 Bedding layer slopes

(1) Bedding Rock Slope

In Figure 7(a), there is a group structure plane in the rock slope II. Its penetrating rate is 100%, and the average distance between two structure planes is 8m. The physical and

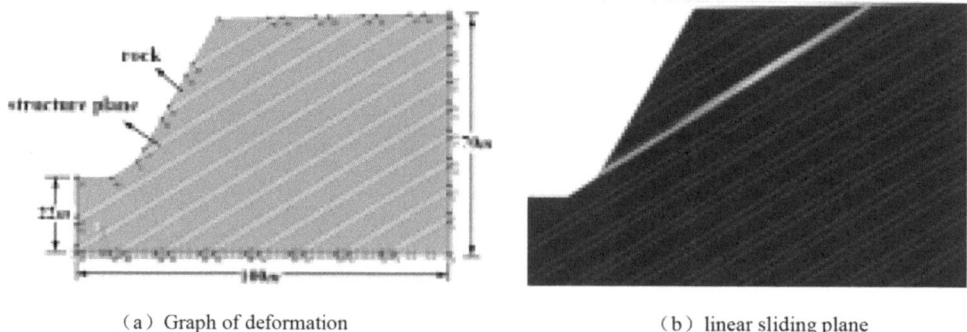

(a) Graph of deformation   (b) linear sliding plane

Figure 7 The bedding rock slope of one group structure plane II.

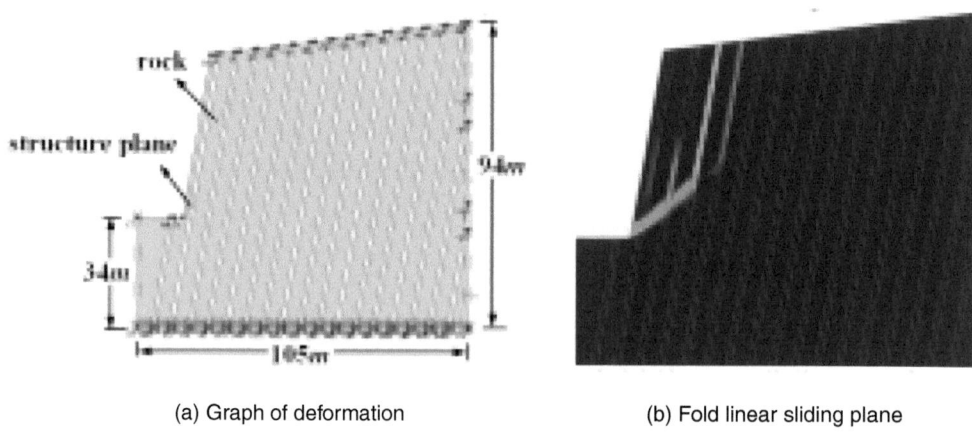

(a) Graph of deformation   (b) Fold linear sliding plane

Figure 8 Buckling rock slope III of one group structure plane.

mechanic parameters of rock and structure plane are shown in Table 8. Safety factor of this slope calculated is 1.011, and corresponding failure surface is shown in Figure 7(b), its failure surface is linear and throughout the whole structure plane.

(2) Buckling rock slope

In Figure 8(a), there is a group structure plane with larger dip angle in the rock slope III, Its penetrating rate is 100%, and the average distance between two structure planes is 5m. The physical and mechanic parameters of rock and structure plane are shown in Table 8. Safety factor of the slope is 1.014, and corresponding failure surface is a fold line, shown in Figure 8(b).

(3) Bidirectional bedding slope

In Figure 9(a), there are two groups different direction structure planes in the rock slope IV, its penetrating rate is 100%, and the average distance between two structure planes is 10m. The dip angle of the first structure plane is 30°, and the second one is 75°. Physical and

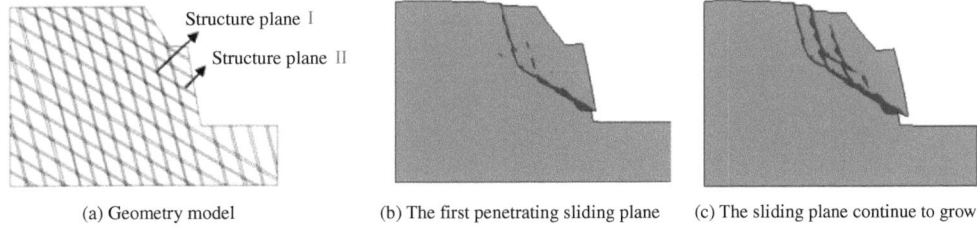

| (a) Geometry model | (b) The first penetrating sliding plane | (c) The sliding plane continue to grow |

Figure 9 The bidirectional bedding slope of two groups structure plane IV.

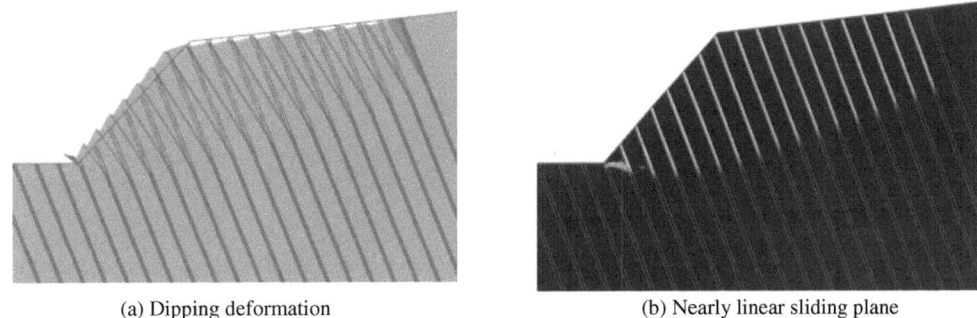

| (a) Dipping deformation | (b) Nearly linear sliding plane |

Figure 10 The toppling rock slope of tilting slope V.

mechanic parameters of the rock and structure plane are shown in Table 8 (Zheng *et al.*, 2011).

Using plane strain simulation, the failure process shown in Figure 9 can be obtained by strength reduction FEM. The main failure surface firstly appears (see fig. 9(b)), then the second and third secondary sliding surfaces appear (see fig .9(c)). The safety factor of the slope is 1.18; when the shape of the main failure surface is known, the safety factor of this slope is 1.17 by the SPENCER method. The two results are very close.

4.1.1.3 Toppling rock slope

When the obliquity of rock formation is larger, toppling failure is easy to appear in counter-tilt layered rock slope. Toppling slope V is shown in Figure 10, in which the counter-tilt layered rock slope generates fold deformation outwards the slope and then results in the toppling instability because of the large obliquity of rock formation.

4.1.1.4 Rock slope with three dimensional wedge body

Wedge body stability in the slope is a typical three dimensional limit equilibrium problems. The wedge body consists of two group structure planes and free surface. Two cases of symmetrical and asymmetrical geometry parameters of rock slope IV are shown in Table 9, and physical and mechanic parameters of the rock and structure plane are shown in Table 8. The soft thickness layer element and the yield criterion of the Mohr-Coulomb

Table 9 Geometric parameters of wedge body.

position	Symmetrical wedge		Asymmetrical wedge	
	Dip direction /°	Dip angle/°	Dip direction /°	Dip angle /°
The left structure plane	115	45	120	40
The right structure plane	245	45	240	60
Top plane	180	10	180	0
Slope plane	180	60	180	60

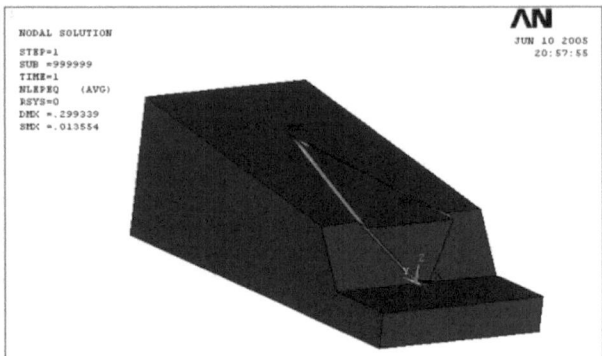

Figure 11 Equivalent plastic strain of symmetry wedge body.

or Mohr-Coulomb equivalent area circle criterion (DP3) is used to simulate the structure planes and calculate the safety factor of slope by strength reduction FEM.

(1) Symmetrical wedge

Three dimensional symmetrical wedge slopes generate sliding failure mainly along the structure plane. Equivalent plastic strain of the symmetrical wedge example is shown in Figure 11. The safety factor of this slope is 1.283 calculated by strength reduction FEM, and 1.293 by the classic method. The calculation error of two safety factors is 1% (Song, 2006; Zheng *et al.*, 2011).

(2) Asymmetrical wedge

Equivalent plastic strain of the asymmetrical wedge example is shown in Figure 12. The safety factor of this slope is 1.60 calculated by strength reduction FEM, and 1.636 by the classic method. The calculation error of two safety factors is 2.2% (Song, 2006; Zheng *et al.*, 2011).

4.1.2 Stability of reservoir slope

4.1.2.1 Stability analysis of slope under seepage

A homogeneous slope is adopted in the calculation, and the height is 10m; the slope angle is 26.57°, the unit weight of soil is γ_{nature}=20.0kN/m3 and $\gamma_{saturation}$=22.0 kN/m3,

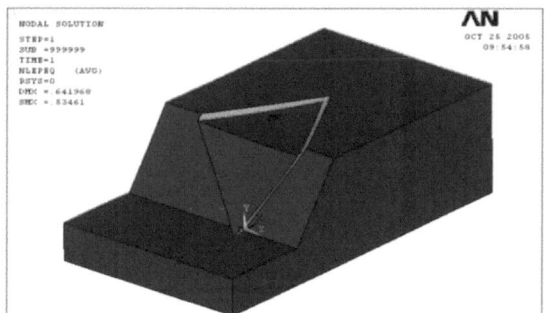

Figure 12 Equivalent plastic strain of asymmetry wedge body.

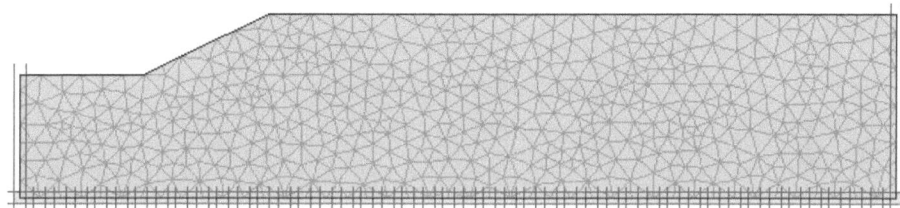

Figure 13 FEM calculation model.

infiltration coefficient is $k_x=k_y=0.001$m/d, the cohesion is $c=20.0$ kPa and the inner friction angle is $\varphi=24°$. The FEM model is shown in Fig. 13. The load of water head and the location of the phreatic surface are calculated through seepage and are shown in Fig. 14 (Tang & Zheng, 2008).

The result of strength reduction FEM is: the safety factor of the slope is 2.003 under calculation condition 1; safety factor is 1.838 under calculation condition 2. The location of the failure surface displayed is shown in Fig. 15. The GEO-SLOPE program is used to testify the calculation result through strength reduction FEM, It shows that the calculational error is within 2%. So adopting strength reduction FEM to calculate the slope stability under seepage can satisfy the requirement of the precision.

4.1.2.2 Stability analysis on the slope under the decline of reservoir water level

Different drawdown rates of reservoir water level and the magnitude of soil permeability coefficient have significant influence on slope stability. A homogeneous soil slope under undrained condition is studied. The unit weight of the soil are γ_{nature} =17.5kN/m3 and $\gamma_{saturation}=19.0$ kN/m3, the cohesion is $c=21.0$kPa, and the internal friction angle is $\varphi=28.0°$. The initial water level in front of the slope is 40m, behind the slope is set as a constant water head h=40 m. Stability analysis model is shown in Fig. 16.

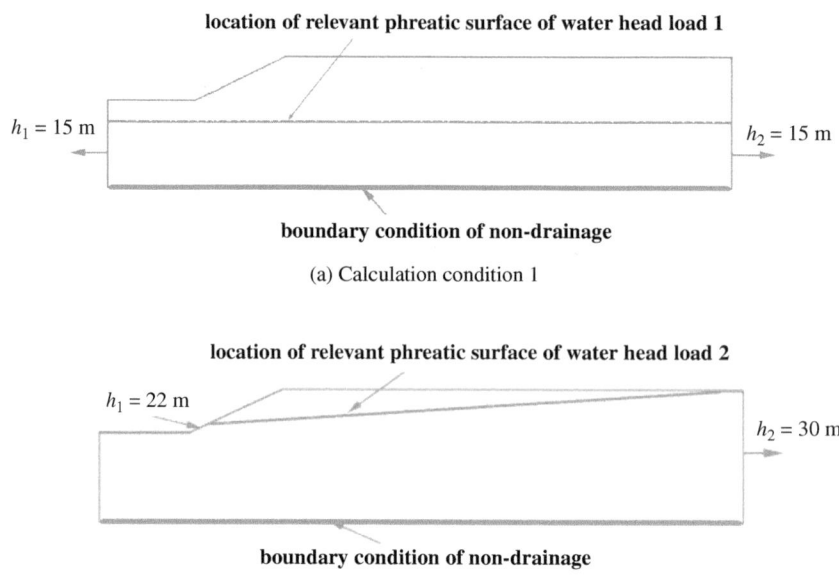

(a) Calculation condition 1

(b) Calculation condition 2

Figure 14 Locations of phreatic surface corresponding to water head loads.

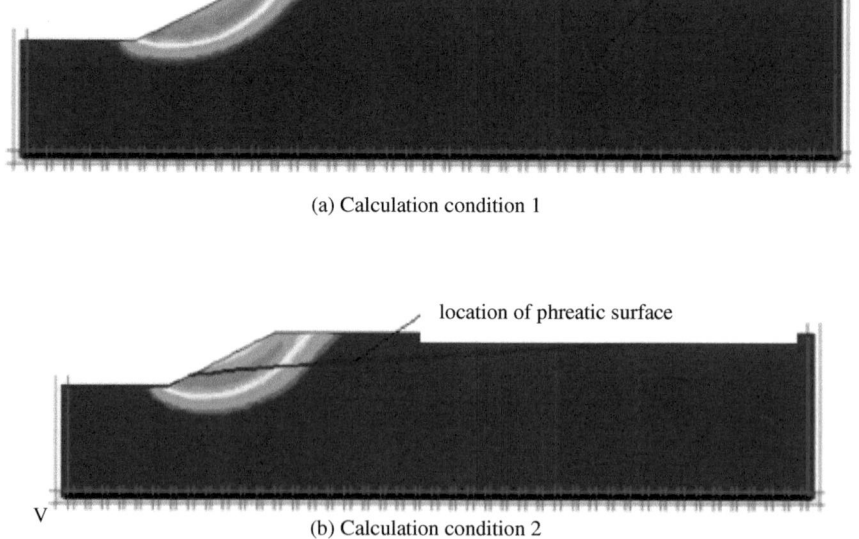

(a) Calculation condition 1

(b) Calculation condition 2

Figure 15 Locations of failure surface and phreatic surface in case 1 and 2.

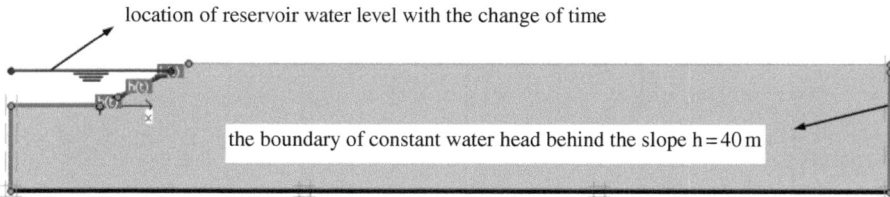

Figure 16 Slope model.

(1) The influence of declining rate of reservoir water level

If the infiltration coefficient is 0.1 m/d, the declining rates of water level in the front are 1,2 and 3 m/d respectively. If the water level declines from 40m by 30m, the stability analysis results are shown in Table 10 and Fig. 17.

Table 10 Calculation results of safety factor under different falling rates of water level.

The height of water level/m	safety factor		
	1 m/d	2 m/d	3 m/d
40	1.624	1.624	1.624
34	1.405	1.403	1.398
28	1.261	1.248	1.245
22	1.138	1.124	1.120
16	1.068	1.052	1.046
10	1.071	1.056	1.053

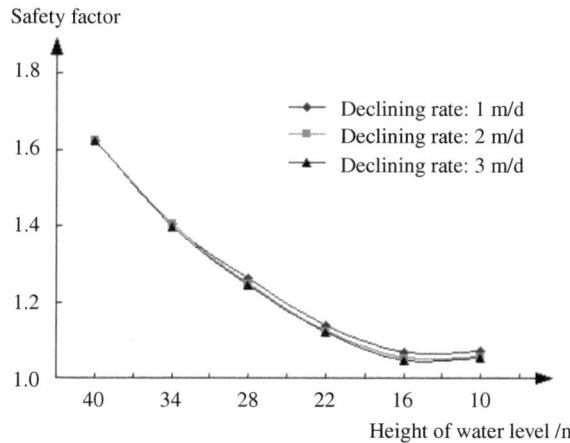

Figure 17 Relation curves of water level and safety factor and safety factor under different declining rates of water level.

(2) Influence of infiltration coefficient of soil

The declining rate of water level is 1 m/d and the water level declines from 40m by 30m. The analysis is conducted if the infiltration coefficients are 0.100, 0.050, 0.010 and 0.005 m/d respectively, and the results of stability analysis are shown in Table 11 and Figure 18.

The results above show that, the higher the declining rate and the lower the infiltration coefficient, the poorer the stability of reservoir slope is. The reason is that the change of the location of phreatic surface inside the slope mass lags behind the change of reservoir water level. For the relatively higher location of phreatic surface, the excess pore water pressure is larger, so the relevant stability of reservoir slope is poorer. The locations of the failure surface and the phreatic surface are shown in Figure 19. In Figure 19(b), the coordinate of the phreatic surface (the infiltration coefficient is smaller) in the y direction of outflow point on the slope surface is obviously higher than that in Figure 19(a) (the infiltration coefficient is larger). It is the concrete manifestation of the "lag effect" of the change of the phreatic surface during the decline of reservoir water level.

Table 11 Calculation results of safety factor under different infiltration coefficients of soil.

The height of water level/m	safety factor			
	0.100 m/d	*0.050 m/d*	*0.010 m/d*	*0.005 m/d*
40	1.624	1.624	1.624	1.624
34	1.405	1.402	1.396	1.392
28	1.261	1.247	1.216	1.213
22	1.138	1.120	1.107	1.084
16	1.068	1.049	1.024	1.012
10	1.071	1.052	0.979	0.950

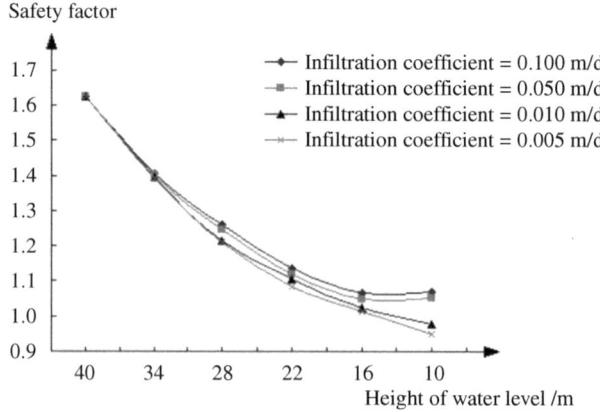

Figure 18 Relation curves of water level under different infiltration coefficients of soil.

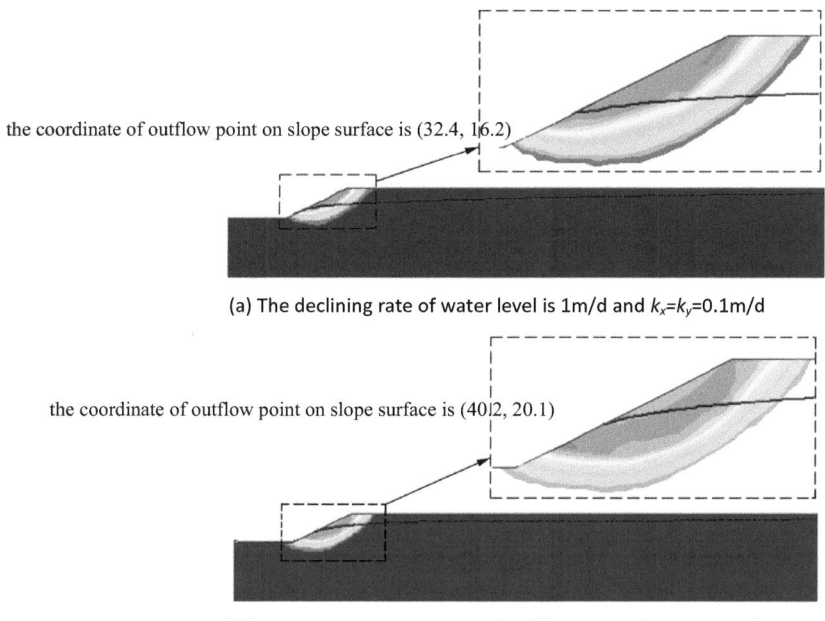

the coordinate of outflow point on slope surface is (32.4, 16.2)

(a) The declining rate of water level is 1m/d and $k_x=k_y=0.1$m/d

the coordinate of outflow point on slope surface is (40.2, 20.1)

(b) The declining rate of water level is 1m/d and $k_x=k_y=0.1$m/d

Figure 19 Locations of failure surface and phreatic surface.

4.1.3 Reasonable length and design of embedded anti-slide pile

Anti-slide piles are widely used in practice because of their strong ability in slope stability, easy construction and comparatively lower costs. The design of anti-slide pile should include two part of the cross section and pile length. At present, only the cross section size can be calculated in the design without the pile length design. The pile length is usually designed pile top extends to the ground. This approach can ensure that the pile does not appear "over the top" failure, nor that the pile is too long to waste. That is reason why the classic limit equilibrium method cannot calculate a reasonable length of pile, while it can be computed by strength reduction FEM, and the goal of safety, economy can be achieved. Generally each large anti-slide pile can save costs by 20% ~ 60%.

4.1.3.1 Determination of a reasonable pile length

The parameters in Table 12 are adopted for the analysis of a typical section of a landslide located within a colluvium deposit (shown in Fig. 20) at Yuhuangge in the Wushan Country. The piles are installed on either side of a road in the plan (Fig. 20). When the pile is installed the downside of the road (position B), and eight lengths of pile are considered: including of 7 m, 9 m, 11 m, 13 m, 15 m, 17 m, 19 m, 21.22 m (full length pile). Fig. 21 shows the positions of sliding surfaces under different pile length conditions. The landslide body, slip strip and an underlying stable rock stratum are represented by domain elements. The embedded anti-slide piles are simulated by beam elements.

Table 12 Physico-mechanical parameters for the analysis of a landslide body at Yuhuangge.

Materials	ρ(kN/m3)	E/MPa	v/l	c/kPa	φ/°
Landslide soil	21.4	30	0.3	34	24.5
Slip soil	20.9	30	0.3	24	18.1
Landslide bed	23.7	1.7×10^3	0.3	200	30
Pile	24.0	29×10^3	0.2	Treated as liner elastic material	

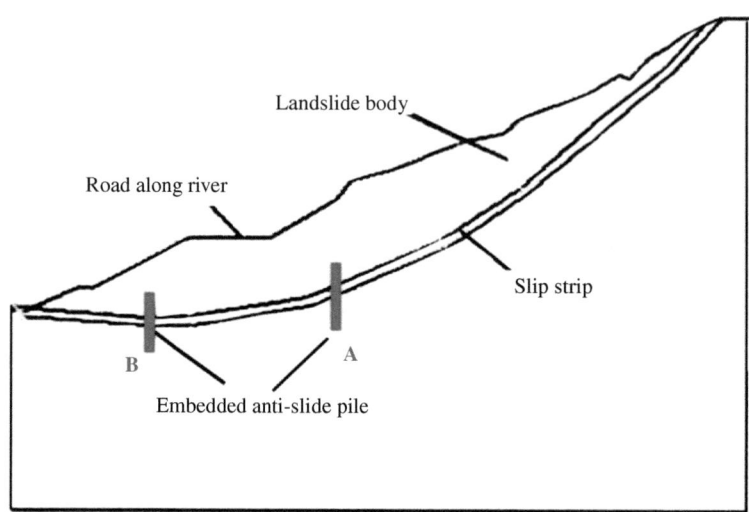

Figure 20 Landslide body, road and anti-slide pile layout over a landslide body at Yuhuangge.

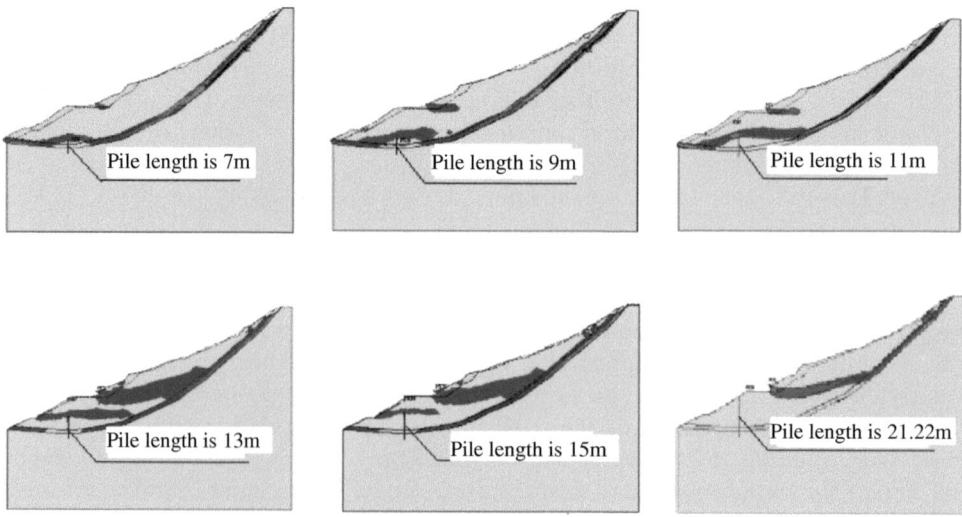

Figure 21 Changes of length and sliding surface positions.

The stability of different pile length is analyzed by strength reduction FEM and the relationships among the pile length, the safety factor and the sliding surface are obtained (Lei, 2004; Zheng, 2009, 2010).

At first, according to the calculated results of strength reduction FEM, the safety factor of the slope under gravity is 1.02, whereas the safety factor calculated by conventional limit equilibrium analysis is 1.04.

(1) The relation between the pile length and the location of sliding surface

Positioning the sliding surface is a very important aspect of slope stability study. Variation in the location of the sliding surface with different pile lengths is shown in Fig. 21. When the embedded pile length is set to 7–11m, a landslide could take place along the sliding surface slipping over the top of the anti-slide piles. When the pile length is 13m, two sliding surfaces developed. The first one is close to the top of the pile with a new sliding surface outcropping on the downhill surface of the road slope, and the other is a secondary sliding surface connected to the major sliding surface. When the pile length is 15m, the pile top is still under the ground surface and sliding surface is same as the one of pile length 13m. When the pile length is 21.22m (full length pile), the pile top reached the slope surface.

(2) The relationship between pile length and the safety factor.

The strength reduction FEM is used to calculate safety factors of different pile lengths. The results are shown in Table 13. The shorter the pile is, the lower the safety factor is. Because the design safety factor is 1.15, and the reasonable pile length should be 9m.

4.1.3.2 The landslide thrust and Internal forces of the anti-slide piles

The strength reduction FEM could be adopted to calculate the thrust behind the pile, but without considering the resistance in front of the pile. The calculated thrust is shown in Table 14.

Table 13 Relationship between pile length and safety factor when the piles are below the road.

Pile length /m	Safety factor	Pile length/m	Safety factor
0	1.02	13	1.19
7	1.13	15	1.19
9	1.15	21.22	1.19
11	1.19	——	——

Table 14 Landslide thrust when the safety factor is 1.15.

Pile length	Landslide thrust on the anti-slide section/ kN	The ratio of landslide thrust to that of full-length pile (%)
9	11405	70.9
15	15237	83.4
19	16167	100.5
22.2	16085	100

Table 15 Comparison of the extreme value of internal force.

Pile length l m	The maximum shear force on anti-slide section (MN)	The maximum force on anchoring section (MN)	Moment/$10^7 N*m$	$S_{oi}(\%)$	$S_{Pi}(\%)$	$M_i(\%)$
9	1.95	4.92	1.09	92.7	51.9	48.8
11	2.39	7.03	1.61	105.4	74.2	71.8
13	2.36	8.28	1.92	104.3	87.3	86.0
15	2.33	9.04	2.12	103.4	95.4	94.7
21.22	2.22	9.48	2.24	100.0	100.0	100.0

The internal forces are usually expressed as moment and shear force. The calculated internal forces and their ratio to the maximum values of internal forces of the full-length pile are shown in Table 15.

4.1.3.3 Comparison of experiment and numerical simulation of embedded anti-slide pile

The large scale complex physical test and numerical simulation are used to study embedded anti-slide piles. The model and its parameters are shown in Fig. 22 and Table 16.

A hydraulic jack is used to input load behind a slope in the experiment. Pile lengths are 1.2m, 1.5m, 1.8m and 2.2m, respectively.

It is loaded gradually in the experiment and numerical simulation. When the pile length is 1.5m, the comparisons of experimental and numerical simulation results are

Table 16 Physical and mechanical parameters.

Length of anti-slide pile /m	c/kPa	$\varphi/^{\circ}$	E/kPa	v/l	$\gamma/kN/m^3$
1.2, 1.5, 1.8, 2.2	33	30.8	1.5×10^7	0.3	20.6

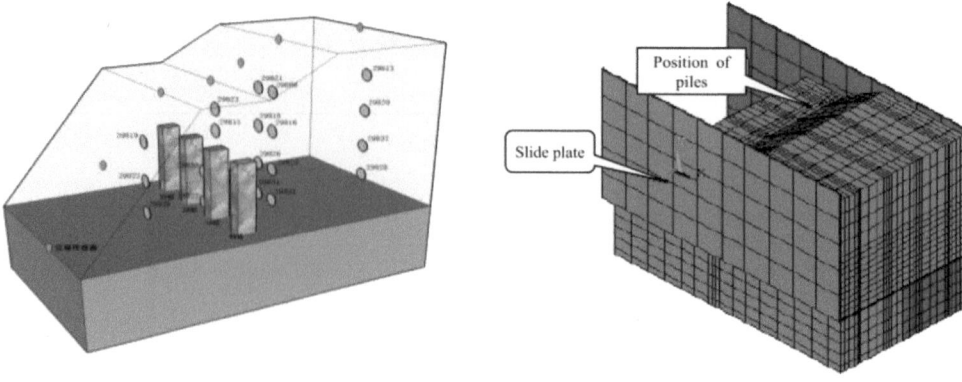

Figure 22 Experiment and numerical simulation model.

Table 17 Comparison of experimental and numerical simulation results when the pile is 1.5m long.

Loading calculation / N	Loading experiment / N	Thrust from numerical calculation / N	Thrust from experiment / N
26 000	31 595	10 650	5 755
52 000	51 903	23 103	15 390
78 000	76 364	40 582	29 754
104 000	105 366	59 021	61 344
130 000	131 310	83 470	87 376

shown in Table 17. The results of the strength reduction FEM are close to experimental results, so it proves the feasibility of the strength reduction FEM for analyzing the thrust on the embedded anti-slide pile (Zheng, 2010).

4.1.3.4 A practical engineering project

The model (Fig. 23) of this practical engineering project is a landslide section at the outlet of a tunnel in Fengjie County, China. The parameters are listed in Table 18.

At first, the reasonable pile length is determined and the calculational results are shown in Table 19. When the embedded anti-slide pile length is 29m, the safety factor of the slope is 1.27, which is greater than the design safety factor of 1.25. So a reasonable pile length is 29m, of which 16m is above the sliding surface and 13 m is below the sliding surface. The full pile length is shortened by 34% (Zheng *et al.*, 2010, 2011).

After determining a reasonable pile length, the thrust and resistance acted on the embedded anti-slide pile can be calculated directly from the strength reduction FEM. The results are shown in Table 20.

The size of the pile section is 3 m×4 m. The stability of slope is analyzed by ANSYS software, and the beam element is used to simulate pile. The calculation parameters are shown in Table 21.

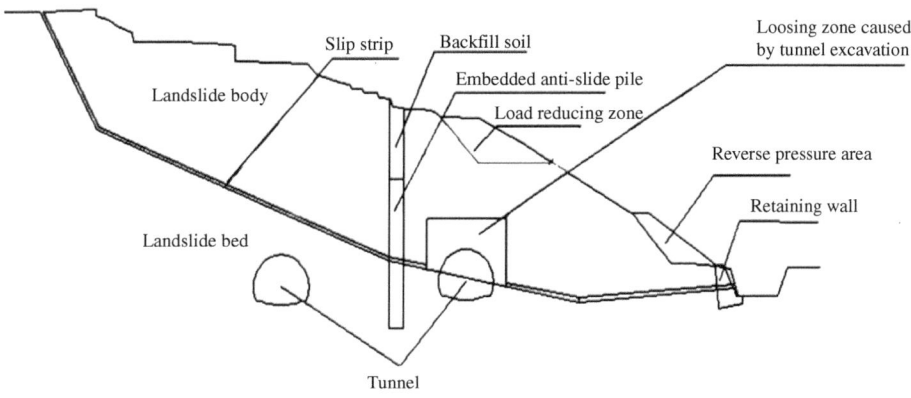

Figure 23 Slope model.

Table 18 Physical and mechanical parameters.

Materials	ρ (kN/m3)	E/MPa	v	c/kPa	φ/°
Landslide body soil	22	10	0.35	28	20.0
Slip strip soil	22	10	0.35	20	17.0
Landslide bed	26.6	8.18×103	0.28	1250	39.1
Pile	24.0	3×104	0.15	Treated as liner elastic material	
Tunnel	25	3×104	0.2	Treated as liner elastic material	

Table 19 Safety factors at different pile lengths from strength reduction FEM.

Pile length	Safety factor
Full-length pile length 44m (28m above the slip surface and 16 m below)	1.35
Pile length 38m (22 m above the slip surface and 16 m below)	1.30
Pile length 29m (16 m above the slip surface and 13 m below)	1.27
Pile length 24m (12 m above the slip surface and 12 m below)	1.24
Pile length 16m (7 m above the slip surface and 9 m below)	1.18

Table 20 Calculation results of thrust and resistance.

Landslide thrust (full length pile length 44m) /kN·m^{-1}	Resistance (full length pile length 44m) /kN·m^{-1}	Designed thrust (full length pile length 44m) /kN·m^{-1}	Landslide thrust (embedded pile length 29m) /kN·m^{-1}	Resistance (embedded pile length 29m) /kN·m^{-1}	Designed thrust (embedded pile length 29 m) /kN·m^{-1}	Ratio of designed thrust of embedded pile to that of full length pile
7 880	3 480	4 400	5 410	2 860	2 550	58 %

Table 21 Physical and mechanical parameters in the calculation.

Materials	Soil unit weight /kN·m^{-3}	Elastic modulus / MPa	Poisson ratio	Cohesion / kPa	Inner friction angle /°
Landslide body soil	21	30	0.3	25.5	24.5
Landslide bed	24	103	0.25	200	30
Pile (C25 concrete)	24	29×103	0.2	Treated as liner elastic material	

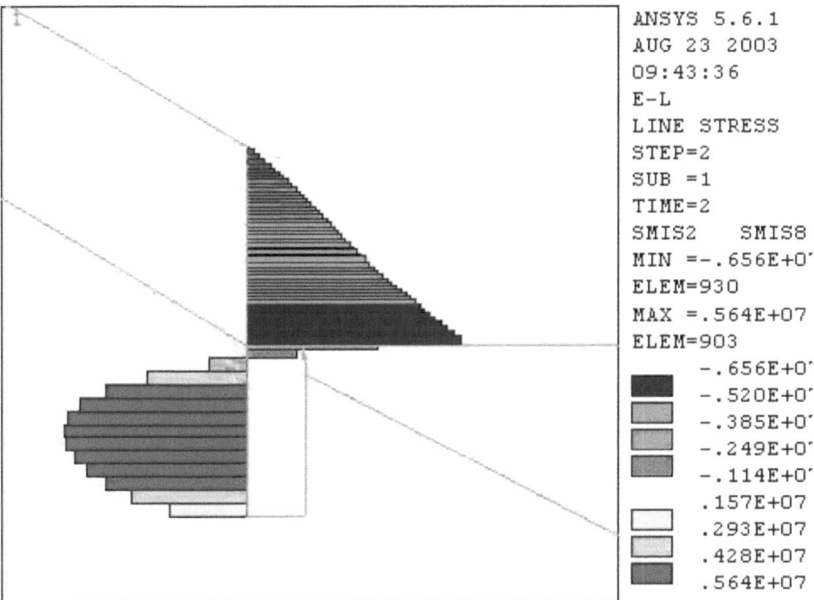

ANSYS 5.6.1
AUG 23 2003
09:43:36
E-L
LINE STRESS
STEP=2
SUB =1
TIME=2
SMIS2   SMIS8
MIN =-.656E+0⁻
ELEM=930
MAX =.564E+07
ELEM=903
      -.656E+0⁻
      -.520E+0⁻
      -.385E+0⁻
      -.249E+0⁻
      -.114E+0⁻
       .157E+07
       .293E+07
       .428E+07
       .564E+07

Figure 24 Shear in the pile without anchoring.

The internal forces of the pile and their distributions are displayed in Fig. 24 and Fig. 25 and the maximum bending moment is 48 100 kN/m and the maximum shear force is 6 560 kN.

4.1.4 Application in the stability analysis of high reinforced slope

4.1.4.1 Failure models of the high reinforced slope

Limit analysis FEM is used to analyze the stability of high reinforced slope. Three failure models are found on the high reinforced slope. The location and state of the failure surface can be obtained, as shown in Figure 26 (Zheng *et al.*, 2011).

1) If the tensile stiffness on the axial direction and the length of reinforced belt are enough, the soil in the front of the slope would loosen and collapse when the slope becomes unstable. It is the interior failure model of slope, as shown in Figure 26 (a).

2) If the length and tensile stiffness on the axial direction of reinforced belt is relatively lower, the upper part of the failure surface will move backward and the entry into the soil mass without reinforced belt. It is the simultaneous failure model of both the interior and the exterior of slope, as shown in Figure 26(b).

3) If the length of reinforced belt reduces to a certain value, failure surface appears in the soil mass without reinforced belt, and reinforced body slides along the bottom. It is the exterior failure model, as shown in Figure 26(c).

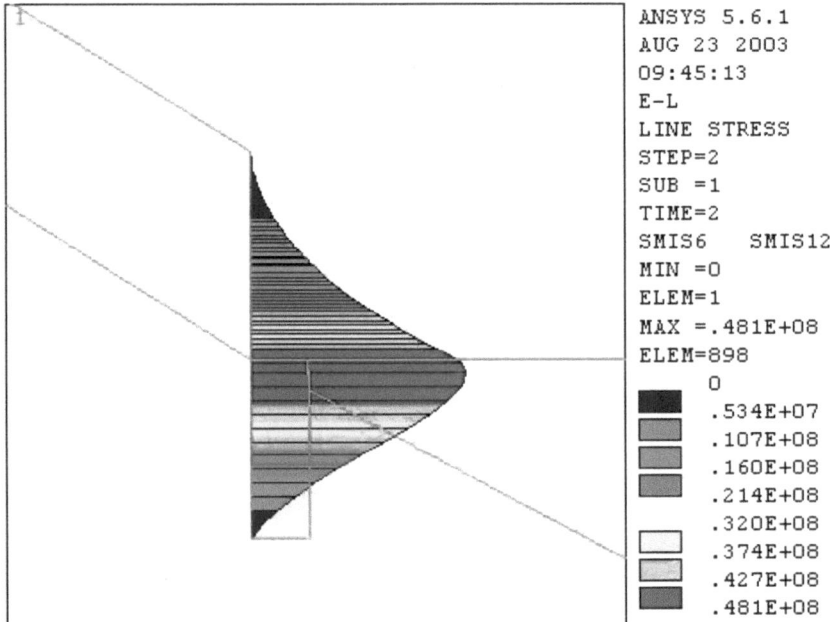

Figure 25 Bending moment in the pile without anchoring.

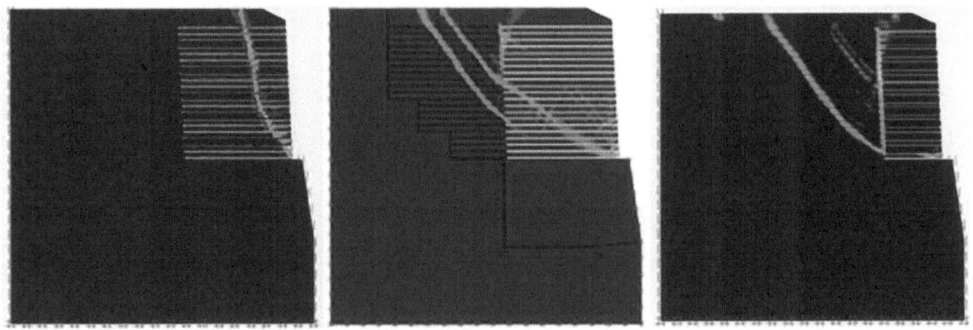

(a) the interior failure   (b) both the interior and the exterior failure   (c) the exterior failure

Figure 26 Three failure models of high slope with reinforced soil.

4.1.4.2 High reinforced slope engineering example

Strength reduction FEM is used in two engineering examples of high reinforced slope for introducing the design process of high reinforced slope.

(1) Example 1

The height of the reinforced soil slope is 60 meters in HeChi airport, China. It consists of four stages. Sidewalk with 3m width is set every 15m along the height, as shown in Figure 27(a). The comprehensive slope gradient of engineering is 1:2.5.

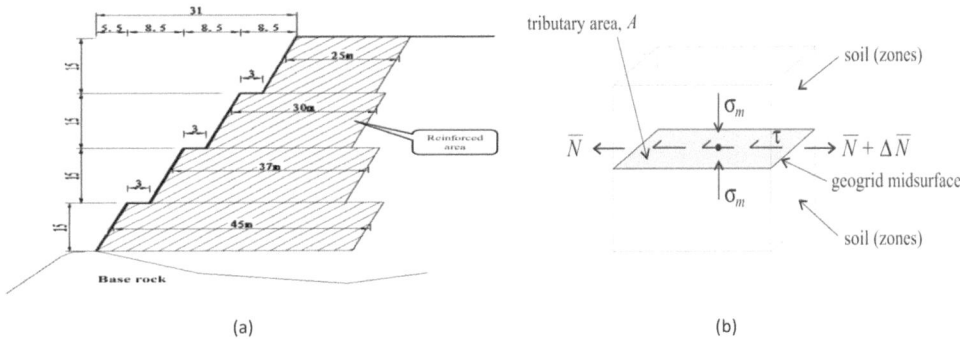

Figure 27 Reinforced earth-retaining wall after construction.

In order to simulate the sliding phenomenon between geo-grid and soil, contact elements is set in the calculation model, which can resist tensile and shear loads but cannot bear bending loads. Assuming that the model can satisfy Mohr-Coulomb yielding criteria, the mechanic relation of the contact elements is shown in Figure 27(b).

$$\tau = \sigma_{\mathrm{m}} \tan \varphi_{\mathrm{inter}} + c_{\mathrm{inter}} \tag{9}$$

The φ_{inter} and c_{inter} of the interface between soil and geo-grid are measured through large-sized direct shear test, φ_{inter} is $17°$ and c_{inter} is 26 kPa. In numerical simulation, the contact between reinforced material and the surrounding soil is simulated through setting the coupling springs in FLAC^{3D} (Song *et al.*, 2010).

The slope engineering is hierarchical construction. As shown in Table 22, the safety factors of each stage are analyzed based on the strength reduction FEM by FLAC^{3D} to ensure the slope safety during the construction. Obviously, the safety factors of each stage satisfy the specification requirements and the safety factor after supporting is 1.39 under gravity. The maximum force of the reinforced belt is 92.54kN/m and its strength reservation safety factor is 2.18, which can satisfy the design requirement under static condition. Figure 28 shows the nephogram of plastic strain after 4th stage construction (Song *et al.*, 2010).

In order to testify to the engineering stability and the reasonability of analysis, monitoring on the horizontal displacement of different stages is conducted. Figure 29 is the monitored displacement and the result of numerical simulation. The maximum of monitored horizontal displacement is at the monitoring point BX59 on the top of the 4th stage. The accumulated horizontal displacement measurement is 280.6mm (observation) and 312mm (numerical simulation) respectively, and the results of numerical

Table 22 Safety factor of each stage.

Stages	The 4th stage	The 3rd stage	The 2nd stage	The 1st stage
Safety factor	1.75	1.58	1.45	1.39

Figure 28 Nephogram of plastic strain after construction of the 4th stage.

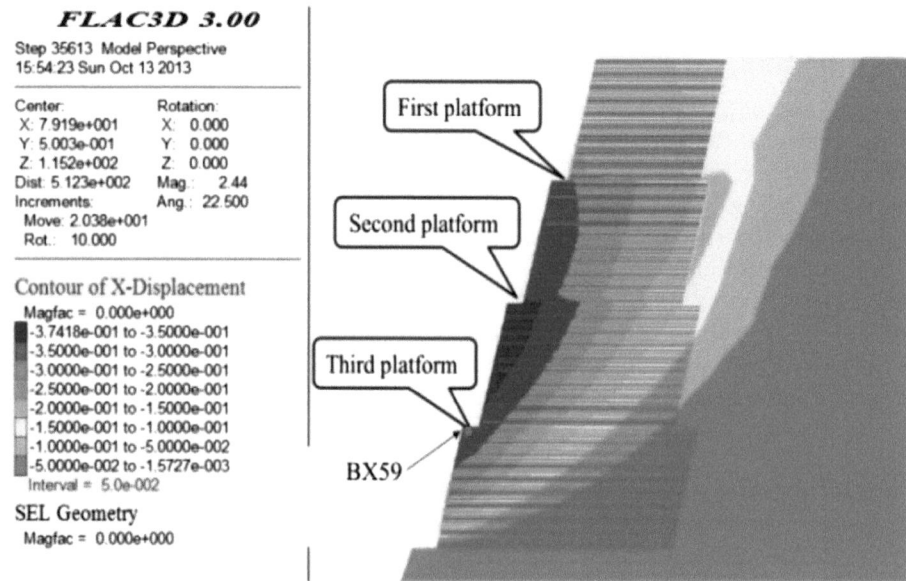

Figure 29 horizontal displacement.

simulation is close to the monitored data. After completion, the horizontal displacement of all the monitoring points is convergent and stable, so the project is stable and reliable. The engineering photograph after the completion of construction is shown in Figure 30.

(2) Example 2

There is collapsed high slope, and the part of the collapsed soil covered on the ancient landslide, as shown in Figure 30. Five grades of retaining wall are constructed with the

Figure 30 As-built drawing.

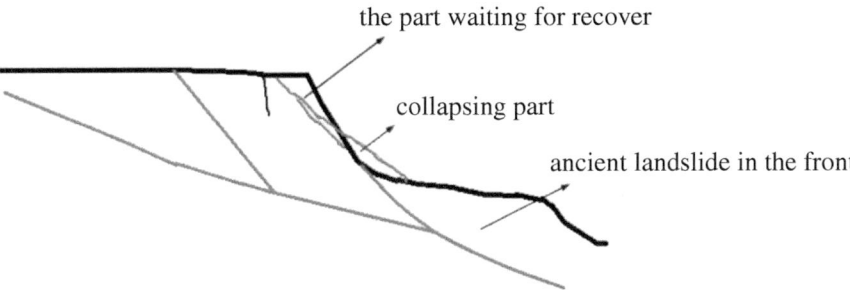

Figure 31 General view of the engineering.

total height is 45m. From top to bottom, the first grade is plain fill, height is 10m, and slope ratio is 1:1.5. Second to fourth grades adopt reinforced retaining wall with geo-grid, for the second to three grade, the retaining wall height is 10m; the geo-grid length is 35m; the interval between geo-grids is 0.5m, slope ratio is 1:0.5; for the fourth grade, the retaining wall height is 15m; the geo-grid length is 45m; the interval between geo-grids is 0.4m, slope ratio is 1:0.5. The retaining wall height of the bottom grade is 10m and adopts a frame structure with pre-stressed anchor cable, as shown in Figure 32. In order to reduce the total height of the retaining wall, two rows of anti-slide piles with anchor cable are set in the bottom of the slope.

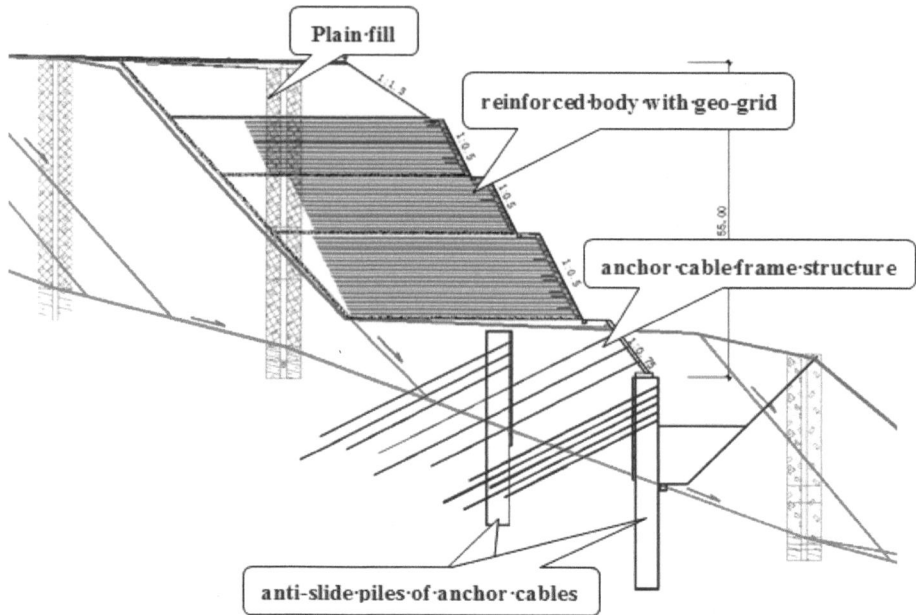

Figure 32 Retaining scheme.

Domain element is used to simulate the frame structure, anchor element is used to simulate the anchor cable, the interval is 4m. The interval between anti-slide piles is 5m. The front row pile has five cables. The back row pile has three cables. The pre-stress acted on each anchor cable is 100kN/m.

Its safety factor is 1.565, which can satisfy the design requirement. The failure surface slides out along the bottom of reinforced body, as shown in Figure 33 (Tang *et al.*, 2014). The designing safety factor is adopted 1.20 to calculate the thrust on the piles. For the back row piles, the thrust behind the pile is 6560kN/m and the resistance in front of pile is 3786kN/m, so the actual bearing thrust on each pile is 2774kN/m. For the front row piles, the thrust behind each pile is 4302kN/m and the resistance in front is 1321kN/m, so the actual bearing thrust on each pile is 2981kN/m (Tang *et al.*, 2014).

4.2 The failure mechanism of tunnel

4.2.1 Introduction

For a long time, stability analysis methods of surrounding rock in the tunnel are unscientific and unreasonable. It is still judged the stability of the tunnel by the empirical criterion of displacements of tunnel perimeter or sizes of plastic zones of surrounding rocks. Displacement of tunnel perimeter is affected by the elastic modulus of surrounding rocks, shape and size of tunnel, and displacements of tunnel perimeter are different at different positions. So it is difficult to get a unified displacement criterion standard. Judging the stability by sizes of plastic zones is superior to

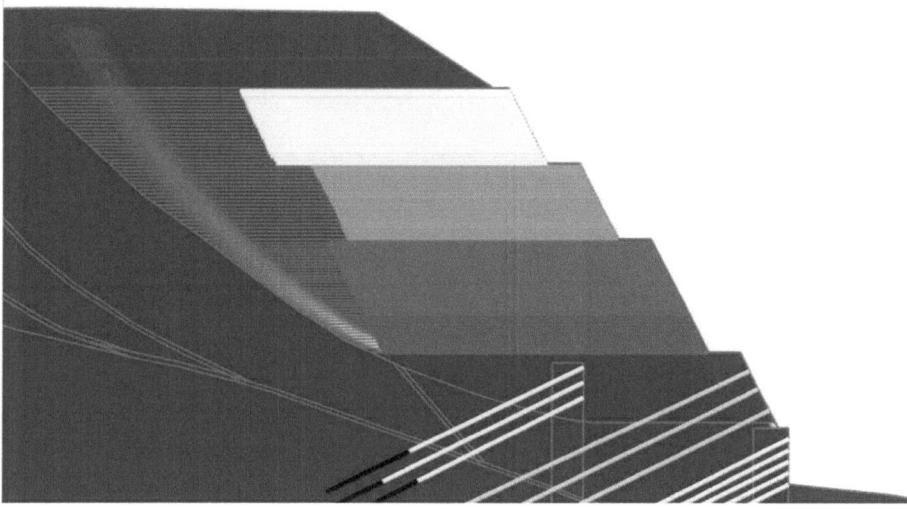

Figure 33 Location of failure surface after reinforced body constructed (post-process without showing the geo-grid units).

displacement criterion. But, plastic zones are affected by Poisson's ratio, shape and size of tunnel, and size of plastic zones calculated by different software are different. The method is also unreliable. So traditional experience methods are unreasonable. The paper puts forward the failure mechanism of tunnel based on the strength reduction FEM, and taking safety factors calculated by strength reduction FEM as the stability criterion. This criterion bases on the strict mechanic foundation, has a universal standard and cannot be affected by other factors (Zhang & Zheng, 2007; Zheng & Qiu, 2008).

4.2.2 The failure mechanism of tunnel

(1) The failure mechanism of deep-buried tunnel
The laboratory model test and numerical analysis method are used to reveal the failure mechanism of a deep-buried tunnel. The length, height and thickness of the model is 400mm, 520mm and 150mm, as shown in Fig. 34. The aggregate material of the sample is sand, and the cementing material is gypsum, cement and talc. These materials are mixed according to a given ratio with water to form the experimental material. A mechanical press is used to impose the load step by step until the tunnel fails. Fig. 35(a) shows the failure surface and failure situate of the model tunnel under 56 kN. Fig. 35(b) is the failure surface simulated by computer under 53 kN. It can be seen that failure occurs on both sides for a deep-buried tunnel (Zheng & Qiu, 2008, 2010; Zheng *et al.*, 2012).

According to the failure characteristics, the displacement or plastic strain of the failure surface mutates when the tunnel failure happens. So the location of the failure surface can be found according to the results of strength reduction FEM, Fig. 36 shows the equivalent plastic strain of surrounding rock when the tunnel fails. The values of

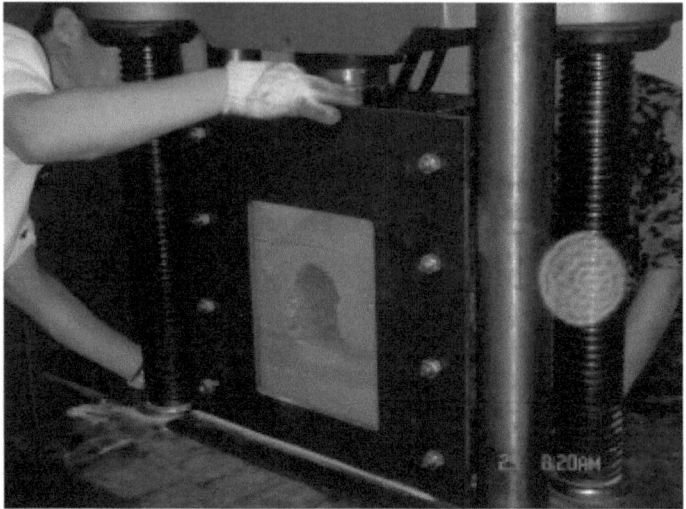

Figure 34 Tunnel model.

(a) The failure surface of model

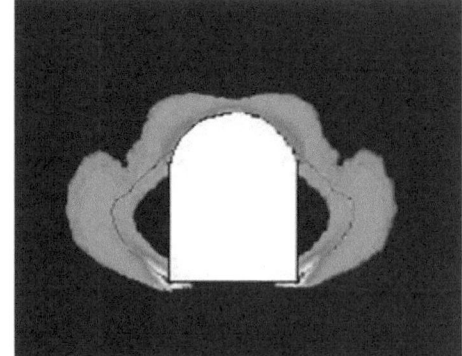

(b) The failure surface by numerical simulation

Figure 35 Failure surface of deep-buried tunnel.

equivalent plastic strain of 5 sections are mapped on Fig. 37(a)-Fig. 37(e) through the program of ANSYS, and the mutation point of the equivalent plastic strain can be found. Connecting each mutation point, the location and the shape of the failure surface can be obtained, as shown by the black line in Fig. 36. The location of the calculated failure surface is very close to that of the failure surface in the model experiment (Zheng *et al.*, 2012). It proves that the failure mechanism studied above is correct and the calculating safety factor by strength reduction FEM is feasible.

(2) The failure mechanism of shallow-buried tunnel

The following is a model experiment of shallow-buried tunnel, whose span is 80mm, height is 120mm and buried depth is 40mm. When the load is 25 kN, obvious cracks

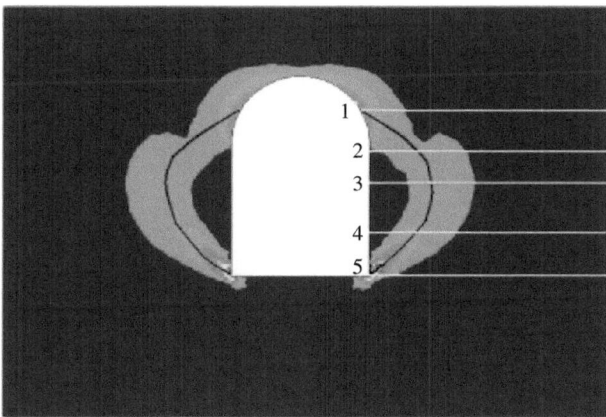

Figure 36 Equivalent plastic strain and failure surface of surrounding rock.

appear on the vault of the tunnel. When the load is increased to 28 kN, two cracks on the vault connect and would collapse, as shown in Figure 38(a). Figure 38 (b) displays the computational failure surface. When the load is 26 kN, the computation is non-convergence and failure happens on the vault. The failure surfaces of Fig. 38(a) and Fig. 38 (b) are very close, and it shows the failure of the surrounding rock of shallow-buried tunnels is in the vault (Zheng & Qiu, 2008).

(3) The failure process of a tunnel under different buried depth

In order to study the failure process of tunnel from shallow-buried to deep-buried, the strength reduction FEM is used to simulate this process. Two tunnels with different cross sections are adopted for the computational models. The width and height of rectangular tunnel are 12m and 5m. The width and height of arched tunnel are 12m and 8m. The computational parameters are as follows: the elastic modulus is 100 MPa; the Poisson ratio is 0.3; the unit weight is 18kN/m³; the cohesion is 0.04 MPa and the internal friction angle is 22°. The failure states and safety factors of rectangular tunnel under different buried depth are shown in Fig. 39 (Zheng *et al.*, 2012).

In Fig. 39(a), the failure surface is arched and extends to the ground surface when the buried depth is 3m, but the arch has not connected, and it shows the shallow-buried failure happens in the vault. In Fig. 39(b), a shallow-buried pressure arch obviously forms in the vault when the buried depth is 9m. The formation of shallow-buried pressure arch is related to the buried depth. In Fig. 39(c), the shallow-buried pressure arch in the vault disappears gradually and deep-buried pressure arch forms when the buried depth is 10m, and it can be defined as the critical boundary between shallow buried and deep buried tunnel. The deep-buried pressure arch is the pressure arch of M. Protodyakonov. In Fig. 39(d) and (e), two failure surfaces form gradually when the buried depth is 12m and 18m, respectively: One is the formed pressure arch of M. Protodyakonov in the vault; another one is the gradually formed lateral failure surface on both sides. According to the equivalent plastic strain shown in Fig. 40, the lateral failure surface forms with the increase of buried depth. Lateral failure surfaces

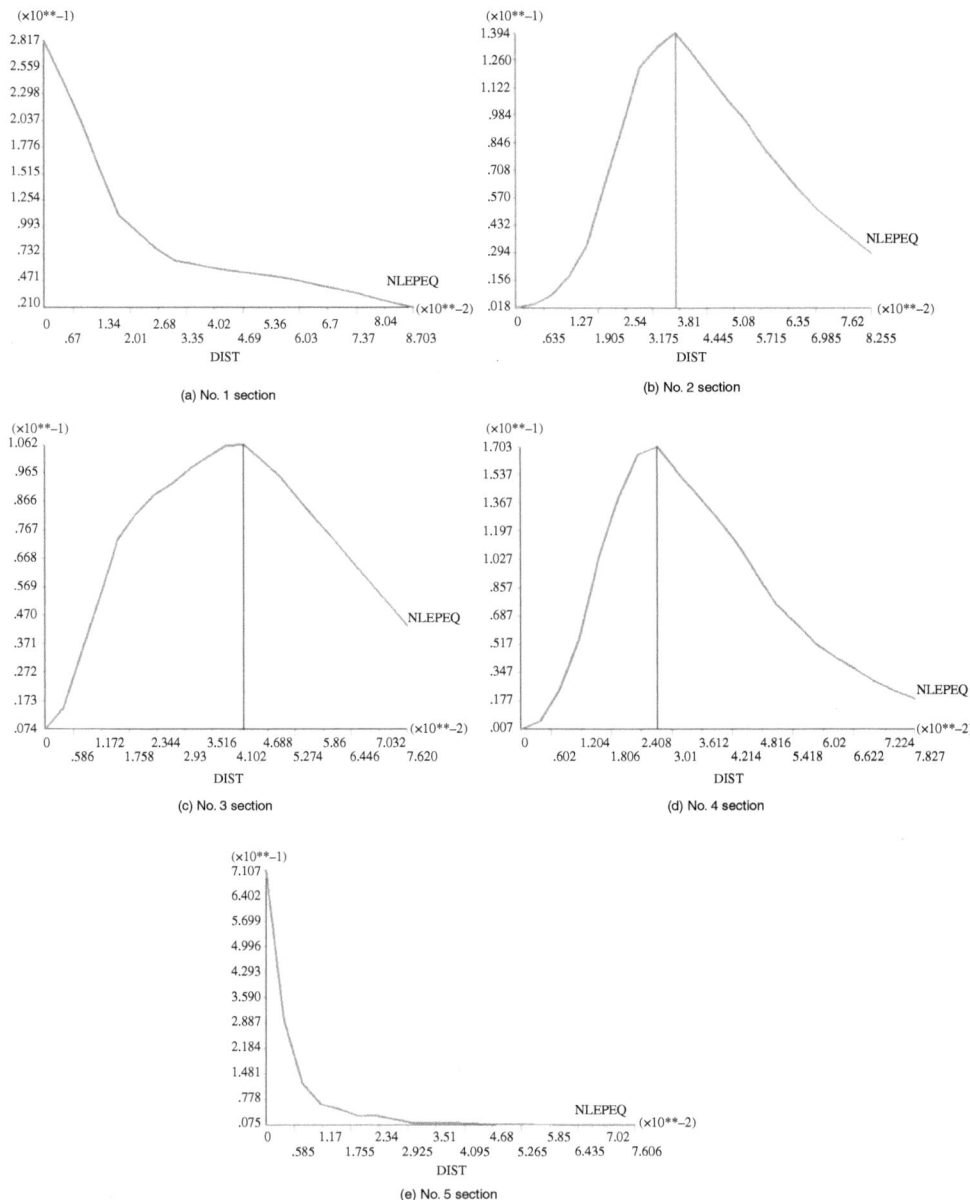

Figure 37 Maps of equivalent plastic strain.

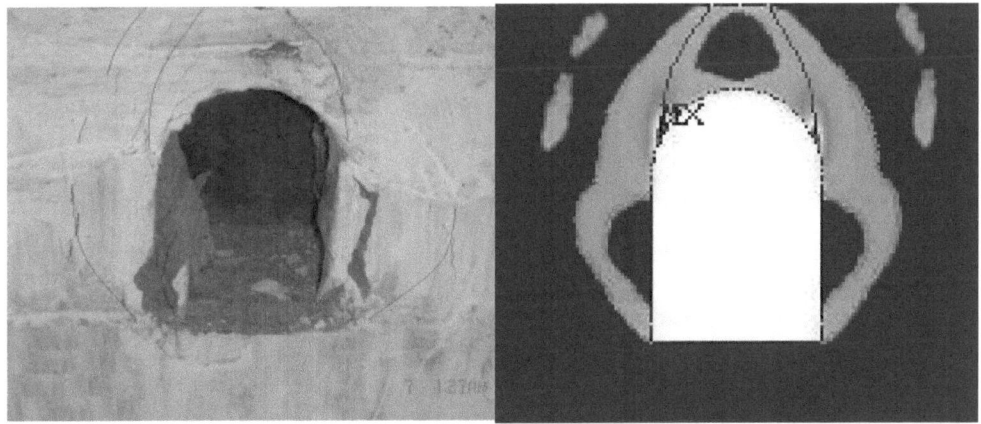

(a) The failure surface of model (pressure is 28 kN)   (b) The failure surface by numerical simulation (pressure is 26 kN)

Figure 38 Failure surface of shallow-buried tunnel.

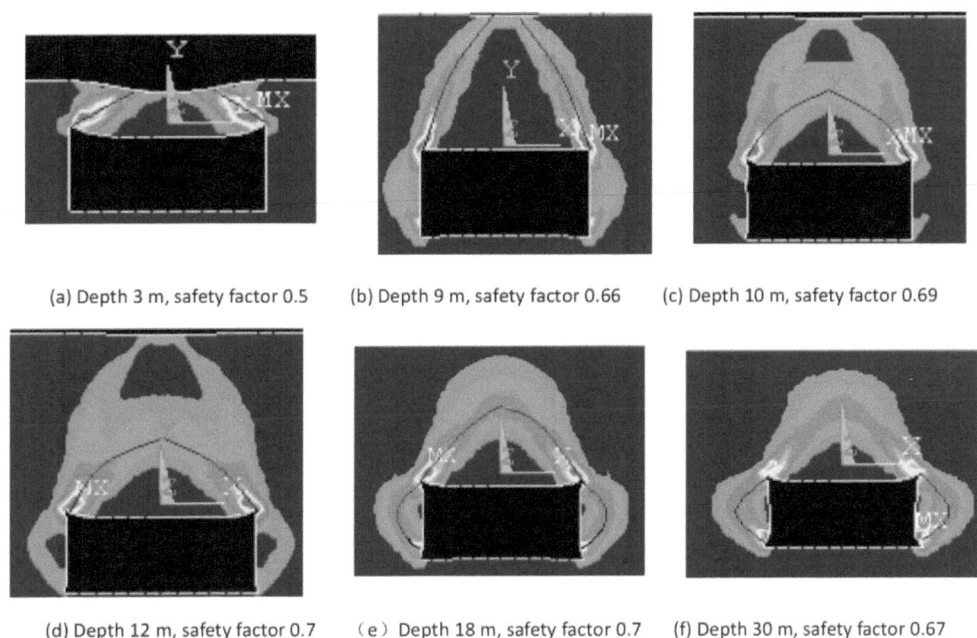

(a) Depth 3 m, safety factor 0.5   (b) Depth 9 m, safety factor 0.66   (c) Depth 10 m, safety factor 0.69

(d) Depth 12 m, safety factor 0.7   （e）Depth 18 m, safety factor 0.7   (f) Depth 30 m, safety factor 0.67

Figure 39 Equivalent plastic strain of the rectangle tunnel.

are finally formed when the depth is 18m, and the safety factor is 0.7. In Fig. 39(f), when the buried depth is 30m, the situation is similar to that of buried depth 18m, and the safety factor is 0.67. It can be concluded from the above that failure happens in the vault and the safety factor increases from 0.5 to 0.7 when the buried depth is between

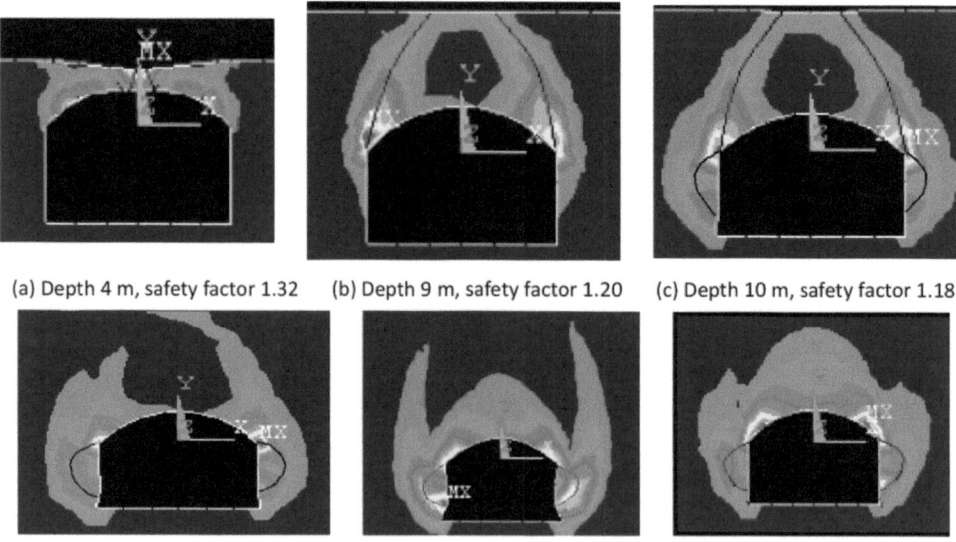

(a) Depth 4 m, safety factor 1.32   (b) Depth 9 m, safety factor 1.20   (c) Depth 10 m, safety factor 1.18

(d) Depth 12 m, safety factor 1.15   (e) Depth 20 m, safety factor 1.07   (f) Depth 30 m, safety factor 1.03

Figure 40 Equivalent plastic strain of the arched tunnel.

3m to 18m, *i.e.* Along with decreasing the depth, the safety factors decrease. When the buried depth is between 18m and 30m, failure happens on both sides and the safety factors decrease from 0.7 to 0.67.

For the arched tunnel, the failure mechanism is basically the same with that of a rectangular tunnel, but the M. Protodyakonov's pressure does not exist because the vault is pressure arch itself. The failure states and safety factors of the arched tunnel under different buried depth are shown in Fig. 40. The results show the larger the depth, the poorer safety is.

(4) The failure mechanism of jointed rock tunnels

The laboratory model test and numerical analysis method are used to reveal the failure mechanism of jointed rock tunnels. The length, height and thickness of model dimension are 400mm, 520mm and 150mm, respectively. The width and height of model tunnel are 80mm and 80mm, respectively. The arch height is 40mm, and the joint obliquity is 30° with the spacing of 40mm (Zheng *et al.*, 2012). In the model test, the mixture of soil and oil is used to simulate joint planes through filling and compacting layer by layer from bottom upwards with the obliquity of 30° and pouring oil on joint planes, the thickness of soil-oil layer 1~2mm. Oil mechanical press of 300t is used to impose load by class. The strength parameters of the experimental soil and the joint planes are determined by direct shear test. The elastic modulus and Poisson's ratio only influence displacement, and do not influence the failure mechanism, so its value is determined by physical and mechanical parameters of experimental material as shown in Table 23.

Press machine is used to impose load on the model top by class till the failure of tunnel. The load in the numerical analysis is increased till the non-convergence of calculation, which shows the failure of the tunnel. The results of the model test and its relevant numerical analysis are shown in Figure 41.

Table 23 Physical and mechanical parameters of experimental material.

Materials	P(kN/m³)	E/MPa	v/1	c /MPa	φ /°
experimental soil	17.8	7E+7	0.32	0.116	21.8
Joint planes	14.0	7E+6	0.36	0.046	14.3

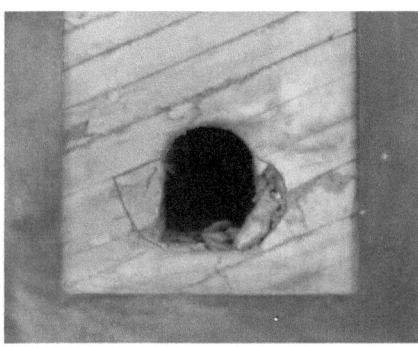 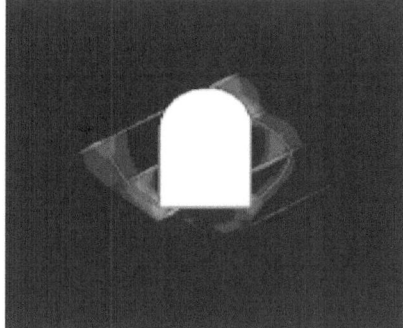

Figure 41 Model experiment and numerical simulation results.

Table 24 Model experiment and numerical simulation results.

Ultimate load of model test /kN	Ultimate load of numerical analysis /kN	Maximum distance between failure surface and tunnel wall in model test /mm		Maximum distance between failure surface and tunnel wall in numerical analysis /mm	
		Left	Right	Left	Right
44	41	58	40	61	43

The ultimate loads obtained by model test and numerical analysis, and the maximum distance between the failure surface and sidewall are listed in Table 24. It shows that the ultimate load and the maximum distance between the failure surface and sidewall are close to the results of numerical simulation. So, the strength reduction FEM is feasible to analyze the failure mechanism of jointed rock tunnel.

4.3 The application in foundation engineering

4.3.1 Ultimate bearing capacity of the plate loading test

The plate loading test is most used to obtain the ultimate bearing capacity of foundation in the world and to check the results of the treatment foundation, but it also has some shortcomings, including of size effects, high expense, time and cost and so on. Step-loading FEM is used to analyze the process of the plate loading test and to obtain the ultimate bearing capacity of foundation.

Table 25 p-s curve of test.

Loads (kPa)	150	270	380	500	610
Settlements (mm)	2.69	5.14	8.12	12.41	16.38
Loads (kPa)	730	840	960	1060	
Settlements (mm)	20.74	24.98	29.16	33.20	

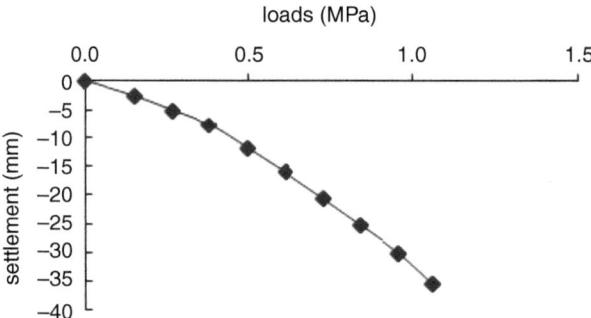

Figure 42 p-s relations.

The shear strength parameters, including c and φ can be obtained by in-situ shear test, the shear test or tri-axial shear test in laboratory. They are the most important parameters in the numerical analysis of the plate loading test. The p-s curves and data of the foundation test in a project are shown in Tab. 25 and Fig. 42. The p-s curve is close to a straight-line before the pressure at 380kPa. When the pressure is about 380kPa, an inflection point appears, and after that the settlement increase more rapidly until the pressure reaches 960kPa. When the pressure is 1.06MPa, the soils around the support plate upheave and it means that the foundation is destroyed. According to the building code, the ultimate bearing capacity of the foundation reaches 960 kPa (Zheng *et al.*, 2007; Zheng, 2012).

Some parameters about foundation by in-situ shear test are as follows: γ=22kN/m³, c=32kPa, φ=30.2°, E=17MPa, υ=0.27, respectively. The numerical analysis of the plate loading test is carried out by step-loading FEM. The FEM model and its meshing are shown in Fig. 43. The process of the plate loading test, as shown in Table 26, is simulated by step-loading FEM and the foundation ultimate bearing capacity is 1.14MPa, which is little bigger than the test value.

The comparison of p-s curves of calculated and the plate loading test are shown in Fig. 44. It shows that FEM results and the plate loading test results are almost identical.

4.3.2 The ultimate bearing capacity of pile foundation

4.3.2.1 Bearing capacity functional and computation

The bearing capacity functional of the tested pile may be expressed by Equation 10.

$$R = R_p(q_{pp}, G_g, M) + R_s(q_{ss}, G_g, M) \tag{10}$$

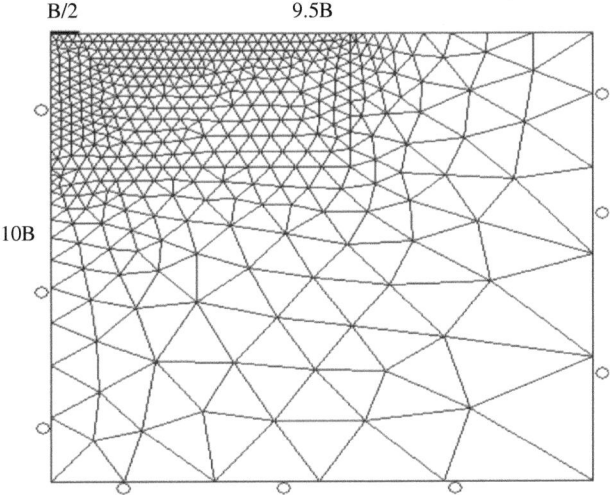

Figure 43 FEM model and meshing.

Table 26 Step-loading process.

Loads step	1	2	3	4	5	6
Loads (MPa)	0.15	0.12	0.11	0.12	0.11	0.12
Loads step	7	8	9	10	11	12
Loads (MPa)	0.11	0.12	0.10	0.07	0.01	0.01

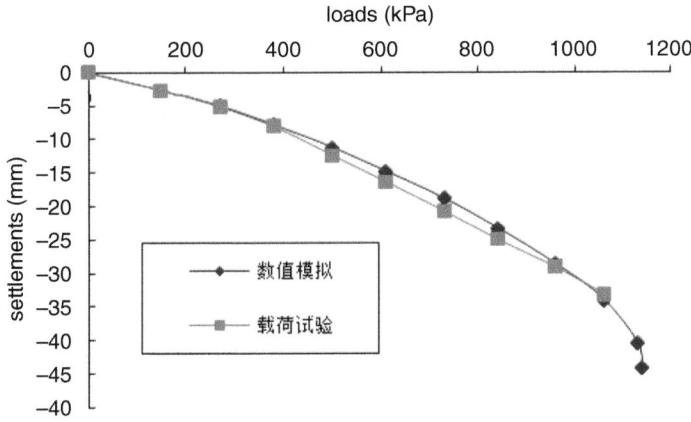

Figure 44 Comparison of p-s curves of calculated and the plate loading test.

Table 27 p-s curve of calculating.

Loads (kPa)	150	270	380	500	610	730
Settlements (mm)	2.68	5.02	7.82	11.23	14.80	18.86
Loads (kPa)	840	960	1060	1130	1140	1150
Settlements (mm)	23.30	28.71	34.11	40.52	44.24	80.64

where R is the vertical load functional of a pile; R_p is the end resistance functional; R_s is the friction resistance functional; q_{pp} is the functional of end resistance; q_{ss} is the functional of friction resistance; G_g is the geometrical characteristic of pile, including of length, section characteristic and size of section; M is the construction method.

Appointed the tested pile, some parameters of Equation 1, for example G_g and M, can be seen as the constants. The q_{pp} and q_{ss}, which are determined by the shear strength function of soil(τ), are mainly influenced functional, so equation (10) can be briefly expressed by equation 11 or equation 12.

$$R = R(\tau) \tag{11}$$

$$R = R_p(\tau_p) + R_s(\tau_s) \tag{12}$$

Where τ is the shear strength function of soils around the pile body; τ_p is the shear strength function of soils under pile toe; τ_s is the shear strength function of soils around the pile side.

Based on Mohr-Coulomb criterion and the strength reduction formulas of geotechnical material parameters (c and φ), the strength reduction factor (F) can be expressed by equation 3.

$$\frac{R}{R'} = F = \frac{c}{c'} = \frac{\tan\varphi}{\tan\varphi'} \tag{13}$$

Where R is the bearing capacity functional of pile; R' is the ultimate bearing capacity functional of pile under P on pile top; c 'φ' are cohesion and internal friction angle while failure. The ultimate load of a pile is as follows:

$$p_u = F \times P \tag{14}$$

where P_u is the ultimate load of pile foundation.

The yield criterion of soils is the Mohr-Coulomb criterion or DP3 criterion, which is suitable for computation in the 3D space problems. Using three judged failure criteria: 1) there are the plastic zone connection of soils around pile body and the infinite movement trend of pile body, 2) the inflexion at the *F-s* curve and the end part *F-s* curve almost paralleling to the s axis; 3) the plastic flow occur in pile tip ground, and the end resistance is quickly decreased, but it is synchronously confined by the resistance of pile shaft and the effective stress of ground upside the pile end, and the bearing capacity of ground will partly increase after happened a certain settlement of pile tip, so the V type inflexion appear on the $F-Q_u$ curve.

The mechanical problem of pile foundation is a 3D space problem. It needs to consider soil-pile interaction and the larger stiffness gradient between pile body and soils, so the finite element model will consider items as follows: 1) A 3D finite element model. The element deformation of soils around the pile is influenced or confine by the near element of soils. In the computing process, computable stabilization of 3D model may be better, and close to the practical condition. 2) Pile-Soil interaction. Contact element is used for simulating pile-soil interaction. For ANSYS software, the computational method and some controls parameters of contact element must be adjusted.

4.3.2.2. Calculation example and comparison

Example 1: Large-diameter pile
The length of pile is 7.2m, the diameter of pile shaft and the pile point cross-section is 0.8m, and concrete strength is C25. The stratum and parameters are as Table 28. Its FEM model is Figure 45. The p-s curves of static loading test and step-loading FEM shown in Figure 46. In static loading test, the settlement of pile top is 208.34mm at $P=4200$kN (maximal loading at the pile top). At $P=3600$kN, the inflexion of P-s curve

Table 28 The geotechnical parameters of ground.

Number	Soil layer name	h /m	E_0 /MPa	c /kPa	φ /o	μ/ l
1	Loam	5.1	6	28.9	18.4	0.3
2	Fine sand	1.6	14.1	4.4	25	0.25
3	Sand gravel	2.3	25.6	5.35	27.5	0.22
4	Gravel	>9	28.5	3.5	30	0.2
5	Contact surface of pile side		2500	9.5	8	0.2
6	Contact surface of pile point		2500	50	8	0.2
7	Ground surface around pile		250	50	30	0.22

Annotation: h was the thickness of soil layer; E_s was compressive modulus; E_0 was Young's modulus; c and φ were respectively cohesion and internal friction angle; μ was Poisson's ratio.

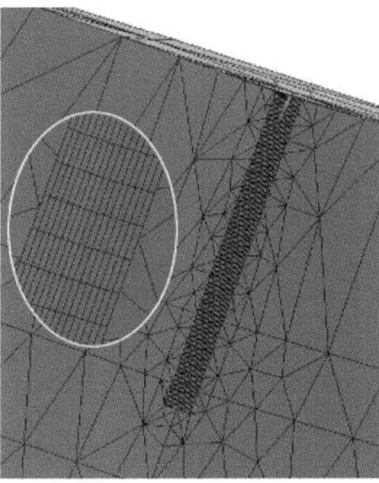

Figure 45 FEM model of pile.

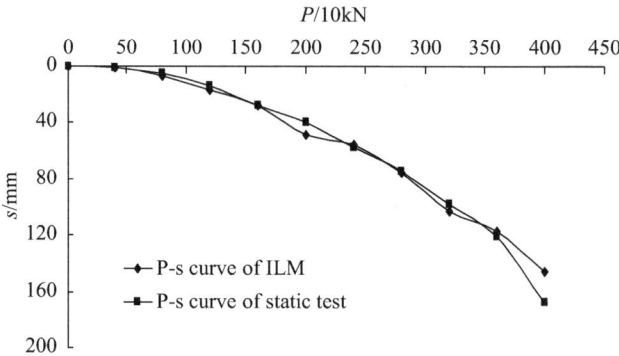

Figure 46 P-s curves of pile.

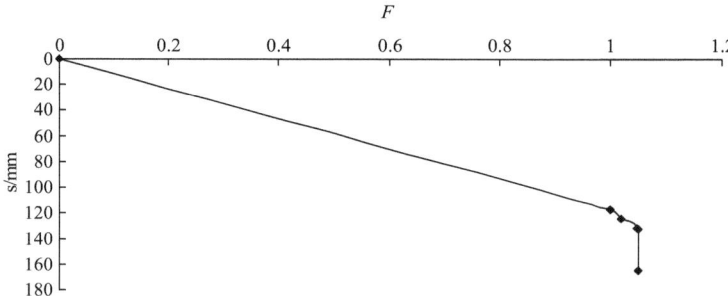

Figure 47 F–s curve.

of static loading test appears, so the ultimate load is estimated as 3600kN and its settlement of pile top is 121.03mm.

The F–s curve and F–Q_u curve are shown in Figures 47 and 48. The ultimate load of this large diameter pile in the example is equal to 3600kN by two P-s curves of the static loading test and ILM, and F=1.047 is computed by strength reduction FEM at 3600kN. Using the formula 14, the ultimate load is 3769.2kN. The difference of double estimated values is 169.2kN, and its error ratio is 4.7% (Dong & Zheng, 2010). If reducing interval value of loading on the pile top in the test and numerical simulation, the error can be eliminated.

Example 2: Vibro-pile foundation
The stratum and its parameters are in Table 29. The construction method is vibro-pile, and the diameter of the pile is 377mm, and the length of the pile is 23m. Its FEM model is Figure 49. The p-s curves of static loading test and step-loading FEM shown in Figure 50. The F–s curve and F–Q_u curve shown in Figures 51 and 52. In static loading test, the maximal loading is 1600kN, and its settlement is 15.05mm. After withdrawing load, the rebound value is 7.51mm and it is 49.9%

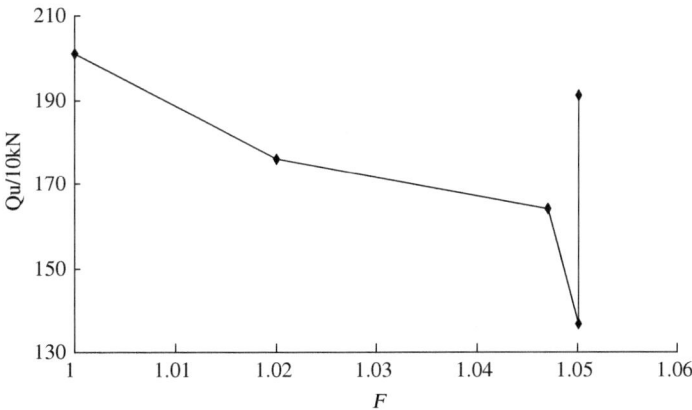

Figure 48 F–Q_u curve.

Table 29 The geotechnical parameters of ground.

Number	Soil layer name	h/m	Es /MPa	E₀ /MPa	c /kPa	φ /°	μ/I
1	Clay	4.6	4.5	8.775	5	5	03
2	Sludge silty-clay	10.4	3.5	5.11	5	5	0.35
3	Silty-sand	14.0	12.8	53.28	5	35	0.22
4	Middling weathering bed rock	30	105	105	800	80	0.3
5	Contact surface			2500	7	8	0.2
6	Ground surface around pile			250	50	35	0.2

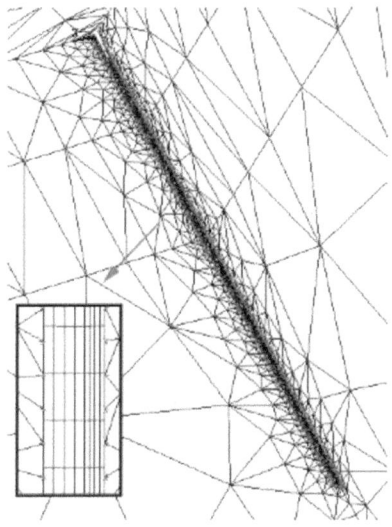

Figure 49 FEM model.

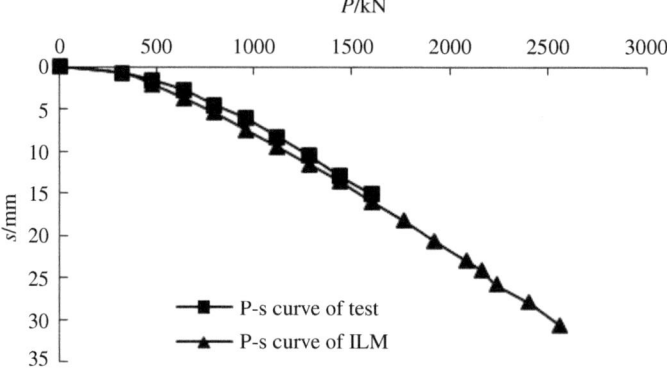

Figure 50 P-s curves.

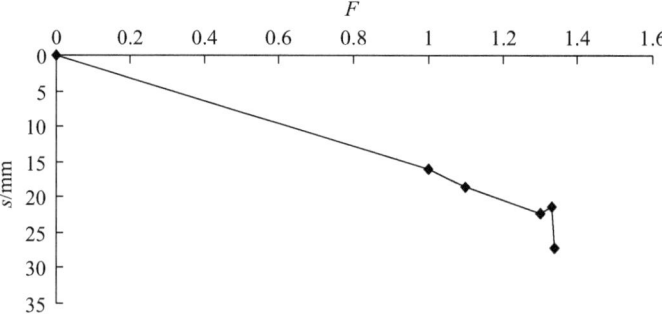

Figure 51 F −s curve.

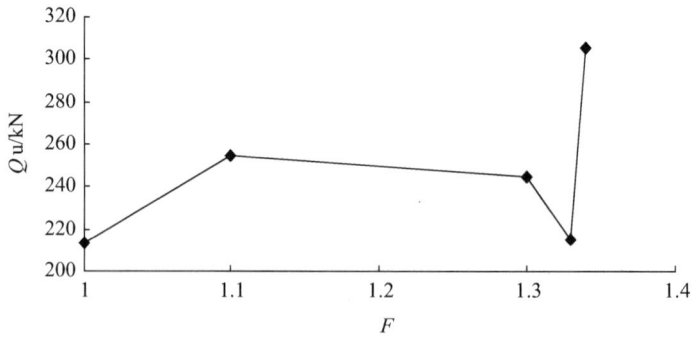

Figure 52 F–Q_u curve.

of full settlement. These show that the design value of the ultimate load (1600kN) is obviously conservative, the real ultimate load and its safety factors are difficult to judge by the existing method.

Analysis result of SRM, the ultimately load is calculated by formula 14 and equal to 2080kN (F=1.3, P=1600kN). The calculated value is less 80kN than 2160kN that be estimated by the computational P-s curve of step-loading FEM, and its error ratio is equal to 3.85% (Dong & Zheng, 2010).

REFERENCES

Dawson E M, Roth W H, & Drescher A. (1999) Slope stability analysis by strength reduction. *Geotechnique*, 49(6), 835–840.

Deng Chujian, He Guojie, & Zheng Yingren. (2006) Studies on Drucker-Prager yield criterions based on M-C yield criterion and application in geotechnical engineering. *Chinese Journal of Geotechnical Engineering*, 28(6), 735–739. (in Chinese)

Dong Tianwen, Zheng Yingren, & Huang Lianzhuang. (2010) Study of Ultimately Loading of Pile Foundation by Strength Reduction Method of No-Linear Limit Analysis of FEM. *Advanced Materials Research*, 12(168), 2537–2542.

Gao Hong, Zheng Yingren, & Feng Xiating. (2007) Study on energy yield criterion of geomaterials. *Chinese Journal of Rock Mechanics and Engineering*, 26(12), 2437–2443. (in Chinese)

Gao Hong, & Zheng Yingren. (2011) Discussion on strength criteria. *Materials Research Innovations*, 2(15), 504–507.

Griffiths D V, & Lane P A. (1999) Slope stability analysis by finite elements. *Geotechnique*, 49(3), 387–403.

Hill Rodney. (1950) *Theory of plasticity mathematics*. Oxford, Oxford University Press.

Lane P A, & Griffiths D V. (2000) Assessment of stability of slopes under drawdown conditions. *Journal of Geotechnical and Geoenvironmental Engineering*, ASCE, 126(5), 443–450.

Lei Wenjie, Zheng Yingren, Feng Xiating, *et al.* (2004) Linit analysis of slope stabilized by deeply buried piles with finite element method. *Chinese Journal of Rock Mechanics and Engineering*, 25(1), 27–33. (in Chinese)

Matsui T, & San K C. (1992) Finite element slope stability analysis by shear strength reduction technique. *Soils and Foundations*, 32(1), 59–70.

Song Yakun, Zheng Yingren, Zhao Shangyi, *et al.* (2006) Application of three-dimensional strength reduction FEM in slope. *Chinese Journal of Underground Space and Engineering*, 5(5), 822–827. (in Chinese)

Song Yakun, Zheng Yingren, Tang Xiaosong, *et al.* (2010) Study on the FEM design of reinforced earth retaining wall with geogrid. *Engineering Sciences*, 8(3), 71–80.

Tang Xiaosong, Zheng Yingren, & Shi Weimin. (2008) Analytic solution of phreatic surface in the slope of reservoir bank. *Engineering Sciences*, 6(3), 2–11.

Tang Xiaosong, Zheng Yingren, & Wang Yongfu. (2014) Application and Analysis of the Reinforced Retaining Wall with Geo-grid, *Proceedings of 2014 International Conference on Mechanics and Civil Engineering*, part of the series AER, ISSN 2352-5401, volume 7, 499–507.

Zienkiewicz O C, Humpheson C, & Lewis R W. (1975) Associated and non-associated visco-plasticity and plasticity in soil mechanics. *Geotechnique*, 25(4), 671–689.

Zhang Liming, Zheng Yingren, Wang Zaiquan, *et al.* (2007) Application of strength reduction finite element method to road tunnels. *Rock and Soil Mechanics*, 28(1), 97–101. (in Chinese)

Zhang Luyu, Shi Weimin, & Zheng Yingren. (2002) The slope stability analysis by FEM under the plane strain condition. *Chinese Journal of Geotechnical Engineering*, 24(4), 487–490. (in Chinese)

Zhang Luyu, Zheng Yingren, & Zhao Shangyi. (2003) The feasibility study of strength reduction method with FEM for calculating safety factors of soil slope stability. *Journal of Hydraulic Engineering*, 24(1), 21–27.

Zhao Shangyi, Zheng Yingren, Shi Weimin, *et al.* (2002) Slope safety factor analysis by strength reduction FEM. *Chinese Journal of Geotechnical Engineering*, 24(3), 343–346. (in Chinese)

Zhao Shangyi, & Zheng Yingren. (2006) Definition and transformation of slope safety factor based on Drucker-Prager criterion. *Chinese Journal of Rock Mechanics and Engineering*, 25 (supp 1), 2730–2734. (in Chinese)

Zhao Shangyi, Zheng Yingren, & Zhang Yufang. (2005) Study on slope failure criterion in strength reduction finite element method. *Rock and Soil Mechanics*, 26(2), 332–336. (in Chinese)

Zheng Yingren. (2012) Development and application of numerical limit analysis for geological materials. *Chinese Journal of Rock Mechanics and Engineering*, 7(31), 1297–1316. (in Chinese).

Zheng Yingren, Deng Chujian, & Zhao Shangyi. (2007) Development of finite element limiting analysis method and its applications in geotechnical engineering. *Engineering Sciences*, 5(3), 10–36. (in Chinese)

Zheng Yingren, & Gao Hong. (2008) Discussion of strength theory for materials. *Journal of Guangxi University*, 33(4), 337–345. (in chinese)

Zheng Yingren, & Kong Liang. (2010) *A Geotechnical plastic mechanics.* Beijing, China Architecture and Building Press. (in Chinese)

Zheng Yingren, & Zhao Shangyi. (2004) Application of strength reduction FEM in soil and rock slope. *Chinese Journal of Rock Mechanics and Engineering*, 23(19), 3381–3388. (in Chinese)

Zheng Yingren, & Zhao Shangyi. (2005) Limit state finite element method for geotechnical engineering analysis and its application. *China Civil Engineering Journal*, 38(1), 91–99. (in Chinese)

Zheng Yingren, Zhao Shangyi, Li Anhong, *et al.* (2011) *FEM limit analysis and its application in slope engineering.* Beijing, China Communications Press. (in Chinese)

Zheng Yingren, Tang Xiaosong, Zhao Shangyi, *et al.* (2009) Strength reduction and step-loading finite element approaches in geotechnical engineering. *Chinese Journal of Rock Mechanics and Geotechnical Engineering*, 1(1), 21–30.

Zheng Yingren, Qiu Chenyu, Zhang Hong, *et al.* (2008) Exploration of stability analysis methods for surrounding rocks of soil tunnel. *Chinese Journal of Rock Mechanics and Engineering*, 27(10), 1968–1980. (in Chinese)

Zheng Yingren, Qiu Chenyu, & Xiao Qiang. (2010) Research on the stability analysis and design of soil tunnel surrounding rock. *Engineering Sciences*, 8(3), 57–70.

Zheng Yingren, Zhao Shangyi, Lei Wenjie, *et al.* (2010) New method of designing anti-slide piles —strength reduction FEM. *Engineering Sciences*, 8(3), 2–11.

Zheng Yingren, Zhu Hehua, Fang Zhengchang, *et al.* (2012) *The stability analysis and design theory of surrounding rock of underground engineering.* Beijing: China Communications Press. (in Chinese)

Chapter 25

Open pit slope design

J.R.L. Read[1] & P.F. Stacey[2]
[1]*John R Read Associates Pty Limited, Brisbane, Australia*
[2]*Stacey Mining Geotechnical Limited, Vancouver, Canada*

Abstract: This chapter highlights what is needed to satisfy best practice with respect to the design of an open pit slope. It is not intended for it to be an instruction manual for geotechnical engineering in open pit mines, but a concise, up-to-date road map that outlines the fundamentals of the slope design process, including how to gather reliable data, how to formulate the design, how to implement the design, how to assess the reliability of the outcome, and how to manage risk.

1 INTRODUCTION

Reliable slopes are essential to the design of an open pit mine, at all scales and at every level of project development. If uncontrolled instabilities do occur, that is, slope failures where displacement has reached a level where it is no longer safe to operate or the intended function of the slope cannot be met, there can be many consequences, principally the following.

- Safety and social consequences
  - loss of life or injury;
  - loss of worker income;
  - loss of worker confidence;
  - loss of corporate credibility, both externally and with shareholders.
- Economic consequences
  - disruption of operations;
  - loss of ore;
  - loss of equipment;
  - increased stripping;
  - cost of cleanup;
  - loss of markets.
- Environmental/regulatory consequences
  - environmental impacts;
  - increased regulation;
  - closure considerations.

If slope instabilities do develop, they must be manageable at all pit scales, from the individual benches to the overall slopes. When managing the failures, it is essential that a degree of stability is ensured to minimize risk. This is required to ensure that the expectations of an economic return held by the various stakeholders in the operation, including the owners, management, the workforce and the regulators, are met. To achieve an optimum economic return, ore recovery must be maximized and waste stripping kept to a minimum throughout the life of the mine. Invariably, these needs result in a compromise that achieves a balance between establishing slope angles that are as steep as possible and yet can be safely and practicably mined.

The question is, how is this balance best achieved? The purpose of this chapter is to answer this question by providing a framework that outlines the fundamentals of the slope design process, including how to gather reliable data, how to formulate the design, how to implement the design, and how to assess the reliability of the outcome.

2 THE SLOPE DESIGN PROCESS

The basic process for the design of a pit slope is summarized in Figure 1. Regardless of size or materials, at any level of project development it essentially involves the following eight steps.

1. Formulation of a geotechnical model for the pit area.
2. Population of the model with relevant and reliable data.
3. Division of the model into geotechnical domains.
4. Subdivision of the domains into design sectors based on slope orientation and/or other factors.
5. Design of the slope elements in the respective sectors of the domains.
6. Assessment of the stability of the resulting slopes in terms of the project acceptance criteria.
7. Definition of implementation and monitoring requirements for the designs.
8. Consideration of the stability requirements for pit closure.

The key elements of these eight steps are outlined in the Sections that follow. Section 3 outlines how the geotechnical model is formulated, with Section 4 presenting how to report and determine the data uncertainties that are incorporated in the model in a manner that is commensurate with each stage of project development (conceptual, pre-feasibility, feasibility, design and construction, and operation, *cf*. Table 1). Section 5 describes the traditional and risk acceptance criteria, which must be set by the owner and applied to the design methods that are outlined in Section 6. The next three sections deal with the implementation of the design (Section 7), how the performance of the design is assessed and the slopes are monitored (Section 8), and how risks are managed (Section 9). The last section (Section 10) presents some final conclusions.

3 THE GEOTECHNICAL MODEL

The geotechnical model is the foundation of the design process. Its objective is to provide the slope design engineer with the information he needs to correctly assess

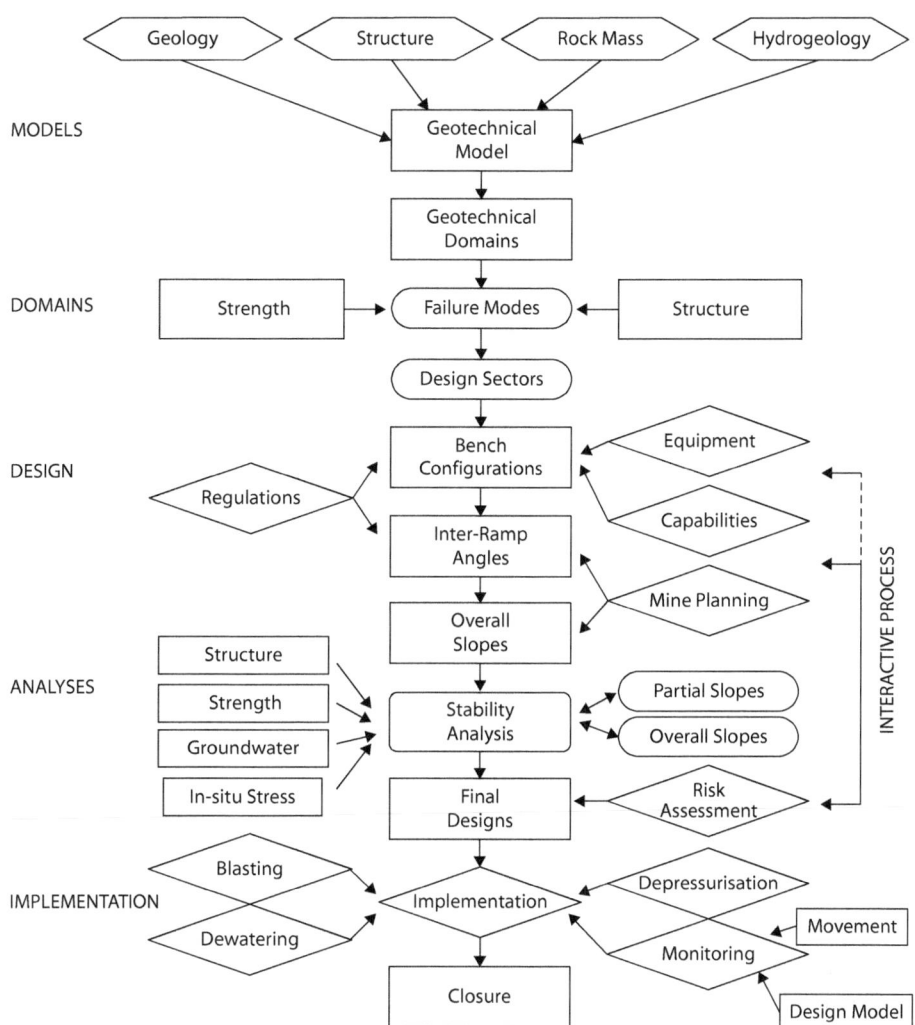

Figure 1 The slope design process.
Source: Read & Stacey, 2009.

the inherently variable properties and characteristics of the natural materials being dealt with. When well prepared, it will enable him to:

- establish robust slope designs;
- reduce uncertainty and improve confidence;
- reduce risk;
- identify the preferred development option;
- identify opportunities.

The model itself is comprised of the four components described below: the geological model; the structural model; the rock mass model; and the hydrogeological model.

3.1 The geological model

3.1.1 Purpose

The geological model presents a 3D distribution of the material types and any alteration variants that will be involved in the pit walls. Its purpose is to link the regional physical geology and the events that lead to the formation of the ore body to a mine-scale description of the setting, distribution and nature of the overburden soils and rock types at the site, including the effects of mineralization (ore versus waste), alteration and weathering.

3.1.2 Physical setting

Pit design too often focuses solely on the mine scale characteristics of the ore body and waste rocks. Before advancing to these details the natural processes that occurred before the deposit was formed must be understood and used to set a framework for the model. Many mines are situated in localities where extreme climatic and related geomorphological and/or geological processes may have had a profound influence not just on the layout of the mining infrastructure, but also on features such as the weathering and alteration profiles and/or the evolution of the deposit.

Many examples can be cited, but two well known extremes are the Ok Tedi mine in Papua New Guinea and the Escondida mine in northern Chile.

1. The Ok Tedi mine in Papua New Guinea, which is located in a mountainous rainforest region where the average annual rainfall is 10 m. The combination of high rainfall and geologically rapid regional uplift and erosion has developed a highly unstable terrain where 40 per cent of the mountain ridges around the mine comprise previous landslide material (Read & Maconochie, 1992). This has been accompanied by tropical weathering and alteration that has affected the strength and stability of the metastable colluvial soils, and the underlying bedrock. All these features required special consideration in the pit wall design.
2. The Escondida mine in northern Chile, which is situated in an arid and geologically mature Tertiary desert landscape where there is no vegetation and rain and/or snow are rare events. However, groundwater fluctuations during regional faulting and uplift since the ore body was emplaced produced a barren leached cap up to 180 m thick above a high-grade sequence of supergene (copper oxide) enriched ore from the deeper primary sulfide ore.

Recognizing such differences, physical regional features that must be incorporated in the model must include at least the following.

- Geographic location.
- Tectonic evolution.
- Climate.
- Geomorphology.
- Topography.
- Drainage system.

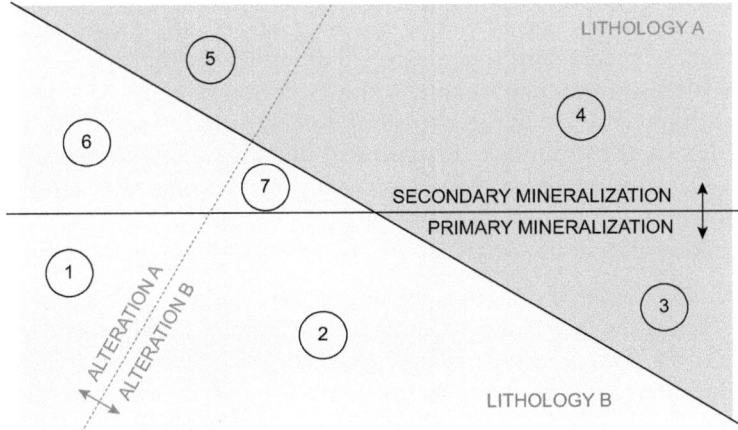

Figure 2 Example of the definition of basic geological units by superimposing the lithology (2 types), the mineralization (2 types), and the alteration (2 types).
Source: Read & Stacey, 2009.

3.1.3 Geotechnical requirements

When the natural regional framework has been established, each rock type at the project site should be subdivided into consistent geological units or domains based on a combination of rock type (lithology), mineralization (ore and waste), and alteration, including all pre- and post-mineralization events.

This requirement is illustrated in Figure 2, using a simple model that includes mineralization and alteration of two different lithologiesto define seven basic geological units.

1. Primary mineralization + Lithology B + Alteration A.
2. Primary mineralization + Lithology B + Alteration B.
3. Primary mineralization + Lithology A + Alteration B.
4. Secondary mineralization + Lithology A +Alteration B.
5. Secondary mineralization + Lithology A +Alteration A.
6. Secondary mineralization + Lithology B + Alteration A.
7. Secondary mineralization + Lithology B + Alteration B.

3.2 The structural model

3.2.1 Purpose

The purpose of the structural model is to describe the orientation and spatial distribution of the structural defects that are likely to influence the stability of the pit slopes. These include major features such as faults and folds that are relatively widely spaced and continuous along strike and dip across the entire mine site, as well as the more closely spaced fabric of joints and faults that typically do not extend for more than two or three benches: all these features should have been identified from outcrop mapping and drilling, and stored in a 3D structural database. They should also have been supported by data that outlines the regional geological setting, describing the tectonic and other geological events that have controlled or influenced the style and shape of the ore body, and the surrounding host rocks from evolution through to mining.

The information contained in the structural model is used to subdivide the rocks at the mine site into a select number of structural domains. Each domain should have distinct boundaries and be characterized internally by a recognizable structural fabric that clearly differentiates it from its neighbors.

The features that should be used to help define each domain include:

- mine-scale contacts marking changes in geology, including changes in lithology (*e.g.*, between igneous and sub-volcanic intrusive rocks and intruded sedimentary rocks), changes in weathering profiles, and changes in alteration styles;
- mine-scale faults that may divide the rocks at the mine site into different structural blocks;
- mine-scale folded structures, with particular emphasis on changes in the orientation of the folds;
- mine-scale metamorphic structures, also with emphasis on changes in the orientation of the structures;
- inter-ramp and bench scale major faults, folds and metamorphic structures;
- bench-scale joints, cleavage and micro-structures such as parasitic or lower-order folds formed on the limbs of any inter-ramp or mine-scale folds.

Figure 3 examples a structural model where the orientation, length and spacing of different sets of major (VIF) and intermediate (Sistemas FT) mine-scale faults have been used to divide the rocks at the mine site into four different structural blocks.

Depending on the nature and application of the model, a number of differing fracture properties need to be defined. The primary properties needed for all models in order to represent the fracture geometry reasonably are:

- fracture orientation distribution;
- fracture spacing distribution;
- fracture intensity;
- spatial variations.

When there are hydraulic and geomechanical applications of the model, a number of other properties need to be defined including:

- fracture aperture;
- fracture transmissivity;
- fracture storativity;
- fracture shear strength properties and stiffnesses.

3.2.2 Blind zones

When the structures that intersect a diamond drill hole are being oriented, the occurrence of structures that have low angles of intersection with the drill hole raises the issue of blind zones, which are common to all structure orientation assessment methods. These zones typically include the loci of the poles of the structures that are parallel to the drill

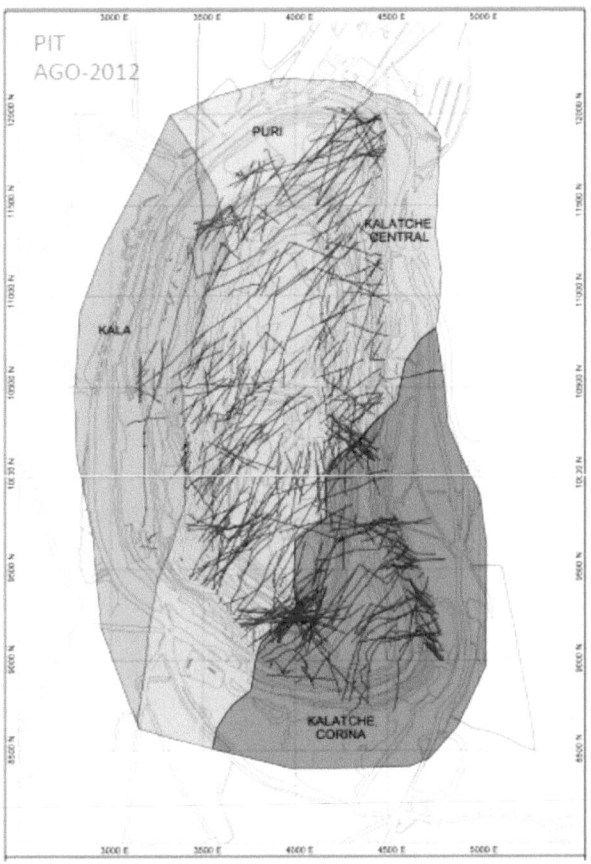

Dominio	Sist	Sistemas VIF Dip (°)	DipDir (°)	Continuidad (m)	Espaciamiento (m)	Sist	Sistemas FT Dip (°)	DipDir (°)	Continuidad (m)	Espaciamiento (m)
Kala	1	64 ± 7	/ 140 ± 11	428	114	1	63 ± 9	/ 163 ± 9	128	34
	2	61 ± 5	/ 291 ± 6	321	61	2	63 ± 7	/ 245 ± 7	96	18
	3	83 ± 3	/ 103 ± 5	206	57	3	75 ± 7	/ 112 ± 7	62	17
						4	77 ± 7	/ 300 ± 7	96	18
kalatche Central	1	73 ± 8	/ 126 ± 9	325	42	1	74 ± 8	/ 128 ± 9	98	13
	2	78 ± 7	/ 300 ± 9	313	58	2	79 ± 6	/ 309 ± 9	94	17
						3	71 ± 6	/ 168 ± 8	98	13
						4	72 ± 7	/ 222 ± 9	96	13
						5	72 ± 8	/ 264 ± 9	94	17
kalatche Corina	1	74 ± 8	/ 282 ± 8	382	77	1	79 ± 7	/ 321 ± 8	197	26
	2	75 ± 7	/ 315 ± 10	656	87	2	77 ± 6	/ 280 ± 9	115	23
	3	77 ± 4	/ 350 ± 8	433	62	3	77 ± 6	/ 353 ± 8	130	19
	4	72 ± 4	/ 53 ± 5	342	41	4	81 ± 4	/ 44 ± 9	103	12
	5	83 ± 3	/ 245 ± 7	356	144	5	80 ± 6	/ 234 ± 9	107	43
						6	81 ± 4	/ 113 ± 9	103	12
Puri	1	69 ± 8	/ 143 ± 8	365	31	1	69 ± 9	/ 135 ± 12	110	9
	2	56 ± 10	/ 207 ± 9	227	22					
	3	76 ± 9	/ 110 ± 6	317	14					

Figure 3 Example structural domains.
Source: Codelco Chile, División Chuquicamata.

hole and therefore are rarely intersected in the drill hole. They can also form perpendicular to the drill hole where an acoustic televiewer is run too quickly in the hole.

All too frequently the occurrence and effects of blind zones are ignored or unrecognized when the structures in an open pit are being modeled. Most commonly they are created when the investigation drill holes along one side of the pit are angled back into the wall. Terzaghi (1965) noted that the only way to overcome their effect is to drill a sufficient number of drill holes so oriented that no structural pole can lie in or near the blind zone of each hole.

An appropriate layout for a single cluster of three holes is for each hole to plunge at approximately 45°, with the orientation of the trace of each hole differing by 120° from that of the other two: a structure of any orientation will be intersected by at least one of these holes at an angle equal to or greater than about 31°.

3.2.3 DFN models

Discrete fracture network (DFN) modeling statistically represents in 3D how the faults and joints recognized by the structural model are spatially distributed within the rockmass behind the pit walls. The DFN model seeks, as illustrated in Figure 4, to represent key elements of the fracture system as discrete objects in space with appropriately defined geometries and properties.

By building geologically realistic models that combine the larger observed deterministic structures with smaller stochastically inferred fractures, DFN models capture both the geometry and connectivity of the fracture network as well as the geometry of the associated intact rock blocks (rock bridges) between the fractures. These features make the DFN model an important tool in helping to visualize how the rock mass deforms and slope failure mechanisms develop, particularly when the failure involves sliding along the major structures and fracture across the rock bridges left between these structures. Other important uses include estimating block size distributions for fragmentation analyses and determining flow conditions in hard rock masses.

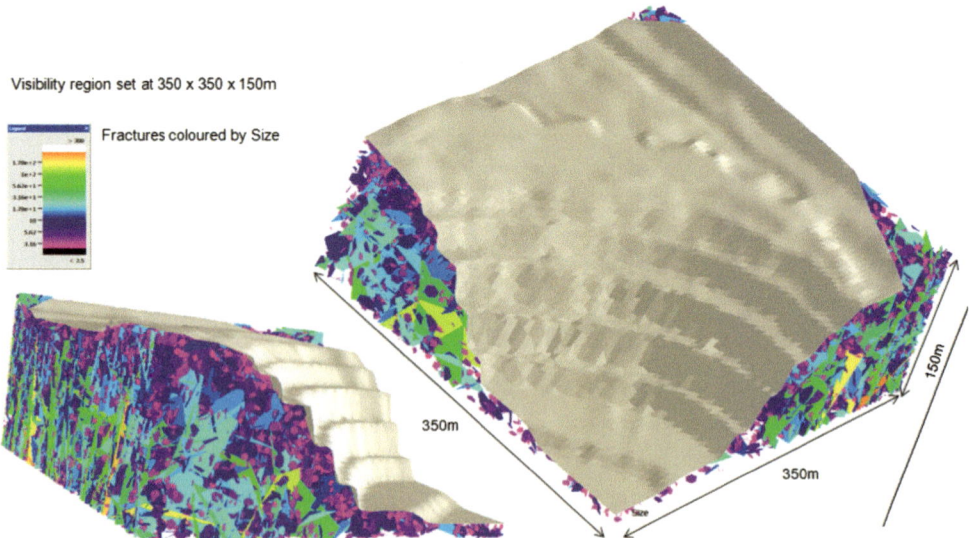

Figure 4 Example DFN model showing cutaway sections of the model behind a pit wall, with color by fracture radius.
Source: Beale & Read, 2014.

3.3 The rock mass model

The purpose of the rock mass model is to compile the engineering properties of the rock mass for use in the stability analyses that will be used to prepare the slope designs at each stage of project development.

When setting out to determine the geotechnical engineering properties of the rock mass a fundamental that must always be considered is that in stronger rocks structure is likely to be the primary control, whereas in weaker rocks strength can be the controlling factor. This means that the rock mass may fail in one of three possible ways.

1. Structurally controlled failure in strong rocks, where the rupture occurs only along the joints, bedding or faults.
2. Failure with partial structural control, where rupture occurs partly through the rock mass and partly through the structures.
3. Failure with limited structural control in weak rocks or in very high slopes, where the rupture occurs predominantly through the rock mass, fundamentally as a result of insufficient capacity (strength).

Because of these differences, when setting out to determine the geotechnical engineering properties of the rock mass the strength of the rock mass and the potential mechanism of failure must be considered and factored into the sampling and testing program. Data representative of the intact pieces of rock, the structures and the rock mass itself will all be required at some stage of the slope design and must be incorporated in the rock mass model.

3.4 The hydrogeological model

The purpose of the hydrogeological model is to present i) the groundwater and ii) the surface water aspects of the hydrogeological regime at the site.

Groundwater pressures acting within the fractures and pore spaces in the rock mass can reduce the effective stress, reducing the shear strength of the rock mass, and thereby increasing the likelihood of slope failures. Potentially, this could lead to a requirement for slope flattening or other remedial measures to compensate for the reduced overall rock mass strength.

Surface water can create saturated conditions and lead to standing water within the pit, which may result in the factors listed below, which are illustrated in Figure 5:

*   loss of access to all or parts of the working mine area;
*   greater use of explosives, or the use of special explosives and increased explosive failures due to wet blast holes;
*   increased equipment wear and inefficient loading;
*   increased damage to tires and inefficient hauling;
*   unsafe working conditions.

It should be noted that groundwater and surface water are the only elements in a slope design that can be readily modified by artificial intervention. Their effects must therefore be fully understood by integrating the geotechnical and hydrological elements during the slope design investigation process, as illustrated in Figure 6.

The cost of 'wet mining'
Wet Haulage
Wet blasting
Wet pit slopes
Wet working conditions

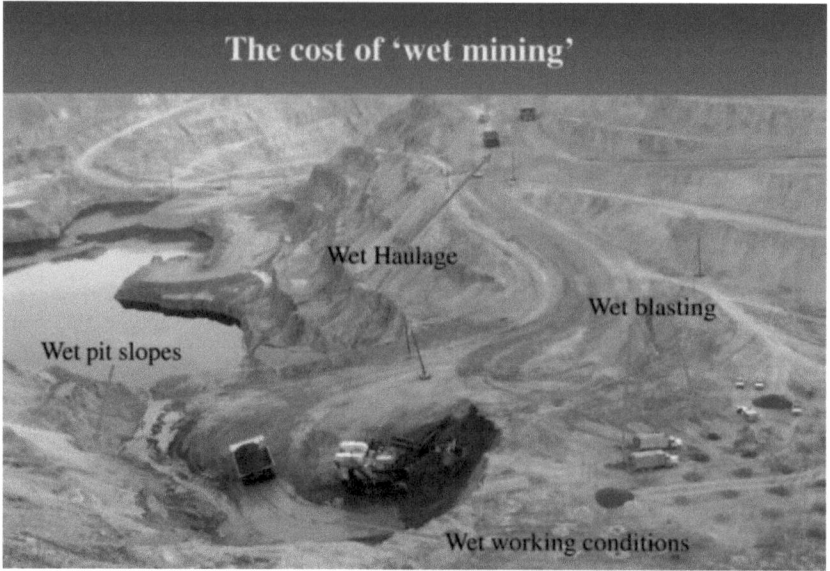

Figure 5 The cost of wet mining.
Source: Beale & Read, 2014.

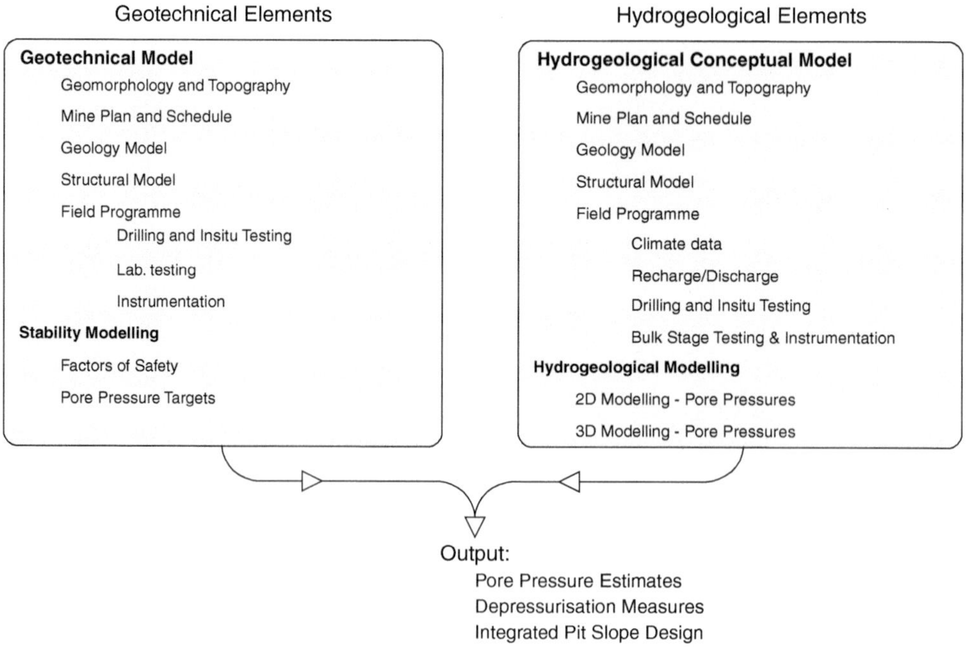

Geotechnical Elements
Geotechnical Model
Geomorphology and Topography
Mine Plan and Schedule
Geology Model
Structural Model
Field Programme
Drilling and Insitu Testing
Lab. testing
Instrumentation
Stability Modelling
Factors of Safety
Pore Pressure Targets
Hydrogeological Elements
Hydrogeological Conceptual Model
Geomorphology and Topography
Mine Plan and Schedule
Geology Model
Structural Model
Field Programme
Climate data
Recharge/Discharge
Drilling and Insitu Testing
Bulk Stage Testing & Instrumentation
Hydrogeological Modelling
2D Modelling - Pore Pressures
3D Modelling - Pore Pressures
Output:
Pore Pressure Estimates
Depressurisation Measures
Integrated Pit Slope Design

Figure 6 Integration of geotechnical and hydrogeological elements in slope design investigations.
Source: Beale & Read, 2014.

4 DATA RELIABILITY

4.1 Introduction

In an open pit mine, data uncertainty stems from the recurrent difficulties geologists, engineering geologists and geotechnical engineers face to correctly predict the inherently variable properties and characteristics of natural materials they are dealing with. Determining the level of uncertainty in the data is one of the most important, if not the most important item in the slope design process. Poor or insufficient data can result in unreliable designs, which are not compensated for by any of the latest and best in slope design stability analyses or by high quality field implementation procedures.

For a pit slope design, the relevant types of uncertainty to be dealt with can be placed into three groups.

1. Parameter uncertainty.
2. Geological uncertainty.
3. Model uncertainty.

- *Parameter uncertainty.* Parameter uncertainty reflects the variability of the elements used to define the various attributes of the rock mass and hydrogeological models; typically values such as the intact strength, friction angle, cohesion, deformation moduli, and hydraulic conductivity and pore pressures.
- *Geological uncertainty.* Geological uncertainty embraces the unpredictability associated with the identification, geometry of and relationships between the different lithologies and structures that constitute the geological and structural models, respectively. It encompasses, for example, uncertainties arising from features such as incorrectly delineated lithological boundaries and major faults, as well as unforeseen geological conditions.
- *Model uncertainty.* Model uncertainty accounts for the unpredictability that surrounds the selection process and the different types of analyses used to formulate the slope design and estimate the reliability of the pit walls. It exists if there is a possibility of obtaining an incorrect result even if exact values are available for all of the parameters used in the analyses.

4.2 Reporting data uncertainty

It is necessary not to just recognize but also to report the reliability of the geotechnical data used in open pit slope designs. These needs were identified and addressed by the sponsors of the industry funded CSIRO Australia Large Open Pit Project (the LOP project). A sponsor panel was formed which recommended that, to provide guidelines for the level of certainty required at each stage of project development, the target levels of effort required at each stage of project development should be matched with the target levels of confidence in the data suggested in the Table 1. In this table the level of work is referenced to the geotechnical characterization needs at each stage of project development, with levels of confidence suggested for each stage of project development for each member of the geotechnical model

Figure 7 illustrates the geotechnical levels of confidence cited in Table 1 relative to the 2004 JORC code (now superseded by the recently introduced 2013 code).

Table 1 Suggested levels of geotechnical effort and target levels of data confidence by project stage.

Project Stage					
Project Level Status	Conceptual	Pre-Feasibility	Feasibility	Design and Construction	Operations
Geotechnical Level Status	Level 1	Level 2	Level 3	Level 4	Level 5
Geotechnical characterization	Pertinent regional information & exploration drill hole data	Assessment & compilation of initial mine scale geotechnical data. Significant program in complex cases	Ongoing assessment & compilation of all new mine scale geotechnical data. Specific programs to address critical issues	Refinement of geotechnical database & 3D model based on initial exposures/ mining	Ongoing maintenance of geotechnical database & 3D model from geotechnical mapping, monitoring & slope performance
Target Levels of Data Confidence in Each Model					
Geology	>50%	50–70%	65–85%	80–90%	>90%
Structural	>20%	40–50%	45–70%	60–75%	>75%
Hydrogeological	>20%	30–50%	40–65%	60–75%	>75%
Rock Mass	>30%	40–65%	60–75%	70–80%	>80%
Geotechnical	>30%	40–60%	50–75%	65–85%	>80%

Source: abridged from Read & Stacey, 2009

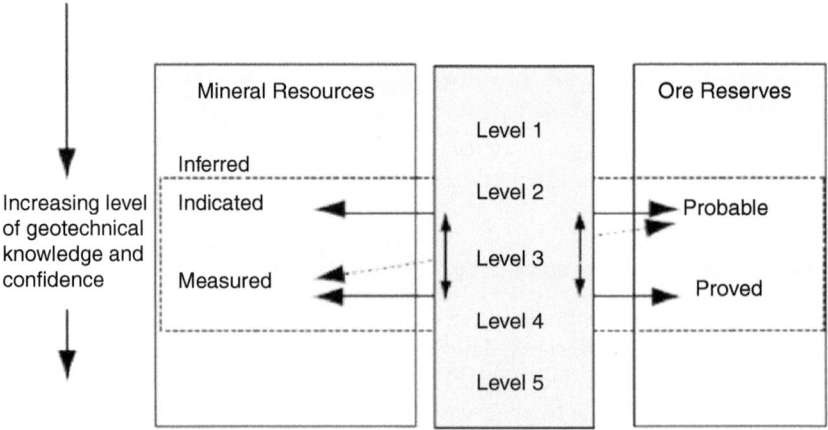

Figure 7 Geotechnical levels of confidence relative to the 2004 JORC code.
Source: Read & Stacey, 2009.

The system outlined in Table 1 has gained widespread acceptance and use across the open pit industry, but it remains subjective. Consequently, there is an ongoing need to develop objective measures of uncertainty that can be confidently used by the slope design engineer, the owner, and the investor to help them assess the reliability of the slope design.

4.3 Determining uncertainty

The system set out in Table 1 has gained widespread acceptance and use across the open pit industry, but it remains subjective. Consequently, there is an ongoing need to develop objective measures of uncertainty that can be confidently used by the slope design engineer, the owner, and the investor to help them assess the reliability of the slope design.

4.3.1 Parameter uncertainty

For parameter uncertainty, this need can be partly met by using the expected value, standard deviation and coefficient of variation of the parameter to assess the inherent uncertainty of the data set. Guided by the measure of the coefficient of variation, where values of less than 10 per cent are considered to be low and values greater than 30 per cent are high, it is customary to accept the expected (average) value as the measure to be used in the design that will be adopted for mining.

However, although the coefficient of variation provides a useful, subjective estimate of performance, it is not a numerical measure of the reliability of the data.

This difficulty can be overcome using a modified Bayesian approach that estimates the expected value of the reliability of the data in a framework of success and failure (Harr, 1996), where the expected value of reliability, E[R], is given by the following equation.

$$E[R] = (S + 1)/(S + F + 2))$$

Where S = the number of successes and F is the number of failures in N trials

The equation can be used in a spreadsheet to provide a straightforward and quick means of checking the estimated reliabilities determined using the expected value, standard deviation and coefficient of variation approach.

An example of a real life case is where, from 24 samples, the designers selected a UCS value of 94 MPa as the feasibility level design value. The standard deviation of the value was 65 MPa and the coefficient of variation 69 per cent, which is well above accepted higher bound of 30 per cent. Using the Bayesian equation the estimated expected value of reliability of the design value of 94 Mpa was 38 per cent, which confirmed the inappropriateness of the value indicated by the high coefficient of variation. Thirty eight per cent is well outside the suggested range of data confidence of 60 to 75 per cent for feasibility level studies and just outside the suggested range of 40 to 65 per cent for pre-feasibility level studies given in Table 1.

4.3.2 Geological uncertainty

In the case of geological uncertainty, through-going fault traces and the boundaries between lithologies and alteration units within geotechnical domains and design sectors are positional, as are the boundaries of the domains and design sectors themselves. Consequently, it is difficult, if not impracticable, to derive probability distributions from measured values that reflect their locations.

Two solutions offer themselves.

1.   To use subjective assessments prepared by competent geologists, engineering geologists and geotechnical engineers, acting individually or as members of a review

panel, as a means of quantifying the uncertainty associated with model geometries and boundaries.

2.   To use generalized plurigaussian simulation to simulate lithologies and structures as a means of quantifying the uncertainty associated with model geometries and boundaries.

Subjective assessments prepared either by an individual 'competent person', or a group of 'competent persons' acting together in a review panel, have become a standard operating procedure in open pit mining. Many aspects of the process raise questions of credibility and defensibility. An overbearing panel member, for example, can disrupt the process of reaching consensus. However, experience over the last 10 to 15 years has shown that a well chosen review panel can work together to produce an effective and balanced opinion of the geotechnical issues under review. The process also recognizes that the judgment and opinion of experienced practitioners are important and of value.

The use of generalized plurigaussian simulations, developed in some detail by the petroleum industry, has been considered by geotechnical practitioners as a means of assessing geological uncertainty. However, it seems that at least for now it is regarded as being too complex a mathematical process to apply to a detailed 3D geological model. Consequently, it would appear that the use of subjective assessments using the judgment and opinions of experienced practitioners as a means of quantifying the uncertainty associated with model geometries and boundaries is likely to remain as the standard operating procedure in open pit mining for the foreseeable future.

5 ACCEPTANCE CRITERIA

The next stage in the slope design process is to apply the data collected and the reliability assigned to them to the iterative design and analysis components of the slope design process, which includes (*cf.* Figure 1):

- the bench configurations;
- the inter-ramp angles;
- the overall slopes;
- the stability analyses;
- the final designs.

5.1 Traditional acceptance criteria

Traditionally, acceptance criteria for open pit mine slopes have been made on the basis of the allowable Factor of Safety (FoS), which is the ratio of the nominal capacity (*C, or resisting forces*) and demand (*D, or driving forces*) of the system.

Over the years other acceptance criteria have been introduced, including the probability of failure (PoF), the consequences of slope displacement on mine operations, and risk.

Factor of safety (FoS) and probability of failure (PoF) criteria typically used across the industry are listed in Table 2.

Table 2 Typical mining industry Factors of Safety (FoS) and probability of failure (PoF).

Slope scale	Consequences of failure	Acceptance criteria[a]		
		FoS (min) (static)	FoS (min) (dynamic)	PoF (max) P[FoS ≤ 1]
Bench	Low–high[b]	1.1	NA	25–50%
Inter-ramp	Low	1.15–1.2	1.0	25%
	Medium	1.2	1.0	20%
	High	1.2–1.3	1.1	10%
Overall	Low	1.2–1.3	1.0	15–20%
	Medium	1.3	1.05	5–10%
	High	1.3–1.5	1.1	≤5%

Source: Read & Stacey, 2009

5.2 Risk acceptance criteria

Traditional acceptance criteria calibrate the performance of the pit slopes, but they do not quantify the risks that may be associated with slope failure.

In slope design, the risks (R) associated with slope failure are defined and quantified as:

$$R = PoF \times (consequences\ of\ failure)$$

Broadly, the consequences of slope failure can be categorized in the following six ways.

1. Fatalities or injuries to personnel, including the costs of industrial and legal action.
2. Damage to equipment and infrastructure, including the costs of replacing equipment and infrastructure.
3. Economic impacts on production, including the costs of:

   - removing failed rock material to the extent that mining can safely continue;
   - slope remediation - the slope may have to be cut back to prevent secondary failures due to steeper upper slopes, or slope support systems may be required;
   - haul road repair and re-access – the haul road and ramp may be damaged and re-access to the mine may need to be considered;
   - equipment re-deployment - the cost of equipment being isolated by the failure and the cost of moving equipment to other parts of the mine unaffected by the failure where it can be used productively should be considered;
   - unrecoverable ore – the loss of a ramp or part of an inter-ramp slope may lead to sterilizing sections of the ore body, at least on a temporary basis.

4. *Force Majeure* (a major economic impact), which should normally equate to failure of an overall slope or loss of medium- to long-term access to ore such that contracts cannot be fulfilled.
5. Industrial action, *i.e.* loss of worker confidence.
6. License to Operate, including public relations aspects such as stakeholder resistance due to social views, environmental impacts arising from the failure, and increased regulatory supervision.

The level of risk that may be accepted by a mining company is an executive management decision. It is likely to be governed by a complex mixture of company culture and attitude to risk, legislative requirements, economics and societal views. The acceptable risk of a fatality from a slope failure is, however, the most sensitive of all the risks, with most companies holding "zero harm" accident policies.

5.3 Voluntary and involuntary risk

Involuntary risks are those to which the average person is exposed without choice, which includes many diseases and general accidents. For voluntary risk, however, only the select few who choose to take part in certain activities are exposed. These include, for example, extreme sports and health-threatening habits such as cigarette smoking and alcohol or drug abuse. Given that social risk acceptance studies have shown that people will accept risk if they perceive the benefit to outweigh the risk, it has been suggested that industrial risk can be regarded as voluntary if and only if the employee has been empowered to consciously accept the risks in order to obtain the reward.

5.4 Consensus risk guidelines

Figure 8 shows the consensus ALARP (as low as reasonably possible) regions on the guideline UK Health and Safety Executive chart. In this chart the upper limit of the ALARP region is defined by a constant risk of 1:1000. For open pit mines it has been suggested that open pit mines slopes should be designed to a fatality risk level between 1:1000 and 1:10 000, within the upper level of the ALARP region.

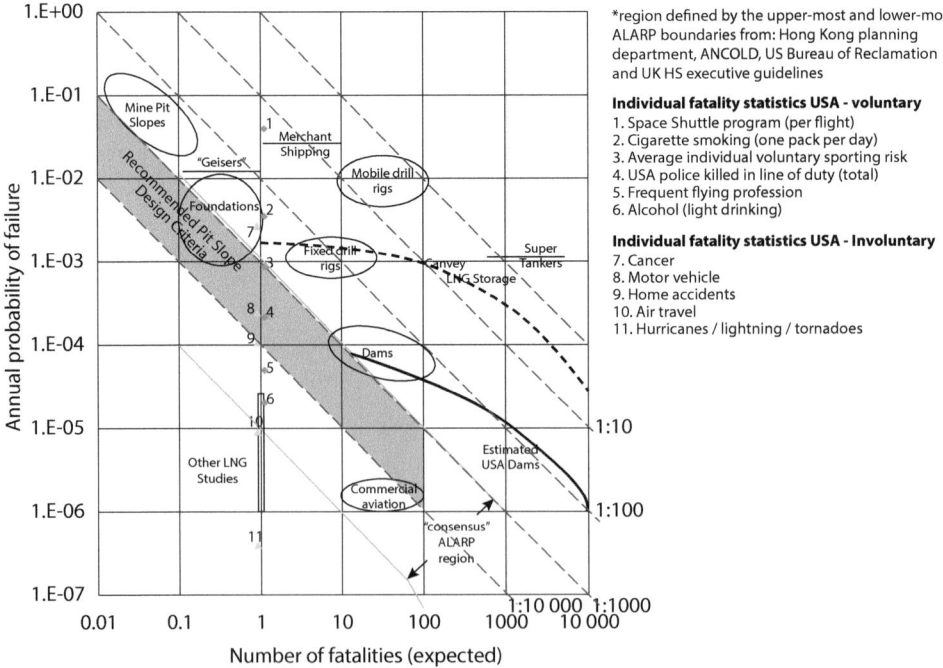

Figure 8 UK Health and Safety Executive consensus ALARP regions.
Source: Read & Stacey, 2009.

Designing mine slopes to the same risk level as that prescribed for dams and other civil engineering structures may seem conservative or even unrealistic. However, contrary to civil engineering structures, this risk level will not be achieved by designing a more conservative slope, but by properly managing the slope to avoid compromising the business plan. In other words, designing flatter slopes does not necessarily help you.

6 SLOPE DESIGN METHODS

6.1 Established methods

The end objective of the slope design process is to enable a safe and economic design for the pit walls at the bench, inter-ramp and overall slope scales within the designated acceptance levels.

The main types of analyses used in slope design studies include the following.

1.   Kinematic analyses, which are based on stereographic projections and are mainly applied to benches.
2.   Limit equilibrium analyses applied to:

   •   structurally controlled failures in bench and inter-ramp slope;
   •   inter-ramp and overall slopes where the stability is controlled by rock mass strength, with or without structural anisotropy.

3.   Numerical analyses using finite element and distinct element methods for the assessment and/or design of the inter-ramp and overall slopes, including deformation assessment.

All of these types of analysis are well described in the public domain and, except for an outline of a recently developed new generation 3D numerical rock slope modeling code (*Slope Model*) directed at reproducing and predicting the behavior of a rock slope based on the interaction of both the rock mass and fractures, and will not be re-described here. However, five design do nots will be emphasized.

1.   **Do not** use the Hoek-Brown strength criterion for determining weak (<R2 and/or GSI≤25) rock mass strengths. A Mohr-Coulomb approach based on suitable, high quality laboratory testing and/or back-analyses is preferred.
2.   **Do not** rely on the Laubscher (1990) rock mass rating criterion (MRMR) to determine the strength of the rock mass in an open pit slope. The MRMR system was designed as an extension of Bieniawski's RMR system (1976 and 1979) for use in underground mining. It contains adjustments to account for the effects of weathering, joint orientation, mining-induced stress, blasting and water and there is no relationship, empirical or other, that enables MRMR determined strengths to be equated with/substituted for the GSI values as used in the Hoek-Brown strength envelope.
3.   At the outset of the project studies, **do not** plunge headlong into performing endless numerical stability analyses. Numerical methods of analysis are capable of modeling many of the complex conditions found in rock slopes, including nonlinear stress-strain behavior, anisotropy and changes in geometry. However, they are not the initiating activity but rather the penultimate activity in the design

process. They should not be performed until there are sufficient reliable data to populate the analytic model, the failure mechanisms perceived for each design sector have been thoroughly evaluated using common sense and limit equilibrium methods, and a need for numerical analyses has been demonstrated.

4. **Do not** rely on moment equilibrium solutions when utilizing limit equilibrium methods of slices codes (*e.g.* Bishop's simplified method of analysis): large scale jointed rock slopes do not form rigid bodies that obey the laws of moment equilibrium. Bishop's simplified method of slices is popular because it is quick and easy to use but, because it is rotational, it rarely approximates the type of geologically controlled failures that occur in large-scale slopes. For such slopes, use only force equilibrium methods (*e.g.* Janbu Modified, Morgenstern & Price, Spencer). Although they will not consider deformations within the sliding body and will ignore out-of-slice forces, they will at least enable you to model irregular failure surfaces such as occur in large-scale slopes. They can also be adapted to model the anisotropic strength of a jointed rock mass and perform step path analyses, as shown in Figures 9 and 10.

5. When calculating pore pressure forces **do not** confuse the phreatic and piezometric surfaces. Phreatic surfaces represent the free ground water level within the slope. In most slopes, this groundwater level will be inclined, indicating groundwater flow. Such conditions require that the pore pressure calculations account for seepage losses. In stability analyses based on any method of slices this requires the determination of the equipotential line passing through the center of the slice base.

If the equipotential line is assumed to be a straight line, the inclination of the phreatic surface and the magnitude of the vertical distance between the phreatic surface and the slice base may be used to estimate the pore pressure head. If the

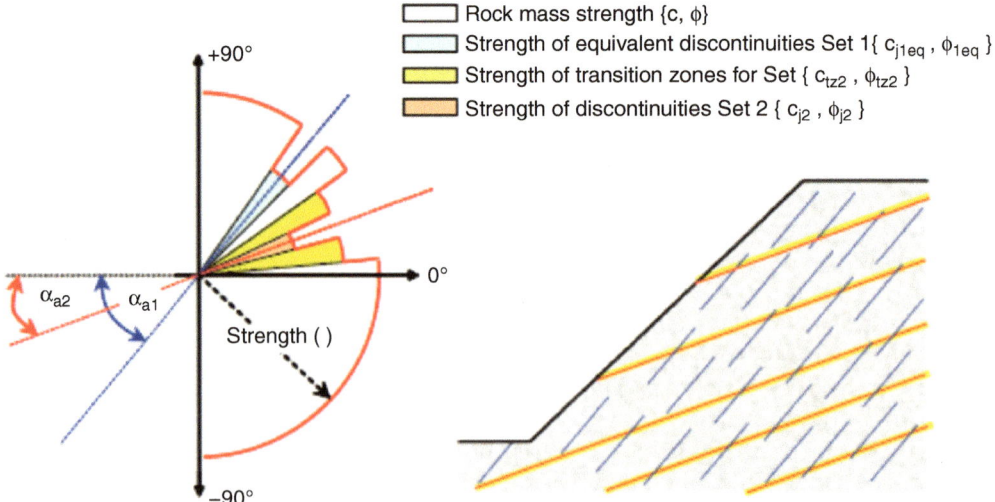

Figure 9 Definition of the directional strength of a rock mass containing two joint sets. *Source: Read & Stacey, 2009.*

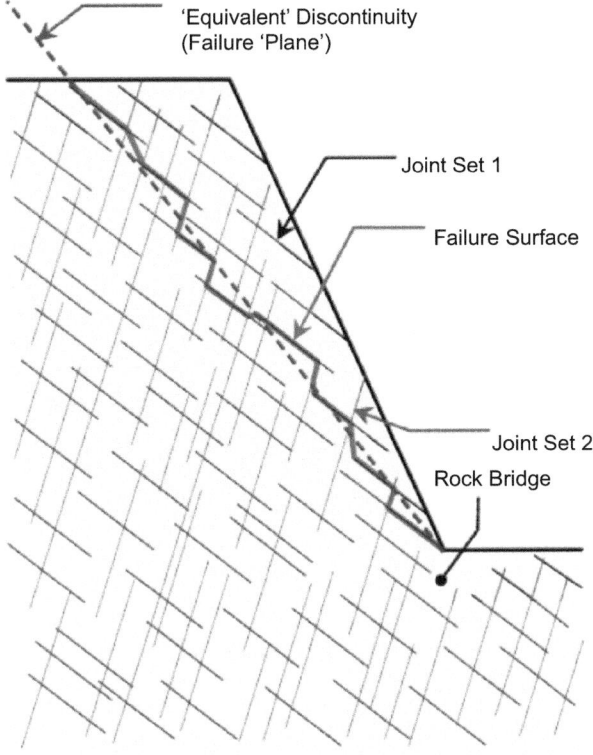

Figure 10 Step path failure surface and "equivalent" discontinuity for a rock slope containing two sets of non persistent discontinuities parallel to the slope.
Source: Read & Stacey, 2009.

piezometric head is known (*i.e.* measured from a piezometer in the slope), the pore pressures should be calculated according to the vertical distance between the base of the slice and the piezometric surface.

Note that, in situations where the phreatic and piezometric surfaces have been confused, if the inclination of the phreatic surface is small (say <5°), the results of the analyses will only be slightly affected. However, for larger angles, the calculated differences will be significantly greater, with the phreatic surfaces always generating lower pore pressures than the piezometric surface.

For numerical methods of analysis the most rigorous method for specifying the distribution of pore pressures within the slopes is to perform a complete flow analysis and use the relevant pore pressures in the stability analyses.

6.2 *Slope model* method of 3D numerical analysis

Continuum (rock mass) and/or discontinuum (fractures and joints) approaches can represent through-going structural defects in a jointed rock mass as interfaces, but do not simulate the brittle fracture that can propagate across the joint fabric within the intact rock bridges between those structures as stress relaxation allows the rock mass to dilate and the rock bridges to separate and move. Instead, empirical friction and cohesion values are applied as "smeared" or "average" values across the rock bridges, which are assumed to behave as a continuum.

This limitation has been overcome by the development of a special 3D code, *Slope Model*, which is a "lattice" model (Beale & Read, 2014, Appendix 6) that achieves high computational efficiency (*e.g.*, 5 to 10 times the execution speed of *PFC3D*). The feature of the code is that it embodies the synthetic rock mass concept (SRM, Pierce *et al.*, 2007). In the SRM model the intact rock is represented by an assemblage of bonded particles numerically calibrated using UCS, modulus and/or Poisson's ratio values to those measured for an intact sample (Potyondy & Cundall, 2004 & Figure 11, lower left). A discrete fracture network (DFN, Figure 11, upper) that captures the geometry and connectivity of the fracture network within the rock bridges is then imported into the particle assembly. In the model the fractures are represented by a smooth joint model (Figure 11, lower right) that allows associated particles to slide through, rather than over, one another and so represent joints that slide and open in the normal way.

Creating and testing the SRM sample illustrated in Figure 11 is essentially a three step process.

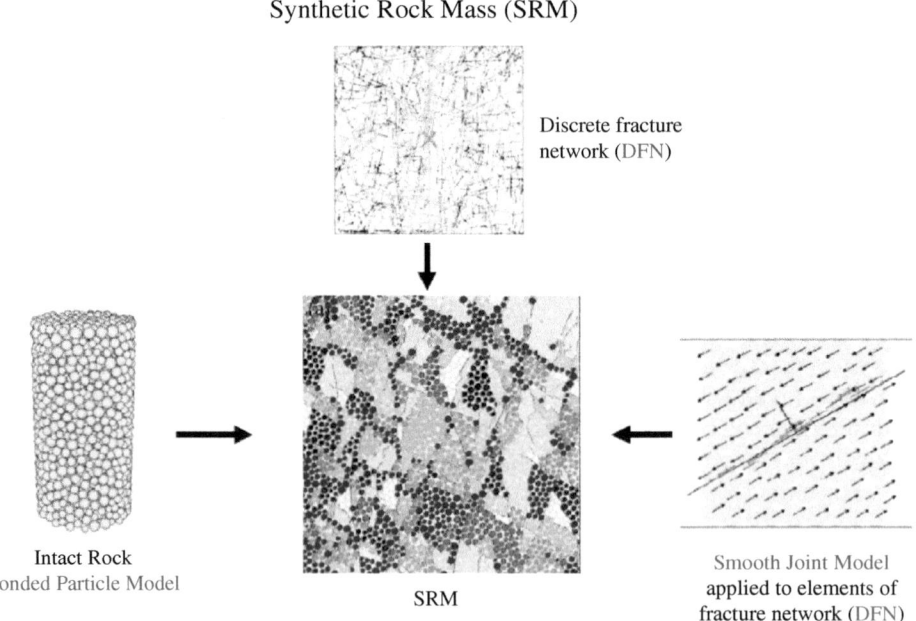

Figure 11 Synthetic rock mass (SRM) assemblage.
Source: Courtesy Itasca Consulting Group, Inc.

1. Creating the particle assembly that represents the intact rock in PFC3D.
2. Generating and importing the DFN that represents the 3D structural pattern of the rock mass into the particle assembly. To produce a DFN that accurately captures these structural properties requires careful measurement in the field, either by hands-on mapping or remote scanning. Regrettably, many mining operations have either reduced or simply do not have the resources to ensure that structural data appropriate for use in DFN simulations is collected. This is a major issue that must be addressed by the owners and management, particularly when the stability of large-scale slopes is involved.
3. Testing. Intermediate stages in preparing the sample involve using the DFN to estimate the average size of the rock bridges that will be modeled and calibrating the micro properties of the synthetic material (*e.g.* particle size distribution and packing, particle and bond stiffness, particle friction coefficients and bond strengths) to the measured properties of the physical material (*e.g.* Young's Modulus, UCS).

Once prepared the 3D failure and deformation characteristics of individual samples can be analyzed under all conditions of stress and stress paths. This is exampled in Figure 12 above that shows the stress-strain response to loading in the x, y & z directions on a 80 m carbonatite sample.

Other advantages of the SRM in rock slope stability analyses include the following.

- It provides a constitutive model (strength envelope) for the material in the rock bridges that is not reliant on either Mohr-Coulomb or Hoek-Brown criteria.
- It provides a strength envelope that honors the strength of the material in the rock bridges at different scales.

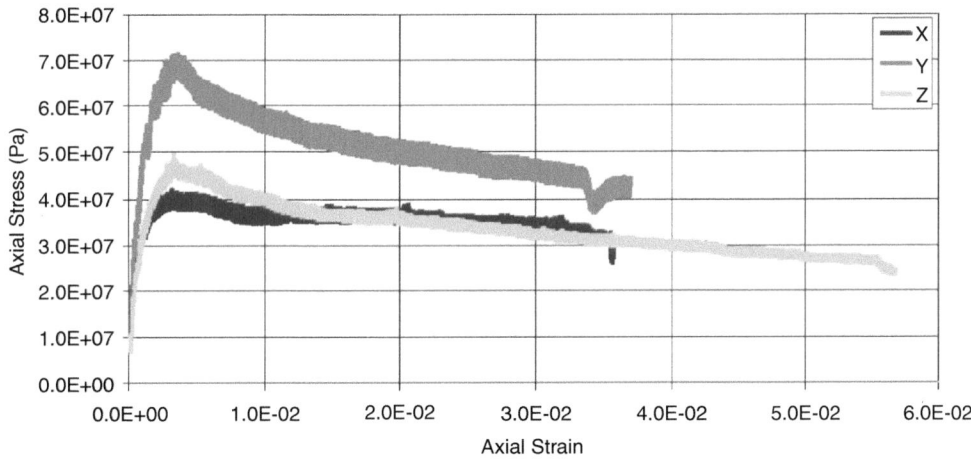

Figure 12 Triaxial stress-strain response to loading in the x, y & z directions on an 80 m carbonatite cube.
Source: Courtesy Itasca Consulting Group, Inc.

- It provides a strength envelope from which the Hoek-Brown parameters can be derived *i.e.* it provides a means of calibrating the Hoek-Brown strength envelope.

7 DESIGN IMPLEMENTATION

7.1 Introduction

The ultimate pit is usually established through a series of interim pit phases that reflect successive cutbacks, a process that allows the final wall design criteria to be confirmed from the operational performance of each preceding phase of pit development. Typically, the process includes requirements for managing water in the pit, controlled blasting, excavation control, scaling, and occasionally slope support to ensure the designs are achieved safely and economically. Of these requirements, controlled blasting and managing water are the most important and are the key to success.

7.2 Controlled blasting

For strong rocks, steep inter-ramp slopes (*e.g.* 50° or more), which are increasingly being introduced in many large open pits as a means of enhancing productivity, and clean bench faces that minimize rock fall hazards, will not be achieved unless the blasting procedures are appropriately controlled.

Unfortunately, no matter whether product prices are high or low, mine operators remain under pressure to minimize mining costs. Inevitably, this introduces conflicts between the interests of production and those of wall stability. Controlled blasting techniques frequently increase the costs on which the operations manager often is judged. However, closer examination will show that even so, steeper, stable walls achieved through controlled blasting will increase the overall profit margin and safety.

In the light of this observation, the following approaches to blasting should always be taken into account.

1. Blasting is the most scientific process in pit operations. It requires high level experienced staff who must be provided with suitable drilling rigs and bench cleaning and bench face scaling equipment that is always maintained in a state of high availability.
2. Fragmentation is the key to achieving high in-pit production and milling targets. This cannot be achieved unless the appropriate production blasting patterns are used together with the appropriate stemming material (*i.e.* crushed rock gravel, not drill cuttings).
3. Close to the pit walls, controlled blasting is essential to minimize wall damage to achieve high safety standards. This can include buffer blasting in weak rocks and trim blasting, with or without pre-splitting in stronger materials. Where benches are stacked, it is preferable to drill the pre-split holes to the full height of the bench face to avoid the formation of small lips at each operating bench level.

4.   Strong QA/QC is required to achieve high quality results. The bench reconciliation procedures outlined below in Section 8.2 (*cf.* Figure 14), supported by QA/QC procedures that ensure i) all loose rock debris has been cleaned from the berms before the blast pattern is set out and drilled, ii) the blast pattern has been accurately positioned, and iii) the blast holes have been drilled at the correct angles and to the correct depths and have been correctly charged, are the basic steps in this process. Additionally, independent QA/QC engineers rather than in-house staff are essential to achieve the required standards.

7.3 Managing water

7.3.1 Surface water

There are two principal goals of a surface water management program related to pit slopes.

1.   To protect the slope from erosion damage caused by surface water.
2.   To shed the water from the slope as quickly as possible in order to minimize the amount of flowing into the pit and/or as recharge to the slope materials. This is particularly important if low permeability, poorly drainable materials occur within the slope.

Surface water may enter the pit from at least the following sources: run-off from incident precipitation onto the slope itself; surface water channels or run-off that may flow onto the slope from the upgradient catchment area above the crest of the wall; gravity drainage, for example from rivers or dams close to the pit crest; and leakage losses from industrial pipes, tanks or other mine facilities.

Typical surface water control measures typically include the following.

- Diversion of any streams that may affect the pit area or may create recharge to alluvial materials that are exposed in the upper pit walls.
- Diversion of runoff from the upgradient catchment area around the crest of the slope.
- Diversion of run-off from the upper benches along drainage channels (toe drains) constructed on catch benches or step-out benches that allow gravity drainage to low points on the pit crest.
- Placement of surface water collection facilities on individual catch benches to shed localized runoff water as quickly as possible.
- Placement of surface water collection facilities below prominent zones of groundwater seepage in the slope.
- Installation of interception trenches, slurry walls or well point cut-off systems to prevent shallow sub-surface seepage into the pit from sources of water near to the pit (*e.g.* rivers or dams).

Uncontrolled surface water flow may lead to (i) erosion and back-cutting of the weaker slope materials, (ii) shallow recharge causing a transient pore pressure rise in the over-break zone creating bench-scale instability and (iii) deeper recharge into the slope creating inter-ramp-scale instability that is often associated with a transient pore pressure rise within larger slopes.

7.3.2 Mine dewatering

A key part in the design of dewatering systems for mines below the water table is the prediction of the potential rate of inflow to the pit and hence the amount of water that will need to be pumped to enable dewatering most commonly to at least one bench below the floor of the pit.

Vertical production pumping wells are the most usual means of dewatering. Projects use a combination of interceptor wells to remove active recharge to the groundwater system at shallower levels, ex-pit wells to remove groundwater flow that would otherwise enter the pit, and in-pit dewatering wells to remove groundwater storage and accelerate drawdown inside the pit shell.

Drainage tunnels installed behind the pit slope or underneath the pit are being increasingly considered by some of the larger open pit mining operators for dewatering. A significant operational advantage of a drainage tunnel is that once the portal is established the drains can be installed and operated from within the tunnel itself, without interfering with mining operations. An obvious potential downside of a tunnel is the up-front cost and time required for planning although, for larger pits, the cost of a tunnel is often comparable with the overall cost of drilling a large number of drains from within the pit for a number of sequential pushbacks. As for all of the drainage and/or depressurisation options, the overall cost must be viewed in terms of the potential benefit of achieving steeper slope angles.

7.3.3 Slope depressurisation

The presence of groundwater can affect open pit mine excavations by changing the effective stress and increasing the resulting pore pressures exerted on the rock mass into which the pit slopes have been excavated.

Increased pore pressures will reduce the shear strength of the rock mass, increasing the likelihood of slope failures and potentially leading to slope flattening or other remedial measures to compensate for the reduced overall rock mass strength.

Of the major factors that control pit slope stability, pore pressure in the slope is the one parameter that can often be readily modified. Other parameters such as lithology, structure and the inherent strength of geological materials (material strength friction and cohesion) cannot normally be changed.

Typical methods of slope depressurisation can be divided into four categories.

1. Natural seepage, which allows pressures to dissipate as a result of seepage to the slope, with no enhanced dewatering/depressurisation measures (*i.e.* passive drainage).
2. Enhanced gravity drainage, by installing gravity flowing drains from the pit slope. These may be horizontal, vertical or inclined (*i.e.* active drainage using gravity flow).
3. Pumped drainage, by installing localized pumping wells or well points, targeting specific units within the slope (*i.e.* active drainage with pumping).
4. Drainage tunnels or galleries installed behind the slope (*i.e.* active drainage that may use a combination of gravity and pumping).

Horizontal drains installed from the pit benches (enhanced gravity drainage) are probably the most commonly used means of active drainage using gravity flow. Typically, hole diameters of 100 mm to 150 mm are used, with 25 mm to 50 mm

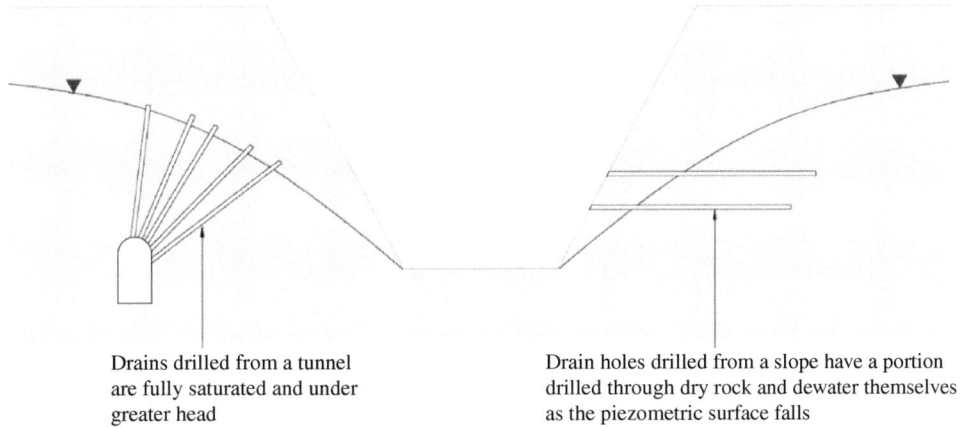

Drains drilled from a tunnel are fully saturated and under greater head

Drain holes drilled from a slope have a portion drilled through dry rock and dewater themselves as the piezometric surface falls

Figure 13 Drain holes from a tunnel.
Source: Beale & Read, 2014.

slotted pipe installed in the drain. However, as for dewatering, drainage tunnels are being increasingly considered to achieve pit slope depressurization.

For depressurization, drilling drains from a tunnel behind the slope offers the following advantages, which are clearly illustrated in Figure 13.

- All drain collars and collection pipes are outside the active open pit mining operation.
- The drains are not affected by subsequent slope pushbacks.
- The drains are under a higher driving hydraulic head and are more efficient than drains drilled from within the pit.
- The drains are saturated throughout their entire length, leading to greater efficiency.

8 PERFORMANCE ASSESSMENT AND SLOPE MONITORING

8.1 Introduction

Slope design is an iterative process whereby:

- design criteria are based on the best available information and a defendable methodology;
- slope designs are implemented according to the established criteria;
- actual geologic conditions, as-built slope geometry and slope behavior are monitored and documented;
- documented versus predicted conditions and behavior are compared and slope design criteria are modified accordingly, completing the cycle.

This process requires systematic monitoring and documentation of the geological conditions and slope performance, and periodic reviews and updates of the design criteria and mine plan as the mine is developed. Techniques for assessing the performance of the slope design and slope monitoring techniques are described in Sections 8.2 and 8.3 below.

8.2 Performance assessment

Systematic documentation and evaluation of the performance of benches is a basic component of any slope assessment program. Benches are the fundamental building blocks of the pit slope, and their geometry and behavior usually control the inter-ramp angles and hence the overall slope design. Additional factors to be taken into account in the inter-ramp slope design include the inter-ramp or bench stack height that can be sustained by the rock mass between the haul roads and/or any intermediate geotechnical safety berms. Systematic monitoring of the piezometric pressures in the inter-ramp and overall slopes is also required to ensure that depressurization targets are being met (*cf.* Section 8.3).

In large open pits advanced digital and laser scanning technologies have become the most widely used means of collecting the information needed to assess the performance of benches and inter-ramp slopes.

For benches, the information collected includes a matrix comprised of design achievement (Fd) and face condition (Fc) factors. The Fd factors include the bench face angle, the bench width and the position of the toe of the bench. The Fc factors include the number of half barrels visible on the face, the degree of intact rock breakage, the openness of joints, the size and quantity of any loose material on the face forming potential rockfall hazards, the shape of the face profile, and the amount damage and/or break-back at the crest.

The total of assigned values for each component in the two factors are then reduced to a factor between 0 and 1 and plotted in the matrix shown in Figure 14. Typical

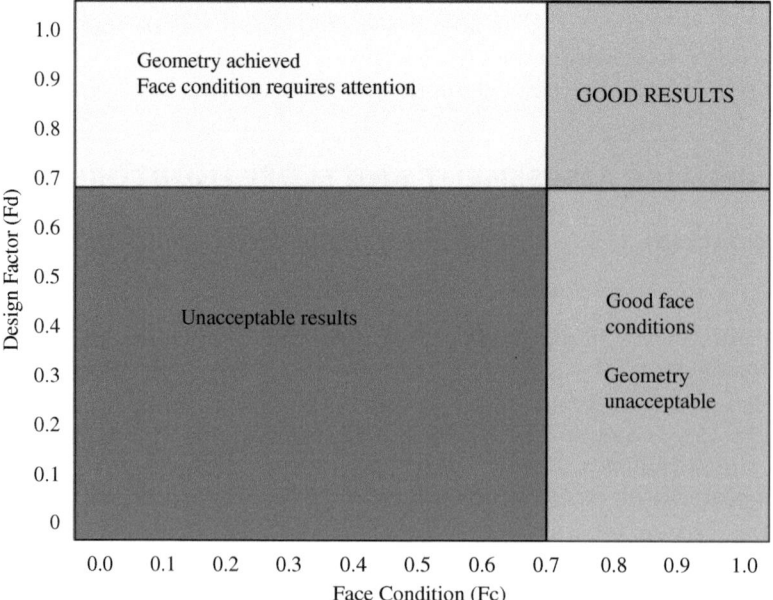

Figure 14 Controlled blast evaluation chart.
Source: Read & Stacey, 2009.

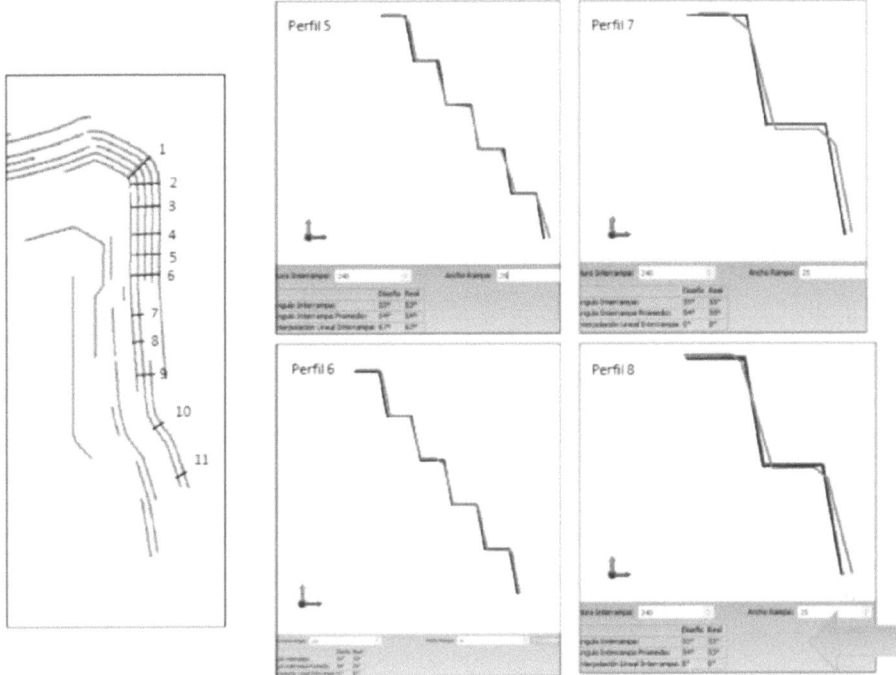

Figure 15 Design and as-constructed inter-ramp slope profiles resulting from 3D laser scanning of the slope.
Source: courtesy AMSA, Chile, Los Pelambres Mine.

assigned factors for each component are given on Table 11.10, Read & Stacey (2009). The system was first introduced at the Chuquicamata mine in Chile, but is now widely used in most of the large open pit mines in Chile, where it is known as "Full Control": at each mine the assigned factors for each component are modified to suit the local conditions.

The Fd information, particularly the information on the shape and position of the toes of each bench in a stack of benches, also provides precise information on whether or not the design inter-ramp slope is being achieved. A typical example is shown in Figure 15: the black linesoutline the design bench profiles and the blue lines the as-constructed profiles.

Additional actions when assessing the performance of the inter-ramp slopes are to look for i) any signs of multi-bench failures involving continuous joints and faults and ii) any signs that the rock mass is not capable of sustaining the bench stack height between the haul roads and/or any intermediate geotechnical safety berms.

Overall slopes are usually limited by the inter-ramp design criteria, the shape of the ore body, haul road access, other mine planning considerations or a combination of factors. Overall slope performance assessments may be limited to documenting the as-built slope geometries to ensure that they are in compliance with the design, and routine monitoring to warn of unanticipated deformations, geological complications or developing adverse pore pressures. If evaluating the overall slope performance based on instrumentation monitoring, it is important to consider the expected response of the slope to mining, and to set thresholds or targets against which performance can be compared.

8.3 Monitoring

The main objectives of a monitoring program can be summarized as follows.

1. Maintaining safe operating conditions to protect personnel and equipment against unexpected slope failures.
2. Providing advance notice of zones of potentially unstable ground so that mine plans can be modified to minimize the impact of slope displacement.
3. Providing geotechnical information for analyzing any slope instability mechanisms that develop, designing appropriate remedial action plans, and conducting future slope designs.
4. Assessing the performance of the implemented slope design.

The golden rule when devising and installing a monitoring system to achieve these objectives is that every instrument installed should be selected and placed to assist in answering a specific question. Following this rule is the key to successful field instrumentation.

Standard monitoring systems and procedures that everyone should be familiar with range from visual observations at the surface to total station survey monitoring of surface prisms, radar monitoring and groundwater monitoring, as listed below.

*   Visual observation from surface walkover inspections.
*   Surface measuring pins and extensometers on observed tension cracks.
*   Real time robotic total station (RTS) survey monitoring of surface prisms.
*   GPS monitoring of surface stations.
*   Surface slope radar monitoring.
*   Subsurface installation of shear strips, time domain reflectometer cables (TDRs), extensometers, and inclinometers.
*   Groundwater pore pressure monitoring from open standpipes and sealed piezometers, particularly strings of vibrating wire piezometers.
*   Microseismic monitoring.

The instruments used to monitor surface displacement in open pits are sophisticated and collectively can provide a real-time 3D record of any surface movements that may be taking place in the walls of the pit

Instruments used to monitor in-ground displacement are, however, less sophisticated. From the list above they typically include:

- shear strips and/or TDRs;
- extensometers;
- inclinometers placed in boreholes to locate subsurface movement after evidence of subsurface deformation has been detected at the surface.

Rarely, if ever, are these systems able to detect in real-time subsurface deformation as it develops and then propagates to the surface.

To augment the current limited sub-surface real time monitoring technology, a subsurface deformation system has been developed in Chile in a joint venture between CSIRO Australia and the University of Chile, with sponsorship from an Australian electronics company and mining companies in Chile. The objectives of the program were to:

- identify when and where the deformation develops;
- Identify the nature and scale of the deformation;
- detect any changes in pore pressure that may accompany the deformation;
- transmit the information wirelessly to the surface in real time.

The technology used is based on subsurface Smart Markers developed for monitoring ore recovery performance in block caves. The markers contain on-board radio transmitters that have an operating life of nine years and currently are used worldwide in 17 different caving operations.

Enhancements of the Smart Markers on-board radio transmitters have enabled the markers to be networked in a chain and to communicate data from below the surface to the surface. Every marker in the chain measures the radio signal strength of communication with the other markers in the chain. Changes in radio signal strength over time indicate:

- movement between markers; or
- changes in the alignment of the markers.

In the open pit chains of markers can be installed in the rock mass behind the pit walls, as illustrated in the left hand schematic in Figure 16.

Any movement of the markers, representing changes in the orientation or distance between the markers caused by subsurface deformation, are reported in real time as the information is "hopped" between markers to the surface reader. The changes in alignment and distance between markers indicate the location and nature/extent of the deformation.

A feature of the system is continuous operation during cutbacks. Although the cutback will remove the markers closest to the pit benches (right hand schematic in Figure 16), the remainder of the system will continue working after the cutback due to the continuing wireless communication between the remainder of the markers in the chain.

Accelerometers and pore pressure sensors have been added to the markers and, together with improvements in movement accuracy and range, the platform will enable:

- real time identification of when and where subsurface deformation develops;
- real time measurement of any pore pressures associated with the deformation;
- transmission of that information to the surface before it propagates to and is observed at the surface.

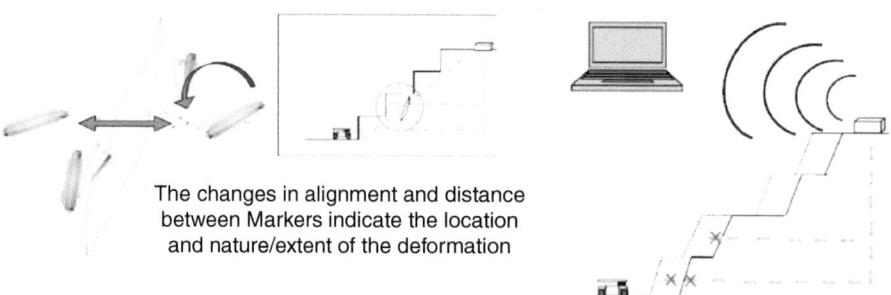

The changes in alignment and distance
between Markers indicate the location
and nature/extent of the deformation

Figure 16 Smart Marker chain installed within the slope to detect subsurface deformation.
Source: Read, 2013.

9 RISK MANAGEMENT

All mining operations are subject to different levels of risk during the life time of a mine. The acceptance of risk levels is the responsibility of the senior management team and their risk advisors who must provide corporate management with a risk-based design that provides sufficient quantitative information to:

- define acceptable risks in terms of safety and economics;
- assess relative risks for different slope configurations;
- benchmark risks against industry norms and the corporate mission statement.

The strategic, operational and technical considerations recommended for managing these risks at all levels of the operation are shown in the flowchart in Figure 17.

The end result of the risk management process should be a hazard management plan that may be a subset of an overarching ground control management plan (GCMP) or may exist as a separate entity. Either way, the plan must include the following information.

1. **An introduction** that contains general statements as to the objective and scope of the plan, plus any relevant historic information used to generate the plan. Because of differences in regulatory requirements, ground conditions and mining methods, this section describes ground control management issues in general terms and outlines the suggested technical content of the plan. Based on information presented in this section, each individual operation should formulate and implement their site specific plan in accordance with local requirements and needs.
2. **A hazard inventory** that describes all geological hazards at the mine site, including rock falls and inter-ramp and overall slope instabilities. The hazards should be communicate to mine personal in a hazard identification plan that is updated weekly so that everyone is aware of the location and scale of any perceived hazards in the pit. Mine personnel should also be asked to provide feedback on any issues that they may observe during their shift so that there is a high awareness of safety practices throughout the operation. This feedback should be formally logged,

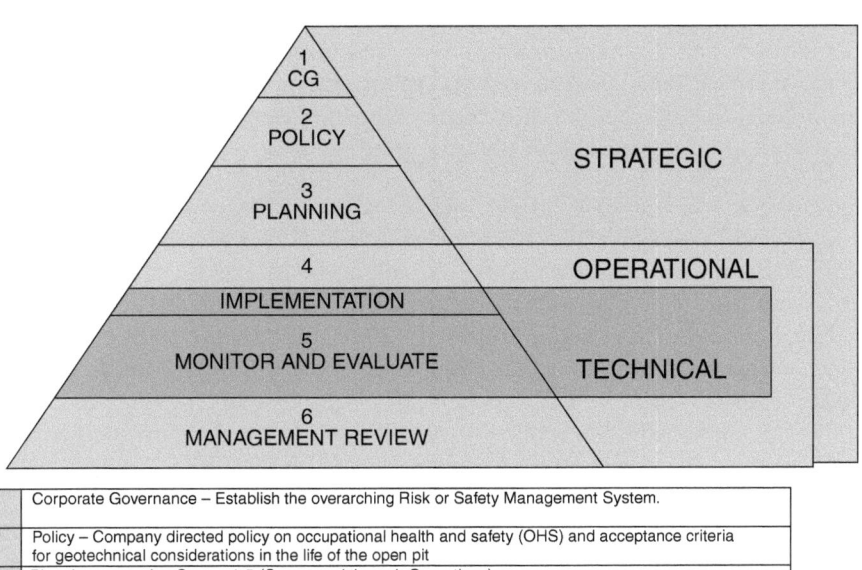

1	Corporate Governance – Establish the overarching Risk or Safety Management System.
2	Policy – Company directed policy on occupational health and safety (OHS) and acceptance criteria for geotechnical considerations in the life of the open pit

3	Planning – covering Stages 1-5 (Conceptual through Operations)
	    3.1 Risk management and generic business continuity management
	    3.2 Legal requirements and compliance with standards
	    3.3 Business objectives, targets, plans, risk benefit analysis, geotechnical model,
	            pit design.

4	Implementation of plans, procedures, standard operating procedures (SOPs) and records at the design, operational and closure/transitional levels
	    4.1 Organisational structure, roles and responsibility
	    4.2 Operational risk management – geotechnical model, pit design, implementation,
	            slope management plan, geotechnical procedures, *e.g.*, mapping and monitoring
	    4.3 Business continuity management
	    4.4 Consultation, communication and reporting
	    4.3 Training and competency
	    4.5 Documentation and data control

5	Monitor and Evaluate – stability management, mine to design, design performance
	    5.1 Monitoring and measurement
	    5.2 Incident investigation, corrective action and preventative action
	    5.3 Records and record management
	    5.4 Audit – internal and external

6	Management Review

Figure 17 Overarching methodology for managing open pit slope failure risks.
Source: Read & Stacey, 2009.

together with the inspection reports from the geotechnical team and the shift supervisors.

3. **A risk reduction plan** that describes the methods that will be utilized to reduce the risks associated with the hazards listed in the hazard inventory.

4. **Trigger action responses (TARPS)** that outline a set of planned responses to all likely events.

5. **Emergency response procedures,** ranging from procedures for managing the impact of low-probability but high-consequence events on the mine to procedures for evacuating personnel and equipment from the mine in cases of emergency. Once developed these procedures, especially the mine evacuation procedures, should be practiced and reviewed at regular intervals.

6.   **Roles and responsibilities** for all employees and contractors at the mine, including the mine manager, to ensure that no work is undertaken without an authorized plan, safe operating procedures are followed, all ground-related hazards are identified and reported to supervisory and/or geotechnical staff, and ground conditions are inspected prior to and during work activities.

The most sensitive of all risks is the risk of a fatality from a slope failure. Most companies hold "zero harm" accident policies, which is a worthy aim but one that accident statistics show is not a reality. To help close this gap it is essential that the GCMP risk reduction plan has in place a hierarchy of controls directed at eliminating or controlling the occurrence and/or impact or consequences of a risk that endangers personnel safety. Monitoring and feedback are also essential parts of the overall risk management process and its adaptation to the management of geotechnical risks in large open pit slopes. Ongoing review is essential to ensure that the geotechnical and slope management plans remain up-to-date, relevant and useful.

10 CONCLUSIONS

To achieve the economic return expected by all of the stakeholders in an open pit mining project, six key elements of the slope design process, outlined below, must be kept in the forefront at each successive development stage of the project.

1.   **Geotechnical Model.** As noted in Section 3, the geotechnical model with its constituent parts, the geological, structural, rock mass and hydrogeological models, is the foundation of the design process. Its objective is to provide the slope design engineer with the information he needs to correctly assess the inherently variable properties and characteristics of the natural materials being dealt with. Without question, the geotechnical model must be in place before the ongoing steps in the design process are initiated.
2.   **Data Reliability.** Determining the level of uncertainty in the data is one of the most important, if not the most important item in the slope design process. Unreliable data will provide unreliable designs, which are not compensated for by any of the latest and best in slope design stability analyses or by high quality field implementation procedures.
3.   **Acceptance Criteria and Risk.** Before the slope designs are accepted they must be aligned with the slope failure criteria that require the walls of the pit to be stable for the required life of the pit, which may extend into closure. The criteria must be specified by the owner and linked to risk levels that reflect executive management's understanding of the company's culture and attitude to risk, legislative requirements, economics, and societal views.
4.   **Design Implementation.** Independently managed QA/QC procedures are required to ensure that the limits blasting and excavation procedures required to implement the wall designs are performed safely and economically. The steep inter-ramp slopes (*e.g.*, 50° or more), with clean bench faces that minimize rock fall hazards sought by many operators any now will not be achieved unless the blasting and face scaling procedures are appropriately controlled.

5. **Performance Assessment and Monitoring.** Systematic monitoring and documentation of the performance of the benches and inter-ramp and overall slopes, supported by periodic reviews and updates of the design criteria and mine plan, must be performed as the mine is developed. Monitoring of the piezometric pressures in the inter-ramp and overall slopes is also required to ensure that depressurization targets are being met.

6. **Risk Management.** The definition and acceptance of risk levels at a mine are the responsibility of the senior management team and their risk advisors, who must provide corporate management with a risk-based design that provides sufficient quantitative information to define acceptable risks in terms of safety and economics, assess relative risks for different slope configurations, and benchmark risks against the corporate mission statement and industry norms.

ACKNOWLEDGMENTS

The information presented in this chapter results from the CSIRO Australia Large Open Pit (LOP) project, an international research and technology transfer project on the stability of rock slopes in open mines supported between 2005 and 2014 by the following mining companies: Anglo American plc; AngloGold Ashanti Limited; Barrick Gold Corporation Pty Limited; BHP Billiton Innovation Pty Limited; BHP Chile Inc; Compañía Vale do Rio Doce (Vale); Compañía Minera Doña Inés de Collahuasi SCM; Corporacion Naciónal del Cobre de Chile; De Beers Group Services (Proprietary) Limited; Debswana Diamond Company; Newcrest Mining Limited; Newmont Mining Services Pty Limited; Newmont USA Limited; Ok Tedi Mining Limited; Technological Resources Pty Limited (RioTinto); Teck Resources Limited; and Xstrata Queensland Limited.

In the 20 years prior to the early 2000s, slopes in "large" open pit mines were mainly in the depth range of 300 m to 500 m and their investigation, design and excavation largely followed procedures enunciated in the 1970s and early 1980s (*cf.* Hoek & Bray, 1981). Since then slope depths have increased toward and beyond 1000 m which, in turn, exposed critical gaps in our knowledge and understanding of the relationships between the strength and deformability of rock masses and the likely mechanisms of failure in such high slopes. This knowledge gap generated a highly diverse literature outlining any number of ideas concerning likely failure mechanisms and a plethora of suggestions of how best to analyze them for stability. Most frequently, the outcome was confusion, facing slope design practitioners with an uncertain choice concerning which design approach and which method of analysis would give them the best outcome. The key question became "do we understand the design and can we quantify its reliability?"

In response to this question, the purpose or theme of the CSIRO Australia LOP project was to address the observed critical gaps and uncertainties in our knowledge base and create more effective ways than previously existed for predicting the reliability of rock slopes in large open pit mines. Ongoing research, which will be sponsored by companies from the original LOP partnership, will ensure that the industry will continue to develop new and improved open pit mining investigation, design and excavation procedures.

REFERENCES

Beale, G. & Read, J. (eds.), 2014. *Guidelines for Evaluating Water in Pit Slope Stability*. CSIRO Publishing, Melbourne.

Bieniawski, Z.T., 1976. Rock mass classification in rock engineering applications. In *Proceedings of the Symposium on Exploration for Rock Engineering*, Johannesburg, vol. 1, pp. 97–106. A.A. Balkema, Rotterdam.

Bieniawski, Z.T., 1979. The geomechanics classification in rock engineering applications. In *Proceedings of 4th Congress of International Society of Rock Mechanics*, Montreux, vol. 2, pp. 41–48. A.A. Balkema, Rotterdam.

Harr, Milton E., 1996. *Reliability-Based Design in Civil Engineering*. Dover Publications, Inc. ISBN: 0-486-69429-1

Hoek, E. & Bray, J.W., 1981. *Rock Slope Engineering*. IMM, London, 358p.

Laubscher, D.H., 1990. A geomechanics classification system for the rating of rock mass in mine design. Journal of the South African Institute of Mining and Metallurgy, 90 (10), 267–273.

Pierce, M., Cundall, P., Potyondy, D. & Ivars, D.M., 2007. A synthetic rock mass model for jointed rock. In *Proceedings of 1st Canada–US Rock Mechanics Symposium* (eds. E. Eberhardt, D. Stead & T. Morrison), 27–31 May, Canada, pp. 341–349. Taylor & Francis, London.

Potyondy, D.O. & Cundall, P.A., 2004. A bonded-particle model for rock. *International Journal of Rock Mechanics and Mining Science and Geomechanics Abstracts*, 41, 1329–1364.

Read, J., 2013. Real time sub-surface monitoring in open pits. *ACG Instrumentation and Slope Monitoring Workshop*, 2013, *International Symposium on Slope Stability in Open Pit Mining and Civil Engineering*, 25–17 September, Brisbane, Australia.

Read, J.R.L. & Maconochie, A.P., 1992. *The Vancouver Ridge Landslide, Ok Tedi Mine, Papua New Guinea*. Proceedings of the 6th International Symposium, Landslides, Christchurch (N.Z.), February 1992, Bell (ed.), pp. 1317–1321. A.A. Balkema, Rotterdam.

Read, J. & Stacey, P. (eds.), 2009. *Guidelines for Open Pit Slope Design*. CSIRO Publishing, Melbourne.

Terzaghi, R. D., 1965. Sources of error in joint surveys. *Geotechnique*, 15, 287–304.

Analysis and Design of Tunnels, Caverns and Stopes

Chapter 26

Q-system: An illustrated guide for tunneling

Nick Barton[1] & Eystein Grimstad[2]
[1]*Nick Barton & Associates, Oslo, Norway*
[2]*Geolog Eystein Grimstad, Oslo, Norway*

Abstract: This paper provides a well-illustrated guide to the workings of the Q-system, with many examples demonstrating its use. Not only rock exposure logging, but also core-logging, and tunnel-logging are illustrated with quantified examples. The Q-system was developed 40 years ago for describing rock mass quality in a quantitative way, using six important parameters and ratings of quality. These were first related to structural geology, in particular the number of joint sets, their roughness, whether there was clay-filling, followed by the effects of water and the stress/strength ratio. A logarithmic-like scale from about 0.001 to 1000 was the result. All the ratings of the key parameters are given in this guide, and include footnotes and a field-logging sheet and examples of its use. Linked to the Q-value and the span or height of the excavation in rock, and also reflecting the final purpose of the excavation, is an updated chart of recommended support and reinforcement for the arch and walls of underground excavations. Both tunnels and caverns are catered for, from roughly 3m to 60m span. In 1993, the S(mr) support was updated by the same authors, replacing mesh reinforced shotcrete with fiber reinforced sprayed concrete or S(fr). The recommended PVC-sleeved (CT) bolts were more resistant to corrosion. The Q-system has always reflected single-shell B+S(fr) concepts of permanent support, as encompassed in the Norwegian Method of Tunneling (NMT). During the 40 years of its use the Q-value has been shown to have empirical relationships to seismic velocity, deformation modulus, and tunnel or cavern deformation. It can also be used for helping to quantify the benefits of high-pressure pre-injection, and to estimate permeability. In addition, the Q-value has been extended for use in TBM prognosis. This is the subject of a separate chapter in this five-volume set.

Keywords: rock mass, classification, tunnels, drill-core, rock support, seismic velocity.

1 INTRODUCTION

Norway is a country with a small population, yet 3,500 km of hydro-power related tunneling, about 180 underground power houses, and some 1,500 km of road and rail tunnels. This has meant that *economic* tunnels, power-houses and also storage caverns, have always been needed, especially prior to the development of North Sea petroleum resources. The Q-system development in 1973 always reflected this, and single-shell tunnel support and reinforcement, meaning shotcrete and rock bolts as final support has been the norm, both before and since Q-system development. About 60% of the first 200-plus case records from which Q was developed were from Scandinavia, and already represented *fifty different rock types*, which is perhaps surprising for those who may focus on the quite frequent pre-Cambrian granites and gneisses, typical in many Norwegian tunnels. Norwegian and Swedish hydro power projects dominated these early cases, giving a wide range of excavation sizes and uses (*i.e.* access tunnels, headrace tunnels, powerhouses). An update of the Q-system support methods, presented by Grimstad & Barton (1993), was based on 1,050 new case records collected between 1986 and 1993. These were deliberately chosen to be independent of Q-system application. They were mostly developed from road tunnel projects, where higher levels of support were generally used. This update specifically replaced S(mr) with S(fr), meaning the replacement of steel mesh with steel fiber-reinforced shotcrete. In 2002, approximately 800 more case records were added, giving further independent measures of S(fr) thickness and bolt spacing. The Q-value had been logged, but was not used in many cases. Some inconsistent results can be noted, including three collapses where Q-recommendations were not used (see Appendix).

2 HOW AND WHY Q WAS DEVELOPED

A question from the Norwegian State Power Board (Statkraft) which was passed to the first author at NGI in 1973, was the following: "Why are Norwegian powerhouses showing such a wide range of deformations?" A lack of quantitative methods for describing rock quality in 1973, besides Deere's RQD from 1964, and the need to consider excavation dimensions, depth and possible stress levels, together with the different support measures used at that time, meant that a new and integrated method was needed. After 6 months of extensive case record study, using a successively updated list of rock mass parameters and constantly updated ratings, the Statkraft question could finally be answered. This 6 months delay saw the development of the Q-system (Barton *et al.*, 1974), which has eventually become one of the main rock mass classification methods used throughout the world of mining and civil engineering. It is often used alongside RQD and RMR (Bieniawski 1989), both of which were developed before Q. Both RMR and Q have made use of RQD, and in the case of Q, the RQD % is used directly, unless it is < 10%. (The minimum used is 10%).

3 CLASSIFICATION METHOD BRIEFLY DESCRIBED

Trial and error using two, three, four and finally six rock mass parameters, with successive adjustment of ratings to get the best fit between rock quality, excavation dimensions, and support quantities, resulted in one of the simplest equations regularly used in rock engineering.

$$Q = \frac{RQD}{J_n} \times \frac{J_r}{J_a} \times \frac{J_w}{SRF} \tag{1}$$

Due to the need for the rock mass classification to fit the case records, the ratings and format of the Q-equation eventually resulted in something resembling a log-scale, with

Table 1 A short summary of the Q-system parameters and the case record back-ground.

Q-parameter definitions	Q case-record back-ground
RQD is the % of *competent* drill-core sticks > 100 mm in length *in a selected domain.* (In tunnel mapping imagine horizontal cores or scan-lines).	The initial database in 1973 was 212 cases of single-shell tunnels and caverns, for hydropower, road, rail, storage, sewage.
Jn = the rating for the number of joint sets (9 for 3 sets, 4 for 2 sets etc.) *in the same domain.*	About 60% of the initial cases were from Scandinavia and about 40% were from Europe, USA, etc.
Jr = the rating for the roughness of the *least favorable* of these joint sets or filled discontinuities, *in the same domain.*	About 50% of the initial cases were from hydropower projects in Norway and Sweden.
Ja = the rating for the degree of alteration or clay filling of the *least favorable* of these joint sets or filled discontinuities, *in the same domain.*	Fifty rock types were initially represented. The majority were igneous and metamorphic rocks, with a smaller number of weak sedimentary rocks.
Jw = the rating for the water inflow and pressure effects, which may cause outwash of discontinuity infillings, *in the same domain.*	Numerous shear zones and faults containing clay, and numerous cases with clay-coated and clay-filled joints were included.
SRF = the rating for faulting, for strength/stress ratios in hard massive rocks, for *squeezing* or for swelling *in soft rock – in the same domain.*	Numerous cases of weathered conditions were also included, with all Q-parameters adversely affected.
(Note: in the 1993 update, three new high-SRF classes related to the observed effects of high stress and extreme support needs were added, specifically for the case of 'spalling' and 'bursting' in initially *massive rock*). See Appendix A1, Table 6 b, L, M and N. Stress-induced fracturing: $\sigma_{\theta(max)} > 0.4\sigma_c$ (However in deep mines with significant numbers of joint sets one should use the original lower SRF values from 1974). Shen & Barton (2017) recently showed that $\sigma_{\theta(max)} = 0.4\sigma_c$ is equal to σ_t/υ, where σ_t = tensile strength and υ = Poisson's ratio. Fracturing due to extensional strain precedes propagation in shear.	In 1993 another 1050 case records were added, mostly from road tunnels. S(mr) was replaced by S(fr) – fiber reinforced shotcrete. S(mr) for tunnel support was totally replaced by S(fr) by 1983 in Norway. These updates provided case records in which the Q-recommendations were not used, ensuring 'independence'. In 2002, approximately 800 more cases of S(fr), RRS and bolt spacing for permanent support were added. The scatter seen in non-Q practice is sometimes wide and includes cases of cave in. See Appendix A4. There are now approximately 2,060 tunneling/cavern cases in total, which lie behind the Q-support-and-reinforcement recommendations for tunnels and caverns.

Q ranging from approximately 0.001 to 1000, from the worst (faulted, squeezing, water-bearing) conditions, to the best (massive dry) conditions. The formal definitions and ratings of the six parameters are tabulated in the Appendix A1. It should be noted that the three pairs of parameters RQD/number of joint sets, joint roughness/joint alteration-filling, and water/stress-strength, *i.e.* $\frac{RQD}{J_n} \times \frac{J_r}{J_a} \times \frac{J_w}{SRF}$ resemble, in very approximate terms: *block size, inter-block shear strength* and *active stress.*

The Q-system is designed to assist in feasibility studies, and to be actively used in detailed site characterization when mapping exposures, interpreting seismic velocities and logging drill-core. It is also used systematically once tunneling begins, since the mapped *rock class* following each blast can be a basis for selecting tunnel reinforcement (bolting) and support (fiber-reinforced shotcrete). Finite element modeling does not answer 'day-to-day (blast-by-blast) questions', so empiricism that works due to its track record is obviously essential. In the following sections we will give photographic examples of core logging, surface-exposure logging, and tunnel logging, so that potential users can get some feel for the method.

The Q-system needs to be used by engineering geologists with some reliable training and experience behind them. The initial assessment naturally involves an evaluation of the degree of jointing, the number of joint sets: *i.e.* the general degree of fracturing and block-size, followed by an assessment of the most adverse Jr/Ja combination, taking into account favorable and unfavorable orientations. What is causing most over-break (*e.g.* Figure 1), and what would happen with no reinforcement or support? Experience is also essential in the determination of the necessary SRF category. Evaluation of this parameter involves knowing the depth or likely stress level in relation to the probable strength of the rock. The degree of stress-induced fracturing, if already occurring, or the

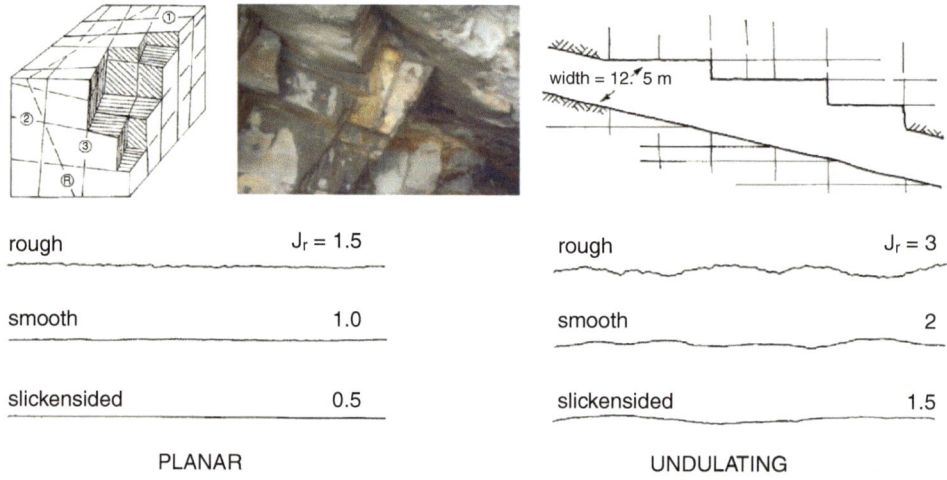

rough	$J_r = 1.5$
smooth	1.0
slickensided	0.5

PLANAR

rough	$J_r = 3$
smooth	2
slickensided	1.5

UNDULATING

width = 12.5 m

Figure 1   Some graphic illustrations of the workings of the Q-parameters, using number of joint sets (Jn) and roughness (Jr). Sufficient numbers of joint sets may or may not cause over-break. When Jn/Jr ≥ 6, over-break becomes extremely likely, even with careful blasting. High Ja obviously assists here.

Figure 2   Contrasting worst (Q ≈ 0.001) and best (Q ≈ 1000) rock mass qualities. The logarithmic appearance of the Q-value scale, stretching over six orders of magnitude, has proved to be a great advantage, and results in simple empirical equations for relating to velocity, modulus, and deformation. The large numerical range appears realistic when considering shear strength and modulus variation.

amount of shearing and clay that is present in the case of fault zones, will each give clues to the appropriate value of SRF. Water inflow is also assessed, with or without the availability of Lugeon or permeability test results in the early stages of logging, and when local measurements are not available.

4 EXAMPLES OF CORE-LOGGING WITH Q

In projects where there are poor exposures due to weathering, the first sight of the rock may be via drill-core. It is strongly advised that a significant number, perhaps most of the bore holes, should be deviated from vertical, because of the frequency of sub-vertical structure which is poorly sampled by vertical holes. In a recent rail tunnel in Norway, all five boreholes were strongly deviated, thereby sampling folded and steeply tilted inter-bedding much more effectively. Vertical holes may give false higher quality, and unrealistically low permeabilities. Deviated holes sample more rock mass.

4.1 Additional advice concerning core logging

Since drill cores are often missing where the rock quality is very poor due to poor recovery (*e.g.* see the plastic containers in Figure 3.1), the rock mass lack-of-quality has to be assessed by other methods, such as seismic velocity or resistivity. Where cores do exist, and there is good recovery, the first four of the Q-parameters may be evaluated with a relatively high degree of accuracy. However, special attention should be addressed to the following:

- Evaluation of the large and medium scale roughness parameter Jr may be difficult when joints are intersecting the borehole at an obtuse angle, due to short samples.
- As water is generally used during drilling, mineral fillings like softer clay minerals may be washed out, making it difficult to evaluate Ja in some cases.
- Joints sub-parallel to the borehole will be under-represented, and will give too high RQD-values and too low Jn-values. So both Q and permeability will be affected.
- RQD is often calculated for every meter. However Jn must usually be estimated for sections of several meters, by observing the core boxes from above and below.
- Water loss or Lugeon tests are often carried out during core drilling, and can form the basis for evaluation of the Jw-value. Since grouting often reduces the permeability, and tends to improve many Q-parameters, there will be an increase in the estimated Q-value in case of logging grouted sections of the rock mass, when logging cored probe holes following pre-injection.
- An estimation of SRF in massive rock can be made based on the height of overburden, using height/steepness of an eventual mountain side. If stress measurements are carried out in boreholes, or experiences from nearby construction sites are available, these should be used so that the probable stress magnitude can be compared with an estimate or measurement of the uniaxial strength of the rock. (Core-disking and subsequent stress-induced or strain-induced fracturing around the tunnel each give clues to stress levels in relation to strength levels, remembering here that an induced stress of about ½ of UCS is already likely to show stress fracturing. (The same applies when $\sigma_{\theta(max)} > \sigma_t/v$).

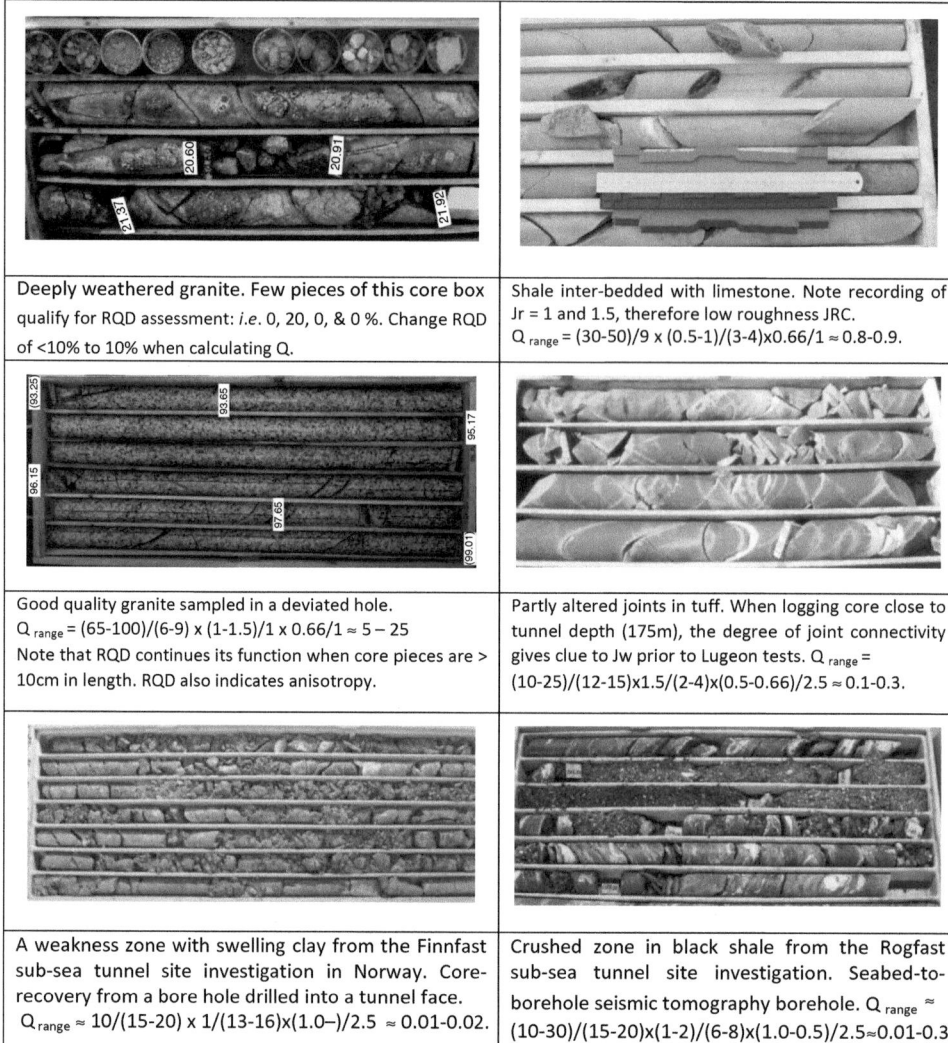

Figure 3 Six contrasting core boxes from road and rail tunnel projects in Norway and Hong Kong. In both countries use of the Q-system for core logging and tunnel logging is required by the authorities. Concerning the two most challenging cases: the weakness zone in the Finnfast tunnel was quite dry both during core drilling and after excavation. There was little water in the drill hole at Rogfast.

5 CHARACTERIZING SURFACE EXPOSURES WITH EXAMPLES

In Nordic countries in particular, where glaciation has exposed a lot of rock, it is possible to gain a good assessment of the higher end of the rock quality scale and likely best tunneling conditions, by observing and mapping surface exposures. When road cuttings are also available, the rock conditions in these better rock-quality terrains can also be readily mapped. However, seismic refraction measurements (next section) and dedicated deviated drilling of low velocity weakness zones will be needed where exposures are absent in flatter and lower areas. These low-relief areas may nevertheless be tunneled under, such as in the case of future high speed railway tunnels. One of these projects is about to start (2016) near Oslo with four double-shield TBM.

More than 300 rock cuttings were Q-logged to obtain rock mass quality input for Q_{TBM} prognoses for these up-coming rail tunnels. However this exposure logging gave data of relevance only to the top five rock classes, and seismic results and core logging of weakness zones was needed to provide approximate information on the lowest rock classes. In Figure 4, some examples of exposure logging using the Q-method are illustrated, using a deliberately wide variation in rock mass quality from the Oslo region.

The massive nature of this motorway rock cutting can be judged by the 2m high 'elk fence'. Class 1 granites. $Q_{range} \approx (90 - 100)/(6-9) \times (1.5-2)/(1-2) \times (0.66-1)/1 \approx 5 - 66$. Massive, abrasive, hard to bore with TBM.	Drammen granite near Lier Tunnels. Classic three joint-set rock mass. Joints in full sun-light have least favourable Jr/Ja combination. $J_n/J_r = 6$ $Q_{range} \approx 100/9 \times 1.5/(1-2) \times 0.66/1 = 6-11$.
Well-jointed shale close to Oslo tunnel portals. Note closely-spaced half-barrels. The shale was interbedded with nodular limestones of higher quality. $Q_{range} \approx (10-20)/9 \times 1/(1-2) \times 0.66/1 \approx 0.4 - 1.5$.	Sheared and clay-bearing hornfels next to granite batholith near Asker along E18 motorway. Examples of three Jr/Ja 'contact' categories. $Q_{range} \approx (10-30)/(9-12) \times (1-1.5)/(4-6) \times (0.33-0.66)/5 \approx 0.01-0.2$.

Figure 4 Rock exposures selected from the Oslo area, mostly connected to tunnels built or planned. Note that in the case of clay-bearing rock, permeability (and water pressure) may be partitioned. High pressures can occur on just one side of a fault zone, until penetrated.

Arguments are sometimes heard in conferences that Norway only has pre-Cambrian granites and gneisses, and therefore excellent tunneling conditions. In fact Norway has some (lower percentage of) extreme tunneling conditions, with quite frequent swelling clay, occasional sand in-rushes, rock bursting where high cover, and some extensively sheared and clay-bearing rock masses, due to regional thrust belts, requiring heavy support, and the actual need of local concrete lining. There are at least ten named collapsed caldera in the geologic history of today's Oslo region.

5.1 Additional advice concerning surface exposure logging

- Near-surface rocks will often be more jointed than the unweathered rock masses at a greater depth. This may especially be the case in schistose rocks, which often have a tendency to disintegrate near the surface. Frequently only the better quality rock masses are exposed at the surface.
- Exposures in the terrain are often well rounded by the ice in Nordic countries and weathered in other countries, reducing the possibility to see joints undisturbed, therefore making reliable description of roughness J_r and joint filling J_a, rather difficult. The parameter RQD will usually be underestimated from natural outcrops, due to weathering or frost damage, while J_n will tend to be over-estimated. However in competent hard rock which has been rounded by ice, RQD will be over-estimated and J_n will be under-estimated, due to erosion and removal of the more jointed materials.
- In weathered rock, the joints may be hidden at the surface. Hence the Q-values relevant to tunnel depth could in some cases be over-estimated. However, depending on rock type, the quality at depth may often be seriously underestimated using surface exposures, and experience is needed to make relevant adjustments for this.
- In high road cuttings or other excavated slopes, the joint surfaces are normally well exposed after blasting, giving a more reliable basis for estimating RQD, J_n, J_r and J_a.
- Rock cuttings excavated in different directions, if sufficiently high, give approximately the same Q-values as in a tunnel, but small cuttings in partly weathered rock should be ignored.
- The water leakage in a tunnel, J_w, will obviously be difficult to predict from field mapping alone. Water loss tests in boreholes and/or empirical data from projects in similar rock masses are necessary to obtain good predictions of the likely water conditions.
- A prediction of the SRF-value may be made based on the topographic features and knowledge of the stress situation in nearby underground openings in the region. High and steep mountain sides often give an anisotropic stress field.
- Geological structures, such as fractures parallel to the mountain side, and sickle shaped exfoliation, are indications of high, anisotropic stresses. The limit for exfoliation in high mountain sides or spalling in a tunnel is dependent on the relation between induced stress and the compressive strength of the rock. In hard rock this limit normally occurs at between 400 and 1100 m of rock cover above the tunnel. This depends on the compressive strength of the intact rock and the gradient of the mountain side.
- Stress measurement within drill holes may be carried out before tunnel excavation in some of the larger tunneling or hydropower projects, and this makes SRF estimation more reliable. Note that stress-induced fracturing referred to above starts when the

ratio of the maximum estimated tangential stress compared to uniaxial strength (σ_θ/σ_c) exceeds about 0.4. (Shen and Barton (2017) showed that fracturing may actually be extensional strain-induced, due to $\sigma_t/\upsilon > 0.4$). This signifies the starting point for considerably increased SRF values. This experience is confirmed in mining and in deep road tunnels.

- Mapping for subsea tunnels is limited to the outcrops which are visible on both sides of the strait or fjord under which the tunnel is planned. For subsea tunnels use of seismic techniques is therefore even more important. Core drilling from the shoreline or islands are carried out. Deviated and steered drill holes up to 1000m long may be used for seabed to borehole seismic tomography. More seldom, because of the expense, core drilling from a ship may be carried out. This will be done for the 27km long planned Rogfast subsea (–390 m) road tunnel, which will be the world's longest sub-sea road tunnel.

6 USING SEISMIC VELOCITY AND Q TO INTERPOLATE BETWEEN BOREHOLES

An empirically-based correlation between the Q-value and the P-wave velocity derived from shallow refraction seismic measurements was developed by Barton, 1995 from trial-and-error lasting several years (Figures 5 and 6). The velocities were based on a large body of data from hard rock sites in Norway and Sweden, thanks to extensive documentation by Sjøgren et al. (1979), using seismic profiles (totaling 113 km) and local profile-oriented core logging results (totaling 2.85 km of core). The initial V_P-Q correlation had the following simple form, and was relevant for *hard rocks with low porosity*, and specifically applied to shallow refraction seismic, *i.e.* 20 to 30m depth, as suggested by Sjøgren.

$$V_p \approx 3.5 + log\,Q\,(\text{km/s}) \tag{2}$$

A more general form of the relation between the Q-value and P-wave velocity is obtained by normalizing the Q-value with the multiplier UCS/100 or $\sigma_c/100$, where the uniaxial compressive strength is expressed in MPa ($Q_c = Q \times \sigma_c/100$). The Q_c form has more general application, as weaker and weathered rock can be included, with a (-ve) correction for porosity.

$$V_p \approx 3.5 + log\,Q_c(\text{km/s}) \tag{3}$$

The derivation of the empirical equations for support pressure (originally in Barton *et al.* (1974)) and for the static deformation modulus (in Barton 1995, 2002) suggest an approximately inverse relationship between *support pressure needs* and rock mass *deformation moduli*. This surprising simplicity is not illogical. However it specifically applies with the mid-range Jr = 2 joint roughness.

7 CHARACTERIZING THE ROCK MASS IN TUNNELS BY INSPECTING EACH TUNNEL ADVANCE

The final role of Q-system logging is to document the rock mass quality of each advance of the tunnel, and thereby assist in the selection of the final support (Sfr) and

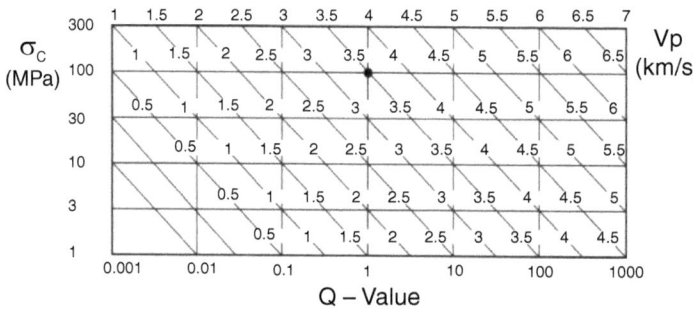

Figure 5 Top: Hard rock, shallow seismic refraction mean trends from Sjøgren *et al.* (1979). The Q-scale was added by Barton (1995), using the hard rock correlation $V_p \approx 3.5 + \log Q$. By remembering Q = 1: $V_p \approx 3.5$ km/s, the Q-Vp approximation to a wide range of qualities is at one's fingertips (*e.g.* for hard, massive rock: Q = 100: $V_p \approx 5.5$ km/s). Bottom: Generalization to include rock with different σ_c values. The results still apply to shallow seismic. The source of this figure is explained in Barton (2006).

reinforcement (B) class. This cannot be done by infrequent finite element modeling or rejection of the empirical method, as suggested by some recent authors in Germany, Austria, Switzerland and Canada. Steel-fiber (or polypropylene) reinforced shotcrete and systematic, corrosion-protected rock bolts (B + Sfr) form the usual Norwegian single-shell tunnel (or cavern) 'lining'. There is infrequent use of rib-reinforced shotcrete (RRS), and occasional cast concrete (CCA) in short sections of bad rock. All of these measures are selected with the help of the Q-support chart (see next section). However, special conditions may demand special measures, so general Q-based methods may be modified when necessary. This will be discussed in the next section.

In Norway, following occasional adverse experiences in the past, what has become known as 'the Owner's half-hour' is allotted to thorough rock mass inspection and characterization when driving the tunnel.

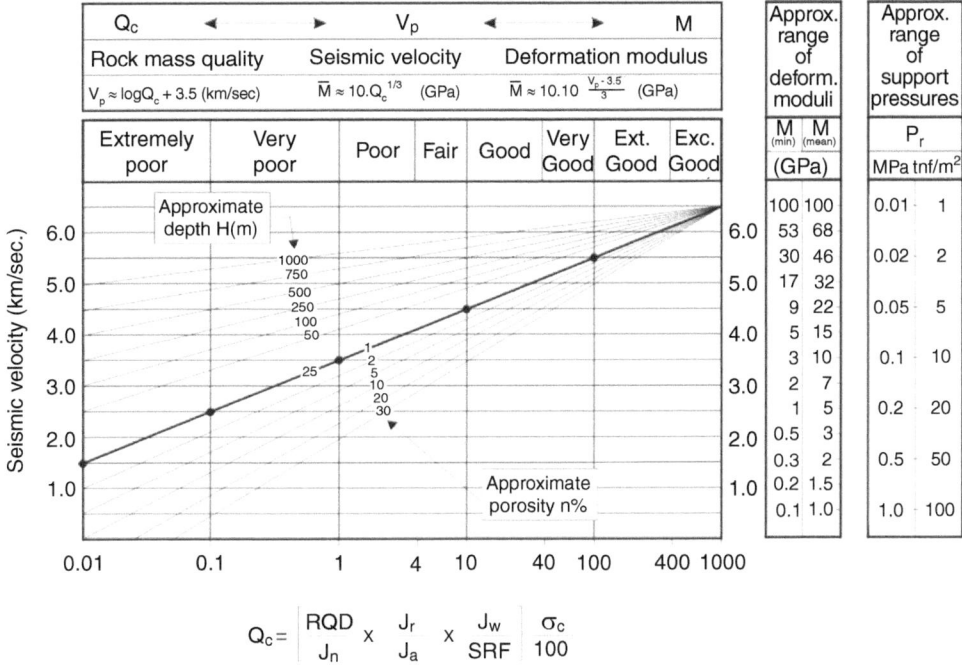

Figure 6 The thick 'central diagonal' line is the same as the sloping line given in Figure 5, and this applies to nominal 25–30m depth shallow seismic refraction results. In practice the nominal 1% (typical hard rock) porosity would be replaced by increased porosity if rock was deeply weathered, and the more steeply sloping lines (below the 'central diagonal') would then suggest the approximate (-ve) correction to V$_P$. Note that very jointed rock with open joints may have lower velocity than saturated soil. The less inclined lines above the 'central diagonal' represent greater depth (50, 100, 250m etc.), and these lines correct V$_P$ for documented stress or depth effects (+ve). These depth-lines were derived from several sets of deep cross-hole seismic tomography, with Q-logging of the respective cores (Barton, 2002). Note the inverse nature of (static) deformation moduli and support pressure shown in the right-hand columns. These derivations are described in Barton (1995) and Barton (2002). For a more detailed treatment of seismic, for example the effects of anisotropy which are accentuated when the rock is dry or above the water table, refer to the numerous cases illustrated and summarized in the text book of Barton (2006).

If conditions permit, this inspection is done before temporary or permanent support and reinforcement operations are commenced, following the last advance of *e.g.* 2 to 5m. This 'half hour' is reserved (and fully costed) so that the engineering geologist representing the contractor, and the engineering geologist representing the owner, can jointly try to come to agreement about the quality (or lack of quality) of the newly exposed rock mass. Having a standard method like the Q-system, and time for discussion, adds to the reliability, and both engineering geologists learn from each other. Shift work for the project engineering geologists is obviously needed when tunneling is progressing during two or more shifts each 24 hours.

The structural-geological (rock-type and joint-set recordings) and Q-logging is of course done following blast-gas displacement, and following scaling ('barring-down') by the contractor. The fact that wet-process shotcreting is used, as opposed to dry-process methods, plus the relatively small number of operatives and vehicles in Norwegian tunnels, means that air quality is generally superior to what is experienced in many other countries. This makes rock mass inspection easier and it is therefore more likely to be correct. Single-shell tunneling demands this reliability. Some examples of Q-logging in tunnels were illustrated in Figure 7.

Because tunnel cross-sections can be quite large, and because full-face blasting is common (and sounder practice) where rock quality allows this, the height of the tunnel arch usually means that rock mass inspection from a hydraulically-lifted and well-lighted cage is imperative. The rock mass quality (especially the lack of quality) is much more likely to be seen when close to the rock surface. Features such as clay-filled discontinuities are less likely to be missed. Geological hammers and a readily available scaling-bar to extend reach and avoid too-frequent moving of the cage, are obvious features of this inspection and decision-making. While some consulting companies may have performed numerical modeling of representative rock mass and tunnel support classes, now is the time to decide which support class and not wait for external decisions. This is important when 40 to 80m per week per face is the typical range of advance rates of single-shell NMT excavations.

There have been two deservedly much-publicized road tunnel rock-falls in the last 20 years in Norway, fortunately with no injuries or fatalities involved. Both have involved incorrect application of the Q-system, with an error in one case of assuming Q = 70, while independent engineering geologists recorded Q = 0.07 after the event, obviously with the benefit of hind-sight, including post-failure observation. In other words there was a 1000:1 error in the Q-estimate, due to failure to recognize a clay-infested section of a sub-sea tunnel, due to inadequate arch inspection and Q-logging routines.

7.1 Effect of orientation of geological structures on Q-value

Over the years many have commented on the *apparent* lack of discontinuity orientation in the derivation and application of the Q-value. Unlike the case in RMR, there is no specific term for an 'orientation rating'. Nevertheless there is the instruction to try to consider the least favorable joint set or discontinuity from the point of view of over-break potential or instability, when selecting the appropriate Jr/Ja ratio. This aspect of Q-logging sometimes requires significantly more experience than required when using RMR, because one needs to visualize the consequences of continuing tunnel advance in case of not providing specific 'feature' support. Prior 3D (*e.g.* 3DEC) modeling of such cases might have been performed.

Figure 8 shows two cases which can be used for illustration. On the left is a small detail from one of Norway's numerous hydropower headrace tunnels. A graphite-coated minor fault strikes sub-parallel to the tunnel axis, while perpendicular to the axis is a set of chlorite coated joints. If considered individually the respective Jr/Ja

<table>
<tr>
<td>

Rock mass classification has to be done close to the crown, using a hydraulically raised cage, mounted on the drill jumbo. It is easy to overlook altered rock and clay when too far below the arch in large tunnels.

</td>
<td>

Inspection of initial shotcrete (Sfr) support must also be carried out from high in the tunnel arch, using a crowbar to check for 'drumminess'. Poorly bonded shotcreted areas due to insufficient jet-washing need repair.

</td>
</tr>
<tr>
<td>

Three joint sets (Jn = 9), planar and rough-surfaced (Jr = 1,5). Sandstone in the Bremanger tunnel in Western Norway. Note that the location with over-break with three well-developed joint sets attracts Q-loggers attention, but may be more jointed than else-where.
$Q_{range} \approx (30\text{-}80)/(6\text{-}9)\text{x}1.5/1\text{x}1/1 \approx 5\text{-}20$.

</td>
<td>

Shallow cavern, with weathering or clay coatings on several of the joint sets. The granite has high RQD (90-100%) and large block sizes. This emphasizes need for additional Q-parameters to reflect the lower quality. Serious over-break due to critical ratio of Jn/Jr ≥ 6.
$Q_{range} \approx (90\text{-}100)/9\text{x}1.5/(2\text{-}4)\text{x}0.66/2.5 \approx 1\text{-}2$.

</td>
</tr>
<tr>
<td>

Stress-induced fracturing in marble in the walls of the Jinping I headrace tunnels, where two large-diameter TBM were eventually removed, due to > 2 km cover. Completion by drill-and-blast: total of four // tunnels.
$Q_{range} \approx (90\text{-}100)/(2\text{-}3)\text{x}(2\text{-}4)/1\text{x}(0.5\text{-}1)/(50\text{-}200) \approx 1$.

</td>
<td>

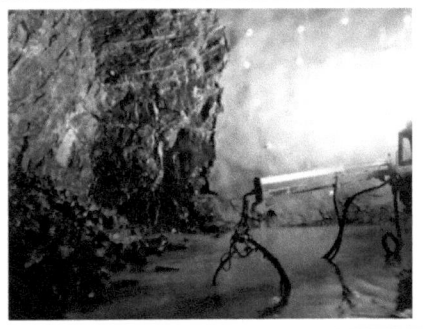

Tunnel face in (pre-injected) shales. Note first layer of S(fr) and permanent (CT) bolts close to previous face. A conservative 3-4m advance. In fact the prior quality of the shales has been improved by 10 MPa pre-injection. $Q_{effective}$ has improved from ≈ 1 to 30.

</td>
</tr>
</table>

Figure 7 Some figures to illustrate tunnel inspection needs following blasting and prior to final support and reinforcement decisions. Some diverse tunnel Q-value estimates are also given.

Figure 8 a and b Some specifically oriented details in a headrace tunnel in granite and in an old road tunnel along the west coast of Norway in massive schist. The discontinuity or joint set most adverse-for-stability gives the appropriate Jr/Ja ratio, and extra support may be long-itudinally extensive.

ratios would be 1.5/3 and 3/4 respectively. The graphite-coated feature follows the tunnel axis for many meters, and is the chief cause of over-break and significant potential instability. Even if the Jr/Ja ratios had been equal, the Jr/Ja = 1.5/3 combination applies in this case, while the more stable perpendicularly oriented feature with Jr/Ja = 3/4, merely contributes to a lower RQD.

In the case of the unsupported portal of the old coastal road tunnel shown in Figure 8b, the smooth sub-vertical joint set with strike sub-parallel to the tunnel axis should have supplied the appropriate Jr/Ja ratio of 1/2. With a local portal rock mass quality of $100/(9 \times 2) \times 1/2 \times 1/2.5 \approx 1$ (note: $2 \times$ Jn and SRF = 2.5 for portals) one would expect B of 1.7 m c/c and 7 cm of S(fr) (see next section), if the tunnel had come under today's Q-support decision-making. However, the tunnel has existed for a long time without support, so the Q-system is seen to be conservative, if correctly applied. Some numerical modelers do not seem to agree here. Of course there are good reasons for this when modelers make all joint sets continuous. This practice needs correction.

8 TUNNEL SUPPORT RECOMMENDATIONS BASED ON Q – SOME HISTORY

The Q-system was originally developed from more than 200 case records, which were mostly Scandinavian or of international origin. The single-shell support methods in the early seventies were B + S(mr), *i.e.* systematic bolting and mesh reinforced shotcrete. The table of support recommendations was based on the location of the case record in 'span-versus-Q' space, as illustrated in Figure 9a, from Barton *et al.* (1974). With the gradual addition of 1050 more case records by Grimstad, the support and

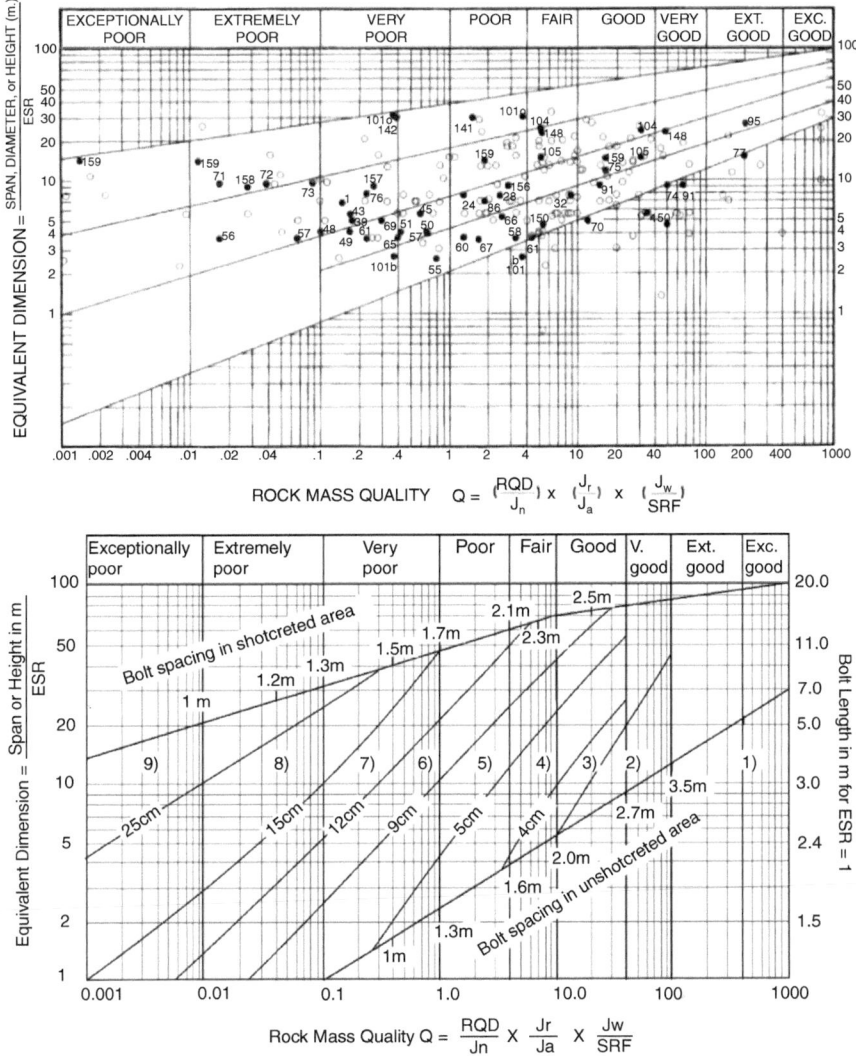

REINFORCEMENT CATEGORIES

1) Unsupported
2) Spot bolting, sb
3) Systematic bolting, B
4) Systematic bolting (and unreinforced shotcrete, 4-10cm, B(+S))
6) Fiber reinforced shotcrete and bolting, 5-9cm, Sfr+B

6) Fiber reinforced shotcrete and bolting, 9-12cm, Sfr+B
7) Fiber reinforced shotcrete and bolting, 12-15cm, Sfr+B
8) Fiber reinforced shotcrete > 15cm, reinforced ribs of shotcrete and bolting, Sft, RRS+B
9) Cast concrete lining, CCA

Figure 9 Top: The Barton *et al.* (1974) Q-based support chart, with each of the (38) boxes having a separate support and reinforcement recommendation. There were 212 case records, and at this time the standard single-shell method was B+S(mr) – *i.e.* bolting and steel mesh reinforced shotcrete. By about 1983 S(mr) had gone out of use as a tunnel support measure in Norway, after the development of robotically-applied wet process S(fr) in 1978/1979. Note that the SPAN (the width of the tunnel or cavern) is divided by a 'tunnel-use' safety requirement number ESR, shown later. Bottom: the Grimstad & Barton (1993) tunnel support and reinforcement chart, which was based on some 1,050 new case records. This chart gave permanent single-shell support and reinforcement, also for large caverns. As will be noted later in Figure 13, some small adjustments to minimum shotcrete thickness were made in 2006, described in Grimstad (2007), based on experiences from 800 new cases assembled in 2002/2003. In addition in Figure 13, the RRS (rib reinforced shotcrete) listed under category #8, is given specific dimensions for a wide range of tunnel sizes and Q-values.

reinforcement recommendations were simplified to the graphic method shown in Figure 9b, from Grimstad & Barton (1993).

In the original Barton *et al.* (1974) version of Q, rock support and rock reinforcement recommendations were 'separated' by the conditional factors RQD/Jn (*i.e.* relative block size) and Jr/Ja (*i.e.* inter-block shear strength). In other words, smaller block sizes (and lower cohesive strength) apparently (and logically) required more S (mr), while lower internal friction apparently (and logically) required closer rock bolt spacing. Later it was discovered (Barton, 2002) that Q, or more specifically Q_c closely resembled the multiplication of 'c' and 'tan φ'. This 'semi-empirical' (*a posteriori*) derivation of the two strength components of a rock mass differs greatly from the *a priori* complex algebra of the Hoek-Brown GSI-based rock mass 'strength criterion' (Barton, 2014), which so many young people use with continuum finite element modeling, obtaining apparent tunnel 'behavior' which they believe to be true.

As discussed later, it is wise to combine empirical methods with numerical methods if one wishes to 'design' tunnel support based on numerical modeling. The often exaggerated 'plastic' zones seen in numerical models, and the numerically modeled deformation need to be viewed with suspicion, and sometimes corrected, by empirical Q-deformation data. A very simple method will be illustrated later. Empirical near-reality is closer than *a priori* modeling.

9 THE COMPONENTS OF Q-SYSTEM BASED SUPPORT: S(FR), CT BOLTS, AND RRS

This section consists of illustrations of some key items of the Q-support recommendations shown in Figure 9, including yesterday's S(mr) and the last thirty years S(fr).

The reality of single-shell NMT-style tunneling, in comparison to double-shell NATM-style tunneling is that each component of support has to be *permanently relied upon*. There is nothing like the *neglect* of the contribution of temporary shotcrete, temporary rock bolts, and temporary steel sets, and reliance on a final concrete lining, as in NATM. Thus more care is taken in the choice and quality of the support and reinforcement components B+S(fr) + (eventual) RRS. Figure 10 (bottom) illustrates application of S(fr). Figure 11 illustrates (in the form of a shortened demo) the workings of the CT bolt. And finally Figure 12 illustrates some of the internal reinforcement details and final appearance of RRS (rib reinforced shotcrete).

It should be noted that the 1993 Q-support chart (shown earlier in Figure 9) suggested the use (at that time) of only 4-5 cm of *unreinforced* sprayed concrete in category 4. The application of *unreinforced* sprayed concrete came to an end during the 1990s, at least in Norway. Furthermore, thickness down to 4 cm is not used any longer in Norway, due to the already appreciated risk of drying out too fast when it is curing. The Q-chart from 1993 (Figure 9) and also an updated 2002/2003 version, indicated a very narrow category 3 consisting of only bolts

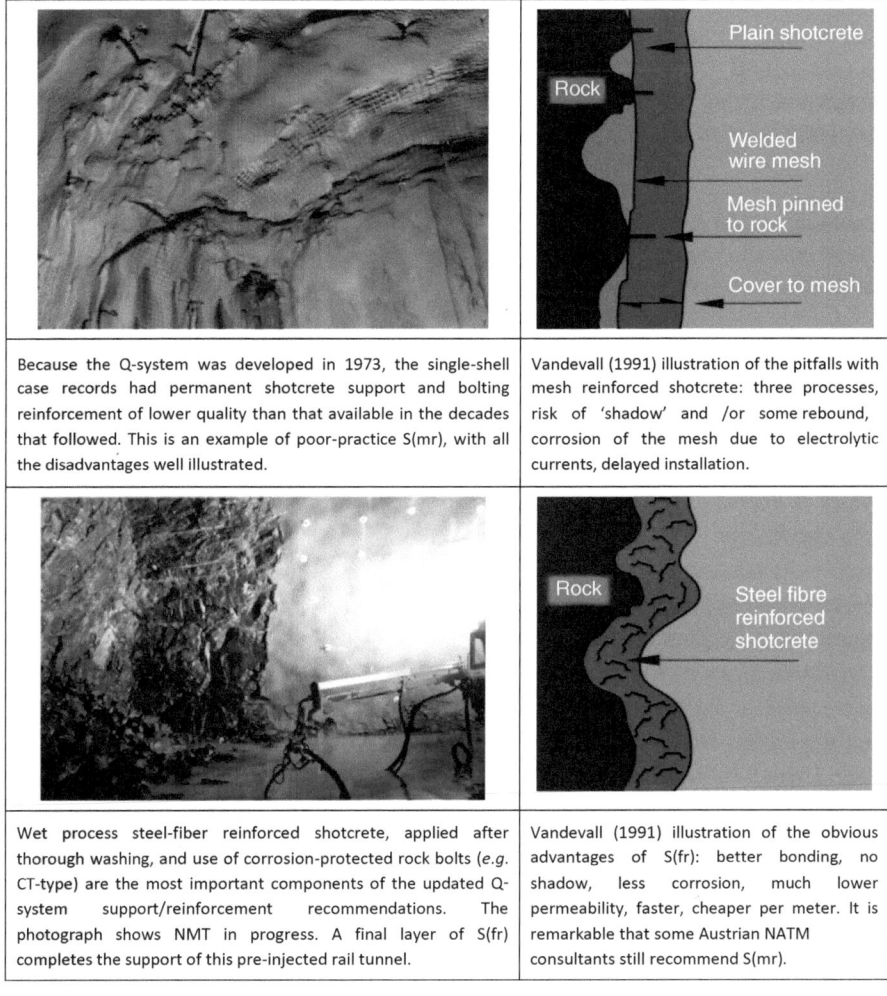

Because the Q-system was developed in 1973, the single-shell case records had permanent shotcrete support and bolting reinforcement of lower quality than that available in the decades that followed. This is an example of poor-practice S(mr), with all the disadvantages well illustrated.	Vandevall (1991) illustration of the pitfalls with mesh reinforced shotcrete: three processes, risk of 'shadow' and /or some rebound, corrosion of the mesh due to electrolytic currents, delayed installation.
Wet process steel-fiber reinforced shotcrete, applied after thorough washing, and use of corrosion-protected rock bolts (*e.g.* CT-type) are the most important components of the updated Q-system support/reinforcement recommendations. The photograph shows NMT in progress. A final layer of S(fr) completes the support of this pre-injected rail tunnel.	Vandevall (1991) illustration of the obvious advantages of S(fr): better bonding, no shadow, less corrosion, much lower permeability, faster, cheaper per meter. It is remarkable that some Austrian NATM consultants still recommend S(mr).

Figure 10 The advantages of S(fr) compared to S(mr) are easily appreciated in these contrasting examples. The sketches from Vandevall's 'Tunnelling the World' (1991) are not-exaggerated.

in a 10m wide tunnel when Q was as high as 10–20. This 'bolt-only' practice is not accepted any longer in Norway for the case of transport tunnels. The category 3 in 1993 and 2002/2003 has been taken away in this newest 2007 chart (Figure 13) which was fine-tuned by Grimstad when still at NGI in 2006. However for less important tunnels with ESR = 1.6 and higher, only spot bolts are still valid. Hence we may distinguish between transport tunnels (road and rail) and head race tunnels, water supply etc. (See later ESR table).

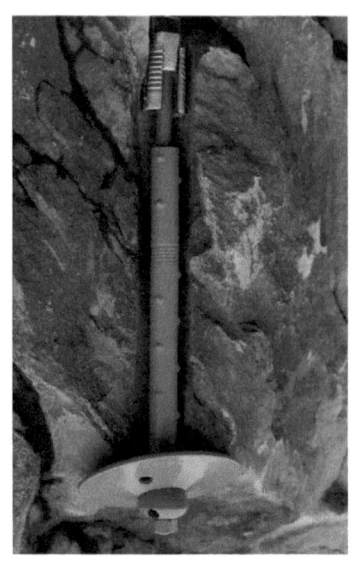
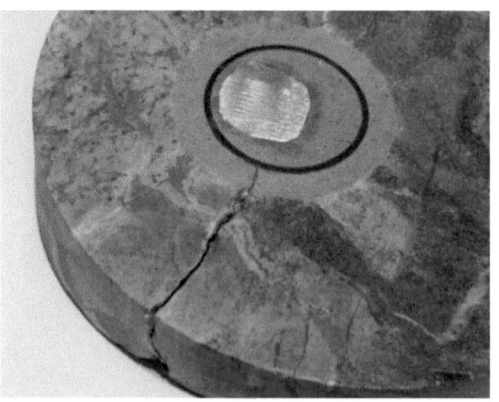

Over-cored CT bolt: Joint/crack deformation next to the bolt (an expected mechanism when installing close to the tunnel face) does not initiate a potential process of corrosion which it might in the case of a conventional bolt without the PVC sleeve.

Figure 11 Because single-shell (NMT) relies on high quality S(fr) and long-life rock bolts, the multi-layer corrosion protection methods developed by Ørsta Stål in the mid-1990s, became an important part of NMT. The left photo shows a blue-colored PVC sleeve: the PVC can also be black or white, and 3, 4 or 5 m long as desired.

10 NMT SINGLE-SHELL TUNNELING CONCEPT SUMMARIZED IN 1992

Shortly before the publication of the updated Q-system tunnel support recommendations by Grimstad & Barton (1993), a multi-company, multi-author group from Norway (Barton, Grimstad, Aas, Opsahl, Bakken, Pedersen and Johansen) from the companies NGI (2), Selmer, Veidekke, Entreprenørservice, NoTeBy and Statkraft, described the main elements of the Norwegian method of tunneling, calling it NMT, in deliberate competition to the much more expensive double-shell NATM. This two-part article in World Tunnelling (Barton *et al.*, 1992) described Q-logging, numerical modeling, tunnel support selection, robotic application of wet process S(fr), support element properties, and the Norwegian tunnel contract system. The initials NMT are now well known after 20 years referencing and inclusion in university courses outside Norway. The initials NMT are helpful for distinguishing it from the very different NATM. (See Barton, 2012a).

10.1 Concerning bolting and fiber types in the Q-recommendations

- The early Q-system nomenclature B_{utg} shown in Figure 14 refers to *untensioned grouted bolts*, which are very stiff. Their use has to be carefully considered when

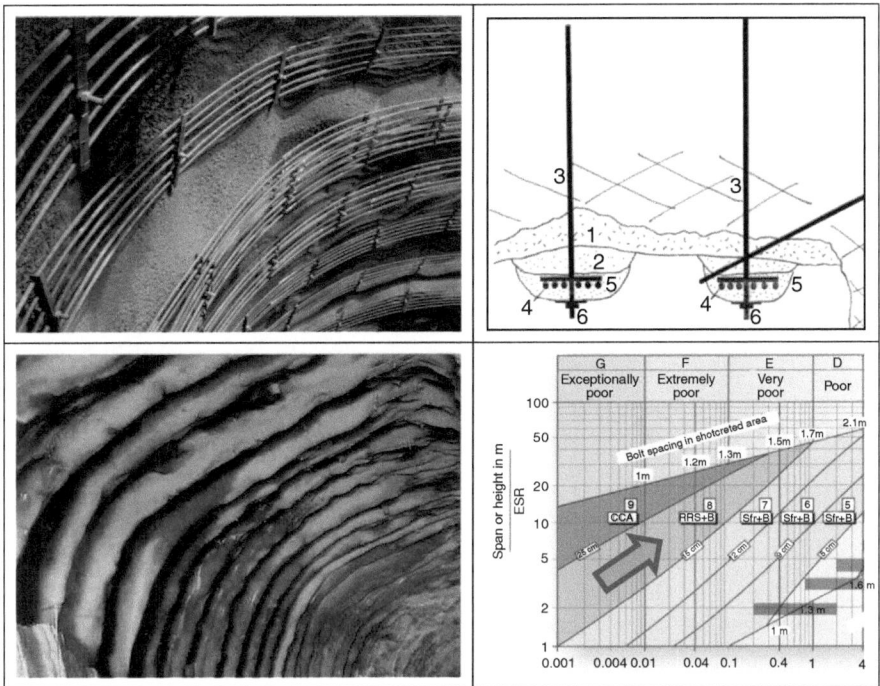

Figure 12 Some details which illustrate the principle of RRS, which is an important component of the Q-system recommendations for stabilizing very poor rock mass conditions. The top left photograph is from an LNS lecture published by NFF, the design sketch is from Barton (1996), the blue arrow shows in which part of the Q-chart the RRS special support-and-reinforcement measure is 'located' (see greater detail in Figure 13). The photograph of completed RRS is from one side of the National Theater station in downtown Oslo, prior to pillar removal beneath only 5m of rock cover and 15m of sand and clay. Final concrete lining followed the RRS for obvious architectural reasons. (Barton, 2012a).

there is early large deformation, spalling or rock burst. Grouting of end-anchored rock bolts too early may increase the adverse effects of spalling, and bolts may also fail in tension in large numbers. It is better to grout the bolts when deformation has slowed down. A highly recommendable alternative is the use of energy/deformation absorbing D-bolts.

- EE-fibers went out of use in Norway in the mid 1990s. Bekaert steel fibers 30-35mm long, among others, are partially being substituted by polypropylene fibers, such as Barchip Kyodo 48mm long, in some sections of the tunneling industry. However it is rather important that these fibers are rough-surfaced to ensure their anchorage and deformation resistance. The desirable decades-long behavior of polypropylene fibers is not yet possible to document, but extensive use in parts of the tunneling industry, such as in subsea road and rail tunnels and when large deformation is expected, is a positive signal.

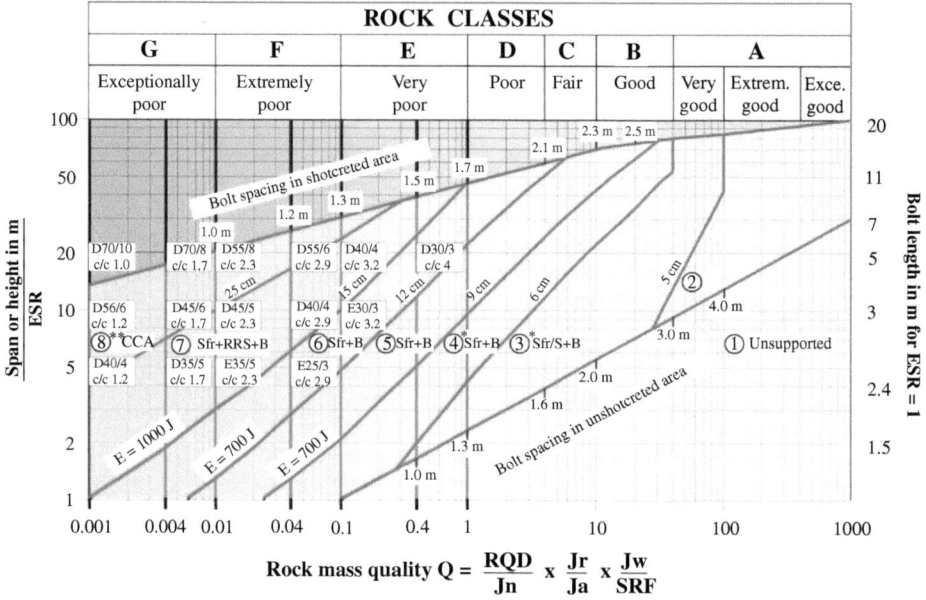

REINFORCEMENT CATEGORIES

1) Unsupported
2) Spot bolting, **sb**
3) Systematic bolting, and unreinforced
   or fibre reinforced shotcrete, 5-6 cm
   Sfr+B+S

4) Fibre reinforced shotcrete and bolting, 6-9cm, **Sfr+B**
5) Fibre reinforced shotcrete and bolting, 9-12 cm, **Sfr (E700)+B**
6) Fibre reinforced shotcrete and bolting,12-15 cm, **Sfr (E700)+B**
7) Fibre reinforced shotcrete > 15 cm +
   reinforced ribs of shotcrete and bolting, **Sfr (E1000)+RRS+B**
8) Cast concrete lining, **CCA or Sfr (E1000)+RRS+B**

The bolts are 20 or 25 mm in diameter
E) Energy absorption in fibre reinforced shotcrete at 25 mm bending during plate testing

| D45/6 | = RRS with totally 6 reinforcement bars in double layer in 45 cm thick ribs with centre to centre (c/c) |
| c/c 1.7 | spacing 1.7 mm. Each box corresponds to Q-values on the left hand side of the box |

*) Up to 10 cm in large spans
) Or **Sfr+RRS+B

Figure 13 The updated Q-support chart first published by Grimstad (2007). The details of RRS dimensioning given in the 'boxes' in the left-hand-side of the Q-support diagram were derived by a combination of empiricism and some specific numerical modeling by a small team of former NGI colleagues. Details of this modeling are given by Grimstad et al. (2002, 2003). Note that each 'box' contains a letter 'D' (double) or a letter 'E' (single) concerning the number of layers of reinforcing bars.(Figure 12a shows both varieties). Following the 'D' or 'E' the 'boxes' show maximum (ridge) thickness in cm (range 30 to 70 cm), and the number of bars in each layer (3 up to 10). The second line in each 'box' shows the c/c spacing of each S(fr) rib (range 4m down to 1m). The 'boxes' are positioned in the Q-support diagram such that the left side corresponds to the relevant Q-value (range 0.4 down to 0.001). Note energy absorption classes E = 1000 Joules (for highest tolerance of deformation), 700 Joules, and 500 Joules in remainder (for when there is lower expected deformation). Note: use S(fr) to form the arch below the steel bars. Note (from Barton et al., 1974) that bolt length (right-side of Figure 13) is estimated from: L = 2 + 0.15 SPAN/ESR (m). For walls L = 2 + 0.15 HEIGHT/ESR (m). For large caverns with eventual cable anchors, the factor 0.15 is replaced respectively by 0.4 (arch) and 0.35 (walls). Note that HEIGHT refers to the full excavation height. See updated ESR values (Table 3).

Table 2 An expanded text to explain the NMT abbreviations in the 'drawers' in Figure 14.

Rock mass characterization *using the six Q-parameters. A relationship between Q and V_P and deformation modulus M is indicated, using Q_o σ_o matrix porosity n% and depth H (m) or stress level.*

Support design measures *consist of none, sb, B, B+S, B+S(fr), RRS, CCA. (Untensioned grouted B_{utg} bolts, tensioned resin end-anchored bolts, and CT bolts). Also may use spilling, drainage, pre-injection, and freezing.*

S(fr) robot technology *using Portland cement, silica fume, plasticizer, super-plasticizer, aggregate and non-alkali (low) accelerator. Steel fiber: EE 20–25 mm (previously), Bekaert 30–35mm/0.5mm. (Today: also PP fiber Barchip 48mm, 0.4/1.4mm).*

Rapid advance *due to wet process S(fr) shotcrete, gives low rebound (4–6%) and improved environment.*

Site investigation *using seismic refraction, radar, cross-hole V_P or E_{dyn} tomogram, or attenuation tomogram. (Note $Q_{seismic}$ = 1/attenuation is numerically close to M GPa).*

Numerical verification of support designs *using codes like UDEC, UDEC-BB, UDEC-Sfr, FLAC, FLAC-3D and 3DEC. Relevant parameters JRC, JCS, φ_r, M, Kn and Ks, c + φ.*

Norwegian tunnel contract system *uses a flexible contract, with unit prices for all possible measures in tender documents: use the motto 'expect the unexpected'.*

Low cost and *less conflicts, permanent single-shell support compared to double-shell NATM.*

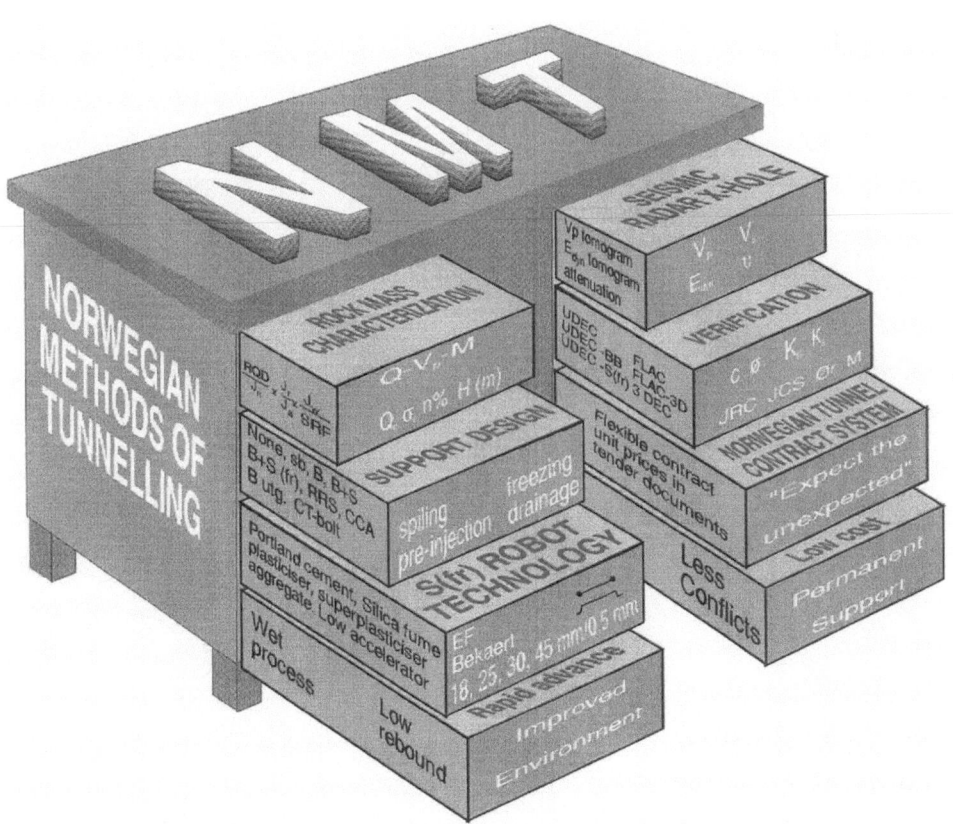

Figure 14 The 'design-and-execute' tunneler's desk-of-drawers, used by Barton (1996) to summarize key elements of NMT for an international readership. Table 2 gives a summary of the 'content' of each drawer, using added connecting text.

11 CONTRASTING SINGLE-SHELL NMT AND DOUBLE-SHELL NATM

The use of steel sets (and lattice girders) is avoided in the practice of single-shell NMT, due to the potential loosening of insufficiently supported rock in the periphery of the excavation. It is difficult to make sufficiently stiff contact between the steel sets and the rock, especially when there is over-break. The results of experiments using different support methods are illustrated in Figure 15. The left-hand diagram shows the results of trial tunnel sections in mudstone (Ward *et al.*, 1983). The five years of monitoring clearly demonstrate the widely different performance of the four different support and reinforcement measures.

In the right-hand diagram, from Barton & Grimstad (1994), the contrasting stiffness of B+S(fr) and steel sets is illustrated in a 'confinement-convergence' diagram, with the implication (and reality) that SRF (loosening variety: see APPENDIX A1) may occur when using steel sets. It should be clear that the early application of S(fr) by shotcrete robot, and the installation of sufficient numbers of permanent corrosion protected rock bolts from the start, as in single-shell NMT, is likely to give a different result from that achieved when using NATM.

In the latter, the commonly used steel sets and mesh-reinforced shotcrete and rock bolts are all considered just as temporary support, and are not 'taken credit for' in the

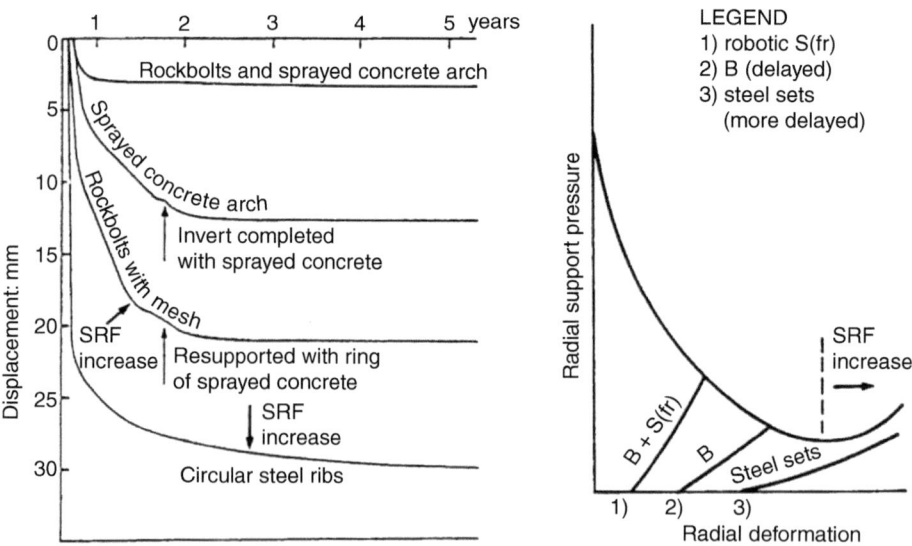

Figure 15 Left: Results of five years of monitoring test-tunnel sections in mudstone, using four different support and reinforcement measures, from Ward *et al.* (1983). The obvious superiority of B + S in relation to steel sets is clear. The last 35 years of B + S(fr), as for instance practiced prior to 2015 in Norway would presumably give an even better result. Right: Representation of the relative stiffness of different support measures, from Barton & Grimstad (1994). SRF may increase due to loosening in the case of steel sets. See Figure 16, which illustrates the implicit difficulty of controlling deformation with steel sets/lattice girders.

Figure 16 Left: An illustration of the challenge of making contact between the excavation periphery and the steel sets, even for the case of limited over-break. In NATM the 'sprayed-in' steel sets, and S(mr) and bolting are considered temporary, and are not included in the design of the final concrete lining. Right: Steel sets are actually a very deformable type of tunnel support. However in squeezing rock as illustrated, the application of RRS might also be a challenge, unless self-boring rock bolts were used to bolt the RRS ribs in the incompetent (over-stressed) rock that is likely to surround the tunnel in such cases.

Figure 17 Left: Illustration of a 'missing component' of support. When the rock mass is significantly jointed and with low cohesion, bolting and mesh alone are clearly inadequate. Right: An illustration of the use of temporary steel sets and mesh reinforced shotcrete, both of which potentially invite increased deformation and possible loss of strength when over-loaded in 'squeezing' conditions. Re-mining with much more robust steel arches is in progress.

design of the final concrete lining. These temporary support measures are assumed to eventually corrode. It is then perhaps not surprising that convergence monitoring is such an important part of NATM, as a degree of loosening seems to be likely when so often using steel sets or lattice girders. The standard procedures involved in NATM are illustrated in Austrian standards, and are reproduced for reference (Figure 18) since so remarkably different to single-shell NMT.

In contrast to the sequences of NATM shown in Figure 18, in NMT the excavation is usually full-face, both for speed and to avoid a very unfavorable top-heading section (as illustrated). This invites the initiation of invert heave if the tunnel is at significant depth and not in hard rock. Final support in NMT is usually B+S(fr), while in double-shell NATM the final support is only the final concrete lining, which also secures the drainage fleece and membrane. The concrete lining is designed to take all ultimate loads from the rock mass. The temporary steel sets, S(mr), and bolting are assumed in the long-term to have corroded and are not featured in the final concrete load-bearing structure. This design philosophy, which is surprising to many, adds to the time and cost of NATM.

Inevitably the cost difference between NATM and NMT is of the order of 1: 3 to 1: 5, but this depends on rock mass quality, and hence on the type and

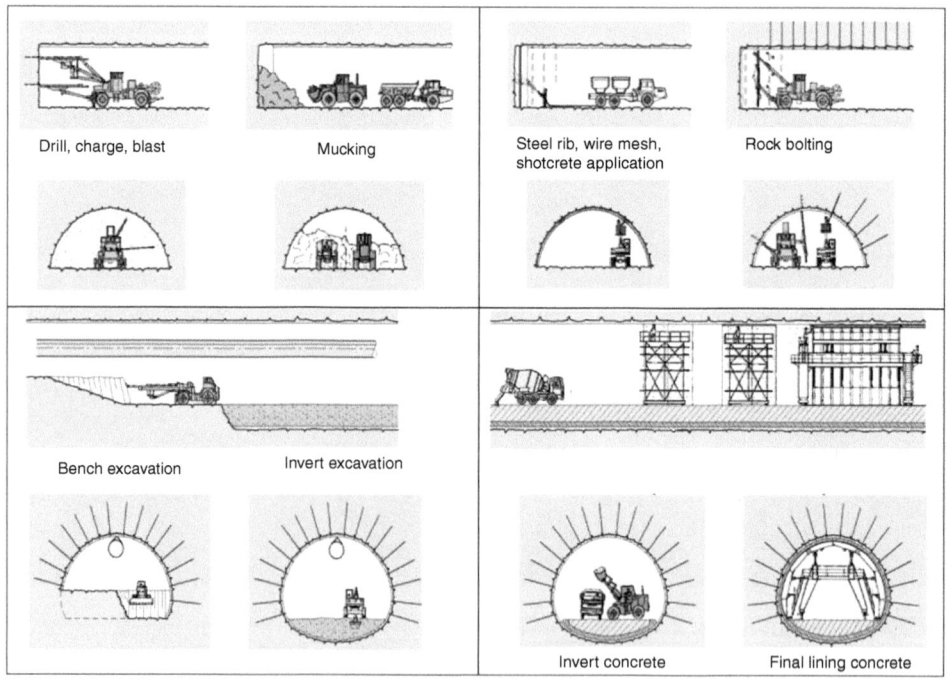

Figure 18 Schematic construction sequence of a typical NATM tunnel, apparently used in both softer and harder rock, from "Austrian Society for Geomechanics (2010). NATM, 'The Austrian Practice of Conventional Tunnelling'. This method has been observed in many countries when Q is 'poor', 'fair', 'good' *i.e.* Q = 1 to 40, where NMT would be eminently suitable and much faster and cheaper.

amount of rock support. The cost difference also depends on differences in labor costs in different countries. There may be a 1:10 difference in the number of tunnel workers involved, and the speed of NATM, including construction of the (3D) membrane and the sometimes locally thick final liner is inevitably much slower than single-shell NMT, due to all the operations, as may be visualized in Figure 18. Those using NATM will often point to poorer rock conditions where NATM tends to be applied. This can be only partly acknowledged, since double-shell NATM procedures are also specified in rock masses of comparable quality to those where NMT would be most applicable. This has been observed in many countries.

Norwegian road and railway tunnels of high standard, (in 2015), with all the technology installed for ventilation, lighting, drainage, safety and communication, cost about 18,000 to 27.000 US $ per meter (road) and about 25,000 to 33,000 US$ per meter (railway), depending on the dimension of the tunnel. International tunneling literature frequently documents 80,000 to 100,000 US per meter for the case of NATM double-shell tunnels.

Concerning tunneling speed, recent Norwegian world records of 164 m and 176 m in best weeks by two different Norwegian contractors, and a 104 m/week project average for 5.8 km in coal-measure rocks, obviously requiring significant rock support and reinforcement, suggests that NMT is a more efficient process. Figure 19 shows the

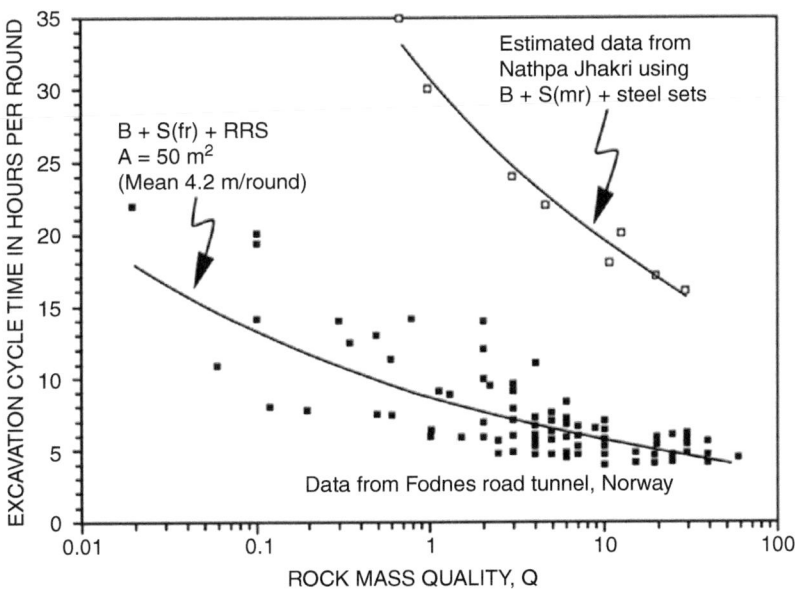

Figure 19 Cycle-time (drilling blast holes, loading with explosives, blasting, waiting for gasses to clear, scaling, geological inspection, mucking, reinforcement and support) as observed by Grimstad in the Fodnes road tunnel, which has a cross-section of 50–55 m². For comparison, the cycle-time for labor intensive temporary support methods in a hydroelectric project in India is also shown in relation to the logged Q-values. A final concrete lining will add to the differences in time and cost of NATM compared to NMT.

source of very fast NMT single-shell tunneling; namely the fast cycle time. This is usually below 10 hours for a wide range of Q-values (*i.e.* 1 to 100), and is as low as 5 to 6 hours at the top end of the rock mass quality scale where support and reinforcement is light or hardly needed.

12 USING THE Q-SYSTEM FOR TEMPORARY SUPPORT PRIOR TO NATM CONCRETE LINING

When the Q-system was first published in 1974, it was designed to provide guidance on suitable permanent support for a variety of tunnel and cavern sizes. By way of a footnote, it was suggested that the Q-system could also be used for guiding *temporary support* selection. The suggested rule-of-thumb was '5Q and 1.5 ESR'. This means a diagonal shift, downwards and to the right, on a Q-support chart, as illustrated by the example in Figure 20. This method has in fact been used systematically by Hong Kong road, rail, and metro authorities for at least 25 years, as the preliminary stage of NATM-style tunneling and station cavern development.

Table 3 (left-side) shows the ESR values recommended in Barton and Grimstad (1994) for various types of excavation. With the world-wide demand for increased safety in the last two decades, the recommended updated areas of the ESR table

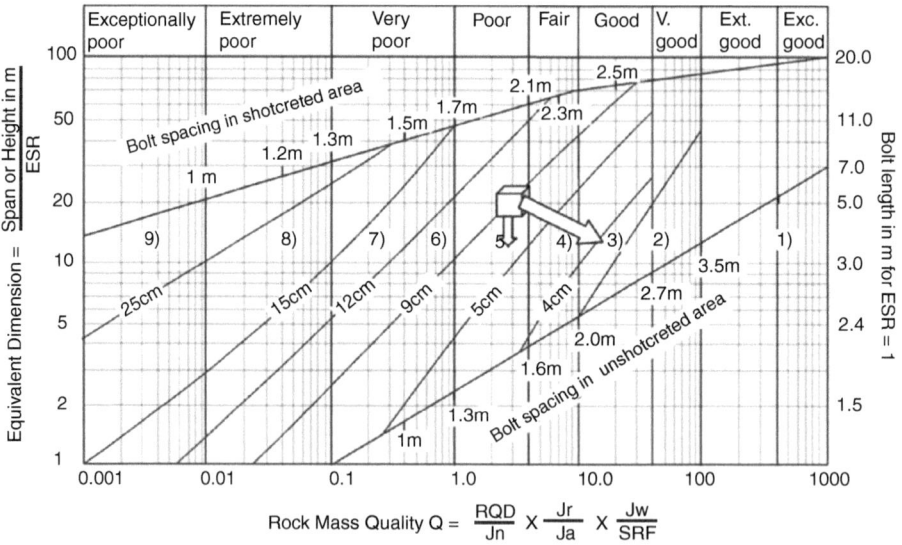

Figure 20 Using the 1993 Q-support chart as illustration, the Barton *et al.* (1974) 'rule-of-thumb' for selecting temporary support is shown for the case of NATM-style tunneling in Hong Kong. Applying 5Q and 1.5 ESR, the span-versus-Q coordinate moves downwards and to the right, ensuring less but sufficient B + S(fr) support while waiting 1 or 2 years for the final concrete lining. The first author has gradually learned to accept this practice when reviewing projects in Hong Kong, but was initially surprised by its widespread use already since the 1990s.

Table 3 On the left the ESR values in use in the nineties (Barton & Grimstad, 1994) are tabulated. Some updates recommended today due to the demand for greater safety are shown on the right (2014). Note the use of italics to emphasize when there has been no change from 1994 to 2014.

Type of Excavation	ESR (1994)	ESR (2014)
A *Temporary mine openings, etc.*	ca. 2–5	*ca. 2 to 5*
B *Permanent mine openings, water tunnels for hydropower (exclude high pressure penstocks), pilot tunnels, drifts and headings for large openings, surge chambers*	1.6–2.0	*1.6 to 2.0*
C *Storage caverns,* water treatment plants, minor road and railway tunnels, *access tunnels*	1.2–1.3	0.9 to 1.1 *Storage caverns 1.2–1.3*
D *Power stations,* major road and railway tunnels, *civil defense chambers, portals, intersections*	0.9–1.1	Major road and rail tunnels 0.5 to 0.8
E *Underground nuclear power stations, railway stations, sports and public facilities, factories, major gas pipeline tunnels*	0.5–0.8	*0.5 to 0.8*

Note: In Barton *et al.* 1974 there was an extra category with few case records: Vertical shafts:
(i) circular ESR ≈ 2.5(?) (ii) rectangular ESR ≈ 2.0(?)

are shown on the right. The ESR table published in 1994 was *correct* (*i.e.* reflecting common practice) in the 1970s and in the 1980s. However, the demand for safety has increased world-wide and also in Norway, particularly in the case of transport tunnels where small falls of rock fragments were occasionally occurring in *minor* road tunnels in the 1970s. (It was of course much less safe under the rock slopes outside the tunnels!). Now there is no tolerance for any rock fragment falls, even in minor transport tunnels. So minor road and railway tunnels should now have ESR = 1. Water treatment plants with a lot of expensive installations and representing a daily working place should have ESR = 0.9–1.1, and are increasingly more important and 'populated' than storage caverns. *Major* road and railway tunnels may need ESR = 0.5–0.8. These suggested updates are tabulated on the right-hand side of 1994 values in Table 3.

13 RELATIVE TIME AND COST IN RELATION TO THE Q-VALUE

As a result of a survey of some 50 km of tunneling mostly in Norway but also in Sweden, Roald produced the two figures of relative time and cost of tunneling in relation to the Q-value shown in Figure 22. These important trends were subsequently published in Barton *et al.* (2001), in which the main topic was rock mass improvement by pre-injection. In fact this was the first exploration of possible improvements in some of the Q-parameters as a result of high pressure pre-injection. A brief discussion of this topic is given near the end of this illustrated guide on the Q-system.

Figure 22 demonstrates the strong influence of the Q-value on tunneling time and cost. This is independently confirmed using the cycle-time changes with Q, as recorded by Grimstad in a Western Norway road tunnel. This was shown in Figure 19. The

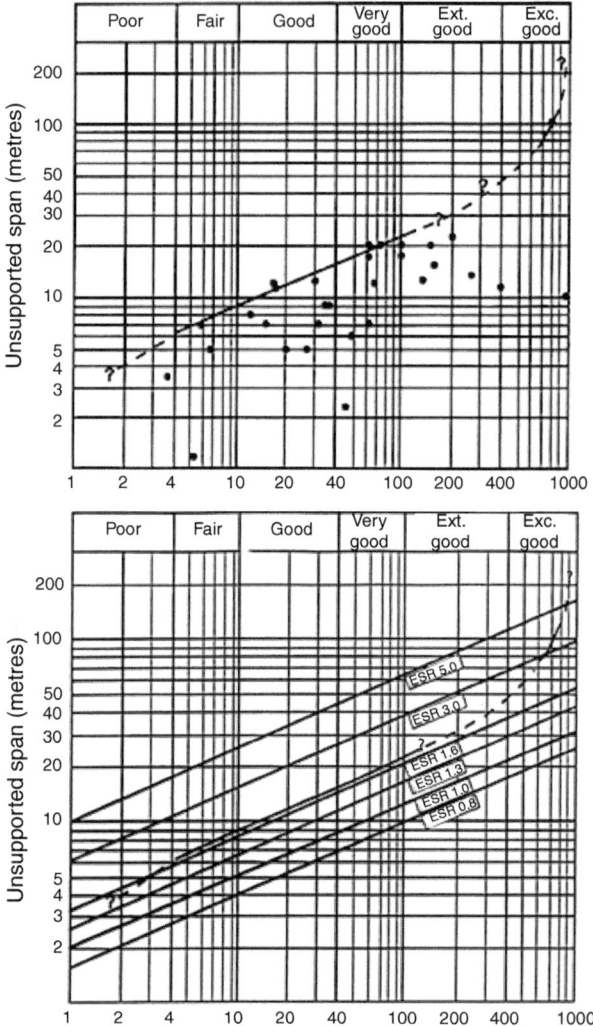

Figure 21 The workings of ESR, for modifying SPAN to equivalent span. The way ESR modifies the equivalent span is shown by the sloping lines, assuming ESR = 1.6 marks the unsupported boundary (for hydropower). (Barton, 1976). Note that 'unsupported span' in the Q-system refers to the width of excavation. In Bieniawski (1989) concerning use of RMR, the 'unsupported span' is the longitudinal distance from the face to the nearest support or reinforcement. These two 'spans' are sometimes confused, due to interest in using the 'stand-up time' chart developed by Bieniawski.

general trends shown in Figure 22 concerning relative cost can also be independently derived by a rigid application of the Q-system recommendations for arch and wall support over the whole range of Q-values, and for a wide range of tunnel spans. With knowledge of Q-values, costs can be derived.

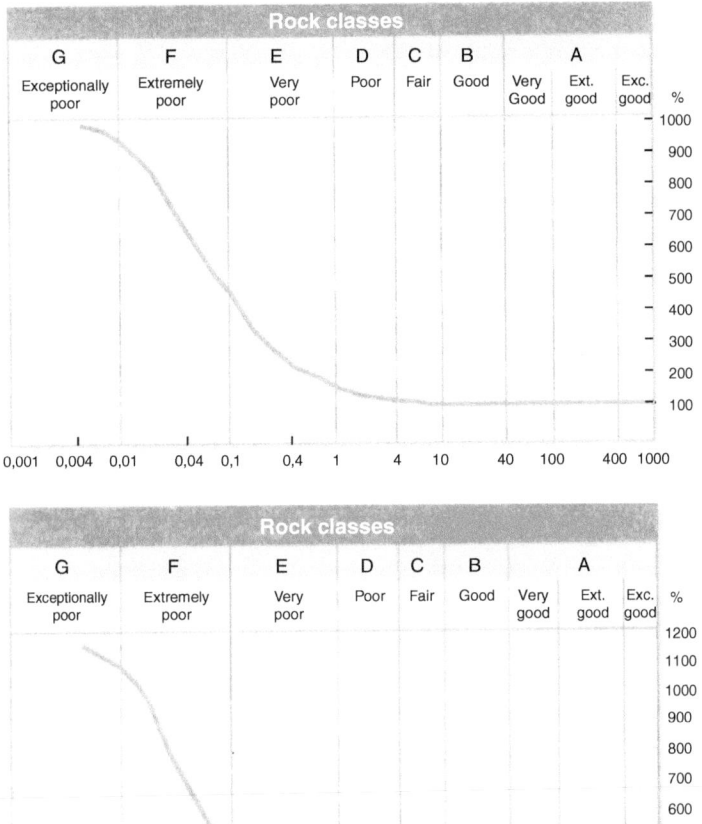

Figure 22 Relative time (top) and cost (bottom) of tunnel construction in relation to Q-value, according to a 50 km survey of Norwegian and Swedish tunnels carried out by Roald, and published as Barton *et al.* (2001).

14 ESTIMATING TUNNEL OR CAVERN DEFORMATION IN RELATION TO Q

It appears that the large numerical range of Q (0.001 to 1000 approx.) referred to in the introduction, helps to allow very simple formulæ for relating the Q-value to parameters of interest to rock engineering performance. The central trend of data in Figure 23 is actually described by the formula $\Delta \approx$ SPAN/Q: surprisingly simple

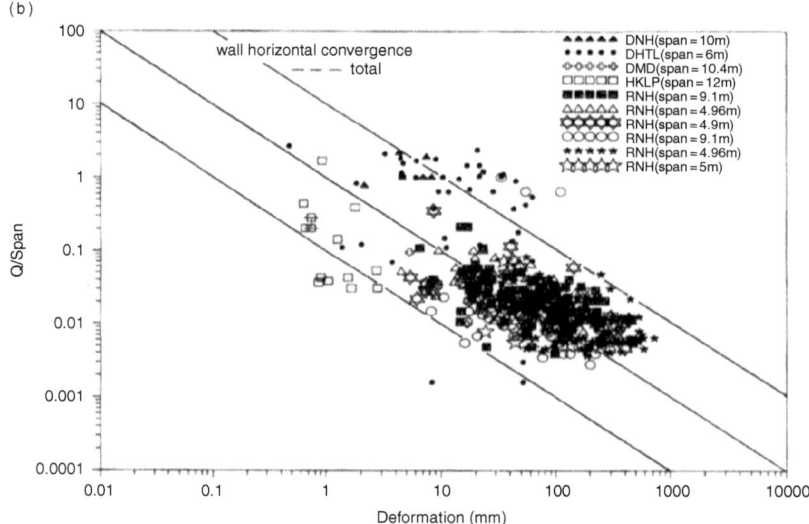

Figure 23   The log-log plotting of Q/span versus deformation was published in Barton *et al.* (1994), with fresh data from the MPBX instrumentation of the top-heading and full 60 m span of the Gjøvik Olympic cavern. Shen & Guo (priv. comm.) later provided similar data for numerous tunnels from Taiwan. When investigated, the central trend of hundreds of data was simply Δ (mm) ≈ SPAN (m) /Q.

and discovered after receiving the data from Taiwan shown in the lower half of Figure 23.

15   GJØVIK CAVERN Q-LOGGING, NMT SINGLE-SHELL B+S(FR) SUPPORT, AND DEFORMATION

The Gjøvik Olympic cavern was a milestone event in Norwegian rock engineering and rock mechanics practice, combining as it did the experience of several of Norway's

Table 4 Empirical equations 4, 5, 6, and 7 derived from Figure 23, with fine-tuning and reduced scatter using the competence factor stress/strength ratios. Examples of application in Indian and Norwegian caverns. Very close approximation to many cases has been found, including metro station caverns in Hong Kong, where over-continuous jointing in numerical models were showing far too large deformations in relation to the measurements. The empirical equations were again giving good results.

$\Delta = \dfrac{SPAN}{Q}$	$\Delta_v = \dfrac{SPAN}{100\,Q}\sqrt{\dfrac{\sigma_v}{\sigma_c}}$	$\Delta_h = \dfrac{HEIGHT}{100Q}\sqrt{\dfrac{\sigma_h}{\sigma_c}}$	$k_o = \left(\dfrac{SPAN}{HEIGHT}\right)^2\left(\dfrac{\Delta_h}{\Delta_v}\right)^2$
Nathpa Jakri power station cavern		**Gjøvik Olympic cavern**	

Nathpa Jakri power station cavern	Gjøvik Olympic cavern
$\Delta_v = \dfrac{20{,}000}{100 \times 3} \times (6/35)^{1/2} = 28$ mm $\Delta_h = \dfrac{50{,}000}{100 \times 3} \times (4/35)^{1/2} = 56$ mm (SPAN = 20m, HEIGHT = 50m, Q = 3, σ_v=4 MPa, σ_h = 6 MPa, σ_c = 35 MPa). (In the middle of the range of MPBX measurements for the arch and walls).	$\Delta_v = \dfrac{60{,}000}{100 \times 10} \times (1/75)^{1/2} = 6.9$ mm (SPAN = 60m, Q_{mean} = 10, σ_v = 1 MPa at 40 m depth, σ_c = 75 MPa) (Almost identical to that measured with nine MPBX, and almost identical to UDEC-BB modelling results).

leading consulting, research institutes and contracting companies. The Q-system was well utilized, as shown in diagrams c) and d) in Figure 24.

The efficient cavern excavation and execution of single-shell NMT-style permanent support, which took just 6 months in 1991, saw the removal of 140,000 m³ of red and grey gneiss. RQD was mostly 60–90%, and the mean UCS was 90 MPa. The 62 × 24 × 90 m raw-cavern dimensions represented a large (almost 100%) jump in the world's largest-span cavern for public use, with capacity for about 5,400 people for artistic events, concerts etc., which both preceded the winter Olympic ice hockey events of 1994 (*i.e.* the grand opening ceremony), and have frequently followed in the years since then.

Four boreholes were used for site investigation, two of them inclined. These holes were used for Q-logging (Figure 26a), seismic tomography (V_P range was 3.5 to 5.5 km/s), with the high velocities due to the surprisingly high 3 to 5 MPa horizontal stress at only 30 to 50m depth.

Representation of the conjugate jointing, the favorably high boundary stresses, the depth-dependent deformation modulus and the eight principal excavation stages used when UDEC-BB modeling the Gjøvik cavern are shown in Figure 25. The modeled vertical deformations above the main arch were approximately 4 and 5 mm depending on the modeled depth of 30 or 50m (relevant to each end of the cavern). The third model shown in Figure 25 had unchanged input data, but included the three Postal Service caverns on the 'right-hand' side (excavation stages # 6, 7 and 8). These caused the central arch vertical deformation to increase to 7 mm. The MPBX-measured results (Figure 24 f) incorporating internal (SINTEF) and external (NGI) extensometers, plus the results of surface leveling (downwards rather than upwards) were 7 to 8 mm.

Note that the assumptions of joint continuity in this model are far different from the 'lazy modeling' that one sees so often, in which all joint sets are made continuous (for

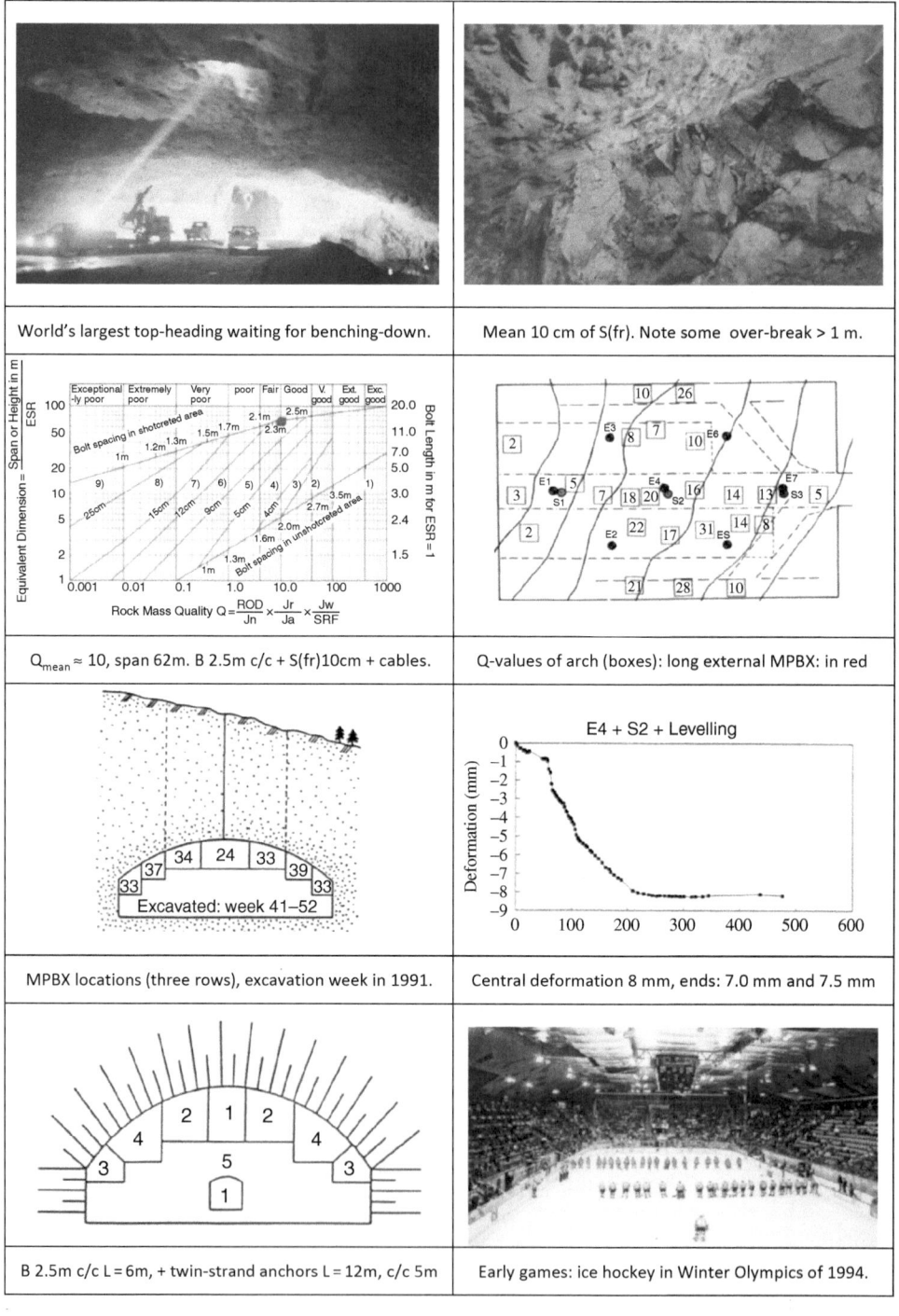

World's largest top-heading waiting for benching-down.

Mean 10 cm of S(fr). Note some over-break > 1 m.

$Q_{mean} \approx 10$, span 62m. B 2.5m c/c + S(fr)10cm + cables.

Q-values of arch (boxes): long external MPBX: in red

MPBX locations (three rows), excavation week in 1991.

Central deformation 8 mm, ends: 7.0 mm and 7.5 mm

B 2.5m c/c L = 6m, + twin-strand anchors L = 12m, c/c 5m

Early games: ice hockey in Winter Olympics of 1994.

Figure 24 Some details of the Gjøvik Olympic cavern. Concept from Jan Rygh, design studies by Fortifikasjon and NoTeBy, design check modeling with UDEC-BB, external MPBX, seismic tomography, stress measurements and Q-logging by NGI, internal MPBX, bolt and cable loads, modeling, research aspects by SINTEF-NTNU. However, most important of all: efficient construction in 6 months using double-access tunnels, by the Veidekke-Selmer JV. The cavern is an example of a drained NMT excavation. Details of the rock engineering and rock mechanics aspects of the project, including the predictive (Class A) numerical modeling, are given in the multi-author paper of Barton *et al.* (1994).

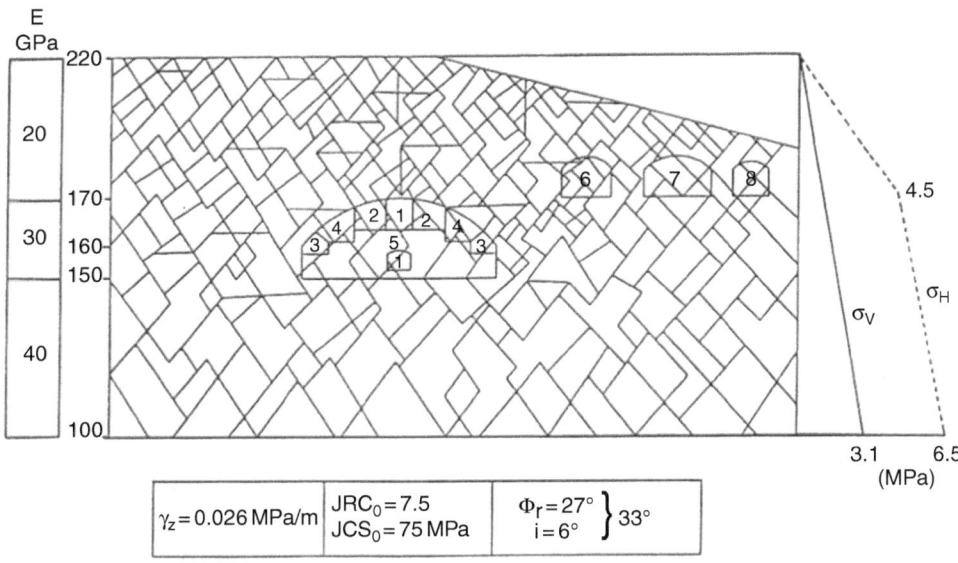

Figure 25 The geometry of the Gjøvik cavern(s), the excavation stages, the depth-dependent deformation moduli, and the joint properties used in UDEC-BB. See *Barton et al. (1994) for details*.

simplicity). Modeled deformations are then much too large, and their authors therefore feel it necessary to criticize empirical methods, when in fact they have grossly over-estimated the deformations and 'plastic zones' due to unrealistic representation of the structural geology.

16 PRE-GROUTING CAN CHANGE EFFECTIVE Q-VALUES

Barton *et al.* (2001) and Barton (2002) suggested, controversially as with most innovations, that several, perhaps most of the Q-parameters could, *in effect* be improved by the typical high pressure 5 to 10 MPa pre-injection of micro-and ultrafine cements-with-microsilica, as regularly practiced in Norway. This suggestion seems to have been proved correct over time, as some others working in dam foundation engineering and mining are also reporting such finds.

The first author systematically Q-logged all the drill-core and analyzed all the permeability measurements for the Jong-Asker and Bærum rail tunnels for JBV (Jernbaneverket). Subsequent experiences suggest that some of the (consultant stipulated) inflow requirements for the first two Jong-Asker tunnels were not stringent enough: due care was taken of the external natural (and built-on) surface environment, but some of the pre-grouting was not sufficiently effective for the inside-the-tunnel environment. Dripping water remained in places, when the least stringent 8 to 16 liters/min/100m inflow criteria were used. However, in the case of the later Bærum rail tunnel of 5 km length, which was systematically injected using more holes and consistently high pressure, a very dry result was

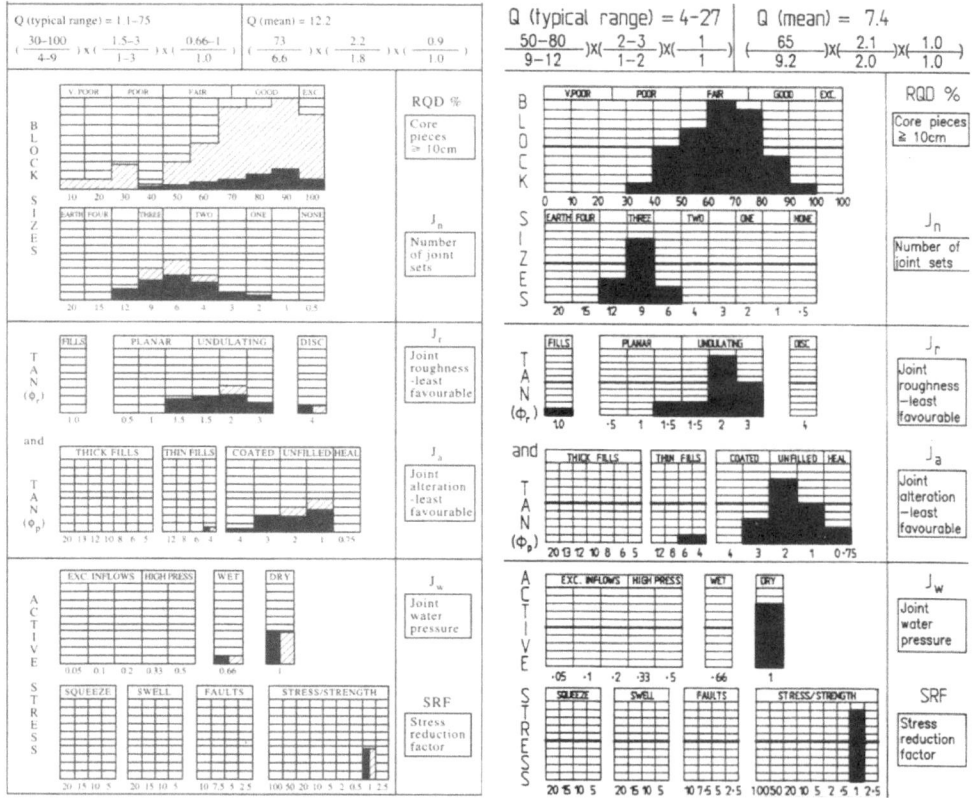

Figure 26 A comparison of three stages of Q-logging. Left: core-logging (cross-hatched) and local cavern-walls logging (black: where no shotcrete cover). Right: Gjøvik cavern top-heading logging. Three different engineering geologists were responsible for this independent Q-histogram logging.

obtained. Several inspections of the pre-injected new rounds of tunnel advance suggested that the consultant's single-shell NMT final support, as seen in Figure 27, was now very conservative. However, this was Q-value based before the possible/probable improvements by pre-grouting.

In relation to the extensive (kilometers) of Q-logged core, logged to depths greater than the tunnel depths, there appeared to have been an improvement in the rock mass quality due to the pre-grouting. Not only was the shotcrete 99.999% dry, but the B+S (fr) which was applied, based on prior Q-based designs by other consultants, seemed to be conservative. This can be concluded just by inspecting the photographs in Figure 27, which are typical of many.

Table 5 shows two hypothetical models for 'before-and-after' Q parameter improvements, to illustrate the possibilities. The pre-grouting models have no relation to the two Bærum tunnel photographs, and were presented by Bàrton (2012b). Reduced tunnel support needs, reduced deformation, increased deformation modulus and

Figure 27 The pre-injected nodular-limestones and shales of the Bærum Tunnel appeared to have increased in Q-class (by two or three classes) due to the effect of high-pressure pre-injection. Top-left: the first 5 cm layer of S(fr) and the permanent CT bolts-and-washers at approximately 1.5 m c/c. Top-right: bolt heads and washers sprayed in with the final 5 cm layer of S(fr). The tunnel now has its completed single-shell NMT support and reinforcement, which appears to be conservative. The quality of the shale/limestone (and igneous dikes elsewhere), appears to have been improved by the high pressure pre-injection.

Table 5 Two hypothetical but not unrealistic 'models' for potential Q-parameter improvement as a result of pre-injection. Barton (2012b).

CONSERVATIVE PRE-INJECTION MODEL	MORE REALISTIC PRE-INJECTION MODEL
RQD increases e.g. 30 to 50%	**RQD increases e.g. 30 to 70%**
Jn reduces e.g. 9 to 6	**Jn reduces e.g. 12 to 4**
Jr increases e.g. 1 to 2	**Jr increases e.g. 1.5 to 2**
(due to sealing of most of set #1)	**(due to sealing of most of set #1)**
Ja reduces e.g. 2 to 1	**Ja reduces e.g. 4 to 1**
(due to sealing of most of set #1)	**(due to sealing of most of set #1)**
Jw increases e.g. 0.5 to 1	**Jw increases e.g. 0.66 to 1**
SRF unchanged e.g. 1.0 to 1.0	**SRF improves e.g. 2.5 to 1.0**
WET CONDITIONS	**WET CONDITIONS**
Before pre-grouting	*Before pre-grouting*
$Q = 30/9 \times 1/2 \times 0.5/1 = 0.8$	$Q = 30/12 \times 1.5/4 \times 0.66/2.5 = 0.2$
$V_p \approx 3.4 \ km/s$	$V_p \approx 2.8 \ km/s$
$E_{mass} \approx 9.3 \ GPa$	$E_{mass} \approx 5.8 \ GPa$
$K \approx 1.3 \times 10^{-7} \ m/s$	$K \approx 5.0 \times 10^{-7} \ m/s$
e.g. for a 10 m tunnel: B 1.6 m c/c, S(fr) 10 cm	e.g. for a 10 m tunnel: B 1.4 m c/c, S(fr) 13 cm
DRY CONDITIONS	**DRY CONDITIONS**
After pre-grouting	*After pre-grouting*
$Q = 50/6 \times 2/1 \times 1/1 = 17$	$Q = 70/4 \times 2/1 \times 1/1 = 35$
$V_p \approx 4.7 \ km/s$	$V_p \approx 5.0 \ km/s$
$E_{mass} \approx 25.7 \ GPa$	$E_{mass} \approx 32.7 \ GPa$
$K \approx 5.9 \times 10^{-9} \ m/s$	$K \approx 2.9 \times 10^{-9} \ m/s$
e.g. for a 10 m tunnel: B 2.4 m c/c	e.g. for a 10 m tunnel: sb (spot bolts)
(Today's conservatism may also demand a single layer of S(fr) due to the demand for lower ESR in the case of transport tunnels).	**(Today's conservatism may also demand a single layer of S(fr) due to the demand for lower ESR in the case of transport tunnels).**

increased seismic velocity as a result of pre-grouting (the latter documented in Barton, 2006) are each suggested. Naturally one expects as a very minimum that a Jw of 0.5 or 0.66 will become 1.0 ('dry') as a result of successful pre-injection. Other parameters seem also to benefit, including the *effective* RQD, and the *effective* Jn. There may also be transfer of lower Jr/Ja ratios to the remaining *uninjected* (probably tightest) joint sets, resulting in higher *effective* Jr/Ja ratios, and therefore to even higher *effective* post-injection Q-values.

In Barton (2002) there is documentation of the measured rotation of all principal permeability tensors (even) as a result of just moderate pressure industrial cement injection at a dam site. This was measured using 3D (multi-borehole) hydro-tomography by Quadros & Correa Filho (1995). So we know that joint sets can become sealed even at low pressures (max. 2 MPa) with non-ideal Portland cement and bentonite. With today's optimized grouting materials and the last 20 years of high pressure (5 to 10 MPa) pre-injection, as often practiced in Norway, it is perhaps time to take credit for the benefits of pre-injection, as already suggested by some tunnel contractors in Norway, who experienced first-hand the benefits of the improved conditions.

17 PERMEABILITY MAY BE RELATED TO 'Q-VALUES', BUT MORE LOGICALLY TO Q_{H2O}

An interesting and pre-injection related 'property' of the Q-value, clearly a result only possible with the big numerical range of Q, is that there is *some evidence* of an inverse relation between Q and the Lugeon value. This is strictly for the case of *clay-free rock masses*. Some theoretical justification is given in Barton (2006) (Chapter 9), based on the interpretation of a Lugeon test (with its 1 MPa over-pressure) as a slightly deforming test. A double Boussinesq 'foundation load' formulation is utilized, plus the cubic law relating flow rates to the cube of aperture, when apertures are large enough (*i.e.*> 0.5mm). For details see Barton (2011, 2012b).

Table 6 shows what to expect in approximate terms in (mostly clay-free) rock masses. Strong deviation from the simple scheme of values tabulated, in one

Table 6 Ball-park estimation of permeability, when clay is absent. Barton (2006). There are valid theoretical arguments linking Lugeon and Q_c, because of deformability effects. $Q_c(= Q \times \sigma_c)$ for the shale (Figure 28) might have increased due to pre-injection from 1 to an equivalent 100, with corresponding property improvements as a result: for example 0.01 Lugeon, and maybe also V_P= 5.5 km/s.

Q_c	0.1	1	10	100
Lugeon	10	1	0.1	0.01
K (m/s)	10^{-6}	10^{-7}	10^{-8}	10^{-9}
V_P (km/s)	2.5	3.5	4.5	5.5

Figure 28 A successfully pre-injected tunnel (a completed section of the Bærum rail tunnel) which demonstrates a '99.999%' dry (non-humid) shotcrete.

direction or the other, would suggest the need to expect clay-filled discontinuities, or weak deformable rock like phyllite, or de-stressed rock, causing lower or higher permeability respectively, despite lower Q-values. The obvious 'over-simplification' of the above inter-related parameters, in particular the inclusion of the inherently complicated *stress and depth dependent permeability* has resulted in the development of a more general 'Q_{H2O}' model for estimating permeability. This has an empirically-developed depth dependence, and Jr/Ja is reversed to a more logical *Ja/Jr*. It is described in Barton (2013a).

18 CONCLUSIONS

1. The Q-value representing rock mass quality or lack of quality, and the Q-system linking Q to recommended single-shell permanent reinforcement and support measures (B + Sfr) has proved its value during its 40 years existence (up to 2015). It has been widely adopted in many countries, as one of the standard empirical tools of rock mass characterization, and as a method for assisting in tunnel and cavern support-class selection in rock engineering. It is most associated with permanent single-shell NMT (Norwegian

Method of Tunnelling). It has also been used for many years to help select temporary support and reinforcement for double-shell tunnels and caverns which will finally have a permanent concrete lining, in the same manner as NATM.

2. The tunnel and cavern reinforcement and support measures were originally based on systematic bolting and mesh-reinforced shotcrete, when the Q-system was first developed in 1974. The development of wet process, robotically applied steel-fiber reinforced shotcrete saw first application in Norway in a hydropower cavern in 1979, and in Holmestrand road tunnel in 1981. The development of multi-layer corrosion protected (CT) bolts soon followed.

3. The Q-system support recommendations were updated in 1993 to reflect the widespread use of B+S(fr) as *single-shell permanent* support. There are about 1250 case records behind the method since 1993, and a further 800 cases since 2002. There have been hundreds of thousands of practical applications, the numbers depending on whether referring to individuals or groups of engineering geologists who apply Q on a 'daily' basis in numerous countries.

4. Besides its widespread use in civil engineering, the mining industry in all the principal mining countries (USA, Canada, Brazil, Peru, Chile, Australia etc.) make active use of the Q-system for support and reinforcement of 'permanent' mine roadways. Q-parameters are also extensively used when differentiating stable, transitional, or caving ground in non-entry stopes. In this application the first four Q-parameters ($Q' = RQD/J_n \times J_r/J_a$) are used, together with stress/strength ratios, stope dimensions, and structural orientation. Stopes may be in need of temporary cable reinforcement. In civil engineering we recommend and need all six Q-parameters. The lack of a faulting term when Q' is used for mining stopes, due to a consulting company's unilateral removal of SRF, and the lack of a Jw term for wet mines are potential weaknesses of the truncated Q' term.

5. The Q-value and its modified form Q_c, obtained by normalizing with UCS/100, has many potential uses in rock engineering. It can be correlated to the seismic P-wave velocity V_P (km/s), to the (static) deformation modulus M or E_{mass} (GPa), to the vertical and horizontal deformation, and it has also been linked tentatively with the Lugeon value of clay-free rock masses. In modified Q_{H2O} form, depth-dependent permeability in the case of clay-bearing or deformable rock seems also to be predictable in approximate terms.

6. In the last 15 years since 2000, the Q-value has been incorporated in a more comprehensive parameter called Q_{TBM}. This has additional machine-rock interaction parameters, and is used as a basis for TBM prognosis. On the basis of numerous (1000 km) of data linking TBM deceleration over time (also seen in world records), the Q-value in the case of poor rock conditions ($Q < 1$) can be shown to explain delays and even stand-stills in fault zones. The gradient of deceleration relates strongly to Q-values. A chapter on this subject will be found in this five-volume publication.

1. Rock Quality Designation		RQD (%)
A	Very poor	0-25
B	Poor	25-50
C	Fair	50-75
D	Good	75-90
E	Excellent	90-100

Notes: i) Where RQD is reported or measured as ≤ 10 (including 0), a nominal value of 10 is used to evaluate Q.

ii) RQD intervals of 5, i.e. 100, 95, 90, etc., are sufficiently accurate.

2. Joint set number		J_n
A	Massive, no or few joints	0.5-1
B	One joint set	2
C	One joint set plus random joints	3
D	Two joint sets	4
E	Two joint sets plus random joints	6
F	Three joint sets	9
G	Three joint sets plus random joints	12
H	Four or more joint sets, random, heavily jointed, 'sugar-cube', etc.	15
J	Crushed rock, earthlike	20

Notes: i) For tunnel intersections, use (3.0 × J_n).

ii) For portals use (2.0 × J_n).

3. Joint roughness number		J_r
a) Rock-wall contact, and b) Rock-wall contact before 10 cm shear		
A	Discontinuous joints	4
B	Rough or irregular, undulating	3
C	Smooth, undulating	2
D	Slickensided, undulating	1.5
E	Rough or irregular, planar	1.5
F	Smooth, planar	1.0
G	Slickensided, planar	0.5

Notes: i) Descriptions refer to small-scale features and intermediate scale features, in that order.

b) No rock-wall contact when sheared		
H	Zone containing clay minerals thick enough to prevent rock-wall contact.	1.0
J	Sandy, gravely or crushed zone thick enough to prevent rock-wall contact	1.0

Notes: ii) Add 1.0 if the mean spacing of the relevant joint set is greater than 3 m.

iii) J_r = 0.5 can be used for planar, slickensided joints having lineations, provided the lineations are oriented for minimum strength.

iv) J_r and J_a classification is applied to the joint set or discontinuity that is least favourable for stability both from the point of view of orientation and shear resistance, τ (where τ ≈ σ_n tan^{-1} (J_r/J_a).)

4. Joint alteration number		ϕ_r approx.	J_a
a) Rock-wall contact (no mineral fillings, only coatings)			
A	Tightly healed, hard, non-softening, impermeable filling, i.e., quartz or epidote.	--	0.75
B	Unaltered joint walls, surface staining only.	25-35°	1.0
C	Slightly altered joint walls. Non-softening mineral coatings, sandy particles, clay-free disintegrated rock, etc.	25-30°	2.0
D	Silty- or sandy-clay coatings, small clay fraction (non-softening).	20-25°	3.0
E	Softening or low friction clay mineral coatings, i.e., kaolinite or mica. Also chlorite, talc, gypsum, graphite, etc., and small quantities of swelling clays.	8-16°	4.0
b) Rock-wall contact before 10 cm shear (thin mineral fillings).			
F	Sandy particles, clay-free disintegrated rock, etc.	25-30°	4.0
G	Strongly over-consolidated non-softening clay mineral fillings (continuous, but < 5 mm thickness).	16-24°	6.0
H	Medium or low over-consolidation, softening, clay mineral fillings (continuous, but < 5 mm thickness).	12-16°	8.0
J	Swelling-clay fillings, i.e., montmorillonite (continuous, but < 5 mm thickness). Value of J_a depends on per cent of swelling clay-size particles, and access to water, etc.	6-12°	8-12
c) No rock-wall contact when sheared (thick mineral fillings).			
KL M	Zones or bands of disintegrated or crushed rock and clay (see G, H, J for description of clay condition).	6-24°	6, 8, or 8-12
N	Zones or bands of silty- or sandy-clay, small clay fraction (non-softening).	--	5.0
OP R	Thick, continuous zones or bands of clay (see G, H, J for description of clay condition).	6-24°	10, 13, or 13-20

5. Joint water reduction factor		approx. water pres. (kg/cm²)	J_w
A	Dry excavations or minor inflow, i.e. < 1 litre/min locally.	< 1	1.0
B	Medium inflow or pressure, i.e. < 5 litre/min locally, occasional outwash of joint fillings.	1-2.5	0.66
C	Large inflow or high pressure in competent rock with unfilled joints.	2.5-10	0.5
D	Large inflow or high pressure, considerable outwash of joint fillings.	2.5-10	0.33
E	Exceptionally high inflow or water pressure at blasting, decaying with time.	> 10	0.2-0.1
F	Exceptionally high inflow or water pressure continuing without noticeable decay.	> 10	0.1-0.05

Notes: i) Factors C to F are crude estimates. Increase J_w if drainage measures installed.

ii) Special problems caused by ice formation are not considered.

iii) For general **characterization** of rock masses distant from excavation influences, the use of J_w = 1.0, 0.66, 0.5, 0.33 etc. as depth increases from say 0-5m, 5-25m, 25-250m to >250m is recommended, assuming that RQD /J_n is low enough (e.g. 0.5-25) for good hydraulic connectivity. This will help to adjust Q for some of the effective stress and water softening effects, in combination with appropriate **characterization** values of SRF. Correlations with depth-dependent static deformation modulus and seismic velocity will then follow the practice used when these were developed.

6. Stress Reduction Factor		SRF
a) Weakness zones intersecting excavation, which may cause loosening of rock mass when tunnel is excavated		
A	Multiple occurrences of weakness zones containing clay or chemically disintegrated rock, very loose surrounding rock (any depth).	10
B	Single weakness zones containing clay or chemically disintegrated rock (depth of excavation ≤ 50 m).	5
C	Single weakness zones containing clay or chemically disintegrated rock (depth of excavation > 50 m).	2.5
D	Multiple shear zones in competent rock (clay-free), loose surrounding rock (any depth).	7.5
E	Single shear zones in competent rock (clay-free), (depth of excavation ≤ 50 m).	5.0
F	Single shear zones in competent rock (clay-free), (depth of excavation > 50 m).	2.5
G	Loose, open joints, heavily jointed or 'sugar cube', etc. (any depth)	5.0

Notes: i) Reduce these values of SRF by 25-50% if the relevant shear zones only influence but do not intersect the excavation. This will also be relevant for characterization.

b) Competent rock, rock stress problems		σ_c/σ_1	σ_θ/σ_c	SRF
H	Low stress, near surface, open joints.	> 200	< 0.01	2.5
J	Medium stress, favourable stress condition.	200-10	0.01-0.3	1
K	High stress, very tight structure. Usually favourable to stability, may be unfavourable for wall stability.	10-5	0.3-0.4	0.5-2
L	Moderate slabbing after > 1 hour in massive rock.	5-3	0.5-0.65	5-50
M	Slabbing and rock burst after a few minutes in massive rock.	3-2	0.65-1	50-200
N	Heavy rock burst (strain-burst) and immediate dynamic deformations in massive rock.	< 2	> 1	200-400

Notes: ii) For strongly anisotropic virgin stress field (if measured): When 5 ≤ σ_1/σ_3 ≤ 10, reduce σ_c to 0.75 σ_c. When σ_1/σ_3 > 10, reduce σ_c to 0.5 σ_c, where σ_c = unconfined compression strength, σ_1 and σ_3 are the major and minor principal stresses, and σ_θ = maximum tangential stress (estimated from elastic theory).

iii) Few case records available where depth of crown below surface is less than span width. Suggest an SRF increase from 2.5 to 5 for such cases (see H).

iv) Cases L, M, and N are usually most relevant for support design of deep tunnel excavations in hard massive rock masses, with RQD /J_n ratios from about 50 to 200.

v) For general characterization of rock masses distant from excavation influences, the use of SRF = 5, 2.5, 1.0, and 0.5 is recommended as depth increases from say 0-5m, 5-25m, 25-250m to >250m. This will help to adjust Q for some of the effective stress effects, in combination with appropriate characterization values of J_w. Correlations with depth - dependent static deformation modulus and seismic velocity will then follow the practice used when these were developed.

c) Squeezing rock: plastic flow of incompetent rock under the influence of high rock pressure		σ_θ/σ_c	SRF
O	Mild squeezing rock pressure	1-5	5-10
P	Heavy squeezing rock pressure	> 5	10-20

Notes: vi) Cases of squeezing rock may occur for depth H > 350 $Q^{1/3}$ according to Singh 1993. Rock mass compression strength can be estimated from SIGMA$_{cm}$ ≈ 5 γ $Q_c^{1/3}$ (MPa) when γ = rock density in t/m³, and Q_c=Q×σ_c/100, Barton, 2000. [29]

d) Swelling rock: chemical swelling activity depending on presence of water		SRF
R	Mild swelling rock pressure	5-10
S	Heavy swelling rock pressure	10-15

APPENDIX A1 Q-parameter definitions and ratings for reference. The Q-logging sheet (following page) is an abbreviated form of these tables used when logging in the field (drill-core, rock exposures, or tunnel advances). See Appendix A4 for details of Jr/Ja, and Appendix A5, Tables A5.1 and A5.2 for specific high stress data related to the highest SRF values in the case of massive rock.

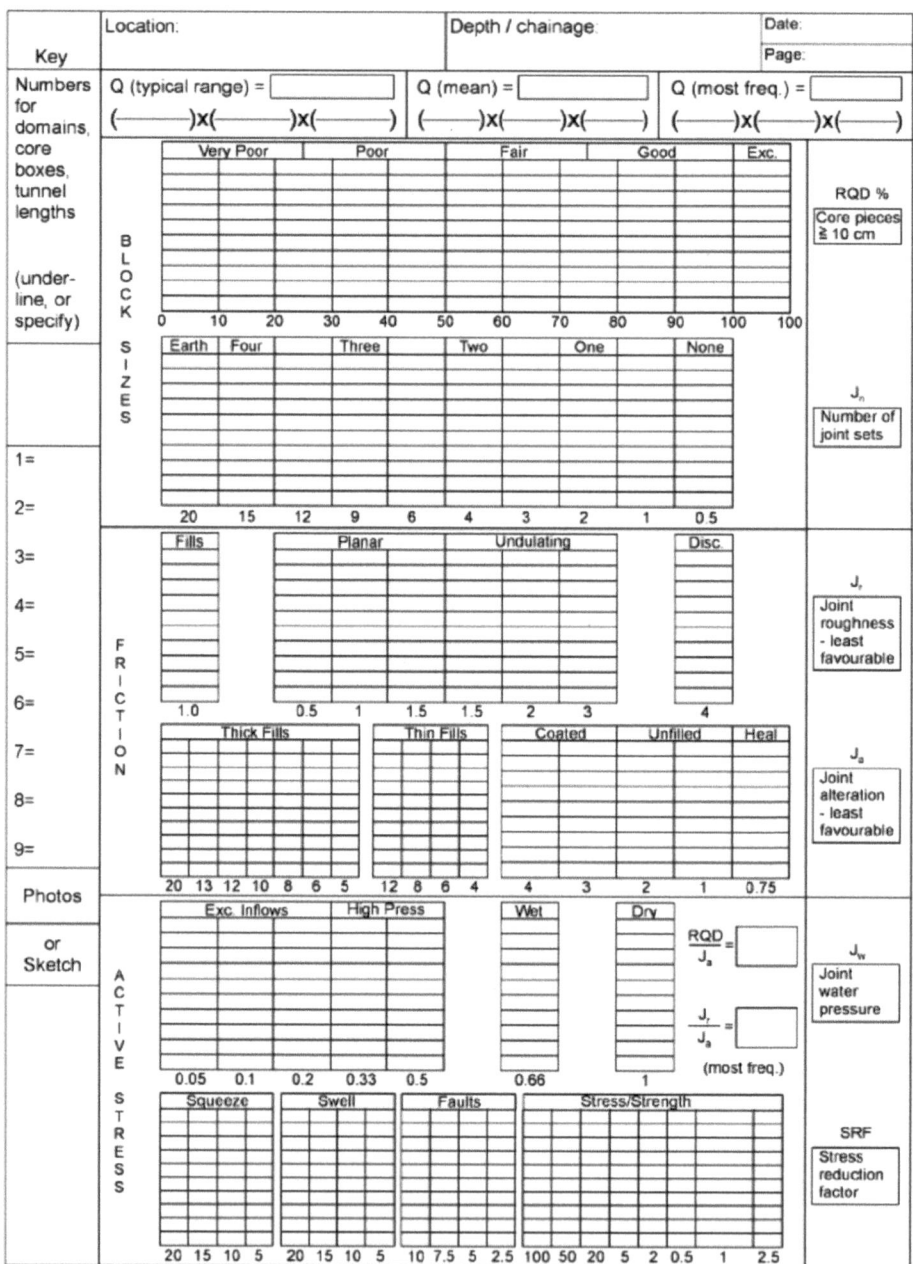

APPENDIX A2 Empty Q-parameter logging sheet. Note (brief) descriptions and ratings above and below each space for recording number of observations. It is convenient to number each row (1 to 9 and result Σ) on the left side, in order to correctly locate (by row) the 5m of core, or 5 m of tunnel being logged e.g. 1= 76.1-81.1m, 2= 81.1−86.1m etc. One can then return to zones (domains, rock types, fault zones) and extract specific observations for separate analysis. Allow five observations per 5m observed (core, exposure, tunnel wall).

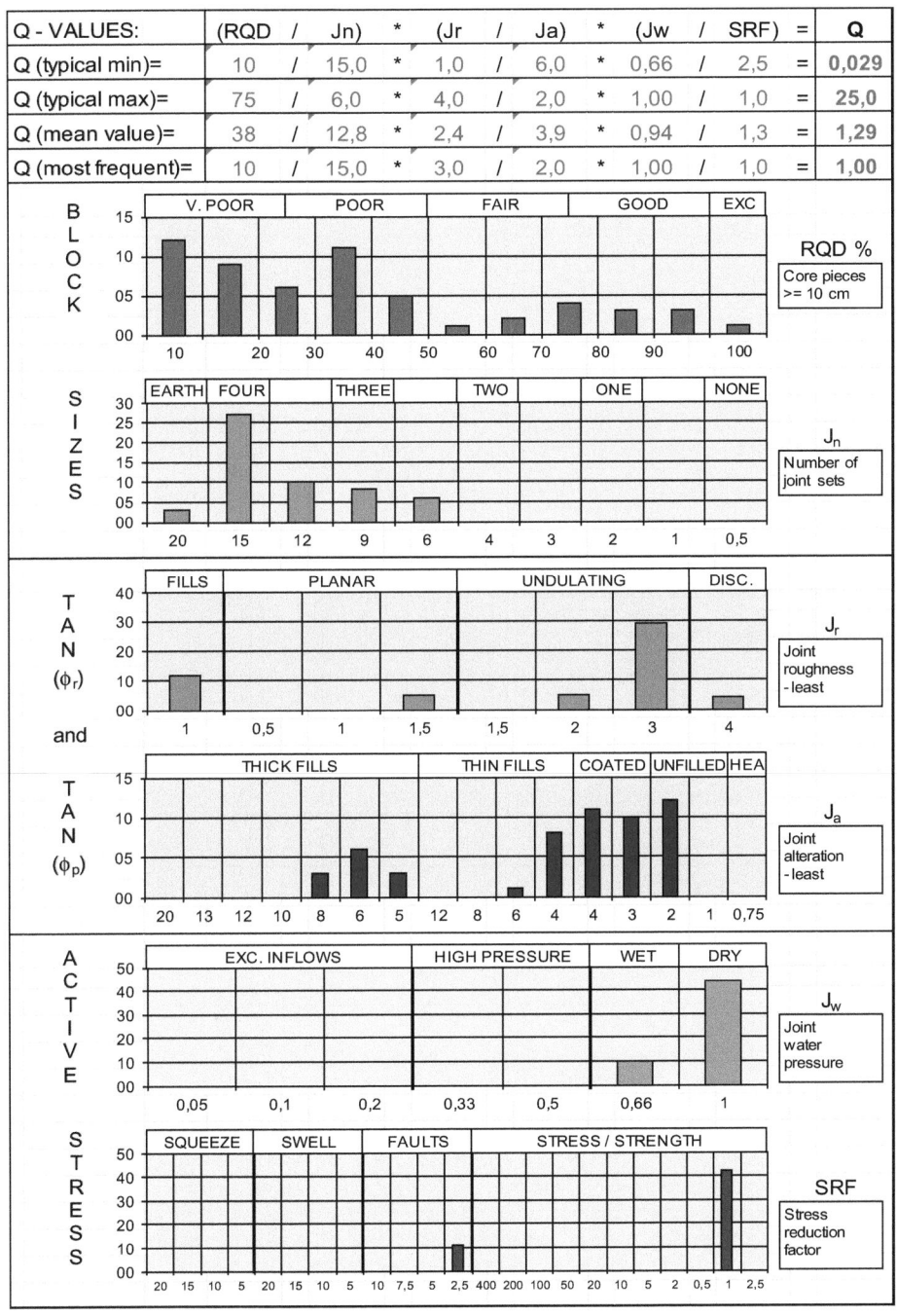

Q - VALUES:	(RQD	/	Jn)	*	(Jr	/	Ja)	*	(Jw	/	SRF)	=	Q
Q (typical min)=	10	/	15,0	*	1,0	/	6,0	*	0,66	/	2,5	=	0,029
Q (typical max)=	75	/	6,0	*	4,0	/	2,0	*	1,00	/	1,0	=	25,0
Q (mean value)=	38	/	12,8	*	2,4	/	3,9	*	0,94	/	1,3	=	1,29
Q (most frequent)=	10	/	15,0	*	3,0	/	2,0	*	1,00	/	1,0	=	1,00

APPENDIX A3 Completed logging sheet with EXCEL calculation of simple Q-statistics. Note weathered, heavily jointed, sheared and clay-bearing nature of this (ore-body) rock mass.

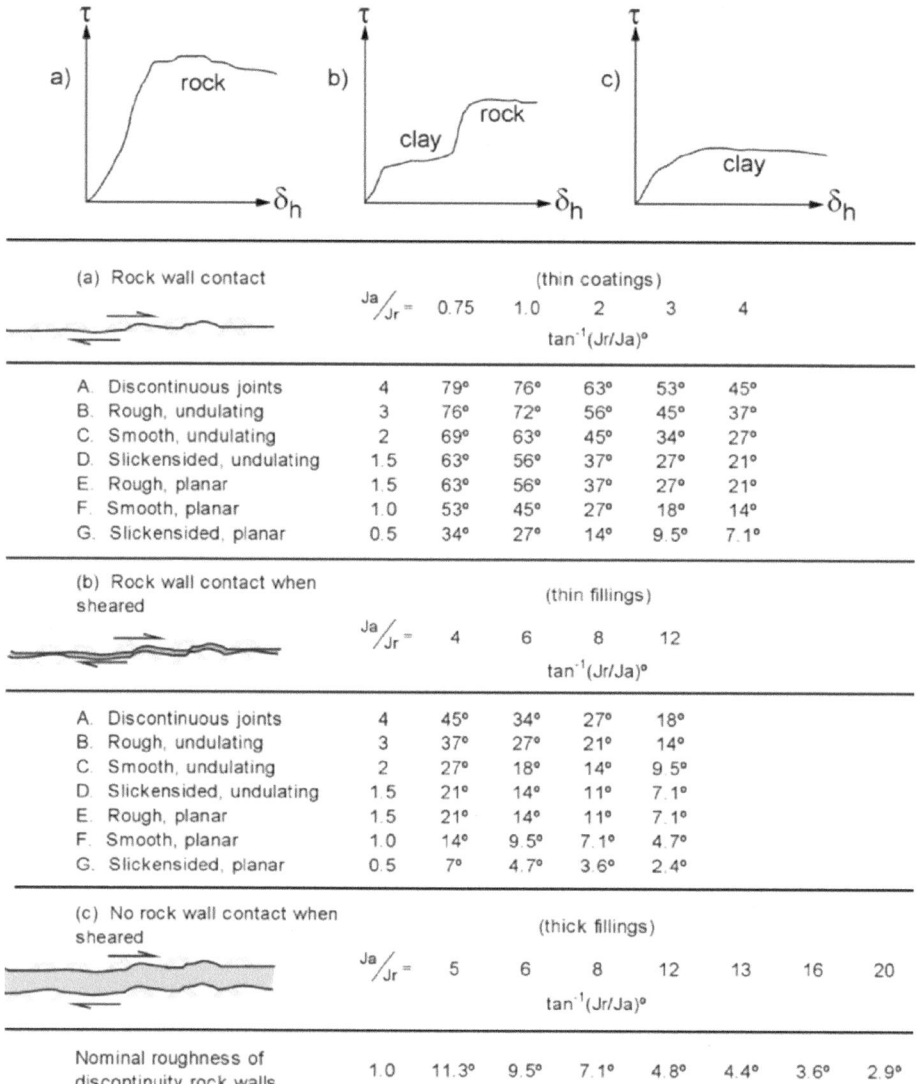

(a) Rock wall contact			(thin coatings)			
$^{Ja}/_{Jr}$ =		0.75	1.0	2	3	4
			$\tan^{-1}(Jr/Ja)°$			
A. Discontinuous joints	4	79°	76°	63°	53°	45°
B. Rough, undulating	3	76°	72°	56°	45°	37°
C. Smooth, undulating	2	69°	63°	45°	34°	27°
D. Slickensided, undulating	1.5	63°	56°	37°	27°	21°
E. Rough, planar	1.5	63°	56°	37°	27°	21°
F. Smooth, planar	1.0	53°	45°	27°	18°	14°
G. Slickensided, planar	0.5	34°	27°	14°	9.5°	7.1°

(b) Rock wall contact when sheared			(thin fillings)		
$^{Ja}/_{Jr}$ =		4	6	8	12
			$\tan^{-1}(Jr/Ja)°$		
A. Discontinuous joints	4	45°	34°	27°	18°
B. Rough, undulating	3	37°	27°	21°	14°
C. Smooth, undulating	2	27°	18°	14°	9.5°
D. Slickensided, undulating	1.5	21°	14°	11°	7.1°
E. Rough, planar	1.5	21°	14°	11°	7.1°
F. Smooth, planar	1.0	14°	9.5°	7.1°	4.7°
G. Slickensided, planar	0.5	7°	4.7°	3.6°	2.4°

(c) No rock wall contact when sheared		(thick fillings)					
$^{Ja}/_{Jr}$ =	5	6	8	12	13	16	20
			$\tan^{-1}(Jr/Ja)°$				

Nominal roughness of discontinuity rock walls	1.0	11.3°	9.5°	7.1°	4.8°	4.4°	3.6°	2.9°

APPENDIX A4 An important feature of the Q-value calculation is the ratio Jr/Ja representing the frictional strength, and closely resembling the inter-block friction coefficient. Therefore tan^{-1} (Jr/Ja) gives a rough indication of the friction angle, with a 'dilational' component seemingly added in the top-left friction angles, and a 'contractile' component subtracted in the bottom-right friction angles. Note the three contact categories, and the symbolic shear strength-displacement diagrams for each case. The above frictional reality was an accidental development of Barton et al. (1974), which was discovered after all Q-parameter ratings were finalized. (Barton, 2012).

REFERENCES

Austrian Society for Geomechanics (2010) NATM: The Austrian Practice of Conventional Tunnelling.

Barton, N., Lien, R. & Lunde, J. (1974) Engineering classification of rock masses for the design of tunnel support. *Rock Mech.* 6(4), 189–236.

Barton, N. (1976) Unsupported underground openings. *Rock Mechanics Discussion Meeting, Befo, Swedish Rock Mechanics Research Foundation*, Stockholm, pp. 61–94.

Barton, N., Grimstad, E., Aas, G., Opsahl, O.A., Bakken, A., Pedersen, L. & Johansen, E.D. (1992) *Norwegian Method of Tunnelling*. WT Focus on Norway, World Tunnelling, June/August 1992.

Barton, N. & Grimstad, E. (1994) The Q-system following twenty years of application in NMT support selection. *43rd Geomechanic Colloquy, Salzburg. Felsbau*, 6/94, 428–436.

Barton, N. (1994) A Q-system case record of cavern design in faulted rock. *5th Int. Rock Mechanics and Rock Engineering Conf., Tunnelling in Difficult conditions*, Torino, Italy, 16.1–16.14.

Barton, N., By, T.L., Chryssanthakis, P., Tunbridge, L., Kristiansen, J., Løset, F., Bhasin, R.K., Westerdahl, H. & Vik, G. (1994) Predicted and measured performance of the 62m span Norwegian Olympic Ice Hockey Cavern at Gjøvik. *Int. J. Rock Mech, Min. Sci. & Geomech. Abstr.* 31(6), 617–641. Pergamon.

Barton, N. (1995) The influence of joint properties in modelling jointed rock masses. Keynote Lecture, *8th ISRM Congress*, Tokyo, Vol. 3, pp. 1023–1032, Balkema, Rotterdam.

Barton, N. (1996) Investigation, design and support of major road tunnels in jointed rock using NMT principles. Keynote Lecture, *IX Australian Tunnelling Conf.* Sydney, pp. 145–159.

Barton, N. (2000) *TBM tunnelling in jointed and faulted rock*. 173p. Balkema, Rotterdam.

Barton, N. (2002) Some new Q-value correlations to assist in site characterization and tunnel design. *Int. J. Rock Mech. & Min. Sci.* 39, 2, 185–216.

Barton, N. (2006) *Rock quality, seismic velocity, attenuation and anisotropy*. Taylor & Francis, UK & Netherlands, 729p.

Barton, N., Buen, B. & Roald, S. (2001) Strengthening the case for grouting. *Tunnels & Tunnelling International*, Dec. 2001: 34–36, and Jan. 2002: 37–39.

Barton, N. (2011) From empiricism, through theory, to problem solving in rock engineering. *ISRM Cong., Beijing*. 6th Müller Lecture. Proceedings, Taylor & Francis, Vol. 1, Qian & Zhou (eds), pp. 3–14.

Barton, N. (2012a) Defining NMT as part of the NATM SCL debate. *TunnelTalk*, Shani Wallace (ed), Sept. 2012, 4p.

Barton, N. (2012b) Assessing pre-injection in tunnelling. *Tunnelling Journal*, Dec. 2011/Jan. 2012, 44–50.

Barton, N. (2013a) Integrated empirical methods for the design of tunnels, shafts and caverns in rock, based on the Q-system. 3rd Int. Symp. on *Tunnels and Shafts in Soil and Rock*, SMIG/Amitos, 17p. Nov. 2013, Mexico City.

Barton, N. (2014) Lessons learned using empirical methods applied in mining. Keynote lecture. *1st. Int. Conf. on Applied Empirical Methods in Mining*. Lima, Peru, 24p.

Bieniawski, Z.T. (1989) *Engineering rock mass classifications: A complete manual for engineers and geologists in mining, civil and petroleum engineering*. 251p. J. Wiley.

Grimstad, E. & Barton, N. (1993) Updating of the Q-System for NMT. *Proc. of Int. Symp. on Sprayed Concrete – Modern Use of Wet Mix Sprayed Concrete for Underground Support*, Fagernes, Norwegian Concrete Association, Oslo, Kompen, Opsahl & Berg (eds), pp. 46–66.

Grimstad, E. (1996) Stability in hard rock affected by high stress and supported by sprayed concrete and rock bolts. *Proc., 2nd Int. Symp. On Sprayed Concrete, Modern use of wet mix sprayed concrete for underground support*. Gol, Norway.

Grimstad, E., Kankes, K., Bhasin, R., Magnussen. A.W. & Kaynia, A. (2002) Rock Mass Q used in designing Reinforced Ribs of Sprayed Concrete and Energy Absorption. *4th Int. Symp. on Sprayed Concrete*, Davos, Switzerland.

Grimstad, E., Kankes, K., Bhasin, R., Magnussen, A.W. & Kaynia, A. (2003) Updating the Q-system for Designing Reinforced Ribs of Sprayed Concrete and General Support. Proceedings, *Underground Construction*, London.

Grimstad, E. (2007) The Norwegian method of tunnelling – A challenge for support design. *XIV European Conference on Soil Machanics and Geotechnical Engineering*. Madrid.

Grimstad, E., Tunbridge, L., Bhasin, R., & Aarset, A. (2008) Measurements of Forces in Reinforced Ribs of Sprayed Concrete. *5th. Int. Conf. on Sprayed Concrete*. Lillehammer, Norway.

Kristiansen, J. & Hansen, S.E. (1993) Cavern Stadium – displacement measurements. *Final Report on the Research Project for the Rock Cavern Stadium*, SINTEF-NGI-Østlandsforskning, Norway.

Løset, F. & Bhasin, R. (1992) Engineering geology – Gjøvik Ice Hockey Cavern. *Research Project: 'Publikumshall I Berg'*, SINTEF-NGI-Østlandsforskning, Norway.

Quadros, E. F. & Correa Filho, D. (1995) Grouting efficiency using directional (3-D) hydraulic tests in Pirapora Dam, Brazil. *Proc. 8th ISRM congress*, Tokyo. Fujii (ed), pp. 823–826.

Shen, B. & Barton, N. (2017) Unstable fracture initiation and propagation in rockbursts and coalbursts. *Int. J. Rock Mech. & Min. Sci.* (in press). Pergamon.

Sjøgren, B., Øfsthus, A. & Sandberg, J. (1979) Seismic classification of rock mass qualities. *Geophys. Prospect.*, 27, 409–442.

Ward, W.H., Todd, P. and Berry, N.S.M. (1983) The Kielder Experimental Tunnel: Final Results. *Geotechnique* 33(3), 275–291.

Vandevall, M. (1990) *Dramix - Tunnelling the World*. NV Bækert S.A, 1991 edition.

Chapter 27

TBM performance, prognosis and risk reduction

Dr. Nick Barton
Nick Barton & Associates, Oslo, Norway

Abstract: World records for drill-and-blast tunneling from Norwegian contractors, bear witness to numerous weeks of more than 100m, and an exceptional 5.8 km in 54 weeks, also from one face. Earlier hard-rock world records using high-powered TBM in Norway, but most frequently and more recently, the records with Robbins TBM through non-abrasive limestones in the USA, provide numbers in meters per day, per week, and per month, which are of course, even more remarkable. Unfortunately there are contrary and undesirable TBM records, which are occasionally recurring events so not records, which see TBM stopped for months or even years in fault zones, or permanently buried in mountains. The many orders of magnitude range of performance suggest the need for better investigations, better choice of TBM, and better facilities for improving the ground ahead of TBM, when probe-drilling indicates that this is essential. Control of water, and improved stand-up behavior in significant weakness zones and faults may demand drainage, which can be unending, and pre-injection. Fortunately there are increasing signs that this is recognized by TBM manufacturers: more guide-holes for drilling pre-injection umbrellas are seen through front-shields nowadays. A little acknowledged fact is that when all hours are included, TBM will generally decelerate as tunnel length and time increases. This is usually seen after improved performance during the learning curve. Deceleration is also a general trend during world-record setting performances. This means that utilization U, equal to the ratio of actual advance rate and penetration rate, AR/PR, is only for specified time intervals, because U is time-dependent. This is rarely quantified by designers, and is therefore a source of risk, by default. Another important item for correct prognosis is the recognition that *reduced* penetration rate PR can sometimes occur when thrust is *increased* by the TBM operator, due to exceptionally resistant rock mass formations. Each of the above, and PR sensitivity to a wide range of cutter forces, UCS and abrasiveness, are provided in the empirical Q_{TBM} method. This method explains variable progress in jointed rock, which is sometimes fast, and also quantifies the likely delays in untreated, or pre-injected, fault zones.

1 INTRODUCTION

During the last 10-15 years, Norwegian contractors have led the world in the fastest drill-and-blast tunneling rates, with 165m and even 176m in single 7 × 24 hour weeks. LNS and Veidekke have had consistent rates of more than 100m/week for several months in specific projects. At the Svea coalmine (one-face) access tunnel, in coal-measure rocks obviously requiring some bolting and shotcreting, LNS achieved 100m per week or more for 32 weeks, and used just 54 weeks to drill-and-blast 5.8 km. The tunnel had a 36 m^2 cross-section. This performance is actually better than many TBM project performances if one considers the whole year of tunneling, but does not appear so impressive in relation to TBM, if shorter time intervals are compared, as typically done with TBM.

TMB have incredible current world records of 172m in 24 hours, 703m in one week, and 2163m in one month. Nevertheless, in the world records for the 3 to 4 m diameter class, the best monthly average is 'only' 1189m, and the overall world record monthly average is 'only' 1352m, found in the 4 to 5m diameter class. The word 'only' is used merely to contrast with the remarkable record best month of 2163m.

Thanks to some detailed TBM world record advance rate statistics provided by Robbins on the internet, it was possible to derive the present (2015) record data shown in Figure 1. The 3 to 6m diameter class shown with the smallest 'cubes' is *the mean of three sets* of data given for 3–4m, 4–5m and 5–6m TBM, based on assumed 24 hours, 168 hours and 720 hours. The 6 to 10m diameter class shown with the larger 'cubes' is *the mean of four sets* of data for 6–7m, 7–8m, 8–9m and 9–10m TBM. This collective averaging helps to see trends more clearly.

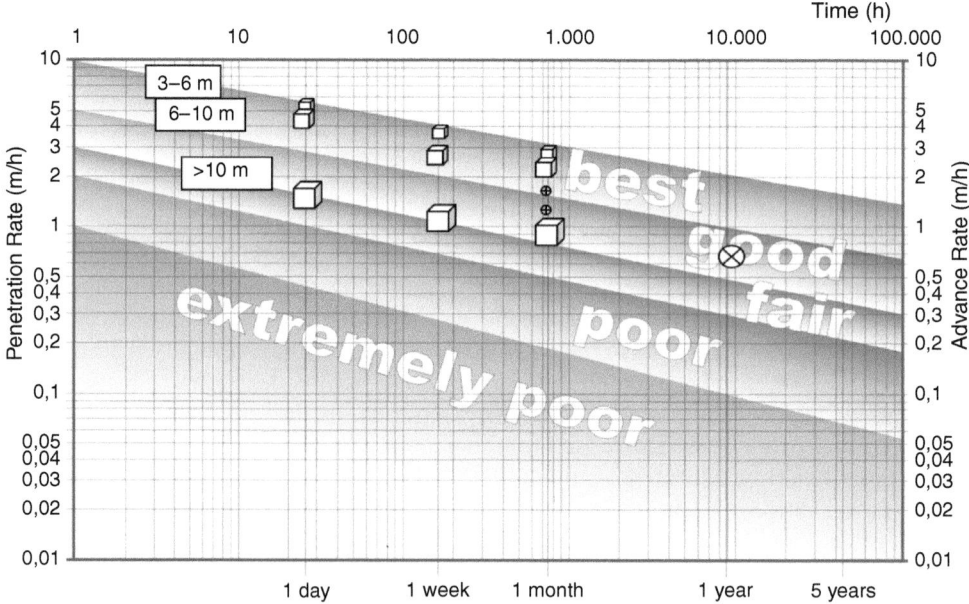

Figure 1 Using a log-log-log plot of PR (penetration rate, left axis only) and AR (advance rate in remainder of plotted area) and time T (total hours), the synthesized **present world-record data** for different sizes of TBM is shown, based on data provided by Robbins. The writer has converted day, week and month records (given in meters) to the form AR (m/hr) by dividing by 24, 168 and 720 hours. Data from 8 countries are represented, chiefly USA and China. The record **mean monthly** data plots at AR = 1.7 m/hr for the 3 to 6m class, and at AR = 1.1 m/hr for the 6 to 10m class, and these results are shown with the two small circles. The larger crossed-circle to the right represents 54 weeks for 5.8 km at the Svea Mine Access Tunnel, achieved during the LNS drill-and-blast record. This was driven in coal-measure rocks and obviously required some shotcreting and rock bolting, due to varied Q-values. Slowest progress was made through a near-surface zone of permafrost.

2 CASE RECORD EVIDENCE OF DECELERATION

There is an all too common habit of reporting utilization (U) of TBM without specifying the time period involved. An estimated average daily utilization is especially an insufficient form of prognosis. Since stand-stills are naturally excluded, the client may get an optimistic view of likely performance. Utilization is estimated from the classic and most used TBM equation:

$$AR = PR \times U \tag{1}$$

where AR = (actual) advance rate in m/hr, and PR = penetration rate (for uninterrupted boring) in m/hr. U is the fraction of time when boring has (or is expected) to actually occur, as seen on the traditional 'pie- or pizza-diagram'. For convenience U is usually expressed (in speech) as a percentage. Note that in Figure 2, U has been expressed as T^m. This is explained in Table 1 and is also shown in Figure 2 (top-right corner).

As illustrated by the world records of Figure 1, and as illustrated by 1000 km of mostly open-gripper case records, summarized in Figure 2 from Barton (2000), there

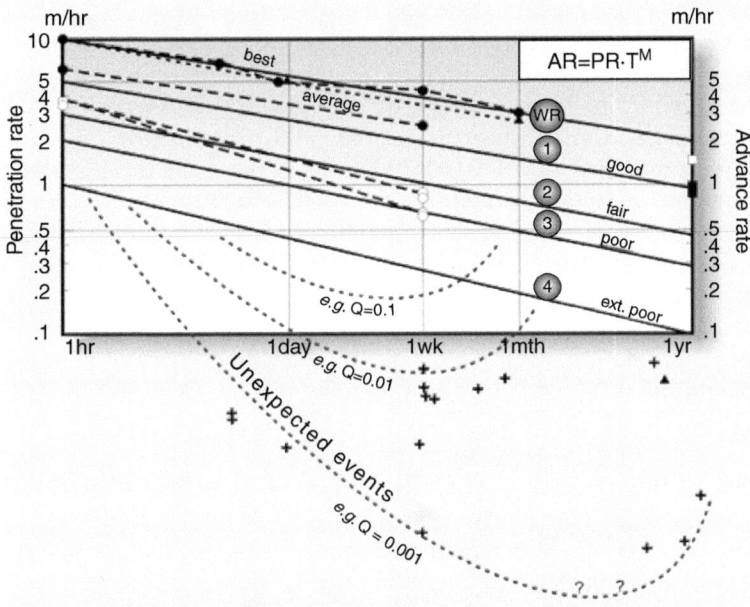

Figure 2 Trends from open-gripper case records representing 145 lengths of well-characterized TBM tunnels, totaling approximately 1000km of tunneling. The performance is represented by log PR – log T – log AR. The PR value applies only to the left axis. T and AR occupy the remainder of the diagram. The five typical 'lines' of performance are the same as shown in Figure 1. The hand-drawn source of this smoothed data was originally given in Barton (2000). In that reference, and in Barton (2013) will be seen numerous sloping red lines (for best performance), and below this numerous and more steeply inclined green lines (for average performance), and below this and even more steeply (adversely) inclined, blue lines (for 'bad-ground' performance). The 'unexpected event' curves (and crosses) in Figure 2, are the low-Q-value-linked worst cases, with (by 2000) three permanent TBM burials in the case of older, poorly equipped TBM.

Table I Deceleration gradients (-m) for the five trends-of-performance lines in Figure 2. A specific 56 km of double-shield performance (two Wirth TBM, two Herrenknecht TBM) is also indicated, but as shown in a later case record, this (optimistic) and at best halving of gradient (-m) may not apply in tough cases, and is hardly evident in the record mean-monthly performances (small circles shown in Figure 1). An EPB machine may apparently almost double these double-shield (harder rock) gradients for obvious reasons related with the greater challenge of semi-continuous face support.

Performance Line	Deceleration Gradient (-) m
# (refer to Figure 2)	(units of LT^{-2})
WR (world records)	−0.13 to −0.17
1, 2, (good, fair)	−0.17, −0.19
3, 4 (poor, extremely poor)	−0.21, −0.25
(trends from 145 cases)	($\Sigma \approx$ 1000 km, mostly open-gripper)
DOUBLE-SHIELD (at Guadarrama)	−0.08 to −0.12 (4 × 14 km)

is actually a time-dependent element in U which is conveniently ignored in a remarkable number of tunnel magazine articles and also in commercial TBM prognoses. Since a client pays for a completed tunnel, a false impression of actual hours (T) is obtained if inevitable standstills, such as in untreated fault zones, are *excluded*. 'Waiting for the train' or broken conveyor belts due to blocky rock may be part of the recorded experience, and cannot be ignored in the prognosis of long TBM tunnels. There are approximately $24 \times 7 \times 50 \approx 8700$ hours of potential three-shifts of work in one year, and during TBM standstills the clock is still running, with the tunnel completion date likely delayed. When U is replaced by T^m, more realistic prognoses are possible. Many TBM projects come in 'late' due to ignorance of this element of time / length. So risk (of cost and time over-run) can be reduced by using T^m in place of a potentially misleading U. A 'monthly utilization of 30%' does not give the correct time for tunnel completion, even if a mean PR of say 3m/hr was quite representative for the whole tunnel.

3 EVIDENCE LINKING Q-VALUES WITH TBM PERFORMANCE

When a TBM tunnel is driven in one predominant rock type, such as the case of 5 km through granites in Malaysia, described by Sundaram & Rafek (1998), there is a surprisingly good correlation of penetration rates (PR) with the Q-value, and with even simpler measures of jointing, such as the volumetric joint count, and even with mean joint spacing. (Other 'necessary' parameters are 'constant' in the continuous granite). The Q-data PR-correlation shown in Figure 4 is based on 2,825m of data analyzed by the above authors, for medium to coarse grained granites with UCS in the range 130 to 246 MPa (mean 182 MPa) – similar to that expected in the four-TBM, total 19km Follobanen Oslo-Ski project, also mostly driven through granites, and granitic gneiss, which will be mentioned later. The authors also found that the *average* Jr/Ja ratio (joint roughness/joint alteration-filling) gave a better correlation of PR to Q than the 'most adverse' Jr/Ja ratio, as traditionally used when selecting suggested tunnel

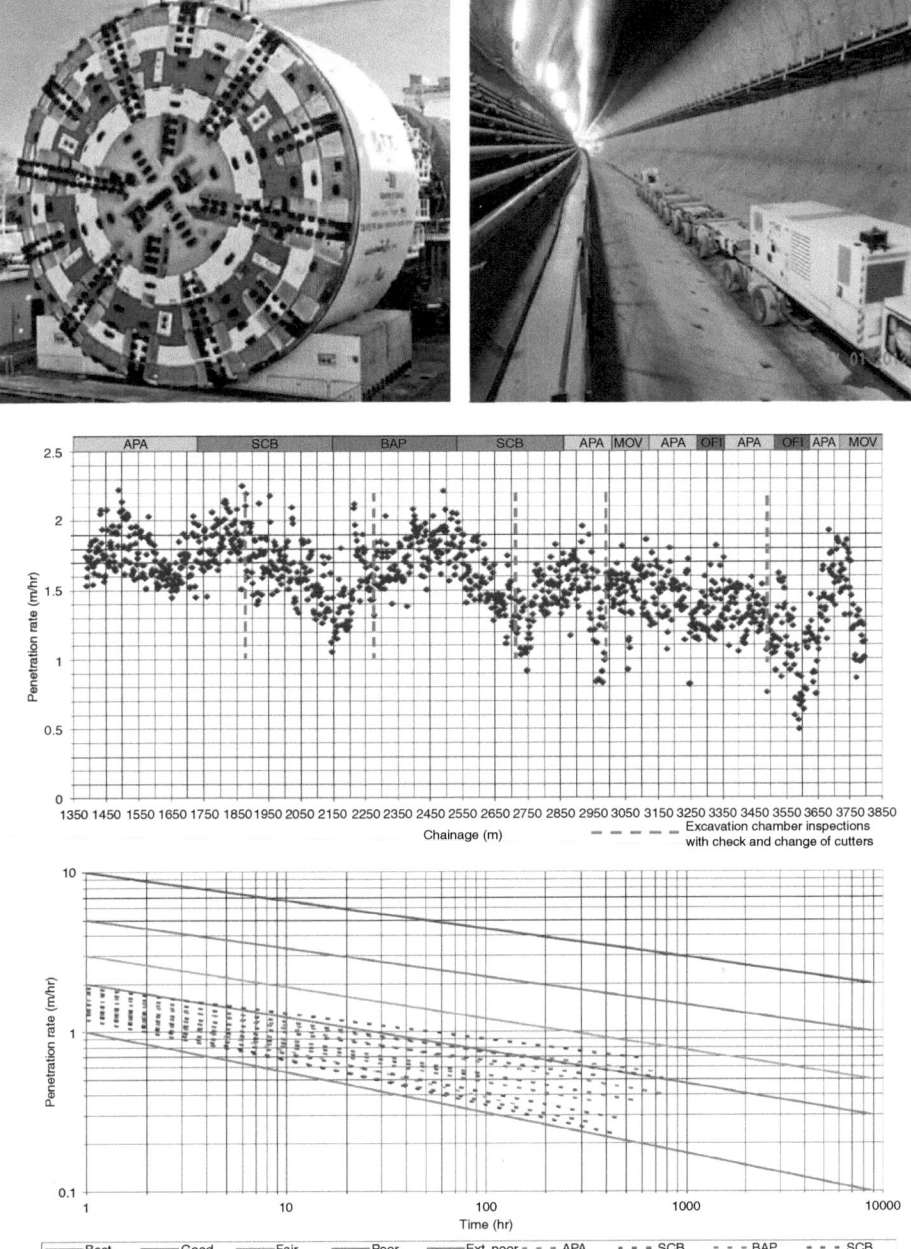

Figure 3 Sparvo Tunnel, driven by the world's (now only) second largest EPB (earth pressure balance) TBM of 15.6 m diameter has twin tunnels of 2.6 km length (Barbieri *et al.* 2013). There were 78 disc cutters due to significant sandstone and conglomerate sections of the tunnels, in addition to the numerous soft ground picks. Note that the range of PR was mostly 1 to 2 m/hr, and due to difficult conditions and use of moderate thrust, the deceleration gradient (-) m varied from (-) 0.16 to (-) 0.31 for both tunnels. However, (-) m was (-) 0.38 during the learning curve, and (-) 0.33 when exiting through bad ground. Due to risk of methane gas, operation was always in closed mode, which of course increases delay and makes (-) m more steeply negative. The mean cutter forces used in the weak sandstone and conglomerate/clay were 16.9 and 10.3 tons.(Pers. comm. M. Tanzini, 2013).

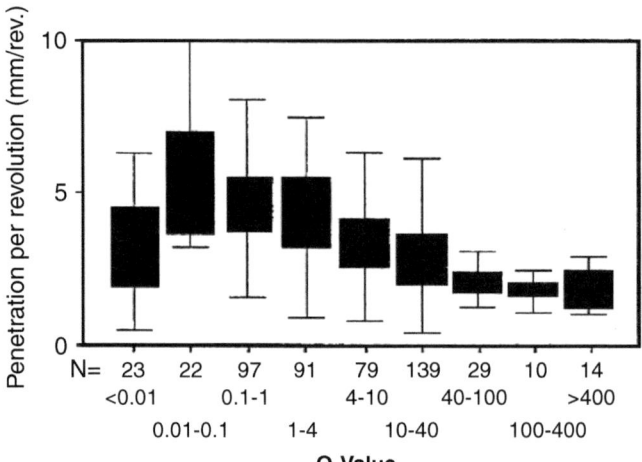

Figure 4 In a project involving only granite, as illustrated here, consistent correlation of penetration rate with Q-values (using mean Jr/Ja) is seen. Sundaram & Rafek (1998). Note the matching PR trend in Figure 5.

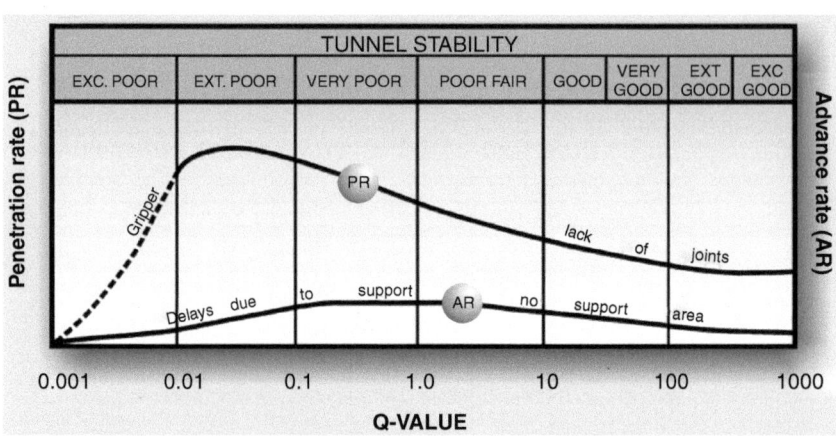

Figure 5 The traditional Q-system adjectives seen at the top of this figure, are clearly not correct for describing TBM performance, as Q-values significantly more than about 30 are adverse for PR, due to lack of joints. Note that the real advance rate AR is likely to be very low at both ends of the Q-scale (Barton, 2000). In a subsequent figure a more representative set of adjectives will be seen, using a new multi-component machine-rock parameter Q_{TBM}, which uses the six Q-parameters for rock mass description, and has extra parameters such as cutter force and rock mass strength.

support and reinforcement for single-shell NMT (Norwegian Method of Tunneling) (Barton *et al.*,1992; Barton & Grimstad, 2014).

When logging more than 300 rock exposures and seven cores drilled through weakness zones as input to Follobanen Oslo-Ski prognoses, the writer also logged all the principle Jr/Ja ratios in the form of Q-histograms. This Q_{TBM} study was described in Barton & Gammelsæter (2010).

4 CUTTER LIFE AND THE EFFECT OF HIGH Q-VALUES AND HIGH STRESS

Although manufacturers of new TBM like to claim that their cross-over machines can tackle all conditions and should be selected instead of drill-and-blast, 'because the tunnel is so long', and nowadays even with the argument 'because the conditions are so bad', this double optimism may not be justified by actual experiences, nor by the numerous older TBM. As we will see later, and as hinted at by the natural *decelerations* seen in Figures 1 and 2, the long tunnel argument is inviting risk, although the ventilation aspect for the long tunnel may weigh heavily in favor of TBM. However, in terms of geotechnical risks, great care is needed before selecting TBM for the long, and probably poorly explored deep tunnel. The prospect of good quality rock may not favor TBM if there is too little jointing and also high cover, as illustrated in Figure 6.

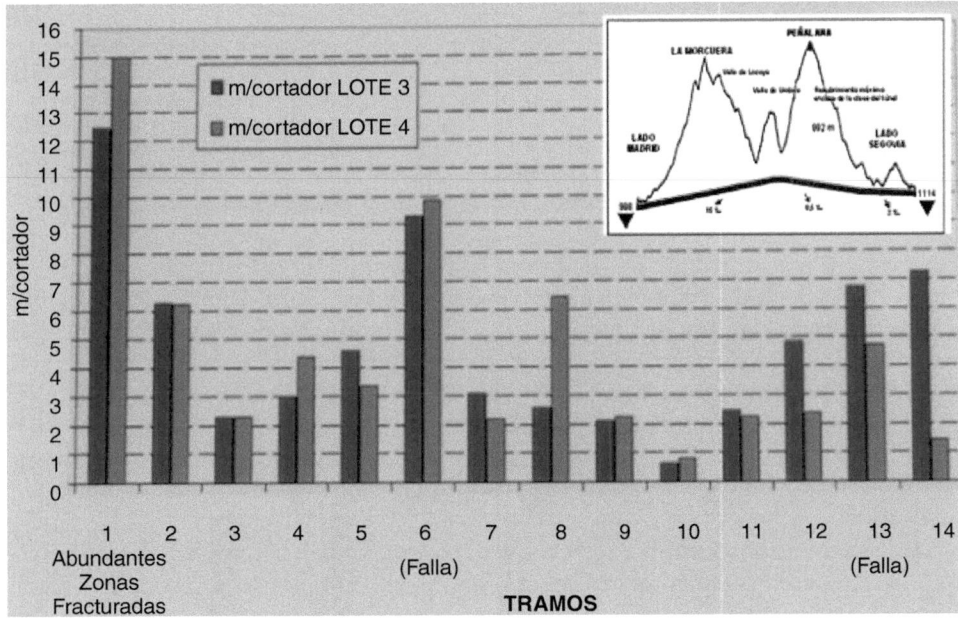

Figure 6 The longitudinal profile for the (4 × 14 km) Guadarrama Tunnels (ADIF, 2005). These were driven in 28 to 33 months by four 'competing' double-shield TBM. The statistics show mean cutter-change frequency (m/cutter) for two of the 14 km lots, with strong correlation to tunnel depth (minimum m/cutter under two mountain ranges) and therefore implied correlation to the level of confining-stress in the predominantly hard and abrasive granites and gneisses. Abundant fracture zones ('zonas fracturadas') and faults ('fallas') give a positive contribution to reduced cutter wear in several locations.

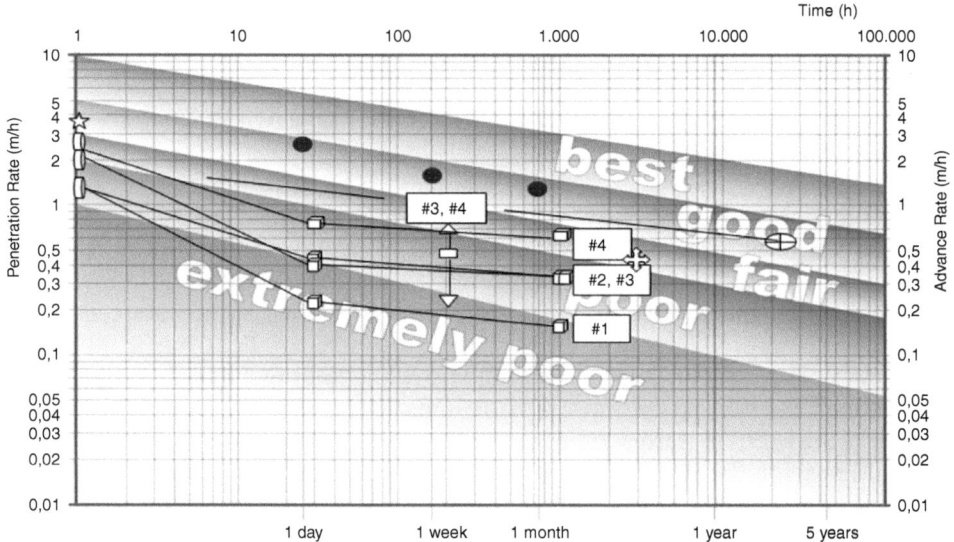

Figure 7 The 'learning curve' performance in the first four months (see the improvement from #1 through #4) of a 5 m diameter and 5 km long double-shield TBM being driven in massive granites with very high RMR (and Q) values. A common feature of 'learning curves' is the initially lower PR and lower AR due to initially poorer utilization: *i.e.* a steeper deceleration gradient (-) m. The cutter change frequency in this 5 km project was typically 2 to 3m/cutter. Rock cover was 200-500 m, half that of the mountainous Guadarrama tunnels shown in Figure 6. The 56 km experience from the four competing TBM at Guadarrama showed a similar mean PR = 2.0 m/hr to this 5km case, yet the general efficiencies of the double-shield method eventually allowed overall performance to reach 'good'(see ellipse with cross beyond the 20,000 hours, 32 months location, over to the right side of this figure).

The adverse nature of massive rock with insufficient jointing, especially when this is combined with high UCS and high abrasiveness is typified by the need to change cutters on average every 1 to 2m, as seen in places in Figure 6, and as occurred in the case illustrated in Figure 7. In practice this means many hours boring with an increasing number of ineffective cutters, because the 5 to 10 (or more) cutters may not be changed until the once-per-24-hours maintenance shift. By then there might be some cutters with 'flats' which consequently have ceased to rotate. So each 24hr period may experience reducing advance rate (AR) if cutter-change needs are significant, a contractor's nightmare.

The very best results for one day, week and month at Guadarrama are shown by the solid circular symbols, high up in Figure 7. They are 62m in 24 hrs, 250m in 1 week, 970m in 1 month. These are well below world records (Figure 1) but nevertheless very good in the circumstances. The cutter life statistics of the two projects described in Figures 6 and 7, actually emphasize the importance of the NTH/NTNU cutter life index CLI parameter, shown in Figure 8, which has been an important part of the writer's prognosis model Q_{TBM} from the start. On a number of occasions, the results of NTNU rock testing and especially CLI results have been requested, where Q_{TBM} is being used at a foreign project. However, there are many other parameters which are also important for the Q_{TBM} calculation.

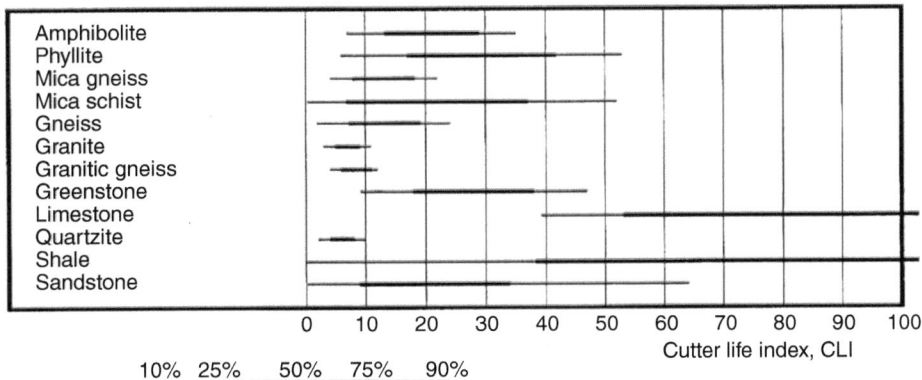

Figure 8 The cutter life index CLI, developed at NTH/NTNU in the 1990s is an important performance indicator, especially when combined with realistic measures of the effect of different degrees of jointing. Naturally for the application of the Q_{TBM} method, Q is needed in preference to other joint description methods. This is especially so when describing faulted rock, which is incorporated in the prognosis rather than excluded as 'special cases', as so often seen in other methods of prognosis.

A combination of four factors: low CLI (as for granite, granitic gneiss, quartzite), high quartz content, high UCS (obviously linked with these rock types) and massive sparsely jointed rock, with for instance Q-values > 100, and RMR > 80 is an inevitable 'recipe' for frequent cutter change statistics. When the above factors are combined with significant depth of cover, the additional confining-pressure acting across the face of the tunnel, and directly adding to the difficulty of chip formation, may cause cutter life to dip below 2m/cutter, and on occasion even below 1m/cutter. Clearly this will be a significant task for the daily/nightly maintenance shift, and besides the time for replacement of say 10 cutters, there will be the added effect that for some of the 10 to15 hours of boring, a number of cutters will have become sub-standard due to excessive wear.

4.1 Valid and invalid critique

While addressing the subject of maintenance shifts, it is unfortunately a fact of life that in the case of double-shield TBM which are convenient for allowing simultaneous PC-element ring assembly and push-off grippers' advance, there will only be the possibility of observing and (very) approximately logging the rock conditions, when the machine has stopped for cutter change. The 'inner climate' with hot cutters and sauna-like conditions at first, are not conductive to easy Q or RMR or NTH/NTNU joint class mapping. The writer has occasionally tried to perform such logging, and has also been a consultant at some smaller-diameter TBM sites where only the smallest engineering geologists get to log the limited data, as seen in a very confined space, and must share their few observations with colleagues (and with the consultants).

It is therefore remarkable that certain authors who will not be named, both in Norway, Italy and the USA, were happy to present other's 'data' showing apparently poor correlation of PR statistics and Q_{TBM}-values, when in reality the only rock mass quality logging was at 15, 20 or 25m intervals (once in each 24 hours) when the relevant TBM were stopped for maintenance, because the rock could not be observed while boring. Worse still, Q was mostly obtained by subsequent estimation from RMR logging, since original (Italian) authors were not at first aware of the new Q_{TBM} method, so they 'retro-actively' estimated Q_{TBM}. Is this a valid basis for over-confident critique (and second-hand use by others in Norway and the USA, on similar dismissive missions?). Most of the more reliable case record data represented by the decelerating lines in Figure 2 were obtained from open-gripper TBM projects, where rock mass conditions were well described on a continuous basis, and not only by the most agile (and smallest) engineering geologists, at much too well-spaced daily-advance intervals, somewhere around midnight. If similar PR scatter had been experienced by the developer, there would obviously have never been Q_{TBM} development or publication for others to criticize. So those with dismissive missions need to evaluate if they are performing valid comparisons, or using other's 'data' in a strictly honest manner.

5 CUTTER THRUST COMPARED TO ROCK MASS STRENGTH

Recent trials with instrumented cutter bearings in Austria, described by Entacher *et al.* (2013) have demonstrated the actual complexity, though logical nature of cutter force distributions. Of necessity one divides net thrust by the number of cutters, to estimate mean thrust per cutter, and then can compare this with a measure of rock strength. The reality, as shown in Figure 9, is that cutter thrust oscillates strongly about the assumed mean, and in addition varies across the face of the tunnel if the resistance to chip and block formation also varies. This of course will be linked to the relative dominance of massive or jointed/foliated rock, and changes of rock strength.

These cutter force oscillations will often be present when the rock mass is frequently varying across the face, which means that comparison of assumed mean cutter force, such as 20, 25 or even 30 tons per cutter, with the assumed chip- or block-formation resistance of the rock mass, is going to be an approximate exercise. Nevertheless it is an obvious advantage if the estimate of resistance of the rock mass is as realistic as possible. This means that the *rock mass* and not just the *rock material* should be used. To base penetration rate prognoses only on rock UCS values is to invite inaccuracies. And to not compare assumed cutter thrust with any measure of rock strength is an invitation to greater lack of reality. The reasons for insisting on this 'thrust/strength' comparison are illustrated in four examples in Figures 10 through 13. Only if the TBM has sufficient thrust in relation to rock mass strength will one obtain the expected result of increased penetration rate with increased thrust, as shown in Figure 10.

A more or less quadratic relation between penetration rate and cutter force is suggested in Figure 10, and this was also seen in high-powered TBM trials in Norway in the 1990s, with thrusts in excess of 30 tnf/cutter. There is often a rapid change of gradient beyond about 20 tnf/cutter, if the effective rock mass strength is finally more easily exceeded.

It seems that the necessity to compare cutter thrust with some measure of rock mass strength more relevant than UCS alone, is not recognized by some who have developed prognosis methods and who also assist TBM manufacturers. Those working for a

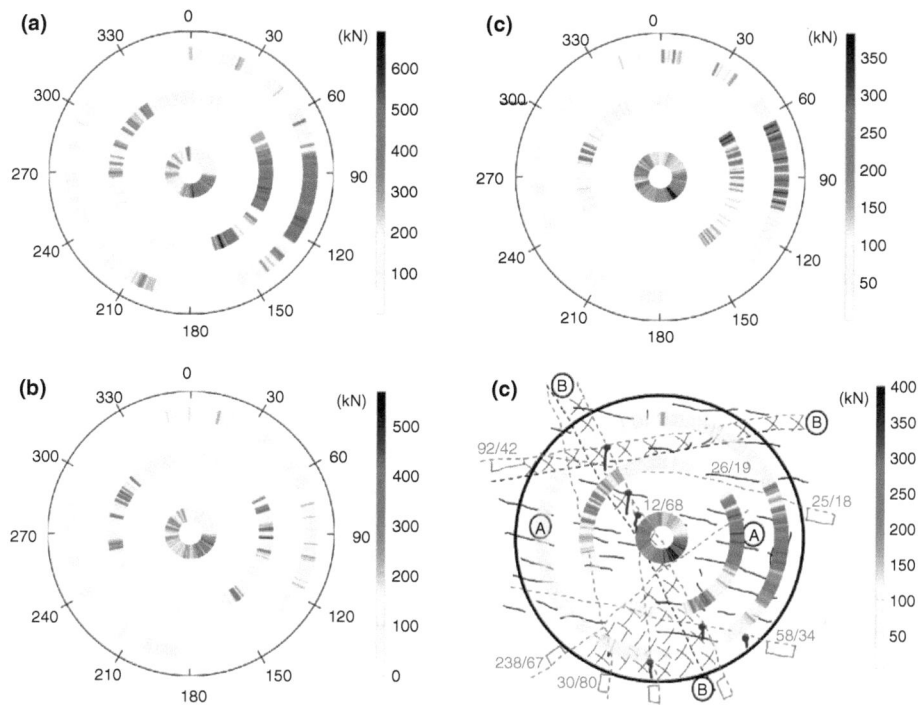

Figure 9 Normal forces monitored during three consecutive cutterhead revolutions (a–c) and the averaged forces of these figures compared with the corresponding geologicalmapping (d). These interesting measurements were made recently at the Koralm tunnel in Austria, and were reported by Entacher *et al.* (2013).

major TBM manufacturer were unaware that cutter thrust increases could be associated with reduced penetration rate, if the rock mass was very strong and sufficiently massive. A further example, with a significant body of data, is that shown in Figure 13. The (NTH) method of prognosis missed the actual TBM behavior, because the metasandstones, perhaps with limited jointing, were stronger than available thrust per cutter.

6 THE DEVELOPMENT OF A TBM PROGNOSIS MODEL CALLED Q_{TBM}

There are two further empirical results that need to be shown, before presenting the Q_{TBM} prognosis model. These concern the essential separation of PR and AR, because it is clearly insufficient to have a prognosis model which only addresses the penetration rate. Figure 14 (also Figure 5) demonstrates that sometimes there may be limited dependence of AR (the actual advance rate) on the penetration rate PR, despite significant variation in the uniaxial compressive strength, from 50 to 150 to 250 MPa for the three rock types. It is clearly the increased need of tunnel support in the weaker rock which

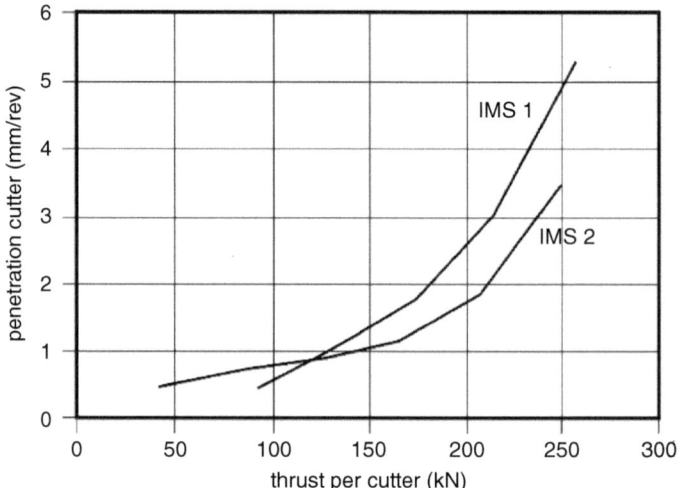

Figure 10 When rock masses are sufficiently jointed and the rock material is not too strong, then increasing the mean thrust on the cutters will often result in the expected increases in penetration rate, as suggested in this data from Hong Kong, where IMS is a local consultant's measure of rock mass class.(Unpredictably, his name is Ian McFeat-Smith). Grandori *et al.* (1995a) quoted this measure of rock mass class in their publication concerning TBM progress through Hong Kong granites.

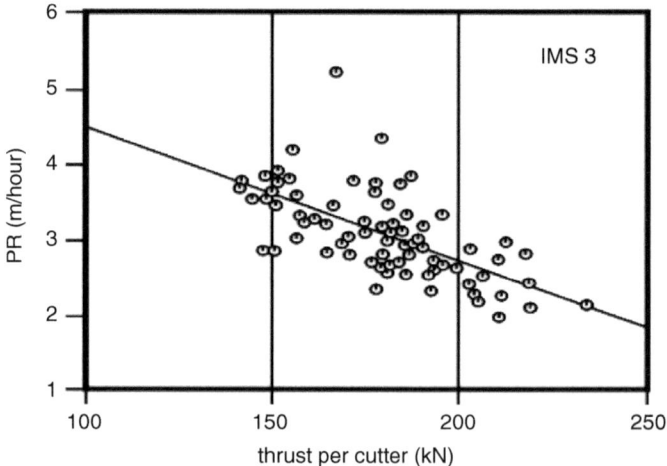

Figure 11 The expected logic of increased PR with increased thrust may break down, as shown here, if the thrust available is insufficient in relation to the effective strength of the rock mass. Too high rock mass strength, including a relative absence of jointing, or too high UCS, would hinder block formation and chip formation. From Grandori *et al.* (1995a).

nearly eliminates the 'initial' advantage of a lower UCS, which gives the best PR result. In faulted rock the problem is often more extreme, with 'too fast' PR occurring using even low thrust, immediately followed by increased support needs, and maybe also delays.

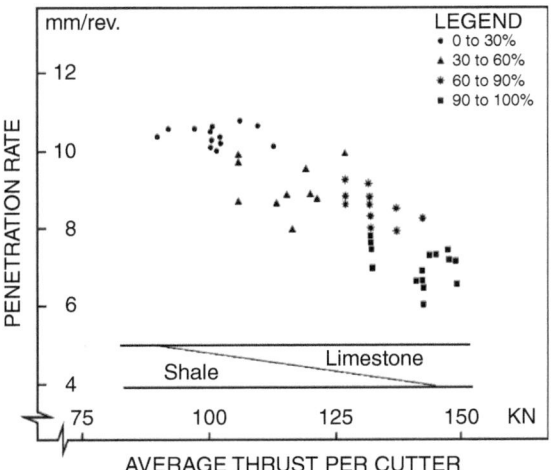

Figure 12 Another example of penetration rate reducing despite increased thrust, from the days when TBM power was more limited. Penetration rate versus thrust per cutter is shown for specified percentages of limestone (σ_c= 130 MPa) and shale (σ_c = 68 MPa), from Nelson *et al.* (1983). A similar situation could arise today with more powerful TBM, if the strongest mixed-face component was even stronger and more massive than this limestone.

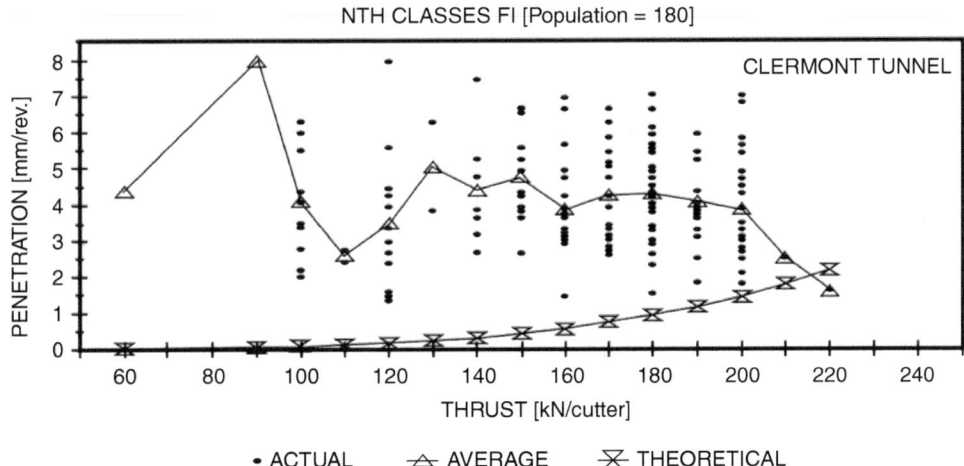

Figure 13 The 'logical' expectation of increased penetration rate with increased thrust may not be experienced if a TBM is underpowered in relation to very hard massive rock. From the Clermont Tunnel in South Africa, back-analyzed by McKelvey *et al.* (1996). The 'theoretical' (and perhaps not sufficiently empirical) NTH prognosis model clearly misrepresented the situation, and gave especially unrealistic predictions of penetration rate when thrust was low, suggesting that the judgment of joint class (F1), or the function of joint classes, was in error.

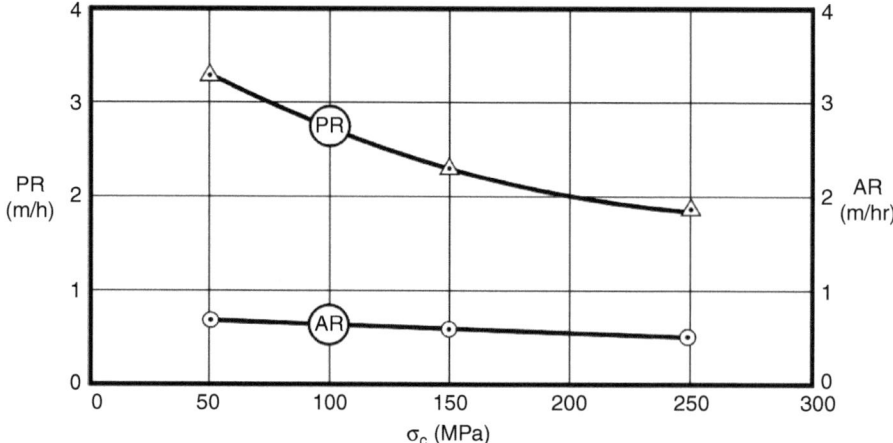

Figure 14 PR and AR data for TBM progress through shale, tillite and sandstone. These curves were derived (in Barton, 2000) from weekly average data reported by Fawcett (1993).

Figure 15 emphasizes the role of jointing, as described by variable Q-values, and the additional influence of UCS reported by Innaurato *et al.* (1991). The recorded PR values for a 3.5m diameter tunnel are given in m/hr at the lower end of each sloping line.

The case-record basis for the development of a TBM prognosis model, detailed stage-by-stage in Barton (2000), later resulted in a user friendly computer program created by Ricardo Abrahão, which Barton *and* Abrahão (2003) termed Q_{TBM}. This indeed employs the Q-system, but modified to an oriented Q_o format. RQD needs to be interpreted with respect to tunnel orientation, and is therefore written as RQD_o. A 'conventional' vertical core can give a false high value of RQD (in relation to the low value in the tunneling direction), if there is a strongly oriented steeply dipping structure such as bedding or foliation. For estimating Q_o, all joint sets are sampled regarding Jr/Ja, unless a particular set is assisting or hindering penetration. It is then allowed to influence the oriented Q_o – value more strongly. Of course a convenient way to gather data is to log rock exposures like recent road cuttings (if available, and not heavily weathered), logging along imaginary horizontal scan-lines. Histogram-based recording of data allows thousands of recordings to be made rapidly. Examples for the Follobanen Oslo-Ski project were given in Barton & Gammelsæter (2010). Some will be shown later.

One of the most important normalized parameters in Q_{TBM} is mean cutter thrust (F, tons) which is normalized by 20 tons. Greater or lesser applied thrust is then compared with SIGMA (rock mass strength estimate $= 5\gamma Q_c^{1/3}$) where $Q_c = Q_0 \times UCS/100$ and γ = density in gm/cm^3. For most conceivable rock masses, SIGMA ranges from 1 to 100 MPa, but in saprolite SIGMA < 1 MPa. The resulting formula for PR (see top-right inset in Figure 16) which is strongly dependent on cutter thrust compared to rock mass strength, has been tested on numerous occasions, both for high-powered TBM with F > 30tnf, and for blind-hole shaft drilling with F as low as 7 to 8 tnf. Realistic values of PR are obtained when the method is correctly used.

In essence we allow the Q-value to assist in determining delays due to support requirements (it therefore effects – where appropriate - the *deceleration gradient -m)*

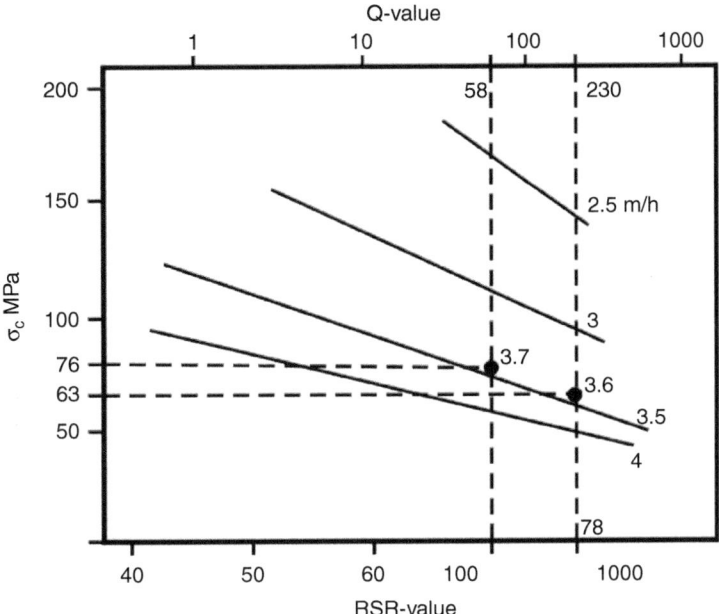

Figure 15 There is a logical correlation between penetration rate (given here in m/hr) and the Q-value. Massive high-Q rock slows progress, and if UCS (σ_c) is also high, an even slower penetration rate is inevitable. Low UCS and low Q-value (but not too low) are positive, as in the left-hand bottom corner. Boring in hard quite high-quality rock masses, will frequently move PR toward the top right-hand corner: *i.e.* combining high Q and high UCS gives lower PR, and possibly frequent cutter change, if the cutter life index CLI is adversely low. From Innaurato *et al.* (1991). (Note that the infrequently used rock mass classification method called RSR, developed just before RMR and Q is not discussed here).

and therefore overall AR. Furthermore we allow the Q-value, and critical *rock-cutter*, and *rock mass-machine* parameters to *also determine* the speed of cutting (therefore effecting slower or faster PR). These dual roles of Q have been criticized, but empirically speaking the method works well. Figure 17 shows an example of a Q_{TBM} data-input screen (explained in detail in Barton & Abrahão, 2003) which shows the parameters used in the simple calculations, most of which appear in the x-axis equation of Figure 16.

7 FAULT AND WEAKNESS ZONES AND THEIR REPRESENTATION IN Q_{TBM}

The fundamental difficulties of tunneling through fault zones, and prognosis-modeling this successfully, will be summarized later, by combining three extremely simple equations. They provide, when combined and presented in terms of time T, a convincing explanation of why so much time can be lost in an unexpected, and therefore usually untreated fault zone. The key to this understanding is that the universal but variable deceleration gradient (-*m*) is strongly linked to low Q-values. Low Q-values and high negative deceleration gradients (meaning low utilization) go hand-in-hand.

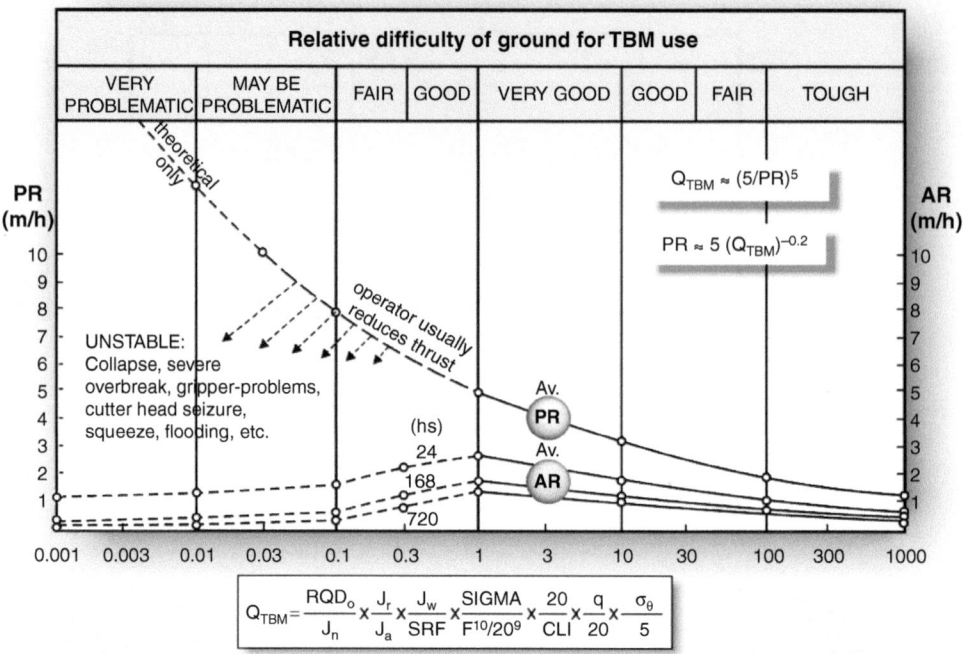

Figure 16 The Q_{TBM} model for TBM prognosis involves an oriented Q_o-value and machine-rock interaction parameters given in normalized form. The Q_{TBM} value is (adversely) increased if CLI (cutter life index) is <20, if q (quartz content %) is > 20, and if the estimated σ_θ (biaxial stress state on tunnel face) is more than 5 MPa (the estimated value at 100 m tunnel depth). Note that the lower set of curves representing AR estimation for 24 hrs, 1 week, 1 month are separated, because of declining utilization. Note the new 'adjectives' (tough, problematic etc) specifically for TBM. It is clear that central Q_{TBM} values of $\approx$ 0.3 to 30 would be ideal for fast progress.

U cannot be independent of time T, as clearly shown in Figures 1 and 2, both for fast and slow TBM tunneling.

Before presenting the three equations, it may be helpful to see how fault zones plot on the *log PR – log T – log AR diagram*, which was used as an introduction to the world records shown in Figure 1. Fault zones in general have a potentially delaying effect on overall tunneling rate. Their low Q-values usually demand heavier local support (or some difficulties with PC-element ring building due to over-break), so a steeper deceleration gradient is usually involved. On the other hand they may often (though not always) be 'easily' bored through using low cutter thrust, but probably need increased torque, since an unusual amount of contact with the cutter-head is occurring. Unless blocks dislodge and jam the cutter-head, as may sometimes happen in unexplored mountainous terrain, the moderate delays may appear as shown in Figure 18. Only one of the steeply inclined lines suggests a delay of (>) 1 week.

Double-shield TBM with push-off liner capabilities may get severely delayed if a fault zone is serious, as over-boring (void development in front of, to one side, or above the cutter-head) can just as easily develop ahead of these machines as ahead of open-

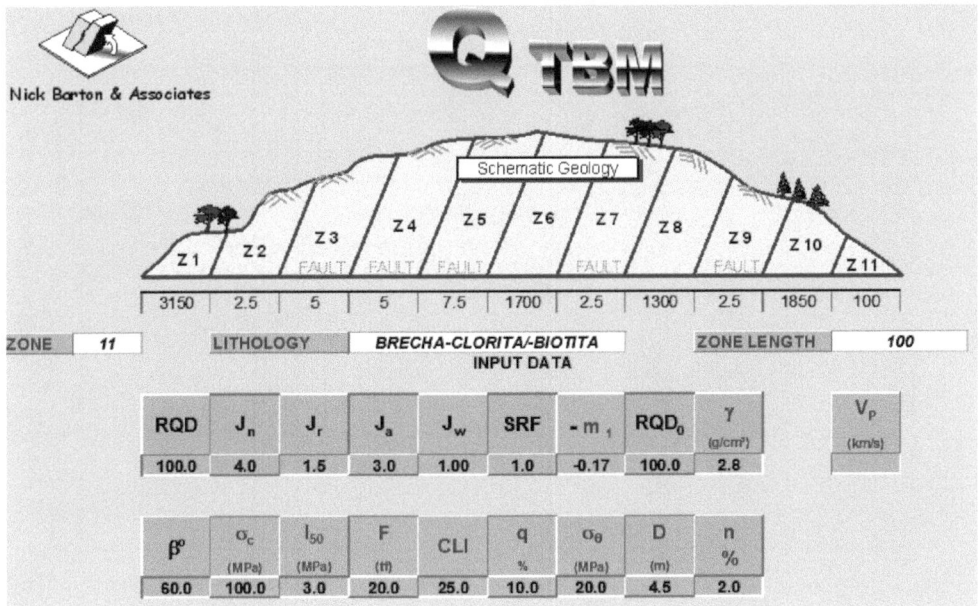

Figure 17 An example of the Q_{TBM} input-data screen showing eleven modeled zones in this case, six of them being assumed fault zones, as viewed in adjacent   deep (mineral-reserves-related) drill-core. The estimated input-data was for a planned TBM tunnel in Chile, successfully constructed some years ago. (Q_{TBM} model by Ricardo Abrahão).

gripper TBM, unless pre-injection in the one case, or spilling has been performed. Facilities for these essential operations are illustrated later in this chapter. When faults are encountered deep below the water-table, and delay TBM progress, inflow of water may occur in an uncontrolled manner and for too long, with groundwater drawdown (and subsidence damage) as a likely result in the case of shallow tunnels beneath towns. Risk analysis should address such consequences and their mitigation. It will not be sufficient to have gasketted PC-elements ready to be installed, while the next 15m of slow advance is negotiated before reaching the element erector. In this sense TBM can be 'too long'. Pre-injection 'would have been' the solution.

7.1 Fault-zone delays explained

We need three basic equations to understand potential delays in fault zones. (The following nomenclature will be used as before: AR = advance rate, PR = penetration rate, U = utilization, expressed as a fraction, for any chosen *total time* period T in hours). Firstly:

$$AR = PR \times U \tag{1}$$

(All TBM must follow this first equation, which was presented at the start of this chapter).

$$U = T^m \tag{2}$$

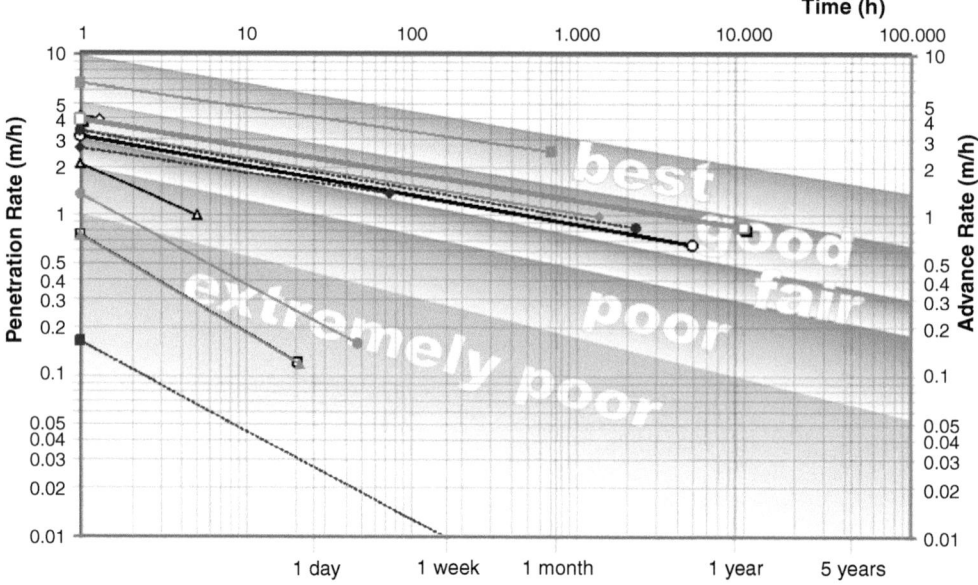

Figure 18 The graphic output concerning the calculated PR and mean AR (both in m/hr) for the eleven separate zones shown in Figure 17. Only one of the faulted rock zones dips below the 'horizon' (T > 1 week) and is therefore predicted to cause trouble even for the double-shield TBM. A weighted mean of about 10,000 hours for 8 km of tunneling was predicted for this double-shield TBM, by combining all of the eleven zones modeled. In practice it took longer due to local labor practices, and due to a long delay out of the starting chamber, where unexplored faulted rock was immediately encountered, and over-boring and cutter-head blockage were intermittently experienced for several weeks.

(Due to the reducing utilization with time, advance rate decelerates, see Figures 1 and 2).

$$T = L/AR \tag{3}$$

(Obviously time needed for advancing length L must be equal to L/AR – in fact this also applies to walking. With *continuous* boring T = L/PR). All readers can agree that these are very simple equations, and also correct equations. But who has seen them combined and employed in prognosis?)

By simple substitution we have the following:

$$T = L / (PR \times T^m)$$

(Here, T appears on both sides of the equation: the final expression for T is therefore:)

$$T = (L/PR)^{1/(1+m)} \tag{4}$$

This is a very important equation for TBM, if one accepts the case record evidence that *(–)m* is strongly related to low Q-values in fault zones and significant weakness zones. It is important because very *negative* (-)m values make the component (1/(1+m)) *too big*.

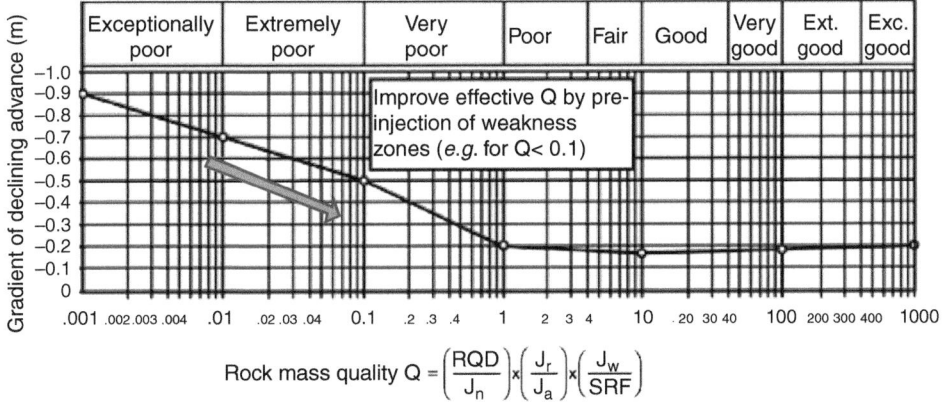

Rock mass quality $Q = \left(\dfrac{RQD}{J_n}\right) \times \left(\dfrac{J_r}{J_a}\right) \times \left(\dfrac{J_w}{SRF}\right)$

Figure 19 An understanding of deceleration tendencies for all TBM (unless conditions improve throughout the length of tunnel) is given by this empirical (a posteriori, not a priori) link between low Q-values and steep deceleration events when passing through (or maybe getting stopped) in significant faults or weakness zones. Pre-grouting (in the blue zone) is the most effective way to prevent such stoppages. It solves other problems as well (such as settlement damage). Double-shield machines may reduce these adverse gradients by as much as one half, at best.

If the fault zone is wide (large L) and PR is low (grippers inefficient, water problems etc.) then L/PR may get too big to tolerate a big component (1/(1+m)) in Equation 4. It is easy (in fact much too easy) to calculate an almost 'infinite' time for a fault zone using this 'theo-empirical' equation. The writer knows of four permanently buried, usually fault-destroyed, occasionally rock-burst destroyed TBM (Pont Ventoux, Dul Hasti, Pinglin, Jinping II). There are certainly many more, and the causes may be related to Equation 4 logic. Fault zones will remain a serious threat to TBM tunneling as we know it, unless the extremely poor rock mass qualities associated with fault zones can be improved by *drainage and pre-grouting*, specifically where Q < 0.1. There are signs that some TBM are being adequately equipped. It has taken many decades.

A steep deceleration gradient demonstrates the adverse nature of these 'unexpected events' (faults), which should alternatively be *anticipated beforehand*, by performing probe drilling during part of each maintenance shift. If highly permeable weakness zones were drained and pre-injected, an effectively increased Q-value (as deduced in *Barton, 2002, 2011/2012*) would cause *(-) m* to reduce to less negative values, as indicated in Figure 19 (see arrow).

8 SOME EXAMPLES OF FAULT ZONE CHALLENGES FOR TBM

In the following illustrated section of this chapter on TBM prognosis and risk (Figures 20 to 23), we can view challenging situations for TBM related with fault zones. The situations illustrated have obviously been a source of delay for contractors, and they hint at the need for more flexibility in detection ahead-of-time, and the ability to pre-treat, so that the inherently adverse properties can be improved before the TBM exposes the problem. Once exposed in the excavation, fault zones, especially those containing

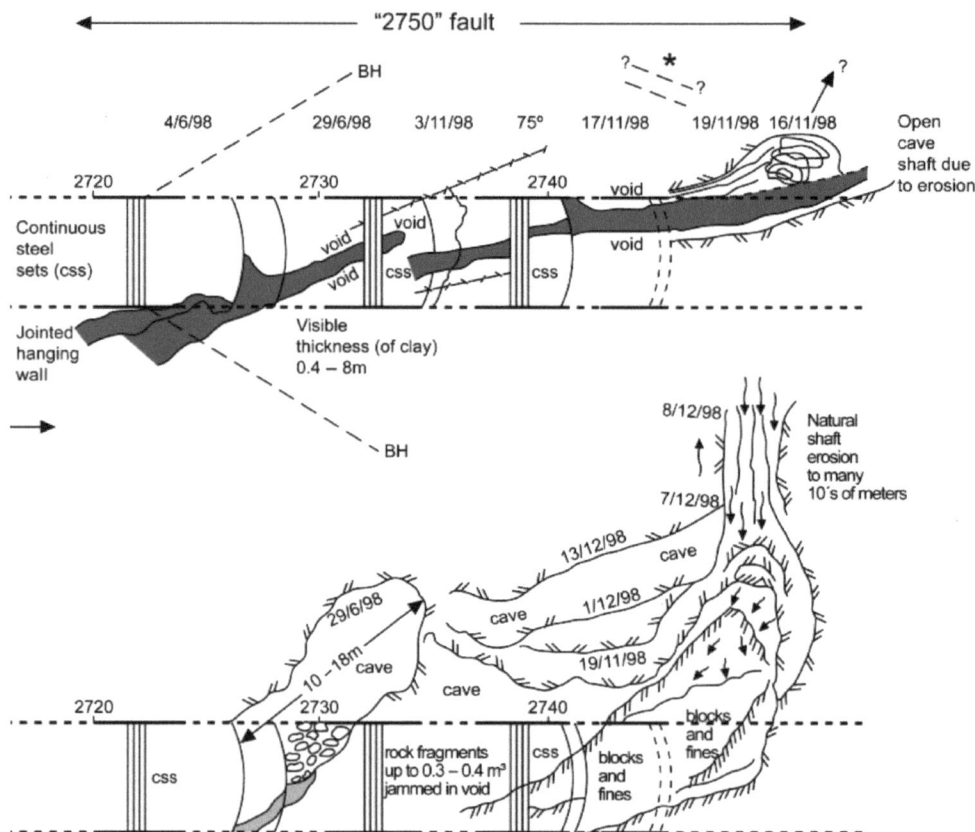

Figure 20 A seemingly minor fault zone with a 1 m thick clay core, but combined with high water pressure on one side, succeeded in delaying this inherited TBM by 5 months, and similar features soon afterwards meant 'drill-and-blast-from-the-other-end' and abandonment of the TBM. Blocks loosened in a steadily eroded 'natural' shaft, due to water inflows, and these blocks which could be heard falling, frequently blocked the cutter-head, preventing rotation. The TBM was not equipped for probe drilling nor for pre-injection. The new contractor inherited a TBM whose limitations were a good illustration of future needs for more versatile TBM, which are now becoming available, as shown later in this section on fault zones. The sketches are from weekly logging by engineering geologist K.G.Holter, each superimposed to see 'progression'.

crushed rock and clay and high water pressures, can present nearly insurmountable problems due to loss of profile, difficulty with gripper operation, difficulty with PC-element ring building, and cutter-head blockage (or over-loading of conveyors) with eroded over-size blocks. In the case of one older TBM, the recesses in the gripper-pads for allowing a 60 cm c/c spacing of circular steel sets proved to be futile, as the stability was so poor that the contractor had placed steel sets flange-to-flange forming a steel-barrel. This was stable except where all the sets were crushed and sheared by the gripper action. With a deforming clay core on one side (Figure 20), the right-side grippers also penetrated into the fault, carrying the steel sets like discarded clothing.

In relation to the 'velocity-confinement' trends seen some figures ahead, in Figure 24, it can be seen that a stress-confined fault met at depth by a TBM, may be unloaded too

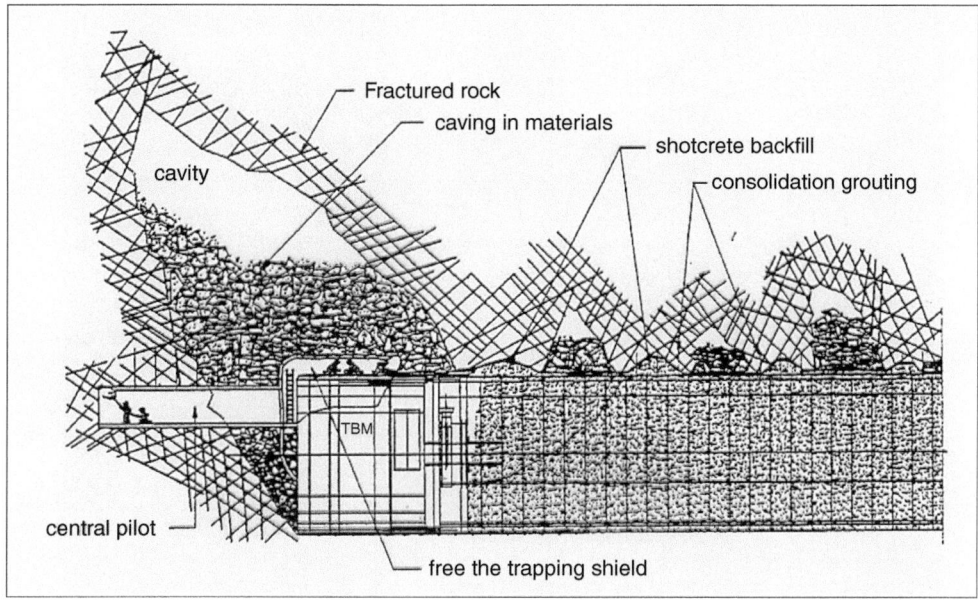

Figure 21 Exceptional problems in faulted meta-sandstones, with the need for a top-heading to release the cutter-head (Shen *et al.*, 1999). Such stoppages occasionally provided the opportunity for replacing cutter-head 'armor'which was worn out every 4 to 5 km, as also at the Guadarrama tunnel project, driven in mostly granites and gneisses. Many smaller delays and large delays like that illustrated, contribute to T and therefore to a steeper (-) m than ever expected by designers or by those offering constant m/month tunneling. TBM are seldom so simple.

much and revert to much more difficult behavior, as if the fault was encountered nearer the surface. (As illustration: follow a constant low Q_c-rock mass quality up toward the surface – the equivalent of unloading.) There are experiences of tunnel-seismic 'illuminating' reflectors well ahead of the face, with known reduced velocities such as to 4.0 km/s compared to higher velocities in the surroundings. Yet even when the contractor is prepared, tunnel collapse occurs. This is probably due to the same undesirable but difficult-to-avoid stress release. Spend one day pre-grouting to avoid this, or risk TBM stand-still.

9 TBM DESIGN ASPECTS FOR TACKLING WEAKNESS ZONES

Some seemingly obvious points about TBM design for more successful penetration of faults and serious weakness zones can be grouped in the following categories.

9.1 Cutter exposure

The cutters should not be 'fully exposed' as this invites blockage when blocks of hard but faulted rock start to be released, for instance due to too high ratios of Jn/Jr (number of joint sets and joint roughness) combined with water pressure and erosion of fines.

Figure 22 At the Pinglin Tunnel project in Taiwan, the use of a pilot tunnel for drainage and pre-injection across to the two main (future road) tunnels was not successful, due to numerous cutter-head blockages and the need for side-access drifts to reach the blocked cutter-head (at least 13 times). In addition the contractor had great difficulty even drilling stable holes for pre-injection, due to the intensely jointed, sheared and clay-coated joints and slickensides in the very hard meta-sandstones. The tunnel in the end was mostly driven by drill-and-blast, due to crushing of one TBM in a major fault zone collapse. Photo: Chris Fong. A 7,000 m³ inrush of clay and rock was witnessed, with a temporary tunnel 'face' moved 100m backwards. There were many fatalities at Pinglin.

When this ratio $Jn/Jr \geq 6$, overbreak and possible over-excavation in front of the cutter-head can occur. (This is illustrated later in Figures 26 and 27). The two TBM illustrated at the top of Figure 23 could be especially susceptible to blockage.

9.2 Cutter 'protection'

with armour plating across the face of the cutter-head is a good way to prevent seizure due to block-fall wedging. However, there is a price to pay which may be experienced when driving long tunnels in hard abrasive rocks: the armour may need in-tunnel replacement at 4 to 5 km intervals. This is obviously not an 'over-night' repair like multiple-cutter change, and requires workers, welders, and lifting-gear access ahead-of-the-cutter-head, if the TBM can be stopped in a stable and preferably dry zone. T is always running, so U reduces.

9.3 Double-shield TBM

may in general have increased utilization U, meaning less steeply inclined deceleration (-) m. They are expected to keep advancing even when grippers cannot be used in weakness zones due to over-break / over-excavation. Continued advance is then achieved by

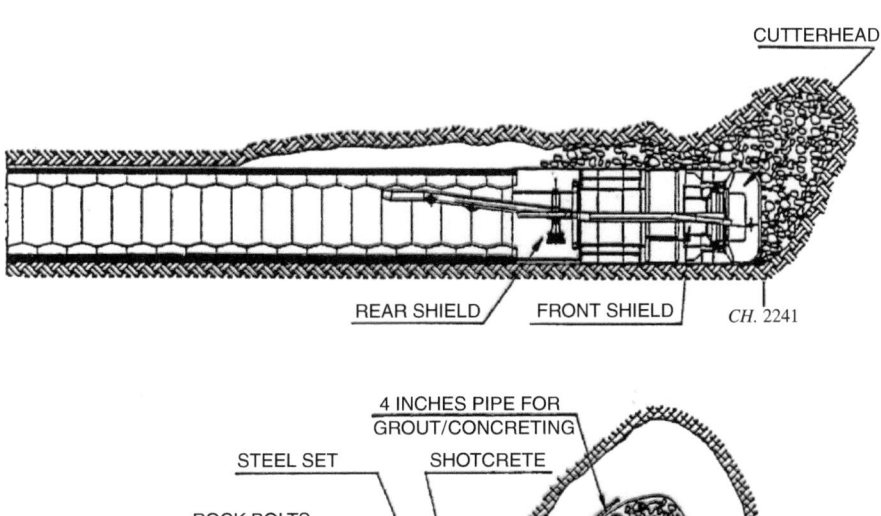

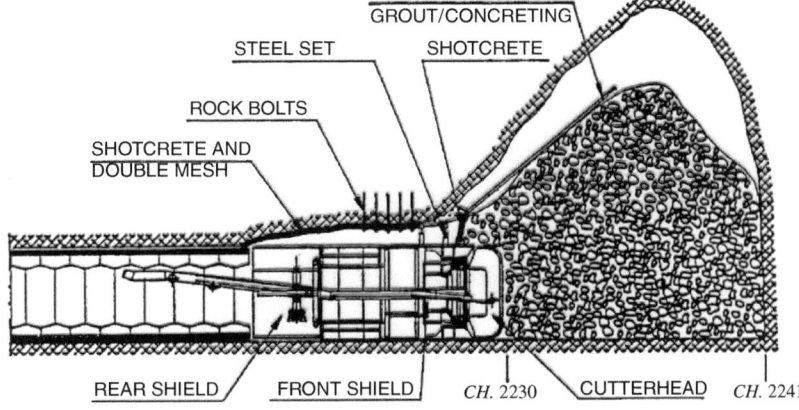

Figure 23 Even the world's most experienced TBM contractor (and manufacturer of TBM) can get stuck in faulted rock, also with double-shield TBM and advanced hexagonal PC-elements. Grandori et al.(1995b). With the benefit of hind-sight, note the adverse situation created by withdrawing the TBM (from ch.2241 to ch.2230). This released the stress on the fault, effectively converting a 'confined V_P' fault-character of say 4 km/s, into a loosened 'unconfined V_P' fault-character of say 2 km/s. A much longer delay and the need for a large-volume post-grouting/concreting was the result.

push-off-liner capabilities, as shown in Figure 26 from a Herrenknecht animation. The photographs are from the Guadarrama Tunnels in Spain. Figure 27 illustrates a double-shield application in a hydropower headrace tunnel, where PC-elements (and therefore push-off-liner capabilities) could be utilized in thirteen specific weakness zones and faults.

In relation to the use of 'nominally unlined' headrace tunnels for economic hydropower, it should be noted that wedges and small blocks that had fallen from some locations in the mostly unlined kilometers were not transported, even by 2.5 m/s water flows in a smooth TBM-driven tunnel. The so-called 'rock trap' just upstream of the pressure shaft, contained only sand and silt and some floating pumice 'pebbles' from an upstream lake. The 'zero velocity' boundary layer ensures transport of nothing larger than rounded, few millimeter size particles. The hydraulic boundary layer phenomenon therefore indirectly provides extra cheap renewable power in Norway and wherever 'nominally unlined' hydropower tunnels are used. There are 3,500 km of such tunnels in Norway, 250 km of them driven by TBM.

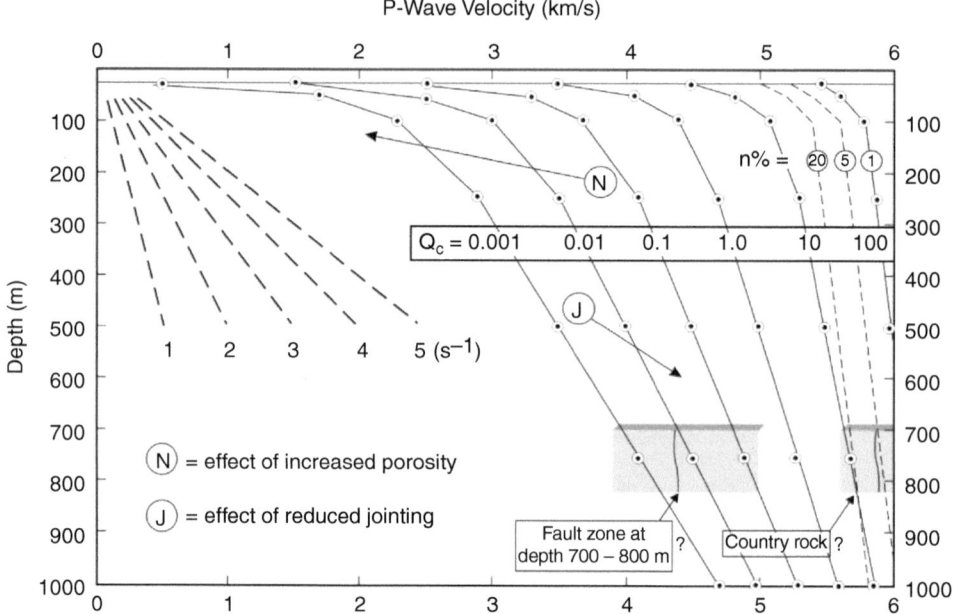

Figure 24 If a highly confined fault zone is remotely sensed ahead of a TBM, its deceptively high P-wave velocity will nevertheless be compromised, when the TBM starts to try to penetrate the zone. Difficult-to-avoid loosening will 'lift' the fault to a near-surface lower rock mass character with V_P perhaps reduced to 2 km/s. Figure 23 is an unnecessarily good example of the possible consequences of loosening. Concerning the sloping dotted lines, note that (s⁻¹) is the unit of velocity gradient, derived from km/s per km. Velocity gradient may be very large close to poor quality weathered rock, hence the severe consequences of allowing loosening. From Barton (2006).

9.4 Pre-injection may be needed

because with insufficient stand-up time, faulted rock can start to over-break and get over-excavated ahead of and to the side of the TBM. The writer has witnessed this several times in different countries (Italy, Kashmir, Chile, Taiwan). An 8 m diameter tunnel can become 11 or 12 m locally, making grippers inoperable until a large void has been filled with concrete or hundreds of sand-cement bags (solution depending on locality). Pre-injection cannot be effectively used if ring-mounting equipment has to be dismantled, as if pre-injection was 'the last resort'. It should always be available. Figures 28 and 29 illustrate some minor (2 to 3 meter high) void formations and post-treatment needs.

9.5 Pre-injection may also be needed

to ensure that water does not flow uncontrolled *into the face and also into the first 5 to 15 m of unlined tunnel,* in the area of the single- or double-shielded TBM, where impervious (gasketted?) tunnel linings are still 'pending'. This could allow a large volume of inflow if one was also advancing more slowly in a faulted area (the –m effect). If this

Figure 25 The upper two photographs are of open (and very cutter-exposed) TBM, showing remarkable similarity in relation to cutter layout, in view of their 35 years difference in dates (#1 mid-seventies: Slemmestad, Oslo, #2 mid-tens: Canada). Performance of both was generally good: one in shales and nodular limestones and igneous dikes, the other in granites. However stoppage in a fault zone was a 'game-changer' (protracted litigation) in the case of the granites. The well-protected cutters in the lower photographs of double-shield machines (#3 Guadarrama, #4 Robbins advertising) are much less likely to be blocked in loosening fault zones.

temporary lack of water control is 'planned' since not solved by pre-injection, and if the tunnel is relatively shallow (say 20 to 200 m) and if the tunnel is passing near to (usually < ½ km, occasionally 1 to 3 km from) built-up areas founded on clay, then groundwater pressure drawdown must be anticipated, with potential for settlement damage. Bolted and gasketted PC-elements may be questionable long-term (100 years) solutions for ensuring no inflow and permanently dry tunnels. This method has not yet been suffi-ciently tested, while (pre-) grouting of rock masses has a much longer track record, in a variety of contexts. Post-grouting in a 'completed' tunnel is famously difficult: a well known Norwegian experience with 15 km of unexplored tunnel (Gardamobanen) need-ing post-injection, after penetrating below two lakes in a rift valley, should be a reminder of this.

Figure 26 Double-shield animation pictures from Herrenknecht, showing push-off-liner when not using gripper (top-left) and PC-element ring assembly when thrusting off grippers (top-right). The two lower photographs show left: thrust off liner (or gripper re-set) and right: PC-element transfer in a northern end Guadarrama high-speed rail tunnel in Spain, near Segovia.

9.6 Pre-injection may also be needed in deep tunnels

in response to probe-drilling evidence that a wet zone is being approached. If MWD also suggests that the rock is heavily jointed, high-pressure injection may be just the measure needed to prevent the very adverse loosening that may occur when a TBM enters such a zone and slows down or stops. The effective 'quality' of a fault zone is reduced when it becomes unconfined: permeability is inevitably increased, and if erosion of finer materials also begins, the long delay may be just the beginning of a sequence of problems that have been known to end in TBM burial. Drill-and-blast 'from the other end' is not so infrequent a decision. It has been used many times, with cases known to the writer in China, Taiwan, Kashmir, Italy etc. Drill-and-blast has even been used to complete a tunnel where two TBM were going to fail their planned meeting by the millennium of 2000. Thus the seriously proposed hybrid-from-the-beginning suggestion described in Barton (2012), meaning drill-and-blast combined with TBM, each where most appropriate.

10 CARE NEEDED WITH AUTOMATIC CHOICE OF TBM 'BECAUSE THE TUNNEL IS SO LONG'

There are many ways to compare TBM and drill-and-blast, such as using over-optimistic 'constant' TBM utilization (say 600 m/month) and conservative (say 40m/week) drill-and-blast progress. The reality will obviously not be so uniform. In Figure 32 an

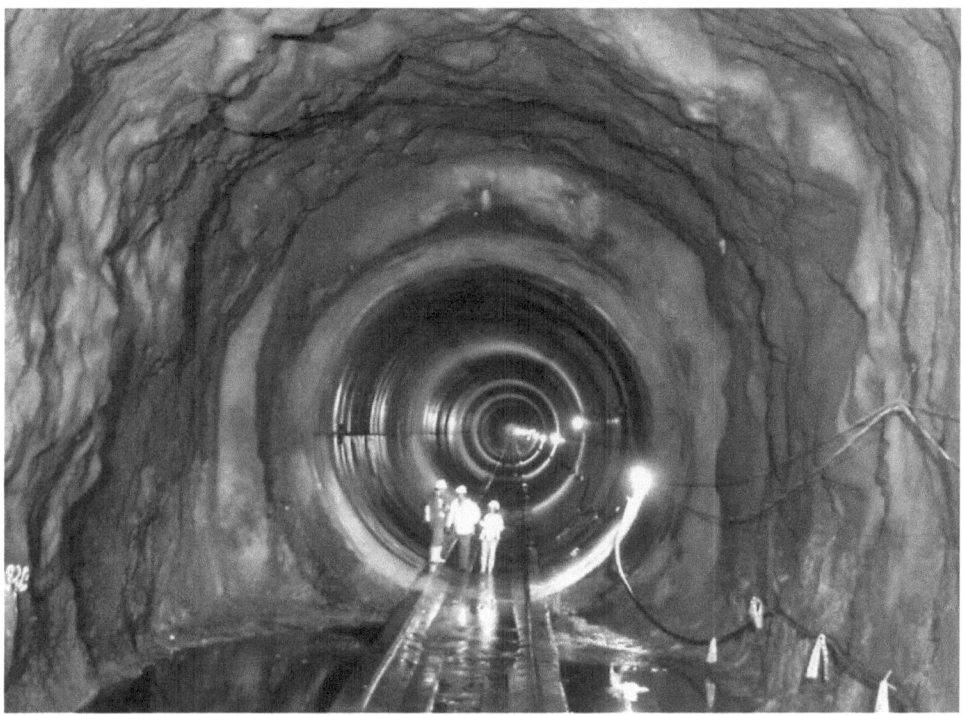

Figure 27 A double shield TBM used to drive a 10 km long headrace tunnel in Ecuador, where inspection during second emptying was led by the writer, with suggestions for local bolting made by our team on behalf of contractor Odebrecht. This project was unusual because the PC-element lining had been used in thirteen specific stretches of bad ground, where the advantages of 'ready-made' support and push-off-liner thrust could be fully utilized if no gripper operation was possible. A serious fault zone nevertheless stopped the machine for several months in one of the thirteen locations. (Photo: Dr. Nghia Trinh).

Figure 28 Void formation ahead of a double-shield TBM, due to adverse ratios of Jn/Jr. There were at least four joint/fracture directions, and the joint/fracture surfaces were sometimes smooth-undulating and slickensided, so Jn/Jr could be as high as 15/1.5. Over-excavation due to unstable ground is a phenomenon that loads conveyor belts with more material than would be consistent with tunnel advance. It can be detected by laser, or by real-time weighing and automatic calculation in relation to the weight expected from measured PR, assuming that only a perfect cylinder of rock is excavated.

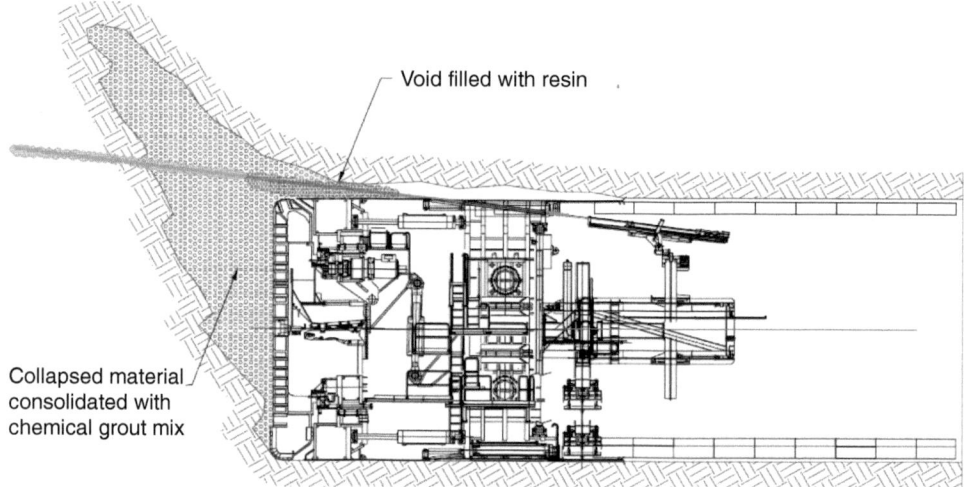

Void filled with resin

Collapsed material
consolidated with
chemical grout mix

Figure 29 Suggested solution for void formation: an after-the-event measure that is actually too late, as a few meters per day may be the limits of advance if the zone of over-excavation continues for several weeks. Note that a tunnel fire can be caused if too large volumes of chemical grout mix are needed for void filling. 'Waiting for the smoke' (to clear), is unexpectedly different from 'waiting for the train' on a pie-diagram. The problem with this TBM was that ring-mounting equipment had to be dismantled before pre-injection drilling could be performed, clearly not the ideal choice of priority if faulted rock is expected. How to do both (build rings and pre-inject) when both were needed was not satisfactorily addressed, so tunneling was delayed unnecessarily.

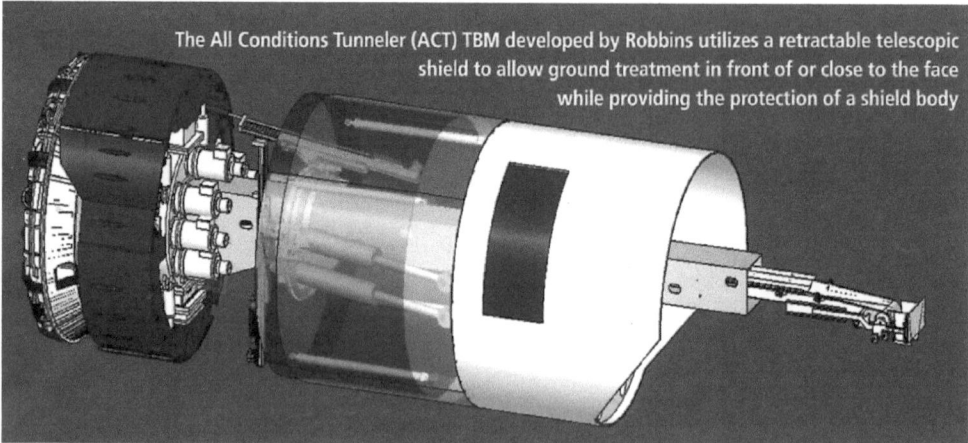

The All Conditions Tunneler (ACT) TBM developed by Robbins utilizes a retractable telescopic shield to allow ground treatment in front of or close to the face while providing the protection of a shield body

Figure 30 One stage closer to the ideal TBM (Robbins: Wallis, 2012) with full acknowledgment of the possible need of pre-injection with minimum delay. Advances in the possibility of pre-treatment are occurring on a regular basis. It seems finally to be acknowledged that rock masses (and hydro-geologies) can exceed the capabilities of 'standard' TBM. Burials and stand-stills have forced this acknowledgment, but it has taken the costly consequences of many decades of stoic optimism, under the sometimes costly motto:'because the tunnel is long we chose TBM'. This decision has enhanced risk of delays more than most other decisions in tunneling history.

Figure 31 Robbins (D. Wallis, 2012) illustrations of the need for pre-injection and its solution with the ACT All Conditions Tunneled illustrated in Figure 30. Water is a remarkably adverse 'partner' in tunneling, but its unwanted presence can be severely curtailed if timely pre-injection is an accepted measure. However pre-injection may be ineffective if too low injection pressures are used. At least 50% of the injection pressure at the pump has been lost, just 1m away from each injection borehole. When flow stops, pressure must not be maintained. The job is complete (Barton, 2011, 2012).

alternative method has been chosen. The D+B curves were derived from single-shell (B + Sfr) cycle-time measurements, performed by Grimstad at a $60m^2$ road tunnel, and recorded over a wide range of Q-values (see Barton & Grimstad, 2014). The TBM prognoses, with the time (T) dependent utilization discussed earlier, was based on the Q-value dependent (–) m values plotted in Figure 19. Application of cutter thrusts was appropriate to the different rock masses. Selection of moderate parameter values allowed the approximation $Q \approx Q_{TBM}$. As may be noted in Figure 32, in one year of tunneling, the predicted mean weekly rate has dropped to about 100 m/week, from more than 200 m/week if the TBM prognosis was on this optimistic short-term basis. As can be

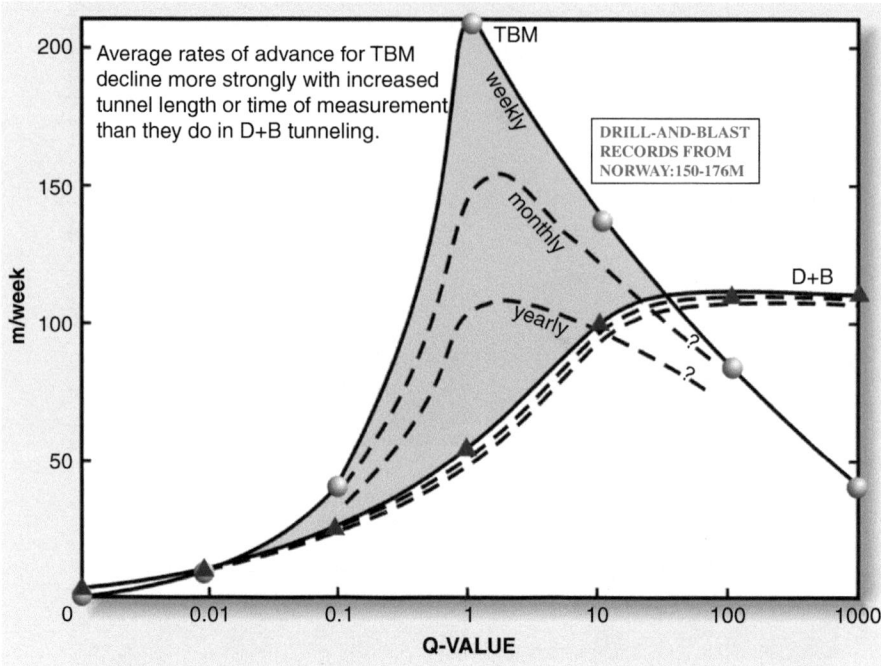

Figure 32 This comparison of TBM and drill-and-blast rates of advance, based on a moderate (>5 km in one year) TBM prognosis, and comparison with Norwegian drill-and-blast cycle-times per Q-value-class measurements, suggests two important things. The longer the tunnel the more the need for central (well jointed but not faulted) rock qualities if the TBM is going to be faster than drill-and-blast. Q-values consistently higher than 100 suggest drill-and-blast superiority, bearing in mind recent records of 150, 165 and 176 m in a 7 x 24 week (high up in the right-hand top quadrant of the figure. LNS has a mean 104 m/wk for 5.8 km in coal-measure rocks, with Q probably mostly 1 to 10, but sometimes 0.1 to 1 needing more support.

imagined, as the TBM tunnel gets longer, and time exceeds 1 year and perhaps even 2 years, more central rock mass qualities will be needed for continued TBM superiority. The more 'extreme-value' qualities will usually favor drill-and-blast.

Whether to choose TBM or drill-and-blast, or a hybrid solution using both, should be based on an honest (not over-optimistic, nor over-pessimistic) assessment of the rock masses likely to be penetrated. In Figure 33, hypothetical 'Q-statistics' have been assembled for a 5 km long tunnel (hardly long enough to choose TBM) and for a 25 km long tunnel with similar 'geology' but somewhat deeper cover. Sampling theory and logic would suggest more 'extreme value' structural geology, hydrogeology and rock strength for the longer tunnel, and of course the likelihood of higher stress if cover is locally greater than in the shorter tunnel. In relation to the empiricism-based prognoses of Figure 32, more 'extreme value' conditions will not be ideal for TBM, especially if the tunnel is likely to take several years to complete.

High rock stresses generally present greater difficulties for TBM because of the maximizing of tangential stress in the less disturbed tunnel wall. When $\sigma_{\theta\ max} > 0.4$ UCS the onset of stress fracturing must be expected. In the Q-system, elevated values of SRF begin at this stress/strength ratio (Barton & Grimstad, 1994). If high cover is

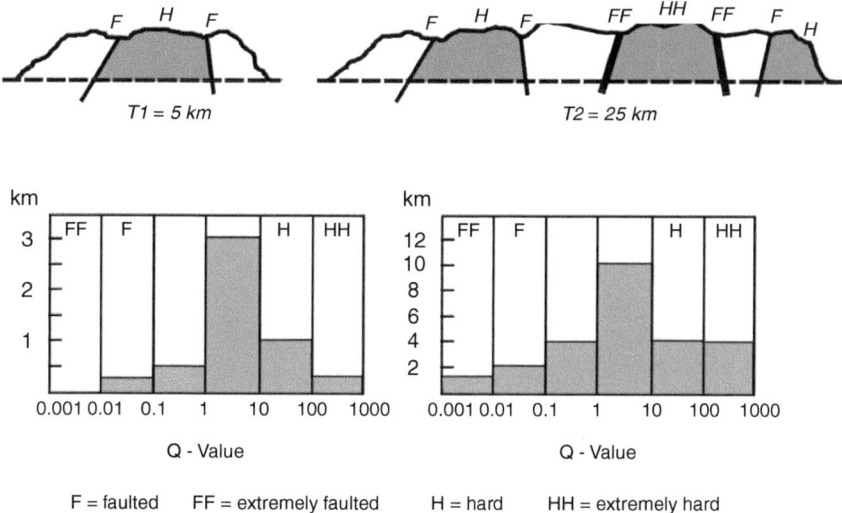

F = faulted   FF = extremely faulted   H = hard   HH = extremely hard

Figure 33 Choosing TBM 'because the tunnel is so long' invites the likelihood of more 'tail distributions' of rock quality, rock hardness, fault-zone severity, rock stress etc. It may be that a given distribution of UCS will be adversely 'intersected' by the distribution of expected maximum (tangential) rock stress, thereby guaranteeing stress-fracturing and possible rock burst problems.

planned, serious consideration of drill-and-blast through such sections should be made, if access for intermediate ventilation can be made. This is one way to reduce risk, as a serious rock burst with TBM carries with it the risk of greater loss of life and much greater material damage than in the case of drill-and-blast.

II AN EXAMPLE OF PROGNOSIS FOR OPEN-GRIPPER AND DOUBLE-SHIELD TBM

This chapter related to TBM prognosis and risk reduction will be concluded with a brief glimpse of a recent application of the Q_{TBM} method in prognoses for possible open-gripper or double-shield TBM for twin tunnels of 8.5 and 9.5 km length. At the time when the studies were performed these two machine-type options remained open. Since 2016, four double-shield TBM will be in operation in this Follobanen project south of Oslo.

The six Q-parameters were recorded on field-logging sheets, with five observations for each 10m of rock mass exposure, which were usually road-cuttings. The rock exposed above the two planned tunnels totaling nearly 19 km length were logged in this way. For the shorter 8.5 km long tunnel, some 200 rock cuttings could be logged in the general neighborhood of the planned tunnel, and logging was of course focused on the same granites and granitic gneisses, some of which were very hard and abrasive and in places also sparsely jointed. Class 1 had mean Q = 150, which is too high for a fast PR, and will likely be associated with high cutter wear.

Five rock classes were identified (Class 1 input data assumptions are represented in Figure 34). In areas where there were no road-cuttings and where there were wet areas or deeper valleys, logging of deviated core and seismic refraction analysis provided the Class

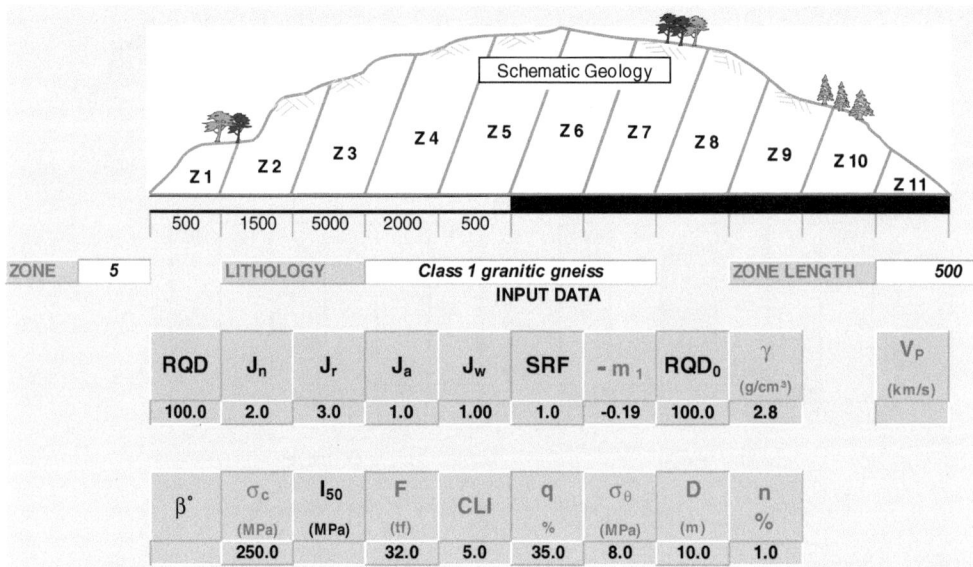

Figure 34 The Q$_{TBM}$ prognosis model input data 'keyboard', which is entered with appropriate Q-parameter numbers for each zone modeled, in this case Class I represented by 500m of massive granite. Slow progress is predicted in this case, despite the high assumed cutter force of 32 tnf. (The Q$_{TBM}$ prognosis model was created by Ricardo Abrahão based on the empirical equations.) developed by Barton, 2000.

6, 7 and 8 Q-parameter input. Velocities were occasionally as low as 2 km/s. Barton & Gammelsæter (2010) have described the process in more detail. The three groupings of weakness zones, with their typical V$_P$ velocities and Q-value interpretation, are shown by stars in Figure 35. The whole range of qualities, with probable differentiation of UCS (due to weathering) are indicated in approximate terms, in Figure 36.

The prognoses using Q$_{TBM}$ were made for both the Follobanen Tunnels south of Oslo, and more complete results can be found in Barton & Gammelsæter (2010). At that time, prognoses were made for tunnels of 7.9 km and 9.6 km length. Respective prognoses were 9.9 and 17.5 months, plus some few months delay for specific low velocity weakness zones. If we average, which is not strictly correct, due to deceleration effects, 9 km might be expected to take from 11 to 16 months, plus some months delays in fault zones. The high end of the above range reflects deceleration effects.

It is understood from recent press releases that both tunnels, going north and south, are now approximately 9 km length due to final layout of the project, which, unusually, sees all four machines assembled in central caverns, and sent on their way 9 km northwards (Sept 2016) and 9 km southwards (probably early 2017). In Figures 37 and 38 only the results predicted for the southern tubes of approx. 8 km length are given. During the time in 2016 when this chapter was finalized, two of four Herrenknecht double-shield TBM have started, under the large contract with Spanish and Italian contractors Acciona and Ghella. The first machines to start are going the 9 km northwards toward Oslo, and boring was initiated in late September 2016. We await the results, with anticipation of four useful new case records.

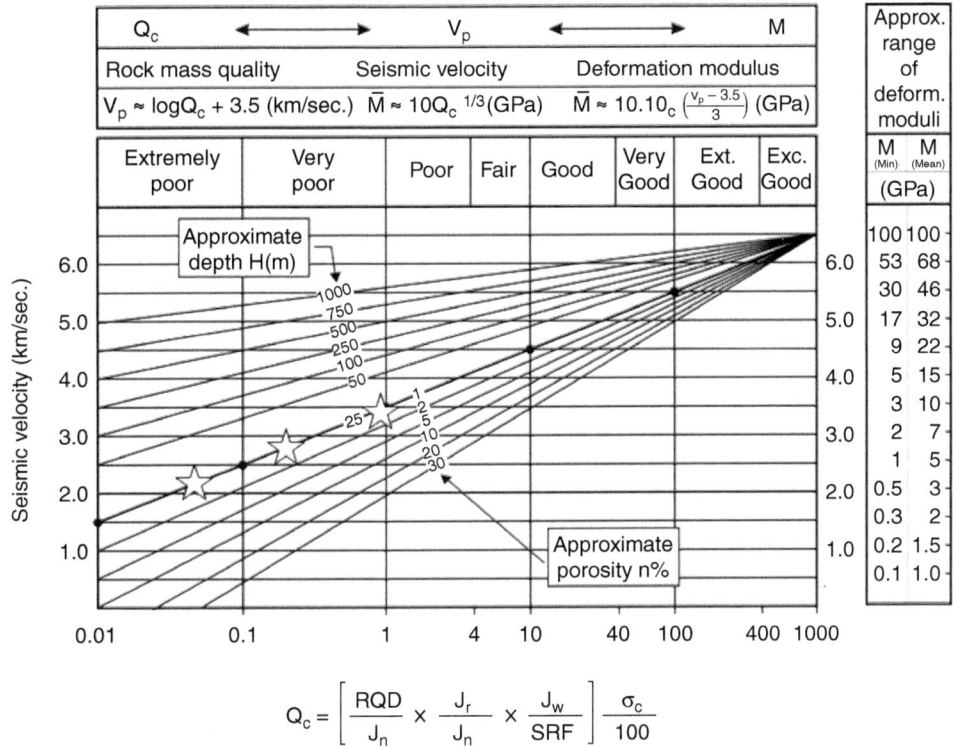

$$Q_c = \left[\frac{RQD}{J_n} \times \frac{J_r}{J_n} \times \frac{J_w}{SRF} \right] \frac{\sigma_c}{100}$$

Figure 35 Inter-relationships between P-wave velocity and Q_c (=Q×UCS/100). Note empirically based correction for depth of measurement (+ve) and porosity (-ve) using diagonal lines. The nominal 25 m depth 'shallow-refraction-seismic' line for hard, low porosity rock with UCS about 100 MPa, has the equation $V_P \approx \log_{10} Q + 3.5$ km/s. (Barton, 1995). Stars show nominal position of Types 1, 2, 3 weakness zones, meaning Class 6, 7 and 8 for this tunnel. An associated declining UCS, Q and Vp are represented by the 'descending stars' in the monogram in Figure 36.

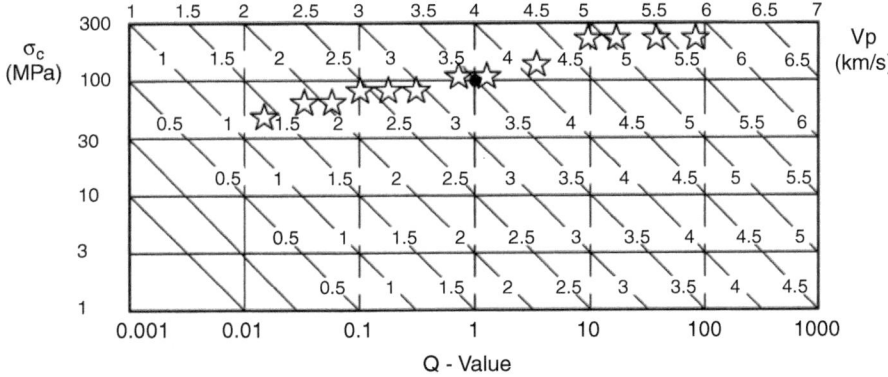

Figure 36 The broken line of 'stars' is designed to follow both the declining UCS and the reducing Q-values, as rock strength reduces, jointing increases, and the weakness zones are approached with lowest V_P. (The black circle is plotted to focus on the easily remembered Q = 1, V_P = 3.5 km/s, UCS = 100 MPa).

Q - VALUES:	(RQD	/	Jn)	*	(Jr	/	Ja)	*	(Jw	/	SRF)	=	Q
Q (typical min)=	75	/	15.0	*	1.0	/	5.0	*	0.50	/	1.0	=	0.500
Q (typical max)=	100	/	4.0	*	4.0	/	1.0	*	1.00	/	1.0	=	100.0
Q (mean value)=	98	/	8.4	*	1.7	/	1.3	*	0.75	/	1.0	=	11.07
Q (most frequent)=	100	/	9.0	*	1.5	/	1.0	*	0.66	/	1.0	=	11.00

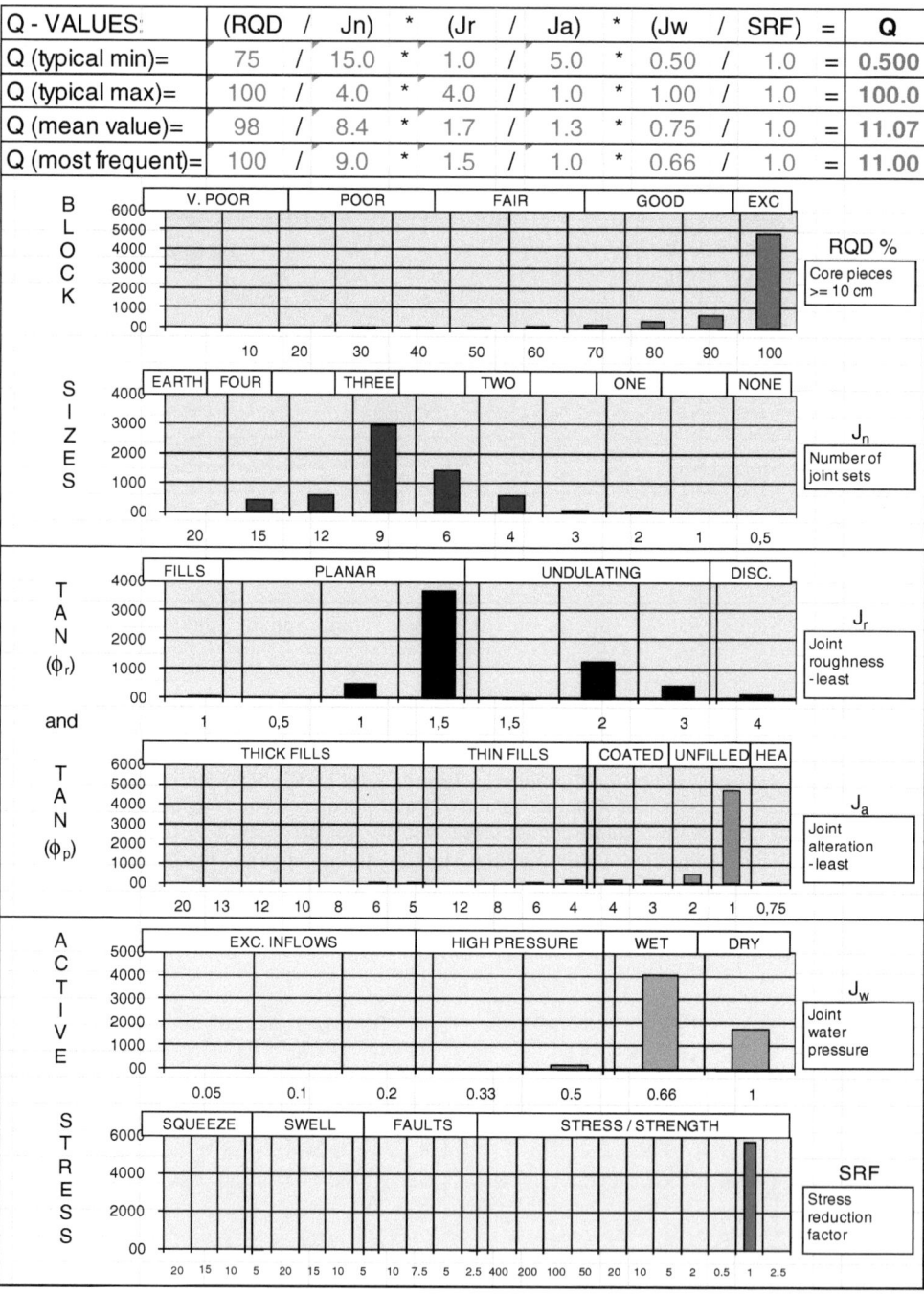

Figure 37 The Q-histogram logging result for the 'best' five rock classes expected in the southern Follobanen tunnels, south of Oslo. Note the large number of observations which were collected for this tunnel (and also for a second longer tunnel) during a few weeks of field work. Logging core from seven deviated boreholes drilled through low-velocity weakness zones are the source of the third set of prognoses shown in Figure 38, indicating that weakness zones might collectively delay the TBM by three months, if all those to be encountered are represented.

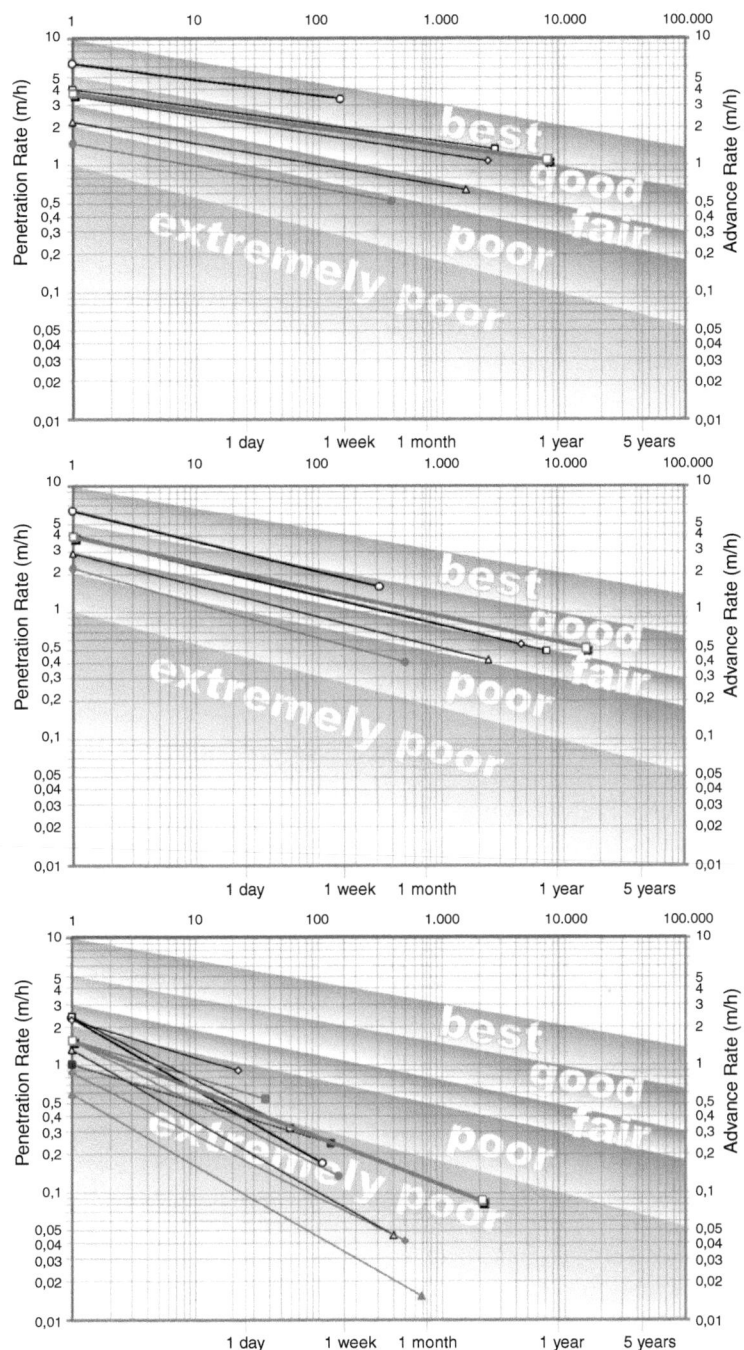

Figure 38 The results of Q_{TBM} prognosis for the 8km long tunnel, southern tunnel made in three parts: Top: double-shield with PC-element liner, showing reduced gradient (-) m due to the assumed efficiencies with 'push-off-liner' capabilities for advance in weakness zones (10 months). Centre: open-gripper, presently without the 'delaying' effect of selective pre-injection (21 months). Bottom: a selection of weakness zones and faults, also not pre-injected as first modeled (3 months). In each case the thickest red line is the predicted weighted-mean performance, with PR (left-end) and AR (right-end).

12 CONCLUSIONS

1   It is misleading to quote a single utilization % for a given TBM project, and worse still to not specify a time interval. It is also misleading to assume that a given tunnel can be driven at a consistent average, say 400 m/month. The advance rate is not a constant in time. Numerous open-gripper case records, and numerous world record results consistently show deceleration with time and tunnel length, if all of time T is included, considering $7 \times 24 \times 52$ hours per year. Deceleration is a natural process that should be a part of realistic TBM prognosis, in preference to denial of its existence. TBM may still be performing very well, despite the deceleration trend. But they get closer to the best drill-and-blast, in long tunnels.

2.   TBM prognosis requires a mix of rock mass description and rock-cutter interaction. So the number of joint sets, their character, the rock UCS, the abrasivity, the quartz content, and the average cutter force, are fundamental requirements for realistic prognosis. In faulted rock with low Q-value, the deceleration is faster, and utilization is often compromised.

3.   A very important component of actual advance rate is the penetration rate. However AR and PR tend to diverge where rock strength is low or where a lot of tunnel support is needed. With double-shield TBM and push-off-liner capabilities, and continuous PC-element lining (if selected), the advance rate is expected to be closer to the penetration rate. If over-excavation (void formation) and ring-building is not compromised, this expectation can be realized. In the case of massive, hard abrasive rock with frequent cutter-change, especially if the tunnel is deep, PR and AR will both be adversely affected.

4.   To avoid over-excavation, compromised ring-building, and unwanted inflows before the liner (and gaskets) are in place, it is necessary to employ high-pressure pre-injection, as if in a drill-and-blast operation, with fast-drilling hydraulic jumbos. This should be available on a continuous basis, and not need disassembly of ring-building equipment. Both are needed as separated facilities. Probe-drilling during the maintenance shift is also a good way to reduce risk.

5.   There are significant numbers of TBM projects that end up with the difficult decision of whether to complete the project by drill-and-blast from the other end. This on its own suggests that the TBM could or should have been used only on the better investigated portion of such projects, for instance the lower-cover section, with drill-and-blast started already from the other end, and deliberately chosen for the less investigated high cover sections. Intermediate access for ventilation is of course an advantage, but stiff-tube negative-pressure reversible ventilation is a way to get round this, and removes blast gasses fastest.

6.   The deliberate selection of both TBM and drill-and-blast may often be a simple matter of common sense, giving schedule advantages and cost savings. This is the preliminary level of hybrid tunneling.

7.   TBM tunneling and drill-and-blast tunneling give quite different performance in hard, massive abrasive rock masses. TBM also exhibit adverse behavior at the lowest end of the rock quality spectrum, in water- and clay-bearing fault zones. Since TBM gradually decelerate with time and tunnel length, even when breaking

records, it is found that central rock qualities are needed for TBM to consistently out-perform drill-and-blast.

8. TBM that are operating in mostly favorable conditions, may record remarkable progress, and are therefore an excellent investment for part or all of many tunneling projects where multiple tunnel cross-sections are not required.

APPENDIX A synthesis of numerous mostly hard rock open-gripper TBM projects, from Barton (1996) and Barton (2000), shows the following expectations for progress, when easy (5 m/hr), moderate (3 m/hr) or tough (2 m/hr) penetration rates are setting the limitations of progress through particular domains in the rock mass. Relevant Q-values might be approx.1–10, 10–30, 30–300 respectively.

Penetration Rate PR m/hr	Daily Progress m/hr	Weekly Progress m/hr	Monthly Progress m/hr	One Year Progress m/hr
5 (GOOD)	3 (72 m/day)	3 (336 m/wk)	1.5 (1040 m/mth)	1.0 (8.6 km)
3 (FAIR)	1.5 (36 m/day)	1.0 (168 m/wk)	0.7 (504 m/mth)	0.5 (4.3 km)
2 (POOR)	1.0 (24 m/day)	0.7 (118 m/wk)	0.5 (360 m/mth)	0.3 (2.6 km)

Assuming: 24 hrs/day, 168 hrs/week, 720 hrs/month, 8640 hrs/year
Note that due to reduced deceleration gradients, at best 50% reduced, double-shield TBM might be able to double the prognoses listed here, at least in the case of the "fair" and "poor" cases.

REFERENCES

ADIF (2005). Guadarrama Base Tunnel. *Ministerio de Fomento*, Madrid, Spain.

Barbieri, G., Collotta, T. & Tanzini, M. (2013) Sparvo Highway Tunnel, experiences from the world's largest EPB shield. In press.

Barton, N., Grimstad, E., Aas, G., Opsahl, O.A., Bakken, A., Pedersen, L. & Johansen, E.D. (1992) Norwegian Method of Tunnelling. *WT Focus on Norway, World Tunnelling*, June/August 1992.

Barton, N. & Grimstad, E. (1994) The Q-system following twenty years of application in NMT support selection. *43rd Geomechanic Colloquy*, Salzburg. Felsbau, 6/94, pp. 428–436.

Barton, N. (1996) Rock mass characterization and seismic measurements to assist in the design and execution of TBM projects. *Proc. of 1996 Taiwan Rock Engineering Symposium*, Keynote Lecture, pp. 1–16.

Barton, N. & Warren, C. (1996) Rock mass classification of chalk marl in the UK Channel Tunnels. *Channel Tunnel Engineering Geology Symposium*, Brighton, September 1995.

Barton, N. (2000) TBM Tunnelling in Jointed and Faulted Rock. *Balkema*, Rotterdam. 173p.

Barton, N. (2001) Are long tunnels faster by TBM? *Proc. Rapid Excavation and Tunnelling Conf. RETC*, San Diego, USA. Soc. Min. Eng., pp. 819–828.

Barton, N. (2002) New applications of Q and the Q-parameters in engineering geology and rock mechanics. *Fjellsprengningsteknikk/Bermekanikk/Geoteknikk*, 40.1–40.15. (in Norwegian).

Barton, N. & Abrahao, R. (2003) Employing the Q_{TBM} prognosis model. *Tunnels and Tunnelling International*, pp. 20–23, December.

Barton, N. & Gammelsæter, B. (2010) Application of the Q-system and Q_{TBM} prognosis to predict TBM tunnelling potential for the planned Oslo-Ski rail tunnels. *Nordic Rock Mechanics Conf.*, Kongsberg, Norway, pp. 56–65.

Barton, N. (2011/2012). Assessing pre-injection in tunnelling. *Tunnelling Journal*, Dec.2011/ Jan. 2012, pp. 44–50.

Barton, N. (2012) Hybrid TBM and Drill-and-Blast from the start. *Tunnelling Journal*, December 2012, 22–32.

Barton, N. (2012) Reducing risk in long deep tunnels by using TBM and drill-and-blast methods in the same project – the hybrid solution. Keynote lecture.*Risk in Underground Construction. Journal of Rock Mechanics and Geotechnical Engineering*, 4(2): 115–126, CSRME, Wuhan. Chinese Academy of Engineering.

Entacher, M., Winter, G. & Galler, R. (2013) Cutter force measurement on tunnel boring machines – Implementation at Koralm tunnel. *Journal of Tunnelling and Underground Space Technology*, 38, 487–496.

Fawcett, D.F. (1993) The effects of rock properties on the economics of full face TBMs. *Comprehensive Rock Engineering.* Hudson *et al.* Ch. 10. 4: pp. 293–311. UK: Pergamon.

Grandori, R., Sem, M., Lembo-Fazio, A. & Ribacchi, R. (1995 a). Tunnelling by double shield TBM in the Hong Kong granite. *Proc. 8th ISRM Congress, Tokyo.* Fuji (ed.) 2: pp. 569–574. Rotterdam: Balkema.

Grandori, R., Jaeger, M., Antonini,F. & Vigl, L. (1995b) Evinos-Mornos Tunnel – Greece. Construction of a 30 km long hydraulic tunnel in less than three years under the most adverse geological conditions. *Proc. of Rapid Excavation and Tunnelling Conf. RETC.* San Francisco, pp.747–767.

Innaurato, N., Mancini, R., Rodena, E. & Zaninetti, A. (1991) Forecasting and effective TBM performances in a rapid excavation of a tunnel in Italy. *Proc. 7th ISRM Congress, Aachen.* Wittke (ed.) 2: pp. 1009–1014. Rotterdam: Balkema.

McKelvey, J.G., Schultz, E.A., Helin, T.A.B. & Blindheim, O.T. (1996) Geotechnical analysis in S. Africa. *World Tunnelling*, November, pp. 377–390.

Nelson, P., O'Rourke, T.D. & Kulhawy, F.H. (1983) Factors affecting TBM penetration rates in sedimentary rocks. *24th US Symposium on Rock Mechanics*, Texas A&M University.

NTH (1994) Full face tunnel boring. [In Norwegian: Fullprofilboring av tunneler.] *Prosjektrapport, anleggsdrift*, Trondheim, Norway, pp. 1–94.

Shen, C.P., Tsai, H.C., Hsieh, Y.S. & Chu, B. (1999) The methodology through adverse geology ahead of Pinglin large TBM. *Proc. RETC. Orlando, FL.* Hilton & Samuelson (eds.) Ch.8: pp. 117–137. Littleton, CO: Soc. for Mining, Metallurgy, and Exploration, Inc.

Sundaram, N.M. and Rafek, A.G. (1998) The influence of rock mass properties in the assessment of TBM performance. *Proc. 8th IAEG congress, Vancouver.* Moore & Hungr (eds.) pp.3553–3559. Rotterdam: Balkema.

Willis, D. (2012) TBM probe drilling and pre-grouting – FIVE THINGS TO KNOW. *Tunnelling Journal*, July 2012, 43–45.

Chapter 28

Design guidelines for open stope support

Y. Potvin
Australian Centre for Geomechanics, The University of Western Australia, Crawley, Australia

Abstract: In this chapter, a description of the most common open stope support practices will be provided. This will be followed by the engineering approach generally used to produce the design parameters associated with open stope support design. Finally, the optimization process that can be implemented to open stope support design will be discussed.

1 INTRODUCTION

Artificial support in the form of rock reinforcement and surface support is widely utilized in underground mines to help maintaining the stability of underground excavations. The main purpose of ground support in underground mines is often to control the risk of rockfalls in mine accesses and as such, to contribute to make the underground infrastructure safer. The role of ground support in open stope mining is somewhat different.

Open stope mining gained popularity world-wide in the late 1970s and early 1980s when the underground mining industry became more mechanized, taking advantage of large equipment to increase productivity. There is a clear link between profitability, productivity and the size of open stopes, large open stopes being more productive at a lesser unit cost per tons. Therefore, mine designers have a strong incentive to make open stopes excavations as large as possible, while accounting for the ground conditions and the level of stresses in which they are to be extracted. It is essential that open stopes remain stable to be profitable. Instability may not only bring safety concerns but also high dilution, which can severely inhibit both productivity and profitability.

The primary role of open stope support is to control dilution. Stope support can also offer an opportunity to extend the size (and therefore the productivity) of stopes for a given ground condition, trading off the additional cost of installing ground support to the gain in productivity of larger supported stopes.

In this chapter, a description of the most common open stope support practices will be provided. This will be followed by the engineering approach generally used to produce the design parameters associated with open stope support design. Finally, the optimization process that can be implemented to open stope support design will be discussed. It is noted that waste or ore pillars, broken ore and mine fill can play very important role in permanent or temporary open stope support, but these techniques will not be covered in this Chapter, which entirely focus on cable bolt support.

2 OPEN STOPE SUPPORT PRACTICES

Open stopes are relatively large excavations and as such, their stabilization is beyond the reach of normal 2.4 to 3 m rockbolts. Consequently, cable bolting is the reinforcement of choice for open stope support.

It is in cut and fill applications that cable bolts have originally gained acceptance in underground mines. The installation of long cable bolts in cut and fill backs had the advantage of covering three to four lifts, reducing the rehabilitation work after each blasts. In cut and fill applications, there is good access to install systematic patterns of cable bolts and the cables are in place before blasting occurred, allowing for an efficient pre-reinforcement of the rock mass.

In open stope applications cable bolts are generally designed to stabilize the crown (roof) or the hanging wall of stopes. Depending on the many possible configurations of open stope layouts, the access for installing cable bolt support can vary from excellent to very limited. The cable bolt pattern must therefore be adapted to be implementable, given the limitations of accesses from which the cable bolt installation can take place.

2.1 Common open stope crown support patterns and specifications

A systematic cable bolt pattern can be installed when the overcut (drilling access) of a stope is completely silled out (excavated for the full stope width, Figure 1 a). This not only facilitates the installation of a systematic ground support pattern (Figure 1b) but also the drilling and blasting activities. The drawback of silling out the overcut is the creation of a potentially large span which could increase the risk of rockfalls and may

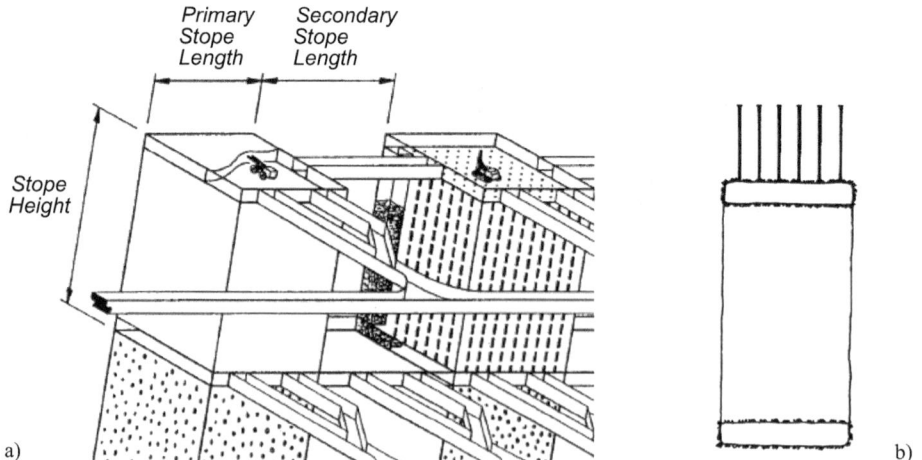

Figure I a) Isometric drawing of open stopes showing the silling out of the overcut (stope of the left) to provide full access for drilling parallel holes for blasting (stope on right). This approach also provides good access to drill and install a systematic cable bolt pattern. b) Cross section of a stope showing a systematic parallel cable bolt patter in the crown of a stope.

have negative consequences on safety. Therefore, this is generally practicable only when the stope overcut is not too wide, the ground conditions are reasonably good and the level of stresses relative to rock strength are not too elevated. The cable bolting pattern should also be conservative and monitoring desirable if people need to enter the area for drilling, charging and blasting activities.

In many cases, the stope overcuts consist of a simple network of few short drifts, designed to enable the execution of the drill and blast strategy for the stope. There are many variances on overcut layouts but conceptually, there are often a drilling drive along the hanging wall, and/or along the footwall which are connected to a footwall haulage drive, using cross-cuts. Three simple examples showing schematic plan view of overcut layouts are given in Figure 2 a to c).

When only partial access from a drives and/or a cross-cuts is available to drill and install cable bolts in the crown, a fan pattern is generally drilled to extend laterally the coverage of the cable bolt support (Figures 3a and b).

In most mining applications, the cable bolts utilized is a seven wires 15.2 mm diameter bolt, generally rated at 25 yielding capacity. Because the reinforcement in the crown may require to support large wedges or significant deadweight, twin strands are recommended in most applications as opposed to single strand. For a very small incremental cost of adding the second strand, the capacity of the reinforcement can be doubled. At least one of the two strands should be use for installing a face-plate. The plate is important in that it serves as the connection between cable bolts and the surface support and prevent deterioration from initiating in between the cable bolts. The modified geometry cable bolts such as bulb cables improve the bond between the cable and the grout and as such, prevent some of the common grout quality control issues common to cable bolting, especially in the case of manual installation. Bulbs also increase the stiffness of the reinforcement which is desirable when dealing with significant dead weight, but not so desirable if high stress and significant deformation is anticipated.

2.2 Common open stope hanging wall support patterns and specifications

As with the support of open stope crowns, the poor access to install cable bolt support in the hanging wall greatly influence the practices and in most applications, the coverage is severely limited. The overcut, sublevels (if the stope has multiple lifts), and undercut generally provides the only accesses to the hanging wall of a stope. A number of cable bolts can be drilled in a fan pattern and installed into the hanging wall from these accesses to effectively reduce the span of the unsupported hanging wall surface, limiting this span to the distance between the bottom of the overcut to the top of the undercut (Figure 4 a). This unsupported span can be further reduced, in some applications, by recessing cable bolts up through the orebody, from the undercut (Figure 4 b).

A variable number of cable bolts can be used when implementing a fan installed from overcut/sublevel/undercut. The general reinforcement principle in this application is to "pin the layers of rock together" and prevent the dilation along the hanging wall discontinuities. The cable bolt fans are also particularly useful to arrest the progression of failure, when the delamination of the hanging wall initiates in the central portion and

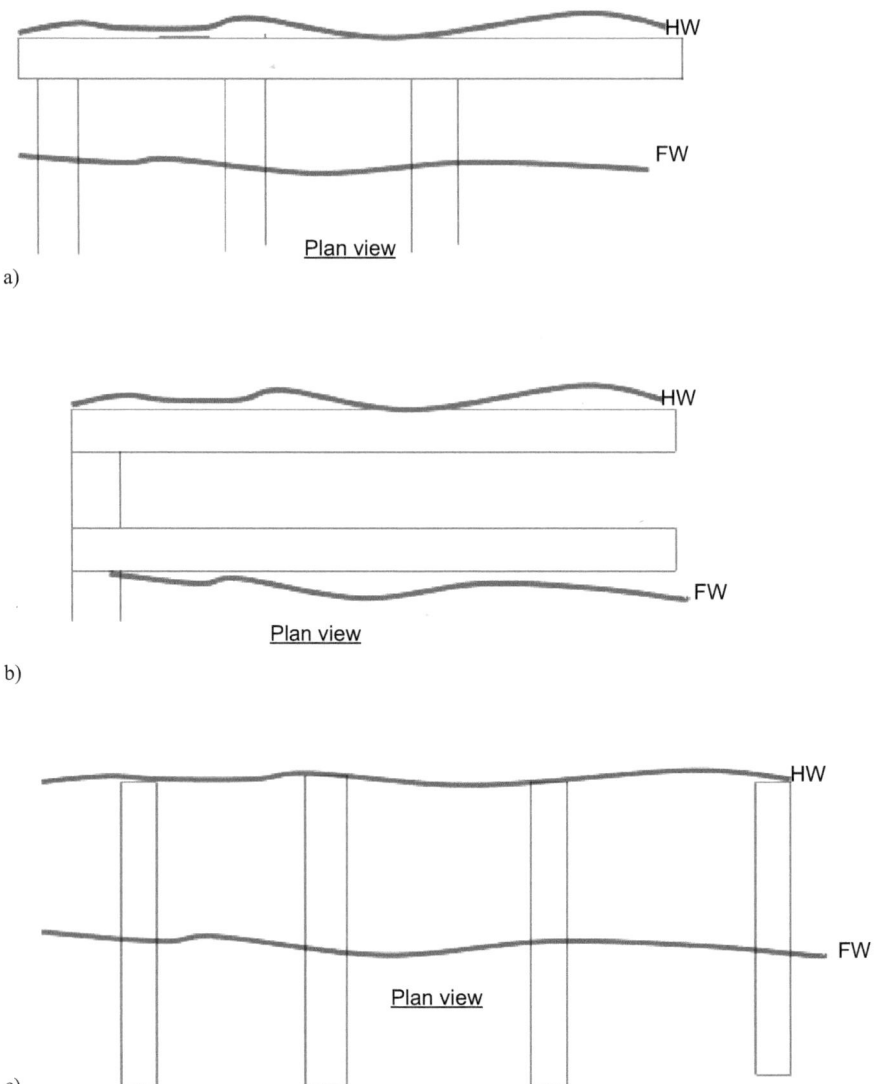

Figure 2 Conceptual plan views of development mining at the crown of open stopes showing the variance in access available for cable bolts drilling and installation. a) hanging wall drive with regular cross cuts for transversal retreat, b) hanging wall and footwall drives with single access cross cut for longitudinal retreat, c) cross cut access only, for transversal retreat.

start propagating upwards toward the overcut. In general hanging wall cables do not take a lot of load from the dead weight of failed rock as it is potentially the case for the crown of the stope. Therefore, using a large number of cable bolts in a fan, or installing twin strands is often not essential. The design of cable bolt layouts needs to be assess on a case by case basis, but it is common to install only 3 to 5 cables in a fan. Perhaps more importantly is the design of a high stiffness reinforcement to prevent the dilation of the

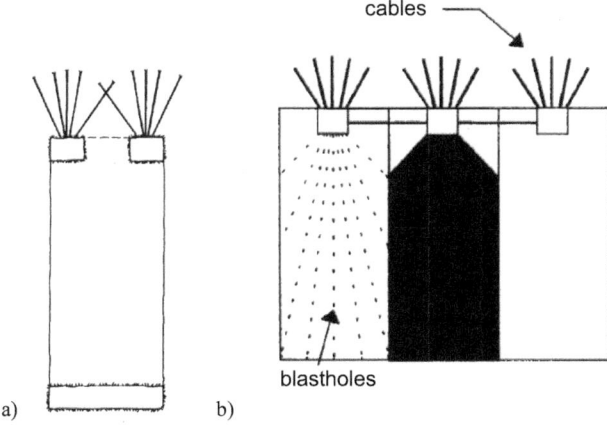

Figure 3 Simple cable bolt fan patterns installed to stabilize the crown of a stope from a) twin drives in a stope and b) in a single cross cut to stabilize the crown of the stope.

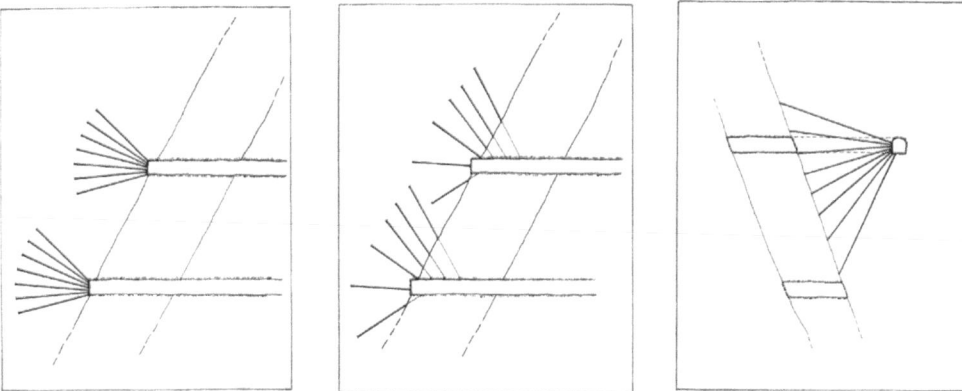

Figure 4 Simple cable bolt fan patterns installed to stabilize the hanging wall of an open stope a) fanned from the overcut, undercut or sub level b) fanned and recessed up to reach further into the hanging wall and c) from a dedicated drive located inside the hanging wall.

rock mass. Modified geometry cables (bulb cables), which are stiffer than the plain strand cables, and plating the cable are likely to produce better results.

The fanned cables are often not required to be very long, since the reinforcement concept is to pin the rock layers together. The objective is to stabilize the first few meters (say approximately 5 meters) of the hanging wall rock mass. Therefore, accounting for the angle of the fanned design, hanging wall cables are generally less than 10 meters long (5 to 7 m are common lengths).

Occasionally, there are accesses readily available in the hanging wall which can be used to install cable bolts or alternatively a dedicated hanging wall support drive can be developed (Figure 4 c), but this latter option is costly. A hanging wall access provides an ideal layout to achieve a systematic pattern and a full support coverage of the hanging wall. As the supporting action will occur from the toe of the cables bolts rather than from the collar, as it is in most applications, in this specific case the plates are not required. However, having a high density of bulbs toward the toe of the cable bolts is of great benefit.

3 ENGINEERING DESIGN OF OPEN STOPE SUPPORT

The engineering design of open stope support is intimately related to the design of open stope dimension and the prevailing ground conditions. In very competent rock masses, large open stopes with no ground support can be successfully mined. When dealing with less competent ground conditions, open stopes need to be smaller to remain stable. In poor rock masses, ground support is often required to enable open stope mining. As mentioned earlier, ground support may provide an opportunity, when mining in a given ground condition, to open larger and more productive stopes.

It is at the pre-feasibility or feasibility study stage, when investigating practicable open stope dimensions, that the design of cable support is first evaluated. At that stage, an empirical approach is most suitable because there are no opportunities to calibrate other design methods. Therefore relying on "pre-calibrated methods" based on the experience acquire in similar conditions elsewhere offers the most robust approach for a preliminary design of open stope and ground support.

3.1 The modified stability graph method for cable bolt support of stope backs

There is ample literature available on the modified stability graph method and it is beyond the scope of this chapter to explain the method in detail. The readers are referred to the original works (Mathews *et al.*, 1980; Potvin, 1988; Nickson, 1992). The method is also summarized in the well-known rock engineering textbooks by Hoek *et al.* (1995), Hutchinson & Diederichs (1996) and Brady & Brown 3rd edition (2004).

The modified stability graph method is used to forecast stable dimensions of open stope excavations, assessing the stability of stope surfaces individually. As such, the stability of the crown is assessed separately to each of the individual stope walls. The stability graph (Figure 5) provides a correlation between the ground condition expressed as a stability number 'N' and the stope surface dimension index called hydraulic radius 'HR', which is calculated by dividing the area of the stope plane being assessed, by its perimeter.

A large stability number reflects good ground condition and a large hydraulic radius reflect a large stope surface. The top left side of the graph represent small stopes in good ground conditions, therefore, very stable stopes, and the bottom right of the graph is the domain of large stopes in poor ground condition, therefore, stopes which will experience caving. From the top left to the bottom right, the stability of stope surfaces transitions from the stable zone, to a zone where cable bolt support is required to a zone where even with support, the surface will not be stable.

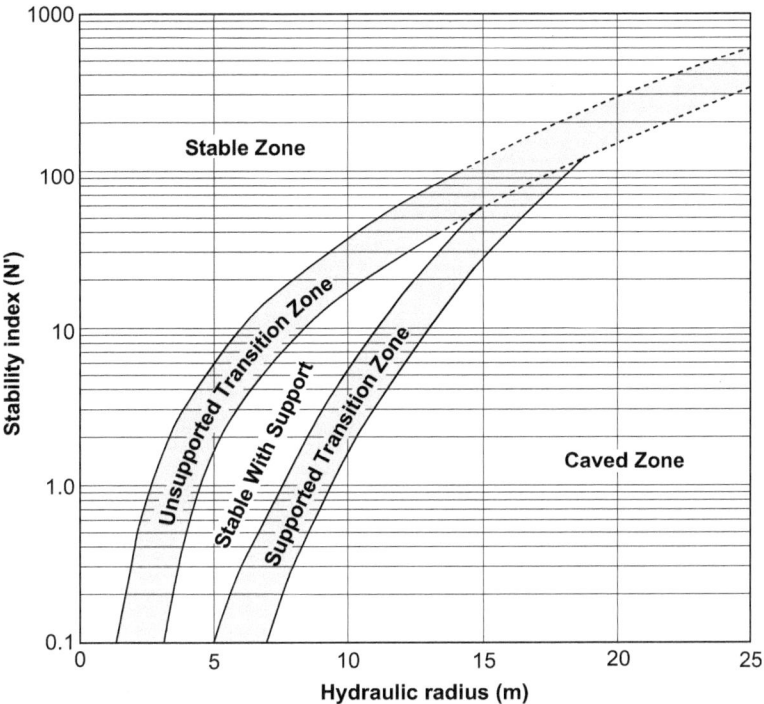

Figure 5 Modified stability graph (after Nickson, 1992).

Typically, at feasibility study, given the ground conditions prevailing in different areas of the mine, stability numbers can be calculated to assess stable stope surface dimensions and whether ground support will be required to stabilize them. For designs plotting within the zone "stable with support", an adequate cable bolt pattern must be selected or alternatively, the stope dimension must be reduced.

The fan patterns in the hanging wall have been discussed previously and often involve 3 to 5 cables bolts, 5 to 7 m long, installed from an ore drive. For stope backs, where systematic pattern cable bolting can be installed (Figure 6 and 7), other empirical graphs have been developed to assist designers on selecting the density of cable bolting required to stabilize them.

In Figure 6, based on 66 case studies, Potvin (1988) related cable bolt densities (# of cable bolts per square meter) to a rock mass block size factor calculated based on Deere's (1964) rock quality designation (RQD) divided by Barton's (1974) joint set number Jn, relative to the previously defined hydraulic radius (HR). The rationale behind this proposal is that the more fractured (smaller block size) is the rock mass relative to the size of the surface, the more cable bolts (higher density pattern) is required to stabilize the stope back.

It is readily observed from Figure 6 that there are three design zones. Moving from the top right of the graph, there is a very conservative zone where the relative block size are large and the cable bolt density is high, moving diagonally toward the bottom left of the graph, there is a conservative zone and a "non entry design zone". The latter

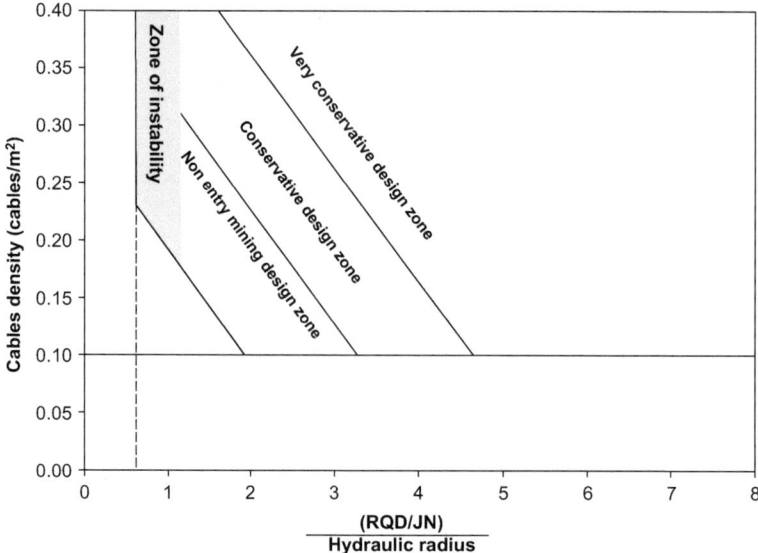

Figure 6 The cable bolt density graph relating the relative block size of the rock mass to the density of the cable bolt pattern (Potvin, 1988).

suggests the least conservative design with lighter cable bolt density and smaller relative block size. It is also noted that the horizontal line shows the minimum practicable cable bolt density used in the 66 case studies and the vertical line indicates the minimum relative block size beyond which the stope surface cannot be stabilized.

These guidelines are very broad and leave the designers with a lot of flexibility into the choice of an appropriate cable-bolting density. It is suggested that the degree of conservatism in selecting a cable bolt density should take into account, amongst other things, where on the stability graph (Figure 5) the stope surface is plotting. For example if a stope surface point plots toward the "supported transition zone" then a more conservative cable bolt design should be applied as it is close to the caving zone. Conversely, when a stope surface plots toward the "unsupported transition zone", then a less conservative approach could be suitable, as the design is already close to the stable zone.

Adding a further 46 case studies to the original 66, Nickson (1992) produced a more specific guideline by combining suggested cable bolt density zones to the stability graph (Figure 7). Eleven zones (A to K, Figure 7) with bolt density varying from 0.1 to 0.32 cable bolt/m^2 are distributed within the "stable with support zone" of the stability graph.

A number of other empirical charts have been proposed to assist in selecting cable bolt lengths (Potvin, 1988; Hutchinson & Diederichs, 1996). Since the cable bolting technique was originally adapted in the early 1980s from cut and fill mining practices, where multiple lifts were pre-reinforced, extra-long cable bolts were often used in open stope mining without really being necessary. As a result, early empirical charts published as a guideline for selecting cable bolt lengths are, in the author's experience, misleading and will not be reproduced here.

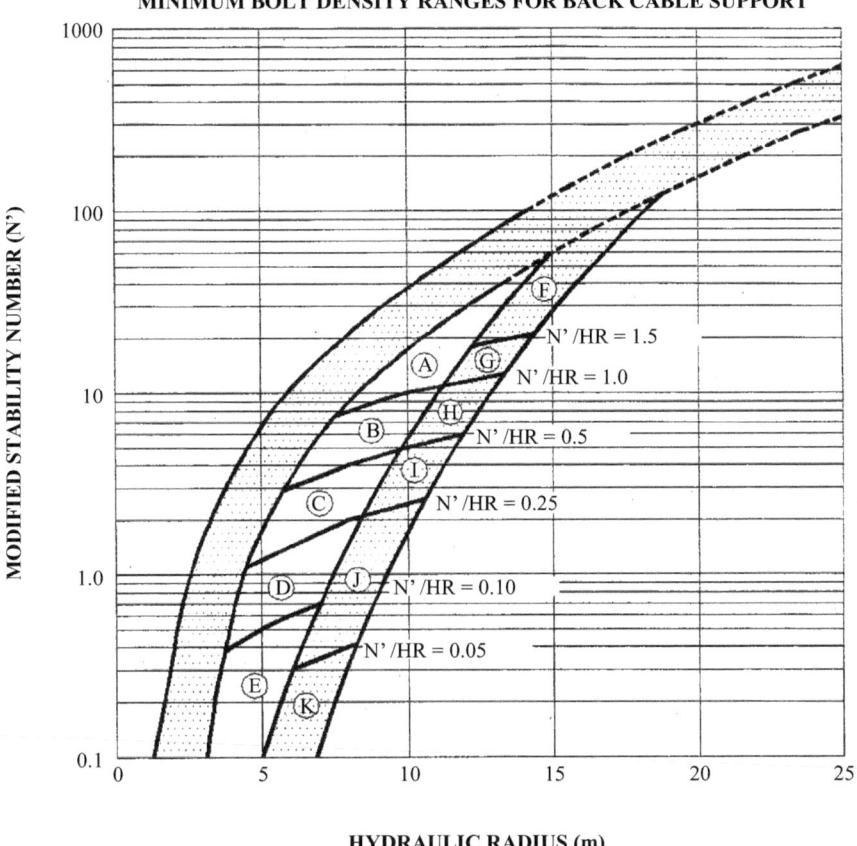

Figure 7 Stability graph with minimum range of cable bolt density zones.

Based on Canadian mining experience, Hadjigeorgiou & Charette (2009) proposed a more recent design chart, which provides a more realistic cable bolt length guideline, based on the span of the excavation (Figure 8).

In reality, the required lengths of cable bolts are intimately related to local geology and structures, which will contribute to produce a potentially unstable zone. Since the critical embedment length of plain strand cable bolts is approximately 2 meters, in theory, cable lengths should exceed the potentially unstable zone by 2 meters, to ensure

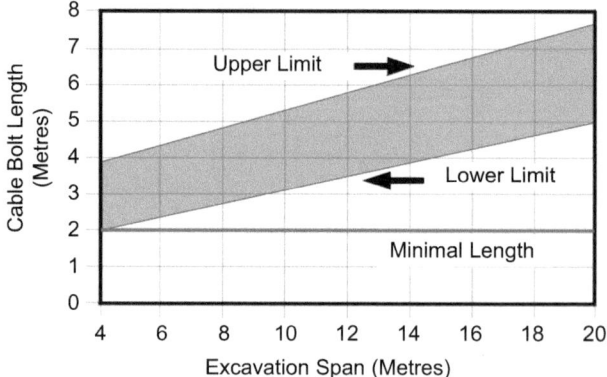

Figure 8 Empirical chart suggesting a range of cable bolt lengths as a function of the excavation spans.

that anchoring in competent ground is achieved. Therefore, in addition to Figure 8, the cable length design should be the result of an attempt by the designer to define a potentially unstable zone, accounting for the geology and structural data, ensuring that cable bolts reach at least 2 meters pass this potentially unstable zone.

3.2 Optimization of open stope ground support design

As mentioned earlier, original ground support patterns are designed at the pre-feasibility or feasibility stages, generally using relatively simple empirical methods. This is because at that stage, there is no opportunity for calibrating more sophisticated methods and also the knowledge of ground conditions is so limited that there is no justification for using complex models. As a mining project progresses toward the construction of the mine, the quantity and quality of geo-mechanics data improves allowing to re-visit the input parameters used in the original empirical design. Some adjustment to the original design may be implemented at that stage, even before the first stope is extracted.

The mining of the first few open stopes provides some of the data essential to initiate a stope optimization process. In particular, the use of the cavity monitoring system (CMS, Miller *et al.* (1992)), when the stope is emptied, generates accurate volumetric data to assess stope stability and the adequacy of the stope design, including the ground support performance. In most cases, mine staff back analyze their design by plotting their own stope surface data on the stability graph and validate the design zones of the graphs against their site experience. Some adjustments of these zones are sometimes implemented.

Other instrumentation, such as borehole extensometers and camera observation boreholes can be installed to better understand the rock mass response and potential failure mechanisms resulting from the open stope mining activities. Instrumented cable bolts are also particularly useful to measure the axial deformation and determine the load distribution along the instrumented cable bolts. This information, in combination with the rock mass response data, can be analyzed to understand which cable bolts in the pattern are doing work and which parts of individual cables are taking load. This can help to achieve a rapid optimization of cable bolt patterns and lengths.

Once a sufficient number of stopes have been extracted and their volume measured, the rock mass properties and behavior are better understood, there is then scope for using more sophisticated tools to design and optimize future open stopes. At this stage, numerical models can be calibrated and validated and add significant value to open stope designs. This process generally begins with the replication by the model of the responses of open stopes which have already been mined out. A good instrumentation program focusing on the early stopes extraction is therefore essential to complete this calibration step. When the numerical model successfully demonstrates that it can reproduce adequately the rock mass deformation surrounding the open stopes and properly capture the failure mechanisms, it can then be utilized to simulate future mining. Different design options, including new support schemes can be trial numerically and optimized before implementation.

The choice of an appropriate model is a function of the rock mass behavior to be simulated and the degree of sophistication of the analyses. Simple elastic models can provide excellent results for simple problems but modeling complicated failure mechanisms will generally require models that can simulate post peak behavior based on continuum or discontinuum theories.

4 CONCLUSION

In open stope applications cable bolts are generally designed to stabilize the crown (roof) or the hanging wall of stopes. Depending on the many possible configurations of open stope layouts, the access for installing cable bolt support can vary from excellent to very limited. The cable bolt pattern must therefore be adapted to be implementable, given the limitations of accesses from which the cable bolt installation can take place.

The engineering design of open stope support is intimately related to the design of open stope dimension and the prevailing ground conditions. In very competent rock masses, large open stopes with no ground support can be successfully mined. When dealing with less competent ground conditions, open stopes need to be smaller to remain stable. In poor rock masses, ground support is often required to enable open stope mining. Ground support may provide an opportunity, when mining in a given ground condition, to open larger and more productive stopes.

It is at the pre-feasibility or feasibility study stage, when investigating practicable open stope dimensions that the design of cable support is first evaluated. At that stage, an empirical approach is most suitable because there are no opportunities to calibrate other design methods. Therefore relying on "pre-calibrated methods" based on the experience acquire in similar conditions elsewhere offers the most robust approach for a preliminary design of open stope and ground support.

Once a sufficient number of stopes have been extracted and their volume measured, the rock mass properties and behavior are better understood, there is then scope for using more sophisticated tools to design and optimize future open stopes.

REFERENCES

Barton, N., Lien, R. & Lunde, J. 1974. Engineering classification of rock masses for the design of tunnel support. *Rock Mechanics* 6(4): 189–236.

Brady, B. H. G. & Brown, E. T. 2004. Rock Mechanics for Underground Mining, 3rd Edition. Dordrecht: Kluwer.

Deere, D. U., 1964. Technical description of rock cores for engineering purposes, *Rock Mechanics and Engineering Geology* 1(1): 17–22.

Hadjigeorgiou, J. & Charette, F. 2009. Guide pratique du soutenement minier. Québec: Association Minière du Québec, 166p.

Hoek, E., Kaiser, P. K. & Bawden, W. F. 1995. Support of Underground Excavations in Hard Rock. Rotterdam: A.A. Balkema, 215p.

Hutchinson, D. J. & Diederichs, M. 1996. Cable bolting in Underground Mines. Richmond: BiTech Publishers, 406p.

Mathews, K. E., Hoek, E., Wyllie, D. C. & Stewart, S. B. V. 1981. Prediction of stable excavation spans for mining at depths below 1000 m in hard rock mines. *Canmet Report* DSS Serial No. OSQ80-00081.

Miller, F., Jacob, D. & Potvin, Y. 1992. Cavity Monitoring System, Update and Applications. *94th CIM Annual General Meeting*, Montreal, Canada.

Nickson, S. D. 1992. Cable support guidelines for underground hard rock mine operations. *M.A. Sc. thesis*, The University of British Columbia, 223p.

Potvin, Y. 1988. Empirical open stope design in Canada. *Ph.D. thesis*. The University of British Columbia, 350p.

The five-volume set *Rock Mechanics and Engineering* consists of the following volumes

Volume 1: Principles
ISBN: 978-1-138-02759-6 (Hardback)
ISBN: 978-1-315-36426-1 (eBook)

Volume 2: Laboratory and Field Testing
ISBN: 978-1-138-02760-2 (Hardback)
ISBN: 978-1-315-36425-4 (eBook)

Volume 3: Analysis, Modeling and Design
ISBN: 978-1-138-02761-9 (Hardback)
ISBN: 978-1-315-36424-7 (eBook)

Volume 4: Excavation, Support and Monitoring
ISBN: 978-1-138-02762-6 (Hardback)
ISBN: 978-1-315-36423-0 (eBook)

Volume 5: Surface and Underground Projects
ISBN: 978-1-138-02763-3 (Hardback)
ISBN: 978-1-315-36422-3 (eBook)